BOUNDARY-LAYER THEORY

McGRAW-HILL SERIES IN MECHANICAL ENGINEERING

JACK P. HOLMAN, *Southern Methodist University*
Consulting Editor

Boundary-Layer Theory

Dr. HERMANN SCHLICHTING

Professor Emeritus at the Engineering University of Braunschweig, Germany
Former Director of the Aerodynamische Versuchsanstalt Göttingen

Translated by

Dr. J. KESTIN

Professor at Brown University in Providence, Rhode Island

Seventh Edition

McGRAW-HILL BOOK COMPANY
New York · St. Louis · San Francisco · Auckland · Bogotá ·
Düsseldorf · Johannesburg · London · Madrid · Mexico · Montreal ·
New Delhi · Panama · Paris · São Paulo · Singapore · Sydney · Tokyo · Toronto

BOUNDARY-LAYER THEORY

234567890 KP KP 7832109

Library of Congress Cataloging in Publication Data

Schlichting, Hermann,
Boundary-layer theory.

(McGraw-Hill series in mechanical engineering)
Translation of Grenzschicht-Theorie.
Bibliography: 17 p.
Includes indexes.
1. Boundary layer. I. Title.
TL574.B6S283   1979   629.132'37   78-17794
ISBN 0-07-055334-3

This book was set in Antiqua. The editor was Frank J. Cerra and the production supervisor was
John F. Harte.

First published in the German language under the title "GRENZSCHICHT-THEORIE" and
Copyright 1951 by G. Braun (vorm. G. Braunsche Hofbuchdruckerei u. Verlag) GmbH, Karlsruhe
First English Edition (Second Edition of the book) published in 1955
Second English Edition (Fourth Edition of the book) published in 1960
Third English Edition (Sixth Edition of the book) published in 1968

Contents

Part A. Fundamental laws of motion for a viscous fluid

Part B.   Laminar boundary layers

List of Tables

List of Tables

Foreword

To the First English Edition

Boundary-layer theory is the cornerstone of our knowledge of the flow of air and other fluids of small viscosity under circumstances of interest in many engineering applications. Thus many complex problems in aerodynamics have been clarified by a study of the flow within the boundary layer and its effect on the general flow around the body. Such problems include the variations of minimum drag and maximum lift of airplane wings with Reynolds number, wind-tunnel turbulence, and other parameters. Even in those cases where a complete mathematical analysis is at present impracticable, the boundary-layer concept has been extraordinarily fruitful and useful.

The development of boundary-layer theory during its first fifty years is a fascinating illustration of the birth of a new concept, its slow growth for many years in the hands of its creator and his associates, its belated acceptance by others, and the subsequent almost exponential rise in the number of contributors to its further development.

The first decade following the classical paper of Prandtl in 1904 brought forth fewer than 10 papers by Prandtl and his students, a rate of about one paper per year. During the past year over 100 papers were published on various aspects of boundary-layer theory and related experiments. The name of H. Schlichting first appears in 1930 with his doctoral thesis on the subject of wake flow. Shortly thereafter Schlichting devoted major effort to the problem of the stability of laminar boundary-layer flow.

My own interest in the experimental aspects of boundary-layer flow began in the late twenties. With the appearance of Schlichting's papers intensive attempts were made to find the amplified disturbances predicted by the theory. For 10 years the experimental results not only failed to confirm this theory but supported the idea that transition resulted from the presence of turbulence in the free air stream as described in a theory set forth by G. I. Taylor. Then on a well-remembered day in August, 1940, the predicted waves were seen in the flow near a flat plate in a wind tunnel of very low turbulence. The theory of stability described in the papers of Tollmien and Schlichting was soon confirmed quantitatively as well as qualitatively.

German periodicals available in the United States after the war referred to a series of lectures by Schlichting on boundary-layer theory which had been published in 1942. This document of 279 pages with 116 figures was not available for some time. An English translation was given limited distribution as NACA Technical Memorandum No. 1217 in 1949. These lectures were completely rewritten to include material previously classified, confidential, or secret from Germany and other countries.

The result was the book of 483 pages and 295 figures published in 1951 in the German language. When this book became known to research workers and educators in the United States, there was an immediate request from several quarters for an English translation, since no comparable book was available in the English language.

The technical content of the present English edition is described in the author's preface. The emphasis is on the fundamental physical ideas rather than on mathematical refinement. Methods of theoretical analysis are set forth along with such experimental data as are pertinent to define the regions of applicability of the theoretical results or to give physical insight into the phenomena.

Aeronautical engineers and research scientists owe a debt of gratitude to Professor Schlichting for this timely review of the present state of boundary-layer theory.

Washington D. C., December 1954   Hugh L. Dryden

Author's Preface to the Seventh (English) Edition

The sixth (English) edition of this book appeared in 1968; it differed very little from the fifth (German) edition of 1965. The first (German) edition of this book was published in 1951. In the time interval between 1951 and 1968 an English edition always followed a German edition. All translations have been prepared by Professor Kestin in an accomplished fashion.

When I decided in 1975 to write a new edition of this book I came to the conclusion that the preceding sequence of a German edition followed by an English edition was no longer practicable. The reason for it was the heavily increased cost of printing. Consequently, I suggested to the two publishing companies, G. Braun in Karlsruhe and McGraw-Hill in New York, to produce a new edition only in the English language. I express my thanks to both Publishers for their consent.

As in the previous editions, I attempted this time also to select for inclusion the most important contributions from among the abundant crop that appeared in the meantime in the field of boundary-layer theory, without, however, altering the basic structure of my book. I hope that the principal thrust of the book remained intact, namely the intent to emphasize and to present theoretical considerations in a form accessible to engineers.

The subdivision of the book into four parts (Fundamental laws of motion of a viscous fluid; Laminar boundary layers; Transition; Turbulent boundary layers) has been retained. Concerning the *additions* I wish to mention a few. Owing to the advent of large electronic computers it became possible to tackle many problems that were considered unsolvable in the past. These include numerical solutions of the Navier-Stokes equations for moderately large Reynolds numbers (Chap. IV), numerical integration of the boundary-layer equations for laminar and turbulent flows (Chap. IX), as well as the explicit numerical integration of the Orr-Sommerfeld equation of the theory of stability of laminar boundary layers (Chap. XVI). Another subject newly taken into account are exact solutions of the Navier-Stokes equations for the non-steady stagnation flow (Chap. V), the theory of the laminar boundary layer of second order (Chap. VII and IX). The sections on the calculation of two-dimensional, incompressible, turbulent boundary layers (Chap. XXII), on the stability of laminar boundary layers with compressibility and heat-transfer effects (Sec. XVIIe), and on losses in cascade flows (Chap. XXV) have been completely revised.

Along with this new material, I feel that I ought to mention the topics which I specifically omitted to include. I do not discuss the effect of chemical reactions on flow processes in boundary layers as they occur in the presence of hypersonic flow. The same applies to boundary layers in magneto-fluid-dynamics, low-density flows and flows of non-Newtonian fluids. I still thought that I ought to refrain from giving an exposition of the statistical theory of turbulence in this edition, as in the previous ones, because nowadays there are available other, good presentations in book form.

Once again, the lists of references have been expanded considerably in many chapters. The number of illustrations increased by about 65, but 20 old ones have been omitted; the number of pages increased by about 70. In spite of this, I hope that the original character of this book has been retained, and that it still can provide the reader with a *bird's-eye view* of this important branch of the physics of fluids.

As I worked on the new manuscript I once more enjoyed the vigorous assistance that I received from several of my professional colleagues. Professor K. Gersten contributed sections on boundary layers of second order to the part on laminar boundary layers (Secs. VIIf and IXj). This is a special field which he successfully worked out in recent years. Professor T. K. Fanneloep contributed the completely reformulated section on the numerical integration of the boundary-layer equations included in Sec. IXi. In the part on turbulent boundary layers, Professor E. Truckenbrodt provided me with a new version of the largest portion of Chapter XXII on two-dimensional and rotationally symmetric boundary layers. Dr. L. M. Mack of the California Institute of Technology was good enough to contribute a new section on the stability of boundary layers in supersonic flow, Sec. XVIIe. Dr. J. C. Rotta thoroughly reviewed Part D on turbulent boundary layers and made many additions to it. For the Russian literature I received much help from Professor Mikhailov. The translation was once again entrusted to Professor J. Kestin's competent pen. I express my sincere thanks to all those gentlemen for their valuable cooperation.

I should also like to repeat my acknowledgement of the help I received from several professional friends when I worked on the fifth (German) edition. Naturally, their contributions have now been retained for the seventh edition. This is the extensive contribution on compressible laminar boundary layers in Chapter XIII written by Dr. F. W. Riegels, Professor K. Gersten's section on thermal boundary layers in Chapter XII and Dr. J. C. Rotta's text on compressible turbulent boundary layers in Chapter XXIII.

I express my thanks to Frau Gerda Wolf, Frau Hilde Kreibohm and Mrs. Leslie Giacin for the careful preparation of the clear copy of the manuscript; Frau Gerda Wolf was also very helpful for me in the library. Messrs. Rotta, Hummel and Starke were kind enough to assist with the reading of the proofs.

Last, but not least, thanks are due to Verlag Braun for their willingness to accede to my wishes and for the pleasing appearance of the book.

Goettingen, August 1978   Hermann Schlichting

Translator's Preface to the Seventh (English) Edition

The present is the fourth edition in the English language of Professor H. Schlichting's "Grenzschicht-Theorie". Once again, the new edition was prepared in close cooperation with the Author whom I visited several times in Goettingen to finalize the contents and the wording. I wish to thank Professor Schlichting for his hospitality and Messrs. McGraw-Hill for partial financial assistance in connexion with these trips.

This time there was no German printed edition and the modifications introduced by the author were transmitted directly to me.

I owe a debt of gratitude to Professor H. E. Khalifa for his help in the task of proof-reading. My wife, Alicia, prepared the authors' and the subject indexes and competently typed them under difficult circumstances. My secretary, Mrs. Giacin in Providence, and Mrs. Kreibohm in Goettingen expertly typed the manuscript; I express to them my sincere thanks for their patience. Both publishers, Messrs. G. Braun of Karlsruhe and Messrs. McGraw-Hill of New York, spared no trouble, as on past occasions, in meeting our wishes regarding the production of the book.

Providence, Rhode Island, August 1978   J. Kestin

From Author's Preface to the First (German) Edition

Since about the beginning of the current century modern research in the field of fluid dynamics has achieved great successes and has been able to provide a theoretical clarification of observed phenomena which the science of classical hydrodynamics of the preceding century failed to do. Essentially three branches of fluid dynamics have become particularly well developed during the last fifty years; they include boundary-layer theory, gas dynamics, and aerofoil theory. The present book is concerned with the branch known as boundary-layer theory. This is the oldest branch of modern fluid dynamics; it was founded by L. Prandtl in 1904 when he succeeded in showing how flows involving fluids of very small viscosity, in particular water and air, the most important ones from the point of view of applications, can be made amenable to mathematical analysis. This was achieved by taking the effects of friction into account only in regions where they are essential, namely in the thin boundary layer which exists in the immediate neighbourhood of a solid body. This concept made it possible to clarify many phenomena which occur in flows and which had previously been incomprehensible. Most important of all, it has become possible to subject problems connected with the occurrence of drag to a theoretical analysis. The science of aeronautical engineering was making rapid progress and was soon able to utilize these theoretical results in practical applications. It did, furthermore, pose many problems which could be solved with the aid of the new boundary-layer theory. Aeronautical engineers have long since made the concept of a boundary layer one of everyday use and it is now unthinkable to do without it. In other fields of machine design in which problems of flow occur, in particular in the design of turbomachinery, the theory of boundary layers made much slower progress, but in modern times these new concepts have come to the fore in such applications as well.

The present book has been written principally for engineers. It is the outcome of a course of lectures which the Author delivered in the Winter Semester of 1941/42 for the scientific workers of the Aeronautical Research Institute in Braunschweig. The subject matter has been utilized after the war in many special lectures held at the Engineering University in Braunschweig for students of mechanical engineering and physics. Dr. H. Hahnemann prepared a set of lecture notes after the first series of lectures had been given. These were read and amplified by the Author. They were subsequently published in mimeographed form by the Office for Scientific Documentation (Zentrale für wissenschaftliches Berichtswesen) and distributed to a limited circle of interested scientific workers.

Several years after the war the author decided completely to re-edit this older compilation and to publish it in the form of a book. The time seemed particularly propitious because it appeared ripe for the publication of a comprehensive book, and because the results of the research work carried out during the last ten to twenty years rounded off the whole field.

The book is divided into four main parts. The first part contains two introductory chapters in which the fundamentals of boundary-layer theory are expounded without the use of mathematics and then proceeds to prepare the mathematical and physical justification for the theory of laminar boundary layers, and includes the theory of thermal boundary layers. The third part is concerned with the phenomenon of transition from laminar to turbulent flow (origin of turbulence), and the fourth part is devoted to turbulent flows. It is now possible to take the view that the theory of laminar boundary layers is complete in its main outline. The physical relations have been completely clarified; the methods of calculation have been largely worked out and have, in many cases, been simplified to such an extent that they should present no difficulties to engineers. In discussing turbulent flows use has been made essentially only of the semi-empirical theories which derive from Prandtl's mixing length. It is true that according to present views these theories possess a number of shortcomings but nothing superior has so far been devised to take their place, nothing, that is, which is useful to the engineer. No account of the statistical theories of turbulence has been included because they have not yet attained any practical significance for engineers.

As intimated in the title, the emphasis has been laid on the theoretical treatment of problems. An attempt has been made to bring these considerations into a form which can be easily grasped by engineers. Only a small number of results has been quoted from among the very voluminous experimental material. They have been chosen for their suitability to give a clear, physical insight into the phenomena and to provide direct verification of the theory presented. Some examples have been chosen, namely those associated with turbulent flow, because they constitute the foundation of the semi-empirical theory. An attempt was made to demonstrate that essential progress is not made through an accumulation of extensive experimental results but rather through a small number of fundamental experiments backed by theoretical considerations.

Braunschweig, October 1950   Hermann Schlichting

Introduction

Towards the end of the 19th century the science of fluid mechanics began to develop in two directions which had practically no points in common. On the one side there was the science of *theoretical hydrodynamics* which was evolved from Euler's equations of motion for a frictionless, non-viscous fluid and which achieved a high degree of completeness. Since, however, the results of this so-called classical science of hydrodynamics stood in glaring contradiction to experimental results — in particular as regards the very important problem of pressure losses in pipes and channels, as well as with regard to the drag of a body which moves through a mass of fluid — it had little practical importance. For this reason, practical engineers, prompted by the need to solve the important problems arising from the rapid progress in technology, developed their own highly empirical science of *hydraulics*. The science of hydraulics was based on a large number of experimental data and differed greatly in its methods and in its objects from the science of theoretical hydrodynamics.

At the beginning of the present century L. Prandtl distinguished himself by showing how to unify these two divergent branches of fluid dynamics. He achieved a high degree of correlation between theory and experiment and paved the way to the remarkably successful development of fluid mechanics which has taken place over the past seventy years. It had been realized even before Prandtl that the discrepancies between the results of classical hydrodynamics and experiment were, in very many cases, due to the fact that the theory neglected *fluid friction*. Moreover, the complete equations of motion for flows with friction (the Navier-Stokes equations) had been known for a long time. However, owing to the great mathematical difficulties connected with the solution of these equations (with the exception of a small number of particular cases), the way to a theoretical treatment of viscous fluid motion was barred. Furthermore, in the case of the two most important fluids, namely water and air, the viscosity is very small and, consequently, the forces due to viscous friction are, generally speaking, very small compared with the remaining forces (gravity and pressure forces). For this reason it was very difficult to comprehend that the frictional forces omitted from the classical theory influenced the motion of a fluid to so large an extent.

In a paper on "Fluid Motion with Very Small Friction", read before the Mathematical Congress in Heidelberg in 1904, L. Prandtl† showed how it was possible to analyze viscous flows precisely in cases which had great practical importance. With

---

† L. Prandtl, Über Flüssigkeitsbewegung bei sehr kleiner Reibung. Proc. Third Intern. Math. Congress, Heidelberg 1904, pp. 484—491; see also L. Prandtl, Gesammelte Abhandlungen zur angewandten Mechanik, Hydro- und Aerodynamik (Collected Works) ed. by W. Tollmien, H. Schlichting and H. Görtler, vol. II pp. 575—584, Springer, Berlin 1961.

the aid of theoretical considerations and several simple experiments, he proved that the flow about a solid body can be divided into two regions: a very thin layer in the neighbourhood of the body (*boundary layer*) where friction plays an essential part, and the remaining region outside this layer, where friction may be neglected. On the basis of this hypothesis Prandtl succeeded in giving a physically penetrating explanation of the importance of viscous flows, achieving at the same time a maximum degree of simplification of the attendant mathematical difficulties. The theoretical considerations were even then supported by simple experiments performed in a small water tunnel which Prandtl built with his own hands. He thus took the first step towards a reunification of theory and practice. This boundary-layer theory proved extremely fruitful in that it provided an effective tool for the development of fluid dynamics. Since the beginning of the current century the new theory has been developed at a very fast rate under the additional stimulus obtained from the recently founded science of aerodynamics. In a very short time it became one of the foundation stones of modern fluid dynamics together with the other very important developments — the aerofoil theory and the science of gas dynamics.

    In more recent times a good deal of attention has been devoted to studies of the mathematical justification of boundary-layer theory. According to these, boundary-layer theory provides us with a first approximation in the framework of a more general theory designed to calculate asymptotic expansions of the solutions to the complete equations of motion. The problem is reduced to a so-called singular perturbation which is then solved by the method of matched asymptotic expansions. Boundary-layer theory thus provides us with a classic example of the application of the method of singular perturbation. A general presentation of perturbation methods in fluid mechanics was prepared by M. Van Dyke†. The basis of these methods can be traced to L. Prandtl's early contributions.

    The boundary-layer theory finds its application in the calculation of the skin-friction drag which acts on a body as it is moved through a fluid: for example the drag experienced by a flat plate at zero incidence, the drag of a ship, of an aeroplane wing, aircraft nacelle, or turbine blade. Boundary-layer flow has the peculiar property that under certain conditions the flow in the immediate neighbourhood of a solid wall becomes reversed causing the boundary layer to separate from it. This is accompanied by a more or less pronounced formation of eddies in the wake of the body. Thus the pressure distribution is changed and differs markedly from that in a frictionless stream. The deviation in pressure distribution from the ideal is the cause of form drag, and its calculation is thus made possible with the aid of boundary-layer theory. Boundary-layer theory gives an answer to the very important question of what shape must a body be given in order to avoid this detrimental separation. Separation can also occur in the internal flow through a channel and is not confined to external flows past solid bodies. Problems connected with the flow of fluids through the channels formed by the blades of turbomachines (rotary compressors and turbines) can also be treated with the aid of boundary-layer theory. Furthermore, phenomena which occur at the point of maximum lift of an aerofoil and which are associated with stalling can be understood only on the basis of boundary-layer

---

† M. Van Dyke, Perturbation methods in fluid mechanics. Academic Press, 1964.

theory. Finally, problems of heat transfer between a solid body and a fluid (gas) flowing past it also belong to the class of problems in which boundary-layer phenomena play a decisive part.

At first the boundary-layer theory was developed mainly for the case of laminar flow in an incompressible fluid, as in this case the phenomenological hypothesis for shearing stresses already existed in the form of Stokes's law. This topic was subsequently developed in a large number of research papers and reached such a stage of perfection that at present the problem of laminar flow can be considered to have been solved in its main outline. Later the theory was extended to include turbulent, incompressible boundary layers which are more important from the point of view of practical applications. It is true that in the case of turbulent flows O. Reynolds introduced the fundamentally important concept of apparent, or virtual turbulent stresses as far back as 1880. However, this concept was in itself insufficient to make the theoretical analysis of turbulent flows possible. Great progress was achieved with the introduction of Prandtl's mixing-length theory (1925) which, together with systematic experiments, paved the way for the theoretical treatment of turbulent flows with the aid of boundary-layer theory. However, a rational theory of fully developed turbulent flows is still nonexistent, and in view of the extreme complexity of such flows it will remain so for a considerable time. One cannot even be certain that science will ever be successful in this task. In modern times the phenomena which occur in the boundary layer of a compressible flow have become the subject of intensive investigations, the impulse having been provided by the rapid increase in the speed of flight of modern aircraft. In addition to a velocity boundary layer such flows develop a thermal boundary layer and its existence plays an important part in the process of heat transfer between the fluid and the solid body past which it flows. At very high Mach numbers, the surface of the solid wall becomes heated to a high temperature owing to the production of frictional heat ("thermal barrier"). This phenomenon presents a difficult analytic problem whose solution is important in aircraft design and in the understanding of the motion of satellites.

The phenomenon of transition from laminar to turbulent flow which is fundamental for the science of fluid dynamics was first investigated at the end of the 19th century, namely by O. Reynolds. In 1914 L. Prandtl carried out his famous experiments with spheres and succeeded in showing that the flow in the boundary layer can also be either laminar or turbulent and, furthermore, that the problem of separation, and hence the problem of the calculation of drag, is governed by this transition. Theoretical investigations into the process of transition from laminar to turbulent flow are based on the acceptance of Reynolds's hypothesis that the latter occurs as a consequence of an instability developed by the laminar boundary layer. Prandtl initiated his theoretical investigation of transition in the year 1921; after many vain efforts, success came in the year 1929 when W. Tollmien computed theoretically the critical Reynolds number for transition on a flat plate at zero incidence. However, more than ten years were to pass before Tollmien's theory could be verified through the very careful experiments performed by H. L. Dryden and his coworkers. The stability theory is capable of taking into account the effect of a number of parameters (pressure gradient, suction, Mach number, transfer of heat) on transition. This theory has found many important applications, among them in the design of aerofoils of very low drag (laminar aerofoils).

Modern investigations in the field of fluid dynamics in general, as well as in the field of boundary-layer research, are characterized by a very close relation between theory and experiment. The most important steps forwards have, in most cases, been taken as a result of a small number of fundamental experiments backed by theoretical considerations. A review of the development of boundary-layer theory which stresses the mutual cross-fertilization between theory and experiment is contained in an article written by A. Betz[†]. For about twenty years after its inception by L. Prandtl in 1904 the boundary-layer theory was being developed almost exclusively in his own institute in Goettingen. One of the reasons for this state of affairs may well have been rooted in the circumstance that Prandtl's first publication on boundary-layer theory which appeared in 1904 was very difficult to understand. This period can be said to have ended with Prandtl's Wilbur Wright Memorial Lecture[°] which was delivered in 1927 at a meeting of the Royal Aeronautical Society in London. In later years, roughly since 1930, other research workers, particularly those in Great Britain and in the U.S.A., also took an active part in its development. Today, the study of boundary-layer theory has spread all over the world; together with other branches, it constitutes one of the most important pillars of fluid mechanics.

The first survey of this branch of science was given by W. Tollmien in 1931 in two short articles in the "Handbuch der Experimentalphysik"[‡]. Shortly afterwards (1935), Prandtl published a comprehensive presentation in "Aerodynamic Theory" edited by W. F. Durand[§]. During the intervening four decades the volume of research into this subject has grown enormously[§]. According to a review published by H. L. Dryden in 1955, the rate of publication of papers on boundary-layer theory reached one hundred *per annum* at that time. Now, some twenty years later, this rate has more than tripled. Like several other fields of research, the theory of boundary layers has reached a volume which is so enormous that an individual scientist, even one working in this field, cannot be expected to master all of its specialized subdivisions. It is, therefore, right that the task of describing it in a modern handbook has been entrusted to several authors[‡]. The historical development of boundary-layer theory has recently been traced by I. Tani[*].

---

† A. Betz, Ziele, Wege und konstruktive Auswertung der Strömungsforschung, Zeitschr. VDI *91*, (1949) 253.

° L. Prandtl, The generation of vortices in fluids of small viscosity (15th Wilbur Wright Memorial Lecture, 1927). J. Roy. Aero. Soc. *31*, 721—741 (1927).

‡ *Cf.* the bibliography on p. 780.

§ L. Prandtl, The mechanics of viscous fluids. Aerodynamic Theory (W. F. Durand, ed.), Vol. 3, 34—208, Berlin, 1935.

§ H. Schlichting, Some developments of boundary-layer research in the past thirty years (The Third Lanchester Memorial Lecture, 1959). J. Roy. Aero. Soc. *64*, 63—80 (1960).

See also: H. Schlichting, Recent progress in boundary-layer research (The 37th Wright Brothers Memorial Lecture, 1973). AIAA Journal *12*, 427—440 (1974).

* I. Tani, History of boundary-layer research. Annual Rev. of Fluid Mechanics *9*, 87—111 (1977).

Part A. Fundamental laws of motion for a viscous fluid

CHAPTER I

Outline of fluid motion with friction

a. Real and perfect fluids

Most theoretical investigations in the field of fluid dynamics are based on the concept of a perfect, i. e. frictionless and incompressible, fluid. In the motion of such a perfect fluid, two contacting layers experience no tangential forces (shearing stresses) but act on each other with normal forces (pressures) only. This is equivalent to stating that a perfect fluid offers no internal resistance to a change in shape. The theory describing the motion of a perfect fluid is mathematically very far developed and supplies in many cases a satisfactory description of real motions, such as e. g. the motion of surface waves or the formation of liquid jets in air. On the other hand the theory of perfect fluids fails completely to account for the drag of a body. In this connexion it leads to the statement that a body which moves uniformly through a fluid which extends to infinity experiences no drag (d'Alembert's paradox).

This unacceptable result of the theory of a perfect fluid can be traced to the fact that the inner layers of a real fluid transmit tangential as well as normal stresses, this being also the case near a solid wall wetted by a fluid. These tangential or friction forces in a real fluid are connected with a property which is called the *viscosity* of the fluid.

Because of the absence of tangential forces, on the boundary between a perfect fluid and a solid wall there exists, in general, a difference in relative tangential velocities, i. e. there is slip. On the other hand, in real fluids the existence of inter-molecular attractions causes the fluid to adhere to a solid wall and this gives rise to shearing stresses.

The existence of tangential (shearing) stresses and the *condition of no slip* near solid walls constitute the essential differences between a perfect and a real fluid. Certain fluids which are of great practical importance, such as water and air, have very small coefficients of viscosity. In many instances, the motion of such *fluids of small viscosity* agrees very well with that of a perfect fluid, because in most cases the shearing stresses are very small. For this reason the existence of viscosity is completely neglected in the theory of perfect fluids, mainly because this introduces a far-reaching simplification of the equations of motion, as a result of which an extensive mathematical theory becomes possible. It is, however, important to stress the fact that even in fluids with very small viscosities, unlike in perfect fluids, the condition of no slip near a solid boundary prevails. This condition of no slip introduces in many cases very large discrepancies in the laws of motion of perfect and real fluids. In particular, the very large discrepancy between the value of drag in a real and a perfect fluid has its physical origin in the condition of no slip near a wall.

This book deals with the motion of fluids of small viscosity, because of the great practical importance of the problem. During the course of the study it will become clear how this partly consistent and partly divergent behaviour of perfect and real fluids can be explained.

b. Viscosity

The nature of viscosity can best be visualized with the aid of the following experiment: Consider the motion of a fluid between two very long parallel plates, one of which is at rest, the other moving with a constant velocity parallel to itself, as shown in Fig. 1.1. Let the distance between the plates be h, the pressure being constant

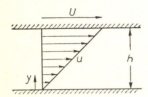

Fig. 1.1. Velocity distribution in a viscous fluid between two parallel flat walls (Couette flow)

throughout the fluid. Experiment teaches that the fluid adheres to both walls, so that its velocity at the lower plate is zero, and that at the upper plate is equal to the velocity of the plate, U. Furthermore, the velocity distribution in the fluid between the plates is linear, so that the fluid velocity is proportional to the distance y from the lower plate, and we have

$$u(y) = \frac{y}{h}\, U\,. \tag{1.1}$$

In order to support the motion it is necessary to apply a tangential force to the upper plate, the force being in equilibrium with the frictional forces in the fluid. It is known from experiments that this force (taken per unit area of the plate) is proportional to the velocity U of the upper plate, and inversely proportional to the distance h. The frictional force per unit area, denoted by τ (frictional shearing stress) is, therefore, proportional to U/h, for which in general we may also substitute du/dy. The proportionality factor between τ and du/dy, which we shall denote by μ, depends on the nature of the fluid. It is small for "thin" fluids, such as water or alcohol, but large in the case of very viscous liquids, such as oil or glycerine. Thus we have obtained the fundamental relation for fluid friction in the form

$$\tau = \mu\, \frac{du}{dy}\,. \tag{1.2}$$

The quantity μ is a property of the fluid and depends to a great extent on its temperature. It is a measure of the *viscosity* of the fluid. The law of friction given by eqn. (1.2) is known as *Newton's law of friction*. Eqn. (1.2) can be regarded as the definition of viscosity. It is, however, necessary to stress that the example considered in Fig. 1.1 constitutes a particularly simple case of fluid motion. A generalization of this simple case is contained in Stokes's law of friction (*cf.* Chap. III). The dimensions of viscosity can be deduced without difficulty from eqn. (1.2)†. The shearing stress is measured in $N/m^2 \equiv Pa$ and the velocity gradient du/dy in sec^{-1}. Hence

$$\mu = \left[\frac{kg}{m\ sec} \right] \equiv Pa\ sec,$$

where the square brackets are used to denote units. The above is not the only, or even the most widely, employed unit of viscosity. Table 1.1 lists the various units together with their conversion factors.

Eqn. (1.2) is related to Hooke's law for an elastic solid body in which case the shearing stress is proportional to the strain

$$\tau = G\gamma \quad \text{with} \quad \gamma = \frac{\partial \xi}{\partial y}\ . \tag{1.3}$$

Here G denotes the modulus of shear, γ the change in angle between two lines which were originally at right angles, and ξ denotes the displacement in the direction of abscissae. Whereas in the case of an elastic solid the shearing stress is proportional to the *magnitude of the strain*, γ, experience teaches that in the case of fluids it is proportional to the *rate of change of strain* $d\gamma/dt$. If we put

$$\tau = \mu \frac{d\gamma}{dt} = \mu \frac{d}{dt}\left(\frac{\partial \xi}{\partial y}\right) = \mu \frac{\partial}{\partial y}\left(\frac{d\xi}{dt}\right),$$

we shall obtain, as before,

$$\tau = \mu \frac{\partial u}{\partial y}$$

because $\xi = ut$. However, this analogy is not complete, because the stresses in a fluid depend on one constant, the viscosity μ, whereas those in an isotropic elastic solid depend on two.

In all fluid motions in which frictional and inertia forces interact it is important to consider the ratio of the viscosity, μ, to the density, ϱ, known as the *kinematic viscosity*, and denoted by ν:

$$\nu = \frac{\mu}{\varrho}\ . \tag{1.4}$$

---

† We shall consistently use in this book the gravitational or engineering system of units; in accordance with international agreement the symbols kp and lbf will be used to denote the respective units of *force;* the corresponding units of *mass* will be denoted by the abbreviations kg and lb respectively. In some tables, the units will be those of the SI system.

Table 1.1. Viscosity conversion factors

a. Absolute viscosity μ

	kp sec/m²	kp hr/m²	Pa sec
kp sec/m²	1	$2\cdot7778 \times 10^{-4}$	$9\cdot8067$
kp hr/m²	3,600	1	$3\cdot5304 \times 10^{4}$
Pa sec	$1\cdot0197 \times 10^{-1}$	$2\cdot8325 \times 10^{-5}$	1
kg/m hr	$2\cdot8325 \times 10^{-5}$	$7\cdot8682 \times 10^{-9}$	$2\cdot7778 \times 10^{-4}$
lbf sec/ft²	$4\cdot8824$	$1\cdot3562 \times 10^{-3}$	$4\cdot7880 \times 10^{1}$
lbf hr/ft²	$1\cdot7577 \times 10^{4}$	$4\cdot8824$	$1\cdot7237 \times 10^{5}$
lb/ft sec	$1\cdot5175 \times 10^{-1}$	$4\cdot2153 \times 10^{-5}$	$1\cdot4882$

kg/m hr	lbf sec/ft²	lbf hr/ft²	lb/ft sec
$3\cdot5316 \times 10^{4}$	$2\cdot0482 \times 10^{-1}$	$5\cdot6893 \times 10^{-5}$	$6\cdot5898$
$127\cdot1 \times 10^{6}$	$7\cdot3734 \times 10^{2}$	$2\cdot0482 \times 10^{-1}$	$2\cdot3723 \times 10^{4}$
1	$2\cdot0885 \times 10^{-2}$	$5\cdot8015 \times 10^{-6}$	$6\cdot7197 \times 10^{-1}$
$0\cdot1724 \times 10^{6}$	$5\cdot8015 \times 10^{-6}$	$1\cdot6115 \times 10^{-9}$	$1\cdot8666 \times 10^{-4}$
$620\cdot8 \times 10^{6}$	1	$2\cdot7778 \times 10^{-4}$	$3\cdot2174 \times 10^{1}$
$5\cdot358 \times 10^{3}$	3,600	1	$1\cdot1583 \times 10^{5}$
	$3\cdot1081 \times 10^{-2}$	$8\cdot6336 \times 10^{-6}$	1

b. Kinematic viscosity ν

	m²/sec	m²/hr	cm²/sec	ft²/sec	ft²/hr
m²/sec	1	3,600	1×10^{4}	$1\cdot0764 \times 10^{1}$	$3\cdot8750 \times 10^{4}$
m²/hr	$2\cdot7778 \times 10^{-4}$	1	$2\cdot778$	$2\cdot9900 \times 10^{-3}$	$1\cdot0764 \times 10^{1}$
cm²/sec (Stokes)	1×10^{-4}	$0\cdot36$	1	$1\cdot0764 \times 10^{-3}$	$3\cdot8750$
ft²/sec	$9\cdot2903 \times 10^{-2}$	$3\cdot3445 \times 10^{2}$	$9\cdot2903 \times 10^{2}$	1	3,600
ft²/hr	$2\cdot5806 \times 10^{-5}$	$9\cdot2903 \times 10^{-2}$	$2\cdot5806 \times 10$	$2\cdot7778 \times 10^{-4}$	1

Table 1.2. Density, viscosity, and kinematic viscosity of water and air in terms of temperature

Temperature	Water			Air at a pressure of 0·099 MPa (14·696 lbf/in²)		
	Density ϱ	Viscosity μ	Kinematic viscosity $\nu \times 10^6$	Density ϱ	Viscosity μ	Kinematic viscosity $\nu \times 10^6$
°C	kg/m³	[μ Pa sec]	[ft²/sec]	kg/m³	[μ Pa sec]	m²/sec
−20	—	—	—	1·39	15·6	11·2
−10	—	—	—	1·34	16·2	12·1
0	999·3	1795	1·80	1·29	16·8	13·0
10	999·3	1304	1·30	1·25	17·4	13·9
20	997·3	1010	1·01	1·21	17·9	14·8
40	991·5	655	0·661	1·12	19·1	17·1
60	982·6	474	0·482	1·06	20·3	19·2
80	971·8	357	0·367	0·99	21·5	21·7
100	959·1	283	0·295	0·94	22·9	24·4

Numerical values: In the case of liquids the viscosity, μ, is nearly independent of pressure and decreases at a high rate with increasing temperature. In the case of gases, to a first approximation, the viscosity can be taken to be independent of pressure but it increases with temperature. The kinematic viscosity, ν, for liquids has the same type of temperature dependence as μ, because the density, ϱ, changes only slightly with temperature, However, in the case of gases, for which ϱ decreases considerably with increasing temperature, ν increases rapidly with temperature. Table 1.2 contains some numerical values of ϱ, μ and ν for water and air.

Table 1.3 contains some additional useful data.

Table 1.3. Kinematic viscosity

Liquid	Temperature °C	$\nu \times 10^6$ [m²/s]
Glycerine	20	680
Mercury	0	0·125
Mercury	100	0·091
Lubricating oil . .	20	400
Lubricating oil . .	40	100
Lubricating oil . .	60	30

c. Compressibility

Compressibility is a measure of the change of volume of a liquid or gas under the action of external forces. In this connexion we can define a *modulus of elasticity*, E, of volume change, by the equation

$$\Delta p = - E \frac{\Delta V}{V_0} . \tag{1.5}$$

Here $\Delta V/V_0$ denotes the relative change in volume brought about by a pressure increase Δp. The compressibility of liquids is very small: e. g. for water $E = 280{,}000 \, \text{lbf/in}^2$ which means that a pressure increase of 1 atm (14.7 lbf/in²) causes a relative change in volume of about 1/20,000, i. e. 0·005 per cent. Other liquids show similar properties so that their compressibility can be neglected in most cases, and flows of liquids can be regarded as incompressible.

In the case of gases, the modulus of elasticity, E, is equal to the initial pressure p_0, if the changes are isothermal, as can easily be deduced from the perfect-gas law†. For air at NTP (atmospheric pressure and ice-point temperature) $E = 14 \cdot 7 \, \text{lbf/in}^2$, which means that air is about 20,000 times more compressible than water. Similar conditions obtain for other gases.

---

† From the perfect gas law it can be deduced that the change in volume, ΔV, caused by a change of pressure Δp, satisfies the relation $(p_0 + \Delta p)(V_0 + \Delta V) = p_0 V_0$. Hence $\Delta p \approx - p_0 \Delta V/V_0$.

In order to answer the question of whether it is necessary to take into account the compressibility of gases in problems of fluid flow it is necessary to consider whether the changes in pressure brought about by the motion of the fluid cause large changes in volume. Instead of considering volumes it is also possible to estimate the change in density, ϱ. Owing to the conservation of mass, we can write: $(V_0 + \Delta V)(\varrho_0 + \Delta \varrho) = V_0 \varrho_0$, so that $\Delta \varrho / \varrho_0 = -\Delta V / V_0$, and eqn. (1.5) can be written as

$$\Delta p = E \frac{\Delta \varrho}{\varrho_0} . \tag{1.5a}$$

Consequently the flow of a gas can be considered incompressible when the relative change in density remains very small, $\Delta \varrho / \varrho_0 \ll 1$. As known from Bernoulli's equation $p + \frac{1}{2} \varrho w^2 = \text{const}$ ($w = $ velocity of flow), the change of pressure, Δp, brought about by the flow is of the order of the dynamic head $q = \frac{1}{2} \varrho w^2$, so that eqn. (1.5a) becomes

$$\frac{\Delta \varrho}{\varrho_0} \approx \frac{q}{E} . \tag{1.6}$$

If, therefore, $\Delta \varrho / \varrho_0$ should be small compared with unity then, as seen from eqn. (1.6), we must also have $q/E \ll 1$. It has thus been proved that flows of gases can be treated as incompressible, with a good degree of approximation, if the dynamic head is small compared with the modulus of elasticity.

The same result can be expressed in a different way if the velocity of sound is introduced into the equation. According to Laplace's equation the velocity of sound is $c^2 = E/\varrho_0$. Hence the condition $\Delta \varrho / \varrho_0 \ll 1$ from eqn. (1.6) can also be written

$$\frac{\Delta \varrho}{\varrho_0} \approx \frac{\varrho_0}{2} \frac{w^2}{E} \approx \frac{1}{2} \left(\frac{w}{c} \right)^2 \ll 1 .$$

The ratio of the velocity of flow, w, to the velocity of sound, c, is known as the Mach number

$$\mathsf{M} = \frac{w}{c} . \tag{1.7}$$

The preceding argument leads to the conclusion that compressibility can be neglected in the treatment of the flow of gases if

$$\frac{1}{2} \mathsf{M}^2 \ll 1 \quad \text{(approximately incompressible),} \tag{1.8}$$

i. e. if the Mach number is small compared with unity, or, in other words, if the flow velocity is small compared with the velocity of sound. In the case of air, with a velocity of sound of about $c = 1100$ ft/sec, the change in density is $\Delta \varrho / \varrho_0 = \frac{1}{2} \mathsf{M}^2 = 0.05$ for a flow velocity $w = 330$ ft/sec. This value can be accepted as the outside limit when a gaseous flow can be considered incompressible.

In what follows we shall often assume the fluid to be incompressible, which will restrict the results to small Mach numbers. However, on several occasions, in particular in Chaps. XII, XIII, and XXIII, our results will be extended to include compressible fluids.

d. The Hagen-Poiseuille equations of flow through a pipe

The elementary law of friction for a simple flow with shear described in Section I b can be applied to the important, and more general, case of flow through a straight pipe of circular cross-section having a constant diameter $D = 2\,R$. The velocity at the wall is zero, because of adhesion, and reaches a maximum on the axis, Fig. 1.2. The velocity remains constant on cylindrical surfaces which are concentric with the axis, and the individual cylindrical laminae slide over each other, the velocity being purely axial everywhere. A motion of this kind is called *laminar*. At a sufficiently large distance from the entrance section the velocity distribution across the section becomes independent of the coordinate along the direction of flow.

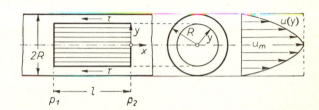

Fig. 1.2. Laminar
flow through a pipe

The fluid moves under the influence of the pressure gradient which acts in the direction of the axis, whereas in sections which are perpendicular to it the pressure may be regarded as constant. Owing to friction individual layers act on each other with a shearing stress which is proportional to the velocity gradient du/dy. Hence, a fluid particle is accelerated by the pressure gradient and retarded by the frictional shearing stress. No additional forces are present, and in particular inertia forces are absent, because along every streamline the velocity remains constant. In order to establish the condition of equilibrium we consider a coaxial fluid cylinder of length l and radius y, Fig. 1.2. The condition of equilibrium in the x-direction requires that the pressure force $(p_1 - p_2)\,\pi\,y^2$ acting on the faces of the cylinder be equal to the shear $2\,\pi\,y\,l\cdot\tau$ acting on the circumferential area, whence we obtain

$$\tau = \frac{p_1 - p_2}{l}\frac{y}{2}\,. \qquad (1.9)$$

In accordance with the law of friction, eqn. (1.2), we have in the present case

$$\tau = -\,\mu\,du/dy$$

because u decreases with y, so that eqn. (1.9) leads to

$$\frac{du}{dy} = -\,\frac{p_1 - p_2}{\mu\,l}\frac{y}{2}$$

and upon integration we find

$$u(y) = \frac{p_1 - p_2}{\mu\,l}\left(C - \frac{y^2}{4}\right).$$

The constant of integration, C, is obtained from the condition of no slip at the wall. Thus $u = 0$ at $y = R$, so that $C = R^2/4$, and finally

$$u(y) = \frac{p_1 - p_2}{4\,\mu\,l}\,(R^2 - y^2)\,. \tag{1.10}$$

The velocity is seen to be distributed parabolically over the radius, Fig. 1.2, and the maximum velocity on the axis becomes

$$u_m = \frac{p_1 - p_2}{4\,\mu\,l}\,R^2\,.$$

The volume Q flowing through a section per unit time can be easily evaluated since the volume of the paraboloid of revolution is equal to $\frac{1}{2} \times$ base area $\times$ height. Hence

$$Q = \frac{\pi}{2}\,R^2\,u_m = \frac{\pi\,R^4}{8\,\mu\,l}\,(p_1 - p_2)\,. \tag{1.11}$$

Eqn. (1.11) states that the volume rate of flow is proportional to the first power of the pressure drop per unit length $(p_1 - p_2)/l$ and to the fourth power of the radius of the pipe. If the mean velocity over the cross-section $\bar{u} = Q/\pi\,R^2$ is introduced, eqn. (1.11) can be rewritten as

$$p_1 - p_2 = 8\,\mu\,\frac{l}{R^2}\,\bar{u}\,. \tag{1.12}$$

Eqn. (1.11) was first deduced by G. Hagen [6] and shortly afterwards by J. Poiseuille [11]. It is known as the Hagen-Poiseuille equation of laminar flow through a pipe.

Eqn. (1.11) can be utilized for the experimental determination of the viscosity, μ. The method consists in the measurement of the rate of flow and of the pressure drop across a fixed portion of a capillary tube of known radius. Thus enough data are provided to determine μ from eqn. (1.11).

The type of flow to which eqns. (1.10) and (1.11) apply exists in reality only for relatively small radii and flow velocities. For larger velocities and radii the character of the motion changes completely: the pressure drop ceases to be proportional to the first power of the mean velocity as indicated by eqn. (1.12), but becomes approximately proportional to the second power of u. The velocity distribution across a section becomes much more uniform and the well-ordered laminar motion is replaced by a flow in which irregular and fluctuating radial and axial velocity components are superimposed on the main motion, so that, consequently, intensive mixing in a radial direction takes place. In such cases Newton's law of friction, eqn. (1.2), ceases to be applicable. This is the case of *turbulent* flow, to be discussed in great detail later in Chap. XX.

e. Principle of similarity; the Reynolds and Mach numbers

The type of fluid motion discussed in the preceding Section was very simple because every fluid particle moved under the influence of frictional and pressure forces only, inertia forces being everywhere equal to zero. In a divergent or convergent channel fluid particles are acted upon by inertia forces in addition to pressure and friction forces.

In the present section we shall endeavour to answer a very fundamental question, namely that concerned with the conditions under which flows of different fluids about two geometrically similar bodies, and with identical initial flow directions display geometrically similar streamlines. Such motions which have geometrically similar streamlines are called *dynamically similar*, or *similar flows*. For two flows about geometrically similar bodies (e. g. about two spheres) with different fluids, different velocities and different linear dimensions, to be similar, it is evidently necessary that the following condition should be satisfied: at all geometrically similar points the forces acting on a fluid particle must bear a fixed ratio at every instant of time.

We shall now consider the important case when only frictional and inertia forces are present. Elastic forces which may be due to changes in volume will be excluded, i. e. it will be assumed that the fluid is incompressible. Gravitational forces will also be excluded so that, consequently, free surfaces are not admitted, and in the interior of the fluid the force of gravity is assumed to be balanced by buoyancy. Under these assumptions the condition of similarity is satisfied only if at all corresponding points the ratio of inertia and friction forces is the same. In a motion parallel to the x-axis the inertia force per unit volume has the magnitude of $\varrho\, Du/Dt$, where u denotes the component of velocity in the x-direction and D/Dt denotes the substantive derivative. In the case of steady flow we can replace it by $\varrho\, \partial u/\partial x \cdot dx/dt = \varrho\, u\, \partial u/\partial x$, where $\partial u/\partial x$ denotes the change in velocity with position. Thus the inertia force per unit volume is equal to $\varrho\, u\, \partial u/\partial x$. For the friction force it is easy to deduce an expression from Newton's law of friction, eqn. (1.2). Considering a fluid particle for which the x-direction coincides with the direction of motion, Fig. 1.3, it is found that the resultant of shearing forces is equal to

$$\left(\tau + \frac{\partial \tau}{\partial y}\, dy\right) dx\, dz - \tau\, dx\, dz = \frac{\partial \tau}{\partial y}\, dx\, dy\, dz\,.$$

Hence the friction force per unit volume is equal to $\partial\tau/\partial y$, or by eqn. (1.2), to $\mu\, \partial^2 u/\partial y^2$.

Consequently, the condition of similarity, i. e. the condition that at all corresponding points the ratio of the inertia to the friction force must be constant, can be written as:

$$\frac{\text{Inertia force}}{\text{Friction force}} = \frac{\varrho\, u\, \partial u/\partial x}{\mu\, \partial^2 u/\partial y^2} = \text{const}\,.$$

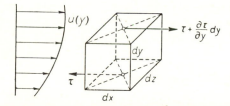

Fig. 1.3. Frictional forces acting on a fluid particle

It is now necessary to investigate how these forces are changed when the magnitudes which determine the flow are varied. The latter include the density ϱ, the viscosity μ, a representative velocity, e. g. the free stream velocity V, and a characteristic linear dimension of the body, e. g. the diameter d of the sphere.

The velocity u at some point in the velocity field is proportional to the free stream velocity V, the velocity gradient $\partial u/\partial x$ is proportional to V/d, and similarly $\partial^2 u/\partial y^2$ is proportional to V/d^2. Hence the ratio

$$\frac{\text{Inertia force}}{\text{Friction force}} = \frac{\varrho\, u\, \partial u/\partial x}{\mu\, \partial^2 u/\partial y^2} = \frac{\varrho\, V^2/d}{\mu\, V/d^2} = \frac{\varrho\, Vd}{\mu}\; .$$

Therefore, the condition of similarity is satisfied if the quantity $\varrho\, V\, d/\mu$ has the same value in both flows. The quantity $\varrho\, V\, d/\mu$, which, with $\mu/\varrho = \nu$, can also be written as $V\, d/\nu$, is a dimensionless number because it is the ratio of the two forces. It is known as the *Reynolds number*, R. Thus two flows are similar when the Reynolds number

$$\mathsf{R} = \frac{\varrho\, Vd}{\mu} = \frac{Vd}{\nu} \tag{1.13}$$

is equal for both. This principle was first enunciated by Osborne Reynolds[12] in connexion with his investigations into the flow through pipes and is known as Reynolds's principle of similarity.

The fact that the Reynolds number is dimensionless can be at once verified directly by considering the dimensions

$$\varrho \left[\frac{\text{lbf sec}^2}{\text{ft}^4}\right], \quad V\left[\frac{\text{ft}}{\text{sec}}\right], \quad d\,[\text{ft}]\,, \quad \mu\left[\frac{\text{lbf sec}}{\text{ft}^2}\right].$$

Hence

$$\frac{\varrho\, Vd}{\mu} = \frac{\text{lbf sec}^2}{\text{ft}^4} \cdot \frac{\text{ft}}{\text{sec}} \cdot \text{ft} \cdot \frac{\text{ft}^2}{\text{lbf sec}} = 1$$

which proves that the Reynolds number is, in fact, dimensionless.

Method of indices: Instead of the consideration of the condition of dynamic similarity, Reynolds's principle can also be deduced by considering dimensions by the method of indices. In this connexion use is made of the observation that all physical laws must be of a form which is independent of the particular system of units employed. In the case under consideration the physical quantities which determine the flow are: the free stream velocity, V, a representative linear dimension of the body, d, as well as the density, ϱ, and the viscosity, μ. We now ask whether there exists a combination of these quantities in the form

$$V^\alpha\, d^\beta\, \varrho^\gamma\, \mu^\delta\,,$$

which would be dimensionless. If F denotes force, L length and T time, a dimensionless combination will be obtained if

$$V^\alpha\, d^\beta\, \varrho^\gamma\, \mu^\delta = \mathsf{F}^0\, \mathsf{L}^0\, \mathsf{T}^0\,.$$

Without restricting the generality of the argument it is permissible to assign the value of unity to one of the four indices $\alpha, \beta, \gamma, \delta$, because any arbitrary power of a dimensionless quantity is also dimensionless. Assuming $\alpha = 1$, we obtain

$$V\, d^\beta\, \varrho^\gamma\, \mu^\delta = \frac{\mathsf{L}}{\mathsf{T}}\, \mathsf{L}^\beta \left(\frac{\mathsf{F}\,\mathsf{T}^2}{\mathsf{L}^4}\right)^\gamma \left(\frac{\mathsf{F}\,\mathsf{T}}{\mathsf{L}^2}\right)^\delta = \mathsf{F}^0\, \mathsf{L}^0\, \mathsf{T}^0\,.$$

Equating the exponents of L, T, and F on both sides of the expression we obtain three equations:

$$F: \qquad \gamma + \quad \delta = 0 \,,$$
$$L: \qquad 1 + \beta - 4\gamma - 2\delta = 0 \,,$$
$$T: \qquad -1 \qquad + 2\gamma + \quad \delta = 0 \,;$$

the solution of which is

$$\beta = 1 \,, \qquad \gamma = 1 \,, \qquad \delta = -1 \,.$$

This shows that there exists a unique dimensionless combination of the four quantities V, d, ϱ and μ, namely the Reynolds number R.

Dimensionless quantities: The reasoning followed in the preceding derivation of the Reynolds number can be extended to include the case of different Reynolds numbers in the consideration of the velocity field and forces (normal and tangential) for flows with geometrically similar boundaries. Let the position of a point in the space around the geometrically similar bodies be indicated by the coordinates x, y, z; then the ratios x/d, y/d, z/d are its dimensionless coordinates. The velocity components are made dimensionless by referring them to the free-stream velocity V, thus u/V, v/V, w/V, and the normal and shearing stresses, p and τ, can be made dimensionless by referring them to the double of the dynamic head, i. e. to ϱV^2 thus: $p/\varrho V^2$ and $\tau/\varrho V^2$. The previously enunciated principle of dynamical similarity can be expressed in an alternative form by asserting that for the two geometrically similar systems with equal Reynolds numbers the dimensionless quantities $u/V, \ldots, p/\varrho V^2$ and $\tau/\varrho V^2$ depend only on the dimensionless coordinates x/d, y/d, z/d. If, however, the two systems are geometrically, but not dynamically, similar, i. e. if their Reynolds numbers are different, then the dimensionless quantities under consideration must also depend on the characteristic quantities V, d, ϱ, μ of the two systems. Applying the principle that physical laws must be independent of the system of units, it follows that the dimensionless quantities $u/V, \ldots, p/\varrho V^2, \tau/\varrho V^2$ can only depend on a dimensionless combination of V, d, ϱ, and μ which is unique, being the Reynolds number $\mathsf{R} = V d \varrho/\mu$. Thus we are led to the conclusion that for the two geometrically similar systems which have different Reynolds numbers and which are being compared, the dimensionless quantities of the field of flow can only be functions of the three dimensionless space coordinates x/d, y/d, z/d and of the Reynolds number R.

The preceding dimensional analysis can be utilized to make an important assertion about the total force exerted by a fluid stream on an immersed body. The force acting on the body is the surface integral of all normal and shearing stresses acting on it. If P denotes the component of the resultant force in any given direction, it is possible to write a dimensionless force coefficient of the form $P/d^2 \varrho V^2$, but instead of the area d^2 it is customary to choose a different characteristic area, A, of the immersed body, e. g. the frontal area exposed by the body to the flow direction which is, in the case of a sphere, equal to $\pi d^2/4$. Hence the dimensionless force coefficient becomes $P/A \varrho V^2$. Dimensional analysis leads to the conclusion that for geometrically similar systems this coefficient can depend only on the dimensionless group formed with V, d, ϱ, and μ, i. e. on the Reynolds number. The component of the resultant force parallel to the undisturbed initial velocity is referred to as the drag D, and the component perpendicular to that direction is called lift, L. Hence the dimensionless coefficients for lift and drag become

$$C_L = \frac{L}{\frac{1}{2}\varrho\,V^2\,A} \quad \text{and} \quad C_D = \frac{D}{\frac{1}{2}\varrho\,V^2\,A}\,, \tag{1.14}$$

if the dynamic head $\frac{1}{2}\varrho\,V^2$ is selected for reference instead of the quantity $\varrho\,V^2$. Thus the argument leads to the conclusion that the dimensionless lift and drag coefficients for geometrically similar systems, i. e. for geometrically similar bodies which have the same orientation with respect to the free-stream direction, are functions of one variable only, namely the Reynolds number:

$$C_L = f_1(\mathsf{R})\,; \qquad C_D = f_2(\mathsf{R})\,. \tag{1.15}$$

It is necessary to stress once more that this important conclusion from Reynolds's principle of similarity is valid only if the assumptions underlying it are satisfied, i. e. if the forces acting in the flow are due to friction and inertia only. In the case of compressible fluids, when elastic forces are important, and for motions with free surfaces, when gravitational forces must be taken into consideration, eqns. (1.15) do not apply. In such cases it is necessary to deduce different similarity principles in which the dimensionless Froude number $\mathsf{F} = V/\sqrt{g\,d}$ (for gravity and inertia) and the dimensionless Mach number $\mathsf{M} = V/c$ (for compressible flows) are included.

The importance of the similarity principle given in eqns. (1.14) and (1.15) is very great as far as the sciences of theoretical and experimental fluid mechanics are concerned. First, the dimensionless coefficients, C_L, C_D, and R are independent of the system of units. Secondly, their use leads to a considerable simplification in the extent of experimental work. In most cases it is impossible to determine the functions $f_1(\mathsf{R})$ and $f_2(\mathsf{R})$ theoretically, and experimental methods must be used.

Supposing that it is desired to determine the drag coefficient C_D for a specified shape of body, e. g. a sphere, then without the application of the principle of similarity it would be necessary to carry out drag measurements for four independent variables, V, d, ϱ, and μ, and this would constitute a tremendous programme of work. It follows, however, that the drag coefficient for spheres of different diameters with different stream velocities and different fluids depends solely on one variable, the Reynolds number. Fig. 1.4 represents the drag coefficient of circular cylinders as a function of the Reynolds number and shows the excellent agreement between experiment and Reynolds's principle of similarity. The experimental points for the drag coefficient of circular cylinders of widely differing diameters fall on a *single* curve. The same applies to points obtained for the drag coefficient of spheres plotted against the Reynolds number in Fig. 1.5. The sudden decrease in the value of the drag coefficient which occurs near $\mathsf{R} = 5 \times 10^5$ in the case of circular cylinders and near $\mathsf{R} = 3 \times 10^5$ in the case of spheres will be discussed, in more detail, later. Fig. 1.6 reproduces photographs of the streamlines about circular cylinders in oil taken by F. Homann [7]. They give a good idea of the changes in the field of flow associated with various Reynolds numbers. For small Reynolds numbers the wake is laminar, but at increasing Reynolds numbers at first very regular vortex patterns, known as Kármán's vortex streets, are formed. At still higher Reynolds numbers, not shown here, the vortex patterns become irregular and turbulent in character.

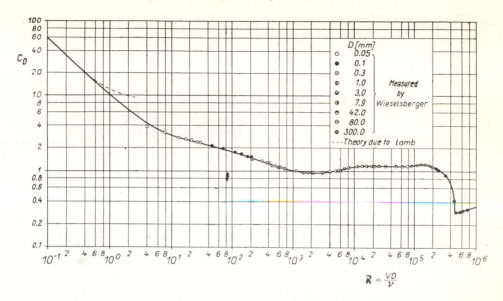

Fig. 1.4. Drag coefficient for circular cylinders as a function of the Reynolds number

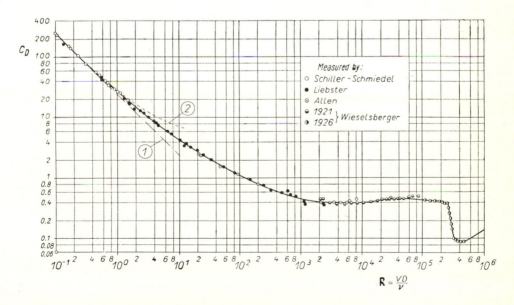

Fig. 1.5. Drag coefficient for spheres as a function of the Reynolds number
Curve (1): Stokes's theory, eqn. (6.10); curve (2): Oseen's theory, eqn. (6.13)

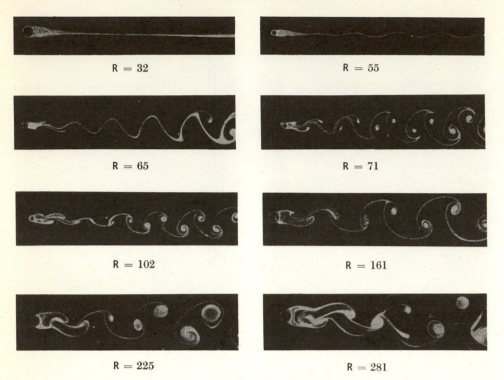

Fig. 1.6. Field of flow of oil about a circular cylinder at varying Reynolds numbers after Homann
[7]; transition from laminar flow to a vortex street in laminar flow. The frequency range for
R = 65 to R = 281 can be taken from Fig. 2.9

In later times such pictures for low Reynolds numbers up to R = 3 were produced
by S. Taneda [14].

It is seen in Fig. 1.4 that the drag coefficient of a circular cylinder reaches a
minimum of $C_D \approx 0.3$ at a Reynolds number between $R = 5 \times 10^5$ and 10^6. A regular
vortex street does not exist in this range of Reynolds numbers. At very high Reynolds
numbers exceeding $R \approx 10^6$, the drag coefficient increases at a considerable rate,
as seen from Fig. 1.7 which is based on the measurements performed by A. Roshko
[13] and G. W. Jones, J. J. Cinotta and R. W. Walker [8]. At $R = 10^7$ the drag
coefficient reaches a value of $C_D \approx 0.55$. According to the preceding authors, a
regular vortex street establishes itself again at $R > 3.5 \times 10^6$.

The drag of spheres has recently also been investigated at very high Reynolds
numbers [1]. Here too, as was the case with the cylinder, the drag coefficient increases
appreciably beyond its minimum at $C_D \approx 0.1$ at about $R = 5 \times 10^5$ attaining $C_D \approx$
0.2 at Reynolds numbers close to $R = 10^7$.

Critical reviews of drag measurements on spheres as a function of the Reynolds
number and the Mach number were prepared by A. B. Bailey and J. Hiatt [1a] as
well as by A. B. Bailey and R. F. Starr [1b].

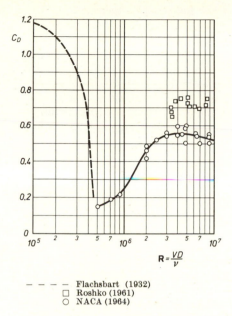

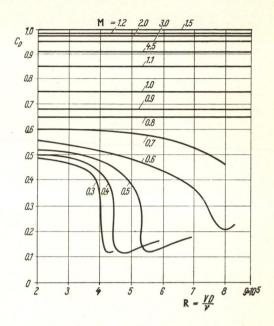

− − − − Flachsbart (1932)
　□　Roshko (1961)
　○　NACA (1964)

Fig. 1.7. Drag coefficient of a circular cylinder at very large Reynolds numbers
and for Mach numbers $M < 0.2$ after the
measurements of A. Roshko [13] and G.
W. Jones and J. J. Walker [8]

Fig. 1.8. Drag coefficient of spheres in terms
of the Reynolds and Mach numbers as measured by A. Naumann [9, 10]

Influence of compressibility: The preceding argument was conducted under the
assumption that the fluid was incompressible, and it was found that the dimensionless
dependent quantities were functions of one dimensionless argument, the Reynolds
number, only. When the fluid is compressible they depend on an additional dimensionless number, the Mach number $M = V/c$ which can be regarded, as shown in Sec. I c,
as a measure of the compressibility of the flowing medium. In the case of such flows,
i. e. when compressibility plays an essential part, the dimensionless coefficients
depend on both parameter R and M. Equation (1.15) is then replaced by

$$C_L = f_1\,(R, M) \; ; \quad C_D = f_2\,(R, M) \; . \qquad\qquad (1.16)$$

An example of such a relationship is given in Fig. 1.8 which shows a plot of the
drag coefficient C_D of spheres in terms of the Reynolds number $R = V\,D/\nu$ and
the Mach number $M = V/c$. The curve for $M = 0.3$ is practically coincident with
that in Fig. 1.5 for incompressible flow which proves that up to $M = 0.3$ the influence
of the Mach number is negligible. On the other hand at higher Mach numbers the
influence is large. In this connexion it is noteworthy that in the range of Reynolds
numbers covered by the diagram, its influence recedes more and more as the Mach
number is increased.

f. Comparison between the theory of perfect fluids and experiment

In the cases of the motion of water and air, which are the most important ones in engineering applications, the Reynolds numbers are very large because of the very low viscosities of these fluids. It would, therefore, appear reasonable to expect very good agreement between experiment and a theory in which the influence of viscosity is neglected altogether, i. e. with the theory of perfect fluids. In any case it seems useful to begin the comparison with experiment by reference to the theory of perfect fluids, if only on account of the large number of existing explicit mathematical solutions.

In fact, for certain classes of problems, such as wave formation and tidal motion, excellent results were obtained with the aid of this theory†. Most problems to be discussed in this book consist in the study of the motion of solid bodies through fluids at rest, or of fluids flowing through pipes and channels. In such cases the use of the theory of perfect fluids is limited because its solutions do not satisfy the condition of no slip at the solid surface which is always the case with real fluids even at very small viscosities. In a perfect fluid there is slip at a wall, and this circumstance introduces, even for small viscosities, such fundamental differences that it is rather surprising to find in some cases (e. g. in the case of very slender, stream-line bodies) that the two solutions display a good measure of agreement. The greatest discrepancy between the theory of a perfect fluid and experiment exists in the consideration of drag. The perfect-fluid theory leads to the conclusion that when an arbitrary solid body moves through an infinitely extended fluid at rest it experiences no force acting in the direction of motion, i. e. that its drag is zero (d'Alembert's paradox). This result is in glaring contradiction to observed fact, as drag is measured on all bodies, even if it can become very small in the case of a stream-line body in steady flow parallel to its axis.

By way of illustration we now propose to make some remarks concerning the flow about a circular cylinder. The arrangement of streamlines for a perfect fluid is given in Fig. 1.9. It follows at once from considerations of symmetry that the resultant force in the direction of motion (drag) is equal to zero. The pressure distribution according to the theory of frictionless motion is given in Fig. 1.10, together with the results of measurements at three values of the Reynolds number. At the leading edge, all measured pressure distributions agree, to a certain extent, with that for a perfect fluid. At the trailing end, the discrepancy between theory and measurement becomes large because of the large drag of a circular cylinder. The pressure distribution at the lowest, subcritical Reynolds number $R = 1{\cdot}9 \times 10^5$ differs most from that given by potential theory. The measurements corresponding to the two largest Reynolds numbers, $R = 6{\cdot}7 \times 10^5$ and $R = 8{\cdot}4 \times 10^6$, are closer to the potential curve than those performed at the lowest Reynolds number. The large variation of pressure distribution with Reynolds number will be discussed in detail in the next chapter. A corresponding pressure-distribution curve around a meridian section of a sphere is reproduced in Fig. 1.11. Here, too, measurements show large differences for the two Reynolds numbers, and, again, the smaller Reynolds number lies in the range

---

† *Cf.* e. g. B. H. Lamb: Hydrodynamics, 6th ed., Dover, New York, 1945.

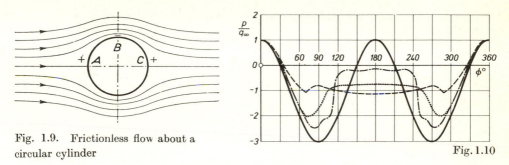

Fig. 1.9. Frictionless flow about a
circular cylinder

Fig. 1.10

Fig. 1.10. Pressure distribution on a circular cylinder in the subcritical and supercritical range of
Reynolds numbers after the measurements of O. Flachsbart [4] and A. Roshko [13]. $q_\infty = \dfrac{1}{2} \varrho\, V^2$
is the stagnation pressure of the oncoming flows

——— frictionless flow
– – – R = 1·9 × 10⁵ } Flachsbart
– · – R = 6·7 × 10⁵ } (1932)
· · · · · R = 8·4 × 10⁶ Roshko (1961)

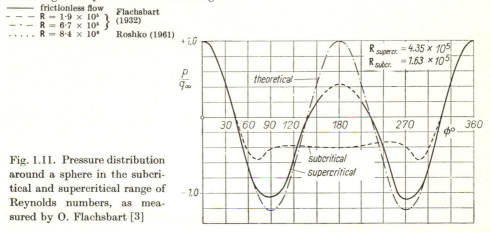

Fig. 1.11. Pressure distribution
around a sphere in the subcritical and supercritical range of
Reynolds numbers, as mea-
sured by O. Flachsbart [3]

of large drag coefficients, whereas the larger value lies in the range of small drag
coefficients, Fig. 1.5. In this case the measured pressure-distribution curve for the
large Reynolds number approximates the theoretical curve of frictionless flow very
well over the greatest part of the circumference.

Considerably better agreement between the theoretical and measured pressure
distribution is obtained for a streamline body in a flow parallel to its axis [5],
Fig. 1.12. Good agreement exists here over almost the whole length of the body,
with the exception of a small region near its trailing end. As will be shown later
this circumstance is a consequence of the gradual pressure increase in the downstream direction.

Although, generally speaking, the theory of perfect fluids does not lead to
useful results as far as drag calculations are concerned, the lift can be calculated from
it very successfully. Fig. 1.13 represents the relation between the lift coefficient and
angle of incidence, as measured by A. Betz [2] in the case of a Zhukovskii aerofoil

of infinite span and provides a comparison with theory. In the range of incidence angles $\alpha = -10°$ to $10°$ the agreement is seen to be good and the small differences can be explained by the influence of friction. The measured and calculated pressure distributions agree very well too, as shown in Fig. 1.14. The discrepancy between theory and measurement displayed in Figs. 1.13 and 1.14 is a consequence of the displacement action of the boundary layer and constitutes a boundary-layer effect of higher order, as will be shown again in Sec. IXj.

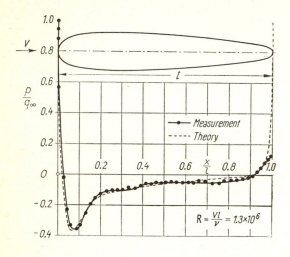

Fig. 1.12. Pressure distribution about a stream-line body of revolution; comparison between theory and measurement, after Fuhrmann [5]

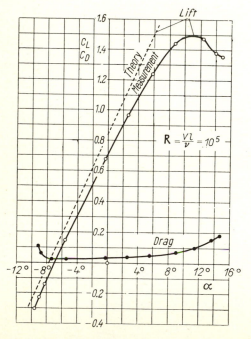

Fig. 1.13. Lift and drag coefficient of a Zhukovskii profile in plane flow, as measured by Betz [2]

Fig. 1.14. Comparison between
the theoretical and measured
pressure distribution for a
Zhukovskii profile at equal lifts,
after A. Betz [2]

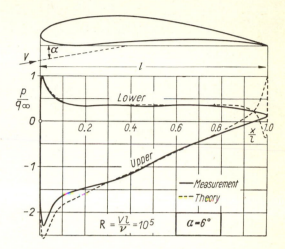

References

[1] Achenbach, E.: Experiments on the flow past spheres at very high Reynolds numbers. JFM *54*, 565—575 (1972).

[1a] Bailey, A.B., and Hiatt, J.: Sphere drag coefficients for a broad range of Mach and Reynolds numbers. AIAA J. *10*, 1436—1440 (1972).

[1b] Bailey, A.B., and Starr, R.F.: Sphere drag at transonic speeds and high Reynolds numbers. AIAA J. *14*, 1631 (1976).

[2] Betz, A.: Untersuchung einer Joukowskischen Tragfläche. ZFM *6*, 173—179 (1915).

[3] Flachsbart, O.: Neuere Untersuchungen über den Luftwiderstand von Kugeln. Phys. Z. *28*, 461—469 (1927).

[4] Flachsbart, O.: Winddruck auf Gasbehälter. Reports of the AVA in Göttingen, IVth Series, 134—138 (1932).

[5] Fuhrmann, G.: Theoretische und experimentelle Untersuchungen an Ballonmodellen. Diss. Göttingen 1910; Jb. Motorluftschiff-Studienges. *V* 63—123 (1911/12).

[6] Hagen, G.: Über die Bewegung des Wassers in engen zylindrischen Röhren. Pogg. Ann. *46*, 423—442 (1839).

[7] Homann, F.: Einfluss grosser Zähigkeit bei Strömung um Zylinder. Forschg. Ing.-Wes. *7*, 1—10 (1936).

[8] Jones, G.W., Cinotta, J.J., and Walker, R.W.: Aerodynamic forces on a stationary and oscillating circular cylinder at high Reynolds numbers. NACA TR R-300 (1969).

[9] Naumann, A.: Luftwiderstand von Kugeln bei hohen Unterschallgeschwindigkeiten. Allgem. Wärmetechnik *4*, 217—221 (1953).

[10] Naumann, A., and Pfeiffer, H.: Über die Grenzschichtströmung am Zylinder bei hohen Geschwindigkeiten. Advances in Aeronautical Sciences (Th. von Kármán, ed.) Vol. *3*, 185—206, London, 1962.

[11] Poiseuille, J.: Récherches expérimentelles sur le mouvement des liquides dans les tubes de très petits diametres. Comptes Rendus *11*, 961—967 and 1041—1048 (1840); *12*, 112—115 (1841); in more detail: Mémoires des Savants Etrangers *9* (1846).

[12] Reynolds, O.: An experimental investigation of the circumstances which determine whether the motion of water shall be direct or sinuous, and of the law of resistance in parallel channels. Phil. Trans. Roy. Soc. *174*, 935—982 (1883) or Scientific Papers II, 51.

[13] Roshko, A.: Experiments on the flow past a circular cylinder at very high Reynolds numbers. JFM *10*, 345—356 (1961); see also: On the aerodynamic drag of cylinders at high Reynolds numbers. Paper presented at the US Japan Research Seminar on Wind Loads on Structures, Univ. of Hawaii, Oct. 1970.

[14] Taneda, S.: Experimental investigation of the wakes behind cylinders and plates at low Reynolds numbers. J. Phys. Soc. Japan *11*, 302—307 (1956).

CHAPTER II

Outline of boundary-layer theory

a. The boundary-layer concept

In the case of fluid motions for which the measured pressure distribution nearly agrees with the perfect-fluid theory, such as the flow past the streamline body in Fig. 1.12, or the aerofoil in Fig. 1.14, the influence of viscosity at high Reynolds numbers is confined to a very thin layer in the immediate neighbourhood of the solid wall. If the condition of no slip were not to be satisfied in the case of a real fluid there would be no appreciable difference between the field of flow of the real fluid as compared with that of a perfect fluid. The fact that at the wall the fluid adheres to it means, however, that frictional forces retard the motion of the fluid in a thin layer near the wall. In that thin layer the velocity of the fluid increases from zero at the wall (no slip) to its full value which corresponds to external frictionless flow. The layer under consideration is called the *boundary layer*, and the concept is due to L. Prandtl [25].

Fig. 2.1. Motion along a thin flat plate, from Prandtl-Tietjens

l = length of plate;
Reynolds number $R = Vl/\nu = 3$

Figure 2.1 reproduces a picture of the motion of water along a thin flat plate in which the streamlines were made visible by the sprinkling of particles on the surface of the water. The traces left by the particles are proportional to the velocity of flow. It is seen that there is a very thin layer near the wall in which the velocity is considerably smaller than at a larger distance from it. The thickness of this boundary layer increases along the plate in a downstream direction. Fig. 2.2 represents diagrammatically the velocity distribution in such a boundary layer at the

plate, with the dimensions across it considerably exaggerated. In front of the
leading edge of the plate the velocity distribution is uniform. With increasing distance
from the leading edge in the downstream direction the thickness, δ, of the retarded
layer increases continuously, as increasing quantities of fluid become affected.
Evidently the thickness of the boundary layer decreases with decreasing viscosity.

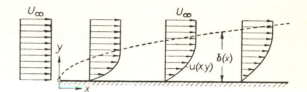

Fig. 2.2. Sketch of boundary
layer on a flat plate in parallel flow at zero incidence

On the other hand, even with very small viscosities (large Reynolds numbers) the
frictional shearing stresses $\tau = \mu \, \partial u / \partial y$ in the boundary layer are considerable
because of the large velocity gradient across the flow, whereas outside the boundary
layer they are very small. This physical picture suggests that the field of flow in the
case of fluids of small viscosity can be divided, for the purpose of mathematical
analysis, into two regions: the thin boundary layer near the wall, in which friction
must be taken into account, and the region outside the boundary layer, where the
forces due to friction are small and may be neglected, and where, therefore, the
perfect-fluid theory offers a very good approximation. Such a division of the field
of flow, as we shall see in more detail later, brings about a considerable simplification
of the mathematical theory of the motion of fluids of low viscosity. In fact, the
theoretical study of such motions was only made possible by Prandtl when he
introduced this concept.

    We now propose to explain the basic concepts of boundary-layer theory with
the aid of purely physical ideas and without the use of mathematics. The mathematical boundary-layer theory which forms the main topic of this book will be discussed
in the following chapters.

    The decelerated fluid particles in the boundary layer do not, in all cases, remain
in the thin layer which adheres to the body along the whole wetted length of the
wall. In some cases the boundary layer increases its thickness considerably in the
downstream direction and the flow in the boundary layer becomes reversed. This
causes the decelerated fluid particles to be forced outwards, which means that
the boundary layer is separated from the wall. We then speak of *boundary-layer
separation*. This phenomenon is always associated with the formation of vortices
and with large energy losses in the wake of the body. It occurs primarily near blunt
bodies, such as circular cylinders and spheres. Behind such a body there exists a region
of strongly decelerated flow (so-called wake), in which the pressure distribution
deviates considerably from that in a frictionless fluid, as seen from Figs. 1.10 and 1.11
in the respective cases of a cylinder and a sphere. The large drag of such bodies can
be explained by the existence of this large deviation in pressure distribution, which
is, in turn, a consequence of boundary-layer separation.

Estimation of boundary-layer thickness: The thickness of a boundary layer which has not separated can be easily estimated in the following way. Whereas friction forces can be neglected with respect to inertia forces outside the boundary layer, owing to low viscosity, they are of a comparable order of magnitude inside it. The inertia force per unit volume is, as explained in Section I e, equal to $\varrho\, u\, \partial u/\partial x$. For a plate of length l the gradient $\partial u/\partial x$ is proportional to U/l, where U denotes the velocity outside the boundary layer. Hence the inertia force is of the order $\varrho\, U^2/l$. On the other hand the friction force per unit volume is equal to $\partial\tau/\partial y$, which, on the assumption of laminar flow, is equal to $\mu\, \partial^2 u/\partial y^2$. The velocity gradient $\partial u/\partial y$ in a direction perpendicular to the wall is of the order U/δ so that the friction force per unit volume is $\partial\tau/\partial y \sim \mu\, U/\delta^2$. From the condition of equality of the friction and inertia forces the following relation is obtained:

$$\mu\, \frac{U}{\delta^2} \sim \frac{\varrho\, U^2}{l}$$

or, solving for the boundary-layer thickness δ†:

$$\delta \sim \sqrt{\frac{\mu\, l}{\varrho\, U}} = \sqrt{\frac{\nu\, l}{U}}\,. \tag{2.1}$$

The numerical factor which is, so far, still undetermined will be deduced later (Chap. VII) from the exact solution given by H. Blasius [4], and it will turn out that it is equal to 5, approximately. Hence for *laminar* flow in the boundary layer we have

$$\delta = 5\,\sqrt{\frac{\nu\, l}{U}}\,. \tag{2.1a}$$

The dimensionless boundary-layer thickness, referred to the length of the plate, l, becomes:

$$\frac{\delta}{l} = 5\,\sqrt{\frac{\nu}{U\, l}} = \frac{5}{\sqrt{R_l}}\,, \tag{2.2}$$

where R_l denotes the Reynolds number related to the length of the plate, l. It is seen from eqn. (2.1) that the boundary-layer thickness is proportional to $\sqrt{\nu}$ and to $\sqrt{l}$. If l is replaced by the variable distance x from the leading edge of the plate, it is seen that δ increases proportionately to $\sqrt{x}$. On the other hand the relative boundary-layer thickness δ/l decreases with increasing Reynolds number as $1/\sqrt{R}$ so that in the limiting case of frictionless flow, with $R \to \infty$, the boundary-layer thickness vanishes.

We are now in a position to estimate the shearing stress τ_0 on the wall, and consequently, the total drag. According to Newton's law of friction (1.2) we have

$$\tau_0 = \mu\, \left(\frac{\partial u}{\partial y}\right)_0,$$

---

† A more rigorous definition of boundary-layer thickness is given at the end of this section.

where subscript 0 denotes the value at the wall, i. e. for $y = 0$. With the estimate $(\partial u/\partial y)_0 \sim U/\delta$ we obtain $\tau_0 \sim \mu\, U/\delta$ and, inserting the value of δ from eqn. (2.1), we have

$$\tau_0 \sim \mu\, U\, \sqrt{\frac{\varrho\, U}{\mu\, l}} = \sqrt{\frac{\mu\, \varrho\, U^3}{l}}\, . \tag{2.3}$$

Thus the frictional stress near the wall is proportional to $U^{3/2}$.

We can now form a dimensionless stress with reference to $\varrho\, U^2$, as explained in Chap. I, and obtain

$$\frac{\tau_0}{\varrho\, U^2} \sim \sqrt{\frac{\mu}{\varrho\, U\, l}} = \frac{1}{\sqrt{R_l}}\, . \tag{2.3a}$$

This result agrees with the dimensional analysis in Chap. I, which predicted that the dimensionless shearing stress could depend on the Reynolds number only.

The total drag D on the plate is equal to $bl\tau_0$ where b denotes the width of the plate. Hence, with the aid of eqn. (2.3) we obtain

$$D \sim b\, \sqrt{\varrho\, \mu\, U^3\, l}\, . \tag{2.4}$$

The laminar frictional drag is thus seen to be proportional to $U^{3/2}$ and $l^{1/2}$. Proportionality to $l^{1/2}$ means that doubling the plate length does not double the drag, and this result can be understood by considering that the downstream part of the plate experiences a smaller drag than the leading portion because the boundary layer is thicker towards the trailing edge. Finally, we can write down an expression for the dimensionless drag coefficient in accordance with eqn. (1.14) in which the reference area A will be replaced by the wetted area bl. Hence eqn. (2.4) gives that

$$C_D \sim \sqrt{\frac{\mu}{\varrho\, U\, l}} = \frac{1}{\sqrt{R_l}}\, . $$

The numerical factor follows from H. Blasius's exact solution, and is 1·328, so that the drag of a plate in parallel laminar flow becomes

$$C_D = \frac{1{,}328}{\sqrt{R_l}}\, . \tag{2.5}$$

The following numerical example will serve to illustrate the preceding estimation: Laminar flow, stipulated here, is obtained, as is known from experiment, for Reynolds numbers Ul/ν not exceeding about 5×10^5 to 10^6. For larger Reynolds numbers the boundary layer becomes turbulent. We shall now calculate the boundary-layer thickness for the flow of air ($\nu = 0{\cdot}144 \times 10^{-3}$ ft^2/sec) at the end of a plate of length $l = 3$ ft at a velocity $U = 48$ ft/sec. This gives $R_l = Ul/\nu = 10^6$ and from eqn. (2.2)

$$\frac{\delta}{l} = \frac{5}{10^3} = 0{\cdot}005; \quad \delta = 0{\cdot}18 \text{ in }.$$

The drag coefficient from eqn. (2.5) is $C_D = 0{\cdot}0013$ i. e. exceedingly small when compared with that for a circular cylinder, Fig. 1.4, because the drag coefficient for a cylinder also includes pressure forces.

Definition of boundary-layer thickness: The definition of the boundary-layer thickness is to a certain extent arbitrary because transition from the velocity in the boundary to that outside it takes place asymptotically. This is, however, of no practical importance, because the velocity in the boundary layer attains a value which is very close to the external velocity already at a small distance from the wall. It is possible to define the boundary-layer thickness as that distance from the wall where the velocity differs by 1 per cent from the external velocity. With this definition the numerical factor in eqn. (2.2) has the value 5. Instead of the boundary-layer thickness, another quantity, the *displacement thickness* δ_1, is sometimes used, Fig. 2.3. It is defined by the equation

$$U \delta_1 = \int_0^\infty (U - u)\, \mathrm{d}y \, . \tag{2.6}$$

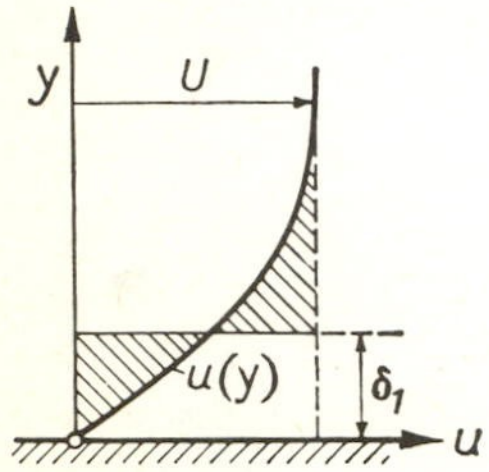

Fig. 2.3.   Displacement thickness δ_1 in a boundary layer

The displacement thickness indicates the distance by which the external stream-lines are shifted owing to the formation of the boundary layer. In the case of a plate in parallel flow and at zero incidence the displacement thickness is about $\frac{1}{3}$ of the boundary-layer thickness δ given in eqn. (2.1a).

b. Separation and vortex formation

The boundary layer near a flat plate in parallel flow and at zero incidence is particularly simple, because the static pressure remains constant in the whole field of flow. Since outside the boundary layer the velocity remains constant the same applies to the pressure because in the frictionless flow Bernoulli's equation remains valid. Furthermore, the pressure remains sensibly constant over the width of the boundary layer at a given distance x. Hence the pressure over the width of the boundary layer has the same magnitude as outside the boundary layer at the same distance, and the same applies to cases of arbitrary body shapes when the pressure outside the boundary layer varies along the wall with the length of arc. This fact is expressed by saying that the external pressure is "impressed" on the boundary layer. Hence in the case of the motion past a plate the pressure remains constant throughout the boundary layer.

The phenomenon of boundary-layer separation mentioned previously is intimately connected with the pressure distribution in the boundary layer. In the boundary layer on a plate no separation takes place as no back-flow occurs.

In order to explain the very important phenomenon of boundary-layer separation let us consider the flow about a blunt body, e. g. about a circular cylinder, as shown in Fig. 2.4. In frictionless flow, the fluid particles are accelerated on the upstream half from D to E, and decelerated on the downstream half from E to F. Hence the pressure decreases from D to E and increases from E to F. When the flow is started up the motion in the first instant is very nearly frictionless, and remains so as long as the boundary layer remains thin. Outside the boundary layer there is a transformation of pressure into kinetic energy along DE, the reverse taking place along EF, so that a particle arrives at F with the same velocity as it had at D. A fluid particle which moves in the immediate vicinity of the wall in the boundary layer remains under the influence of the same pressure field as that existing outside, because the external pressure is impressed on the boundary layer. Owing to the large friction forces in the thin boundary layer such a particle consumes so much of its kinetic

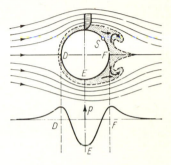

Fig. 2.4. Boundary-layer separation and vortex formation on a
circular cylinder (diagrammatic)

S = point of separation

energy on its path from D to E that the remainder is too small to surmount the "pressure hill" from E to F. Such a particle cannot move far into the region of increasing pressure between E and F and its motion is, eventually, arrested. The external pressure causes it then to move in the opposite direction. The photographs reproduced in Fig. 2.5 illustrate the sequence of events near the downstream side of a round body when a fluid flow is started. The pressure increases along the body contour from left to right, the flow having been made visible by sprinkling aluminium dust on the surface of the water. The boundary layer can be easily recognized by reference to the short traces. In Fig. 2.5a, taken shortly after the start of the motion, the reverse motion has just begun. In Fig. 2.5b the reverse motion has penetrated a considerable distance forward and the boundary layer has thickened appreciably. Fig. 2.5c shows how this reverse motion gives rise to a vortex, whose size is increased still further in Fig. 2.5d. The vortex becomes separated shortly afterwards and moves downstream in the fluid. This circumstance changes completely the field of flow in the wake, and the pressure distribution suffers a radical change, as compared with frictionless flow. The final state of motion can be inferred from Fig. 2.6. In the eddying region behind the cylinder there is considerable suction, as seen from the pressure distribution curve in Fig. 1.10. This suction causes a large pressure drag on the body.

At a larger distance from the body it is possible to discern a regular pattern of vortices which move alternately clockwise and counterclockwise, and which is known as a Kármán vortex street [20], Fig. 2.7 (see also Fig. 1.6). In Fig. 2.6 a vortex moving in a clockwise direction can be seen to be about to detach itself from the body before joining the pattern. In a further paper, von Kármán [21] proved that such vortices are generally unstable with respect to small disturbances parallel

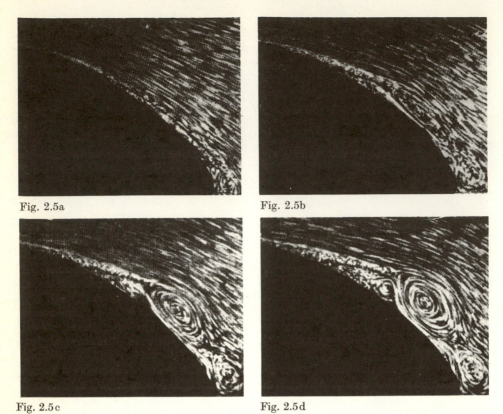

Fig. 2.5a

Fig. 2.5b

Fig. 2.5c

Fig. 2.5d

Fig. 2.5a, b, c, d.   Development of boundary-layer separation with time, after Prandtl-Tietjens [27]. See also Fig. 15.5

to themselves. The only arrangement which shows neutral equilibrium is that with $h/l = 0\cdot281$ (Fig. 2.8). The vortex street moves with a velocity u, which is smaller than the flow velocity U in front of the body. It can be regarded as a highly idealized picture of the motion in the wake of the body. The kinetic energy contained in the velocity field of the vortex street must be continually created, as the body moves through the fluid. On the basis of this representation it is possible to deduce an expression for the drag from the perfect-fluid theory. Its magnitude per unit length of the cylindrical body is given by

$$D = \varrho\, U^2\, h \left[2\cdot83\, \frac{u}{U} - 1\cdot12 \left(\frac{u}{U} \right)^2 \right].$$

The width h, and the velocity ratio u/U must be known from experiment.

More recent experimental investigations due to W. W. Durgin and others [13] established that in an accelerating vortex street the ratio of the longitudinal to the transverse spacing of the vortices changes considerably. As a result, the regular arrangement of vortices is transformed into a turbulent wake.

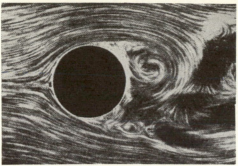

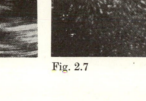

Fig. 2.6   Fig. 2.7

Fig. 2.6. Instantaneous photograph of flow with complete boundary-layer separation in the wake of a circular cylinder, after Prandtl-Tietjens [27]

Fig. 2.7. Kármán vortex street, from A. Timme [38]

Fig. 2.8. Streamlines in a vortex street $(h/l = 0.28)$. The fluid is at rest at infinity, and the vortex street moves to the left

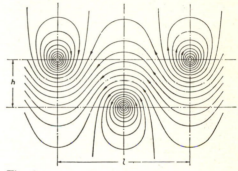

Fig. 2.8

Circular cylinder. The frequency with which vortices are shed in a Kármán vortex street behind a circular cylinder was first extensively measured by H. Blenk, D. Fuchs and L. Liebers [5]. A regular Kármán street is observed only in the range of Reynolds numbers VD/ν from about 60 to 5000. At lower Reynolds numbers the wake is laminar and has the form visible in the first two photographs of Fig. 1.6; at higher Reynolds numbers there is complete turbulent mixing. Measurements show that in the regular range given above, the dimensionless frequency,

$$\frac{nD}{V} = \mathsf{S}, \qquad \text{(Strouhal number)}$$

also known as the Strouhal number [37], depends only on the Reynolds number. This relationship is shown plotted in Fig. 2.9 which is based on measurements performed by A. Roshko [32]; see also [15]. The experimental points which were obtained with cylinders of different diameters D and at different velocities V arrange themselves well on a single curve. At the higher Reynolds numbers the Strouhal number remains approximately constant at $\mathsf{S} = 0.21$. This value of S, as seen from Fig. 2.9, prevails up to a Reynolds number $\mathsf{R} = 2 \times 10^5$, that is in the subcritical range (see also Fig. 1.4). At higher Reynolds numbers, say around $\mathsf{R} = 10^6$, a regular vortex street does not exist. According to A. Roshko [31], such a regular street re-appears at extremely large Reynolds numbers ($\mathsf{R} > 3 \times 10^6$) when the Strouhal

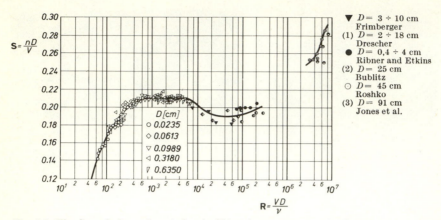

Fig. 2.9. The Strouhal number, S, for the Kármán vortex street in the flow past a circular cylinder in terms of the Reynolds number, R. Measurements performed by A. Roshko [31, 32], H. S. Ribner, B. Etkins and K. K. Nelly [30], E. F. Relf and L. F. G. Simmons [28] as well as G. W. Jones et al. ([8] of Chap. I). In the range $R = 3 \times 10^5$ to 3×10^6 (supercritical regime with very low drag, Fig 1.4) the Kármán vortex street is no loger regular. It is only at $R > 4 \times 10^6$ that a regular pattern forms again; its Strouhal number is now higher at $S = 0.26$ to 0.30 compared with S ≈ 0.20 at $R = 10^3$ to 3×10^5

number assumes values around $S = 0.27$. In this connexion the paper by P. W. Bearman [3a] may also be consulted. When the diameters of the cylinders are small and the velocities are moderate, the resulting frequencies lie in the acoustic range. For example, the familiar "aeolian tones" emited by telegraph wires are the result of these phenomena. At a velocity of $V = 10$ m/sec (30·48 ft/sec) and a wire of 2 mm (0·079 in) in diameter, the frequency becomes $n = 0.21 \, (10/0.002) = 1050$ sec^{-1}, and the corresponding Reynolds number $R \approx 1200$.

Flat plate at zero incidence. The fact that a regular vortex street establishes itself, among others, behind slender bodies as well as in compressible streams has only been established recently by H. J. Heinemann et al. [18]. The photograph of Fig. 2.10 shows such a regular vortex street behind a flat plate at zero incidence for a Mach number $M_\infty = 0.61$. The diagram in Fig. 2.11 contains a plot of the Strouhal number, $S = nd/V$, formed with the plate thickness, d, in terms of the Mach number, but only for the subsonic range $M = 0.2$ to 0.85. The diagram proves that here too $S \approx 0.2$, as was the case for the circular cylinder in Fig. 2.9. The corresponding Reynolds numbers, referred to the length of the plate, are in the range $R = Vl/\nu = 3 \times 10^5$ to 8×10^5 in which the flow is laminar.

Two papers by C. C. Lin. [22] and U. Domm [11] concern themselves with the theory of the Kármán vortex street. The formation of a vortex pair behind a flat plate in cross-flow at right angles to it has been investigated theoretically by E. Wedemeyer [38a], whereas T. Sarpkaya [33b] conducted theoretical and experimental studies for a plate arranged at a large angle of attack (see Fig. 4.2); in this connexion

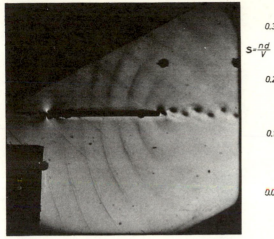

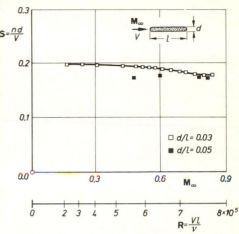

Fig. 2.10. Von Kármán vortex street behind a flat plate at zero incidence at a Mach number $M = 0.61$ and a Reynolds number $R = Vl/\nu = 6.5 \times 10^5$ after H. J. Heinemann et al. [18]. Length of plate $l = 60$ mm, thickness ratio $d/l = 0.05$. Exposure time approx. 20 nanosec $(20 \times 10^{-9}$ sec)

Fig. 2.11. Strouhal number $S = nd/V$ in terms of the Mach number for the vortex street behind a flat plate at zero incidence, after H. J. Heinemann et al. [18]

an earlier paper by L. Rosenhead [32a] may also be consulted. The reader may also be interested to look up the text of a remark made by L. Prandtl on the occasion of a lecture by K. Friedrichs („Bemerkung über die ideale Strömung um einen Körper bei verschwindender Zähigkeit" Lectures on aerodynamics and allied subjects, Aachen 1929, Springer, Berlin 1930, pp. 51, 52).

Separation. The boundary-layer theory succeeds in this manner, i.e. with the aid of the explanation of the phenomenon of separation, in throwing light on the occurrence of pressure or form drag in addition to viscous drag. The danger of boundary-layer separation exists always in regions with an adverse pressure gradient and the likelihood of its occurrence increases in the case of steep pressure curves, i.e. behind bodies with blunt ends. The preceding argument explains also why the experimental pressure distribution shown in Fig. 1.11 for the case of a slender streamline body differs so little from that predicted for frictionless flow. The pressure increase in the downstream direction is here so gradual that there is no separation. Consequently, there is no appreciable pressure drag and the total drag consists mainly of viscous drag and is, therefore, small.

The streamlines in the boundary layer near separation are shown diagrammatically in Fig. 2.12. Owing to the reversal of the flow there is a considerable thickening of the boundary layer, and associated with it, there is a flow of boundary-layer material into the outside region. At the point of separation one streamline inter-

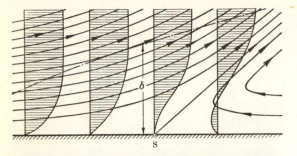

S = point of separation

Fig. 2.12. Diagrammatic representation of flow in the boundary layer near a point of separation

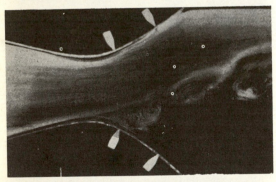

Fig. 2.13. Flow with separation in a highly divergent channel, from Prandtl-Tietjens [27]

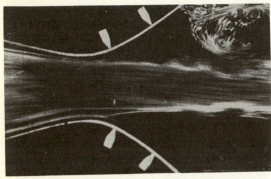

Fig. 2.14. Flow with boundary-layer suction on upper wall of highly divergent channel

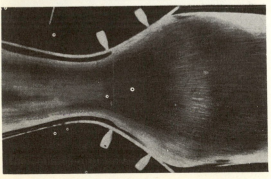

Fig. 2.15. Flow with boundary-layer suction on both walls of highly divergent channel

sects the wall at a definite angle, and the point of separation itself is determined by the condition that the velocity gradient normal to the wall vanishes there:

$$\left(\frac{\partial u}{\partial y}\right)_{wall} = 0 \quad \text{(separation)} . \tag{2.7}$$

The precise location of the point of separation can be determined only with the aid of an exact calculation, i. e. by the integration of the boundary-layer equations.

Separation, as described for the case of a circular cylinder, can also occur in a highly divergent channel, Fig. 2.13. In front of the throat the pressure decreases in the direction of flow, and the flow adheres completely to the walls, as in a frictionless fluid. However, behind the throat the divergence of the channel is so large that the boundary layer becomes separated from both walls, and vortices are formed. The stream fills now only a small portion of the cross-sectional area of the channel. However, separation is prevented if boundary-layer suction is applied at the wall (Figs. 2.14 and 2.15).

The photographs in Figs. 2.16 and 2.17† prove that the adverse pressure gradient together with friction near the wall determine the process of separation which is independent of such other circumstance as e. g. the curvature of the wall. The first picture shows the motion of a fluid against a wall at right angles to it (plane stagnation flow). Along the streamline in the plane of symmetry which leads to the stagnation point there is a considerable pressure increase in the direction of flow. No separation, however, occurs, because no wall friction is present. There is no separation near the wall, either, because here the flow in the boundary layer takes place in the direction of decreasing pressure on both sides of the plane of symmetry. If now a thin wall is placed along the plane of symmetry at right angles to the first wall, Fig. 2.17, the new boundary layer will show a pressure increase in the direction of flow. Consequently, separation now occurs near the plane wall. The incidence of separation is often rather sensitive to small changes in the shape of the solid body, particularly when the pressure distribution is strongly affected by this change in shape. A very instructive example is given in the pictures of Fig. 2.18 which show photographs of the flow field about a model of a motor vehicle (the Volkswagen delivery van), [23, 35]. When the nose was flat giving it an angular shape (a), the flow past the fairly sharp corners in front caused large suction followed by a large pressure increase along the side walls. This led to complete separation and to the formation of a wide wake behind the body. The drag coefficient of the vehicle with this angular shape had a value of $C_D = 0.76$. The large suction near the front end and the separation along the side walls were eliminated when the shape was changed by adding the round nose shown at (b). Simultaneously, the drag coefficient became markedly smaller and had a value of $C_D = 0.42$. Further research on such vehicles have been performed by W. H. Hucho [19] for the case of a non-symmetric stream.

---

† Fig. 2.16. and 2.17. have been taken from the paper "Strömungen in Dampfkesselanlagen" by H. Foettinger, Mitteilungen der Vereinigung der Groß-Kesselbesitzer, No. 73, p. 151 (1939).

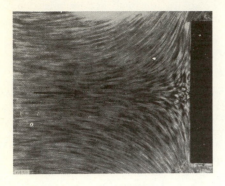 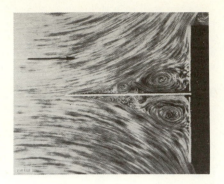

Fig. 2.14. Free stagnation flow without separation, as photographed by Foettinger

Fig. 2.15. Decelerated stagnation flow with separation, as photographed by Foettinger

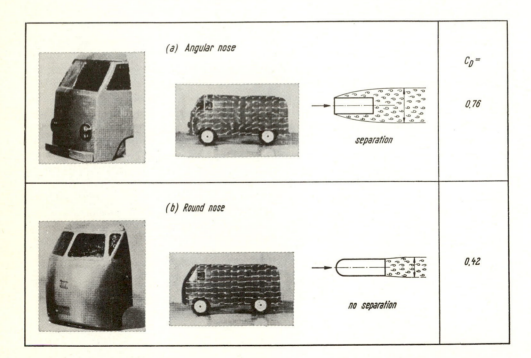

Fig. 2.16. Flow about a model of a motor vehicle (Volkswagen delivery van), after E. Moeller [23]. a) Angular nose with separated flow along the whole of the side wall and large drag coefficient ($C_D = 0\cdot76$); b) Round nose with no separation and small drag coefficient ($C_D = 0\cdot42$)

Separation is also important for the lifting properties of an aerofoil. At small incidence angles (up to about 10°) the flow does not separate on either side and closely approximates frictionless conditions. The pressure distribution for such a case ("sound" flow, Fig. 2.19a) was given in Fig. 1.14. With increasing incidence there is danger of separation on the suction side of the aerofoil, because the pressure increase becomes steeper. For a given angle of incidence, which is about 15°, separation finally occurs. The separation point is located fairly closely behind the leading edge. The wake, Fig. 2.19b, shows a large "dead-water" area. The frictionless, lift-creating flow pattern has become disturbed, and the drag has become very large. The beginning of separation nearly coincides with the occurrence of maximum lift of the aerofoil.

Structural aerodynamics. Flow around land-based bluff bodies, such as structures and buildings, is considerably more complex than flow around streamlined bodies and aircraft. The principal cause of complication is the presence of the ground and the shear created in the turbulent wind as a consequence. The interaction between the incident shear flow and the structure produces coexisting static and dynamic loads [8, 9, 10]. The fluctuating forces produced by vortex formation and shedding can induce oscillations in the structures at their natural frequencies.

The flow patterns observed on a detached rectangular building is shown schematically in Fig. 2.20. In front of the building there appears a bound vortex which arises from the interaction of the boundary layer in the sheared flow ($dV/dz > 0$) and the ground. There is, furthermore, strong vortex shedding from the sharp corners of the building and a complex wake is created behind it. So far no theoretical methods have been developed to cope with this extremely complicated flow pattern. It is, therefore, necessary to resort to wind-tunnel studies with the aid of adequately scaled models.

a)

b)

Fig. 2.19a.b.   Flow around an aerofoil,
after Prandtl-Tietjens [27]. a) 'sound' flow,
b) flow with separation

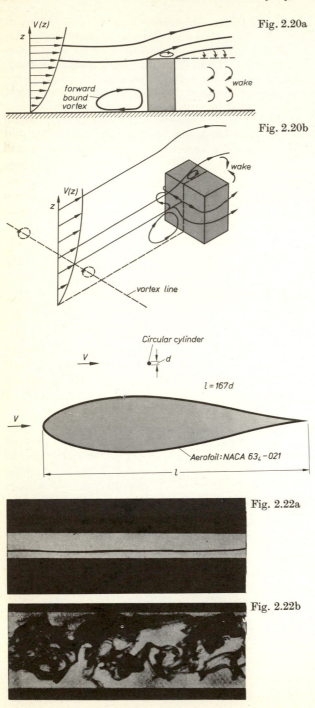

Fig. 2.20a

Fig. 2.20b

Fig. 2.21.

Fig. 2.22a

Fig. 2.22b

Fig. 2.20. Overall view of flow pattern (schematic) around a rectangular structure [34]. a) Side view with forward bound vortex in the stagnation zone and a separated roof boundary layer; b) upwind face and vortex shedding from the the windward corner of the roof

Fig. 2.21. Aerofoil and circular cylinder drawn in such relation to each other as to produce the same drag in parallel flows (parallel to axis of symmetry of aerofoil) of the same velocity. *Aerofoil:* Laminar aerofoil NACA $63_4 - 021$ with laminar boundary layer. Drag coefficient $c_{D_0} = 0\cdot006$ at $R_l = 10^6$ to 10^7, Fig. 17.14. *Circular cylinder:* Drag coefficient $c_D = 1\cdot0$ at $R_d = 10^4$ to 10^5; Fig. 1.4. Thus the ratio of the chord of the aerofoil, l, to the diameter, d, of the cylinder is $l/d = 1\cdot0/0\cdot006 = 167$

Fig. 2.22. The Reynolds dye experiment. Flow in water made visible by the injection of a dye, after W. Dubs [12]; a) laminar flow, $R = 1150$; b) turbulent flow, $R = 2520$

To conclude this section, we wish to discuss a particularly telling example of how effectively it is possible to reduce the drag of a body in a stream when the separation of the boundary layer is completely eliminated and when, in addition, the body itself is given a shape which is conducive to low resistance. Fig 2.21 illustrates the effect of a favorable shape (streamline body) on drag: a symmetric aerofoil and a circular cylinder (thin wire) have been drawn here to a relative scale which assures equal drag in streams of equal velocity. The cylinder has a drag coefficient $C_D \approx 1$ with respect to its frontal area (see also Fig. 1.4). On the other hand, the drag coefficient of the aerofoil, referred to its cross-sectional area, has the very low value of $C_D = 0{\cdot}006$. The extremely low drag of the aerofoil is achieved as a result of a carefully chosen profile which assures that the boundary layer remains laminar over almost the whole of its wetted length (laminar aerofoil). In this connexion, Chap. XVII and, especially, Fig. 17.14, should be consulted.

c. Turbulent flow in a pipe and in a boundary layer

Measurements show that the type of motion through a circular pipe which was calculated in Section Id, and in which the velocity distribution was parabolic, exists only at low and moderate Reynolds numbers. The fact that in the laminar motion under discussion fluid laminae slide over each other, and that there are no radial velocity components, so that the pressure drop is proportional to the first power of the mean flow velocity, constitutes an essential characteristic of this type of flow. This characteristic of the motion can be made clearly visible by introducing a dye into the stream and by discharging it through a thin tube, Fig. 2.22. At the moderate Reynolds numbers associated with laminar flow the dye is visible in the form of a clearly defined thread extending over the whole length of the pipe, Fig. 2.22a. By increasing the flow velocity it is possible to reach a stage when the fluid particles cease to move along straight lines and the regularity of the motion breaks down. The coloured thread becomes mixed with the fluid, its sharp outline becomes blurred and eventually the whole cross-section becomes coloured, Fig. 2.22b. On the axial motion there are now superimposed irregular radial fluctuations which effect the mixing. Such a flow pattern is called *turbulent*. The dye experiment was first carried out by O. Reynolds [29], who ascertained that the transition from the laminar to the turbulent type of motion takes place at a definite value of the Reynolds number (critical Reynolds number). The actual value of the critical Reynolds number depends further on the details of the experimental arrangement, in particular on the amount of disturbance suffered by the fluid before entering the pipe. With an arrangement which is as free from disturbances as possible critical Reynolds numbers $(\bar{u}d/\nu)_{crit}$ exceeding 10^4 can be attained ($\bar{u}$ = denotes the mean velocity averaged over the cross-sectional area). With a sharp-edged entrance the critical Reynolds number becomes approximately

$$\left(\frac{\bar{u}\,d}{\nu}\right)_{crit} = \mathsf{R}_{crit} \approx 2300 \quad \text{(pipe)}\,. \tag{2.8}$$

This value can be regarded as the lower limit for the critical Reynolds number below which even strong disturbances do not cause the flow to become turbulent.

In the turbulent region the pressure drop becomes approximately proportional to the square of the mean flow velocity. In this case a considerably larger pressure difference is required in order to pass a fixed quantity of fluid through the pipe, as compared with laminar flow. This follows from the fact that the phenomenon of turbulent mixing dissipates a large quantity of energy which causes the resistance to flow to increase considerably. Furthermore, in the case of turbulent flow the velocity distribution over the cross-sectional area is much more even than in laminar flow. This circumstance is also to be explained by turbulent mixing which causes an exchange of momentum between the layers near the axis of the tube and those near the walls. Most pipe flows which are encountered in engineering appliances occur at such high Reynolds numbers that turbulent motion prevails as a rule. The laws of turbulent motion through pipes will be discussed in detail in Chap. XX.

In a way which is similar to the motion through a pipe, the flow in a boundary layer along a wall also becomes turbulent when the external velocity is sufficiently large. Experimental investigations into the transition from laminar to turbulent flow in the boundary layer were first carried out by J. M. Burgers [6] and B. G. van der Hegge Zijnen [17] as well as by M. Hansen [16]. The transition from laminar to turbulent flow in the boundary layer becomes most clearly discernible by a sudden and large increase in the boundary-layer thickness and in the shearing stress near the wall. According to eqn. (2.1), with l replaced by the current coordinate x, the dimensionless boundary-layer thickness $\delta/\sqrt{\nu\,x/U_\infty}$ becomes constant for laminar flow, and is, as seen from eqn. (2.1a), approximately equal to 5. Fig. 2.23 contains a plot of this dimensionless boundary-layer thickness against the Reynolds number $U_\infty\,x/\nu$. At $R_x > 3\cdot2 \times 10^5$ a very sharp increase is clearly visible, and

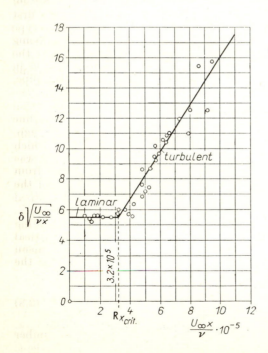

Fig. 2.23. Boundary-layer thickness plotted against the Reynolds number based on the current length x along a plate in parallel flow at zero incidence, as measured by Hansen [16]

an identical phenomenon is observed in a plot of wall shearing stress. The sudden increase in these quantities denotes that the flow has changed from laminar to turbulent. The Reynolds number R_x based on the current length x is related to the Reynolds number $R_\delta = U_\infty\,\delta/\nu$ based on the boundary-layer thickness through the equation

$$R_\delta = 5\,\sqrt{R_x}\;.$$

as seen from eqn. (2.1a). Hence to the critical Reynolds number

$$R_{x\ crit} = \left(\frac{U_\infty\,x}{\nu}\right)_{crit} = 3{\cdot}2\cdot 10^5 \quad (\text{plate})$$

there corresponds $R_{\delta\ crit} \approx 2800$. The boundary layer on a plate is laminar near the leading edge and becomes turbulent further downstream. The abscissa x_{crit} of the point of transition can be determined from the known value of $R_{x\ crit}$. In the case of a plate, as in the previously discussed pipe flow, the numerical value of R_{crit} depends to a marked degree on the amount of disturbance in the external flow, and the value $R_{x\ crit} = 3{\cdot}2 \times 10^5$ should be regarded as a lower limit. With exceptionally disturbance-free external flow, values of $R_{x\ crit} = 10^6$ and higher have been attained.

A particularly remarkable phenomenon connected with the transition from laminar to turbulent flow occurs in the case of blunt bodies, such as circular cylinders or spheres. It will be seen from Figs. 1.4 and 1.5 that the drag coefficient of a circular cylinder or a sphere suffers a sudden and considerable decrease near Reynolds numbers $V\,D/\nu$ of about 5×10^5 or 3×10^5 respectively. This fact was first observed on spheres by G. Eiffel [14]. It is a consequence of transition which causes the point of separation to move downstream, because, in the case of a turbulent boundary layer, the accelerating influence of the external flow extends further due to turbulent mixing. Hence the point of separation which lies near the equator for a laminar boundary layer moves over a considerable distance in the downstream direction. In turn, the dead area decreases considerably, and the pressure distribution becomes more like that for frictionless motion (Fig. 1.11). The decrease in the dead-water region considerably reduces the pressure drag, and that shows itself as a jump in the curve $C_D = f(R)$. L. Prandtl [26] proved the correctness of the preceding reasoning by mounting a thin wire ring at a short distance in front of the equator of a sphere. This causes the boundary layer to become artificially turbulent at a lower Reynolds number and the decrease in the drag coefficient takes place earlier than would otherwise be the case. Figs. 2.24 and 2.25 reproduce photographs of flows which have been made visible by smoke. They represent the subcritical pattern with a large value of the drag coefficient and the supercritical pattern with a small dead-water area and a small value of the drag coefficient. The supercritical pattern was achieved with Prandtl's tripping wire. The preceding experiment shows in a convincing manner that the jump in the drag curve of a circular cylinder and sphere can only be interpreted as a boundary-layer phenomenon. Other bodies with a blunt or rounded stern, (e. g. elliptic cylinders) display a type of relationship between drag coefficient and Reynolds number which is substantially similar. With increasing slenderness the jump in the curve becomes progressively less pronounced. For a streamline body, such as that shown in Fig. 1.12 there is no jump, because no appreciable separation occurs; the very gradual pressure increase on the back of such bodies can be overcome by the boundary layer without separation. As we shall also see later in greater detail, the pressure distribution in the external flow exerts a decisive influence on the position of the transition point. The boundary layer is laminar in the region of pressure decrease, i. e. roughly from the leading edge to the point of minimum pressure, and becomes turbulent, in most cases, from that point onward throughout the region of pressure increase. In this connexion it is important to state that separation can only be avoided in regions of increasing pressure when the flow in the boundary layer is turbulent. A laminar boundary layer,

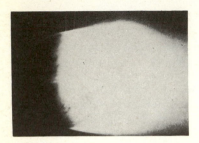

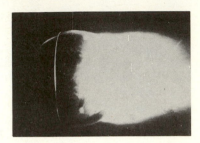

Fig. 2.24. Flow past a sphere at a subcritical Reynolds number; from Wieselsberger [39]

Fig. 2.25. Flow past a sphere at a supercritical Reynolds number; from Wieselsberger [39]. The supercritical flow pattern is achieved by the mounting of a thin wire ring (tripping wire)

as we shall see later, can support only a very small pressure rise so that separation would occur even with very slender bodies. In particular, this remark also applies to the flow past an aerofoil with a pressure distribution similar to that in Fig. 1.14. In this case separation is most likely to occur on the suction side. A smooth flow pattern around an aerofoil, conducive to the creation of lift, is possible only with a turbulent boundary layer. Summing up it may be stated that the small drag of slender bodies as well as the lift of aerofoils are made possible through the existence of a turbulent boundary layer.

Boundary-layer thickness: Generally speaking, the thickness of a turbulent boundary layer is larger than that of a laminar boundary layer owing to greater energy losses in the former. Near a smooth flat plate at zero incidence the boundary layer increases downstream in proportion to $x^{0.8}$ (x = distance from leading edge). It will be shown later in Chap. XXI that the boundary-layer thickness variation in turbulent flow is given by the equation

$$\frac{\delta}{l} = 0.37 \left(\frac{U_\infty l}{\nu}\right)^{-1/5} = 0.37\,(\mathsf{R}_l)^{-1/5} \tag{2.9}$$

which corresponds to eqn. (2.2) for laminar flow. Table 2.1 gives values for the boundary-layer thickness calculated from eqn. (2.9) for several typical cases of air and water flows.

Table 2.1. Thickness of boundary layer, δ, at trailing edge of flat plate at zero incidence in parallel turbulent flow

U_∞ = free stream velocity; l = length of plate; ν = kinematic viscosity

	U_∞ [ft/sec]	l [ft]	$R_l = \dfrac{U_\infty l}{\nu}$	δ [in]
Air $\nu = 150 \times 10^{-6}$ ft²/sec	100	3	$2\cdot0 \times 10^6$	0·73
	200	3	$4\cdot0 \times 10^6$	0·64
	200	15	$2\cdot0 \times 10^7$	2·30
	500	25	$8\cdot3 \times 10^7$	2·90
	750	25	$1\cdot25 \times 10^8$	2·68
Water $\nu = 11 \times 10^{-6}$ ft²/sec	5	5	$2\cdot3 \times 10^6$	1·19
	10	15	$1\cdot35 \times 10^7$	2·52
	25	150	$3\cdot4 \times 10^8$	13·1
	50	500	$2\cdot3 \times 10^9$	29·8

Methods for the prevention of separation: Separation is mostly an undesirable phenomenon because it entails large energy losses. For this reason methods have been devised for the artificial prevention of separation. The simplest method, from the physical point of view, is to move the wall with the stream in order to reduce the velocity difference between them, and hence to remove the cause of boundary-layer formation, but this is very difficult to achieve in engineering practice. However, Prandtl† has shown on a *rotating circular cylinder* that this method is very effective. On the side where the wall and stream move in the same direction separation is completely prevented. Moreover, on the side where the wall and stream move in opposite directions, separation is slight so that on the whole it is possible to obtain a good experimental approximation to perfect flow with circulation and a large lift.

Another very effective method for the prevention of separation is *boundary-layer suction*. In this method the decelerated fluid particles in the boundary layer are removed through slits in the wall into the interior of the body. With sufficiently strong suction, separation can be prevented. Boundary-layer suction was used on a circular cylinder by L. Prandtl in his first fundamental investigation into boundary-layer flow. Separation can be almost completely eliminated with suction through a slit at the back of the circular cylinder. Instances of the effect of suction can be seen in Figs. 2.14 and 2.15 on the example of flows through a highly divergent channel. Fig. 2.13 demonstrates that without suction there is strong separation. Fig. 2.14 shows how the flow adheres to the one side on which suction is applied, whereas from Fig. 2.15 it is seen that the flow completely fills the channel cross-section when the suction slits are put into operation on both sides. In the latter case the streamlines assume a pattern which is very similar to that in frictionless flow. In later years suction was successfully used in aeroplane wings to increase the lift. Owing to suction on the upper surface near the trailing edge, the flow adheres

---

† Prandtl-Tietjens: Hydro- and Aerodynamics. Vol. II, Tables 7, 8 and 9.

to the aerofoil at considerably larger incidence angles than would otherwise be the case, stalling is delayed, and much larger maximum-lift values are achieved [36].

After having given a short outline of the fundamental physical principles of fluid motions with very small friction, i. e. of the boundary-layer theory, we shall proceed to develop a rational theory of these phenomena from the equations of motion of viscous fluids. The description will be arranged in the following way: We shall begin in Part A by deriving the general Navier-Stokes equations from which, in turn, we shall derive Prandtl's boundary-layer equations with the aid of the simplifications which can be introduced as a consequence of the small values of viscosity. This will be followed in Part B by a description of the methods for the integration of these equations for the case of laminar flow. In Part C we shall discuss the problem of the origin of turbulent flow, i. e. we shall discuss the process of transition from laminar to turbulent flow, treating it as a problem in the stability of laminar motion. Finally, Part D will contain the boundary-layer theory for completely developed turbulent motions. Whereas the theory of laminar boundary layers can be treated as a deductive sequence based on the Navier-Stokes differential equations for viscous fluids, the same is not, at present, possible for turbulent flow, because the mechanism of turbulent flow is so complex that it cannot be mastered by purely theoretical methods. For this reason a treatise on turbulent flow must draw heavily on experimental results and the subject must be presented in the form of a semi-empirical theory.

References

[1] Achenbach, E.: Experiments on the flow past spheres at very high Reynolds numbers. JFM *54*, 565—575 (1972).

[2] Berger, E., and Wille, R.: Periodic flow phenomena. Annual Review of Fluid Mech. *4*, 313—340 (1972).

[3] Berger, E.: Bestimmung der hydrodynamischen Grössen einer Kármánschen Wirbelstrasse aus Hitzdrahtmessungen bei kleinen Reynolds-Zahlen. ZFW *12*, 41—59 (1964).

[3a] Bearman, P.W.: On the vortex shedding from a circular cylinder in the critical Reynolds number range. JFM *37*, 577—585 (1969).

[4] Blasius, H.: Grenzschichten in Flüssigkeiten mit kleiner Reibung. Diss. Göttingen 1907; Z. Math. u. Phys. *56*, 1—37 (1908); Engl. transl. in NACA TM 1256.

[5] Blenk, H., Fuchs, D., and Liebers, L.: Über die Messung von Wirbelfrequenzen. Luftfahrtforschung *12*, 38—41 (1935).

[6] Burgers, J.M.: The motion of a fluid in the boundary layer along a plane smooth surface. Proc. First International Congress for Applied Mechanics, Delft, 113—128 (1924).

[7] Chang, P.K.: Separation of flow. Pergamon Press, Washington D.C., 1970.

[8] Cermak, J.E.: Application of fluid mechanics to wind engineering — A Freeman Scholar lecture. Trans. ASME Fluids Engineering *97*, Ser. I, 9—38 (1975); see also: Laboratory simulation of the atmospheric boundary layer. AIAA J. *9*, 1746—1754 (1971).

[8a] Cermak, J.E.: Aerodynamics of buildings. Annual Review of Fluid Mech. *8*, 75—106 (1976).

[9] Cermak, J.E., and Sadeh, W.Z.: Wind-tunnel simulation of wind loading on structures. Meeting Preprint 1417, ASCE National Structural Engineering Meeting, Baltimore, Maryland, 19—23 April, 1971.

[10] Davenport, A.G.: The relationship of wind structure to wind loading. Proc. Conference on Wind Effects on Buildings and Structures, National Physical Laboratory, Teddington, Middlesex, Great Britain, 26—28 June 1963, Her Majesty's Stationary Office, London, Vol. *I*, 54—112 (1965).

[11] Domm, U.: Ein Beitrag zur Stabilitätstheorie der Wirbelstrassen unter Berücksichtigung endlicher und zeitlich wachsender Wirbelkerndurchmesser. Ing.-Arch. *22*, 400 — 410 (1954).

[12] Dubs, W.: Über den Einfluss laminarer und turbulenter Strömung auf das Röntgenbild von Wasser und Nitrobenzol. Helv. phys. Acta *12*, 169—228 (1939).

[13] Durgin, W.W., and Karlsson, S.K.F.: On the phenomenon of vortex street breakdown. Klasse 1914, 177—190; see also Coll. Works *II*, 597—608.

[14] Eiffel, G.: Sur la résistance des sphères dans l'air en mouvement. Comptes Rendus *155*, 1597 (1912).

[14a] Försching, H.W.: Aeroelastische Probleme an Hochbaukonstruktionen in freier Windumströmung. Vulkan-Verlag, Essen, Haus der Technik, Part 347, 3—18 (1976).

[15] Frimberger, R.: Experimentelle Untersuchungen an der Kármánschen Wirbelstrasse. ZFW *5*, 355—359 (1957).

[16] Hansen, M.: Die Geschwindigkeitsverteilung in der Grenzschicht an der längsangeströmten ebenen Platte. ZAMM *8*, 185—199 (1928); NACA TM 585 (1930).

[17] van der Hegge Zijnen, B.G.: Measurements of the velocity distribution in the boundary layer along a plane surface. Thesis Delft 1924.

[18] Heinemann, H.J., Lawaczeck, O., and Bütefisch, K.A.: Kármán vortices and their frequency determination in the wakes of profiles in the sub- and transonic regime. Symposium Transsonicum II Göttingen, Sept. 1975. Springer Verlag, 1976, pp. 75—82; see also: AGARD-Conference Proc. No. 177, Unsteady Phenomena in Turbomachinery (1975).

[19] Hucho, W.H.: Einfluss der Vorderwagenform auf Widerstand, Giermoment und Seitenkraft von Kastenwagen. ZFW *20*, 341—351 (1972).

[20] von Kármán, Th.: Über den Mechanismus des Widerstandes, den ein bewegter Körper in einer Flüssigkeit erzeugt. Nachr. Ges. Wiss. Göttingen, Math. Phys. Klasse 509—517 (1911) and 547—556 (1912); see also Coll. Works *I*, 324—338.

[21] von Kármán, Th., and Rubach, H.: Über den Mechanismus des Flüssigkeits- und Luftwiderstandes. Phys. Z. *13*, 49—59 (1912); see also Coll. Works *I*, 339—358.

[22] Lin, C.C.: On periodically oscillating wakes in the Oseen approximation. R. v. Mises Anniversary Volume, Studies in Mathematics and Mechanics. Academic Press, New York, 1950, 170—176.

[23] Möller, E.: Luftwiderstandsmessungen am Volkswagen-Lieferwagen. Automobiltechnische Z. *53*, 1—4 (1951).

[23a] Novak, I.: Strouhal number of bodies and their systems (in Russian). Strojnicky Casopis *26*, 72—89 (1975).

[24] Owen, M., Griffin, S., and Ramberg, E.: The vortex-street wakes of vibrating cylinders. JFM *66*, 553—576 (1974).

[25] Prandtl, L.: Über Flüssigkeitsbewegung bei sehr kleiner Reibung. Proc. 3rd Intern. Math. Congr. Heidelberg 1904, 484—491. Reprinted in: Vier Abhandlungen zur Hydrodynamik und Aerodynamik, Göttingen, 1927; see also Coll. Works *II*, 575—584; Engl. transl. NACA TM 452 (1928).

[26] Prandtl, L.: Der Luftwiderstand von Kugeln. Nachr. Ges. Wiss. Göttingen, Math. Phys. Klasse, 1914, 177—190; see also Coll. Works *II*, 597—608.

[27] Prandtl, L., and Tietjens, O.: Hydro- und Aeromechanik (based on Prandtl's lectures). Vol. *I* and *II*, Berlin, 1929 and 1931; Engl. transl. by L. Rosenhead (Vol. *I*) and J.P. den Hartog (Vol. II), New York, 1934.

[28] Relf, E.F., and Simmons, L.F.G.: The frequencies of eddies generated by the motion of circular cylinders through a fluid. ARC RM 917, London (1924).

[29] Reynolds, O.: An experimental investigation of the circumstances which determine whether the motion of water shall be direct or sinuous, and of the law of resistance in parallel channels. Phil. Trans. Roy. Soc. *174*, 935—982 (1883); see also Scientific Papers *2*, 51.

[30] Ribner, H.S., Etkins, B., and Nelly, K.K.: Noise research in Canada: Physical and bioacoustic. Proc. First Int. Congress Aero. Sci. Madrid, Pergamon Press, London, Vol. *I*, 393—411 (1959).

[31] Roshko, A.: Experiments on the flow past a circular cylinder at very high Reynolds number. JFM *10*, 345—356 (1961).

[32] Roshko, A.: On the development of turbulent wakes from vortex streets. NACA Rep. 1191 (1954).

[32a] Rosenhead, L.: The formation of vortices from a surface of discontinuity. Proc. Roy. Soc. A *134*, 170 (1931).

[33] Rubach, H.: Über die Entstehung und Fortbewegung des Wirbelpaares bei zylindrischen Körpern. Diss. Göttingen 1914; VDI-Forschungsheft 185 (1916).

[33a] Sarpkaya, T.: An inviscid model of two-dimensional vortex shedding for transient and asymptotically steady flow over an inclined plate. JFM *68*, 109—128 (1975).

[34] Sadeh, W. Z., and Cermak, J. E.: Turbulence effect on wall pressure fluctuations. J. Eng. Mech. Div. ASCE *98*, No. EM 6, Proc. Paper 9445, 189—198 (1972).

[35] Schlichting, H.: Aerodynamische Untersuchungen an Kraftfahrzeugen. Rep. Techn. Hochschule Braunschweig, 130—139 (1954).

[36] Schrenk, O.: Versuche mit Absaugeflügeln. Luftfahrtforschung *XII*, 10—27 (1935).

[37] Strouhal, V.: Über eine besondere Art der Tonerregung. Ann. Phys. und Chemie, New Series *5*, 216—251 (1878).

[38] Timme, A.: Über die Geschwindigkeitsverteilung in Wirbeln. Ing.-Arch. *25*, 205—225 (1957).

[38a] Wedemeyer, E.: Ausbildung eines Wirbelpaares an den Kanten einer Platte. Ing.-Arch. *30*, 187—200 (1961).

[39] Wieselsberger, C.: Der Luftwiderstand von Kugeln. ZFM *5*, 140—144 (1914).

CHAPTER III

Derivation of the equations of motion of a compressible viscous fluid

(Navier-Stokes equations)†

a. Fundamental equations of motion and continuity applied to fluid flow

We shall now proceed to derive the equations of motion of a compressible, viscous, Newtonian fluid. In the general case of three-dimensional motion, the flow field is specified by the velocity vector

$$w = i\,u + j\,v + k\,w$$

where u, v, w are the three orthogonal components, by the pressure p, and by the density ϱ, all conceived as functions of the coordinates x, y, z, and time t. For the determination of these five quantities there exist five equations: the continuity equation (conservation of mass), the three equations of motion (conservation of momentum) and the thermodynamic equation of state $p = f(\varrho)$.‡

The equation of continuity expresses the fact that for a unit volume there is a balance between the masses entering and leaving per unit time, and the change in density. In the case of non-steady flow of a compressible fluid this condition leads to the equation:

$$\frac{D\varrho}{Dt} + \varrho \operatorname{div} w = \frac{\partial \varrho}{\partial t} + \operatorname{div}(\varrho\,w) = 0\,, \tag{3.1}$$

whereas for an incompressible fluid, with $\varrho = \text{const}$, the equation of continuity assumes the simplified form

$$\operatorname{div} w = 0\,. \tag{3.1a}$$

The symbol $D\varrho/Dt$ denotes here the substantive derivative which consists of the local contribution (in non steady flow) $\partial \varrho / \partial t$, and the convective contribution (due to translation), $w \cdot \operatorname{grad} \varrho$.

---

† In the Sixth Edition this chapter has been revised by the Translator at the Author's invitation.
‡ If the equation of state contains temperature as an additional variable, a further equation is supplied by the principle of the conservation of energy in the form of the First Law of Thermodynamics; *cf.* Chap. XII.

The equations of motion are derived from Newton's Second Law, which states that the product of mass and acceleration is equal to the sum of the external forces acting on the body. In fluid motion it is necessary to consider the following two classes of forces: forces acting throughout the mass of the body (gravitational forces) and forces acting on the boundary (pressure and friction). If $F = \varrho\, g$ denotes the gravitational force per unit volume (g = vector of acceleration due to gravity) and P denotes the force on the boundary per unit volume, then the equations of motion can be written in the following vector form

$$\varrho \frac{Dw}{Dt} = F + P \tag{3.2}$$

with

$$F = i\,X + j\,Y + k\,Z \qquad \text{body force} \tag{3.3}$$

and

$$P = i\,P_x + j\,P_y + k\,P_z \qquad \text{surface force}. \tag{3.4}$$

The symbol Dw/Dt denotes here the substantive acceleration which, like the substantive derivative of density, consists of the local contribution (in non-steady flow) $\partial w/\partial t$, and the convective contribution (due to translation) $dw/dt = (w{\cdot}\mathrm{grad})\,w$ †

$$\frac{Dw}{Dt} = \frac{\partial w}{\partial t} + \frac{dw}{dt}\ .$$

The body forces are to be regarded as given external forces, but the surface forces depend on the *rate* at which the fluid is *strained* by the velocity field present in it. The system of forces determines a *state of stress*, and it is now our task to indicate the relationship between stress and rate of strain, noting that it can only be given empirically. In our present derivation we shall restrict attention to *isotropic, Newtonian fluids* for which it may be assumed that this relation is a linear one. All gases and many liquids of interest in boundary-layer theory, in particular water, belong to this class. A fluid is said to be isotropic when the relation between the components of stress and those of the rate of strain is the same in all directions; it is said to be Newtonian when this relation is linear, that is when the fluid obeys Stokes's law of friction. In the case of isotropic, elastic solid bodies, experiment teaches that the state of stress depends on the magnitude of strain itself, most engineering materials obeying Hooke's linear law which is somewhat analogous to Stokes's law. Whereas the relation between stress and strain for an isotropic elastic solid involves two constants which characterize the properties of a given material (e. g. elastic modulus and Poisson's ratio), the relation between stress and rate of strain in an isotropic fluid involves a single constant (the viscosity, μ) as long as relaxation phenomena do not occur within it, as we shall see in Sec. IIIe.

---

† In order to express the vector $(w{\cdot}\mathrm{grad})\,w$ in an arbitrary system of coordinates, the following general relation should be used

$$(w{\cdot}\mathrm{grad})\,w = \mathrm{grad}\ \tfrac{1}{2}\,w^2 - w \times \mathrm{curl}\ w\ ,$$

where $w^2 = w \cdot w$.

b. General stress system in a deformable body

In order to write down expressions for the surface forces acting on the boundary, let us imagine a small parallepiped of volume $dV = dx\,dy\,dz$ isolated instantaneously from the body of the fluid, Fig. 3.1, and let its lower left-hand vertex coincide with the point x, y, z. On the two faces of area $dy \cdot dz$ which are perpendicular to the x-axis there act two resultant stresses (vectors $=$ surface force per unit area):

$$\boldsymbol{p_x} \quad \text{and} \quad \boldsymbol{p_x} + \frac{\partial \boldsymbol{p_x}}{\partial x}\,dx \quad \text{respectively .} \tag{3.5}$$

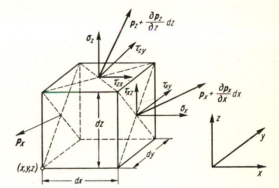

Fig. 3.1. Derivation of the expressions for the stress tensor of an inhomogeneous stress system and of its symmetry in the absence of a volumetric distribution of local moments

(Subscript x denotes that the stress vector acts on an elementary plane which is perpendicular to the x-direction.) Similar terms are obtained for the faces $dx \cdot dz$ and $dx \cdot dy$ which are perpendicular to the y- and z-axes respectively. Hence the three net components of the surface force are:

$$\text{plane} \perp \text{direction } x: \quad \frac{\partial \boldsymbol{p_x}}{\partial x} \cdot dx \cdot dy \cdot dz$$

$$\text{,, \quad ,, \quad ,, \quad } y: \quad \frac{\partial \boldsymbol{p_y}}{\partial y} \cdot dx \cdot dy \cdot dz$$

$$\text{,, \quad ,, \quad ,, \quad } z: \quad \frac{\partial \boldsymbol{p_z}}{\partial z} \cdot dx \cdot dy \cdot dz .$$

and the resultant surface force $\boldsymbol{P}$ per unit volume is, therefore, given by

$$\boldsymbol{P} = \frac{\partial \boldsymbol{p_x}}{\partial x} + \frac{\partial \boldsymbol{p_y}}{\partial y} + \frac{\partial \boldsymbol{p_z}}{\partial z} . \tag{3.6}$$

The quantities $\boldsymbol{p_x}$, $\boldsymbol{p_y}$, $\boldsymbol{p_z}$ are vectors which can be resolved into components perpendicular to each face, i. e., into normal stresses denoted by σ with a suitable subscript indicating the direction, and into components parallel to each face, i. e. into shearing stresses denoted by τ. The symbol for a shearing stress will be provided with two subscripts: the first subscript indicates the axis to which the face is perpendicular, and the second indicates the direction to which the shearing stress is parallel. With this notation we have

$$\left.\begin{array}{l} p_x = i\,\sigma_x + j\,\tau_{xy} + k\,\tau_{xz} \\ p_y = i\,\tau_{yx} + j\,\sigma_y + k\,\tau_{yz} \\ p_z = i\,\tau_{zx} + j\,\tau_{zy} + k\,\sigma_z\,. \end{array}\right\} \tag{3.7}$$

The stress system is seen to require nine scalar quantities for its description. These nine quantities form a *stress tensor*. The set of nine components of the stress tensor is sometimes called the stress matrix:

$$II = \begin{pmatrix} \sigma_x & \tau_{xy} & \tau_{xz} \\ \tau_{yx} & \sigma_y & \tau_{yz} \\ \tau_{zx} & \tau_{zy} & \sigma_z \end{pmatrix}. \tag{3.8}$$

The stress tensor and the corresponding matrix are symmetric, which means that two shearing stresses with subscripts which differ only in their order are equal. This can be demonstrated with reference to the equations of motion of an element of fluid. In general, its motion can be separated into an instantaneous translation and an instantaneous rotation, and only the latter needs to be considered for our purpose. Denoting the instantaneous angular acceleration of the element by $\dot{\omega}\,(\dot{\omega}_x,\ \dot{\omega}_y,\ \dot{\omega}_z)$, we can write for the rotation about the y-axis that

$$\dot{\omega}_y\,\mathrm{d}I_y = (\tau_{xz}\,\mathrm{d}y\,\mathrm{d}z)\,\mathrm{d}x - (\tau_{zx}\,\mathrm{d}x\,\mathrm{d}y)\,\mathrm{d}z = (\tau_{xz} - \tau_{zx})\,\mathrm{d}V$$

where $\mathrm{d}I_y$ is the elementary moment of inertia about the y-axis. Now the moment of inertia, $\mathrm{d}I$, is proportional to the fifth power of the linear dimensions of the parallelepiped, whereas its volume, $\mathrm{d}V$, is proportional to their third power. On contracting the element to a point, we notice that the left-hand side of the preceding equation vanishes faster than the right-hand side. Hence, ultimately,

$$\tau_{xy} - \tau_{yx} = 0$$

if $\dot{\omega}_y$ is not to become infinitely large. Analogous equations can be written for the remaining two axes, and the symmetry of the stress tensor can thus be demonstrated. It is clear from the argument that the stress tensor would cease to be symmetric if the fluid developed a local moment which was proportional to its volume, $\mathrm{d}V$. The latter may occur, for example, in an electrostatic field.

Owing to the fact that

$$\tau_{xy} = \tau_{yx}; \qquad \tau_{xz} = \tau_{zx}; \qquad \tau_{yz} = \tau_{zy}, \tag{3.9}$$

the stress matrix (3.8) contains only six different stress components and becomes symmetrical with respect to the principal diagonal:

$$II = \begin{pmatrix} \sigma_x & \tau_{xy} & \tau_{xz} \\ \tau_{xy} & \sigma_y & \tau_{yz} \\ \tau_{xz} & \tau_{yz} & \sigma_z \end{pmatrix}. \tag{3.10}$$

The surface force per unit volume can be calculated from eqns. (3.6), (3.7), and (3.10) and becomes

$$P = i \left(\frac{\partial \sigma_x}{\partial x} + \frac{\partial \tau_{xy}}{\partial y} + \frac{\partial \tau_{xz}}{\partial z} \right) \cdots \cdots \text{comp. } x$$

$$+ j \left(\frac{\partial \tau_{xy}}{\partial x} + \frac{\partial \sigma_y}{\partial y} + \frac{\partial \tau_{yz}}{\partial z} \right) \cdots \cdots \text{comp. } y \qquad (3.10\,\text{a})$$

$$+ k \left(\frac{\partial \tau_{xz}}{\partial x} + \frac{\partial \tau_{yz}}{\partial y} + \frac{\partial \sigma_z}{\partial z} \right) \cdots \cdots \text{comp. } z \, .$$

$$\underbrace{\qquad}_{\substack{\text{face} \\ yz}} \quad \underbrace{\qquad}_{\substack{\text{face} \\ zx}} \quad \underbrace{\qquad}_{\substack{\text{face} \\ xy}}$$

Introducing the expression (3.10a) into the equation of motion (3.2), and resolving into components we have:

$$\varrho \, \frac{Du}{Dt} = X + \left(\frac{\partial \sigma_x}{\partial x} + \frac{\partial \tau_{xy}}{\partial y} + \frac{\partial \tau_{xz}}{\partial z} \right)$$

$$\varrho \, \frac{Dv}{Dt} = Y + \left(\frac{\partial \tau_{xy}}{\partial x} + \frac{\partial \sigma_y}{\partial y} + \frac{\partial \tau_{yz}}{\partial z} \right) \qquad (3.11)$$

$$\varrho \, \frac{Dw}{Dt} = Z + \left(\frac{\partial \tau_{xz}}{\partial x} + \frac{\partial \tau_{yz}}{\partial y} + \frac{\partial \sigma_z}{\partial z} \right) .$$

If the fluid is "frictionless" all shearing stresses vanish; only the normal stresses remain in the equation, and they are, moreover, equal. Their negative is defined as the pressure at the point x, y, z in the fluid:

$$\tau_{xy} = \tau_{xz} = \tau_{yz} = \quad 0$$

$$\sigma_x = \sigma_y = \sigma_z = - p \, .$$

In such a *hydrostatic stress system*, the fluid pressure is equal to the arithmetical mean of the normal stresses taken with a negative sign. Since measurements which lead to the establishment of the thermodynamic equation of state are performed under such conditions, the fluid being at rest, this pressure is identical with the thermodynamic pressure in the equation of state. It is convenient to introduce the arithmetical mean of the three normal stresses — their sum being called the *trace* of the stress tensor — as a useful numerical quantitiy in the case of a *viscous fluid* in a state of motion also. It is still called the pressure, but its relation to the thermodynamic pressure requires further investigation. Although it then ceases to be equal to a particular stress which is normal to the surface, it has the property of being invariant with respect to transformations of the system of coordinates, as it is an invariant of the stress tensor, being defined as

$$\tfrac{1}{3} \left(\sigma_x + \sigma_y + \sigma_z \right) = - p \, . \qquad (3.12)$$

We shall see in Sec. IIIe that it remains equal to the thermodynamic pressure in the absence of relaxation.

The system of the three equations (3.11) contains the six stresses $\sigma_x, \sigma_y, \sigma_z,$ $\tau_{xy}, \tau_{xz}, \tau_{yz}$. The next task is to determine the relation between them and the strains so as to enable us to introduce the velocity components u, v, w into eqn. (3.11). Before giving this relation in Sec. IIId we shall investigate the system of strains in greater detail.

c. The rate at which a fluid element is strained in flow

When a continuous body of fluid is made to flow, every element in it is, generally speaking, displaced to a new position in the course of time. During this motion elements of fluid become strained, and since the motion of the fluid is completely determined when the velocity vector w is given as a function of time and position, $w = w(x,y,z,t)$, there exist kinematic relations between the components of the rate of strain and this function. The rate at which an element of fluid is strained depends on the *relative* motion of two points within it. We, therefore, consider the two neighbouring points A and B which are shown in Fig. 3.2. Owing to the presence of the velocity field, point A will be displaced to A' in time dt by a distance $s = w \, dt$; since, however, the velocity at B, imagined at a distance dr from A, is different, point B will move to B' displaced from B by $s + ds = (w + dw) \, dt$. More explicitly, if the components of velocity have the values u, v, w at A, then, at the neighbouring point B, the velocity components will be given to first order by the Taylor-series expansions

$$u + du = u + \frac{\partial u}{\partial x} \, dx + \frac{\partial u}{\partial y} \, dy + \frac{\partial u}{\partial z} \, dz$$

$$v + dv = v + \frac{\partial v}{\partial x} \, dx + \frac{\partial v}{\partial y} \, dy + \frac{\partial v}{\partial z} \, dz \tag{3.13}$$

$$w + dw = w + \frac{\partial w}{\partial x} \, dx + \frac{\partial w}{\partial y} \, dy + \frac{\partial w}{\partial z} \, dz .$$

Thus, the relative motion of point B with respect to A is described by the following matrix of nine partial derivatives of the local velocity field

$$\begin{pmatrix} \dfrac{\partial u}{\partial x} & \dfrac{\partial u}{\partial y} & \dfrac{\partial u}{\partial z} \\[2mm] \dfrac{\partial v}{\partial x} & \dfrac{\partial v}{\partial y} & \dfrac{\partial v}{\partial z} \\[2mm] \dfrac{\partial w}{\partial x} & \dfrac{\partial w}{\partial y} & \dfrac{\partial w}{\partial z} \end{pmatrix} \tag{3.13a}$$

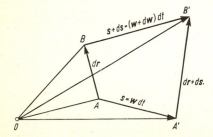

Fig. 3.2.   Relative displacement

It is convenient to rearrange the expressions for the relative velocity components du, dv, dw from eqn. (3.13) to the form

$$du = (\dot{\varepsilon}_x \ dx + \dot{\varepsilon}_{xy} \ dy + \dot{\varepsilon}_{xz} \ dz) + (\eta \ dz - \zeta \ dy)$$

$$dv = (\dot{\varepsilon}_{yx} \ dx + \dot{\varepsilon}_y \ dy + \dot{\varepsilon}_{yz} \ dz) + (\zeta \ dx - \xi \ dz) \qquad (3.14)$$

$$dw = (\dot{\varepsilon}_{zx} \ dx + \dot{\varepsilon}_{zy} \ dy + \dot{\varepsilon}_z \ dz) + (\xi \ dy - \eta \ dx),$$

it being easy to verify that the new symbols have the following meanings

$$\dot{\varepsilon}_{ij} = \begin{pmatrix} \dot{\varepsilon}_x & \dot{\varepsilon}_{xy} & \dot{\varepsilon}_{xz} \\ \dot{\varepsilon}_{yx} & \dot{\varepsilon}_y & \dot{\varepsilon}_{yz} \\ \dot{\varepsilon}_{zx} & \dot{\varepsilon}_{zy} & \dot{\varepsilon}_z \end{pmatrix} = \begin{pmatrix} \dfrac{\partial u}{\partial x} & ; & \dfrac{1}{2}\left(\dfrac{\partial v}{\partial x} + \dfrac{\partial u}{\partial y}\right); & \dfrac{1}{2}\left(\dfrac{\partial w}{\partial x} + \dfrac{\partial u}{\partial z}\right); \\ \dfrac{1}{2}\left(\dfrac{\partial u}{\partial y} + \dfrac{\partial v}{\partial x}\right); & \dfrac{\partial v}{\partial y} & ; & \dfrac{1}{2}\left(\dfrac{\partial w}{\partial y} + \dfrac{\partial v}{\partial z}\right); \\ \dfrac{1}{2}\left(\dfrac{\partial u}{\partial z} + \dfrac{\partial w}{\partial x}\right); & \dfrac{1}{2}\left(\dfrac{\partial v}{\partial z} + \dfrac{\partial w}{\partial y}\right); & \dfrac{\partial w}{\partial z} \end{pmatrix}$$

$$(3.15\,a)$$

and

$$\xi = \frac{1}{2}\left(\frac{\partial w}{\partial y} - \frac{\partial v}{\partial z}\right); \quad \eta = \frac{1}{2}\left(\frac{\partial u}{\partial z} - \frac{\partial w}{\partial x}\right); \quad \zeta = \frac{1}{2}\left(\frac{\partial v}{\partial x} - \frac{\partial u}{\partial y}\right). \quad (3.15\,b)$$

It is noted that the matrix $\dot{\varepsilon}_{ij}$ is symmetric, so that

$$\dot{\varepsilon}_{yx} = \dot{\varepsilon}_{xy}; \qquad \dot{\varepsilon}_{xz} = \dot{\varepsilon}_{zx}; \qquad \dot{\varepsilon}_{zy} = \dot{\varepsilon}_{yz}, \qquad (3.15\,c)$$

and that ξ, η, ζ are related to the components of the vector

$$\boldsymbol{\omega} = \operatorname{curl} \boldsymbol{w} \qquad (3.15\,d)$$

Each of the new terms can be given a kinematic interpretation, and we now proceed to obtain it.

Since we concentrate our attention on the immediate neighbourhood of point A, and since interest is centred on the motion of B relative to A, we shall place point A at the origin, and interpret dx, dy, dz as the coordinates of point B in a Cartesian system of coordinates. In this manner, the expressions in eqns. (3.14) will define a field of relative velocities in which the components du, dv, dw are linear functions of the space coordinates. In order to understand the meaning of the different terms in the matrix (3.15a) and in eqns. (3.15b), we proceed to interpret them one by one.

The diagram in Fig. 3.3 represents the field of relative velocities when all terms except $\partial u/\partial x$ vanish on the assumption that $\partial u/\partial x > 0$. The relative velocity of any point B with respect to A is now

$$du = \left(\frac{\partial u}{\partial x}\right) dx,$$

and the field consists of planes $x = \text{const}$ which displace themselves uniformly with a velocity which is proportional to the distance dx away from the plane $x = 0$. An elementary parallelepiped with A and B at its vertices placed in such a velocity field will be distorted in extension, its face BC receding from AD with an increasing

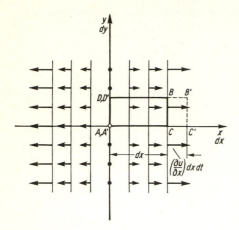

Fig. 3.3. Local distortion of fluid element when $\partial u/\partial x > 0$ with all other terms being equal to zero; uniform extension in the x-direction

velocity. Thus $\dot{\varepsilon}_x$ represents the rate of *elongation* in the x-direction suffered by the element. Similarly, the additive terms $\dot{\varepsilon}_y = \partial v/\partial y$ and $\dot{\varepsilon}_z = \partial w/\partial z$ describe the rate of elongation in the y- and z-directions, respectively.

It is now easy to visualize the distortion imparted to a fluid element by the simultaneous action of all three diagonal elements of matrices (3.13a) or (3.15a). The element expands in all three directions, and the change in the length of its three sides produces a change in volume at a relative rate

$$\dot{e} = \frac{\left\{dx + \frac{\partial u}{\partial x}\, dx\, dt\right\}\left\{dy + \frac{\partial v}{\partial y}\, dy\, dt\right\}\left\{dz + \frac{\partial w}{\partial z}\, dz\, dt\right\} - dx\, dy\, dz}{dx\, dy\, dz\, dt}$$

$$= \frac{\partial u}{\partial x} + \frac{\partial v}{\partial y} + \frac{\partial w}{\partial z} = \text{div } \boldsymbol{w}\, , \tag{3.16}$$

to first order in the derivatives. During this distortion, however, the shape of the element, described by the angles at its vertices, remains unchanged, since all right angles continue to be that way. Thus $\dot{e}$ describes the local, instantaneous *volumetric dilatation* of a fluid element. When the fluid is incompressible, $\dot{e} = 0$, as must be expected. In a compressible fluid the continuity equation (3.1) shows that

$$\dot{e} = \text{div } \boldsymbol{w} = -\frac{1}{\varrho}\frac{D\varrho}{Dt}\, , \tag{3.17}$$

that is that the volumetric dilatation, the relative change in volume, is equal to the negative of the relative rate of change in the local density.

The relative velocity field presents a different appearance when one of the off-diagonal terms of matrix (3.13a), for example $\partial u/\partial y$, has a non-vanishing, say positive, value. The corresponding field, sketched in Fig. 3.4, is one of pure shear strain. A rectangular element of fluid centred on A now distorts into a parallelogram as indicated in the diagram. The original right angle at A changes at a rate measured by the angle $\gamma_{xy} = [(\partial u/\partial y)\, dy\, dt]/dy$, that is at a rate $\partial u/\partial y$. When both $\partial u/\partial y$

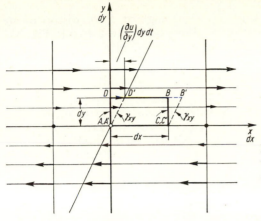

Fig. 3.4. Local distortion of fluid element when $\partial u/\partial y > 0$ with all other terms being equal to zero; uniform shear deformation.

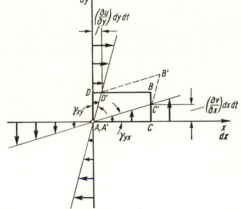

Fig. 3.5. Local distortion of fluid element when

$$\dot{\varepsilon}_{xy} = \dot{\varepsilon}_{yx} = \tfrac{1}{2}\{(\partial u/\partial y) + (\partial v/\partial x)\} > 0$$

with all other terms being equal to zero; distortion in shape. (The diagram has been drawn for $\partial u/\partial y = \partial v/\partial x$)

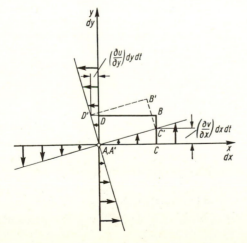

Fig. 3.6. Local distortion of fluid element when

$$\zeta = \tfrac{1}{2}\{(\partial v/\partial x) - (\partial u/\partial y)\} \neq 0 \; ;$$

instantaneous rigid-body rotation

and $\partial v/\partial x$ have positive nonvanishing values, the right angle at A will distort owing to the superposition of two motions, the state of affairs being illustrated in Fig. 3.5. It is clear that the right angle at A now distorts at twice the rate

$$\dot{\varepsilon}_{yx} = \dot{\varepsilon}_{xy} = \frac{1}{2}\left(\frac{\partial u}{\partial y} + \frac{\partial v}{\partial x}\right)$$

described by two of the off-diagonal terms of matrix (3.15a). In general, the three off-diagonal terms $\dot{\varepsilon}_{xy} = \dot{\varepsilon}_{yx}$, $\dot{\varepsilon}_{xz} = \dot{\varepsilon}_{zx}$, and $\dot{\varepsilon}_{zy} = \dot{\varepsilon}_{yz}$ describe the rate of distortion of a right angle located in a plane normal to the axis the index of which does not appear as a subscript. The distortion is volume-preserving and affects only the shape of the element.

Circumstances are again different in the particular case when $\partial u/\partial y = -\partial v/\partial x$ illustrated in Fig. 3.6. From the preceding considerations and from the fact that now $\dot{\varepsilon}_{xy} = 0$ we can infer at once that the right angle at A remains undistorted. This is also clear from the diagram which shows that the fluid element rotates with respect to the reference point A. *Instantaneously*, this rotation occurs without distortion and can be described as a rigid-body rotation. The instantaneous angular velocity of this rotation is

$$\frac{(\partial v/\partial x)\,dx\,dt}{dx\,dt} = \frac{\partial v}{\partial x} \quad \text{or} \quad = -\frac{\partial u}{\partial y}\ .$$

It is now easy to see that the component ζ of $\frac{1}{2}$ curl $\boldsymbol{w}$ from eqn. (3.15b), known as the vorticity of the velocity field, represents the angular velocity of this instantaneous rigid-body rotation, and that

$$\frac{1}{2}\left(\frac{\partial v}{\partial x} - \frac{\partial u}{\partial y}\right) \neq 0\ .$$

In the more complex case when $(\partial v/\partial x) \neq -(\partial u/\partial y)$, the element of fluid rotates and its shape is distorted simultaneously. We can still interpret the term

$$\dot{\varepsilon}_{xy} = \dot{\varepsilon}_{yx} = \frac{1}{2}\left(\frac{\partial u}{\partial y} + \frac{\partial v}{\partial x}\right)$$

as describing the rate of distortion in shape, the term

$$\zeta = \frac{1}{2}\left(\frac{\partial v}{\partial x} - \frac{\partial u}{\partial y}\right)$$

describing the rate at which the element of fluid participates in a rigid-body rotation.

The linearity of eqns. (3.13) or of the entirely equivalent eqns. (3.14) signifies that the most general case arises by a superposition of the simple cases just described. Therefore, if attention is fixed on two neighbouring points A and B in a body of fluid which sustains a continuous velocity field $\boldsymbol{w}(x,y,z)$, the motion of an element of fluid surrounding these two points can be uniquely decomposed into four component motions:

(a) A pure translation described by the velocity components u, v, w of $\boldsymbol{w}$.

(b) A rigid-body rotation described by the components ξ, η, ζ of $\frac{1}{2}$ curl $\boldsymbol{w}$.

(c) A volumetric dilatation described by $e = \operatorname{div} \boldsymbol{w}$, the linear dilatations in the direction of the axes being described by $\dot{\varepsilon}_x, \dot{\varepsilon}_y$ and $\dot{\varepsilon}_z$, respectively.

(d) A distortion in shape described by the components $\dot{\varepsilon}_{xy}$ etc with mixed indices.

Only the last two motions produce an intrinsic deformation of a fluid element surrounding the reference point A, the first two causing a mere, general, displacement of its location.

The elements of matrix (3.15a) constitute the components of a symmetric tensor known as the *rate-of-strain tensor*; its mathematical properties are analogous to those of the equally symmetric stress tensor. It is known from the theory of elasticity [3, 7] or from general considerations of tensor algebra [11] that with every symmetric tensor it is possible to associate three mutually orthogonal *principal axes* which determine three mutually orthogonal principal planes that is a privileged Cartesian system of coordinates. In this system of coordinates, the stress vector or the instantaneous motion in any one of the principal planes is normal to it, that is, parallel to one of the axes. When such a special system of coordinates is used, the matrices (3.10) or (3.15a) retain their diagonal terms only. Denoting the values of the respective components by symbols with bars, we would be dealing with the matrices

$$
\begin{pmatrix} \bar{\sigma}_x & 0 & 0 \\ 0 & \bar{\sigma}_y & 0 \\ 0 & 0 & \bar{\sigma}_z \end{pmatrix} \quad \text{and} \quad \begin{pmatrix} \bar{\dot{\varepsilon}}_x & 0 & 0 \\ 0 & \bar{\dot{\varepsilon}}_y & 0 \\ 0 & 0 & \bar{\dot{\varepsilon}}_z \end{pmatrix} \tag{3.18}
$$

It should, finally, be remembered that such a transformation of coordinates does not affect the sum of the diagonal terms, so that

$$
\sigma_x + \sigma_y + \sigma_z = \bar{\sigma}_x + \bar{\sigma}_y + \bar{\sigma}_z , \tag{3.19a}
$$

and

$$
\dot{\varepsilon}_x + \dot{\varepsilon}_y + \dot{\varepsilon}_z = \bar{\dot{\varepsilon}}_x + \bar{\dot{\varepsilon}}_y + \bar{\dot{\varepsilon}}_z \quad (= \dot{e} = \operatorname{div} \boldsymbol{w}) , \tag{3.19b}
$$

because they constitute invariants of the tensors, as already intimated earlier. Viewed in such two systems of coordinates (both denoted by bars), an element

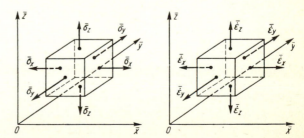

Fig. 3.7. Principial axes for stress and rate of strain

of fluid is stressed in three mutually perpendicular directions, and its faces are displaced instantaneously also in three mutually perpendicular directions, as suggested by Figs. 3.7 a and b. This does not, of course, mean that there exist no shearing stresses in other planes or that the shape of the element remains undistorted.

d. Relation between stress and rate of deformation

It should, perhaps, be stressed once more that the equations which relate the surface forces to the flow field must be obtained by a perceptive interpretation of experimental results and that our interest is restricted to isotropic and Newtonian fluids. The considerations of the preceding section provided us with a useful mathematical framework which allows us now to state the requirements suggested by experiments in a somewhat more precise form.

When the fluid is at rest, it develops a uniform field of hydrostatic stress (negative pressure $-p$) which is identical with the thermodynamic pressure. When the fluid is in motion, the equation of state still determines a pressure at every point ("principle of local state" [4]), and it is convenient to consider the deviatoric normal stresses

$$\sigma_x' = \sigma_x + p \; ; \quad \sigma_y' = \sigma_y + p \; ; \quad \sigma_z' = \sigma_z + p \; ; \tag{3.20}$$

together with the unchanged shearing stresses. The six quantities so obtained constitute a symmetric stress tensor the existence of which is due to the motion because at rest all its components vanish identically. From what has been said before it follows that the components of this deviatoric tensor are created solely by the components of the rate-of-strain tensor, that is to the exclusion of the components u, v, w of velocity as well as of the components ξ, η, ζ of vorticity. This is equivalent to saying that the instantaneous translation [component motion (a)] as well as the instantaneous rigid-body rotation [component motion (b)] of an element of fluid produce no surface forces on it in addition to the existing components of hydrostatic pressure. The preceding statement, evidently, merely represents a precise local formulation of what we expect to observe when a finite body of fluid performs a general motion which is indistinguishable from that of an equivalent rigid body. We thus conclude that the expressions for the components $\sigma_x', \sigma_y', \ldots, \tau_{zx}$ of the deviatoric stress tensor can contain in them only the velocity gradients $\partial u/\partial x, \ldots, \partial w/\partial z$ in appropriate combinations which we now proceed to determine. These relations are postulated to be linear; they must remain unchanged by a rotation of the system of coordinates or by an interchange of axes to ensure isotropy. Isotropy also requires that at every point in the continuum, the principal axes of the stress tensor must coincide with the principal axes of the rate-of-strain tensor, for, otherwise, a preferred direction would be introduced. The simplest way to achieve our aim is to select an arbitrary point in the continuum and to imagine that the local system of coordinates $\bar{x}, \bar{y}, \bar{z}$ has been provisionally so chosen as to coincide with the three common principal axes of the two tensors. The components of the velocity field in this system of coordinates are denoted by $\bar{u}, \bar{v}, \bar{w}$.

It is now clear that isotropy can be secured only if each one of the three normal stresses $\overline{\sigma}_x', \overline{\sigma}_y', \overline{\sigma}_z'$ is made to depend on the component of rate of strain the direction of which conicides with it and on the sum of the three, each with a different factor of proportionality. Thus we record, directly in terms of the space-derivatives, that

$$
\bar{\sigma}_x{}' = \lambda \left(\frac{\partial \bar{u}}{\partial \bar{x}} + \frac{\partial \bar{v}}{\partial \bar{y}} + \frac{\partial \bar{w}}{\partial \bar{z}} \right) + 2 \mu \frac{\partial \bar{u}}{\partial \bar{x}}
$$

$$
\bar{\sigma}_y{}' = \lambda \left(\frac{\partial \bar{u}}{\partial \bar{x}} + \frac{\partial \bar{v}}{\partial \bar{y}} + \frac{\partial \bar{w}}{\partial \bar{z}} \right) + 2 \mu \frac{\partial \bar{v}}{\partial \bar{y}} \qquad (3.21)
$$

$$
\bar{\sigma}_z{}' = \lambda \left(\frac{\partial \bar{u}}{\partial \bar{x}} + \frac{\partial \bar{v}}{\partial \bar{y}} + \frac{\partial \bar{w}}{\partial \bar{z}} \right) + 2 \mu \frac{\partial \bar{w}}{\partial \bar{z}} \; .
$$

The quantities u, v, w and ξ, η, ζ do not appear in these expressions for the reasons just explained. In each expression, the last term represents the appropriate rate of linear dilatation, that is, in essence, a change in shape, and the first term represents the volumetric dilatation, that is the rate of change in volume, in essence, a change in density. The factors 2 in the last terms are not essential, being merely convenient to facilitate the interpretation, as we shall see later. The factors of proportionality, μ and λ, two in all, must be the same in each of the three preceding equations to secure isotropy. It is easy to see that an interchange between any two axes, that is an interchange of any of the three pairs of quantities: $(\bar{u}, \bar{x})$, $(\bar{v}, \bar{y})$, $(\bar{w}, \bar{z})$ leaves the set of relations invariant, as they must be in an isotropic medium. Moreover, the preceding is the only combination of spatial gradients which possesses the required properties. If the reader cannot see this directly, he may consult a more rigorous proof in a treatise on tensor algebra (or e. g. [11] p. 89).

The relations in eqns. (3.21) can be re-written to apply in an arbitrary system of coordinates by performing a general rotation with the aid of the appropriate linear transformation formulae. We shall refrain from putting down the detailed steps because, though tedious if performed directly, they are quite straightforward. They become simple if tensor calculus is used. The approriate direct formulae may be found in refs. [3, 6, 7], whereas their tensorial counterparts are given in ref. [11]. Such a derivation would show that

$$
\sigma_x{}' = \lambda \operatorname{div} \boldsymbol{w} + 2 \mu \frac{\partial u}{\partial x}
$$

$$
\sigma_y{}' = \lambda \operatorname{div} \boldsymbol{w} + 2 \mu \frac{\partial v}{\partial y} \qquad (3.22\,\mathrm{a})
$$

$$
\sigma_z{}' = \lambda \operatorname{div} \boldsymbol{w} + 2 \mu \frac{\partial w}{\partial z} \; ;
$$

$$
\tau_{xy} = \tau_{yx} = \mu \left(\frac{\partial v}{\partial x} + \frac{\partial u}{\partial y} \right)
$$

$$
\tau_{yz} = \tau_{zy} = \mu \left(\frac{\partial w}{\partial y} + \frac{\partial v}{\partial z} \right) \qquad (3.22\,\mathrm{b})
$$

$$
\tau_{zx} = \tau_{xz} = \mu \left(\frac{\partial u}{\partial z} + \frac{\partial w}{\partial x} \right) \; ;
$$

where div w has been used for brevity. The reader may notice the regularity with which the indices x, y, z, the components u, v, w, and the coordinates x, y, z are permuted†.

Applying these equations to the simple case represented in Fig. 1.1, we recover eqn. (1.2) and so confirm that the preceding more general relation reduces to Newton's law of friction in the case of simple shear and does, therefore, constitute its proper generalization. At the same time, we identify the factor μ with the viscosity of the fluid, amply discussed in Sec Ib, and, incidentally, justify the factor 2 previously inserted into eqns. (3.21). The physical significance of the second factor, λ, requires further discussion, but we note that it plays no part in an incompressible fluid when div $w = 0$; it then disappears from the equations altogether, and so is seen to be important for compressible fluids only.

e. Stokes's hypothesis

Although the problem that we are about to discuss has arisen more than a century and a half ago, the physical interpretation of the second factor, λ, in eqns. (3.21) or (3.22a, b) and for flows in which div w does not vanish identically, is still being disputed, even though the *value* which should be given to it *in the working equations* is not. This numerical value is determined with the aid of a hypothesis advanced by G. G. Stokes in 1845 [13]. Without, for the moment, concerning ourselves with the physical reasons which justify *Stokes's hypothesis*, we first state that according to it, it is necessary to assume

$$3\,\lambda + 2\,\mu = 0\,, \quad \text{or} \quad \lambda = -\,\frac{2}{3}\,\mu\,. \tag{3.23}$$

This relates the value of the factor λ to the viscosity, μ, of the compressible fluid and reduces the number of properties which characterize the field of stresses in a flowing compressible fluid from two to one, that is to the same number as is required for an incompressible fluid.

Substituting this value into eqns. (3.22a), we obtain the normal components of deviatoric stress:

$$\left.\begin{aligned}
\sigma_x{}' &= -\,\frac{2}{3}\,\mu \operatorname{div} w + 2\,\mu\,\frac{\partial u}{\partial x} \\[2mm]
\sigma_y{}' &= -\,\frac{2}{3}\,\mu \operatorname{div} w + 2\,\mu\,\frac{\partial v}{\partial y} \\[2mm]
\sigma_z{}' &= -\,\frac{2}{3}\,\mu \operatorname{div} w + 2\,\mu\,\frac{\partial w}{\partial z}\,,
\end{aligned}\right\} \tag{3.24}$$

---

† The above set of six equations can be contracted to a single one in Cartesian-tensor notation (with Einstein's summation convention):

$$\sigma_{ij}{}' = \lambda\,\delta_{ij}\,\frac{\partial v_k}{\partial x_k} + \mu\left(\frac{\partial v_i}{\partial x_j} + \frac{\partial v_j}{\partial x_i}\right), \quad (i, j, k = 1, 2, 3)$$

where the Kronecker delta $\delta_{ij} = 0$ for $i \neq j$ and $\delta_{ij} = 1$ for $i = j$.

the shearing stresses remaining unchanged. Making use of eqns. (3.20), we obtain the so-called *constitutive equation* for an isotropic, Newtonian fluid

$$
\left.
\begin{aligned}
\sigma_x &= -p - \frac{2}{3}\,\mu\,\text{div}\,\boldsymbol{w} + 2\,\mu\,\frac{\partial u}{\partial x} \\[2mm]
\sigma_y &= -p - \frac{2}{3}\,\mu\,\text{div}\,\boldsymbol{w} + 2\,\mu\,\frac{\partial v}{\partial y} \\[2mm]
\sigma_z &= -p - \frac{2}{3}\,\mu\,\text{div}\,\boldsymbol{w} + 2\,\mu\,\frac{\partial w}{\partial z}
\end{aligned}
\right\}
\tag{3.25a}
$$

$$
\left.
\begin{aligned}
\tau_{xy} &= \tau_{yx} = \mu\left(\frac{\partial v}{\partial x} + \frac{\partial u}{\partial y}\right) \\[2mm]
\tau_{yz} &= \tau_{zy} = \mu\left(\frac{\partial w}{\partial y} + \frac{\partial v}{\partial z}\right) \\[2mm]
\tau_{zx} &= \tau_{xz} = \mu\left(\frac{\partial u}{\partial z} + \frac{\partial w}{\partial x}\right)
\end{aligned}
\right\}
\tag{3.25b}
$$

in its final form, noting that p represents the local thermodynamic pressure[†].

Regarded as a pure hypothesis, or even guess, eqn. (3.23) can certainly be accepted on the ground that the working equations which result from the substitution of eqns. (3.25a, b) into (3.11) have been subjected to an unusually large number of experimental verifications, even under quite extreme conditions, as the reader will concede after having studied this book. Thus, even if it should not represent the state of affairs exactly, it certainly constitutes an excellent approximation.

Since the deviatoric components are the only ones which arise in motion, they represent those components of stress which produce dissipation in an isothermal flow, there being further dissipation in a temperature field due to thermal conduction, Chap. XII. Furthermore, since the factor λ occurs only in the normal components $\sigma_x{}', \sigma_y{}', \sigma_z{}'$ which also contain the thermodynamic pressure, eqns. (3.20), it becomes clear that the physical significance of λ is connected with the mechanism of dissipation when the volume of the fluid element is changed at a finite rate as well as with the relation between the total stress tensor and thermodynamic pressure.

f. Bulk viscosity and thermodynamic pressure

We now revert to the general discussion, without necessarily accepting the validity of Stokes's hypothesis, but confine it to the case when no shearing stresses are involved, because their physical significance and origin is clear. Consequently,

---

† In the compact tensorial notation we would write

$$
\sigma_{ij} = -p\,\delta_{ij} + \mu\left(\frac{\partial v_i}{\partial x_j} + \frac{\partial v_j}{\partial x_i} - \frac{2}{3}\,\delta_{ij}\,\frac{\partial v_k}{\partial x_k}\right) \quad (i, j, k = 1, 2, 3)\,.
$$

we consider a fluid system, say the sphere shown in Fig. 3.8a which is subjected to a uniform normal stress, $\bar{\sigma}$, on its boundary. In the absence of motion $\bar{\sigma}$ is obviously equal and opposite in sign to the thermodynamic pressure, p. Taking the sum of the three equations (3.21) and utilizing eqns. (3.20), we find that

$$\bar{\sigma} = - p + (\lambda + \frac{2}{3} \mu) \operatorname{div} \boldsymbol{w} , \tag{3.26}$$

and notice that our equations reflect this fact, as already pointed out earlier. Now, the question poses itself as to what this relation should be in a general flow field.

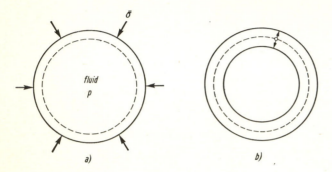

Fig. 3.8.   Quasistatic compression and oscillatory motion of a spherical mass of fluid

When the system is compressed quasistatically and reversibly, we again recover the previous case because then $\operatorname{div} \boldsymbol{w} \to 0$ asymptotically. We note that in such cases the rate at which work is performed in a thermodynamically reversible process per unit volume becomes

$$\dot{W} = p \operatorname{div} \boldsymbol{w} \tag{3.26a}$$

which is the same as

$$\dot{W} = p \, \frac{\mathrm{d}V}{\mathrm{d}t} \tag{3.26b}$$

in the notation customary in thermodynamics.

When $\operatorname{div} \boldsymbol{w}$ is finite, and the fluid is compressed, expanded or made to oscillate, at a finite rate, equality between $\bar{\sigma}$ and $- p$ persists only if the coefficient

$$\mu' = \lambda + \frac{2}{3} \mu \tag{3.27}$$

vanishes identically (Stokes's hypothesis); otherwise it does not. If $\mu' \neq 0$, the oscillatory motion of a spherical system, Fig. 3.8b, would produce dissipation, even if the temperature remained constant throughout the bulk of the gas. The same would be true in the case of expansion or compression at a finite rate. For this reason, the coefficient μ' is called the *bulk viscosity* of the fluid: it represents that property, like the shear viscosity μ for deformation in shape, which is responsible for energy dissipation in a fluid of uniform temperature during a change in volume at a finite rate. The bulk viscosity would thus constitute a second property of a compressible, isotropic, Newtonian fluid needed to determine its constitutive equation and would have to be measured in addition to μ. It is evident that

$$\mu' = 0 \quad \text{implies} \quad p = -\bar{\sigma}$$
$$\mu' \neq 0 \quad \text{implies} \quad p \neq -\bar{\sigma} \,.$$

Thus the acceptance of Stokes's hypothesis is equivalent to the assumption that the thermodynamic pressure p is equal the one-third of the invariant sum of normal stresses even in cases when compression or expansion proceeds at a finite rate. Furthermore, it is also equivalent to the assumption that the oscillatory motion of a large spherical system would be reversible if it were isothermal. More detailed considerations in terms of the concepts of thermodynamics as it applies to irreversible processes in continuous systems can be found in the works of J. Meixner [8], I. Prigogine [12] and S. R. de Groot and P. Mazur [1].

In order to determine under what conditions the bulk viscosity of a compressible fluid vanishes, it is necessary to have recourse to experiment or to the methods of statistical thermodynamics which permit us to calculate transport coefficients from first principles. The direct measurement of bulk viscosity is very difficult to perform, and no definitive results are in existence. Statistical methods for dense gases or liquids have not yet been developed to a point which would allow us to make a complete statement on the subject. It appears, however, that the bulk viscosity vanishes identically in gases of low density, that is under conditions when only binary collisions of molecules need to be taken into account. In dense gases, the numerical value of bulk viscosity appears to be very small. This means that eqns. (3.26a,b) continue to describe the work in a continuous system in the absence of shear to an excellent degree of approximation and that dissipation at constant temperature, even in the general case, occurs only through the intervention of the deviatoric stresses. Thus, once again, we are led to Stokes's hypothesis and so to eqn. (3.26). This conclusion does not extend to fluids which are capable of undergoing relaxation processes by virtue of a local departure from a state of chemical equilibrium [1,8]. Such relaxation processes occur, for example, when a chemical reaction can take place, or, in gases of complex structure, when a comparatively slow transfer of energy between the translational and rotational degrees of freedom on the one hand, and the vibrational degrees of freedom on the other, becomes possible. Thus when relaxation processes are possible, the thermodynamic pressure is no longer equal to one-third of the trace of the stress tensor.

It is sometimes argued that the adoption of Stokes's hypothesis, that is the supposition that the bulk viscosity of Newtonian fluid vanishes, does not accord with our intuitive feeling that a sphere of fluid whose boundary oscillates so that there is a cyclic sequence of compression and expansion, Fig. 3.8b, would dissipate no energy. This would, indeed, be the case, as is easily seen from the preceding argument, because the dissipative part of the stress field vanishes under such conditions. It must, however, not be forgotten that such a conclusion is valid only if the temperature of the sphere of gas were to be kept constant during the oscillation throughout the whole volume. Normally this is impossible. Consequently, an oscillating sphere of gas will soon develop a temperature field and energy will be dissipated down the existing temperature gradients [5].

g. The Navier-Stokes equations

With the aid of eqns. (3.20) the non-viscous pressure terms can be separated in the equation of motion (3.11) so that they become

$$\varrho \frac{Du}{Dt} = X - \frac{\partial p}{\partial x} + \left(\frac{\partial \sigma_x'}{\partial x} + \frac{\partial \tau_{xy}}{\partial y} + \frac{\partial \tau_{xz}}{\partial z} \right)$$

$$\varrho \frac{Dv}{Dt} = Y - \frac{\partial p}{\partial y} + \left(\frac{\partial \tau_{xy}}{\partial x} + \frac{\partial \sigma_y'}{\partial y} + \frac{\partial \tau_{yz}}{\partial z} \right) \qquad (3.28)$$

$$\varrho \frac{Dw}{Dt} = Z - \frac{\partial p}{\partial z} + \left(\frac{\partial \tau_{xz}}{\partial x} + \frac{\partial \tau_{yz}}{\partial y} + \frac{\partial \sigma_z'}{\partial z} \right).$$

Introducing the constitutive relation from eqns. (3.24) we obtain the resultant surface force in terms of the velocity components, e. g. for the x-direction we obtain with the aid of eqn. (3.10a):

$$P_x = \frac{\partial \sigma_x}{\partial x} + \frac{\partial \tau_{xy}}{\partial y} + \frac{\partial \tau_{xz}}{\partial z} = -\frac{\partial p}{\partial x} + \frac{\partial \sigma_x'}{\partial x} + \frac{\partial \tau_{xy}}{\partial y} + \frac{\partial \tau_{xz}}{\partial z}$$

$$P_x = -\frac{\partial p}{\partial x} + \frac{\partial}{\partial x} \left[2\,\mu \frac{\partial u}{\partial x} - \frac{2}{3} \mu \operatorname{div} w \right] + \frac{\partial}{\partial y} \left[\mu \left(\frac{\partial u}{\partial y} + \frac{\partial v}{\partial x} \right) \right] + \frac{\partial}{\partial z} \left[\mu \left(\frac{\partial w}{\partial x} + \frac{\partial u}{\partial z} \right) \right]$$

and corresponding expressions for the y- and z-components. In the general case of a compressible flow, the viscosity μ must be regarded as dependent on the space coordinates, because μ varies considerably with temperature (Tables 1.2 and 12.1), and the changes in velocity and pressure together with the heat due to friction bring about considerable temperature variations. The temperature dependence of viscosity μ(T) must be obtained from experiments (*cf.* Sec. XIII a).

If these expressions are introduced into the fundamental equations (3.11), we obtain

$$\varrho \frac{Du}{Dt} = X - \frac{\partial p}{\partial x} + \frac{\partial}{\partial x} \left[\mu \left(2\frac{\partial u}{\partial x} - \frac{2}{3} \operatorname{div} w \right) \right] + \frac{\partial}{\partial y} \left[\mu \left(\frac{\partial u}{\partial y} + \frac{\partial v}{\partial x} \right) \right] + \frac{\partial}{\partial z} \left[\mu \left(\frac{\partial w}{\partial x} + \frac{\partial u}{\partial z} \right) \right]$$

$$\varrho \frac{Dv}{Dt} = Y - \frac{\partial p}{\partial y} + \frac{\partial}{\partial y} \left[\mu \left(2\frac{\partial v}{\partial y} - \frac{2}{3} \operatorname{div} w \right) \right] + \frac{\partial}{\partial z} \left[\mu \left(\frac{\partial v}{\partial z} + \frac{\partial w}{\partial y} \right) \right] + \frac{\partial}{\partial x} \left[\mu \left(\frac{\partial u}{\partial y} + \frac{\partial v}{\partial x} \right) \right]$$

$$\varrho \frac{Dw}{Dt} = Z - \frac{\partial p}{\partial z} + \frac{\partial}{\partial z} \left[\mu \left(2\frac{\partial w}{\partial z} - \frac{2}{3} \operatorname{div} w \right) \right] + \frac{\partial}{\partial x} \left[\mu \left(\frac{\partial w}{\partial x} + \frac{\partial u}{\partial z} \right) \right] + \frac{\partial}{\partial y} \left[\mu \left(\frac{\partial v}{\partial z} + \frac{\partial w}{\partial y} \right) \right].$$

$$(3.29 \text{a, b, c})\dagger$$

These very well known differential equations form the basis of the whole science of fluid mechanics. They are usually referred to as the Navier-Stokes equations.

---

† In indicial notation:

$$\varrho \left(\frac{\partial v_i}{\partial t} + v_j \frac{\partial v_i}{\partial x_j} \right) = X_i - \frac{\partial p}{\partial x_i} + \frac{\partial}{\partial x_j} \left\{ \mu \left(\frac{\partial v_i}{\partial x_j} + \frac{\partial v_j}{\partial x_i} - \frac{2}{3} \delta_{ij} \frac{\partial v_k}{\partial x_k} \right) \right\} \quad (i, j, k = 1, 2, 3) \,.$$

It is necessary to include here the equation of continuity which, as seen from eqn. (3.1), assumes the following form for compressible flow:

$$\frac{\partial \varrho}{\partial t} + \frac{\partial (\varrho\, u)}{\partial x} + \frac{\partial (\varrho\, v)}{\partial y} + \frac{\partial (\varrho\, w)}{\partial z} = 0 \,. \tag{3.30}$$

The above equations do not give a complete description of the motion of a compressible fluid because changes in pressure and density effect temperature variations, and principles of thermodynamics must, therefore, once more enter into the considerations. From thermodynamics we obtain, in the first place, the characteristic equation (equation of state) which combines pressure, density, and temperature, and which for a perfect gas has the form

$$p - \varrho\, R\, T = 0 \,, \tag{3.31}$$

with R denoting the gas constant and T denoting the absolute temperature. Secondly, if the process is not isothermal, it is further necessary to make use of the energy equation which draws up a balance between heat and mechanical energy (First Law of Thermodynamics), and which furnishes a differential equation for the temperature distribution. The energy equation will be discussed in greater detail in Chap. XII. The final equation of the system is given by the empirical viscosity law $\mu(T)$, its dependence on pressure being, normally, neglected. In all, if the forces X, Y, Z are considered given, there are seven equations for the seven variables $u, v, w, p, \varrho, T, \mu$.

For isothermal processes these reduce to five equations (3.29 a, b, c), (3.30) and (3.31) for the five unknowns u, v, w, p, ϱ.

Incompressible flow: The above system of equations becomes further simplified in the case of incompressible fluids ($\varrho = $ const) even if the temperature is not constant. First, as already shown in eqn. (3.1a), we have div $\boldsymbol{w} = 0$. Secondly, since temperature variations are, generally speaking, small in this case, the viscosity may be taken to be constant†.

The equation of state as well as the energy equation become superfluous as far as the calculation of the field of flow is concerned. The field of flow can now be considered independently from the equations of thermodynamics. The equations of motion (3.29 a, b, c) and (3.30) can be simplified and, if the acceleration terms are written out fully, they assume the following form:

$$\left.\begin{aligned}
\varrho\left(\frac{\partial u}{\partial t} + u\,\frac{\partial u}{\partial x} + v\,\frac{\partial u}{\partial y} + w\,\frac{\partial u}{\partial z}\right) &= X - \frac{\partial p}{\partial x} + \mu\left(\frac{\partial^2 u}{\partial x^2} + \frac{\partial^2 u}{\partial y^2} + \frac{\partial^2 u}{\partial z^2}\right) \\[2mm]
\varrho\left(\frac{\partial v}{\partial t} + u\,\frac{\partial v}{\partial x} + v\,\frac{\partial v}{\partial y} + w\,\frac{\partial v}{\partial z}\right) &= Y - \frac{\partial p}{\partial y} + \mu\left(\frac{\partial^2 v}{\partial x^2} + \frac{\partial^2 v}{\partial y^2} + \frac{\partial^2 v}{\partial z^2}\right) \\[2mm]
\varrho\left(\frac{\partial w}{\partial t} + u\,\frac{\partial w}{\partial x} + v\,\frac{\partial w}{\partial y} + w\,\frac{\partial w}{\partial z}\right) &= Z - \frac{\partial p}{\partial z} + \mu\left(\frac{\partial^2 w}{\partial x^2} + \frac{\partial^2 w}{\partial y^2} + \frac{\partial^2 w}{\partial z^2}\right)
\end{aligned}\right\} \tag{3.32 a, b, c}$$

$$\frac{\partial u}{\partial x} + \frac{\partial v}{\partial y} + \frac{\partial w}{\partial z} = 0 \,. \tag{3.33}$$

---

† This condition is more nearly satisfied in gases than in liquids.

With known body forces there are four equations for the four unknowns u, v, w, p.

If vector notation is used the simplified Navier-Stokes equations for incompressible flow, eqns. (3.32 a, b, c), can be shortened to

$$\varrho \, \frac{\mathbf{D}w}{\mathbf{D}t} = F - \mathrm{grad} \, p + \mu \, \nabla^2 \, w \, , \tag{3.34}$$

where the symbol ∇^2 denotes the Laplace operator, $\nabla^2 = \partial^2/\partial x^2 + \partial^2/\partial y^2 + \partial^2/\partial z^2$. The above Navier-Stokes equations differ from Euler's equations of motion by the viscous terms $\mu \, \nabla^2 \, w$.

The solutions of the above equations become fully determined physically when the boundary and initial conditions are specified. In the case of viscous fluids the condition of no slip on solid boundaries must be satisfied, i. e., on a wall both the normal and tangential components of the velocity must vanish:

$$v_n = 0 \, , \qquad v_t = 0 \text{ on solid walls} \, . \tag{3.35}$$

The equations under discussion were first derived by M. Navier [9] in 1827 and by S. D. Poisson [10] in 1831, on the basis of an argument which involved the consideration of intermolecular forces. Later the same equations were derived without the use of any such hypotheses by B. de Saint Venant [14] in 1843 and by G. G. Stokes [13] in 1845. Their derivations were based on the same assumption as made here, namely that the normal and shearing stresses are linear functions of the rate of deformation, in conformity with the older law of friction, due to Newton, and that the thermodynamic pressure is equal to one-third of the sum of the normal stresses taken with an opposite sign.

Since the hypothesis of linearity is evidently completely arbitrary, it is not a priori certain that the Navier-Stokes equations give a true description of the motion of a fluid. It is, therefore, necessary to verify them, and that can only be achieved by experiment. In this connexion it should, in any case, be noted that the enormous mathematical difficulties encountered when solving the Navier-Stokes equations have so far prevented us from obtaining a single analytic solution in which the convective terms interact in a general way with the friction terms. However, known solutions, such as laminar flow through a circular pipe, as well as boundary-layer flows, to be discussed later, agree so well with experiment that the general validity of the Navier-Stokes equations can hardly be doubted.

Cylindrical coordinates: We shall now transform the Navier-Stokes equations to cylindrical coordinates for future reference. If r, ϕ, z denote the radial, azimuthal, and axial coordinates, respectively, of a three-dimensional system of coordinates, and v_r, v_ϕ, v_z denote the velocity components in the respective directions, then the transformation of variables [3, 11] for the case of incompressible fluid flow, eqns. (3.33) and (3.34), leads to the following system of equations:

$$\varrho \left(\frac{\partial v_r}{\partial t} + v_r \frac{\partial v_r}{\partial r} + \frac{v_\phi}{r} \frac{\partial v_r}{\partial \phi} - \frac{v_\phi^2}{r} + v_z \frac{\partial v_r}{\partial z} \right) =$$

$$= F_r - \frac{\partial p}{\partial r} + \mu \left(\frac{\partial^2 v_r}{\partial r^2} + \frac{1}{r} \frac{\partial v_r}{\partial r} - \frac{v_r}{r^2} + \frac{1}{r^2} \frac{\partial^2 v_r}{\partial \phi^2} - \frac{2}{r^2} \frac{\partial v_\phi}{\partial \phi} + \frac{\partial^2 v_r}{\partial z^2} \right) \tag{3.36 a}$$

$$\varrho\left(\frac{\partial v_\phi}{\partial t} + v_r\frac{\partial v_\phi}{\partial r} + \frac{v_\phi}{r}\frac{\partial v_\phi}{\partial \phi} + \frac{v_r\,v_\phi}{r} + v_z\frac{\partial v_\phi}{\partial z}\right) =$$

$$= F_\phi - \frac{1}{r}\frac{\partial p}{\partial \phi} + \mu\left(\frac{\partial^2 v_\phi}{\partial r^2} + \frac{1}{r}\frac{\partial v_\phi}{\partial r} - \frac{v_\phi}{r^2} + \frac{1}{r^2}\frac{\partial^2 v_\phi}{\partial \phi^2} + \frac{2}{r^2}\frac{\partial v_r}{\partial \phi} + \frac{\partial^2 v_\phi}{\partial z^2}\right) \quad (3.36\,\text{b})$$

$$\varrho\left(\frac{\partial v_z}{\partial t} + v_r\frac{\partial v_z}{\partial r} + \frac{v_\phi}{r}\frac{\partial v_z}{\partial \phi} + v_z\frac{\partial v_z}{\partial z}\right) =$$

$$= F_z - \frac{\partial p}{\partial z} + \mu\left(\frac{\partial^2 v_z}{\partial r^2} + \frac{1}{r}\frac{\partial v_z}{\partial r} + \frac{1}{r^2}\frac{\partial^2 v_z}{\partial \phi^2} + \frac{\partial^2 v_z}{\partial z^2}\right) \quad (3.36\,\text{c})$$

$$\frac{\partial v_r}{\partial r} + \frac{v_r}{r} + \frac{1}{r}\frac{\partial v_\phi}{\partial \phi} + \frac{\partial v_z}{\partial z} = 0. \quad (3.36\,\text{d})$$

The stress components assume the form

$$\begin{aligned}
\sigma_r &= -p + 2\mu\frac{\partial v_r}{\partial r}\,; & \tau_{r\phi} &= \mu\left[r\frac{\partial}{\partial r}\left(\frac{v_\phi}{r}\right) + \frac{1}{r}\frac{\partial v_r}{\partial \phi}\right] \\
\sigma_\phi &= -p + 2\mu\left(\frac{1}{r}\frac{\partial v_\phi}{\partial \phi} + \frac{v_r}{r}\right)\,; & \tau_{\phi z} &= \mu\left(\frac{\partial v_\phi}{\partial z} + \frac{1}{r}\frac{\partial v_z}{\partial \phi}\right) \\
\sigma_z &= -p + 2\mu\frac{\partial v_z}{\partial z}\,; & \tau_{rz} &= \mu\left(\frac{\partial v_r}{\partial z} + \frac{\partial v_z}{\partial r}\right).
\end{aligned} \right\} \quad (3.37)$$

Curvilinear coordinates: It is often useful to employ a curvilinear system of coordinates which is adapted to the shape of the body. In the case of two-dimensional flow along a curved wall, it is possible to select a coordinate system whose abscissa, x, is measured along the wall, the ordinate, y, being measured at right angles to it, Fig. 3.9. Thus the curvilinear net consists of curves which are parallel to the wall

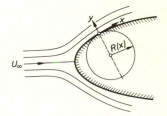

Fig. 3.9. Two-dimensional boundary layer along a curved wall

and of straight lines perpendicular to them. The corresponding velocity components are denoted by u and v, respectively. The radius of curvature at position x is denoted by $R(x)$; it is positive for walls which are convex outwards, and negative when the wall is concave. The appropriate form of the complete Navier-Stokes equations has been derived by W. Tollmien [15]. They are:

$$\frac{\partial u}{\partial t} + \frac{R}{R+y} u \frac{\partial u}{\partial x} + v \frac{\partial u}{\partial y} + \frac{v\,u}{R+y} = -\frac{R}{R+y} \frac{1}{\varrho} \frac{\partial p}{\partial x} +$$

$$+ \nu \left\{ \frac{R^2}{(R+y)^2} \frac{\partial^2 u}{\partial x^2} + \frac{\partial^2 u}{\partial y^2} + \frac{1}{R+y} \frac{\partial u}{\partial y} - \frac{u}{(R+y)^2} + \right. \tag{3.38a}$$

$$\left. + \frac{2R}{(R+y)^2} \frac{\partial v}{\partial x} - \frac{R}{(R+y)^3} \frac{dR}{dx} v + \frac{Ry}{(R+y)^3} \frac{dR}{dx} \frac{\partial u}{\partial x} \right\};$$

$$\frac{\partial v}{\partial t} + \frac{R}{R+y} u \frac{\partial v}{\partial x} + v \frac{\partial v}{\partial y} - \frac{u^2}{R+y} = -\frac{1}{\varrho} \frac{\partial p}{\partial y} + \nu \left\{ \frac{\partial^2 v}{\partial y^2} - \frac{2R}{(R+y)^2} \frac{\partial u}{\partial x} + \right.$$

$$+ \frac{1}{R+y} \frac{\partial v}{\partial y} + \frac{R^2}{(R+y)^2} \frac{\partial^2 v}{\partial x^2} - \frac{v}{(R+y)^2} + \tag{3.38b}$$

$$\left. + \frac{R}{(R+y)^3} \frac{dR}{dx} u + \frac{Ry}{(R+y)^3} \frac{dR}{dx} \frac{\partial v}{\partial x} \right\};$$

$$\frac{R}{R+y} \frac{\partial u}{\partial x} + \frac{\partial v}{\partial y} + \frac{v}{R+y} = 0. \tag{3.38c}$$

The stress components are

$$\left.\begin{aligned} \sigma_x &= -p + 2\mu \left(\frac{R}{R+y} \frac{\partial u}{\partial x} + \frac{v}{R+y} \right) \\ \sigma_y &= -p + 2\mu \frac{\partial v}{\partial y} \\ \tau_{xy} &= \mu \left(\frac{\partial u}{\partial y} - \frac{u}{R+y} + \frac{R}{R+y} \frac{\partial v}{\partial x} \right), \end{aligned}\right\} \tag{3.39}$$

and the vorticity [see eqn. (4.5)] becomes

$$\omega = \frac{1}{2} \left(\frac{R}{R+y} \frac{\partial v}{\partial x} - \frac{\partial u}{\partial y} - \frac{1}{R+y} u \right). \tag{3.40}$$

References

[1] de Groot, S.R., and Mazur, P.: Non-equilibrium thermodynamics. North-Holland Publ. Co., 1962.
[2] Föppl, A.: Vorlesungen über technische Mechanik. Vol. 5, Teubner, Leipzig, 1922.
[3] Hopf, L.: Zähe Flüssigkeiten. Contribution to: Handbuch der Physik, Vol. *VII* (H. Geiger and K. Scheel, ed.), Berlin, 1927.
[4] Kestin, J.: A course in thermodynamics. Vol. *1*, Blaisdell, 1966.
[5] Kestin, J.: Etude thermodynamique des phénomènes irréversibles. Rep. No. 66—7, Lab. d'Aérothermique, Meudon, 1966.
[6] Lamb, H.: Hydrodynamics. 6th ed., Cambridge, 1957; also Dover, 1945.
[7] Love, A.E.H.: The mathematical theory of elasticity. 4th ed., Cambridge Univ. Press, 1952.
[8] Meixner, J., and Reik, H.G.: Thermodynamik der irreversiblen Prozesse. Contribution to Handbuch der Physik, Vol. *III/2* (S. Flügge, ed.), Springer, 1959, pp. 413—523.

[9] Navier, M.: Mémoire sur les lois du mouvement des fluides. Mém. de l'Acad. de Sci. *6*, 389—416 (1827).

[10] Poisson, S.D.: Mémoire sur les équations générales de l'équilibre et du mouvement des corps solides élastiques et des fluides. J. de l'Ecole polytechn. *13*, 139—186 (1831).

[11] Prager, W.: Introduction to mechanics of continua. Ginn & Co., 1961.

[12] Prigogine, I.: Etude thermodynamique des phénomènes irréversibles, Dunod-Desoer, 1947.

[13] Stokes, G.G.: On the theories of internal friction of fluids in motion. Trans. Cambr. Phil. Soc. *8*, 287—305 (1845).

[14] de St. Venant, B.: Note à joindre un mémoire sur la dynamique des fluides. Comptes Rendus *17*, 1240—1244 (1843).

[15] Tollmien, W.: Grenzschichttheorie. Handbuch der Exper.-Physik, Vol. *IV*, Part. 1, 241—287 (1931).

CHAPTER IV

General properties of the Navier-Stokes equations

Before passing on to the integration of the Navier-Stokes equations in the following chapters, it now seems pertinent to discuss some of their general properties. In doing so we shall restrict ourselves to incompressible viscous fluids.

a. Derivation of Reynolds's principle of similarity from the Navier-Stokes equations

Until the present day no general analytic methods have become available for the integration of the Navier-Stokes equations. Furthermore, solutions which are valid for all values of viscosity are known only for some particular cases, e. g. for Poiseuille flow through a circular pipe, or for Couette flow between two parallel walls, one of which is at rest, the other moving along its own plane with a constant velocity (see Fig. 1.1). For this reason the problem of calculating the motion of a viscous fluid was attacked by first tackling limiting cases, that is, by solving problems for very large viscosities, on the one hand, and for very small viscosities on the other, because in this manner the mathematical problem is considerably simplified. However, the case of moderate viscosities cannot be interpolated between these two extremes.

Even the limiting cases of very large and very small viscosities present great mathematical difficulties so that research into viscous fluid motion proceeded to a large extent by experiment. In this connexion the Navier-Stokes equations furnish very useful hints which point to a considerable reduction in the quantity of experimental work required. It is often possible to carry out experiments on *models*, which means that in the experimental arrangement a geometrically similar model of the actual body, but reduced in scale, is investigated in a wind tunnel, or other suitable arrangement. This always raises the question of the *dynamic similarity* of fluid motions which is, evidently, intimately connected with the question of how far results obtained with models can be utilized for the prediction of the behaviour of the full-scale body.

As already explained in Chap. I, two fluid motions are dynamically similar if, with geometrically similar boundaries, the velocity fields are geometrically similar, i. e., if they have geometrically similar streamlines.

This question was answered in Chap. I for the case in which only inertia and viscous forces take part in the process. It was found there that for the two motions the Reynolds numbers must be equal (Reynolds's principle of similarity). This conclusion was drawn by estimating the forces in the stream; we now propose to deduce it again directly from the Navier-Stokes equations.

The Navier-Stokes equations express the condition of equilibrium, namely that for each particle there is equilibrium between body forces (weight), surface forces and inertia forces. The surface forces consist of pressure forces (normal forces) and friction forces (shear forces). Body forces are important only in cases when there is a free surface or when the density distribution is inhomogeneous. In the case of a homogeneous fluid in the absence of a free surface there is equilibrium between the weight of each particle and its hydrostatic buoyancy force, in the same way as at rest. Hence in the motion of a homogeneous fluid, in the absence of a free surface, body forces can be cancelled if pressure is taken to mean the difference between that in motion and at rest. In the following argument we shall restrict our attention to cases for which this assumption is true because they are the most important ones in applications. Thus the Navier-Stokes equations will now contain only forces due to pressure, viscosity, and inertia.

Under these assumptions and conventions the Navier-Stokes equations for an incompressible fluid, restricted to steady flow and in vector form, simplify to

$$\varrho \, (\boldsymbol{w} \cdot \text{grad}) \, \boldsymbol{w} = - \, \text{grad} \, p + \mu \, \nabla^2 \, \boldsymbol{w} \, . \tag{4.1}\dagger$$

This differential equation must be independent of the choice of the units for the various physical quantities, such as velocity, pressure, etc., which appear in it.

We now consider flows about two geometrically similar bodies of different linear dimensions in streams of different velocities, e. g., flows past two spheres in which the densities and viscosities may also be different. We shall investigate the condition for dynamic similarity with the aid of the Navier-Stokes equations. Evidently, dynamic similarity will prevail if with a suitable choice of the units of length, time, and force, the Navier-Stokes eqn. (4.1) is so transformed that it becomes identical for the two flows with geometrically similar boundaries. Now, it is possible to free oneself from the fortuitously selected units if dimensionless quantities are introduced into eqn. (4.1). This is achieved by selecting certain suitable characteristic magnitudes in the flow as our units, and by referring all others to them. Thus e. g., the free-stream velocity and the diameter of the sphere can be selected as the respective units of velocity and length.

Let V, l, and p_1 denote these characteristic reference magnitudes. If we now introduce into the Navier-Stokes eqn. (4.1) the dimensionless ratios

$$\text{velocity} \quad W = \frac{w}{V} \, ,$$

$$\text{lengths} \quad X = \frac{x}{l} \, , \quad Y = \frac{y}{l} \, , \quad Z = \frac{z}{l} \, ,$$

$$\text{pressure} \quad P = \frac{p}{p_1} \, ,$$

---

† See footnote on p. 48.

we obtain

$$\varrho \frac{V^2}{l} (W \cdot \text{grad}) W = - \frac{p_1}{l} \text{grad } P + \frac{\mu V}{l^2} \nabla^2 W .$$

or, dividing by $\varrho V^2/l$:

$$(W \cdot \text{grad}) W = - \frac{p_1}{\varrho V^2} \text{grad } P + \frac{\mu}{\varrho V l} \nabla^2 W . \qquad (4.2)\dagger$$

The fluid motions under consideration can become similar only if the solutions expressed in terms of the respective dimensionless variables are identical. This requires that for both motions the respective dimensionless Navier-Stokes equations differ only by a factor common to all terms. The quantity $p_1/\varrho V^2$ represents the ratio of pressure to the double of the dynamic head and is unimportant for the dynamic similarity of the two motions because in incompressible flow a change in pressure causes no change in volume. The second factor $\varrho V l/\mu$ is, however, very important and must assume the same value for both motions if they are to be dynamically similar. Hence dynamic similarity is assured if for the two motions

$$\frac{\varrho_1 V_1 l_1}{\mu_1} = \frac{\varrho_2 V_2 l_2}{\mu_2} .$$

This principle was discovered by Osborne Reynolds when he investigated fluid motion through pipes and is, therefore, known as the *Reynolds principle of similarity*. The dimensionless ratio

$$\frac{\varrho V l}{\mu} = \frac{V l}{\nu} = \mathsf{R} \qquad (4.3)$$

is called the Reynolds number. Here the ratio of the dynamic viscosity μ, to the density ϱ, denoted by $\nu = \mu/\varrho$, is the kinematic viscosity of the fluid, introduced earlier. Summing up we can state that flows about geometrically similar bodies are dynamically similar when the Reynolds numbers for the flows are equal.

Thus Reynolds's similarity principle has been deduced once more, this time from the Navier-Stokes equations, having been previously derived first from an estimation of forces and secondly from dimensional analysis.

b. Frictionless flow as "solutions" of the Navier-Stokes equations

It may be worth noting, parenthetically, that the solutions for incompressible *frictionless* flows may also be regarded as exact solutions of the Navier-Stokes equations, because in such cases the frictional terms vanish identically. In the case of incompressible, frictionless flows the velocity vector can be represented as the gradient of a potential:

$$w = \text{grad } \Phi ,$$

where the potential Φ satisfies the Laplace equation

$$\nabla^2 \Phi = 0 .$$

We then also have grad $(\nabla^2 \Phi) = \nabla^2 (\text{grad } \Phi) = 0$, that is, $\nabla^2 w = 0$.

---

† See footnote on p. 48.

Thus the frictional terms in eqn. (4.1) vanish identically for potential flows, but generally speaking both boundary conditions (3.35) for the velocity cannot then be satisfied simultaneously. If the normal component must assume prescribed values along a boundary, then, in potential flow, the tangential component is thereby determined so that the no slip condition cannot be satisfied at the same time. For this reason one cannot regard potential flows as physically meaningful solutions of the Navier-Stokes equations, because they do not satisfy the prescribed boundary conditions. There exists, however, an important exception to the preceding statement which occurs when the solid wall is in motion and when this condition does not apply. The simplest particular case is that of flow past a rotating cylinder when the potential solution does constitute a meaningful solution to the Navier-Stokes equations, as explained in greater detail on p. 80. The reader may refer to two papers, one by G. Hamel [4] and one by J. Ackeret [1], for further details.

The following sections will be restricted to the consideration of plane (two-dimensional) flows because for such cases only is it possible to indicate some general properties of the Navier-Stokes equations, and, on the other hand, plane flows constitute by far the largest class of problems of practical importance.

c. The Navier-Stokes equations interpreted as vorticity transport equations

In the case of two-dimensional non-steady flow in the x, y-plane the velocity vector becomes

$$w = i\,u(x,y,t) + j\,v(x,y,t)\,.$$

and the system of equations (3.32) and (3.33) transforms into

$$
\left.
\begin{aligned}
\frac{\partial u}{\partial t} + u\frac{\partial u}{\partial x} + v\frac{\partial u}{\partial y} &= \frac{1}{\varrho}X - \frac{1}{\varrho}\frac{\partial p}{\partial x} + \nu\left(\frac{\partial^2 u}{\partial x^2} + \frac{\partial^2 u}{\partial y^2}\right) \\
\frac{\partial v}{\partial t} + u\frac{\partial v}{\partial x} + v\frac{\partial v}{\partial y} &= \frac{1}{\varrho}Y - \frac{1}{\varrho}\frac{\partial p}{\partial y} + \nu\left(\frac{\partial^2 v}{\partial x^2} + \frac{\partial^2 v}{\partial y^2}\right) \\
\frac{\partial u}{\partial x} + \frac{\partial v}{\partial y} &= 0\,.
\end{aligned}
\right\}
\quad\text{(4.4\,a, b, c)}
$$

which furnishes three equations for u, v, and p.

We now introduce the vector of vorticity, curl w, which reduces to the one component about the z-axis for two-dimensional flow:

$$\frac{1}{2}\,\text{curl}\,w = \omega_z = \omega = \frac{1}{2}\left(\frac{\partial v}{\partial x} - \frac{\partial u}{\partial y}\right)\,. \tag{4.5}$$

Frictionless motions are irrotational so that curl $w = 0$ in such cases. Eliminating pressure from eqns. (4.4a, b) we obtain

$$\frac{\partial \omega}{\partial t} + u\frac{\partial \omega}{\partial x} + v\frac{\partial \omega}{\partial y} = \nu\left(\frac{\partial^2 \omega}{\partial x^2} + \frac{\partial^2 \omega}{\partial y^2}\right) \tag{4.6}$$

or, in shorthand form

$$\frac{D\omega}{Dt} = \nu\,\nabla^2\,\omega\,. \tag{4,7}$$

This equation is referred to as the *vorticity transport*, or *transfer*, *equation*. It states that the substantive variation of vorticity, which consists of the local and convective terms, is equal to the rate of dissipation of vorticity through friction. Eqn. (4.6), together with the equation of continuity (4.4c), form a system of two equations for the two velocity components u and v.

Finally, it is possible to transform these two equations with two unknowns into one equation with one unknown by introducing the stream function $\psi(x, y)$. Putting

$$u = \frac{\partial \psi}{\partial y} ; \quad v = -\frac{\partial \psi}{\partial x} , \tag{4.8}$$

we see that the continuity equation is satisfied automatically. In addition the vorticity from eqn. (4.5) becomes

$$\omega = -\tfrac{1}{2} \nabla^2 \psi , \dagger \tag{4.9}$$

and the vorticity transport equation (4.6) becomes

$$\frac{\partial \nabla^2 \psi}{\partial t} + \frac{\partial \psi}{\partial y} \frac{\partial \nabla^2 \psi}{\partial x} - \frac{\partial \psi}{\partial x} \frac{\partial \nabla^2 \psi}{\partial y} = \nu \nabla^4 \psi . \tag{4.10}$$

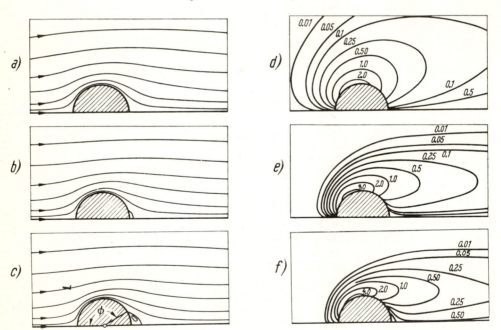

Fig. 4.1   Patterns of motion in a viscous flow past a sphere at different Reynolds numbers $R = VD/\nu$ derived from the vorticity transport equation (4.10) by V. G. Jenson [5].

a, b, c, Patterns of streamlines; d, e, f, Distribution of vorticity $\omega D/V = \text{const}$

a, d	$R = 5$,	$C_D = 8\cdot0$,	no separation
b, e	$R = 20$,	$C_D = 2\cdot9$,	separation at $\phi = 171°$
c, f	$R = 40$,	$C_D = 1\cdot9$,	separation at $\phi = 148°$

† This equation is sometimes called the Poisson equation for the streamfunction ψ.

In this form the vorticity transport equation contains only one unknown, ψ. The left-hand side of eqn. (4.10) contains, as was the case with the Navier-Stokes equations, the inertia terms, whereas the right-hand side contains the frictional terms. It is a fourth-order partial differential equation in the stream function ψ. Its solution in general terms is, again, very difficult, owing to its being non-linear.

V. G. Jenson [5] found a solution to the vorticity transport equation (4.10) for the case of a sphere by numerical integration. The resulting patterns of streamlines for different Reynolds numbers are seen plotted in Fig. 4.1 which also contains diagrams of the distribution of vorticity in the flow field. The smallest Reynolds number included, $R = 5$ in Figs. 4.1a and 4.1d, corresponds to the case when the viscous forces by far outweigh the inertia forces and the resulting flow can be described as creeping motion, Sec. IVd and Chapter VI. In this case the whole flow field is rotational and the patterns of streamlines forward and aft are nearly identical. As the Reynolds number is increased the sphere develops on its rear a separated region with back-flow and the intensity of vorticity is progressively more concentrated near the downstream portion of the sphere, whereas in the forward portion the flow becomes nearly irrotational. The flow patterns under consideration which have been deduced from the Navier-Stokes equation, allow us to recognize the characteristic changes which take place in the stream as the Reynolds number is made to increase, even if at the highest Reynolds number reached, $R = 40$ in Figs. 4.1c and 4.1f, the boundary layer pattern has not yet had a chance to develop fully.

Very extensive experimental investigations of the wake behind a circular cylinder in the range of Reynolds numbers $5 < R < 40$ are described in two papers by M. Coutanceau and R. Bouard [1c, 1d] who covered both steady and unsteady flows.

The development of very efficient electronic computers in modern times has made it possible to solve the Navier-Stokes equations for flow past geometrically simple bodies by purely numerical methods. In order to do this, the differential equations are replaced by difference equations. The numerical techniques used for this purpose will be explained in Sec. IX1. Without discussing this matter here in any depth, we quote one interesting result. Figure 4.2 shows the flow past a rectangular plate placed at right angles to the stream calculated by J. E. Fromm and F. H. Harlow [3]. At the back of the plate there forms a vortex street similar to that behind a circular cylinder shown in Figs. 1.6 and 2.7. Figure 4.2a shows an experimentally determined pattern of streamlines, whereas Fig. 4.2b represents the calculated field, both for a Reynolds number $Vd/\nu = 6000$. The agreement between the two patterns is remarkably good, in spite of the fact that in this range of Reynolds numbers the flow acquires an oscillatory character, Fig. 1.6. The earliest attempts to obtain such numerical solutions to the Navier-Stokes equations can be traced to A. Thom [6] who performed such calculations for a circular cylinder at the low Reynolds numbers $R = 10$ to 20. Later, the calculations were carried to $R = 100$ [2]. As the Reynolds number increases, the degree of difficulty of such numerical integrations increases steeply. In this connexion it is worth consulting the comprehensive summary by A. Thom and C. J. Apelt [7], as well as the work of C. J. Apelt [1a] and D. N. de G. Allen and R. V. Southwell [1b] and of H. B. Keller and H. Takami [5a].

Fig. 4.2a

Fig. 4.2. Pattern of streamlines behind a rectangular flat plate ($H/d = 1.6$) placed at right angle to the flow at a Reynolds number $\mathsf{R} = V\,H/\nu = 6000$, after J. E. Fromm and F. H. Harlow [3]. ($H =$ height of plate, $d =$ thickness of plate)
a) streamline pattern determined experimentally,
b) streamline pattern calculated by numerical integration of the Navier-Stokes equation for $\mathsf{T} = t\,V/H = 2.78$ ($t =$ time from start of motion). Numerical integration performed on an IBM 7090 computer

Fig. 4.2b

d. The limiting case of very large viscosity
(very small Reynolds number)

In very slow motions or in motions with very large viscosity the viscous forces are considerably greater than the inertia forces because the latter are of the order of the velocity squared, whereas the former are linear with velocity. To a first approximation it is possible to neglect the inertia terms with respect to the viscous terms so that from eqn. (4.10) we obtain

$$\nabla^4\,\psi = 0 \,. \tag{4.11}$$

This is, now, a linear equation which is considerably more amenable to mathematical treatment than the complete eqn. (4.10). Flows described by eqn. (4.11) proceed with very small velocities and are sometimes called *creeping motions*. The omission of the inertia terms is permissible from the mathematical point of view because the order of the equation is not thereby reduced, so that with the simplified differential eqn. (4.11) it is possible to satisfy as many boundary conditions as with the full eqn. (4.10).

Creeping motions can also be regarded as solutions of the Navier-Stokes equations in the *limiting case of very small Reynolds numbers* ($R \to 0$), because the Reynolds number represents the ratio of inertia to friction forces.

Solutions of eqn. (4.11) for the creeping motion of a viscous fluid were found by G. G. Stokes in the case of a sphere and by H. Lamb in the case of a circular cylinder. Stokes's solution can be applied to the falling of particles of mist in air, or to the motion of small spheres in a very viscous oil, when the velocities are so small that inertia forces can be neglected with good accuracy. Furthermore, the *hydrodynamic theory of lubrication*, i. e. the theory of the motion of lubricating oil in the very narrow channel between the journal and bearing uses this simplified equation of motion as its starting point. In the latter case it will be observed that if the velocities are not very small, the very small clearance heights, and the relatively large viscosity of the oil, ensure that the viscous forces are much larger than the inertia forces. However, apart from the theory of lubrication, the field of application of the theory of creeping motion is fairly insignificant.

e. The limiting case of very small viscous forces
(very large Reynolds numbers)

From the point of view of practical applications the second extreme case, namely that of very small viscous forces in eqn. (4.10) compared with the inertia forces, is of far greater importance. Since the two most important fluids, namely water and air, have very small viscosities, the case under consideration occurs, generally speaking, already at moderately high velocities. This is the *limiting case of very large Reynolds numbers* ($R \to \infty$). In this case the process of mathematical simplification of the differential eqn. (4.10) requires a considerable amount of care. It is not permissible simply to omit the viscous terms, i. e., the right-hand side of eqn. (4.10). This would reduce the order of the equation from four to two, and the solution of the simplified equation could not be made to satisfy the full boundary conditions of the original equation. The problem which was outlined in the preceding sentences belongs essentially to the realm of *boundary-layer theory*. We now propose to discuss briefly the general statements which can be made about the solutions of the Navier-Stokes equations for the special case of small viscous forces as compared with the inertia forces, that is in the limiting case of very large Reynolds numbers.

The following analogy may serve to illustrate the character of the solutions of the Navier-Stokes equations for the limiting case of very small viscosity, i. e., of very small friction terms, as compared with the inertia terms. The temperature distribution $\theta(x, y)$ about a hot body in a fluid stream is described by the following differential equation, Chap. XII:

$$\varrho\, c \left(\frac{\partial \theta}{\partial t} + u\, \frac{\partial \theta}{\partial x} + v\, \frac{\partial \theta}{\partial y} \right) = k \left(\frac{\partial^2 \theta}{\partial x^2} + \frac{\partial^2 \theta}{\partial y^2} \right) . \qquad (4.12)$$

Here ϱ, c, and k denote the density, specific heat, and conductivity of the fluid respectively; θ is the difference between the local temperature and that at a very large distance from the body, where the temperature, T, is constant and equal to T_∞, i. e., $\theta = T - T_\infty$. The velocity field $u(x, y)$ and $v(x, y)$ in eqn. (4.12) is assumed to be known. The temperature distribution on the boundaries of the body defined by $T_0 \gtreqless T_\infty$ is prescribed and in the simplest case it is constant with respect to space and time but, generally speaking, it varies with both. From the physical point of view eqn. (4.12) represents the heat balance for an elementary volume. The left-hand side represents the quantity of heat exchanged by convection, whereas the right-hand side is the quantity of heat exchanged by conduction. The frictional heat generated in the fluid is neglected. If $T_0 > T_\infty$ the problem is that of determining the temperature field around a hot body which is cooled. By inspection it is seen that eqn. (4.12) is of the same form as eqn. (4.6) for the vorticity ω. In fact they become identical if the vorticity is replaced by the temperature difference and the kinematic viscosity ν by the ratio $k/\varrho\, c$ known as the thermal diffusivity. The boundary condition $\theta = 0$ at a large distance from the body corresponds to the condition $\omega = 0$ for the undisturbed parallel stream also at a large distance from the body. Hence we may expect that the solutions of the two equations, i. e. the distribution of vorticity and that of temperature around the body will be similar in character.

Now, the temperature distribution around the body may be perceived intuitively, to a certain extent. In the limiting case of zero velocity (fluid at rest) the influence of the heated body will extend uniformly on all sides. With very small velocities the fluid around the body will still be affected by it in all directions. With increasing velocity of flow, however, it is clearly seen that the region affected by the higher temperature of the body shrinks more and more into a narrow zone in the immediate vicinity of the body, and into a tail of heated fluid behind it, Fig. 4.3.

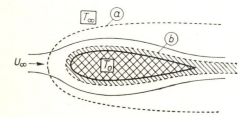

Fig. 4.3. Analogy between temperature and vorticity distribution in the neighbourhood of a body placed in a stream of fluid

a), b) Limits of region of increased temperature
a) for small velocities
b) for large velocities of flow

The solution of eqn. (4.12) must, as mentioned, be of a character similar to that for vorticity. At small velocities (viscous forces large compared with inertia forces) there is vorticity in the whole region of flow around the body. On the other hand for large velocities (viscous forces small compared with inertia forces), we may expect a field of flow in which vorticity is confined to a small layer along the surface of the body and in a wake behind the body, whereas the rest of the field of flow remains, practically speaking, free from vorticity (see Fig. 4.1). It is, therefore, to be expected that in the limiting case of very small viscous forces, i. e. at large Reynolds numbers, the solutions of the Navier-Stokes equations are so constituted as to permit a subdivision of the field of flow into an external region which is free from vorticity, and a thin layer near the body together with a wake behind it. In the first region the flow may be expected to satisfy the equations of frictionless flow, the potential flow theory being used for its evaluation, whereas in the second region vorticity is inherent, and, therefore, the Navier-Stokes equations must be used for its evaluation. Viscous forces are important, i. e. of the same order of magnitude as inertia forces, only in the second region known as the *boundary layer*. This concept of a boundary layer was introduced into the science of fluid mechanics by L. Prandtl at the beginning of the present century: it has proved to be very fruitful. The subdivision of the field of flow into the frictionless external flow and the essentially viscous boundary-layer flow permitted the reduction of the mathematical difficulties inherent in the Navier-Stokes equations to such an extent that it became possible to integrate them for a large number of cases. The description of these methods of integration forms the subject of the boundary-layer theory presented in the following chapters.

From a numerical analysis of the available solutions of the Navier-Stokes equations it is also possible to show directly that in the limiting case of very large Reynolds numbers there exists a thin boundary layer in which the influence of viscosity is concentrated. We shall revert to this topic in Chap. V.

The previously discussed limiting case in which viscous forces heavily outweigh inertia forces (*creeping motion*, i. e., very small Reynolds number) results in a considerable mathematical simplification of the Navier-Stokes equations. By omitting the inertia terms their order is not reduced, but they become linear. The second limiting case, when inertia forces outweigh viscous forces (*boundary layer*, i. e. very large Reynolds numbers) presents greater mathematical difficulties than creeping motion. For, if we simply substitute $\nu = 0$ in the Navier-Stokes equations (3.32), or in the stream-function equation (4.10), we thereby suppress the derivatives of the highest order and with the simpler equation of lower order it is impossible to satisfy simultaneously all boundary conditions of the complete differential equations. However, this does not signify that the solutions of such an equation, simplified by the elimination of viscous terms, lose their physical meaning. Moreover, it is possible to prove that this solution agrees with the complete solution of the full Navier-Stokes equations almost everywhere in the limiting case of very large Reynolds numbers. The exception is confined to a thin layer near the wall — the boundary layer. Thus, the complete solution of the Navier-Stokes equations can be thought of as consisting of two solutions, the so-called "outer" solution which is obtained with the aid of Euler's equations of motion, and a so-called "inner" or boundary-layer solution which is valid only in the thin layer adjacent to the wall. The "inner" solution satisfies the so-called boundary-layer equations which are deduced from the Navier-Stokes equations by coordinate stretching and passage to the limit $\mathsf{R} \to \infty$, as will be shown in Chap. VII. The outer and inner solutions must be matched to each other by exploiting the condition that there must exist an overlapping region in which both solutions are valid.

f. Mathematical illustration of the process of going to the limit $R \to \infty$ †

Since the preceding argument constitutes one of the fundamental principles of boundary-layer theory, it may be worth while to illustrate the basic ideas by quoting a mathematical analog which was first given by L. Prandtl*.

Let us consider the damped vibrations of a point-mass described by the differential equation

$$m \frac{d^2x}{dt^2} + k \frac{dx}{dt} + c\,x = 0. \tag{4.13}$$

Here m denotes the vibrating mass, c the spring constant, k the damping factor, x the length coordinate measured from the position of equilibrium, and t the time. The initial conditions are assumed to be

$$x = 0 \quad \text{at} \quad t = 0. \tag{4.14}$$

In analogy with the Navier-Stokes equations for the case when the kinematic viscosity, ν, is very small, we consider here the limiting case of very small mass m, because this too causes the term of the highest order in eqn. (4.13) to become very small.

The complete solution of eqn. (4.13) subject to the initial condition (4.14) has the form

$$x = A \{\exp(-c\,t/k) - \exp(-k\,t/m)\}; \quad m \to 0, \tag{4.15}$$

where A is a free constant whose value can be determined with reference to a second initial condition.

If we put $m = 0$ in eqn. (4.13), we are led to the simplified equation

$$k \frac{dx}{dt} + c\,x = 0, \tag{4.16}$$

which is of first order, and whose solution is

$$x_0(t) = A \exp(-c\,t/k). \tag{4.17}$$

This solution is identical with the first term of the complete solution due to the felicitous choice of the adjustable constant. However, this solution cannot be made to satisfy the initial condition (4.14); it thus represents a solution for large values of the time, t ("outer" solution). The solution for small values of time ("inner" solution) satisfies another differential equation which can also be derived from eqn. (4.13). In order to achieve this, the independent variable t is "stretched" in that a new "inner" variable

$$t^* = t/m \tag{4.18}$$

is introduced. In this manner, eqn. (4.13) is transformed to

$$\frac{d^2x}{dt^{*2}} + k \frac{dx}{dt^*} + m\,c\,x = 0. \tag{4.19}$$

In the limit $m = 0$, we deduce the differential equation

$$\frac{d^2x}{dt^{*2}} + k \frac{dx}{dt^*} = 0 \tag{4.20}$$

which governs the "inner" solution. The solution now is

$$x_i(t^*) = A_1 \exp(-k\,t^*) + A_2. \tag{4.21}$$

---

† I am indebted to Professor Klaus Gersten for the revised version of this section.
* L. Prandtl, Anschauliche und nuetzliche Mathematik. Lectures delivered at Goettingen University in the Winter-Semester of 1931/32.

In spite of the simplification, the differential equation (4.20) is one of second degree; it can be made to satisfy the initial condition (4.14) by the choice

$$A_1 = - A_2. \tag{4.22}$$

The value of constant A_2 follows from the matching to the "outer" solution, eqn. (4.17). In an overlapping range, that is for moderate values of time, the solutions in eqns. (4.17) and (4.21) must agree. Thus we must have

$$\lim_{t^* \to \infty} x_i (t^*) = \lim_{t \to 0} x_0 (t) \tag{4.23}$$

or, in words: The "outer" limit of the "inner" solution must be equal to the "inner" limit of the "outer" solution. Condition (4.23) leads at once to

$$A_2 = A, \tag{4.24}$$

and so to the inner solution

$$x_i(t^*) = A \left\{ 1 - \exp\left(- k\, t^*\right) \right\}. \tag{4.25}$$

The same form can be obtained from the complete solution from eqn. (4.15) by expanding the first term for small values of t and retaining the first term only, that is by putting

$$\lim_{t \to 0} \exp\left(- c\, t/k\right) = 1. \tag{4.26}$$

The two solutions, the outer solution from eqn. (4.17) and the inner solution from eqn. (4.25), together form the complete solution on condition that each is used in its proper range of validity. At finite t, eqn. (4.15) tends to the outer solution for $m \to 0$, whereas at constant t^* eqn. (4.15) tends to the inner solution. The partial solutions give the complete, composite solution which is valid in the entire range of values of t by adding them together, remembering that the common term from eqn (4.23) must be included only once, that is subtracted from the sum according to the prescription

$$x(t) = x_0(t) + x_i(t^*) - \lim_{t^* \to \infty} x_i (t^*) = x_0 (t) + x_i(t^*) - \lim_{t \to 0} x_0(t). \tag{4.27}$$

A graphical representation of the complete solution from eqn. (4.15) is shown in Fig. 4.4 for the case when $A > 0$. Curve (a) corresponds to the outer solution (4.17). Curves (b), (c) and (d) represent solutions of the complete differential equation (4.13) with m decreasing from (b) to (d).

If we now compare this example with the Navier-Stokes equations, we conclude that the complete equation (4.13) is analogous to the Navier-Stokes equations for a viscous fluid, whereas the simplified equation (4.16), corresponds to Euler's equations for an ideal fluid. The initial

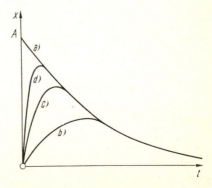

Fig. 4.4. Solutions of the vibration equation (4.13). (a) Solution of the simplified equation (4.14), $m = 0$; (b), (c), (d) represent solutions of the complete differential equation (4.13) with various values of m. When m is very small, solution (d) acquires boundary-layer character

condition (4.14) plays a part which is similar to the no-slip condition of a real fluid. The latter *can* be satisfied by the solutions of the Navier-Stokes equations but not by those of Euler's equations. The slowly-varying solution is analogous to the frictionless solution (potential flow) which fails to satisfy the no-slip condition. The fast-varying solution represents the counterpart of the boundary-layer solution which is determined by the presence of viscosity; it differs from zero only in a narrow zone near the wall (boundary layer). It is to be noted that the second boundary condition (no slip at the wall) can only be satisfied if this boundary-layer solution is added, thus making the whole solution physically real.

This simple example exhibits the same mathematical features as those discussed in the preceding chapter. It is, namely, not permissible simply to omit the viscous terms in the Navier-Stokes equations when performing the process of going over to the limit for very small viscosity (very large Reynolds number). This can only be done in the integral solution itself.

We shall demonstrate later in greater detail that it is not necessary to retain the full Navier-Stokes equations for the process of finding the limit for $R \to \infty$. For the sake of mathematical simplification it will prove possible to omit certain terms in it, particularly certain small viscous terms. It is, however, important to note that not all viscous terms can be neglected, as this would depress the order of the Navier-Stokes equations.

References

[1] Ackeret, J.: Über exakte Lösungen der Stokes-Navier-Gleichungen inkompressibler Flüssigkeiten bei veränderten Grenzbedingungen. ZAMP *3*, 259—271 (1952).

[1a] Apelt, C. J.: The steady flow of a viscous fluid past a circular cylinder at Reynolds numbers 40 and 44. British ARC RM 3175 (1961).

[1b] Allen, D. N. De G., and Southwell, R. V.: Relaxation methods applied to determine the motion, in two dimensions, of a viscous fluid past a fixed cylinder. Quart. J. Mech. Appl. Math. *8*, 129—145 (1955).

[1c] Coutanceau, M., and Bouard, R.: Experimental determination of the main features of the viscous flow in the wake of a circular cylinder in uniform translation. Part 1. Steady flow. JFM *79*, 231—256 (1977).

[1d] Coutanceau, M., and Bouard, R.: Experimental determination of the main features of the viscous flow in the wake of a circular cylinder in uniform translation. Part 2. Unsteady flow. JFM *79*, 257—272 (1977).

[2] Dennis, S.C.R., and Gau-Zu Chang: Numerical solutions for steady flow past a circular cylinder at Reynolds numbers up to 100. JFM *42*, 471—489 (1970).

[3] Fromm, J.E., and Harlow, F.H.: Numerical solutions of the problem of vortex street development. Phys. of Fluids *6*, 975—982 (1963); see also: AIAA Selected Reprints, Computational Fluid Dynamics (C.K. Chu, ed.), 82—89 (1968) and AGARD Lecture Series *34* (1971).

[4] Hamel, G.: Über die Potentialströmung zäher Flüssigkeiten. ZAMM *21*, 129—139 (1941).

[5] Jenson, V.G.: Viscous flow round a sphere at low Reynolds numbers (< 40). Proc. Roy. Soc. London A *249*, 346—366 (1959).

[5a] Keller, H.B., and Takami, H.: Numerical studies of steady viscous flow about cylinders. Numerical solutions of non-linear differential equations. Proc. Adv. Symp. at Univ. of Wisconsin, Madison, 1966 (D. Greenspan, ed.), J. Wiley & Sons, New York, 1966, pp. 115—140.

[6] Thom, A.: Flow past circular cylinders at low speeds. Proc. Roy. Soc. London A *141*, 651—669 (1933).

[7] Thom, A., and Apelt, C. J.: Field computations in engineering and physics. Van Nostrand, London, 1961.

CHAPTER V

Exact solutions of the Navier-Stokes equations

In general, the problem of finding exact solutions of the Navier-Stokes equations presents insurmountable mathematical difficulties. This is, primarily, a consequence of their being non-linear, so that the application of the principle of superposition, which serves so well in the case of frictionless potential motions, is excluded. Nevertheless, it is possible to find exact solutions in certain particular cases, mostly when the quadratic convective terms vanish in a natural way. In this chapter we shall devote our attention to the discussion of several exact solutions. Incidentally, it will be shown that in the case of small viscosity many of the exact solutions have a *boundary-layer structure* which means that the influence of viscosity is confined to a thin layer near the wall.

A comprehensive review of solutions of the Navier-Stokes equations has been given by R. Berker [4].

a. Parallel flow

Parallel flows constitute a particularly simple class of motions. A flow is called parallel if only one velocity component is different from zero, all fluid particles moving in one direction. For example: if the components v and w are zero everywhere, it follows at once from the equation of continuity that $\partial u/\partial x \equiv 0$, which means that the component u cannot depend on x. Thus for parallel flow we have

$$u = u(y,z,t); \quad v \equiv 0; \quad w \equiv 0 . \tag{5.1}$$

Further, it also follows immediately from the Navier-Stokes equations (3.32) for the y- and z-directions† that $\partial p/\partial y = 0$, and $\partial p/\partial z = 0$, so that the pressure depends only on x. In addition, in the equation for the x-direction all convective terms vanish. Hence

$$\varrho \, \frac{\partial u}{\partial t} = -\frac{\mathrm{d}p}{\mathrm{d}x} + \mu \left(\frac{\partial^2 u}{\partial y^2} + \frac{\partial^2 u}{\partial z^2} \right) \tag{5.2}$$

which is a linear differential equation for $u(y, z, t)$.

---

† In the following argument the term "pressure" denotes the difference between the total pressure and the hydrostatic pressure (pressure at rest). This causes the body forces to cancel, as they are in equilibrium with the hydrostatic pressure.

1. Parallel flow through a straight channel and Couette flow. A very simple solution of equation (5.2) is obtained for the case of steady flow in a channel with two parallel flat walls, Fig. 5.1. Let the distance between the walls be denoted by $2\,b$, so that eqn. (5.2) can be written

$$\frac{dp}{dx} = \mu\,\frac{d^2u}{dy^2} \tag{5.3}$$

with the boundary condition: $u = 0$ for $y = \pm\,b$. Since $\partial p/\partial y = 0$ the pressure gradient in the direction of flow is constant, as seen from eqn. (5.3). Thus $dp/dx = \text{const}$ and the solution is

$$u = -\,\frac{1}{2\,\mu}\,\frac{dp}{dx}\,(b^2 - y^2)\,. \tag{5.4}$$

The resulting velocity profile, Fig. 5.1, is parabolic.

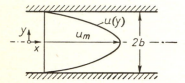

Fig. 5.1. Parallel flow with parabolic velocity distribution

Another simple solution of eqn. (5.3) is obtained for the so-called Couette flow between two parallel flat walls, one of which is at rest, the other moving in its own plane with a velocity U, Fig. 5.2. With the boundary conditions

$$y = 0:\quad u = 0\,;\qquad y = h:\quad u = U$$

we obtain the solution

$$u = \frac{y}{h}\,U - \frac{h^2}{2\,\mu}\,\frac{dp}{dx}\,\frac{y}{h}\left(1 - \frac{y}{h}\right) \tag{5.5}$$

which is shown in Fig. 5.2. In particular for a vanishing pressure gradient we have

$$u = \frac{y}{h}\,U\,. \tag{5.5a}$$

This particular case is known as simple Couette flow, or simple shear flow. The general case of Couette flow is a superposition of this simple case over the flow between two flat walls. The shape of the velocity profile is determined by the dimensionless pressure gradient

$$P = \frac{h^2}{2\,\mu\,U}\left(-\,\frac{dp}{dx}\right)\,.$$

For $P > 0$, i. e., for a pressure decreasing in the direction of motion, the velocity is positive over the whole width of the channel. For negative values of P the velocity

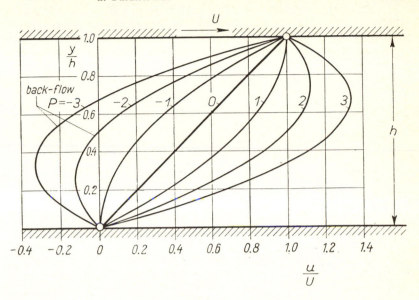

Fig. 5.2. Couette flow between two parallel flat walls

$P > 0$, pressure decrease in direction of wall motion; $P < 0$, pressure increase; $P = 0$, zero pressure gradient

over a portion of the channel width can become negative, that is, *back-flow* may occur near the wall which is at rest, and it is seen from Fig. 5.2 that this happens when $P < -1$. In this case the dragging action of the faster layers exerted on fluid particles in the neighbourhood of the wall is insufficient to overcome the influence of the adverse pressure gradient. This type of Couette flow with a pressure gradient has some importance in the hydrodynamic theory of lubrication. The flow in the narrow clearance between journal and bearing is, by and large, identical with Couette flow with a pressure gradient (*cf.* Sec. VI c).

2. The Hagen-Poiseuille theory of flow through a pipe. The flow through a straight tube of circular cross-section is the case with rotational symmetry which corresponds to the preceding case of two-dimensional flow through a channel. Let the x-axis be selected along the axis of the pipe, Fig. 1.2, and let y denote the radial coordinate measured from the axis outwards. The velocity components in the tangential and radial directions are zero; the velocity component parallel to the axis, denoted by u, depends on y alone, and the pressure is constant in every cross-section. Of the three Navier-Stokes equations in cylindrical coordinates, eqns. (3.36), only the one for the axial direction remains, and it simplifies to

$$\mu \left(\frac{d^2 u}{dy^2} + \frac{1}{y} \frac{du}{dy} \right) = \frac{dp}{dx} \tag{5.6}$$

the boundary condition being $u = 0$ for $y = R$. The solution of eqn. (5.6) gives the velocity distribution

$$u(y) = -\frac{1}{4\mu} \frac{dp}{dx} (R^2 - y^2) \tag{5.7}$$

where $-\mathrm{d}p/\mathrm{d}x = (p_1 - p_2)/l = \text{const}$ is the pressure gradient, to be regarded as given. Solution (5.7), which was obtained here as an exact solution of the Navier-Stokes equations, agrees with the solution in eqn. (1.10) which was obtained in an elementary way. The velocity over the cross-section is distributed in the form of a paraboloid of revolution. The maximum velocity on the axis is

$$u_m = \frac{R^2}{4\,\mu}\left(-\frac{\mathrm{d}p}{\mathrm{d}x}\right).$$

The mean velocity $\bar{u} = \frac{1}{2}\,u_m$, that is,

$$\bar{u} = \frac{R^2}{8\,\mu}\left(-\frac{\mathrm{d}p}{\mathrm{d}x}\right), \tag{5.8}$$

and the volume rate of flow becomes

$$Q = \pi\,R^2\,\bar{u} = \frac{\pi\,R^4}{8\,\mu}\left(-\frac{\mathrm{d}p}{\mathrm{d}x}\right). \tag{5.9}$$

The laminar flow described by the above solution occurs in practice only as long as the Reynolds number $R = \bar{u}\,d/\nu$ (d = pipe diameter) has a value which is less than the so-called critical Reynolds number, in spite of the fact that the above formulae constitute an exact solution of the Navier-Stokes equations for arbitrary values of $\mathrm{d}p/\mathrm{d}x$, R, and μ, or hence, of $\bar{u}$, R, and μ. According to experiments

$$\left(\frac{\bar{u}\,d}{\nu}\right)_{crit} = R_{crit} = 2300$$

approximately. For $R > R_{crit}$ the flow pattern is entirely different and becomes *turbulent*. We shall discuss this type of flow in greater detail in Chap. XX.

The relation between the pressure gradient and the mean velocity of flow is normally represented in engineering applications by introducing a *resistance coefficient of pipe flow*, λ. This coefficient is defined by setting the pressure gradient proportional to the dynamic head, i. e., to the square of the mean velocity of flow, according to the equation †

$$-\frac{\mathrm{d}p}{\mathrm{d}x} = \frac{\lambda}{d}\,\frac{\varrho}{2}\,\bar{u}^2. \tag{5.10}$$

Introducing the expression for $\mathrm{d}p/\mathrm{d}x$ from eqn. (5.9) we obtain

$$\lambda = \frac{2\,d}{\varrho\,\bar{u}^2}\,\frac{8\,\mu\,\bar{u}}{R^2} = \frac{32\,\mu}{\varrho\,\bar{u}\,R}$$

that is

$$\lambda = \frac{64}{R} \tag{5.11}$$

with

$$R = \frac{\varrho\,\bar{u}\,d}{\mu} = \frac{\bar{u}\,d}{\nu}. \tag{5.12}$$

---

† This quadratic law which assumes $\mathrm{d}p/\mathrm{d}x \sim \bar{u}^2$ fits turbulent flow very well. It is retained for laminar flow, although in that range $\mathrm{d}p/\mathrm{d}x \sim \bar{u}$. Thus for laminar flow λ ceases to be a constant.

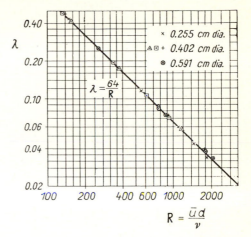

Fig. 5.3. Laminar flow through pipe; resistance coefficient, λ, plotted against Reynolds number (measured by Hagen), from Prandtl-Tietjens

Here R denotes the Reynolds number calculated for the pipe diameter and mean velocity of flow. The laminar equation for pressure loss in pipes, eqn. (5.11), is in excellent agreement with experimental results for the laminar range, as seen from Fig. 5.3 which reproduces experimental points measured by G. Hagen [10]. From this it is possible to infer that the Hagen-Poiseuille parabolic velocity distribution represents a solution of the Navier-Stokes equations which is in agreement with experimental results [22]. It is also possible to indicate an exact solution of the Navier-Stokes equations for the case of a pipe with a circular annular cross-section [20]. The problem of laminar and turbulent flow through pipes with excentric annular cross-sections was discussed theoretically in ref. [38] which also contains experimental results.

3. **The flow between two concentric rotating cylinders.** A further example which leads to a simple exact solution of the Navier-Stokes equations is afforded by the flow between two concentric rotating cylinders, both of which move at different but steady rotational speeds. We shall denote the inner and outer radii by r_1, and r_2 respectively, and similarly, the two angular velocities by ω_1, and ω_2. The Navier-Stokes equations (3.36) for plane polar coordinates reduce to

$$\varrho \frac{u^2}{r} = \frac{dp}{dr} \; ; \tag{5.13}$$

and

$$\frac{d^2u}{dr^2} + \frac{d}{dr}\left(\frac{u}{r}\right) = 0 \tag{5.14}$$

with u denoting the circumferential velocity. The boundary conditions are $u = r_1 \omega_1$ for $r = r_1$ and $u = r_2 \omega_2$ for $r = r_2$. The solution of (5.14) which satisfies these requirements is

$$u(r) = \frac{1}{r_2^2 - r_1^2}\left[r(\omega_2 r_2^2 - \omega_1 r_1^2) - \frac{r_1^2 r_2^2}{r}(\omega_2 - \omega_1)\right] . \tag{5.15}$$

Equation (5.13) determines the radial pressure distribution resulting from the motion.

The case when the inner cylinder is at rest, while the outer cylinder rotates, has some practical significance. In this instance the torque transmitted by the outer cylinder to the fluid becomes

$$M_2 = 4\,\pi\,\mu\,h\,\frac{r_1^{\,2}\,r_2^{\,2}}{r_2^{\,2} - r_1^{\,2}}\,\omega_2\,, \tag{5.16}$$

where h is the height of the cylinder. The moment M_1 with which the fluid acts on the inner cylinder has the same magnitude. The arrangement under consideration has been used occasionally for the determination of viscosity. The angular velocity of the external cylinder and the moment acting on the inner cylinder are measured, so that the viscosity can be evaluated with the aid of eqn. (5.16).

We now propose to indicate the velocity distributions in the annulus between the two cylinders for two particular cases. In Case I, the inner cylinder rotates with the outer one at rest; in Case II, the inner cylinder does not move, but the outer one rotates. Both flows are called Couette flow. Denoting the ratio of the two radii by $\varkappa = r_1/r_2$, the width of the annulus by $s = r_2 - r_1$, and the current relative radius by $x = r/r_2$, Fig. 5.4, we find

$$\text{(I)}\quad \frac{u}{u_1} = \frac{\varkappa}{1 - \varkappa^2}\cdot\frac{1 - x^2}{x}\qquad\text{(inner rotating; outer at rest),}\tag{5.16a}$$

and

$$\text{(II)}\quad \frac{u}{u_2} = \frac{\varkappa}{1 - \varkappa^2}\left(\frac{x}{\varkappa} - \frac{\varkappa}{x}\right)\quad\text{(inner at rest; outer rotating).}\tag{5.16b}$$

Here, $u_1 = r_1\omega_1$ is the peripheral velocity of the inner cylinder, and $u_2 = r_2\omega_2$ is that for the outer cylinder. Figure 5.4 represents the two velocity distributions in terms of the dimensionless distance from the inner cylinder

$$\frac{x'}{s} = \frac{r - r_1}{s}\,.$$

It is noteworthy that the velocity varies strongly with the ratio $\varkappa = r_1/r_2$ of the two radii in Case I, whereas for Case II it is almost independent of it. When $\varkappa = r_1/r_2 \to 1$, both cases tend to the linear velocity distribution of Couette flow, as it occurred between two flat plates in the case represented in Fig. 1.1. The equation of Case II yields the same limit for $r_1 = 0$, i. e. for $\varkappa = 0$ when no inner cylinder is present. In this case, the fluid rotates inside the outer cylinder as a rigid body. Hence it is seen that Case II yields a linear velocity distribution for the two symptotic cases $\varkappa = 0$ and $\varkappa = 1$. This behavior makes it easy to understand why the velocity distributions for the other, intermediate values of $\varkappa$ differ so little from a straight line.

In the particular case of a single cylinder rotating in an infinite fluid ($r_2 \to \infty$, $\omega_2 = 0$) eqn. (5.15) gives $u = r_1^{\,2}\,\omega_1/r$, and the torque transmitted by the fluid to the cylinder becomes $M_1 = 4\,\pi\,\mu\,h\,r_1^{\,2}\,\omega_1$. The velocity distribution in the fluid is the same as that around a line vortex of strength $\Gamma_1 = 2\,\pi\,r_1^{\,2}\,\omega_1$ in frictionless flow, or

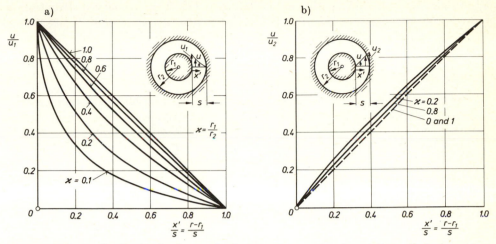

Fig. 5.4. Velocity distribution in the annulus between two concentric, rotating cylinders as calculated with the aid of eqns. (5.15 a, b).
a) Case I: inner cylinder rotating; outer cylinder at rest, $\omega_2 = 0$
b) Case II: inner cylinder at rest, $\omega_1 = 0$; outer cylinder rotating
r_1 = radius of inner cylinder, r_2 = radius of outer cylinder

$$u = \frac{\Gamma_1}{2 \pi r} \, .$$

It is seen, therefore, that the case of frictionless flow in the neighbourhood of a vortex line constitutes a solution of the Navier-Stokes equations (*cf.* Sec. IV b). In this connexion it may be instructive to mention an example of an exact *non-steady* solution of the Navier-Stokes equations, namely that which describes the process of decay of a vortex through the action of viscosity. The distribution of the tangential velocity component u with respect to the radial distance r and time t is given by

$$u(r,t) = \frac{\Gamma_0}{2 \pi r} \left\{ 1 - \exp\left(- r^2/4 \, v \, t\right) \right\}$$

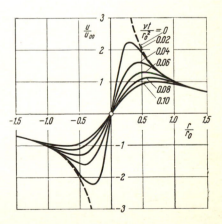

Fig. 5.5.   Velocity distribution at varying times in the neighbourhood of a vortex filament caused by the action of viscosity

Γ_0 = circulation of the vortex filament at time $t = 0$ when viscosity begins to act; $u_{00} = \Gamma_0/2 \pi r_0$

as derived by C. W. Oseen [21] and G. Hamel [11]. This velocity distribution is represented graphically in Fig. 5.5 Here Γ_0 denotes the circulation of the vortex filament at time $t = 0$, i. e. at the moment when viscosity is assumed to begin its action. An experimental investigation of this process was undertaken by A. Timme [40]. K. Kirde [17] made an analytic study of the case when the velocity distribution in the vortex differs from that imposed by potential theory.

4. The suddenly accelerated plane wall; Stokes's first problem. We now proceed to calculate some *non-steady* parallel flows. Since the convective acceleration terms vanish identically, the friction forces interact with the local acceleration. The simplest flows of this class occur when motion is started impulsively from rest. We shall begin with the case of the flow near a flat plate which is suddenly accelerated from rest and moves in its own plane with a constant velocity U_0. This is one of the problems which were solved by G. Stokes in his celebrated memoir on pendulums [35]†. Selecting the x-axis along the wall in the direction of U_0, we obtain the simplified Navier-Stokes equation

$$\frac{\partial u}{\partial t} = \nu \frac{\partial^2 u}{\partial y^2} \, . \tag{5.17}$$

The pressure in the whole space is constant, and the boundary conditions are:

$$\left. \begin{array}{ll} t \leq 0: \quad u = 0 \quad \text{for all } y\,; \\ t > 0: \quad u = U_0 \quad \text{for } y = 0; \quad u = 0 \quad \text{for } y = \infty\,. \end{array} \right\} \tag{5.18}$$

The differential equation (5.17) is identical with the equation of heat conduction which describes the propagation of heat in the space $y > 0$, when at time $t = 0$ the wall $y = 0$ is suddenly heated to a temperature which exceeds that in the surroundings. The partial differential equation (5.17) can be reduced to an ordinary differential equation by the substitution

$$\eta = \frac{y}{2 \sqrt{\nu t}} \, . \tag{5.19}$$

If we, further, assume

$$u = U_0 f(\eta) \, , \tag{5.20}$$

we obtain the following ordinary differential equation for $f(\eta)$:

$$f'' + 2 \eta f' = 0 \tag{5.21}$$

with the boundary conditions $f = 1$ at $\eta = 0$ and $f = 0$ at $\eta = \infty$. The solution is

$$u = U_0 \operatorname{erfc} \eta \, , \tag{5.22}$$

where

$$\operatorname{erfc} \eta = \frac{2}{\sqrt{\pi}} \int_\eta^\infty \exp\left(-\eta^2\right) \mathrm{d}\eta = 1 - \operatorname{erf} \eta = 1 - \frac{2}{\sqrt{\pi}} \int_0^\eta \exp\left(-\eta^2\right) \mathrm{d}\eta;$$

---

† Some authors refer to this problem as the 'Rayleigh problem'; there is no justification for this designation because the problem can be found fully discussed and solved in ref. [35].

the *complementary error function*, erfc η, has been tabulated[†]. The velocity distribution is represented in Fig. 5.6, and it may be noted that the velocity profiles for varying times are 'similar', i. e., they can be reduced to the same curve by changing the scale along the axis of ordinates. The complementary error function which appears in eqn. (5.22) has a value of about 0·01 at $\eta = 2\cdot0$. Taking into account the definition of the thickness of the boundary layer, δ, we obtain

$$\delta = 2\,\eta_\delta\,\sqrt{\nu\,t} \approx 4\,\sqrt{\nu\,t}\;. \tag{5.23}$$

It is seen to be proportional to the square root of the product of kinematic viscosity and time.

This problem was generalized by E. Becker [3] to include more general rates of acceleration as well as the cases involving suction or blowing or the effect of compressibility.

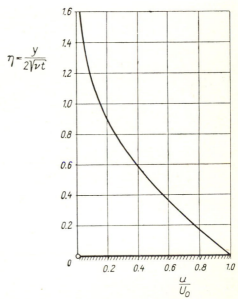

$$\eta = \frac{y}{2\sqrt{\nu t}}$$

Fig. 5.6. Velocity distribution above a suddenly accelerated wall

$$\frac{u}{U_o}$$

5. Flow formation in Couette motion. The substitution (5.19) which leads to eqn. (5.21) does not, in general, lead to a solution of the so-called heat conduction equation (5.17) if more complicated boundary conditions are imposed. Since eqn. (5.17) is linear, solutions for it can be obtained by the use of the Laplace transformation and by more direct methods developed in connexion with the study of the conduction of heat in solids. Many results obtained, e. g., for the temperature variation in an infinite or semi-infinite solid, can be directly transposed and used for the solution of problems in viscous flow. Thus the preceding problem in which the formation of the boundary layer near a suddenly accelerated wall has been investigated can also be solved for the case when the wall moves in a direction parallel to another flat wall at rest and at a distance h from it. This is the problem of flow formation in Couette motion, i. e.,

[†] See e. g. Sheppard, "The Probability Integral", British Assoc. Adv. Sci.: Math. Tables vol. vii (1939) and Works Project Administration "Tables of the Probability Function", New York, 1941.

the problem of how the velocity profile varies with time tending asymptotically to the linear distribution shown in Fig. 1.1. The differential equation is the same as before, eqn. (5.17), but with modified boundary conditions which now are:

$$t \leq 0; \; u = 0 \quad \text{for all } y, \text{ if } 0 \leq y \leq h;$$
$$t > 0: \; u = U_0 \text{ for } y = 0; \; u = 0 \text{ for } y = h \, .$$

The solution of eqn. (5.17) which satisfies the boundary and initial conditions can be obtained in the form of a series of complementary error functions

$$\frac{u}{U_0} = \sum_{n=0}^{\infty} \text{erfc} \, [2 \, n \, \eta_1 + \eta] - \sum_{n=0}^{\infty} \text{erfc} \, [2 \, (n+1) \, \eta_1 - \eta]$$

$$= \text{erfc} \, \eta - \text{erfc} \, (2 \, \eta_1 - \eta) + \text{erfc} \, (2 \, \eta_1 + \eta) - \text{erfc} \, (4 \, \eta_1 - \eta) + \text{erfc} \, (4 \, \eta_1 + \eta) - \ldots + \ldots$$

$$(5.24)$$

where $\eta_1 = h/2 \sqrt{\nu \, t}$ denotes the dimensionless distance between the two walls. The solution is represented in Fig. 5.7. The early profiles are still approximately similar and remain so, as long as the boundary layer has not spread to the stationary wall. The succeeding velocity profiles are no longer "similar" and tend asymptotically to the linear distribution of the steady state.

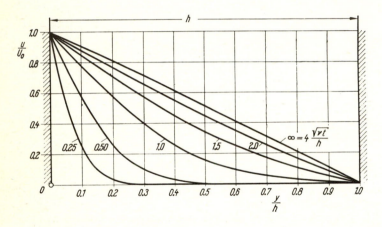

Fig. 5.7. Flow formation in Couette motion

Exact solutions for non-steady Couette flow were derived by J. Steinheuer [33] for the case when one of the walls is at rest in a steady flow and is then suddenly accelerated to a given, constant velocity. To do this, it is necessary to solve eqn. (5.17), which is identical with the one-dimensional heat conduction equation, by means of a Fourier series. A special case in this class of solutions is that when the moving wall is suddenly stopped so that it represents the decay of Couette flow.

6. Flow in a pipe, starting from rest. The acceleration of a fluid in a pipe is closely related to the preceding examples. Suppose that the fluid in an infinitely long pipe of circular cross-section is at rest for $t < 0$. At the instant $t = 0$ a pressure gradient dp/dx, which is constant in time, begins to act along it. The fluid will begin to move under the influence of viscous and inertia forces, and the velocity profile will approach asymptotically the parabolic distribution in Hagen-Poiseuille flow. The solution of this problem which leads to a differential equation involving Bessel functions was given by F. Szymanski [37]. The velocity profile is drawn in Fig. 5.8 for various instants. It is noteworthy that in the early stages the velocity near the axis is approximately constant over the radius and that viscosity makes itself felt in a narrow layer near the wall. The influence of viscosity reaches the pipe centre only in the later stages of motion, and the velocity profile tends asymptotically to the parabolic distribution for steady flow. The corresponding solution for an annular circular cross-section was given by W. Mueller [20].

The analogous case when the pressure gradient is removed instantly was solved by W. Gerbers [9].

The acceleration of a fluid over the whole length of pipe discussed here must be carefully distinguished from the acceleration of a fluid in the inlet portions of a pipe in steady flow. The rectangular velocity profile which exists in the entrance section is gradually transformed as the fluid progresses through the pipe with x increasing, and tends, under the influence of viscosity, to assume the Hagen-Poiseuille parabolic distribution. Since here $\partial u/\partial x \neq 0$ the flow is not one-dimensional, and the velocity depends on x, as well as on the radius. This problem was discussed by H. Schlichting [30], who gave the solution for two-dimensional flow through a straight channel, and by L. Schiller [29], and B. Punnis [24] for axially symmetrical flow (circular pipe); see also Secs. IX i and XI b.

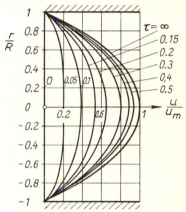

Fig. 5.8.   Velocity profile in a circular pipe during acceleration, as given by F. Szymanski [37]; $\tau = \nu\,t/R^2$

7. The flow near an oscillating flat plate; Stokes's second problem. In this section we propose to discuss the flow about an infinite flat wall which executes linear harmonic oscillations parallel to itself and which was first treated by G. Stokes [35] and later by Lord Rayleigh [25]. Let x denote the coordinate parallel to the direction of motion and y the coordinate perpendicular to the wall. Owing to the condition of no slip at the wall, the fluid velocity at it must be equal to that of the wall. Supposing that this motion is given by

$$y = 0: \qquad u(0,t) = U_0 \cos nt \qquad\qquad (5.25)$$

we find that the fluid velocity $u(y,t)$ is the solution of eqn. (5.17), together with the boundary condition (5.25), which, as already mentioned, is known from the theory of heat conduction. For the case under consideration

$$u(y,t) = U_0\,e^{-ky}\cos(nt - ky) . \qquad\qquad (5.26)$$

It is easy to verify that eqn. (5.26) is the required solution if

$$k = \sqrt{\frac{n}{2\nu}} .$$

Putting $\eta = ky = y \sqrt{n/2\,\nu}$ we have

$$u(y,t) = U_0\, e^{-\eta} \cos{(nt - \eta)}\,. \tag{5.26 a}$$

The velocity profile $u(y,t)$ thus has the form of a damped harmonic oscillation, the amplitude of which is $U_0\, e^{-y\sqrt{n/2\nu}}$, in which a fluid layer at a distance y has a phase lag $y \sqrt{n/2\,\nu}$ with respect to the motion of the wall. Fig. 5.9 represents this motion for several instants of time. Two fluid layers, a distance $2\,\pi/k = 2\,\pi\sqrt{2\,\nu/n}$ apart, oscillate in phase. This distance can be regarded as a kind of wave length of the motion: it is sometimes called the *depth of penetration* of the viscous wave. The layer which is carried by the wall has a thickness of the order $\delta \sim \sqrt{\nu/n}$ and decreases for decreasing kinematic viscosity and increasing frequency†.

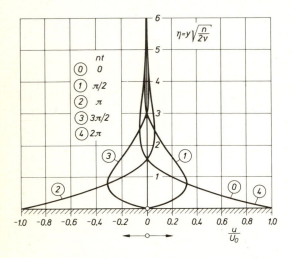

Fig. 5.9. Velocity distribution in the neighbourhood of an oscillating wall (Stokes's second problem)

8. A general class of non-steady solutions. A general class of non-steady solutions of the Navier-Stokes equations which possess boundary-layer character is obtained in the special case when the velocity components are independent of the longitudinal coordinate, x. The system of equations (3.32), written for plane flow, assumes the form

$$\frac{\partial u}{\partial t} + v\,\frac{\partial u}{\partial y} = -\frac{1}{\varrho}\,\frac{\partial p}{\partial x} + \nu\,\frac{\partial^2 u}{\partial y^2} \tag{5.27 a}$$

$$\frac{\partial v}{\partial t} = -\frac{1}{\varrho}\,\frac{\partial p}{\partial y} \tag{5.27 b}$$

$$\frac{\partial v}{\partial y} = 0\,. \tag{5.27 c}$$

---

† The solution in eqn. (5.26 a) represents also the temperature distribution in the earth which is caused by the periodic fluctuation of the temperature on the surface, say, from day to day or over the seasons in a year.

If we now prescribe a constant velocity $v_0 < 0$ at the wall (suction), we notice that eqn. (5.27c) is satisfied immediately by a flow for which $v = v_0$ and that the pressure p becomes independent of y simultaneously. Accordingly, we put $-(1/\varrho)\,(\partial p/\partial x) = dU/dt$, where $U(t)$ denotes the free-stream velocity at a very large distance from the wall, and hence obtain the following differential equation for $u(y, t)$:

$$\frac{\partial u}{\partial t} + v_0 \frac{\partial u}{\partial y} = \frac{dU}{dt} + v \frac{\partial^2 u}{\partial y^2}\,. \tag{5.28}$$

According to J. T. Stuart [32] there exists an exact solution of eqn. (5.28) for the arbitrary external velocity

$$U(t) = U_0\,[1 + f(t)]\,. \tag{5.29}$$

This solution is

$$u(y, t) = U_0\,[\zeta(y) + g(y, t)] \tag{5.30}$$

where

$$\zeta(y) = 1 - e^{\frac{y v_0}{v}}\,. \tag{5.31}$$

Substituting the last three equations into eqn. (5.28), we are led to a partial differential equation for the unknown function $g(y, t) = g(\eta, t)$; this has the form

$$\frac{\partial g}{\partial T} - 4 \frac{\partial g}{\partial \eta} = f'(T) + 4 \frac{\partial^2 g}{\partial \eta^2}\,, \tag{5.32}$$

and the boundary conditions are:

$$\eta = 0:\ g = 0\,; \quad \eta = \infty:\ g = f\,.$$

The following non-dimensional variables have been introduced in the preceding:

$$\eta = \frac{y(-v_0)}{v}\,; \quad T = \frac{t\,v_0{}^2}{4\,v}\,. \tag{5.33}$$

Solutions of (5.32) have been obtained by J. Watson [41] who employed Laplace transformations and who restricted himself to several special forms of the function $f(t)$. Generally speaking, the following external flows, $U(t)$, have been included:

a) damped and undamped oscillations,
b) step-like change from one value of velocity to another,
c) linear increase from one value to another.

In the special case when the external flow is independent of time, $f(t) = 0$, equation (5.32) leads to the simple solution $g(\eta, T) = 0$. This causes the velocity profile from eqn. (5.30) to become identical with the asymptotic suction profile given later in eqn. (14.6).

b. Other exact solutions

The preceding examples on one-dimensional flows were very simple, because the convective acceleration which renders the equations non-linear vanished identically everywhere. We shall now proceed to examine some exact solutions in which these terms are retained, so that non-linear equations will have to be considered. We shall, however, restrict ourselves to steady flows.

9. Stagnation in plane flow (Hiemenz flow). The first simple example of this type of flow, represented in Fig. 5.10, is that leading to a stagnation point in plane,

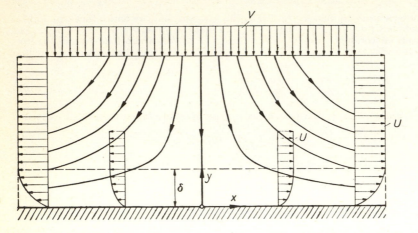

Fig. 5.10.   Stagnation in plane flow

i. e., two-dimensional flow. The velocity distribution in frictionless potential flow
in the neighbourhood of the stagnation point at $x = y = 0$ is given by

$$U = a\,x\,; \qquad V = -\,a\,y\,,$$

where a denotes a constant. This is an example of a plane potential flow which arrives from the y-axis and impinges on a flat wall placed at $y = 0$, divides into two
streams on the wall and leaves in both directions. The viscous flow must adhere to
the wall, whereas the potential flow slides along it. In potential flow the pressure is
given by Bernoulli's equation. If p_0 denotes the stagnation pressure, and p is the
pressure at an arbitrary point, we have in potential flow

$$p_0 - p = \tfrac{1}{2}\,\varrho\,(U^2 + V^2) = \tfrac{1}{2}\,\varrho\,a^2(x^2 + y^2)\,.$$

For viscous flow, we now make the assumptions

$$u = x\,f'(y)\,; \qquad v = -\,f(y)\,, \tag{5.34}$$

and

$$p_0 - p = \tfrac{1}{2}\,\varrho\,a^2\,[x^2 + F(y)]\,. \tag{5.35}$$

In this way the equation of continuity (4.4 c) is satisfied identically, and the two
Navier-Stokes equations of plane flow (4.4 a, b) are sufficient to determine the functions $f(y)$ and $F(y)$. Substituting eqns. (5.34) and (5.35) into eqn. (4.4 a, b) we obtain two ordinary differential equations for f and F:

$$f'^2 - f\,f'' = a^2 + \nu\,f''' \tag{5.36}$$

and

$$f\,f' = \tfrac{1}{2}\,a^2\,F' - \nu\,f''\,. \tag{5.37}$$

The boundary conditions for f and F are obtained from $u = v = 0$ at the wall, where $y = 0$, and $p = p_0$ at the stagnation point, as well as from $u = U = a\,x$ at a large distance from the wall. Thus

$$y = 0: \quad f = 0 \; ; \quad f' = 0 \; ; \quad F = 0 \; ; \qquad y = \infty : \quad f' = a \; .$$

Eqns. (5.36) and (5.37) are the two differential equations for the functions $f(y)$ and $F(y)$ which determine the velocity and pressure distribution. Since $F(y)$ does not appear in the first equation, it is possible to begin by determining $f(y)$ and then to proceed to find $F(y)$ from the second equation. The non-linear differential equation (5.36) cannot be solved in closed terms. In order to solve it numerically it is convenient to remove the constants a^2 and v by putting

$$\eta = \alpha\,y \; ; \quad f(y) = A\,\phi(\eta) \; .$$

Thus

$$\alpha^2\,A^2(\phi'^2 - \phi\,\phi'') = a^2 + v\,A\,\alpha^3\,\phi''' \; ,$$

where the prime now denotes differentiation with respect to η. The coefficients of the equation become all identically equal to unity if we put

$$\alpha^2\,A^2 = a^2 \; ; \quad v\,A\,\alpha^3 = a^2$$

or

$$A = \sqrt{v\,a} \; ; \quad \alpha = \sqrt{\frac{a}{v}}$$

so that

$$\eta = \sqrt{\frac{a}{v}}\,y; \qquad f(y) = \sqrt{a\,v}\,\phi(\eta) \; . \tag{5.38}$$

The differential equation for $\phi(\eta)$ now has the simple form

$$\phi''' + \phi\,\phi'' - \phi'^2 + 1 = 0 \tag{5.39}$$

with the boundary conditions

$$\eta = 0: \quad \phi = 0 \; , \quad \phi' = 0; \qquad \eta = \infty : \quad \phi' = 1 \; .$$

The velocity component parallel to the wall becomes

$$\frac{u}{U} = \frac{1}{a}\,f'(y) = \phi'(\eta) \; .$$

The solution of the differential equation (5.39) was first given in a thesis by K. Hiemenz [12] and later improved by L. Howarth [14]. It is shown in Fig. 5.11 (see also Table 5.1). The curve $\phi'(\eta)$ begins to increase linearly at $\eta = 0$ and tends asymptotically to unity. At approximately $\eta = 2\cdot4$ we have $\phi' = 0\cdot99$, i. e. the final value is reached there with an accuracy of 1 per cent. If we consider the corresponding distance from the wall, denoted by $y = \delta$, as the boundary layer, we have

$$\delta = \eta_\delta \sqrt{\frac{v}{a}} = 2\cdot4 \sqrt{\frac{v}{a}} \; . \tag{5.40}$$

Table 5.1. Functions occurring in the solution of plane and axially symmetrical flow with stagnation point. Plane case from L. Howarth [14]; axially symmetrical case from N. Froessling [8]

plane				axially symmetrical			
$\eta = \sqrt{\dfrac{a}{\nu}}\, y$	ϕ	$\dfrac{d\phi}{d\eta} = \dfrac{u}{U}$	$\dfrac{d^2\phi}{d\eta^2}$	$\sqrt{2}\cdot\zeta = \sqrt{\dfrac{2a}{\nu}}\, z$	ϕ	$\dfrac{d\phi}{d\zeta} = \dfrac{u}{U}$	$\dfrac{d^2\phi}{d\zeta^2}$
0	0	0	1·2326	0	0	0	1·3120
0·2	0·0233	0·2266	1·0345	0·2	0·0127	0·1755	1·1705
0·4	0·0881	0·4145	0·8463	0·4	0·0487	0·3311	1·0298
0·6	0·1867	0·5663	0·6752	0·6	0·1054	0·4669	0·8910
0·8	0·3124	0·6859	0·5251	0·8	0·1799	0·5833	0·7563
1·0	0·4592	0·7779	0·3980	1·0	0·2695	0·6811	0·6283
1·2	0·6220	0·8467	0·2938	1·2	0·3717	0·7614	0·5097
1·4	0·7967	0·8968	0·2110	1·4	0·4841	0·8258	0·4031
1·6	0·9798	0·9323	0·1474	1·6	0·6046	0·8761	0·3100
1·8	1·1689	0·9568	0·1000	1·8	0·7313	0·9142	0·2315
2·0	1·3620	0·9732	0·0658	2·0	0·8627	0·9422	0·1676
2·2	1·5578	0·9839	0·0420	2·2	0·9974	0·9622	0·1175
2·4	1·7553	0·9905	0·0260	2·4	1·1346	0·9760	0·0798
2·6	1·9538	0·9946	0·0156	2·6	1·2733	0·9853	0·0523
2·8	2·1530	0·9970	0·0090	2·8	1·4131	0·9912	0·0331
3·0	2·3526	0·9984	0·0051	3·0	1·5536	0·9949	0·0202
3·2	2·5523	0·9992	0·0028	3·2	1·6944	0·9972	0·0120
3·4	2·7522	0·9996	0·0014	3·4	1·8356	0·9985	0·0068
3·6	2·9521	0·9998	0·0007	3·6	1·9769	0·9992	0·0037
3·8	3·1521	0·9999	0·0004	3·8	2·1182	0·9996	0·0020
4·0	3·3521	1·0000	0·0002	4·0	2·2596	0·9998	0·0010
4·2	3·5521	1·0000	0·0001	4·2	2·4010	0·9999	0·0006
4·4	3·7521	1·0000	0·0000	4·4	2·5423	0·9999	0·0003
4·6	3·9521	1·0000	0·0000	4·6	2·6837	1·0000	0·0001

Hence again, as before, the layer which is influenced by viscosity is small at low kinematic viscosities and proportional to $\sqrt{\nu}$. The pressure gradient $\partial p/\partial y$ becomes proportional to $\varrho\, a\, \sqrt{\nu a}$ and is also very small for small kinematic viscosities.

It is, further, worth noting that the dimensionless velocity distribution u/U and the boundary-layer thickness from eqn. (5.40) are independent of x, i. e., they do not vary along the wall.

The type of flow under consideration does not occur near a plane wall only, but also in two-dimensional flow past any cylindrical body, provided that it has a blunt nose near the stagnation point. In such cases the solution is valid for a small neighbourhood of the stagnation point, if the portion of the curved surface can be replaced by its tangent plane near the stagnation point.

The non-steady flow pattern which results upon the superposition of an arbitrary, time-dependent transverse motion of the plane was studied by J. Watson [42]. The special case of a harmonic transverse motion was solved earlier by M. B. Glauert ([14] in Chap. XV).

9a. Two-dimensional non-steady stagnation flow. The case of non-steady, two-dimensional flow studied by N. Rott [28a] constitutes a generalization of the preceding case. We consider the case of two-dimensional stagnation flow depicted in Fig. 5.10 and bounded by a wall at $y = 0$. We assume that the velocity at a large distance from the wall is directed towards the wall, and that the wall itself performs a harmonic motion in its own plane. In the resulting flow pattern, the velocity remains steady at a large distance ($y \to \infty$), whereas near the wall it acquires a non-steady pattern of the same kind as that near the oscillating wall of Fig. 5.9 (Stokes's second problem). According to [28a], it is possible to integrate the nonsteady Navier-Stokes equation (4.4a, b, c) by assuming

$$u(x, y, t) = a\, x\, \phi'(\eta) + b\, g(\eta) \exp(i\omega t) \qquad (5.40\,\mathrm{a})$$

$$v(y) = -(a\,\nu)^{1/2}\, \phi(\eta),$$

in the same way as was done in eqn. (5.34). As far as the pressure is concerned, we put

$$p = p_0 - (1/2)\, \varrho\, a^2\, x^2 - \varrho\, \nu\, a\, F(\eta). \qquad (5.40\,\mathrm{c})$$

Here, $\eta = y(a/\nu)^{1/2}$ denotes the dimensionless distance from the wall from eqn. (5.38), b is the constant amplitude of the wall oscillating in its own plane, and ω is the circular frequency of this oscillation.

The preceding assumptions (5.40a, b, c) are introduced into the Navier-Stokes equations (4.4a, b, c), and the problem is reduced to solving the following system of equations:

$$\phi''' + \phi\,\phi'' - \phi'^2 + 1 = 0, \qquad (5.40\,\mathrm{d})$$

$$g'' + g'\,\phi - g\,(\phi' + i\,k) = 0, \qquad (5.40\,\mathrm{e})$$

$$\phi\,\phi' = F' - \phi''. \qquad (5.40\,\mathrm{f})$$

Here $k = \omega/a$ denotes the dimensionless frequency of the wall oscillation. The differential equations (5.40d) and (5.40e) result from the non-steady Navier-Stokes equation in the x-direction, eqn. (4.4a), when the velocity component u is represented as the sum of a steady term, ϕ', and an unsteady term, g, as was done in eqn. (5.40a). The function $\phi(\eta)$ satisfies the boundary conditions

$$\phi(0) = \phi'(0) = 0 \quad \text{and} \quad \phi'(\infty) = 1.$$

A comparison between eqns. (5.39) and (5.40d) shows that this function is identical with the well-known solution of the steady-state problem due to Hiemenz. The function $g(\eta)$ satisfies the boundary conditions

$$g(0) = 1 \quad \text{and} \quad g(\infty) = 0.$$

It is seen from eqns. (5.40d) and (5.40e) that in this case the steady component is independent of the superimposed non-steady component. The differential equation (5.40e) for the non-steady contribution g of the x-component of the velocity can be easily solved, because the function $\phi(\eta)$, Table 5.1, is known. Further details concerning this problem can be found in [28a]. The reader may also consult the papers by M. Glauert, [14] in Chap. XV, and J. Watson, [65] in Chap. XV.

10. Stagnation in three-dimensional flow. In a similar way it is possible to obtain an exact solution of the Navier-Stokes equations for the three-dimensional case of flow with stagnation, i. e., for the axisymmetrical case. A fluid stream impinges on a wall at right angles to it and flows away radially in all directions. Such a case occurs in the neighbourhood of a stagnation point of a body of revolution in a flow parallel to its axis.

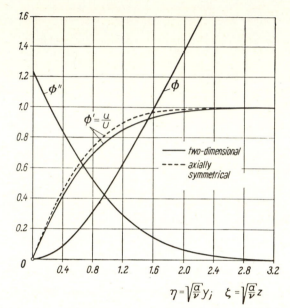

$$\eta = \sqrt{\tfrac{a}{\nu}}\, y; \quad \zeta = \sqrt{\tfrac{a}{\nu}}\, z$$

Fig. 5.11. Velocity distribution of plane and rotationally symmetrical flow at a stagnation point

To solve the problem we shall use cylindrical coordinates r, ϕ, z, and we shall assume that the wall is at $z = 0$, the stagnation point is at the origin and that the flow is in the direction of the negative z-axis. We shall denote the radial and axial components in frictionless flow by U and W respectively, whereas those in viscous flow will be denoted by $u = u(r,z)$, and $w = w(r,z)$. In accordance with eqn. (3.36) the Navier-Stokes equation for rotational symmetry can be written as

$$\left.\begin{aligned}
u\,\frac{\partial u}{\partial r} + w\,\frac{\partial u}{\partial z} &= -\frac{1}{\varrho}\frac{\partial p}{\partial r} + \nu\left(\frac{\partial^2 u}{\partial r^2} + \frac{1}{r}\frac{\partial u}{\partial r} - \frac{u}{r^2} + \frac{\partial^2 u}{\partial z^2}\right), \\[2mm]
u\,\frac{\partial w}{\partial r} + w\,\frac{\partial w}{\partial z} &= -\frac{1}{\varrho}\frac{\partial p}{\partial z} + \nu\left(\frac{\partial^2 w}{\partial r^2} + \frac{1}{r}\frac{\partial w}{\partial r} + \frac{\partial^2 w}{\partial z^2}\right), \\[2mm]
\frac{\partial u}{\partial r} + \frac{u}{r} + \frac{\partial w}{\partial z} &= 0,
\end{aligned}\right\} \tag{5.41}$$

because $v_\phi = 0$ and $\partial/\partial\phi = 0$, and we have put $v_r = u$ and $v_z = w$.

The boundary conditions are

$$z = 0 : \quad u = 0 , \quad w = 0 ; \quad z = \infty : \quad u = U . \tag{5.41a}$$

For the frictionless case we can write

$$U = a r ; \quad W = - 2 a z , \tag{5.42}$$

where a is a constant. It is seen at once that such a solution satisfies the equation of continuity. Denoting once more the stagnation pressure by p_0, we find the pressure in ideal flow:

$$p_0 - p = \tfrac{1}{2} \varrho (U^2 + W^2) = \tfrac{1}{2} \varrho \, a^2 (r^2 + 4 z^2) .$$

In the case of viscous flow we assume the following form of the solutions for the velocity and pressure distributions

$$u = r f'(z) ; \quad w = - 2 f(z) , \tag{5.43}$$

$$p_0 - p = \tfrac{1}{2} \varrho \, a^2 [r^2 + F(z)] . \tag{5.44}$$

It can be easily verified that a solution of the form (5.43) satisfies the equation of continuity identically, whereas the equations of motion lead to the following two equations for $f(z)$ and $F(z)$:

$$f'^2 - 2 f f'' = a^2 + v f''' , \tag{5.45}$$

$$2 f f' = \tfrac{1}{4} a^2 F' - v f'' . \tag{5.46}$$

The boundary conditions for $f(z)$ and $F(z)$ follow from eqn. (5.41a), and are

$$z = 0 : \quad f = f' = 0 , \quad F = 0 ; \quad z = \infty : \quad f' = a .$$

As before, the first of the two equations for f and F can be freed of the constants a^2 and v by a similarity transformation, which is identical with that in the plane case, thus

$$\zeta = \sqrt{\frac{a}{v}}\, z ; \quad f(z) = \sqrt{a v}\ \phi(\zeta) .$$

The differential equation for $\phi(\zeta)$ simplifies to

$$\phi''' + 2 \phi \phi'' - \phi'^2 + 1 = 0 \tag{5.47}$$

with the boundary conditions

$$\zeta = 0 : \quad \phi = \phi' = 0 ; \quad \zeta = \infty : \quad \phi' = 1 .$$

The solution of eqn. (5.47) was first given by F. Homann [13] in the form of a power series. The plot of $\phi' = u/U$ is given in Fig. 5.11 together with the plane case, and the values for ϕ' given in Table 5.1 have been taken from a paper by N. Froessling [8].

11. Flow near a rotating disk. A further example of an exact solution of the Navier-Stokes equations is furnished by the flow around a flat disk which rotates about an axis perpendicular to its plane with a uniform angular velocity, ω, in a fluid otherwise at rest. The layer near the disk is carried by it through friction and is thrown outwards owing to the action of centrifugal forces. This is compensated by particles which flow in an axial direction towards the disk to be in turn carried and ejected centrifugally. Thus the case is seen to be one of fully three-dimensional flow, i. e., there exist velocity components in the radial direction, r, the circumferential direction, ϕ, and the axial direction, z, which we shall denote respectively by u, v, and w. An axonometric representation of this flow field is shown in Fig. 5.12. At first the calculation will be performed for the case of an infinite rotating plane. It will then be easy to extend the result to include a disk of finite diameter $D = 2\,R$, on condition that the edge effect is neglected.

Taking into account rotational symmetry as well as the notation for the problem we can write down the Navier-Stokes equations (3.36) as:

$$\left.\begin{array}{l}
u\dfrac{\partial u}{\partial r} - \dfrac{v^2}{r} + w\dfrac{\partial u}{\partial z} = -\dfrac{1}{\varrho}\dfrac{\partial p}{\partial r} + \nu\left\{\dfrac{\partial^2 u}{\partial r^2} + \dfrac{\partial}{\partial r}\left(\dfrac{u}{r}\right) + \dfrac{\partial^2 u}{\partial z^2}\right\}, \\[3mm]
u\dfrac{\partial v}{\partial r} + \dfrac{uv}{r} + w\dfrac{\partial v}{\partial z} = \qquad\quad \nu\left\{\dfrac{\partial^2 v}{\partial r^2} + \dfrac{\partial}{\partial r}\left(\dfrac{v}{r}\right) + \dfrac{\partial^2 v}{\partial z^2}\right\}, \\[3mm]
u\dfrac{\partial w}{\partial r} \qquad + w\dfrac{\partial w}{\partial z} = -\dfrac{1}{\varrho}\dfrac{\partial p}{\partial z} + \nu\left\{\dfrac{\partial^2 w}{\partial r^2} + \dfrac{1}{r}\dfrac{\partial w}{\partial r} + \dfrac{\partial^2 w}{\partial z^2}\right\}, \\[3mm]
\qquad\qquad \dfrac{\partial u}{\partial r} + \dfrac{u}{r} + \dfrac{\partial w}{\partial z} = 0.
\end{array}\right\} \quad (5.48)$$

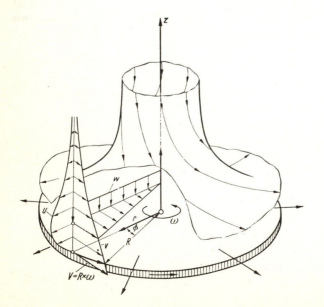

Fig. 5.12. Flow in the neighbourhood of a disk rotating in a fluid at rest

Velocity components: u-radial, v-circumferential, w-axial. A layer of fluid is carried by the disk owing to the action of viscous forces. The centrifugal forces in the thin layer give rise to secondary flow which is directed radially outward

The no-slip condition at the wall gives the following boundary conditions:

$$z = 0 : \quad u = 0, \quad v = r\,\omega, \quad w = 0, \left.\begin{matrix} \\ \\ \end{matrix}\right\} \quad (5.49)$$
$$z = \infty: \quad u = 0, \quad v = 0.$$

We shall begin by estimating the thickness, δ, of the layer of fluid 'carried' by the disk [23]. It is clear that the thickness of the layer of fluid which rotates with the disk owing to friction decreases with the viscosity and this view is confirmed when compared with the results of the preceding examples. The centrifugal force per unit volume which acts on a fluid particle in the rotating layer at a distance r from the axis is equal to $\varrho\, r\, \omega^2$. Hence for a volume of area $dr \cdot ds$ and height, δ, the centrifugal force becomes: $\varrho\, r\, \omega^2\, \delta\, dr\, ds$. The same element of fluid is acted upon by a shearing stress τ_w, pointing in the direction in which the fluid is slipping, and forming an angle, say θ, with the circumferential velocity. The radial component of the shearing stress must now be equal to the centrifugal force, and hence

$$\tau_w \sin \theta\, dr\, ds = \varrho\, r\, \omega^2\, \delta\, dr\, ds$$

or

$$\tau_w \sin \theta = \varrho\, r\, \omega^2\, \delta\,.$$

On the other hand the circumferential component of the shearing stress must be proportional to the velocity gradient of the circumferential velocity at the wall. This condition gives

$$\tau_w \cos \theta \sim \mu\, r\, \omega/\delta\,.$$

Eliminating τ_w from these two equations we obtain

$$\delta^2 \sim \frac{\nu}{\omega} \tan \theta.$$

If it is assumed that the direction of slip in the flow near the wall is independent of the radius, the thickness of the layer carried by the disk becomes

$$\delta \sim \sqrt{\frac{\nu}{\omega}}\,,$$

which is identical with the result obtained in the case of the oscillating wall on p. 94. Further, we can write for the shearing stress at the wall

$$\tau_w \sim \varrho\, r\, \omega^2\, \delta \sim \varrho\, r\, \omega\, \sqrt{\nu\,\omega}\,.$$

The torque, which is equal to the product of shearing stress at the wall, area and arm becomes

$$M \sim \tau_w R^3 \sim \varrho\, R^4\, \omega\, \sqrt{\nu\,\omega}\,, \qquad (5.50)$$

R denoting the radius of the disk.

In order to integrate the system of equations (5.48) it is convenient to introduce a dimensionless distance from the wall, $\zeta \sim z/\delta$, thus putting

$$\zeta = z\, \sqrt{\frac{\omega}{\nu}}\,. \qquad (5.51)$$

Further, the following assumptions are made for the velocity components and pressure

$$u = r\,\omega\,F(\zeta); \qquad v = r\,\omega\,G(\zeta); \qquad w = \sqrt{\nu\,\omega}\,H(\zeta) \qquad (5.52)$$

$$p = p\,(z) = \varrho\,\nu\,\omega\,P(\zeta)\,.$$

Inserting these equations into eqns. (5.48) we obtain a system of four simultaneous ordinary differential equations for the functions F, G, H, and P:

$$\left.\begin{aligned} 2\,F + H' &= 0 \\ F^2 + F'\,H - G^2 - F'' &= 0 \\ 2\,F\,G + H\,G' - G'' &= 0 \\ P' + H\,H' - H'' &= 0\,. \end{aligned}\right\} \qquad (5.53)$$

The boundary conditions can be calculated from eqn. (5.49) and are:

$$\zeta = 0 \;:\; F = 0\,, \quad G = 1\,, \quad H = 0\,, \quad P = 0$$
$$\zeta = \infty \;:\; F = 0\,, \quad G = 0\,.$$

The first solution of the system of eqns. (5.53) by an approximate method was given by a method of numerical integration†. They are plotted in Fig. 5.13. The starting values of the solution indicated in Table 5.2 were given by E. M. Sparrow and J. L. Gregg [32].

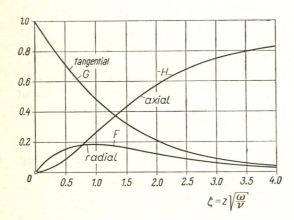

Fig. 5.13.   Velocity distribution near a disk rotating in a fluid at rest

In the case under discussion, just as in the example involving a stagnation point, the velocity field is the first to be evaluated from the equation of continuity and the equations of motion parallel to the wall. The pressure distribution is found subsequently from the equation of motion perpendicular to the wall.

---

† This solution was obtained in the form of a power series near $\zeta = 0$ and an asymptotic series for large values of ζ which were then joined together for moderate values of ζ.

Table 5.2.   Values of the functions needed for the description of the flow of a disk rotating in a fluid at rest calculated at the wall and at a large distance from the wall, as calculated by E. M. Sparrow and J. L. Gregg [32]

$\zeta = z \sqrt{\dfrac{\omega}{\nu}}$	F'	$-G'$	$-H$	P
0	0·510	0·6159	0	0
∞	0	0	0·8845	0·3912

It is seen from Fig. 5.13 that the distance from the wall over which the peripheral velocity is reduced to half the disk velocity is $\delta_{0\cdot5} \approx \sqrt{\nu/\omega}$. It is to be noted from the solution that when $\delta \approx \sqrt{\nu/\omega}$ is small, the velocity components u and v have appreciable values only in a thin layer of thickness $\sqrt{\nu/\omega}$. The velocity component w, normal to the disk is, at any rate, small and of the order $\sqrt{\nu\,\omega}$. The inclination of the relative streamlines near the wall with respect to the circumferential direction, if the wall is imagined at rest and the fluid is taken to rotate at a large distance from the wall, becomes

$$\tan \phi_0 = - \left(\frac{\partial u/\partial z}{\partial v/\partial z}\right)_{z=0} = - \frac{F'(0)}{G'(0)} = \frac{0\cdot510}{0\cdot616} = 0\cdot828,$$

or

$$\phi_0 = 39\cdot6° .$$

Although the calculation is, strictly speaking, applicable to an infinite disk only, we may utilize the same results for a finite disk, provided that its radius R is large compared with the thickness δ of the layer carried with the disk. We shall now evaluate the turning moment of such a disk. The contribution of an annular disk element of width dr on radius r is $dM = - 2\,\pi\,r\,dr\,r\,\tau_{z\phi}$, and hence the moment for a disk wetted on one side becomes

$$M = - 2\pi \int_0^R r^2\, \tau_{z\phi}\, dr .$$

Here $\tau_{z\phi} = \mu\,(\partial v/\partial z)_0$ denotes the circumferential component of the shearing stress. From eqn. (5.52) we obtain

$$\tau_{z\phi} = \varrho\,r\,\nu^{1/2}\,\omega^{3/2}\,G'\,(0) .$$

Hence the moment for a disk wetted on both sides becomes

$$2\,M = - \pi\,\varrho\,R^4\,(\nu\omega^3)^{1/2}\,G'\,(0) = 0\cdot616\,\pi\,\varrho\,R^4\,(\nu\omega^3)^{1/2} . \tag{5.54}$$

It is customary to introduce the following dimensionless moment coefficient,

$$C_M = \frac{2\,M}{\tfrac{1}{2}\,\varrho\,\omega^2\,R^5} . \tag{5.55}$$

This gives

$$C_M = -\frac{2\pi G'(0)\,\nu^{1/2}}{R\,\omega^{1/2}},$$

or, defining a Reynolds number based on the radius and tip velocity,

$$R = \frac{R^2\,\omega}{\nu}$$

and introducing the numerical value $-2\pi G'(0) = 3{\cdot}87$, we obtain finally

$$C_M = \frac{3{\cdot}87}{\sqrt{R}}.$$

$$(5.56)$$

Fig. 5.14 shows a plot of this equation, curve (1), and compares it with measurements [39]. For Reynolds numbers up to about $R = 3 \times 10^5$ there is excellent agreement between theory and experiment. At higher Reynolds numbers the flow becomes turbulent, and the respective case is considered in Chap. XXI.
Curves (2) and (3) in Fig. 5.14 are obtained from the turbulent flow theory. Older measurements, carried out by G. Kempf [16] and W. Schmidt [31], show tolerable agreement with theoretical results. Prior to these solutions, D. Riabouchinsky [26], [27] established empirical formulae for the turning moment of rotating disks which were based on very careful measurements. These formulae showed very good agreement with the theoretical equations discovered subsequently.

The quantity of liquid which is pumped outwards as a result of the centrifuging action on the one side of a disk of radius R is

$$Q = 2\pi R \int_{z=0}^{\infty} u \, dz.$$

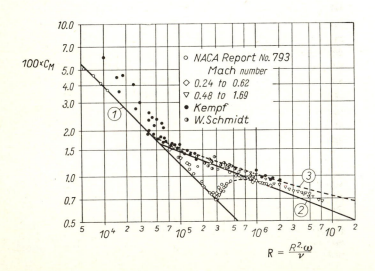

Fig. 5.14. Turning moment on a rotating disk; curve (1) from eqn. (5.56), *laminar*; curves (2) and (3) from eqns. (21.30) and (21.33), *turbulent*

Calculation shows that

$$Q = 0{\cdot}885\,\pi\,R^2\,\sqrt{\nu\,\omega} = 0{\cdot}885\,\pi\,R^3\,\omega\,\mathrm{R}^{-1/2}\ . \tag{5.57}$$

The quantity of fluid flowing towards the disk in the axial direction is of equal magnitude. It is, further, worthy of note that the pressure difference over the layer carried by the disk is of the order $\varrho\,\nu\,\omega$, i. e., very small for small viscosities. The pressure distribution depends only on the distance from the wall, and there is no radial pressure gradient.

A generalised form of the preceding problem has been studied by M. G. Rogers and G. N. Lance [28] who assumed that the fluid moves with an angular velocity $\Omega = s\,\omega$ at infinity. With this assumption, the second equation (5.53) becomes modified to

$$F^2 + F'\,H - G^2 - F'' + s^2 = 0\ ,$$

and the second boundary condition for the function $G(\zeta)$ must be replaced by $G(\infty) = s$. In this connexion a comparison should be made with the case of rotating flow over a fixed disk given in Sec. XIa. Numerical solutions for rotation in the same sense ($s > 0$) can be found in [26]. When the rotations are in opposite senses ($s < 0$), physically meaningful solutions can be obtained for $s < -0{\cdot}2$ only if uniform suction at right angles to the disk is admitted.

The problem of a rotating disk in a housing is discussed in Chap. XXI.

It is particularly noteworthy that the solution for the rotating disk as well as the solutions obtained for the flow with stagnation are, in the first place, exact solutions of the Navier-Stokes equations and, in the second, that they are of a *boundary-layer type*, in the sense discussed in the preceding chapter. In the limiting case of very small viscosity these solutions show that the influence of viscosity extends over a very small layer in the neighbourhood of the solid wall, whereas in the whole of the remaining region the flow is, practically speaking, identical with the corresponding ideal (potential) case. These examples show further that the boundary layer has a thickness of the order $\sqrt{\nu}$. The one-dimensional examples of flow discussed previously display the same boundary-layer character. In this connexion the reader may wish to consult a paper by G. K. Batchelor [2] which discusses the solution of the Navier-Stokes equations for the case of two co-axial, rotating disks placed at a certain distance apart, as well as a paper by K. Stewartson [34]. An extension of the preceding solution to the case of uniform suction is due to J. T. Stuart ([92] in Chap. XIV) and to E. M. Sparrow and J. L. Gregg (see p. 3 in [32]). The latter contains also an analysis of the case with homogeneous blowing. The limiting case of very vigorous blowing was discussed by H. K. Kuiken [18].

12. Flow in convergent and divergent channels. A further class of exact solutions of the Navier-Stokes equations can be obtained in the following way: Let it be assumed that the family of straight lines passing through a point in a plane constitute the streamlines of a flow. Let the velocity differ from line to line, which means that it is assumed to be a function of the polar angle ϕ. The rays along which the velocity vanishes can then be regarded as the solid walls of a convergent or a divergent channel. The continuity equations can be satisfied by assuming that the velocity along every ray is inversely proportional to the distance from the origin. Hence the radial velocity u has the form $u \sim F(\phi)/r$, or, if F is to be dimensionless,

$$u = \frac{v}{r} F(\phi) \, .$$

The peripheral velocity vanishes everywhere. Introducing this form into the Navier-Stokes equations written in polar coordinates, eqn. (3.36), and eliminating pressure from the equations in the r and ϕ directions, we obtain the following ordinary differential equation for $F(\phi)$:

$$2\,F F' + 4 F' + F''' = 0 \, .$$

Integrating once, we are led to the equation

$$F^2 + 4 F + F'' + K = 0 \, . \tag{5.58}$$

The constant K denotes the radial pressure gradient at the walls, $K = -(1/\varrho)\,(\partial p/\partial r)\,(r^3/v^2)$, where we have $F = 0$ for $\phi = \alpha$ and $\phi = -\alpha$, as well as $F' = 0$ for $\phi = 0$. The solution of eqn. (5.58) was given by G. Hamel [11]. The function F can be expressed explicitly as an elliptic function of ϕ.

We shall now briefly sketch the character of the solution refraining from discussing the details of the derivation. The graph in Fig. 5.15 shows a family of velocity profiles for a convergent and a divergent channel for different Reynolds numbers plotted on the basis of the numerical calculations performed by K. Millsaps and K. Pohlhausen [19]. The velocity distribution for the convergent and for the divergent channel differ markedly from each other. In the latter case, they also differ markedly for different Reynolds numbers. In a *convergent* channel the velocity distribution for the highest Reynolds number (R = 5000) remains nearly constant over a large centre-portion and decreases steeply to zero near the walls; thus it exhibits in this case a clear "boundary-layer character".

In a *divergent* channel the shape of the velocity profiles is markedly affected by the Reynolds number. Each of these velocity distributions is more curved at the centerline than the parabola that characterizes flow through a channel with parallel walls. The velocity distribution for the largest Reynolds number, curve 7, is distinguished by the fact that it shows two regions of back-flow. Thus, the velocity vanishes at four points. Since the wall could place itself at any one of these points, it is possible to envisage this velocity distribution at an included angle of 10° with two symmetric regions of back-flow or at an included angle of 6·9° but with a single, asymmetric region of back-flow. Such asymmetric velocity distributions are actually observed, and the back-flow signals the start of separation.

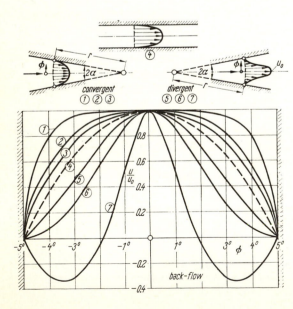

Fig. 5.15.   Velocity distribution in a convergent and a divergent channel after G. Hamel [11] and K. Millsaps and K. Pohlhausen [19]

Included angle $2\,\alpha = 10°$
Reynolds number $R = u_0\,r/v$

Convergent channel	Divergent channel
Curve 1: R = 5000	Curve 5: R = 684
Curve 2: R = 1342	Curve 6: R = 1342
Curve 3: R = 684	Curve 7: R = 5000

Curve 4 refers to a channel with parallel walls (Poiseuille's parabolic velocity distribution, *cf.* Fig. 5.1)

In the paper referred to above, G. Hamel has set himself the problem of calculating all three-dimensional flows whose streamlines are identical with those of a potential flow. The solution consisted of streamlines in the shape of logarithmic spirals. The case of radial flow considered here, and the case of potential vortex-flow, discussed in Sec. V 3, constitute particular examples of this general solution.

The preceding example of an exact solution exhibits once more the *boundary-layer character* of the flow. In particular, in the case of a convergent channel, the existence of a thin layer near the wall is confirmed together with the fact that the influence of viscosity is concentrated in it. Further, the calculation confirms that the boundary-layer thickness increases as $\sqrt{\nu}$ here too. The divergent case exhibits an additional phenomenon, that of back-flow, and, resulting from it, separation. This is an essential property of all boundary-layer flows and we shall discuss it later in greater detail on the basis of the equations of boundary-layer flow. Its existence is fully confirmed by experiment.

The cases of two-dimensional and axi-symmetrical flow through channels with small angles of divergence have been investigated earlier by H. Blasius [5] from first principles, i. e., with the aid of the Navier-Stokes equations. In this connexion it was shown that laminar flow can support only a very small pressure increase without the incidence of separation. The condition for the avoidance of back-flow at the wall in a divergent tube of radius $R(x)$ was found to be $dR/dx < 12/R$. (condition for separation), where $R = \bar{u} \, d/\nu$ denotes the Reynolds number referred to the mean velocity of flow through the channel and to its diameter. In more modern times M. Abramowitz [1] extended these calculations for divergent channels, and discovered that the point of separation moves downstream from the channel entrance as the Reynolds number is increased and as the angle of divergence is decreased.

13. Concluding remark. This example concludes the discussion of exact solution of the Navier-Stokes equations and the next topic will deal with approximate solutions. In the previous description an exact solution meant a solution of the Navier-Stokes equations in which all its terms were taken into account, provided that they did not vanish identically for the problem. In the following chapter we shall concern ourselves with approximate solutions of the Navier-Stokes equations, that is, with solutions which are obtained when small terms are neglected in the differential equations themselves. As already mentioned in Chap. IV, the two limiting cases of very large and very small viscosity are of particular importance. In very slow, or so-called creeping motion, viscous forces are very large compared with inertia forces, and in boundary-layer motion they are very small. Whereas in the first case it is permissible to omit the inertia terms completely, no such simplification is possible in boundary-layer theory, because if the viscous terms are simply disregarded the physically essential condition of no slip at the solid boundary cannot be met.

K. W. Mangler [6] developed a general theory for the solution of the Navier-Stokes equations for the case of two-dimensional laminar flow at very high Reynolds numbers, that is for flows in which the effect of viscosity is included, and which possess boundary-layer character. In Prandtl's boundary-layer theory (see also Chap. VII for details) the contour of the solid body in the stream is prescribed and the effect of viscosity is accounted for only in the thin layer adjoining the wall. By contrast, the new procedure is an indirect one. Instead of the contour of the real body, the theory prescribes an appropriate form for the so-called displacement contour which surrounds the body. The displacement contour makes an allowance for the displacement effect exerted on the external flow and on the wake. This permits us to determine the external, frictionless flow about the displacement contour; the next step consists in the computation of the flow field in the frictional layer with the aid of an asymptotic treatment of the Navier-Stokes equations for very large Reynolds numbers which yields, finally, the real shape of the body. The remarkable feature of this new procedure consists in the fact that the boundary-layer calculation can be carried beyond the point of separation. This is in contrast with Prandtl's boundary-layer theory which can be applied at most as far as the point of separation only. Furthermore, the new theory succeeds in some cases even with the evaluation of the complex flow patterns which exist in the region of back-flow behind the point of separation as well as that in the region of re-attachment.

The brief mention of an alternative theory to that given by Prandtl must suffice here. The boundary-layer theory expounded in the remainder of this book is based on Prandtl's line of thought.

References

[1] Abramowitz, M.: On backflow of a viscous fluid in a diverging channel. J. Math. Phys. *28*, 1—21 (1949).

[2] Batchelor, G.K.: Note on a class of solutions of the Navier-Stokes equations representing steady non-rotationally symmetric flow. Quart. J. Mech. Appl. Math. *4*, 29—41 (1951).

[3] Becker, E.: Eine einfache Verallgemeinerung der Rayleigh-Grenzschicht. ZAMP *11*, 146—152 (1960).

[4] Berker, R.: Intégration des équations du mouvement d'un fluide visqueux incompressible. Contribution to: Handbuch der Physik (S. Flügge, ed.) *VIII/2*, 1—384, Berlin, 1963.

[5] Blasius, H.: Laminare Strömung in Kanälen wechselnder Breite. Z. Math. u. Physik *58*, 225 (1910).

[6] Catherall, D., and Mangler, K. W.: The integration of the two-dimensional laminar boundary-layer equations past the point of vanishing skin friction. JFM *26*, 163—182 (1966).

[7] Cochran, W. G.: The flow due to a rotating disk. Proc. Cambr. Phil. Soc. *30*, 365—375 (1934).

[7a] Florent, P. and Peube, J.L.: Écoulement laminaire d'un fluide visqueux incompressible entre deux disques poreux. J. Mécanique *14*, 435—459 (1975).

[8] Frössling, N.: Verdunstung, Wärmeübertragung und Geschwindigkeitsverteilung bei zweidimensionaler und rotationssymmetrischer laminarer Grenzschichtströmung. Lunds. Univ. Arsskr. N. F. Afd. 2, *35*, No. 4 (1940).

[9] Gerbers, W.: Zur instationären, laminaren Strömung einer inkompressiblen zähen Flüssigkeit in kreiszylindrischen Rohren. Z. angew. Physik *3*, 267—271 (1951).

[10] Hagen, G.: Über die Bewegung des Wassers in engen zylindrischen Rohren. Pogg. Ann. *46*, 423—442 (1839).

[11] Hamel, G.: Spiralförmige Bewegung zäher Flüssigkeiten. Jahresber. Dt. Mathematiker-Vereinigung *25*, 34—60 (1916).

[12] Hiemenz, K.: Die Grenzschicht an einem in den gleichförmigen Flüssigkeitsstrom eingetauchten geraden Kreiszylinder. Thesis Göttingen 1911. Dingl. Polytech. J.*326*, 321 (1911).

[13] Homann, F.: Der Einfluss grosser Zähigkeit bei der Strömung um den Zylinder und um die Kugel. ZAMM *16*, 153—164 (1936); Forschg. Ing.-Wes. *7*, 1—10 (1936).

[14] Howarth, L.: On the calculation of the steady flow in the boundary layer near the surface of a cylinder in a stream. ARC RM 1632 (1935).

[15] von Kármán, Th.: Über laminare und turbulente Reibung. ZAMM *1*, 233—252 (1921); NACA TM 1092 (1946); see also: Coll. Works *II*, 70—97.

[16] Kempf, G.: Über Reibungswiderstand rotierender Scheiben. Vorträge auf dem Gebiet der Hydro- und Aerodynamik, Innsbruck Congr. 1922; Berlin, 1924, 168.

[17] Kirde, K.: Untersuchungen über die zeitliche Weiterentwicklung eines Wirbels mit vorgegebener Anfangsverteilung. Ing.-Arch. *31*, 385—404 (1962).

[18a] Mellor, G. L., Chapple, P. J. and Stokes, V. K.: On the flow between a rotating and a stationary disk. JFM *31*, 95—112 (1968).

[18] Kuiken, H.K.: The effect of normal blowing on the flow near a rotating disk of infinite extent. JFM *47*, 789—798 (1971).

[19] Millsaps, K., and Pohlhausen, K.: Thermal distribution in Jeffrey-Hamel flows between nonparallel plane walls. JAS *20*, 187—196 (1953).

[20] Müller, W.: Zum Problem der Anlaufströmung einer Flüssigkeit im geraden Rohr mit Kreisring- und Kreisquerschnitt. ZAMM *16*, 227—238 (1936).

[21] Oseen, C.W.: Ark. f. Math. Astron. och. Fys. *7* (1911); Hydromechanik, Leipzig, 1927, p. 82.

[22] Poiseuille, J.: Recherches expérimentelles sur le mouvement des liquides dans les tubes de très petits diamètres. Comptes Rendus *11*, 961—967 and 1041—1048 (1840); *12*, 112 (1841); in more detail: Memoires des Savants Etrangers *9* (1846).

[23] Prandtl, L.: Führer durch die Strömungslehre. 6th ed., 500, 1965; Engl. transl. Blackie and Son, London, 1952.

[24] Punnis, B.: Zur Berechnung der laminaren Einlaufströmung im Rohr. Diss. Göttingen 1947.

[25] Rayleigh, Lord: On the motion of solid bodies through viscous liquid. Phil. Mag. *21*, 697—711 (1911); also Sci. Papers *VI*, 29.

[26] Riabouchinsky, D.: Bull. de l'Institut Aerodyn. de Koutchino, *5*, 5—34 Moscow (1914); see also J. Roy. Aero. Soc. *39*, 340—348 and 377—379 (1935).

[27] Riabouchinsky, D.: Sur la résistance de frottement des disques tournant dans un fluide et les équations integrales appliquées à ce problème. Comptes Rendus *233*, 899—901 (1951).

[27a] Roberts, S. M. and Shipman, J. S.: Computing of the flow between a rotating and a stationary disk. JFM *73*, 53—63 (1976).

[28] Rogers, M. G., and Lance, G. N.: The rotationally symmetric flow of a viscous fluid in the presence of an infinite rotating disk. JFM *7*, 617—631 (1960).

[28a] Rott, N.: Unsteady viscous flow in the vicinity of a stagnation point. Quart. Appl. Math. *13*, 444—451 (1955/56).

[29] Schiller, L.: Untersuchungen über laminare und turbulente Strömung. VDI-Forschungsheft 248 (1922).

[29a] Schobeiri, M. T.: Näherungslösungen der Navier-Stokes'schen Differentialgleichung für eine zweidimensionale stationäre Laminarströmung konstanter Viskosität in konvexen und konkaven Diffusoren und Düsen. ZAMP *27*, 9—21 (1976).

[30] Schlichting, H.: Laminare Kanaleinlaufströmung. ZAMM *14*, 368—373 (1934).

[31] Schmidt, W.: Ein einfaches Messverfahren für Drehmomente. Z. VDI *65*, 441—444 (1921).

[32] Sparrow, E. M., and Gregg, J. L.: Mass transfer, flow and heat transfer about a rotating disk. Transactions ASME, J. Heat Transfer *82*, 294—302 (1960).

[33] Steinheuer, J.: Eine exakte Lösung der instationären Couette-Strömung. Proc. Scientific Soc. of Braunschweig *XVII*, 154—164 (1965).

[34] Stewartson, K.: On the flow between two rotating coaxial disks. Proc. Cambr. Phil. Soc. *49*, 333—341 (1953).

[35] Stokes, G. G.: On the effect of the internal friction of fluids on the motion of pendulums. Cambr. Phil. Trans. *IX*, 8 (1851); Math. and Phys. Papers, Cambridge, *III*, 1—141 (1901).

[36] Stuart, J. T.: A solution of the Navier-Stokes and energy equations illustrating the response of skin friction and temperature of an infinite plate thermometer to fluctuations in the stream velocity. Proc. Roy. Soc. London A *231*, 116—130 (1955).

[37] Szymanski, F.: Quelques solutions exactes des équations de l'hydrodynamique de fluide visqueux dans le cas d'un tube cylindrique. J. de math. pures et appliquées, Series 9, *11*, 67 (1932); see also Proc. Intern. Congr. Appl. Mech. Stockholm *I*, 249 (1930).

[38] Tao, L. N., and Donovan, W. F.: Through-flow in concentric and excentric annuli of fine clearance with and without relative motion of the boundaries. Trans. ASME *77*, 1291—1301 (1955).

[39] Theodorsen, Th., and Regier, A.: Experiments on drag of revolving discs, cylinders, and streamline rods at high speeds. NACA Rep. 793 (1944).

[40] Timme, A.: Über die Geschwindigkeitsverteilung in Wirbeln. Ing.-Arch. *25*, 205—225 (1957).

[41] Watson, J.: A solution of the Navier-Stokes equations illustrating the response of a laminar boundary layer to a given change in the external stream velocity. Quart. J. Mech. Appl. Math. *11*, 302—325 (1958).

[42] Watson, J.: The two-dimensional laminar flow near the stagnation point of a cylinder which has an arbitrary transverse motion. Quart. J. Mech. Appl. Math. *12*, 175—190 (1959).

CHAPTER VI

Very slow motion

a. The differential equations for the case of very slow motion

In this chapter we propose to discuss some approximate solutions of the Navier-Stokes equations which are valid in the limiting case when the viscous forces are considerably greater than the inertia forces. Since the inertia forces are proportional to the square of the velocity whereas the viscous forces are only proportional to its first power, it is easy to appreciate that a flow for which viscous forces are dominant is obtained when the velocity is very small, or, speaking more generally, when the Reynolds number is very small. When the inertia terms are simply omitted from the equations of motion the resulting solutions are valid approximately for $R \ll 1$. This fact can also be deduced from the dimensionless form of the Navier-Stokes equations, eqns. (4.2), where the inertia terms are seen to be multiplied by a factor $R = \varrho \, V \, l/\mu$ compared with the viscous terms. In this connexion we may remark that in each particular case it is necessary to examine in detail the quantities with which this Reynolds number is to be formed. However, apart from some special cases, motions at very low Reynolds numbers, sometimes also called *creeping motions*, do not occur too often in practical applications†.

It is seen from eqns. (3.34) that when the inertia terms are neglected the incompressible Navier-Stokes equations assume the form

$$\text{grad } p = \mu \nabla^2 \boldsymbol{w} \,, \tag{6.1}$$

$$\text{div } \boldsymbol{w} = 0 \,, \tag{6.2}$$

or, in extended form

$$\left.
\begin{aligned}
\frac{\partial p}{\partial x} &= \mu \left(\frac{\partial^2 u}{\partial x^2} + \frac{\partial^2 u}{\partial y^2} + \frac{\partial^2 u}{\partial z^2} \right) \\[2mm]
\frac{\partial p}{\partial y} &= \mu \left(\frac{\partial^2 v}{\partial x^2} + \frac{\partial^2 v}{\partial y^2} + \frac{\partial^2 v}{\partial z^2} \right) \\[2mm]
\frac{\partial p}{\partial z} &= \mu \left(\frac{\partial^2 w}{\partial x^2} + \frac{\partial^2 w}{\partial y^2} + \frac{\partial^2 w}{\partial z^2} \right)
\end{aligned}
\right\} \tag{6.3}$$

$$\frac{\partial u}{\partial x} + \frac{\partial v}{\partial y} + \frac{\partial w}{\partial z} = 0 \,. \tag{6.4}$$

---

† In the case of a sphere falling in air ($\nu = 160 \times 10^{-6}$ ft²/sec) we obtain e. g. $R = V \, d/\nu = 1$, when the diameter $d = 0 \cdot 04$ in ($= 0 \cdot 00333$ ft) and the velocity $V = 0 \cdot 048$ ft/sec.

This system of equations must be supplemented with the same boundary conditions as the full Navier-Stokes equations, namely those expressing the absence of slip in the fluid at the walls, i. e. the vanishing of the normal and tangential components of velocity:

$$v_n = 0 , \quad v_t = 0 \quad \text{at walls} . \tag{6.5}$$

An important characteristic of creeping motion can be obtained at once from eqn. (6.1), when the divergence of both sides is formed and when it is noticed that the operations div and ∇^2 on the right-hand side may be performed in the reverse order. Thus, with eqn. (6.2) we have

$$\text{div grad } p = \nabla^2 p = 0 . \tag{6.6}$$

The pressure field in creeping motion satisfies the potential equation and the pressure $p(x,y,z)$ is a potential function.

   The equations for *two-dimensional* creeping motion become particularly simple in form with the introduction of the stream function ψ defined by $u = \partial\psi/\partial y$ and $v = -\partial\psi/\partial x$. As explained in Chap. IV, and as seen from eqns. (6.3), when pressure is eliminated from the first two equations, the stream function must satisfy the equation

$$\nabla^4 \psi = 0 .$$

The stream function of plane creeping motion is thus a bipotential (biharmonic) function.

   In the remaining sections of this chapter we propose to discuss three examples of creeping motion: 1. Parallel flow past a sphere; 2. The hydrodynamic theory of lubrication; 3. The Hele-Shaw flow.

b. Parallel flow past a sphere

   The oldest known solution for a creeping motion was given by G. G. Stokes who investigated the case of parallel flow past a sphere [17]. We shall now describe the result of his calculations without going into the mathematical details of the theory. We shall base our description on that given by L. Prandtl [12]. The solution of eqns. (6.3) and (6.4) for the case of a sphere of radius R, the centre of which coincides with the origin, and which is placed in a parallel stream of uniform velocity U_∞, Fig. 6.1, along the x-axis can be represented by the following equations for the pressure and velocity components:

$$\left.\begin{array}{l}
u = U_\infty \left[\dfrac{3}{4} \dfrac{Rx^2}{r^3} \left(\dfrac{R^2}{r^2} - 1 \right) - \dfrac{1}{4} \dfrac{R}{r} \left(3 + \dfrac{R^2}{r^2} \right) + 1 \right] \\[3mm]
v = U_\infty \dfrac{3}{4} \dfrac{Rxy}{r^3} \left(\dfrac{R^2}{r^2} - 1 \right) \\[3mm]
w = U_\infty \dfrac{3}{4} \dfrac{Rxz}{r^3} \left(\dfrac{R^2}{r^2} - 1 \right) \\[3mm]
p - p_\infty = -\dfrac{3}{2} \dfrac{\mu U_\infty Rx}{r^3}
\end{array}\right\} \tag{6.7}$$

where $r^2 = x^2 + y^2 + z^2$ has been introduced for the sake of brevity. It is easy to verify that these expressions satisfy eqns. (6.3) and (6.4) and that the velocity vanishes at all points on the surface of the sphere. The pressure on the surface becomes

$$p - p_\infty = -\frac{3}{2}\,\mu\,\frac{x}{R^2}\,U_\infty\,. \tag{6.7a}$$

The maximum and minimum of pressure occurs at points P_1 and P_2, respectively, their values being

$$p_{1,2} - p_\infty = \pm\,\frac{3}{2}\,\frac{\mu\,U_\infty}{R} \tag{6.7b}$$

The pressure distribution along a meridian of the sphere as well as along the axis of abscissae, x, is shown in Fig. 6.1. The shearing-stress distribution over the sphere can also be calculated from the above formulae. It is found that the shearing stress has its largest value at point A where $\tau = \frac{3}{2}\,\mu\,U_\infty/R$ and is equal to the pressure rise at P_1 or pressure decrease at P_2. Integrating the pressure distribution and the shearing stress over the surface of the sphere we obtain the total drag

$$D = 6\pi\,\mu\,R\,U_\infty\,. \tag{6.8}$$

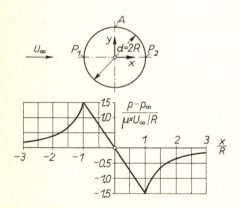

Fig. 6.1.   Pressure distribution around a sphere in parallel uniform flow

This is the very well known *Stokes equation* for the drag of a sphere. It can be shown that one third of the drag is due to the pressure distribution and that the remaining two thirds are due to the existence of shear. It is further remarkable that the drag is proportional to the first power of velocity. If a drag coefficient is formed by referring the drag to the dynamic head $\frac{1}{2}\,\varrho\,U_\infty^2$ and the frontal area, as is done in the case of higher Reynolds numbers, or if we put

$$D = C_D\,\pi\,R^2\,(\tfrac{1}{2}\,\varrho\,U_\infty^2)\,, \tag{6.9}$$

then

$$C_D = \frac{24}{R}\,; \quad R = \frac{U_\infty\,d}{\nu}\,. \tag{6.10}$$

A comparison between Stokes's equation and experiment was given in Fig. 1.5 from which it is seen that is applies only to cases when R < 1. The pattern of streamlines in front of and behind the sphere must be the same, as by reversing the direction of free flow, i. e., by changing the sign of velocity components in eqns. (6.3) and (6.4) the system is transformed into itself. The streamlines in viscous flow past a sphere are shown in Fig. 6.2. They were drawn as they would appear to an observer in front of whom the sphere is dragged with a constant velocity U_∞. The sketch contains also velocity profiles at several cross-sections. It is seen that the sphere drags with it a very wide layer of fluid which extends over about one diameter on both sides. At very high Reynolds numbers this boundary layer becomes very thin.

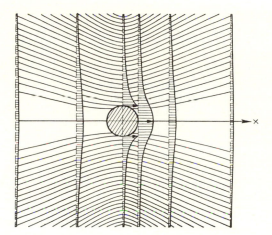

Fig. 6.2. Streamlines and velocity distribution in Stokes' solution for a sphere in parallel flow

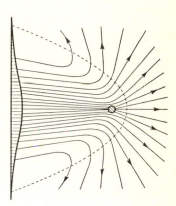

Fig. 6.3. Streamlines in the flow past a sphere from Oseen's solution

Oseen's improvement: An improvement of Stokes's solution was given by C. W. Oseen [11], who took the inertia terms in the Navier-Stokes equations partly into account. He assumed that the velocity components can be represented as the sum of a constant and a perturbation term. Thus

$$u = U_\infty + u' ; \quad v = v' ; \quad w = w' , \tag{6.11}$$

where u', v' and w' are the perturbation terms, and as such, small with respect to the free stream velocity U_∞. It is to be noted, however, that this is not true in the immediate neighbourhood of the sphere. With the assumption (6.11) the inertia terms in the Navier-Stokes eqns. (3.32) are decomposed in two groups, e. g.:

$$U_\infty \frac{\partial u'}{\partial x} , \quad U_\infty \frac{\partial v'}{\partial x} , \dots \quad \text{and} \quad u' \frac{\partial u'}{\partial x} , \quad u' \frac{\partial v'}{\partial x} , \dots$$

The second group is neglected as it is small of the second order compared with the first group. Thus we obtain the following equations of motion from the Navier-Stokes equations:

$$\varrho\, U_\infty \frac{\partial u'}{\partial x} + \frac{\partial p}{\partial x} = \mu\,\nabla^2 u'$$

$$\varrho\, U_\infty \frac{\partial v'}{\partial x} + \frac{\partial p}{\partial y} = \mu\,\nabla^2 v'$$

$$\varrho\, U_\infty \frac{\partial w'}{\partial x} + \frac{\partial p}{\partial z} = \mu\,\nabla^2 w' \qquad (6.12)$$

$$\frac{\partial u'}{\partial x} + \frac{\partial v'}{\partial y} + \frac{\partial w'}{\partial z} = 0.$$

The boundary conditions are the same as for the Navier-Stokes equations, but the Oseen equations are linear as was the case with the Stokes equations.

The pattern of streamlines is now no longer the same in front of and behind the sphere. This can be recognized if reference is made to eqns. (6.12), because if we change the sign of the velocities and of the pressure, the equations do not transform into themselves, whereas the Stokes equations (6.3) did. The streamlines of the Oseen equations are plotted in Fig. 6.3, and the observer is again assumed to be at rest with respect to the flow at a large distance from the sphere; it is imagined that the sphere is dragged with a constant velocity U_∞. The flow in front of the sphere is very similar to that given by Stokes, but behind the sphere the streamlines are closer together which means that the velocity is larger than in the former case. Furthermore, behind the sphere some particles follow its motion as is, in fact, observed experimentally at large Reynolds numbers.

The improved expression for the drag coefficient now becomes

$$C_D = \frac{24}{\mathsf{R}} \left(1 + \frac{3}{16}\,\mathsf{R} \right) ; \qquad \mathsf{R} = \frac{U_\infty d}{\nu} . \qquad (6.13)$$

Experimental results show, Fig. 1.5, curve (2), that Oseen's equation is applicable up to $\mathsf{R} = 5$ approximately.

c. The hydrodynamic theory of lubrication

The phenomena which take place in oil lubricated bearings afford another example of flow in which viscous forces are predominant. From the practical point of view these phenomena are very important. At high velocities the clearance between two machine elements which are in relative motion (e. g. journal and bearing) is filled by an oil stream in which extremely large pressure differences may be created. As a consequence, the revolving journal is lifted somewhat by the oil film and metallic contact between the moving parts is prevented. The essential features of this type of motion can be understood on the example of a slide block or slipper moving on a plane guide surface, Fig. 6.4, it being important that they are inclined at a small angle, δ, to each other. We shall assume that the sliding surfaces are very large in a transverse direction with respect to the motion so that the problem is one in two dimensions†. In order to obtain a steady-state problem let us assume that the block is at rest and that the plane guide is forced to move with a constant velocity U with respect to it. The x-axis is assumed in the direction of motion, and the y-axis is at right angles to the plane of the guide. The height $h(x)$ of the wedge between the block and the guide is assumed to be very small compared with the length l of the block.

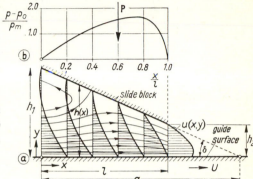

Fig. 6.4. Lubrication in a bearing: a) Flow in wedge between slide block and plane guide surface; b) Pressure distribution over block, $a/l = 1\cdot 57$

This motion is a more general example of that considered in Section V 1, i. e. of the motion between two parallel flat walls with a pressure gradient. The essential difference consists in the fact that here the two walls are inclined at an angle to each other. For this reason the convective acceleration $u\,\partial u/\partial x$ is evidently different from zero. An estimation of the viscous and inertia forces shows immediately that, in spite of that, in all cases of practical importance, the viscous forces are predominant. The largest viscous term in the equation of motion for the x-direction is equal to $\mu\,\partial^2 u/\partial y^2$. Hence we can make the following estimate:

$$\frac{\text{Inertia force}}{\text{Viscous force}} = \frac{\varrho\,u\,\partial u/\partial x}{\mu\,\partial^2 u/\partial y^2} = \frac{\varrho\,U^2/l}{\mu\,U/h^2} = \frac{\varrho\,Ul}{\mu}\cdot\left(\frac{h}{l}\right)^2.$$

The inertia forces can be neglected with respect to the viscous forces if the reduced Reynolds number

$$\mathsf{R}* = \frac{Ul}{\nu}\left(\frac{h}{l}\right)^2 \ll 1\,, \tag{6.14}$$

or, by way of numerical example:

$$U = 40 \text{ ft/sec};\qquad\qquad l = 4 \text{ in} = 0\cdot 333 \text{ ft}$$
$$\nu = 4\times 10^{-4} \text{ ft}^2/\text{sec};\quad h = 0\cdot 008 \text{ in}.$$

This leads to a value of the Reynolds number referred to the length of the block of $U\,l/\nu = 25{,}000$, whereas the reduced Reynolds number $\mathsf{R}* = 0\cdot 1$.

---

† The two-dimensional theory was first formulated by O. Reynolds, *cf.* Phil. Trans.Roy. Soc. (1886), Pt. I, see also Ostwalds Klassiker No. 218, p. 39.

The differential equations of creeping motion, eqns. (6.3), can be further simplified for the case under consideration. The equation for the y-direction can be omitted altogether because the component v is very small with respect to u. Further, in the equation for the x-direction $\partial^2 u/\partial x^2$ can be neglected with respect to $\partial^2 u/\partial y^2$, because the former is smaller than the latter by a factor of the order $(h/l)^2$. The pressure distribution must satisfy the condition that $p = p_0$ at both ends of the slipper. Compared with the case of flow between parallel sliding walls, the pressure gradient in the direction of motion, $\partial p/\partial x$, is no longer constant, but the very small pressure gradient in the y-direction can be neglected. With these simplifications the differential equations (6.3) reduce to

$$\frac{dp}{dx} = \mu \frac{\partial^2 u}{\partial y^2} , \tag{6.15}$$

and the equation of continuity in differential form can be replaced by the condition that the volume of flow in every section must be constant:

$$Q = \int_0^{h(x)} u \, dy = \text{const} . \tag{6.16}$$

The boundary conditions are:

$$
\begin{aligned}
y = 0: & \quad u = U; & x = 0: & \quad p = p_0 \\
y = h: & \quad u = 0 ; & x = l: & \quad p = p_0 .
\end{aligned} \tag{6.17}
$$

The solution of eqn. (6.15) which satisfies the boundary conditions (6.17) is similar to eqn. (5.5), namely

$$u = U \left(1 - \frac{y}{h} \right) - \frac{h^2 p'}{2 \mu} \frac{y}{h} \left(1 - \frac{y}{h} \right) \tag{6.18}$$

where $p' = dp/dx$ denotes the pressure gradient, which must be determined in such a way as to satisfy the continuity equation (6.16), and the boundary conditions for pressure. Inserting (6.18) into (6.16) we first obtain

$$Q = \frac{Uh}{2} - \frac{h^3 p'}{12 \mu} ,$$

or, solving for p':

$$p' = 12 \mu \left(\frac{U}{2 h^2} - \frac{Q}{h^3} \right) . \tag{6.19}$$

Hence by integration

$$p(x) = p_0 + 6 \mu U \int_0^x \frac{dx}{h^2} - 12 \mu Q \int_0^x \frac{dx}{h^3} . \tag{6.20}$$

Inserting the condition $p = p_0$ at $x = l$ we obtain the value

$$Q = \tfrac{1}{2} U \int_0^l \frac{dx}{h^2} \left/ \int_0^l \frac{dx}{h^3} \right. . \tag{6.21}$$

Thus the mass flow is known when the shape of the wedge is given as the function $h(x)$. Eqn. (6.19) gives the pressure gradient, and eqn. (6.20) gives the pressure distribution over the slipper.

The quantities

$$b_1(x) = \int_0^x dx/h^2 \quad \text{and} \quad b_2(x) = \int_0^x dx/h^3 , \tag{6.22}$$

which appear in eqn. (6.20) depend only on the geometrical shape of the gap between the slider and the plane. Their ratio

$$c(x) = b_1(x)/b_2(x) \tag{6.23}$$

which has the dimension of a length plays an important part in the theory of lubrication; its value for the whole channel,

$$H = c(l) = (\int_0^l dx/h^2)/(\int_0^l dx/h^3) , \tag{6.24}$$

is sometimes called the *characteristic thickness*. With its aid, the equation of continuity (6.21) can be contracted to

$$Q = \tfrac{1}{2} U H , \tag{6.25}$$

from which its physical interpretation is evident. The pressure can now be written

$$p(x) = p_0 + 6 \mu U b_1(x) - 12 \mu Q b_2(x) , \tag{6.26}$$

and the pressure gradient becomes

$$p' = \frac{6 \mu U}{h^2} \left(1 - \frac{H}{h}\right), \tag{6.27}$$

which shows that the pressure has a maximum or a minimum at a place where the channel thickness is equal to its characteristic value, $h = H$.

Often it is desirable to maintain a positive excess of pressure $p - p_0$, and the preceding equation can be used to derive the condition for it. Assuming that $p - p_0 = 0$ at $x = 0$ and that the thickness H is placed at $x = x_H$, we must have

$$\left. \begin{array}{l} h(x) > H \text{ for } 0 < x < x_H, \text{ implying } p' > 0 \\ h(x) < H \text{ for } x_H < x < l, \text{ implying } p' < 0 . \end{array} \right\} \tag{6.28}$$

These conditions lead to a wedge-like shape which is convergent in the direction of flow and which admits local both positive and negative gradients dh/dx. Since H depends on the shape of the whole channel, the direction of the pressure gradient at a section cannot be determined from dh/dx at the section alone unlike in potential flow.

In the case of a wedge with flat faces for which $h(x) = \delta(a - x)$, where a and δ are constants, see Fig. 6.4, we obtain finally

$$Q = U \delta \frac{a(a-l)}{2 a-l} ,$$

and for the pressure distribution

$$p(x) = p_0 + 6\,\mu U \frac{x\,(l-x)}{h^2\,(2\,a-l)}\;.\tag{6.29}$$

The relations become somewhat simpler if the shape of the channel is described by the gap widths h_1 and h_2 at inlet and exit, respectively, see Fig. 6.4. The characteristic width now becomes equal to the harmonic mean

$$H = \frac{2\,h_2\,h_1}{h_1 + h_2}\;,\tag{6.30}$$

and the condition for positive pressure excess, eqn. (6.28), now requires that the channel should be convergent. In this notation, the pressure distribution is given by

$$p(x) = p_0 + 6\,\mu\,U \frac{l}{h_1{}^2 - h_2{}^2} \cdot \frac{(h_1 - h)\,(h - h_2)}{h^2}\;,\tag{6.31}$$

and the resultant of the pressure forces can be computed by integration, when we obtain

$$P = \int_0^l p\,\mathrm{d}x = \frac{6\,\mu\,U\,l^2}{(k-1)^2\,h_2{}^2}\left[\ln k - \frac{2\,(k-1)}{k+1}\right],\tag{6.32}$$

with $k = h_1/h_2$. The resultant of the shearing stresses can be calculated in a similar manner:

$$F = -\int_0^l \mu\left(\frac{\mathrm{d}u}{\mathrm{d}y}\right)_0 \mathrm{d}x = \frac{\mu\,U\,l}{(k-1)\,h_2}\left[4\ln k - \frac{6\,(k-1)}{k+1}\right].\tag{6.33}$$

It is interesting to note [9] that the resultant pressure force possesses a maximum for $k = 2\cdot2$ approximately, when its value is

$$P_{max} \approx 0\cdot16\,\frac{\mu\,U\,l^2}{h_2{}^2}$$

and when

$$F = F_1 \approx 0\cdot75\,\frac{\mu\,U\,l}{h_2}\;.$$

The coefficient of friction F/P is proportional to h_2/l and can be made very small.

The coordinates of the centre of pressure, x_c, can be shown to be equal to

$$x_c = \tfrac{1}{2}\,l\left[\frac{2\,k}{k-1} - \frac{k^2 - 1 - 2\,k\ln k}{(k^2 - 1)\ln k - 2\,(k-1)^2}\right].\tag{6.34}$$

For small angles of inclination between block and slide ($k \approx 1$), the pressure distribution from eqn. (6.29) is nearly parabolic, the characteristic thickness and centre of pressure being very nearly at $x = \tfrac{1}{2}\,l$. Putting $h_m = h(\tfrac{1}{2}\,l)$ we can find that the pressure difference becomes

$$p_m = \mu\,U \frac{l^2}{(2\,a-l)\,h_m{}^2}\;.\tag{6.35}$$

If we compare this result with that for creeping motion past a sphere in eqn. (6.7b), we notice that in the case of the slipper the pressure difference is greater by a factor $(l/h_m)^2$. Since l/h_m is of the order of 500 to 1000 ($l = 4$, $h_m = 0 \cdot 004$ to $0 \cdot 008$ in), the prevailing pressures are seen to assume very large values†. The occurrence of such high pressures in slow viscous motion is a peculiar property of the type of flow encountered in lubrication. At the same time it is recognized that the angle formed between the two solid surfaces is an essential feature of the flow.

The pressure and velocity distribution, and the shape of streamlines for the case of a plane slipper are given in Fig. 6.4. It will be noticed that back-flow occurs in the region of pressure rise near the wall at rest, just as was the case with the channel in Fig. 5.2, when the pressure increased in the direction of wall motion. W. Froessel [5] calculated the pressure distribution and thrust supported by a slipper of finite width as well as by a spherical slipper and confirmed these calculations by experiment.

In many cases when the width of the slipper is finite, the assumption made earlier that the flow is one-dimensional is insufficient, and the existence of a component w in the z-direction must be taken into account; here z is perpendicular to the plane of the sketch in Fig. 6.4. The equation preceding eqn. (6.19) must now be supplemented by

$$Q_z = \int_0^h w \, dy = \tfrac{1}{2} \, h \, W - \frac{h^3}{12 \, \mu} \cdot \frac{\partial p}{\partial z} \,, \tag{6.36}$$

and the equation of continuity becomes

$$\frac{\partial}{\partial x} \int_0^h u \, dy + \frac{\partial}{\partial z} \int_0^h w \, dy = 0 \tag{6.37}$$

or

$$\frac{\partial}{\partial x} \left(h^3 \, \frac{\partial p}{\partial x} \right) + \frac{\partial}{\partial z} \left(h^3 \, \frac{\partial p}{\partial z} \right) = 6 \, \mu \left[\frac{\partial}{\partial x} \, (h \, U) + \frac{\partial}{\partial z} \, (h \, W) \right], \tag{6.38}$$

which is known as *Reynolds's* equation of lubrication. Here W denotes the component of the velocity of the boundary in the z-direction at a given x.

In the case of a journal and bearing there must be eccentricity between them in order to create a wedge of variable height which is essential if a thrust is to be created. The relevant theory, based on the preceding principles, as well as on exact two-dimensional theory, was developed in great detail by A. Sommerfeld [16], L. Guembel [6] and G. Vogelpohl [20, 21]. Figure 6.5 shows the pressure distribution in the narrow gap between journal and bearing; it possesses a very pronounced maximum near to the narrowest section of the lubrication wedge. Hence, a significant contribution to the load-bearing capacity is made by that portion of the gap which is *convergent* in the direction of rotation of the journal. The resultant of the pressure forces in this distribution balances the load on the bearing. It has also been

---

† Numerical example: $U = 10$ m/sec; $\mu = 0 \cdot 04$ kg/m sec; $l = 0 \cdot 1$ m; $a = 2\,l = 0 \cdot 2$ m; $h_m = 0 \cdot 2$ mm. Hence $\mu \, U/(2 \, a - l) = 1 \cdot 33$ N/m²; $p_m = 1 \cdot 33 \times 500^2 = 0 \cdot 33$ MPa ($= 3 \cdot 3$ bar).

extended to include the case of bearings with finite width [1, 9], when it was found that the decrease in thrust supported by such a bearing is very considerable due to the sidewise decrease in the pressure. Most theoretical calculations have been conducted under the assumption of constant viscosity. In reality heat is evolved through friction and the temperature of the lubricating oil is increased. Since the viscosity of oil decreases rapidly with increasing temperature (Table 1.2), the thrust also decreases greatly. In more recent times F. Nahme [10] extended the hydrodynamic theory of lubrication to include the effect of the variation of viscosity with temperature (*cf.* Chap. XII).

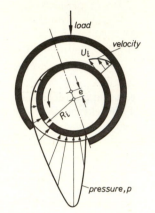

Fig. 6.5. Pressure distribution on the surface of a journal which rotates eccentrically in a bearing (schematic)

e = eccentricity of journal in bearing

With large velocities and high temperatures (low viscosity), the reduced Reynolds number R^* from eqn. (6.14) can assume values near or exceeding unity. This means that inertia forces become comparable with viscous forces and the validity of the theory may be questioned. It is possible to improve the theory, and to extend it to higher Reynolds numbers, by a step-by-step procedure. The neglected inertia terms can be calculated from the first approximation and introduced as external forces so that a second approximation is obtained. This procedure corresponds to Oseen's improved solution for the flow past a sphere. Such calculations have been performed by W. Kahlert [8], who found that the inertia corrections in the case of a plane slipper or circular bearing do not exceed 10 per cent. of the solution under consideration for values of up to $R^* = 5$, approximately. A comparison between theoretical and experimental results is contained in a book by G. Vogelpohl [22], and in an earlier paper [21].

Turbulence. The modern tendency to increase the loads and hence also the peripheral velocities of bearings has led to a situation in which the inertia forces that nowadays occur in the lubricating film begin to play an important part in the process. Under certain conditions, this causes the laminar Couette flow to become unstable, and leads to turbulence.

As early as 1923, G. I. Taylor [18] investigated the case of a bearing in which the journal rotates concentrically in the bushing so that the lubricating gap is one of constant thickness. The instability and the transition to turbulence is governed by the dimensionless Taylor number

$$T = \frac{U_i d}{\nu} \sqrt{\frac{d}{R_i}} . \tag{6.39}$$

Here R_i and U_i denote, respectively, the radius and the peripheral velocity of the concentric journal ($e = 0$) and d is the width of the gap.

After the onset of instability, the flow in the gap develops regularly spaced, cellular vortices which alternately rotate in opposite directions. The axes of these vortices coincide with the circumferential direction, as shown schematically in Figs. 17.32 and in the photograph of Fig. 17.33. In a certain range of Taylor numbers, the flow in the Taylor vortices remains laminar. Transition to turbulent flow occurs at values of the Taylor number which considerably exceed the limit of stability. The three regimes of flow (as will be repeated in Sec. XVIIf and in Fig. 17.34) are characterized as follows:

$$T < 41 \cdot 3 \qquad \text{laminar Couette flow};$$
$$41 \cdot 3 < T < 400 \qquad \text{laminar flow with cellular Taylor vortices};$$
$$T > 400 \qquad \text{turbulent flow}.$$

When the flow becomes unstable, the torque acting on the rotating cylinder increases steeply, because the kinetic energy stored in the secondary flow must be compensated by work.

The same flow phenomena, generally speaking, occur when the bearing is loaded and the gap width varies circumferentially, but the details of the flow become more complex. Attempts have been made to calculate the turbulent flow in a gap of a bearing with the aid of Prandtl's mixing length [Chap. XIX, eqn. (19.7)]. The set of these problems has attracted a wide circle of investigators, such as D. F. Wilcock [19], V. N. Constantinescu [2, 3, 4]. E. A. Saibel and N. A. Macken [14, 15] have written two general accounts that contain numerous literature references.

d. The Hele-Shaw flow

Another remarkable solution of the three-dimensional equations of creeping motion, eqns. (6.3) and (6.4), can be obtained for the case of flow between two parallel flat walls separated by a small distance $2h$. If a cylindrical body of arbitrary cross-section is inserted between the two plates at right angles so that it completely fills the space between them, the resulting pattern of streamlines is identical with that in potential flow about the same shape. H. S. Hele-Shaw [7] used this method to obtain experimental patterns of streamlines in potential flow about arbitrary bodies. It is easy to prove that the solution for creeping motion from eqns. (6.3) and (6.4) possesses the same streamlines as the corresponding potential flow.

We select a system of coordinates with its origin in the centre between the two plates, and make the x, y-plane parallel to the plates, the z-axis being perpendicular to them. The body is assumed to be placed in a stream of velocity U_∞ parallel to the x-axis. At a large distance from the body the velocity distribution is parabolic, as in the motion in a rectangular channel which was considered in Section V.1. Hence

$$x = \infty: \qquad u = U_\infty \left(1 - \frac{z^2}{h^2} \right), \qquad v = 0, \qquad w = 0 .$$

A solution of eqns. (6.3) and (6.4) can be written as:

$$\left. \begin{array}{l} u = u_0(x, y) \left(1 - \frac{z^2}{h^2} \right) ; \qquad v = v_0(x, y) \left(1 - \frac{z^2}{h^2} \right) ; \qquad w \equiv 0 \\[3mm] p = -\frac{2\mu}{h^2} \int\limits_{x_0}^{x} u_0(x, y) \, dx = -\frac{2\mu}{h^2} \int\limits_{y_0}^{y} v_0(x, y) \, dy \end{array} \right\} \quad (6.39)$$

where $u_0(x, y)$, $v_0(x, y)$ and $p_0(x, y)$ denote the velocity and pressure distribution of the two-dimensional potential flow past the given body. Thus u_0, v_0 and p_0 satisfy the equations

$$
\left.
\begin{aligned}
u_0 \frac{\partial u_0}{\partial x} + v_0 \frac{\partial u_0}{\partial y} &= -\frac{1}{\varrho}\frac{\partial p_0}{\partial x} \\[2mm]
u_0 \frac{\partial v_0}{\partial x} + v_0 \frac{\partial v_0}{\partial y} &= -\frac{1}{\varrho}\frac{\partial p_0}{\partial y} \\[2mm]
\frac{\partial u_0}{\partial x} + \frac{\partial v_0}{\partial y} &= 0 \;.
\end{aligned}
\right\}
\qquad (6,40\,\mathrm{a,\ b,\ c})
$$

First we notice at once from the solution (6.39) that the equation of continuity and the equation of motion in the z-direction are satisfied. The fact that the equations of motion in the x- and y-directions are also satisfied follows from the potential character of u_0 and v_0. The functions u_0 and v_0 satisfy the condition of irrotationality

$$\partial u_0/\partial y - \partial v_0/\partial x = 0 \;,$$

so that the potential equations $\nabla^2 u_0 = 0$ and $\nabla^2 v_0 = 0$, where $\nabla^2 = \partial^2/\partial x^2 + \partial^2/\partial y^2$, are satisfied.

The first two equations (6.3) reduce to $\partial p/\partial x = \mu\,\partial^2 u/\partial z^2$ and $\partial p/\partial y = \mu\,\partial^2 v/\partial z^2$; they are, however, satisfied, as seen from eqns. (6.39). Thus eqns. (6.39) represent a solution of the equations for creeping motion. On the other hand the flow represented by eqns. (6.39) has the same streamlines as potential flow about the body, and the streamlines for all parallel layers $z = $ const are congruent. The condition of no slip at the plates $z = \pm h$ is seen to be satisfied by eqn. (6.39), but the condition of no slip at the surface of the body is not satisfied.

The ratio of inertia to viscous forces in Hele-Shaw motion, just as in the case of the motion of lubricating oil, is given by the reduced Reynolds number

$$\mathsf{R}^{*} = \frac{U_\infty L}{\nu}\left(\frac{h}{L}\right)^2 \ll 1 \;,$$

where L denotes a characteristic linear dimension of the body in the x, y-plane. If R^{*} exceeds unity the inertia terms become considerable and the motion deviates from the simple solution (6.39).

The solution given by eqn. (6.39) can be improved in the same manner as Stokes's solution for a sphere or the solution for very slow flow. The inertia terms are calculated from the first approximation and introduced into the equations as external forces, and an improved solution results. This was carried out by F. Riegels [13] for the case of Hele-Shaw flow past a circular cylinder.

For $\mathsf{R}^{*} > 1$ the streamlines in the various layers parallel to the walls cease to be congruent. The slow particles near the two plates are deflected more by the presence of the body than the faster particles near the centre. This causes the streamlines to appear somewhat blurred and the phenomenon is more pronounced at the rear of the body than in front of it, Fig. 6.6.

Solutions in the case of creeping motion are inherently restricted to very small Reynolds numbers. In principle it is possible to extend the field of application to larger Reynolds numbers by successive approximation, as mentioned previously. However, in all cases the calculations become so complicated that it is not practicable to carry out more than one step in the approximation. For this reason it is not possible to reach the region of moderate Reynolds numbers from this direction. To all intents and purposes the region of moderate Reynolds numbers in which the inertia and viscous forces are of comparable magnitude throughout the field of flow has not been extensively investigated by analytic means.

It is, therefore, the more useful to have the possibility of integrating the Navier-Stokes equation for the other limiting case of very large Reynolds numbers. Thus we are led to the boundary-layer theory which will form the subject of the following chapters.

Fig. 6.6. Hele-Shaw flow past circular cylinder at $R^* = 4$, after Riegels [13]

References

[1] Bauer, K.: Einfluss der endlichen Breite des Gleitlagers auf Tragfähigkeit und Reibung. Forschg. Ing.-Wes. *14*, 48—62 (1943).

[2] Constantinescu, V.N.: Analysis of bearings operating in turbulent regime. Trans. ASME, Series D, J. Basic Eng. *84*, 139—151 (1962).

[3] Constantinescu, V.N.: On the influence of inertia forces in turbulent and laminar self-acting films. Trans. ASME, Series F, J. Lubrication Technology *92*, 473—481 (1970).

[4] Constantinescu, V.N.: On gas lubrication in turbulent regime. Trans. ASME, Series D, J. Basic Eng. *86*, 475—482 (1964).

[5] Frössel, W.: Reibungswiderstand und Tragkraft eines Gleitschuhes endlicher Breite. Forschg. Ing.-Wes. *13*, 65—75 (1942).

[6] Gümbel, L., and Everling, E.: Reibung und Schmierung im Maschinenbau, Berlin, 1925.

[7] Hele-Shaw, H.S.: Investigation of the nature of surface resistance of water and of stream motion under certain experimental conditions. Trans. Inst. Nav. Arch. *XI*, 25 (1898); see also Nature *58*, 34 (1898) and Proc. Roy. Inst. *16*, 49 (1899).

[8] Kahlert, W.: Der Einfluss der Trägheitskräfte bei der hydrodynamischen Schmiermittel-theorie. Diss. Braunschweig 1947; Ing.-Arch. *16*, 321—342 (1948).

[9] Michell, A.G.M.: Z. Math. u. Phys. *52*, S. 123 (1905); see also Ostwald's Klassiker No. 218.

[10] Nahme, F.: Beiträge zur hydrodynamischen Theorie der Lagerreibung. Ing.-Arch. *11*, 191—209 (1940).

[11] Oseen, C.W.: Über die Stokes'sche Formel und über eine verwandte Aufgabe in der Hydro-dynamik. Ark. f. Math. Astron. och Fys. *6*, No. 29 (1910).

[12] Prandtl, L.: The mechanics of viscous fluids. In W.F. Durand: Aerodynamic Theory *III*, 34—208 (1935).

[13] Riegels, F.: Zur Kritik des Hele-Shaw-Versuches. Diss. Göttingen 1938; ZAMM *18*, 95—106 (1938).

[14] Saibel, E.A., and Macken, N.A.: The fluid mechanics of lubrication. Annual Review of Fluid Mech. (M. Van Dyke, ed.) *5*, 185—212 (1973).

[15] Saibel, E.A., and Macken, N.A.: Non-laminar behavior in bearings. Critical review of the literature. Trans. ASME, Series F, J. Lubrication Technology *96*, 174—181 (1974).

[16] Sommerfeld, A.: Zur hydrodynamischen Theorie der Schmiermittelreibung. Z. Math. u. Physik *50*, 97 (1904); also Ostwald's Klassiker No. 218, p. 108, and: Zur Theorie der Schmiermittelreibung. Z. Techn. Phys. *2*, 58 (1921); also Ostwald's Klassiker No. 218, p. 181.
[17] Stokes, G. G.: On the effect of internal friction of fluids on the motion of pendulums. Trans. Cambr. Phil. Soc. *9*, Part II, 8—106 (1851) or Coll. Papers *III*, 55.
[18] Taylor, G.I.: Stability of a viscous liquid contained between two rotating cylinders. Phil. Trans. A *223*, 289—293 (1923).
[19] Wilcock, D. F.: Turbulence in high-speed journal bearings. Trans. ASME *72*, 825 (1950).
[20] Vogelpohl, G.: Beiträge zur Kenntnis der Gleitlagerreibung. VDI-Forschungsheft 386 (1937).
[21] Vogelpohl, G.: Ähnlichkeitsbeziehungen der Gleitlagerreibung und untere Reibungsgrenze. Z. VDI *91*, 379 (1949).
[22] Vogelpohl, G.: Betriebssichere Gleitlager. Berechnungsverfahren für Konstruktion und Betrieb. Vol. *1*, Springer-Verlag, 2nd. ed., Berlin, 1967.

Part B. Laminar boundary layers

CHAPTER VII

Boundary-layer equations for two-dimensional incompressible flow; boundary layer on a plate

a. Derivation of boundary-layer equations for two-dimensional flow

We now proceed to examine the second limiting case, namely that of very small viscosity or very large Reynolds number. An important contribution to the science of fluid motion was made by L. Prandtl [21] in 1904 when he clarified the essential influence of viscosity in flows at high Reynolds numbers and showed how the Navier-Stokes equations could be simplified to yield approximate solutions for this case. We shall explain these simplifications with the aid of an argument which preserves the physical picture of the phenomenon, and it will be recalled that in the bulk of the fluid inertia forces predominate, the influence of viscous forces being vanishingly small.

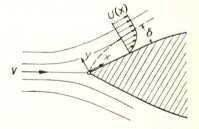

Fig. 7.1. Boundary-layer flow along a wall

For the sake of simplicity we shall consider two-dimensional flow of a fluid with very small viscosity about a cylindrical body of slender cross-section, Fig. 7.1. With the exception of the immediate neighbourhood of the surface, the velocities are of the order of the free-stream velocity, V, and the pattern of streamlines and the velocity distribution deviate only slightly from those in frictionless (potential) flow. However, detailed investigations reveal that, unlike in potential flow, the fluid does not slide over the wall, but adheres to it. The transition from zero velocity at the wall to the full magnitude at some distance from it takes place in a very thin layer, the so-called boundary layer. In this manner there are two regions to consider, even if the division between them is not very sharp:

1. A very thin layer in the immediate neighbourhood of the body in which the velocity gradient normal to the wall, $\partial u/\partial y$, is very large (*boundary layer*). In this region the very small viscosity μ of the fluid exerts an essential influence in so far as the shearing stress $\tau = \mu(\partial u/\partial y)$ may assume large values.

2. In the remaining region no such large velocity gradients occur and the influence of viscosity is unimportant. In this region the flow is frictionless and potential.

In general it is possible to state that the thickness of the boundary layer increases with viscosity, or, more generally, that it decreases as the Reynolds number increases. It was seen from several exact solutions of the Navier-Stokes equations presented in Chap. V that the boundary-layer thickness is proportional to the square root of kinematic viscosity:

$$\delta \sim \sqrt{\nu}\,.$$

When making the simplifications to be introduced into the Navier-Stokes equations it is assumed that this thickness is very small compared with a still unspecified linear dimension, L, of the body:

$$\delta \ll L\,.$$

In this way the solutions obtained from the boundary-layer equations are asymptotic and apply to very large Reynolds numbers.

We shall now proceed to discuss the simplification of the Navier-Stokes equations, and in order to achieve it, we shall make an estimate of the order of magnitude of each term. In the two-dimensional problem shown in Fig. 7.1 we shall begin by assuming the wall to be flat and coinciding with the x-direction, the y-axis being perpendicular to it. We now rewrite the Navier-Stokes equations in dimensionless form by referring all velocities to the free-stream velocity, V, and by referring all linear dimensions to a characteristic length, L, of the body, which is so selected as to ensure that the dimensionless derivative, $\partial u/\partial x$, does not exceed unity in the region under consideration. The pressure is made dimensionless with ϱV^2, and time is referred to L/V. Further, the expression

$$\mathsf{R} = \frac{VL\varrho}{\mu} = \frac{VL}{\nu}$$

denotes the Reynolds number which is assumed very large. Under these assumptions, and retaining the same symbols for the dimensionless quantities as for their dimensional counterparts, we have from the Navier-Stokes equations for plane flow, eqns. (3.32) or (4.4):

continuity:
$$\frac{\partial u}{\partial x} + \frac{\partial v}{\partial y} = 0\,. \tag{7.1}$$
$$\;\;1\qquad\;\;1$$

direction x:
$$\frac{\partial u}{\partial t} + u \frac{\partial u}{\partial x} + v \frac{\partial u}{\partial y} = -\frac{\partial p}{\partial x} + \frac{1}{R}\left(\frac{\partial^2 u}{\partial x^2} + \frac{\partial^2 u}{\partial y^2}\right), \qquad (7.2)$$

$$1 \qquad 1 \;\; 1 \qquad \delta \frac{1}{\delta} \qquad\qquad \delta^2 \quad 1 \qquad \frac{1}{\delta^2}$$

direction y:
$$\frac{\partial v}{\partial t} + u \frac{\partial v}{\partial x} + v \frac{\partial v}{\partial y} = -\frac{\partial p}{\partial y} + \frac{1}{R}\left(\frac{\partial^2 v}{\partial x^2} + \frac{\partial^2 v}{\partial y^2}\right) \qquad (7.3)$$

$$\delta \qquad 1 \;\; \delta \qquad \delta \;\; 1 \qquad\qquad \delta^2 \;\; \delta \qquad \frac{1}{\delta}$$

The boundary conditions are: absence of slip between the fluid and the wall, i. e. $u = v = 0$ for $y = 0$, and $u = U$ for $y \to \infty$.

With the assumptions made previously the dimensionless boundary-layer thickness, δ/L, for which we shall retain the symbol δ, is very small with respect to unity, $(\delta \ll 1)$.

We shall now estimate the order of magnitude of each term in order to be able to drop small terms and thus to achieve the desired simplification of the equations. Since $\partial u/\partial x$ is of the order 1, we see from the equation of continuity that equally $\partial v/\partial y$ is of the order 1, and hence, since at the wall $v = 0$, that in the boundary layer v is of the order δ. Thus $\partial v/\partial x$ and $\partial^2 v/\partial x^2$ are also of the order δ. Further $\partial^2 u/\partial x^2$ is of the order 1. (The orders of magnitude are shown in eqns. (7.1) to (7.3) under the individual terms.)

We shall, further, assume that the non-steady acceleration $\partial u/\partial t$ is of the same order as the convective term $u\, \partial u/\partial x$ which means that very sudden accelerations, such as occur in very large pressure waves, are excluded. In accordance with our previous argument some of the viscous terms must be of the same order of magnitude as the inertia terms, at least in the immediate neighbourhood of the wall, and in spite of the smallness of the factor $1/R$. Hence some of the second derivatives of velocity must become very large near the wall. In accordance with what was said before this can only apply to $\partial^2 u/\partial y^2$ and $\partial^2 v/\partial y^2$. Since the component of velocity parallel to the wall increases from zero at the wall to the value 1 in the freestream across the layer of thickness δ, we have

$$\frac{\partial u}{\partial y} \sim \frac{1}{\delta} \quad \text{and} \quad \frac{\partial^2 u}{\partial y^2} \sim \frac{1}{\delta^2},$$

whereas $\partial v/\partial y \sim \delta/\delta \sim 1$ and $\partial^2 v/\partial y^2 \sim 1/\delta$. If these values are inserted into eqns. (7.2) and (7.3), it follows from the first equation of motion that the viscous forces in the boundary layer can become of the same order of magnitude as the inertia forces only if the Reynolds number is of the order $1/\delta^2$:

$$\frac{1}{R} = \delta^2. \qquad (7.4)$$

The first equation, that of continuity, remains unaltered for very large Reynolds numbers. The second equation can now be simplified by neglecting $\partial^2 u/\partial x^2$ with respect to $\partial^2 u/\partial y^2$. From the third equation we may infer that $\partial p/\partial y$ is of the order δ. The pressure increase across the boundary layer which would be obtained by integrating the third equation, is of the order δ^2, i. e. very small. Thus the pressure in a direction normal to the boundary layer is practically constant; it may be assumed equal to that at the outer edge of the boundary layer where its value is determined by the frictionless flow. The pressure is said to be "impressed" on the boundary layer by the outer flow. It may, therefore, be regarded as a known function as far as boundary-layer flow is concerned, and it depends only on the coordinate x, and on time t.

At the outer edge of the boundary layer the parallel component u becomes equal to that in the outer flow, $U(x,t)$. Since there is no large velocity gradient here, the viscous terms in eqn. (7.2) vanish for large values of R, and consequently, for the outer flow we obtain

$$\frac{\partial U}{\partial t} + U \frac{\partial U}{\partial x} = -\frac{1}{\varrho} \frac{\partial p}{\partial x}, \tag{7.5}$$

where again the symbols denote dimensional quantities.

In the case of steady flow the equation is simplified still further in that the pressure depends only on x. We shall emphasize this circumstance by writing the derivative as dp/dx, so that

$$U \frac{dU}{dx} = -\frac{1}{\varrho} \frac{dp}{dx}. \tag{7.5a}$$

This may also be written in the usual form of Bernoulli's equation

$$p + \tfrac{1}{2} \varrho U^2 = \text{const} . \tag{7.6}$$

The boundary conditions for the external flow are nearly the same as for frictionless flow. The boundary-layer thickness is very small and the transverse velocity component v is very small at the edge of the boundary layer ($v/V \sim \delta/L$). Thus potential non-viscous flow about the body under consideration in which the perpendicular velocity component is vanishingly small near the wall offers a very good approximation to the actual external flow. The pressure gradient in the x-direction in the boundary layer can be obtained by simply applying the Bernoulli equation (7.5a) to the streamline at the wall in the known potential flow.

Summing up, we are now in a position to write down the simplified Navier-Stokes equations, known as *Prandtl's boundary-layer equations*. We return again to dimensional quantities, and obtain:

$$\frac{\partial u}{\partial x} + \frac{\partial v}{\partial y} = 0, \tag{7.7}$$

$$\frac{\partial u}{\partial t} + u \frac{\partial u}{\partial x} + v \frac{\partial u}{\partial y} = -\frac{1}{\varrho} \frac{\partial p}{\partial x} + \nu \frac{\partial^2 u}{\partial y^2}, \tag{7.8}$$

with the boundary conditions

$$y = 0: \quad u = v = 0; \quad y = \infty: \quad u = U(x,t). \tag{7.9}$$

The potential flow $U(x,t)$ is to be considered known; it determines the pressure distribution with the aid of eqn. (7.5). In addition, a suitable boundary-layer flow must be prescribed over the whole x, y region under consideration for the instant $t = 0$.

In the case of *steady flow* the above system of equations simplifies to

$$\frac{\partial u}{\partial x} + \frac{\partial v}{\partial y} = 0 \tag{7.10}$$

$$u\frac{\partial u}{\partial x} + v\frac{\partial u}{\partial y} = -\frac{1}{\varrho}\frac{dp}{dx} + \nu\frac{\partial^2 u}{\partial y^2} \tag{7.11}$$

with the boundary conditions

$$y = 0: \quad u = 0, \quad v = 0; \quad y = \infty: \quad u = U(x). \tag{7.12}$$

It is necessary to prescribe, in addition, a velocity profile at the initial section, $x = x_0$, say, by indicating the function $u(x_0, y)$. The problem is thus seen to reduce itself to the calculation of the further change of a given velocity profile with a given potential motion.

The mathematical simplification achieved on the preceding pages is considerable: it is true that, as distinct from the case of creeping motion, the non-linear character of the Navier-Stokes equation has been preserved, but of the three original equations for u, v, and p of the two-dimensional flow problem, one, the equation of motion normal to the wall, has been dropped completely. Thus the number of unknowns has been reduced by one. There remains a system of two simultaneous equations for the two unknowns u and v. The pressure ceased to be an unknown function and can now be evaluated from the potential flow solution for the body with the aid of the Bernoulli equation. Further, one viscous term in the remaining equation of motion has also been dropped.

Finally, we shall note that the estimation of the boundary-layer thickness in eqn. (7.4) showed that

$$\frac{\delta}{L} \sim \frac{1}{\sqrt{R}} = \sqrt{\frac{\nu}{VL}}. \tag{7.13}$$

The fact that $\delta \sim \sqrt{\nu}$, inferred from the exact solutions of the Navier-Stokes equations, is thereby confirmed. The numerical coefficient, still missing in eqn. (7.13), will turn out to be equal to 5 for the case of a flat plate at zero incidence, when L will mean the distance from its leading edge.

The preceding derivations were related to a flat plate, but there is no difficulty in extending them to the case of a curved wall [26]. When this is done, it is found that equations (7.10) to (7.12) continue to be applicable on condition that the curvature does not change abruptly as would be the case with sharp edges.

The present argument assumed at the outset that the viscosity affects the flow essentially only in a very thin layer. It should be recorded, however, that attempts have been made to derive the boundary-layer equations from the Navier-Stokes equations in a purely mathematical way, that is without the adoption of physically plausible concepts [24].

b. The separation of a boundary layer

It is already possible to draw some important conclusions from the preceding deliberations, i. e. without first discussing the question of the methods of integration.

The first important question to answer is to determine the circumstances under which some of the retarded fluid in the boundary layer can be transported into the main stream or, in other words, to find when *separation* of the flow from the wall may occur. When a region with an adverse pressure gradient exists along the wall, the retarded fluid particles cannot, in general, penetrate too far into the region of increased pressure owing to their small kinetic energy. Thus the boundary layer is deflected sideways from the wall, separates from it, and moves into the main stream, Fig. 7.2. In general the fluid particles behind the point of separation follow the pressure gradient and move in a direction opposite to the external stream.

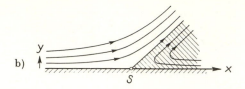

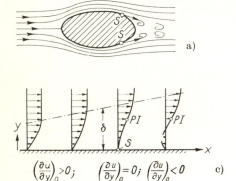

$$\left(\frac{\partial u}{\partial y}\right)_0 > 0; \qquad \left(\frac{\partial u}{\partial y}\right)_0 = 0; \qquad \left(\frac{\partial u}{\partial y}\right)_0 < 0 \qquad c)$$

Fig. 7.2. Separation of the boundary layer. a) Flow past a body with separation (S = point of separation). b) Shape of streamlines near point of separation. c) Velocity distribution near the point of separation (PI = point of inflexion)

The point of separation is defined as the limit between forward and reverse flow in the layer in the immediate neighbourhood of the wall, or

$$\text{point of separation}: \left(\frac{\partial u}{\partial y}\right)_{y=0} = 0 . \qquad (7.14)\dagger$$

In order to answer the question of whether and where separation occurs, it is necessary, in general, first to integrate the boundary-layer equations. Generally speaking, the boundary-layer equations are only valid as far as the point of separation. A short distance downstream from the point of separation the boundary-layer becomes so thick that the assumptions which were made in the derivation of the boundary-layer equations no longer apply. In the case of bodies with blunt sterns the separated boundary layer displaces the potential flow from the body by an appreciable distance and the pressure distribution impressed on the boundary layer must be determined by experiment, because the external flow depends on the phenomena connected with separation.

The fact that separation in steady flow occurs only in decelerated flow ($dp/dx > 0$) can be easily inferred from a consideration of the relation between the pressure gradient dp/dx and the velocity distribution $u(y)$ with the aid of the boundary-layer

---

† The velocity profile at the point of separation is seen to have a perpendicular tangent at the wall. The velocity profiles downstream from the point of separation will show regions of reversed flow near the wall, Fig. 7.2c.

equations. From eqn. (7.11) with the boundary conditions $u = v = 0$ we have at $y = 0$

$$\mu \left(\frac{\partial^2 u}{\partial y^2} \right)_{y=0} = \frac{dp}{dx} \tag{7.15}$$

and, further, after differentiation with respect to y:

$$\left(\frac{\partial^3 u}{\partial y^3} \right)_{y=0} = 0 . \tag{7.16}$$

In the immediate neighbourhood of the wall the curvature of the velocity profile depends only on the pressure gradient, and the curvature of the velocity profile at the wall changes its sign with the pressure gradient. For flow with decreasing pressure (accelerated flow, $dp/dx < 0$) we have from eqn. (7.15) that $(\partial^2 u/\partial y^2)_{wall} < 0$ and, therefore, $\partial^2 u/\partial y^2 < 0$ over the whole width of the boundary layer, Fig. 7.3. In the region of pressure increase (decelerated flow, $dp/dx > 0$) we find $(\partial^2 u/\partial y^2) > 0$. Since, however, in any case $\partial^2 u/\partial y^2 < 0$ at a large distance from the wall, there must exist a point for which $\partial^2 u/\partial y^2 = 0$. This is a point of inflexion† of the velocity profile in the boundary layer, Fig. 7.4.

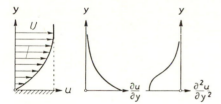

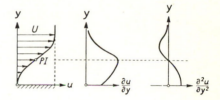

Fig. 7.3. Velocity distribution in a boundary layer with pressure decrease

Fig. 7.4. Velocity distribution in a boundary layer with pressure increase; PI = point of inflexion

It follows that in the region of retarded potential flow the velocity profile in the boundary layer always displays a point of inflexion. Since the velocity profile at the point of separation and with a zero tangent must have a point of inflexion, it follows that separation can only occur when the potential flow is retarded.

c. A remark on the integration of the boundary-layer equations

In order to integrate the boundary-layer equations, whether in the non-steady case, eqns. (7.7) and (7.8), or in the steady case, eqns. (7.10) and (7.11), it is often convenient to introduce a stream function $\psi(x, y, t)$ defined by

$$u = \frac{\partial \psi}{\partial y} , \qquad v = -\frac{\partial \psi}{\partial x} , \tag{7.17}$$

---

† The existence of a point of inflexion in the velocity profile in the boundary layer is important for its stability (transition from laminar to turbulent flow), see Chap. XVI.

so that the equation of continuity is thereby satisfied. Introducing this assumption into eqn. (7.7) we have

$$\frac{\partial^2 \psi}{\partial y\, \partial t} + \frac{\partial \psi}{\partial y}\, \frac{\partial^2 \psi}{\partial x\, \partial y} - \frac{\partial \psi}{\partial x}\, \frac{\partial^2 \psi}{\partial y^2} = -\frac{1}{\varrho}\, \frac{\partial p}{\partial x} + \nu\, \frac{\partial^3 \psi}{\partial y^3} \, , \tag{7.18}$$

which is a partial differential equation of the third order for the stream function. The boundary conditions require the absence of slip at the wall, or $\partial \psi/\partial y = \partial \psi/\partial x = 0$ at the wall. Further, the initial condition at $t = 0$ prescribes the velocity distribution $u = \partial \psi/\partial y$ over the whole region. If this equation for the stream function is compared with the complete Navier-Stokes equations (4.10), it is seen that the boundary-layer assumptions have reduced the order of the equation from four to three.

d. Skin friction

When the boundary-layer equations are integrated, the velocity distribution can be deduced, and the position of the point of separation can be determined. This, in turn, permits us to calculate the viscous drag (skin friction) around the surface by a simple process of integrating the shearing stress at the wall over the surface of the body. The shearing stress at the wall is

$$\tau_0 = \mu \left(\frac{\partial u}{\partial y}\right)_{y=0} \, .$$

The viscous drag for the case of two-dimensional flow becomes

$$D_f = b \int\limits_{s=0}^{l} \tau_0 \cos \phi \, \mathrm{d}s \, , \tag{7.19}$$

where b denotes the height of the cylindrical body; ϕ is the angle between the tangent to the surface and the free-stream velocity U_∞, and s is the coordinate measured along the surface, Fig. 7.5. The process of integration is to be performed

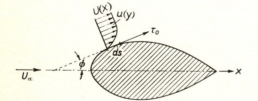

Fig. 7.5.   Illustrating the calculation of skin friction

over the whole surface, from the stagnation point at the leading edge to the trailing edge, assuming that there is no separation. Since $\cos \phi \, \mathrm{d}s = \mathrm{d}x$, where x is measured parallel to the free-stream velocity, we can also write

$$D_f = b \, \mu \int\limits_{x=0}^{l} \left(\frac{\partial u}{\partial y}\right)_{y=0} \mathrm{d}x \, , \tag{7.20}$$

and the integration, as before, is to be extended over the whole wetted surface from the leading to the trailing edge. In order to calculate the skin friction it is necessary to know the velocity gradient at the wall, which can be achieved only through the integration of the differential equations of the boundary layer. If separation occurs before the trailing edge, eqn. (7.20) is valid only as far as the point of separation. Furthermore, if the laminar boundary layer transforms into a turbulent one, eqn. (7.20) applies only as far as the point of transition. Behind the point of transition there is turbulent friction, to be discussed in Chap. XXII.

If separation exists, the pressure distribution differs considerably from that in the ideal case of frictionless, potential flow and pressure, or form drag, results. Thus the boundary-layer theory explains the fact that, in addition to skin friction, there is also form drag, but its magnitude cannot be calculated with the aid of the boundary-layer theory in a simple manner. A rough estimate will, however, be given in Chap. XXV.

e. The boundary layer along a flat plate

In the succeeding chapter we shall deduce a number of general properties of the differential equations of the boundary layer. However, before doing that it seems opportune to consider now a specific example and so to gain greater familiarity with the equations. The simplest example of the application of the boundary-layer equations is afforded by the flow along a very thin flat plate. Historically this was the first example illustrating the application of Prandtl's boundary-layer theory;

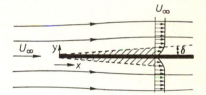

Fig. 7.6. The boundary layer along a flat plate at zero incidence

it was discussed by H. Blasius [2] in his doctor's thesis at Goettingen. Let the leading edge of the plate be at $x = 0$, the plate being parallel to the x-axis and infinitely long downstream, Fig. 7.6. We shall consider steady flow with a free-stream velocity, U_∞, which is parallel to the x-axis. The velocity of potential flow is constant in this case, and, therefore, $dp/dx \equiv 0$. The boundary-layer equations (7.10) to (7.12) become

$$\frac{\partial u}{\partial x} + \frac{\partial v}{\partial y} = 0 , \tag{7.21}$$

$$u \frac{\partial u}{\partial x} + v \frac{\partial u}{\partial y} = \nu \frac{\partial^2 u}{\partial y^2} , \tag{7.22}$$

$$y = 0 : \quad u = v = 0 ; \qquad y = \infty : \quad u = U_\infty . \tag{7.23}$$

Since the system under consideration has no preferred length it is reasonable to suppose that the velocity profiles at varying distances from the leading edge are similar to each other, which means that the velocity curves $u(y)$ for varying distances x can be made identical by selecting suitable scale factors for u and y†. The scale factors for u and y appear quite naturally as the free-stream velocity, U_∞ and the boundary-layer thickness, $\delta(x)$, respectively. It will be noted that the latter increases with the current distance x. Hence the principle of similarity of velocity profiles in the boundary layer can be written as $u/U_\infty = \phi(y/\delta)$, where the function ϕ must be the same at all distances x from the leading edge.

We can now estimate the thickness of the boundary layer. From the exact solutions of the Navier-Stokes equations considered previously (Chap. V) it was found, e. g. in the case of a suddenly accelerated plate, that $\delta \sim \sqrt{\nu t}$, where t denoted the time from the start of the motion. In relation to the problem under consideration we may substitute for t the time which a fluid particle consumes while travelling from the leading edge to the point x. For a particle outside the boundary layer this is $t = x/U_\infty$, so that we may put $\delta \sim \sqrt{\nu x/U_\infty}$. We now introduce the new dimensionless coordinate $\eta \sim y/\delta$ so that

$$\eta = y\sqrt{\frac{U_\infty}{\nu x}}. \tag{7.24}$$

The equation of continuity, as already discussed in Sec. VIId, can be integrated by introducing a stream function $\psi(x,y)$. We put

$$\psi = \sqrt{\nu x\, U_\infty} f(\eta), \tag{7.25}$$

where $f(\eta)$ denotes the dimensionless stream function. Thus the velocity components become:

$$u = \frac{\partial \psi}{\partial y} = \frac{\partial \psi}{\partial \eta}\frac{\partial \eta}{\partial y} = U_\infty f'(\eta), \tag{7.26}$$

the prime denoting differentiation with respect to η. Similarly, the transverse velocity component is

$$v = -\frac{\partial \psi}{\partial x} = \frac{1}{2}\sqrt{\frac{\nu U_\infty}{x}} (\eta f' - f). \tag{7.27}$$

Writing down the further terms of eqn. (7.22), and inserting, we have

$$-\frac{U_\infty^2}{2x}\eta f' f'' + \frac{U_\infty^2}{2x}(\eta f' - f) f'' = \nu \frac{U_\infty^2}{x\nu} f'''.$$

After simplification, the following ordinary differential equation is obtained:

$$f f'' + 2 f''' = 0 \quad \text{(Blasius's equation)}. \tag{7.28}$$

As seen from eqns. (7.23), as well as (7.26) and (7.27), the boundary conditions are:

$$\eta = 0: \quad f = 0, \quad f' = 0; \quad \eta = \infty: \quad f' = 1. \tag{7.29}$$

---

† The problem of *affinity* or *similarity* of velocity profiles will be considered from a more general point of view in Chap. VIII. The more exact theory shows that the region immediately behind the leading edge must be excluded; see p. 141.

In this example both partial differential equations (7.21) and (7.22) have been transformed into an *ordinary* differential equation for the stream function by the similarity transformation, eqns. (7.24) and (7.25). The resulting differential equation is non-linear and of the third order. The three boundary conditions (7.29) are, therefore, sufficient to determine the solution completely.

The analytic evaluation of the solution of the differential equation (7.28) is quite tedious. H. Blasius obtained this solution in the form of a series expansion around $\eta = 0$ and an asymptotic expansion for η very large, the two forms being matched at a suitable value of η. The resulting procedure was described in detail by L. Prandtl [22]. Subsequent to that, L. Bairstow [1] and S. Goldstein [13] solved the same equation but with the aid of a slightly modified procedure. Somewhat earlier, C. Toepfer [27] solved the Blasius equation (7.28) numerically by the application of the method of Runge and Kutta. The same equation was solved again, this time with an increased accuracy, by L. Howarth [16]; the numerical values of f, f' and f'' quoted in Table 7.1 have been taken from his paper. In this connexion, the reader may also consult a new method of integration devised by D. Meksyn [19].

The variation of the longitudinal component $u/U_\infty = f'(\eta)$ is seen plotted in Fig. 7.7. Comparing it with the profile near a stagnation point, Fig. 5.10, we see that the velocity profile on a flat plate possesses a very small curvature at the wall and turns rather abruptly further from it in order to reach the asymptotic value. At the wall itself the curve has a point of inflexion, since for $y = 0 : \partial^2 u/\partial y^2 = 0$.

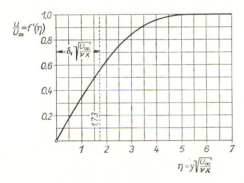

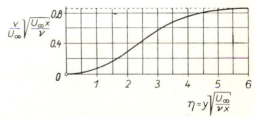

Fig. 7.7. Velocity distribution in the boundary layer along a flat plate, after Blasius [2]

Fig. 7.8. The transverse velocity component in the boundary layer along a flat plate

The transverse component of the velocity in the boundary layer, given by eqn. (7.27), is represented in Fig. 7.8. It is worth noting here that at the outer edge of the boundary layer, i. e. for $\eta \to \infty$ this component differs from zero; we have

$$v_\infty = 0 \cdot 8604\, U_\infty \sqrt{\frac{\nu}{x\, U_\infty}}\;.$$

This means that at the outer edge there is a flow outward which is due to the fact that the increasing boundary-layer thickness causes the fluid to be displaced from the wall as it flows along it. There is no boundary-layer separation in the present case, as the pressure gradient is equal to zero.

J. Steinheuer [25] published a systematic review of the solutions to Blasius's equation. In particular, he provided a discussion of the character of the solutions in the integration range where $\eta < 0$ in the presence of a variety of boundary conditions. It turns out that there exist three sets of solutions which differ from each other by their asymptotic behavior at $\eta \to -\infty$. Apart from the laminar boundary layer on a flat plate, the solutions which can be given a physically meaningful interpretation include laminar flow between two parallel streams of which the two-dimensional half-jet is a special case (see Sec. IX h), laminar flow with suction or blowing at right angles (see Sec. XIV b), as well as the laminar boundary layer formed over a wall moving parallel to the stream in the same or in the opposite direction.

Skin friction: The skin friction can be easily determined from the preceding data. From eqn. (7.19) we obtain for one side of the plate

$$D = b \int_{x=0}^{l} \tau_0 \, \mathrm{d}x , \tag{7.30}$$

where b is the width and l is the length of the plate. Now the local shearing stress at the wall is given by

$$\tau_0(x) = \mu \left(\frac{\partial u}{\partial y} \right)_{y=0} = \mu \, U_\infty \sqrt{\frac{U_\infty}{\nu x}} f''(0) = \alpha \, \mu \, U_\infty \sqrt{\frac{U_\infty}{\nu x}} , \tag{7.31}$$

with $f''(0) = \alpha = 0 \cdot 332$ from Table 7.1. Hence the dimensionless shearing stress becomes:

$$\frac{1}{2} c'_f = \frac{\tau_0(x)}{\varrho \, U_\infty^2} = 0 \cdot 332 \sqrt{\frac{\nu}{U_\infty x}} = \frac{0 \cdot 332}{\sqrt{R_x}} . \tag{7.32}$$

Consequently, from eqn. (7.30), the skin friction of one side becomes

$$D = \alpha \, \mu \, b \, U_\infty \sqrt{\frac{U_\infty}{\nu}} \int_{x=0}^{l} \frac{\mathrm{d}x}{\sqrt{x}} = 2 \, \alpha \, b \, U_\infty \sqrt{\mu \, \varrho \, l \, U_\infty} ,$$

and for a plate wetted on both sides:

$$2 \, D = 4 \alpha \, b \, U_\infty \sqrt{\mu \, \varrho \, l \, U_\infty} = 1 \cdot 328 \, b \, \sqrt{U_\infty^3 \, \mu \, \varrho \, l} . \tag{7.33}$$

It is remarkable that the skin friction is proportional to the power $\frac{3}{2}$ of velocity whereas in creeping motion there was proportionality to the first power of velocity. Further, the drag increases with the square root of the length of the plate. This can be interpreted as showing that the downstream portions of the plate contribute proportionately less to the total drag than the portions near the leading edge,

Table 7.1.   The function $f(\eta)$ for the boundary layer along a flat plate at zero incidence, after
L. Howarth [16]

$\eta = y\sqrt{\dfrac{U_\infty}{\nu x}}$	f	$f' = \dfrac{u}{U_\infty}$	f''
0	0	0	0·33206
0·2	0·00664	0·06641	0·33199
0·4	0·02656	0·13277	0·33147
0·6	0·05974	0·19894	0·33008
0·8	0·10611	0·26471	0·32739
1·0	0·16557	0·32979	0·32301
1·2	0·23795	0·39378	0·31659
1·4	0·32298	0·45627	0·30787
1·6	0·42032	0·51676	0·29667
1·8	0·52952	0·57477	0·28293
2·0	0·65003	0·62977	0·26675
2·2	0·78120	0·68132	0·24835
2·4	0·92230	0·72899	0·22809
2·6	1·07252	0·77246	0·20646
2·8	1·23099	0·81152	0·18401
3·0	1·39682	0·84605	0·16136
3·2	1·56911	0·87609	0·13913
3·4	1·74696	0·90177	0·11788
3·6	1·92954	0·92333	0·09809
3·8	2·11605	0·94112	0·08013
4·0	2·30576	0·95552	0·06424
4·2	2·49806	0·96696	0·05052
4·4	2·69238	0·97587	0·03897
4·6	2·88826	0·98269	0·02948
4·8	3·08534	0·98779	0·02187
5·0	3·28329	0·99155	0·01591
5·2	3·48189	0·99425	0·01134
5·4	3·68094	0·99616	0·00793
5·6	3·88031	0·99748	0·00543
5·8	4·07990	0·99838	0·00365
6·0	4·27964	0·99898	0·00240
6·2	4·47948	0·99937	0·00155
6·4	4·67938	0·99961	0·00098
6·6	4·87931	0·99977	0·00061
6·8	5·07928	0·99987	0·00037
7·0	5·27926	0·99992	0·00022
7·2	5·47925	0·99996	0·00013
7·4	5·67924	0·99998	0·00007
7·6	5·87924	0·99999	0·00004
7·8	6·07923	1·00000	0·00002
8·0	6·27923	1·00000	0·00001
8·2	6·47923	1·00000	0·00001
8·4	6·67923	1·00000	0·00000
8·6	6·87923	1·00000	0·00000
8·8	7·07923	1·00000	0·00000

because they lie in the region where the boundary layer is thicker and where, consequently, the shearing stress at the wall is smaller. Introducing, as usual, a dimensionless drag coefficient by the definition

$$c_f = \frac{2\,D}{\frac{1}{2}\,\varrho\,A\,U_\infty{}^2}\,,$$

where $A = 2\,b\,l$ denotes the wetted surface area, we obtain from eqn. (7.33) the
formula:

$$\boxed{c_f = \frac{1{,}328}{\sqrt{R_l}}} \tag{7.34}$$

Here $R_l = U_\infty\,l/\nu$ denotes the Reynolds number formed with the length of the
plate and the free-stream velocity. This law of friction on a plate first deduced by
H. Blasius, is valid only in the region of laminar flow, i. e. for $R_l = U_\infty\,l/\nu < 5 \times 10^5$
to 10^6. It is represented in Fig. 21.2 as curve (1). In the region of turbulent motion,
$R_l > 10^6$, the drag becomes considerably greater than that given in eqn. (7.34).

Boundary-layer thickness: It is impossible to indicate a boundary-layer thickness
in an unambiguous way, because the influence of viscosity in the boundary layer
decreases asymptotically outwards. The parallel component, u, tends asymptotically
to the value U_∞ of the potential flow (the function $f'(\eta)$ tends asymptotically to 1).
If it is desired to define the boundary-layer thickness as that distance for which
$u = 0{\cdot}99\,U_\infty$, then, as seen from Table 7.1, $\eta \approx 5{\cdot}0$. Hence the boundary-layer
thickness, as defined here, becomes

$$\delta \approx 5{\cdot}0\,\sqrt{\frac{\nu\,x}{U_\infty}}\,. \tag{7.35}$$

A physically meaningful measure for the boundary layer thickness is the *displacement thickness* δ_1, which was already introduced in eqn. (2.6), Fig. 2.3. The displacement thickness is that distance by which the external potential field of flow
is displaced outwards as a consequence of the decrease in velocity in the boundary
layer. The decrease in volume flow due to the influence of friction is $\int\limits_{y=0}^{\infty} (U_\infty - u)\,dy$,
so that for δ_1 we have the definition

$$U_\infty\,\delta_1 = \int\limits_{y=0}^{\infty} (U_\infty - u)\,dy\,,$$

or

$$\delta_1 = \int\limits_{y=0}^{\infty} \left(1 - \frac{u}{U_\infty}\right) dy\,. \tag{7.36}$$

With u/U_∞ from eqn. (7.26) we obtain

$$\delta_1 = \sqrt{\frac{\nu\,x}{U_\infty}} \int\limits_{\eta=0}^{\infty} [1 - f'(\eta)]\,d\eta = \sqrt{\frac{\nu\,x}{U_\infty}}\,[\eta_1 - f(\eta_1)]\,,$$

where η_1 denotes a point outside the boundary layer. Using the value $f(\eta)$ from Table 7.1 we obtain $\eta_1 - f(\eta_1) = 1.7208$ and hence

$$\delta_1 = 1.7208 \sqrt{\frac{\nu x}{U_\infty}} \quad \text{(displacement thickness).} \quad (7.37)$$

The distance $y = \delta_1$ is shown in Fig. 7.7. This is the distance by which the stream-lines of the external potential flow are displaced owing to the effect of friction near the wall. The boundary-layer thickness, δ, given in eqn. (7.35), over which the potential velocity is attained to within 1 per cent. is, in round figures, three times larger than the displacement thickness.

We may at this point evaluate the *momentum thickness* δ_2 which will be used later. The loss of momentum in the boundary layer, as compared with potential flow, is given by $\varrho \int_0^\infty u(U_\infty - u)\,dy$, so that a new thickness can be defined by

$$\varrho\, U_\infty^2\, \delta_2 = \varrho \int_{y=0}^{\infty} u\,(U_\infty - u)\,dy\ ,$$

or

$$\delta_2 = \int_{y=0}^{\infty} \frac{u}{U_\infty} \left(1 - \frac{u}{U_\infty}\right) dy\ . \quad (7.38)$$

Numerical evaluation for the plate at zero incidence gives:

$$\delta_2 = \sqrt{\frac{\nu x}{U_\infty}} \int_{\eta=0}^{\infty} f'\,(1-f')\,d\eta\ ,$$

or

$$\delta_2 = 0.664 \sqrt{\frac{\nu x}{U_\infty}} \quad \text{(momentum thickness).} \quad (7.39)$$

It is necessary to remark here that near the leading edge of the plate the boundary-layer theory ceases to apply, since there the assumption $|\partial^2 u/\partial x^2| \ll |\partial^2 u/\partial y^2|$ is not satisfied. The boundary-layer theory applies only from a certain value of the Reynolds number $R = U_\infty x/\nu$ onwards. The relationship near the leading edge can only be found from the full Navier-Stokes equations because it involves a singularity at the leading edge itself. An attempt to carry out such a calculation was made by G. F. Carrier and C. C. Lin [5] as well as by B. A. Boley and M. B. Friedman [3].

Experimental investigations: Measurements to test the theory given on the preceding pages were carried out first by J. M. Burgers [4] and B. G. van der Hegge Zijnen [15], and subsequently by M. Hansen [14]. Particularly careful and com-prehensive measurements were reported later by J. Nikuradse [20]. It was found that the formation of the boundary layer is greatly influenced by the shape of the leading edge as well as by the very small pressure gradient which may exist in the external flow. J. Nikuradse introduced careful corrections for these possible effects, when he carried out his measurements on a plate in a stream of air. The velocity distribution in the laminar boundary layer has been plotted from Nikuradse's measurements in Fig. 7.9 for several distances from the leading edge. The similarity

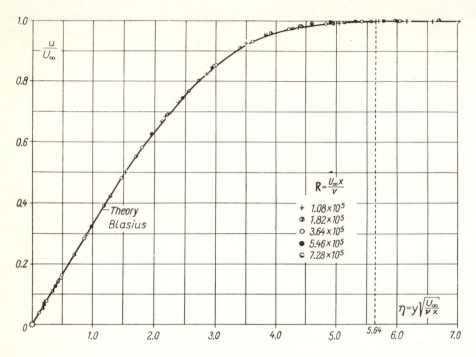

Fig. 7.9. Velocity distribution in the laminar boundary layer on a flat plate at zero incidence, as measured by Nikuradse [20]

of the velocity profiles at various distances x from the leading edge predicted by the theory is confirmed by these measurements. The shape of the velocity profile agrees equally well with that calculated with the aid of the theory. The relation between the dimensionless boundary-layer thickness $\delta \sqrt{U_\infty/\nu x}$ and the Reynolds number formed with the current length, x, was already plotted in Fig. 2.19. This dimensionless thickness remains constant as long as the boundary layer is laminar, and its numerical value is nearly that given in eqn. (7.35). At large Reynolds numbers $U_\infty x/\nu$ the boundary layer ceases to be laminar and transition to turbulent motion takes place. This fact can be recognized in Fig. 2.19 by noticing the marked increase in the thickness of the boundary layer as the distance from the leading edge is increased. According to the measurements performed by B. G. van der Hegge Zijnen and M. Hansen transition from laminar to turbulent flow takes place at $U_\infty x/\nu = 300,000$. This corresponds to a value of the Reynolds number referred to the displacement thickness, $U_\infty \delta_1/\nu = 950$. More recent measurements, to be discussed in Chap. XVI, have demonstrated that the value of this 'critical' Reynolds number can become considerably larger in an air stream which is made very free from disturbance. In this way it is possible to reach values of up to about $U_\infty x/\nu = 3 \times 10^6$.

$$C_f' = \frac{\tau_0}{\frac{1}{2}\varrho U_\infty{}^2}$$

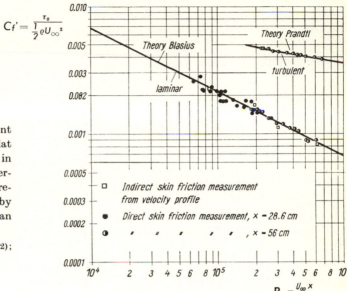

Fig. 7.10. Local coefficient of skin friction on a flat plate at zero incidence in incompressible flow, determined from direct measurement of shearing stress by Liepmann and Dhawan [6, 18]

Theory: laminar from eqn. (7.32); turbulent from eqn. (21.12)

The laminar law of friction on a flat plate was also subjected to careful experimental verification. The local shearing stress at the wall can be determined indirectly from the slope of the velocity profile at the wall together with eqn. (7.31). In recent times H. W. Liepmann and S. Dhawan [18] measured the shearing stress directly from the force acting on a small portion of the plate which was arranged so that it could move slightly with respect to the main plate. The results of their very careful measurements are seen reproduced in Fig. 7.10, which shows a plot of the local coefficient of skin friction $c_f' = \tau_0/\frac{1}{2}\varrho\,U_\infty{}^2$, against the Reynolds number $R_x = U_\infty x/\nu$. In the range of $R_x = 2 \times 10^5$ to 6×10^5 both laminar and turbulent flows are possible. It can be seen that direct and indirect measurements are in excellent agreement with each other. Measurements in the laminar range give a striking confirmation of Blasius's eqn. (7.32) from which $c_f' = 0.664/\sqrt{R_x}$. In the turbulent range there is also good agreement with Prandtl's theoretical formula which will be deduced in Chap. XXI, eqn. (21.12).

The complete agreement between theoretical and experimental results which exists for the velocity distribution and the shearing stress in a laminar boundary layer on a flat plate at zero incidence that has been brought into evidence in Figs. 7.9 and 7.10 for the range $R_x > 10^5$ unequivocally demonstrates the validity of the boundary-layer approximations from the physical point of view. In spite of this, certain mathematicians have expended much effort to create a "mathematical proof" for the validity of these simplifications; in this connexion consult the work of H. Schmidt and K. Schroeder [24].

f. Boundary layer of higher order†

The boundary-layer equations have been obtained in Sec. VII a of this chapter by a process of estimating orders of magnitude of individual terms in the complete equations of motion. The boundary-layer equations can, however, also be derived with the aid of a more general theory. In order to obtain asymptotic expansions of the solutions of the Navier-Stokes equations for large Reynolds numbers, it is possible to establish a perturbation scheme in which

$$\varepsilon = \frac{1}{\sqrt{\overline{R}}} = \frac{1}{\sqrt{\dfrac{U_\infty\,R_0}{\nu}}} \tag{7.40}$$

is chosen as the perturbation parameter. This leads to a so-called singular perturbation scheme and results in the separation of the required asymptotic expansion of the solution into an outer expansion (external flow) and an inner expansion (boundary-layer flow). With the aid of the method of matched asymptotic expansions it thus becomes possible to derive an asymptotic expansion of the complete solution.

The first term of such an asymptotic expansion is precisely the solution of the boundary-layer equations. Moreover, the continuation of the perturbation calculation allows us to compute further terms of the expansion and so to extend the classical theory of boundary layers due to Prandtl. We thus create a boundary-layer theory of higher order. The second terms of the expansion are of particular practical importance because we can look upon them as corrections to the classical theory which represent boundary-layer effects of second order.

Extensive presentations of boundary-layer theory of higher order were published by M. Van Dyke [9], K. Gersten [10], and K. Gersten and J. F. Gross [12]. In addition, reference [8] contains a detailed exposition of the method of matched asymptotic expansions. The basic ideas of this method can be traced to L. Prandtl; they have been made plausible with reference to a simple mathematical example in Sec. IVf.

In what follows, we give a brief description of the theory of asymptotic solutions for large Reynolds numbers as it applies to a two-dimensional, incompressible flow. The main purpose of this argument is to find an extension of Prandtl's boundary-layer theory and to derive the boundary-layer equations of higher order. Details of the derivations can be found in the treatise of M. Van Dyke [7].

The starting point is constituted by the Navier-Stokes equations written with reference to a curvilinear, rectangular system of coordinates in Sec. IVf, Fig. 3.6. All lengths are measured in units of a convenient length R_0, for example the radius of curvature at the stagnation point. Velocities are referred to U_∞ and the overpressures are referred to $\varrho\,U_\infty^2$. The geometrical shape is described by the local radius of curvature, $R(x)$, and the dimensionless curvature of the surface is

$$K(x) = R_0/R(x). \tag{7.41}$$

Outer expansions: In order to solve the system of equations (3.35), we assume the following asymptotic expansions:

$$\left.\begin{aligned}
u(x, y, \varepsilon) &= U_1(x, y) + \varepsilon\,U_2(x, y) + \ldots\\[4pt]
v(x, y, \varepsilon) &= V_1(x, y) + \varepsilon\,V_2(x, y) + \ldots\\[4pt]
p(x, y, \varepsilon) &= P_1(x, y) + \varepsilon\,P_2(x, y) + \ldots.
\end{aligned}\right\} \tag{7.42}$$

These forms are substituted into eqns. (3.35) and the terms are ordered by the powers of ε. In this manner, we obtain a sequence of systems of equations for the first-order solution $U_1(x, y)$,

---

† I owe this section to Professor K. Gersten.

$V_1(x, y)$, $P_1(x, y)$, for the second-order solution $U_2(x, y)$, $V_2(x, y)$, $P_2(x, y)$, etc. Up to solutions of the second order, terms proportional to ε^2, that is the frictional terms in the Navier-Stokes equations, remain unaccounted for. Thus, solutions of first and second order correspond to inviscid flows or even to potential flows when only fields with a uniform oncoming velocity are studied.

The solutions of *first order* satisfy the boundary conditions

$$
\left.
\begin{aligned}
y = 0: \quad & V_1(x, 0) = 0 \\
y \to \infty: \quad & U_1^2 + V_1^2 = 1\,.
\end{aligned}
\right\}
\tag{7.43}
$$

The solution of the potential-flow equations $U_1(x, y)$, $V_1(x, y)$ gives the velocity $U_1(x, 0)$ at the wall, and Bernoulli's equation leads to the wall pressure

$$
P_1(x, 0) = \frac{1}{2} - \frac{1}{2} U_1^2(x, 0)\,.
\tag{7.44}
$$

The solutions of *second order* satisfy the boundary conditions

$$
\left.
\begin{aligned}
y = 0: \quad & V_2(x, 0) = \frac{1}{\varepsilon} \frac{d}{dx} [U_1(x, 0), \delta_1(x)] \\
y \to \infty: \quad & U_2^2 + V_2^2 = 0\,,
\end{aligned}
\right\}
\tag{7.45}
$$

where $\delta_1(x)$ denotes the displacement thickness defined in an analogous way as that in eqn. (7.36); see also eqn. (7.51).

The solution of the potential equation leads again to the distribution of the parallel velocity components at the wall, $U_2(x, 0)$, and to the pressure

$$
P_2(x, 0) = - U_1(x, 0) \cdot U_2(x, 0)\,.
\tag{7.46}
$$

The resulting solutions do not, generally speaking, satisfy the no-slip condition at the wall and for this reason they are not valid near it; they are given the name "outer solutions" or "outer asymptotic expansions".

Inner expansions: In order to obtain solutions valid near the wall, it is necessary to apply a special procedure. Instead of the distance, y, from the wall, we introduce a new, stretched coordinate

$$
N = y/\varepsilon\,.
\tag{7.47}
$$

This so-called inner variable was so selected as to prevent the disappearance of some of the viscous terms in the equations of first order in the coordinate system x, N.

For the solutions near the wall (in the boundary layer), we again assume asymptotic expansions, viz.

$$
\left.
\begin{aligned}
u(x, y, \varepsilon) &= u_1(x, N) + \varepsilon\, u_2(x, N) + \cdots, \\
v(x, y, \varepsilon) &= \varepsilon\, v_1(x, N) + \varepsilon^2\, v_2(x, N) + \cdots, \\
p(x, y, \varepsilon) &= p_1(x, N) + \varepsilon\, p_2(x, N) + \cdots;
\end{aligned}
\right\}
\tag{7.48}
$$

Substitution into the system of equations (3.35) and ordering according to powers of ε, yields the following systems of equations.

Boundary-layer equations of first order:

$$
\left.
\begin{aligned}
\frac{\partial u_1}{\partial x} + \frac{\partial v_1}{\partial N} &= 0\,, \\
u_1 \frac{\partial u_1}{\partial x} + v_1 \frac{\partial u_1}{\partial N} &= - \frac{\partial p_1}{\partial x} + \frac{\partial^2 u_1}{\partial N^2}\,, \\
0 &= - \frac{\partial p_1}{\partial N}\,,
\end{aligned}
\right\}
\tag{7.49}
$$

with the boundary conditions

$$N = 0: \qquad u_1 = 0,\ v_1 = 0, \left.\vphantom{\begin{matrix}a\\a\end{matrix}}\right\} \tag{7.50}$$
$$N \to \infty: \qquad u_1 = U_1(x, 0).$$

These are exactly Prandtl's boundary-layer equations, eqns. (7.10) and (7.11) transformed to coordinates x, N. In addition $p_1(x) = p_1(x, 0)$.

The solution $u_1(x, N)$ allows us to compute the displacement thickness δ_1 defined as

$$\delta_1 = \varepsilon \int\limits_0^\infty \left(1 - \frac{u_1(x, N)}{U_1(x, 0)}\right) \mathrm{d}N. \tag{7.51}$$

The equations of first order, eqns. (7.49), do not contain the Reynolds number explicitly. It follows that $u_1(x, N)$ and $v_1(x, N)$ must also be independent of the Reynolds number. This proves that the location of the point of laminar separation is independent of the Reynolds number.

Boundary-layer equations of second order:

$$\frac{\partial u_2}{\partial x} + \frac{\partial v_2}{\partial N} = K\left(N\frac{\partial u_1}{\partial x} - v_1\right),$$

$$u_1\frac{\partial u_2}{\partial x} + u_2\frac{\partial u_1}{\partial x} + v_1\frac{\partial u_2}{\partial N} + v_2\frac{\partial u_1}{\partial N} + \frac{\partial p_2}{\partial x} - \frac{\partial^2 u_2}{\partial N^2} =$$

$$= K\left(N\frac{\partial^2 u_1}{\partial N^2} + \frac{\partial u_1}{\partial N} - N v_1\frac{\partial u_1}{\partial N} - u_1 v_1\right),$$

$$\frac{\partial p_2}{\partial N} = K u_1^2,$$
$$\left.\vphantom{\begin{matrix}a\\a\\a\\a\\a\\a\\a\end{matrix}}\right\} \tag{7.52}$$

with the boundary conditions

$$N = 0:\ u_2 = 0,\ v_2 = 0,$$
$$N \to \infty:\ u_2 = U_2(x, 0) - K\, U_1(x, 0)\, N, \left.\vphantom{\begin{matrix}a\\a\\a\end{matrix}}\right\} \tag{7.53}$$
$$p_2 = P_2(x, 0) + K\, U_1^2(x, 0)\, N.$$

The outer boundary conditions (i. e. for $N \to \infty$) of the inner solutions as well as the inner boundary conditions of the outer solutions (e. g. eqn. (7.45) for $V_2(x, 0)$) follow from the matching of the inner and outer solutions; see also [7].

The system of equations (7.52), (7.53) for the second-order boundary layer too does not contain the Reynolds number explicitly. However, it contains solutions of first order and is more extensive than the first-order system, but it consists of linear differential equations. For this reason, it is possible, in turn, to separate the whole solution into a sum of partial solutions. It has become customary to split the solution into a curvature term and into a displacement term, but we shall not pursue this discussion any further here.

Due to the fact that the curvature of the wall is accounted for in the second-order theory, there appears a pressure gradient in the direction normal to the wall. For this reason, the pressure at the wall becomes different from that which is impressed on the boundary layer by the outer flow. Integrating across the boundary layer, we obtain the pressure coefficient at the wall in the form

$$\frac{1}{2}\, c_{pw} = p(x, 0, \varepsilon) =$$
$$= P_1(x, 0) + \varepsilon\, \{P_2(x, 0) + K \int\limits_0^\infty [U_1^2(x, 0) - u_1^2(x, N)]\, \mathrm{d}N\} + O\,(\varepsilon^2). \left.\vphantom{\begin{matrix}a\\a\\a\\a\end{matrix}}\right\} \tag{7.54}$$

The pressure at the wall exceeds the impressed pressure when the wall is convex $(K > 0)$.

The distribution of the local shearing stress to second order is

$$\frac{1}{2} c_f' = \frac{\tau_0(x)}{\varrho \, U_\infty^2} = \varepsilon \left(\frac{\partial u_1}{\partial N}\right)_{N=0} + \varepsilon^2 \left(\frac{\partial u_2}{\partial N}\right)_{N=0} + O\left(\varepsilon^3\right). \tag{7.55}$$

The boundary layer of second order also reacts on the outer flow. The paper by K. Gersten [11] contains a calculation of the displacement thickness to second order.

Example: Flat plate at zero incidence. In the case of an impermeable flat plate at zero incidence, the displacement thickness δ_1 is calculated with the aid of eqn. (7.37). According to eqn. (7.45), the boundary condition for the outer flow is

$$V_2(x, 0) = 0.8604/\sqrt{x}, \tag{7.56}$$

where the length of the plate has been chosen as a reference. The solution of the two-dimensional potential equation subject to this boundary condition yields

$$\left.\begin{aligned}
U_2(x, y) &= -\frac{0.8604}{r} \sqrt{\frac{r-x}{2}} \\[2mm]
V_2(x, y) &= \frac{0.8604}{r} \sqrt{\frac{r+x}{2}},
\end{aligned}\right\} \tag{7.57}$$

where

$$r = \sqrt{x^2 + y^2}. \tag{7.58}$$

The associated streamlines are parabolae whose foci are at the origin and whose vertices lie on the x-axis. It follows that in this particular case the velocity $U_2(x, 0)$ at the wall vanishes, and the solution of the system of equations (7.52) and (7.53) is the trivial solution. We conclude, therefore, that in the case of the flat plate the second-order correction to skin friction vanishes. Nevertheless, we must not draw the conclusion that the second-order drag coefficient also vanishes. This is due to the fact that the second-order external flow described by eqn. (7.57) contributes a momentum term. This can be identified by calculating the integral of momentum over the whole plate when it will be discovered that this contribution is equivalent to an increase in drag. Such calculations have been carried out by I. Imai [17] who found that the drag coefficient of a flat plate is given by

$$c_f = \frac{1.328}{\sqrt{R_l}} + \frac{2.326}{R_l}, \tag{7.59}$$

where $2.326 = \pi \times (0.8604)^2$. The correction (the second term) in eqn. (7.59) amounts to 5.5% at $R_l = 10^3$, decreasing to 0.2% at $R_l = 10^6$, compared to the first term.

The fact that the second term in eqn. (7.59) does not represent skin friction is explained by the observation that the singular character of the flow at the leading edge induces a pressure drag. Presumably, at the leading edge there arises an infinite overpressure which contributes a finite force in spite of the vanishingly small plate thickness. In this connexion a comparison with the case of the parabola of Sec. IXj should be made.

Strictly speaking, the preceding analysis of flow past a flat plate is restricted to a semi-infinite plate. In the case of a finite length, the shearing stress becomes modified at a certain distance upstream of the trailing edge. However, Prandtl's boundary-layer equations, being parabolic, cannot account for this "trailing-edge effect".

According to K. Stewartson [25a], it is possible to master such trailing-edge effects, or, generally speaking, the effects which are expressed as singularities (e. g. leading edge, trailing edge, separation) by Prandtl's equations, through a generalization of Prandtl's concept of the boundary layer. This is done by the introduction of the idea of "multistructured" boundary layers or the "triple-deck" concept.

For the case of a flat plate, again, K. Stewartson [25a] and A. F. Messiter [18b] find that the skin-friction coefficient is given by

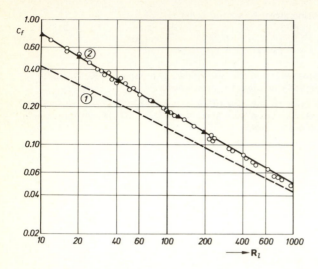

Fig. 7.11. Skin-friction coefficient of a flat plate of finite length at zero incidence

(1) Theory after H. Blasius, eqn. (7.34)
(2) Theory after A. F. Messiter [18 b], eqn. (7.60)
▲ Theory after Dennis (solution of Navier-Stokes equations)
○ Experiments after Z. Janour [30]

$$c_f = \frac{1 \cdot 328}{\sqrt{R_l}} + \frac{2 \cdot 668}{(R_l)^{7/8}} \qquad (7.60)$$

Here, the trailing edge has been accounted for, but not the displacement effect.

The diagram in Fig. 7.11, reproduced from the work of R. E. Melnik and R. Chow [18a], shows that the values of c_f computed with the aid of eqn. (7.60) agree very well with the results obtained from the complete Navier-Stokes equations as well as with those of measurements down to $R_l = 10$. At $R_l = 40$ eqn. (7.60) leads to $c_f = 0 \cdot 316$ which is less than 2% in excess of the exact value $c_f = 0 \cdot 311$.

Section IXj will return to the discussion of exact solutions of boundary-layer equations of second order.

References

[1] Bairstow, L.: Skin friction. J. Roy. Aero. Soc. *19*, 3 (1925).
[2] Blasius, H.: Grenzschichten in Flüssigkeiten mit kleiner Reibung. Z. Math. Phys. *56*, 1—37 (1908). Engl. transl. in NACA TM 1256.
[3] Boley, B.A., and Friedman, M.B.: On the viscous flow around the leading edge of a flat plate. JASS *26*, 453—454 (1959).
[4] Burgers, J.M.: The motion of a fluid in the boundary layer along a plane smooth surface. Proc. First Intern. Congr. of Appl. Mech., Delft 1924 (C.B. Biezeno and J.M. Burgers, ed.) Delft, 1925, pp. 113—128.
[5] Carrier, G.F., and Lin, C.C.: On the nature of the boundary layer near the leading edge of a flat plate. Quart. Appl. Math. *VI*, 63—68 (1948).
[6] Dhawan, S.: Direct measurements of skin friction. NACA Rep. 1121 (1953).
[7] Van Dyke, M.: Higher approximations in boundary layer theory. Part 1: General analysis. JFM *14*, 161—177 (1962). Part 2: Application to leading edges. JFM *14*, 481—495 (1962). Part 3: Parabola in uniform stream. JFM *19*, 145—159 (1964).
[8] Van Dyke, M.: Perturbation methods in fluid mechanics. Academic Press, New York, 1964.
[9] Van Dyke, M.: Higher-order boundary layer theory. Annual Review of Fluid Mech. *I*, 265—292 (1969).

[10] Gersten, K.: Grenzschichteffekte höherer Ordnung. Anniversary volume commemorating Professor H. Schlichting's 65th anniversary (Sept. 30, 1972). Rep. 72/5 Inst. f. Strömungsmech. Techn. Univ. at Braunschweig, 29—53 (1972).

[11] Gersten, K.: Die Verdrängungsdicke bei Grenzschichten höherer Ordnung. ZAMM 54, 165—171 (1974).

[12] Gersten, K., and Gross, J. F.: Higher-order boundary layer theory. Fluid Dynamics Transactions (1975).

[13] Goldstein, S.: Concerning some solutions of the boundary layer equations in hydrodynamics. Proc. Cambr. Phil. Soc. 26, 1—30 (1930); see also: Modern developments in fluid dynamics, Vol. I, 135, Oxford, 1938.

[14] Hansen, M.: Die Geschwindigkeitsverteilung in der Grenzschicht an einer eingetauchten Platte. ZAMM 8, 185—199 (1928); NACA TM 585 (1930).

[15] Van der Hegge-Zijnen, B. G.: Measurements of the velocity distribution in the boundary layer along a plane surface. Thesis, Delft 1924.

[16] Howarth, L.: On the solution of the laminar boundary layer equations. Proc. Roy. Soc. London A 164, 547—579 (1938).

[17] Imai, I.: Second approximation to the laminar boundary layer flow over a flat plate. JAS 24, 155—156 (1957).

[18] Liepman, H. W., and Dhawan, S.: Direct measurements of local skin friction in low-speed and high-speed flow. Proc. First US Nat. Congr. Appl. Mech. 869 (1951).

[18a] Melnik, R. E., and Chow, R.: Asymptotic theory of two-dimensional trailing edge flows. Grumman Research Department Rep. RE-510 (1975).

[18b] Messiter, A. F.: Boundary layer flow near the trailing edge of a flat plate. SIAM J. Appl. Math. 18, 241—257 (1970).

[19] Meksyn, D.: New methods in laminar boundary layer theory. London, 1961.

[20] Nikuradse, J.: Laminare Reibungsschichten an der längsangeströmten Platte. Monograph. Zentrale f. wiss. Berichtswesen, Berlin, 1942.

[21] Prandtl, L.: Über Flüssigkeitsbewegung bei sehr kleiner Reibung. Proc. Third Intern. Math. Congr. Heidelberg 1904. Reprinted in: Vier Abhandlungen zur Hydro- und Aerodynamik. Göttingen, 1927; NACA TM 452 (1928); see also: Coll. Works II, 575—584 (1961).

[22] Prandtl, L: The mechanics of viscous fluids. In W. F. Durand: Aerodynamic Theory III, 34—208 (1935).

[23] Rotta, J. C.: Grenzschichttheorie zweiter Ordnung für ebene und achsensymmetrische Hyperschallströmung. ZFW 15, 329—334 (1967).

[24] Schmidt, H., and Schröder, K.: Laminare Grenzschichten. Ein kritischer Literaturbericht. Part I: Grundlagen der Grenzschichttheorie. Luftfahrtforschung 19, 65—97 (1942).

[25] Steinheuer, J.: Die Lösungen der Blasiusschen Grenzschichtdifferentialgleichung. Proc. Wiss. Ges. Braunschweig XX, 96—125 (1968).

[25a] Stewartson, K.: Multistructured boundary layers on flat plates and related bodies. Adv. Appl. Mech. 14, 146—239, Academic Press, New York, 1974.

[26] Tollmien, W.: Grenzschichttheorie. Handbuch der Exper.-Physik IV, Part I, 241—287 (1931).

[27] Töpfer, C.: Bemerkungen zu dem Aufsatz von H. Blasius: Grenzschichten in Flüssigkeiten mit kleiner Reibung. Z. Math. Phys. 60, 397—398 (1912).

[28] Weyl, H.: Concerning the differential equations of some boundary layer problems. Proc. Nat. Acad. Sci. Washington 27, 578—583 (1941).

[29] Weyl, H.: On the differential equations of the simplest boundary layer problems. Ann. Math. 43, 381—407 (1942).

[30] Janour, Z.: Resistance of a flat plate at low Reynolds numbers. NACA TM 1316 (1951).

CHAPTER VIII

General properties of the boundary-layer equations

Before passing to the calculation of further examples of boundary-layer flow in the next chapter, we propose first to discuss some general properties of the boundary-layer equations. In doing so we shall confine our attention to steady, two-dimensional, and incompressible boundary layers.

Although the boundary-layer equations have been simplified to a great extent, as compared with the Navier-Stokes equations, they are still so difficult from the mathematical point of view that not very many general statements about them can be made. To begin with, it is important to notice that the Navier-Stokes equations are of the elliptic type with respect to the coordinates, whereas Prandtl's boundary-layer equations are parabolic. It is a consequence of the simplifying assumptions in boundary-layer theory that the pressure can be assumed constant in a direction at right angles to the boundary layer, whereas along the wall the pressure can be regarded as being "impressed" by the external flow so that it becomes a given function. The resulting omission of the equation of motion perpendicular to the direction of flow can be interpreted physically by stating that a fluid particle in the boundary layer has zero mass, and suffers no frictional drag, as far as its motion in the transverse direction is concerned. It is, therefore, clear that with such fundamental changes introduced into the equations of motion we must anticipate that their solutions will exhibit certain mathematical singularities, and that agreement between observed and calculated phenomena cannot always be expected.

a. Dependence of the characteristics of a boundary layer on the Reynolds number†

The assumptions which were made in the derivation of the boundary-layer equations are satisfied with an increasing degree of accuracy as the Reynolds number increases. Thus boundary-layer theory can be regarded as a process of *asymptotic integration* of the Navier-Stokes equations at very large Reynolds numbers*. This statement leads us now to a discussion of the relationship between the Reynolds number and the characteristics of a boundary layer on our individual body under consideration. It will be recalled that in the derivation of the boundary-layer equations

---

† *Cf.* Secs. VIIf and IXj.

* The argument contained in this section was already discussed in Sec. VIIf on higher-order approximations. The amplification is given here for the sake of better understanding.

dimensionless quantities were used; all velocities were referred to the free-stream velocity U_∞, all lengths having been reduced with the aid of a characteristic length of the body, L. Denoting all dimensionless magnitudes by a prime, thus $u/U_\infty = u'$, $\ldots, x/L = x', \ldots$, we obtain the following equations for the steady, two-dimensional case:

$$u' \frac{\partial u'}{\partial x'} + v' \frac{\partial u'}{\partial y'} = U' \frac{dU'}{dx'} + \frac{1}{R} \frac{\partial^2 u'}{\partial y'^2}, \tag{8.1}$$

$$\frac{\partial u'}{\partial x'} + \frac{\partial v'}{\partial y'} = 0, \tag{8.2}$$

$$y' = 0: \quad u' = v' = 0; \quad y' = \infty: \quad u' = U'(x');$$

see also eqs. (7.10) to (7.12). Here R denotes the Reynolds number formed with the aid of the reference quantities

$$R = \frac{U_\infty L}{\nu}.$$

It is seen from eqns. (8.1) and (8.2) that the boundary-layer solution depends on one parameter, the Reynolds number R, if the shape of the body, and, hence, the potential motion $U'(x')$ are given.   By the use of a further transformation it is possible to eliminate the Reynolds number also from eqns. (8.1) and (8.2). If we put

$$v'' = v' \sqrt{R} = \frac{v}{U_\infty} \sqrt{\frac{U_\infty L}{\nu}} \tag{8.3}†$$

and

$$y'' = y' \sqrt{R} = \frac{y}{L} \sqrt{\frac{U_\infty L}{\nu}}, \tag{8.4}†$$

eqns. (8.1) and (8.2) transform into:

$$u' \frac{\partial u'}{\partial x'} + v'' \frac{\partial u'}{\partial y''} = U' \frac{dU'}{dx'} + \frac{\partial^2 u'}{\partial y''^2}, \tag{8.5}$$

$$\frac{\partial u'}{\partial x'} + \frac{\partial v''}{\partial y''} = 0, \tag{8.6}$$

with the boundary conditions: $u' = 0$ and $v'' = 0$ at $y'' = 0$ and $u' = U'$ at $y'' = \infty$. These equations do not now contain the Reynolds number, so that the solutions of this system, i. e. the functions $u'(x', y'')$ and $v''(x', y'')$, are also independent of the Reynolds number. A variation in the Reynolds number causes an affine transformation of the boundary layer during which the ordinate and the velocity in the transverse direction are multiplied by $R^{-1/2}$. In other words, for a given body the dimensionless velocity components u/U_∞ and $(v/U_\infty) \cdot (U_\infty L/\nu)^{1/2}$ are functions of the dimensionless coordinates x/L and $(y/L) \cdot (U_\infty L/\nu)^{1/2}$; the functions, moreover, do not depend on the Reynolds number any longer.

The practical importance of this *principle of similarity with respect to Reynolds number* consists in the fact that for a given body shape it suffices to find the solution to the boundary-layer problem only once in terms of the above dimensionless variables.

---

† This transformation is identical with that implied in eqns. (7.47) and (7.48).

Such a solution is valid for any Reynolds number, provided that the boundary layer is laminar. In particular, it follows further that the position of the point of separation is independent of the Reynolds number. The angle which is formed between the streamline through the point of separation and the body, Fig. 7.2, simply decreases in the ratio $1/R^{1/2}$ as the Reynolds number increases.

Moreover, the fact that separation does take place is preserved when the process of passing to the limit $R \to \infty$ is carried out. Thus, in the case of body shapes which exhibit separation, the boundary-layer theory presents a totally different picture of the flow pattern than the frictionless potential theory, even in the limit of $R \to \infty$. This argument confirms the conclusion which was already emphatically stressed in Chap. IV, namely that the process of passing to the limit of frictionless flow must not be performed in the differential equations themselves; it may only be undertaken in the integral solution, if physically meaningful results are to be obtained.

b. 'Similar' solutions of the boundary-layer equations

A second, and very important, question arising out of the solution of boundary-layer equations, is the investigation of the conditions under which two solutions are 'similar'. We shall define here 'similar' solutions as those for which the component u of the velocity has the property that two velocity profiles $u(x, y)$ located at different coordinates x differ only by a scale factor in u and y. Therefore, in the case of such 'similar' solutions the velocity profiles $u(x, y)$ at all values of x can be made congruent if they are plotted in coordinates which have been made dimensionless with reference to the scale factors. Such velocity profiles will also sometimes be called *affine*. The local potential velocity $U(x)$ at section x is an obvious scale factor for u, because the dimensionless $u(x)$ varies with y from zero to unity at all sections. The scale factor for y denoted by $g(x)$, must be made proportional to the local boundary-layer thickness. The requirement of 'similarity' is seen to reduce itself to the requirement that for two arbitrary sections, x_1 and x_2, the components $u(x, y)$ must satisfy the following equation

$$\frac{u\{x_1, [y/g(x_1)]\}}{U(x_1)} = \frac{u\{x_2, [y/g(x_2)]\}}{U(x_2)}. \tag{8.7}$$

The boundary layer along a flat plate at zero incidence considered in the preceding chapter possessed this property of 'similarity'. The free-stream velocity U_∞ was the scale factor for u, and the scale factor for y was equal to the quantity $g = \sqrt{v\,x/U_\infty}$ which was proportional to the boundary-layer thickness. All velocity profiles became identical in a plot of u/U_∞ against $y/g = y\sqrt{U_\infty/v\,x} = \eta$, Fig. 7.7. Similarly, the cases of two- and three-dimensional stagnation flow, Chap. V, afforded examples of solutions which proved to be 'similar' in the present sense.

The quest for 'similar' solutions is particularly important with respect to the mathematical character of the solution. In cases when 'similar' solutions exist it is possible, as we shall see in more detail later, to reduce the system of partial differential equations to one involving ordinary differential equations, which, evidently, constitutes a considerable mathematical simplification of the problem. The boundary layer along a flat plate can serve as an example in this respect also.

It will be recalled that with the *similarity transformation* $\eta = y \sqrt{U_\infty / \nu x}$, eqn. (7.24), we obtained an ordinary differential equation, eqn. (7.28), for the stream function $f(\eta)$, instead of the original partial differential equations.

We shall now concern ourselves with the types of potential flows for which such 'similar' solutions exist. This problem was discussed in great detail first by S. Goldstein [4], and later by W. Mangler [9]. The point of departure is to consider the boundary-layer equations for plane steady flow, eqns. (7.10) and (7.11) together with eqn. (7.5a), which can be written as

$$\frac{\partial u}{\partial x} + \frac{\partial v}{\partial y} = 0 \; ,$$

$$u \frac{\partial u}{\partial x} + v \frac{\partial u}{\partial y} = U \frac{dU}{dx} + \nu \frac{\partial^2 u}{\partial y^2} \; , \qquad \left.\begin{array}{}\\\\\\\end{array}\right\} \qquad (8.8)$$

the boundary conditions being $u = v = 0$ for $y = 0$, and $u = U$ for $y = \infty$. The equation of continuity is integrated by the introduction of the stream function $\psi(x, y)$ with

$$u = \frac{\partial \psi}{\partial y} \; , \qquad v = - \frac{\partial \psi}{\partial x} \; .$$

Thus the equation of motion becomes

$$\frac{\partial \psi}{\partial y} \frac{\partial^2 \psi}{\partial x \partial y} - \frac{\partial \psi}{\partial x} \frac{\partial^2 \psi}{\partial y^2} = U \frac{dU}{dx} + \nu \frac{\partial^3 \psi}{\partial y^3} \; , \qquad (8.9)$$

with the boundary conditions $\partial \psi / \partial x = 0$ and $\partial \psi / \partial y = 0$ for $y = 0$, and $\partial \psi / \partial y = U$ for $y = \infty$. In order to discuss the question of 'similarity', dimensionless quantities are introduced, as was done in Sec. VIII a. All lengths are reduced with the aid of a suitable reference length, L, and all velocities are made dimensionless with reference to a suitable velocity, U_∞. As a result the Reynolds number

$$R = \frac{U_\infty L}{\nu}$$

appears in the equation. Simultaneously the y-coordinate is referred to the dimensionless scale factor $g(x)$, so that we put

$$\xi = \frac{x}{L} \; , \qquad \eta = \frac{y \sqrt{R}}{L \, g(x)} \; . \qquad (8.10)\dagger$$

---

† The transformation

$$\eta^* = \frac{1}{2 A} \ln{(1 + 2 A \eta)},$$

proposed by F. Schultz-Grunow [6a, 15a], makes it possible to reduce several problems involving self-similar solutions to that of the flat plate at zero incidence. If $A = \delta/2 R$ is chosen as the curvature parameter, the transformations can be applied to flows along longitudinally curved walls with blunt or sharp leading edges as well as with blowing or suction (Chapt. XIV). The preceding transformation is exact to second order in curvature which means that all terms of the order A have been included.

The factor $\sqrt{R}$ for the ordinate already appeared in eqn. (8.4). The stream function is made dimensionless by the substitution

$$f(\xi, \eta) = \frac{\psi(x, y) \sqrt{R}}{L U(x) g(x)}. \tag{8.11}$$

Consequently, the velocity components become

$$u = \frac{\partial \psi}{\partial y} = U \frac{\partial f}{\partial \eta} = U f',$$

$$-\sqrt{R}\, v = \sqrt{R}\, \frac{\partial \psi}{\partial x} = L f \frac{d}{dx}(U g) + U g \left(\frac{\partial f}{\partial \xi} - L \frac{g'}{g} \eta f'\right),$$

$$\left. \vphantom{\frac{\partial \psi}{\partial x}} \right\} \tag{8.12}$$

where the prime in f' denotes differentiation with respect to η, and with respect to x in g'. It is now seen directly from eqn. (8.12) that the velocity profiles $u(x, y)$ are similar in the previously defined sense, when the stream function f depends only on the one variable η, eqn. (8.10), so that the dependence of f on ξ is cancelled. In this case, moreover, the partial differential equation for the stream function, eqn. (8.9), must reduce itself to an ordinary differential equation for $f(\eta)$. If we now proceed to investigate the conditions under which this reduction of eqn. (8.9) takes place, we shall obtain the condition which must be satisfied by the potential flow $U(x)$ for such 'similar' solutions to exist.

If we introduce now the dimensionless variables from eqns. (8.10) and (8.11) into eqn. (8.9), we obtain the following differential equation for $f(\xi, \eta)$:

$$f''' + \alpha f f'' + \beta (1 - f'^2) = \frac{U}{U_\infty} g^2 \left(f' \frac{\partial f'}{\partial \xi} - f'' \frac{\partial f}{\partial \xi}\right) \tag{8.13}$$

where α and β are contractions for the following functions of x:

$$\alpha = \frac{L g}{U_\infty} \frac{d}{dx}(U g); \qquad \beta = \frac{L}{U_\infty} g^2 U', \tag{8.14}$$

and where $U' = dU/dx$. The boundary conditions for eqn. (8.13) are $f = 0$ and $f' = 0$ for $\eta = 0$ and $f' = 1$ for $\eta = \infty$.

'Similar' solutions exist only when f and f' do not depend on ξ, i. e. when the right-hand side of eqn. (8.13) vanishes. Simultaneously the coefficients α and β on the left-hand side of eqn. (8.13) must be independent of x, i. e., they must be constant. This latter condition, combined with eqn. (8.14), furnishes two equations for the potential velocity $U(x)$ and the scale factor $g(x)$ for the ordinate, so that they can be evaluated. Hence, if similar solutions of boundary-layer flow are to exist, the stream function $f(\eta)$ must satisfy the following ordinary differential equation:

$$f''' + \alpha f f'' + \beta (1 - f'^2) = 0 \tag{8.15}$$

with the boundary conditions

$$\eta = 0: \quad f = 0, \quad f' = 0; \quad \eta = \infty: \quad f' = 1. \tag{8.16}$$

This equation was first given by V. M. Falkner and S. W. Skan [2], and its solutions were later studied in detail by D. R. Hartree [6]. We shall revert to this point in the succeeding chapter.

It remains now to determine from eqn. (8.14) the conditions for $U(x)$ and $g(x)$. From (8.14) we obtain first

$$2\,\alpha - \beta = \frac{L}{U_\infty}\frac{d}{dx}(g^2\,U)$$

and hence, if $2\,\alpha - \beta \neq 0$,

$$\frac{U}{U_\infty}g^2 = (2\alpha - \beta)\,\frac{x}{L}\,.\tag{8.17}$$

Further from (8.14) we have

$$\alpha - \beta = \frac{L}{U_\infty}\,g\,g'\,U$$

and hence

$$(\alpha - \beta)\,\frac{U'}{U} = \frac{L}{U_\infty}\,g^2\,U'\,\frac{g'}{g} = \beta\,\frac{g'}{g}$$

so that upon integration

$$\left(\frac{U}{U_\infty}\right)^{(\alpha-\beta)} = K\,g^{\,\beta},\tag{8.18}$$

where K is a constant. The elimination of g from eqns. (8.17) and (8.18) yields the velocity distribution of the potential flow

$$\frac{U}{U_\infty} = K^{\frac{2}{2\alpha-\beta}}\left[(2\alpha-\beta)\,\frac{x}{L}\right]^{\frac{\beta}{2\alpha-\beta}}\tag{8.19}$$

and

$$g = \sqrt{(2\alpha-\beta)\,\frac{x}{L}}\left(\frac{U}{U_\infty}\right)^{-\frac{1}{2}}.\tag{8.20}$$

It will be recalled that the case $2\,\alpha - \beta = 0$ has been excluded.

As seen from eqn. (8.14) the result is independent of any common factor of α and β, as it can be included in g. Therefore as long as $\alpha \neq 0$ it is permissible to put $\alpha = +1$ without loss of generality. It is, furthermore, convenient to introduce a new constant m to replace β by putting

$$m = \frac{\beta}{2-\beta}\,,\tag{8.21}$$

as in this way the physical meaning of the solution will become clearer. Hence

$$\beta = \frac{2m}{m+1}$$

so that, with $\alpha = 1$, the velocity distribution of the potential flow and the scale factor g for the ordinate become

$$\frac{U}{U_\infty} = K^{(1+m)} \left(\frac{2}{1+m} \frac{x}{L} \right)^m , \tag{8.22}$$

$$g = \sqrt{\frac{2}{m+1} \frac{x}{L} \frac{U_\infty}{U}} , \tag{8.23}$$

and the transformation equation (8.10) for the ordinate is

$$\eta = y \sqrt{\frac{m+1}{2} \frac{U}{\nu x}} . \tag{8.24}$$

It is thus concluded that similar solutions of the boundary-layer equations are obtained when the velocity distribution of the potential flow is proportional to a power of the length of arc, measured along the wall from the stagnation point. Such potential flows occur, in fact, in the neighbourhood of the stagnation point of a wedge whose included angle is equal to $\pi \beta$, as shown in Fig. 8.1. It is easy to verify with the aid of potential theory that we have here

$$U(x) = C\, x^m , \tag{8.25}$$

where C is a constant. The relationship between the wedge angle factor β and the exponent m is exactly that given in eqn. (8.21).

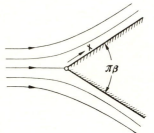

Fig. 8.1. Flow past a wedge. In the neighbourhood of the leading edge the potential velocity distribution is $U(x) = C x^m$

Particular cases for $\alpha = 1$: (a) For $\beta = 1$ we have $m = 1$, and eqn. (8.22) becomes $U(x) = a\,x$. This is the case of two-dimensional *stagnation flow*, which was considered in Sec. Vb 9, and which led to an exact solution of the Navier-Stokes equations. With $\alpha = 1$, and $\beta = 1$, the differential equation (8.15) transforms into eqn. (5.39) which was already considered earlier. The transformation equation for the ordinate, eqn. (8.24), becomes identical with the already familiar equation (5.38), if we put $U/x = a$.

(b) For $\beta = 0$ we have $m = 0$, hence $U(x)$ is constant and equal to U_∞. This is the case of a *flat plate at zero incidence*. It follows from eqn. (8.24) that $\eta = y \sqrt{U_\infty/2\,\nu\,x}$. This value differs only by a factor $\sqrt{2}$ from that introduced in eqn. (7.24). Correspondingly the differential equation $f''' + ff'' = 0$ which follows from eqn. (8.15) differs by a factor 2 in the second term from eqn. (7.28) which was solved earlier. The two equations become identical when transformed to identical definitions of η.

Solution for different values of m will be considered later in Chap. IX.

The case $\alpha = 0$: The case $\alpha = 0$ which has, so far, been left out of account, leads, as is easily inferred from eqn. (8.19), to potential flows $U(x)$ which are proportional to $1/x$ for all values of β. Depending on the sign of U this is the case of a two-dimensional sink or source, and can also be interpreted as flow in a divergent or convergent channel with flat walls. This type of flow will also be considered in greater detail in Chap. IX.

The second case excluded earlier, namely that when $2\alpha - \beta = 0$, leads to 'similar' solutions with $U(x)$ proportional to e^{px}, where p is a positive or negative constant. We shall, however, refrain from discussing this case.

The problem of the existence of similar solutions involving non-steady boundary layers was discussed by H. Schuh [15]; the same problem in relation to compressible boundary layers will be discussed in Sec. XIII d.

d. Transformation of the boundary-layer equations into the heat-conduction equation

R. von Mises [10] published in 1927 a remarkable transformation of the boundary-layer equations. This transformation exhibits the mathematical character of the equations even more clearly than the original form. Instead of the Cartesian coordinates x and y, von Mises introduced the stream function ψ, together with the length coordinate x as independent variables. Substituting

$$u = \frac{\partial \psi}{\partial y}, \qquad v = -\frac{\partial \psi}{\partial x}$$

into eqns. (7.10) and (7.11), as well as introducing the new coordinates $\xi = x$ and $\eta = \psi$ instead of x and y, we obtain

$$\frac{\partial u}{\partial x} = \frac{\partial u}{\partial \xi}\frac{\partial \xi}{\partial x} + \frac{\partial u}{\partial \eta}\frac{\partial \eta}{\partial x} = \frac{\partial u}{\partial \xi} - v\frac{\partial u}{\partial \psi},$$

$$\frac{\partial u}{\partial y} = \frac{\partial u}{\partial \xi}\frac{\partial \xi}{\partial y} + \frac{\partial u}{\partial \eta}\frac{\partial \eta}{\partial y} = 0 + u\frac{\partial u}{\partial \psi}.$$

Hence, from eqn. (7.10), it follows that

$$u\frac{\partial u}{\partial \xi} + \frac{1}{\varrho}\frac{dp}{d\xi} = v\,u\frac{\partial}{\partial \psi}\left(u\frac{\partial u}{\partial \psi}\right).$$

Introducing, further, the 'total head'

$$g = p + \tfrac{1}{2}\varrho\,u^2, \tag{8.26}$$

where the small quantity $\tfrac{1}{2}\varrho\,v^2$ can be neglected, we obtain, reverting to the symbol x for ξ:

$$\frac{\partial g}{\partial x} = v\,u\frac{\partial^2 g}{\partial \psi^2}. \tag{8.27}$$

We may also put

$$u = \sqrt{\frac{2}{\varrho}\,[g - p(x)]}.$$

Equation (8.27) is a differential equation for the total pressure $g(x, \psi)$, and its boundary conditions are

$$g = p(x) \text{ for } \psi = 0 \quad \text{and} \quad g = p(x) + \frac{1}{2}\varrho U^2 = \text{const for } \psi = \infty .$$

In order to represent the flow in the physical plane x, y, it is necessary to transform from ψ to y with the aid of the equation

$$y = \int \frac{d\psi}{u} = \sqrt{\frac{\varrho}{2}} \int\limits_{\psi=0} \frac{d\psi}{\sqrt{g - p(x)}} .$$

Equation (8.27) is related to the heat-conduction equation. The differential equation for the one-dimensional case, e. g. for a bar, is given by

$$\frac{\partial T}{\partial t} = a \frac{\partial^2 T}{\partial x^2} , \tag{8.28}$$

where T denotes the temperature, t denotes the time, and a is the thermal diffusivity, see Chap. XII. However, the transformed boundary-layer equation, unlike eqn. (8.28), is non-linear, because the thermal diffusivity is replaced by νu, which depends on the independent variable x, as well as on the dependent variable g.

At the wall, $\psi = 0$, $u = 0$, $g = p$, eqn. (8.27) exhibits an unpleasant singularity. The left-hand side becomes $\partial g/\partial x = dp/dx \neq 0$. On the right-hand side we have $u = 0$, and, therefore, $\partial^2 g/\partial \psi^2 = \infty$. This circumstance is disturbing when numerical methods are used, and is intimately connected with the singular behaviour of the velocity profile near the wall. A detailed discussion of eqn. (8.27) was given by L. Prandtl [11], who had deduced the transformation a long time before the paper by R. von Mises appeared, without, however, publishing it†, cf. [1, 12, 16].

H. J. Luckert [8] applied eqn. (8.27) to the example of the boundary layer on a flat plate in order to test its practicability. L. Rosenhead and H. Simpson [13] gave a critical discussion of the preceding publication.

e. The momentum and energy-integral equations for the boundary layer

A complete calculation of the boundary layer for a given body with the aid of the differential equations is, in many cases, as will be seen in more detail in the next chapter, so cumbersome and time-consuming that it can only be carried out with the aid of an electronic computer (see also Sec. IX i). It is, therefore, desirable to possess at least approximate methods of solution, to be applied in cases when an exact solution of the boundary-layer equations cannot be obtained with a reasonable amount of work, even if their accuracy is only limited. Such approximate methods can be devised if we do not insist on satisfying the differential equations for every fluid particle. Instead, the boundary-layer equation is satisfied in a stratum near the wall and near the region of transition to the external flow by satisfying the boundary

---

† See footnote on p. 79 of ref. [11] and the letter of L. Prandtl to ZAMM 8, 249 (1928).

conditions, together with certain compatibility conditions. In the remaining region of fluid in the boundary layer only a mean over the differential equation is satisfied, the mean being taken over the whole thickness of the boundary layer. Such a mean value is obtained from the momentum equation which is, in turn, derived from the equation of motion by integration over the boundary-layer thickness. Since this equation will be often used in the approximate methods, to be discussed later, we shall deduce it now, writing it down in its modern form. The equation is known as the *momentum-integral equation* of boundary-layer theory, or as von Kármán's integral equation [7].

We shall restrict ourselves to the case of steady, two-dimensional, and incompressible flow, i. e., we shall refer to eqns. (7.10) to (7.12). Upon integrating the equation of motion (7.10) with respect to y, from $y = 0$ (wall) to $y = h$, where the layer $y = h$ is everywhere outside the boundary layer, we obtain:

$$\int_{y=0}^{h} \left(u \frac{\partial u}{\partial x} + v \frac{\partial u}{\partial y} - U \frac{dU}{dx} \right) dy = - \frac{\tau_0}{\varrho} . \tag{8.29}$$

The shearing stress at the wall, τ_0, has been substituted for $\mu (\partial u/\partial y)_0$, so that eqn. (8.29) is seen to be valid both for laminar and turbulent flows, on condition that in the latter case u and v denote the time averages of the respective velocity components. The normal velocity component, v, can be replaced by $v = - \int_0^y (\partial u/\partial x) dy$, as seen from the equation of continuity, and, consequently, we have

$$\int_{y=0}^{h} \left(u \frac{\partial u}{\partial x} - \frac{\partial u}{\partial y} \int_0^y \frac{\partial u}{\partial x} dy - U \frac{dU}{dx} \right) dy = - \frac{\tau_0}{\varrho} .$$

Integrating by parts, we obtain for the second term

$$\int_{y=0}^{h} \left(\frac{\partial u}{\partial y} \int_0^y \frac{\partial u}{\partial x} dy \right) dy = U \int_0^h \frac{\partial u}{\partial x} dy - \int_0^h u \frac{\partial u}{\partial x} dy ,$$

so that

$$\int_0^h \left(2u \frac{\partial u}{\partial x} - U \frac{\partial u}{\partial x} - U \frac{dU}{dx} \right) dy = - \frac{\tau_0}{\varrho} ,$$

which can be contracted to

$$\int_0^h \frac{\partial}{\partial x} [u(U-u)] dy + \frac{dU}{dx} \int_0^h (U-u) dy = \frac{\tau_0}{\varrho} . \tag{8.29a}$$

Since in both integrals the integrand vanishes outside the boundary layer, it is permissible to put $h \to \infty$.

We now introduce the displacement thickness, δ_1, and the momentum thickness, δ_2, which have already been used in Chap. VII. They are defined by

$$\delta_1\, U = \int_{y=0}^{\infty} (U-u)\, dy \qquad \text{(displacement thickness)}, \qquad (8.30)$$

and

$$\delta_2\, U^2 = \int_{y=0}^{\infty} u(U-u)\, dy \qquad \text{(momentum thickness)} . \qquad (8.31)$$

It will be noted that in the first term of the eqn. (8.29a), differentiation with respect to x, and integration with respect to y, may be interchanged as the upper limit h is independent of x. Hence

$$\boxed{\frac{\tau_0}{\varrho} = \frac{d}{dx}\, (U^2\, \delta_2) + \delta_1\, U\, \frac{dU}{dx}} . \qquad (8.32)$$

This is the *momentum-integral equation for two-dimensional, incompressible boundary layers*. As long as no statement is made concerning τ_0, eqn. (8.32) applies to laminar and turbulent boundary layers alike. This form of the momentum integral equation was first given by H. Gruschwitz [5]. It finds its application in the approximate theories for laminar and turbulent boundary layers (Chaps. X, XI and XXII).

Using a similar approach, K. Wieghardt [17] deduced an *energy-integral equation* for laminar boundary layers. This equation is obtained by multiplying the equation of motion by u and then integrating from $y = 0$ to $y = h > \delta(x)$. Substituting, again, v from the equation of continuity we obtain

$$\varrho \int_0^h \left[u^2\, \frac{\partial u}{\partial x} - u\, \frac{\partial u}{\partial y} \left(\int_0^y \frac{\partial u}{\partial x}\, dy \right) - u\, U\, \frac{dU}{dx} \right] dy = \mu \int_0^h u\, \frac{\partial^2 u}{\partial y^2}\, dy .$$

The second term can be transformed by integrating by parts:

$$\int_0^h \left[u\, \frac{\partial u}{\partial y} \left(\int_0^y \frac{\partial u}{\partial x}\, dy \right) \right] dy = \frac{1}{2} \int_0^h (U^2 - u^2)\, \frac{\partial u}{\partial x}\, dy ,$$

whereas by combining the first with the third term we have

$$\int_0^h \left[u^2\, \frac{\partial u}{\partial x} - u\, U\, \frac{dU}{dx} \right] dy = \frac{1}{2} \int_0^h u\, \frac{\partial}{\partial x}\, (u^2 - U^2)\, dy .$$

Finally, upon integrating the right-hand side by parts, we obtain

$$\frac{1}{2}\, \varrho\, \frac{d}{dx} \int_0^{\infty} u(U^2 - u^2)\, dy = \mu \int_0^{\infty} \left(\frac{\partial u}{\partial y} \right)^2 dy . \qquad (8.33)$$

The upper limit of integration could here, too, be replaced by $y = \infty$, because the integrands become equal to zero outside the boundary layer. The quantity $\mu\, (\partial u/\partial y)^2$ represents the energy, per unit volume and time, which is transformed into heat by friction (dissipation, *cf.* Chap. XII). The term $\frac{1}{2}\, \varrho\, (U^2 - u^2)$ on the left-hand side represents the loss in mechanical energy (kinetic and pressure energy) taking place in the boundary layer as compared with the potential flow. Hence the term $\frac{1}{2}\varrho \int_0^\infty u(U^2-u^2)\,dy$ represents the flux of dissipated energy, and the left-hand side represents the rate of change of the flux of dissipated energy per unit length in the x-direction.

If, in addition to the displacement and momentum thickness from eqns. (8.30) and (8.31) respectively, we introduce the dissipation-energy thickness, δ_3, from the definition

$$U^3\,\delta_3 = \int_0^\infty u(U^2-u^2)\,dy \qquad \text{(energy thickness)}, \qquad (8.34)$$

we can rewrite the energy-integral equation (8.33) in the following simplified form:

$$\frac{d}{dx}(U^3\,\delta_3) = 2\,\nu \int_0^\infty \left(\frac{\partial u}{\partial y}\right)^2 dy \qquad (8.35)$$

which represents the *energy-integral equation for two-dimensional, laminar boundary layers in incompressible flow* †.

In order to visualize the displacement thickness, the momentum thickness, and the energy-dissipation thickness, it is convenient to calculate them for the simple case of linear velocity distribution, as shown in Fig. 8.2. In this case we find:

displacement thickness	$\delta_1 = \tfrac{1}{2}\,\delta$	
momentum thickness	$\delta_2 = \tfrac{1}{6}\,\delta$	
energy thickness	$\delta_3 = \tfrac{1}{4}\,\delta.$	

The extension of the preceding approximate method to axially symmetrical boundary layers will be discussed in Chap. XI. Approximate methods for thermal boundary layers are treated in Sec. XIIg; those for compressible and non-steady boundary layers will be given in Sec. XIIId and Chap. XV, respectively.

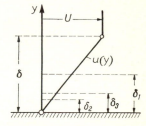

Fig. 8.2. Boundary layer with linear velocity distribution

δ — boundary-layer thickness
δ_1 — displacement thickness
δ_2 — momentum thickness
δ_3 — energy thickness

---

† In the case of turbulent flows, the energy-integral equation assumes the form

$$\frac{d}{dx}(U^3\,\delta_3) = 2 \int_0^\infty \frac{\tau}{\varrho}\,\frac{\partial u}{\partial y}\,dy\,.$$

References

[1] Betz, A.: Zur Berechnung des Überganges laminarer Grenzschichten in die Aussenströmung. Fifty years of boundary-layer research (W. Tollmien and H. Görtler, ed.), Braunschweig, 1955, 63—70.

[2] Falkner, V. M., and Skan, S. W.: Some approximate solutions of the boundary layer equations. Phil. Mag. *12*, 865—896 (1931); ARC RM. 1314 (1930).

[3] Geis, Th.: Ähnliche Grenzschichten an Rotationskörpern. Fifty years of boundary layer research (W. Tollmien and H. Görtler, ed.), Braunschweig, 1955, 294—303.

[4] Goldstein, S.: A note on the boundary layer equations. Proc. Cambr. Phil. Soc. *35*, 338—340 (1939).

[5] Gruschwitz, E.: Die turbulente Reibungsschicht in ebener Strömung bei Druckabfall und Druckanstieg. Ing.-Arch. *2*, 321—346 (1931).

[6] Hartree, D. R.: On an equation occurring in Falkner and Skan's approximate treatment of the equations of the boundary layer. Proc. Cambr. Phil. Soc. *33*, Part II, 223—239 (1937).

[6a] Holt, M.: Basic developments in fluid dynamics. Contribution of F. Schultz-Grunow and W. Breuer, 377—436, New York, 1965.

[7] von Kármán, Th.: Über laminare und turbulente Reibung. ZAMM *1*, 233—253 (1921). Engl. transl. in NACA TM 1092; see also Coll. Works *II*, 70—97, London 1956.

[8] Luckert, H. J.: Über die Integration der Differentialgleichung einer Gleitschicht in zäher Flüssigkeit. Diss. Berlin 1933, reprinted in: Schriften des Math. Seminars, Inst. f. angew. Math. der Univ. Berlin *1*, 245 (1933).

[9] Mangler, W.: Die "ähnlichen" Lösungen der Prandtlschen Grenzschichtgleichungen. ZAMM *23*, 241—251 (1943).

[10] von Mises, R.: Bemerkungen zur Hydrodynamik. ZAMM *7*, 425—431 (1927).

[11] Prandtl, L.: Zur Berechnung der Grenzschichten. ZAMM *18*, 77—82 (1938); see also Coll. Works *II*, 663—672, J. Roy. Aero. Soc. *45*, 35—40 (1941), and NACA TM 959 (1940).

[12] Riegels, F., and Zaat, J.: Zum Übergang von Grenzschichten in die ungestörte Strömung. Nachr. Akad. Wiss. Göttingen, Math. Phys. Klasse, 42—45 (1947).

[13] Rosenhead, L., and Simpson, J. H.: Note on the velocity distribution in the wake behind a flat plate placed along the stream. Proc. Cambr. Phil. Soc. *32*, 285—291 (1936).

[14] Schröder, K.: Verwendung der Differenzenrechnung zur Berechnung der laminaren Grenzschicht. Math. Nachr. *4*, 439—467 (1951).

[15] Schuh, H.: Über die "ähnlichen" Lösungen der instationären laminaren Grenzschichtgleichung in inkompressibler Strömung. Fifty years of boundary-layer research (W. Tollmien and H. Görtler, ed.), Braunschweig, 1955, 147—152.

[15a] Schultz-Grunow, F., and Henseler, H.: Ähnliche Grenzschichtlösungen zweiter Ordnung für Strömungs- und Temperaturgrenzschichten an longitudinal gekrümmten Wänden mit Grenzschichtbeeinflussung. Wärme- und Stoffübertragung *1*, 214—219 (1968).

[16] Tollmien, W.: Über das Verhalten einer Strömung längs einer Wand am äusseren Rand ihrer Reibungsschicht. Betz Anniversary Volume, 218—224 (1945).

[17] Wieghardt, K.: Über einen Energiesatz zur Berechnung laminarer Grenzschichten. Ing.-Arch. *16*, 231—242 (1948).

CHAPTER IX

Exact solutions of the steady-state boundary-layer equations in two-dimensional motion

The present chapter will deal with some exact solutions of the boundary-layer equations. A solution will be considered exact when it is a complete solution of the boundary-layer equations, irrespective of whether it is obtained analytically or by numerical methods. On the other hand, Chap. X will deal with approximate solutions, i. e. with solutions which are obtained from integral relations, such as the momentum and energy-integral equations described in the preceding chapter, rather than from differential equations.

There are in existence only comparatively few exact analytical solutions, and we shall discuss them first. Generally speaking, the process of obtaining analytical solutions of the boundary-layer equations encounters considerable mathematical difficulties, as already illustrated with the example of a flat plate. The differential equations are non-linear in most cases so that, again generally speaking, they can be solved only by power-series expansions or by numerical methods. Even for the physically simplest case of the boundary layer on a flat plate at zero incidence with incompressible flow no closed-form analytic solution has been discovered so far.

In the case of two-dimensional motion, the boundary-layer equations and their boundary conditions are given by eqns. (7.10) to (7.12):

$$\frac{\partial u}{\partial x} + \frac{\partial v}{\partial y} = 0 , \tag{9.1}$$

$$u \frac{\partial u}{\partial x} + v \frac{\partial u}{\partial y} = U \frac{dU}{dx} + \nu \frac{\partial^2 u}{\partial y^2} , \tag{9.2}$$

$$y = 0: \quad u = 0 , \quad v = 0 ; \quad y = \infty : \quad u = U(x) . \tag{9.3}$$

In addition, a velocity profile $u(0, y)$ must be given at an initial section, say, at $x = 0$.

In most cases it is convenient to integrate the equation of continuity by the introduction of a stream function $\psi(x, y)$, so that

$$u = \frac{\partial \psi}{\partial y} ; \quad v = -\frac{\partial \psi}{\partial x} .$$

Consequently the stream function must satisfy the following equation [see also eqn. (7.18)]:

$$\frac{\partial \psi}{\partial y} \frac{\partial^2 \psi}{\partial x \partial y} - \frac{\partial \psi}{\partial x} \frac{\partial^2 \psi}{\partial y^2} = U \frac{dU}{dx} + \nu \frac{\partial^3 \psi}{\partial y^3} , \tag{9.4}$$

with the boundary conditions $\partial\psi/\partial y = 0$ and $\partial\psi/\partial x = 0$ at the wall, $y = 0$, and $\partial\psi/\partial y = U(x)$ at $y = \infty$.

a. Flow past a wedge

The 'similar' solutions discussed in Chap. VIII constitute a particularly simple class of solutions $u(x,y)$ which have the property that the velocity profiles at different distances, x, can be made congruent with suitable scale factors for u and y. The system of partial differential equations (9.1) and (9.2) is now reduced to one ordinary differential equation. It was proved in Chap. VIII that such similar solutions exist when the velocity of the potential flow is proportional to a power of the length coordinate, x, measured from the stagnation point, i. e. for

$$U(x) = u_1 \, x^m \, .$$

From eqn. (8.24) it follows that the transformation of the independent variable y, which leads to an ordinary differential equation, is:

$$\eta = y \sqrt{\frac{m+1}{2} \frac{U}{\nu x}} = y \sqrt{\frac{m+1}{2} \frac{u_1}{\nu}} \; x^{\frac{m-1}{2}} \, . \tag{9.5}$$

The equation of continuity is integrated by the introduction of a stream function, for which we put

$$\psi(x,y) = \sqrt{\frac{2}{m+1}} \sqrt{\nu \, u_1} \; x^{\frac{m+1}{2}} \, f(\eta) \, ,$$

as seen from eqns. (8.11) and (8.23). Thus the velocity components become

$$\left. \begin{array}{l} u = u_1 \, x^m \, f'(\eta) = U \, f'(\eta) \, , \\[2mm] v = -\sqrt{\dfrac{m+1}{2} \, \nu \, u_1 \, x^{m-1}} \left\{ f + \dfrac{m-1}{m+1} \, \eta \, f' \right\} \, . \end{array} \right\} \tag{9.6}$$

Introducing these values into the equation of motion (9.1), dividing by $m \, u_1 \, x^{2m-1}$, and putting, as in eqn. (8.21),

$$m = \frac{\beta}{2-\beta} \, , \qquad \frac{2m}{m+1} = \beta \, , \tag{9.7}$$

we obtain the following differential equation for $f(\eta)$:

$$f''' + f f'' + \beta(1 - f'^2) = 0 \, . \tag{9.8}$$

It will be recalled that it was already given as eqn. (8.15), and that its boundary conditions are

$$\eta = 0 : \quad f = 0 \, , \quad f' = 0 \, ; \qquad \eta = \infty : \quad f' = 1 \, .$$

Equation (9.8) was first deduced by V. M. Falkner and S. W. Skan, and its solutions were later investigated in detail by D. R. Hartree (see References to Chap. VIII). The solution is represented in Fig. 9.1. In the case of accelerated flow ($m > 0, \beta > 0$)

the velocity profiles have no point of inflexion, whereas in the case of decelerated flow $(m < 0,\ \beta < 0)$ they exhibit a point of inflexion. Separation occurs for $\beta = -0\cdot199$, i. e. for $m = -0\cdot091$. This result shows that the laminar boundary layer is able to support only a very small deceleration without separation occuring.

K. Stewartson [64] gave a detailed analysis of the manifold of solutions of eqn. (9.8). According to this analysis, in the range of increasing pressures $(-0.199 < \beta < 0)$ there exists a further solution, that is, in addition to the one discovered by Hartree. The additional solution leads to a velocity profile with back-flow (cf. Chap. X f).

The potential flow given by $U(x) = u_1\, x^m$ exists in the neighbourhood of the stagnation point on a wedge, Fig. 8.1, whose included angle β, is given by eqn. (9.7). Two-dimensional stagnation flow, as well as the boundary layer on a flat plate at zero incidence, constitute particular cases of the present solutions, the former for $\beta = 1$ and $m = 1$, the latter for $\beta = 0$ and $m = 0$.

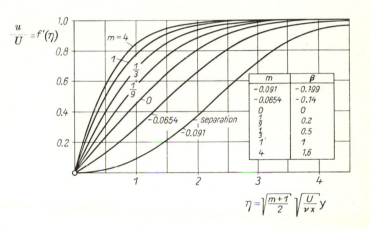

Fig. 9.1. Velocity distribution in the laminar boundary layer in the flow past a wedge given by $U(x) = u_1\, x^m$. The exponent m and the wedge angle β (Fig. 8.1) are connected through eqn. (9.7)

m	β
-0.091	-0.199
-0.0654	-0.14
0	0
$\tfrac{1}{9}$	0.2
$\tfrac{1}{3}$	0.5
1	1
4	1.6

The case $\beta = \tfrac{1}{2}$, $m = \tfrac{1}{3}$ is worthy of attention. In this case the differential equation for $f(\eta)$ becomes: $f''' + f\,f'' + \tfrac{1}{2}(1 - f'^2) = 0$; it transforms into the differential equation of rotationally symmetrical flow with stagnation point, eqn. (5.47), i. e., $\phi''' + 2\,\phi\,\phi'' + 1 - \phi'^2 = 0$ for $\phi(\zeta)$, if we put $\eta = \zeta\sqrt{2}$ and $df/d\eta = d\phi/d\zeta$. This means that the calculation of the boundary layer in the rotationally symmetrical case can be reduced to the calculation of two-dimensional flow past a wedge whose included angle is $\pi\,\beta = \pi/2$.

The relationship between the two-dimensional and rotationally symmetrical boundary layers will be further discussed, in a more general form, in Chap. XI.

If the similarity variable η defined in eqn. (9.5) were replaced by the independent variable $\overline{\eta} = y\,\sqrt{U(x)/\nu\, x}$, the differential equation for the function $f'(\overline{\eta}) = u/U$ would change its form to

$$f''' + \frac{m+1}{2}\, f\, f'' + m(1 - f'^2) = 0 \,.\tag{9.8a}$$

This equation transforms into that for a flat plate, eqn. (7.28), in the special case when $m = 0$. The solutions of the Falkner-Skan equation (9.8) have been discussed in detail in [61].

According to J. Steinheuer [63], an interesting extension of the solution of the Falkner-Skan equation (9.8) which is valid for retarded flows ($\beta < 0$) in cases when velocity distributions possessing a velocity excess ($f'(\eta) > 1$) with a maximum near the wall are admitted. In such cases, the limit $f'(\eta) = 1$ for $\eta \to \infty$ is attained asymptotically "from above" rather than "from below", as was the case so far. Such solutions can be interpreted physically as corresponding to a laminar wall-jet produced in an external stream with a positive pressure gradient, $dp/dx > 0$. Reference [63] demonstrates that the limiting case of these solutions, obtained when the maximum velocity excess tends to infinity, transforms into the well-known self-similar solution of a pure wall-jet in the absence of an external velocity — a case treated by M. B. Glauert (see [40] in Chap. XI)— when we put $\beta = -2$.

A particularly detailed monograph on exact, self-similar solutions for laminar boundary layers in two-dimensional and rotationally symmetric arrangements, inclusive of the associated thermal boundary layers (see Chap. XII),was published by C. F. Dewey and J. F. Gross [14]. Their considerations include the effects of compressibility (see Chap. XIII) with and without heat transfer, relate to varying values of the Prandtl number, and include some cases of suction and blowing.

K. K. Chen and P. A. Libby [9] carried out an extensive investigation of boundary layers which are characterized by small departures from the self-similar wedge-flow boundary layers of the Falkner-Skan type. Evidently, such boundary layers are no longer self-similar.

b. Flow in a convergent channel

The case of potential flow given by the equation

$$U(x) = -\frac{u_1}{x} \tag{9.9}$$

is related to flows past a wedge, and also leads to 'similar' solutions. With $u_1 > 0$ it represents two-dimensional motion in a convergent channel with flat walls (sink).

The volume of flow for a full opening angle 2π and for a stratum of unit height is $Q = 2\pi u_1$ (Fig. 9.2). Introducing the similarity transformation

$$\eta = y \sqrt{\frac{U}{-x\nu}} = \frac{y}{x}\sqrt{\frac{u_1}{\nu}} = \frac{y}{x}\sqrt{\frac{Q}{2\pi\nu}} \tag{9.10}$$

as well as the stream function

$$\psi(x,y) = -\sqrt{\nu u_1}\, f(\eta)$$

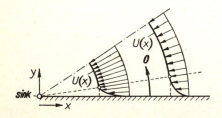

Fig. 9.2.   Flow in a convergent channel

we obtain the velocity components

$$u = U f'; \quad v = -\sqrt{\nu u_1} \frac{\eta}{x} f' . \tag{9.11}$$

Substituting into eqn. (9.2) we obtain the differential equation for the stream function

$$f''' - f'^2 + 1 = 0 . \tag{9.12}$$

The boundary conditions follow from eqn. (9.3) and are: $f' = 0$ at $\eta = 0$, and $f' = 1$ and $f'' = 0$ at $\eta = \infty$. This is also a particular case of the class of 'similar' solutions considered in Chap. VIII. Equation (9.12) is obtained from the more general differential equation (8.15) for the case of 'similar' boundary layers, if we put $\alpha = 0$, and $\beta = + 1$. The example under consideration is one of the rare cases when the solution of the boundary-layer equation can be obtained analytically in closed form.

First, upon multiplying eqn. (9.12) by f'' and integrating once, we have

$$f''^2 - \tfrac{2}{3} (f' - 1)^2 (f' + 2) = a ,$$

where a is a constant of integration. Its value is zero, as $f' = 1$ and $f'' = 0$ for $\eta \to \infty$. Thus

$$\frac{df'}{d\eta} = \sqrt{\frac{2}{3} (f' - 1)^2 (f' + 2)}$$

or

$$\eta = \sqrt{\frac{3}{2}} \int_0^{f'} \frac{df'}{\sqrt{(f' - 1)^2 (f' + 2)}} ,$$

where the additive constant of integration is seen to be equal to zero in view of the boundary condition $f' = 1$ at $\eta = \infty$. The integral can be expressed in closed form as follows:

$$\eta = \sqrt{2} \left\{ \tanh^{-1} \frac{\sqrt{2 + f'}}{\sqrt{3}} - \tanh^{-1} \sqrt{\frac{2}{3}} \right\} ,$$

or, solving for $f' = u/U$:

$$f' = \frac{u}{U} = 3 \tanh^2 \left(\frac{\eta}{\sqrt{2}} + 1 \cdot 146 \right) - 2 . \tag{9.13}$$

Here we have substituted $\tanh^{-1} \sqrt{\frac{2}{3}} = 1 \cdot 146$. Introducing the polar angle $\theta = y/x$, as well as $Q = 2 \pi r U$ ($r =$ radial distance from the sink), we can replace η from eqn. (9.10) by

$$\eta = \theta \sqrt{\frac{U r}{\nu}} = \frac{y}{x} \sqrt{\frac{U r}{\nu}} . \tag{9.14}$$

The velocity distribution given by eqn. (9.13) is represented in Fig. 9.3. At $\eta = 3$, approximately, the boundary layer merges with the potential flow. Hence the boundary-layer thickness becomes $\delta = 3 x \sqrt{\nu/U r}$; it decreases, as in other examples, as $1/\sqrt{R}$.

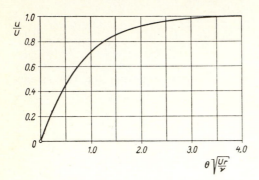

The preceding solution was first obtained by K. Pohlhausen [50]. It will be recalled from Sec. V b 12 on p. 107 that the flow through a divergent channel discussed by G. Hamel constitutes an exact solution of the Navier-Stokes equations. The diagram in Fig. 5.15 contained some numerical results pertaining to this solution. In this connexion, a paper by B. L. Reeves and C. J. Kippenhan [52] may usefully be consulted.

c. Flow past a cylinder; symmetrical case (Blasius series)

The class of 'similar' solutions of the boundary-layer equations considered so far is comparatively narrow. Apart from the examples of the flat plate, stagnation flow, flow past a wedge, and flow in a convergent channel which have already been described, few additional solutions can be obtained. We shall now consider the general case of the boundary layer on a cylindrical body placed in a stream which is perpendicular to its axis. The method of solution was first given by H. Blasius [4]; it was developed further by K. Hiemenz [39] and L. Howarth [40]. It is necessary to distinguish here two cases depending on whether the cylinder is symmetrical about an axis which is parallel to the stream at a large distance from the body or not. We shall refer to these two cases as to the symmetrical and asymmetrical case, respectively.

In either case the velocity of the potential flow is assumed to have the form of a power series in x, where x denotes the distance from the stagnation point measured along the contour. The velocity profile in the boundary layer is also represented as a similar power series in x, where the coefficients are assumed to be functions of the coordinate y, measured at right angles to the wall (Blasius series). L. Howarth succeeded in finding a substitution for the velocity profile which confers universal validity on the y-dependent coefficients. In other words, by a suitable assumption regarding the power series, its coefficients have been made independent of the particulars of the cylindrical body, so that the resulting functions could be evaluated and presented in the form of tables. Thus the calculation of the boundary layer for a given shape becomes very simple if use is made of the tables, provided that the tabulation extends over a sufficiently large number of terms of the series.

The usefulness of Blasius's method is, however, severely restricted by the fact that, precisely in the most important case of very slender body-shapes, a large number of terms is required; in fact, their number is so large that it ceases to be practicable to tabulate them all with a reasonable amount of numerical work. This is caused by the circumstance that in the case of slender body-sections, e. g. in the case of an ellipse, placed in a stream parallel to its major axis, or in the case of an aerofoil, the potential velocity near the stagnation point in the neighbourhood of the leading edge increases steeply at first and then varies very slowly over a considerable distance downstream. A function of this type cannot be well represented by a power series with a small number of terms. In spite of this limitation Blasius's method is of great fundamental importance because, in cases when its convergence is insufficient to reach the point of separation, it can be used to calculate analytically and with great accuracy the initial portion of the boundary layer near the stagnation point. The calculation can then be continued with the aid of a suitable numerical integration method, such as, for example, the one described in Sec. IXi.

We shall now very briefly describe the procedure that is followed for the calculation of a boundary layer with the aid of a Blasius series. A more detailed account can be found in the earlier editions of this book [57a]. However, the numerical results for the circular cylinder are given more fully.

We consider the *symmetric case* and assume that the potential flow is given in the form of the series

$$U(x) = u_1 x + u_3 x^3 + u_5 x^5 + \ldots . \tag{9.15}$$

The coefficients u_1, u_3, ... depend only on the shape of the body and are to be considered known. The continuity equation is satisfied by the introduction of a stream-function $\psi (x, y)$. In analogy with eqn. (9.15) it is plausible also to adopt a power series in x, its coefficients being treated as functions of y. The choice of the particular form of the power series is governed by the desire to render the functions of y contained in it independent of the coefficients u_1, u_3, u_5, ... which describe the potential flow. In this manner, the functions of y become universal and can be calculated once and for all.

The distance from the wall is made dimensionless by assuming[†]

$$\eta = y \sqrt{\frac{u_1}{\nu}} . \tag{9.16}$$

This leads to the form

$$\psi = \sqrt{\frac{\nu}{u_1}} \{u_1 \, x \, f_1(\eta) + 4 \, u_3 \, x^3 \, f_3(\eta) + 6 \, u_5 \, x^5 \, f_5(\eta) + \ldots\} \tag{9.17}$$

for the stream-function with the aid of which it now becomes possible to determine the appropriate series for the velocity components $u = \partial\psi/\partial y$ and $v = -\partial\psi/\partial x$. Substituting these expressions into the equation of motion (9.2) we compare coeffi-

---

[†] This form is obtained from the Blasius eqn. (7.24) by substituting in it the first term (9.15), i. e. $u_1 x$ for U_∞. This brings with it the disadvantage that it does not make an allowance for the increase in the boundary-layer thickness in the downstream direction.

cients and thus obtain a system of ordinary differential equations for the functions $f_1, f_3, \ldots$ The first two equations turn out to be

$$\left.\begin{array}{l} f_1'^2 - f_1 f_1'' = 1 + f_1''' \\ 4 f_1' f_3 - 3 f_1'' f_3 - f_1 f_3'' = 1 + f_3'''. \end{array}\right\} \qquad (9.18)$$

In these, differentiation with respect to η is denoted by primes. The associated boundary conditions are

$$\left.\begin{array}{l} \eta = 0: \ f_1 = f_1' = 0; \ \ f_3 = f_3' = 0 \\[2mm] \eta = \infty: \ f_1' = 1; \qquad f_3' = \dfrac{1}{4}. \end{array}\right\} \qquad (9.19)$$

All differential equations for the functional coefficients are of the third order, and only the first one, that for f_1, is non-linear; it is identical with the equation for two-dimensional stagnation flow, eqn. (5.39), discussed in Chap. V. All remaining equations are linear and their coefficients are expressed in terms of the functions associated with the preceding terms. The functions f_1 and f_3 have been calculated already by K. Hiemenz [39], and their first derivatives are represented graphically in Fig. 9.4. The function f_1' for the velocity distribution was reproduced earlier in Fig. 5.10 and Table 5.1 (when it was denoted by ϕ'). The higher-order functions can be found in the earlier editions [57a].

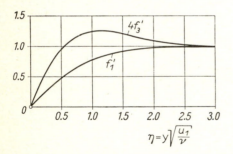

Fig. 9.4.   The functions f_1' and f_3' which appear in the Blasius power series.

Example: Circular cylinder. We shall now apply the method outlined in the preceding paragraphs to the case of a circular cylinder. In order to be definite, we base the calculation on the pressure distribution obtained from potential theory, although in the literature the problem was frequently solved with the aid of an experimentally determined pressure distribution. The ideal velocity distribution in non-viscous, irrotational flow past a circular cylinder of radius R and free-stream velocity U_∞ parallel to the x-axis is given by

$$u(x) = 2\,U_\infty \sin x/R = 2\,U_\infty \sin \phi\,, \qquad (9.20)$$

where ϕ is the angle measured from the stagnation point. Expanding $\sin x/R$ into a series and comparing it with that in eqn. (9.15), we find that

$$u_1 = 2\,\frac{U_\infty}{R}\,; \ \ u_3 = -\frac{2}{3!}\,\frac{U_\infty}{R^3}\,, \ \ldots \ \text{and } \eta = \frac{y}{R}\sqrt{\frac{2\,U_\infty\,R}{\nu}}\,.$$

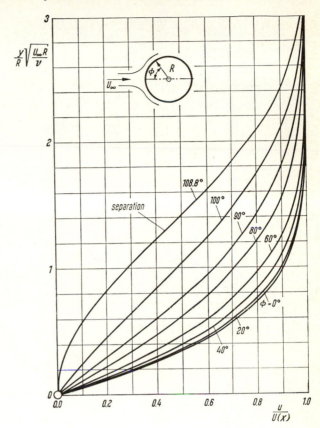

Fig. 9.5. Velocity distribution in the boundary layer on a circular cylinder

ϕ — angle measured from stagnation point

The velocity profiles for different values of ϕ are seen plotted in Fig. 9.5 which is based on a series for the velocity, u, carried as far as the term in x^{11}. The velocity profiles for $\phi > 90°$ possess a point of inflexion because they lie in the region of increasing pressure.

The distribution of shearing stress $\tau_0 = \mu(\partial u/\partial y)_0$ is plotted in Fig. 9.6. The position of the point of separation results from the condition that $\tau_0 = 0$, and is given by

$$\phi_S = 108 \cdot 8°$$

Fig. 9.6. Variation of shearing stress at the wall over the circumference of a circular cylinder for a laminar boundary layer

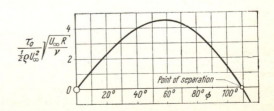

If the power series were terminated at x^9, the point of separation would turn out to be at $\phi_S = 109 \cdot 6°$. Better accuracy can nowadays be obtained with numerical methods, see Secs. IXi and Xc3.

The accuracy of this calculation based on a power series can be tested for speed of convergence of the omitted portion of the series by invoking the *conditions of compatibility* at the wall. According to eqn. (7.15), we must have

$$U \frac{\mathrm{d}U}{\mathrm{d}x} = -\nu \left(\frac{\partial^2 u}{\partial y^2} \right)_{y=0}. \tag{9.21}$$

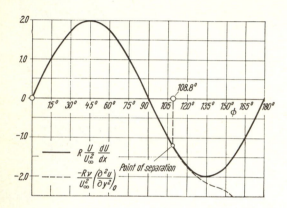

Fig. 9.7. Verification of the first compatibility condition from eqn. (9.21) for the laminar boundary layer on a circular cylinder from Fig. 9.5. The first compatibility condition is satisfied approximately as far as some point beyond separation

Figure 9.7 compares the curvature of the velocity profiles measured at the wall with its exact value represented by $U \,\mathrm{d}U/\mathrm{d}x$. The agreement is good for a distance x beyond the point of separation. We may, therefore, conclude that the Blasius series terminating at the term x^{11} satisfies the compatibility condition on a circular cylinder up to a point which lies beyond the point of separation. It does not, however, necessarily follow that the truncated series represents the velocity profile with good accuracy.

As already mentioned, in the case of more slender body-shapes considerably more terms of the Blasius series are required, if it is desired to obtain the velocity profiles as far as the point of separation. However, the evaluation of further functional coefficients is hindered by considerable difficulties. These are due not only to the fact that for every additional term in the series the number of differential equations to be solved increases, but also, and even more forcibly, the difficulties are due to the need to evaluate the functions for the lower power terms with ever increasing accuracy, if the functions for the higher power terms are to be sufficiently accurate.

L. Howarth [40] extended the present method to include the asymmetrical case, but the tabulation of the functional coefficients was not carried beyond those corresponding to the power x^2. N. Froessling [23] carried out an extension of this method to the rotationally symmetrical case which will be considered in Chap. XI.

Measurements of the pressure distribution around a circular cylinder were reported by K. Hiemenz in his thesis presented to Goettingen University [39]. They were made the basis of his boundary-layer calculations. His measurements showed separation at $\phi_S = 81°$, whereas the calculation indicated $\phi_S = 82°$. Later O. Flachsbart published extensive experimental data on the pressure distribution, Fig. 1.10, which point to a large influence of the Reynolds number. For values of the Reynolds number *below the critical* the pressure minimum occurs already near $\phi = 70°$, and the pressure is nearly constant over the whole downstream portion of the cylinder. For Reynolds numbers *above the critical* the pressure minimum shifts to $\phi = 90°$ approximately, in agreement with the potential-flow theory and, on the whole, the pressure distribution departs less from that given by the potential theory than in the previous case. Between these values, i. e. near a critical Reynolds number of approximately $U_\infty D/\nu = 3 \times 10^5$, the drag coefficient of the circular cylinder decreases abruptly (Fig. 1.4), and this phenomenon indicates that the boundary layer has become turbulent (see Sec. XVIIIf).

The laminar boundary layer on a circular cylinder was also investigated by A. Thom [67], at a Reynolds number $U_\infty D/\nu = 28{,}000$ and by A. Fage [16] in the range $U_\infty D/\nu = 1 \cdot 0$ to $3 \cdot 3 \times 10^5$. A paper by L. Schiller and W. Linke [54] contains some considerations concerning pressure drag and skin friction in the region of Reynolds numbers below the critical. In the range of Reynolds numbers from about 60 to about 5000 there exists behind the cylinder a vortex street which shows a regular, periodic structure (Figs. 2.7 and 2.8). The frequency at which vortices are shed in this so-called von Kármán vortex street has been investigated by H. Blenk, D. Fuchs and H. Liebers, and, more recently, by A. Roshko (see Chap. II).

d. Boundary layer for the potential flow given by $U(x) = U_0 - ax^n$

A further family of solutions of the boundary-layer equations was found by L. Howarth [41] and I. Tani [66]. These solutions relate to the potential flow given by

$$U(x) = U_0 - ax^n \quad (n = 1, 2, 3 \ldots),$$
(9.22)

which, evidently, constitutes a generalized form of the flow along a flat plate (see Sec. VIIe), and becomes identical with it when we put $a = 0$. In the simplest case with $n = 1$, which was treated by L. Howarth, the flow can be interpreted as that which occurs in a channel which consists of a portion with parallel walls (velocity U_0) followed by either a convergent $(a < 0)$ or a divergent $(a > 0)$ section†. This is another example of a boundary layer for which the velocity profiles are not similar. L. Howarth introduced the new independent variable

$$\eta = \frac{1}{2} y \sqrt{\frac{U_0}{\nu x}}$$
(9.23)

which is identical with that used in the flat plate solution at zero incidence. He assumed further

$$x^* = \frac{ax}{U_0}$$

$(x^* < 0$, accelerated flow; $x^* > 0$, decelerated flow). It is now possible to stipulate a power series

---

† When equation (9.22) is written in the form $U(x) = U_0(1 - x/L)$ for $n = 1$, it can also be interpreted as representing the potential flow along a flat wall which starts at $x = 0$ and which abuts on to another infinite wall at right angles to it at $x = L$. It is of the same type as the case of decelerated stagnation flow shown in Fig. 2.17, the stagnation point being at $x = L$.

in x^* for the stream function in a manner similar to the case of the cylinder, Sec. IX c, the coefficients being functions of y:

$$\psi(x,y) = \sqrt{U_0\, \nu x}\; \{f_0(\eta) - (8\,x^*)\, f_1(\eta) + (8\,x^*)^2\, f_2(\eta) - + \ldots\}. \tag{9.24}$$

Hence the velocity of flow becomes

$$u = \tfrac{1}{2} U_0\, \{f_0'(\eta) - (8\,x^*)\, f_1'(\eta) + (8\,x^*)^2\, f_2'(\eta) - + \ldots\}. \tag{9.25}$$

Introducing these values into the equations of motion (9.2) and comparing coefficients we obtain a system of ordinary differential equations for the functions $f_0(\eta)$, $f_1(\eta)$, ... The first three of these are:

$$f_0''' + f_0\, f_0'' = 0\,,$$
$$f_1''' + f_0\, f_1'' - 2\, f_0'\, f_1' + 3\, f_0''\, f_1 = -1\,,$$
$$f_2''' + f_0\, f_2'' - 4\, f_0'\, f_2' + 5\, f_0''\, f_2 = -\tfrac{1}{8} + 2\, f_1'^2 - 3\, f_1\, f_1''\,,$$

with the boundary conditions

$$\eta = 0\;:\quad f_0 = f_0' = 0\;:\quad f_1 = f_1' = 0\;;\quad f_2 = f_2' = 0\;;$$
$$\eta = \infty:\quad f_0' = 2\;;\qquad\qquad f_1' = \tfrac{1}{4}\;;\qquad\quad f_2' = 0\,.$$

Only the first equation is non-linear, and it is identical with that for a flat plate at zero incidence [‡]. All remaining equations are linear and contain only the function f_0 in the homogeneous portion, whereas the non-homogeneous terms are formed with the aid of the remaining functions f_ν. L. Howarth solved the first seven differential equations (up to and including f_6), and calculated tables for them.

The series (9.25) converges well with these values of f_ν in the range $-0.1 \leq x^* \leq +0.1$. In the case of decelerated flow ($x^* > 0$) the point of separation is at $x^* = 0.12$ approximately, but for the slightly extended range of values the convergence of the series (9.25) is no longer assured. In order to reach the point of separation, L. Howarth used a numerical procedure for the continuation of the solution. Velocity profiles for several values of x^* for both accelerated

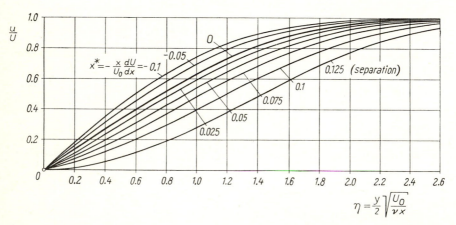

Fig. 9.8.   Velocity distribution in the laminar boundary layer for the potential flow given by $U(x) = U_0 - ax$, after Howarth [41]

[‡] The independent variable η in the above equations differs from that in Chap. VII by a factor $\tfrac{1}{2}$.

and decelerated flow are seen plotted in Fig. 9.8. It should be noted that all profiles in decelerated flow have a point of inflexion. D. R. Hartree [38] repeated these calculations and obtained good agreement with L. Howarth. The case for $a/U_0 = 0.125$ was calculated more accurately by D. C. F. Leigh [44] who used an electronic digital computer for the purpose and who paid special attention to the region of separation. The value of the form factor at the point of separation itself was found to be $x^* = 0.1198$.

The method employed by L. Howarth was extended by I. Tani [66] to include the cases corresponding to $n \geq 1$ (with $a > 0$). However, I. Tani did not publish any tables of the functional coefficients but confined himself to reporting the final result for $n = 2, 4$ and 8. In his case, too, the poor convergence of the series did not permit him to determine the point of separation with sufficient accuracy and he found himself compelled to use L. Howarth's numerical continuation scheme.

e. Flow in the wake of flat plate at zero incidence

The application of the boundary-layer equations is not restricted to regions near a solid wall. They can also be applied when a stratum in which the influence of friction is dominating exists in the interior of a fluid. Such a case occurs, among others, when two layers of fluid with different velocities meet, for instance, in the wake behind a body, or when a fluid is discharged through an orifice. We shall consider three examples of this type in the present and in the succeeding sections, and we shall return to them when considering turbulent flow.

As our first example we shall discuss the case of flow in the wake of a flat plate at zero incidence, Fig. 9.9. Behind the trailing edge the two velocity profiles coalesce into one profile in the wake. Its width increases with increasing distance, and its mean velocity decreases. The magnitude of the depression in the velocity curve is directly connected with the drag on the body. On the whole, however, as we shall see later, the velocity profile in the wake, at a large distance from the body, is independent of the shape of the body, except for a scale factor. On the other hand the velocity profile very close to the body is, evidently, determined by the boundary layer on the body, and its shape depends on whether or not the flow has separated.

The momentum equation can be used to calculate the drag from the velocity profile in the wake. For this purpose we draw a rectangular control surface $AA_1 B_1 B$,

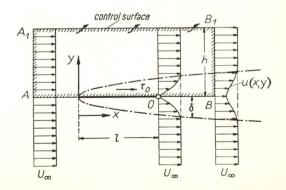

Fig. 9.9. Application of the momentum equation in the calculation of the drag on a flat plate at zero incidence from the velocity profile in the wake

as shown in Fig. 9.9. The boundary $A_1 B_1$, parallel to the plate, is placed at such a distance from the body that it lies everywhere in the region of undisturbed velocity, U_∞. Furthermore, the pressure is constant over the whole of the control surface, so that pressure forces do not contribute to the momentum. When calculating the flux of momentum across the control surface it is necessary to remember that, owing to continuity, fluid must leave through the boundary $A_1 B_1$; the quantity of fluid leaving through $A_1 B_1$ is equal to the difference between that entering through $A_1 A$ and leaving through $B_1 B$. The boundary AB contributes no term to the momentum in the x-direction because, owing to symmetry, the transverse velocity vanishes along it. The momentum balance is given in tabular form on the next page, and in it the convention is followed that inflowing masses are considered positive, and outgoing masses are taken to be negative. The width of the plate is denoted by b. The total flux of momentum is equal to the drag D on a flat plate wetted on one side. Thus we have

$$D = b \, \varrho \int\limits_{y=0}^{\infty} u(U_\infty - u) \, \mathrm{d}y \,. \tag{9.26}$$

Integration may be performed from $y = 0$ to $y = \infty$ instead of to $y = h$, because for $y > h$ the integrand in eqn. (9.26) vanishes. Hence the drag on a plate wetted on both sides becomes

$$2\,D = b \, \varrho \int\limits_{-\infty}^{+\infty} u(U_\infty - u) \, \mathrm{d}y \,. \tag{9.27}$$

This equation applies to any symmetrical cylindrical body and not only to a flat plate. It is to be remembered that in the more general case the integral over the profile in the wake must be taken at a sufficiently distant section, and one across which the static pressure has its undisturbed value. Since near a plate there are no pressure differences either in the longitudinal or in the transverse direction, eqn. (9.27) applies to any distance behind the plate. Furthermore, eqn. (9.27) may be applied to any section x of the boundary layer, when it gives the drag on the portion of the plate between the leading edge and that section. The physical meaning of the integral in eqn. (9.26) or (9.27) is that it represents the loss of momentum due to friction. It is identical with the integral in eqn. (8.31) which defined the *momentum thickness* δ_2, so that eqn. (9.26) can be given the alternative form

$$D = b \, \varrho U_\infty^2 \, \delta_2 \,. \tag{9.28}$$

We shall now proceed to calculate the velocity profile in the wake, in particular, at a large distance x behind the trailing edge of the flat plate. The calculation must be performed in two steps: 1. Through an expansion in the downstream direction from the leading to the trailing edge, i. e. by a calculation which involves the continuation of the Blasius profile on the plate near the trailing edge, and 2. Through an expansion in the upstream direction. The latter is a kind of asymptotic integration for a large distance behind the plate and is valid irrespective of the shape of the body. It will be necessary here to make the assumption that the velocity difference in the wake

$$u_1(x, y) = U_\infty - u(x, y) \tag{9.29}$$

Cross-section	Rate of flow	Momentum in direction x
A B	0	0
A A$_1$	$b \int\limits_0^h U_\infty \, dy$	$\varrho \, b \int\limits_0^h U_\infty{}^2 \, dy$
B B$_1$	$- b \int\limits_0^h u \, dy$	$- \varrho \, b \int\limits_0^h u^2 \, dy$
A$_1$ B$_1$	$- b \int\limits_0^h (U_\infty - u) \, dy$	$- \varrho \, b \int\limits_0^h U_\infty \, (U_\infty - u) \, dy$
Σ = Control surface	Σ Rate of flow $= 0$	Σ Momentum flux = Drag

is small compared with U_∞, so that quadratic and higher terms in u_1 may be neglected. The procedure makes use of a method of continuing a known solution. The calculation starts with the profile at the trailing edge, calculated with the aid of Blasius's method, and we shall refrain from further discussing it here. The asymptotic expansion in the upstream direction was calculated by W. Tollmien [69]. Since it is typical for problems of flow in the wake, and since we shall make use of it in the more important turbulent case, we propose to devote some time to an account of it.

As the pressure term is equal to zero, the boundary-layer equation (9.2) combined with eqn. (9.29) gives

$$U_\infty \frac{\partial u_1}{\partial x} = \nu \, \frac{\partial^2 u_1}{\partial y^2} , \tag{9.30}$$

where the quadratic terms in u_1 and v_1 have been omitted. The boundary conditions are:

$$y = 0 : \quad \frac{\partial u_1}{\partial y} = 0 \; ; \; y = \infty : \; u_1 = 0 .$$

The partial differential equation can, here too, be transformed into an ordinary differential equation by a suitable transformation. Similarly to the assumption (7.24) in Blasius's method for the flat plate we put

$$\eta = y \sqrt{\frac{U_\infty}{\nu \, x}} ,$$

and, in addition, we assume that u_1 is of the form

$$u_1 = U_\infty \, C \left(\frac{x}{l} \right)^{-\frac{1}{2}} g(\eta) , \tag{9.31}$$

where l is the length of the plate, Fig. 9.9.

The power $- \tfrac{1}{2}$ for x in eqn. (9.31) is justified on the ground that the momentum integral which gives the drag on the plate in eqn. (9.27) must be independent of x.

Hence, omitting quadratic terms in u_1, the drag on a plate wetted on both sides, as given in eqn. (9.27), is transformed to

$$2\,D = b\,\varrho\,U_\infty \int\limits_{y=-\infty}^{+\infty} u_1\,\mathrm{d}y \,.$$

Substituting eqn. (9.31) we obtain

$$2\,D = b\,\varrho\,U_\infty{}^2\,C\,\sqrt{\frac{\nu\,l}{U_\infty}} \int\limits_{-\infty}^{+\infty} g(\eta)\,\mathrm{d}\eta \,. \tag{9.32}$$

Introducing, further, the assumption (9.31) into (9.30), and dividing through by $C\,U_\infty{}^2 \cdot (x/l)^{-1/2}\,x^{-1}$, we obtain the following differential equation for $g(\eta)$:

$$g'' + \tfrac{1}{2}\,\eta\,g' + \tfrac{1}{2}\,g = 0 \tag{9.33}$$

with the boundary conditions

$$g' = 0 \quad \text{at} \quad \eta = 0 \quad \text{and} \quad g = 0 \quad \text{at} \quad \eta = \infty \,.$$

Integrating once, we have

$$g' + \tfrac{1}{2}\,\eta\,g = 0 \,,$$

where the constant of integration vanishes on account of the boundary condition at $\eta = 0$. Repeated integration gives the solution

$$g = \exp\left(-\tfrac{1}{4}\,\eta^2\right) \,. \tag{9.34}$$

Here the constant of integration appears in the form of a coefficient and can be made equal to unity without loss of generality, as the velocity distribution function u_1 from eqn. (9.31) still contains a free coefficient C. This constant C is determined from the condition that the drag calculated from the loss of momentum, eqn. (9.32), must be equal to that on the plate, eqn. (7.33).

First we notice that

$$\int\limits_{-\infty}^{+\infty} g(\eta)\,\mathrm{d}\eta = \int\limits_{-\infty}^{+\infty} \exp\left(-\tfrac{1}{4}\,\eta^2\right)\mathrm{d}\eta = 2\,\sqrt{\pi}\,,$$

so that from eqn. (9.32) we have

$$2\,D = 2\,\sqrt{\pi}\,C\,b\,\varrho\,U_\infty{}^2\sqrt{\frac{\nu\,l}{U_\infty}} \,.$$

On the other hand, from eqn. (7.33) we can write down the skin friction on a plate wetted on both sides in the form:

$$2\,D = 1{\cdot}328\,b\,\varrho U_\infty{}^2\,\sqrt{\frac{\nu\,l}{U_\infty}} \,.$$

Hence $2\,C\,\sqrt{\pi} = 1{\cdot}328$ and $C = 0{\cdot}664/\sqrt{\pi}$, and the final solution for the velocity difference in the wake of a flat plate at zero incidence becomes

$$\frac{u_1}{U_\infty} = \frac{0\cdot664}{\sqrt{\pi}}\left(\frac{x}{l}\right)^{-1/2}\exp\left\{-\frac{1}{4}\,\frac{y^2\,U_\infty}{x\,\nu}\right\}. \tag{9.35}$$

The velocity distribution given by this asymptotic equation is represented in Fig. 9.10. It is remarkable that the velocity distribution is identical with Gauss's error-distribution function. As assumed at the beginning, eqn. (9.35) is valid only at great distances from the plate. W. Tollmien verified that it may be used at about $x > 3\,l$. Fig. 9.11 contains a plot from which the whole velocity field can be inferred.

The flow in the wake of a plate as well as in that behind any other body is, in most cases, turbulent. Even in the case of small Reynolds numbers, say $R_l < 10^6$, when the boundary layer remains laminar as far as the trailing edge, the flow in the wake still becomes turbulent, because the velocity profiles in the wake, all of which possess a point of inflexion, are extremely unstable. In other words, even with comparatively small Reynolds numbers the wake becomes turbulent. Turbulent wakes will be discussed in Chap. XXIV.

f. The two-dimensional laminar jet

The efflux of a jet from an orifice affords a further example of motion in the absence of solid boundaries to which it is possible to apply the boundary-layer theory. We propose to discuss the two-dimensional problem so that we shall assume

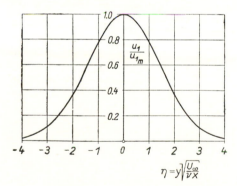

Fig. 9.10. Asymptotic velocity distribution in the laminar wake behind a flat plate, from eqn. (9.35)

Fig. 9.11. Velocity distribution in the la- ➤ minar wake behind a flat plate at zero incidence

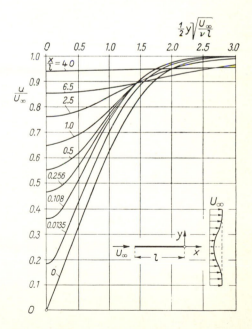

that the jet emerges from a long, narrow slit and mixes with the surrounding fluid. This problem was solved by H. Schlichting [56] and W. Bickley [3]. In practice, in this case, as in the previous ones, the flow becomes turbulent. We shall, however, discuss here the laminar case in some detail, since the turbulent jet, which will be considered later, can be analyzed mathematically in an identical way.

The emerging jet carries with it some of the surrounding fluid which was originally at rest because of the friction developed on its periphery. The resulting pattern of streamlines is shown in Fig. 9.12. We shall adopt a system of coordinates with its origin in the slit and with its axis of abscissae coinciding with the jet axis.

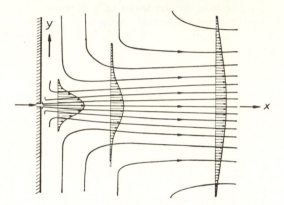

Fig. 9.12.   The laminar two-dimensional free jet

The jet spreads outwards in the downstream direction owing to the influence of friction, whereas its velocity in the centre decreases in the same direction. For the sake of simplicity we shall assume that the slit is infinitely small, but in order to retain a finite volume of flow as well as a finite momentum, it is necessary to assume an infinite fluid velocity in the slit. The pressure gradient dp/dx in the x-direction can here, as in the previous example, be neglected, because the constant pressure in the surrounding fluid impresses itself on the jet. Consequently, the total momentum in the x-direction, denoted by J, must remain constant and independent of the distance x from the orifice. Hence

$$J = \varrho \int\limits_{-\infty}^{+\infty} u^2 \, dy = \text{const} .\qquad(9.36)$$

It is possible to make a suitable assumption regarding the velocity distribution if it is considered that the velocity profiles $u(x,y)$, just as in the case of a flat plate at zero incidence, are most probably similar, because the problem as a whole possesses no characteristic linear dimension. We shall assume, therefore, that the velocity u is a function of y/b, where b is the width of the jet, suitably defined. We shall also assume that b is proportional to x^q. Accordingly we can write the stream function in the form

$$\psi \sim x^p \, f\left(\frac{y}{b}\right) = x^p \, f\left(\frac{y}{x^q}\right) .$$

The two unknown exponents p and q will be determined from the following conditions:

1. The flux of momentum in the x-direction is independent of x, according to eqn. (9.36).
2. The acceleration terms and the friction term in eqn. (9.2) are of the same order of magnitude.

This gives two equations for p and q:

$$2p - q = 0 \text{ and } 2p - 2q - 1 = p - 3q ,$$

and hence,

$$p = \tfrac{1}{3} ; \quad q = \tfrac{2}{3} .$$

Consequently, the assumptions for the independent variable and for the stream function can be written as

$$\eta = \frac{1}{3 \, \nu^{1/2}} \frac{y}{x^{2/3}} ; \quad \psi = \nu^{1/2} \, x^{1/3} \, f(\eta) ,$$

if suitable constant factors are included. Therefore, the velocity components are given by the following expressions:

$$\left.\begin{array}{l} u = \dfrac{1}{3 \, x^{1/3}} \, f'(\eta); \\[2mm] v = -\tfrac{1}{3} \, \nu^{1/2} \, x^{-2/3} \, (f - 2\eta f') . \end{array}\right\} \tag{9.37}$$

Introducing these values into the differential equation (9.2), and equating the pressure term to zero, we obtain the following differential equation for the stream function $f(\eta)$:

$$f'^2 + f f'' + f''' = 0 \tag{9.38}$$

with the boundary conditions $v = 0$ and $\partial u/\partial y = 0$ at $y = 0$, and $u = 0$ at $y = \infty$. Thus

$$\eta = 0 : \quad f = 0, \quad f'' = 0; \quad \eta = \infty: \quad f' = 0 . \tag{9.39}$$

The solution of eqn. (9.38) is unexpectedly simple. Integrating once we have

$$f f' + f'' = 0 .$$

The constant of integration is zero because of the boundary conditions at $\eta = 0$, and the resulting differential equation of the second order could be integrated immediately if the first term contained the factor 2. This can be achieved by the following transformation:

$$\xi = \alpha \eta ; \quad f = 2 \alpha F(\xi) ,$$

where α is a free constant, to be determined later. Thus the above equation transforms into

$$F'' + 2 F F' = 0 \tag{9.40}$$

and the dash now denotes differentiation with respect to ξ. The boundary conditions are

$$\xi = 0 : \quad F = 0 ; \qquad \xi = \infty : \quad F' = 0 , \tag{9.41}$$

and the equation can be integrated once more to give

$$F' + F^2 = 1 , \tag{9.42}$$

where the constant of integration was made equal to 1. This follows if we put $F'(0) = 1$, which is permissible without loss of generality because of the free constant α in the relation between f and F. Equation (9.42) is a differential equation of Riccati's type and can be integrated in closed terms. We obtain

$$\xi = \int_{0}^{F} \frac{\mathrm{d}F}{1 - F^2} = \frac{1}{2} \ln \frac{1 + F}{1 - F} = \tanh^{-1} F .$$

Inverting this equation we obtain

$$F = \tanh \xi = \frac{1 - \exp(-2\xi)}{1 + \exp(-2\xi)} . \tag{9.43}$$

Since, further, $\mathrm{d}F/\mathrm{d}\xi = 1 - \tanh^2 \xi$, the velocity distribution can be deduced from eqn. (9.37) and is

$$u = \tfrac{2}{3} \alpha^2 \, x^{-1/3} \, (1 - \tanh^2 \xi) . \tag{9.44}$$

The velocity distribution from eqn. (9.37) is seen plotted in Fig. 9.13.

It now remains to determine the constant α, and this can be done with the aid of condition (9.36) which states that the momentum in the x-direction is constant. Combining eqns. (9.44) and (9.36) we obtain

$$J = \tfrac{4}{3} \varrho \, \alpha^3 \, \nu^{1/2} \int_{0}^{\infty} (1 - \tanh^2 \xi)^2 \, \mathrm{d}\xi = \tfrac{16}{9} \varrho \, \alpha^3 \, \nu^{1/2} . \tag{9.45}$$

We shall assume that the flux of momentum, J, for the jet is given. It is proportional to the excess in pressure with which the jet leaves the slit. Introducing the *kinematic momentum* $J/\varrho = K$, we have from eqn. (9.45)

$$\alpha = 0 \cdot 8255 \left(\frac{K}{\nu^{1/2}} \right)^{\frac{1}{3}}$$

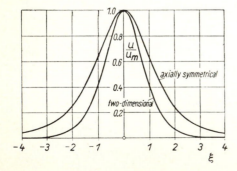

Fig. 9.13. Velocity distribution in a two-dimensional and circular free jet from eqns. (9.44) and (11.15) respectively. For the two-dimensional jet $\xi = 0 \cdot 275 \, K^{1/3} \, y/(\nu x)^{2/3}$, and for the circular jet $\xi = 0 \cdot 244 \, K'^{1/2} \, y/\nu x$. K and K' denote the kinematic momentum J/ϱ

and, hence, for the velocity distribution

$$u = 0\cdot 4543 \left(\frac{K^2}{\nu x}\right)^{\frac{1}{3}} (1 - \tanh^2 \xi) ,$$

$$v = 0\cdot 5503 \left(\frac{K\nu}{x^2}\right)^{\frac{1}{3}} [2 \xi (1 - \tanh^2 \xi) - \tanh \xi] , \qquad (9.46)$$

$$\xi = 0\cdot 2752 \left(\frac{K}{\nu^2}\right)^{\frac{1}{3}} \frac{y}{x^{2/3}} .$$

The transverse velocity at the boundary of the jet is

$$v_\infty = - 0\cdot 550 \left(\frac{K\nu}{x^2}\right)^{\frac{1}{3}} , \qquad (9.47)$$

and the volume-rate of discharge per unit height of slit becomes $Q = \varrho \int_{-\infty}^{+\infty} u \, dy$, or

$$Q = 3\cdot 3019 \, (K \nu x)^{1/3} . \qquad (9.48)$$

The volume-rate of discharge increases in the downstream direction, because fluid particles are carried away with the jet owing to friction on its boundaries. It also increases with increasing momentum.

The corresponding rotationally symmetrical case in which the jet emerges from a small circular orifice will be discussed in Chap. XI. The problem of the two-dimensional laminar compressible jet emerging from a narrow slit was solved by S. I. Pai [49] and M. Z. Krzywoblocki [42].

Measurements performed by E. N. Andrade [1] for the two-dimensional laminar jet confirm the preceding theoretical argument very well. The jet remains laminar up to R = 30 approximately, where the Reynolds number is referred to the efflux velocity and to the width of the slit. The case of a two-dimensional and that of a circular turbulent jet is discussed in Chap. XXIV. A comprehensive review of all problems involving jets can be found in S. I. Pai's book [49].

g. Parallel streams in laminar flow

We shall now briefly examine the layer between two parallel, laminar streams which move at different velocities, and so provide a further example of the applicability of the boundary-layer equations. The formulation of the problem is seen illustrated in Fig. 9.14: Two initially separated, undisturbed, parallel streams which move with the velocities U_1 and U_2, respectively, begin to interact through friction. It is possible to assume that the transition from the velocity U_1 to velocity U_2 takes place in a narrow zone of mixing and that the transverse velocity component, v, is everywhere small compared with the longitudinal velocity, u. Consequently, the boundary-layer equation (9.1) can be used to describe the flow in the zones I and II, and the pressure term may be omitted.

In a manner analogous to that employed for the boundary layer on a flat plate (Sec. VIIe), it is possible to obtain the ordinary differential equation

$$f f'' + 2 f''' = 0 . \tag{9.49}$$

by introducing the dimensionless transverse coordinate $\eta = y \sqrt{U_1/\nu x}$ and the stream function $\psi = \sqrt{\nu U_1 x} \, f$. Assuming that $u/U = f'$, we are led to the boundary conditions

and

$$\left.\begin{array}{l} \eta = +\infty : \quad f' = 1 \\[2mm] \eta = -\infty : \quad f' = \dfrac{U_2}{U_1} = \lambda . \end{array}\right\} \tag{9.50}$$

At the plane of separation at $y = 0$ we must have

$$\eta = 0 : \quad f = 0 . \tag{9.51}$$

because $\psi = 0$ there. The solution of the differential equation (9.49) subject to the boundary conditions (9.50) and (9.51) cannot be obtained in closed form, and a numerical method must be employed. It is possible to obtain exact numerical solutions by the use of asymptotic expansions for $\eta \to -\infty$ and $\eta \to +\infty$ together with a series expansion about $\eta = 0$; several such solutions were provided by R. C. Lock [45]. The problem was first solved by numerical integration by M. Lessen [44a] starting with an asymptotic expansion for $\eta \to -\infty$.

The diagram in Fig. 9.14 presents velocity profiles for $\lambda = U_2/U_1 = 0$ and 0.5. An improved numerical solution was published by W. J. Christian [10]. This special case of the interaction between a wide, homogeneous jet and an adjoining mass of quiescent air is often described by the term "plane half-jet".

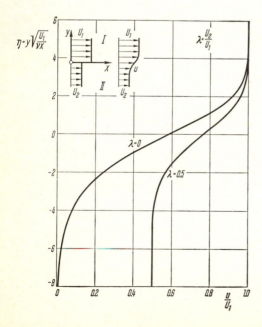

Fig. 9.14. Velocity distribution in the zone between two interacting parallel streams, after R. C. Lock [45]

According to the investigation carried out by J. Steinheuer [63], as mentioned in Chap. VII, these solutions belong to a special group of solutions of Blasius's equation (9.49). The values of $f'(0)$ and $f''(0)$ recently calculated for various values of the velocity ratio λ can be found in that reference. In addition, the displacement of the zero streamline has also been calculated. This occurs as a result of the circumstance that the normal velocity components, $v = -\partial \psi / \partial x \sim (\eta f' - f)$ for $\eta \to \pm \infty$ are not equal at the two edges of the boundary layer.

R. C. Lock [45] studied, in addition, the case when the two half-jets differ in their densities and viscosities, and not only in their velocities. An example of such a case is the flow of air over a water surface. The solution now depends on the parameter $\varkappa = \varrho_2 \mu_2 / \varrho_1 \mu_1$ in addition to λ. Lock provided several exact solutions as well as solutions which were based on the momentum integral equation. An approximate method was also conceived by O. E. Potter [51].

The problem of the compressible half-jet was studied by D. R. Chapman [7]. Compressible flows of this type play a certain part in the calculation of separated, free shear layers in wakes [8, 13].

h. Flow in the inlet length of a straight channel

As a further example of two-dimensional flow in the boundary layer, we shall now consider the case of flow in the inlet length of a straight channel with flat parallel walls. At a large upstream distance from the inlet the velocity distribution is assumed to be uniform and parabolic over the width of the channel, as indicated in Chap. V. We shall assume that the velocity in the inlet section is uniformly distributed over its width, $2\,a$, and that its magnitude is U_0. Owing to viscous friction, boundary layers will be formed on both walls, and their width will increase in the downstream direction. At the beginning, i. e. at small distances from the inlet section, the boundary layers will grow in the same way as they would along a flat plate at zero incidence. The resulting velocity profile will consist of two boundary-layer profiles on the two walls joined in the centre by a line of constant velocity. Since the volume of flow must be the same for every section, the decrease in the rate of flow near the walls which is due to friction must be compensated by a corresponding increase in the rate of flow near the axis. Thus the boundary layer is formed under the influence of an accelerated external flow, as distinct from the case of the flat plate. At larger distances from the inlet section the two boundary layers gradually merge into each other, and finally the velocity profile is transformed asymptotically into the parabolic distribution of Poiseuille flow.

This process can be analysed mathematically in one of two ways. First, the integration can be performed in the downstream direction so that the boundary-layer growth is calculated for an accelerated external stream. Secondly, it is possible to analyse the progressive deviation of the profile from its asymptotic parabolic distribution, i. e. integration can proceed in the upstream direction. Having obtained both solutions, say in the form of series expansions, we can retain a sufficient number of terms in either of them and join the two solutions at a section where both are still applicable. In this way the flow for the whole inlet length is obtained. The method which was first used by H. Schlichting [57], will now be outlined in brief.

We assume a system of coordinates whose axis of abscissae coincides with that of the channel, Fig. 9.15. For the expansion in the upstream direction we shall measure the ordinate y from the centre-line of the channel, whereas for the expansion in the downstream direction the ordinate y' will be measured from one of the walls. The inlet velocity will be denoted by U_0, and that in the central stream by $U(x)$.

We begin by writing the equation of continuity:

$$\int_{y'=0}^{a} u \, \mathrm{d}y = U_0 \, a . \tag{9.52}$$

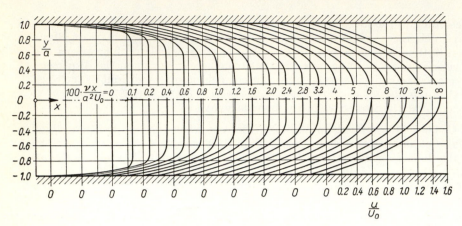

Fig. 9.15.   Velocity distribution for laminar flow in the inlet section of a channel

Introducing the displacement thickness δ_1 from eqn. (8.30) we can write

$$\int_0^a (U - u)\, dy = U\, \delta_1,$$

and with the aid of eqn. (9.52) we can write

$$U(x) = U_0\, \frac{a}{a - \delta_1} = U_0 \left[1 + \frac{\delta_1}{a} + \left(\frac{\delta_1}{a}\right)^2 + \cdots \right]. \tag{9.53}$$

. Near the inlet section the boundary layer develops in the same way as on a flat plate at zero incidence in unaccelerated flow, so that from eqn. (7.37) we have

$$\frac{\delta_1}{a} = 1.72\, \sqrt{\frac{\nu\, x}{a^2\, U_0}} = 1.72\, \varepsilon = K_1\, \varepsilon,$$

where

$$\varepsilon = \sqrt{\frac{\nu\, x}{a^2\, U_0}} \tag{9.54}$$

is the characteristic dimensionless inlet length. Equation (9.53) can also be written as

$$U(x) = U_0\, [1 + K_1\, \varepsilon + K_2\, \varepsilon^2 + \cdots] \tag{9.55}$$

with $K_1 = 1.72$. In this manner the velocity outside the boundary layer has been developed in powers of $\sqrt{x}$. The value of K_1 is known from Blasius's solution for the flat plate, but the remaining coefficients $K_2, K_3, \ldots$ are unknown, as they depend on the boundary layer which has not yet been determined.

In the series expansion from the upstream direction we assume $u = u_0(y) - u'(x, y)$, where $u_0(y)$ is the parabolic velocity distribution, i. e. $u_0(y) = \frac{3}{2}\, U_0 (1 - y^2/a^2)$, and u' is an additional velocity whose higher orders may be neglected in the first approximation.

Figure 9.15 gives an indication of the change in the velocity profile over the inlet length. It is seen that the parabolic profile is formed at about $\nu\, x/a^2\, U_0 = 0.16$, so that the actual inlet length is $l_E = 0.16 a (U_0 a/\nu) = 0.04 (2a) \cdot R$ where R denotes the Reynolds number referred to the width of the channel. For example at $R = 2000$ to 5000 the inlet length extends over 80 to 200 channel widths. Consequently, the flow does not become fully developed at all if the channel is short or if the Reynolds number is comparatively large.

An approximate method of calculation for the two-dimensional case which is based on the momentum equation (see Chap. X), as well as numerous experimental results which reach into the turbulent region, have been reported in two papers by H. Hahnemann and L. Ehret [36] and [37]. The flow in the entrance of a pipe has been studied by L. Schiller [55].

The details of a calculation which develops the solution from the downstream direction upwards can be found in [57].

The problem of the development of the flow pattern in the inlet length of a channel was examined critically by M. Van Dyke [71] when he formulated his second-order theory, see Secs. VIIf and IXm. He drew attention to the fact that the solution displayed in Fig. 9.15 represents a first-order solution and is valid only for very large Reynolds numbers. For this reason it is found to show certain deviations at low Reynolds numbers from the corresponding numerical solution of the full Navier-Stokes equations.

i. The method of finite differences†

Modern methods (digital computers). In recent years a large number of numerical methods has been developed for solving the laminar boundary-layer equations. These methods fall mainly under the heading of *implicit* finite-difference procedures and represent a development of a numerical procedure first formulated by Fluegge-Lotz and Blottner [21]. The methods referred to are accurate and fast but require access to a digital computer. The choice of method in a given case depends on the nature of the problem considered, but is also a matter of personal preference. For a review of existing methods the reader is referred to a survey article by Blottner [5].

The method proposed herein is chosen for its simplicity and its wide range of possible applications; it differs from the early methods in that transformed (similarity-type) variables are used and the step sizes are allowed to vary in the streamwise and normal direction. Some of the advantages of using transformed variables are: (a) the growth in the domain of calculation associated with the increasing boundary-layer thickness is largely eliminated; (b) the boundary-layer profiles are smoother and vary more slowly in the transformed plane allowing larger step sizes to be used; and (c) the finite-difference formulation becomes virtually identical for compressible and incompressible plane and axially symmetric boundary-layer flows. The use of variable step size in the normal direction makes it possible to calculate turbulent as well as laminar flows with only minor changes in formulation.

Special classes of laminar flows characterized by boundary layers with different length scales (e. g. large blowing rates), can also be handled with greater accuracy.

The boundary-layer equations considered are

$$\frac{\partial}{\partial x}(r^j\, u) + \frac{\partial}{\partial y}(r^j\, v) = 0\,, \tag{9.56}$$

$$u\,\frac{\partial u}{\partial x} + v\,\frac{\partial u}{\partial y} = -\frac{1}{\varrho}\,\frac{dp}{dx} + \frac{\partial}{\partial y}\left[\left(\nu + \frac{\varepsilon_t}{\varrho}\right)\frac{\partial u}{\partial y}\right], \tag{9.57}$$

---

† I am indebted to Professor T. I. Fanneloep of the Institute of Technology in Trondheim, who kindly provided me with the following presentation.

where $j = 0$ (plane flow) or $j = 1$ (flow with axial symmetry). The boundary conditions are $u = v = 0$ at $y = 0$ and $u = U(x)$ at $y = \delta$. For turbulent flows u and v are the appropriate mean velocities and ε_t represents a suitably defined eddy viscosity, see for instance A. M. O. Smith and T. Cebeci [61]. For laminar flows $\varepsilon_t = 0$. The transformation of eqns. (9.56) and (9.57) to dimensionless variables incorporates both the Blasius and the Mangler transformation (see also H. Goertler [33, 34]) and is defined as follows:

$$\xi = \frac{1}{\nu} \int_0^x U(x)\,\mathrm{d}x, \tag{9.58a}$$

$$\eta = y\,U(x) \left/ \left(2\,\nu \int_0^x U(x)\,\mathrm{d}x\right)^{1/2}\right. . \tag{9.58b}$$

The continuity equation is satisfied by the stream function

$$u\,r^j = \frac{\partial \psi}{\partial y}\;;\quad v\,r^j = -\frac{\partial \psi}{\partial x}\,. \tag{9.59}$$

The corresponding dimensionless form, $f(\xi, \eta)$, of the stream function expressed in terms of the independent variables ξ, η is defined by

$$\psi(x,\,y) = \nu\sqrt{2\,\xi}\,f(\xi, \eta). \tag{9.60}$$

It can now be shown that $f(\xi, \eta)$ satisfies the partial differential equation

$$(N\,f_{\eta\eta})_\eta + f\,f_{\eta\eta} + \beta(1 - f_\eta^2) = 2\,\xi\,(f_\eta\,f_{\eta\xi} - f_\xi\,f_{\eta\eta}). \tag{9.61}$$

Here

$$N = 1 + \varepsilon_t/\varrho\nu,$$

and ε_t is the eddy viscosity from eqn. (19.2). The subscripts denote partial differentiation, and the quantity

$$\beta(\xi) = 2\,\frac{U'(x)}{U^2(x)} \int_0^x U(x)\,\mathrm{d}x \tag{9.62}$$

is the only one determined by the flow. The boundary conditions for f are

$$\eta = 0;\quad f = 0;\quad f_\eta = 0 \text{ and } \eta = \infty;\quad f_\eta = 1. \tag{9.63}$$

Finite-difference equations of second order can be solved (by matrix inversion routines) much more efficiently than third (or higher) order equations. It is of interest, therefore, to reduce equations (9.61) to second order. To this end the variable $F = f_\eta$ is introduced and eqn. (9.61) is rewritten as

$$[N\,F_\eta]_\eta + f\,F_\eta + \beta(1 - F^2) = 2\,\xi\,[F\,F_\xi - f_\xi\,F_\eta]. \tag{9.64}$$

This equation now contains two unknown functions, f and F, but these are related by the simple expression

$$f(\xi, \eta) = \int_0^\eta F(\xi, \eta)\,\mathrm{d}\eta + f(\xi, 0). \tag{9.65}$$

In the absence of suction or blowing the boundary conditions are

$$F(\xi, 0) = f(\xi, 0) = 0, \quad F(\xi, \infty) = 1. \tag{9.66}$$

Near the leading edge of a cusped body and in the stagnation region of a blunt body, equations (9.64) and (9.65) reduce to true similarity form. The corresponding similar solutions can be used, therefore, as initial values for the step-by-step finite-difference method. The method presented here solves the partial differential equation (9.64) and (with small modifications) also the associated similar equations required as initial values. The method is thus self-starting and requires no additional input.

Finite-difference quotients: The domain of calculation in the (ξ, η)-plane can be represented by a semi-infinite strip bounded by the wall $\eta = 0$, the edge of the boundary layer $\eta = \eta_e$, with η_e suitably defined, and the initial line $\xi = \xi_0$ where the solution is presumed to be known.

This strip is completely covered by a grid with lines drawn parallel to the ξ and η coordinates as illustrated in Fig. 9.16. The step size $\Delta\xi$ represents the distance between two successive grid lines $\xi = $ constant; it is presumed to be small but is otherwise unspecified. The corresponding step sizes $\Delta\eta$ in the η-direction are specified to vary in geometric progression. The ratio between two successive grid lines, η_n and η_{n+1}, is denoted by $K = 1 + k$ where $|k|$ varies from 0 to 0·05 in typical cases. Each nodal point is identified by a double index m, n which defines its position ξ_m, η_n according to

$$\xi_m = \xi_0 + \sum_{j=1}^{m} \Delta\xi_j, \quad \eta_n = \frac{\Delta\eta_1}{k}(K^{n-1} - 1). \tag{9.67}$$

In writing the finite-difference quotients it is convenient to introduce the mean of two successive $\Delta\eta$-values

$$\overline{\Delta\eta}_n = \frac{1}{2}(\Delta\eta_{n+1} + \Delta\eta_n). \tag{9.68}$$

In the step-by-step calculations the solution is considered known at ξ_m and all preceding grid lines, and the variables F and f are sought at ξ_{m+1}.

Central-difference approximations to the derivatives F_η and $F_{\eta\eta}$ at ξ_{m+1} are obtained by expanding $F_{m+1,\ n+1}$ and $F_{m+1,\ n-1}$, respectively, in a Taylor series

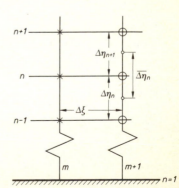

Fig. 9.16. Variable step size finite-difference grid for the calculation of laminar and turbulent boundary layers
× known values,
○ unknown values

centered at $(m+1, n)$. The two expressions are thereupon combined in such a way that terms of order $\Delta\eta^2$ are eliminated. The corresponding difference quotients can be given the form (index $m+1$ omitted):

$$\frac{\partial F_n}{\partial \eta} = \frac{1}{2\,\Delta\eta_n}\{L_3\,F_{n-1} + L_2\,F_n + L_1\,F_{n+1}\} + O\,(\Delta\eta^2, k\,\Delta\eta),\qquad(9.69)$$

where

$$L_1 = K^{-1},\quad L_2 = (K^2-1)\,L_1,\quad L_3 = -K;$$

$$\frac{\partial^2 F_n}{\partial \eta^2} = \frac{P_1}{\Delta\eta_n^2}\{P_4\,F_{n-1} - P_3\,F_n + P_2\,F_{n+1}\} + O\,(\Delta\eta^2, k\,\Delta\eta),\qquad(9.70)$$

where

$$P_1 = {}^1\!/_2\,(1+K),\quad P_2 = L_1,\quad P_3 = 2\,P_1\,P_2,\quad P_4 = 1.$$

Equations (9.69) and (9.70) reduce to the standard form for central differences when $K=1$.

For the ξ-derivatives in equation (9.64) a simple backward difference formula is used

$$F_\xi = \frac{F_{m+1,\,n} - F_{m,\,n}}{\Delta\xi} + O\,(\Delta\xi).\qquad(9.71)$$

The larger truncation error which appears here is balanced by the iterative scheme proposed for solving the difference equation. The non-linear terms in equation (9.64) have to be replaced by linearized difference quotients. The terms fF_η and FF_ξ may serve as examples and they are written as

$$F\,F_\xi = F^i\,(F_\xi)_{m+1,\,n} \text{ and } f^i\,(F_\eta)_{m+1,\,n},\qquad(9.72)$$

where $(F_\xi)_{m+1,n}$ and $(F_\eta)_{m+1,n}$ are given by (9.71) and (9.69), respectively. The unknown coefficients F^i and f^i are set equal to the known values $F_{m,n}$ and $f_{m,n}$ in the first iteration $(i=0)$ and later updated by $F^i, f^i\,(i=1,2,3,\ldots)$ in the second and further successive iterations. Experience suggests that the term F^2 should be approximated by

$$F^2_{m+1,\,n} = 2\,F_m\,F_{m+1,\,n} - F^2_{m,\,n}.\qquad(9.73)$$

The linearized finite-difference quotients given above are substituted into the differential equation (9.73) and the result is multiplied through by $\Delta\xi$ to give a difference equation. This is written as follows

$$A_n\,F_{m+1,\,n-1} + B_n\,F_{m+1,\,n} + C_n\,F_{m+1,\,n+1} = D_n,\qquad(9.74)$$

where

$$A_n = \frac{\Delta\xi}{\Delta\eta_n^2}\,N^i\,P_1\,P_4 + \frac{\Delta\xi}{2\,\Delta\eta_n}\,L_3\,[f^i + (N_\eta)^i + 2\,\xi\,(f_\xi)^i],\qquad(9.75\mathrm{a})$$

$$B_n = -\frac{\Delta\xi}{\Delta\eta_n^2}\,P_1\,P_3\,N^i + \frac{\Delta\xi}{2\,\Delta\eta_n}\,L_2\,[f^i + (N_\eta)^i + 2\,\xi\,(f_\xi)^i\,L_2] - \qquad(9.75\mathrm{b})$$
$$- 2\,F_{m,\,n}\,(\Delta\xi \times \beta + \xi),$$

$$C_n = \frac{\Delta\xi}{\Delta\eta_n^2} N^i P_1 P_2 + \frac{\Delta\xi}{2 \Delta\eta_{|n}} L_1 [f^i + (N_\eta)^i + 2 \xi (f_\xi)^i], \qquad (9.75\,c)$$

$$D_n = -\beta (1 - F_{m, n}^2) - 2 \xi F_{m.n}^2 . \qquad (9.75\,d)$$

In equations (9.75), ξ and β are evaluated at $(m+1)$, and only the variables with superscript i are updated through successive iterations. To speed-up the iteration process the terms $(f_\xi)^i$ can be kept constant (equal to the value at the previous station) until initial convergence is achieved.

Method of solution: Equations (9.74) represent a set of N—1 simultaneous algebraic equations for the unknown $F_{m+1,n}$ $(n = 2, 3, \ldots, N)$. At each level n three unknown quantities appear, namely $F_{m+1, n-1}$, $F_{m+1,n}$ and $F_{m+1, n+1}$, but since $F_{m+1,1}$ and $F_{m+1,N}$ are known from the boundary conditions, the total number of equations equals the number of unknowns. The set of algebraic equations can be written in so-called three-diagonal matrix form. Matrices of this type where off-diagonal elements vanish outside the three-diagonal band can be inverted by a simple and direct method well suited for digital computers. To end this equation (9.74) is rewritten in "standard form" (subscripts $(m+1)$ omitted)

$$A_n F_{n-1} + B_n F_n + C_n F_{n+1} = D_n; \; 2 \leqslant n \leqslant N - 1. \qquad (9.74\,b)$$

The boundary conditions are

$$F_1 = 0 \quad \text{and} \quad F_N = 1, \qquad (9.76)$$

where $n = 1$ denotes the wall and $n = N$ the edge of the boundary layer. It is assumed now that a solution exists† in the form

$$F_n = E_n F_{n+1} + G_n. \qquad (9.77)$$

The boundary condition $F_1 = 0$ and the requirement that equation (9.77) should remain valid independently of the step size $\Delta\eta$ leads to

$$E_1 = 0, \, G_1 = 0. \qquad (9.78)$$

A direct consequence of equation (9.77) is that

$$F_{n-1} = E_{n-1} F_n + G_{n-1}. \qquad (9.79)$$

When the preceding expressions are substituted into eqn. (9.74b), the following relation is obtained

$$F_n = \frac{-C_n}{B_n + A_n E_{n-1}} F_{n+1} + \frac{D_n - A_n G_{n-1}}{B_n + A_n E_{n-1}} . \qquad (9.80)$$

A comparison between equations (9.77) and (9.80) shows that

$$E_n = \frac{-C_n}{B_n + A_n E_{n-1}} , \quad G_n = \frac{D_n - A_n G_{n-1}}{B_n + A_n E_{n-1}} \qquad (9.81)$$

By means of equation (9.81) and the condition (9.78), it becomes possible to compute

---

† For a justification see R. D. Richtmeyer [53].

E_n and G_n for successive values of n starting with $n = 2$ for all grid points between the wall and the edge of the boundary layer.

Since F_{n+1} for $n = N-1$ is known from equation (9.76), it becomes possible to evaluate all unknowns F_n by means of equation (9.77) while traversing the boundary layer from the edge to the wall through decreasing values of n, i. e. for $n = N-1$, $N-2$, ..., 2. This completes the calculation of $F_n (\equiv F_{m+1,n})$ in one iteration. Once $F_{m+1,n}$ has been determined, the corresponding solution for $f_{m+1,n}$ can be found by direct numerical integration of equation (9.65). The trapezoidal rule suffices for this purpose.

The calculated values $F_{m+1,n}$ and $f_{m+1,n}$ are used to determine new and improved values of the coefficients A_n, B_n, C_n, which in turn leads to new and improved values of $F_{m+1,n}$ and $f_{m+1,n}$. The process is terminated when the results of two successive iterations agree to within a specified tolerance, typically of order 10^{-5}. The convergence is usually rapid, three to four iterations being adequate in most cases with step sizes Δx in the range $0{\cdot}01$ to $0{\cdot}05$.

In certain problems it becomes necessary to allow for boundary-layer growth by increasing N (or η_e) as the calculations proceed downstream. The boundary-layer edge is defined by the requirement that the difference $F_N - F_{N-1}$ should be less than a specified value, typically of order 10^{-4}. The growth, in terms of the present variables, is usually very modest even for cases involving separation.

A variable of primary interest in the calculation is the stress at the wall; its value can be determined with good accuracy from the five point formula

$$\left(\frac{\partial F}{\partial \eta}\right)_w = \frac{\Gamma_1}{\Delta \eta_1} \{\Gamma_2 F_2 + \Gamma_3 F_3 + \Gamma_4 F_4 + \Gamma_5 F_5\}, \tag{9.82}$$

where

$$\left.\begin{aligned}
&\Gamma_1 = -K^{-3}, \quad \Gamma_2 = -(1 + K + K^2 + K^3), \\
&\Gamma_3 = -\frac{\Gamma_2 (1 + K + K^2)}{K^2 (1 + K)^2}, \\
&\Gamma_4 = -\frac{\Gamma_1 \Gamma_2}{1 + K + K^2}, \quad \Gamma_5 = \frac{\Gamma_1}{\Gamma_2}.
\end{aligned}\right\} \tag{9.83}$$

Initial values: When using tabulated similar solutions as starting values, extensive interpolation is required whenever variable step sizes $\Delta \eta_n$ are used. It is more convenient and efficient also to generate the similarity solution by finite differences through successive iterations. The equation to be solved is obtained from equation (9.64), and can be written in linearized form as

$$N_{i-1} F_i'' + (f_{i-1} + N_{i-1}') F_i' + \beta (1 - F_{i-1} F_i) = 0, \tag{9.84}$$

with

$$f_i = \int_0^\eta F_i \, d\eta. \tag{9.85}$$

The indices $(i, i-1)$ indicate the iteration for which the variable is evaluated and $(\eta)'$ denotes $d/d\eta$. Variables with index $i-1$ are considered known (initially by guessing a solution which satisfies the boundary conditions), whereas those with index i are to be found in the i-th or current iteration. The difference quotients (9.69) and (9.70) are now substituted into equation (9.84). The result is a difference equation which can be written in the standard form of equation (9.74), with coefficients

$$A_n = N_{i-1}\, P_1\, P_4 + \frac{1}{2}\, L_3\, \overline{\Delta\eta}_n\, (f_{i-1} - N'_{i-1}), \tag{9.86a}$$

$$B_n = -N_{i-1}\, P_1\, P_3 + \frac{1}{2}\, L_2\, \overline{\Delta\eta}_n\, (f_{i-1} + N'_{i-1}) - \overline{\Delta\eta}_n^2\, \beta\, F_{i-1}, \tag{9.86b}$$

$$C_n = N_{i-1}\, P_1\, P_2 + \frac{1}{2}\, L_1\, \overline{\Delta\eta}_n\, (f_{i-1} + N'_{i-1}), \tag{9.86c}$$

$$D_n = -\overline{\Delta\eta}_n^2\, \beta. \tag{9.86d}$$

A linear variation in F suffices as an initial guess, F_0, and the corresponding value of f is determined from equation (9.85). The coefficients A_n, B_n, C_n and D_n are calculated next and the corresponding values of E_n and G_n are determined across the boundary layer. The recurrence relation (9.77) and the boundary conditions (9.78) are then used to determine the new iterate, F_i, across the boundary layer. The process is repeated until the difference between successive iterates becomes smaller than the specified tolerance. The number of iterations required is typically of order 8 to 12. The method is simpler than the usual "shooting" method used for two-point boundary-value problems and it converges in many cases where the latter method fails, for instance for very large blowing rates.

Applications: The finite-difference method presented here is intended as a practical engineering tool. Greater accuracy can be achieved with a more elaborate procedure, but this in turn leads to greater complexity in formulation and programming and to an increased demand for computer time and capacity. The computing time and accuracy depend for all difference methods on the step size used in the calculations. It is of interest to examine the accuracy of the present method in a few cases for which very accurate solutions are known. The cases considered are Howarth's linearly retarded flow (cf. Sec. IX d) and the circular cylinder with a pressure distribution according to potential theory and according to the experiments of Hiemenz (cf. Sec. X c). The results for a "normal" step size and a "small" step size are tabulated below. From the calculated results only the location of the separation points are shown.

Case Considered	Present results	Exact
Linearly retarded flow	(1) $x_s{}^* = 0{\cdot}1227$ (2) $x_s{}^* = 0{\cdot}1210$	$x_s{}^* = 0{\cdot}1199$ (Howarth) or $x_s{}^* = 0{\cdot}1198$ (Leigh)[44] or $x_s{}^* = 0{\cdot}1203$ (Schoenauer)
Circular cylinder (Potential flow)	(1) $\phi_s = 106{\cdot}13°$ (2) $\phi_s = 105{\cdot}01°$	$\phi_s = 104{\cdot}5°$ (Schoenauer) (cf. Sec. X c)
Circular cylinder (Hiemenz press. data)	(1) $\phi_s = 80{\cdot}98°$ (2) $\phi_s = 80{\cdot}08°$	$\phi_s = 80{\cdot}0°C$ (Jaffe and Smith)[42] (interpolated)

(1) "Normal" step size: $\Delta\xi = 0{\cdot}01$, $\Delta\eta = 0{\cdot}05$; (2) "Small" step size: $\Delta\xi = 0{\cdot}001$, $\Delta\eta = 0{\cdot}025$

The computing time with the "normal" step size is typically 5 to 10 seconds on the UNIVAC 1108 computer. The accuracy with the small step size is seen to be better but at the expense of a twenty-fold increase in computer time. For engineering calculations the coarser grid should suffice; it requires running times of the order 10 seconds in case of practical interest such as the laminar boundary layer of an aerofoil. Improved economy can be achieved by varying the step size as the calculation proceeds, that is using the fine mesh only in the critical region near separation.

A summary account of numerical methods in fluid mechanics is given in the lecture notes of Smolderen [65].

j. Boundary layer of second order†

The second-order equations, eqns. (7.52) and (7.53) for flow in a boundary layer were derived in Sec. VIIf. This system of linear partial differential equations can be solved if the first-order solutions $u_1(x, N)$ and $v_1(x, N)$ are known, and if the functions $K(x)$, $U_2(x, 0)$ and $P_2(x, 0)$ are suitably prescribed.

It follows that the calculation of a second-order boundary layer on a given body in a stream requires that the following steps should be taken:

(a) Calculation of the potential flow (external flow of first order) about the body with the boundary conditions $V_1(x, 0) = 0$. The solution yields $U_1(x, 0)$.

(b) Calculation of the first-order boundary layer for given $U_1(x, 0)$, that is, determination of the solution of the system of equations (7.49). In particular, from the solution $u_1(x, N)$, $v_1(x, N)$ we calculate the function $V_2(x, 0)$ with the aid of eqn. (7.45).

(c) Calculation of the second-order external flow for the boundary conditions $V_2(x, 0)$ and zero velocity at large distance from the body in accordance with eqn. (7.45). The solution provides us with $U_2(x, 0)$ and $P_2(x, 0)$.

In what follows, we shall assume that these steps have already been taken. We shall concentrate on more detailed second-order calculation for several particular cases.

Symmetric stagnation flow: This type of flow was analyzed in detail by M. Van Dyke (see also Chap. VII, [7]). It is assumed that the expressions for the external flow of first and second order on a convex wall at the stagnation point ($K = 1$ at $x = 0$) have been found and yield

$$U(x, 0) = U_{11} x + \varepsilon U_{21} x + O(\varepsilon^2), \tag{9.87}$$

where U_{11} and U_{21} are constants which depend on the shape of the body. According to eqn. (7.48), we make the following assumption for the inner solution:

$$u(x, y, \varepsilon) = U_{11} x f'(\eta) + \varepsilon [\sqrt{U_{11}} x F'_c(\eta) + U_{21} x F'_d(\eta)] + O(\varepsilon^2), \tag{9.88}$$

$$v(x, y, \varepsilon) = -\varepsilon U_{11} f(\eta) - \varepsilon^2 [F_c(\eta) - \eta f(\eta) + U_{21} F_d(\eta)/\sqrt{U_{11}}] + O(\varepsilon^3). \tag{9.89}$$

---

† I am indebted to Professor K. Gersten for the exposition in this section.

Here, the new variable is defined as

$$\eta = \sqrt{U_{11}}\, N = \sqrt{U_{11}}\, y/\varepsilon . \dagger \qquad (9.90)$$

Substituting these forms into the boundary-layer equations of first and second order, we obtain

$$f''' + f f'' + 1 - f'^2 = 0, \qquad (9.91)$$

$$F_c''' + f F_c'' - 2 f' F_c' + f'' F_c = \eta \,(f f'' - f'^2 + 2) + 0 \cdot 6479, \qquad (9.92)$$

$$F_d''' + f F_d'' - 2 f' F_d' + f'' F_d = - 2, \qquad (9.93)$$

with the boundary conditions

$$\left.\begin{array}{l} \eta = 0: \quad f = 0;\ f' = 0;\ F_c = 0;\ F_c' = 0,\, F_d = 0,\, F_d' = 0, \\[4pt] \eta \to \infty: \quad f' = 1;\ F_c'' = -1;\ F_d' = 1. \end{array}\right\} \qquad (9.94)$$

The first equation determines the first-order boundary layer which is identical with eqn. (5.39) for stagnation flow at a flat plate. The two succeeding equations determine the second-order boundary layer. The solution has been split into two parts, the partial solution due to curvature (subscript c) and the partial solution due to the displacement effect (subscript d). The latter is induced by the external flow of second order with the velocity $U_2(x, 0) = U_{21} x$, as determined in step (c) above. For F_d we obtain the following simple solution

$$F_d = \frac{1}{2}\,(f + \eta f'). \qquad (9.95)$$

The skin-friction coefficient follows from eqn. (7.55). Inserting the numerical values

$$f''(0) = 1 \cdot 2326;\ F_c''(0) = - 1 \cdot 9133;\ F_d''(0) = 1 \cdot 8489, \qquad (9.96)$$

we find that

$$\tfrac{1}{2}\, c_f' = \varepsilon\, x\, \sqrt{U_{11}}\, \{1 \cdot 2326\, U_{11} - \varepsilon\,(1 \cdot 9133\, \sqrt{U_{11}} - 1 \cdot 8489\, U_{21})\} + O\,(\varepsilon^3) \quad (9.97)$$

According to eqn. (7.54), the pressure coefficient is

$$c_{pw} = 1 - U_{11}^2\, x^2\, \{1 - \varepsilon\,(1 \cdot 8805/\sqrt{U_{11}} - 2\, U_{21}/U_{11}) + O\,(\varepsilon^2)\}. \qquad (9.98)$$

The formulae for the pressure and skin-friction coefficient are universal. The missing numerical values of the constants U_{11} and U_{21} depend only on the shape of the body. In all known examples, U_{21} has turned out to be negative. This signifies that the skin-friction coefficient near the stagnation point on a convex wall decreases due to higher-order boundary-layer effects (curvature and displacement); the opposite is true about the pressure coefficient at the wall.

---

† This equation is seen to be identical with eqn. (9.13) of Sec. IX c if it is noted that the coordinates x, y in it denote lengths, whereas here they have been referred to the characteristic length R_0 (radius of curvature of the body at the stagnation point) and made dimensionless. When comparing the velocity distribution of the external flows from eqns. (9.81) and (9.15), we have $\eta = \sqrt{U_{11}}\, N = y\, \sqrt{u_1/\nu}$.

Parabola in a symmetric stream: The second-order boundary layer on a parabola in a symmetric stream was calculated by M. Van Dyke (see also Chap. VII, [7]). In the neighbourhood of stagnation, we have

$$U_{11} = 1 \text{ and } U_{21} = -0\cdot61. \tag{9.100}$$

In the case of the parabola we have at our disposal a numerical solution of the complete Navier-Stokes equations due to R. T. Davis [11] and can use it for a direct evaluation of the improvement made by the second-order theory. Figure 9.17 shows a plot of the skin-friction coefficient from (9.97) at a stagnation point of a parabola in terms of the Reynolds number formed with the radius of curvature at the vertex. It follows from eqn. (9.97) that

$$c'_f/2\,\varepsilon\,x = 1\cdot2326 - 3\cdot04\,\varepsilon + O\,(\varepsilon^2), \tag{9.101}$$

which is equivalent to

$$c'_f/2\,\varepsilon\,x = \frac{1\cdot2326}{1 + 2\cdot47\,\varepsilon} + O(\varepsilon^2). \tag{9.102}$$

Curve II in Fig. 9.17 is a plot of this relation, whereas Curve I depicts the first-order solution. Curve III has been plotted with the results of R. T. Davis's numerical solution. The considerable improvement effected by the second-order theory in the lower range of Reynolds numbers is clearly visible. In addition, the diagrams give an unambiguous indication that the second-order theory allows us to identify the range of validity of first-order theory. If an error of up to 2% is to be tolerated, it follows that first-order theory applies at Reynolds numbers in excess of $R = 1\cdot5 \times 10^5$. Similar comparisons based on R. T. Davis's numerical solutions reveal that the lower limit of validity for the second-order theory is at $R = 100$ for a 2% tolerance.

Figure 9.18 gives diagrams of static pressure and skin-friction distributions along the contour of a parabola at zero incidence, both evaluated with the aid of second-order theory. For purposes of comparison, the diagrams contain distributions calculated with the aid of first-order boundary-layer theory ($R \to \infty$). Both pressure distributions start with $c_p = 1$ at the stagnation point. Frictionless flow ($R \to \infty$) gives

$$c_p = \frac{1}{1 + 2\,x^*}, \tag{9.102}$$

where $x^* = x'/R_0$ denotes the dimensionless distance from the vertex of the parabola and measured along the centerline; see also Fig. 9.18. For $R = 100$ we find from eqn. (9.98) for the neighbourhood of the stagnation point that

$$c_p = 1 - x^2\,(1 - 3\cdot10\,\varepsilon) + O(\varepsilon^2). \tag{9.103}$$

This is equivalent to

$$c_p = \frac{1}{1 + 1\cdot38\,x^*}, \tag{9.104}$$

where $x = (2\,x^*)^{1/2}$ near the stagnation point. As expected, the higher-order corrections decrease in the downstream direction, particularly also due to the decreasing curvature in that direction. At about $x^* = 2$, the higher-order effects vanish to all intents and purposes. Similar conclusions apply to the skin-friction coefficient which, however, displays at the stagnation point the largest second-order correction.

Fig. 9.17. Local skin-friction coefficient in the neighbourhood of a stagnation point of a parabola in terms of the Reynolds number $R = U_\infty R/\nu$

(1) First-order boundary-layer theory, $R \to \infty$;
(2) Second-order boundary-layer theory, eqn. (9.101), after K. Gersten ([10] of Chap. VII)
(3) Numerical solution of the Navier-Stokes equations after R. T. Davis [11]
(4) $R = 0$, Stokesian flow

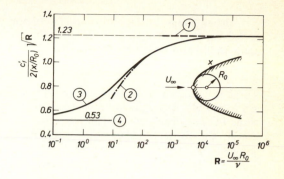

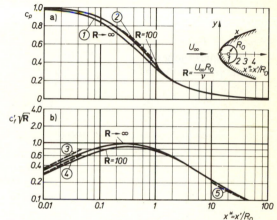

(1) c_p for nonviscous flow, $R \to \infty$, eqn. (9.102)
(2) c_p for $R = 100$; eqn. (9.104)
(3) c'_f $R^{1/2} = 3\cdot486\, x^{*1/2}$; stagnation point; $R \to \infty$; eqn. (9.100); $\varepsilon = 0$
(4) c'_f $R^{1/2} = 2\cdot63\, x^{*1/2}$; stagnation point; $R \to \infty$; eqn. (9.100); $\varepsilon = 0\cdot1$
(5) c'_f $R^{1/2} = 0\cdot664\, x^{*-1/2}$; flat plate

Fig. 9. 18. a) Static pressure distribution and b) distribution of shearing stress around the contour of a parabola at zero incidence. The curves for $R = 100$ correspond to second-order boundary-layer theory; the curves $R \to \infty$ correspond to first-order theory

Whereas the pressure coefficient is increased by higher-order effects, the skin-friction is reduced thereby. It follows that the pressure drag of a parabola increases, whereas the skin friction decreases for Reynolds numbers decreasing from $R \to \infty$.

For the pressure drag of a parabola of width b (exclusive of base drag), we find [11] that

$$c_D = \frac{D}{\frac{1}{2}\,\varrho\,U_\infty^2\,R_0\,b} = \pi + 5\cdot2\,\varepsilon + O(\varepsilon^2). \qquad (9.105)$$

Thus, at $R = 100$ the pressure drag exceeds that in an inviscid flow by 16%.

The fact that the pressure drag increases as a result of the operation of second-order boundary-layer effects points to the possibility that such drag should appear in the framework of a second-order theory also for a flat plate at zero incidence, as already intimated in Sec. VIIf.

Other shapes: Second-order effects for half-bodies have been investigated by L. Devan [12]. The results are similar to those for the parabola. The coefficients for eqns. (9.97) and (9.98) are

$$U_{11} = 1.5; \quad U_{21} = -0.62.$$

Further solutions of the boundary-layer equations (7.52) and (7.53) of second order are available, as might have been expected, for cases which lead to self-similar solutions in first order, Sec. VII b. In the case of flows whose first-order external flows are of the form $U_1(x, 0) \sim x^m$ the second-order theory also leads to self-similar solutions if

$$K(x) \sim x^{(m-1)/2}; \quad U_2(x, 0) \sim x^n. \tag{9.106}$$

Further details concerning the effects of second order can be found in Chap. VII as well as in [31, 33]. The latter contain indications about second-order effects in the presence of suction, blowing, heat transfer and compressibility. Second-order effects acquire increasing importance for high Mach numbers and in the presence of blowing. In this connexion consult [24, 25, 47, 48, 59].

References

[1] Andrade, E.N.: The velocity distribution in a liquid-into-liquid jet. The plane jet. Proc. Phys. Soc. London 51, 784—793 (1939).

[2] Baxter, D.C., and Flügge-Lotz, I.: The solution of compressible laminar boundary layer problems by a finite difference method. Part II: Further discussion of the method and computation of examples. Techn. Rep. 110, Div. Eng. Mech. Stanford Univ. (1957); short version: ZAMP 9b, 81—96 (1958).

[3] Bickley, W.: The plane jet. Phil. Mag. Ser. 7, 23, 727—731 (1939).

[4] Blasius, H.:, Grenzschichten in Flüssigkeiten mit kleiner Reibung. Z. Math. u. Phys. 56, 1—37 (1908); Engl. transl. in NACA TM 1256.

[5] Blottner, F.G.: Finite difference methods of solution of the boundary-layer equations. AIAA J. 8, 193—205 (1970).

[5a] Blottner, F. G.: Investigation of some finite difference techniques for solving the boundary layer equations. Comp. Math. Appl. Mech. Eng. 6, 1—30 (1975).

[6] Cebeci, T., and Smith, A.M.O.: A finite difference method for calculating compressible laminar and turbulent boundary layers. Trans. ASME, J. Basic Eng. 92, 523—535 (1970).

[7] Chapman, D.R.: Laminar mixing of a compressible fluid. NACA TN 1800 (1949).

[8] Chapman, D.R.: Theoretical analysis of heat transfer in regions of separated flow. NACA TN 3792 (1956).

[9] Chen, K.K., and Libby, P.A.: Boundary layers with small departures from the Falkner-Skan profile. JFM 33, 243—282 (1968).

[10] Christian, W.J.: Improved numerical solution of the Blasius problem with three point boundary condition. JASS 28, 911—912 (1961).

[11] Davis, R.T.: Numerical solution of the Navier-Stokes equations for symmetric laminar incompressible flow past a parabola. JFM 51, 417—433 (1972).

[12] Devan, L.: Second order incompressible laminar boundary layer development on a two-dimensional semi-infinite body. Ph. D. Thesis, Univ. of California at Los Angeles, 1964.

[13] Denison, M.R., and Baum, E.: Compressible free shear layer with finite initial thickness. AIAA J. 1, 342—349 (1963).

[14] Dewey, C.F., and Gross, F.: Exact similar solutions of the laminar boundary layer equations. Advances in Heat Transfer Vol. 4, Academic Press, New York, 1967, 317—446.

[15] Evans, H.L.: Laminar boundary layer theory. Addison-Wesley Publishing Company, London, 1968.

[16] Fage, A.: The airflow around a circular cylinder in the region where the boundary separates from the surface. Phil. Mag. 7, 253 (1929).

[17] Fage, A., and Falkner, V.M.: Further experiments on the flow around a circular cylinder. ARC RM 1369 (1931).

[18] Falkner, V.M.: A further investigation of solution of boundary layer. ARC RM 1884 (1939).

[19] Falkner, V.M.: Simplified calculation of the laminar boundary layer. ARC RM 1895 (1941).

[20] Fanneloep, T., and Flügge-Lotz, I.: The compressible boundary layer along a wave-shaped wall. Ing.-Arch. 33, 24—35 (1963).

[21] Flügge-Lotz, I., and Blottner, F.G.: Computation of the compressible laminar boundary layer flow including displacement thickness interaction using finite difference methods. Stanford Univ. Div. Eng. Mech. Tech. Rep. 131 (1962). Shortened version in Journal de Mécanique 2, 397—423 (1963).

[22] Flügge-Lotz, I.: The computation of the laminar compressible boundary layer. Dep. Mech. Eng. Stanford Univ., Rep. R. 352—30—7 (1954).

[23] Frössling, N.: Verdunstung, Wärmeübergang und Geschwindigkeitsverteilung bei zwei-dimensionaler und rotationssymmetrischer laminarer Grenzschichtströmung. Lunds. Univ. Arsskr. N. F. Avd. 2, 36, No. 4 (1940); see also NACA TM 1432.

[24] Gersten, K., and Gross, J.F.: The second-order boundary-layer along a circular cylinder in supersonic flows. Int. J. Heat Mass Transfer 16, 2241—2260 (1973).
The leading edge of a swept cylinder. Int. J. Heat Mass Transfer 16 (1972).

[25] Gersten, K., Gross, J.F., and Börger, G.: Die Grenzschicht höherer Ordnung an der Stau-linie eines schiebenden Zylinders mit starkem Ausblasen. Z. Flugwiss. 20, 330—341 (1972).

[26] Goldstein, S.: On the two-dimensional steady flow of a viscous fluid behind a solid body. Proc. Roy. Soc. London A 142, 545—562 (1933).

[27] Goldstein, S. (ed.): Modern developments in fluid dynamics, Vol. I, 105. Clarendon Press, Oxford, 1938.

[28] Goldstein, S.: On laminar boundary layer flow near a position of separation. Quart. J. Mech. Appl. Math. 1, 43—69 (1948).

[29] Görtler, H.: Ein Differenzenverfahren zur Berechnung laminarer Grenzschichten. Ing.-Arch. 16, 173—187 (1948).

[30] Görtler, H.: Einfluss einer schwachen Wandwelligkeit auf den Verlauf der laminaren Grenz-schichten. Parts I and II. ZAMM 25/27, 233—244 (1947) and 28, 13—22 (1948).

[31] Görtler, H.: Zur Approximation stationärer laminarer Grenzschichtströmungen mit Hilfe der abgebrochenen Blasiusschen Reihe. Arch. Math. 1, No. 3, 235—240 (1949).

[32] Görtler, H.: Reibungswiderstand einer schwach gewellten längsangeströmten Platte. Arch. Math. 1, 450—453 (1949).

[33] Görtler, H.: Eine neue Reihenentwicklung für laminare Grenzschichten. ZAMM 32, 270—271 (1952).

[34] Görtler, H.: A new series for the calculation of steady laminar boundary layer flows. J. Math. Mech. 6, 1—66 (1957).

[35] Görtler, H., and Witting, H.: Zu den Tanischen Grenzschichten. Österr. Ing.-Archiv 11, 111—122 (1957).

[36] Hahnemann, H., and Ehret, L.: Der Druckverlust der laminaren Strömung in der Anlauf-strecke von geraden, ebenen Spalten. Jb. dt. Luftfahrtforschung I, 21—36 (1941).

[37] Hahnemann, H., and Ehret, L.: Der Strömungswiderstand in geraden, ebenen Spalten unter Berücksichtigung der Einlaufverluste. Jb. dt. Luftfahrtforschung I, 186—207 (1942).

[38] Hartree, D.R.: A solution of the laminar boundary layer equation for retarded flow. ARC RM 2426 (1949).

[39] Hiemenz, K.: Die Grenzschicht an einem in den gleichförmigen Flüssigkeitsstrom einge-tauchten geraden Kreiszylinder. Thesis Göttingen 1911; Dingl. Polytechn. J. 326, 321 (1911).

[40] Howarth, L.: On the calculation of steady flow in the boundary layer near the surface of a cylinder in a stream. ARC RM 1632 (1935).

[41] Howarth, L.: On the solution of the laminar boundary layer equations. Proc. Roy. Soc. London A 164, 547—579 (1938).

[42] Jaffe, N.A., and Smith, A.M.O.: Calculation of laminar boundary layers by means of a differential-difference method. Progress in Aerospace Sciences, Vol. 12 (D. Küchemann, ed.), Pergamon Press, 1972.

[42a] Keller, H. B.: Numerical methods in boundary layer theory. Ann. Rev. Fluid Mech. (M. van Dyke, ed.) 10, 417—433 (1978).

[43] Krzywoblocki, M.Z.: On steady, laminar two-dimensional jets in compressible viscous gases far behind the slit. Quart. Appl. Math. 7, 313 (1949).

[44] Leigh, D.C.F.: The laminar boundary layer equation: A method of solution by means of an automatic computer. Proc. Cambr. Phil. Soc. *51*, 320—332 (1955).

[44a] Lessen, M.: On the stability of the laminar free boundary layer between parallel streams. NACA Rep. 979 (1950); see also Sc. D. Thesis, MIT (1948).

[45] Lock, R.C.: The velocity distribution in the laminar boundary layer between parallel streams. Quart. J. Mech. Appl. Math. *4*, 42—63 (1951).

[46] Mills, R.H.: A note on some accelerated boundary layer velocity profiles. JAS *5*, 325 (1938).

[47] Papenfuss, H.D.: Higher-order solutions for the incompressible, three-dimensional boundary-layer flow at the stagnation point of a general body. Archives of Mechanics (Warsaw) *26*, 459—478 (1974).

[48] Papenfuss, H.D.: Mass-transfer effects on the three-dimensional second order boundary-layer flow at the stagnation point of blunt bodies. Mech. Res. Comm. *1*, 285—290 (1974).

[49] Pai, S.I.: Fluid dynamics of jets. D. Van Nostrand Company, New York, 1954.

[50] Pohlhausen, K.: Zur näherungsweisen Integration der Differentialgleichung der Grenzschicht. ZAMM *1*, 252—268 (1921).

[51] Potter, O.E.: Laminar boundary layers at the interface of co-current parallel streams. Quart. J. Mech. Appl. Math. *10*, 302 (1957).

[52] Reeves, B.L., and Kippenhan, C.J.: A particular class of similar solutions of the equations of motion and energy of a viscous fluid. JASS *29*, 38—47 (1962).

[53] Richtmeyer, R.D.: Difference methods for initial value problems. Interscience, New York, 1957.

[54] Schiller, L., and Linke, W.: Druck- und Reibungswiderstand des Zylinders bei Reynoldsschen Zahlen 500 bis 40000. ZFM *24*, 193—198 (1933).

[55] Schiller, L.: Die Entwicklung der laminaren Geschwindigkeitsverteilung (im Kreisrohr) und ihre Bedeutung für die Zähigkeitsmessungen. ZAMM *2*, 96—106 (1922).

[56] Schlichting, H.: Laminare Strahlenausbreitung. ZAMM *13*, 260—263 (1933).

[57] Schlichting, H.: Laminare Kanaleinlaufströmung. ZAMM *14*, 368—373 (1934).

[57a] Schlichting, H.: Grenzschichttheorie. Engl. transl. by Kestin, J.: Boundary-layer theory. 6th ed., McGraw-Hill, New York, 1968.

[58] Schroeder, K.: Ein einfaches numerisches Verfahren zur Berechnung der laminaren Grenzschicht. FB 1741 (1943); later expanded and reprinted in Math. Nachr. *4*, 439—467 (1951).

[59] Schultz-Grunow, F., and Henseler, H.: Ähnliche Grenzschichtlösungen zweiter Ordnung für Strömungs- und Temperaturgrenzschichten an longitudinal gekrümmten Wänden mit Grenzschichtbeeinflussung. Wärme- und Stoffübertragung *1*, 214—219 (1968).

[60] Smith, A.M.O., and Clutter, D.W.: Solution of the incompressible boundary layer equations. AIAA J. *1*, 2062—2071 (1963).

[61] Smith, A.M.O., and Cebeci, T.: Numerical solution of the turbulent boundary-layer equations. McDonnell-Douglas Rep. No. DAC 33735 (1967).

[63] Steinheuer, J.: Similar solutions for the laminar wall jet in a decelerating outer flow. AIAA J. *6*, 2198—2200 (1968).

[64] Stewartson, K.: Further solutions of the Falkner-Skan equation. Proc. Cambr. Phil. Soc. *50*, 454—465 (1954).

[65] Smolderen, E.: Numerical methods in fluid dynamics. AGARD Lecture Ser. No. 48 (1972).

[66] Tani, I.: On the solution of the laminar boundary layer equations. J. Phys. Soc. Japan *4*, 149—154 (1949). See also: Fifty years of boundary-layer research (W. Tollmien and H. Görtler, ed.), Braunschweig, 193—200 (1955).

[67] Thom, A.: The laminar boundary layer of the front part of a cylinder. ARC RM 1176 (1928); see also ARC RM 1194 (1929).

[68] Tifford, A.N.: Heat transfer and frictional effects in laminar boundary layers. Part 4: Universal series solutions. WADC Techn. Rep. 53—288 (1954).

[69] Tollmien, W.: Grenzschichten. Handbuch der Exper. Physik *IV*, Part 1, 241—287 (1931).

[70] Ulrich, A.: Die ebene laminare Reibungsschicht an einem Zylinder. Arch. Math. *2*, 33—41 (1949).

[71] Van Dyke, M.: Entry flow in a channel. JFM *44*, 813—823 (1970).

[72] Witting, H.: Über zwei Differenzenverfahren der Grenzschichttheorie. Arch. Math. *4*, 247—256 (1953).

[73] Anonymous: Interpolation and allied tables. Prepared by H.M. Nautical Almanac Office. H.M. Stationary Office (1956).

CHAPTER X

Approximate methods for the solution of the two-dimensional, steady boundary-layer equations

Introductory remark: The examples of *exact* solutions of the boundary-layer equations which have been discussed in the preceding chapters have shown that the mathematical difficulties associated with *analytic* solutions for them are considerable. In particular, the general problem involving the flow of fluid round a body of arbitrary shape cannot be completely solved with the aid of the analytical methods presented thus far. The *numerical* or step-by-step methods (see Sec. IX i) allow us to solve most problems with a tolerable amount of work if a fast digital computer is available. For this reason, the approximate methods for the solution of our boundary-layer equations developed in earlier times, that is before the advent of computers, do not enjoy the same importance now as they did then. Nevertheless, we propose to give here an outline of these approximate methods, because they are well-suited to the generation of a quick outline of a solution even in more complex cases; in this connexion the summary by E. Truckenbrodt [24] will be found helpful.

This chapter deals with approximate methods for laminar boundary layers only. Analogous methods for turbulent boundary layers (*cf.* Chap. XXII) have retained their special importance up to this day.

All approximate methods are *integral* methods which do not attempt to satisfy the boundary-layer equations for every streamline; instead, the equations are satisfied only on an average extended over the thickness of the boundary layer. All approximate methods are based on the momentum and energy equations of boundary-layer theory known to us from Sec. VIII e. All these methods can be traced to two papers, one due to Th. von Kármán [7], and the other to K. Pohlhausen [15]. Before proceeding to apply the method to the general cases of two-dimensional and axially-symmetrical boundary layers with pressure gradients, we shall consider first the essential features of the method as applied to the flat plate at zero incidence. This example is particularly simple in that the pressure gradient vanishes along the whole plate. Moreover, we shall have the opportunity of assessing the power of the approximate method, at least in a particular case, and to compare it with the exact solution which is already known from Chap. VII.

a. Application of the momentum equation to the flow past a flat plate at zero incidence

Applying the momentum equation to the fluid within the control surface shown in Fig. 10.1, we can derive the statement that the flux of momentum through the control surface, considered fixed in space, is equal to the skin friction on the plate $D(x)$ from the leading edge ($x=0$) to the current section at x. The application of the momentum equation to this particular case has already been discussed in Sec. IXf. It was then found, eqn. (9.26), that the drag of a plate wetted on one side is given by

$$D(x) = b\,\varrho \int\limits_{y=0}^{\infty} u\,(U_\infty - u)\,\mathrm{d}y\,, \tag{10.1}$$

where the integral is to be taken at section x. On the other hand the drag can be expressed as an integral of the shearing stress τ_0 at the wall, taken along the plate:

$$D(x) = b \int\limits_{0}^{x} \tau_0(x)\,\mathrm{d}x\,. \tag{10.2}$$

Upon comparing eqns. (10.1) and (10.2) we obtain

$$\tau_0(x) = \varrho\,\frac{\mathrm{d}}{\mathrm{d}x} \int\limits_{y=0}^{\infty} u\,(U_\infty - u)\,\mathrm{d}y\,. \tag{10.3}$$

This equation can be also deduced in a purely formal way from the boundary-layer equation (7.22) by first integrating the equation of motion in the x-direction with respect to y from $y = 0$ to $y = \infty$. Equation (10.3) is, finally, obtained without difficulty if the velocity component v is eliminated with the aid of the equation of continuity, and if it is noticed that $\mu\,(\partial u/\partial y)_{y=0} = \tau_0$.

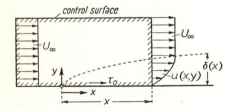

Fig. 10.1.   Application of the momentum equation to the flow past a flat plate at zero incidence

Introducing the momentum thickness, δ_2, defined by eqn. (8.31), we have

$$U_\infty{}^2\,\frac{\mathrm{d}\delta_2}{\mathrm{d}x} = \frac{\tau_0}{\varrho}\,. \tag{10.4}$$

The momentum equation in its form (10.4) represents a particular case of the general momentum equation of boundary-layer theory as given in eqn. (8.32), being valid for the case of a flat plate at zero incidence. Its physical meaning expresses the fact that the shearing stress at the wall is equal to the loss of momentum in the boundary layer, because in the example under consideration there is no contribution from the pressure gradient.

So far eqn. (10.4) introduced no additional assumptions, as will be the case with the approximate method, but before discussing this matter it might be useful to note a relation between τ_0 and δ_2, which is obtained from eqn. (10.4) by introducing the exact value for τ_0 from eqn. (7.32). Putting $\tau_0/\varrho\, U_\infty^2 = \alpha\sqrt{\nu/U_\infty\, x}$ with $\alpha = 0\cdot332$, we have

$$\delta_2 = \int_0^x \frac{\tau_0}{\varrho\, U_\infty^2}\, dx = 2\,\alpha\sqrt{\frac{\nu\, x}{U_\infty}}$$

and hence

$$\delta_2 = 2\,\frac{\tau_0}{\varrho\, U_\infty^2}\, x\,. \tag{10.5}$$

With reference to eqn. (10.3) or (10.4) we can now perform an approximate calculation of the boundary layer along a flat plate at zero incidence. The essence of the approximate method consists in assuming a suitable expression for the velocity distribution $u(y)$ in the boundary layer, taking care that it satisfies the important boundary conditions for $u(y)$, and that it contains, in addition, one free parameter, such as a suitably chosen boundary-layer thickness which is finally determined with the aid of the momentum equation (10.3).

In the particular case of a flat plate at zero incidence now being considered it is possible to take advantage of the fact that the velocity profiles are similar. Hence we put

$$\frac{u}{U_\infty} = f\left(\frac{y}{\delta(x)}\right) = f(\eta)\,, \tag{10.6}$$

where $\eta = y/\delta(x)$ is the dimensionless distance from the wall referred to the boundary-layer thickness. The similarity of velocity profiles is here accounted for by assuming that $f(\eta)$ is a function of η only, and contains no additional free parameter. The function f must vanish at the wall ($\eta = 0$) and tend to the value 1 for large values of η, in view of the boundary conditions for u. When using the approximate method, it is expedient to place the point at which this transition occurs at a finite distance from the wall, or in other words, to assume a finite boundary-layer thickness $\delta(x)$, in spite of the fact that all exact solutions of the boundary-layer equations tend asymptotically to the potential flow associated with the particular problem The boundary-layer thickness has no physical significance in this connexion, being only a quantity which it is convenient to use in the computation.

Having assumed the velocity profile in eqn. (10.6), we can now proceed to evaluate the momentum integral (10.3), and we obtain

$$\int_{y=0}^\infty u\,(U_\infty - u)\, dy = U_\infty^2\,\delta(x)\int_{\eta=0}^1 f(1-f)\, d\eta\,. \tag{10.7}$$

The integral in eqn. (10.7) can now be evaluated provided that a specific assumption is made for $f(\eta)$. Putting

$$\alpha_1 = \int_0^1 f(1-f)\, d\eta \tag{10.8}$$

for short, we have

$$\int_{y=0}^\infty u\,(U_\infty - u)\, dy = U_\infty^2\,\delta_2 = \alpha_1\,\delta\, U_\infty^2\,,$$

or

$$\delta_2 = \alpha_1\,\delta\,. \tag{10.9}$$

The value of the displacement thickness δ_1 from eqn. (8.30) will now also be calculated as it will be required later. Putting

$$\alpha_2 = \int_0^1 (1 - f)\, d\eta \,, \tag{10.10}$$

we obtain

$$\delta_1 = \alpha_2\, \delta \,. \tag{10.11}$$

Furthermore, the viscous shearing stress at the wall is given by

$$\frac{\tau_0}{\varrho} = \nu \left(\frac{\partial u}{\partial y}\right)_{y=0} = \frac{\nu\, U_\infty}{\delta} \cdot f'(0) = \beta_1\, \frac{\nu\, U_\infty}{\delta} \,, \tag{10.12}$$

where

$$\beta_1 = f'(0) \,. \tag{10.13}$$

Introducing these values into the momentum equation (10.4), we obtain

$$\delta\, \frac{d\delta}{dx} = \frac{\beta_1}{\alpha_1}\, \frac{\nu}{U_\infty} \,.$$

Integration from $\delta = 0$ at $x = 0$ gives the first result for the approximate theory
in the form

$$\delta(x) = \sqrt{\frac{2\beta_1}{\alpha_1}}\, \sqrt{\frac{\nu x}{U_\infty}} \,. \tag{10.14}$$

Hence the shearing stress at the wall from eqn. (10.12) becomes

$$\tau_0(x) = \sqrt{\frac{\alpha_1\,\beta_1}{2}}\, \mu\, U_\infty \sqrt{\frac{U_\infty}{\nu x}} \,. \tag{10.15}$$

Finally, the total drag on a plate wetted on both sides can be written as
$2\,D = 2\,b \int_0^l \tau_0\, dx$, i. e.

$$2\,D = 2\,b \sqrt{2\,\alpha_1\beta_1}\, \sqrt{\mu\, \varrho\, l\, U_\infty^3} \,, \tag{10.16}$$

and from eqns. (10.11) and (10.14) we obtain the displacement thickness

$$\delta_1 = \alpha_2 \sqrt{\frac{2\beta_1}{\alpha_1}}\, \sqrt{\frac{\nu x}{U_\infty}} \,. \tag{10.17}$$

A comparison of the approximate expressions for the boundary-layer thickness,
for the shearing stress at the wall, and for drag with the respective formulae of
the accurate theory, eqns. (7.37), (7.31) and (7.33), shows that the use of the integral
momentum equation leads in all cases to a perfectly correct formulation of the
equations. In other words, the dependence of these quantities on the current length, x,
the free-stream velocity, U_∞, and the coefficient of kinematic viscosity, ν, is correctly
deduced. Furthermore, the relation between momentum thickness and shearing
stress at the wall given by eqn. (10.5) can also be deduced from the approximate
calculation, as is easily verified. The still-unknown coefficients α_1, α_2 and β_1 can only

be calculated if a specific assumption regarding the velocity profile is made, i. e. if the function $f(\eta)$ from eqn. (10.6) is given explicitly.

When writing down an expression for $f(\eta)$, it is necessary to satisfy certain boundary conditions for $u(y)$, i. e. for $f(\eta)$. At least the no-slip condition $u = 0$ at $y = 0$ and the condition of continuity when passing from the boundary-layer profile to the potential velocity, $u = U$ at $y = \delta$, must be satisfied. Further conditions might include the continuity of the tangent and curvature at the point, where the two solutions are joined. In other words, we may seek to satisfy the conditions $\partial u/\partial y = 0$ and $\partial^2 u/\partial y^2 = 0$ at $y = \delta$. In the case of a plate the condition that $\partial^2 u/\partial y^2 = 0$ at $y = 0$ is also of importance, and it can be seen from eqn. (7.15) that it is satisfied by the exact solution.

Numerical examples:

We now propose to test the usefulness of the preceding approximate method with the aid of several examples. The quality of the result depends to a great extent on the assumption which is made for the velocity function (10.6). In any case, as already mentioned, the function $f(\eta)$ must vanish at $\eta = 0$ in view of the no-slip condition at the wall. Moreover, for large values of η we must have $f(\eta) = 1$. If only a rough approximation is desired, the transition to the value $f(\eta) = 1$ may occur with a discontinuous first derivative. For a better approximation, continuity in $df/d\eta$ may be postulated. Independently of the particular assumption for $f(\eta)$ the quantities

$$\delta_1 \sqrt{\frac{U_\infty}{\nu x}}\,, \quad \delta_2 \sqrt{\frac{U_\infty}{\nu x}}\,, \quad \frac{\tau_0}{\mu\, U_\infty} \sqrt{\frac{\nu x}{U_\infty}}\,, \quad c_f \sqrt{\frac{U_\infty l}{\nu}}$$

must be pure numbers. They can be easily calculated from eqns. (10.8) to (10.17).

Fig. 10.2. Velocity distribution in the boundary layer on a flat plate at zero incidence

(1) Linear approximation
(2) Cubic approximation from Table 10.1

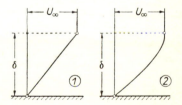

Table 10.1 contains results of several calculations with alternative velocity-distribution functions. The first two functions are illustrated with the aid of Fig. 10.2. The linear function satisfies only the conditions $f(0) = 0$ and $f(1) = 1$, whereas the cubic function satisfies in addition the conditions $f'(1) = 0$ and $f''(0) = 0$; finally, a fourth degree polynomial can be made to satisfy the additional condition $f''(1) = 0$. The sine function satisfies the same boundary conditions as the polynomial of fourth degree, except for $f''(1) = 0$. The polynomials of third and fourth degree and the sine-function lead to values of shearing stress at the wall which are in error by less than 3 per cent and may be considered entirely adequate. The values of the displacement thickness δ_1 show acceptable agreement with the corresponding exact values.

Table 10.1.   Results of the calculation of the boundary layer for a flat plate at zero incidence based on approximate theory

	Velocity distribution $u/U = f(\eta)$	α_1	α_2	β_1	$\delta_1\sqrt{\dfrac{U_\infty}{\nu\,x}}$	$\dfrac{\tau_0}{\mu\,U_\infty}\sqrt{\dfrac{\nu\,x}{U_\infty}}$	$c_f\left(\dfrac{U_\infty l}{\nu}\right)^{\frac{1}{2}}$	$\dfrac{\delta_1}{\delta_2} = H_{12}$
1	$f(\eta) = \eta$	$\dfrac{1}{6}$	$\dfrac{1}{2}$	1	1·732	0·289	1·155	3·00
2	$f(\eta) = \tfrac{3}{2}\,\eta - \tfrac{1}{2}\,\eta^3$	$\dfrac{39}{280}$	$\dfrac{3}{8}$	$\dfrac{3}{2}$	1·740	0·323	1·292	2·70
3	$f(\eta) = 2\eta - 2\eta^3 + \eta^4$	$\dfrac{37}{315}$	$\dfrac{3}{10}$	2	1·752	0·343	1·372	2·55
4	$f(\eta) = \sin\left(\dfrac{\pi}{2}\,\eta\right)$	$\dfrac{4-\pi}{2\,\pi}$	$\dfrac{\pi-2}{\pi}$	$\dfrac{\pi}{2}$	1·741	0·327	1·310	2·66
5	exact	—	—	—	1·721	0·332	1·328	2·59

$$\delta_2\sqrt{\frac{U_\infty}{\nu\,x}} = \frac{2\,\tau_0}{\mu\,U_\infty}\sqrt{\frac{\nu x}{U_\infty}}\,; \qquad c_f\left(\frac{U_\infty l}{\nu}\right)^{\frac{1}{2}} = 2\,\delta_2\sqrt{\frac{U_\infty}{\nu\,x}}$$

It is seen that the approximate method leads to satisfactory results in the case of a flat plate at zero incidence, and the extraordinary simplicity of the calculation is quite remarkable, compared with the complexity of the exact solution.

b. The approximate method due to Th. von Kármán and K. Pohlhausen for two-dimensional flows

We now propose to develop the approximate method of the preceding section so that it can be applied to the general problem of a two-dimensional boundary layer with pressure gradient. The method in its original form was first indicated by K. Pohlhausen [15]. The succeeding description of the method is based on its more modern form as developed by H. Holstein and T. Bohlen [5]. We now choose, as before, a system of coordinates in which x denotes the arc measured along the wetted wall and where y denotes the distance from the wall. The basic equation of the momentum theory is obtained by integrating the equation of motion with respect to y from the wall at $y = 0$ to a certain distance $h(x)$ which is assumed to be outside the boundary layer for all values of x. With this notation the momentum equation has the form already given in (8.32), namely

$$U^2\frac{d\delta_2}{dx} + (2\,\delta_2 + \delta_1)\,U\,\frac{dU}{dx} = \frac{\tau_0}{\varrho}. \tag{10.18}$$

This equation gives an ordinary differential equation for the boundary-layer thickness, as was the case with the flat plate in the preceding section, provided that a suitable form is assumed for the velocity profile. This allows us to calculate the momentum thickness, the displacement thickness, and the shearing stress at the wall. In choosing a suitable velocity function it is necessary to take into account the same considerations as before, namely those regarding the no-slip condition at the wall, as well as the requirements of continuity at the point where this solution is joined to the potential solution. Furthermore, in the presence of a pressure gradient the function must admit the existence of profiles with and without a point of inflexion corresponding to their occurrence in regions of negative or positive pressure gradients. In order to be in a position to calculate the point of separation with the aid of the approximate method the existence of a profile with zero gradient at the wall $(\partial u/\partial y)_{y=0}=0$ must also be possible. On the other hand functions postulating similarity of velocity profiles for various values of x may no longer be prescribed. Following K. Pohlhausen we assume a polynomial of the fourth degree for the velocity function in terms of the dimensionless distance from the wall $\eta = y/\delta(x)$, i. e. we put

$$\frac{u}{U} = f(\eta) = a\,\eta + b\,\eta^2 + c\,\eta^3 + d\,\eta^4 \tag{10.19}$$

in the range $0 \leq \eta \leq 1$, whereas for $\eta > 1$ we assume simply $u/U = 1$. We further demand, as before, that the boundary layer should join the potential flow at the finite distance from the wall $y = \delta(x)$.

In order to determine the four free constants, $a\ b,\ c,\ d$, we shall prescribe the following four boundary conditions

$$
\begin{aligned}
&y = 0: &&\nu\,\frac{\partial^2 u}{\partial y^2} = \frac{1}{\varrho}\frac{dp}{dx} = -\,U\,\frac{dU}{dx}\;; \\
&y = \delta: \quad u = U\;; &&\frac{\partial u}{\partial y} = 0,\quad \frac{\partial^2 u}{\partial y^2} = 0\,.
\end{aligned}
\right\} \tag{10.20}
$$

As seen from eqns. (7.10) to (7.12), they are all satisfied by the exact solution.

These requirements are sufficient to determine the constants $a,\ b,\ c,\ d$, because the no-slip condition at the wall is implicit in eqn. (10.19). The first condition which is satisfied by all exact solutions, as seen from eqn. (7.15), is of particular importance. It determines the curvature of the velocity profile near the wall and makes sure that there is no point of inflexion in the velocity profile in regions of decreasing pressure. Furthermore, regions of increasing pressure contain points of inflexion as required by the exact solution in Chap. VII., Figs. 7.3 and 7.4. Introducing the dimensionless quantity

$$\varLambda = \frac{\delta^2}{\nu}\frac{dU}{dx}\,, \tag{10.21}$$

we obtain the following expressions for the coefficients in eqn. (10.19):

$$a = 2 + \frac{\varLambda}{6}\;; \qquad b = -\frac{\varLambda}{2}\;; \qquad c = -2 + \frac{\varLambda}{2}\;; \qquad d = 1 - \frac{\varLambda}{6}\;;$$

and hence for the velocity profile:

$$\frac{u}{U} = F(\eta) + \varLambda\,G(\eta) = (2\,\eta - 2\,\eta^3 + \eta^4) + \frac{\varLambda}{6}\,(\eta - 3\,\eta^2 + 3\,\eta^3 - \eta^4), \tag{10.22}$$

where

$$F(\eta) = 2\,\eta - 2\,\eta^3 + \eta^4 = 1 - (1 - \eta)^3\,(1 + \eta)\,,$$
$$G(\eta) = \tfrac{1}{6}\,(\eta - 3\,\eta^2 + 3\,\eta^3 - \eta^4) = \tfrac{1}{6}\,\eta(1 - \eta)^3\,. \qquad \Big\} \qquad (10.23)$$

It is easily recognized that the velocity profiles expressed in terms of $\eta = y/\delta(x)$ constitute a one-parameter family of curves, the dimensionless quantity Λ being a shape factor. The dimensionless quantity Λ which may also be written as

$$\Lambda = \frac{\delta^2}{\nu}\,\frac{dU}{dx} = -\frac{dp}{dx}\,\frac{\delta}{\mu U/\delta}$$

can be interpreted physically as the ratio of pressure forces to viscous forces. In order to obtain a quantity to which real physical significance can be ascribed, it would be necessary to replace δ in the above definition by a linear quantity which itself possesses physical significance, such as the momentum thickness. This will be done later in this section.

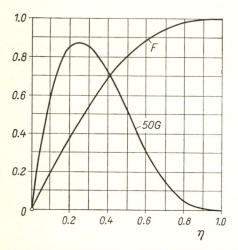

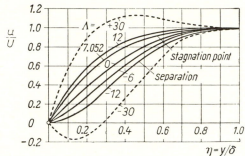

Fig. 10.3. The functions $F(\eta)$ and $G(\eta)$ for the velocity distribution in the boundary layer from eqns. (10.22) and (10.23)

Fig. 10.4. The one-parameter family of velocity profiles from eqn. (10.22)

The two functions $F(\eta)$ and $G(\eta)$ defined by eqn. (10.23), which together compose the velocity-distribution function given in eqn. (10.22), are seen plotted in Fig. 10.3. Velocity profiles for various values of Λ are shown in Fig. 10.4. The profile which corresponds to $\Lambda = 0$ is obtained when $dU/dx = 0$, i. e. for the boundary layer with no pressure gradient (flat plate at zero incidence), or for a point where the velocity of the potential flow passes through a minimum or a maximum. In this case the velocity profile becomes identical with the fourth-degree polynomial used for the flat plate in the preceding section. The profile at separation with $(\partial u/\partial y)_0 = 0$, i. e. with $a = 0$, occurs for $\Lambda = -12$. It will be shown later that the profile at the stagnation point corresponds to² $\Lambda = 7{\cdot}052$. For $\Lambda > 12$ values $u/U > 1$ occur in the boundary layer, but this must be excluded in steady flow. Since behind the point of separation the present calculation based, as it is, on the boundary-layer concept, loses significance, the shape factor is seen to be restricted to the range $-12 \leq \Lambda \leq +12$, see Fig. 10.4.

Before proceeding to calculate the boundary-layer thickness $\delta(x)$ from the momentum theorem, it is now convenient to calculate the momentum thickness, δ_2, the displacement thickness, δ_1, and the viscous shearing stress at the wall, τ_0, with the aid of the approximate velocity profile in the same way as was done for the flat plate at zero incidence in the preceding section. Thus we obtain from eqns. (8.33) and (8.31), together with eqn. (10.22),

$$\frac{\delta_1}{\delta} = \int_{\eta=0}^{1} [1 - F(\eta) - \Lambda\,G(\eta)]\,\mathrm{d}\eta\,,$$

$$\frac{\delta_2}{\delta} = \int_{\eta=0}^{1} [F(\eta) + \Lambda\,G(\eta)]\,[1 - F(\eta) - \Lambda\,G(\eta)]\,\mathrm{d}\eta\,.$$

Computing the definite integrals with the aid of the values of $F(\eta)$ and $G(\eta)$ from eqn. (10.23), we have

$$\frac{\delta_1}{\delta} = \frac{3}{10} - \frac{\Lambda}{120}\,; \quad \frac{\delta_2}{\delta} = \frac{1}{63}\left(\frac{37}{5} - \frac{\Lambda}{15} - \frac{\Lambda^2}{144}\right). \tag{10.24}$$

Similarly, the viscous stress at the wall, $\tau_0 = \mu\,(\partial u/\partial y)_{y=0}$, is given by

$$\frac{\tau_0\,\delta}{\mu\,U} = 2 + \frac{\Lambda}{6}\,. \tag{10.25}$$

In order to determine the still-unknown shape factor $\Lambda(x)$ and, hence, the function $\delta(x)$ from eqn. (10.21), it is now necessary to refer to the momentum equation (10.18). Multiplying by $\delta_2/\nu\,U$ we can represent it in the following dimensionless form:

$$\frac{U\delta_2\delta_2'}{\nu} + \left(2 + \frac{\delta_1}{\delta_2}\right)\frac{U'\delta_2{}^2}{\nu} = \frac{\tau_0\,\delta_2}{\mu\,U}\,, \tag{10.26}$$

in which the boundary-layer thickness δ does not appear explicitly; this circumstance is not surprising, because it constitutes only a fortuitous quantity associated with the approximate method of calculation and has no particular physical meaning. On the other hand eqn. (10.26) contains the really important physical quantities, viz. the displacement thickness, δ_1, the momentum thickness, δ_2, and the shearing stress at the wall τ_0. It is, therefore, natural to begin with the calculation of δ_2 from the momentum equation (10.26) and to deduce δ from it with the aid of eqn. (10.24). Following H. Holstein and T. Bohlen [5] it is convenient to introduce for this purpose a second shape factor

$$K = \frac{\delta_2{}^2}{\nu}\frac{\mathrm{d}U}{\mathrm{d}x} \tag{10.27}$$

which is connected with the momentum thickness in the same way as the first shape factor, Λ, was connected with the boundary-layer thickness, δ, in eqn. (10.21). In addition we shall put

$$Z = \frac{\delta_2{}^2}{\nu} , \tag{10.28}$$

so that

$$K = Z \frac{dU}{dx} . \tag{10.29}$$

It is seen from eqns. (10.21), (10.27) and (10.24) that the shape factors Λ and K satisfy the universal relation

$$K = \left(\frac{37}{315} - \frac{1}{945} \Lambda - \frac{1}{9072} \Lambda^2 \right)^2 \Lambda . \tag{10.30}$$

Denoting

$$H_{12} = \frac{\delta_1}{\delta_2} = \frac{\frac{3}{10} - \frac{1}{120} \Lambda}{\frac{37}{315} - \frac{1}{945} \Lambda - \frac{1}{9072} \Lambda^2} = f_1(K) , \tag{10.31}†$$

and

$$\frac{\tau_0 \delta_2}{\mu U} = \left(2 + \frac{1}{6} \Lambda \right) \left(\frac{37}{315} - \frac{1}{945} \Lambda - \frac{1}{9072} \Lambda^2 \right) = f_2(K) \tag{10.32}$$

for the sake of brevity, and substituting, K and Z from eqns. (10.27) and (10.28) respectively, together with $f_1(K)$ and $f_2(K)$ from eqns. (10.31) and (10.32), we obtain, further, from the momentum equation (10.26) together with $\delta_2 \, \delta_2'/\nu = \frac{1}{2} dZ/dx$, the relation

$$\frac{1}{2} U \frac{dZ}{dx} + [2 + f_1(K)] K = f_2(K) . \tag{10.33}$$

Finally, we introduce the additional abbreviation

$$2 f_2(K) - 4 K - 2 K f_1(K) = F(K) \tag{10.34}$$

or, written out fully,

$$F(K) = 2 \left(\frac{37}{315} - \frac{1}{945} \Lambda - \frac{1}{9072} \Lambda^2 \right) \left[2 - \frac{116}{315} \Lambda + \left(\frac{2}{945} + \frac{1}{120} \right) \Lambda^2 + \frac{2}{9072} \Lambda^3 \right], \tag{10.35}$$

where the relation between Λ and K was given in eqn. (10.30). With all these abbreviations and substitutions the momentum equation (10.33) can now be rewritten in the very condensed form

$$\boxed{\frac{dZ}{dx} = \frac{F(K)}{U} ; \quad K = Z \, U' .} \tag{10.36}$$

---

† The quantity $H_{12} = \delta_1/\delta_2$ is also regarded as a shape factor; it is of particular importance for the turbulent boundary layer, cf. Chap. XXII. Its value for laminar boundary layers ranges from about 2·3 to 3·5, cf. Table 10.2; it assumes values from about 1·3 to 2·2 in the case of turbulent boundary layers. At the point of transition H_{12} increases considerably, cf. Fig. 16.5.

This is a non-linear differential equation of the first order for $Z = \delta_2{}^2/\nu$ as a function of the current length coordinate, x. The fact that the form of the function $F(K)$ is very complex does not constitute a real difficulty insofar as the solution of eqn. (10.36) is concerned, because it is a universal function, i. e. one which is independent of the shape of the body and it can, therefore, be calculated once and for all. The functions $K(\Lambda)$ from eqn. (10.30), as well as $f_1(K)$, $f_2(K)$, and $F(K)$ from eqns. (10.31), (10.32), and (10.35), respectively, are given in Table 10.2. The auxiliary function $F(K)$ is represented graphically in Fig. 10.6.

Solution of the differential equation for momentum thickness: Concerning the solution of eqn. (10.36) it is possible to make the following remarks: The calculation should begin at the stagnation point $x = 0$, where $U = 0$ and dU/dx is finite and different from zero, unless the body possesses there a sharp cusped edge with zero angle. The initial slope of the integral curve dZ/dx would become infinite at the upstream stagnation point were it not for the fact that $F(K)$ vanishes there simultaneously. Thus the function $F(K)$ is seen to have a physically meaningful initial value. The zero of $F(K)$ occurs for values of Λ for which the second bracketed term on the right-hand side of eqn. (10.35) vanishes. Thus

$$F(K) = 0 \text{ for } K = K_0 = 0 \cdot 0770, \text{ or for } \Lambda = \Lambda_0 = 7 \cdot 052 \ .$$

Hence $\Lambda = 7 \cdot 052$ is the value of the first shape factor at the stagnation point, as already mentioned. In this manner the initial slope of the integral curve at the upstream stagnation point is seen to be of the indeterminate form $\frac{0}{0}$ (singular point of eqn. (10.36)), but its value can be easily computed by a simple process of going over to the limit. We obtain

$$Z_0 = \frac{K_0}{U_0{}'} = \frac{0 \cdot 0770}{U_0{}'} \ ; \quad \left(\frac{dZ}{dx}\right)_0 = - 0 \cdot 0652 \, \frac{U_0{}''}{U_0{}'^2} \ . \tag{10.36a}$$

Here the subscript 0 refers to the upstream stagnation point. With these initial values the equation can be conveniently integrated, e. g. by the method of isoclines. Figure 10.5 illustrates the use of this method as applied to a symmetrical aerofoil at zero incidence. The calculation begins with the values $\Lambda_0 = 7 \cdot 052$ and $K_0 = 0 \cdot 0770$ at the leading-edge stagnation point, and becomes completed upon reaching the point of separation with $\Lambda = - 12$ and $K = - 0 \cdot 1567$. The velocity function $U(x)$, together with its first derivative dU/dx, is given by the potential-flow solution. The value of d^2U/dx^2 is only required at the leading edge for the initial slope of the integral curve.

The procedure used in the computation may be summarized as follows:

1. The potential flow function $U(x)$, together with its derivative dU/dx, are given in terms of the arc length.

2. Integration of eqn. (10.36) gives $Z(x)$ and the second shape factor $K(x)$ so that the momentum thickness $\delta_2(x)$ can be calculated from equation (10.27), and the position of the point of separation may be found subsequently.

Table 10.2.　Auxiliary functions for the approximate calculation of laminar boundary layers, after Holstein and Bohlen [5]

Λ	K	$F(K)$	$f_1(K) = \dfrac{\delta_1}{\delta_2} = H_{12}$	$f_2(K) = \dfrac{\delta_2 \tau_0}{\mu U}$
15	0·0884	−0·0658	2·279	0·346
14	0·0928	−0·0885	2·262	0·351
13	0·0941	−0·0914	2·253	0·354
12	0·0948	−0·0948	2·250	0·356
11	0·0941	−0·0912	2·253	0·355
10	0·0919	−0·0800	2·260	0·351
9	0·0882	−0·0608	2·273	0·347
8	0·0831	−0·0335	2·289	0·340
7·8	0·0819	−0·0271	2·293	0·338
7·6	0·0807	−0·0203	2·297	0·337
7·4	0·0794	−0·0132	2·301	0·335
7·2	0·0781	−0·0051	2·305	0·333
7·052	0·0770	0	2·308	0·332
7	0·0767	0·0021	2·309	0·331
6·8	0·0752	0·0102	2·314	0·330
6·6	0·0737	0·0186	2·318	0·328
6·4	0·0721	0·0274	2·323	0·326
6·2	0·0706	0·0363	2·328	0·324
6	0·0689	0·0459	2·333	0·321
5	0·0599	0·0979	2·361	0·310
4	0·0497	0·1579	2·392	0·297
3	0·0385	0·2255	2·427	0·283
2	0·0264	0·3004	2·466	0·268
1	0·0135	0·3820	2·508	0·252
0	0	0·4698	2·554	0·235
− 1	−0·0140	0·5633	2·604	0·217
− 2	−0·0284	0·6609	2·647	0·199
− 3	−0·0429	0·7640	2·716	0·179
− 4	−0·0575	0·8698	2·779	0·160
− 5	−0·0720	0·9780	2·847	0·140
− 6	−0·0862	1·0877	2·921	0·120
− 7	−0·0999	1·1981	2·999	0·100
− 8	−0·1130	1·3080	3·085	0·079
− 9	−0·1254	1·4167	3·176	0·059
−10	−0·1369	1·5229	3·276	0·039
−11	−0·1474	1·6257	3·383	0·019
−12	−0·1567	1·7241	3·500	0
−13	−0·1648	1·8169	3·627	−0·019
−14	−0·1715	1·9033	3·765	−0·037
−15	−0·1767	1·9820	3·916	−0·054

3. The variation of the first shape factor $\Lambda(x)$ is obtained from eqn. (10.30) and Table 10.2.

4. The displacement thickness, δ_1, and the shearing stress at the wall, τ_0, are found from eqn. (10.31) and (10.32), respectively, together with the values in Table 10.2.

5. The boundary-layer thickness $\delta(x)$ follows from eqn. (10.24).

6. Finally, the velocity distribution is found from eqn. (10.22).

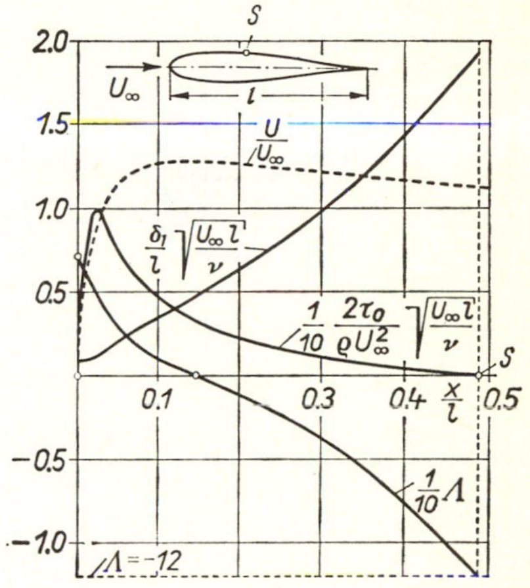

Fig. 10.5. Example of the calculation of the boundary layer by the approximate method due to Pohlhausen and Holstein-Bohlen [8]. Solution of the differential equation (10.36) by the methods of isoclines for the symmetrical Zhukovskii aerofoil J 015 at an incidence angle $\alpha = 0$. See also Fig. 10.12

S = point of separation

A. Walz [25] pointed out that eqn. (10.36) can be reduced to a simple quadrature by the introduction of a further approximation without any appreciable loss of accuracy. He found that the function $F(K)$ can be approximated quite closely by the straight line

$$F(K) = a - b\,K\,.$$

With $a = 0\cdot470$ and $b = 6$ the approximation is particularly close between the stagnation point and the point of maximum velocity (Fig. 10.6) .In this manner eqn. (10.36) reduces to

$$U\,\frac{\mathrm{d}Z}{\mathrm{d}x} = a - b\,K$$

or, substituting the original values for Z and K,

$$\frac{\mathrm{d}}{\mathrm{d}x}\left(\frac{U\delta_2{}^2}{\nu}\right) = a - (b-1)\frac{U\delta_2{}^2}{\nu}\frac{1}{U}\frac{\mathrm{d}U}{\mathrm{d}x}\,.$$

This differential equation for $U\,\delta_2{}^2/\nu$ can be integrated explicitly to

$$\frac{U\delta_2{}^2}{\nu} = \frac{a}{U^{b-1}}\int_0^x U^{b-1}\,\mathrm{d}x\,,$$

or, using the numerical values for a and b given earlier:

$$\frac{U\delta_2{}^2}{\nu} = \frac{0 \cdot 470}{U^5} \int_{x=0}^{x} U^5 \, dx \,.$$

(10.37)

Thus the solution of eqn. (10.36) is seen to reduce to a simple quadrature. An analogous quadrature will be used in Chap. XXII for the solution of the equations of turbulent flow.

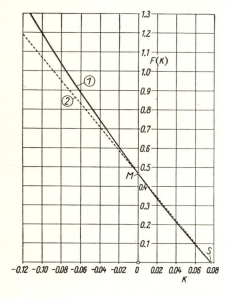

Fig. 10.6. The auxiliary function $F(K)$ for the calculation of laminar boundary layer by the method of Holstein and Bohlen [5]

(1) using eqn. (10.35);
(2) linear approximation $F(K) = 0 \cdot 470 - 6\,K$;
S = stagnation point;
M = velocity maximum

c. Comparison between the approximate and exact solutions

1. Flat plate at zero incidence. It is easy to see from eqn. (10.22) that the Pohlhausen approximation becomes equivalent to Example 3 in Table 10.1 for the case of a flat plate at zero incidence. This case can also be obtained directly from eqn. (10.36), where $U(x) = U_\infty$, $U' \equiv 0$ and hence $K = \Lambda \equiv 0$, so that eqn. (10.36) gives $dZ/dx = F(0)/U_\infty = 0 \cdot 4698/U_\infty$. Taking into account that $Z = 0$ at $x = 0$ it follows that $Z = 0 \cdot 4698\, x/U_\infty$, or $\delta_2 = 0 \cdot 686 \sqrt{\nu\, x/U_\infty}$ in agreement with Table 10.1. Table 10.1 contains exact and approximate values of the boundary-layer parameters for the purpose of comparison. It is seen that agreement is very satisfactory.

2. Two-dimensional stagnation flow. The exact solution of the problem of two-dimensional stagnation flow for which $U(x) = U' \cdot x$, was given in Sec. V 9. The exact values of displacement thickness, momentum thickness and shearing stress at the wall, calculated with the aid of that theory, are given in Table 10.3.

Table 10.3.   Comparison of exact and approximate values of the boundary-layer parameters for the case of *two-dimensional stagnation flow*

	$\delta_1 \sqrt{\dfrac{U'}{\nu}}$	$\delta_2 \sqrt{\dfrac{U'}{\nu}}$	$\dfrac{\tau_0}{\mu\,U} \sqrt{\dfrac{\nu}{U'}}$	$H_{12} = \dfrac{\delta_1}{\delta_2}$
Approximate method due to K. Pohlhausen	0·641	0·278	1·19	2·31
exact solution	0·648	0·292	1·233	2·21

In the approximate method we have $Z_0 = K_0/U'$, and from eqn. (10.36) it follows that the momentum thickness is given by $\delta_2 \sqrt{U'/\nu} = \sqrt{K_0} = \sqrt{0\cdot0770} = 0\cdot278$. It is seen from eqn. (10.31) that the displacement thickness is approximated by $\delta_1 \sqrt{U'/\nu} = f_1(K_0) \sqrt{K_0} = 0\cdot641$ and eqn. (10.32) gives $\tau_0/\mu\,U \cdot \sqrt{\nu/U'} = f_2(K_0)/\sqrt{K_0} = 0\cdot332/0\cdot278 = 1\cdot19$ for the shearing stress at the wall. The agreement between the approximate and exact values is here also completely satisfactory.

3. Flow past a circular cylinder. A comparison of the result of the approximate calculation for a circular cylinder with the solution due to Hiemenz (Sec. IX c) was given by K. Pohlhausen [15] in his original paper. He used Hiemenz's experimental pressure distribution function for the circular cylinder and compared the results with Hiemenz's solution which takes into account only the first three terms of the Blasius series. Hiemenz's solution indicates that separation occurs at an angle $\phi = 82\cdot0°$, whereas Pohlhausen's approximate value was $\phi = 81\cdot5°$. However, the approximate method leads to values for the boundary-layer thickness near the point of separation which are considerably larger than the values obtained by Hiemenz. On the other hand it must be realized that such a comparison is not conclusive, because a Blasius series containing only three terms is in itself inadequate to represent the solution near the point of separation.

We now propose to give a comparison between a set of calculations obtained with the aid of Pohlhausen's approximate method and numerical calculations which have been performed with great accuracy on a digital computer programmed to solve the differential equations directly. The example chosen for comparison is that of a circular cylinder in the presence of a free-stream velocity computed from potential theory, the boundary-layer velocities having been calculated with a Blasius series containing terms up to x^{11} (Sec. IX c). This comparison shows that the power-series method gives very high accuracy up to the immediate vicinity of the point of separation, However, at the point of separation itself, the series broken off at the term x^{11} becomes inaccurate. Figure 10.7 shows a plot of the boundary-layer parameters, displacement thickness, δ_1, momentum thickness, δ_2, and wall shearing stress, τ_0. It is seen that the recent numerical calculations performed by W. Schoenauer [20] show somewhat different trends in the vicinity of the point of separation as far as the variations in the displacement and momentum thicknesses as well as in the shearing stress are concerned, and predict an earlier point of separation. W. Schoenauer found that the separation angle is at $\phi_S = 104 \cdot 5°$ as against $\phi_S = 109 \cdot 5°$ obtained with the aid of the Pohlhausen approximation and $\phi_S = 108 \cdot 8°$ suggested by the series expansion continued up to the term x^{11}. A comparison between the velocity distributions, Fig. 10.8, leads to the conclusion that there exists almost perfect agreement between the exact solution and the approximation in the range of angles $0 < \phi < 90°$, that is in the range of accelerated external flow. By contrast, downstream of the pressure minimum the discrepancies increase very fast on approaching the point of separation.

No general criterion regarding the admissibility of the approximation has been given so far, and it seems that it will be difficult to obtain. Judging by the above and similar calculations as well as by experimental results it appears, however, to be reasonably certain that Pohlhausen's approximate method leads to very satisfactory results in regions of accelerated potential flow. Similarly, it may be

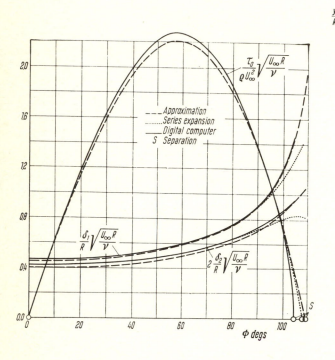

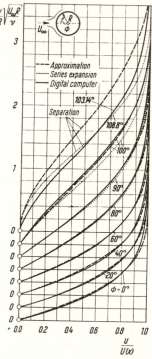

Fig. 10.7. Comparison of Pohlhausen's approximate solution with the exact solution for the case of a circular cylinder

δ_1 = displacement thickness;
δ_2 = momentum thickness;
τ_0 = shearing stress at the wall

Fig. 10.8. Comparison of Pohlhausen's approximate solution with the exact solution for the case of a circular cylinder; velocity profiles

stated that in regions of retarded potential flow the approximate solution becomes somewhat inaccurate as the point of separation is approached. The position of the point of separation can only be calculated with a certain degree of uncertainty, particularly in cases when the point of separation is situated comparatively far behind the point of minimum pressure†‡.

From the assumption that the velocity profiles constitute a one-parameter family it necessarily follows that the point of separation is determined solely by the value of this parameter. It was, however, shown by I. Tani [22] that the position of the point of separation depends, in addition, on the pressure gradient of the external flow.

d. Further examples

In this section we propose to summarize some examples illustrating the calculation of boundary layers by the preceding approximate methods which were first given in a paper by H. Schlichting and A. Ulrich [19]. The first set of examples is concerned with elliptical cylinders whose major axes are parallel to the direction of the stream. The ratio of the major to the minor axis of the cylinders ranged over $a/b = 1, 2, 4, 8$ and the potential velocity-distribution functions are seen plotted in Figs. 10.9. The value of the velocity maximum is $U_m/U = 1 + b/a$. The characteristic parameter of the boundary layer, namely the displacement thickness, δ_1, the shape factor, Λ, and the shearing stress at the wall, τ_0, are seen plotted in Fig. 10.10. The results for the flat plate at zero incidence have been plotted in the same figure for the purpose of comparison. In the case of a circular cylinder separation occurs at $x/l' = 0.609$, i. e. at $\phi = 109.5°$, as already mentioned, $(2\,l' = $ circumference) and moves downstream as the ellipse becomes more slender. The position of the point of separation is marked in the velocity profile plots in Fig. 10.9. The results for an ellipse of $a/b = 8$ differ only very little from those for a flat plate at zero incidence. Fig. 10.11 contains velocity profiles for the boundary layer on an elliptic cylinder with $a/b = 4$. Calculations concerning elliptic cylinders whose minor axes are parallel to the direction of the stream as well as ellipsoids of revolution may be found in a paper by J. Pretsch [17].

---

† For example: G. B. Schubauer [21] measured the position of the point of minimum pressure on an elliptical cylinder of slenderness $a : b = 2.96 : 1$ placed in a stream parallel to the major axis. He found that it was located at $x/b = 1.3$ and that separation took place at $x/b = 1.99$. A calculation based on Pohlhausen's approximation showed very good agreement with measurements for velocity profiles up to the point of minimum pressure but it predicted no separation. D. Meksyn [13] developed a method of computation which leads to a value of $x/b = 2.02$ for the point of separation in the above example. In his method, the boundary-layer equations are transformed into ordinary differential equations which are related to Falkner and Skan's equation (9.8).

‡ Here it may be worth mentioning that approximate integration by the method of isoclines in connexion with Pohlhausen's approximation fails in the region of large pressure gradients which occur for $\Lambda > 12$ ($K > 0.095$), because the plot of K against Λ turns at this point (Table 10.2) and cannot, therefore, be continued beyond $K = 0.095$. Moreover, for $\Lambda > 12$ the velocity profiles become unacceptable as they contain points for which $u/U > 1$ (Fig. 10.4). These difficulties are obviated when eqn. (10.37) is used.

A further example is shown in Fig. 10.12 which contains results for a symmetrical Zhukovskii aerofoil at zero incidence. The point of minimum pressure is at $x/l' = 0.141$ which is very far forward on the aerofoil. The pressure rise at the rear is very gradual so that the point of separation lies very far downstream of the point of minimum pressure, i. e. $x/l' = 0.470$. Since the Zhukovskii aerofoil has a cusped trailing edge the potential velocity at the trailing edge is different from zero. For details of additional systematic boundary-layer calculations concerning an extensive series of Zhukovskii aerofoils with different thickness and camber ratios and at different angles of incidence, reference may be made to a paper by K. Bussmann and A. Ulrich [2].

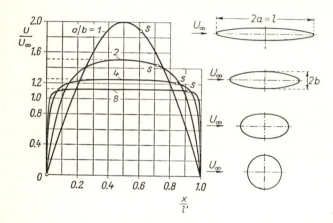

Fig. 10.9. Potential velocity distribution function on elliptical cylinders of slenderness $a/b = 1, 2, 4, 8$, the direction of the stream being parallel to the major axis

S = position of point of separation

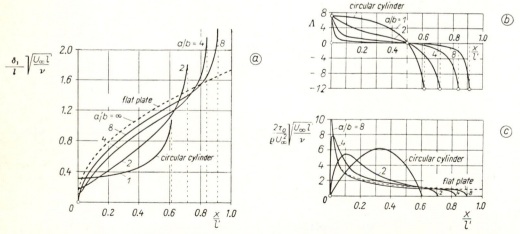

Fig. 10.10. Results of the calculation of boundary layers on elliptical cylinders of slenderness $a/b = 1, 2, 4, 8$, Fig. 10.9. a) displacement thickness of the boundary layer, b) shape factor c) shearing stress at the wall, $2 l'$ = circumference of the ellipse; $a/b = 1$ circular cylinder; $a/b = \infty$ flat plate

A review of the very numerous approximate methods which have been proposed so far is contained in the collective book entitled "Laminar Boundary Layers"
[18] and edited by L. Rosenhead.

In an effort to improve the accuracy of the calculation of laminar boundary
layers, many authors replaced the preceding single-parameter methods by one employing *two parameters*. This is achieved when the energy integral equation is satisfied in addition to the momentum integral equation (see e. g. K. Wieghardt [27]).
Two-parameter methods have been extensively developed by L. G. Loitsianskii and
his coworkers [8, 9, 10, 11, 12, 14]. Since such two-parameter methods are very
complex, and since their accuracy is difficult to assess, modern authors favour *exact
numerical* methods employed in conjunction with large electronic computers; their
principles have been outlined in Sec. IX i.

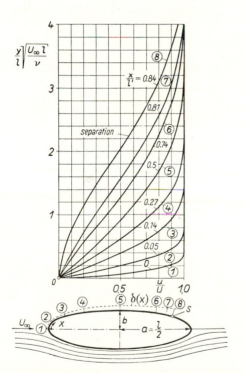

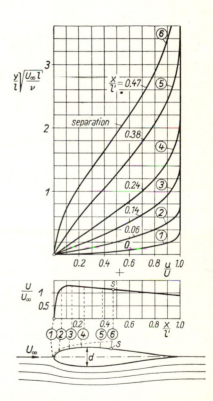

Fig. 10.11. Velocity profiles in the laminar
boundary layer on an elliptical cylinder.
Ratio of axes $a/b = 4$

Fig. 10.12. Velocity profiles in the laminar
boundary layer and potential velocity function for a Zhukovskii aerofoil J 015 of thickness ratio $d/l = 0.15$ at an angle of incidence
$\alpha = 0$

e. Laminar flow with adverse pressure gradient; separation

Flows with adverse pressure gradients (retarded flows) are of great practical importance. In this connexion it is always desired to *avoid separation* from the wall, because this phenomenon is associated with large energy losses. The flow about an aerofoil is a case in point. Owing to the fact that on the suction side the pressure must increase to its free-stream value at the trailing edge, the flow is always likely to separate. The flow in a divergent channel (diffuser) affords another example. The object in using this shape of channel is to convert kinetic energy into pressure energy, and if the angle of divergence is made too large, separation may occur.

Theoretical investigations on the behaviour of the boundary layer in the vicinity of the point of separation have been carried out by S. Goldstein [4] and B. S. Stratford [21a]. *Cf.* the review by S. N. Brown and K. Stewartson [1].

Observations show that a laminar boundary layer which separates from a wall frequently becomes reattached to it, having first become turbulent. This leads to the creation of a laminar separation bubble, Fig. 10.13b, which places itself between the separation point S and the reattachment point R. The fluid in the bubble performs a circulatory motion. According to Fig. 10.13a, the pressure distribution along the wall can be represented, in simplified fashion, by a constant value between the point of separation S and point T of largest thickness followed by a linear increase from T to the point of reattachment R. Phenomena of this kind have been described in detail by I. Tani [23]. More recent experimental investigations into the nature of laminar separation bubbles have been performed by A. D. Young et al. [28] as well as by M. Gaster [4a] and J. L. Van Ingen [6]. For theoretical contributions see [2b, 3a, 5a].

It will now be shown with the aid of several examples that a laminar flow can only support very small adverse pressure gradients without separation. Adverse pressure gradients which exist in practical applications would, therefore, almost always lead to separation if the flow were laminar. The circumstance that real flows can support considerable rates of pressure increase in a large number of cases without separation is due to the fact that the flow is mostly *turbulent*. It will be seen later that turbulent flows are capable of overcoming much larger adverse pressure gradients without separation. The best known examples include the cases of flow past circular cylinders and spheres, when separation occurs much further upstream in laminar than in turbulent flow. In practice when adverse pressure gradients exist the flow is almost always turbulent because, in addition, the existence of an adverse pressure gradient favours the transition from laminar to turbulent flow. It is, nevertheless, useful to clarify some of the fundamental relations associated with the prevention of separation on the example of laminar flow, in particular, because laminar flows are much more readily amenable to mathematical treatment than is the case with turbulent flows.

There are several methods of preventing separation. The simplest of them consists in arranging for the adverse pressure gradients to remain below the limit for which separation

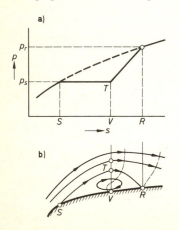

Fig. 10.13. Separation bubble in a laminar boundary layer after I. Tani [23]. a) Shape of bubble (schematic); b) Pressure distribution in bubble along the wall (schematic). The pressure between S and V in the bubble remains constant at p_s; further downstream the pressure increases to p_r

S = point of separation
R = point of reattachment
$V - T$ = height of bubble

does occur. A numerical example will serve to make this idea clear. Another possibility consists in controlling the boundary layer, e. g. by suction or by injecting fluid into it, or by addition of an aerofoil at a point where its presence favourably affects the boundary layer in critical regions. These methods will be discussed more fully in Chap. XIV.

Following L. Prandtl [16] we shall show how it is possible to estimate the permissible magnitude of the adverse pressure gradient for which separation is just prevented. The argument will be based on the von Kármán-Pohlhausen approximation discussed in Sec. X b. It will be assumed that the boundary layer is acted upon by the pressure distribution determined by the free-stream potential flow up to a point which lies very close to the point of separation, such as point 0 in Fig. 10.14. Starting with this point it will be assumed that the pressure gradient is such that the shape of the velocity profile remains unchanged proceeding downstream, or that, in other words, the shape factor Λ remains constant; since at separation $\Lambda = -12$ a value of $\Lambda = -10$ will be chosen. As seen from Table 10.2 this leads to a definite value for the second shape factor, namely $K = -0.1369$, so that $F(K) = 1.523$. Using these values it is seen from eqns. (10.28) and (10.29) that prevention of separation implies the following relationship between the velocity $U(x)$ of potential flow and the momentum thickness $\delta_2(x)$:

$$\frac{\delta_2^2}{\nu} = Z = \frac{0.1369}{-U'(x)}.$$

It follows that $dZ/dx = 0.1369\, U''/U'^2$, or

$$U \frac{dZ}{dx} = 0.1369 \frac{UU''}{U'^2} = 0.1369\, \sigma\,, \qquad (10.38)$$

where

$$\sigma = \frac{U\, U''}{U'^2}\,. \qquad (10.39)$$

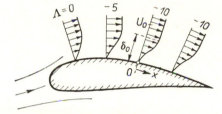

Fig. 10.14. Development of boundary layer in the case when laminar separation is prevented

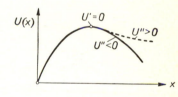

Fig. 10.15. Potential velocity function for a laminar boundary layer with and without separation

On the other hand the succeeding velocity profiles are given by the momentum equation (10.36) for $x > 0$, or

$$U \frac{dZ}{dx} = F(K) = 1.523\,, \qquad (10.40)$$

where the numerical value for $F(K)$ which corresponds to $\Lambda = -10$ has been inserted. From eqns. (10.38) and (10.40) it follows that the value of the shape factor remains constant at $\Lambda = -10$ if $0.1369\, \sigma = 1.523$, or if

$$\sigma = \frac{U\, U''}{U'^2} = 11.13 \approx 11\,, \qquad (10.41)$$

$$\sigma > 11: \text{no separation}; \qquad \sigma < 11: \text{separation}\,. \qquad (10.41a)$$

The preceding argument shows that the boundary layer can support the adverse pressure gradients if $\sigma > 11$, whereas $\sigma < 11$ implies separation. If σ remains constant at $\sigma = 11$, with $\Lambda = -10$, the boundary layer remains on the verge of separation.

Qualitatively it is at once possible to make the following statement regarding the shape of the potential velocity function $U(x)$ which leads to no separation. In view of eqn. (10.41)

$$U'' > 0$$

is a *necessary condition* for a retarded flow ($U' < 0$) to adhere to the wall. In other words, the magnitude of the adverse pressure gradient must decrease in the flow direction. Fig. 10.15. Thus separation will always occur if the function $U(x)$ is curved downwards behind its maximum ($U'' < 0$). In the opposite case, when the velocity function curves upwards ($U'' > 0$), separation may be obviated. Even the limiting case of $U'' = 0$, i. e. the case of a velocity which decreases linearly with the length of arc, always leads to separation. This latter remark agrees with the result found in Sec. IX d; it was concerned with the boundary layer associated with a potential flow velocity which decreased linearly, and the solution of the differential equations was quoted from a paper by L. Howarth. The *sufficient condition* for the absence of separation is given by

$$U'' > 11 \ U'^2/U .$$

We shall now proceed to calculate the potential flow and the variation of boundary-layer thickness which are associated with the limiting case of $\sigma = 11$, when the boundary layer remains on the verge of separation. From eqn. (10.41) we have

$$\frac{U''}{U'} = 11 \times \frac{U'}{U}$$

or, upon integrating: $\ln U' = 11 \ln U + \ln (-C_1')$, i. e. $U'/U^{11} = -C_1'$, where C_1' denotes the constant of integration. Repeated integration gives

$$\frac{1}{10} U^{-10} = C_1' x + C_2 . \tag{10.46}$$

For $x = 0$ we should have $U(x) = U_0$, so that $C_2 = \frac{1}{10}U_0^{-10}$. Putting further $C_1' U_0^{10} = C_1$, we obtain from eqn. (10.41)

$$U(x) = \frac{U_0}{(1 + 10 \ C_1 \ x)^{1/10}} . \tag{10.43}$$

Equation (10.43) represents the potential velocity for which separation can just be avoided. The constant C_1 can be determined from the value of the boundary-layer thickness δ_0 at the origin $x = 0$. We have $\Lambda = U' \delta^2/\nu = -10$ or $\delta = \sqrt{10 \ \nu/(-U')}$. From eqn. (10.43) we obtain

$$U' = - \frac{C_1 \ U_0}{(1 + 10 \ C_1 \ x)^{11/10}} ,$$

and hence

$$\delta = \sqrt{\frac{10 \ \nu}{C_1 \ U_0}} (1 + 10 \ C_1 \ x)^{11/20} .$$

From $\delta = \delta_0$ at $x = 0$ we have $C_1 = 10 \ \nu/U_0 \ \delta_0^2$, which gives the final solution for the potential flow and the variation of boundary-layer thickness

$$U(x) = U_0 \left(1 + 100 \ \frac{\nu x}{U_0 \ \delta_0^2}\right)^{-0 \cdot 1} ; \tag{10.44}$$

$$\delta(x) = \delta_0 \left(1 + 100 \ \frac{\nu x}{U_0 \ \delta_0^2}\right)^{0 \cdot 55} . \tag{10.45}$$

It is seen that the magnitude of the permissible deceleration (decrease in velocity) is very small, being proportional to $x^{-0 \cdot 1}$. Its value is very nearly realized for the case of constant velocity along the flat plate at zero incidence. In the present case the increase in boundary-layer thickness, δ, is proportional to $x^{0 \cdot 55}$; this value also differs but little from the case of a flat plate at zero incidence for which $\delta \sim x^{0 \cdot 5}$.

By way of a further example of retarded flow we shall consider the flow through a divergent channel whose walls are straight. This case is corollary to the case of the boundary layer in a divergent channel treated in Sec. IX b. The flow is seen sketched in Fig. 10.16, where x denotes the radial distance from the source at 0. The wall is assumed to begin at $x = a$ where the entrance velocity of the potential stream is put equal to U_0. The potential flow is given by

$$U(x) = U_0 \, \frac{a}{x} \; ; \quad U'(x) = - U_0 \, \frac{a}{x^2} \; ; \quad U''(x) = 2 \, U_0 \, \frac{a}{x^3} \, . \tag{10.46}$$

Computing the value of the quantity σ from eqn. (10.41), which is decisive for separation, we obtain here $\sigma = 2$. Applying the criterion given in eqn. (10.41a) we conclude that separation occurs in all cases irrespective of the magnitude of the angle of divergence. This example shows very clearly that a laminar stream has only a very limited capacity for supporting an adverse pressure gradient without separation.

According to a calculation performed by K. Pohlhausen [15] the point of separation occurs at $x_s/a = 1 \cdot 21$ and is seen to be independent of the angle of divergence.

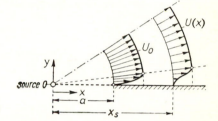

Fig. 10.16.   Laminar boundary layer in a divergent channel. Separation occurs at $x_s/a = 1 \cdot 21$ independently of the angle of divergence

The preceding conclusions apply only as long as the displacement effect of the boundary layer may be neglected. However, this is not the case when the angle of divergence is small. When this angle is small, the boundary layers fill the whole channel cross-section after a certain inlet length (cf. Sec. XI i) and the flow goes over asymptotically to that discussed in Sec. V 12 under the heading of channel flow. When the included angle does not exceed a certain value which depends on the Reynolds number, there is no separation.

Recently, S. N. Brown and K. Stewartson [1] published a summary review on separation in which the mathematical question centered on the singularity which occurs in the differential equations at the critical point has been emphasized. See also the work of S. Goldstein [4]. A more physically inspired review of this problem area has recently been published by J. C. Williams III [29], and by P. K. Chang [2c].

References

[1] Brown, S. N., and Stewartson, K.: Laminar separation. Annual Review of Fluid Mech. *1*, 45—72 (1969).
[2] Bussmann, K., and Ulrich, A.: Systematische Untersuchungen über den Einfluss der Profilform auf die Lage des Umschlagpunktes. Preprint Jb. dt. Luftfahrtforschung 1943 in: Techn. Berichte *10*, No. 9 (1943); NACA TM 1185 (1947).
[2a] Chan, Y. Y.: Loitsianskii's method for boundary layers with suction and injection. AIAA J. *7*, 562—563 (1969).
[2b] Briley, W. R. and McDonald, D. H.: Numerical prediction of incompressible separation bubbles. JFM *69*, 631—656 (1975).
[2c] Chang, P. K.: Separation of flow. Pergamon Press, New York, 1970.
[3] Glauert, M. B., and Lighthill, M. J.: The axisymmetric boundary layer on a long thin cylinder. Proc. Roy. Soc. A *230*, 188—203 (1955).
[3a] Crimi, P., and Reeves, B. L.: Analysis of leading edge separation bubbles on airfoils. AIAA J. *14*, 1548—1555 (1976).
[4] Goldstein, S.: On laminar boundary layer flow near a point of separation. Quart. J. Mech. Appl. Math. *1*, 43—69 (1948).

[4a] Gaster, M.: The structure and behaviour of laminar separation bubbles. AGARD Conf. Proc. *4*, 819—854 (1966).

[5] Holstein, H., and Bohlen, T.: Ein einfaches Verfahren zur Berechnung laminarer Reibungsschichten, die dem Näherungsverfahren von K. Pohlhausen genügen. Lilienthal-Bericht S. 10, 5—16 (1940).

[5a] Horton, H. P.: A semi-empirical theory for the growth and bursting of laminar separation bubbles. Aero. Res. Council, Current Paper No. 107 (1967).

[6] Van Ingen, J. L.: On the calculation of laminar separation bubbles in two-dimensional incompressible flow. AGARD Conf. Proc. Flow Separation, No. 168, 11—1 to 11—16 (1975).

[7] von Kármán, Th.: Über laminare und turbulente Reibung. ZAMM *1*, 233—252 (1921); NACA 1092 (1946); see also Coll. Works *II*, 70—97 (1956).

[8] Kotschin, N. J., and Loitsianskii, L. G.: Über eine angenäherte Methode der Berechnung der Laminargrenzschicht. Dokl. Akad. Nauk, SSSR *36*, No. 9 (1942); see also: An approximate method of calculating the laminar boundary layer. Comptes Rendus (Doklady) de l'Académie des Sciences de l'URSS *46*, 262—266 (1942).

[9] Loitsianskii, L. G.: Laminarnyi pogranichnyi sloi. Fizmatgiz Moscow. Germ. transl. by H. Limberg: Laminare Grenzschichten. Akademie-Verlag, Berlin, 1967.

[10] Loitsianskii, L. G.: Mekhanika zhidkostei i gazov. Nauka, Moscow, 1973.

[11] Loitsianskii, L. G.: Universal'nye uravnenia i parametricheskie priblizhenia v teorii laminarnykh pogranichnykh sloev. Prikl. Mat. i Mekh. *XXIX*, No. 1 (1965). See also: The universal equations and parametric approximations in the theory of laminar boundary layers. J. Appl. Math. Mech. (PMM) *29*, 70—87 (1965).

[12] Loitsianskii, L. G.: Sur la méthode paramétrique de la théorie de la couche limite laminaire. Proc. 11th Intern. Congress Appl. Mech., Munich 1966 (H. Görtler, ed.), Springer Verlag, Berlin, 1966, 722—728.

[13] Meksyn, D.: Integration of the boundary layer equations. Proc. Roy. Soc. A *237*, 543—559 (1956).

[14] Ozerova, E. F., and Simuni, L. M.: Approximate two-parameter solution of the equation for steady-state laminar boundary layers (in Russian). Trudy Leningr. Polyt. Inst. No. 313 (1970).

[15] Pohlhausen, K.: Zur näherungsweisen Integration der Differentialgleichung der laminaren Reibungsschicht. ZAMM *1*, 252—268 (1921).

[16] Prandtl, L.: The mechanics of viscous fluids. In W. F. Durand (ed.): Aerodynamic Theory *III*, 34—208 (1935).

[17] Pretsch, J.: Die laminare Reibungsschicht an elliptischen Zylindern und Rotationsellipsoiden bei symmetrischer Anströmung. Luftfahrtforschung *18*, 397—402 (1941).

[18] Rosenhead, L. (ed.): Laminar boundary layers. Clarendon Press, Oxford, 1963.

[19] Schlichting, H., and Ulrich, A.: Zur Berechnung des Umschlages laminar-turbulent. Jb. dt. Luftfahrtforschung *I*, 8—35 (1942); see also: Lilienthal-Bericht S 10, 75—135 (1940).

[20] Schönauer, W.: Ein Differenzenverfahren zur Lösung der Grenzschichtgleichung für stationäre, laminare, inkompressible Strömung. Ing.-Arch. *33*, 173—189 (1964).

[21] Schubauer, G. B.: Airflow in a separating laminar boundary layer. NACA Rep. 527 (1935).

[21a] Stratford, B. S.: Flow in the laminar boundary layer near separation. ARC, RM 3002, 1—27 (1957).

[22] Tani, I.: On the solution of the laminar boundary layer equations. Fifty years of boundary layer research (H. Görtler, ed.), Braunschweig, 1955, 193—200.

[23] Tani, I.: Low speed flows involving bubble separation. Progress in Aeronautical Sciences *5*, 70—103 (1964).

[24] Truckenbrodt, E.: Näherungslösungen der Strömungsmechanik und ihre physikalische Deutung. Nineteenth Prandtl Memorial Lecture. ZFW *24*, 177—188 (1976).

[25] Walz, A.: Ein neuer Ansatz für das Geschwindigkeitsprofil der laminaren Reibungsschicht. Lilienthal-Bericht 141, 8—12 (1941).

[26] Watson, E. J., and Preston, J. H.: An approximate solution of two flat plate boundary layer problems. ARC RM 2537 (1951).

[27] Wieghardt, K.: Über einen Energiesatz zur Berechnung laminarer Grenzschichten. Ing.-Arch. *16*, 231—242 (1948).

[28] Young, A. D., and Horton, H. P.: Some results of investigations of separation bubbles. AGARD Conf. Proc. Flow Separation 4, Part II, 779—818 (1966).

[29] Williams, J. C. III: Incompressible boundary layer separation. Annual Review of Fluid Mech. *9*, 113—144 (1977).

CHAPTER XI

Axially symmetrical and three-dimensional boundary layers

In the discussion of boundary layers in the preceding chapter we have considered exclusively two-dimensional cases for which the velocity components depended on only two space coordinates. At the same time the velocity component in the direction of the third space coordinate did not exist. The general three-dimensional case of a boundary layer in which the three velocity components depend on all three co-ordinates has, so far, been hardly elaborated because of the enormous mathematical difficulties associated with the problem. We shall describe the first attempts in this direction at the end of the present chapter.

On the other hand the mathematical difficulties encountered in the study of axially symmetrical boundary layers are considerably fewer and hardly exceed those in the two-dimensional case. Axially symmetrical boundary layers occur, e. g., in flows past axially symmetrical bodies; the axially symmetrical jet also belongs under this heading. Two examples, that of the rotating disk and axially symmetrical flow with stagnation, have already been discussed in the chapter on exact solutions of the Navier-Stokes equations.

We shall begin the present chapter with a discussion of some further examples of steady axially symmetrical flows which can be solved with the aid of the differential equations, and will continue with the extension of the approximate procedure which was explained in the preceding chapter to include the axially symmetrical case. Further, we shall discuss the principal features of three-dimensional boundary layers. Non-steady axially symmetrical boundary layers will be considered in Chap. XV together with non-steady two-dimensional examples.

a. Exact solutions for axially symmetrical boundary layers

1. Rotation near the ground. In Chap. V we have considered the case of flow in the neighbourhood of a disk which rotates in a fluid at rest. The case of motion near a stationary wall, when the fluid at a large distance above it rotates at a constant angular velocity, is closely connected with it, Fig. 11.1. This example was studied by U. T. Boedewadt [9]. One of the essential effects in the example of the disk which rotates in a fluid at rest consists in the fact that in the thin layer near the wall the fluid is thrown outwards owing to the existence of centrifugal forces. The fluid which is forced outwards in a radial direction is replaced by a fluid stream in the axial direction. In the case under consideration, in which the fluid rotates over the wall, there is a similar effect but its sign is reversed: the particles which rotate at a large distance from the wall are in equilibrium under the influence of the centrifugal force which is balanced by a radial pressure gradient. The peripheral velocity of the particles near the wall is reduced, thus decreasing materially the centrifugal force, whereas the radial pressure gradient directed towards the axis remains the same. This set of circumstances causes the particles near the wall to flow radially inwards, and for reasons of continuity that motion must be compensated by an axial flow upwards, as shown in Fig. 11.1. A superimposed field of flow of this nature which occurs in the boundary layer and whose direction deviates from that in the external flow is quite generally referred to as a *secondary flow*. It was first discovered by E. Gruschwitz [45] when he analyzed the flow in a curved channel, see also E. Becker [6].

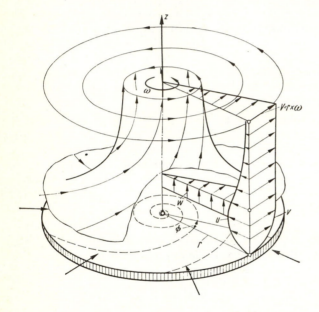

Fig. 11.1. Rotation of flow near the ground

Velocity components: u — radial; v — tangential; w — axial. Owing to friction, the tangential velocity suffers a deceleration in the neighbourhood of the disk at rest. This gives rise to a secondary flow which is directed radially inwards

The secondary flow which accompanies rotation near a solid wall and which has been described in the preceding paragraph can be clearly observed in a teacup: after the rotation has been generated by vigorous stirring and again after the flow has been left to itself for a short while, the radial inward flow field near the bottom will be formed. Its existence can be inferred from the fact that tea leaves settle in a little heap near the centre at the bottom.

In order to formulate the mathematical problem, we shall assume cylindrical polar coordinates r, ϕ, z, the stationary wall being at $z = 0$, see Fig. 11.1. The fluid at a large distance from the wall will be assumed to rotate like a rigid body, with a constant angular velocity ω. We shall denote the velocity component in the radial direction by u, that in the tangential direction by v, the axial component being denoted by w. For reasons of axial symmetry the derivatives with respect to ϕ may be dropped from the Navier-Stokes equations. The solution which we are about to find will be an exact solution of the Navier-Stokes equations, just as was that for the rotating disk, because the terms which are neglected in the boundary-layer equations vanish here on their own accord. By eqn. (3.36) we can write down the Navier-Stokes equations as

$$u \frac{\partial u}{\partial r} + w \frac{\partial u}{\partial z} - \frac{v^2}{r} = -\frac{1}{\varrho} \frac{\partial p}{\partial r} + \nu \left\{ \frac{\partial^2 u}{\partial r^2} + \frac{\partial}{\partial r} \left(\frac{u}{r} \right) + \frac{\partial^2 u}{\partial z^2} \right\}, \qquad (11.1\,a)$$

$$u \frac{\partial v}{\partial r} + w \frac{\partial v}{\partial z} + \frac{u v}{r} = \nu \left\{ \frac{\partial^2 v}{\partial r^2} + \frac{\partial}{\partial r} \left(\frac{v}{r} \right) + \frac{\partial^2 v}{\partial z^2} \right\}, \qquad (11.1\,b)$$

$$u \frac{\partial w}{\partial r} + w \frac{\partial w}{\partial z} = -\frac{1}{\varrho} \frac{\partial p}{\partial z} + \nu \left\{ \frac{\partial^2 w}{\partial r^2} + \frac{1}{r} \frac{\partial w}{\partial r} + \frac{\partial^2 w}{\partial z^2} \right\}, \qquad (11.1\,c)$$

$$\frac{\partial u}{\partial r} + \frac{u}{r} + \frac{\partial w}{\partial z} = 0. \qquad (11.1\,d)$$

The boundary conditions are

$$\left. \begin{array}{llll} z = 0 : & u = 0; & v = 0 ; & w = 0 \\[2mm] z = \infty : & u = 0; & v = r \omega . \end{array} \right\} \qquad (11.2)$$

It is convenient to introduce the dimensionless coordinate

$$\zeta = z \sqrt{\frac{\omega}{\nu}} \qquad (11.3)$$

in place of z, as in the case of the rotating disk (Sec. V 11). We assume that the velocity components have the form

$$u = r \omega F(\zeta); \qquad v = r \omega G(\zeta); \qquad w = \sqrt{\nu \omega} H(\zeta). \qquad (11.4)$$

The radial pressure gradient can be computed for the frictionless flow at a large distance from the wall from the condition: $(1/\varrho) \cdot (\partial p/\partial r) = V^2/r$, or, with $V = r \omega$,

$$\frac{1}{\varrho} \frac{\partial p}{\partial r} = r \omega^2 . \qquad (11.5)$$

In the framework of the boundary-layer theory it is assumed that the same pressure gradient acts in the viscous layer near the wall. Introducing eqns. (11.4) and (11.5) into eqns. (11.1a, b, d), we obtain a system of ordinary differential equations which is analogous to that in Sec. V 11:

$$\left. \begin{array}{ll} F^2 - G^2 + H F' - F'' + 1 = 0 \\[2mm] 2 G F + H G' - G'' \qquad\qquad = 0 \\[2mm] 2 F + H' \qquad\qquad\qquad\quad = 0 \end{array} \right\} \qquad (11.6)$$

with the boundary conditions

$$\left. \begin{array}{llll} \zeta = 0 : & F = 0; & G = 0; & H = 0; \\[2mm] \zeta = \infty : & F = 0; & G = 1 . \end{array} \right\} \qquad (11.7)$$

The pressure gradient in the z-direction may be assumed equal to zero, as such an assumption is compatible with boundary-layer theory. Alternatively, it can be calculated from eqn. (11.1c), after the principal solution had been obtained, which then results in an exact solution of the Navier-Stokes equations.

The system of equations (11.6) with the boundary conditions (11.7) was first solved by U. T. Boedewadt [9] in a very laborious way by means of a power series expansion at $\zeta = 0$ and an asymptotic expansion for $\zeta = \infty$. Recently this solution was corrected by J. E. Nydahl [81a] in an unpublished paper. The values of the functions F, G, H according to Nydahl are given in Table 11.1 and in Fig. 11.2. The horizontal velocity. i.e. the resultant of u and v, is also shown plotted in a polar diagram in Fig. 11.3. The angle between the horizontal velocity component and the peripheral direction depends only on the height, and the vectors in Fig. 11.3 indicate this direction for varying heights. The deviation from the peripheral direction prescribed at a large height is largest near the ground and has a value of 50·6° inwards. The largest deviation of 7·4° outwards occurs for $\zeta = 4\cdot63$ so that the largest angular

Table 11.1. The functions for the velocity distribution for the case of rotation over a stationary wall, after J. E. Nydahl [81a]

ζ	F	G	H
0.0	0.0000	0.0000	0.0000
0.5	−0.3487	0.3834	0.1944
1.0	−0.4788	0.7354	0.6241
1.5	−0.4496	1.0134	1.0987
2.0	−0.3287	1.1924	1.4929
2.5	−0.1762	1.2721	1.7459
3.0	−0.0361	1.2714	1.8496
3.5	0.0663	1.2182	1.8308
4.0	0.1227	1.1413	1.7325
4.5	0.1371	1.0640	1.5995
5.0	0.1210	1.0016	1.4685
5.5	0.0878	0.9611	1.3632
6.0	0.0499	0.9427	1.2944
6.5	0.0162	0.9407	1.2620
7.0	−0.0084	0.9530	1.2585
7.5	−0.0223	0.9693	1.2751
8.0	−0.0268	0.9857	1.3004
8.5	−0.0243	0.9991	1.3264
9.0	−0.0179	1.0078	1.3477
9.5	−0.0102	1.0119	1.3617
10.0	−0.0033	1.0121	1.3683
10.0	0.0018	1.0099	1.3689
11.0	0.0047	1.0065	1.3654
11.5	0.0057	1.0031	1.3601
12.0	0.0052	1.0003	1.3546
12.5	0.0038	0.9984	1.3500
∞	0.0000	1.0000	1.3494

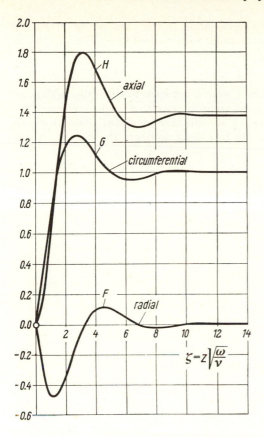

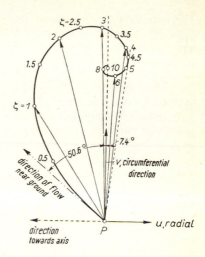

Fig. 11.3. Rotation near a solid wall, after Boedewadt. Vector representation of the horizontal velocity component

◄

Fig. 11.2. Rotation near a solid wall, after Boedewadt [9]. Velocity distribution in the boundary layer from eqn. (11.4); see also Table 11.1

difference, i. e. that between the ground and that at $\zeta = 4.63$, is 58°. It is further remarkable that the axial velocity component w does not depend on the distance r from the axis but only on the distance from the ground. The motion at all points is upwards with $w > 0$. As already mentioned, this is caused by the inward flow near the ground, consequent upon the decrease in the centrifugal forces. In any case, as seen from Fig. 11.2, this is compensated by a radial flow outwards at a greater height, but on the whole, the radial flow inwards predominates. The total volume flowing towards the axis taken over a cylinder of radius R around the z-axis is

$$Q = 2\pi R \int_{z=0}^{\infty} u \, dz = 2\pi R^2 \sqrt{\omega \nu} \int_0^{\infty} F(\zeta) \, d\zeta = -\pi R^2 \sqrt{\omega \nu} \, H(\infty) \, .$$

Inserting the numerical value of $H(\infty)$ from Table 11.1 we obtain

$$Q = -1.387 \pi R^2 \sqrt{\omega \nu} \, . \tag{11.8}$$

The volume of flow in the positive z-direction is of equal magnitude. The largest upward motion occurs at $\zeta = 3.1$, where $w = 1.85\sqrt{\omega \nu}$. It is also worth noting that the boundary layer extends considerably higher than in the example with the disk rotating in a fluid at rest (Sec. Vb). If the *boundary-layer thickness* δ is defined as the height for which the deviation of the peripheral velocity is equal to 2 per cent , we shall obtain $\delta = 8\sqrt{\nu/\omega}$ as against $\delta = 4\sqrt{\nu/\omega}$ for the stationary fluid.

The example of the motion of a vortex source between two parallel walls considered by G. Vogelpohl [120] is related to some extent to the present case. For very small Reynolds numbers the velocity distribution deviates little from the parabolic curve of Poiseuille flow. For large Reynolds numbers the velocity profile approaches a rectangular distribution, and a boundary layer is seen to be forming. The corresponding case of turbulent flow was discussed by C. Pfleiderer [85]. In this connexion the paper by E. Becker [6] may also be consulted.

Similar phenomena can be found in swirling flow through a conical funnel-like channel investigated by K. Garbsch [32]. The potential flow is generated by a sink of strength Q placed at the vertex of the cone and a potential vortex of strength Γ placed along the axis, Fig. 11.4. The solution to the boundary-layer equations is obtained by an iterative procedure which is said to lead to a good approximation with a small number of steps only. Two particular cases of such flows have also been investigated with the aid of approximate methods, and they will be mentioned in Chap. X: A. M. Binnie and D. P. Harris [7] studied pure sink flow ($\Gamma = 0$), and G. I. Taylor [111] and J. C. Cooke [17] studied pure vortex flow ($Q = 0$). In the latter case, as shown in Fig. 11.4, the flow forms a boundary layer on the wall of the conical channel. The flow field in the boundary layer develops a velocity component w in the direction of the cone generators, whereas the frictionless core, being a pure swirl, possesses only tangential velocity components v. The secondary flow in the boundary layer transports some fluid towards the vertex. The reader may further wish to study a related paper by H. E. Weber [121].

2. The circular jet. We shall now indicate H. Schlichting's [97] solution for the laminar circular jet which is analogous to the one for a two-dimensional jet given in Sec. IXg. The subject of the investigation is, thus, a jet which leaves a small circular opening and mixes with the surrounding fluid. In most practical cases the circular jet is also turbulent. The turbulent circular jet will be considered in Chap. XXIV, but since it leads to a differential equation which is identical with that for the laminar case we shall discuss the latter in some greater detail.

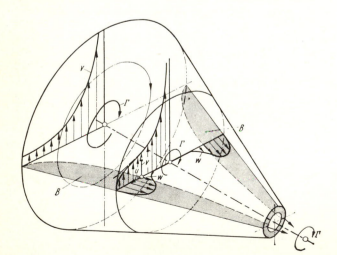

Fig. 11.4.   Swirling flow in a convergent conical channel, after G. I. Taylor [111]

B = boundary layer on the wall of the conical channel with secondary flow towards the vertex

The pressure can here be regarded constant, as in the two-dimensional case. The system of coordinates will be selected with its x-axis in the axis of the jet, the radial distance being denoted by y. The axial and radial velocity components will be denoted by u and v, respectively. Owing to the assumption of a constant pressure the flux of momentum in the direction of x is constant once more:

$$J = 2\pi \varrho \int_0^\infty u^2\, y\, dy = \text{const.} \tag{11.9}$$

In the adopted system of coordinates the equation of motion in the direction of x, under the usual boundary-layer simplifications, together with the equation of motion, can be written as

$$u \frac{\partial u}{\partial x} + v \frac{\partial u}{\partial y} = \nu \frac{1}{y} \frac{\partial}{\partial y} \left(y \frac{\partial u}{\partial y} \right), \tag{11.10a}$$

$$\frac{\partial u}{\partial x} + \frac{\partial v}{\partial y} + \frac{v}{y} = 0, \tag{11.10b}$$

and the boundary conditions are

$$\left.\begin{aligned} y = 0 \; : \quad & v = 0; \quad \frac{\partial u}{\partial y} = 0; \\ y = \infty : \quad & u = 0. \end{aligned}\right\} \tag{11.11}$$

As before, the velocity profiles $u(x, y)$ can be assumed similar. The width of the jet will be taken to be proportional to x^n, it being further assumed that

$$\psi \sim x^p\, F(\eta) \qquad \text{with} \qquad \eta = \frac{y}{x^n}.$$

In order to determine the exponents p and n we can use the same two conditions as in the two-dimensional case. First the momentum from eqn. (11.9) must be independent of x, and secondly, the inertia and frictional terms in eqn. (11.10a) must be of the same order of magnitude. Hence

$$u \sim x^{p-2n}, \quad \frac{\partial u}{\partial x} \sim x^{p-2n-1}, \quad \frac{\partial u}{\partial y} \sim x^{p-3n}, \quad \frac{1}{y} \frac{\partial}{\partial y} \left(y \frac{\partial u}{\partial y} \right) \sim x^{p-4n}.$$

Thus the following two equations for p and n result:

$$2p - 4n + 2n = 0; \qquad 2p - 4n - 1 = p - 4n,$$

so that $p = n = 1$. Consequently, we may now put

$$\psi = \nu x\, F(\eta) \qquad \text{and} \qquad \eta = \frac{y}{x},$$

from which it follows that the velocity components are

$$u = \frac{\nu}{x} \frac{F'}{\eta} ; \qquad v = \frac{\nu}{x} \left(F' - \frac{F}{\eta} \right). \tag{11.12}$$

Inserting these values into eqn. (11.10a), we obtain the following equation for the stream function

$$\frac{FF'}{\eta^2} - \frac{F'^2}{\eta} - \frac{FF''}{\eta} = \frac{d}{d\eta}\left(F'' - \frac{F'}{\eta}\right),$$

which can be integrated once to give

$$F F' = F' - \eta\, F'' . \tag{11.13}$$

The boundary conditions are $u = u_m$ and $v = 0$ for $y = 0$. It follows that $F' = 0$ and $F = 0$ for $\eta = 0$. Since u is an even function of η, F'/η must be even, F' odd and F even. Because of $F(0) = 0$ the constant term in the expansion of F in powers of η must vanish, which determines one constant of integration. The second constant of integration, which will be denoted by γ, can be evaluated as follows: If $F(\eta)$ is a solution of eqn. (11.13), then $F(\gamma\,\eta) = F(\xi)$ is also a solution. A particular solution of the differential equation

$$F\,\frac{dF}{d\xi} = \frac{dF}{d\xi} - \xi\,\frac{d^2F}{d\xi^2}$$

which satisfies the boundary condition $\xi = 0$: $F = 0$, $F' = 0$, is given by

$$F = \frac{\xi^2}{1 + \frac{1}{4}\,\xi^2} . \tag{11.14}$$

Hence we obtain from eqn. (11.12)

$$u = \frac{\nu}{x}\,\gamma^2\,\frac{1}{\xi}\,\frac{dF}{d\xi} = \frac{\nu}{x}\,\frac{2\,\gamma^2}{\left(1 + \frac{1}{4}\,\xi^2\right)^2},$$

$$v = \frac{\nu}{x}\,\gamma\left(\frac{dF}{d\xi} - \frac{F}{\xi}\right) = \frac{\nu}{x}\,\gamma\,\frac{\xi - \frac{1}{4}\,\xi^3}{\left(1 + \frac{1}{4}\,\xi^2\right)^2} .$$

Here $\xi = \gamma\,y/x$, and the constant of integration γ can now be determined from the given value of momentum.

From eqn. (11.9) we obtain for the momentum of the jet

$$J = 2\,\pi\,\varrho\int_0^\infty u^2\,y\,dy = \frac{16}{3}\,\pi\,\varrho\,\gamma^2\,\nu^2 .$$

Finally, the above results can be expressed in a form to contain only the kinematic viscosity, ν, and the *kinematic momentum* $K' = J/\varrho$. Thus

$$u = \frac{3}{8\pi}\,\frac{K'}{\nu\,x}\,\frac{1}{\left(1 + \frac{1}{4}\,\xi^2\right)^2} , \tag{11.15}$$

$$v = \frac{1}{4}\,\sqrt{\frac{3}{\pi}}\,\frac{\sqrt{K'}}{x}\,\frac{\xi - \frac{1}{4}\,\xi^3}{\left(1 + \frac{1}{4}\,\xi^2\right)^2} , \tag{11.16}$$

$$\xi = \sqrt{\frac{3}{16\pi}}\,\frac{\sqrt{K'}}{\nu}\,\frac{y}{x} . \tag{11.17}$$

Figure 11.5 represents a streamline pattern calculated from the preceding equations. The longitudinal velocity u is shown plotted together with that for the two-dimensional jet in Fig. 9.13.

The volume of flow $Q = 2\pi \int\limits_0^\infty u\, y\, dy$ (volume per second), which increases with the distance from the orifice owing to the flow from the surroundings, is represented by the simple equation

$$Q = 8\pi\,\nu\,x\,. \tag{11.18}$$

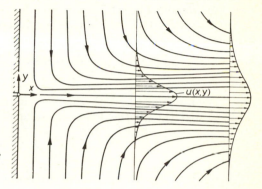

Fig. 11.5.   Streamline pattern for a circular laminar jet

This equation should be compared with eqn. (9.48) for the two-dimensional jet. It is seen that, unexpectedly, the volume of flow at a given distance from the orifice is independent of the momentum of the jet, i. e., independent of the excess of pressure under which the jet leaves the orifice. A jet which leaves under a large pressure difference (large velocity) remains narrower than one leaving with a smaller pressure difference (small velocity). The latter carries with it comparatively more stationary fluid, namely in a manner to make the volume of flow at a given distance from the orifice equal to that in a faster jet, provided that the kinematic viscosity is the same in both cases.

H. B. Squire [105, 106] was able to find solutions to the boundary-layer equations as well as to the complete Navier-Stokes equations and to make a comparison between them for the case of a conical jet which possesses an additional, radial velocity component in its annular orifice. In this latter class of radial jets the velocities are also inversely proportional to the distance from the orifice. The theory can be extended to turbulent flows by replacing the kinematic viscosity with the apparent kinematic viscosity of turbulent flow, which in this case remains constant, see Chap. XXIV. The case when a jet impinges at right angles on a wall and is spread along it was solved by M. B. Glauert [40], who included plane as well as axially symmetrical, and laminar as well as turbulent flows.

The corresponding case of a compressible circular laminar jet was evaluated by M. Z. Krzywoblocki [61] and D. C. Pack [83]. In the subsonic regime, the density on the axis of the jet is larger, and the temperature is smaller than on its boundary. These differences are inversely proportional to the square of the distance from the orifice. According to H. Goertler [43], the case when a weak swirl is superimposed on the jet can also be treated mathematically, and the effect of the swirling motion present in the orifice can be traced in the downstream direction. It turns out that the swirl decreases faster with the distance from the orifice than the jet velocity on the axis.

3. The axially symmetric wake. The flow in an axially symmetric wake, such as occurs downstream of an axially symmetric body placed in a stream parallel to its axis, can also be described with the aid of the system of equations (11.10a, b). The solution is quite analogous to that for the two-dimensional case which was described in detail in Sec. IX f. Let U_∞ denote the oncoming velocity and let $u(x, y)$ be the flow velocity in the wake. We assume, as was done in eqn. (9.29), that the velocity difference in the wake,

$$u_1(x,y) = U_\infty - u(x,y) \tag{11.19}$$

is very small compared with U_∞ far downstream. Consequently, we shall neglect quadratic terms in u_1. With this simplification it is possible to deduce from eqns. (11.10a) and (11.19) the following differential equation for u_1:

$$U_\infty \frac{\partial u_1}{\partial x} = \frac{\nu}{y} \frac{\partial}{\partial y}\left(y\, \frac{\partial u_1}{\partial y}\right). \tag{11.20}$$

The analytic form to be assumed for the dependence of the velocity difference $u_1(x,y)$ on the axial coordinate, x, and on the radial coordinate, y, can be discovered from the condition that the drag evaluated from the momentum of the wake must become independent of x at large distances downstream of the body. This leads to the relation

$$D = 2\,\pi\varrho\, U_\infty \int\limits_{y=0}^{\infty} u_1 \cdot y \ \ dy = \text{const}, \tag{11.21}$$

which is satisfied by the form

$$u_1 = CU_\infty \frac{f(\eta)}{x}, \tag{11.22}$$

where

$$\eta = \tfrac{1}{2}\, y\, \sqrt{\frac{U_\infty}{\nu x}}. \tag{11.23}$$

This form is analogous to that in eqn. (9.31) for the two-dimensional problem. Substituting eqns. (11.22) and (11.23) into eqn. (11.20), we obtain a differential equation for $f(\eta)$. This is

$$(\eta f')' + 2\,\eta^2 f' + 4\,\eta f = 0, \tag{11.24}$$

and the boundary conditions are

$$f' = 0 \quad \text{at} \quad \eta = 0 \quad \text{and} \quad f = 0 \quad \text{at} \quad \eta = \infty.$$

It is easy to verify that the solution of eqn. (11.24) has the form of an exponential,

$$f(\eta) = \exp\left(-\eta^2\right), \tag{11.25}$$

this form, too, being analogous to that in eqn. (9.34) for the two-dimensional case. Hence, the velocity difference turns out to be

$$u_1(x,y) = \frac{C}{x} U_\infty \exp\left(-\tfrac{1}{4} \frac{U_\infty y^2}{\nu x}\right).$$

The value of the constant C must be determined from the drag with the aid of eqn. (11.21); its value is

$$C = \frac{\pi}{32} c_D \mathsf{R} d ,$$

where c_D denotes the drag coefficient referred to the frontal area of the body, and $\mathsf{R} = U_\infty d/\nu$. Hence we obtain

$$\frac{u_1(x,y)}{U_\infty} = \frac{\pi c_D}{32}\left(\frac{d}{x}\mathsf{R}\right)\exp\left(-\eta^2\right). \tag{11.26}$$

The plot of the velocity difference from eqn. (11.26) is the same as that in Fig. 9.10. Experimental data can be found in F. R. Hama's work [45a].

4. Boundary layer on a body of revolution. The flow of a viscous fluid past a body of revolution when the stream is parallel to its axis is of great practical importance. The boundary-layer equations have been adapted to this case by E. Boltze [10]. Assuming a curvilinear system of coordinates (Fig. 11.6), we denote by x the current length measured along a meridian from the stagnation point, y denoting the coordinate at right angles to the surface. The contour of the body of revolution will be specified by the radii $r(x)$ of the sections taken at right angles to the axis. We assume that there are no sharp corners so that d^2r/dx^2 does not assume extremely large values. The velocity components parallel and normal to the wall will be denoted by u and v, respectively, and the potential flow will be given by $U(x)$. According to Boltze the boundary-layer equations will then assume the form:

$$\left.\begin{aligned}
\frac{\partial u}{\partial t} + u\frac{\partial u}{\partial x} + v\frac{\partial u}{\partial y} &= -\frac{1}{\varrho}\frac{\partial p}{\partial x} + \nu\frac{\partial^2 u}{\partial y^2}, \\[2ex]
\frac{\partial(ur)}{\partial x} + \frac{\partial(vr)}{\partial y} &= 0 ,
\end{aligned}\right\} \tag{11.27 a, b}$$

with the boundary conditions:

$$y = 0: \quad u = v = 0; \quad y = \infty: \quad u = U(x,t). \tag{11.28}$$

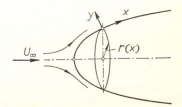

Fig. 11.6. Boundary layer near a body of revolution. System of coordinates

The equation of motion in the x-direction is seen to remain unchanged compared with two-dimensional flow. An order-of-magnitude estimate of terms in the equation of motion in the y-direction shows that the pressure gradient normal to the wall $\partial p/\partial y \sim u^2/r \sim 1$. Consequently the pressure difference across the boundary layer is of the order of the boundary layer thickness δ, and it is again possible to assume that the pressure gradient of the potential stream, $\partial p/\partial x$, is impressed on the boundary layer.

We shall limit the considerations of this chapter to the case of steady flow. In order to integrate eqns. (11.27 a, b) for the axially symmetrical case it is once more possible to introduce a stream function $\psi(x,y)$ given by:

$$u = \frac{1}{r} \frac{\partial (\psi r)}{\partial y} = \frac{\partial \psi}{\partial y} ,$$

$$v = - \frac{1}{r} \frac{\partial (\psi r)}{\partial x} = - \frac{\partial \psi}{\partial x} - \frac{1}{r} \frac{dr}{dx} \psi .$$

$$\left.\vphantom{\begin{array}{c}1\\1\\1\\1\end{array}}\right\} \quad (11.29\,\mathrm{a, b})\dagger$$

This transforms eqn. (11.27 a) into

$$\frac{\partial \psi}{\partial y} \frac{\partial^2 \psi}{\partial x \, \partial y} - \left(\frac{\partial \psi}{\partial x} + \frac{1}{r} \frac{dr}{dx} \psi \right) \frac{\partial^2 \psi}{\partial y^2} = U \frac{dU}{dx} + v \frac{\partial^3 \psi}{\partial y^3} , \qquad (11.30)$$

with the boundary conditions

$$y = 0 : \quad \psi = 0 , \quad \frac{\partial \psi}{\partial y} = 0 ; \quad y = \infty : \quad \frac{\partial \psi}{\partial y} = U(x) .$$

We now proceed to give a brief account of the methods used to calculate the boundary layer on a body of revolution. A more detailed account can be found in an earlier edition of this book [101]. The numerical results for a sphere, however, will be discussed in more complete detail. The boundary layer on a *body of revolution of arbitrary shape* can be determined by the same method as that used in Sec. IX c for the case of a cylinder of arbitrary cross-section (two-dimensional problem). The velocity of the potential flow, $U(x)$, is expanded into a power series in x and the stream-function, ψ, is assumed to be represented by a similar series in x, with coefficients depending on the wall distance y (Blasius series). Following N. Froessling [29] it is found that here also the coefficient-functions of y can be so arranged as to become independent of the parameters of any particular problem. In this manner the functions can be calculated once and applied universally.

$\dagger$ The equation of continuity can also be satisfied by an alternative stream function $\bar\psi$, such that

$$u = \frac{1}{r} \frac{\partial \bar\psi}{\partial y} ; \quad v = - \frac{1}{r} \frac{\partial \bar\psi}{\partial x} .$$

This form of the stream function was used by E. Boltze when he calculated non-steady axially symmetrical boundary layers, as described in Sec. XV b 2.

The body contour is given by the series

$$r(x) = r_1\, x + r_3\, x^3 + r_5\, x^5 + \ldots, \tag{11.31}$$

the potential flow being defined by the series

$$U(x) = u_1\, x + u_3\, x^3 + u_5\, x^5 + \ldots. \tag{11.32}$$

The distance from the wall is represented by the dimensionless coordinate

$$\eta = y\,\sqrt{\frac{2\,u_1}{\nu}}\,, \tag{11.33}$$

and in analogy with eqn. (11.32), the stream-function is represented by the Blasius series

$$\psi(x,\, y) = \sqrt{\frac{\nu}{2\,u_1}}\,\{u_1\, x\, f_1(\eta) + 2\, u_3\, x^3\, f_3(\eta) + \ldots\}. \tag{11.34}$$

Substituting eqns. (11.31), (11.32) and (11.35) with (11.36) into eqn. (11.30) and comparing terms, we obtain a set of differential equations for the functions $f_1, f_3, \ldots$. The first equation is

$$f_1''' = -\, f_1\, f_1'' + \frac{1}{2}\,(f_1'^2 - 1), \tag{11.35}$$

where differentiation with respect to η is denoted by primes. The boundary conditions are:

$$\eta = 0: \quad f_1 = f_1' = 0; \quad \eta = \infty: \quad f_1' = 1. \tag{11.36}$$

The first equation of the set is non-linear and identical with that for three-dimensional stagnation flow which was considered in Sec. V 10†. A plot of f_1' is shown in Fig. 5.10, where $f_1' = \phi'$. The equations for the terms in x^3 and x^5 have been solved by N. Froessling [29]. The succeeding ten functions of the term x^7 have been evaluated by F. W. Scholkemeyer [102].

Example: *Sphere.*

In a manner analogous to that employed for a circular cylinder in Sec. IXc, we can use the preceding scheme to solve the case of the sphere. The current radius for a sphere of radius R is given by

$$r(x) = R \sin x/R, \tag{11.37}$$

and the velocity distribution at the surface of the sphere we have

$$U(x) = \frac{3}{2}\, U_\infty \sin x/R = \frac{3}{2}\, U_\infty \sin \phi, \tag{11.38}$$

where ϕ denotes the central angle measured from the stagnation point. Comparing

---

† The equation for $f_1(\eta)$ transforms into eqn. (5.47) for $\phi(\zeta)$, if it is noticed that $\eta = \zeta\,\sqrt{2}$ and $df_1/d\eta = d\phi/d\zeta$.

the two series expansions for $\sin (x/R)$ in eqns. (11.37) and (11.38), we determine the coefficients of eqn. (11.3) as follows

$$u_1 = \frac{3}{2} \frac{U_\infty}{R}; \quad u_3 = -\frac{1}{4} \frac{U_\infty}{R^3}; \quad \ldots \text{ and } \quad \eta = \frac{y}{R} \sqrt{\frac{3 U_\infty R}{\nu}}.$$

The resulting velocity distributions for various values of the angle ϕ are seen plotted in Fig. 11.7; for these graphs the velocity u has been computed up to the term x^7. The velocity profiles for $\phi > 90°$ exhibit a point of inflexion because they are associated with the range of pressure increase.

In connexion with the problem at hand, we can repeat our previous remarks concerning the general practicability of applying a Blasius series. The calculation of the fundamental coefficients beyond the term x^7 involves an unacceptable amount of computation, and furthermore, the calculation of slender bodies requires considerably more terms. All this puts a very severe limitation on this method. For further results concerning spheres, reference should be made to the succeeding section.

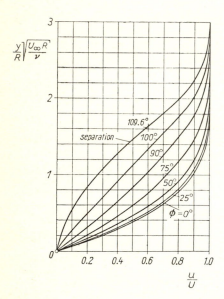

Fig. 11.7. Velocity distribution in the boundary layer on a sphere

Transverse curvature. We have stated repeatedly that the equation of motion (11.27a) of an axially symmetric flow has the same form as that for the two-dimensional case only on condition that the boundary-layer thickness is everywhere much smaller than the radius of the contour of the body ($\delta \ll r$). This condition is not satisfied in the case of a long but thin cylinder or, for that matter, in the case of any long and slender body of revolution. The boundary layer on such a body grows downstream and its thickness becomes comparable with the radius eventually. This brings into evidence the essentially three-dimensional nature of the boundary layer on a body of revolution which results from the comparatively large curvature of the surface of the body in the transverse direction.

R. A. Seban and R. Bond [95] treated the case of a slender cylinder, of radius $r_0 = a = \text{const}$, placed in a uniform axial stream. The same problem was studied by H. R. Kelly [60] who introduced certain numerical corrections. M. B. Glauert and M. J. Lighthill [41] obtained solutions by the application of Pohlhausen's approximate method (see Sec. XIb) and of an asymptotic series expansion. The flow along the generators of a cylinder of arbitrary cross-section was worked out by J. C. Cooke [18] who employed a Blasius series as well as Pohlhausen's approximate procedure.

The more general case of a compressible, axially symmetric boundary layer on a body of revolution whose contour is a function of the longitudinal coordinate, x, in particular, the cases of a circular cylinder and a sphere, were studied by R. F. Probstein and D. Elliot [88]. It turned out that the transverse curvature has the same effect on such flows with a pressure gradient as a supplementary, favourable pressure gradient. As a result, the shearing stress is increased and separation is delayed.

b. Approximate solutions for axially symmetric boundary layers

1. Approximate solutions for boundary layers on bodies which do not rotate. The approximate method for the solution of the differential equations of boundary-layer flow for two-dimensional steady problems which was presented in detail in Chap. X can be extended to the case of axially symmetrical flow. An approximate method for the calculation of boundary layers on bodies of revolution in axial flow was first indicated by C. B. Millikan [75]. Pohlhausen's approximate method of calculation which was described in Chap. X and which is based on a polynomial of 4th degree was extended by S. Tomotika [116, 117] to include a body of revolution.

The following account of the method as applied to bodies of revolution is based on the work of F. W. Scholkemeier [102], who used the modern version of the momentum equation in a similar way to that employed by H. Holstein and T. Bohlen for the case of two-dimensional flow. The momentum equation for the axially symmetrical case is obtained in the same way as that used in Sec. VIIId for the two-dimensional case. Starting with eqns. (11.27a, b) we obtain†

$$U^2 \frac{d\delta_2}{dx} + (2\,\delta_2 + \delta_1)U \frac{dU}{dx} + U^2 \frac{\delta_2}{r} \frac{dr}{dx} = \frac{\tau_0}{\varrho} . \qquad (11.39)$$

---

† The definitions employed for the displacement thickness, δ_1, and the momentum thickness, δ_2, of a boundary layer on a body of revolution are the same as in the two-dimensional case, eqns. (8.30) and (8.31), with y denoting the coordinate at right angles to the wall. Sometimes, however, slightly different definitions have been used [122]:

$$\text{displacement thickness: } \delta_1\,U = \int_0^\infty (U-u)\left(1 + \frac{y}{r}\right) dy ,$$

$$\text{momentum thickness: } \delta_2\,U^2 = \int_0^\infty u\,(U-u)\left(1 + \frac{y}{r}\right) dy .$$

The factor $(1 + y/r)$ takes into account the circumstance that the velocity u at a distance y from the wall is associated with the volume flow which passes through a strip of width dy. This volume is larger by a factor $(1 + y/r)$ than that which passes through a flat area of width $2\,\pi\,r$.

The significance of $r(x)$ may be inferred from Fig. 11.6. Retracing the steps of Sec. Xb we obtain the following differential equation for the quantity $Z = \delta_2{}^2/\nu$:

$$\frac{1}{2} U \frac{dZ}{dx} + [2 + f_1(K)] K + \frac{1}{r} \frac{dr}{dx} \frac{U}{U'} K = f_2(K) .$$

The quantities $K, f_1(K), f_2(K)$ have the same meaning as in the two-dimensional case, eqns. (10.27), (10.31) and (10.32). Introducing $F(K)$ as before, eqn. (10.34), we have

$$\frac{dZ}{dx} = \frac{1}{U}\left\{ F(K) - 2 K \frac{1}{r} \frac{dr}{dx} \frac{U}{U'} \right\} ; \quad K = ZU' . \tag{11.40}$$

It is easy to see that the substitution

$$\bar{Z} = r^2 Z \tag{11.41}$$

transforms the preceding equation to the form

$$\frac{d\bar{Z}}{dx} = \frac{r^2}{U} F(K) ; \quad K = \frac{\bar{Z}U'}{r^2} . \tag{11.42}$$

This form is preferable to that in eqn. (11.40) because it does not contain the derivative dr/dx.

The point of separation is again at $\Lambda = -12$, i. e. at $K = -0.1567$, but at the stagnation point the values of the shape factors Λ and K are now different. If the body of revolution has a blunt nose, we have at $x = 0$, i. e. at the upstream stagnation point,

$$\lim_{x \to 0} \left(\frac{1}{r} \frac{dr}{dx} \frac{U}{U'} \right) = 1 .$$

With this value the terms in the bracket in eqn. (11.40) reduce to $F(K) - 2 K$. By following the same argument as in the two-dimensional case it is found that the initial value of K at the stagnation point is determined by the condition $F(K) - 2 K = 0$, or, explicitly

$$\Lambda_0 = +4.716 ; \quad K_0 = 0.05708 .$$

Hence the initial values of the integral curve (11.40) at the stagnation point become

$$\left. \begin{array}{c} Z_0 = \dfrac{K_0}{U'_0} = \dfrac{0.05708}{U'_0} , \\[2ex] \left(\dfrac{dZ}{dx} \right)_0 = 0 . \end{array} \right\} \tag{11.43}$$

The initial slope is zero for a body of revolution, because for reasons of symmetry we must have $U_0'' = 0$ at the stagnation point. The method of direct integration described in Sec. Xb can be extended to the case of axially symmetrical bodies, as shown by N. Rott and L. F. Crabtree [93]. Equation (10.37) for the momentum thickness is now replaced by

$$\frac{U \delta_2{}^2}{\nu} = \frac{0.470}{r^2 U^5} \int_0^x r^2 U^5 \, dx . \tag{11.44}$$

Some numerical examples have been calculated by F. W. Scholkemeier [102] in his thesis presented to the Engineering University at Braunschweig as well as in the paper by J. Pretsch [87], already quoted. S. Tomotika [117] calculated the boundary layer on a sphere for a range of Reynolds numbers using both potential and measured pressure distributions. A comparison with measurement is given by A. Fage [27], and further results of measurements are contained in a paper by W. Moeller [76].

In this connexion it is useful to mention A. Michalke's [74] theoretical and experimental investigations on a rotationally symmetrical nozzle.

2. Flow in the entrance of a pipe. In this connexion it may be worth drawing attention to another axially-symmetrical boundary-layer problem, namely that associated with laminar flow in the inlet portion of a pipe. Strictly speaking, this is not a problem in boundary-layer theory but it has been solved with the aid of methods similar to the ones now being considered. The initially rectangular velocity distribution in the entrance section of the pipe $(x=0)$ is gradually transformed into a parabolic, Poiseuille, distribution by the action of viscous forces as sections further downstream are considered. The analogous two-dimensional problem, namely laminar flow in the inlet portion of a rectangular channel, has already been considered in Sec. IXi on the basis of the differential equations of boundary layer flow. The approximate method due to L. Schiller [96] is based on an equation which expresses the condition of equilibrium between momentum, pressure drop and viscous drag in a manner similar to the momentum equation discussed earlier. The velocity profiles in the inlet portion of the pipe are approximated by a constant velocity near the axis of the tube combined with two tangent portions of a parabola near the wall, so that at the wall the velocity becomes equal to zero. At the inlet section the width of the parabolic portions is zero and increases downstream until they coalesce into a single parabola at a definite distance from the entrance. This distance constitutes the theoretical initial length, and its magnitude, as calculated by L. Schiller, is given by $x \, \nu/R^2 \, \bar{u} = 0\cdot115$. Measurements performed by J. Nikuradse, Fig. 11.8, show good agreement with Schiller's theory for about a third of the initial length near the entrance (about $x \, \nu/R^2 \, \bar{u} = 0\cdot04$). The actual transition to a parabolic velocity profile appears to proceed more slowly than implied in the approximate calculation. Owing to the acceleration imparted to the fluid near the center the pressure drop in the entrance increases compared with that of a developed flow. The additional pressure drop at the entrance is $\Delta p = 1\cdot16 \, \varrho \, \bar{u}^2/2$. An approximate solution to this problem was also given by H. L. Langhaar [65].

The flow in the entrance region of a pipe was studied by B. Punnis [89] in the year 1947 and, more recently, by E. M. Sparrow et al. [106a].

The flow becomes much more complex in axially symmetrical flows in cases when there exists a tangential (whirl) component, in addition to the longitudinal component, which decays in the downstream direction. This problem was investigated by L. Talbot [110] and L. Collatz and H. Goertler [14]. Assuming that the whirl component of velocity is small compared with the axial velocity of Hagen and Poiseuille, it is possible to compute the former by formulating a boundary-value problem in relation to a linear differential equation of second order whose first eigenvalues have been evaluated. According to Talbot, the whirl component

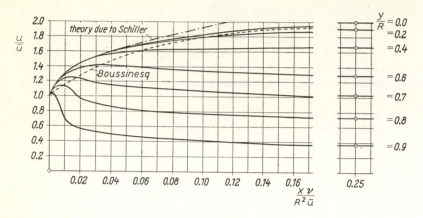

Fig. 11.8.   Velocity distribution in the inlet portion of a pipe for the laminar case; measurements performed by Nikuradse and quoted from Prandtl-Tietjens vol. II. Theory due to Schiller [96]

has practically decayed at a distance of 40 pipe radii when the Reynolds number has a value of $R = 10^3$. This is in good agreement with experimental results.

3. Boundary layers on rotating bodies of revolution. The simplest example of a boundary layer on a rotating body is that considered in Sec. Vb 11, namely the problem of a disk rotating in a fluid at rest. The fluid particles which rotate with the boundary layer are thrown outwards owing to the existence of centrifugal forces ('centrifuging') and are replaced by particles flowing towards the boundary layer in an axial direction. The case of a disk of radius R rotating with an angular velocity ω in an axial stream of velocity U_∞ affords a simple extension of the previous problem. In the latter case the flow is governed by two parameters: the Reynolds number and the rotation parameter, $U_\infty/R\omega$, which is given by the ratio of free-stream to tip velocity. An exact solution to the problem under consideration was given by Miss D. M. Hannah [46]† and A. N. Tifford [113] for the case of laminar flow; H. Schlichting and E. Truckenbrodt [98] provided an approximate solution. E. Truckenbrodt [119] investigated the case of turbulent flow. Figure 11.9 contains a plot of the torque coefficient, $C_M = M/\frac{1}{2}\varrho\,\omega^2\,R^5$, in terms of the Reynolds number and rotation parameter, $U_\infty/R\omega$, obtained from such calculations. Here M denotes the torque on the leading side of the disk only. When the disk rotates we may still assume that separation occurs at the edge of the disk. The 'stagnant' fluid behind the disk partly rotates with the disk and contributes little to the torque. Any such contribution has been left out of account in C_M in Fig. 11.9. It is seen that the torque increases rapidly with U_∞ at constant angular velocity.

---

† Actually ref. [38] solves a related problem in which the external field is that due to a source at infinity.

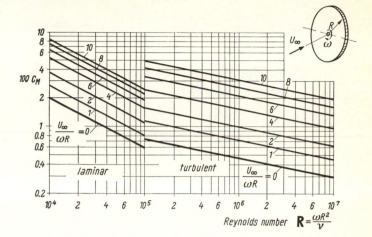

Fig. 11.9.
Moment coefficient on
a rotating disk in axial
flow, after Schlichting
and Truckenbrodt
[98, 119]

$C_M = M/\frac{1}{2} \varrho \, \omega^2 \, R^5$;
$M =$ torque on leading side
of disk

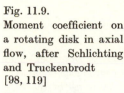

The flow in a circular box provided with a rotating lid shows a marked resemblance to that between two rotating disks mentioned in Sec. V b 11. The case of the flow inside the box was investigated in detail by D. Grohne [44] who discovered two peculiar features in it: First, the flow in the friction-free core in the interior of the box can only be determined by taking into account the influence of the boundary layers which form on the wall, in contrast to normal cases when one naturally assumes that the influence of the flow in a boundary layer results at most in a displacement. Secondly, the boundary layers are unusual in that they join each other. Similarly, in the arrangement consisting of a rotating channel investigated by H. Ludwieg [68], it is possible to discern two regions of flow when the speed of rotation is sufficiently high, namely a frictionless core and boundary layers which form on the side walls and which give rise to a secondary flow. The theory leads to a large increase of the drag coefficient which is due to rotation, and this fact has been confirmed by experiment.

Blunt bodies, such as e. g. a sphere or a slender body of revolution, placed in axial streams, show a marked influence of rotation on drag, as evidenced by the measurements performed by C. Wieselsberger [123], and S. Luthander and A. Rydberg [69]. Fig. 11.10 contains a plot of the drag coefficient of a rotating sphere in terms of the Reynolds number. It is seen that the critical Reynolds number, for which the drag coefficient decreases abruptly, depends strongly on the rotation parameter $U_\infty/R\omega$, and the same is true of the position of the point of separation. The effect of rotary motion on the position of the line of laminar separation on a sphere is described by the graph in Fig. 11.11; the data for it have been computed by N. E. Hoskin [50]. When the rotation parameter has attained the value $\Omega = \omega R/U_\infty = 5$, the line of separation will have moved by about 10° in the upstream direction, as compared with a sphere at rest. The physical reason for this behaviour is connected with the centrifugal forces acting on the fluid particles rotating with the body in its boundary layer. The centrifugal forces have the same effect as an additional pressure gradient directed towards the plane of the equator.

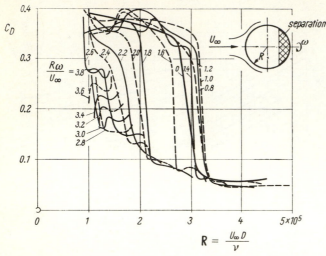

Fig. 11.10. Drag coefficients of a rotating sphere in axial flow in terms of the Reynolds number R and rotation parameter $\Omega = \omega R/U_\infty$ as measured by Luthander and Rydberg [69]

$$R = \frac{U_\infty D}{\nu}$$

A theoretical explanation of the very complex three-dimensional effects in the boundary layer of rotating bodies of revolution in axial flow is contained in the papers by H. Schlichting [99], E. Truckenbrodt [118] and O. Parr [84]; these authors employed the approximate method explained earlier. It is true that the boundary layer of a rotating body of revolution in axial flow still retains its axial symmetry, but owing to the rotation there appears a peripheral velocity component in addition to that in the meridional direction. For this reason, the calculation for such a boundary layer must introduce a momentum equation in the circumferential direction (y-direction) in addition to that in the meridional direction (x-direction). Assuming that the angular velocity of the body is ω, and denoting the coordinate at right angles to the wall by y, we can write the two equations of momentum in the form

$$U^2 \frac{d\delta_{2x}}{dx} + U \frac{dU}{dx}\left(2\,\delta_{2x} + \delta_{1x}\right) + \frac{1}{r}\frac{dr}{dx}\left(U^2\,\delta_{2x} + w_0^2\,\delta_{2x}\right) = \frac{\tau_{x0}}{\varrho}, \qquad (11.45)$$

$$\frac{w_0}{r^3}\frac{d}{dx}\left(U\,r^3\,\delta_{2xz}\right) = -\frac{\tau_{z0}}{\varrho}. \qquad (11.46)$$

The components of the shearing stress at the wall are then given by

$$\tau_{x0} = \mu\left(\frac{\partial u}{\partial y}\right)_0 ; \quad \tau_{z0} = \mu\left(\frac{\partial w}{\partial y}\right)_0 , \qquad (11.47)$$

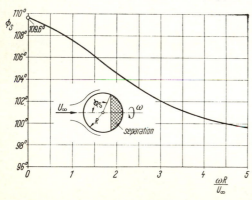

Fig. 11.11. Position of line of laminar separation on a sphere rotating in axial stream, after N. E. Hoskin [50]

and the displacement and momentum thicknesses are defined as

$$
\delta_{1x} = \int\limits_{y=0}^{\infty} \left(1 - \frac{u}{U} \right) dy \; ; \qquad \delta_{2x} = \int\limits_{y=0}^{\infty} \frac{u}{U} \left(1 - \frac{u}{U} \right) dy \; ;
$$

$$
\delta_{2z} = \int\limits_{y=0}^{\infty} \left(\frac{w}{w_0} \right)^2 dy \; ; \quad \delta_{2xz} = \int\limits_{y=0}^{\infty} \frac{u}{U} \cdot \frac{w}{w_0} \, dy \; .
$$

(11.48)

In the preceding equations, the local peripheral velocity $w_0 = r \, \omega$ has been chosen as a reference velocity for the azimutal component, $w(x, z)$. The preceding equations make it possible to perform calculations for laminar as well as for turbulent flows, it being necessary to introduce different expressions for the shearing stress at the wall in the latter case (see ref. [84] and Sec. XXII c). In some of the cases, it proved possible to evaluate the drag coefficient in addition to the turning moment, the former decreasing as the parameter $\omega R/U_\infty$ is increased. In this connexion, the papers by C. R. Illingworth [54] and S. T. Chu and A. N. Tifford [13] may also be studied. The approximate procedure conceived by H. Schlichting [98] was extended for compressible flows by J. Yamaga [125]. The preceding investigations have been extended for laminar as well as for turbulent flows by theoretical and experimental investigations described in several papers by Japanese authors [29a, 30, 31, 79, 80].

Problems connected with laminar flow about a sphere rotating in a fluid at rest have been discussed by L. Howarth [51] and S. D. Nigam [81]. An extension to the case involving ellipsoids of revolution was provided by B. S. Fadnis [26]. Near the poles, the flow is the same as on a rotating disk and near the equator it is like the one on a rotating cylinder. The accompanying secondary stream causes fluid particles to flow into the boundary layer near the poles, and out of it at the equator. The rate of this secondary flow increases with increasing slenderness, the equatorial area and speed of rotation remaining constant. However, the phenomena in the plane of the equator where the two boundary layers impinge on each other and are thrown outwards can no longer be analyzed with the aid of boundary-layer theory, *cf.* W. H. H. Banks [5a].

Further theoretical and experimental investigations of this problem have been later undertaken by O. Sawatzki [94] and by P. Dumargue et al. [21a]. Reference [94] describes measurements of the torque exerted on a rotating sphere in the range of Reynolds number $2 \cdot 10^5 < R < 1 \cdot 5 \times 10^6$ which goes far beyond the laminar regime. The investigation of Ref. [21a] included the visualization of the spiral streamlines near the wall on a sphere and on cones of various included angles as they occur in laminar flow.

It has been observed that in axial turbomachines there may, under certain circumstances, appear an extended zone of dead fluid in the whirl behind the row of stationary blades and near the hub. This phenomenon was described in great detail by K. Bammert and H. Klaeukens [5]. The origin of this dead-water area is connected with the radial increase in pressure in the outward direction which is due to the whirl. Owing to the whirl the axial pressure increase near the hub in the bladeless annulus behind the guides is much greater than at the outer wall. The influence of the boundary layer is here only secondary. Attention may, further, be drawn to an investigation due to K. Bammert and J. Schoen [4] concerning the flow through a rotating hollow shaft. It is observed that a funnel-like free surface is formed at the exit owing to the interaction between centrifugal and viscous forces.

c. Relation between axially symmetrical and two-dimensional boundary layers; Mangler's transformation

The preceding considerations demonstrate that the calculation of an axially symmetrical boundary layer is, generally speaking, more difficult than that of a two-dimensional boundary layer. That this is the case can be appreciated if it is remembered that the flow field in a two-dimensional boundary layer, say on a cylinder in cross-flow, depends only on the potential velocity distribution, $U(x)$. By contrast, when an axially symmetrical boundary layer is studied, for example that on a rotating body of revolution, it is found that the contour $r(x)$ of the body enters explicitly into the corresponding equations. The present section is devoted to a more detailed investigation into the relation between two-dimensional and axially symmetric boundary layers.

In steady flow the boundary-layer equations for two-dimensional flow and for axially symmetrical flow are given by eqns. (7.10), (7.11) and (11.27 a, b), respectively. The latter refer to a curvilinear system of coordinates with x denoting the current arc length and y denoting the distance from the wall in a direction normal to it. The respective velocity components are denoted by u and v, and the magnitudes with a bar refer to the two-dimensional case. With these symbols, we have for the two-dimensional case:

$$\bar{u}\,\frac{\partial \bar{u}}{\partial \bar{x}} + \bar{v}\,\frac{\partial \bar{u}}{\partial \bar{y}} = \bar{U}\,\frac{d\bar{U}}{d\bar{x}} + \nu\,\frac{\partial^2 \bar{u}}{\partial \bar{y}^2}\,, \qquad \frac{\partial \bar{u}}{\partial \bar{x}} + \frac{\partial \bar{v}}{\partial \bar{y}} = 0\,; \qquad (11.49)$$

for the axially symmetrical case

$$u\,\frac{\partial u}{\partial x} + v\,\frac{\partial u}{\partial y} = U\,\frac{dU}{dx} + \nu\,\frac{\partial^2 u}{\partial y^2}\,, \qquad \frac{\partial(r\,u)}{\partial x} + \frac{\partial(r\,v)}{\partial y} = 0\,. \qquad (11.50)$$

Here $r(x)$ denotes the distance of a point on the wall from the axis of symmetry. The first equations of both systems are identical, the difference being only in the appearance of the radius $r(x)$ in the equation of continuity.

It seems thus reasonable to inquire whether it is possible to indicate a transformation which would permit the use of the solutions of the two-dimensional case to derive solutions of the axially symmetrical case. Such a general relationship between two-dimensional and axially symmetrical boundary layers has been discovered by W. Mangler [72]. It reduces the calculation of the laminar boundary layer for an axially symmetrical body to that on a cylindrical body. The given body of revolution is associated with an ideal potential velocity distribution for a cylindrical body, the function being easily calculated from the contour and the potential velocity distribution of the body of revolution. Mangler's transformation is also valid for compressible boundary layers, as well as for thermal boundary layers in laminar flow. We shall, however, consider it here only in relation to incompressible flow.

According to Mangler, the equations which transform the coordinates and the velocities of the axially symmetrical problem to those of the equivalent two-dimensional problem are as follows:

$$\bar{x} = \frac{1}{L^2}\int_0^x r^2(x)\,dx\,; \quad \bar{y} = \frac{r(x)}{L}\,y\,;$$

$$\bar{u} = u\,; \quad \bar{v} = \frac{L}{r}\left(v + \frac{r'}{r}\,y\,u\right)\,;$$

$$\bar{U} = U\,, \qquad\qquad\qquad\qquad (11.51)$$

where L denotes a constant length. Remembering that

$$\frac{\partial f}{\partial x} = \frac{r^2}{L^2} \frac{\partial f}{\partial \bar{x}} + \frac{r'}{r} \bar{y} \frac{\partial f}{\partial \bar{y}} ; \qquad \frac{\partial f}{\partial y} = \frac{\partial f}{\partial \bar{y}} \frac{r}{L} ,$$

it is easy to verify that the system of equations (11.50) transforms into eqns. (11.49) by the use of the substitutions (11.51).

The boundary layer on a body of revolution $r(x)$ having the ideal potential velocity distribution $U(x)$ can be evaluated by computing the two-dimensional boundary layer for a velocity distribution $\bar{U}(\bar{x})$, where $U = \bar{U}$ and $\bar{x}$ and x are related by eqns. (11.51). Having calculated the velocity components $\bar{u}$, and $\bar{v}$ for the two-dimensional boundary layer it is possible to determine the components u and v of the axially symmetrical boundary layer with the aid of the transformation equations (11.51).

The method may be better understood with the aid of the following example. We shall consider rotationally symmetrical stagnation flow, for which

$$r(x) = x ; \qquad U(x) = u_1 x .$$

Hence, from eqn. (11.51), we have

$$\bar{x} = \frac{x^3}{3 L^2} , \text{ and consequently } x = \sqrt[3]{3 L^2 \bar{x}} .$$

The potential flow of the associated two-dimensional flow becomes

$$\bar{U}(\bar{x}) = u_1 \sqrt[3]{3 L^2 \bar{x}} ,$$

so that $\bar{U}(\bar{x}) = C \bar{x}^{\frac{1}{3}}$, where C denotes a constant. The associated two-dimensional flow belongs to the class of wedge flows discussed in Sec. IX a and is given by $U = C x^m$, with $m = \frac{1}{3}$ for the present example. From eqn. (9.7) we find the wedge angle $\beta = 2 m/(m+1) = \frac{1}{2}$. The associated two-dimensional flow is that past a wedge with an angle $\pi \beta = \pi/2$. The fact that axially symmetrical stagnation flow can be reduced to the case of flow past a wedge whose angle is $\pi/2$ was stated in Sec. IX a and is now confirmed.

d. Three-dimensional boundary layers

Until now we have restricted ourselves almost exclusively to the consideration of two-dimensional and axially symmetrical problems. Problems of two-dimensional and of axially symmetrical flow have this in common that the prescribed potential flow depends only on one space coordinate, and the two velocity components in the boundary layer depend on two space coordinates each. In the case of a three-dimensional boundary layer the external potential flow depends on two coordinates in the wall surface and the flow within the boundary layer possesses all three velocity components which, moreover, depend on all three space coordinates in the general case. The flow about a disk rotating in a fluid at rest (Sec. V b) and rotation in the neighbourhood of a fixed wall (Sec. XI a) constitute examples of three-dimensional boundary layers, apart from being exact solutions of the Navier-Stokes equations.

If the streamlines of the potential motion are straight lines which either converge or diverge then, essentially, the flow differs from a two-dimensional pattern only in that there is a change in the boundary-layer thickness. On the other hand, if the potential motion is curved the pressure gradient across the streamlines of the potential flow impressing itself upon the boundary layer gives rise to additional influences, such as secondary flow: outside the boundary layer the transverse pressure gradient is balanced with the centrifugal force, but within it the centrifugal forces are decreased because of the decreased velocities and, consequently, the pressure gradient causes mass to flow inwards, i. e. towards the concave side of the potential streamlines. The rotation of air over a fixed wall affords an example of this behaviour and illustrates the existence of a flow inwards.

A further example of secondary flow is afforded by the motion on the sidewall of the channel formed by turbine or compressor blades or by a deflector. The boundary layer which forms on the wall develops a secondary flow from the pressure side of one blade to the suction side of the next one owing to the curvature of the streamlines in the external flow field. The secondary flow caused by the sidewall is further affected by the boundary layer on the blades themselves causing the flow pattern through a turbine or compressor stage to become very complex. This presents a very difficult problem to boundary-layer theory because the three-dimensional nature of the flow is essential to it. For a long time problems of this kind had been studied by experimental means only [47].

1. The boundary layer on a yawed cylinder. Another important case of a three-dimensional boundary layer is that of an aeroplane wing, whose leading edge is not perpendicular to the stream, as in the case of swept-back wings and yawed wings. It is known from experience that on the suction side considerable quantities of the fluid move towards the receding end, the phenomenon having a very detrimental effect on the aerodynamic behaviour of the wing.

In two-dimensional motion through a boundary layer, the geometrical shape of the body influences the field of flow only indirectly, i. e. through the velocity distribution of the potential flow which alone enters the calculation. By contrast, three-dimensional boundary layers are affected by both: by the external velocity distribution and by the geometrical shape directly. For example, in the case of a body of revolution the variation of the radius with distance expressed by the function $R(x)$ appears explicitly in the differential equations, see eqn. (11.27 b).

For the purpose of establishing the boundary-layer equations we shall confine ourselves to the simplest case of a plane wall or to a curved wall which is developable into a plane (Fig. 11.12). Let x and z denote the coordinates in the wall surface, y denoting (as previously) the coordinate which is perpendicular to the wall. The velocity vector of potential flow V will be assumed to have the components $U(x,z)$ and $W(x,z)$, so that in the steady-state case the pressure distribution in the potential stream is given by

$$p + \tfrac{1}{2} \varrho \, [U^2 + W^2] = \text{const} . \tag{11.52}$$

If we now perform the same estimation, under the assumption of very large Reynolds numbers, relative to the three-dimensional Navier-Stokes equations (3.32), as explained in detail in Sec. VII a in relation to the two-dimensional case, we shall reach the conclusion that in the frictional terms of the equations for the x- and z-directions, respectively, it is possible to neglect the derivatives with respect to the coordinates which are parallel to the wall as against the derivative with respect to the coordinate at right angles to it. Regarding the equation in the y-direction we again obtain the result that $\partial p/\partial y$ is very small and may be neglected. Thus the pressure is seen to depend on x and z alone, and is impressed on the boundary layer by the potential flow. The estimation further shows that, generally speaking, none of the convective terms may be omitted. The three-dimensional boundary-layer equations are, then, as follows:

$$\left.\begin{array}{c} u\dfrac{\partial u}{\partial x}+v\dfrac{\partial u}{\partial y}+w\dfrac{\partial u}{\partial z}=-\dfrac{1}{\varrho}\dfrac{\partial p}{\partial x}+\nu\dfrac{\partial^2 u}{\partial y^2} \\[3mm] u\dfrac{\partial w}{\partial x}+v\dfrac{\partial w}{\partial y}+w\dfrac{\partial w}{\partial z}=-\dfrac{1}{\varrho}\dfrac{\partial p}{\partial z}+\nu\dfrac{\partial^2 w}{\partial y^2} \\[3mm] \dfrac{\partial u}{\partial x}+\dfrac{\partial v}{\partial y}+\dfrac{\partial w}{\partial z}=0\,, \end{array}\right\} \qquad (11.53\text{a, b, c})$$

with the following boundary conditions:

$$y=0:\quad u=v=w=0\,;\qquad y=\infty:\quad u=U\,;\quad w=W\,. \qquad (11.54)$$

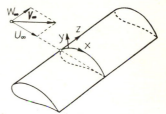

Fig. 11.12. System of coordinates for a three-dimensional boundary layer

The pressure gradients $\partial p/\partial x$ and $\partial p/\partial z$ are known from the potential flow in accordance with eqn. (11.52). This is a system of three equations for u, v, and w. For $W\equiv 0$ and $w=0$ the system transforms into the familiar system of equations (7.10) and (7.11) for two-dimensional boundary-layer flow.

Up to the present time no exact solutions of this general system of equations for three-dimensional flow have been found, apart from the examples which we have mentioned previously. Th. Geis [33, 34] investigated the special class of flows which lead to similar solutions. In analogy with wedge flows, the velocity profiles are now similar in the direction of each of the two axes of coordinates, and this allows us to transform the system (11.53) into a set of ordinary differential equations.

A particular case of three-dimensional boundary-layer flow which is considerably more amenable to numerical calculation is that where the potential flow depends on x but not on z, i. e. when

$$U=U(x)\,;\qquad W=W(x)\,. \qquad (11.55)$$

These conditions apply in the case of a yawed cylinder and also, approximately, in the case of a yawed wing at zero lift. The system of equations (11.53 a, b, c) is simplified in that there is no dependence on z. With $W = W_\infty = \text{const}$ and taking into account that $- (1/\varrho) \cdot (\partial p/\partial x) = U \cdot (dU/dx)$, we obtain

$$
\left.
\begin{aligned}
u \frac{\partial u}{\partial x} + v \frac{\partial u}{\partial y} &= U \frac{dU}{dx} + \nu \frac{\partial^2 u}{\partial y^2} \\[2mm]
u \frac{\partial w}{\partial x} + v \frac{\partial w}{\partial y} &= \nu \frac{\partial^2 w}{\partial y^2} \\[2mm]
\frac{\partial u}{\partial x} + \frac{\partial v}{\partial y} &= 0 ,
\end{aligned}
\right\}
\tag{11.56}
$$

with the same boundary conditions as before. In this particular case the system is reducible in the sense that it is possible to calculate u and v from the first and last equation, the solution being identical with that for a two-dimensional case, and subsequently, to complete the calculation of w from the second equation, which is, moreover, linear in w. This renders such cases really simple. Incidentally, it might be noted that the equation for the velocity component w is identical with that for the temperature distribution in a two-dimensional boundary layer when the Prandtl number is equal to unity (see Chap. XII).

Specializing the system (11.56) still further for the case when $U(x) = U_\infty = \text{const}$, we obtain the example of the flat plate in yaw but at zero incidence. In this case the pressure term in the first equation vanishes, and the second equation becomes identical with the first when w is replaced by u. Thus the solutions $u(x, y)$ and $w(x, y)$ become proportional, $w(x, y) = \text{const} \cdot u(x, y)$, or

$$
\frac{w}{u} = \frac{W_\infty}{U_\infty} .
$$

This means that in the case of a yawed flat plate the resultant of the velocity in the boundary layer which is parallel to the wall is also parallel to the potential flow at all points. The fact that the plate is yawed is seen to have no influence on the formation of the boundary layer (independence principle).

When the flow in the boundary layer on a yawed flat plate becomes *turbulent*, the right-hand sides of the first two equations (11.56) must be supplemented with the terms due to turbulent Reynolds stresses (Chap. XIX). Then, the two equations can no longer be transformed into each other by the substitution of u for w and vice versa. Consequently, the streamlines in the boundary layer cease to be parallel to the flow direction in the free stream, as can be verified by direct experiment [3]. In addition, ref. [3] has established that the displacement thickness of a turbulent boundary layer on a yawed plate grows somewhat faster in the downstream direction than is the case with an unyawed plate. This again demonstratese the inapplicability of the independence principle to turbulent boundary layers.

The calculation of the three-dimensional boundary layer on a yawed cylinder, eqns. (11.56), can be carried out by a method similar to that used in the case of two-dimensional flow about a cylinder whose axis is at right angles to the stream (Sec. IX c), i. e. by assuming a series expansion with respect to the length of arc, x, measured from the stagnation point. For a symmetrical cylinder we may put

$$U(x) = u_1\, x + u_3\, x^3 + \ldots, \qquad W(x) = W_\infty = \text{const}.$$

It is further assumed that the velocity components $u(x, y)$ and $v(x, y)$ of this flow (in which the stagnation points lie on a definite line) may also be expressed with the aid of a series in x with coefficients depending on y (Blasius series), the flow pattern being independent of the coordinate z measured along the generatrix of the cylinder. Thus, putting

$$\eta = y\,\sqrt{\frac{u_1}{v}}\,, \tag{11.57}$$

we obtain

$$
\left.
\begin{aligned}
u(x, y) &= u_1\, x\, f'_1(\eta) + 4\, u_3\, x^3\, f'_3(\eta) + \ldots\,; \\[2mm]
v(x, y) &= -\sqrt{\frac{v}{u_1}}\,\{u_1\, f_1(\eta) + 12\, u_3\, x^2\, f_3(\eta) + \ldots\}\,; \\[2mm]
w(x, y) &= W_\infty \left\{g_0(\eta) + \frac{u_3}{u_1}\, x^2\, g_2(\eta) + \ldots\right\}.
\end{aligned}
\right\} \tag{11.58a, b, c}
$$

The functions $f_1, f_3, \ldots$ satisfy the differential equations (9.18). The computation of the component w was first given by W. R. Sears [103]. It was later considereably extended by H. Goertler [42]. The functions $g_0, g_2, \ldots$ satisfy the differential equations

$$
\left.
\begin{aligned}
g_0'' + f_1\, g_0' &= 0\,, \\
g_2'' + f_1\, g_2' - 2\, f_1'\, g_2 &= -12\, f_3\, g_0'
\end{aligned}
\right\} \tag{11.59a, b}
$$

whose boundary conditions are

$$
\begin{aligned}
\eta = 0 &\;:\; g_0 = 0\,, \quad g_2 = 0\,, \ldots\,, \\
\eta = \infty &\;:\; g_0 = 1\,, \quad g_2 = 0\,, \ldots\,.
\end{aligned}
$$

As indicated by L. Prandtl [86] the equation for g_0 can be solved by direct integration, the result being

$$g_0(\eta) = \frac{\displaystyle\int_0^\eta \left\{\exp\left(-\int_0^\eta f_1\, d\eta\right)\right\} d\eta}{\displaystyle\int_0^\infty \left\{\exp\left(-\int_0^\eta f_1\, d\eta\right)\right\} d\eta} \tag{11.60}$$

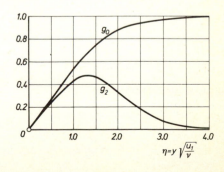

Fig. 11. 13. Laminar boundary layer on a yawed cylinder. The functions g_0 and g_2 for the velocity component w along the axis of the cylinder, eqn. (11.58c). At the stagnation line we have $w/W_\infty = g_0(\eta)$.

where f_1 denotes the solution for the two-dimensional stagnation-point flow according to eqn. (5.39) and Table 5.1; here $f_1(\eta) \equiv \Phi(\eta)$. The functions g_0 and g_2 which appear in the differential eqns. (11.61a, b) are seen plotted in Fig. 11.13. A tabulation for both functions can be found in [101] Chap. XI and in [42].

Approximate method. L. Prandtl [72] laid down a programme for obtaining approximate solutions with the aid of the momentum theorem, i. e. in a way which is similar to that used in Sec. XI b. In particular, the set of equations (11.45) to (11.48) transforms into that for a yawed cylinder when it is assumed formally that $r = \mathrm{const}$ and when the azimuthal momentum thickness δ_{2xz} is represented by the formula

$$\delta_{2xz} = \int\limits_{y=0}^{\infty} \frac{u}{U}\left(1 - \frac{w}{W_\infty}\right) \mathrm{d}y \ .$$

A procedure which is based on these equations was published by W. Dienemann [21].

A similar approximate method was used by J. M. Wild [124] for the solution of the problem of the yawed cylinder. Figure 11.14 represents the pattern of streamlines for a yawed elliptic cylinder of slenderness ratio 6 : 1, placed at an angle of incidence to the stream. The lift coefficient has a value of 0·47. The arrows shown in the sketch, indicate the direction of flow of the velocity component parallel to the wall in its immediate neighbourhood, i. e. the value

$$\lim_{y \to 0} (w/u) \ .$$

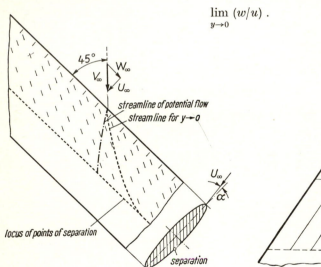

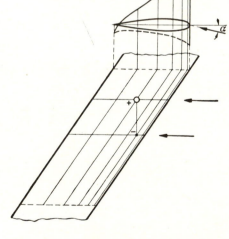

Fig. 11.14.   Boundary-layer flow about a yawed elliptical cylinder with lift, after J. M. Wild [124]

Fig. 11.15.   Explanation of origin of cross-flow on a yawed wing at an angle of incidence. Curves of constant pressure (isobars) on the suction side of the wing. Near the leading edge on the upper surface of the wing there is a sharp pressure gradient at right angles to the main stream and towards the receding end causing cross-flow

The respective streamline is shown as a broken line, and the potential streamline is seen plotted for comparison. It is noticeable that the flow direction in the boundary layer is turned by a large angle towards the receding end of the cylinder. This circumstance is very important when flow patterns on yawed wings are observed with the aid of tufts.

Swept wings. The existence of cross-flow which occurs in the boundary layer of a yawed cylinder is important for the aerodynamic properties of swept wings. When yawed or swept-back wings operate at higher lift values the pressure on the suction side near the leading edge shows a considerable gradient towards the receding tip, the effect being due to the rearward shift of the aerofoil sections of the wing. This phenomenon can be inferred from Fig. 11.15 which shows the isobars on the suction side of a yawed wing. The fluid particles which become decelerated in the boundary layer have a tendency to travel in the direction of this gradient, and a cross-flow in the direction of the receding tip results. As demonstrated by measurements performed by R. T. Jones [58] and W. Jacobs [55], the boundary layer on the receding portion thickens, the effect leading to premature separation. In aircraft equipped with swept-back wings separation begins at the receding portion, i. e. near the ailerons, and causes the dreaded one-winged stall to occur. It is possible to avoid this kind of separation, and hence to prevent one-winged stalling, by equipping the wing with a 'boundary-layer fence' which consists of a sheet-metal wall placed on the suction side in the forward portion of the wing, thus preventing cross-flow. An aircraft with swept-back wings and a boundary-layer fence on each half of the wing is shown in Fig. 11.16. W. Liebe [66] reported on the improvement in wing characteristics which can be attained by these means. A paper by M. J. Queijo, B. M. Jaquet and W. D. Wolhart [90] describes extensive measurements on models provided with 'boundary-layer fences'. The papers by J. Black [8] and D. Kuechemann [64] contain more details

Fig. 11.16.   Jet fighter De Havilland D. H. 110 with swept-back wings and a boundary-layer fence at edge of each aileron; from W. Liebe [66]

concerning the very complex flow patterns in boundary layers on swept wings. Experimental results obtained by A. Das [20] indicate that a boundary-layer fence causes a considerable improvement in the flow on its inner side in addition to that on the outer side. Drag coefficients of fences accommodated on flat plates have been investigated by K. G. R. Raju et al. [91a].

The case with $W = $ const, eqn. (11.55), is not the only one which has been given attention. H. G. Loos [67] studied the case of flow past a flat plate when the free stream is described by $U = $ const, $W = a_0 + a_1 x$, whereas A. G. Hansen and H. Z. Herzig [48] considered the generalized case with

$$U = \text{const}; \qquad W = \sum_n a_n x^n .$$

Since such external flows are not irrotational, the velocity in the boundary layer can become larger than that in the free stream. The excess in velocity is due to the secondary flow in the boundary layer which transfers into it fluid particles from regions of higher energy. It sometimes also happens that the initial velocity profiles in the principal flow direction show regions of back-flow which, nevertheless, do not signify separation; they usually disappear further downstream. This type of behaviour can also be explained as being due to a transfer of energy by the secondary flow. The reader will recognize from the preceding example that the definition of separation is beset with difficulties when three-dimensional boundary layers are being considered. This is due to the fact that the relation between back-flow and shearing stress has ceased to be as simple as in the two-dimensional case [49, 77]. A separation of terms identical with the one encountered in connexion with the free stream described by eqn. (11.55) can be successfully achieved, according to L. E. Fogarty [28], when considering an infinitely long wing which is made to rotate about a vertical axis (helicopter rotor). It is found that the rotary motion does not affect the chordwise velocity component and so the incidence of separation remains unaffected. Rotation merely causes the appearance of slight radial velocity components.

A further special case of the general problem described by eqns. (11.53) and (11.54) which is amenable to calculation occurs when the external flow consists of a two-dimensional basic pattern on which there is superimposed a weak disturbance of the kind described by

$$U(x,z) = U_0(x) + U_1(x,z) , \qquad U_1 \ll U_0 ,$$
$$W(x,z) = \qquad\quad W_1(x,z) , \qquad W_1 \ll U_0 .$$

The boundary-layer flow can then, equally, be separated into a two-dimensional basic pattern with a weak perturbation superimposed on it. The requisite differential equations can, once more, be uncoupled by linearization. Examples of this kind were given by A. Mager [70, 71] and H. S. Tan [110a].

2. Boundary layers on other bodies. Three-dimensional boundary-layer flows become even more complicated in cases when the external flow cannot be represented simply by the superposition of two components. The latter case occurs, for example, on a yawed body of revolution. In arrangements of this kind, the direction of velocity components in the boundary layer deviates considerably from that in the free stream at the point; in other words, a strong secondary-flow field is generated. An idea of the very complicated three-dimensional flow pattern in such boundary layers is conveyed by the photograph shown in Fig. 11.17b; it was taken by E. A. Eichelbrenner and A. Oudart [22] on the upper side of a yawed ellipsoid of revolution, the flow pattern having been made visible by streaks of dye issuing from drillings on the surface of the body. The photograph shows, in particular, that the pattern of a three-dimensional boundary layer which exists in an adverse pressure gradient is markedly different from that in a two-dimensional boundary layer. The principal difference is this: in the two-dimensional case, the fluid in the boundary layer is generally forced into the external flow if the pressure gradient is sufficiently strong, thus causing separation from the wall, (cf. Fig. 7.2b); in the three-dimensional case the fluid particles can escape sideways along the wall. The photograph in Fig. 11.17b clearly exhibits this type of behaviour: the streaks in the neighbourhood of the rear stagnation point, i. e. in the region of a strong adverse pressure gradient (see also Fig. 11.17a), are clearly seen to be deflected sideways; they do, however, remain clinging to the surface. The streamlines at the surface which are shown in Fig. 11.17c and which have been obtained by calculation show good qualitative agreement with

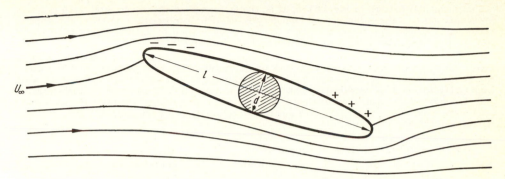

Fig. 11.17a

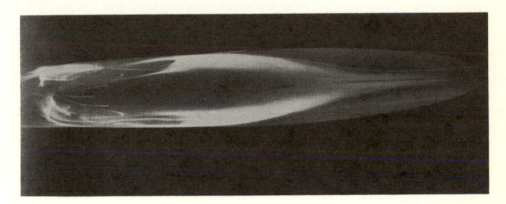

Fig. 11.17b

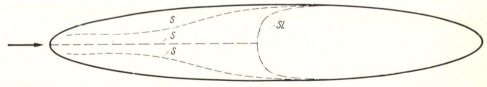

Fig. 11.17c

Figs. 11.17a,b,c. Three-dimensional, laminar boundary layer on upper side of ellipsoid of revolution of aspect ratio $l/d = 6$ and with a $10°$ yaw, after E. A. Eichelbrenner and A. Oudart [21, 22]

a) Schematic diagram of streamlines viewed sideways

b) Photograph of the upper side of the ellipsoid of revolution in the water channel of the ONERA in Chatillon-sous-Bagneux (Paris). Reynolds number $U_\infty l/\nu = 2 \times 10^4$. The flow pattern was made visible by streaks of dye issuing from the surface of the body. The streamlines in the layers in the neighbourhood of the rear stagnation point show a marked deflexion sideways. This corresponds to separation in a three-dimensional boundary layer

c) Streamlines at the wall, marked S, obtained by calculation, and theoretical separation line, marked SL; there is satisfactory qualitative agreement with the pattern the photograph of which is shown in b) the experimental pattern in Fig. 11.17b. It is, therefore, not at all easy to establish a criterion for separation in a three-dimensional boundary layer, if proper weight is given to this type of behaviour. At this point, we wish to draw the reader's attention to the investigations on yawed cones due to W. J. Rainbird, R. S. Crabbe and L. S. Jurewicz [91].

It appears to be possible to attempt a theoretical analysis of three-dimensional boundary layers with the aid of a scheme suggested by L. Prandtl [86] who proposed to introduce a curvilinear system of coordinates in which the potential lines and streamlines of the free stream would play the part of coordinates. This programme was carried out by E. A. Eichelbrenner and A. Oudart [22] when they calculated the laminar case mentioned earlier. It has already been mentioned that good qualitative agreement resulted, as shown in Fig. 11.17c. See also R. Timman [114].

The method of calculation proposed by L. Prandtl [86] was recently developed numerically by W. Geissler [35, 36, 37]. Figure 11.18 illustrates the results referring to the three-dimensional boundary layer on a yawed ellipsoid of revolution. In addition to the potential lines and streamlines of the external flow, Figure 11.18a shows the separation line S; the latter has a course similar to that in Fig. 11.17. Figures 11.18b and 11.18c represent the velocity distribution in the boundary layer at various stations on a particular potential line.

The laminar boundary layer on a *yawed* rotating circular cone in a supersonic stream was earlier investigated by R. Sedney [104], whereas J. C. Martin [73] investigated the Magnus effect on bodies of revolution at a small angle of incidence.

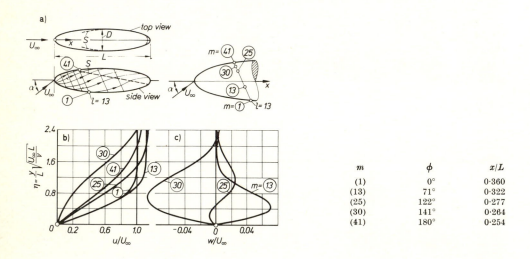

m	ϕ	x/L
(1)	0°	0·360
(13)	71°	0·322
(25)	122°	0·277
(30)	141°	0·264
(41)	180°	0·254

Fig. 11.18. Velocity distribution in the three-dimensional boundary-layer on an ellipsoid of revolution of axis ratio $L/D = 4$ at an angle of incidence $\varkappa = 15°$, after W. Geissler [36, 37]. a) System of potential lines and streamlines in outer flow; S = separation line. b) Primary flow velocity profiles, u/U_∞, in the direction of the outer flow streamlines. c) Secondary flow velocity profiles, w/U_∞, at right angles to the direction of the outer flow streamlines. The velocity profiles are given for potential line $l = (13)$ at different stations m with azimuth angle ϕ and station x as per table above ($\phi = 0°$ — windward symmetry)

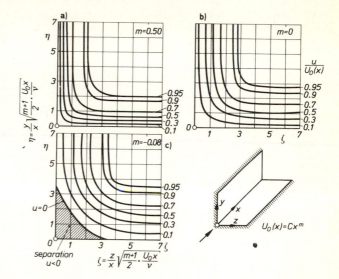

Fig. 11.19. Velocity distribution in the laminar boundary layer along a corner, after Vasanta Ram [92]. Line of constant velocity in the boundary, $u/U_0(x) =$ const. Free-stream velocity $U_0(x) = C\,x^m$

Another important example of a three-dimensional boundary layer can be found in the corner formed by two mutually perpendicular planes in a stream parallel to their line of intersection. This flow configuration was investigated theoretically by Vasanta Ram [92]. The external velocity at far distance has been assumed to be of Hartree's type, i. e. given by

$$U_0(x) = C\,x^m.$$

It is recalled from Sec. IXa that this type of external stream leads to similar velocity profiles in the boundary layer. This feature continues to hold in the case of flow in a corner. Some of the results of these studies are given in Fig. 11.19; this shows the velocity distributions in the corner for three different values of the pressure parameter m. A comparison between the distributions for different values of m demonstrates that the boundary layer in a corner thickens appreciably in the presence of a pressure increase in the external flow.

Experimental observations [82, 39] suggest that the flow in the corner separates earlier than that on the portions of the walls at a larger distance from it, even in the presence of small adverse pressure gradients. This physically understandable mode of behavior is fully confirmed by these theoretical results. On a flat plate separation occurs at $m = -0\cdot091$ (see Fig. 9.1), separation in a right-angled corner occurs as early as for $m = -0\cdot05$. At $m = -0\cdot08$, Fig. 11.19, the flow in the neighbourhood of the corner displays a separation region with reverse flow ($u < 0$). By contrast, at a large distance no reverse flow occurs. M. Zamir and A. D. Young [126, 127] carried out extensive experiments on the laminar boundary layer along a right-angled corner at zero incidence. See also S. G. Rubin [93a].

An extension of Pohlhausen's method to rotating bodies was given by G. Jungclaus [49]; he applied it to the investigation of relative motion through a curved channel which is important in the theory of centrifugal pumps. The theory leads to predictions regarding separation which are in good agreement with measurements.

In conclusion, attention may be drawn to the calculation of the boundary layer on two mutually perpendicular flat plates at zero incidence performed by G. F. Carrier [12] and K. Gersten [38]. The same problem with supersonic flow and heat transfer has been dealt with by M. Z. von Krzywoblocki who makes use of G. F. Carrier's earlier work; cf. H. A. Dwyer [21b].

The so-called "quarter-plate problem" is closely related to the above. In it, an investigation is made of the flow along a flat plate at zero incidence which possesses a side edge parallel to the stream in addition to the leading edge. Assumptions for the theoretical treatment of this problem are due to K. Stewartson and L. Howarth [108], and to K. Stewartson [109]. The side edge causes the appearance of a supplementary secondary motion in the boundary layer which produces, among others, an increase in the shearing stress. This result agrees with the measurements performed by J. W. Elder [25] on a plate of finite width. However, the flow at the side edge, like that in the immediate neighbourhood of the leading edge, is not yet completely understood.

Many three-dimensional boundary-layer flows are so physically complex that they will, most probably, remain inaccessible to a numerical treatment for a long time. An example of this type is illustrated with the aid of Fig. 11.20. This depicts schematically the three-dimensional boundary layer which forms in the neighbourhood of a squat cylindrical body (small height compared to length) provided with a blunt nose. The body is placed on a flat plate. On the plate, and at a large distance from the cylinder, the boundary layer is two-dimensional. As the cylinder is approached, and *outside the plane of symmetry*, there forms in the boundary layer† a region of secondary flow in which the velocity profiles are skewed, rather like those in Fig.

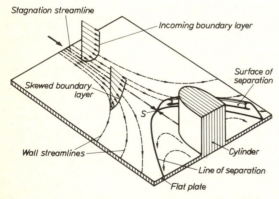

Fig. 11.20 (schematic). Three-dimensional boundary layer formed in the corner between a squat cylindrical body placed on a flat plate, after J. P. Johnston [56, 57]. The streamlines are curved outside the plane of symmetry; as a result, there forms in the boundary layer a secondary flow, and the velocity profiles become skewed. The boundary layer in the plane of symmetry separates at point S in the stagnation region. The flow forms a surface of separation; see also Fig. 11.21

---

† We describe this type of flow in conjunction with our discussion on laminar boundary layers even though such flows are turbulent in most cases, because their character is, fundamentally speaking, the same.

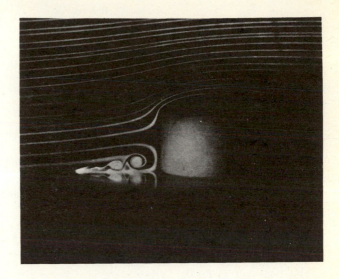

Fig. 11.21. Separation of a three-dimensional boundary layer in stagnation flow; squat cylinder mounted on a flat plate, after Thwaites [112a]; see also Fig. 11.20

11.1. The boundary layer in the plane of symmetry must overcome a strong pressure increase in the stagnation region of the cylinder. This leads to separation at point S, in a manner similar to that in the decelerated stagnation flow of Fig. 2.15.

In the neighbourhood of this separation region, the flow develops a separated vortex sheet which curls up around the cylinder at its base like a horseshoe. The photograph of Fig. 11.21 was taken with the injection of smoke to provide visualization. Here it is possible to notice that in addition to the primary vortex which rotates clockwise ahead of the cylinder there forms an additional, smaller vortex which rotates in the same sense. The primary vortex differs fundamentally from its two-dimensional analog in that it is not formed by the same fluid particles; on the contrary, it accepts continuously fresh material from the upstream direction and discharges material, also continuously, into the zone of separation confined inside the vortex sheet. It is very difficult to master such a complex flow pattern by calculation, particularly because it is predominantly turbulent in most cases, even though it can be laminar near the wall.

Contemporary research investigations into separation in three-dimensional flows are extensively discussed in the AGARD Conference Proceedings No. 168 [1].

Summary papers on three-dimensional boundary layers were given by W. R. Sears [103a], F. K. Moore [78]. J. C. Cooke and M. G. Hall [19], as well as by H. Schlichting [100].

References

[1] AGARD Conference Proceedings No. 168 on "Flow Separation" (1975) containing 42 contributions.

[2] Andrade, E.N., and Tsien, H.S.: The velocity distribution in a liquid-into-liquid jet. Proc. Phys. Soc. London *49*, 381—391 (1937).

[3] Ashkenas, H., and Riddell, F. R.: Investigation of the turbulent boundary layer on a yawed flat plate. NACA TN 3383 (1955).

[4] Bammert, K., and Schoen, J.: Die Strömung von Flüssigkeiten in rotierenden Hohlwellen. Z. VDI *90*, 81—87 (1948).

[5] Bammert, K., and Kläukens, H.: Nabentotwasser hinter Leiträdern von axialen Strömungsmaschinen. Ing.-Arch. *17*, 367—380 (1949).

[5a] Banks, W. H. H.: The boundary layer on a rotating sphere. Quart. J. Mech. Appl. Math. *18*, 443—454 (1965).

[6] Becker, E.: Berechnung der Reibungsschichten mit schwacher Sekundärströmung nach dem Impulsverfahren. ZFW *7*, 163—175 (1959); see also: Mitt. Max-Planck-Institut für Strömungsforschung No. 13 (1956) and ZAMM-Sonderheft 3—8 (1956); Diss. Göttingen 1954.

[7] Binnie, A.M., and Harris, D.P.: The application of boundary layer theory to swirling liquid flow through a nozzle. Quart. J. Mech. Appl. Math. *3*, 89—106 (1950).

[8] Black, J.: A note on the vortex patterns in the boundary layer flow of a swept-back wing. J. Roy. Aero. Soc. *56*, 279—285 (1952).

[9] Bödewadt, U.T.: Die Drehströmung über festem Grund. ZAMM *20*, 241—253 (1940).

[10] Boltze, E.: Grenzschichten an Rotationskörpern. Diss. Göttingen 1908.

[11] Burgers, J.M.: Some considerations on the development of boundary layer in the case of flows having a rotational component. Kon. Akad. van Wetenschappen, Amsterdam *45*, No. 1—5, 13—25 (1941).

[12] Carrier, G.F.: The boundary layer in a corner. Quart. Appl. Math. *4*, 367—370 (1946).

[13] Chu, S.T., and Tifford, A.N.: The compressible laminar boundary layer on a rotating body of revolution. JAS *21*, 345—346 (1954).

[14] Collatz, L., and Görtler, H.: Rohrströmung mit schwachem Drall. ZAMP *5*, 95—110 (1954).

[15] Cooke, J.C.: The boundary layer of a class of infinite yawed cylinders. Proc. Cambr. Phil. Soc. *46*, 645—648 (1950).

[16] Cooke, J.C.: Pohlhausen's method for three-dimensional laminar boundary layers. Aero. Quart. *3*, Part I, 51—60 (1951).

[17] Cooke, J.C.: On Pohlhausen's method with application to a swirl problem of Taylor. JAS *19*, 486—490 (1952).

[18] Cooke, J.C.: The flow of fluids along cylinders. Quart. J. Mech. Appl. Math. *10*, 312—321 (1957).

[19] Cooke, J.C., and Hall, M.G.: Boundary layers in three dimensions. Progress in Aeronautical Sciences *2*, 221—282, Pergamon Press, London, 1962.

[19a] Crabtree, L. F., Küchemann, D., and Sowerby, L.: Three-dimensional boundary layers. Chapter in: L. Rosenhead (ed.): Laminar boundary layers. Clarendon Press, Oxford, 1963, p. 409—491.

[20] Das, A.: Untersuchungen über den Einfluss von Grenzschichtzäunen auf die aerodynamischen Eigenschaften von Pfeil- und Deltaflügeln. Diss. Braunschweig 1959; ZFW *7*, 227—242 (1959).

[21] Dienemann, W.: Berechnung des Wärmeüberganges an laminar umströmten Körpern mit konstanter und ortsveränderlicher Wandtemperatur. Diss. Braunschweig 1951; ZAMM *33*, 89—109 (1953); see also JAS *18*, 64—65 (1951).

[21a] Dumarque, P., Laghoviter, G., and Daguenet, M.: Détermination des lignes de courant pariétales sur un corps de révolution tournant autour de son axe dans un fluide au repos. ZAMP *26*, 325—336 (1975).

[21b] Dwyer, H. A.: Solution of a three-dimensional boundary-layer flow with separation. AIAA J., *6*, 1336—1342 (1968).

[22] Eichelbrenner, E.A., and Oudart, A.: Méthode de calcul de la couche limite tridimensionnelle. Application à un corps fuselé incliné sur le vent. ONERA-Publication No. 76, Chatillon, 1955.

[23] Eichelbrenner, E.A.: Décollement laminaire en trois dimensions sur un obstacle fini. ONERA-Publication No. 89, Chatillon, 1957.
[24] Eichelbrenner, E.A.: Three-dimensional boundary layers. Annual Review of Fluid Mech. 5, 339—360 (1973).
[25] Elder, J.W.: The flow past a flat plate of finite width. JFM 9, 133—153 (1960).
[26] Fadnis, B.S.: Boundary layer on rotating spheroids. ZAMP V, 156—163 (1954).
[27] Fage, A.: Experiments on a sphere at critical Reynolds-numbers. ARC RM 1766 (1936).
[28] Fogarty, L.E.: The laminar boundary layer on a rotating blade. JAS 18, 247—252 (1951).
[29] Frössling, N.: Verdunstung, Wärmeübergang und Geschwindigkeitsverteilung bei zwei-dimensionaler und rotationssymmetrischer laminarer Grenzschichtströmung. Lunds. Univ. Arsskr. N. F. Avd. 2, 35, No. 4 (1940).
[29a] Furuya, Y., and Nakamura, I.: Velocity profiles in the skewed boundary layers on some rotating bodies in axial flow. J. Appl. Mech. 37, 17—24 (1970).
[30] Furuya, Y., Nakamura, K., and Kawachi, H.: The experiment on the skewed boundary layer on a rotating body. Bulletin of JSME 9, 702—710 (1966).
[31] Furuya, Y., and Nakamura, I.: An experimental investigation of the skewed boundary layer on a rotating body (2nd Report). Bulletin of JSME 11, 107—246 (1968).
[32] Garbsch, K.: Über die Grenzschicht an der Wand eines Trichters mit innerer Wirbel- und Radialströmung. Fifty years of boundary-layer research (W. Tollmien and H. Görtler, ed.), Braunschweig, 1955, 471—486; see also: ZAMM-Sonderheft 11—17 (1956).
[33] Geis, Th.: Ähnliche Grenzschichten an Rotationskörpern. Fifty years of boundary-layer research, (W. Tollmien, and H. Görtler, ed.), Braunschweig, 1955, 204—303.
[34] Geis, Th.: „Ähnliche" dreidimensionale Grenzschichten. J. Rat. Mech. Analysis 5, 643—686 (1956).
[35] Geissler, W.: Berechnung der Potentialströmung um rotationssymmetrische Rümpfe, Ringprofile und Triebwerkseinläufe. ZFW 20, 457—462 (1972).
[36] Geissler, W.: Berechnung der dreidimensionalen laminaren Grenzschicht an angestellten Rotationskörpern mit Ablösung. AVA-Bericht 74 A 19 (1974); Ing.-Arch. 43, 413—425 (1974).
[37] Geissler, W.: The three-dimensional laminar boundary layer over a body of revolution at incidence and with separation. AVA-Bericht 74 A 08 (1974); AIAA J. 12, 1743—1745 (1974).
[38] Gersten, K.: Corner interference effects. AGARD Rep. 299 (1959).
[39] Gersten, K.: Die Grenzschichtströmung in einer rechtwinkligen Ecke. ZAMM 39, 428—429 (1959).
[40] Glauert, M.B.: The wall jet. JFM I, 625—643 (1956).
[41] Glauert, M.B., and Lighthill, M.J.: The axisymmetric boundary layer on a long thin cylinder. Proc. Roy. Soc. London A 230, 188—203 (1955).
[42] Görtler, H.: Die laminare Grenzschicht am schiebenden Zylinder. Arch. Math. 3, Fasc. 3, 216—231 (1952).
[43] Görtler, H.: Decay of swirl in an axially symmetrical jet far from the orifice. Revista Math. Hisp.-Amer. IV, Ser. 14, 143—178 (1954).
[44] Grohne, D.: Zur laminaren Strömung in einer kreiszylindrischen Dose mit rotierendem Deckel. ZAMM-Sonderheft 17—20 (1956).
[45] Gruschwitz, E.: Turbulente Reibungsschichten mit Sekundärströmung. Ing.-Arch. 6, 355—365 (1935).
[45a] Hama, F.R., and Peterson, L.F.: Axisymmetric laminar wake behind a slender body of revolution. JFM 76, 1—15 (1976).
[46] Hannah, D.M.: Forced flow against a rotating disc. ARC RM 2772 (1952).
[47] Hansen, A.G., Herzig, H.Z., and Costello, G.R.: A visualization study of secondary flows in cascades. NACA TN 2947 (1953).
[48] Hansen, A.G., and Herzig, H.Z.: Cross flows in laminar incompressible boundary layers. NACA TN 3651 (1956).
[49] Hayes, W.D.: The three-dimensional boundary layer. NAVORD Rep. 1313 (1951).
[50] Hoskin, N.E.: The laminar boundary layer on a rotating sphere. Fifty years of boundary layer research (W. Tollmien and H. Görtler, ed.), Braunschweig, 1955, 127—131.
[51] Howarth, L.: Note on the boundary layer on a rotating sphere. Phil. Mag. VII, 42, 1308—1315 (1951).
[52] Howarth, L.: The boundary layer in three-dimensional flow. Part I. Phil. Mag. VII, 42, 239—243 (1951).

[53] Howarth, L.: The boundary layer in three-dimensional flow. Part II: The flow near a stagnation point. Phil. Mag. VII, *42*, 1433—1440 (1951).

[54] Illingworth, C.R.: The laminar boundary layer of a rotating body of revolution. Phil. Mag. *44*, 351—389 (1953).

[55] Jacobs, W.: Systematische Sechskomponentenmessungen an Pfeilflügeln. Ing.-Arch. *18*, 344—362 (1950).

[56] Johnston, J.P.: On the three-dimensional turbulent boundary layer generated by secondary flow. Trans. ASME, Series D, J. Basic Eng. *82*, 233—248 (1960).

[57] Johnston, J.P.: The turbulent boundary layer at a plane of symmetry in a three-dimensional flow. Trans. ASME, Series D, J. Basic Eng. *82*, 622—628 (1960).

[58] Jones, R.T.: Effects of sweep-back on boundary layer and separation. NACA Rep. 884 (1947).

[59] Jungclaus, G.: Grenzschichtuntersuchungen in rotierenden Kanälen und bei scherenden Strömungen. Mitt. Max-Planck-Institut für Strömungsforschung No. 11, Göttingen (1955).

[60] Kelly, H.R.: A note on the laminar boundary layer on a circular cylinder in axial incompressible flow. JAS *21*, 634 (1954).

[61] Krzywoblocki, M.Z.: On steady, laminar round jets in compressible viscous gases far behind the mouth. Österr. Ing.-Arch. *3*, 373—383 (1949).

[62] Krzywoblocki, M.Z.: On the boundary layer in a corner by use of the relaxation method. GANITA *VII*, No. 2, 77—112 (1956).

[63] Küchemann, D.: Aircraft shapes and their aerodynamics for flight at supersonic speeds. Advances in Aeronautical Sciences *3*, 221—252 (1962).

[64] Küchemann, D.: The effect of viscosity on the type of flow on swept wings. Proc. Symposium Nat'l. Phys. Lab. (NPL) 1955.

[65] Langhaar, H.: Steady flow in the transition length of a straight tube. J. Appl. Mech. *9*, A 55—A 58 (1942).

[66] Liebe, W.: Der Grenzschichtzaun. Interavia *7*, 215—217 (1952).

[67] Loos, H.G.: A simple laminar boundary layer with secondary flow. JAS *22*, 35—40 (1955).

[68] Ludwieg, H.: Die ausgebildete Kanalströmung in einem rotierenden System. Ing.-Arch. *19*, 296—308 (1951).

[69] Luthander, S., and Rydberg, A.: Experimentelle Untersuchungen über den Luftwiderstand bei einer um eine mit der Windrichtung parallele Achse rotierenden Kugel. Phys. Z. *36*, 552—558 (1935).

[70] Mager, A.: Three-dimensional laminar boundary layer with small cross-flow. JAS *21*, 835—845 (1954).

[71] Mager, A.: Thick laminar boundary layer under sudden perturbation. Fifty years of boundary layer research (W. Tollmien and H. Görtler, ed.), Braunschweig, 1955, 21—33.

[71a] Mager, A.: Three-dimensional boundary layers. Princeton University Series. High Speed Aerodynamics and Jet Propulsion. Princeton University Press. Vol. IV, 286—394 (1964).

[72] Mangler, W.: Zusammenhang zwischen ebenen und rotationssymmetrischen Grenzschichten in kompressiblen Flüssigkeiten. ZAMM *28*, 97—103 (1948).

[73] Martin, J.C.: On the Magnus effects caused by the boundary-layer displacement thickness on the bodies of revolution at small angles of attack. JAS *24*, 421—429 (1957).

[74] Michalke, A.: Theoretische und experimentelle Untersuchung einer rotationssymmetrischen laminaren Düsengrenzschicht. Ing.-Arch. *31*, 268—279 (1962).

[75] Millikan, C.B.: The boundary layer and skin friction for a figure of revolution. Trans. ASME *54*, 29—43 (1932).

[76] Möller, W.: Experimentelle Untersuchungen zur Hydrodynamik der Kugel. Phys. Z. *39*, 57—80 (1938).

[77] Moore, F.K.: Three-dimensional laminar boundary layer flow. JAS *20*, 525—534 (1953).

[78] Moore, F.K.: Three-dimensional boundary layer theory. Advances in Appl. Mech. *IV*, 159—228 (1956).

[79] Nakamura, I.: The laminar boundary layer on a spinning body of arbitrary shape in axial flow. Research Bulletin No. XVI, *16*, 31—45 (1972).

[80] Nakamura, I., Yamashita, S., and Furuya, Y.: The thick turbulent boundary layers on rotating cylinders in axial flow. Second Intern. Symposium Fluid Machinery and Fluidics, Tokyo, Sept. 1972.

[81] Nigam, S.D.: Note on the boundary layer on a rotating sphere. ZAMP *5*, 151—155 (1954).

[81a] Nydahl, J.E.: Heat transfer for the Bödewadt problem. Dissertation, Colorado State University, Fort Collins, Colorado 1971.

[82] Oman, R.: The three-dimensional laminar boundary layer along a corner. Sc. D. Thesis, MIT, Cambridge, Mass., 1959.

[83] Pack, D.C.: Laminar flow in an axially symmetrical jet of compressible fluid, far from the orifice. Proc. Cambr. Phil. Soc. *50*, 98—104 (1954).

[84] Parr, O.: Untersuchungen der dreidimensionalen Grenzschicht an rotierenden Drehkörpern bei axialer Anströmung. Diss. Braunschweig 1962; Ing.-Arch. *32*, 393—413 (1963); see also: Die Strömung um einen axial angeströmten rotierenden Drehkörper. Jb. Schiffbautechn. Ges. *53*, 260—271 (1959), and: Flow in the three-dimensional boundary layer on a spinning body of revolution. AIAA J. *2*, 362—363 (1964).

[85] Pfleiderer, C.: Untersuchungen auf dem Gebiet der Kreiselradmaschinen. VDI-Forschungsheft No. 295 (1927).

[86] Prandtl., L.: Über Reibungsschichten bei dreidimensionalen Strömungen. Betz-Festschrift 1945, 134—141, or Coll. Works *2*, 679—686 (1961).

[87] Pretsch, J.: Die laminare Reibungsschicht an elliptischen Zylindern und Rotationsellipsoiden bei symmetrischer Anströmung. Luftfahrtforschung *18*, 397—402 (1941).

[88] Probstein, R.F., and Elliot, D.: The transverse curvature effect in compressible axially symmetric laminar boundary-layer flow. JAS *23*, 208—224 (1956).

[89] Punnis, B.: Zur Berechnung der laminaren Einlaufströmung im Rohr. Diss. Göttingen 1947.

[90] Queijo, M.J., Jaquet, B.M., and Wolhart, W.D.: Wind-tunnel investigation at low speed of the effects of chordwise wing fences and horizontal-tail position on the static longitudinal stability characteristics of an airplane model with a 35° swept-back wing. NACA Rep. 1203 (1954).

[91] Rainbird, W.J., Crabbe, R.S., and Jurewicz, L.S.: The flow separation about cones at incidence. Nat. Res. Council Canada, DMENAE Quart. Bull. 1963 (2).

[91a] Raju, K.G.R., Loeser, J., and Plate, E.J.: Velocity profiles and fence for a turbulent boundary layer along smooth and rough plates. JFM *76*, 383—399 (1976).

[92] Ram, Vasanta: Ähnliche Lösungen für die Geschwindigkeits- und Temperaturverteilung in der inkompressiblen laminaren Grenzschicht entlang einer rechtwinkligen Ecke. Ein theoretischer Beitrag zum Problem der Interferenz von Grenzschichten. Diss. Braunschweig 1966; Jb. WGL 156—178 (1966).

[93] Rott, N., and Crabtree, L.F.: Simplified laminar boundary layer calculation for bodies of revolution and for yawed wings. JAS *19*, 553—565 (1952).

[93a] Rubin, S.G.: Incompressible flow along a corner. JFM *26*, 97—110 (1966).

[94] Sawatzki, O.: Strömungsfeld um eine rotierende Kugel. Acta Mech. *9*, 159—214 (1970).

[95] Seban, R.A., and Bond, R.: Skin friction and heat-transfer characteristics of a laminar boundary layer on a cylinder in axial incompressible flow. JAS *18*, 671—675 (1951).

[96] Schiller, L.: Untersuchungen über laminare und turbulente Strömung. Forschg. Ing.-Wes. Heft 428, (1922); ZAMM *2*, 96—106 (1922); Phys. Z. *23*, 14 (1922).

[97] Schlichting, H.: Laminare Strahlausbreitung. ZAMM *13*, 260—263 (1933).

[98] Schlichting, H., and Truckenbrodt, E.: Die Strömung an einer angeblasenen rotierenden Scheibe. ZAMM *32*, 97—111 (1952).

[99] Schlichting, H.: Die laminare Strömung um einen axial angeströmten rotierenden Drehkörper. Ing.-Arch. *21*, 227—244 (1953).

[100] Schlichting, H.: Three-dimensional boundary layer flow. Lecture at the IXth Convention of the International Association for Hydraulic Research at Dubrovnik/Jugoslavia, Sept. 1961. Proc. Neuvième Assemblée Générale de l'Association Internationale de Recherches Hydrauliques, Dubrovnik, 1262—1290; see also DFL-Rep. 195 (1961).

[101] Schlichting, H.: Grenzschichttheorie. 5th ed., G. Braun Verlag, Karlsruhe, 1965.

[102] Scholkemeier, F.W.: Die laminare Reibungsschicht an rotationssymmetrischen Körpern. Diss. Braunschweig 1943. Shortened version in Arch. Math. *1*, 270—277 (1949).

[103] Sears, W.R.: Boundary layer of yawed cylinders. JAS *15*, 49—52 (1948).

[103a] Sears, W.R.: Boundary layers in three-dimensional flow. Appl. Mech. Rev. *7*, 281—285 (1954).

[104] Sedney, R.: Laminar boundary layer on a spinning cone at small angles of attack in a supersonic flow. JAS *24*, 430—436 (1957).

[105] Squire, H.B.: The round laminar jet. Quart. J. Mech. Appl. Math. *4*, 321—329 (1951).

[106] Squire, H.B.: Radial jets. Fifty years of boundary-layer research (W. Tollmien and H. Görtler, ed.), Braunschweig, 1955, 47—54.

[106a] Sparrow, E.M., Lin, S., and Lundgren, T.S.: Flow development in the hydrodynamic entrance region of tubes and ducts. Phys. Fluids *7*, 338—347 (1964).

[107] Steinheuer, T.: Three-dimensional boundary layers on rotating bodies and in corners. AGARDograph No. 97, Part 2, 567—611 (1965).

[108] Stewartson, K., and Howarth, L.: On the flow past a quarter infinite plate using Oseen's equations. JFM 7, 1—21 (1960).

[109] Stewartson, K.: Viscous flow past a quarter infinite plate. JAS 28, 1—10 (1961).

[110] Talbot, L.: Laminar swirling pipe flow. J. Appl. Mech. 21, 1—7 (1954).

[110a] Tan, S.: On laminar boundary layer over a rotating blade. JAS 20, 780—781 (1953).

[111] Taylor, G.I.: The boundary layer in the converging nozzle of a swirl atomizer. Quart. J. Mech. Appl. Math. 3, 129—139 (1950).

[112] Tetervin, N.: Boundary-layer momentum equations for three-dimensional flow. NACA TN 1479 (1947).

[113] Tifford, A.N., and Chu, S.T.: On the flow around a rotating disc in a uniform stream. JAS 19, 284—285 (1952).

[114] Timman, R.: The theory of three-dimensional boundary layers. Boundary layer effects in aerodynamics. Proc. of a Symposium held at NPL, London, 1955.

[115] Timman, R., and Zaat, J.A.: Eine Rechenmethode für dreidimensionale laminare Grenzschichten. Fifty years of boundary-layer research (W. Tollmien and H. Görtler, ed.), Braunschweig, 1955, 432—445.

[116] Tomotika, S.: Laminar boundary layer on the surface of a sphere in a uniform stream. ARC RM. 1678 (1935).

[117] Tomotika, S., and Imai, I.: On the transition from laminar to turbulent flow in the boundary layer of a sphere. Rep. Aero. Res. Inst. Tokyo 13, 389—423 (1938); and Tomotika, S.: Proc. Phys. Math. Soc. Japan 20 (1938).

[118] Truckenbrodt, E.: Ein Quadraturverfahren zur Berechnung der Reibungsschicht an axial angeströmten rotierenden Drehkörpern. Ing.-Arch. 22, 21—35 (1954).

[119] Truckenbrodt, E.: Die turbulente Strömung an einer angeblasenen rotierenden Scheibe. ZAMM 34, 150—162 (1954).

[120] Vogelpohl, G.: Die Strömung der Wirbelquelle zwischen ebenen Wänden mit Berücksichtigung der Wandreibung. ZAMM 24, 289—294 (1944).

[121] Weber, H.E.: The boundary layer inside a conical surface due to swirl. J. Appl. Mech. 23, 587—592 (1956).

[122] Wieghardt, K.: Einige Grenzschichtmessungen an Rotationskörpern. Schiffstechnik 3, 102—103 (1955/56).

[123] Wieselsberger, C.: Über den Luftwiderstand bei gleichzeitiger Rotation des Versuchskörpers. Phys. Z. 28, 84—88 (1927).

[124] Wild, J.M.: The boundary layer of yawed infinite wings. JAS 16, 41—45 (1949).

[125] Yamaga, J.: An approximate solution of the laminar boundary layer on a rotating body of revolution in uniform compressible flow. Proc. 6th Japan. Nat. Congr. Appl. Mech., 295—298 (1956).

[126] Young, A.D.: Some special boundary-layer problems (20th Prandtl Memorial Lecture). ZFW 1, 401—414 (1977).

[127] Zamir, M., and Young, A.D.: Experimental investigation of the boundary layer in a streamwise corner. Aero. Quart. 21, 313—339 (1970).

CHAPTER XII

Thermal boundary layers in laminar flow †

a. Derivation of the energy equation

The transfer of heat between a solid body and a liquid or gaseous flow is a problem whose consideration involves the science of fluid motion. On the physical motion of the fluid there is superimposed a flow of heat and, generally speaking, the two fields interact. In order to determine the temperature distribution it is necessary to combine the equations of motion with those of heat conduction. It is intuitively evident that the temperature distribution around a hot body in a fluid stream will often have the same character as the velocity distribution in boundary-layer flow. For example, if we imagine a solid body which is placed in a fluid stream and which is heated so that its temperature is maintained above that of the surroundings then it is clear that the temperature of the stream will increase only over a thin layer in the immediate neighbourhood of the body and over a narrow wake behind it, Fig. 4.2. The major part of the transition from the temperature of the hot body to that of the colder surroundings takes place in a thin layer in the neighbourhood of the body which, in analogy with flow phenomena, may be termed the thermal boundary layer. It is evident that flow phenomena and thermal phenomena interact to a high degree.

To begin the investigation of such phenomena, it is necessary to establish the energy balance for a fluid element in motion and to consider it in addition to the equations of motion. For an incompressible fluid the energy balance is determined by the internal energy, the conduction of heat, the convection of heat with the stream and the generation of heat through friction. In a compressible fluid there is an additional term due to the work of expansion (or compression) when the volume is changed. In all cases radiation may also be present, but its contribution is small at moderate temperatures, and we shall neglect it completely. We shall now proceed to establish this energy balance on the basis of the First Law of Thermodynamics, considering an elementary volume $\Delta V = dx\, dy\, dz$ of mass $\Delta M = \varrho \Delta V$ as it flows along its path. The quantity of heat dQ added to the volume during an element of time dt serves to increase its internal energy by an amount dE_T and to perform work dW. Hence

$$\frac{dQ}{dt} = \underbrace{\frac{dE_T}{dt}}_{\text{energy}} + \underbrace{\frac{dW}{dt}}_{\text{work}} \qquad [\text{J/sec}]\ddagger . \qquad (12.1)$$

---

† I am indebted to Professor K. Gersten for the revised version of this chapter.

‡ Here and in what follows, we employ the Joule (1 Joule = 1 Newtonmeter or 1 J = 1 Nm) as the unit of work and energy).

The term dE_T/dt represents a substantial derivative which consists of a local and a convective contribution.

If the transfer of heat by radiation is neglected, then it can occur only through conduction. According to Fourier's law, the heat flux q [J/m^2 sec] per unit area A and time is proportional to the temperature gradient, so that

$$\frac{1}{A}\frac{dQ}{dt} = q = -k\frac{\partial T}{\partial n}. \tag{12.2}$$

Here k [J/m sec deg] denotes the thermal conductivity of the fluid. The negative sign signifies that the heat flux is reckoned as positive in the direction of the temperature gradient. Hence, the amount of heat transferred into volume ΔV through surface elements which are normal to the x-direction (Fig. 12.1) is equal to $-(k\partial T/\partial x)\,dydz$. By contrast, the amount leaving the volume is given by $[(k\partial T/\partial x) + (\partial/\partial x)\,(k\partial T/dx)dx]\,dy\,dz$. Thus, the amount of heat added by conduction during time dt to a volume ΔV can be written

$$dQ = dt\,\Delta V\left\{\frac{\partial}{\partial x}\left(k\frac{\partial T}{\partial x}\right) + \frac{\partial}{\partial y}\left(k\frac{\partial T}{\partial y}\right) + \frac{\partial}{\partial z}\left(k\frac{\partial T}{\partial z}\right)\right\}. \tag{12.3}$$

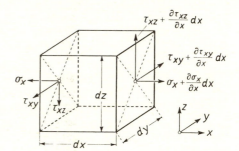

Fig. 12.1. Frictional stresses on a fluid element

The change in the total energy, dE_T, consists of a change $dE = \varrho\Delta V\,de$ in the internal energy and a change in kinetic energy by an amount $d\{\frac{1}{2}\varrho\,\Delta V\,(u^2+v^2+w^2)\}$, if the change in the potential energy due to a displacement in the gravitational field is neglected. Hence

$$\frac{dE_T}{dt} = \varrho\,\Delta V\left\{\frac{de}{dt} + \frac{1}{2}\frac{d}{dt}\,(u^2+v^2+w^2)\right\}. \tag{12.4}$$

In order to determine the work performed, we consider first the contribution from the component σ_x of stress. According to Fig. 12.1, we see that the work per unit time is then

$$dW_{\sigma_x} = -\,dy\,dz\left\{-u\sigma_x + \left(u + \frac{\partial u}{\partial x}\,dx\right)\left(\sigma_x + \frac{\partial\sigma_x}{\partial x}\,dx\right)\right\}$$

$$= -\Delta V\frac{\partial}{\partial x}\,(u\,\sigma_x). \tag{12.5}$$

The negative sign is added in order to follow the sign convention of eqn. (12.1) according to which work added to the fluid from the outside is negative. The total work performed by the normal and shearing stresses per unit time can now be written as

$$dW = - \Delta V \left\{ \frac{\partial}{\partial x} (u\,\sigma_x + v\,\tau_{xy} + w\,\tau_{xz}) + \right.$$
$$+ \frac{\partial}{\partial y} (u\,\tau_{yx} + v\,\sigma_y + w\,\tau_{yz}) +$$
$$\left. + \frac{\partial}{\partial z} (u\,\tau_{zx} + v\,\tau_{zy} + w\,\sigma_z) \right\}. \qquad (12.6)$$

Here $\sigma_x, \sigma_y, \ldots, \tau_{zy}$ denote the normal and shearing stresses introduced earlier in eqns. (3.20) and (3.25). Substituting eqns. (12.3), (12.4) and (12.6) into eqn. (12.1), and performing a number of obvious simplifications, including those introduced by eqn. (3.11), we obtain, after some calculation, the following energy equations of the flow:

$$\varrho \frac{de}{dt} + p \, \mathrm{div}\, \boldsymbol{w} = \frac{\partial}{\partial x} \left(k\, \frac{\partial T}{\partial x} \right) + \frac{\partial}{\partial y} \left(k\, \frac{\partial T}{\partial y} \right) + \frac{\partial}{\partial z} \left(k\, \frac{\partial T}{\partial z} \right) + \mu\, \Phi. \qquad (12.7)$$

Here Φ represents the dissipation function given by

$$\Phi = 2 \left\{ \left(\frac{\partial u}{\partial x} \right)^2 + \left(\frac{\partial v}{\partial y} \right)^2 + \left(\frac{\partial w}{\partial z} \right)^2 \right\} + \left(\frac{\partial v}{\partial x} + \frac{\partial u}{\partial y} \right)^2 +$$
$$+ \left(\frac{\partial w}{\partial y} + \frac{\partial v}{\partial z} \right)^2 + \left(\frac{\partial u}{\partial z} + \frac{\partial w}{\partial x} \right)^2 - \frac{2}{3} \left(\frac{\partial u}{\partial x} + \frac{\partial v}{\partial y} + \frac{\partial w}{\partial z} \right)^2. \qquad (12.8)$$

Equation (12.7) enjoys general validity, but in most practical cases it is possible to simplify it still further. In doing so, it is necessary carefully to distinguish between the case of a perfect gas and that of an incompressible fluid. The thermodynamic properties of the latter *do not* constitute a limiting case of the properties of the former. In fact, the variation in the internal energy of a perfect gas is $de = c_v\, dT$, whereas that of its enthalpy is $dh = c_p dT$. The corresponding variations for an incompressible fluid are $de = c\, dT$ and $dh = c\, dT + (1/\varrho)dp$.

Employing the equation of continuity (3.1), we obtain in the case of a perfect gas that

$$\mathrm{div}\, \boldsymbol{w} = \frac{\partial u}{\partial x} + \frac{\partial v}{\partial y} + \frac{\partial w}{\partial z} = - \frac{1}{\varrho} \frac{D\varrho}{Dt}. \qquad (12.9)$$

With the aid of this equation and of

$$c_p\, dT = c_v\, dT + d \left(\frac{p}{\varrho} \right) \qquad (12.10)$$

we can specialize eqn. (12.7) to the form

$$\varrho\, c_p\, \frac{dT}{dt} = \frac{dp}{dt} + \left\{ \frac{\partial}{\partial x} \left(k\, \frac{\partial T}{\partial x} \right) + \frac{\partial}{\partial y} \left(k\, \frac{\partial T}{\partial y} \right) + \right.$$
$$\left. + \frac{\partial}{\partial z} \left(k\, \frac{\partial T}{\partial z} \right) \right\} + \mu\, \Phi. \qquad (12.11)$$

Here c_p [J/kg deg] represents the specific heat at constant pressure per unit mass. In general, c_p depends on temperature. In the case of a constant thermal conductivity, we obtain the simpler form

$$\varrho \, c_p \, \frac{dT}{dt} = \frac{dp}{dt} + k \left(\frac{\partial^2 T}{\partial x^2} + \frac{\partial^2 T}{\partial y^2} + \frac{\partial^2 T}{\partial z^2} \right) + \mu \, \Phi \, . \tag{12.12}$$

In the case of an incompressible fluid, we have div $w = 0$, and eqn. (12.7) together with $de = c \, dT$ yields

$$\varrho \, c \, \frac{dT}{dt} = k \left(\frac{\partial^2 T}{\partial x^2} + \frac{\partial^2 T}{\partial y^2} + \frac{\partial^2 T}{\partial z^2} \right) + \mu \, \Phi \, . \tag{12.13}$$

b. Temperature increase through adiabatic compression; stagnation temperature

The temperature changes brought about by the dynamic pressure variation in a compressible flow are important for its heat balance. In particular, it appears useful to compare the temperature differences which result from the heat due to friction with those caused by compression. For this reason we shall first evaluate the temperature increase due to compression in a frictionless fluid stream: If the velocity varies along a streamline the temperature must vary also. In order to simplify the argument it is permissible to assume that the process is adiabatic and reversible because the small value of conductivity and the high rate of change in the thermodynamic properties of state will, in general, prevent any appreciable exchange of heat with the surroundings. In particular we propose to calculate the temperature increase $(\Delta T)_{ad} = T_0 - T_\infty$ which occurs at the stagnation point of a body in a stream and which is due to compression from p_∞ to p_0, Fig. 12.2.

Fig. 12.2.   Calculation of the temperature increase at stagnation point due to adiabatic compression $(\Delta T)_{ad} = T_0 - T_\infty$

For the case of zero heat conduction in frictionless flow the energy equation (12.11) gives the following relation between temperature and pressure along a streamline (coordinate s)

$$\varrho \, c_p \, w \, \frac{dT}{ds} = w \, \frac{dp}{ds} \, ,$$

where $w(s)$ denotes the velocity along a streamline. Dividing by ϱw and integrating along a streamline we obtain

$$c_p \, (T - T_\infty) = \int_{s_\infty}^{s} \frac{1}{\varrho} \frac{dp}{ds} \, ds = \int_{p_\infty}^{p} \frac{dp}{\varrho} \, .$$

Table 12.1.   Physical constants

$(1 \text{ J} = 1 \text{ Nm}; \quad 1 \text{ kJ/kg deg} = 10^3 \text{ m}^2/\text{sec}^2 \text{ deg})$

Substance	Temperature		Specific heat	Thermal conductivity	Thermal diffusivity	Viscosity	Kinematic viscosity	Prandtl number
	t	T	c_p	k	$a \times 10^6$	$\mu \times 10^6$	$\nu \times 10^6$	P
	[°C]	[K]	[kJ/kg K]	[J/m sec K]	[m²/sec]	[kg/m sec = Pa s]	[m²/sec]	[—]
Water (atmosph.)	20	293	4·183	0·598	0·143	1 000	1·006	7·03
	40	313	4·179	0·627	0·151	654	0·658	4·35
	60	333	4·191	0·650	0·159	470	0·478	3·01
	80	353	4·199	0·670	0·164	354	0·364	2·22
	100	373	4·215	0·681	0·169	276	0·294	1·75
Mercury	20	293	0·138	9.3	5	1 560	0·115	0·023
Lubr. oil	20	293	1·84	0·145	0·088	796 000	892	10 100
	40	313	1·92	0·143	0·084	204 000	231	2 750
	60	333	2·00	0·141	0·081	71 300	82	1 020
	80	353	2·10	0·140	0·078	31 500	37	471
Air (atmosph.)	−50	223	1·006	0·0205	13·1	14·6	9·5	0·72
	0	273	1·006	0·0242	19·2	17·1	13·6	0·71
	+50	323	1·006	0·0278	26·2	19·6	18·6	0·71
	100	373	1·009	0·0310	33·6	21·8	23·8	0·71
	200	473	1·028	0·0368	49·7	25·9	35·9	0·71
	300	573	1·048	0·0430	69·0	29·6	49·7	0·72

In an analogous manner, the complete Navier-Stokes equations (3.26) lead to the Bernoulli equation when viscosity is neglected in them and when an integral along a streamline is taken:

$$\frac{w^2}{2} + \int \frac{dp}{\varrho} = \text{const},$$

so that the temperature increase

$$T - T_\infty = \frac{1}{2\,c_p}\,(w_\infty{}^2 - w^2), \tag{12.14a}$$

and, in particular, the temperature increase at the stagnation point $(w = 0)$ due to adiabatic compression becomes

$$T_0 - T_\infty = (\Delta T)_{ad} = \frac{w_\infty{}^2}{2\,c_p}. \tag{12.14b}$$

Here w_∞ denotes the free-stream velocity (Fig. 12.2). The temperature T_0 assumed by the fluid when the velocity is reduced to zero is known as the *stagnation temperature*, sometimes also referred to as the *total temperature*. The difference $(\Delta T)_{ad} = T_0 - T_\infty$ between the stagnation and the free-stream temperature will here be called the *adiabatic temperature increase*.

Equation (12.14a), which is also known as the compressible Bernoulli equation, has been deduced on the assumption that the flow in the stream is reversible, i. e. that the entropy remains constant along a streamline. In actual fact eqn. (12.14a) is more general than this argument would suggest, as it applies to any one-dimensional stream, such as the flow through a slender nozzle, on condition that there is no external exchange of heat but irrespective of whether the entropy remains constant or not. The equation can be shown to be approximately true along a streamline in steady three-dimensional flow†. For air with $c_p = 1.006$ kJ/kg deg the adiabatic temperature increase at a velocity of $w_\infty = 100$ m/sec has a value of

$$(\Delta T)_{ad} = \frac{100^2}{2 \times 1006} = 4{\cdot}97\,{}^\circ\mathrm{C}\,.$$

The adiabatic temperature increase calculated for air from eqn. (12.14b) is shown plotted in Fig. 12.3. The specific heat, conductivity, and other thermal properties for a number of substances are listed in Table 12.1.

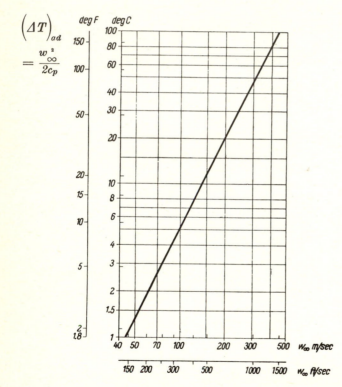

Fig. 12.3. Adiabatic temperature increase at stagnation point for air from eqn. (12.14b)
($c_p = 0{\cdot}24$ Btu/lbf R)

---

† Cf. e. g. eqn. (10) on p. 761 of "Modern Developments in Fluid Dynamics: High-Speed Flow" edited by L. Howarth, Clarendon Press, Oxford 1953.

c. Theory of similarity in heat transfer

In motions where temperature differences bring about differences in density it is necessary to include buoyancy forces in the equations of motion (3.26) and to treat them as body forces impressed on the liquid or gas, their magnitude being

$$X = \varrho \, g_x, \qquad Y = \varrho \, g_y, \qquad Z = \varrho \, g_z,$$

where g_x, g_y, g_z denote the components of the vector of gravitational acceleration $\mathbf{g}$. Generally speaking, the density is a function of pressure p and temperature T. If the latter do not deviate too much from their reference values p_∞ or T_∞, it is possible to use the expansion

$$\varrho = \varrho_\infty + \left(\frac{\partial \varrho}{\partial T}\right)_{p, \, T=\infty} (T - T_\infty) + \left(\frac{\partial \varrho}{\partial p}\right)_{T, \, p=p_\infty} (p - p_\infty), \tag{12.15}$$

or

$$\varrho = \varrho_\infty - \varrho_\infty \, \beta \, (T - T_\infty) + \frac{\gamma}{c_\infty^2} (p - p_\infty). \tag{12.16}$$

Here β denotes the coefficient of thermal expansion at temperature T_∞, γ is the ratio of the two specific heats, and c_∞ is the speed of sound of the fluid.

The last term can be neglected in flows which are affected by gravitation. This means, generally speaking, that the dependence of density on pressure can be ignored. Further, if we subtract the term $p = \varrho \, \mathbf{g}$ which is generated by the static field, we obtain from eqns. (3.26) and (3.27) the following modified form of the Navier-Stokes equations for the steady flow of a compressible fluid of constant viscosity:

$$\frac{\partial(\varrho u)}{\partial x} + \frac{\partial(\varrho v)}{\partial y} + \frac{\partial(\varrho w)}{\partial z} = 0, \tag{12.17}$$

$$\left.\begin{aligned}
\varrho \left(u \frac{\partial u}{\partial x} + v \frac{\partial u}{\partial y} + w \frac{\partial u}{\partial z}\right) &= -\frac{\partial p}{\partial x} + \varrho \, g_x \, \beta \, \theta + \mu \left[\varDelta^2 u + \frac{1}{3} \frac{\partial}{\partial x} \operatorname{div} w\right] \\
\varrho \left(u \frac{\partial v}{\partial x} + v \frac{\partial v}{\partial y} + w \frac{\partial v}{\partial z}\right) &= -\frac{\partial p}{\partial y} + \varrho \, g_y \, \beta \, \theta + \mu \left[\varDelta^2 v + \frac{1}{3} \frac{\partial}{\partial y} \operatorname{div} w\right] \\
\varrho \left(u \frac{\partial w}{\partial x} + v \frac{\partial w}{\partial y} + w \frac{\partial w}{\partial z}\right) &= -\frac{\partial p}{\partial z} + \varrho \, g_z \, \beta \, \theta + \mu \left[\varDelta^2 w + \frac{1}{3} \frac{\partial}{\partial z} \operatorname{div} w\right].
\end{aligned}\right\} \tag{12.18}$$

In addition it is necessary to consider the energy equation (12.12), also under the assumption of constant properties:

$$\varrho \, c_p \left(u \frac{\partial T}{\partial x} + v \frac{\partial T}{\partial y} + w \frac{\partial T}{\partial z}\right) = k \left(\frac{\partial^2 T}{\partial x^2} + \frac{\partial^2 T}{\partial y^2} + \frac{\partial^2 T}{\partial z^2}\right) +$$

$$+ u \frac{\partial p}{\partial x} + v \frac{\partial p}{\partial y} + w \frac{\partial p}{\partial z} + \mu \, \Phi. \tag{12.19}$$

Here the dissipation function, Φ, is given by eqn. (12.8). For perfect gases the equation of state can be written as

$$\frac{p}{\varrho} = R \, T. \tag{12.20}$$

In the general case of a compressible medium, eqns. (12.17) to (12.20) form a system of six simultaneous equations for the six variables: u, v, w, p, ϱ, T †. For incompressible media (liquids) the last equation as well as the terms $u\,\partial p/\partial x$ etc. which represent compression work vanish. In this case there are five equations for u, v, w, p, T.

It is necessary to emphasize that the symbol p does not denote the same physical quantity in eqns. (12.18), (12.19) and (12.20). Whereas in the last two equations p stands for the thermodynamic property, the symbol p in eqns. (12.18) represents the difference between the actual pressure and the static pressure of the medium at rest when its density is ϱ_∞ (cf. remark concerning fluids without free surfaces in Sec. IV a). In the cases treated in detail in the literature so far, the pressure term has been included either only in eqns. (12.18) — the case of free flows — or in the pair of equations (12.19) and (12.20) for compressible flows.

Before proceeding to indicate solutions of the above equations, which we shall discuss in the succeeding sections, we propose, first, to examine them from the point of view of the *principle of similarity* [106]. In this way we shall discover the dimensionless groups on which the solutions must depend. We begin by introducing dimensionless quantities into eqns. (12.18) and (12.19) in the same manner as in Sec. IV a, when Reynolds's similarity principle was deduced from the Navier-Stokes equations. All lengths will be referred to a representative length l, the velocities will be made dimensionless with reference to the free-stream velocity U_∞, the density with respect to ϱ_∞, and the pressure will be referred to $\varrho_\infty\,U_\infty^2$. The temperature in the energy equation will be made dimensionless with reference to the temperature difference $(\Delta T)_0 = T_w - T_\infty$ between the wall and the fluid at a large distance from the body; thus $\theta^* = (T - T_\infty)/(\Delta T)_0$. Denoting all dimensionless quantities by a star we obtain from eqns. (12.18) and (12.19) for the equation of motion in the x-direction and for the energy equation in the two-dimensional case with $g_x = -g^*\cos\alpha$:

$$\varrho^*\left(u^*\frac{\partial u^*}{\partial x^*} + v^*\frac{\partial u^*}{\partial y^*}\right) = -\frac{\partial p^*}{\partial x^*} + \frac{g\,\beta\,(\Delta T)_0\,l}{U_\infty^2}\,\varrho^*\,\theta^*\cos\alpha + \frac{\mu}{\varrho_\infty\,U_\infty l}\left(\frac{\partial^2 u^*}{\partial x^{*2}} + \frac{\partial^2 u^*}{\partial y^{*2}}\right),$$

$$(12.21)$$

$$\varrho^*\left(u^*\frac{\partial\theta^*}{\partial x^*} + v^*\frac{\partial\theta^*}{\partial y^*}\right) = \frac{k}{\varrho_\infty\,c_p\,U_\infty l}\left(\frac{\partial^2\theta^*}{\partial x^{*2}} + \frac{\partial^2\theta^*}{\partial y^{*2}}\right) +$$

$$+ \frac{U_\infty^2}{c_p\,(\Delta T)_0}\left(u^*\frac{\partial p^*}{\partial x^*} + v^*\frac{\partial p^*}{\partial y^*}\right) + \frac{\mu\,U_\infty}{\varrho_\infty\,c_p\,l\,(\Delta T)_0}\,\Phi^*. \quad (12.22)$$

The dimensionless dissipation function is here given by

$$\Phi^* = 2\left[\left(\frac{\partial u^*}{\partial x^*}\right)^2 + \cdots\right] + \cdots.$$

---

† Since the viscosity μ was assumed constant the above system is valid only for moderate changes in temperature. In the case of large temperature differences in gases (over 50° C or 90° F), or moderate ones (over 10° C or 18° F) in liquids, μ must be taken to vary with temperature. In this case the equation of motion retains the form (3.29). The six equations under consideration must be supplemented by the empirical viscosity law $\mu(T)$, eqn. (13.3), and, in all, we have a system of seven simultaneous equations for the seven functions $u, v, w, p, \varrho, T, \mu$.

It is recognized that the solutions of eqns. (12.21) and (12.22) depend on the following five dimensionless groups:

$$R = \frac{\varrho_\infty U_\infty l}{\mu} \;\; ; \;\; \frac{g \, \beta \, (\Delta T)_0 \, l}{U_\infty^2} \;\; ; \;\; \frac{k}{\varrho_\infty \, c_p \, U_\infty l} \;\; ; \;\; \frac{U_\infty^2}{c_p \, (\Delta T)_0} \;\; ; \;\; \frac{\mu \, U_\infty}{\varrho_\infty \, c_p \, l \, (\Delta T)_0} \; .$$

The *first* group is the already familiar Reynolds number. The fourth and fifth groups differ only by the factor R, so that, in all, there are only *four independent dimensionless quantities*. The *second* group can be represented as

$$\frac{g \, \beta \, l \, (\Delta T)_0}{U_\infty^2} = \frac{g \, \beta \, l^3 \, (\Delta T)_0}{\nu^2} \; \frac{\nu^2}{U_\infty^2 \, l^2} = G \, \frac{1}{R^2} \; .$$

This gives the Grashof number

$$G = \frac{g \, \beta \, l^3 \, (\Delta T)_0}{\nu^2} \; . \tag{12.23}$$

The *third* quantity can be written as

$$\frac{k}{\varrho_\infty \, c_p \, U_\infty l} = \frac{a}{U_\infty l} = \frac{a}{\nu} \; \frac{\nu}{U_\infty l} = \frac{1}{P} \, \frac{1}{R} \; , \tag{12.24}$$

where

$$a = \frac{k}{\varrho_\infty \, c_p} \tag{12.25}$$

is the *thermal diffusivity* [m²/sec or ft²/sec] and

$$P = \frac{\nu}{a} = \frac{\mu \, c_p}{k}$$

is the dimensionless Prandtl number. It will be noted that it depends only on the properties of the medium. For air $P = 0.7$ approximately and for water at 20 °C $P = 7$ approximately, whereas for oils it is of the order of 1000† owing to their large viscosity (see also Table 12.1). The *fourth* dimensionless quantity leads directly to the temperature increase through adiabatic compression as calculated in eqn. (12.14b). We have

$$E = \frac{U_\infty^2}{c_p (\Delta T)_0} = 2 \, \frac{(\Delta T)_{ad}}{(\Delta T)_0} \quad \text{(Eckert number)}, \tag{12.26‡}$$

where E is known as the dimensionless Eckert number. The quantity $E = U_\infty^2/c_p (\Delta T)_0$

---

† In heat-transfer theory the Péclet number

$$P_e = \frac{U_\infty l}{a},$$

is sometimes used. It is related to the Prandtl number by the equation $P_e = P \, R$.

‡ The ratio of the two temperature differences has, so far, not received a separate name. Following a suggestion by Professor E. Schmidt it has been proposed in an earlier edition to call it after Professor E. R. G. Eckert, and to give it the name of the Eckert number, E.

can be retained in incompressible flow also, but the interpretation with reference to adiabatic compression ceases to be valid. It is now possible to conclude that frictional heat and heat due to compression are important for the calculation of the temperature field when the free-stream velocity U_∞ is so large that the adiabatic temperature increase is of the same order of magnitude as the prescribed temperature difference between the body and the stream.

If this prescribed temperature difference is of the same order of magnitude as the absolute temperature of the free stream, which is, for example, the case with a rocket at very high altitude, the Eckert number becomes equivalent to the Mach number, as seen from the following calculation: from the equation of state of a perfect gas

$$\frac{p_\infty}{\varrho_\infty} = R\,T_\infty = T_\infty\,(c_p - c_v) = c_p\,T_\infty\,\frac{\gamma-1}{\gamma}\,,$$

with $c_p/c_v = \gamma$. Hence the velocity of sound

$$c_\infty{}^2 = \gamma\,p_\infty/\varrho_\infty = T_\infty\,c_p\,(\gamma-1)\,.$$

Now

$$\mathsf{E} = \frac{U_\infty{}^2}{c_p\,(\varDelta T)_0} = \frac{U_\infty{}^2}{c_p\,T_\infty}\,\frac{T_\infty}{(\varDelta T)_0} = (\gamma-1)\,\frac{U_\infty{}^2}{c_\infty{}^2}\,\frac{T_\infty}{(\varDelta T)_0} = (\gamma-1)\,\mathsf{M}^2\,\frac{T_\infty}{(\varDelta T)_0}\,,$$

so that

$$\mathsf{E} = (\gamma-1)\,\mathsf{M}^2\,\frac{T_\infty}{(\varDelta T)_0}\,, \tag{12.27}$$

where $\mathsf{M} = U_\infty/c_\infty$ is the Mach number. The work of compression and that due to friction become important when the free-stream velocity is comparable with that of sound, and when the prescribed temperature difference becomes of the order of the absolute temperature of the free stream; this occurs in practice in the flight of rockets at very high altitudes.

The preceding dimensional analysis leads to the conclusion that the solutions of the above system of equations for the velocity and temperature fields depend on the following four dimensionless groups:

$$
\left.
\begin{array}{lll}
\text{Reynolds number} & \mathsf{R} = \dfrac{U_\infty\,l}{\nu} & \\[3mm]
\text{Prandtl number} & \mathsf{P} = \dfrac{\nu}{a} = \dfrac{\mu\,c_p}{k} & \\[3mm]
\text{Grashof number} & \mathsf{G} = \dfrac{g\,\beta\,(\varDelta T)_0\,l^3}{\nu^2} & \\[3mm]
\text{Eckert number} & \mathsf{E} = \dfrac{U_\infty{}^2}{c_p(\varDelta T)_0}\,.
\end{array}
\right\} \tag{12.28}
$$

If $(\varDelta T)_0 \approx T_\infty$ the Eckert number is determined by the Mach number in accordance with eqn. (12.27). The problem of determining the dimensionless groups which govern flows with heat transmission is treated in a paper by P. Fischer [36].

In most applications we do not require to know all the details of the temperature and velocity field, but we wish, in the first place, to know the quantity of heat exchanged between the body and the stream. This quantity can be expressed with the aid of a coefficient of heat transfer, α, which is defined either as a local quantity or as a mean quantity over the surface of the body under consideration.

The coefficient of heat transfer is referred to the difference between the temperature of the wall and that of the fluid, the latter being taken at a large distance from the wall. If $q(x)$ denotes the quantity of heat exchanged per unit area and time ($=$ heat flux) at a point x, then according to *Newton's law of cooling* it is assumed that

$$q(x) = \alpha(x) \times (T_w - T_\infty) = \alpha(x)\,(\Delta T)_0 . \tag{12.29}$$

The coefficient of heat transfer has the dimension [J/m² sec deg]. At the boundary between a solid body and a fluid the transfer of heat is due solely to conduction. In accordance with Fourier's law the absolute value of the heat flux is, eqn. (12.2),

$$q(x) = -\,k\left(\frac{\partial T}{\partial n}\right)_{n=0} . \tag{12.30}$$

Comparing eqns. (12.29) and (12.30), and introducing dimensionless quantities, we obtain a local dimensionless coefficient of heat transfer which is known as the Nusselt number N [91]:

$$N\,(x) = \frac{\alpha(x)\,l}{k} = -\left(\frac{\partial T^*}{\partial n^*}\right)_{n^*=0} = -\,\frac{l}{(\Delta T)_0}\left(\frac{\partial T}{\partial n}\right)_{n=0} .$$

Thus the heat flux becomes

$$q = \frac{k}{l}\,N\,(T_w - T_\infty) = \frac{k}{l}\,N\,(\Delta T)_0 . \tag{12.31}$$

In accordance with the preceding argument it is to be expected that the velocity field and the temperature field as well as the local dimensionless coefficient of heat transfer must depend on the dimensionless groups considered previously. Thus

$$\left.\begin{aligned}
\frac{w}{U_\infty} &= f_1\,(s^*;\ \mathsf{R},\ \mathsf{P},\ \mathsf{G},\ \mathsf{E}) \\[4pt]
\frac{T - T_\infty}{(\Delta T)_0} &= f_2\,(s^*;\ \mathsf{R},\ \mathsf{P},\ \mathsf{G},\ \mathsf{E}) \\[4pt]
\mathsf{N} &= f_3\,(s^*;\ \mathsf{R},\ \mathsf{P},\ \mathsf{G},\ \mathsf{E})\ .
\end{aligned}\right\} \tag{12.32}$$

The second equation states that similar processes are also characterized by the fact that for them the ratio $T_\infty/(\Delta T)_0$ must have the same value (*cf.* [36]). The symbol s^* denotes here the three dimensionless space coordinates. If a mean value of the coefficient of heat transfer is formed by integrating over the whole surface, the space coordinate will cease to appear and

$$\mathsf{N}_m = f(\mathsf{R},\ \mathsf{P},\ \mathsf{G},\ \mathsf{E}) \tag{12.32a}$$

for geometrically similar surfaces.

When special solutions are considered then, in most cases, one or more of the dimensionless groups will disappear as the problem will only seldom be of this most general nature. As seen from eqn. (12.27) the temperature field and, hence, the coefficient of heat transfer depend on the Eckert number only when the temperature differences are large (50 to 100° C or 100 to 200° F) and when, simultaneously, the velocities are very large and of the order of the velocity of sound. With moderate velocities the temperature and velocity fields depend on the Eckert number when temperature differences are small (several degrees). Further, even with moderate velocities, the buoyancy forces in eqn. (12.21) caused by temperature differences are small compared with the inertia and friction forces. In such cases the problem ceases to depend on the Grashof number. Such flows are called *forced flows*. Hence, for *forced convection*

$$N_m = f(R, P) \quad \text{(forced convection)}.$$

The Grashof number becomes important only at very small velocities of flow, particularly if the motion is caused by buoyancy forces, such as in the stream which rises along a heated vertical plate. Such flows are called *natural*, and we refer to the problem as one in *natural convection*. In such cases the flow becomes independent of the Reynolds number, and

$$N_m = f(G, P) \quad \text{(natural convection)}.$$

Examples of problems in forced flow are given in Secs. e to g of the present chapter; examples of problems in natural convection are contained in Sec. h.

d. Exact solutions for the problem of temperature distribution in a viscous flow

We shall now proceed to solve several particular problems of temperature distribution. The examples to be discussed will be selected from the large number of possible cases on the ground of mathematical simplicity. We shall begin by discussing several cases of exact solutions, as given by H. Schlichting [101], just as we have begun with the discussion of examples of exact solutions of the equations of flow with friction in Chap. V. For the case of incompressible two-dimensional flow with constant properties the system of equations for the velocity and temperature distribution in steady flow along a horizontal x, z-plane we obtain from eqns. (12.17) to (12.19):

$$\frac{\partial u}{\partial x} + \frac{\partial v}{\partial y} = 0, \tag{12.33a}$$

$$\varrho \left(u \frac{\partial u}{\partial x} + v \frac{\partial u}{\partial y} \right) = -\frac{\partial p}{\partial x} + \mu \left(\frac{\partial^2 u}{\partial x^2} + \frac{\partial^2 u}{\partial y^2} \right), \tag{12.33b}$$

$$\varrho \left(u \frac{\partial v}{\partial x} + v \frac{\partial v}{\partial y} \right) = -\frac{\partial p}{\partial y} + \mu \left(\frac{\partial^2 v}{\partial x^2} + \frac{\partial^2 v}{\partial y^2} \right), \tag{12.33c}$$

$$\varrho c \left(u \frac{\partial T}{\partial x} + v \frac{\partial T}{\partial y} \right) = k \left(\frac{\partial^2 T}{\partial x^2} + \frac{\partial^2 T}{\partial y^2} \right) + \mu \Phi, \tag{12.34}$$

Fig. 12.4. Velocity and temperature distribution in Couette flow. a) Velocity distribution. b) Temperature distribution with heat generated by friction when the temperatures of both walls are equal. c) Temperature distribution with heat generated by friction for the case when the lower wall is non-conducting

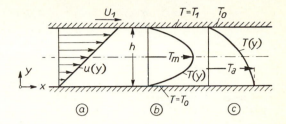

where

$$\Phi = 2\left[\left(\frac{\partial u}{\partial x}\right)^2 + \left(\frac{\partial v}{\partial y}\right)^2\right] + \left(\frac{\partial v}{\partial x} + \frac{\partial u}{\partial y}\right)^2.$$

1. Couette flow. A particularly simple exact solution of this system is obtained for *Couette flow*, i. e. for the case of flow between two parallel flat walls of which one is at rest, the other moving with a constant velocity U_1 in its own plane, Fig. 12.4. The solution of the equations of motion in the absence of a pressure gradient in the x-direction is

$$u(y) = U_1 \frac{y}{h}; \qquad v \equiv 0; \qquad p = \text{const}.$$

A very simple solution for the temperature distribution is obtained when it is postulated that the temperature is constant along the wall, the boundary conditions being

$$y = 0: \quad T = T_0; \qquad y = h: \quad T = T_1. \tag{12.35a}$$

In this case the dissipation function reduces to the simple expression $\Phi = (\partial u/\partial y)^2$, and the equation for temperature distribution becomes consequently

$$\varrho\, c\left(u\,\frac{\partial T}{\partial x} + v\,\frac{\partial T}{\partial y}\right) = k\left(\frac{\partial^2 T}{\partial x^2} + \frac{\partial^2 T}{\partial y^2}\right) + \mu\left(\frac{\partial u}{\partial y}\right)^2. \tag{12.35b}$$

With the boundary conditions (12.35a), the above equation has a solution which is independent of x. Since, with $v = 0$, the term $v\,\partial T/\partial y$ on the left-hand side also vanishes, all the convective terms on the left-hand side of eqn. (12.34) become equal to zero. The resulting temperature distribution is, therefore, due solely to the generation of heat through friction and to conduction in the transverse direction. From eqn. (12.35b) we obtain

$$k\,\frac{d^2 T}{dy^2} = -\mu\left(\frac{du}{dy}\right)^2, \tag{12.35c}$$

and, substituting du/dy, we have

$$k\,\frac{d^2 T}{dy^2} = -\mu\,\frac{U_1^2}{h^2}.$$

The solution of this equation which satisfies conditions (12.35a) is

$$\frac{T - T_0}{T_1 - T_0} = \frac{y}{h} + \frac{\mu \, U_1{}^2}{2 \, k \, (T_1 - T_0)} \frac{y}{h} \left(1 - \frac{y}{h}\right).$$

The dimensionless parameter

$$\frac{\mu \, U_1{}^2}{k \, (T_1 - T_0)}$$

can also be written as

$$\frac{\mu \, U_1{}^2}{k \, (T_1 - T_0)} = \frac{\mu \, c_p}{k} \, \frac{U_1{}^2}{c_p \, (\Delta T)_0} = \mathsf{P} \cdot \mathsf{E} \, ,$$

if we put $T_1 - T_0 = (\Delta T)_0$. It is seen that it can be expressed in terms of the Prandtl number and the Eckert number from eqn. (12.28). In the case under consideration, i. e. when there is no convection of heat, the temperature distribution is seen to depend on the product $\mathsf{P} \times \mathsf{E}$. If, finally, the abbreviation $\eta = y/h$ is introduced, the following very simple equation for temperature distribution is obtained:

$$\frac{T - T_0}{T_1 - T_0} = \eta + \frac{1}{2} \, \mathsf{P} \cdot \mathsf{E} \, \eta \, (1 - \eta) \, . \tag{12.36}$$

The temperature distribution consists of a linear term which is the same as in the case of a fluid at rest with no frictional heat generated. Superimposed on it there is a parabolic distribution which is due to the heat generated through friction. The temperature distribution for various values of the product $\mathsf{P} \times \mathsf{E}$ is seen plotted in Fig. 12.5. It is worthy of note that for a given value of the temperature difference of the two walls $T_1 - T_0 > 0$ heat flows from the upper wall to the fluid only as long as the velocity U_1 of the upper wall does not exceed a certain value. A reversal of the direction of the flow of heat at the upper plate occurs when the temperature gradient at it changes sign. It is seen from eqn. (12.36) that $(dT/dy)_{y=h} = 0$ for $\mu \, U_1{}^2/2 \, k = T_1 - T_0$. Hence the following rule applies to the direction of heat flow at the upper wall:

Heat from upper wall $\rightarrow$ fluid (cooling of upper wall):

$$\frac{\mu \, U_1{}^2}{2 \, k} < T_1 - T_0 \quad \text{or} \quad \mathsf{P} \cdot \mathsf{E} < 2 \, .$$

Heat from fluid $\rightarrow$ upper wall (heating of upper wall):

$$\frac{\mu \, U_1{}^2}{2 \, \lambda} > T_1 - T_0 \quad \text{or} \quad \mathsf{P} \cdot \mathsf{E} > 2 \, . \tag{12.37}$$

This simple example shows that the generation of heat due to friction exerts a large effect on the process of cooling and that at high velocities the warmer wall may become heated instead of being cooled. This effect is of fundamental importance for the consideration of cooling at high velocities. It will recur in the problems connected with thermal boundary layers and will be discussed later.

In the case when the two walls in Couette flow have equal temperatures ($T_1 = T_0$), eqn. (12.36) leads to a simple parabolic temperature distribution which is symmetrical with respect to the mean axis

$$T(y) - T_0 = \frac{\mu \, U_1{}^2}{2 \, k} \, \frac{y}{h} \left(1 - \frac{y}{h}\right).$$

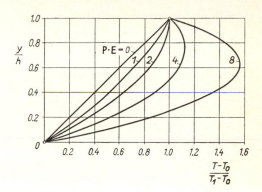

Fig. 12.5. Temperature distribution in Couette flow for various temperatures of both walls with heat generated by friction (T_0 = temperature of the lower wall, T_1 = temperature of the upper wall)

This distribution is seen plotted in Fig. 12.4b. The highest temperature T_m created by frictional heat occurs in the centre and has a value given by

$$T_m - T_0 = \frac{\mu U_1^2}{8 k}. \qquad (12.38)$$

In the case of compressible flow for which the above solution remains valid provided that the viscosity may be assumed to be independent of temperature, eqn. (12.38) can be put in the following dimensionless form

$$\frac{T_m - T_0}{T_0} = \frac{\gamma - 1}{8} \, \mathsf{P} \cdot \mathsf{M}^2, \qquad (12.38\,\mathrm{a})$$

where $\mathsf{M} = U_1/c_0$ denotes the Mach number and c_0 is the velocity of sound at temperature T_0. It is remarkable that the maximum temperature does not depend on the distance between the walls. The quantity of heat generated by friction is distributed evenly between the stationary and the moving wall.

The temperature distribution in the present example is important for the flow in the clearance between a journal and its bearing and was discussed in detail by G. Vogelpohl [143]. The flow in the clearance is laminar in view of the small dimensions of the latter and of the high viscosity of the oil. The temperature rise due to friction becomes considerable even at moderate velocities, as shown by the following example: Viscosity of oil at moderate temperature (say 30° C) from Table 12.1: $\mu = 0.4$ kg/m sec; conductivity of oil $k = 0.14$ J/m sec deg. Hence from eqn. (12.38) with $U_1 = 5$ m/sec: $T_m - T_0 = 9$ deg C, and for $U_1 = 10$ m/sec: $T_m - T_0 = 36\,°$C. The temperature rise in the lubricating oil is so large that its dependence on temperature becomes important. R. Nahme [90] extended the preceding solution to the case of temperature-dependent viscosity and found that the velocity distribution at right angles to the walls ceases to be linear.

A further important solution for the temperature distribution from eqn. (12.34) is obtained when it is postulated that all the heat due to friction is transferred to one of the walls only, whereas no heat transfer takes place at the other wall (adiabatic wall). Let it be assumed that the lower wall is insulated, so that the boundary conditions for temperature become:

$$y = h: \quad T = T_0; \qquad y = 0: \quad \frac{dT}{dy} = 0. \tag{12.39}$$

The solution of eqn. (12.34) with the above boundary conditions is

$$T(y) - T_0 = \mu \, \frac{U_1^2}{2k} \left(1 - \frac{y^2}{h^2}\right); \tag{12.40}$$

it is seen plotted in Fig. 12.4c. Thus the temperature increase of the lower wall is given by

$$T(0) - T_0 = T_a - T_0 = \mu \, U_1^2/2k. \tag{12.41}$$

The value T_a is called the *adiabatic wall temperature* as already mentioned; it is equal to the reading on a thermometer in the form of a flat plate. Upon comparing eqns. (12.41) and (12.38) it is seen that the highest temperature rise in the centre of the channel for the case of equal wall temperatures is equal to one quarter of the adiabatic wall temperature rise

$$T_a - T_0 = 4(T_m - T_0). \tag{12.42}$$

The criterion for cooling in the case of different wall temperatures given in eqn. (12.37) can be simplified if the adiabatic wall temperature T_a is introduced. We then have

$$T_1 - T_0 \gtrless T_a - T_0: \quad \left.\begin{array}{c} \text{cooling} \\ \text{heating} \end{array}\right\} \text{ of upper wall}. \tag{12.43}$$

H. M. de Groff [48] generalized the preceding solution for Couette motion to include the case when the viscosity of the fluid depends on temperature. The further extension to a compressible fluid was given by C.R. Illingworth [58] and A.J.A. Morgan [87].

2. Poiseuille flow through a channel with flat walls. A further and very simple exact solution for temperature distribution is obtained in the case of two-dimensional flow through a channel with parallel flat walls. Using the symbols explained in Fig. 12.6 we note with Poiseuille that the velocity distribution is parabolic:

Fig. 12.6. Velocity and temperature distribution in a channel with flat walls with frictional heat taken into account

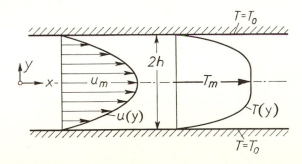

$$u(y) = u_m \left(1 - \frac{y^2}{h^2} \right).$$

Assuming, again, equal temperatures of the walls, i. e. $T = T_0$ for $y = \pm h$, we obtain from eqn. (12.35c)

$$k \frac{d^2 T}{dy^2} = - \frac{4 \mu (u_m)^2}{h^4} y^2$$

the solution of which is

$$T(y) - T_0 = \frac{1}{3} \frac{\mu (u_m)^2}{k} \left[1 - \left(\frac{y}{h} \right)^4 \right]. \tag{12.44}$$

The temperature distribution is represented by a parabola of the fourth degree, Fig. 12.6, and the maximum temperature rise in the centre of the channel is

$$T_m - T_0 = \frac{1}{3} \frac{\mu (u_m)^2}{k}. \tag{12.45}$$

An extension of the solution to the case of temperature-dependent viscosity was given by H. Hausenblas [53]. The corresponding solution for a circular pipe was given by U. Grigull [47].

A further exact solution for the thermal boundary layer can be obtained for the flow in a convergent and a divergent channel already considered in Sec. V 12. The solution for the velocity field due to O. Jeffery and G. Hamel quoted in that section was utilized by K. Millsaps and K. Pohlhausen [86] in order to solve the thermal problem. The temperature distribution across the channel is seen plotted in Fig. 12.7 for different Prandtl numbers. Owing to the dissipation of energy which is particularly large near the wall, the resulting temperature profiles acquire a pronounced "boundary-layer appearance". In fact, boundary-layer-like appearance becomes more pronounced as the Prandtl number increases. The velocity distribution u/u_0 from Fig. 5.15 has been plotted in Fig. 12.7 to provide a comparison.

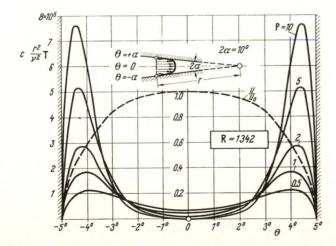

Fig. 12.7. Temperature distributions in a convergent channel of included angle $2\alpha = 10°$ at varying Prandtl numbers P, after K. Millsaps and K. Pohlhausen [86]. Reynolds number R = 1342 Velocity distribution from Fig. 5.15.

Owing to the dissipation of energy which is particularly large near the wall, the resulting temperature profiles acquire a pronounced "boundary-layer appearance"

Rotating disk: Chap. V, in particular Figs. 5.12 and 5.13, contained a solution to the flow problem around an infinitely large disk rotating in a fluid at rest. This was a solution of the system of eqns. (5.53). In order to determine the temperature field in the neighbourhood of a heated, rotating disk, it is necessary to expand the preceding system of equations by including the equation for the temperature distribution itself (energy equation). Such calculations have been performed by K. Millsaps and K. Pohlhausen [86a] who also included in them the effects of dissipation. Several additional solutions of the Navier-Stokes equations together with the energy equation which lead to similar solutions were given by B. L. Reeves and Ch. J. Kippenhan [97].

e. Boundary-layer simplifications

It has already been stated on several occasions that it is intuitively evident that in many cases the temperature field around a hot body in a fluid stream is of the *boundary-layer type*. This means that the temperature field which spreads from the body extends essentially only over a narrow zone in the immediate neighbourhood of the surface, whereas the regions at a larger distance from it are not materially affected by the higher body temperature. In particular this is the case when the conductivity, k, is small, as for gases and liquids. In such cases there is a very steep temperature gradient at right angles to the wall and the heat flux due to conduction is of the same order of magnitude as that due to convection only across a thin layer near the wall. On the other hand, it is to be expected that the temperature increase near an unheated body in a fluid stream flowing at a high Reynolds number, and which is due to the generation of frictional heat, is important only in the thin boundary layer, because the quantity of mechanical energy which is transformed into heat through friction is significant only there. Hence it may be expected that in conjunction with the velocity boundary layer there will be formed a thermal boundary layer across which the temperature gradient is very large. It is, therefore, possible to take advantage of this fact and to introduce into the energy equation, which governs the temperature distribution, simplifications of a similar nature to those introduced earlier into the equations of motion (Chap. VII).

Dimensionless forms of the equations of motion and energy were given in Sec. c of this chapter where a representative velocity, U_∞, a representative length, l, as well as a representative temperature difference, $(\Delta T)_0$, were used to render the relevant quantities dimensionless. For the sake of simplicity we shall restrict ourselves to the two-dimensional case with constant fluid properties and we shall choose the x-axis along the direction of the main stream. Under these assumptions the equation of motion in the x-direction and the energy equation, eqns. (12.21) and (12.22), can be written in the following form:

$$\varrho \left(u \frac{\partial u}{\partial x} + v \frac{\partial u}{\partial y} \right) = -\frac{\partial p}{\partial x} + \frac{G}{R^2} \; \theta \cos \alpha + \frac{1}{R} \left(\frac{\partial^2 u}{\partial x^2} + \frac{\partial^2 u}{\partial y^2} \right), \qquad (12.46a)$$

$$1 \; 1 \quad \delta_s \frac{1}{\delta_s} \qquad 1 \qquad\qquad\qquad \delta_s^2 \;\; 1 \quad\; \frac{1}{\delta_s^2}$$

$$\varrho \left(u\, \frac{\partial \theta}{\partial x} + v\, \frac{\partial \theta}{\partial y}\right) = \frac{1}{\mathsf{P\,R}} \left(\frac{\partial^2 \theta}{\partial x^2} + \frac{\partial^2 \theta}{\partial y^2}\right) + \mathsf{E} \left(u\, \frac{\partial p}{\partial x} + v\, \frac{\partial p}{\partial y}\right) + \mathsf{E}\, \frac{1}{\mathsf{R}}\, \varPhi \,. \qquad (12.46\,\mathrm{b})$$

$$\begin{array}{cccccccc} 1 \;\; 1 & \delta_S \dfrac{1}{\delta_\mathsf{T}} & \delta_T{}^2 & 1 & \dfrac{1}{\delta_T{}^2} & 1\;\;1 & \delta_S \delta_S & \delta_S{}^2 \dfrac{1}{\delta_S{}^2} \end{array}$$

The stars have now been omitted as superfluous. The orders of magnitude of the various terms in the two equations which have been estimated with the aid of the velocity boundary-layer equation (7.2) are shown above. The essential result of the previous estimation was that the viscous forces were of the same order of magnitude as the inertia forces only if the velocity boundary-layer thickness, δ_S, satisfied the condition

$$\left(\frac{\delta_S}{l}\right)^2 \sim \frac{1}{\mathsf{R}}\,. \qquad (12.47)$$

As a consequence it proved to be possible to neglect $\partial^2 u/\partial x^2$ against $\partial^2 u/\partial y^2$ in the first equation of motion and the second equation of motion dropped out altogether. This was connected with the fact that the transverse pressure gradient $\partial p/\partial y \sim \delta_S$, so that the pressure could be assumed to depend on x alone. As seen from eqn. (12.46a) the body force which is due to the buoyancy of the hotter fluid particles, i. e. to their thermal expansion, is of the same order of magnitude as the inertia and viscous forces if

$$\mathsf{G} \approx \mathsf{R}^2$$

which occurs only with very small velocities and considerable temperature differences.

It is now possible to make a similar estimation of terms in the energy equation. The dimensionless group

$$\frac{k}{\varrho_\infty c_p U_\infty l} = \frac{k}{c_p \mu}\, \frac{1}{\mathsf{R}} = \frac{1}{\mathsf{P}}\, \frac{1}{\mathsf{R}}\,, \qquad (12.54)$$

i. e. the multiplicative factor of the thermal conduction terms, is also a small quantity as far as liquids and gases are concerned if the Reynolds number is large, because the Prandtl number for gases is of the order of 1, and for liquids it ranges from 10 to 1000. Hence it is seen that the conduction terms can become of the same order of magnitude as the convection terms only if $\partial \theta/\partial y$ is very large, i. e., only if in the vicinity of the surface of the body there is a layer with a steep transverse temperature gradient: the *thermal boundary layer*. The order of magnitude of the convectional and viscous terms can now be estimated. It is shown under the equation and the symbol δ_T denotes the thickness of the thermal boundary layer†. The term $\partial^2 \theta/\partial x^2$ can be neglected against $\partial^2 \theta/\partial y^2$ and the conduction term becomes of the same order of magnitude as the convectional term only if the thickness of the thermal

---

† Since the Prandtl number may vary from fluid to fluid by several orders of magnitude (see Table 12.1), the present estimate cannot be expected to hold in the two limits $\mathsf{P} \to 0$ or $\mathsf{P} \to \infty$. In such cases better estimates result from the solutions given in eqns. (12.58) and (12.62a).

boundary layer is of the order

$$\left(\frac{\delta_T}{l}\right)^2 \sim \frac{1}{R} f_1(\mathsf{P}) .$$ (12.48)

In view of the previously obtained estimation for the thickness of the velocity boundary layer $\delta_S \sim 1/\sqrt{R}$, we obtain

$$\frac{\delta_T}{\delta_S} = f_2\left(\frac{x}{l}, \mathsf{P}, \mathsf{E}, \frac{\mathsf{G}}{\mathsf{R}^2}\right)$$ (12.49)

It follows that the ratio of the thicknesses of the two boundary layers is independent of the Reynolds number. If energy dissipation through friction and the buoyancy forces are omitted, the ratio of the two boundary-layer thicknesses becomes dependent on a *single* characteristic number — the Prandtl number. In this case it is possible to give a very good physical interpretation of the Prandtl number, as will be shown in Sec. XII f 4 in more detail.

Estimating the remaining terms in the energy equation it is concluded that in the expression for the dissipation function only the term $(\partial u/\partial y)^2$ remains significant, and

$$\Phi = \left(\frac{\partial u}{\partial y}\right)^2 \sim \frac{1}{\delta_S^2} .$$

The heat due to friction is seen to be important only if

$$\mathsf{E} = \frac{U_\infty^2}{c_p\,(\Delta T)_0} \sim 1 .$$

In the case of gases the heat generated by friction becomes important only if the temperature rise due to adiabatic compression is of the same order of magnitude as the difference in temperature between the body and the fluid. The same remark applies to the work of compression.

Reverting to dimensional quantities and taking into account the dependence of viscosity on temperature, we obtain the following simplified equations for two-dimensional compressible fluid flow:

$$\frac{\partial(\varrho\,u)}{\partial x} + \frac{\partial\,(\varrho v)}{\partial y} = 0 ,$$ (12.50a)

$$\varrho\left(u\frac{\partial u}{\partial x} + v\frac{\partial u}{\partial y}\right) = \frac{\partial}{\partial y}\left(\mu\frac{\partial u}{\partial y}\right) - \frac{dp}{dx} + \varrho\,g_x\,\beta\,(T - T_\infty) ,$$ (12.50b)

$$\varrho\,c_p\left(u\frac{\partial T}{\partial x} + v\frac{\partial T}{\partial y}\right) = k\frac{\partial^2 T}{\partial y^2} + \mu\left(\frac{\partial u}{\partial y}\right)^2 + u\frac{dp}{dx} ,$$ (12.50c)

$$\frac{p}{\varrho} = R\,T , \qquad \mu = \mu\,(T) .$$ (12.50d; e)

Since in the framework of boundary-layer theory the pressure may be regarded as a given, impressed force, we have here a system of five simultaneous equations for the five unknowns ϱ, u, v, T, μ.

Regarding the differences in the significance of p in eqn. (12.50b) on the one hand and in eqn. (12.50d) n the other, we refer the reader to the remark made in Sec. XII c just after eqns. (12.17) to (12.20).

For the incompressible case ($\varrho = \varrho_\infty = $ const) and for constant viscosity these equations reduce to

$$\frac{\partial u}{\partial x} + \frac{\partial v}{\partial y} = 0 \; , \tag{12.51 a}$$

$$\varrho_\infty \left(u \frac{\partial u}{\partial x} + v \frac{\partial u}{\partial y} \right) = \mu \frac{\partial^2 u}{\partial y^2} - \frac{dp}{dx} - \varrho_\infty \, g_x \, \beta \, (T - T_\infty), \tag{12.51 b}$$

$$\varrho_\infty \left(u \frac{\partial T}{\partial x} + v \frac{\partial T}{\partial y} \right) = k \frac{\partial^2 T}{\partial y^2} + \mu \left(\frac{\partial u}{\partial y} \right)^2, \tag{12.51 c}$$

giving three equations for $u, v,$ and T.

f. General properties of thermal boundary layers

1. Forced and natural flows. The differential equations for the velocity and thermal boundary layer, eqns. (12.51 b) and (12.51 c), are very similar in structure They differ only in the last two terms in the equation of motion and in the last term in the temperature equation. In the general case the velocity field and the temperature field mutually interact which means that the temperature distribution depends on the velocity distribution and, conversely, the velocity distribution depends on the temperature distribution. In the special case when buoyancy forces may be disregarded, and when the properties of the fluid may be assumed to be independent of temperature, mutual interaction ceases, and the velocity field no longer depends on the temperature field, although the converse dependence of the temperature field on the velocity field still persists. This happens at large velocities (large Reynolds numbers) and small temperature differences, such flows being termed *forced* (*cf.* p. 276). The process of heat transfer in such flows is described as *forced convection.* Flows in which buoyancy forces are dominant are called *natural*, the respective heat transfer being known as *natural convection.* This case occurs at very small velocities of motion in the presence of large temperature differences. The state of motion which accompanies natural convection is evoked by buoyancy forces in the gravitational field of the earth, the latter being due to density differences and gradients. For example, the field of motion which exists outside a vertical hot plate belongs to this class. Forced flows can be subdivided into those with moderate and those with high velocities depending on whether the heat due to friction and compression need or need not be taken into account. In both cases the temperature field depends on the field of flow. At moderate velocities, when the heat due to friction and compression may be neglected, the dependence of the temperature field on the velocity field is governed solely by the Prandtl number. To each *single* velocity field there corresponds a singly infinite family of temperature distributions with the Prandtl number as its parameter. At high velocities work due both to friction and compression must be included. Whether this is necessary or not depends on the Eckert number $\mathsf{E} = 2 (\varDelta T)_{ad}/(\varDelta T)_0$, i. e. on whether it is comparable with unity. In other words, the work due to friction and compression must be taken into account when the temperature increase due to friction and compression is comparable with the temperature difference prescribed as a boundary condition (temperature difference between body and fluid). If the prescribed temperature difference is of the order of the mean absolute temperature, the work due to friction and compression becomes important only if the velocity of flow is comparable with that of sound.

It is important to note that the temperature equation is linear, unlike the equation of motion. This leads to considerable simplifications in the process of integrating, and superposition of known solutions becomes possible.

2. Adiabatic wall. Finally it is necessary to mention that the variety of possible sets of boundary conditions is much greater for the temperature field than for the velocity field. The temperature on the surface of the body may be constant or variable but, moreover, it is also possible to encounter problems for which the heat flux is prescribed. In view of eqn. (12.30), this means that the temperature gradient at the wall may appear as a boundary condition. The so-called *adiabatic wall* constitutes a particular example of the latter class of cases, since it must be postulated that there is no heat flux from the wall to the fluid, i. e., the boundary condition at the wall is

$$\left(\frac{\partial T}{\partial n}\right)_{n=0} = 0 \qquad \text{(adiabatic wall)} .$$

This case can be visualized by imagining that the wall of the body is perfectly insulated against heat flow. The heat generated by the fluid through friction serves to heat the wall until the condition $(\partial T/\partial n)_{n=0} = 0$ is reached. Thus the temperature of the wall which we may also call the *adiabatic wall temperature* becomes higher than that of the fluid at some distance from it. Such conditions are satisfied in practice when a so-called plate thermometer is used, i. e. when the temperature of a fluid stream is measured with the aid of a flat plate which is placed parallel to the stream†. The excess temperature on the plate constitutes the error of the plate thermometer. The error must be deducted in order to obtain the true temperature of the moving fluid. This difference is sometimes called the *kinetic temperature*.

3. Analogy between heat transfer and skin friction. For boundary-layer flow there exists a remarkable relationship between heat transfer and skin friction which, in its simplest form, was discoverd by O. Reynolds [98] in 1874. For this reason, this relation is known as the Reynolds analogy.

It has been shown in Sec. VIIIa that *all* solutions of the two-dimensional boundary-layer equations for an incompressible fluid have the form

$$\frac{u}{U_\infty} = f_1\left(\frac{x}{l}, \frac{y}{l}\sqrt{\mathsf{R}}\right), \qquad (12.52\,\mathrm{a})$$

---

† For this reason in older textbooks the problem of an adiabatic wall was referred to as the plate-thermometer problem.

$$\frac{v}{U_\infty} \sqrt{R} = f_2 \left(\frac{x}{l}, \frac{y}{l} \sqrt{R} \right) \quad \text{with} \quad R = \frac{U_\infty l}{v}.$$ (12.52 b)

If the work of compression as well as the evolution of heat through dissipation can be neglected, the same reasoning shows that *all* solutions of equations (12.51 c) which describe the thermal boundary layer, must be of the form:

$$\theta^* = \frac{T - T_\infty}{T_w - T_\infty} = f_3 \left(\frac{x}{l}, \frac{y}{l} \sqrt{R}, P \right).$$ (12.52 c)

Hence, the heat flux from eqn. (12.30) can be written

$$q = - k \left(\frac{\partial T}{\partial y} \right)_{y=0} = \frac{k}{l} (T_w - T_\infty) \sqrt{R} \, \bar{f}_3 \left(\frac{x}{l}, P \right);$$ (12.52 d)

this gives the Nusselt number

$$N = \frac{q \, l}{k (T_w - T_\infty)} = \sqrt{R} \, \bar{f}_3 \left(\frac{x}{l}, P \right).$$ (12.53)

This very important relation states that for *all* laminar boundary layers — always on the assumption that compression work and frictional heat are negligible — the Nusselt number is proportional to the square root of the Reynolds number. Instead of the general relation between the Nusselt and the Reynolds numbers implied by eqn. (12.32) on the basis of the full Navier-Stokes equations, the boundary-layer simplifications lead to this special, more explicit relation.

Equation (12.52 a) allows us to write down the following formula for the local shearing stress:

$$\tau_w = \mu \left(\frac{\partial u}{\partial y} \right)_{y=0} = \frac{\mu \, U_\infty \sqrt{R}}{l} \, \bar{f}_1 \left(\frac{x}{l} \right)$$ (12.53 a)

and to form with it the local skin-friction coefficient

$$c_f' = \frac{\tau_0}{\frac{1}{2} \varrho \, U^2{}_\infty} = \frac{1}{\sqrt{R}} \, \bar{f}_1 \left(\frac{x}{l} \right).$$ (12.53 b)

On combining eqn. (12.53 b) with (12.53), we obtain the general relation

$$\boxed{N = \frac{1}{2} c_f' \, R \, f \left(\frac{x}{l}, P \right)}.$$ (12.54)

As already stated, this most general form of Reynolds's analogy is valid for *all* laminar boundary layers.

In particular, if there exists a class of *similar* solutions, namely that given by external flows of the form $U(x) = u_1 \, x^m$, then the considerations of Sec. IX a allow us to write

$$\frac{u}{U\,(x)} = F_1\left(y\,\sqrt{\frac{U\,(x)}{\nu\,x}}\right) \tag{12.54a}$$

and

$$\frac{v}{U\,(x)}\,\sqrt{\frac{x\,U\,(x)}{\nu}} = F_2\left(y\,\sqrt{\frac{U\,(x)}{\nu\,x}}\right). \tag{12.54b}$$

It follows immediately from the temperature equation that

$$\theta* = \frac{T - T_\infty}{T_w - T_\infty} = F_3\left(y\,\sqrt{\frac{U\,(x)}{\nu\,x}}\,,\,\mathsf{P}\right). \tag{12.54c}$$

In analogy with eqn. (12.53), the local Nusselt number formed with the coordinate x assumes the form

$$\mathsf{N}_x = \frac{\alpha\,x}{k} = \sqrt{\mathsf{R}_x} \cdot F\,(m,\,\mathsf{P})\,, \tag{12.55}$$

where

$$\mathsf{R}_x = \frac{x\,U\,(x)}{\nu}\,.$$

The function $F\,(m,\,\mathsf{P})$ will be discussed in more detail in Sec. XII g 2 (see eqn. (12.87) and Fig. 12.14). Thus between the local skin-friction coefficient

$$c'_{fx} = \frac{\tau_0}{\frac{1}{2}\,\varrho[U(x)]^2} = \frac{2}{\sqrt{\mathsf{R}_x}}\,\overline{F}_1(m) \tag{12.55a}$$

and the Nusselt number there exists the relation

$$\mathsf{N}_x = \tfrac{1}{2}\,c'_{fx}\,\mathsf{R}_x \cdot \overline{F}(m,\,\mathsf{P})\,. \tag{12.56}$$

The simplest type of flow, that on a flat plate at zero incidence, is characterized by the value $m = 0$ and by the fact that eqns. (12.51b) and (12.51c) for the velocity field and the temperature field, respectively, become completely analogous if the Prandtl number has the value of unity. In this case, the solutions themselves acquire identical algebraic forms, and we have

$$F_3\left(y\,\sqrt{\frac{u_1}{\nu\,x}}\,,\,1\right) \equiv F_1\left(y\,\sqrt{\frac{u_1}{\nu\,x}}\right) \quad \text{(if } m = 0\text{)}. \tag{12.56a}$$

Consequently,

$$\overline{F}\,(0,1) = 1\,,$$

and eqn. (12.56) simplifies to

$$\mathsf{N}_x = \tfrac{1}{2}\,c_{fx}'\,\mathsf{R}_x \quad (m = 0\,,\;\mathsf{P} = 1)\,, \tag{12.56b}$$

when applied to a flat plate. This is the simplest form of the Reynolds analogy; it was, as already stated, first discovered by O. Reynolds himself.

The preceding argument is applicable, so far, only to laminar, incompressible flows at constant wall temperature and on condition that energy dissipation may be neglected. Nevertheless, the preceding results can be extended to include other cases, such as that of a flat plate with frictional heat (see eqn. (12.81) and footnote on p. 299) or that with compression work (see Sec. XIIIc). It is particularly noteworthy that the Reynolds analogy can be recovered in turbulent flows where it plays an essential part in the calculation of heat-transfer rates (*cf.* Chap. XXIII).

4. Effect of Prandtl number. The considerations of this chapter convince us that the Prandtl number constitutes that parameter whose value is decisive for the extent of the thermal boundary layer and, therefore, for the rate at which heat is transferred in forced or free convection. According to its definition

$$P = \frac{\nu}{a} \,,$$

the Prandtl number is equal to the ratio of two quantities: one of them (viscosity) characterizes the fluid's transport properties with respect to the transport of momentum, the other (thermal diffusivity) doing the same for the transport of heat. If the fluid possesses a particularly large viscosity, it can be stated loosely that its ability to transport momentum is large. Consequently, the destruction of momentum introduced by the presence of a wall (no-slip condition) extends far into the fluid and the velocity boundary layer is comparatively large. Similar statements can be made with respect to the thermal boundary layer. It is, therefore, understandable that the Prandtl number serves as a direct measure for the ratio of the thicknesses of the two layers in forced flow, as already demonstrated in eqn. (12.49). The special case when $P = 1$ (already discussed) corresponds to flows for which the two boundary layers are approximately equal in extent; they are exactly equal along a flat plate at zero incidence when its temperature is uniform. In addition to this, the two limiting cases when the Prandtl number is either very large or very small are also worthy of attention; they are represented schematically in Fig. 12.8

Very small Prandtl numbers: It is clear from Fig. 12.8 that in the case of very small Prandtl numbers, such as occur in molten metals (for example in mercury), it

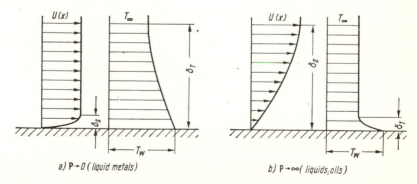

Fig. 12.8. Comparison between the temperature and velocity fields for boundary layers with very small and with very large values of Prandtl number

is possible to disregard the velocity boundary layer in the calculation of the thermal boundary layer. Consequently, the velocity components $u(x, y)$ and $v(x, y)$ can be replaced by $U(x)$ and $V(x, y) = - (dU/dx) y$, respectively, the approximation for V stemming from the continuity equation applied at the wall. The energy equation (12.51c) then assumes the particularly simple form

$$U(x) \frac{\partial T}{\partial x} - y \frac{dU}{dx} \frac{\partial T}{\partial y} = a \frac{\partial^2 T}{\partial y^2} \quad (\text{P} \to 0). \tag{12.57}$$

The temperature field has thus become independent of the velocity field in the boundary layer.

Introducing the similarity parameter

$$\eta = y \frac{U(x)}{2 \sqrt{a \int_0^x U(x) \, dx}}, \tag{12.57a}$$

we can transform the partial differential equation for temperature distribution into an ordinary one. This, in turn, leads to the following universal expression for the Nusselt number

$$\mathsf{N}_x = \frac{a \, x}{l} = \frac{x \, U(x)}{\sqrt{\pi \, \nu \int_0^x U(x) \, dx}} \, \mathsf{P}^{1/2} \quad (\text{P} \to 0). \tag{12.57b}$$

Equations (12.59a) and (12.59b) are special cases of this general equation.

In the case of a flat plate ($U(x) = U_\infty = \text{const}$) with a uniform wall temperature T_w, we obtain the same differential equation as that encountered in another connexion in Chap. V, eqn. (5.17). Its solution is

$$T - T_\infty = (T_w - T_\infty) \left(1 - \frac{2}{\sqrt{\pi}} \int_0^\eta \exp(-\eta^2) \, d\eta \right), \tag{12.58}$$

where

$$\eta = \tfrac{1}{2} y \sqrt{\frac{U_\infty}{a \, x}}$$

According to eqn. (12.31), the corresponding Nusselt number is

$$\mathsf{N}_x = \frac{\alpha \, x}{k} = \sqrt{\frac{U_\infty x}{\pi \, a}} = \frac{1}{\sqrt{\pi}} \sqrt{\mathsf{R}_x \, \mathsf{P}} \quad (\text{flat plate}, \quad \text{P} \to 0). \tag{12.59a}$$

In the case of stagnation-point flow ($U(x) = u_1 \, x$), it follows that

$$\mathsf{N}_x = \sqrt{\frac{2}{\pi}} \sqrt{\mathsf{R}_x \, \mathsf{P}} \quad (\text{stagnation flow}, \ \text{P} \to 0), \tag{12.59b}$$

where

$$\mathsf{R}_x = U \, x / \nu .$$

Very large Prandtl numbers: The second limiting case when $P \to \infty$ was solved for the first time many years ago by M. A. Levêque [76]. He introduced the very reasonable assumption that the whole of the temperature field is confined inside that zone of the velocity field where the longitudinal velocity component, u, is still proportional to the transverse distance y. The same circumstances can also occur at intermediate values of the Prandtl number in cases when the thermal boundary layer starts with a temperature jump at the wall at $x = x_0$ (cf. Fig. 12.17) inside a developed velocity boundary layer. Accordingly, in the energy equation, eqn. (12.51 c), we suppose that the velocity distribution in the velocity boundary layer is represented by $u = (\tau_0/\mu)\, y$. It can then be verified that, in accordance with refs. [76] and [68a] (see also refs. [111] and [112]), the substitution

$$\eta = \frac{y \sqrt{\dfrac{\tau_0}{\mu}}}{\left\{ 9\,a \displaystyle\int_{x_0}^{x} \dfrac{\tau_0}{\mu}\,dx \right\}^{1/3}} \tag{12.60}$$

transforms the energy equation into the following ordinary differential equation:

$$\frac{d^2 T}{d\eta^2} + 3\,\eta^2\,\frac{dT}{d\eta} = 0 . \tag{12.61}$$

Here x_0 denotes the coordinate at which the temperature jump at the wall has been placed, it being remembered that the effect of frictional heat has been neglected. The solution of this ordinary differential equation can be expressed in closed form in terms of the incomplete gamma functions. Performing the required calculation, we would obtain the Nusselt number

$$N_x = \frac{\alpha\,(x - x_0)}{k} = \frac{x - x_0}{0 \cdot 8930} \sqrt{\frac{\tau_0}{\mu}} \left\{ 9\,a \int_{x_0}^{x} \sqrt{\frac{\tau_0}{\mu}}\,dx \right\}^{-1/3} \quad (P \to \infty) \ (12.61\,\text{a})\dagger$$

or

$$N_x = 0 \cdot 5384\ P^{1/3}\, \frac{x - x_0}{\nu} \sqrt{\frac{\tau_0}{\varrho}} \left\{ \int_{x_0}^{x} \sqrt{\frac{\tau_0}{\varrho}}\,\frac{dx}{\nu} \right\}^{-1/3} \quad (P \to \infty). \quad (12.61\,\text{b})\dagger$$

In the particular case of a flat plate at zero incidence with $x_0 = 0$, we substitute from eqn. (7.31)

$$\tau_0 = 0 \cdot 332\ \mu\, U_\infty \sqrt{\frac{U_\infty}{\nu\, x}}$$

and obtain

$$N_x = 0 \cdot 339\ P^{1/3} \sqrt{R_x} \quad (\text{flat plate}, \quad P \to \infty). \tag{12.62\,a}$$

It is shown in Fig. 12.14 that this equation represents a very good approximation

---

$\dagger\ 0 \cdot 8930 = (\tfrac{1}{3})!$, $\ 0 \cdot 5384 = 9^{-1/3}[(\tfrac{1}{3})!]^{-1}$, where $z! = \Gamma\,(z + 1)$.

even in the case of moderately-valued Prandtl numbers. At the stagnation point, the corresponding equation is

$$N_x = 0{\cdot}661 \; P^{1/3} \; \sqrt{R_x} \quad \text{(stagnation point, } \quad P \to \infty).\qquad (12.62\,\text{b})$$

Analogous, simple asymptotic formulae can also be established for the case of free convection on a vertical flat plate, [73], see also eqns. (12.118a) and (12.118b).

g. Thermal boundary layers in forced flow

In the present section we shall consider several examples of thermal boundary layers in forced flow. In solving these problems, use will be made of the simplified thermal boundary-layer equations. Just as in the case of a velocity boundary layer, the general problem of evaluating the thermal boundary layer for a body of arbitrary shape proves to be extremely difficult, so that we shall begin with the simpler example of the flat plate at zero incidence.

1. Parallel flow past a flat plate at zero incidence. We shall assume that the x-axis is placed in the plane of the plate in the direction of flow, the y-axis being at right angles to it and to the flow, with the origin at the leading edge. The boundary-layer equations for incompressible flow and constant properties (i. e. independent of temperature) have been given in eqns. (12.51a, b, c): assuming that the buoyancy forces are equal to zero as well as that $dp/dx = 0$ [18, 94], we obtain

$$\frac{\partial u}{\partial x} + \frac{\partial v}{\partial y} = 0 \;, \qquad\qquad (12.63\,\text{a})$$

$$\varrho \left(u\,\frac{\partial u}{\partial x} + v\,\frac{\partial u}{\partial y} \right) = \mu\,\frac{\partial^2 u}{\partial y^2} \;, \qquad\qquad (12.63\,\text{b})$$

$$\varrho\,c_p \left(u\,\frac{\partial T}{\partial x} + v\,\frac{\partial T}{\partial y} \right) = k\,\frac{\partial^2 T}{\partial y^2} + \mu \left(\frac{\partial u}{\partial y} \right)^2 . \qquad\qquad (12.63\,\text{c})$$

The boundary conditions are:

$$y = 0 \;:\quad u = v = 0 \;; \qquad T = T_w \quad \text{or} \quad \partial T/\partial y = 0$$

$$y = \infty \;:\quad u = U_\infty \;; \qquad T = T_\infty \;.$$

The velocity field is independent of the temperature field so that the two flow equations (12.63a, b) can be solved first and the result can be employed to evaluate the temperature field. An important relationship between the velocity distribution and the temperature distribution can be obtained immediately from eqns. (12.63b) and (12.63c). If the heat of friction $\mu\,(\partial u/\partial y)^2$ may be neglected in eqn. (12.63c), the two equations, (12.63b) and (12.63c), become identical if T is replaced by u in the second equation and if, in addition, the properties of the fluid satisfy the equation

$$\frac{\mu}{\varrho} = \frac{k}{c_p\,\varrho} \quad \text{or} \quad \nu = a, \quad \text{i. e.} \quad P = 1 \;.$$

If the frictional heat is neglected then a temperature field exists only if there is a difference in temperature between the wall and the external flow, e.g., if $T_w - T_\infty > 0$ (cooling). Hence it follows that for a flat plate at zero incidence in parallel flow and at small velocities the temperature and velocity distributions are identical provided that the Prandtl number is equal to unity:

$$\frac{T - T_w}{T_\infty - T_w} = \frac{u}{U_\infty} \qquad (\mathsf{P} = 1) . \tag{12.64}$$

This result corresponds to eqn. (12.52) which led us to the formulation of the important Reynolds analogy between heat transfer and skin friction.

H. Blasius introduced new variables for the solution of the flow equations, see eqns. (7.24) and (7.25), (ψ is the stream function):

$$\eta = y \sqrt{\frac{U_\infty}{\nu x}} ; \quad \psi = \sqrt{\nu x U_\infty} f(\eta) .$$

Hence

$$u = U_\infty f'(\eta) ; \quad v = \frac{1}{2} \sqrt{\frac{\nu U_\infty}{x}} (\eta f' - f) .$$

The differential equation for $f(\eta)$, eqn. (7.28) becomes

$$f f'' + 2 f''' = 0 ,$$

with the boundary conditions: $\eta = 0 : f = f' = 0 ; \quad \eta = \infty : f' = 1$. The solution of these equations was given in Chap. VII, Table 7.1.

Including the effect of frictional heat, as seen from eqn. (12.63c), the temperature distribution $T(\eta)$ is given by the equation

$$\frac{d^2 T}{d\eta^2} + \frac{\mathsf{P}}{2} f \frac{dT}{d\eta} = -2\mathsf{P} \frac{U_\infty^2}{2 c_p} f''^2 . \tag{12.65}$$

It is convenient to represent the general solution of eqn. (12.65) by the super-position of two solutions of the form:

$$T(\eta) - T_\infty = C \theta_1(\eta) + \frac{U_\infty^2}{2 c_p} \theta_2(\eta) . \tag{12.66}$$

Here $\theta_1(\eta)$ denotes the general solution of the homogeneous equation and $\theta_2(\eta)$ denotes a particular solution of the non-homogeneous equation. It is, further, convenient to choose the boundary conditions for $\theta_1(\eta)$ and $\theta_2(\eta)$ so as to make $\theta_1(\eta)$ the solution of the cooling problem with a prescribed temperature difference between the wall and the external stream, $T_w - T_\infty$, with $\theta_2(\eta)$ giving the solution for the adiabatic wall. Thus $\theta_1(\eta)$ and $\theta_2(\eta)$ satisfy the following equations:

$$\theta_1'' + \tfrac{1}{2} \mathsf{P} f \theta_1' = 0 , \tag{12.67}$$

with $\theta_1 = 1$ at $\eta = 0$ and $\theta_1 = 0$ at $\eta = \infty$, and

$$\theta_2'' + \tfrac{1}{2}\, \mathsf{P}\, f\, \theta_2' = -\,2\,\mathsf{P}\, f''^2 \tag{12.68}$$

with $\theta_2' = 0$ at $\eta = 0$ and $\theta_2 = 0$ at $\eta = \infty$. The value $\theta_2(0)$ permits us to evaluate the constant C from eqn. (12.66) in a manner to satisfy the boundary condition $T = T_w$ for $\eta = 0$. This yields

$$C = T_w - T_\infty - \frac{U_\infty^2}{2c_p}\, \theta_2(0)\,. \tag{12.68a}$$

Cooling problem: The solution of eqn. (12.67) was first given by E. Pohlhausen [94]. It can be written as

$$\theta_1(\eta, \mathsf{P}) = \int_{\xi=\eta}^{\infty} [f''(\xi)]^{\mathsf{P}}\, \mathrm{d}\xi \; \Big/ \; \int_{\xi=0}^{\infty} [f''(\xi)]^{\mathsf{P}}\, \mathrm{d}\xi\,. \tag{12.69}$$

Hence for $\mathsf{P} = 1$: $\theta_1(\eta) = 1 - f'(\eta) = 1 - u/U_\infty$, and for $\mathsf{P} = 1$ the temperature distribution becomes identical with the velocity distribution in accordance with eqn. (12.64). The temperature gradient at the wall, as calculated from eqn. (12.69), with $f''(0) = 0 \cdot 332$, becomes:

$$-\left(\frac{\mathrm{d}\theta_1}{\mathrm{d}\eta}\right)_0 = a_1(\mathsf{P}) = (0 \cdot 332)^{\mathsf{P}} \; \Big/ \; \int_0^{\infty} [f''(\xi)]^{\mathsf{P}}\, \mathrm{d}\xi\,. \tag{12.70}$$

The constant a_1 is seen to depend solely on the Prandtl number, $a_1(\mathsf{P})$. Some values calculated by E. Pohlhausen are reproduced in Table 12.2. They can be interpolated with good accuracy from the formula

$$a_1 = 0 \cdot 332 \sqrt[3]{\mathsf{P}} \qquad (0 \cdot 6 < \mathsf{P} < 10)\,. \tag{12.71a}$$

For very small Prandtl numbers, eqn. (12.59a) gives

$$a_1 = 0 \cdot 564 \sqrt[2]{\mathsf{P}} \qquad (\mathsf{P} \to 0)\,, \tag{12.71b}$$

whereas for very large Prandtl numbers eqn. (12.62a) leads to

$$a_1 = 0 \cdot 339 \sqrt[3]{\mathsf{P}} \qquad (\mathsf{P} \to \infty)\,. \tag{12.71c}$$

The temperature distribution calculated from eqn. (12.69) is shown plotted in Fig. 12.9. As already mentioned, the curve for $\mathsf{P} = 1$ gives also the velocity distribution. For values of $\mathsf{P} > 1$ the thermal boundary layer is thinner than the velocity boundary layer. For example, for oil with a Prandtl number $\mathsf{P} = 1000$ the thermal boundary layer is only one tenth of the velocity boundary layer.

Adiabatic wall: The solution of eqn. (12.68) can be obtained by the method of 'variation of the parameter'. It is

$$\theta_2(\eta, \mathsf{P}) = 2\, \mathsf{P} \int_{\xi=\eta}^{\infty} [f''(\xi)]^{\mathsf{P}} \left(\int_0^{\xi} [f''(\tau)]^{2-\mathsf{P}}\, \mathrm{d}\tau \right) \mathrm{d}\xi\,. \tag{12.72}$$

Table 12.2. Dimensionless coefficient of heat transfer, a_1, and dimensionless adiabatic wall temperature, b, for a flat plate at zero incidence, from eqns. (12.70) and (12.75)

P	0·6	0·7	0·8	0·9	1·0	1·1	7·0	10·0	15·0
a_1	0·276	0·293	0·307	0·320	0·332	0·344	0·645	0·730	0·835
b	0·770	0·835	0·895	0·950	1·000	1·050	2·515	2·965	3·535

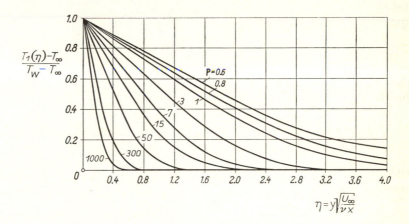

Fig. 12.9. Temperature distribution on a heated flat plate at zero incidence, with small velocity plotted for various Prandtl numbers P (frictional heat neglected)

For $P = 1$ we have:

$$\theta_2(\eta) = 1 - f'^2(\eta) \,. \tag{12.73}$$

The temperature which is assumed by the wall owing to frictional heat, the *adiabatic wall temperature* T_a, is thus, by eqns. (12.66) and (12.72):

$$T_{2w} - T_\infty = T_a - T_\infty = \frac{U_\infty{}^2}{2\,c_p}\, b(\mathsf{P}) \tag{12.74}$$

with

$$b(\mathsf{P}) = \theta_2(0,\ \mathsf{P})\,, \tag{12.75}$$

from eqn. (12.72). For a constant Prandtl number the adiabatic wall temperature is proportional to the adiabatic temperature rise $U_\infty{}^2/2\,c_p$ which was plotted in Fig. 12.3. Some numerical values of the factor $b(\mathsf{P})$ are given in Table 12.2; for moderate Prandtl numbers these values may be interpolated with sufficient accuracy from the formula $b = \sqrt{\mathsf{P}}$. The values for larger Prandtl numbers can be inferred from Fig. 12.10. In the limiting case, we have [84]

$$b(\mathsf{P}) = 1\!\cdot\!9\ \mathsf{P}^{1/3} \qquad (\mathsf{P} \to \infty)\,.$$

It is remarkable that for $\mathsf{P} = 1$ we have exactly $b = 1$. Thus, for a gas with $\mathsf{P} = 1$ flowing in a parallel stream with velocity U_∞ past a flat plate at zero incidence the temperature rise due to frictional heat is equal to the adiabatic temperature, i. e. to that which occurs from velocity U_∞ to zero. The adiabatic wall temperature [16, 20] measured at various Reynolds numbers $U_\infty x/\nu$ is seen plotted in Fig. 12.11. The agreement is very good in the laminar region. At the point of transition from laminar to turbulent flow in the boundary layer the temperature increases suddenly. The temperature distribution for an adiabatic wall represented non-dimensionally is

$$\frac{T_2(\eta) - T_\infty}{T_{2w} - T_\infty} = \frac{T_2(\eta) - T_\infty}{T_a - T_\infty} = \frac{\theta_2(\eta, \mathsf{P})}{b(\mathsf{P})} ,$$

and is seen plotted in Fig. 12.12 for various values of the Prandtl number. From eqns. (12.74) and (12.75), we obtain that the constant C from eqn. (12.68a) is

$$C = (T_w - T_\infty) - (T_a - T_\infty) = T_w - T_a .$$

The general solution for a prescribed temperature difference between the wall and the free stream, $T_w, - T_\infty$, eqn. (12.66), is thus

$$T(\eta) - T_\infty = [(T_w - T_\infty) - (T_a - T_\infty)] \theta_1(\eta, \mathsf{P}) + \frac{U_\infty^2}{2 c_p} \theta_2(\eta, \mathsf{P}) , \quad (12.76)$$

with $T_a - T_\infty$ from eqn. (12.74). The dimensionless temperature distribution becomes

$$\frac{T - T_\infty}{T_w - T_\infty} = [1 - \tfrac{1}{2} \mathsf{E}\, b\, (\mathsf{P})] \theta_1(\eta, \mathsf{P}) + \tfrac{1}{2} \mathsf{E}\, \theta_2(\eta, \mathsf{P}) . \quad (12.76a)$$

It is shown plotted in Fig. 12.13 for various values of the Eckert number $\mathsf{E} = U_\infty^2/c_p(T_w - T_\infty)$, from eqn. (12.28). For $b \times \mathsf{E} > 2$ the boundary layer near the wall is warmer than the wall itself owing to the generation of frictional heat. In such cases the wall will not be cooled by the stream of air flowing past it.

Heat transfer. As seen from eqn. (12.2) the heat flux from plate to fluid at station x has the value $q(x) = - k(\partial T/\partial y)_{y=0}$ or

$$q(x) = - k \sqrt{\frac{U_\infty}{\nu x}} \left(\frac{\mathrm{d}T}{\mathrm{d}\eta}\right)_{\eta=0} . \quad (12.77)$$

The rate of heat transfer per unit time for both sides of a plate (length l, width b) is $Q = 2b \int_0^l q(x)\, \mathrm{d}x$, so that

$$Q = 4\, b\, k \sqrt{\frac{U_\infty l}{\nu}} \left(-\frac{\mathrm{d}T}{\mathrm{d}\eta}\right)_0 . \quad (12.78)$$

a) *Neglecting frictional heat*: In this case $T(\eta) - T_\infty = (T_w - T_\infty) \theta_1(\eta)$ by eqn. (12.69) with $(\mathrm{d}T/\mathrm{d}\eta)_0 = - a_1(T_w - T_\infty)$. With a_1 from eqn. (12.71a) we have

Fig. 12.10. Adiabatic wall temperature T_a of a flat plate at zero incidence with velocity U_∞ for various values of the Prandtl number; after E. Eckert and O. Drewitz [15] as well as D. Meksyn [84]. For large Prandtl numbers, according to D. Meksyn [84], we have $b = 1 \cdot 9\ P^{1/3}$

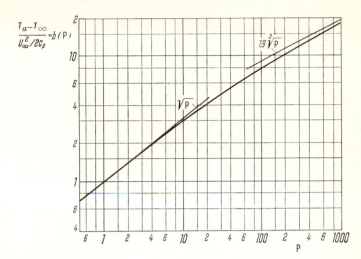

Fig. 12.11. Measurement of adiabatic wall temperature on a flat plate in a parallel air stream at zero incidence in a laminar and turbulent boundary layer, after Eckert and Weise [20]; theory for laminar flow and $P = 0 \cdot 7$

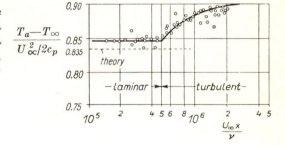

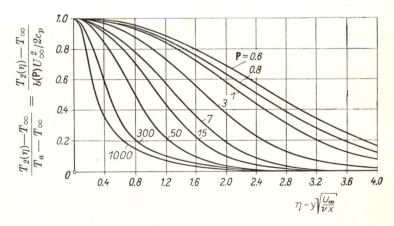

Fig. 12.12. Temperature excess in the laminar boundary layer on a flat plate at zero incidence in a parallel stream with high velocity in the *absence of heating* for various Prandtl numbers (adiabatic wall)

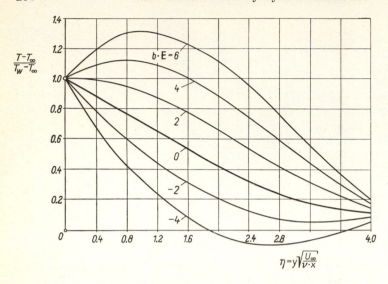

$$\eta = y\sqrt{\frac{U_\infty}{\nu \cdot x}}$$

Fig. 12.13.   Temperature distribution in a laminar boundary layer on a heated ($E > 0$) and cooled ($E < 0$) flat plate at zero incidence in a parallel stream for the case of a laminar boundary layer and with frictional heat accounted for as calculated from eqn. (12.76). Prandtl number $P = 0.7$ (air). The temperature of the wall is maintained constant at T_w. Curve $b \times E = 0$ for zero frictional heat; curve $b \times E = 2$ corresponds to an adiabatic wall; $E = U_\infty^2/c_L(T_w - T_\infty)$; $b = 0.835$. For $b \times E > 2$ the hot wall ceases to be cooled by the stream of cooler air, since the 'heat cushion' provided by frictional heat prevents cooling

$$\left(\frac{dT}{d\eta}\right)_0 = -0.332 \sqrt[3]{P}\,(T_w - T_\infty)$$

so that
$$q(x) = 0.332\, k \sqrt[3]{P}\, \sqrt{\frac{U_\infty}{\nu x}}\,(T_w - T_\infty)\,,$$

$$Q = 1.328\, b\, k \sqrt[3]{P}\, \sqrt{R_l}\,(T_w - T_\infty)\,. \qquad\qquad \left.\vphantom{\begin{array}{c}1\\1\\1\end{array}}\right\} \quad (12.79)$$

Introducing dimensionless coefficients in the form of the Nusselt number from eqn. (12.31) instead of the local and total heat flux, respectively

$$q(x) = \frac{k}{x}\, N_x\,(T_w - T_\infty) \quad \text{and} \quad Q = 2\,b\,l\,\frac{k}{l}\, N_m\,(T_w - T_\infty)\,,$$

we have

$$N_x = 0.564\ \sqrt{P}\ \sqrt{R_x} \quad \text{for } P \to 0\,,$$

$$N_x = 0.332\ \sqrt[3]{P}\ \sqrt{R_x} \quad \text{for } 0.6 < P < 10\,, \qquad \left.\vphantom{\begin{array}{c}1\\1\\1\\1\\1\end{array}}\right\} \quad (12.79\text{a})$$

$$N_x = 0.339\ \sqrt[3]{P}\ \sqrt{R_x} \quad \text{for } P \to \infty\,.$$

The total heat flux yields the mean Nusselt number

$$\frac{\mathsf{N}_m}{\sqrt{\mathsf{R}_l}} = 2\,\frac{\mathsf{N}_x}{\sqrt{\mathsf{R}_x}}\,. \tag{12.79b}$$

The case of turbulent flow can be approximated by the equations

$$\mathsf{N}_x = 0\cdot0296\,\sqrt[3]{\mathsf{P}}\cdot\mathsf{R}_x{}^{0\cdot8}\quad\text{(turbulent)}\,, \tag{12.79c}$$

$$\mathsf{N}_m = 0\cdot037\,\sqrt[3]{\mathsf{P}}\cdot\mathsf{R}_l{}^{0\cdot8}\quad\text{(turbulent)}\,, \tag{12.79d}$$

which we quote here for completeness, but without proof. The preceding formulae for the rate of heat transfer are in good agreement with the measurements due to F. Elias [31], A. Edwards and B. N. Furber [27] and J. Kestin, P. F. Maeder and H. E. Wang [66].

b) *With frictional heat*: In this case with $T(\eta)$ from eqn. (12.76) we obtain

$$\left(\frac{\mathrm{d}T}{\mathrm{d}\eta}\right)_0 = -\,a_1\,(T_w - T_a) = -\,0\cdot332\,\sqrt[3]{\mathsf{P}}\,(T_w - T_a)\,,$$

where T_a is the adiabatic wall temperature. It is identical with the wall temperature in the thermometer problem and follows from the equation

$$T_a - T_\infty = b(\mathsf{P})\,\frac{U_\infty{}^2}{2\,c_p} = \sqrt{\mathsf{P}}\,\frac{U_\infty{}^2}{2\,c_p}\,. \tag{12.80}$$

Here $b(\mathsf{P})$ can be taken from Table 12.2. Introducing the Mach number $\mathsf{M} = U_\infty/c_\infty$ from (12.27), T_a may also be taken from

$$T_a = T_\infty\left(1 + \frac{\gamma-1}{2}\,\mathsf{M}^2\right)\quad\text{for}\quad\mathsf{P}=1\,.$$

Thus we obtain the following expressions for the local and total heat flux from eqns. (12.77) and (12.78) respectively

$$\left.\begin{aligned} q(x) &= 0\cdot332\,k\,\sqrt[3]{\mathsf{P}}\,\sqrt{\frac{U_\infty}{\nu\,x}}\,(T_w - T_a)\,,\\[2mm] Q &= 1\cdot328\,b\,k\,\sqrt[3]{\mathsf{P}}\,\sqrt{\mathsf{R}_l}\,(T_w - T_a)\,. \end{aligned}\right\} \tag{12.81}$$

It now ceases to be useful to base the coefficient of heat transfer $\alpha(x)$ on the temperature difference $(T_w - T_\infty)$ from eqn. (12.29) or to define the Nusselt number as in eqn. (12.31) because the heat flux is no longer proportional to that temperature difference†.

---

† E. Eckert and W. Weise [17] have, therefore, suggested to introduce a Nusselt number N^* based on the difference $(T_w - T_a)$. We might then expect to obtain as a first approximation, even in compressible flow, the same formulae for N^* as in eqn. (12.79a, b). If, on the other hand, we retain the Nusselt number based on $(T_w - T_\infty)$ then eqn. (12.81) leads to the following expressions instead of (12.79a):

$$\mathsf{N}_x = 0\cdot332\,\sqrt[3]{\mathsf{P}}\,\sqrt{\mathsf{R}_x}\,[1 - \tfrac{1}{2}\,\mathsf{E}\,b(\mathsf{P})]\,,$$

i. e. $\mathsf{N}_x = 0$ for $b\,\mathsf{E} = 2$ and $\mathsf{N}_x < 0$ for $b\,\mathsf{E} > 2$, see Fig. 12.13.

The cooling action of a stream of fluid on a wall is considerably reduced because of the heat generated by friction. In the absence of frictional heat, heat will flow from the plate to the fluid $(q > 0)$ as long as $T_w > T_\infty$ but in actual fact, if frictional heat is present, a flow of heat persists only if $T_w > T_a$, eqn. (12.81). Taking into account the value deduced for T_a, we obtain the condition that heat flows from wall to fluid (upper sign) or in the reverse direction (lower sign), if

$$T_w - T_\infty \gtrless \sqrt{P}\ \frac{U_\infty^2}{2\,c_p} \ . \tag{12.82}$$

A numerical example may serve to illustrate the significance of eqn. (12.82): In a stream of air flowing at $U_\infty = 200$ m/sec, $P = 0\cdot7$, $c_p = 1\cdot006$ kJ/kg deg we obtain $\sqrt{P}\ U_\infty^2/2\,c_p = 16$ deg C. The wall will begin to be cooled when

$$T_w - T_\infty > 16°\,\mathrm{C} \ .$$

If the temperature difference between wall and stream is smaller than this value the wall will pick up a portion of the heat generated by friction. In particular this is the case when the temperature of the wall and stream are equal.

An equation for the rate of heat transferred from a flat plate at zero incidence but with variable material properties was derived by H. Schuh [110]. The temperature field on a plate placed in a stream with a linear temperature distribution was studied in ref. [128].

2. Additional similar solutions of the equations for thermal boundary layers.
In the case of a flat plate at zero incidence, the velocity and the temperature profiles turned out to be similar among themselves. This means that the distributions at different distances x along the plate could be made congruent by a suitable stretching in the y-direction. Since it is known that there exist velocity boundary layers other than those on a flat plate for which this is true (e. g. the wedge profiles discussed in Chap. IX), it appears useful to study the possibility of the existence of additional similar solutions of the energy equation. This problem was investigated in detail in ref. [135]. At the present time, we shall start with the class of velocity boundary layers on wedges and will assume that the external flow is of the form $U(x) = u_1\,x^m$. In an analogous manner, we stipulate that the wall-temperature distribution also satisfies a power law, say one of the form $T_w(x) - T_\infty = (\Delta T)_0 = T_1\,x^n$. Walls of constant temperature are included as the case $n = 0$, and the value $n = (1-m)/2$ corresponds to a constant heat flux q. Introducing the similarity variable

$$\eta = y\ \sqrt{\frac{U(x)}{v\,x}}\ ,$$

we obtain the familiar equations (9.8a) for the velocity $u = U(x) \cdot f'(\eta)$, or

$$f''' + \frac{m+1}{2}\,f\,f'' + m\,(1 - f'^2) = 0 \ . \tag{12.83}$$

The dimensionless temperature

$$\theta = \frac{T - T_\infty}{T_w - T_\infty}$$

then satisfies the equation

$$\theta'' + \frac{m+1}{2} \, \mathsf{P} f \, \theta' - n \, \mathsf{P} f' \, \theta = - \, \mathsf{PE} \, x^{2m-n} f''^2 \,, \tag{12.84}$$

and the solution must satisfy the boundary conditions

$$\eta = 0: \qquad \theta = 1; \qquad \eta = \infty: \quad \theta = 0 \,.$$

Here $\mathsf{E} = u_1^2/c_p \, T_1$ represents the appropriate form of the Eckert number for the problem.

It is clear from eqn. (12.84) that its right-hand side vanishes in the absence of frictional heat and that *all* solutions are then of the similar type. However, if frictional heat is included, similar solutions are restricted to that combination of parameters for which the right-hand side becomes independent of x. This occurs when $2\,m - n = 0$, that is, when there exists a firm coupling between the velocity distribution in the external flow and the temperature distribution along the wall. According to this result, the case of a constant temperature leads to similar solutions only on a flat plate ($m=n=0$). On the other hand, if the condition $2\,m - n = 0$ is satisfied, then for every pair of values of m and P there exists one definite value E_a for which there is no flow of heat ($\theta'(0) = 0$). In this case, the temperature distribution along the wall, once again known as the adiabatic wall-temperature distribution T_a, is given by

$$\frac{T_a - T_\infty}{U^2/2\,c_p} = 2\,c_p \, \frac{T_1 \, x^{2m}}{u_1^2 \, x^{2m}} = \frac{2}{\mathsf{E}_a} = b\,(m, \mathsf{P}) \,. \tag{12.85}$$

Numerical values for the function $b\,(m, \mathsf{P})$ have been computed by E. A. Brun [7]. In the particular case when $m = 0$, the numerical values of Table 12.2 are recovered.

When the effect of dissipative heat is neglected, we obtain the simpler equation

$$\theta'' + \frac{m+1}{2} \, \mathsf{P} f \, \theta' - n \, \mathsf{P} f' \, \theta = 0 \,, \tag{12.86}$$

whose solutions for different values of the parameters m, n, and P have been published by a number of authors [79, 121, 32, 33, 89, 140]. E. R. G. Eckert [19] has demonstrated that for $n = 0$, the local Nusselt number is given by the equation

$$\frac{\mathsf{N}_x}{\sqrt{\mathsf{R}_x}} = F\,(m, \mathsf{P}) = \left\{ \int_0^\infty \exp\left[- \, \mathsf{P} \, \sqrt{\frac{m+1}{2}} \int_0^\eta f(\eta) \, \mathrm{d}\eta \right] \mathrm{d}\eta \right\}^{-1} \,. \tag{12.87}$$

Here

$$\mathsf{N}_x = \frac{\alpha\,x}{k} = - \sqrt{\frac{U\,(x) \cdot x}{\nu}} \, \theta'\,(0) = - \sqrt{\mathsf{R}_x} \, \theta'\,(0) \,. \tag{12.88}$$

The function $F\,(m, \mathsf{P})$ is seen plotted in Fig. 12.14 on the basis of the numerical data provided by H. L. Evans [33]. In addition, the asymptotes for very small and very large Prandtl numbers, eqns. (12.57) and (12.61a), respectively, have also

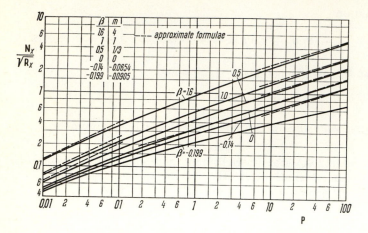

Fig. 12.14. Local Nusselt number as a function of the Prandtl number and of the flow parameter m for flows whose free-stream velocity is distributed according to the law $U(x) = u_1\, x^m = u_1\, x^{\beta/(2-\beta)}$ (wedge flow) but for a constant wall temperature and in the absence of dissipation

Asymptotic approximations for $P \to 0$

$$\frac{N_x}{\sqrt{R_x}} = \sqrt{\frac{2}{\pi(2-\beta)}}\; P^{1/2}$$

Asymptotic approximations for $P \to \infty$ and $\beta \neq -0.199$ according to eqn. (12.61 a), and for $P \to \infty$ and $\beta = -0.199$:

$$\frac{N_x}{\sqrt{R_x}} = 0.224\; P^{1/4}$$

Approximation for intermediate Prandtl numbers and $\beta = 0$, according to eqn. (12.71 a).

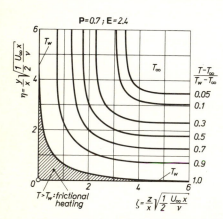

Fig. 12.15. Temperature distribution along a heated wall ($T_w > T_\infty$) in a right-angled corner in a laminar boundary layer with a constant external velocity U_∞ (inclusive of dissipation), after Vasanta Ram [144].

Lines of constant temperature for $P = 0.7$ and $E = 2.4$. The local temperature exceeds the wall temperature ($T > T_w$) in the hatched region; consequently, in that region heat flows fluid $\to$ wall in spite of the fact that the wall temperature exceeds the free-stream temperature. The reason for this process lies in dissipation. Eckert number $E = U_\infty^2/c_p\,(T_w - T_\infty)$

been indicated (see also [119]). For the flat plate ($m=0$) the earlier relations from eqns. (12.59a) and (12.62a) are, naturally, recovered. The case of stagnation flow ($m=1$) leads to eqns. (12.59b) and (12.62b). In the special case of a separation profile ($m = -0.091$) it becomes necessary to adopt a different asymptotic approximation for $P \to \infty$, as shown in [32].

The thermal boundary layer associated with the three-dimensional velocity boundary layer on a rectangular corner at zero incidence is also of the self-similar type when the external velocity distribution is of the Hartree class given by $U(x) = C x^m$. The velocities as well as the temperature distributions for this case have been worked out in a thesis by Vasanta Ram (ref. [92] in Chap. XI). Figure 11.19 gives an idea of the velocity distribution for different values of the pressure-gradient parameter m. The diagram in Fig. 12.15 supplements the preceding one in that it contains an example of the associated temperature distribution. For a uniform external flow with $U(x) = U_\infty = $ const and in the case of a hotter (i. e. cooled) wall for which $T_w > T_\infty$ the solution nevertheless exhibits a zone near the corner itself, shaded in the figure, in which $(T - T_\infty)/(T_w - T_\infty) > 0$, that is in which $T > T_w$. This zone occurs when dissipation is included and corresponds to a condition where the local fluid temperature exceeds the wall temperature. Thus, locally, the heat flux is reversed and proceeds from the fluid to the wall in spite of the fact that at a large distance from the wall the temperature of the fluid is lower than that of the wall, $T_\infty < T_w$. The physical reason for this seemingly anomalous behavior is rooted in the increased local rate of heating due to dissipation which occurs near the corner. Phenomena of this kind are important in the hypersonic flow regime. The resulting large increases in temperature which occur in such cases can cause burning of the surface of the body in the stream (*cf.* Sec. XIII e).

3. Thermal boundary layers on isothermal bodies of arbitrary shape. N. Froessling [39] carried out calculations on the temperature distribution in the laminar boundary layer about a body of arbitrary shape for the two-dimensional and axially symmetrical cases. In his calculations, in which friction and compression work were neglected throughout, he assumed a power series for the potential velocity distribution around the body expanded in terms of the length of arc (Blasius series), similar to Sec. IX c, i. e. of the form:

$$U = u_1 x + u_3 x^3 + u_5 x^5 + \cdots .$$

The velocity distribution in the boundary layer is assumed to have the form:

$$u(x, y) = u_1 x f_1(y) + u_3 x^3 f_3(y) + \cdots .$$

Correspondingly, the assumption for the temperature distribution was of the form:

$$T(x, y) = T_0 + x T_1(y) + x^3 T_3(y) + \cdots .$$

In a manner similar to that for the velocity boundary layer in Sec. IX c it is found that the functions $T_1(y)$, $T_3(y)$, ... satisfy ordinary differential equations which include the functions $f_1, f_3, \ldots$ of the velocity distribution. In this case, however, the functions $T_1, T_3, \ldots$ also depend on the Prandtl number. The first auxiliary functions $T_r(y)$

for the two-dimensional and axially symmetrical case were evaluated numerically for a Prandtl number of 0·7. The method under consideration is somewhat cumbersome by its nature, as was the case with the velocity boundary layer, particularly for slender body forms when a large number of terms in the power expansion is required, as shown in [28].

Numerous solutions for self-similar thermal boundary layers inclusive of the effects of blowing and suction can be found in [34, 44, 134, 10].

In the special case when $P = 1$, and when the heat due to friction is neglected, the differential equation for the temperature distribution in the boundary layer around an arbitrary cylinder is identical with that for the transverse velocity component (velocity component in the direction of the generatrix of the yawed cylinder). This can be seen upon comparing eqns. (12.63c) and (11.58). The relation, which has already been discussed in Sec. XId, was utilized by L. Goland [46] for the evaluation of the temperature distribution in the boundary layer around a cylinder of special form.

In the neighbourhood of a stagnation point, where the velocity distribution is represented by $U(x) = u_1 x$ with $m = \beta = 1$, the Nusselt number defined in eqn. (12.87) can be represented by the equation

$$\frac{N_x}{\sqrt{R_x}} = F(P, 1) = A(P),\qquad(12.89)$$

on condition that energy dissipation is neglected. The character of the function $A(P)$ emerges from Fig. 12.14 and Table 12.3. In the former, the curve for $\beta = 1$ corresponds to the function A. For a circular cylinder we put $U(x) = U_\infty \sin(x/R)$, so that $u_1 = 4 U_\infty/D$. Hence

$$\frac{N_x}{\sqrt{R_x}} = \frac{\alpha x}{k} \sqrt{\frac{D\nu}{4 U_\infty x^2}} = \frac{1}{2} \frac{\alpha D}{k} \sqrt{\frac{\nu}{U_\infty D}} = \frac{1}{2} \frac{N_D}{\sqrt{R_D}} = A(P).\qquad(12.90)$$

The above expression agrees reasonably well with the measurements performed by E. Schmidt and K. Wenner [107] at lower Reynolds numbers, see Fig. 12.16. It appears, however, that the ratio $N_D/\sqrt{R_D}$ depends systematically on the Reynolds number; this systematic influence is not accounted for by the theory. For example at $R = 1·7 \times 10^5$, the measured value exceeds the theoretical one at the stagnation point by 10 to 15%. We shall revert to this point in Sec. XIIg 7 where it will be shown that the discrepancy is due to the effect of varying free-stream turbulence produced by a change in the Reynolds number.

When performing numerical calculations on thermal boundary layers, it is found that *approximate methods* are much simpler than the preceding exact methods. Such approximate methods are based on equations modelled on the integral momentum equation for the calculation of velocity boundary layers which was considered in detail in Chap. X. Neglecting frictional heat and the effects of compressibility it is possible to integrate the energy equation (12.51c) from $y = 0$ to $y = \infty$ and so to obtain the *heat-flux equation*

Table 12.3.   The constant A in the equation for the calculation of the coefficient of heat transfer in the neighbourhood of a stagnation point, after H. B. Squire [131]

P	0·6	0·7	0·8	0·9	1·0	1·1	7·0	10	15
A	0·466	0·495	0·521	0·546	0·570	0·592	1·18	1·34	1·54

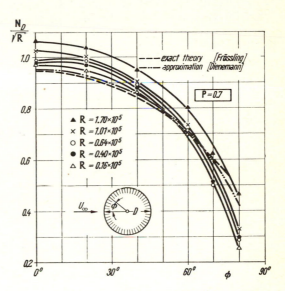

Fig. 12.16. Local rate of heat transfer around a circular cylinder. Comparison between theory and experiment. Nusselt number N_D and Reynolds number R_D referred to cylinder diameter $D = 100$ mm. Measurements performed by E. Schmidt and K. Wenner [107]. Theory due to N. Froessling [39] and W. Dienemann [11]. Systematic influence of Reynolds number due to varying free-stream turbulence, Sec. XII g 7

$$\frac{\mathrm{d}}{\mathrm{d}x} \int_0^\infty [u\,(T - T_\infty)]\,\mathrm{d}y = -\,a \left(\frac{\partial T}{\partial y}\right)_{y=0} \tag{12.91}$$

where $a = k/\varrho\, c_p$ is the thermal diffusivity introduced in eqn. (12.25). The preceding equation, sometimes also called the energy-integral equation†, is quite analogous to the momentum-integral equation (8.32) for the velocity boundary layer.

From among the numerous procedures which are available for the solution of the heat flux equation (12.91), we propose to describe that due to H. B. Squire [132] in some detail because it is particularly simple and because it is a natural continuation of Pohlhausen's approximate method for the solution of the velocity boundary layer described in Chap. X. In order to evaluate the integral on the left-hand side of eqn. (12.91) we introduce the variables $\eta = y/\delta$ for the velocity boundary layer and

_____

† Not to be confused with the energy-integral equation (8.35).

$\eta_T = y/\delta_T$ for the thermal layer. We denote, further, their ratio by $\varDelta = \delta_T/\delta$, and we assume that the velocity and temperature distributions, respectively, have the forms

$$u = U(x) \left[2\,\eta - 2\,\eta^3 + \eta^4\right] = U(x)\,F(\eta)\;, \tag{12.92a}$$

$$T - T_\infty = (T_W - T_\infty)\left[1 - 2\,\eta_T + 2\,\eta_T^3 - \eta_T^4\right] = (T_W - T_\infty)\,L(\eta_T)\;. \tag{12.92b}$$

The velocity distribution stipulated here corresponds to the Pohlhausen assumption in eqn. (10.23) and the form of the temperature distribution function is so selected as to ensure identical velocity and temperature distributions for $\delta_T = \delta$, as required by the Reynolds analogy for a flat plate at $P = 1$, eqn. (12.64). On substituting eqns. (12.92 a, b) into eqn. (12.91), we obtain

$$\frac{d}{dx}\left\{\delta_T \cdot U \cdot H(\varDelta)\right\} = 2\,\frac{a}{\delta_T}\;. \tag{12.93}$$

Here $H(\varDelta)$ is a universal function of $\varDelta = \delta_T/\delta$ which turns out to be given by

$$H = \int\limits_0^\infty F(\eta) \cdot L(\eta_T) \cdot d\eta_T\;. \tag{12.94}$$

Performing the indicated integrations, we obtain

$$H(\varDelta) = \frac{2}{15}\,\varDelta - \frac{3}{140}\,\varDelta^3 + \frac{1}{180}\,\varDelta^4 \qquad\qquad \text{for } \varDelta < 1$$

and $\qquad H(\varDelta) = \frac{3}{10} - \frac{3}{10}\frac{1}{\varDelta} + \frac{2}{15}\frac{1}{\varDelta^2} - \frac{3}{140}\frac{1}{\varDelta^4} + \frac{1}{180}\frac{1}{\varDelta^5} \quad \text{for } \varDelta > 1\,.$

Some numerical values of the function $H(\varDelta)$, calculated by W. Dienemann[11], have been listed in Table 12.4.

Table 12.4.   Numerical values of the function $H(\varDelta)$

$\varDelta$	0·7	0·8	0·9	1·0	1·2	1·4
H	0·0873	0·0980	0·1080	0·1175	0·1345	0·1492

The integration of eqn. (12.93) yields

$$(\delta_T\,U \cdot H)^2 = 4\,a\int\limits_0^x U \cdot H \cdot dx\;. \tag{12.95}$$

The velocity boundary-layer thickness δ can be evaluated with the aid of eqn. (10.37) when it is remembered from eqn. (10.24)† that $\delta/\delta_2 = 315/37$. Thus

$$\delta^2 = 34\,\frac{\nu}{U^6}\int\limits_{x=0}^x U^5\,dx\;. \tag{12.96}$$

_____

† For the sake of simplicity the calculation is based throughout on the flat-plate relations ($\varDelta = 0$).

Upon dividing eqn. (12.95) by eqn. (12.96), we obtain

$$\Delta^2 \cdot H(\Delta) = \frac{4}{34} \frac{1}{P} \frac{U^4 \int_0^x U H \cdot dx}{H \int_0^x U^5 \, dx} . \tag{12.97}$$

Since $H(\Delta)$ is a known function, Table 12.4, the preceding equation can be used to determine $\Delta(x)$. The calculation is best performed by successive approximations, starting with the initial assumption that $\Delta = \text{const.}$ Hence we obtain

$$\Delta^2 \cdot H(\Delta) = \frac{4}{34} \frac{1}{P} \frac{U^4 \int_0^x U \, dx}{\int_0^x U^5 \, dx} . \tag{12.97a}$$

The resulting value of Δ is now introduced into the left-hand side of eqn. (12.97) thus leading to an improved value of Δ. In general, two steps in the iteration are found to be sufficient.

The local rate of heat transfer becomes

$$q(x) = -k \left(\frac{\partial T}{\partial y} \right)_0 = 2 \left(T_W - T_\infty \right) \frac{k}{\delta_T} ,$$

and hence the local Nusselt number referred to a characteristic length l is

$$\mathsf{N}_x = \frac{q(x)}{T_W - T_\infty} \cdot \frac{l}{k} = 2 \frac{l}{\delta_T} . \tag{12.98}$$

The steps to be taken to evaluate the thermal boundary layer, and in particular, to determine the variation of the Nusselt number along a body of prescribed shape are thus the following ones:

    1.   evaluate $\Delta(x)$ from eqns. (12.97) and (12.97a)

    2.   evaluate $\delta(x)$ from eqn. (12.96)

    3.   steps 1 and 2 give $\delta_T(x)$; finally, the local Nusselt number follows from eqn. (12.98).

Flat plate at zero incidence: The preceding approximate method will now be compared with the exact solution in the case of a flat plate at zero incidence. Inserting $U(x) = U_\infty$ into eqn. (12.97), we obtain

$$\Delta^2 H(\Delta) = \frac{4}{34} \cdot \frac{1}{P} .$$

The expression $\Delta = P^{-1/3}$ constitutes an approximation to the solution of this equation which is in error by no more than 5 per cent. as compared with the exact solution. The boundary-layer thickness from eqn. (12.96) is

$$\delta = 5 \cdot 83 \sqrt{\nu \, x / U_\infty} .$$

Hence the local Nusselt number referred to the current length x along the plate, eqn. (12.98), becomes

$$N_x = 0.343 \; \sqrt[3]{P} \; R_x^{1/2}, \tag{12.99}$$

whereas the exact solution, eqn. (12.79a), showed the numerical coefficient to be equal to 0.332.

Alternative approximate procedures for the calculation of the thermal boundary layer on bodies of arbitrary shapes have been indicated by E. Eckert [19] and by E. Eckert and J. N. B. Livingood [23, 25]; the latter require a somewhat larger amount of numerical work, but their accuracy is improved. In this connexion the papers by W. Dienemann [11], H. J. Merk [85], M. B. Skopets [118] and A. G. Smith and D. B. Spalding [119] may be useful to the reader. In contrast with H. B. Squire's, the latter procedures make use of the results of the theory of similar thermal boundary layers outlined in the preceding section. This improves the accuracy of the calculation. The various approximate methods have been examined critically and compared with each other in a paper by D. B. Spalding and W. M. Pun [122]; their accuracy has been judged by performing comparisons with the exact solution for the circular cylinder provided by N. Froessling. According to these studies, the methods due to H. J. Merk [85] and A. G. Smith and D. B. Spalding [119] turn out to be relatively the most accurate in spite of their simplicity. The latter reference shows that at a Prandtl number of $P = 0.7$, the similar wedge profiles satisfy with good accuracy the relation

$$\frac{U(x)}{\nu} \frac{d(\delta_T^2)}{dx} = 46.72 - 2.87 \frac{\delta_T^2}{\nu} \frac{dU}{dx} . \tag{12.100}$$

This equation is exact for $\beta = 0$ (plate) and $\beta = 1$ (stagnation point). If it is supposed that eqn. (12.100) enjoys universal validity, it is possible immediately to write down a simple integral formula for δ_T, namely

$$\left(\frac{\delta_T}{l}\right)^2 \frac{U_\infty l}{\nu} = \frac{46.72}{\left(\frac{U(x)}{U_\infty}\right)^{2.87}} \int_0^{x/l} \left(\frac{U(x)}{U_\infty}\right)^{1.87} d\left(\frac{x}{l}\right) \quad \text{(for } P = 0.7\text{)} . \tag{12.101}$$

Here U_∞ and l denote constant reference values. This equation corresponds to eqn. (10.37) which was derived by A. Walz for the momentum thickness. The local Nusselt number is again determined by eqn. (12.98). At the stagnation point we obtain

$$\left(\frac{\delta_T}{l}\right)^2 \frac{U_\infty l}{\nu} = \frac{16.28}{\left[\frac{d(U/U_\infty)}{d(x/l)}\right]} \quad \text{(for } P = 0.7\text{)} . \tag{12.102}$$

Drastically simplified procedures for the calculation of thermal boundary layers on plates and rotationally symmetric bodies were indicated by H. J. Allen and B. C. Look [1], and by E. Eckert and W. Weise [17].

E. Eckert and O. Drewitz [18] performed calculations on the temperature distribution in the boundary layer allowing for the effects of compressibility and frictional heat. In general, in gaseous motion the work of compression is of the same order of magnitude as that dissipated through viscosity. It is then no longer possible to reduce the equation for temperature distribution to a differential equation of the first order, as was the case with the flat plate, and this circumstance renders the calculation much more difficult. In particular, the preceding authors made detailed calculations for the thermal boundary layers associated with the wedge flows which correspond to $U(x) = u_1 x^m$ and whose velocity boundary layers, calculated earlier by D. R. Hartree, have been discussed in Sec. IX a. The thermal boundary layer for wedge flow is also discussed in a paper by A. N. Tifford [139].

4. Thermal boundary layers on walls with an arbitrary temperature distribution. With the exception of the similar solutions for wedge flows discussed in Sec. XII g 2, all thermal boundary layers discussed so far were calculated on the assumption that the temperature difference between wall and free stream which creates the heat flux remained constant. The calculation of the temperature field and of the rate of heat transfer in the presence of a temperature $T_w(x)$ which is distributed along the wall presents many difficulties. In many cases, these are created by the fact that the local heat flux is by no means determined solely by the local temperature difference $T_w(x) - T_\infty$. It proves to be strongly influenced by the "past history" of the boundary layer.

The extension of the expansion in a Blasius series to include arbitrary distributions of wall temperature was worked out by C. R. Guha and C. S. Yih [38] as well as by N. Froessling [40]. The special case when the velocity profiles in the boundary layer can be represented in the form of a power law and when the temperature distribution along the wall can be represented as a power series was investigated by D. R. Davies and D. E. Bourne [9]. Approximate procedures for the calculation of thermal boundary layer along non-isothermal walls were elaborated by the following authors: D. R. Chapman and W. M. Rubesin [8], J. Klein and M. Tribus [80], P. L. Donoughe and N. B. Livingood [12], M. J. Lighthill [133], H. Schuh [111], G. S. Ambrok [2], D. B. Spalding [120], E. Eckert, J. P. Hartnett and R. Birkeback [26], B. Le Fur [74, 75] and H. Schlichting [102]. The scheme provided by H. B. Squire, and discussed earlier can also be extended to include non-isothermal walls [133]. In most cases, the authors neglect the evolution of frictional heat when incompressible flows are being studied.

Since the differential equation for the thermal boundary layer is linear, it is possible to write down the general solution to the problem in the form of a linear combination of certain standard solutions. Such a standard solution is obtained by considering a wall, Fig. 12.17, whose temperature is equal to that of the free stream, T_∞, from $x = 0$ to $x = x_0$. At $x = x_0$ the wall temperature is suddenly changed to a value T_S, producing the step function sketched in Fig. 12.17. If the solution to this problem is denoted by

$$\theta(x, y, x_0) = \frac{T(x, y, x_0) - T_\infty}{T_S - T_\infty} , \qquad (12.103)$$

then, for an arbitrary temperature distribution $T_S(x)$, we have

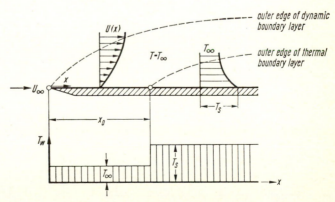

outer edge of dynamic boundary layer

outer edge of thermal boundary layer

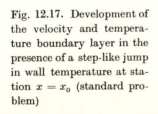

Fig. 12.17. Development of the velocity and temperature boundary layer in the presence of a step-like jump in wall temperature at station $x = x_0$ (standard problem)

$$T\,(x,y) - T_\infty = \int_0^x \theta\,(x,y,x_0)\,\mathrm{d}T_S\,(x_0)\,.\tag{12.104}$$

In a similar manner, the heat flux $q(x)$ can be computed from the known distribution
$$q(x,x_0) = q^*(x,x_0)\,(T_S - T_\infty)$$
for the standard problem of Fig. 12.17. In this case

$$q\,(x) = \int_0^x q^*\,(x,x_0)\,\mathrm{d}T_S\,(x_0)\,.\tag{12.105}$$

Here, eqns. (12.104) and (12.105) contain Stieltjes integrals. When the distributions $T_S(x)$ at $x > 0$ are continuous, it is possible to simplify the above expression to

$$T\,(x,y) - T_\infty = \int_0^x \theta\,(x,y,x_0)\,\frac{\mathrm{d}T_S}{\mathrm{d}x_0}\,\mathrm{d}x_0\,,\tag{12.106}$$

and to write an analogous version of eqn. (12.105). Referring to eqn. (12.61 b), it is now possible to obtain the following expression for the heat flux along a wall provided with a varying temperature $T_S(x)$ which was first discovered by M. J. Lighthill [80]. This is

$$q\,(x) = 0\cdot5384\left(\frac{\varrho\,\mathsf{P}}{\mu^2}\right)^{\frac13}\sqrt{\tau_0\,(x)}\left\{\int_0^x\left(\int_{x_0}^x\sqrt{\tau_0\,(z)}\,\mathrm{d}z\right)^{-\frac13}\mathrm{d}T_S\,(x_0)\right\}.\tag{12.107}$$

Strictly speaking, the preceding equation is valid only for the asymptotic case when $\mathsf{P} \to \infty$. According to M. J. Lighthill [80], a greater measure of agreement in the range $0 < \beta < 1$ can be secured when the factor $0\cdot5384$ is replaced by $0\cdot487$. The same equation, but with a factor of $0\cdot523$, was also obtained by H. W. Liepmann [78] who employed a different mode of reasoning.

As indicated in the caption to Fig. 12.14, the general, asymptotic approximation is not valid for the separation profile. It is, therefore, clear that Lighthill's equation (12.107) must break down at the point of separation. An improvement in the computation of the rate of heat transfer was indicated by D. B. Spalding [120]. According to this method, the distribution of the heat flux for the standard problem of Fig. 12.17 must be obtained by iteration from the following two equations:

$$q_{n+1}^*\,(x,x_0) = \frac{q_{n+1}\,(x,x_0)}{T_S - T_\infty} = \left(\frac{\varrho\mathsf{P}}{\mu^2}\right)^{1/3}\sqrt{\tau_0\,(x)}\left[\int_{x_0}^x\sqrt{\tau_0\,(x)}\,F\,(\chi_n)\,\mathrm{d}x\right]^{-\frac13}\tag{12.108}$$

with

$$\chi_{n+1}\,(x,x_0) = -\,\frac{k}{\tau_0\,(x)}\,\frac{\mathrm{d}p}{\mathrm{d}x}\,\frac{1}{q_{n+1}^*\,(x,x_0)}\,.\tag{12.109}$$

The function $F(\chi)$ was taken over from the known class of similar solutions and is given in ref[120]; a few numerical values have been reproduced here in Table 12.5. The iteration starts with $F(\chi_0) = 6\cdot4$ and leads to $q_1^*(x,x_0)$, from eqn. (12.108). This, in turn, allows us to calculate $\chi_1(x,x_0)$ from eqn. (12.109) and to insert it once again into eqn. (12.108), and so on. Unfortuna-

Table 12.5.   Values of the function $F(\chi)$ for the calculation of a thermal boundary layer on a nonisothermal wall; after D. B. Spalding [120]

χ	-4	-3	-2	-1	0	$+1$	$+2$	$+3$
$F(\chi)$	$3\cdot5$	$3\cdot8$	$4\cdot3$	$5\cdot1$	$6\cdot4$	$8\cdot5$	$11\cdot6$	$15\cdot8$

tely, this method, too, fails at the point of separation because the function $\chi(x,x_0)$ becomes infinite as $\tau_0 \to 0$.

A comparatively accurate method which, in addition, makes an allowance for frictional heat, has been indicated by B. Le Fur [74, 75]. This method was extended to include compressible flows.

5. Thermal boundary layers on rotationally symmetric and rotating bodies. The calculation of rotationally symmetric thermal boundary layers presents no particular difficulties because the energy equation is the same as for the two-dimensional case. Consequently, most methods which have been evolved for two-dimensional problems can be extended to apply to rotationally symmetric surfaces, see, for example, [1, 17, 111]. Furthermore, the rotationally symmetric case can be reduced to a two-dimensional one by the application of the Mangler transformation [71], Sec. XI c.

Thermal boundary layers on rotating rotationally symmetric surfaces have been investigated in a number of publications. Solutions for a disk which rotates in still air (cf. Sec. V 11) are contained in [129, 130, 51]; the corresponding problem concerning a rotating sphere (cf. Sec. XI b 2) was solved by S. N. Singh [117].

A. N. Tifford and S. T. Chu [141] examined the case of a rotating disk placed in an axial stream, whereas the problem of a sphere rotating in an axial stream forms the subject of a study by J. Siekmann [116]. Additional solutions for rotating bodies can be found in refs. [3] and [138]. A generally valid approximate method for the study of thermal boundary layers on bodies rotating in an axial stream was developed by Y. Yamaga [145] who based himself on H. Schlichting's procedure mentioned in Chap. XI (ref. [99] of Chap. XI).

6. Measurements on cylinders and other body shapes. Measurements on the coefficient of heat transfer by forced convection, mostly from circular cylinders, can be found in papers by R. Hilpert [56] and E. Schmidt and K. Wenner [107]. R. Hil-

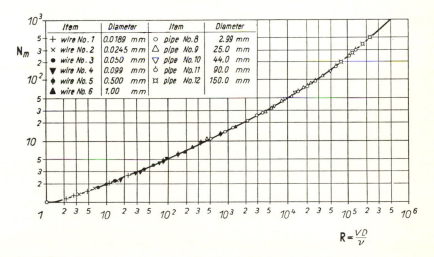

$$R = \frac{VD}{\nu}$$

Fig. 12.18. The Nusselt number N_m in terms of the Reynolds number R for circular cylinders after R. Hilpert [56]. Surface temperature 100° C approx. A comparison with the measurements due to J. Kestin and P. F. Maeder [67] suggests that Hilpert's measurements were performed in a stream of 0·9% turbulence intensity

pert performed measurements on circular cylinders in a cross-flow of air covering
a very wide range of Reynolds numbers. Figure 12.18 contains a plot of the mean Nusselt
number N_m taken for the whole circumference of the cylinder against the Reynolds
number R. Both N_m and R are based on the diameter of the cylinder. As a first
crude approximation it can be assumed that N_m is proportional to $R^{1/2}$ as confirmed
by the theoretical calculations for the flat plate at zero incidence, eqn. (12.79a, b),
and for the flow near a stagnation point, eqn. (12.90), in laminar flow.

The local coefficient of heat transfer varies considerably over the surface of
cylinders and other bodies; measurements on circular cylinders due to E. Schmidt
and K. Wenner [107] are shown in Fig. 12.19. It is seen that in the laminar boundary
layer the coefficient of heat transfer decreases with distance from the stagnation
point and reaches a minimum in the neighbourhood of the point of separation. In
the flow behind the point of separation its value is about equal to that at the leading
edge in the laminar layer. Similar work is reported in refs. [72] and [99].

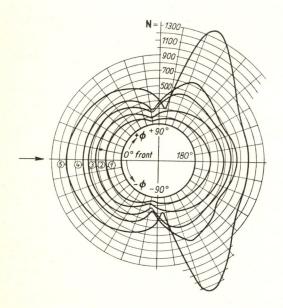

Fig. 12.19. Local coefficient of heat
transfer for a circular cylinder at
varying Reynolds numbers as measured by E. Schmidt and K. Wenner
[107]. Curves (1) and (2) refer to the
region below the critical Reynolds
number, curves (3) and (4) were measured in the critical range, and curve
(5) above the critical range

(1) R = 39,800, (2) R = 101,300,
(3) R = 170,000, (4) R = 257,600,
(5) R = 426,000

A comparison between the measured values for the forward portion of the cylinder
from Fig. 12.19, i. e. for the portion where the flow is laminar, and theoretical
calculations has already been given in Fig. 12.16 The theoretical curves were based
on the actual, measured velocity distribution in the external flow. As is known,
near the forward stagnation point the latter agrees very well with that given by
potential theory. The agreement, as already stated, is satisfactory. D. Johnson and
J. P. Hartnett [62] performed measurements of heat-transfer rates on a circular
cylinder with blowing (sweat-cooling). E. Eckert and W. Weise [17, 20] published
the results of their measurements on the mean and local adiabatic wall temperatures

on unheated cylinders in parallel and in cross-flow covering a range of air velocities nearly up to sonic. In the case of a stream parallel to the axis of the cylinder they obtained a mean value $(T_a - T_\infty) \, 2 \, c_p / U_\infty^2 = 0 \cdot 84$ which was independent of the Mach number, in good agreement with the value from eqn. (12.80) for a flat plate. In cross-flow they obtained a value between $0 \cdot 6$ and $0 \cdot 8$ which was also reasonably independent of the Mach number. A summary of recent work on the total heat-transfer rate from a circular cylinder is contained in a paper by V. T. Morgan [88].

R. Eichhorn, E. Eckert, and A. D. Anderson [30] measured the rate of heat transferred along a circular cylinder placed in an axial stream, the surface temperature of the cylinder being variable. They obtained good agreement with theoretical calculations when allowance was made for the effect of the curvature of the surface. Reviews of current papers which concern themselves with heat transfer problems are published from time to time in the *International Journal of Heat and Mass Transfer*.

The thermal boundary layer can be conveniently made visible with the aid of interferometric photographs. Figure 12.20 represents the flow past a turbine cascade. The shift in the lines is a measure of the difference between the local density and that at a reference state (e. g., with respect to the undisturbed stream). The changes in density in the region of potential flow are due mainly to pressure changes, but in the boundary layer the heat due to friction contributes greatly to the change in density. Upon close examination it is possible to discern in Fig. 12.20 sudden sharp kinks in the lines. These are due to the considerable additional change in density produced by frictional heat. Thus the kinks trace the outer edge of the thermal boundary layer. In natural convection it is even easier to render the boundary layer visible as it is possible to use a Schlieren method for this purpose, first described by E. Schmidt [105]), *cf*. p. 314.

Fig. 12.20. Thermal boundary layer on a turbine cascade, made visible with the aid of the interferometer method, after E. Eckert. Angle of flow at inlet $\beta_1 = 40°$; solidity $l/t = 2 \cdot 18$; Reynolds number $R = 1 \cdot 97 \times 10^5$

The shift of the interferometer lines is proportional to the change in density. The sudden kinks in the lines near the wall show the outer edge of the thermal boundary layer, since the heat of friction produces a large change in density in that region

7. Effect of free-stream turbulence. In all previous considerations concerning laminar boundary layers it was tacitly implied that the external stream was also laminar. However, in the overwhelming majority of cases, particularly during wind-tunnel tests, the external stream carries with it a certain degree of turbulence which means that at every point in it the velocity fluctuates, changing its magnitude and direction. When the velocity is steady on the average, there are superimposed on it three fluctuating velocity components whose time averages over sufficiently long intervals of time vanish. The effect of such fluctuations on the velocity boundary layer will be examined in greater detail in Chap. XV which deals with non-steady boundary layers. In the present section we shall examine the effect of such free-stream oscillations, particularly those due to turbulence, on thermal boundary layers and on rates of heat transfer.

It is recognized that there exists a difficulty in providing an unequivocal description of such fluctuating streams. Since turbulence involves stochastic fluctuations, strictly speaking, no two turbulent streams can ever be similar. However, it is found by experiment that certain average properties of the oscillatioris are adequate to describe them. These are: the intensity of turbulence, T, defined in Sec. XVI d 1, and the scale of turbulence, L, defined in Sec. XVIII d. It is found, further, that in cases when the scale of turbulence is small compared with the dimensions of the body, which occurs in most cases in practice, the degree of turbulence alone suffices to characterize the flow. It is, therefore, to be expected that the Nusselt number for geometrically similar, isothermal bodies which are placed in fluctuating, parallel, isothermal streams, depends on the turbulence intensiy, T, in addition to its dependence on the Prandtl and Reynolds numbers. Thus, for the local or the mean Nusselt number we may write, respectively,

$$N_x = f_1(R, P, T) ,$$ (12.110a)
$$N_m = f_2(R, P, T) ,$$ (12.110b)

instead of the earlier relations in eqns. (12.32) and (12.32a).

An increase in the intensity of turbulence of the free stream must produce two essentially different effects. First, as will be shown in Chap. XVI, an increase in intensity causes earlier transition to turbulence in the boundary layer and hence an increase in the rate of heat transfer which is characteristic of a turbulent as compared with a laminar boundary layer. This effect will be discussed in more detail in Chap. XVI. In addition, there exists a second effect which can become particularly pronounced in the presence of a laminar boundary layer. The diagram in Fig. 12.21 depicts the variation of the local Nusselt number on a ci᷉ ⸱ular cylinder at different values of the Reynolds number and of the intensity of turbulence of the external stream according to measurements performed by J. Kestin, P. F. Maeder and H. H. Sogin [64]. These measurements have been compared in the diagram with N. Froessling's [39] theoretical calculations which correspond to the case of a turbulence-free external stream. These results are quite similar to those reproduced in Fig. 12.19. It is noted that the effect is remarkably high, a turbulence intensity of about 2·5% producing an increase in the local heat flux by something like 80%.

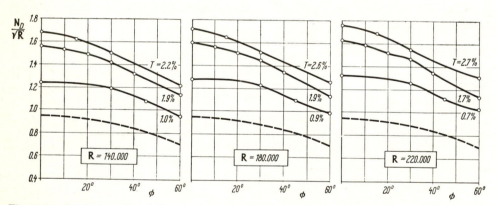

Fig. 12.21. Variation of local Nusselt number N_D on a circular cylinder with turbulence intensity T and angular coordinate ϕ, after J. Kestin, P. F. Maeder and H. H. Sogin [64] (Values of intensity of turbulence T approximate only) – – – Theory after N. Froessling [39]

In more recent times L. Kayalar [63] investigated both theoretically and experimentally the influence of turbulence intensity on the transfer of heat from a circular cylinder. The experimental result is shown in Fig. 12.22. These measurements also show that for intensities between T = 1 and 5% there occurs a steep increase

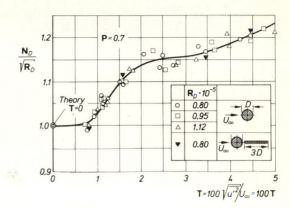

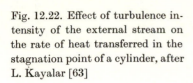
Fig. 12.22. Effect of turbulence intensity of the external stream on the rate of heat transferred in the stagnation point of a cylinder, after L. Kayalar [63]

in the Nusselt number as the turbulence intensity increases, even though the increase is not as large as that in Fig. 12.21. L. Kayalar attempted to explain this phenomenon theoretically. He assumes that the stagnation stream (see Fig. 5.9), characterized by streamlines which are concave outwards, develops a system of stationary, counter-rotating vortices whose axes are aligned with the principal flow direction, rather like those on a concave wall shown in Fig. 17.32 b (Goertler vortices). As a result, the flow becomes highly three-dimensional in the boundary layer which explains the increase in the heat-transfer rate. In this connection, relevant considerations are contained in the papers by H. Goertler [45], H. Schlichting [103], J. Kestin [65], E. A. Brun et al. [4], G. W. Lowery and R. J. Vachon [82] and J. Kestin and L. N. Persen [68a]. See also ref. [118] on p. 549.

Unexpectedly, however, the preceding effect is absent on a flat plate at zero incidence. Measurements performed by J. Kestin, P. F. Maeder and H. E. Wang [68] on a flat plate show no sensitivity to free-stream turbulence in the laminar range. The same result was obtained by A. Edwards and N. Furber [27]. Such results suggest that external turbulence affects the local heat transfer only in the presence of a pressure gradient. The experiments quoted in ref. [67] provide a certain confirmation of such a supposition. By imposing a pressure gradient artificially on a flat plate, it was found possible to increase the local Nusselt number by increasing the turbulence intensity. A qualitative explanation of this behavior can be obtained with the aid of C. C. Lin's theory described in Chap. XV, as pointed out in ref. [68]. The effect of free-stream turbulence on heat transfer has been studied also in references [5, 42, 43, 54, 83, 100, 113, 136]. A modern summary can be found in [88].

h. Thermal boundary layers in natural flow (free convection)

Motions which are caused solely by the density gradients created by temperature differences are termed 'natural' as distinct from those 'forced' on the stream by external causes. Such a natural flow exists around a vertical hot plate or around a horizontal hot cylinder. Natural flows also display, in most cases, a boundary-layer structure, particularly if the viscosity and conductivity of the fluid are small. A comprehensive review of the field was prepared by A. J. Ede [28].

In the case of a *vertical hot plate*, the pressure in each horizontal plane is equal to the gravitational pressure and is thus constant. The only cause of motion is furnished by the difference between weight and buoyancy in the gravitational field of the earth. The equation of motion is obtained from eqns. (12.51 a, b, c) with $dp/dx = 0$ and $\beta = 1/T_\infty$. Neglecting frictional heat we have

$$\frac{\partial u}{\partial x} + \frac{\partial v}{\partial y} = 0, \tag{12.111}$$

$$u \frac{\partial u}{\partial x} + v \frac{\partial u}{\partial y} = \nu \frac{\partial^2 u}{\partial y^2} + g \frac{T_w - T_\infty}{T_\infty} \theta, \tag{12.112}$$

$$u \frac{\partial \theta}{\partial x} + v \frac{\partial \theta}{\partial y} = a \frac{\partial^2 \theta}{\partial y^2}. \tag{12.113}$$

Here $a = k/\varrho\, c_p$ is the thermal diffusivity and $\theta = (T - T_\infty)/(T_w - T_\infty)$ is the dimensionless local temperature. In a theoretical investigation concerning the experimentally determined temperature and velocity field of a case involving natural convection on a vertical hot plate, due to E. Schmidt and W. Beckmann [104], E. Pohlhausen demonstrated that if a stream function is introduced by putting $u = \partial\psi/\partial y$ and $v = -\partial\psi/\partial x$, then the resulting partial differential equation for ψ can be reduced to an ordinary differential equation by the similarity transformation

$$\eta = c\, \frac{y}{\sqrt[4]{x}}; \qquad \psi = 4\, \nu\, c\, x^{3/4} \zeta(\eta)$$

where

$$c = \sqrt[4]{\frac{g\,(T'_w - T_\infty)}{4\, \nu^2\, T_\infty}}. \tag{12.114}$$

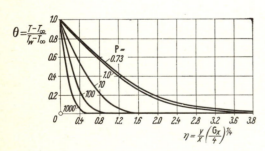

$$\theta = \frac{T - T_\infty}{T_w - T_\infty}$$

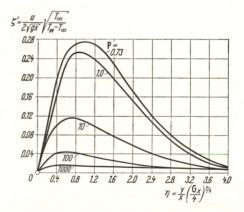

$$\zeta' = \frac{u}{2\sqrt{gx}} \sqrt{\frac{T_\infty}{T_w - T_\infty}}$$

Fig. 12.23. Temperature distribution in the laminar boundary layer on a hot vertical flat plate in natural convection. Theoretical curves, for $P = 0.73$, after E. Pohlhausen [94] and S. Ostrach [93]

$$G_x = \frac{g\, x^2}{\nu^2}\, \frac{T_w - T_\infty}{T_\infty} = \text{Grashof number}$$

Fig. 12.24. Velocity distribution in the laminar boundary layer on a hot vertical flat plate in natural convection (see also Fig. 12.23)

The velocity components now become

$$u = 4\,\nu\,x^{1/2}\,c^2\,\zeta'\;;\qquad v = \nu\,c\,x^{-1/4}\,(\eta\,\zeta' - 3\,\zeta)\,,$$

and the temperature distribution is determined by the function $\theta(\eta)$. Equations (12.112), (12.113) and (12.114) lead to the following differential equations

$$\zeta''' + 3\,\zeta\,\zeta'' - 2\,\zeta'^2 + \theta = 0\,,\qquad \theta'' + 3\,\mathrm{P}\,\zeta\,\theta' = 0\,,\qquad \text{(12.115 a, b)}$$

with the boundary conditions $\zeta = \zeta' = 0$ and $\theta = 1$ at $\eta = 0$ and $\zeta' = 0$, $\theta = 0$ at $\eta = \infty$. Figures 12.23 and 12.24 illustrate the solutions of these equations for various values of P. Figures 12.25 and 12.26 contain a comparison between the calculated velocity and temperature distribution and those measured by E. Schmidt and W. Beckmann [104]. The agreement is seen to be very good. It is seen, further, that the velocity and thermal boundary-layer thickness are proportional to $x^{1/4}$.

Fig. 12.25. Temperature distribution in the laminar boundary layer on a hot vertical flat plate in natural convection in air, as measured by E. Schmidt and W. Beckmann [104]; x = distance from the lower edge of the plate

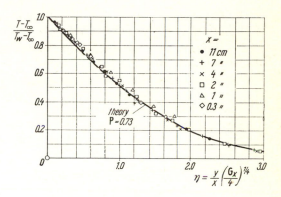

Fig. 12.24. Velocity distribution in the laminar boundary layer on a vertical plate in natural convection in air as measured by E. Schmidt and W. Beckmann [104]

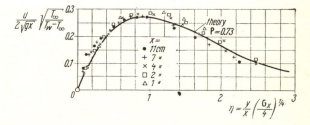

Heat transfer: The quantity of heat $q(x) = -\,k(\partial T/\partial y)_0$ transferred per unit time and area from the plate to the fluid at section x becomes

$$q(x) = -\,k\,c\,x^{-1/4}\left(\frac{d\theta}{d\eta}\right)_0 (T_w - T_\infty)\,,$$

with $(\partial\theta/\partial\eta)_0 = -\,0.508$ for $\mathrm{P} = 0.733$. The total heat transferred by a plate of length l and width b is $Q = b \int\limits_0^l q(x)\,\mathrm{d}x$, and hence

$$Q = \tfrac{4}{3} \times 0 \cdot 508\, b\, l^{3/4}\, c\, k\, (T_w - T_\infty)\,.$$

The mean Nusselt number defined by $Q = b\, k\, \mathsf{N}_m (T_w - T_\infty)$ thus becomes $\mathsf{N}_m = {}$ $= 0 \cdot 677\, c\, l^{3/4}$, or, inserting the value of c from eqn. (12.114):

$$\mathsf{N}_m = 0 \cdot 478\, (\mathsf{G})^{1/4}\,, \tag{12.116}$$

where

$$\mathsf{G} = \frac{g\, l^3\, (T_w - T_\infty)}{\nu^2\, T_\infty} \tag{12.117}$$

is the Grashof number. It can also be written as $\mathsf{G} = g\, l^3\, \beta\, (T_w - T_\infty)/\nu^2$ in the case of liquids.

The diagram in Fig. 12.27 gives a comparison between theoretical results on free convection with measurements on heated vertical cylinders and flat plates performed by E. R. G. Eckert and T. W. Jackson [22]. When the product $\mathsf{G}\mathsf{P} < 10^8$, the flow is laminar, and for $\mathsf{G}\mathsf{P} > 10^{10}$ the flow is turbulent. The agreement between theory and experiment is excellent.

E. Pohlhausen's calculations have been extended by H. Schuh [109] to the case of large Prandtl numbers such as exist in oils.

The case of very small Prandtl numbers is treated in a paper by E. M. Sparrow and J. L. Gregg [126]. The limiting cases when $\mathsf{P} \to 0$ and $\mathsf{P} \to \infty$ were examined by E. J. Le Fevre [73], according to whom we may write

$$\frac{\mathsf{N}_m}{(\mathsf{G}\mathsf{P}^2)^{1/4}} = 0 \cdot 800 \quad (\mathsf{P} \to 0)\,, \tag{12.118a}$$

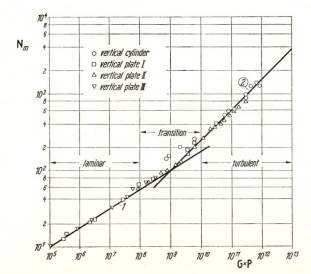

Fig. 12.27. Average Nusselt number for free convection on vertical plates and cylinders, after E. R. G. Eckert and T. W. Jackson [22]

Curve (1) laminar:
$\mathsf{N}_m = 0 \cdot 555\,(\mathsf{G}\mathsf{P})^{1/4}$; $\mathsf{G}\mathsf{P} < 10^9$

Curve (2) turbulent:
$\mathsf{N}_m = 0 \cdot 0210\,(\mathsf{G}\mathsf{P})^{2/5}$; $\mathsf{G}\mathsf{P} > 10^9$

Table 12.6. Coefficients of heat transfer on a heated vertical plate in natural convection (laminar), according to refs. [93, 94, 109, 126]

P	0	0·003	0·008	0·01	0·02	0·03	0·72	0·73	1	2	10	100	1000	∞
$\dfrac{N_m}{(GP)^{1/4}}$	0	0·182	0·228	0·242	0·280	0·305	0·516	0·518	0·535	0·568	0·620	0·653	0·665	0·670

$$\frac{N_m}{(GP)^{1/4}} = 0.670 \qquad (P \to \infty) \,. \tag{12.118b}$$

Some numerical values for intermediate Prandtl numbers are contained in Table 12.6. Calculations with a temperature-dependent viscosity were performed by T. Hara [50]. The effect of suction or blowing on the rate of heat transfer from a vertical plate in natural convection is described in refs. [29, 124]. Additional classes of similar solutions in natural flows were discussed by K. T. Yang [146]. Thus, temperature distributions on the surface of the plate of the form $T_w - T_\infty = T_1 x^n$ also produce similar solutions, but the differential equation (12.115) is now replaced by

$$\zeta''' + (n+3)\,\zeta\,\zeta'' - 2\,(n+1)\,\zeta'^2 + \theta = 0 \,, \tag{12.119a}$$

$$\theta'' + P\,(n+3)\,\zeta\,\theta' - 4\,P\,n\,\zeta'\,\theta = 0. \tag{12.119b}$$

Solutions to these equations were found by E. M. Sparrow and J. L. Gregg [127]. Reference [125] discusses similar solutions in the simultaneous presence of free and forced convection. In such cases, the velocity of the external stream must be proportional to x^m (wedge flow) and the temperature distribution on the plate must be proportional to x^{2m-1} .

Measurements on a vertical hot plate in oil performed by H. H. Lorenz [81] gave the value $N_m = 0.555 \ (G \times P)^{1/4}$ which constitutes very satisfactory agreement with theoretical calculations if it is considered that the theory does not take into account the dependence of viscosity on temperature, which is important precisely in the case of oils.

The laminar thermal boundary layer around heated bodies in natural convection can be conveniently made visible with the aid of a Schlieren method devised by E. Schmidt [105]. A parallel beam of light is passed through the boundary layer in a direction parallel to the plate and produces shadows on a screen placed at a large distance from the body. The density gradient in the air at right angles to the surface causes the rays of light to be deflected outwards. The deflexion is largest at points where the density gradient is steep, i. e. near the body. With a sufficiently large distance between screen and body the space taken up by the heated layer remains dark so that in the Schlieren picture the shadow of the body is surrounded by a shadow due to the thermal boundary layer. The rays of light which are deflected out of the temperature field create an illuminated zone around the dark shadow. The outer edge of this zone of light is formed by the rays which just skirt the surface; consequently their deflexion is proportional to the density

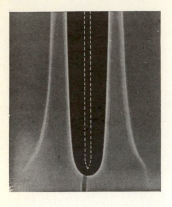

Fig. 12.28. Schlieren photograph of the thermal boundary layer on a heated vertical flat plate, after E. Schmidt [105]

gradient at the surface, i. e. to the local coefficient of heat transfer. Figure 12.28 represents a Schlieren photograph taken on a heated vertical flat plate. The contour of the plate is shown by a broken white line. It is easy to recognize on the shadow that the boundary-layer thickness increases as $x^{1/4}$. The edge of the zone of light shows that the local coefficient of heat transfer is proportional to $x^{-1/4}$. The picture in Fig. 12.29 gives an interferogram for the same type of boundary layer; it was obtained by E. R. G. Eckert and E. Soehngen [13].

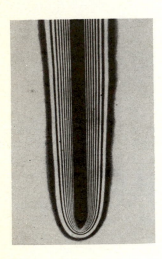

Fig. 12.29. Interferogram of a thermal boundary layer on a vertical heated flat plate, after E. R. G. Eckert and E. Soehngen [13]

Other shapes: The motion due to natural convection around a horizontal heated circular cylinder was treated in an analogous way by R. Hermann [55]. He found for $P = 0.7$ a mean heat transfer coefficient $N_m = 0.372\, G^{1/4}$, where G is based on the diameter. Measurements in air performed by K. Jodlbauer [61] gave $N_m = 0.395\, G^{1/4}$ at $G = 10^5$ which shows a satisfactory agreement with theory. Measurements on vertical cylinders [142] in water and ethylene glycol gave ($P \times G =$

$= 2 \times 10^8$ to 4×10^{10}): $N = 0{\cdot}726 \, (P \times G)^{1/4}$ for laminar flow, and $(P \times G = 4 \times 10^{10}$ to 9×10^{11}): $N = 0{\cdot}0674 \, (G \times P^{1{\cdot}29})^{-1/3}$ for turbulent flow.

For the sphere J.I. Shell [115] calculated $N_m = 0{\cdot}429 \, G^{1/4}$, which was confirmed by measurements in air. Summaries of recent work on natural convection are contained in refs. [65, 96].

References

[1] Allen, H. J., and Look, B. C.: A method for calculating heat transfer in the laminar flow regions of bodies. NACA Rep. 764 (1943).
[2] Ambrok, G. S.: The effect of surface temperature variability on heat exchange in laminar flow in a boundary layer. Soviet Phys. Techn. Phys. *2*, 738—748 (1957). Translation of Zh. Tekh. Fiz. *27*, 812—821 (1957).
[3] Bjorklund, G. S., and Kays, W. M.: Heat transfer between concentric rotating cylinders. J. Heat Transfer *81*, 175—186 (1959).
[4] Brun, E. A., Diep, A., and Kestin, J.: Sur un nouveau type des tourbillons longitudinaux dans l'écoulement autour d'un cylindre. C. R. Acad. Sci. *263*, 742 (1966).
[5] Büyüktür, A. R., Kestin, J., and Maeder, P. F.: Influence of combined pressure gradient and turbulence on the transfer of heat from a plate. Int. J. Heat Mass Transfer *7*, 1175—1186 (1964).
[6] Ten Bosch, M.: Die Wärmeübertragung. Berlin, 1936.
[7] Brun, E. A.: Selected combustion problems. Vol. *II*, 185—198, AGARD, Pergamon Press, London, 1956.
[8] Chapman, D. R., and Rubesin, M. W.: Temperature and velocity profiles in the compressible, laminar boundary layer with arbitrary distribution of surface temperature. JAS *16*, 547—565 (1949).
[9] Davies, D. R., and Bourne, D. E.: On the calculation of heat and mass transfer in laminar and turbulent boundary layers. I. The laminar case. Quart. J. Mech. Appl. Math. *9*, 457—467 (1956); see also Quart. J. Mech. Appl. Math. *12*, 337—339 (1959).
[10] Dewey, C. F., and Gross, J. F.: Exact similar solution of the laminar boundary-layer equations. Advances in Heat Transfer *4*, 317—446 (1967).
[11] Dienemann, W.: Berechnung des Wärmeüberganges an laminar umströmten Körpern mit konstanter und ortsveränderlicher Wandtemperatur. Diss. Braunschweig 1951; ZAMM *33*, 89—109 (1953); see also JAS *18*, 64—65 (1951).
[12] Donoughe, P. L., and Livingood, J. N. B.: Exact solutions of laminar boundary layer equations with constant property values for porous wall with variable temperature. NACA Rep. 1229 (1955).
[12a] Driest, E. R. van: Convective heat transfer in gases. Princeton University Series, High Speed Aerodynamics and Jet Propulsion, Vol. *V*, 339—427 (1959).
[13] Eckert, E. R. G., and Drake, R. M.: Heat and mass transfer. McGraw-Hill, New York, 1959.
[14] Eckert, E.: Einführung in den Wärme- und Stoffaustausch. 3rd ed., Berlin, 1966.
[15] Eckert, E., and Drewitz, O.: Der Wärmeübergang an eine mit großer Geschwindigkeit längsangeströmte Platte. Forschg. Ing.-Wes. *11*, 116—124 (1940).
[16] Eckert, E.: Temperaturmessungen in schnell strömenden Gasen. Z. VDI *84*, 813—817 (1940).
[17] Eckert, E., and Weise, W.: Die Temperatur unbeheizter Körper in einem Gasstrom hoher Geschwindigkeit. Forschg. Ing.-Wes. *12*, 40—50 (1941).
[18] Eckert, E., and Drewitz, O.: Die Berechnung des Temperaturfeldes in der laminaren Grenzschicht schnell angeströmter unbeheizter Körper. Luftfahrtforschung *19*, 189—196 (1942).
[19] Eckert, E.: Die Berechnung des Wärmeüberganges in der laminaren Grenzschicht umströmter Körper. VDI-Forschungsheft *416* (1942).
[20] Eckert, E., and Weise, W.: Messung der Temperaturverteilung auf der Oberfläche schnell angeströmter unbeheizter Körper. Forschg. Ing.-Wes. *13*, 246—254 (1942).
[21] Eckert, E. R. G., and Soehngen, E.: Distribution of heat transfer coefficients around circular cylinders in cross-flow at Reynolds numbers from 20 to 500. Trans. ASME *74*, 343—347 (1952).

[22] Eckert, E. R. G., and Jackson, T. W.: Analysis of turbulent free convection boundary layer on a flat plate. NACA Rep. 1015 (1951).

[23] Eckert, E. R. G., and Livingood, J. N. B.: Method for calculation of laminar heat transfer in air flow around cylinders of arbitrary cross-section (including large temperature differences and transpiration cooling). NACA Rep. 1118 (1953).

[24] Eckert, E. R. G., and Diaguila, A. J.: Experimental investigation of free-convection heat transfer in vertical tube at large Grashof numbers. NACA Rep. 1211 (1955).

[25] Eckert, E. R. G., and Livingood, J. N. B.: Calculations of laminar heat transfer around cylinders of arbitrary cross-section and transpiration cooled walls with application to turbine blade cooling. NACA Rep. 1220 (1955).

[26] Eckert, E. R. G., Hartnett, J. P., and Birkeback, R.: Simplified equations for calculating local and total heat flux to non-isothermal surface. JAS 24, 549−551 (1957).

[27] Edwards, A., and Furber, B. N.: The influence of free stream turbulence on heat transfer by convection from an isolated region of a plane surface in parallel air flow. Proc. Inst. Mech. Eng. 170, 941 (1956).

[28] Ede, A. J.: Advances in free convection. Advances in Heat Transfer, Acad. Press, 4, 1−64 (1967).

[29] Eichhorn, R.: The effect of mass transfer on free convection. J. Heat Transfer 32, 260−263, (1960).

[30] Eichhorn, R., Eckert, E. R. G., and Anderson, A. D.: An experimental study of the effects of nonuniform wall temperature on heat transfer in laminar and turbulent axisymmetric flow along a cylinder. J. Heat Transfer 82, 349−359 (1960).

[31] Elias, F.: Der Wärmeübergang einer geheizten Platte an strömende Luft. Abhandl. Aerodyn. Inst. TH Aachen, Heft 9 (1930); ZAMM 9, 434−453 (1929) and 10, 1−14 (1930).

[32] Evans, H. L.: Mass transfer through laminar boundary layers. 3a. Similar solution to the b-equation when B = 0 and $\sigma \geqslant 0.5$. Int. J. Heat Mass Transfer 8, 26−41 (1961).

[33] Evans, H. L.: Mass transfer through laminar boundary layers. 7. Further similar solutions to the b-equation for the case B = 0. Int. J. Heat Mass Transfer 5, 35−37 (1962).

[34] Evans, H. L.: Laminar boundary layer theory. Addison-Wesley Publishing Company, Reading, Mass., 1968.

[35] Fage, A., and Falkner, V. M.: Relation between heat transfer and surface friction for laminar flow. ARC RM 1408 (1931).

[36] Fischer, P.: Ähnlichkeitsbedingungen für Strömungsvorgänge mit gleichzeitigem Wärmeübergang. ZAMM 48, T 122−T 125 (1963).

[37] Frick, C. W., and McCullough, G. B.: A method for determining the rate of heat transfer from a wing or streamlined body. NACA Rep. 830 (1945).

[38] Fritzsche, A. F., Bodnarescu, M., Kirscher, O., and Esdorn, H.: Probleme der Wärmeübertragung. VDI-Forschungsheft 450 (1955).

[39] Frössling, N.: Verdunstung, Wärmeübergang und Geschwindigkeitsverteilung bei zweidimensionaler und rotationssymmetrischer Grenzschichtströmung. Lunds Univ. Arssk., N. F. Avd. 2, 36, No. 4 (1940); see also NACA TM 1432; see also Lunds Univ. Arssk., N. F. Avd. 2, 154, No. 3 (1958).

[40] Frössling, N.: Calculating by series expansion of the heat transfer in laminar, constant property boundary layers at non-isothermal surfaces. Archiv för Fysik 14, 143−151 (1958).

[41] Frössling, N.: Problems of heat transfer across laminar boundary layers. Theory and fundamental research in heat transfer. Proc. Ann. Meeting of the American Soc. of Mech. Engrs. (J. A. Clark, ed.), Pergamon Press, 181−202, 1963.

[42] Giedt, W. H.: Investigation of variation of point unit heat transfer coefficient around a cylinder normal to an airstream. Trans. ASME 71, 375−381 (1949).

[43] Giedt, W. H.: Effect of turbulence level of incident air stream on local heat transfer and skin friction on a cylinder. JAS 18, 725−730, 766 (1951).

[44] Gersten, K., and Körner, H.: Wärmeübergang unter Berücksichtigung der Reibungswärme bei laminaren Keilströmungen mit veränderlicher Temperatur und Normalgeschwindigkeit entlang der Wand. Intern. J. Heat Mass Transfer 11, 655−673 (1968).

[45] Görtler, H.: Über eine Analogie zwischen Instabilitäten laminarer Grenzschichtströmungen an konkaven Wänden und an erwärmten Wänden. Ing.-Arch. 28, 71−78 (1959).

[46] Goland, L.: A theoretical investigation of heat transfer in the laminar flow regions of airfoils. JAS 17, 436−440 (1950).

[47] Grigull, U.: Wärmeübertragung in laminarer Strömung mit Reibungswärme. Chemie-Ingenieur-Technik 480−483 (1955).

[47a] Grigull, U.: Technische Thermodynamik. 3rd ed., 194 p., Berlin, 1977.

[48] De Groff, H.M.: On viscous heating. JAS 23, 395—396 (1956).

[49] Guha, C.R., and Yih, C.S.: Laminar convection of heat from two-dimensional bodies with variable wall temperatures. Proc. 5th Midw. Conf. Fluid Mech. 29—40 (1957).

[50] Hara, T.: Heat transfer by laminar free convection about a vertical flat plate with large temperature difference. Bull. JSME 1, 251—254 (1958).

[51] Hartnett, J.P.: Heat transfer from a non-isothermal disk rotating in still air. J. Appl. Mech. 26, 672—673 (1959).

[52] Hassan, H.A.: On heat transfer to laminar boundary layers. JASS 26, 464 (1959).

[53] Hausenblas, H.: Die nicht isotherme Strömung einer zähen Flüssigkeit durch enge Spalten und Kapillarröhren. Ing.-Arch. 18, 151—166 (1950).

[54] Van Der Hegge-Zijnen, B.G.: Heat transfer from horizontal cylinders to a turbulent air flow. Appl. Sci. Res. A 7, 205—223 (1957).

[55] Hermann, R.: Wärmeübertragung bei freier Strömung am waagerechten Zylinder in zweiatomigen Gasen. VDI-Forschungsheft 379 (1936).

[56] Hilpert, R.: Wärmeabgabe von geheizten Drähten und Rohren im Luftstrom. Forschg. Ing.-Wes. 4, 215—224 (1933).

[57] Howarth, L.: Velocity and temperature distribution for a flow along a flat plate. Proc. Roy. Soc. London A 154, 364—377 (1936).

[58] Illingworth, C.R.: Some solutions of the equations of flow of a viscous compressible fluid. Proc. Cambr. Phil. Soc. 46, 469—478 (1950).

[59] Imai, I.: On the heat transfer to constant property laminar boundary layer with power function free stream velocity and wall temperature distributions. Quart. Appl. Math. 16, 33—45 (1958).

[60] Jakob, M.: Heat transfer, I and II. McGraw-Hill, New York, 1949 and 1957.

[61] Jodlbauer, K.: Das Temperatur- und Geschwindigkeitsfeld um ein geheiztes Rohr bei freier Konvektion. Forschg. Ing.-Wes. 4, 157—172 (1933).

[62] Johnson, D.V., and Hartnett, J.P.: Heat transfer from a cylinder in crossflow with transpiration cooling. J. Heat Transfer 85, 173—179 (1963).

[63] Kayalar, L.: Experimentelle und theoretische Untersuchungen über den Einfluß des Turbulenzgrades auf den Wärmeübergang in der Umgebung des Staupunktes eines Kreiszylinders. Diss. Braunschweig 1968; Forschg. Ing.-Wes. 35, 157—167 (1969).

[64] Kestin, J., Maeder, P.F., and Sogin, H.H.: The influence of turbulence on the transfer of heat to cylinders near the stagnation point. ZAMP 12, 115—132 (1961).

[65] Kestin, J.: The effect of free-stream turbulence on heat transfer rates. Advances in Heat Transfer (Th. Irvine and J.P. Harnett, ed.) Acad. Press, Vol. 3, 1—32 (1966).

[66] Kestin, J., Maeder, P.F., and Wang, H.E.: On boundary layers associated with oscillating streams. Appl. Sci. Res. A 10, 1 (1961).

[67] Kestin, J., and Maeder, P.F.: Influence of turbulence on transfer of heat from cylinders. NACA TN 4018 (1954).

[68] Kestin, J., Maeder, P.F., and Wang, H.E.: Influence of turbulence on the transfer of heat from plates with and without a pressure gradient. Int. J. Heat Mass Transfer 3, 133—154 (1961).

[68a] Kestin, J., and Persen, L.N.: The transfer of heat across a turbulent boundary layer at very high Prandtl numbers. Int. J. Heat Mass Transfer 5, 355—371 (1962).

[69] Klein, J., and Tribus, M.: Forced convection from non-isothermal surfaces. Heat Transfer Symposium, Engineering Research Institute, Univ. of Michigan, Aug. 1952.

[70] Knudsen, J.G., and Katz, D.L.: Fluid dynamics and heat transfer. McGraw-Hill, New York, 1958.

[71] Ko, S.Y.: Calculation of local heat transfer coefficients on slender surfaces of revolution by the Mangler transformation. JAS 25, 62—63 (1958).

[72] Kroujilin, G.: The heat transfer of a circular cylinder in a transverse airflow in the range of Re = 6000 — 425000. Techn. Physics USSR 5, 289—297 (1938).

[73] Le Fevre, E.J.: Laminar free convection from a vertical plane surface. Mech. Eng. Res. Lab., Heat 113, Gt. Britain, 1956.

[74] Le Fur, B.: Nouvelle méthode de résolution par itération des équations dynamiques et thermiques de la couche limite laminaire. Publ. Sci. et Techn. du Ministère de l'Air, No. 383 (1962).

[75] Le Fur, B.: Convection de la chaleur en régime laminaire dans le cas d'un gradient de pression et d'une température de paroi quelquonques, le fluide étant à propriétés physiques constantes. Int. J. Heat Mass Transfer *1*, 68—80 (1960).

[76] Levèque, M.A.: Les lois de la transmission de chaleur par convection. Ann. Mines *13*, 201—239 (1928).

[77] Levy, S.: Heat transfer to constant property laminar boundary layer flows with power-function free-stream velocity and wall temperature variation. JAS *19*, 341—348 (1952).

[78] Liepmann, H.W.: A simple derivation of Lighthill's heat transfer formula. JFM *3*, 357—360 (1958).

[79] Lietzke, A.F.: Theoretical and experimental investigation of heat transfer by laminar natural convection between parallel plates. NACA Rep. 1223 (1955).

[80] Lighthill, M.J.: Contributions to the theory of heat transfer through a laminar boundary layer. Proc. Roy. Soc. London A *202*, 359—377 (1950).

[81] Lorenz, H.H.: Die Wärmeübertragung an einer ebenen senkrechten Platte an Öl bei natürlicher Konvektion. Z. Techn. Physik *362* (1934).

[82] Lowery, G.W., and Vachon, R.J.: The effect of turbulence on heat transfer from heated cylinders. Int. J. Heat Mass Transfer *18*, 1229—1242 (1975).

[83] Maisel, D.S., and Sherwood, T.K.: Evaporation of liquids into turbulent gas streams. Chem. Eng. Progr. *46*, 131—138 (1950).

[84] Meksyn, D.: Plate thermometer. ZAMP *11*, 63—68 (1960).

[85] Merk, H.J.: Rapid calculations for boundary layer heat transfer using wedge solutions and asymptotic expansions. JFM *5*, 460—480 (1959).

[86] Millsaps, K., and Pohlhausen, K.: Thermal distribution in Jeffery-Hamel flows between non-parallel plane walls. JAS *20*, 187—196 (1953).

[86a] Millsaps, K., and Pohlhausen, K.: Heat transfer by laminar flow from a rotating plate.

[87] Morgan, A.J.A.: On the Couette flow of a compressible viscous, heat conducting, perfect gas. JAS *24*, 315—316 (1957).

[88] Morgan, V.T.: The overall convection heat transfer from smooth circular cylinders. Advances in Heat Transfer *11*, 199—265 (1975).

[89] Morgan, G.W., Pipkin, A.C., and Warner, W.H.: On heat transfer in laminar boundary layer flows of liquids having a very small Prandtl number. JAS *25*, 173—180 (1958).

[90] Nahme, R.: Beiträge zur hydrodynamischen Theorie der Lagerreibung. Ing.-Arch. *11*, 191—209 (1940).

[91] Nusselt, W.: Das Grundgesetz des Wärmeüberganges. Ges. Ing. *38*, 477 (1915).

[92] Oldroyd, J.G.: Calculations concerning theoretical values of boundary layer thickness and coefficients of friction and heat transfer for steady two-dimensional flow in an incompressible boundary layer with main stream velocity $U \sim x^m$ or $U \sim e^x$. Phil. Mag. *36*, 587—600 (1945).

[93] Ostrach, S.: An analysis of laminar free-convection flow and heat transfer about a flat plate parallel to the direction of the generating body force. NACA Rep. 1111 (1953).

[94] Pohlhausen, E.: Der Wärmeaustausch zwischen festen Körpern und Flüssigkeiten mit kleiner Reibung und kleiner Wärmeleitung. ZAMM *1*, 115—121 (1921).

[95] Prandtl, L.: Eine Beziehung zwischen Wärmeaustausch und Strömungswiderstand in Flüssigkeiten. Phys. Z. *11*, 1072—1078 (1910); see also Coll. Works *II*, 585—596 (1961).

[96] Raithby, G.D., and Hollands, K.G.T.: A general method of obtaining approximate solutions to laminar and turbulent free convection problems. Advances in Heat Transfer *11*, 265—315 (1975).

[97] Reeves, B.L., and Kippenhan, Ch.J.: On a particular class of similar solutions of the equations of motion and energy of a viscous fluid. JASS *29*, 38—47 (1962).

[98] Reynolds, O.: On the extent and action of the heating surface for steam boilers. Proc. Manchester Lit. Phil. Soc. *14*, 7—12 (1874).

[99] Richardson, E.G.: The aerodynamic characteristics of a cylinder having a heated boundary layer. Phil. Mag. *23*, 681—692 (1937).

[100] Sato, K., and Sage, B.H.: Thermal transfer in turbulent gas streams; Effect of turbulence on macroscopic transport from spheres. Trans. ASME *80*, 1380—1388 (1958).

[101] Schlichting, H.: Einige exakte Lösungen für die Temperaturverteilung in einer laminaren Strömung. ZAMM *31*, 78—83 (1951).

[102] Schlichting, H.: Der Wärmeübergang an einer längsangeströmten Platte mit veränderlicher Wandtemperatur. Forschg. Ing.-Wes. *17*, 1—8 (1951).

[103] Schlichting, H.: A survey on some recent research investigations on boundary layers and heat transfer. J. Appl. Mech. *38*, 289—300 (1971).

[104] Schmidt, E., and Beckmann, W.: Das Temperatur- und Geschwindigkeitsfeld von einer Wärme abgebenden, senkrechten Platte bei natürlicher Konvektion. Forschg. Ing.-Wes. *1*, 391—404 (1930).

[105] Schmidt, E.: Schlierenaufnahmen der Temperaturfelder in der Nähe wärmeabgebender Körper. Forschg. Ing.-Wes. *3*, 181—189 (1932).

[106] Schmidt, E.: Einführung in die technische Thermodynamik und in die Grundlagen der chemischen Thermodynamik, 10th ed. Berlin, 1963.

[107] Schmidt, E., and Wenner, K.: Wärmeabgabe über den Umfang eines angeblasenen geheizten Zylinders. Forschg. Ing.-Wes. *12*, 65—73 (1941).

[108] Schmidt, E.: Thermische Auftriebsströmungen und Wärmeübergang. Vierte Ludwig-Prandtl-Gedächtnisvorlesung, ZFW *8*, 273—284 (1960).

[109] Schuh, H.: Einige Probleme bei freier Strömung zäher Flüssigkeiten. Göttinger Monographien Bd. B, Grenzschichten, 1946.

[110] Schuh, H.: Über die Lösung der laminaren Grenzschichtgleichung an einer ebenen Platte für Geschwindigkeits- und Temperaturfeld bei veränderlichen Stoffwerten und für das Diffusionsfeld bei höheren Konzentrationen. ZAMM *25/27*, 54—60 (1947).

[111] Schuh, H.: Ein neues Verfahren zur Berechnung des Wärmeüberganges in ebenen und rotationssymmetrischen laminaren Grenzschichten bei konstanter und veränderlicher Wandtemperatur. Forschg. Ing.-Wes. *20*, 37—47(1954); see also: Schuh, H.: A new method for calculating laminar heat transfer on cylinders of arbitrary cross-section and on bodies of revolution at constant and variable wall temperature. KTH Aero. TN 33 (1953).

[112] Schuh, H.: On asymptotic solutions for the heat transfer at varying wall temperatures in a laminar boundary layer with Hartree's velocity profiles. JAS *20*, 146—147 (1953).

[113] Seban, R. A.: The influence of free-stream turbulence on the local transfer from cylinders. Trans. ASME Ser. C, J. Heat Transfer *82*, 101—107 (1960).

[114] Shao Wen Yean: Heat transfer in laminar compressible boundary layer on a porous flat plate with fluid injection. JAS *16*, 741—748 (1949).

[115] Shell, J. I.: Die Wärmeübergangszahl von Kugelflächen. Bull. Acad. Sci. Nat. Belgrade *4*, 189 (1938).

[116] Siekmann, J.: The calculation of the thermal laminar boundary layer on a rotating sphere. ZAMP *13*, 468—482 (1962); see also AGARD Rep. 283 (1960).

[117] Singh, S. N.: Heat transfer by laminar flow from a rotating sphere. Appl. Sci. Res. A *9*, 197—205 (1960).

[118] Skopets, M. B.: Approximate method for integrating the equations of a laminar boundary layer in an incompressible gas in the presence of heat transfer. Soviet Phys. Techn. Phys. *4*, 411—419 (1959). Translation of Zh. Tekh. Fiz *29*, 461—471 (1959).

[119] Smith, A. G., and Spalding, D. B.: Heat transfer in a laminar boundary layer with constant fluid properties and constant wall temperature. J. Roy. Aero. Soc. *62*, 60—64 (1958).

[120] Spalding, D. B.: Heat transfer from surfaces of non-uniform temperature. JFM *4*, 22—32 (1958).

[121] Spalding, D. B., and Evans, H. L.: Mass transfer through laminar boundary layers. 3. Similar solutions to the b-equation. Int. J. Heat Mass Transfer *2*, 314—341 (1961).

[122] Spalding, D. B., and Pun, W. M.: A review of methods for predicting heat transfer coefficients for laminar uniform-property boundary layer flows. Int. J. Heat Mass Transfer *5*, 239—250 (1962).

[123] Sparrow, E. M.: The thermal boundary layer on a non-isothermal surface with non-uniform free stream velocity. JFM *4*, 321—329 (1958).

[124] Sparrow, E. M., and Cess, R. D.: Free convection with blowing or suction. J. Heat Transfer *83*, 387—389 (1961).

[125] Sparrow, E. M., Eichhorn, R., and Gregg, J. L.: Combined forced and free convection in a boundary layer flow. Physics of Fluids *2*, 319—328 (1959).

[126] Sparrow, E. M., and Gregg, J. L.: Details of exact low Prandtl number boundary layer solutions for forced and for free convection. NASA Memo. 2-27-59 E (1959).

[127] Sparrow, E. M., and Gregg, J. L.: Similar solutions for free convection from a non isothermal vertical plate. Trans. ASME *80*, 379—386 (1958).

[128] Sparrow, E. M., and Gregg, J. L.: The effect of a non isothermal free stream on boundary layer heat transfer. J. Appl. Mech. *26*, 161—165 (1959).

[129] Sparrow, E.M., and Gregg, J.L.: Heat transfer from a rotating disk to fluids of any Prandtl number. J. Heat Transfer *81*, 249—251 (1959).
[130] Sparrow, E.M., and Gregg, J.L.: Mass transfer, flow, and heat transfer about a rotating disk. J. Heat Transfer *82*, 294—302 (1960).
[131] Squire, H.B.: Section of: Modern Developments in Fluids Dynamics (S. Goldstein, ed.), Oxford, *II*, 623—627 (1938).
[132] Squire, H.B.: Heat transfer calculation for aerofoils. ARC RM 1986 (1942).
[133] Squire, H.B.: Note on the effect of variable wall temperature on heat transfer. ARC RM 2753 (1953).
[134] Stewart, W.E., and Prober, R.: Heat transfer and diffusion in wedge flows with rapid mass transfer. Int. J. Heat Transfer *5*, 1149—1163 (1962).
[135] Stojanovic, D.: Similar temperature boundary layers. JASS *26*, 571—574 (1959).
[136] Sugawara, S., Sato, T., Komatsu, H., and Osaka, H.: The effect of free stream turbulence on heat transfer from a flat plate. NACA TM 1441 (1958).
[137] Sutera, S.P.: Vorticity amplification in stagnation point flow and its effect on heat transfer. JFM *21*, 513—534 (1965).
[138] Tien, C.L.: Heat transfer by laminar flow from a rotating cone. J. Heat Transfer *82*, 252—253 (1960).
[139] Tifford, A.N.: The thermodynamics of the laminar boundary layer of a heated body in a high speed gas flow field. JAS *12*, 241—251 (1945).
[140] Tifford, A.N., and Chu, S.T.: Heat transfer in laminar boundary layers subject to surface pressure and temperature distributions. Proc. Second Midwestern Conf. Fluid Mech. 1949, 363—377 (1949).
[141] Tifford, A.N., and Chu, S.T.: On the flow and temperature field in forced flow against a rotating disc. Proc. Second U.S. Nat. Congr. Appl. Mech. 1955, 793—800 (1955).
[142] Touloukian, Y.S., Hawkins, G.A., and Jakob, M.: Heat transfer by free convection from heated vertical surfaces to liquids. Trans. ASME *70*, 13—23 (1948).
[143] Vogelpohl, G.: Der Übergang der Reibungswärme von Lagern aus der Schmierschicht in die Gleitflächen. — Temperaturverteilung und thermische Anlaufstrecke in parallelen Schmierschichten bei Erwärmung durch innere Reibung. VDI-Forschungsheft 425 (1949).
[144] Vasanta Ram: Ähnliche Lösungen für die Geschwindigkeits- und Temperaturverteilung in der inkompressiblen laminaren Grenzschicht entlang einer rechtwinkligen Ecke. Ein theoretischer Beitrag zum Problem der Interferenz von Grenzschichten. Diss. Braunschweig 1966; Jb. WGL 1966, 156—178 (1967).
[145] Yamaga, J.: An approximate solution of the laminar flow heat-transfer in a rotating axially symmetrical body surface in a uniform incompressible flow. J. Mech. Lab. Japan *2*, No. 1, 1—14 (1956).
[146] Yang, K.T.: Possible similarity solutions for laminar free convection on vertical plates and cylinders. J. Appl. Mech. *27*, 230—236 (1960).

CHAPTER XIII

Laminar boundary layers in compressible flow†

a. Physical considerations

The development of the theory of boundary-layer flow in compressible streams was stimulated by the progress in aeronautical engineering and, in recent times, by the development of rockets and artificial satellites. When flight velocities of the order of multiples of the velocity of sound are attained, the work of compression and energy dissipation produces considerable increases in temperature and forces us always to include the thermal boundary layer in the analysis, because the two boundary layers strongly interact each with the other. At a velocity w_∞, the temperature rise due to adiabatic compression attains a value of

$$(\Delta T)_{ad} = \frac{w_\infty^2}{2\,c_p} \tag{13.1}$$

as is known form eqn. (12.14b). Here c_p denotes the specific heat of the gas per unit mass. Since $\gamma\, p_\infty/\rho_\infty = (\gamma - 1)c_p\,T_\infty$, we may also write

$$\frac{(\Delta T)_{ad}}{T_\infty} = \frac{\gamma - 1}{2}\,\mathsf{M}_\infty^2\,, \tag{13.2}$$

where the Mach number is defined as $\mathsf{M}_\infty = w_\infty/c_\infty$. The rise in temperature through friction in the boundary layer is of the same order of magnitude as the rise due to adiabatic compression, as already mentioned in Chap. XII, and as will be shown in greater detail later in this chapter.

The numerical evaluation of eqns. (13.1) and (13.2) for air conceived of as a perfect gas is represented graphically in Fig. 13.1 (in which $c_p = 1{\cdot}006$ kJ/kg deg and $\gamma = 1{\cdot}4$ have been assumed). It is seen from it that at a flight velocity of $w_\infty = 2$ km/sec, which corresponds to a Mach number of $\mathsf{M}_\infty = 6$, the temperature rise of the gas stream attains a value of $\Delta T = 2000$ deg C. This temperature rise increases rapidly as the flight velocity is increased. However, a gas at high temperature changes its physical properties in comparison with the corresponding perfect gas. In *real gases* there occur the processes of dissociation and ionization (formation of a plasma). Consequently, the absorption of energy associated with such processes causes the temperature rise in a real gas to be smaller than it would be in a perfect gas. At an

---

† I am indebted to Dr. F. W. Riegels who contributed a revised version of this chapter to the previous edition; in particular, Dr. Riegels formulated the presentation of the extended Illingworth-Stewartson transformation contained in Sec. XIII d 1.

orbital velocity of a satellite of $w_\infty \approx 8$ km/sec, the temperature rise even in a real gas is still of the order of 10,000 deg C. The range of Mach numbers $M_\infty > 6$, in which there exist large differences between the behaviour of a real as opposed to a perfect gas, is given the name of *hypersonic flow*. The occurrence of chemical reactions (ionization, dissociation) which set in behind a shock wave or in the boundary layer on a solid body in a hypersonic stream by virtue of the existence of a high temperature, considerably complicates the task of analyzing the flow. For this reason, we shall restrict our considerations to that range of Mach numbers in which the fluid can still be assumed to obey the perfect-gas law; in air, this corresponds to a range of $M_\infty < 6$. In modern times much attention has been given to the study of boundary-layer flows at hypersonic velocities and in the presence of chemical reactions. For details, the reader is referred to the book by W. H. Dorrance [29].

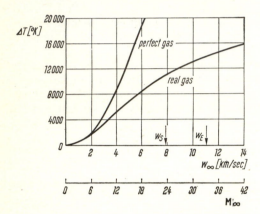

Fig. 13.1. Temperature rise in air in terms of the flight velocity, w_∞, and the Mach number, M_∞. The curve labelled "perfect gas" was calculated with the aid of eqns. (13.1) and (13.2). The velocity $w_S = 7 \cdot 9$ km/sec is that of an artificial satellite in orbit, and $w_E = 11 \cdot 2$ km/sec represents the escape velocity of a satellite from the earth

Even in the range of supersonic Mach numbers ($M_\infty < 6$ in air), the temperature rise in the gaseous stream is high enough to force us to take into account the effect of temperature on the properties of the gas, in particular, on its viscosity. The kinematic viscosity of most gases, and of air among them, increases considerably as the temperature is increased.

In the case of air, as shown by E. R. van Driest [30], it is possible to use an interpolation formula based on D. M. Sutherland's theory of viscosity. This can be written

$$\frac{\mu}{\mu_0} = \left(\frac{T}{T_0}\right)^{\frac{3}{2}} \frac{T_0 + S_1}{T + S_1} \tag{13.3}$$

where μ_0 denotes the viscosity at the reference temperature T_0, and S_1 is a constant which for air assumes the value

$$S_1 = 110 \text{ K}.$$

The preceding relation between the viscosity μ of air and the temperature, T, is seen plotted as curve (1) in Fig. 13.2. Since the relation (13.3) is still too complicated, it is customary to approximate it in theoretical calculations by the simpler power law

$$\frac{\mu}{\mu_0} = \left(\frac{T}{T_0}\right)^{\omega} \quad \text{with} \quad 0 \cdot 5 < \omega < 1 . \tag{13.4}$$

The curves corresponding to $\omega = 0.5, 0.75$ and 1.0 are also shown plotted in Fig. 13.2. It is seen from the graph that Sutherland's formula (13.3) can be approximated at high temperatures by adopting values of ω between 0.5 and 0.75, whereas at lower temperatures the value $\omega = 1.0$ appears to be adequate. The specific heat, c_p, and the Prandtl number, P, can both be assumed to be constant with a satisfactory degree of approximation, even at large temperature differences, as seen from Table 12.1.

Sometimes, the viscosity law $\mu(T)$ is assumed to be of the form

$$\frac{\mu}{\mu_0} = b \, \frac{T}{T_0} \, , \tag{13.4a}$$

where the constant b serves to achieve a better approximation to the more exact Sutherland formula (13.3) in the neighbourhood of a desired temperature range (*cf.* Sec. XIIId).

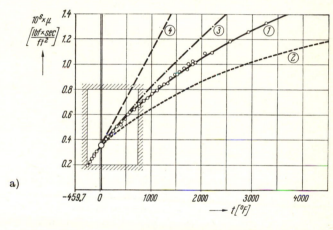

a)

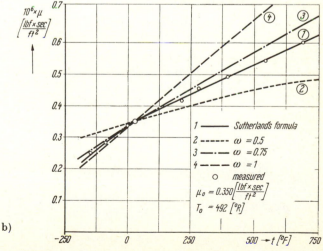

b)

Fig. 13.2. The dynamic viscosity, μ, of air in terms of the temperature T

Curve (1) Measurements and interpolation formula (13.3) based on Sutherland's equation. Curves (2), (3), and (4) power laws, eqn. (13.4), with different values of the exponent ω

The phenomena under consideration become, naturally, very complicated because of the interaction between the velocity and the thermal boundary layers. Compared with incompressible flow there are at least four additional quantities which must be taken into account in the calculation of compressible boundary layers:

1. the Mach number
2. the Prandtl number
3. the viscosity function $\mu(T)$
4. boundary condition for temperature distribution
    (heat transfer or adiabatic wall).

It is clear that the large number of additional parameters, compared with incompressible flow, causes the number of cases likely to occur in practice to become almost intractable as a consequence.

Comprehensive reviews of the numerous papers concerned with compressible boundary layers were given by G. Kuerti [57] and A. D. Young [106]. Details of special mathematical methods employed by various authors have been discussed by N. Curle [26] and K. Stewartson [96]. Problems of compressible turbulent layers are discussed in Chap. XXIII.

b. Relation between the velocity and the temperature fields

In the case of two-dimensional flow, and irrespective of the shape of the body, there exists a remarkably simple relation between the fields of velocity and temperature. In the particular case when $\mathsf{P} = 1$, the integration of the differential equations becomes much easier. The corresponding proposition was first used by A. Busemann [10] and L. Crocco [20] when they calculated the compressible boundary layer on a plate. It can be stated simply by asserting that irrespective of the form of the viscosity function $\mu(T)$, the temperature T depends solely on the velocity component u taken parallel to the wall, i. e. $T = T(u)$. Thus curves of constant velocity ($u = $ const) are identical with the isotherms ($T = $ const).

This remarkable theorem can be easily deduced from the boundary-layer equations. Neglecting the buoyancy forces but taking into account the temperature dependence of the properties μ and k, we can rewrite the boundary-layer equations (12.50a, b, c) as:

$$\frac{\partial(\varrho\,u)}{\partial x} + \frac{\partial(\varrho\,v)}{\partial y} = 0\,, \tag{13.5}$$

$$\varrho\left(u\,\frac{\partial u}{\partial x} + v\,\frac{\partial u}{\partial y}\right) = -\frac{dp}{dx} + \frac{\partial}{\partial y}\left(\mu\,\frac{\partial u}{\partial y}\right), \tag{13.6}$$

$$\varrho\,c_p\left(u\,\frac{\partial T}{\partial x} + v\,\frac{\partial T}{\partial y}\right) = u\,\frac{dp}{dx} + \frac{\partial}{\partial y}\left(k\,\frac{\partial T}{\partial y}\right) + \mu\left(\frac{\partial u}{\partial y}\right)^2, \tag{13.7}$$

$$p = \varrho\,R\,T\,. \tag{13.8}$$

The pressure gradient, as was the case with incompressible flow, is now also determined by the frictionless external flow:

$$\frac{dp}{dx} = -\varrho_1 \, U \, \frac{dU}{dx} = \varrho_1 \, c_p \, \frac{dT_1}{dx} \tag{13.9}$$

with $\varrho_1(x)$ and $T_1(x)$ denoting the density and temperature, respectively, at the outer edge of the boundary layer. Since $\partial p/\partial y = 0$ at any point x along the flow, the temperature and density satisfy the relation

$$\varrho(x, y) \cdot T(x, y) = \varrho_1(x) \cdot T_1(x) . \tag{13.10}$$

Making the assumption in eqns. (13.5) to (13.7) that the temperature depends on the single variable u, i. e. that

$$T = T(u) ,$$

we can deduce from eqn. (13.7) that

$$\varrho \, c_p \, T_u \left(u \, \frac{\partial u}{\partial x} + v \, \frac{\partial u}{\partial y} \right) = u \, \frac{dp}{dx} + \frac{\partial}{\partial y} \left(k T_u \, \frac{\partial u}{\partial y} \right) + \mu \left(\frac{\partial u}{\partial y} \right)^2 ,$$

where differentiation with respect to u is denoted by the subscript, so that $T_u = dT/du$. Eliminating the left-hand side with the aid of eqn. (13.6), we have

$$c_p \, T_u \left[-\frac{dp}{dx} + \frac{\partial}{\partial y} \left(\mu \, \frac{\partial u}{\partial y} \right) \right] = u \, \frac{dp}{dx} + T_u \, \frac{\partial}{\partial y} \left(k \, \frac{\partial u}{\partial y} \right) + (T_{uu} \, k + \mu) \left(\frac{\partial u}{\partial y} \right)^2$$

or

$$-\frac{dp}{dx} \, (c_p \, T_u + u) + T_u \left[c_p \, \frac{\partial}{\partial y} \left(\mu \, \frac{\partial u}{\partial y} \right) - \frac{\partial}{\partial y} \left(k \, \frac{\partial u}{\partial y} \right) \right] = (T_{uu} \, k + \mu) \left(\frac{\partial u}{\partial y} \right)^2 .$$

Introducing the Prandtl number $\mathsf{P} = \mu \, c_p/k$, which may be assumed independent of temperature as far as gases are concerned (*cf.* Table 12.1), we obtain

$$-\frac{dp}{dx} \, (c_p \, T_u + u) + c_p \, \frac{\mathsf{P}-1}{\mathsf{P}} \, T_u \, \frac{\partial}{\partial y} \left(\mu \, \frac{\partial u}{\partial y} \right) = (T_{uu} \, k + \mu) \left(\frac{\partial u}{\partial y} \right)^2 .$$

It is clear from this form that $T = T(u)$ is a solution of the system of equations (13.5) to (13.7) if, simultaneously,

$$\frac{dp}{dx} = 0 : \mathsf{P} = 1 \quad \text{and} \quad T_{uu} = -\frac{\mu}{k} = -\frac{1}{c_p}. \tag{13.11}$$

or, if

$$\frac{dp}{dx} \neq 0 : \text{and if, in addition, } T_u = 0 \text{ at } y = 0 . \tag{13.11a}$$

This proves our proposition.

The actual function which describes the relation between temperature and velocity is obtained by integration. Thus from eqn. (13.11) we have the general solution

$$T(u) = -\frac{u^2}{2 \, c_p} + C_1 \, u + C_2 .$$

The constants of integration C_1 and C_2 can now be determined from the boundary conditions. For $dp/dx \neq 0$ we have $C_1 = 0$.

1. Adiabatic wall

The boundary conditions are

$$y = 0 : u = 0 ; \quad \frac{\partial T}{\partial y} = 0 , \quad \text{and hence} \quad \frac{dT}{du} = 0 .$$

$$y = \infty : u = U ; \quad T = T_1 .$$

Here $T_1(x)$ denotes the temperature at the outer edge of the boundary layer, and the solution becomes

$$T = T_1 + \frac{1}{2 c_p} (U^2 - u^2) . \tag{13.12}$$

Consequently the adiabatic wall temperature $T = T_a$ for $u = 0$ is given by

$$T_a = T_1 + \frac{U^2}{2 c_p} . \tag{13.12a}$$

Introducing the Mach number $M = U/c_1$ where $c_1{}^2 = (\gamma - 1) c_p T_1$ we can re-write eqn. (13.12a) in the form

$$T_a = T_1 \left(1 + \frac{\gamma - 1}{2} M^2 \right) , \quad (P = 1) . \tag{13.12b}$$

The quantity $T_a - T_1$ represents the temperature increase of an adiabatic wall which is due to frictional heat. It is independent of the exponent of the viscosity function.

2. Heat transfer (flat plate, $dp/dx = 0$)

We assume that the temperature of the wall is kept constant and equal to T_w. Thus the boundary conditions are

$$y = 0 : u = 0 , \quad T = T_w ; \quad y = \infty : u = U_\infty , \quad T = T_\infty ,$$

which gives the solution

$$\frac{T - T_w}{T_\infty} = \left(1 - \frac{T_w}{T_\infty} \right) \frac{u}{U_\infty} + \frac{U_\infty{}^2}{2 c_p T_\infty} \frac{u}{U_\infty} \left(1 - \frac{u}{U_\infty} \right) . \tag{13.13}$$

Expressing it in terms of the Mach number $M_\infty = U_\infty/c_\infty$, we obtain

$$\frac{T - T_w}{T_\infty} = \left(1 - \frac{T_w}{T_\infty} \right) \frac{u}{U_\infty} + \frac{\gamma - 1}{2} M_\infty{}^2 \frac{u}{U_\infty} \left(1 - \frac{u}{U_\infty} \right) . \tag{13.13a}$$

In the limiting case when $M_\infty \to 0$, eqn. (13.13a) assumes the form of eqn. (12.64) which was obtained earlier for incompressible flow.

The relation between the velocity and temperature distribution given in eqn. (13.13) is seen plotted in Fig. 13.3. The direction of heat flow can be deduced at once from the temperature gradient at the wall. Since $(\partial u/\partial y)_w > 0$, the direction of the heat flow is determined by the gradient $(dT/du)_w$ at the wall. In fact, we can deduce from eqn. (13.13) that

$$\frac{U_\infty}{T_\infty}\left(\frac{dT}{du}\right)_w = 1 - \frac{T_w}{T_\infty} + \frac{U_\infty^2}{2\,c_p\,T_\infty}\,, \qquad (13.14)$$

so that for $(dT/du)_w < 0$ there is a flow of heat from the wall to the fluid, and conversely, for $(dT/du)_w > 0$ heat flows from the fluid to the wall. In this manner

$$T_w - T_\infty \gtreqless \frac{U_\infty^2}{2\,c_p} \quad \text{or} \quad \frac{T_w - T_\infty}{T_\infty} \gtreqless \frac{\gamma-1}{2}\,\mathsf{M}_\infty^2 : \qquad (13.15)$$

Heat flux wall $\rightleftarrows$ fluid, valid for $\mathsf{P} = 1$.

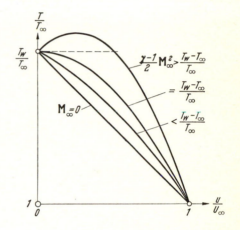

Fig. 13.3.   Relationship between velocity and temperature distribution for the compressible laminar boundary layer on a flat plate including frictional heat, from eqn. (13.13)

Prandtl   number $\mathsf{P} = 1$. T_w = wall   temperature;
T_∞ = free-stream temperature.   For
$$\tfrac{1}{2}\,(\gamma-1)\,\mathsf{M}^2 > (T_w - T_\infty)/T_\infty$$
we have $(\partial T/\partial y)_{wall} > 0$, and heat is transferred to the wall owing to the large quantity of heat generated by friction, although $T_w > T_\infty$

c. The flat plate at zero incidence

The boundary layer on a flat plate at zero incidence has been studied extensively in numerous publications, and we propose to begin with a more detailed discussion of this case. First we shall deduce the relation between the velocity and temperature distribution on a flat plate from the preceding general proposition.

In the case of an *adiabatic wall* (flat-plate thermometer) we substitute $T_1 = T_\infty$ and $U = U_\infty$ into eqn. (13.12), so that the temperature distribution in the boundary layer on a flat plate becomes

$$T = T_\infty + \frac{1}{2\,c_p}\,(U_\infty^2 - u^2)\,, \qquad (13.16)$$

and the adiabatic wall temperature, eqns. (13.12a, b), is

$$T_a = T_\infty + \frac{U_\infty^2}{2\,c_p} = T_\infty\left(1 + \frac{\gamma-1}{2}\,\mathsf{M}_\infty^2\right)\ (\mathsf{P}=1)\,, \qquad (13.17)$$

which follows with $\mathsf{M}_\infty = U_\infty/c_\infty$, and $c_\infty^2 = (\gamma-1)\,c_p\,T_\infty$.   It is worth noting that the temperature of a wall in compressible flow given by eqn. (13.17) is identical with that for an incompressible fluid from eqn. (12.80) provided that in the former case $P = 1$. H. W. Emmons and J. G. Brainerd [34] have shown that in the case of Prandtl numbers which differ from unity the deviations in wall temperature caused by compressibility effects, as compared with the incompressible equation (12.80), are only very slight. Thus the adiabatic-wall temperature equation

$$T_a = T_\infty + \sqrt{P}\, \frac{U_\infty^2}{2 c_p} = T_\infty \left(1 + \sqrt{P}\, \frac{\gamma - 1}{2}\, M_\infty^2\right) \tag{13.18}$$

remains valid for compressible flows with a very good degree of approximation. For air, with $\gamma = 1.4$ and $P = 0.71$, we obtain

$$T_a = T_\infty \left(1 + 0.170\, M_\infty^2\right) . \tag{13.18 a}$$

The resulting dependence of the adiabatic-wall temperature on the Mach number has been represented graphically by the plot in Fig. 13.4. For example, at a Mach number $M_\infty = 1$ the wall becomes heated by $45°$ C (or $80°$ F) in round figures. At $M_\infty = 3$, the temperature increase becomes as high as $400°$ C (or $720°$ F), and at $M_\infty = 5$, it is as much as $1200°$ C (or $2200°$ F).

It has now become customary to write eqn. (13.18) in the more general form

$$T_a = T_\infty + r\, \frac{U_\infty^2}{2 c_p} = T_\infty \left(1 + r\, \frac{\gamma - 1}{2}\, M_\infty^2\right) . \tag{13.19}$$

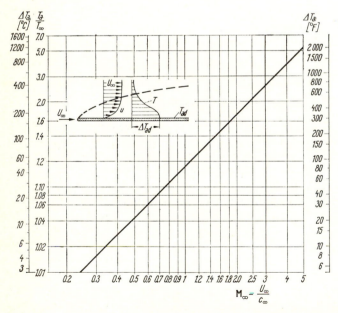

Fig. 13.4. Increase in the temperature of a flat wall owing to frictional heat when the wall is *adiabatic*, in terms of the Mach number, for air, from eqn. (13.18a)

Prandtl number, $P = 0.7$;   Adiabatic-wall temperature $= T_a$;   External temperature $= T_\infty$;   Increase in wall temperature, $(\varDelta T)_a = T_a - T_\infty$; $T_\infty = 273°$ K ($492°$ R)

The *recovery factor*, r, then represents the ratio of the frictional temperature increase of the plate, $(T_a - T_\infty)$, to that due to adiabatic compression,

$$\varDelta T_a = \frac{U_\infty{}^2}{2 c_p},$$

from eqn. (12.14). On comparing eqns. (13.18) and (13.19) it is seen that the recovery factor has the value

$$r = \sqrt{P} \qquad \text{(laminar)}, \qquad (13.19\,\mathrm{a})$$

hence for air

$$r = \sqrt{0.71} = 0.84 \qquad \text{(laminar)}. \qquad (13.19\,\mathrm{b})$$

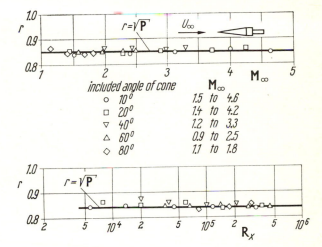

Fig. 13.5. Measured recovery factors, r, for laminar boundary layers on cones at supersonic velocities for different Mach numbers and Reynolds numbers, after G. R. Eber [32]; comparison with theoretical values from eqn. (13.19a)

The diagrams in Fig. 13.5 represent the results of measurements on the recovery factor in the case of laminar boundary layers on cones in supersonic streams, performed by G.R. Eber [32]. The numerical value $r = P^{1/2}$ is seen to be confirmed by these measurements. Similar results follow from measurements performed on various cones and a paraboloid performed by B. des Clers and J. Sternberg [27] and R. Scherrer [89].

Velocity and temperature distributions in the absence of heat transfer: Two papers by W. Hantzsche and H. Wendt [44, 46] and a paper by L. Crocco [21] contain explicit formulae for the calculation of the velocity and temperature distribution in a number of specific cases. Figure 13.6 contains plots of the velocity distribution in the boundary layer for several Mach numbers. It represents Crocco's calculations for a boundary layer on an *adiabatic flat plate* on the assumption of a viscosity law with $\omega = 1$ and for $P = 1$. The distance, y, from the wall has been made dimensionless with reference to $\sqrt{\nu_\infty\, x / U_\infty}$ where ν_∞ denotes the kinematic viscosity in the external flow. It is seen that for increasing Mach numbers there is a considerable thickening of the boundary layer and that for very large Mach numbers the velocity distribution is approximately linear over its whole thickness.

The temperature distribution is also shown in Fig. 13.6, and it is seen that the frictional increase in the temperature in the boundary layer assumes large values for large Mach numbers. The paper by W. Hantzsche and H. Wendt [44], quoted earlier, contains calculations for $P = 0\cdot 7$ (air) for the case of a heat-conducting plate. It is shown that the velocity distribution u/U_∞ plotted in terms of $y\sqrt{U_\infty/x\ \nu_w}$ deviates considerably from that for $P = 1$ when the Mach number assumes larger values. The velocity profiles shown in Fig. 13.6 can be made nearly to coincide when the distance from the wall, y, is made dimensionless with reference to $\sqrt{\nu_w\,x/U_\infty}$, Fig. 13.7, where ν_w denotes the kinematic viscosity of the air at the wall. This circumstance denotes physically that the increase in boundary-layer thickness with Mach number (at constant Reynolds number) is mainly due to the increase in volume which is associated with the increase in the temperature of the air near the wall. This fact was first noticed by A. N. Tifford [98].

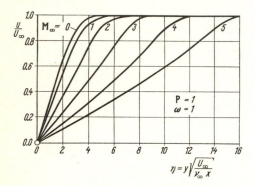

In this method of plotting, the curves for different Mach numbers have been made nearly to coincide. It is possible to conclude from this that the large increase in the boundary-layer thickness with Mach number is mainly due to the increase in volume which is associated with the increase in temperature of the air near the wall

▼

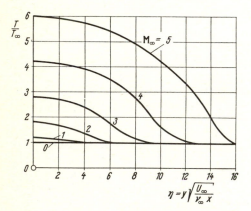

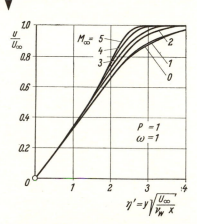

Fig. 13.6. Velocity and temperature distribution in compressible, laminar boundary layer on adiabatic flat plate, after Crocco [21]

Prandtl number $P = 1$, $\omega = 1$, $\gamma = 1\cdot 4$. Distance from wall referred to $\sqrt{\nu_\infty\,x/U_\infty}$

Fig. 13.7. Velocity distributions in the laminar boundary layer on an adiabatic flat plate at zero incidence; data identical with those in Fig. 13.6. The distance from the wall is referred to $\sqrt{\nu_w\,x/U_\infty}$. For $\omega = 1$, we have $\sqrt{\nu_w/\nu_\infty} = T_w/T_\infty$

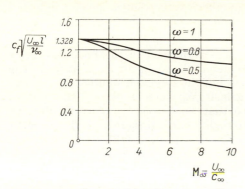

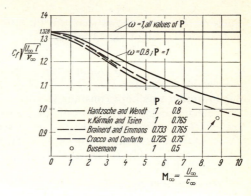

Fig. 13.8. Coefficient of skin friction on *adiabatic* flat plate with compressible, laminar boundary layer. $P = 1$, $\gamma = 1\cdot4$ (air), after Hantzsche and Wendt [44]

Fig. 13.9. Coefficient of skin friction for *adiabatic* flat plate at zero incidence with compressible, laminar boundary layer, after Rubesin and Johnson [88]

Adiabatic coefficient of skin friction: The coefficient of skin friction for an adiabatic wall, as calculated by W. Hantzsche and H. Wendt, has been plotted in terms of the Mach number in Fig. 13.8. For $\omega = 1$ the product $c_f \sqrt{R}$ is independent of the Mach number, but for different values of ω the coefficient of skin friction decreases with increasing Mach number, the rate of decrease being larger for smaller values of ω. Figure 13.9 contains a comparison between the values of the coefficient of skin friction for an adiabatic flat plate obtained by several authors, i. e. for different values of the Prandtl number, P, and of the exponent in the viscosity

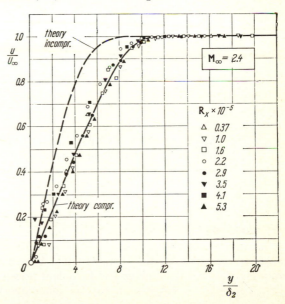

Fig. 13.10. Measurements of the velocity distribution in an *adiabatic, laminar* boundary layer in supersonic flow, after R. M. O'Donnell [28]. Mach number $M_\infty = 2\cdot4$. Theory from ref. [13]

function. The plot shows that the Prandtl number exerts a much smaller influence on the coefficient of skin friction than the exponent ω.

Figure 13.10 shows the results of measurements on compressible boundary layers performed by R. M. O'Donnell [28]. They were performed in the boundary layer of very long circular cylinders of small diameter placed in an axial stream. The Mach number was kept constant at $\mathsf{M}_\infty = 2\cdot4$ but the Reynolds number was varied. The velocity distribution has been plotted against y/δ_2, where δ_2 denotes the momentum thickness from eqn. (13.75). It is seen that the velocity profiles at different distances from the leading edge are similar to each other and there is good agreement with the theory due to D. R. Chapman and M. W. Rubesin [13].

Velocity and temperature distributions in the presence of heat transfer: In the general case, *with heat transfer present*, the relation between the velocity and temperature distribution can be deduced from eqn. (13.13a). When $\mathsf{P} = 1$, it can be written

$$\frac{T}{T_\infty} = 1 + \frac{\gamma - 1}{2}\, \mathsf{M}_\infty{}^2\left[1 - \left(\frac{u}{U_\infty}\right)^2\right] + \frac{T_w - T_{ad}}{T_\infty}\left[1 - \frac{u}{U_\infty}\right] \quad (\mathsf{P} = 1), \quad (13.20)$$

where T_{ad} is given by eqn. (13.17). The preceding equation can be extended to Prandtl numbers differing from unity by the introduction of the recovery factor, when we obtain

$$\frac{T}{T_\infty} = 1 + r\,\frac{\gamma - 1}{2}\, \mathsf{M}_\infty{}^2\left[1 - \left(\frac{u}{U_\infty}\right)^2\right] + \frac{T_w - T_{ad}}{T_\infty}\left[1 - \frac{u}{U_\infty}\right]. \quad (13.21)$$

In this equation, the adiabatic wall temperature, T_{ad}, should be calculated from eqn. (13.18), but it must be realized that this is only an approximation. The direction in which heat is transferred can be deduced from eqn. (13.21) and written

$$T_w - T_\infty \gtrless \sqrt{\mathsf{P}}\;\frac{U_\infty^2}{2\,c_p} \;:\; \text{Heat: wall} \rightleftarrows \text{gas}\,, \quad (13.22)$$

in complete agreement with eqn. (12.82) for incompressible flow.

The second paper by W. Hantzsche and H. Wendt [46] contains numerous examples for the case of the heat-conducting wall. Some results are seen plotted in Fig. 13.11. They refer to the case when the temperature of the walls is reduced by cooling to that in the free stream ($T_w = T_\infty$). A comparison of the velocity distributions in Figs. 13.11 and 13.6 shows that the boundary layer on a heat-conducting wall is considerably smaller than on an adiabatic one. The temperature profiles show that in the case under consideration the highest temperature increase in the boundary layer attains a value of about 20 per cent of that due to adiabatic compression irrespective of the Mach number.

Since for $\omega = 1$ the coefficient of skin friction is independent of the Mach number (Fig. 13.8), the rate at which heat is transferred becomes equal to that in an incompressible stream, eqn. (12.81). A survey of heat-transfer coefficients and recovery factors for laminar and turbulent flow at high Mach numbers can be found in a paper by J. Kaye [55]. In this connexion ref. [105] may also be mentioned.

The case when the temperature varies along the wall, i. e. when $T_w = T_w(x)$, has been studied by D. R. Chapman and M. W. Rubesin [13] on the assumption of a viscosity function $\mu/\mu_0 = b\,T/T_0$. The analysis shows that the local heat flux (quantity of heat transferred per unit area and time) cannot be determined from the temperature difference $T_w(x) - T_a$ alone but that it depends to a large degree on the previous "history" of the boundary layer, i. e. on the conditions which prevail upstream of the section under consideration. The local Nusselt number loses its significance in cases when the wall temperature varies along the flow, because its use implies that the local heat flux is proportional to $T_w - T_\infty$, eqn. (12.31), or, taking into account the heat generated by friction, that it is proportional to $T_w - T_a$.

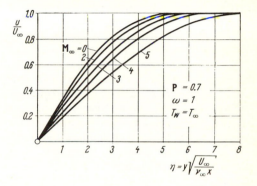

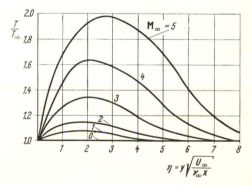

Fig. 13.11. Velocity and temperature distribution in compressible *laminar* boundary layer on flat plate at zero incidence *with heat transfer*, after Hantzsche and Wendt [44]

Wall temperature = free stream temperature, $T_w = T_\infty$; P = 0·7, $\omega = 1$; $\gamma = 1·4$

Calculations concerning compressible boundary layers on flat plates which are based on the momentum-integral equation (Chap. X) have been performed by Th. von Kármán and H. S. Tsien [53]; see also Fig. 13.9. Approximate solutions for the flat plate were also published by F. Bouniol and E. A. Eichelbrenner [7], D. Coles [17], L. Crocco [22] and R. J. Monaghan [75]. Solutions for the equations of laminar boundary layers with variable properties were given by L. L. Moore [77] and G. B. W. Young and E. Janssen [108].

d. Boundary layer with non-zero pressure gradient

1. Exact solutions. The calculations concerning boundary layers with non-zero pressure gradients are more difficult than those concerning flat plates, owing to the large number of independent variables. L. Crocco [21] discovered quite early a transformation which simplifies the task of integrating the equations for the cases when either (1) $P = 1$, and the viscosity function $\mu(T)$ is arbitrary, or (2) when the Prandtl number has an arbitrary value but $\mu/T = \text{const}$ (i.e. when $\omega = 1$). In the special cases of an adiabatic wall with $P = 1$ and $\omega = 1$, L. Howarth [48], C. R. Illingworth [70] and K. Stewartson [94] discovered a transformation which reduces the compressible boundary-layer equations to almost the same form as that valid for incompressible flow.

1.1. The Illingworth-Stewartson transformation. We now proceed to derive the Illingworth-Stewartson transformation by the use of a slightly modified method compared with ref. [94], and without, at first, restricting the argument to adiabatic walls. Furthermore, we shall suppose that the Prandtl number, P, may have an arbitrary, though constant, value. The viscosity law, $\mu(T)$, will be assumed linear as in eqn. (13.4a), and the subscript 0 for viscosity as well as for the other parameters will refer to stagnation conditions in the external stream. The constant b serves to provide an improved approximation to the more exact Sutherland equation (13.3) in the meighbourhood of the desired temperature. If the latter is chosen at the wall temperature, T_w, assumed constant, then, according to eqns. (13.3) and (13.4a), it is necessary to put

$$b = \sqrt{\frac{T_w}{T_0} \cdot \frac{T_0 + S_1}{T_w + S_1}} \, . \tag{13.23}$$

The Illingworth-Stewartson transformation introduces two new coordinates by the definitions

$$\tilde{x} = \int_0^x b \, \frac{p_1 c_1}{p_0 c_0} \, \mathrm{d}x \tag{13.24}$$

and

$$\tilde{y} = \frac{c_1}{c_0} \int_0^y \frac{\varrho}{\varrho_0} \, \mathrm{d}y \, . \tag{13.25}$$

Here, c denotes the velocity of sound, and the subscript 1 refers to conditions in the external flow (at the outer edge of the boundary layer) at station x. Now,

$$c_1^2 = (\gamma - 1) \, c_p \, T_1 \quad \text{and} \quad c_0^2 = (\gamma - 1) \, c_p \, T_0 \, . \tag{13.26}$$

Since, however, T_1 depends only on x, we also have that $c_1 = c_1(x)$; further, since $p_1 = p_1(x)$, we find that $\tilde{x} = \tilde{x}(x)$ is a function of x alone. By contrast, $\tilde{y}$ depends on both y and x, because the density, ϱ, in the boundary layer depends on y. We may also invert these relations and note that

$$x = x(\tilde{x}) \quad \text{but} \quad y = y(\tilde{x}, \tilde{y}) \, .$$

The succeeding derivation aims at expressing the boundary-layer equations (13.5) and (13.6) in terms of the new coordinates $\tilde{x}$ and $\tilde{y}$. The continuity equation (13.5) is satisfied identically by the introduction of the stream function $\psi(x,y)$ defined through its derivatives

$$\frac{\partial \psi}{\partial y} = \frac{\varrho}{\varrho_0} u \quad \text{and} \quad \frac{\partial \psi}{\partial x} = -\frac{\varrho}{\varrho_0} v \; . \tag{13.27}$$

Regarding ψ as a function of $\tilde{x}$ and $\tilde{y}$, we find that

$$\frac{\partial \psi}{\partial x} = \frac{d\tilde{x}}{dx} \frac{\partial \psi}{\partial \tilde{x}} + \frac{\partial \tilde{y}}{\partial x} \frac{\partial \psi}{\partial \tilde{y}} = \frac{bp_1 c_1}{p_0 c_0} \frac{\partial \psi}{\partial \tilde{x}} + \frac{\partial \tilde{y}}{\partial x} \frac{\partial \psi}{\partial \tilde{y}}$$

and

$$\frac{\partial \psi}{\partial y} = \frac{\partial \tilde{y}}{\partial y} \frac{\partial \psi}{\partial \tilde{y}} = \frac{c_1 \varrho}{c_0 \varrho_0} \frac{\partial \psi}{\partial \tilde{y}} \; , \tag{13.28}$$

since $\partial \tilde{x}/\partial y = 0$. Hence, for example

$$u = \frac{\varrho_0}{\varrho} \frac{\partial \psi}{\partial y} = \frac{c_1}{c_0} \frac{\partial \psi}{\partial \tilde{y}} \; . \tag{13.29}$$

Further, after a calculation during which terms in $\partial \tilde{y}/\partial x$ are found to cancel, we can show that

$$u \frac{\partial u}{\partial x} + v \frac{\partial u}{\partial y} = \left(\frac{c_1}{c_0}\right)^3 \frac{p_1 b}{p_0} \left[\frac{\partial \psi}{\partial \tilde{y}} \frac{\partial^2 \psi}{\partial \tilde{y} \partial \tilde{x}} - \frac{\partial \psi}{\partial \tilde{x}} \frac{\partial^2 \psi}{\partial \tilde{y}^2} + \frac{1}{c_1} \frac{dc_1}{d\tilde{x}} \left(\frac{\partial \psi}{\partial \tilde{y}}\right)^2 \right] .$$

Along the external flow, assumed isentropic, the stagnation enthalpy remains constant, that is,

$$h_1 = c_p T_1 + \tfrac{1}{2} u_1^2 = c_p T_0 \; ; \quad h = c_p T + \tfrac{1}{2} u^2 \; , \tag{13.30)†}$$

or, in view of (13.26),

$$c_1^2 + \frac{1}{2} (\gamma - 1) u_1^2 = c_0^2 \; . \tag{13.31}$$

It follows that

$$\frac{1}{c_1} \frac{dc_1}{dx} = -\frac{1}{2} (\gamma - 1) \frac{u_1}{c_1^2} \frac{du_1}{dx} \; , \tag{13.32}$$

so that, finally,

$$u \frac{\partial u}{\partial x} + v \frac{\partial u}{\partial y} = \left(\frac{c_1}{c_0}\right)^3 \frac{p_1 b}{p_0} \left(\frac{\partial \psi}{\partial \tilde{y}} \frac{\partial^2 \psi}{\partial \tilde{y} \partial \tilde{x}} - \frac{\partial \psi}{\partial \tilde{x}} \frac{\partial^2 \psi}{\partial \tilde{y}^2} \right) -$$
$$- \frac{1}{2} (\gamma - 1) \frac{u^2}{c_1^2} u_1 \frac{du_1}{dx} \tag{13.33}$$

because

$$\frac{d\tilde{x}}{dx} = b \frac{p_1 c_1}{p_0 c_0} \; .$$

---

† In the present section we find it simpler to denote the external velocity by the symbol u_1 instead of U as in the past.

The viscous term in the equation of motion can be transformed with the aid of eqn. (13.4a) and the perfect-gas law $p = p_1 = \varrho R T$ to yield

$$\frac{1}{\varrho} \frac{\partial}{\partial y} \left(\mu \frac{\partial u}{\partial y} \right) = \frac{\nu_0 b p_1}{p_0} \left(\frac{c_1}{c_0} \right)^3 \frac{\partial^3 \psi}{\partial \tilde{y}^3} . \tag{13.34}$$

Finally, according to eqns. (13.9) and (13.10), we have

$$-\frac{1}{\varrho} \frac{dp}{dx} = -\frac{T}{\varrho_1 T_1} \frac{dp_1}{dx} = \frac{T}{T_1} u_1 \frac{du_1}{dx} ,$$

or, by introducing the dimensionless temperature function (relative stagnation-enthalpy difference), defined by

$$S = \frac{c_p T + \frac{1}{2} u^2}{c_p T_0} - 1 = \frac{\hat{h} + \frac{1}{2} u^2}{h_0} - 1 , \tag{13.35}$$

and by making use of eqn. (13.26), we obtain that

$$-\frac{1}{\varrho} \frac{dp}{dx} = \left\{ (1 + S) \left(\frac{c_0}{c_1} \right)^2 - \frac{\gamma - 1}{2} \frac{u^2}{c_1^2} \right\} u_1 \frac{du_1}{dx} . \tag{13.36}$$

Here $\hat{h}$ denotes the local, as distinct from the stagnation enthalpy. Introducing the expression in eqns. (13.33), (13.34), and (13.36) into eqn. (13.6) imagined divided by ϱ, we derive:

$$\frac{\partial \psi}{\partial \tilde{y}} \frac{\partial^2 \psi}{\partial \tilde{y} \partial \tilde{x}} - \frac{\partial \psi}{\partial \tilde{x}} \frac{\partial^2 \psi}{\partial \tilde{y}^2} = (1 + S) \left(\frac{c_0}{c_1} \right)^5 \frac{p_0}{p_1 b} u_1 \frac{du_1}{dx} + \nu_0 \frac{\partial^3 \psi}{\partial \tilde{y}^3} . \tag{13.37}$$

Putting

$$\tilde{u}_1 = \frac{c_0}{c_1} u_1 , \tag{13.38}$$

we find that

$$\frac{du_1}{dx} = \frac{1}{c_0} \left(\tilde{u}_1 \frac{dc_1}{dx} + c_1 \frac{d\tilde{u}_1}{d\tilde{x}} \right) ,$$

and hence, with the aid of eqns. (13.31) and (13.32), we see that

$$u_1 \frac{du_1}{dx} = \left(\frac{c_1}{c_0} \right)^5 \frac{b p_1}{p_0} \tilde{u}_1 \frac{d\tilde{u}_1}{d\tilde{x}} . \tag{13.39}$$

As a last step we define

$$\tilde{u} = \frac{\partial \psi}{\partial \tilde{y}} \quad \text{and} \quad \tilde{v} = -\frac{\partial \psi}{\partial \tilde{x}} , \tag{13.40}$$

substitute eqns. (13.39) and (13.40) into eqn. (13.37), and derive the transformed equation of motion

$$\tilde{u} \frac{\partial \tilde{u}}{\partial \tilde{x}} + \tilde{v} \frac{\partial \tilde{u}}{\partial \tilde{y}} = \tilde{u}_1 \frac{d\tilde{u}_1}{d\tilde{x}} (1 + S) + \nu_0 \frac{\partial^2 \tilde{u}}{\partial \tilde{y}^2} . \tag{13.41}$$

This transformed equation differs from the corresponding boundary-layer equation of incompressible flow merely by the factor $(1 + S)$ which multiplies the pressure term.

In order to transform the energy equation, we multiply eqn. (13.6) by u and add eqn. (13.7). Remembering that the Prandtl number is

$$P = \frac{\mu \, c_p}{k},$$

we obtain

$$\varrho \, u \, \frac{\partial}{\partial x} \left(c_p \, T + \frac{1}{2} \, u^2 \right) + \varrho \, v \, \frac{\partial}{\partial y} \left(c_p \, T + \frac{1}{2} \, u^2 \right) =$$

$$= \frac{\partial}{\partial y} \left\{ \mu \, \frac{\partial}{\partial y} \left(\frac{c_p \, T}{P} + \frac{1}{2} \, u^2 \right) \right\}. \tag{13.42}$$

Making use of the temperature function S from eqn. (13.35), we can transform this into

$$\varrho \, u \, \frac{\partial S}{\partial x} + \varrho \, v \, \frac{\partial S}{\partial y} = \frac{\partial}{\partial y} \left\{ \mu \left[\frac{1}{P} \, \frac{\partial S}{\partial y} + \frac{P-1}{P} \, \frac{\partial}{\partial y} \left(\frac{u^2}{2 \, c_p \, T_0} \right) \right] \right\}. \tag{13.43}$$

As was done for eqn. (13.28), we express the partial derivatives with respect to x and y by those with respect to $\tilde{x}$ and $\tilde{y}$, note that $\mu = b \, \mu_0 \, P_1 \, \varrho_0 / p_0 \, \varrho$ and make use of the definitions (13.40) to obtain

$$\tilde{u} \, \frac{\partial S}{\partial \tilde{x}} + \tilde{v} \, \frac{\partial S}{\partial \tilde{y}} = \nu_0 \left\{ \frac{1}{P} \, \frac{\partial^2 S}{\partial \tilde{y}^2} + \frac{P-1}{P} \, \frac{\partial^2}{\partial \tilde{y}^2} \left(\frac{u^2}{2 \, c_p \, T_0} \right) \right\}. \tag{13.44}$$

Equations (13.26) and (13.30) together with the relation

$$\frac{u}{u_1} = \frac{\tilde{u}}{\tilde{u}_1} \tag{13.45}$$

yield

$$\frac{u^2}{2 \, c_p \, T_0} = \frac{\frac{1}{2} (\gamma - 1) \, \mathsf{M}_1{}^2}{1 + \frac{1}{2} (\gamma - 1) \mathsf{M}_1{}^2} \left(\frac{\tilde{u}}{\tilde{u}_1} \right)^2. \tag{13.46}$$

Here, $\mathsf{M}_1 = u_1/c_1$ is the Mach number of the external flow. Since

$$\frac{\partial}{\partial \tilde{y}} = \frac{\partial y}{\partial \tilde{y}} \, \frac{\partial}{\partial y} + \frac{\partial x}{\partial \tilde{y}} \, \frac{\partial}{\partial x} = \frac{\partial y}{\partial \tilde{y}} \, \frac{\partial}{\partial y},$$

the factor of $(\tilde{u}/\tilde{u}_1)^2$ in eqn. (13.46) can be put in front of the operator $\partial^2/\partial y^2$ in eqn. (13.44), so that the transformed energy equation acquires the form:

$$\tilde{u} \, \frac{\partial S}{\partial \tilde{x}} + \tilde{v} \, \frac{\partial S}{\partial \tilde{y}} = \nu_0 \left\{ \frac{1}{P} \, \frac{\partial^2 S}{\partial \tilde{y}^2} + \right.$$

$$\left. + \frac{P-1}{P} \, \frac{\frac{1}{2} (\gamma - 1) \, \mathsf{M}_1{}^2}{1 + \frac{1}{2} (\gamma - 1) \mathsf{M}_1{}^2} \, \frac{\partial^2}{\partial \tilde{y}^2} \left[\left(\frac{\tilde{u}}{\tilde{u}_1} \right)^2 \right] \right\}. \tag{13.47}$$

Equations (13.41) and (13.47) together with the continuity equation

$$\frac{\partial \tilde{u}}{\partial \tilde{x}} + \frac{\partial \tilde{v}}{\partial \tilde{y}} = 0 \,, \tag{13.48}$$

which is a direct consequence of eqn. (13.40), now constitute the new set of boundary-layer equations.

The system of equations (13.5), (13.6), (13.7) was subject to the boundary conditions

$$y = 0 : u = v = 0 \quad \text{and} \quad \frac{\partial T}{\partial y} = 0 \quad \text{or} \quad T = T_w \,,$$

the latter depending on whether the wall is adiabatic or isothermal, together with

$$y = \infty : \quad u = u_1(x); \quad T = T_1(x) \,.$$

It is easy to see that these boundary conditions transform as follows:

$$\tilde{y} = 0 : \quad \tilde{u} = \tilde{v} = 0 \text{ with } \frac{\partial S}{\partial \tilde{y}} = 0 \quad \text{or} \quad S = S_w$$

together with

$$\tilde{y} = \infty : \quad \tilde{u} = \tilde{u}_1(\tilde{x}); \quad S = 0 \,.$$
$$\tag{13.49}$$

Limiting cases: If $\mathsf{P} = 1$ then $S = 0$ is a special solution of the energy equation (13.47). Together with eqn. (13.30), it leads to the relation between temperature and velocity for an adiabatic wall discovered earlier as eqn. (13.12). In this case, eqn. (13.41) assumes the "incompressible" form of eqn. (9.1) *exactly*.

Along a flat plate we have $dp/dx = 0$ which implies that $d\tilde{u}_1/d\tilde{x} = 0$ as well. Then for $\mathsf{P} = 1$ we discover that $S = S_w(1 - \tilde{u}/\tilde{u}_1)$ with S_w equal to a constant constitutes a special solution of (13.47), as is easily verified by substitution. Confronting eqns. (13.45) and (13.30), we recover the relation between temperature and velocity first indicated as eqn. (13.13), remembering that u_1 must be written for U_∞ and T_1 for T_∞.

1.2. Self-similar solutions. The Illingworth-Stewartson transformation has been used to derive exact solutions and to formulate a large number of approximate procedures. Self-similar solutions play an important part within the class of exact solutions. In the context of incompressible flows, we considered that a solution belonged to this group if the velocity profiles $u(x, y)$ at two different stations x could be made congruent by the application of a single scale factor each for u and y (Sec. VIII b). It was then shown that such similar solutions existed in the presence of a definite group of external flows $u_1(x)$. In cases of this kind, the partial differential equation for the stream function reduced to an ordinary differential equation which is considerably easier to solve than the former.

Making use of a number of studies, for example [48, 49, 50], T. Y. Li and H. T. Nagamatsu [60, 61] demonstrated in a number of praiseworthy investigations that such similar solutions exist in the case of compressible boundary layers as well. As far as the velocity boundary layer is concerned, here too, similarity extends to the longitudinal velocity component, u; with respect to the thermal layer, similarity relates to the stagnation enthalpy $h = c_p T + \frac{1}{2} u^2$ which we encountered earlier in eqn. (13.35) in the form of a "temperature function". The system of partial differential equations for u, v, and T reduces in such cases to two coupled ordinary differential equations containing the stream function and the stagnation enthalpy.

Similar solutions for compressible boundary layers constitute exact solutions of the system of equations and are, therefore, intrinsically very important. Perhaps even more importantly, solutions of this kind are employed as touchstones against which the accuracy of approximate procedures can be judged. For these reasons, we now propose roughly to sketch the line of reasoning which leads to similar solutions starting with the Illingworth-Stewartson transformation. We shall conclude this topic with a number of numerical results. We shall postulate the validity of the viscosity law from eqn. (13.4a) so that $\omega = 1$ and $\mathsf{P} = 1$ are implied. In the case of boundary layers *with heat transfer*, an arbitrary, but constant wall temperature, T_w, will be assumed, so that S_w will become a constant. In problems involving an *adiabatic* wall, the stagnation enthalpy is given by eqn. (13.12):

$$c_p T + \frac{1}{2} u^2 = c_p T_1 + \frac{1}{2} u_1{}^2 = c_p T_0 \, ,$$

and remains constant over the boundary-layer thickness, implying $S = 0$ (*cf.* also end of preceding section). In this case, the similarity of the stagnation-enthalpy profiles assumes a trivial form.

Employing the stream function ψ, we rewrite eqns. (13.41) and (14.47) in the form:

$$\frac{\partial \psi}{\partial \tilde{y}} \frac{\partial^2 \psi}{\partial \tilde{y} \, \partial \tilde{x}} - \frac{\partial \psi}{\partial \tilde{x}} \frac{\partial^2 \psi}{\partial \tilde{y}^2} = \tilde{u}_1 \frac{d\tilde{u}_1}{d\tilde{x}} (1 + S) + v_0 \frac{\partial^3 \psi}{\partial \tilde{y}^3} \tag{13.50}$$

and

$$\frac{\partial \psi}{\partial \tilde{y}} \frac{\partial S}{\partial \tilde{x}} - \frac{\partial \psi}{\partial \tilde{x}} \frac{\partial S}{\partial \tilde{y}} = v_0 \frac{\partial^2 S}{\partial \tilde{y}^2} \, . \tag{13.51}$$

The similarity variable is introduced with the aid of the following assumptions:

$$\left. \begin{aligned} \psi &= A \, \tilde{x}^q \, \tilde{u}_1{}^r \times f(\eta) \, , \\ \tilde{y} &= B \, \tilde{x}^s \, \tilde{u}_1{}^t \times \eta \, , \\ S &= S(\eta) \end{aligned} \right\} \tag{13.52}$$

where A, B, r, s, t play the parts of undetermined constants, $f(\eta)$ is an unknown stream function, and $S(\eta)$ is the temperature function defined in eqn. (13.35), now conceived to be a function of η alone.

Equations (13.50) and (13.51) are now transformed to the coordinates $\tilde{x}$ and η, and in the resulting expressions it is demanded that the terms in $\tilde{x}$ must disappear. In this manner we obtain ordinary differential equations for the functions $f(\eta)$ and $S(\eta)$. Such calculations have been performed by T. Y. Li and H. T. Nagamatsu [60] who found that there existed four classes of solutions for $\tilde{u}_1(\tilde{x})$. Following this work, C. B. Cohen [16] demonstrated that three of these classes can be reduced to the common form

$$\tilde{u}_1 = K \, \tilde{x}^m \tag{13.53}$$

(K and m are constants). The fourth case

$$\tilde{u}_1 = K' \exp (K'' x)$$

is of marginal interest only, and we shall ignore it henceforth.

We now wish to determine the form of the external flow, $u_1 = u_1(x)$, which corresponds to the law (13.53) expressed there in the transformed variable $\tilde{x}$. It follows from eqns. (13.38) and (13.31) that

$$c_1{}^2 = \frac{c_0{}^4}{c_0{}^2 + \frac{1}{2}(\gamma - 1)\tilde{u}_1{}^2} . \tag{13.54}$$

The external flow is isentropic, and hence

$$\frac{p_1}{p_0} = \left(\frac{\varrho_1}{\varrho_0}\right)^\gamma \quad \text{so that with} \quad \frac{c_1{}^2}{c_0{}^2} = \frac{p_1/p_0}{\varrho_1/\varrho_0}$$

it follows that

$$\frac{p_1}{p_0} = \left(\frac{c_1}{c_0}\right)^{2\gamma/(\gamma-1)} \tag{13.55}$$

Thus, taking account of eqns. (13.24), (13.53), and (13.54), we derive that

$$dx = \frac{1}{b}\left[1 + \frac{\gamma-1}{2\,c_0{}^2}\,K^2\,\tilde{x}^{2m}\right]^{(3\gamma-1)/2(\gamma-1)} d\tilde{x} . \tag{13.56}$$

The resulting differential equation can be solved in closed form only for particular values of m. If we choose

$$m = m_0 = \frac{\gamma-1}{3-5\gamma} , \tag{13.57}$$

and so

$$\frac{3\gamma-1}{2(\gamma-1)} = -1 - \frac{1}{2\,m_0} ,$$

we obtain by integration that

$$x = \frac{\tilde{x}}{b\left[1 + \frac{\gamma-1}{2\,c_0{}^2}\,K^2\,\tilde{x}^{\,2(\gamma-1)/(3-5\gamma)}\right]^{(3-5\gamma)/2(\gamma-1)}} . \tag{13.58}$$

Invoking eqns. (13.53) and (13.54), we deduce that

$$u_1 = \frac{c_1}{c_0}\,\tilde{u}_1 = K\,b^{\,(\gamma-1)/(3-5\gamma)}\,x^{\,(\gamma-1)/(3-5\gamma)} = K'\,x^{\,(\gamma-1)/(3-5\gamma)} . \tag{13.59}$$

It is seen that in this particular case, the external flow $u_1(x)$ is also a power in x, the exponent, moreover, being the same as for $\tilde{u}_1(\tilde{x})$. Inserting the values for γ which apply to monatomic, diatomic and polyatomic gases into the exponent of x from eqn. (13.59), we can establish the following table:

Gas	γ	$m_0 = \dfrac{\gamma - 1}{3 - 5\gamma}$
monatomic	$\dfrac{5}{3} = 1\cdot67$	$-\dfrac{1}{8}$
diatomic or linear polyatomic, rigid	$\dfrac{7}{5} = 1\cdot40$	$-\dfrac{1}{10}$
polyatomic nonlinear, rigid	$\dfrac{1}{3} = 1\cdot33$	$-\dfrac{1}{11}$

The external flow which secures similarity turns out to be decelerated for all three classes of gases.

When an arbitrary value of m is assumed, it is possible to integrate eqn. (13.56), generally speaking, only by a series expansion. This, however, no longer leads to a simple power function for the relation $u_1 = u_1(x)$. The diagram in Fig. 13.12 contains graphs of the velocity distribution $u_1(x)$ together with $\tilde{u}_1(\tilde{x}) = K\,\tilde{x}^m$ for the values of $m = -1, 0, +\frac{1}{2}, +1$ together with $K = 1$, $b = 1$ and $\gamma = 7/5$.

We now apply the transformation (13.52) in a particular form, assuming

$$A = B = \sqrt{2\,\nu_0/(m+1)}\;;\quad q = r = s = \tfrac{1}{2}\;;\quad t = -\tfrac{1}{2}\,,$$

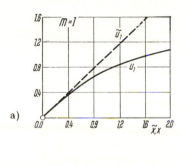

a)

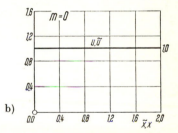

b)

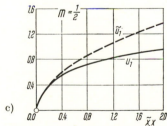

c)

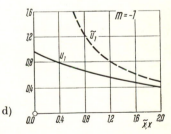

d)

Fig. 13.12. Velocity distributions for similar solutions according to the Illingworth-Stewartson transformation in the case when $\gamma = 7/5$. $\tilde{u}_1 = K\,\tilde{x}^m$ according to eqn. (13.53); $u_1 = K'\,x^m$ according to eqn. (13.59)
Relation between x and $\tilde{x}$ is given in eqn. (13.56)
Relation between m and γ is given in eqn. (13.57)

which gives
$$\psi = f(\eta) \sqrt{\frac{2\,\nu_0}{m+1}\, \tilde{u}_1\, \tilde{x}}; \quad S = S(\eta)\,,$$

(13.60)

$$\eta = \tilde{y} \sqrt{\frac{(m+1)\, \tilde{u}_1}{2\,\nu_0\, \tilde{x}}}\,,$$

and write down the transformed boundary-layer equations (13.50) and (13.51) in the form of the following two ordinary differential equations:

$$\left.\begin{aligned} f''' + ff'' &= \beta(f'^2 - 1 - S)\,, \\ S'' + fS' &= 0\,, \end{aligned}\right\}$$

(13.61)

in which primes denote differentiation with respect to η. The parameter β, in the same way as earlier in eqn. (9.7), is defined by

$$\beta = \frac{2\,m}{m+1}\,;$$

it characterizes the pressure gradient of the external stream.

Remembering that
$$\tilde{u} = \frac{\partial \psi}{\partial \tilde{y}} = \frac{\partial \psi}{\partial \eta}\, \frac{\partial \eta}{\partial \tilde{y}}\,,$$

we conclude with the aid of eqn. (13.60) that f' constitutes a dimensionless form of the longitudinal velocity component in the boundary layer, because

$$f' = \frac{\tilde{u}}{\tilde{u}_1} = \frac{u}{u_1}\,.$$

(13.62)

Since $y = 0$, or $y = \infty$ implies $\eta = 0$ and $\eta = \infty$, respectively, the boundary conditions for the system (13.61) must be written

$$\left.\begin{aligned} \eta = 0\!: f = f' = 0; &\quad \eta = \infty\!: f' = 1 \\ S = S_w &\quad\quad\quad S = 0\,. \end{aligned}\right\}$$

(13.63)

In the case of an *adiabatic wall*, the second equation (13.61) is satisfied identically, and it is necessary to solve the single equation

$$f''' + ff'' = \beta(f'^2 - 1)\,.$$

This equation, however, is identical with that for incompressible wedge flows given earlier as eqn. (9.8); it is remembered that D. R. Hartree studied its solutions for different values of β. He then discovered that for all values of $\beta < -0.199$ $(m < -0.0904)$ there is separation. Consequently, among the specific values of m displayed in the table above, the first two, namely $m = -1/8$ and $m = -1/10$ also lead to separation when the wall is adiabatic.

When the wall permits the *transfer of heat*, it becomes necessary to solve the system of equations (13.61). Since the wall temperature, T_w, can be prescribed in an arbitrary manner, it will be found that the solutions depend on the parameter

$$S_w = \frac{T_w}{T_0} - 1 \,,$$

in addition to their dependence on β. Solutions for a large number of values of these two parameters have been worked out by T. Y. Li and H. T. Nagamatsu [61] as well as by C. B. Cohen and E. Reshotko [16a].

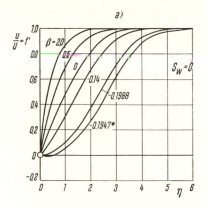

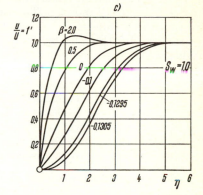

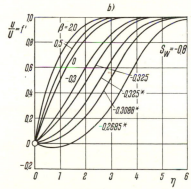

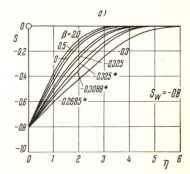

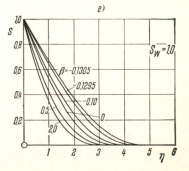

Fig. 13.13.   Velocity and enthalpy distributions in compressible, laminar boundary layers with pressure gradient and heat transfer, after C. B. Cohen and E. Reshotko [16a], and in conformity with eqns. (13.62) and (13.35)

Prandtl Number: $P = 1$; $\omega = 1$. $U(x) = u_1(x)$ denotes the velocity of the external flow. a), b), c) velocity distributions; d), e) enthalpy distributions; a) $S_w = 0$; $T_w = T_0$; adiabatic wall; b), d) $S_w = -0.8$; $T_w = 0.2$ T_0 cooled wall; c), e) $S_w = 1.0$; $T_w = 2$ T_0; heated wall

It is worthy of note that the system (13.61) subject to the boundary conditions (13.63) yields two physically sensible solutions when $\beta < 0$ (this is also true in the case of an adiabatic wall, cf. Sec. IXa). According to the views expressed by C. B. Cohen and E. Reshotko [16a], the one of the two solutions which sets in in an experiment is determined by the initial conditions which establish the pressure field acting on the developing boundary layer.

The diagrams in Figs. 13.13a, b, c represent the *velocity* distribution $u/u_1 = f'$ in terms of the dimensionless transverse distance η for various values of the temperature parameter S_w and of β. The particular values of S_w chosen in the diagram correspond to the following cases (taken in order): an adiabatic wall with $T_w = T_0$; a cooled wall with $T_w = 0\cdot2\,T_0$ (transfer of heat from wall to fluid). In the case of multiple solutions, the locus of lower values of f' for a given value of β has been distinguished by an asterisk. It is noteworthy that in the case of a heated wall and a favorable pressure gradient ($\beta > 0$, Fig. 13.13c) the velocity in the boundary layer can exceed the external velocity , u, in a certain range away from the wall. The reason for it can be found in the large increase in volume imparted to the gas in the boundary layer owing to strong heating. The gas of lower density is accelerated by the external pressure forces more strongly than that in the external flow in spite of its being decelerated by viscous stresses.

Figures 13.13d, e represent the *enthalpy distribution*, S, in the boundary layer in accordance with eqn. (13.35) for $T_w = 0\cdot2\,T_0$ and $T_w = 2\,T_0$, respectively. It is seen that the pressure gradient exerts a considerably stronger influence on the velocity profiles than on the enthalpy profiles.

Figures 13.14a, b, c illustrate the *variation in shearing stress*. This can be calculated with the aid of eqns. (13.10), (13.26), (13.29) together with the isentropic law $p_1/p_0 = (\varrho_1/\varrho_0)^\nu$. It is given by

$$\tau = \mu\,\frac{\partial u}{\partial y} = b\,\mu_0\,\tilde{u}_1 \left(\frac{T_1}{T_0}\right)^{(2\gamma-1)/(\gamma-1)} \sqrt{\frac{m+1}{2}\,\frac{\tilde{u}_1}{\nu_0\,\tilde{x}}}\,f''(\eta)\,. \tag{13.64}$$

The figures contain plots of $f''(\eta)$ for different values of the parameters β and S_w. When the external flow is accelerated ($\beta > 0$), the largest shearing stress occurs at the wall itself ($\eta = 0$); when the flow is decelerated ($\beta < 0$), this maximum moves away from the wall and places itself further from it as the pressure rise is increased, that is for larger absolute values of the negative value of β. Introducing the local skin-friction coefficient

$$c_f' = \frac{\tau_w}{\tfrac{1}{2}\,\varrho_w\,u_1^2} \tag{13.65}$$

as well as the Reynolds number

$$\mathsf{R}_w = \frac{u_1\,x}{\nu_w}\,,$$

where the subscript w refers to the wall, we find that

$$c_f'\,\sqrt{\mathsf{R}_w} = f_w''\,\sqrt{2\,(m+1)\,\frac{x}{\tilde{x}}\,\frac{d\tilde{x}}{dx}}\,. \tag{13.66}$$

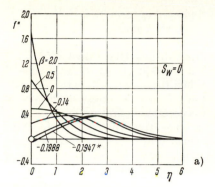

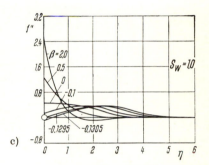

Fig. 13.14. Distribution of shearing stresses in compressible, laminar boundary layers with pressure gradient and heat transfer, after C. B. Cohen and E. Reshotko [16a], and in conformity with eqns. (13.64) Prandtl number $P = 1$; $\omega = 1$
$a: S_w = 0$; $T_w = T_0$; adiabatic wall.
$b: S_w = -0.8$; $T_w = 0.2\,T_0$; cooled wall.
$c: S_w = 1.0$; $T_w = 2\,T_0$; heated wall

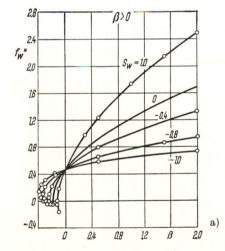

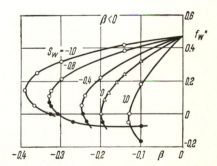

Fig. 13.15. Local skin-friction coefficient in compressible, laminar boundary layers with pressure gradient and heat transfer, after C. B. Cohen and E. Reshotko [16a], and in conformity with eqn. (13.66)
Prandtl number $P = 1$; $\omega = 1$. $S_w = 0$; adiabatic wall. $S_w < 0$; cooled wall. $S_w > 0$; heated wall

The values of f_w'' for different values of S_w are seen plotted in terms of β in Fig. 13.15. It is recognized that a change in pressure gradient exerts a much stronger influence on f_w'', and hence on the shearing stress at the wall, when the wall is heated ($S_w > 0$) than when the latter is cooled ($S_w < 0$). In the range of negative values of β there exist two values of τ_w for each value of β. This is a consequence of the existence of two solutions in this range, as mentioned earlier. When the wall is adiabatic ($S_w = 0$), the lower branch of the curve yields negative values of shearing stress which points to reverse flow. When the wall is heated ($S_w > 0$) it is possible to find sufficiently small values of $\beta - \beta_{min}$ for which both values of f_w'' are negative, that is for which the flow has reversed its direction. In the case of a cooled wall ($S_w < 0$), both values of f_w'' can be positive, that is both can represent non-separated flow patterns. It is seen, finally, that separation ($f_w'' = 0$) moves in the direction of smaller pressure rises as the temperature of the wall is increased.

In order to transform from the variable η to the physical distance y, it is necessary to utilize eqns. (13.8), (13.10), (13.24), (13.25) and (13.52). It is then found that

$$y = b \, \frac{\mathrm{d}x}{\mathrm{d}\tilde{x}} \, \sqrt{\frac{2}{m+1} \, \frac{v_0 \, \tilde{x}}{\tilde{u}_1}} \int\limits_0^\eta \frac{T}{T_0} \, \mathrm{d}\eta \ . \tag{13.67}$$

The factor ahead of the integral is computed from eqn. (13.53), and the functional relation between x and $\tilde{x}$ must be taken from eqn. (13.56). According to eqns. (13.46) and (13.62), the integrand is

$$\frac{T}{T_0} = 1 + S - \frac{u^2}{2 \, c_p \, T_0} = 1 + S(\eta) - \frac{\frac{1}{2}(\gamma - 1) \, \mathsf{M}_1{}^2}{1 + \frac{1}{2}(\gamma - 1) \mathsf{M}_1{}^2} \, [f'(\eta)]^2 \ . \tag{13.68}$$

T. Y. Li and H. T. Nagamatsu [61] have succeeded in deriving expressions for similar solutions without having to employ the Illingworth-Stewartson transformation. W. Mangler [71] indicated substitutions which can be used for an exact calculation with an arbitrary pressure distribution.

E. Reshotko and I. E. Beckwith [86] published an exact calculation for the three-dimensional boundary layer on a yawed circular cylinder with heat transfer and at an arbitrary value of the Prandtl number.

2. Approximate methods. The numerous approximate methods which have been devised for the calculation of compressible laminar boundary layers are based, in most cases, on the momentum-integral and energy-integral equations of the boundary layer. The reader will recall that this was also the case with incompressible boundary layers. All these approximate methods have the common feature that they lead to much more involved procedures than the ones applicable to incompressible flows which were discussed in Chap. X. The number of alternative procedures is much larger in the case of compressible than incompressible boundary layers as would be expected from the increased number of variables. In this connexion the surveys written by A. D. Young [106] and M. Morduchow [79] may be consulted. A more recent summary was prepared by N. Curle [26].

When discussing approximate methods, it is necessary to make a clear distinction between those as are applicable only to *adiabatic walls* and the ones that are valid in the presence of *heat transfer* as well.

Methods which are restricted to *adiabatic walls* include an early one developed by L. Howarth [48] and later modified somewhat by H. Schlichting [90], both for the case when $P = 1$; further, one should mention the procedures developed by E. Gruschwitz [43], by J. A. Zaat [110] and I. Fluegge-Lotz and A. F. Johnson [36], all valid for arbitrary values of the Prandtl number. In the special case when $dp/dx = 0$, the last-mentioned method can be modified to include the transfer of heat. All procedures are based on the assumptions that $\omega = 1$.

In the course of the last years, work proceeded mainly on the solution of problems with *heat transfer*. From among the procedures which are restricted to $P = 1$, it is necessary to mention those due to M. Morduchow [79], C. B. Cohen and E. Reshotko [16a], R. J. Monaghan [76] and G. Poots [85], all of them further limited to the assumption that $\omega = 1$. The second and third method on this list serve to determine the momentum thickness, skin-friction coefficient and coefficient of heat transfer, whereas the first and fourth also permit the determination of velocity and temperature profiles. When the numbers P and ω each differ little from unity, it is possible to resort to the method developed by R. E. Luxton and A. D. Young [68].

The method due to N. Curle [24, 25] and G. M. Lilley [65] are valid for arbitrary values of the Prandtl number, but are based on the *viscosity law* given by eqns. (13.4a) and (13.23) with a constant b which may, at most, depend on x; this means that the function $\mu(T)$ is linear. The work of N. Curle [24] calculates the characteristic parameters of the velocity boundary layer with an adverse pressure gradient and assumes a foreknowledge of the temperature field, but admits that the wall temperature may vary. Another paper by the same author [25] makes it possible to calculate the rate of heat transfer on condition that the distribution of the shearing stress at the wall is known. G. M. Lilley determines the shearing stress at the wall and the coefficient of heat transfer in the presence of a temperature which varies along the wall. The methods worked out by I. Ginzel [41] and D. N. Morris and J. W. Smith [80] are valid for *arbitrary viscosity laws*, the latter also including a variable wall temperature.

The results obtained with the aid of a conventional approximate calculation can be improved considerably by the application of a procedure developed by K. T. Yang [104].

Momentum-integral and energy-integral equations: We begin our account by the derivation of the momentum-integral and energy-integral equations for compressible boundary layers, because they form the starting point of most approximate procedures. In order to achieve this we refer to the fundamental equations for compressible, laminar boundary layers as recorded in eqns. (13.5) to (13.8). Introducing the local enthalpy

$$\hat{h} = c_p \, T \,, \tag{13.69}$$

and the stagnation enthalpy

$$h = \hat{h} + \tfrac{1}{2} \, u^2 = c_p \, T + \tfrac{1}{2} \, u^2 \tag{13.70}$$

we can rewrite the energy equation (13.7) in the form:

$$\varrho \left(u \frac{\partial \hat{h}}{\partial x} + v \frac{\partial \hat{h}}{\partial y} \right) = u \frac{dp}{dx} + \mu \left(\frac{\partial u}{\partial y} \right)^2 + \frac{\partial}{\partial y} \left(\frac{\mu}{P} \frac{\partial \hat{h}}{\partial y} \right). \tag{13.71}$$

The boundary conditions are

a)   with heat transfer:

$$\left. \begin{array}{l} y = 0: \quad u = v = 0; \quad T = T_{\mathrm{W}}; \\ y = \infty: \quad u = U(x); \quad \hat{h} = \hat{h}_1(x); \end{array} \right\} \tag{13.72a}$$

b)   for an adiabatic wall

$$\left. \begin{array}{l} y = 0: \quad u = v = 0; \quad \dfrac{\partial \hat{h}}{\partial y} = 0, \\[2mm] y = \infty: \quad u = U(x); \quad \hat{h} = \hat{h}_1(x). \end{array} \right\} \tag{13.72b}$$

Equations (13.5), (13.6), (13.8) and (13.71) together with the boundary conditions (13.72) constitute a system of four equations for the variables u, v, ϱ and $\hat{h}$. The pressure $p(x)$ is known from Bernoulli's equation and is given by eqn. (13.9); it remains constant over the thickness of the boundary layer, i. e. $\partial p/\partial y = 0$. Since the pressure remains constant across the layer, we have at every point

$$\frac{\hat{h}}{\hat{h}_1} = \frac{T}{T_1} = \frac{\varrho_1}{\varrho}, \tag{13.73}$$

where $\hat{h}_1$, T_1, ϱ_1 denote the values of enthalpy, temperature, and density, respectively, at the outer edge of the boundary layer.

We now introduce a displacement thickness, a momentum thickness and an energy-dissipation thickness in the same way as in incompressible flow and several additional quantities defined with the aid of enthalpy. In this connexion the former parameters are so defined as to reduce to the respective quantities for incompressible flow, eqns. (8.30), (8.31), and (8.34), when $\varrho = \mathrm{const}$ is substituted in the definitions. Denoting the boundary-layer thickness of the velocity layer by δ, we introduce the definitions:

$$\delta_1 = \int_0^\delta \left(1 - \frac{\varrho \, u}{\varrho_1 \, U} \right) dy \qquad \text{(displacement thickness)}, \tag{13.74}$$

$$\delta_2 = \int_0^\delta \frac{\varrho \, u}{\varrho_1 \, U} \left(1 - \frac{u}{U} \right) dy \qquad \text{(momentum thickness)}, \tag{13.75}$$

$$\delta_3 = \int_0^\delta \frac{\varrho \, u}{\varrho_1 \, U} \left(1 - \frac{u^2}{U^2} \right) dy \qquad \text{(energy-dissipation thickness)}, \tag{13.76}$$

$$\delta_H = \int_0^\delta \frac{\varrho \, u}{\varrho_1 \, U} \left(\frac{\hat{h}}{\hat{h}_1} - 1 \right) dy \qquad \text{(enthalpy thickness)}, \tag{13.77}$$

$$\delta_u = \int_0^\delta \left(1 - \frac{u}{U}\right) dy \qquad \text{(velocity thickness)} . \qquad (13.78)$$

It is easy to verify from eqns. (13.73), (13.74), (13.77) and (13.78) that the parameters δ_1, δ_H and δ_u satisfy the relation

$$\delta_1 - \delta_u = \delta_H . \qquad (13.79)$$

Integrating the momentum equation (13.6) and the energy equation (13.71) over y, in the same way as was done for incompressible flow, we can obtain the momentum-integral and energy-integral equation for compressible flow. Taking into account that

$$\frac{1}{\varrho_1} \frac{d\varrho_1}{dx} = -\frac{(1-M^2)}{U} \frac{dU}{dx} ,$$

we obtain the *momentum-integral equation* in the form

$$\boxed{\frac{d\delta_2}{dx} + \frac{\delta_2}{U} \frac{dU}{dx}\left(2 + \frac{\delta_1}{\delta_2} - M^2\right) = \frac{\mu_w}{\varrho_1 U^2}\left(\frac{\partial u}{\partial y}\right)_w .} \qquad (13.80)$$

The equation for *mechanical energy* is obtained by first multiplying eqn. (13.6) by the velocity component u and then integrating with respect to y. Making use of the continuity equation and performing a number of simplifications, we obtain

$$\frac{1}{2} \frac{d}{dx}(\varrho_1 U^3 \delta_3) + \varrho_1 U^2 \frac{dU}{dx}(\delta_1 - \delta_u) = \int_0^\delta \mu \left(\frac{\partial u}{\partial y}\right)^2 dy . \qquad (13.81)$$

On the left-hand side of this equation we discover the mechanical work of the flow, the term on the right-hand side representing the dissipation. In incompressible flow, the second term on the left-hand side vanishes because then, with $\varrho = \text{const}$, we find that $\delta_H = 0$. As a result, eqn. (13.81) transforms into eqn. (8.35).

The equation for the increase in enthalpy — habitually known as the energy equation for short — is obtained as a result of the integration of eqn. (13.71) over y. Thus

$$\frac{d}{dx}(\varrho_1 \hat{h}_1 U \delta_H) + \varrho_1 U^2 \frac{dU}{dx} \delta_H = -\left(\frac{\mu}{P} \frac{\partial \hat{h}}{\partial y}\right)_w + \int_0^\delta \mu \left(\frac{\partial u}{\partial y}\right)^2 dy . \qquad (13.82)$$

The left-hand side of this equation represents the change in the enthalpy of the stream, whereas the terms on the right-hand side describe its changes due to the transfer of heat at the wall (subscript w) and to its generation through dissipation. Noting that eqn. (13.81) describes the loss in mechanical energy, whereas eqn. (13.82) describes the gain in enthalpy, we can obtain an equation which describes the increase in *total enthalpy* in the x-direction by forming their difference. This yields

$$\frac{d}{dx}\left[\varrho_1 U \left(\hat{h}_1 \delta_H - \frac{1}{2} U^2 \delta_3\right)\right] = -\left(\frac{\mu}{P} \frac{\partial \hat{h}}{\partial y}\right)_w . \qquad (13.83)$$

Introducing the stagnation enthalpy, $\hat{h}$, per unit mass from eqn. (13.70), we transform eqn. (13.83) to the form

$$\frac{\mathrm{d}}{\mathrm{d}x} \int_0^\delta \varrho\, u\, (\hat{h} - \hat{h}_1)\, \mathrm{d}y = - \left(\frac{\mu}{\mathsf{P}}\, \frac{\partial \hat{h}}{\partial y} \right)_w . \tag{13.84}$$

The left-hand side of this equation represents the increase in the stagnation enthalpy of the stream in the x-direction, whereas the right-hand side describes the quantity of heat added or subtracted by the wall.

Integrating eqn. (13.83) with respect to x, we obtain

$$\frac{1}{2}\, U^2\, \delta_3 = \hat{h}_1\, \delta_H + \frac{1}{\varrho_1\, U} \int_0^x \left(\frac{\mu}{\mathsf{P}}\, \frac{\partial \hat{h}}{\partial y} \right)_w \mathrm{d}x . \tag{13.85}$$

The right-hand side of eqn. (13.85) vanishes for an *adiabatic wall* since then $(\partial \hat{h}/\partial y)_w = 0$. It is now convenient to introduce the velocity of sound $c_1^2 = \gamma R T_1$ which corresponds to the state at the edge of the boundary layer. Since then

$$\hat{h}_1 = c_p\, T_1 = \frac{c_1{}^2}{\gamma - 1}$$

it follows that

$$\delta_H = \frac{1}{2}\, (\gamma - 1)\, \mathsf{M}^2\, \delta_3 , \tag{13.86}$$

in view of eqn. (13.85), where $\mathsf{M} = U/c_1$ denotes the local Mach number at the outer edge of the boundary layer. Taking into account the relations (13.79), (13.81) and (13.86) we obtain the final form of the *energy-integral equation:*

$$\frac{\mathrm{d}\delta_3}{\mathrm{d}x} + \frac{\delta_3}{U}\, \frac{\mathrm{d}U}{\mathrm{d}x}\, [3 - (2 - \gamma)\, \mathsf{M}^2] = \frac{2}{\varrho_1\, U^3} \int_0^\delta \mu \left(\frac{\partial u}{\partial y} \right)^2 \mathrm{d}y . \tag{13.87}$$

Equations (13.80) and (13.87) represent, respectively, the integral forms of the momentum and energy equation of a compressible, laminar boundary layer on an adiabatic wall. They constitute the basis for further calculation in approximate procedure devised, among others, by Gruschwitz. For incompressible flow, that is in the limit when $\mathsf{M} \to 0$, eqns. (13.80) and (13.87) transform into the momentum-integral and the energy-integral equations (8.32) and (8.35), respectively.

The approximate procedure due to Gruschwitz: In what follows, we shall pursue only one of the very numerous approximate procedures, namely that which was devised by E. Gruschwitz [43]. It is applicable to adiabatic walls with $\omega = 1$, but for arbitrary Prandtl numbers. As far as the amount of numerical work is concerned, this procedure is still relatively simple. It has the added advantage that it goes over into the scheme devised by K. Pohlhausen and H. Holstein and T. Bohlen, both described earlier in detail in Chap. X, when a passage to the limit of incompressible flow is performed.

We shall refrain from discussing this procedure in detail here, but the interested reader will be able to find it in Chap. XIII of the Sixth Edition of this book. We confine our attention to the results of such calculation performed for an aerofoil in a subsonic stream at a relatively high Mach number†.

The potential pressure distributions for the suction side of the aerofoil in question are seen plotted in Fig. 13.16 for three Mach numbers: $M_\infty = 0$; $0\cdot6$; and $0\cdot8$ and for an angle of incidence $\alpha = 0$. The diagrams also include plots of the temperature T_1 outside the boundary layer. The results of the calculation are shown in Figs. 13.17 and 13.18. The graph in Fig. 13.17 shows the variation of the momentum thickness δ_2, the displacement thickness δ_1, as well as the shearing stress τ_w along the suction side. As the Mach number is increased, the point of laminar separation moves slightly forward. The variation of the momentum thickness and shearing stress depends only little on the Mach number, whereas the displacement thickness δ_1 increases considerably as the Mach number is increased. Finally, Fig. 13.18 displays the velocity and temperature distributions at several positions along the contour of the aerofoil. The velocity profiles do not change much with the Mach number, but the temperature profiles show large increases in the wall temperature with increasing Mach number. This is to be expected, because the wall was assumed to be adiabatic. The adiabatic-wall temperatures, T_a, are also shown plotted in Fig. 13.16. An integration procedure for an adiabatic wall and with the further restrictions that the Prandtl number is equal to unity and that the viscosity is proportional to temperature was indicated by N. Rott and L. F. Crabtree [87].

The case of a circular cone in an axial supersonic stream calculated by W. Hantzsche and H. Wendt [45] constitutes an example of an axially symmetrical boundary layer. The boundary layer on a yawed circular cone in a supersonic stream was considered by F. K. Moore [78], whereas the case of a slightly yawed cone rotating in a supersonic stream was treated by R. Sedney [93]. Additional calculations for rotating bodies were performed by S. T. Chu and A. N. Tifford [15] and by J. Yamaga [103].

The proposition due to Mangler, described in Sec. XIc, permits us to reduce the calculation of axially symmetrical boundary layers on arbitrary bodies of revolution to that in two-dimensional flow. It remains valid in the field of compressible fluid flow.

R. M. Inman [51] analyzed the case of compressible Couette flow and calculated the skin-friction coefficient for an adiabatic as well as for a heat-transmitting wall, but on the simplified assumption that the viscosity is proportional to temperature. I. E. Beckwith [5] demonstrated that it is possible to perform approximate calculations of compressible boundary layers on arbitrary, three-dimensional bodies on condition that the components of the secondary stream should be small compared with those of the main flow.

---

† The author is indebted to Mr. F. Moser for working out this example. Since Gruschwitz's method does not lead to reasonable temperatures, the present temperature profiles have been calculated on the basis of ref. [36].

Figs. 13.16. to 13.18. Laminar boundary layer in compressible subsonic flow for the suction
side of the NACA 8410 aerofoil on the assumption of an adiabatic wall. Angle of incidence $\alpha = 0°$.
Mach number $M_\infty = U_\infty/c_\infty$; Prandtl number $P = 0 \cdot 725$. Calculation based on the approximate method by E. Gruschwitz [43]
S = point of separation

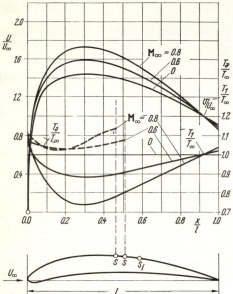

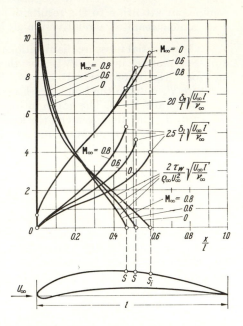

Fig. 13.16. Potential velocity distribution
U/U_∞ at the outer edge of the boundary
layer, the corresponding temperature distribution T_1/T_∞, and the variation in
the adiabatic wall temperature, T_a/T_∞

Fig. 13.17. Momentum thickness δ_2, displacement thickness δ_1, and shearing stress τ_w
for different Mach numbers

e. Interaction between shock wave and boundary layer

When a solid body is placed in a stream whose velocity is high, or when it flies
through air with a high velocity, local regions of supersonic velocity can be
formed in its neighbourhood. The transition from supersonic velocity to subsonic
velocity against the adjoining adverse pressure gradient will usually take place
through a shock wave. On crossing the very thin shock wave, the pressure, density,
and temperature of the fluid change at extremely high rates. The rates of change
are so high that the transition can be regarded as being discontinuous, except for
the immediate neighbourhood of the wall. The existence of shock waves is of fundamental importance for the drag of the body as they often cause the boundary
layer to separate. The theoretical calculation of shock waves and associated flow
fields is very difficult, and we do not propose to discuss this topic here. Experiments
show that the processes of shock and boundary-layer formation interact strongly

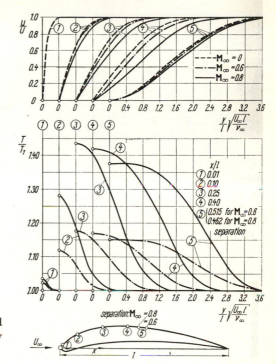

Fig. 13.18. a) Velocity distributions and b) temperature distributions in the boundary layer at different Mach numbers

with each other. This leads to phenomena of great complexity because the behaviour of the boundary layer depends mainly on the Reynolds number, whereas the conditions in a wave are primarily dependent on the Mach number. The earliest systematic investigations in which these two influences were clearly separated have been put to hand a long time ago. J. Ackeret, F. Feldmann and N. Rott [1], H. W. Liepmann [63], G. E. Gadd, W. Holder and J. D. Regan [38] varied in their experiments the Reynolds and Mach numbers independently of each other and so succeeded in providing some clarification of this complex interaction. The most important results obtained in the above three investigations are described in this section. We must, however, add that a complete understanding of these complex phenomena has eluded us to this day.

The pressure increase along the boundary layer must ultimately be the same as that in the external flow because the streamline which separates the two regions must become parallel to the contour of the body after the shock. In the boundary layer, by its nature, the particles near the wall move with subsonic velocities but shock waves can only occur in supersonic streams. It is, therefore, clear that a shock wave which originates in the external stream cannot reach right up to the wall, and it follows that the pressure gradient parallel to the wall must be much more gradual in the neighbourhood of the wall than in the external stream. Near the point where the shock wave reaches towards the wall, the rates of change of $\partial u/\partial x$ and

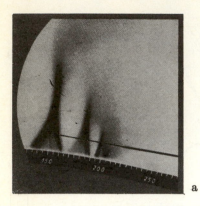

a

b

Fig. 13.19. Schlieren photograph of shock wave; direction of flow from left to right, after Ackeret, Feldmann and Rott [1]; a) laminar boundary layer; multiple λ-shock, M = 1·92, $R_{\delta_2} = 390$; b) turbulent boundary layer; normal shock, M = 1·28, $R_{\delta_2} = 1159$

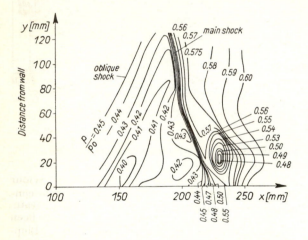

Fig. 13.20. Isobars in a shock region in laminar flow (λ-shock), after Ackeret, Feldmann and Rott [1]

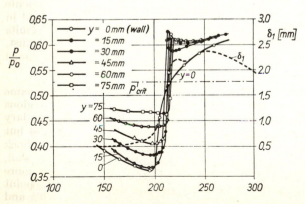

Fig. 13.21. Turbulent boundary layer in a shock region; pressure distribution at various distances from wall, after Ackeret, Feldmann and Rott [1]

$\partial u/\partial y$ become of the same order of magnitude, and transverse pressure gradients can also occur there. Both conditions render the well-known assumptions of boundary-layer theory invalid.

The appearance of the shock wave is fundamentally different depending on whether the boundary layer is laminar or turbulent, Fig. 13.19. A short distance ahead of the point where the essentially perpendicular shock wave impinges on a laminar boundary layer, there appears a short leg forming a so-called λ-shock, Fig. 13.19a. In general, when the boundary layer is turbulent, the normal shock does not split and no λ-shocks are formed, Fig. 13.19b. An *oblique* shock which impinges on a laminar boundary layer from the outside becomes reflected from it in the form of a fan of expansion waves, Fig. 13.30a. However, when the boundary layer is turbulent, the reflexion appears in the form of a more concentrated expansion wave (Fig. 13.30b).

The plot of isobaric curves in Fig. 13.20 and the pressure curves in Fig. 13.21 show that the rate of pressure increase along a laminar or a turbulent boundary layer is more gradual than in the external stream. This flattening of the pressure gradient in the boundary layer is described by stating that the pressure distribution "diffuses" near the wall. It is observed that diffusion is much more pronounced for a laminar than for a turbulent boundary layer. The difference between laminar and turbulent shock diffusions can also be recognized from Fig. 13.22 which represents the pressure variation along a flat plate placed parallel to a supersonic stream. The measurements were performed by H. W. Liepmann, A. Roshko and S. Dhawan [64]. The pressure plots have been taken near the point on the plate where the oblique shock produced by a wedge interacts with the boundary layer. The pressure gradient is considerably steeper for the turbulent than for the laminar boundary layer. The width of diffusion is equal to about 100 δ in the case of interaction with

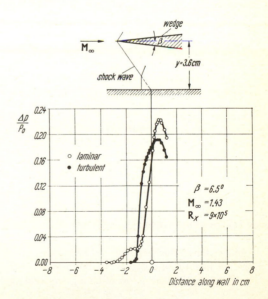

Fig. 13.22. Pressure distribution along a flat plate at supersonic velocity in the neighbourhood of the region of reflexion of a shock wave from laminar and turbulent boundary layers, after Liepmann, Roshko and Dhawan [64]

boundary-layer thickness: laminar $\delta \approx 0\cdot7$ mm (0·028 in), turbulent $\delta \approx 1\cdot4$ mm (0·056 in)

a laminar boundary layer, but decreases to about 10 δ for a turbulent boundary layer; the symbol δ denotes here the boundary-layer thickness in the shock region. The higher degree of diffusion which is characteristic of laminar boundary layers can be understood if it is noted that the subsonic region of flow extends further away from the wall in a laminar than in a turbulent boundary layer.

Irrespective of whether separation does or does not occur, the boundary-layer thickness increases ahead of the point of arrival of the shock wave. The pressure increase at the outer edge of the boundary layer, and hence also inside the boundary layer, corresponds to the curved streamline which is convex in the direction of the wall and which separates the external from the boundary-layer flow. Even in the domain of influence of the expansion waves which appear in the reflexion of an oblique shock wave, the slight decrease in pressure in the boundary layer, Fig 13.22, corresponds to the fact that the curvature of the dividing streamline is concave towards the wall. A laminar boundary layer which has not separated can support only very small pressure rises because the external flow impresses on it the pressure gradient exclusively through viscous forces. A non-separated turbulent boundary layer can take up much larger pressure gradients because now the turbulent mixing motion aids the process. Both laminar and turbulent boundary layers are in a position to support the large pressure increases of strong shocks if they separate. In particular,

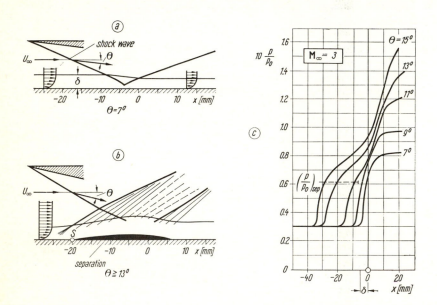

Fig. 13.23.   Reflexion of a shock wave from a turbulent boundary layer on a flat wall, after S. M. Bogdonoff and C. E. Kepler [6].   Boundary-layer thickness ahead of shock wave $\delta \approx 3$ mm ($0\cdot 12$ in). a) Weak shock, deflexion angle $\theta = 7°$. Reflexion similar to that in frictionless flow; no boundary-layer separation; b) Strong shock; deflexion angle $\theta \geq 13°$. Reflexion in form of a system of compression and expansion waves; boundary-layer separation; c) Pressure distributions at different deflexion angles θ. Separation occurs at $p_{sep}/p_\infty = 2$ approx.

Fig. 13.24. Schlieren photograph of the flow past an aerofoil. Shock-wave and boundary-layer interaction. Case (2): Laminar boundary layer with separation ahead of the shock, but re-attaching behind shock. $M = 0.84$, $R = 8.45 \times 10^5$, after Liepmann [63]

in the turbulent case the dead-water vortex between the separated boundary layer and the wall can create considerable velocities which carry the inner edge of the boundary layer against the pressure rise by the action of viscosity. The sketch in Fig. 13.23 shows how the boundary layer and the dead-water region thicken ahead of the front and become thinner behind it. Finally, as shown in Fig. 13.23, the boundary layer re-attaches itself completely. The same phenomenon is also visible in Fig. 13.24.

The sketches in Fig. 13.23 reproduce the results of some measurements performed by S. M. Bogdonoff and C. E. Kepler [6] in connexion with their investigations into the reflexion of oblique shock waves from a flat wall carrying a turbulent boundary layer at a Mach number $M_\infty = 3$ in the external stream. The sketches in Figs. 13.23a and b show the reflexions of a weak and a strong shock, respectively, their strength being regulated by the magnitude of the deflexion angle θ. When the shock is weak ($\theta = 7°$), the reflected shock presents a pattern which would be expected on the basis of ideal-flow theory, and the boundary layer does not separate. When the strength of the shock is increased ($\theta = 13°$), the reflected pattern contains a system of compression and expansion waves. The boundary layer exhibits a large local thickening which leads to separation. The boundary layer is thicker behind the reflected than ahead of the incident shock. The corresponding pressure distri-

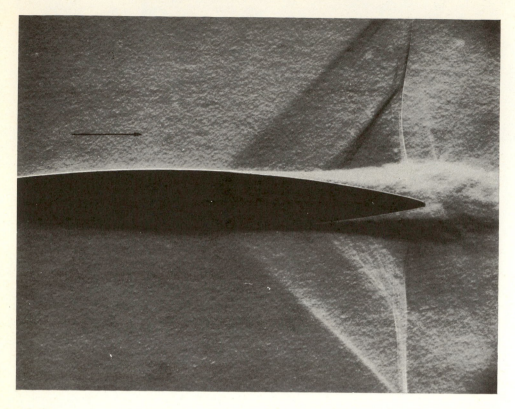

Fig. 13.25.   Schlieren photograph of the flow past an aerofoil. Shock-wave and boundary-layer interaction. Case (3): Laminar boundary layer with separation behind shock. $M = 0.90$, $R = 8.74 \times 10^5$, after Liepmann [63]

butions along the wall are shown plotted in Fig. 13.23 c for different deflexion angles (and hence different shock strengths). Separation occurs for $\theta > 9°$. The pressure rise which leads to separation is independent of the deflexion angle and has a value of about $p_{sep}/p_\infty = 2$.

The incidence of transition and separation in the neighbourhood of an impinging shock wave are governed principally by the Reynolds number of the boundary layer and by the Mach number of the external stream. When the shock is weak and the Reynolds number is very small, the boundary layer remains laminar throughout. Increasing the Reynolds number at a fixed, small Mach number, causes transition to occur at the point of impingement. When the shock is strong (large Mach number) and the Reynolds number is small, the laminar boundary layer will separate ahead of the shock front owing to pressure diffusion; it may also undergo transition ahead of the shock front. When the Reynolds number is large enough, transition in the boundary layer occurs ahead of the shock, whether the boundary layer has

Fig. 13.26. Schlieren photograph of the flow past an aerofoil. Shock-wave and boundary-layer interaction. Case (4): Boundary layer turbulent ahead of shock, no separation. $M = 0.85$, $R = 1.69 \times 10^6$, after Liepmann [63]

separated or not. According to observations made by A. Fage and R. Sargent [35], turbulent boundary layers do not separate when the pressure ratio p_2/p_1 is smaller than 1·8, which corresponds to a Mach number $M_\infty < 1.3$ for a normal shock wave. Further experimental results on the interaction between shock waves and boundary layers can be found in the publications by W. A. Mair [69], N. H. Johannesen [52], O. Bardsley and W. A. Mair [3], and J. Lukasiewicz and J. K. Royle [67].

During more recent times attempts have been made to describe the interaction between a laminar boundary layer and a shock wave by *theoretical* means. Such attempts were mostly unsuccessful because, generally speaking, the assumptions of boundary-layer theory break down near a shock wave. In some studies numerical integrations were based on the Navier-Stokes equations. A review of the current state of this complex of problems can be found in the papers by J. D. Murphy [81a], R. W. MacCormack [14a], J. M. Klineber [56a], J. C. Carter [14b], and J. D. Murphy et al. [81b].

The various effects of shocks impinging on a boundary layer will now be illustrated with reference to Schlieren photographs. As pointed out by A.D. Young [106], it is possible to distinguish the following cases:

(1) The approaching boundary layer is laminar and remains so beyond the shock without separation.

(2) The approaching boundary layer is laminar, but separates ahead of the shock because of the adverse pressure gradient and then returns to the surface in either a laminar or turbulent state, Fig. 13.24†.

(3) The approaching boundary layer is laminar, separates completely from the surface ahead of the shock, and does not re-attach itself to the surface, Fig. 13.25; the shock is normal and sprouts a λ-limb.

(4) The approaching boundary layer is turbulent and does not separate from the surface, Fig. 13.26.

(5) The approaching boundary layer is turbulent and separates from the surface, Figs. 13.27 and 13.28.

Figures 13.30a, b illustrate the reflexion of an oblique shock wave from a laminar and a turbulent boundary layer, respectively.

A long time ago, A. Busemann [10] published observations on boundary-layer separation in supersonic flow. Supersonic tunnels are usually equipped with a diffuser which serves to recover pressure from the high wind velocity. These diffusers are made in the shape of convergent-divergent channels through which the stream flows with an adverse pressure gradient in both the convergent and the divergent portion. A. Busemann observed that at all Mach numbers separation did not depend on either the angle of convergence or on the angle of divergence, but that it could always be associated with the adverse pressure gradient. In this connexion it should be realized that the change in the character of the flow which occurs at higher Mach numbers is linked with changed conditions for the adverse pressure gradients.

The considerations concerning the behaviour of boundary layers on aerofoils in the transonic regime that follow refer essentially to turbulent boundary layers which will be studied later in Chaps. XXII and XXIII. Since, however, transition plays a part in these processes, we shall insert them here, even though the transition process itself will also be discussed later, namely in Chaps. XVI and XVII.

The interaction between the boundary layer and the external flow is particularly strong in the transonic regime. Figure 13.27 after G.L. Loving [66] contains an experimental contribution consisting of measurements of pressure distributions on an aerofoil in flight (large Reynolds number) compared to that in a wind tunnel (reduced Reynolds number). The boundary layers are turbulent over the greater part of the wetted perimeter in either case. The two pressure distributions, the one in the wind tunnel and the one in flight, agree satisfactorily at $M_\infty = 0{\cdot}75$ (Fig. 13.27a) for which the pressure distribution is still subcritical, even though the Reynolds numbers differ by a factor of 10. However, at $M_\infty = 0{\cdot}85$ (Fig. 13.27b) which generates local

---

† Thanks are due to Professor H.W. Liepmann of the California Institute of Technology for his permission to use the photographs in Figs. 13.24 to 13.26 and Figs. 13.29 to 13.30 and for his kindness in supplying the original prints for publication.

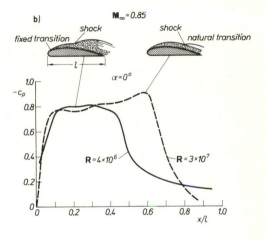

Fig. 13.27. Effect of Reynolds number on pressure distribution on the upper side of an aerofoil in the transonic velocity range, after G. L. Loving [66]; comparison between flight test (full scale) at $R = 3 \times 10^7$ and wind-tunnel test at $R = 4 \times 10^6$. Natural transition in free flight; tripped transition in wind-tunnel test. a) Subcritical velocity distribution for $M_\infty = 0.75$; lift coefficient $c_L \approx 0.3$; satisfactory agreement between free-flight test at $R = 3 \times 10^7$ and wind-tunnel test at $R = 4 \times 10^6$. b) Supercritical pressure distribution for $M_\infty = 0.85$; lift coefficient $c_L \approx 0.34$; large deviation between free-flight test at $R = 3 \times 10^7$ and wind-tunnel test at $R = 4 \times 10^6$

——————— wind tunnel (transition fixed)

– – – – flight full scale

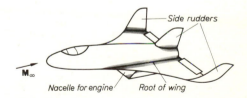

Fig. 13.28. Corners in parallel, hypersonic flow

▓▓▓▓ Corners with considerable overheating

supersonic regions on the aerofoil, the two pressure distributions differ considerably. The shock wave (and the point of separation linked with it) is located much further downstream at the larger Reynolds number $R = 3 \times 10^7$ of free flight than at the wind-tunnel Reynolds number of $R = 4 \times 10^6$. The physical explanation of this circumstance is connected, presumably, with the fact that the boundary layer at the lower Reynolds number of the wind-tunnel experiment is considerably thicker (compared to the aerofoil thickness) and therefore displaces the shock wave and the point of separation triggered by it further upstream. It is thus possible to conclude that

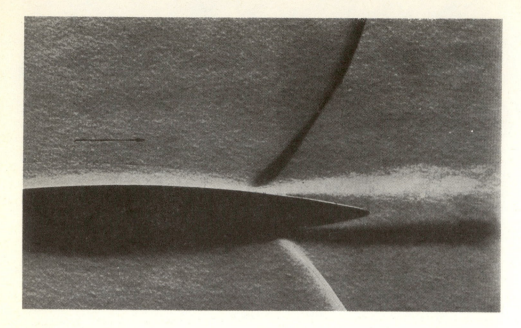

Fig. 13.29.   Schlieren photograph of the flow past an aerofoil. Shock-wave and boundary-layer interaction. Case (5): Turbulent boundary layer with strong separation behind shock. $M = 0.90$, $R = 1.75 \times 10^6$, after Liepmann [63]

the influence of Reynolds number on the boundary layer — and hence also on the shock wave as well as on the associated point of separation — is quite considerable in transonic flow. As a result, the value of the Reynolds number has a much greater effect on all aerodynamic characteristics of an aerofoil in the *transonic range of Mach numbers* than either in subsonic or in the purely supersonic regime. For this reason it is necessary to exercise utmost caution when test results from wind tunnels in the transonic range are used to predict behaviour in flight. Further experimental results on this topic can be found in [27a, 84, 91].

   Very extensive recent experimental results in this field were communicated by J. J. Kacprzynski [53].

   Heat-transfer problems in the *hypersonic* range occur during re-entry of space vehicles and ballistic rockets into the terrestrial atmosphere. In cases when air resistance is utilized to provide deceleration as the moving body approaches the surface of the earth, a large proportion of the energy so dissipated is imparted to the body in the form of heat. Such processes take place in the boundary layer in the hypersonic regime whether it is laminar or turbulent. J. C. Rotta [87a] published a summary article describing the problems that arise in this connexion in two-dimensional and rotationally symmetric bodies.

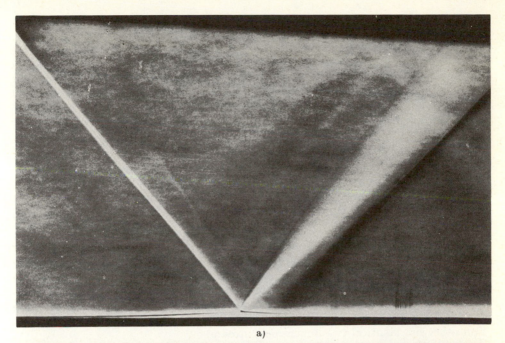

a)

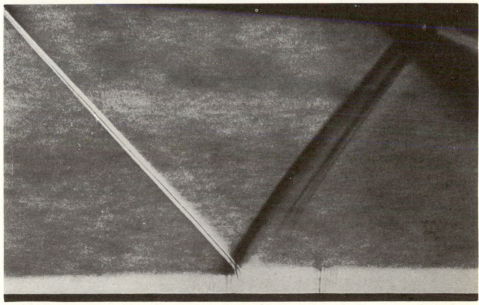

b)

Fig. 13.30a,b.   Reflexion of oblique shock from flat plate with boundary layer, after Liepmann, Roshko and Dhawan [64]. a) Laminar boundary layer; b) Turbulent boundary layer

Yet another important problem of interaction between boundary layer and shock wave occurs in *hypersonic corner flow* at zero incidence. The flow is accompanied by intense heating in the corner caused by the very much larger rate of dissipation in the corner compared to the dissipation in the neighbouring two-dimensional flow. A hint in that direction is visible in Fig. 12.15. It was shown there that even in incompressible flow along a rectangular corner with the wall being at a temperature exceeding that of the free stream there exists a heat flux transferring heat from the fluid to the wall. By contrast, at a large distance from the corner, the flow of heat takes place in the reverse direction.

Scientists became aware of the above problem only recently, namely in connexion with the flight tests in the range of Mach numbers $M = 3$ to 6 on the American experimental aeroplane X-15. Reports on this phenomenon were published by R. D. Neumann [82, 99]. Figure 13.28 reminds the reader that such corner configurations exist at the root of the wing, at the side fins, at the engine pods or at the air inlet in air-breathing engines.

More recent experimental investigations on hypersonic corner flows were performed by K. Kipke and D. Hummel [56] at the very large Mach numbers of $M = 12$ to 16; they measured the pressure distribution and the local rate of heat transfer in the corner and discovered the extraordinarily complex structure of the zone of interaction of the shock wave and boundary layer. The flow develops strong separation zones and the local heat-transfer rates were as much as tenfold larger than those in an otherwise identical two-dimensional stream.

Theoretical investigations into the problem of interaction between a shock wave and a laminar boundary layer are very numerous. We shall be satisfied with the mention of the following few authors only: E. A. Mueller [81], D. Meksyn [74], M. Honda [47] and J. Appleton and H. J. Davies [2]. Particular attention should be given to the paper by N. Curle [24]. This study contains an investigation of the effect of heat transfer on the pressure rise on a flat plate as well as the description of an approximate method of calculating the boundary layer for arbitrary wall temperatures and Prandtl numbers. The method takes advantage of the experimental results due to G. E. Gadd [39] which demonstrated that at the point of separation the velocity gradient dU/dx differs from zero but the velocity U itself remains nearly constant. With this simplification, it becomes possible to integrate the equations subject to the assumption of a relation between the unknown pressure gradient and the increase in the boundary-layer thickness. It turns out that the pressure coefficient at the point of separation is independent of the temperature at the wall but that the spread in the extent of the zone of interaction is proportional to T_w. Consequently, the pressure rise at separation becomes inversely proportional to T_w.

Assuming that p differs little from p_0, and that we have, approximately

$$c_p \approx \frac{2}{\gamma M_0^2} \left(\frac{p}{p_0} - 1 \right) \approx 1 - \frac{U^2(x)}{U_0^2} , \qquad (13.88)$$

N. Curle calculated a function $F(X)$ which is reproduced here in Table 13.1. The following abbreviations have there been introduced:

Table 13.1.   The function $F(X)$ for the pressure distribution along a flat plate in the neighbour-hood of a shock wave, in accordance with eqns. (13.89) and (13.90), after N. Curle [24]

X	$F(X)$	$F'(X)$	X	$F(X)$	$F'(X)$
$-7 \cdot 03$	$0 \cdot 02$	$0 \cdot 0103$	$0 \cdot 70$	$0 \cdot 40$	$0 \cdot 0885$
$-5 \cdot 12$	$0 \cdot 05$	$0 \cdot 0237$	$1 \cdot 86$	$0 \cdot 50$	$0 \cdot 0828$
$-4 \cdot 09$	$0 \cdot 08$	$0 \cdot 0351$	$3 \cdot 21$	$0 \cdot 60$	$0 \cdot 0645$
$-3 \cdot 14$	$0 \cdot 12$	$0 \cdot 0479$	$5 \cdot 03$	$0 \cdot 70$	$0 \cdot 0465$
$-2 \cdot 21$	$0 \cdot 17$	$0 \cdot 0612$	$7 \cdot 61$	$0 \cdot 80$	$0 \cdot 0323$
$-1 \cdot 32$	$0 \cdot 23$	$0 \cdot 0736$	$11 \cdot 75$	$0 \cdot 90$	$0 \cdot 0174$
$-0 \cdot 55$	$0 \cdot 29$	$0 \cdot 0832$	$15 \cdot 52$	$0 \cdot 95$	$0 \cdot 0101$
0	$0 \cdot 338$	$0 \cdot 0900$	$23 \cdot 33$	$1 \cdot 00$	$0 \cdot 0042$
			∞	$1 \cdot 03$	0

Fig. 13.31 a.

$\mathbf{M}_0 = 3$; $\mathbf{R} = 4 \cdot 2 \times 10^5$

$+$ heated wall $T_w = 1 \cdot 25\, T_0$

$\times$ zero heat transfer $T_w = T_0$

$\bigcirc$ cooled wall $T_w = 0 \cdot 88\, T_0$

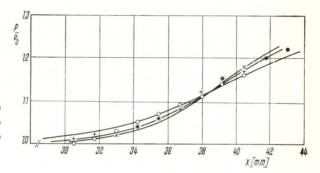

Fig. 13.31 b.

$\mathbf{M}_0 = 2 \cdot 7$; $\mathbf{R} = 1 \cdot 5 \times 10^5$

$\times$ zero heat transfer $T_w = T_0$

$\bigcirc$ heated wall $T_w = 1 \cdot 5\, T_0$

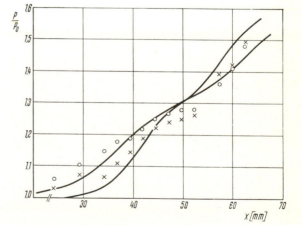

Fig. 13.31. Laminar boundary layer in the zone of interaction with a shock wave: pressure distribution on a flat plate placed at zero incidence in a supersonic stream at different wall temperatures, T_w. Full lines: theory due to N. Curle [24]

$$F = 0.4096 \, (M_0{}^2 - 1)^{1/4} \, R^{1/4} \left(1 - \frac{U^2(x)}{U_0{}^2}\right), \qquad (13.89)$$

$$X = 1.820 \, (M_0{}^2 - 1)^{1/4} \, R^{1/4} \left(\frac{T_w}{T_1}\right)_{sep}^{-1} \left(\frac{x}{x_{sep}} - 1\right). \qquad (13.90)$$

The subscript *sep* refers to the point of separation, the subscript 0 describes the state upstream of the shock wave, and subscript 1 denotes the state at the edge of the boundary layer.

The pressure coefficient at separation turns out to have the form

$$c_{p \, sep} = \frac{2}{\gamma \, M_0{}^2} \left(\frac{p_{sep}}{p_0} - 1\right) = 0.825 \, (M_0{}^2 - 1)^{-1/4} \, R^{-1/4}, \qquad (13.91)$$

where $R = U \, x/\nu_1$ and p_0, M_0 denote the pressure and Mach number, respectively, upstream of the shock wave.

The diagrams in Fig. 13.31 a and b contain a comparison between the theoretical curves and the results of measurements performed by G. E. Gadd and J. L. Attridge [40]. Both theory and experiment lead to the conclusion that the pressure ahead of the zone of separation is higher when the wall is heated than when it is adiabatic. A comparison between the two diagrams, each of which corresponds to a different temperature at the wall, convinces us that this effect becomes more pronounced as the temperature is increased.

Numerical solutions which contain the zone of interaction between a separated laminar boundary layer and a frictionless supersonic stream were performed by V. N. Vatsa and S. D. Bertke [101], as well as by O. R. Burggraf [9], G. S. Settles, S. M. Bogdonoff and I. E. Vas [93a].

References

[1] Ackeret, J., Feldmann, F., and Rott, N.: Untersuchungen an Verdichtungsstössen und Grenzschichten in schnell bewegten Gasen. Report No. 10 of the Inst. of Aerodynamics ETH Zürich 1946; see also NACA TM 1113 (1947).
[2] Appleton, J. P., and Davies, H. J.: A note on the interaction of a normal shock wave with a thermal boundary layer. JAS *25*, 722—723 (1958).
[3] Bardsley, O., and Mair, W. A.: Separation of the boundary layer at a slightly blunt leading edge in supersonic flow. Phil. Mag. *43*, 338, 344—352 (1952).
[4] Barry, F. W., Shapiro, H. A., and Neumann, E. P.: Some experiments on the interaction of shock waves with boundary layers on a flat plate. J. Appl. Mech. *17*, 126—131 (1950).
[5] Beckwith, I. E.: Similarity solutions for small cross flows in laminar compressible boundary layers. NASA TR R 107, 1—67 (1961).
[6] Bogdonoff, S. M., and Kepler, C. E.: Separation of a supersonic turbulent boundary layer. JAS *22*, 414—424 (1955).
[7] Bouniol, F., and Eichelbrenner, E. A.: Calcul de la couche limite laminaire compressible. Méthode rapide applicable au cas de la plaque plane. La Récherche Aéron. *28* (1952).

[8] Bradfield, W.S., Decoursin, D.G., and Blumer, C.B.: The effect of leading-edge bluntness on a laminar supersonic boundary layer. JAS 21, 373—382 and 398 (1954).

[9] Burggraf, O.R.: Asymptotic theory of separation and reattachment of a laminar boundary layer on a compression ramp. AGARD Conf. Proc. Flow Separation, No. 168, 10/1—10/9 (1975).

[10] Busemann, A.: Gasströmung mit laminarer Grenzschicht entlang einer Platte. ZAMM 15, 23—25 (1935).

[10a] Busemann, A.: Die achsensymmetrische kegelige Überschallströmung. Luftfahrtforschung 19, 137—144 (1942).

[11] Busemann, A.: Das Abreissen der Grenzschicht bei Annäherung an die Schallgeschwindigkeit. Jb. Luftfahrtforschung I, 539—541 (1940).

[12] Byran, L.F.: Experiments on aerodynamic cooling. Report of the Inst. of Aerodynamics ETH Zürich, No. 18, 1951.

[13] Chapman, D.R., and Rubesin, M.W.: Temperature and velocity profiles in the compressible laminar boundary layer with arbitrary distribution of surface temperature. JAS 16, 547—565 (1949).

[14] Charwat, A.F., and Redekopp, L.G.: Supersonic interference flow along the corner of intersecting wedges. AIAA J. 5, 480—488 (1967).

[14a] MacCormack, R.W.: Numerical solution of the interaction of shock wave with a laminar boundary layer. Proceedings 2nd Intern. Conf. on Numerial Methods in Fluid Dynamics. Lecture Notes in Physics 8, Springer Verlag, 1971.

[14b] Carter, J.E.: Solutions for laminar boundary layers with separation and reattachment. AIAA Paper 74—583 (1974).

[15] Chu, S.T., and Tifford, A.N.: The compressible laminar boundary layer on a rotating body of revolution. JAS 21, 345—346 (1954).

[16] Cohen, C.B.: Similar solutions of compressible laminar boundary layer equations. JAS 21, 281—282 (1954).

[16a] Cohen, C.B., and Reshotko, E.: The compressible laminar boundary layer with heat transfer and arbitrary pressure gradient. NACA Rep. 1294 (1956).

[17] Coles, D.: Measurements of turbulent friction on a smooth flat plate in supersonic flow. JAS 21, 433—448 (1954).

[18] Cope, W.F., and Hartree, D.R.: The laminar boundary layer in a compressible flow. Phil. Trans. Roy. Soc. A 241, 1—69 (1948).

[19] Crabtree, L.F.: The compressible laminar boundary layer on a yawed infinite wing. Aero. Quart. 5, 85—100 (1954).

[20] Crocco, L.: Sulla trasmissione del calore da una lamina piana a un fluido scorrente ad alta velocità. L'Aerotecnica 12, 181—197 (1932).

[21] Crocco, L.: Sullo strato limite laminare nei gas lungo una lamina piana. Rend. Mat. Univ. Roma V 2, 138—152 (1941).

[22] Crocco, L.: Lo strato laminare nei gas. Mon. Sci. Aer. Roma (1946).

[23] Crocco, L., and Cohen, C.B.: Compressible laminar boundary layer with heat transfer and pressure gradient. Fifty years of boundary layer research (W. Tollmien and H. Görtler, ed.), Braunschweig, 1955, 280—293; see also NACA Rep. 1294 (1956).

[24] Curle, N.: The effects of heat transfer on laminar boundary layer separation in supersonic flow. Aero. Quart. 12, 309—336 (1961).

[25] Curle, N.: Heat transfer through a compressible laminar boundary layer. Aero. Quart. 13, 255—270 (1962).

[26] Curle, N.: The laminar boundary layer equations. Clarendon Press, Oxford, 1962.

[27] Des Clers, B., and Sternberg, J.: On boundary layer temperature recovery factors. JAS 19, 645—646 (1952).

[27a] Delery, J., Chattot, J.J., and Le Balleur, J.C.: Interaction visqueuse avec décollement en écoulement transsonique. AGARD Conf. Proc. Flow Separation, No. 168, 27—1 to 27—13 (1975).

[28] O'Donnell, R.M.: Experimental investigation at Mach number of 2·41 of average skin friction coefficients and velocity profiles for laminar and turbulent boundary layers assessment of probe effects. NACA TN 3122 (1954).

[29] Dorrance, W.H.: Viscous hypersonic flow. Theory of reacting hypersonic boundary layers. McGraw-Hill, New York, 1962.

[30] Van Driest, E.R.: Investigation of laminar boundary layer in compressible fluids using the Crocco-Method. NACA TN 2597 (1952).

[31] Van Driest, E.R.: The problem of aerodynamic heating. Aero. Eng. Review *15*, 26—41 (1956).

[32] Eber, G.R.: Recent investigations of temperature recovery and heat transmission on cones and cylinders in axial flow in the NOL Aeroballistics Wind Tunnel. JAS *19*, 1—6 (1952).

[33] Eichelbrenner, E.A.: Méthodes de calcul de la couche limite laminaire bidimensionelle en régime compressible. Office National d'Etudes et de Récherche Aéronautiques (ONERA), Paris, Publication No. *83* (1956).

[34] Emmons, H.W., and Brainerd, J.G.: Temperature effects in a laminar compressible fluid boundary layer along a flat plate. J. Appl. Mech. *8*, A 105 (1941) and J. Appl. Mech. *9*, I (1942).

[35] Fage, A., and Sargent, R.: Shock wave and boundary layer phenomena near a flat plate surface. Proc. Roy. Soc. A *190*, 1—20 (1947).

[36] Flügge-Lotz, I., and Johnson, A.F.: Laminar compressible boundary layer along a curved insulated surface. JAS *22*, 445—454 (1955).

[37] Gadd, G.E.: Some aspects of laminar boundary layer separation in compressible flow with no heat transfer to the wall. Aero. Quart. *4*, 123—150 (1953).

[38] Gadd, G.E., Holder, D.W., and Regan, J.D.: An experimental investigation of the interaction between shock waves and boundary layers. Proc. Roy. Soc. A *226*, 227—253 (1954).

[39] Gadd, G.E.: An experimental investigation of heat transfer effects on boundary layer separation in supersonic flow. JFM *2*, 105—122 (1957).

[40] Gadd, G.E., and Attridge, J.L.: A note on the effects of heat transfer on the separation of laminar boundary layer. ARC CP 569 (1961).

[41] Ginzel, I.: Ein Pohlhausen-Verfahren zur Berechnung laminarer kompressibler Grenzschichten. ZAMM *29*, 6—8 (1949); Ginzel, I.: Ein Pohlhausen-Verfahren zur Berechnung laminarer kompressibler Grenzschichten an einer geheizten Wand. ZAMM *29*, 321—337 (1949).

[42] Green, J.E.: Interactions between shock waves and turbulent boundary layers. Progress in Aerospace Sciences (D. Küchemann, ed.), *11*, 235—340 (1970).

[43] Gruschwitz, E.: Calcul approché de la couche limite laminaire en écoulement compressible sur une paroi non-conductrice de la chaleur. Office National d'Etudes et de Récherche Aéronautiques (ONERA), Paris, Publication No. *47* (1950).

[44] Hantzsche, W., and Wendt, H.: Zum Kompressibilitätseinfluss bei der laminaren Grenzschicht der ebenen Platte. Jb. dt. Luftfahrtforschung *I*, 517—521 (1940).

[45] Hantzsche, W., and Wendt, H.: Die laminare Grenzschicht an einem mit Überschallgeschwindigkeit angeströmten nicht angestellten Kreiskegel. Jb. dt. Luftfahrtforschung *I*, 76—77 (1941).

[46] Hantzsche, W., and Wendt, H.: Die laminare Grenzschicht an der ebenen Platte mit und ohne Wärmeübergang unter Berücksichtigung der Kompressibilität. Jb. dt. Luftfahrtforschung *I*, 40—50 (1942).

[47] Honda, M.: A theoretical investigation of the interaction between shock waves and boundary layers. JAS *25*, 667—678 (1958).

[48] Howarth, L.: Concerning the effect of compressibility on laminar boundary layers and their separation. Proc. Roy. Soc. London A *194*, 16—42 (1948).

[49] Illingworth, C.R.: The laminar boundary layer associated with retarded flow of a compressible fluid. ARC RM 2590 (1946).

[50] Illingworth, C.R.: Steady flow in the laminar boundary layer of a gas. Proc. Roy. Soc. A *199*, 533—558 (1949).

[51] Inman, R.M.: A note on the skin-friction coefficient of a compressible Couette flow. JASS *26*, 182 (1959).

[52] Johannesen, N.H.: Experiments on two-dimensional supersonic flow in corners and over concave surfaces. Phil. Mag. *43*, 340, 568—580 (1952).

[53] Kacprzynski, J.J.: Viscous effects in transonic flow past airfoils. ICAS Paper No. 74—19, Ninth Congress of the International Council of the Aeronautical Sciences Haifa, Israel, August 1974.

[54] von Kármán, Th., and Tsien, H.S.: Boundary layer in compressible fluids. JAS *5*, 227—232 (1938); see also: von Kármán, Th.: Report on Volta Congress, Rome 1935; see also Coll. Works *III*, 313—325.

[55] Kaye, J.: Survey of friction coefficients, recovery factors and heat transfer coefficients for supersonic flow. JAS *21*, 117—129 (1954).

[56] Kipke, K., and Hummel, D.: Untersuchungen an längsangeströmten Eckenkonfigurationen im Hyperschallbereich. ZFW 23, 417—429 (1975).

[56a] Klineber, J.M., and Steger, J.L.: Numerical calculation of laminar boundary-layer separation. NASA TN 7732 (1974).

[57] Kuerti, G.: The laminar boundary layer in compressible flow. Advances in Appl. Mech. 11, 21—92 (1951).

[58] Lees, L.: On the boundary layer equations in hypersonic flow and their approximate solution. JAS 20, 143—145 (1953).

[59] Lees, L.: Influence of the leading-edge shock wave on the laminar boundary layer at hypersonic speeds. JAS 23, 594—600 and 612 (1956).

[60] Li, T.Y., and Nagamatsu, H.T.: Similar solutions of compressible boundary layer equations. JAS 20, 653—655 (1953).

[61] Li, T.Y., and Nagamatsu, H.T.: Similar solutions of compressible boundary-layer equations. JAS 22, 607—616 (1955).

[62] Libby, P.A., and Morduchow, M.: Method for calculation of compressible boundary layer with axial pressure gradient and heat transfer. NACA TN 3157 (1964).

[63] Liepmann, H.W.: The interaction between boundary layer and shock waves in transonic flow. JAS 13, 623—637 (1946).

[64] Liepmann, H.W., Roshko, A., and Dhawan, S.: On reflection of shock waves from boundary layers. NACA Rep. 1100 (1952).

[65] Lilley, G.M.: A simplified theory of skin friction and heat transfer for a compressible laminar boundary layer. Coll. Aero. Cranfield, Note No. 93 (1959).

[66] Loving, G.L.: Wind-tunnel-flight correlation of shock induced separated flow. NASA TND 3580 (1966).

[67] Lukasiewicz, J., and Royle, J.K.: Boundary layer and wake investigation in supersonic flow. ARC RM 2613 (1952).

[68] Luxton, R.E., and Young, A.D.: Generalised methods for the calculation of the laminar compressible boundary layer characteristics with heat transfer and non-uniform pressure distribution. ARC RM 3233 (1962).

[69] Mair, W.A.: Experiments on separation of boundary layers on probes in a supersonic airstream. Phil. Mag. 43, 342, 695—716 (1952).

[70] Mangler, W.: Zusammenhang zwischen ebenen und rotationssymmetrischen Grenzschichten in kompressiblen Flüssigkeiten. ZAMM 28, 97—103 (1948).

[71] Mangler, W.: Ein Verfahren zur Berechnung der laminaren Grenzschicht mit beliebiger Druckverteilung und Wärmeübergang für alle Mach-Zahlen. ZFW 4, 63—66 (1956).

[72] Maydew, R.C., and Pappas, C.C.: Experimental investigation of the local and average skin friction in the laminar boundary layer on a flat plate at a Mach-number of 2·4. NACA TN 2740 (1952).

[73] Meksyn, D.: Integration of the boundary layer equations for a plane in a compressible fluid. Proc. Roy. Soc. London A 195, 180—188 (1948).

[74] Meksyn, D.: The boundary layer equations of compressible flow separation. ZAMM 38, 372—379 (1958).

[75] Monaghan, R.J.: An approximate solution of the compressible laminar boundary layer on a flat plate. ARC RM 2760 (1949).

[76] Monaghan, R.J.: Effects of heat transfer on laminar boundary layer development under pressure gradients in compressible flow. ARC RM 3218 (1961).

[77] Moore, L.L.: A solution of the laminar boundary layer equations for a compressible fluid with variable properties, including dissociation. JAS 19, 505—518 (1952).

[78] Moore, F.K.: Three-dimensional laminar boundary layer flow. JAS 20, 525—534 (1953).

[79] Morduchow, M.: Analysis and calculation by integral methods of laminar compressible boundary layer with heat transfer and with and without pressure gradient. NACA Rep. 1245 (1955).

[80] Morris, D.N., and Smith, J.W.: The compressible laminar boundary layer with arbitrary pressure and surface temperature gradients. JAS 20, 805—818 (1953). See also: Morris, D.N., and Smith, J.W.: Ein Näherungsverfahren für die Integration der laminaren kompressiblen Grenzschichtgleichungen. ZAMM 34, 193—194 (1954).

[81] Müller, E.A.: Theoretische Untersuchungen über die Wechselwirkung zwischen einem einfallenden schwachen Verdichtungsstoß und der laminaren Grenzschicht in einer Überschallströmung. Fifty years of boundary-layer research (W. Tollmien and H. Görtler, ed.), Braunschweig, 1955, 343—363.

[81a] Murphy, J.D.: A critical evaluation of analytical methods for predicting laminar boundary layer, shock-wave interaction. NASA TN D-7044 (1971).

[81b] Murphy, J.D., Presley, L.L., and Rose, W.C.: On the calculation of supersonic separating and reattaching flows. AGARD Conf. Proc. Flow Separation, No. 168, 22—1 to 22—12 (1975).

[82] Neumann, R.D.: Special topics in hypersonic flow. AGARD Lecture Series No. 42, *1*, 7—1 to 7—64 (1972).

[83] Pai, S.I., and Shen, S.F.: Hypersonic viscous flow over an inclined wedge with heat transfer. Fifty years of boundary-layer research (W. Tollmien and H. Görtler, ed.), Braunschweig, 1955, 112—121.

[84] Pearcey, H.H., Osborne, J., and Haines, A.B.: The interaction between local effects at the shock and rear separation — a source of significant scale effects in windtunnel tests on airfoils and wings. AGARD Conf. Proc. No. 35, 11—1 to 11—23 (1968).

[85] Poots, G.: A solution of the compressible laminar boundary layer equations with heat transfer and adverse pressure gradient. Quart. J. Mech. Appl. Math. *13*, 57—84 (1960).

[86] Reshotko, E., and Beckwith, I.E.: Compressible laminar boundary layer over a yawed infinite cylinder with heat transfer and arbitrary Prandtl number. NACA Rep. 1379, 1—49 (1958).

[87] Rott, N., and Crabtree, L.F.: Simplified laminar boundary layer calculations for bodies of revolution and for yawed wings. JAS *19*, 553—565 (1952).

[87a] Rotta, J.C.: Wärmeübergangsprobleme bei hypersonischen Grenzschichten. Jb. WGLR 1962, 190—196 (1963).

[88] Rubesin, M.W., and Johnson, H.A.: A critical review of skin friction and heat transfer solutions of the laminar boundary layer of a flat plate. Trans. ASME *71*, 383—388 (1949).

[88a] Ryzhov, O. S.: Viscous transonic flows. Ann. Rev. Fluid Mech. (M. van Dyke, ed.) *10*, 65—92 (1978).

[89] Scherrer, R.: Comparison of theoretical and experimental heat transfer characteristics of bodies of revolution of supersonic speeds. NACA Rep. 1055 (1951).

[90] Schlichting, H.: Zur Berechnung der laminaren Reibungsschicht bei Überschallgeschwindigkeit. Abh. der Braunschweigischen Wiss. Gesellschaft *3*, 239—264 (1951).

[91] Stanewsky, E., and Little, B.H.: Separation and reattachment in transonic airfoil flow. J. Aircraft *8*, 952—958 (1971).

[92] Stainback, P.C.: An experimental investigation at a Mach number 4·95 of flow in the vicinity of a 90° interior corner aligned with the free stream velocity. NASA TN 184 (1960).

[93] Sedney, R.: Laminar boundary layer on a spinning cone at small angles of attack in a supersonic flow. JAS *24*, 430—436, 455 (1957).

[93a] Settles, G. S., Bogdonoff, S. M., and Vas, I. E.: Incipient separation of a supersonic boundary layer at high Reynolds numbers. AIAA J. *14*, 50—56 (1976).

[94] Stewartson, K.: Correlated compressible and incompressible boundary layers. Proc. Roy. Soc. A *200*, 84—100 (1949).

[95] Stewartson, K.: On the interaction between shock waves and boundary layers. Proc. Cambr. Phil. Soc. *47*, 545—553 (1951).

[96] Stewartson, K.: The theory of laminar boundary layers in compressible fluids. Oxford, 1964.

[97] Tani, I.: On the approximate solution of the laminar boundary layer equations. JAS *21*, 487—495 (1954).

[98] Tifford, A.N.: Simplified compressible laminar boundary layer theory. JAS *18*, 358—359 (1951).

[99] Toll, T.A., and Fischel, G.: The X-15 project-results and new research. Astronautics and Aeronautics, *2*, 25—32 (1964).

[100] Watson, R.D., and Weinstein, L.M.: A study of hypersonic corner flow interactions. AIAA J. *9*, 1280—1286 (1971).

[101] Werle, M.J., Polak, A., Vatsa, V.N., and Bertke, S.D.: Finite difference solutions for supersonic separated flows. AGARD Conf. Proc. Flow Separation, No. 168, 8—1 to 8—12 (1975).

[102] West, J.E., and Korgegi, R.H.: Supersonic interaction in the corner of intersecting wedges at high Reynolds numbers. AIAA J. *10*, 652—656 (1972).

[103] Yamaga, J.: An approximate solution of the laminar boundary layer on a rotating body of revolution in uniform compressible flow. Proc. 6th Japan Nat. Congr. Appl. Mech. Univ. Kyoto, Japan, 295—298 (1956).

[104] Yang, K.T.: An improved integral procedure for compressible laminar boundary layer analysis. J. Appl. Mech. *28*, 9—20 (1961).
[105] Young, A.D.: Section on "Boundary Layers" in: Modern developments in fluid mechanics. High speed flow (L. Howarth, ed.), *I*, 375—475, Clarendon Press, Oxford, 1953.
[106] Young, A.D.: Skin friction in the laminar boundary layer of a compressible flow. Aero. Quart. *I*, 137—164 (1949).
[107] Young, A.D.: Boundary layers and skin friction in high speed flow. J. Roy. Aero. Soc. *55*, 285—302 (1951).
[108] Young, G.B.W., and Janssen, E.: The compressible boundary layer. JAS *19*, 229—236, 288 (1952).
[109] Young, A.D., and Harris, H.D.: A set of similar solutions of the compressible laminar boundary layer equations for the flow over a flat plate with unsteady wall temperature. ZFW *15*, 295—301 (1967).
[110] Zaat, J.A.: A one-parameter method for the calculation of laminar compressible boundary layer flow with a pressure gradient. Nat. Luchtv. Lab. Amsterdam, Rep. F *141* (1953).

CHAPTER XIV

Boundary-layer control in laminar flow†

a. Methods of boundary-layer control

There are in existence several methods which have been developed for the purpose of artificially controlling the behaviour of the boundary layer. The purpose of these methods is to affect the whole flow in a desired direction by influencing the structure of the boundary layer. As early as in his first paper published in 1904, L. Prandtl described several experiments in which the boundary layer was controlled. He intended to prove the validity of his fundamental ideas by suitably designed experiments and achieved quite remarkable results in this way. Figure 14.1 shows the flow past a circular cylinder with suction applied on one side of it through a small slit. On the suction side the flow adheres to the cylinder over a considerably larger portion of its surface and separation is avoided; the drag is reduced appreciably, and simultaneously a large cross-force is induced owing to the lack of symmetry in the flow pattern.

Fig. 14.1. Flow past a circular cylinder with suction on one side, after Prandtl

As demonstrated in Chap. X, a laminar boundary layer can support only very small adverse pressure gradients without the occurence of separation. In the case of turbulent flow the danger of separation is intrinsically reduced, compared with laminar flow, because owing to the turbulent mixing motion there is a continuous flow of momentum from the external flow towards the wall. Nevertheless, even in turbulent flow it is often desirable to prevent separation by adopting suitable *boundary-layer control* measures. The problem of boundary-layer control has become very important for a time, in particular in the field of aeronautical engineering;

---

† Professor Dr. W. Wuest assisted in the preparation of the new version of this chapter for the Fifth Edition of this book.

in actual applications it is often necessary to prevent separation in order to reduce drag and to attain high lift. Several methods of controlling the boundary layer have been developed experimentally, and also on the basis of theoretical considerations [6, 75, 76]. These can be classified as follows:

1. Motion of the solid wall
2. Acceleration of the boundary layer (blowing)
3. Suction
4. Injection of a different gas (binary boundary layers)
5. Prevention of transition to turbulent flow by the provision of suitable shapes (laminar aerofoils)
6. Cooling of the wall.

Methods 1 to 4 will be discussed in this chapter. Methods 5 and 6 will be described in Chap. XVII in connexion with the consideration of the theory of transition from laminar to turbulent flow.

The treatise entitled "Boundary-Layer and Flow Control" [44] by G. V. Lachmann contains a summary of the subject of boundary-layer control according to the state of research at the time; compare also P. K. Chang [12a]. Until the end of the Second World War, the problems under consideration were worked on almost exclusively in Germany; the corresponding achievements have been reported on by A. Betz [9]. The development of this subject which took place in other countries since the end of the Second World War has been summarized in ref. [44] as well as in [27, 36, 65, 104].

The present chapter will principally deal with the problems of control in laminar boundary layers. Problems related to turbulent boundary layers will be studied in Sec. XXIIb 6.

Fig. 14.2. Flow past a rotating cylinder

1. Motion of the solid wall. The most obvious method of avoiding separation is to attempt to prevent the formation of a boundary layer. Since a boundary layer owes its existence to the difference between the velocity of the fluid and that of the solid wall, it is possible to eliminate the formation of a boundary layer by attempting to suppress that difference, i. e. by causing the solid wall to move with the stream. The simplest way of achieving such a result involves the rotation of a circular cylinder. Figure 14.2 shows the flow pattern which exists about a rotating cylinder placed in a stream at right angles to its axis. On the upper side, where the flow and the cylinder move in the same direction, separation is completely eliminated. Furthermore, on the lower side where the direction of fluid motion is opposite to that of the solid wall, separation is developed only incompletely. On the whole, the flow pattern which exists in this case approximates very closely the pattern of frictionless flow past a circular cylinder with circulation. The stream exerts a considerable force on the cylinder at right angles to the mean flow direction, and this is sometimes referred to as the Magnus effect. This effect can be seen, e. g., when a tennis ball is 'sliced' in play. Attempts were also made to utilize the occurrence of lift on rotating cylinders for the propulsion of ships (Flettner's rotor [1]). With the exception of rotating cylinders, the idea of moving the solid wall with the stream can be realized only at the cost of very great complications as far as shapes other than cylindrical are concerned, and consequently, this method has not found much practical application. Nevertheless, A. Favre [26] made a thorough experimental investigation of the influence of a moving boundary on an aerofoil. A portion of the upper surface of the aerofoil was formed into an endless belt which moved over two rollers so that the return motion occurred in the interior of the model. The arrangement proved very effective for the avoidance of separation, and yielded very high maximum lift coefficients ($C_{Lmax} = 3.5$) at high angles of incidence ($\alpha = 55°$). The laminar boundary layer for a flat plate moving in its rear part with the stream has been calculated by E. Truckenbrodt [100].

2. Acceleration of the boundary layer (blowing). An alternative method of preventing separation consists in supplying additional energy to the particles of fluid which are being retarded in the boundary layer. This result can be achieved by discharging fluid from the interior of the body with the aid of a special blower (Fig. 14.3a), or by deriving the required energy directly from the main stream. This latter effect can be produced by connecting the retarded region to a region of higher pressure through a slot in the wing (slotted wing, Fig. 14.3b). In either case additional energy is imparted to the particles of fluid in the boundary layer near the wall. When fluid is discharged, say in the manner shown in Fig. 14.3a, it is mandatory to pay careful attention to the shape of the slit in order to prevent the jet from dissolving into vortices at a short distance behind the exit section. Later experiments performed in France [64] have made it very attractive to apply blowing at the trailing edge of an aerofoil in order to increase its maximum lift. Attempts considerably to increase the maximum lift of a flap wing through blowing in the slot have also met with success (cf. Sec. XIIb 6).

In the case of the slotted wing [7], shown in Fig. 14.3b, the effect is produced as follows: The boundary layer formed on the forward slat A — B is carried into the main stream before separation occurs, and from point C onwards a new boundary layer is formed. Under favourable conditions this new boundary layer will reach the trailing edge D without separation. In this way it is possible to relegate separation to considerably larger angles of incidence, and to achieve much larger lifts. Fig. 14.4 shows a polar diagram (lift coefficient plotted against drag coefficient) for a wing section with and without forward slat and flap. The phenomena in the slot formed by the flap near the trailing edge are, in principle, the same as those at the forward slat. The gain in lift is seen to be very considerable.

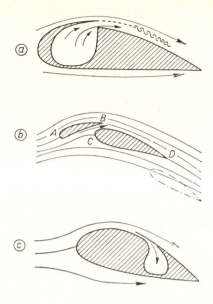

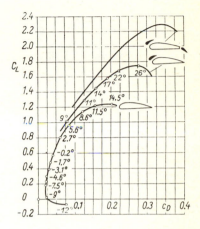

Fig. 14.3. Different arrangements for boundary-layer control.   a) discharge of fluid,
b) slotted wing,   c) suction

Fig. 14.4.   Polar diagram of a
wing with forward slat and flap

A review of recent work on control through blowing is contained in ref. [13].

3. Suction. The effect of suction consists in the removal of decelerated fluid particles from the boundary layer before they are given a chance to cause separation, Fig. 14.3c. A new boundary layer which is again capable of overcoming a certain adverse pressure gradient is allowed to form in the region behind the slit. With a suitable arrangement of the slits and under favourable conditions separation can be prevented completely. Simultaneously, the amount of pressure drag is greatly reduced owing to the absence of separation. The application of suction, which was first tried by L. Prandtl (Fig. 14.1), was later widely used in the design of aircraft wings. By applying suction, considerably greater pressure increases on the upper side of the aerofoil (i. e. lower absolute pressure) are obtained at large angles of incidence, and, consequently, much larger maximum lift values. O. Schrenk [85] investigated a large number of different arrangements of suction slits and their effect on maximum lift; see also ref. [104].

In more recent times suction was also applied to reduce drag. By the use of suitable arrangements of suction slits it is possible to shift the point of transition in the boundary layer in the downstream direction; this causes the drag coefficient to decrease, because laminar drag is substantially smaller than turbulent drag, Fig. 14.9. The effect of the *delay in transition* caused by suction is to reduce the boundary-layer thickness which then becomes less prone to turning turbulent [3]. Furthermore, the velocity profiles in a boundary layer with suction, being fuller (Fig. 14.6), have forms which are less likely to induce turbulence compared with those in laminar boundary layers without suction and of equal thickness. Problems connected with the phenomenon of transition, and in particular those associated with suction, will be discussed more fully in Chap. XVII.

4. Injection of a different gas. The injection of a light gas, which is different from that in the external stream, through a porous wall into the boundary layer reduces the rate at which heat is exchanged between the wall and the stream [34]. This is the most important one of the effects produced this way, and for this reason, an arrangement of this kind is often used to provide thermal protection at high supersonic velocities. Injection creates a gaseous mixture in the boundary layer, and to the processes of momentum and heat transfer there is added the process of mass transfer by diffusion. Generally speaking, the thermal diffusion must not be neglected with respect to the diffusion along concentration gradients. Similar processes arise when a liquid film evaporates at the wall or when the material of the wall itself melts or sublimates. The latter process is described by the term *ablation*; we shall revert to it in Sec. XIVc.

5. Prevention of transition by the provision of suitable shapes. Laminar aerofoils. Transition from laminar to turbulent flow can also be delayed by the use of suitably shaped bodies. The object, as in the case of suction, is to reduce frictional drag by causing the point of transition to move downstream. It has been established that the location of the point of transition in the boundary layer is strongly influenced by the pressure gradient in the external stream. With a decrease in pressure, transition occurs at much higher Reynolds numbers than with pressure increase. A decrease in pressure has a highly stabilizing effect on the boundary layer, and the opposite is true of an increase in pressure along the stream. This circumstance is utilized in modern low-drag aerofoils. The desired result is achieved by displacing the section of maximum thickness far rearwards. In this manner a large portion of the aerofoil remains under the influence of a pressure which decreases downstream and a laminar boundary layer is maintained. We shall revert to this question in Chap. XVII.

6. Cooling of the wall. In a certain range of supersonic Mach numbers it is possible completely to stabilize the boundary layer by the application of cooling at the wall (*cf.* Sec. XVII e). Cooling can also be applied in order to reduce the thickness of the boundary layer, and this possibility may become important, e. g. when gases of very low density are made to flow through the nozzles of wind tunnels, because otherwise the very thick boundary layers would unacceptably reduce the useful area of the test section.

The method of boundary-layer control by suction, together with the prevention of transition on laminar aerofoils, have the greatest practical importance among all the methods discussed previously. For this reason various mathematical methods for the calculation of the influence of suction on boundary-layer flow have been developed, and we now propose to review them briefly.

b. Boundary-layer suction

1. Theoretical results

1.1. Fundamental equations. It is simplest to begin the mathematical study of the laminar boundary layer with suction by first considering the case with continuous suction which may be imagined realized with the aid of a porous wall. The usual system of coordinates will be adopted, the x-axis being along the wall, and the y-axis being at right angles to it, Fig. 14.5. Suction will be accounted for by prescribing a non-zero normal velocity component $v_0(x)$ at the wall; in the case of

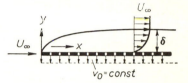

Fig. 14.5. Flat plate with homogeneous suction at zero incidence

suction we shall put $v_0 < 0$, making $v_0 > 0$ for discharge. It will be assumed that the quantity of fluid removed from the stream is so small that only fluid particles in the immediate neighbourhood of the wall are sucked away. This is equivalent to saying that the ratio of suction velocity $v_0(x)$ to free-stream velocity U_∞ is very small, say $v_0/U_\infty = 0.0001$ to 0.01†. The condition of no slip at the wall is retained with suction present, as well as the expression $\tau_0 = \mu(\partial u/\partial y)_0$ for the shearing stress at the wall. The quantity of fluid removed, Q, will be expressed through a dimensionless volume coefficient by putting

$$Q = c_Q \, A \, U_\infty, \tag{14.1}$$

where A denotes the wetted area. For the flat plate $Q = b \int_0^l [-v_0(x)] \, dx$ and $A = b \, l$ so that consequently

$$c_Q = \frac{1}{l U_\infty} \int_0^l [-v_0(x)] \, dx, \tag{14.2}$$

and for the case of uniform suction, $v_0 = \text{const}$,

$$c_Q = \frac{-v_0}{U_\infty}. \tag{14.2a}$$

---

† In order to ensure that a flow with suction, or blowing, at the wall satisfies the simplifying conditions which form the basis of boundary-layer theory, it is necessary to limit the velocity v_0 at the wall to a magnitude of the order of $U_\infty R^{-1/2}$, where $R = U_\infty l/\nu$ and l denotes a characteristic dimension of the solid body placed in the flow. At $R = 10^6$ this condition gives $v_0 \sim 0.001 \, U_\infty$. When the suction velocity is of such a small order of magnitude, it is possible to neglect the loss of mass or "sink-effect" on the external potential flow. In other words, the potential flow may be assumed to remain unaffected by such blowing or suction applied at the surface of the solid body.

Assuming incompressible two-dimensional flow we have the following differential equations

$$\frac{\partial u}{\partial x} + \frac{\partial v}{\partial y} = 0 \; ,$$

$$u\frac{\partial u}{\partial x} + v\frac{\partial u}{\partial y} = -\frac{1}{\varrho}\frac{dp}{dx} + \nu\frac{\partial^2 u}{\partial y^2} \; ,$$

\right\} \quad (14.3)

with the boundary conditions

$$y = 0 : \quad u = 0, \quad v = v_0(x) \; ;$$

$$y = \infty : \quad u = U(x) \; .$$

\right\} \quad (14.4)

Evidently, the integration of the above system of equations for the general case of arbitrary body shape, implying an arbitrary velocity function $U(x)$, presents no fewer difficulties than does the case with no suction.

Nevertheless, the qualitative effect of suction on separation can be estimated with the aid of the preceding equations even without integration. Along the streamline at the wall ($y = 0$), equations (14.3) and (14.4) yield

$$\nu\left(\frac{\partial^2 u}{\partial y^2}\right)_{y=0} = \frac{1}{\varrho}\frac{dp}{dx} + v_0\left(\frac{\partial u}{\partial y}\right)_{y=0} \; . \qquad (14.5)$$

It is seen that in a region of adverse pressure gradient ($dp/dx > 0$), the superposition of suction ($v_0 < 0$) reduces the curvature of the velocity profile at the wall. According to the arguments advanced in Chap. VII, this signifies that the point of separation is displaced rearwards. Now, in accordance with the theory which will be given in Chap. XVII, this has the additional effect of stabilizing the laminar boundary layer. These two effects produced by suction, namely avoidance of separation and the relegation of the point of laminar-turbulent transition to higher Reynolds numbers, have been confirmed by the results of experiments.

A summary of methods used for the calculation of boundary layers with suction was published by W. Wuest [108].

1.2. Exact solutions. The method of using a power-series expansion in terms of the length of arc for the potential velocity (Blasius series) described in Sec. IX c can, in principle, be applied in this case as well. However, just as in the case without suction, the resulting computations become very laborious [75]. Reasonable simple solutions can be obtained only in the case of a flat plate at zero incidence.

Flat plate: A surprisingly simple solution can be obtained in the case of a *flat plate at zero incidence* with *uniform suction*, Fig. 14.5. The system of differential equations now reduces to

$$\frac{\partial u}{\partial x} + \frac{\partial v}{\partial y} = 0 \; ; \qquad (14.5\,\mathrm{a})$$

$$u\frac{\partial u}{\partial x} + v\frac{\partial u}{\partial y} = \nu\frac{\partial^2 u}{\partial y^2} \; , \qquad (14.5\,\mathrm{b})$$

with the boundary conditions: $u = 0$, $v = v_0 = \text{const} < 0$ for $y = 0$, and $u = U_\infty$ for $y = \infty$. It can be seen at once that this system possesses a particular solution for which the velocity is independent of the current length x [52, 78]. Putting $\partial u/\partial x \equiv 0$ we see from the equation of continuity that $v(x, y) = v_0 = \text{const}$. Hence the equation of motion becomes $v_0 \, \partial u/\partial y = \nu \, \partial^2 u/\partial y^2$, with the solution

$$u(y) = U_\infty \left[1 - \exp\left(v_0 y/\nu\right)\right] ; \quad v(x, y) = v_0 < 0 . \qquad (14.6)$$

It is worth noting that this simple solution is even an exact solution of the complete Navier-Stokes equations. The displacement thickness and the momentum thickness are

$$\delta_1 = \frac{\nu}{-v_0} ; \qquad \delta_2 = \frac{1}{2} \frac{\nu}{-v_0} ; \qquad (14.7); \ (14.8)$$

and the shearing stress at the wall $\tau_0 = \mu (\partial u/\partial y)_0$ becomes simply

$$\tau_0 = \varrho(-v_0) \, U_\infty , \qquad (14.9)$$

and is independent of viscosity. The velocity distribution is seen plotted in Fig. 14.6, curve I. Curve II, drawn for the purpose of comparison, represents the Blasius velocity distribution without suction. It should be noted that the suction profile is fuller. The solution thus discovered can be realized on a flat plate at zero incidence with uniform suction only at some distance from the leading edge, even if suction is applied from the leading edge onwards. The boundary layer, evidently, begins to grow from zero thickness at the leading edge and continues downstream tending asymptotically to the value given in eqn. (14.7). The velocity profile attains the simple form given by eqn. (14.6) only asymptotically, i. e. from the practical point of view after a certain initial length. For these reasons the preceding particular solution may be regarded as the *asymptotic suction profile*.

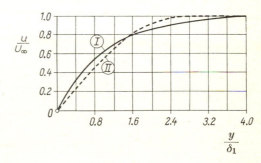

Fig. 14.6. Velocity distribution in the boundary layer on a flat plate at zero incidence

I. uniform suction; 'asymptotic suction profile'
II. no suction; 'Blasius profile'

A more detailed investigation into the flow in the *initial length*, i. e., before the asymptotic state has been reached, was carried out by R. Iglisch [40] who has shown that the asymptotic state is reached after a length of about

$$\left(\frac{-v_0}{U_\infty}\right)^2 \frac{U_\infty x}{\nu} = 4 \quad \text{or} \quad c_Q \sqrt{R_x} = 2 .$$

The velocity profiles in the initial length are not similar among themselves. They are practically identical with those for the case with no suction at short distances

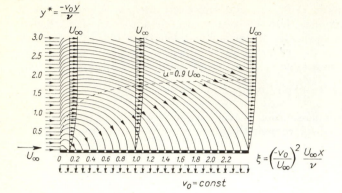

$$\xi = \left(\frac{-v_0}{U_\infty}\right)^2 \frac{U_\infty x}{\nu}$$

Fig. 14.7. Flat plate with uniform suction; pattern of streamlines

from the leading edge (Blasius profile, Fig. 7.7). The pattern of streamlines in the initial length is seen drawn in Fig. 14.7, and the velocity profiles are seen plotted in Fig. 14.8. The way in which the boundary-layer thickness increases from zero at the leading edge to its asymptotic value given in eqn. (14.7) is described by the values in Table 14.1 which have been taken from R. Iglisch's paper.

Particular interest is attached to the *saving in drag* caused by preserving laminar flow with the aid of suction, and, therefore, to the law of friction for the

Table 14.1. Dimensionless boundary-layer thickness δ_1 and shape factor δ_1/δ_2 for the velocity profiles in the initial length on a flat plate at zero incidence with uniform suction, after R. Iglisch [40]

$$\xi = \left(\frac{-v_0}{U_\infty}\right)^2 \frac{U_\infty x}{\nu} \; ; \quad \begin{array}{l} \delta_1 \text{ — displacement thickness;} \\ \delta_2 \text{ — momentum thickness} \end{array}$$

ξ	$\dfrac{-v_0\,\delta_1}{\nu}$	$\dfrac{\delta_1}{\delta_2}$
0	0	2·59
0·005	0·114	2·53
0·02	0·211	2·47
0·045	0·303	2·43
0·08	0·381	2·39
0·125	0·450	2·35
0·18	0·511	2·31
0·245	0·566	2·28
0·32	0·614	2·25
0·405	0·658	2·23
0·5	0·695	2·21
0·72	0·761	2·17
0·98	0·812	2·14
1·28	0·853	2·11
2·0	0·911	2·07
2·88	0·948	2·05
5·12	0·983	2·01
8·0	0·996	2·00
∞	1	2

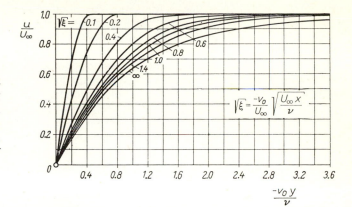

Fig. 14.8. Flat plate with uniform suction; velocity profiles over initial length, after Iglisch [40]

The curve $\xi = \infty$ corresponds to the 'asymptotic suction profile' of eqn. (14.6)

case with suction. This is seen plotted in Fig. 14.9. In the case of very large Reynolds numbers $U_\infty l/\nu$, when the major portion of the plate falls within the region of the asymptotic solution, the drag is given by the simple equation (14.9), whence we can obtain the local drag coefficient

$$c'_{f\infty} = \frac{\tau_0}{\frac{1}{2}\varrho\,U_\infty{}^2} = 2\,\frac{-v_0}{U_\infty} = 2\,c_Q\,. \tag{14.10}\dagger$$

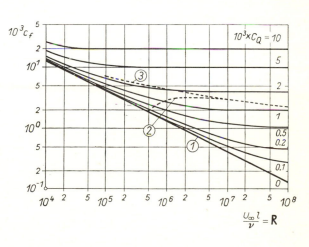

Fig. 14.9. Drag coefficients for the flat plate at zero incidence with uniform suction

$c_Q = (-v_0)/U_\infty$ = volume coefficient of suction

Curves (1), (2) and (3) refer to no suction

(1) laminar
(2) transition from laminar to turbulent
(3) fully turbulent

---

† This drag is completely independent of viscosity. With $D = \tau_0\,b\,l$ and $Q = (-v_0)\,b\,l$ we find from eqn. (14.9) that

$$D = \varrho\,Q\,U_\infty\,.$$

This is the drag due to sinking, i.e. the drag experienced by a body which is placed in a frictionless stream of velocity U_∞ and which 'swallows' a quantity Q of fluid. The above expression can be deduced very simply by the application of the momentum theorem (cf. Prandtl-Tietjens, Hydro- u. Aeromechanik, vol. II, 1931, p. 140, Engl. transl. by J. P. den Hartog, 1934.

The drag coefficient is larger for small Reynolds numbers, because the shearing stress is greater over the front portion of the plate, i. e. that which falls within the initial region and where the boundary layer is thinner than further down-stream. The drag on a plate with a turbulent boundary layer with no suction is shown plotted in Fig. 14.9 for the purpose of comparison. It will be discussed more fully in Chap. XXI. The saving in drag can be deduced from this diagram only if the value of the smallest volume coefficient of suction which is capable of ensuring laminar conditions in the boundary layer at large Reynolds numbers is known. This problem will be investigated in Chap. XVII, together with the phenomenon of transition. It will then be shown that there exists a curve of 'most favourable suction'; it can be seen plotted in Fig. 17.19. It will be noticed that the reduction in drag through suction is very considerable and that the required intensity of suction is very small, as it corresponds to values of the order $c_Q = 10^{-4}$. A solution for the flat plate with uniform suction in a compressible stream was found by H. G. Lew and J. B. Fanucci [47]; the same problem for cylindrical bodies of arbitrary cross-section was solved by W. Wuest [107].

J. M. Kay [41a] undertook to verify these theoretical results for the flat plate at zero incidence with the aid of experiments. The assumption that uniform suction begins at the leading edge, which formed the basis of Iglisch's theoretical calculations, was not satisfied in the test plate. The latter, moreover, had a portion near the leading edge completely devoid of suction. Figure 14.10 shows a comparison between the measured and calculated displacement thickness and momentum thickness respectively. The asymptotic values from eqns. (14.7) and (14.8) are seen to have been confirmed by the measurements. Figure 14.11 shows a comparison between theory and measurement for various values of ξ; the measurements have been performed by M. R. Head [35]. Again, the agreement is very satisfactory. Measurements performed by P. A. Libby, L. Kaufmann and R. P. Harrington [48] confirm, in addition, the strong stabilizing effect caused by suction (increase in the critical Reynolds number), as will be reported more fully in Sec. XVIIc. The large decrease in the skin friction which results from the preservation of laminar flow when suction is applied, and which is shown in Fig. 14.9, was confirmed by measurements performed by M. Jones and M. R. Head [41], and A. Raspet [70].

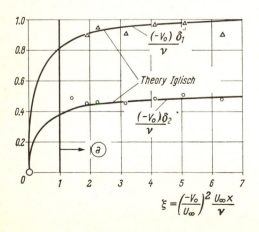

Fig. 14.10. Laminar boundary layer on a flat plate at zero incidence with uniform suction. Displacement thickness δ_1 and momentum thickness δ_2 have been measured by J. M. Kay [41a]. Theoretical curves after R. Iglisch [40], Table 14.1

a = section at which suction begins

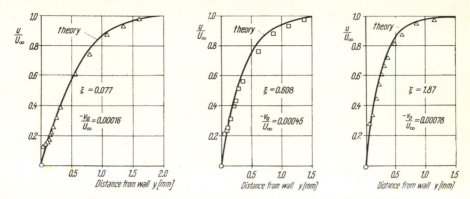

Fig. 14.11. Velocity distribution in the laminar boundary layer on an aerofoil with suction applied through its porous surface. Measurements performed by M. R. Head [35]; comparison with the theory due to R. Iglisch [40]

Boundary layer with pressure gradient: Additional *exact solutions* of the boundary-layer equations (14.3) and (14.4) are known only for flow patterns which can be associated with similar velocity profiles. The class of similar solutions discussed in Chap. VIII can be extended to include boundary layers with suction and blowing. When the velocity in the external stream can be described by the function $U(x) = u_1 x^m$ and when the suction velocity $v_0(x)$ is proportional to $x^{(1/2)(m-1)}$, we recover from the boundary-layer equations the already familiar ordinary differential equation for the stream function $f(\eta)$, first derived by Falkner and Skan, namely the familiar equation (9.8):

$$f''' + ff'' + \beta(1-f'^2) = 0$$

in which η has been defined in eqn. (9.5). That this is so can be inferred by inspection from eqn. (9.5). In the present case, the stream function $f(\eta)$ has a value which is different from zero at the wall when $\eta = 0$. This value is positive in the case of suction and negative for blowing.

The particular case for $m = 0$ which corresponds to a flat with a suction velocity

$$v_0(x) = -\tfrac{1}{2} C \sqrt{\frac{\nu\, U_\infty}{x}}\;; \qquad \begin{array}{l} C > 0 : \text{suction} \\ C < 0 : \text{blowing} \end{array} \qquad (14.11)$$

was investigated by H. Schlichting and K. Bussmann [79, 80]. The resulting velocity profiles for several values of the volume coefficient have been plotted in Fig. 14.12. It is worth noting that all velocity profiles for the case of discharge have points of inflexion with $\partial^2 u/\partial y^2 = 0$. This fact is important for the study of transition (Chap. XVI). Similar velocity profiles are also obtained in the case of two-dimensional stagnation flow with a velocity function $U(x) = u_1 x$ with suction, provided that $v_0 = $ const. This case was also investigated in the paper by H. Schlichting and K. Bussmann already quoted.

Extensive tables for boundary layers on a plate with suction ($m = 0$) covering a wide range of values of the parameter C were calculated by H. W. Emmons and D. C. Leigh [22] as well as by J. Steinheuer (*cf.* Chap. VII). For cases when $m \neq 0$ there exist additional numerical solutions extending over a wide range of values of the parameters [57]. The diagram in Fig. 14.13 presents the relation between the shearing stress at the wall which is proportional to $f''(0)$, the suction velocity — which is proportional to $f(0)$ — and the parameter β of the external flow. The position of the point of separation is determined by the parameter for which $\tau_0 = 0$, that is by the condition that $f''(0) = 0$. It is clear from Fig. 14.13 that separation can be eliminated by vigorous suction even in strongly decelerated flows (e. g. when $\beta = -1$, i. e. when $m = -\tfrac{1}{3}$).

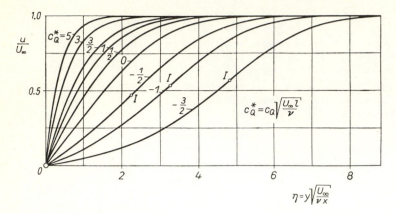

Fig. 14.12.   Velocity distribution in the boundary layer on a flat plate at zero incidence with suction and discharge according to the law $v_0(x) \sim 1/\sqrt{x}$ from eqn. (14.11), after H. Schlichting and K. Bussmann [79]

$C = c_Q^* =$ reduced volume coefficient of suction. $c_Q^* > 0$: suction; $c_Q^* < 0$: discharge; $I =$ point of inflexion

When the mass flow in blowing is made large, it is observed that the corresponding numerical calculation becomes difficult, because the velocity profile acquires a kink. This detail was first discovered by J. Pretsch [69] who derived it from a consideration of the asymptotic solution. The asymptotic behaviour of the preceding similar solutions for large suction velocities has been investigated by E. J. Watson [102].

The solutions for the external flow corresponding to $U(x) = u_1 x^m$ form the basis for a series of further investigations aimed at discovering additional exact solutions for laminar boundary layers with suction and blowing:

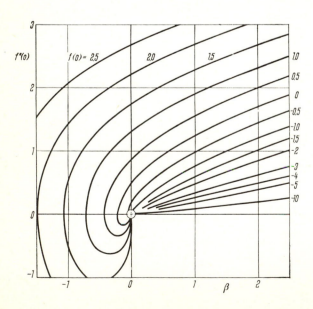

Fig. 14.13. Relation between shearing stress at the wall, τ_0, and the suction velocity, v_0, for laminar boundary layers with suction when the external velocity is $U(x) = u_1 x^m$, after K. Nickel [57].

The position of the point of separation is determined by the condition that $\tau_0 = 0$, i.e. by $f''(0) = 0$. Note that

$$\tau_0/\varrho\, U^2 = \sqrt{\frac{m+1}{2}}\ \sqrt{\frac{\nu}{U x}}\ f''(0)\ ,$$

$$v_0/U = -\sqrt{\frac{m+1}{2}}\ \sqrt{\frac{\nu}{U x}}\ f(0)\ .$$

$f(0) > 0$ denotes suction, $f(0) < 0$ denotes blowing, and

$$\beta = \frac{2\,m}{m+1}\ ,\ \text{see eqn. (9.7)}$$

a) K. D. P. Sinhar [86] studied the case of an infinitely long, yawed cylinder with suction. The velocity distribution along the stream was assumed to be proportional to x^m. The investigation has some bearing on the control of the boundary layer on swept wings.

b) When the temperature of the fluid being blown out is different from that in the external flow, the boundary layer will develop a temperature profile; the resulting thermal boundary layer was calculated in refs. [55] and [111]. The knowledge of the temperature distribution in the boundary layer is of particular importance for the problem of cooling. It turns out that cooling by means of blowing by means of the coolant through a porous wall, so-called transpiration cooling, is much more effective than cooling the wall on the inside. In this connexion the papers by B. Brown [11, 12], P. L. Donoughe and J. N. B. Livingood [19] and W. Wuest [109] may be consulted.

c) The cooling problem becomes very important at high velocities of flow. G. M. Low [51] found solutions for the case of compressible flow over an isothermal flat plate; see also [45, 110].

Compressible boundary layers with suction: It was shown by A. D. Young [112] and H. G. Lew [45] that an asymptotic solution exists also for the case of *compressible flow* along a flat plate at zero incidence in the presence of homogeneous suction. This can be done as follows: According to eqns. (13.5) and (13.6), the continuity and momentum equations can be written

$$\frac{d(\varrho\, v)}{dy} = 0 \qquad\qquad (14.12)$$

and

$$\varrho\, v\, \frac{du}{dy} = \frac{d}{dy}\left(\mu\, \frac{du}{dy}\right). \qquad\qquad (14.13)$$

It follows from eqn. (14.12) that

$$\varrho\, v = \varrho_0\, v_0 = \text{const},$$

and, consequently, we obtain from eqn. (14.13) that

$$\frac{du}{u - U_\infty} = \frac{v_0\, \varrho_0}{\mu}\, dy. \qquad\qquad (14.14)$$

Assuming the validity of the viscosity law $\mu/\mu_\infty = C T/T_\infty$, we see that $\varrho\, \mu = \varrho_0\, \mu_0$. Thus eqn. (14.14) integrates to

$$u(y) = U_\infty\left\{1 - \exp\left(\frac{v_0\, \varrho_\infty\, y_1}{\mu_0}\right)\right\} \qquad\qquad (14.15)$$

where

$$y_1 = \int_0^y \frac{\varrho}{\varrho_\infty}\, dy. \qquad\qquad (14.16)$$

The preceding relations are valid for an arbitrary value of the Prandtl number. The shearing stress at the wall is now

$$\tau_0 = \varrho_0(-v_0)\, U_\infty, \qquad\qquad (14.17)$$

in agreement with eqn. (14.9). When $P = 1$ and the wall is adiabatic, it is possible further to specialize eqns. (14.15) and (14.16). The calculation yields the following explicit expression:

$$\frac{v_0\, \varrho_\infty\, y}{\mu_0} = \left(\frac{T_a}{T_\infty} - 1\right)\left\{\frac{1}{2}\left(\frac{u}{U_\infty}\right)^2 + \frac{u}{U_\infty}\right\} - \ln\left(1 - \frac{u}{U_\infty}\right), \qquad\qquad (14.18)$$

($P = 1$; adiabatic wall). When the flow is incompressible, we have $T_a = T_\infty$, and eqn. (14.18) reduces to eqn. (14.6).

1.3. Approximate solutions. In the general case of an arbitrary body shape and an arbitrary law of suction we must resort to approximate methods based on the momentum equation; they were described in Chap. X. The momentum equation for the case with suction is obtained in exactly the same way as before, except that it is now necessary to take into account the fact that the normal component of the velocity at the wall differs from zero. Performing the same calculation as in Sec. VIIIe, we find that the equation for the normal component of velocity at a distance $y = h$ from the wall now becomes

$$v_h = v_0 - \int_0^h \frac{\partial u}{\partial x} \, \mathrm{d}y \, .$$

The calculation is continued in exactly the same way as in Sec. VIIIe, and leads finally to the following momentum equation for the boundary layer with suction

$$\frac{\mathrm{d}}{\mathrm{d}x} (U^2 \, \delta_2) + \delta_1 \, U \, \frac{\mathrm{d}U}{\mathrm{d}x} - v_0 \, U = \frac{\tau_0}{\varrho} \, , \tag{14.19}$$

the energy-integral equation, according to K. Wieghardt [103], assuming the form:

$$\frac{\mathrm{d}}{\mathrm{d}x} (U^3 \, \delta_3) - v_0 \, U^2 = 2 \int_0^\infty \frac{\tau}{\varrho} \, \frac{\partial u}{\partial y} \, \mathrm{d}y \, . \tag{14.20}$$

The additional terms, as compared with eqns. (8.32) and (8.35), represent the change in momentum or energy, respectively, due to suction at the wall.

Equation (14.19) was used by L. Prandtl [67] to make a simple estimate of the suction velocity which is just sufficient to prevent separation. Assuming that the velocity profiles along the whole length are identical with that at the point of separation, i. e. that for which $\tau_0 = \mu (\partial u/\partial y)_0 = 0$, and that, as assumed by Pohlhausen, $\varLambda = -12$, we can deduce from eqn. (10.22) that the velocity is given by

$$u = U \left\{ 6 \left(\frac{y}{\delta} \right)^2 - 8 \left(\frac{y}{\delta} \right)^3 + 3 \left(\frac{y}{\delta} \right)^4 \right\} \, .$$

The displacement and momentum thickness follow from eqn. (10.24), and are, respectively

$$\delta_1 = \frac{2}{5} \, \delta \; ; \qquad \delta_2 = \frac{4}{35} \, \delta \, ,$$

so that

$$\delta_1 + 2 \, \delta_2 = \frac{22}{35} \, \delta \, .$$

Substituting this value into eqn. (14.19) and taking into account that $\mathrm{d}\delta_2/\mathrm{d}x = 0$, because of the assumption of constant boundary-layer thickness, we have

$$v_0 = \frac{22}{35} \, \delta \, \frac{\mathrm{d}U}{\mathrm{d}x} \, . \tag{14.20a}$$

Further, by eqn. (14.5) we have at $y = 0$:

$$v_0 \left(\frac{\partial u}{\partial y}\right)_0 = U \frac{dU}{dx} + v \left(\frac{\partial^2 u}{\partial y^2}\right)_0. \tag{14.21}$$

In the case under consideration $(\partial u/\partial y)_0 = 0$ and $(\partial^2 u/\partial y^2)_0 = 12\, U/\delta^2$. Hence we obtain from eqn. (14.21) that

$$\delta = \sqrt{\frac{12\, v}{(-\, dU/dx)}}, \tag{14.22}$$

and from eqns. (14.20a) and (14.22) that

$$v_0 = -\, 2{\cdot}18 \sqrt{v \left(-\frac{dU}{dx}\right)}. \tag{14.23}$$

This velocity of suction is seen to be just sufficient to prevent separation all along the wall. Taking as an example the case of flow past a circular cylinder of radius R with $dU/dx = -\, 2\, U_\infty/R$ at the downstream stagnation point, and applying eqn. (14.23), we obtain that the volume coefficient which must be used to prevent separation is given by

$$c_Q \sqrt{\frac{U_\infty R}{v}} = 2{\cdot}18 \sqrt{2} = 3{\cdot}08\,.$$

H. Schlichting [77, 81] indicated an approximate method for the calculation of the boundary layer on a body of arbitrary shape, with arbitrary suction $v_0(x)$ applied. The method is similar to the Kármán-Pohlhausen method and is based on the use of the momentum equation. T. P. Torda [98] made improvements in this method. Papers by L. Trilling [99], B. Thwaites [39, 96], and F. Ringleb [74] contain descriptions of procedures suitable for arbitrary pressure distributions as well as arbitrary distributions of the suction velocity. K. Wieghardt [103] extended them to the case of axially symmetrical bodies and J.T. Stuart [88a] solved the case of a rotating disk. E. Truckenbrodt [101] developed an approximate method which is suitable in the two-dimensional as well as in the axially symmetrical case and which excels the other methods in its outstanding simplicity. The whole problem has been reduced here to the solution of an ordinary differential equation of the first order. The equation becomes identical with that given by A. Walz (cf. Sec. Xb) in the limiting case of zero suction, i. e. when the wall is impermeable. The results of calculations for a Zhukovskii aerofoil, performed with the aid of this method, are represented graphically in Fig. 14.14. It is seen that the point of separation moves towards the trailing edge as the intensity of suction increases, and that no separation occurs at all from a certain suction intensity onwards.

R. Eppler [23] worked out an approximate method for the calculation of laminar and turbulent boundary layers with suction which is well-suited for pro- gramming on a digital computer. Corresponding approximate methods for com- pressible boundary layers with suction and blowing have been developed in refs. [49, 55, 111], with particular attention given to the associated problem in heat transfer which is so important for cooling. Approximate methods, at least for flat plates, are also available for the calculation of turbulent boundary layers with blowing and suction [14, 20, 76]; they all make use of Prandtl's mixing length hypothesis (*cf.* sec. XIXb). W. Pechau [60a] published an approximate procedure for the calculation of compressible, laminar boundary layers with an arbitrary external stream and arbitrary distribution of suction, but for the special case when the wall is adiabatic and when the Prandtl number has the value $P = 1$.

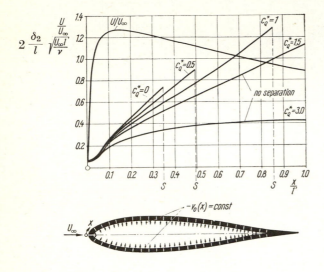

Fig. 14.14. Laminar boundary layer on a symmetrical Zhukovskii aerofoil with uniform suction; $v_0(x) =$ const, angle of incidence $\alpha = 0$, as calculated by E. Trucken-brodt [101]

$\delta_2 =$ momentum thickness; $l' =$ half perimeter length; $c_Q{}^* = c_Q \sqrt{U_\infty l/\nu} =$ reduced volume coefficient of suction. With increasing suction, i. e. for $c_Q{}^* > 1 \cdot 12$ no separation occurs at all

2. Experimental results on suction. As early as 1904 L. Prandtl published photographs of flow patterns which demonstrated that suction causes the flow to adhere to the wall even in the case of non-streamlined blunt bodies, such as circular cylinders, in which there would otherwise be strong eddy formation. Figs. 2.14 and 2.15 show the effect of suction on the flow in a divergent channel. Under normal conditions, Fig. 2.13, the flow in a rapidly divergent channel separates violently from the wall, whereas suction applied through two slits on either side causes the flow completely to adhere to it, Fig. 2.15.

When suction is applied to a wing, it is necessary to discern two distinct problems which might arise:

1. It may be desired to increase the maximum lift by delaying separation.

2. It may be desirable to maintain laminar flow and to avoid transition in order to reduce skin friction. We propose to give a short account of the considerations connected with these two problems.

2.1. Increase in lift. An increase in the maximum lift of an aerofoil can be achieved with the aid of suction and blowing when the boundary layer is either laminar or turbulent. An account of some recent investigations will be given in Sec. XXIIb 6 which treats turbulent boundary layers. At this point, we shall, at first, confine ourselves to the mention of some older experimental results. Extensive experimental material concerning the increase in the lift coefficient due to suction was collected at the end of the twenties and at the beginning of the thirties in the course of a programme of research instituted at the Aerodynamische Versuchs-anstalt in Goettingen under the direction of O. Schrenk. The effect of suction is to preserve the potential flow pattern at higher angles of incidence than would otherwise be the case. O. Schrenk published a comprehensive review of this work in ref. [84]. The scope of these experiments reached such a degree of advancement [85] that at the end of the thirties the Institute in Goettingen was in a position to build two experimental aeroplanes in which suction was applied for the improvement of performance. A detailed description of these experimental aeroplanes was given by J. Stueper [93]. Photographs of the flow field on the wings of one of these experimental aeroplanes are shown in Fig. 14.15. The effect of suction, which was applied in the slit between the wing and the flap, can be inferred clearly from the behaviour of the tufts which are visible in the photographs: Without suction (Fig. 14.15a), the flow is completely separated from the flap; it is brought back completely (Fig. 14.15b) when suction is turned on. A. Gerber [30] investigated systematically certain aspects of suction, such as the best shape of slits, the velocity distribution near the slit, the pressure distribution around it, etc.

More recently, in Great Britain [58, 59] and in the U.S.A. [88], extensive experimental investigations have been carried out into the effect of suction on thin

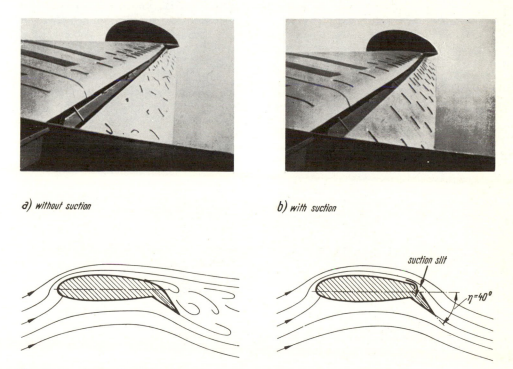

a) without suction   b) with suction

Fig. 14.15.   Flow about the wing of the Goettingen experimental aeroplane; the flap is in the down position; the two photographs represent flow without and with suction. a) without suction: the flow is detached from the flap, b) with suction: the flow adheres to the flap

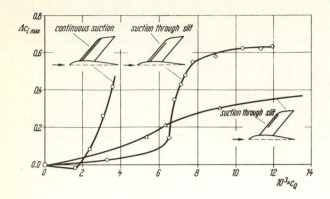

Fig. 14.16. Increase in the maximum lift of a swept-back wing
by suction. Comparison between
continuous suction and suction
applied through slits, as measured by E. D. Poppleton [66]

Reynolds number $R = 1\cdot3 \times 10^6$;
relative width of slits $s/l = 0\cdot004$

aerofoils. Since at high angles of incidence thin aerofoils develop a sharp negative-
pressure peak near the nose on the upper side, it is necessary to apply suction there.
In this connexion it is important to know whether to apply suction through a porous
wall (uniform suction) or through a system of slits. The diagram in Fig. 14.16 shows
a comparison between the results of continuous suction and suction applied through
slits on a swept-back wing as measured by E. D. Poppleton [66]; see also ref. [38].
It is clear that the same increase in the lift coefficient can be obtained with a much
reduced mass flow when continuous suction is used. The diagram in Fig. 14.17
contains information on the most favourable position of the suction zone at the
nose. The measurements carried out on an 8% thick symmetrical aerofoil seem
to indicate that continuous suction is most effective when it is confined to the upper
side of the wing and when it extends over a region of $0\cdot15\,l$ approximately. The
minimum mass flow required to avoid separation depends on the position and the

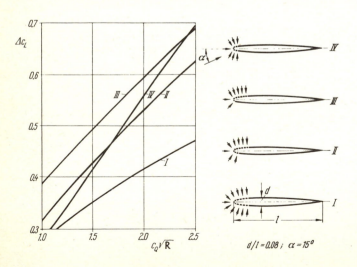

Fig. 14.17.   Effect on increase in lift coefficient of
changing the position of the
porous suction surface for
an 8% thick aerofoil at an
angle of incidence of $\alpha = 15°$

extent of the porous surface and, even more significantly, on the Reynolds number. This, of course, is a very important consideration when results of model experiments are applied to full-scale arrangements. Some data on the dependence of the mass flow on the Reynolds number are shown in Fig. 14.18. They are based on measurements performed by N. Gregory and W. S. Walker [32] on a thin symmetrical aerofoil. The graph shows the minimum volume flow of suction required to avoid separation for a fixed angle of incidence of $\alpha = 14°$ plotted in terms of the Reynolds number. Several curves of $c_Q \sqrt{R} = \text{const}$, which were obtained from the theory of purely laminar flow, have also been plotted for comparison.

2.2. Decrease in drag. An experimental proof of the fact that it is possible to maintain laminar conditions in the boundary layer with the aid of suction was first given by H. Holstein [37], and shortly afterwards by J. Ackeret, M. Ras

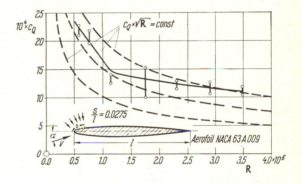

Fig. 14.18. Minimum suction volume required for the prevention of separation as a function of the Reynolds number for an angle of incidence of $\alpha = 14°$, after Gregory and Walker [32]

and W. Pfenninger [3]. W. Pfenninger [61] carried out extensive experiments on the problem of reducing drag by the application of suction through which laminar flow is maintained. Figure 14.19 reproduces some of his results, obtained with a thin aerofoil which was provided with a large number of suction slits. The graph in Fig. 14.19a shows the optimum values of the skin-friction coefficient plotted in terms of the Reynolds number. It is seen that there is a large saving in drag, even if the power consumption of the suction pump is debited against it. The graph shows, further, that, at moderate values of the lift coefficient, even at large Reynolds numbers, the values of the skin-friction coefficient are not much higher than those for a flat plate at zero incidence. Moreover, Fig. 14.19b demonstrates that these low values persist over a considerable range of values of the lift coefficient, c_L. Further, the experiments demonstrated that the decrease in the drag effected by maintaining a laminar boundary layer with the aid of suction depends largely on a careful shaping of slits. If this precaution is not taken, the flow may be so much affected by the presence of the slits that transition to turbulent flow occurs readily; in this connexion see also N. Gregory [33]. In an American paper [10], the possibility of using continuous suction through a porous wall to maintain laminar flow to considerably larger Reynolds numbers (of the order of $R = 20 \times 10^6$) was carefully investigated. In this case too, substantial reductions in drag were achieved, allowing for the mechanical work required to maintain it.

When an attempt is made to preserve a laminar boundary layer either by suction, or, as already mentioned, merely by proper shaping, it is very important to have a good knowledge of the potential velocity distribution. In either case it is necessary to arrange for the pressure to decrease over as large a portion of the section as possible. Very extensive experiments on this subject were carried out by S. Goldstein [31] and his collaborators. The calculations led to the determination of the shape of the section of the aerofoil which would produce a prescribed potential

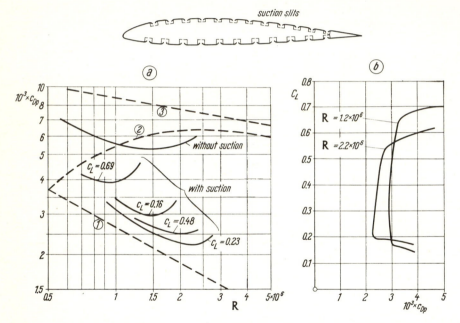

Fig. 14.19.   Decrease in the drag of an aerofoil in which transition is delayed by suction through a large number of slits, after W. Pfenninger [61]. The energy consumption of the pump has been included in the drag coefficient

a) Optimum values of the drag coefficient in terms of the Reynolds number R

Curves (1), (2) and (3) without suction; (1) Flat plate, laminar; (2) Flat plate, transitional; (3) Flat plate, fully developed turbulent

b) Polar diagrams for two different Reynolds numbers. The extremely low drag coefficients exist for an increased range of values of the lift coefficient c_L

velocity distribution. In order to obtain aerofoils which maintain a laminar boundary layer as far as the trailing edge it was suggested to use shapes showing a decrease in pressure (an increase in velocity) over the whole length, and only displaying an abrupt pressure increase at one position, as shown in Fig. 14.20. If the slits are arranged at the point of pressure jump, as suggested by Griffith [73], it is possible to secure a laminar boundary layer on thick aerofoils as far as the slit and separation is prevented behind it. B. Regenscheit [71, 72], and B. Thwaites [94] proposed to 'regulate' the lift on very thick aerofoils by varying the intensity of suction and so to obtain a lift which is independent of the angle of incidence. In more recent times there were many proposals to use the air sucked away from the boundary layer for the purpose of increasing the thrust of a jet aircraft [87].

The papers by F. X. Wortmann [105] and W. Pfenninger [62, 63] report on more recent results concerning the design of laminar aerofoils and of the delay of transition on swept-back wings.

A comprehensive review of problems concerned with aircraft construction and boundary-layer control has been given recently by G. V. Lachmann [43], and C. R. Pankhurst [59]. A paper by M. H. Smith [88] contains a comprehensive list of references.

The process of transition from laminar to turbulent flow in the boundary layer with suction will be studied in detail in Sec. XVII c.

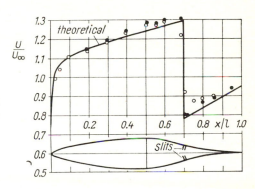

Fig. 14.20.   Theoretical and experimental velocity distribution on a symmetrical aerofoil with suction after Goldstein [31]

$c_L = 0$; $R = 3.85 \times 10^6$

o   = measurement without suction;
•   = measurement with suction

c. Injection of a different gas (Binary boundary layers)

1. Theoretical results.

1.1. The fundamental equations. When a space-vehicle returns to the denser layers of the atmosphere, the stagnation effect which is produced at the nose or in the boundary layers along the walls gives rise to very high temperatures. In order to reduce the quantity of heat transferred to the vehicle to small proportions, it is possible to inject a light gas or a fluid through a porous wall. The light gas or the vapourizing fluid thus create a thin film along the walls. A similar effect can also be produced if the material of the wall (e.g. graphite, glass, or a synthetic material) is allowed to sublimate thus reducing its thickness (ablation). In all such cases, boundary layers are formed in which two or more gases mix with one another by diffusion.

In a streaming gaseous mixture, every component i moves with a mean velocity $\mathbf{w}_i$ which differs from species to species. In order to describe the velocity field, it is convenient to introduce a *mean* or *barycentric* velocity $\mathbf{w} = \Sigma \varrho_i \mathbf{w}_i / \Sigma \varrho_i$, where $\Sigma \varrho_i = \varrho$ denotes the total density. The departure of a velocity, $\mathbf{w}_i$, from the barycentric velocity, $\mathbf{w}$, is known as the diffusion velocity, $\mathbf{W}_i$, of the species, so that

$$\mathbf{w}_i = \mathbf{W}_i + \mathbf{w}.$$

Owing to the definition of w, we must have $\Sigma \, \varrho_i \, \mathbf{W}_i = 0$, and for each component i we may write the law of mass conservation in the form

$$\operatorname{div}\,(\varrho_i \, \mathbf{w}_i) = \operatorname{div}\,\{\varrho_i \, (\mathbf{w} + \mathbf{W}_i)\} = 0 \, . \tag{14.24}$$

Upon summing over all components, we obtain the continuity equation

$$\operatorname{div}\,(\varrho \, \mathbf{w}) = 0 \tag{14.25}$$

which has the familar form of eqn. (3.1).

In the absence of external fields, the diffusive flow is driven, essentially, by concentration gradients as well as by thermal diffusion which produces a flow of masses in the presence of a temperature gradient. In the case of a binary mixture, we may write the law of diffusion in the form

$$c_1 \, \mathbf{W}_1 = - \, D_{12} \, (\operatorname{grad} c_1 + k_T \operatorname{grad} \ln T) \, , \tag{14.26}$$

where D_{12} denotes the coefficient of binary diffusion, k_T is the thermal diffusion ratio, and $c_1 = \varrho_1/\varrho$ is the mass concentration of the first gas, assumed to be the one which emerges from the wall. The coefficient of binary diffusion depends only little on concentration and is affected by temperature in the same way as the kinematic viscosity. The thermal diffusion ratio, k_T, depends essentially on concentration and is frequently approximated by the rather crude relation

$$k_T = \alpha \, c_1 \, (1 - c_1) \tag{14.27}$$

due to Onsager, Furry and Jones. Here, the coefficient of thermal diffusion, α, is assumed to be a constant for every specific combination of gases.

Inserting eqn. (14.26) into the law of mass conservation, eqn. (14.24), written for the first component, and taking into account eqn. (14.25), we obtain

$$\varrho \left(u \, \frac{\partial c_1}{\partial x} + v \, \frac{\partial c_1}{\partial y} \right) = \operatorname{div}\,\{\,\varrho \, D_{12} \, (\operatorname{grad} c_1 + k_T \operatorname{grad} \ln T)\,\} \, .$$

We may now introduce the normal boundary-layer simplifications into the right-hand side of this equation thus neglecting terms in $\partial/\partial x$ with respect to those in $\partial/\partial y$. In this manner we obtain the *concentration equation*

$$\varrho \left(u \, \frac{\partial c_1}{\partial x} + v \, \frac{\partial c_1}{\partial y} \right) = \frac{\partial}{\partial y} \left\{ \varrho \, D_{12} \left(\frac{\partial c_1}{\partial y} + k_T \, \frac{\partial \ln T}{\partial y} \right) \right\} \, . \tag{14.28}$$

A corresponding equation is valid for the second component; however, this second equation becomes trivial when the modified form of eqn. (14.28) is used because $c_1 + c_2 = 1$. For this reason, the second equation is replaced by the continuity equation (14.25).

The momentum equations for a gas mixture are identical with those for a single gas and are written

$$\varrho \left(u \, \frac{\partial u}{\partial x} + v \, \frac{\partial u}{\partial y} \right) = - \, \frac{dp}{dx} + \frac{\partial}{\partial y} \left(\mu \, \frac{\partial u}{\partial y} \right) , \tag{14.29}$$

$$\frac{\partial p}{\partial y} = 0 \, , \tag{14.30}$$

where now ϱ and μ depend on concentration in addition to their familiar dependence on temperature.

The energy equation for a gaseous mixture must be formulated with due regard being paid to the normal thermal conduction, to the transfer of heat by diffusion, and to that by thermal diffusion. Restricting our considerations to perfect gases, we introduce the mixture enthalpy

$$h = c_1 \, h_1 + c_2 \, h_2 \, . \tag{14.31}$$

Since the derivation is lengthy, we merely quote the result in which the boundary-layer approximations have already been introduced:

$$\varrho\, c_p \left(u\, \frac{\partial T}{\partial x} + v\, \frac{\partial T}{\partial y}\right) = \frac{\partial}{\partial y}\left(k\, \frac{\partial T}{\partial y}\right) + u\, \frac{dp}{dx} + \mu \left(\frac{\partial u}{\partial y}\right)^2$$

$$+ \frac{RT\, k_T}{c_1\, c_2} \left\{ \frac{\partial}{\partial y}\left[\varrho\, D_{12}\left(\frac{\partial c_1}{\partial y} + k_T\, \frac{\partial \ln T}{\partial y}\right)\right]\right\} \qquad (14.32)$$

$$+ \varrho\, D_{12} \left\{\frac{\partial c_1}{\partial y} + k_T\, \frac{\partial \ln T}{\partial y}\right\} \frac{\partial}{\partial y}\left\{(h_1 - h_2) + \frac{RT\, k_T}{c_1\, c_2}\right\}.$$

Here R stands for the universal gas constant. If thermal diffusion is neglected, the underlined terms are deleted. In the derivation of this equation use has been made of Onsager's principle according to which the coefficient of the concentration gradient in the heat-flux vector is the same as that of the temperature gradient in the mass flux.

The boundary conditions for velocity and temperature are the same as in boundary layers consisting of single gases. These must be supplemented with two new boundary conditions for the concentration. At a large distance from the wall, there is present only the external gas which means that the concentration c_1 of the species emanating from the wall vanishes at $y = \infty$. The second boundary condition must be prescribed at the wall. In most cases it is permissible to assume that the external gas cannot pass through the wall, that is, that the diffusion velocity of the external gas is equal and opposite in sign to the injection velocity, v_w, at the wall. Since

$$c_2\, W_2 = (1 - c_1)\, W_2 = c_1\, W_1,$$

and in view of eqn. (14.26), we obtain the condition that

$$v_w = \left\{\frac{D_{12}}{1 - c_1}\, (\text{grad } c_1 + k_T\, \text{grad } \ln T)\right\}_w. \qquad (14.33)$$

Equations (14.25), (14.28), (14.29), and (14.32) constitute a system of four equations for the four quantities: u, v, T, and c_1.

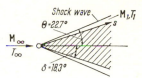

Fig. 14.21 Binary boundary layer on a cone in laminar flow with supersonic velocity $M_\infty = 12{\cdot}9$ in the presence of the injection of helium into air, after W. Wuest [110]. Velocity distribution, u, temperature distribution, T, and concentration distribution, c_1, for different ratios of wall temperature, T_w, to external temperature, T_1. Injection velocity:

$$v/u_1 = 0{\cdot}2\,(\varrho_1/\varrho_w)/\sqrt{2\, R_s/3}\ ;$$
$$\eta = \sqrt{3\, R_s/2}\ (y/s);\qquad R_s = u_1\, s/\nu_1;$$
$$M_\infty = 12{\cdot}9;\qquad T_\infty = -\,50°C;$$
$$M_1 = 5;\qquad T_1 = 1023°C.$$

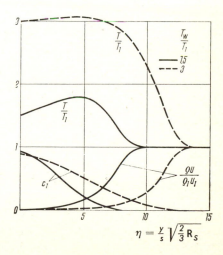

$$\eta = \frac{y}{s}\sqrt{\frac{2}{3}R_s}$$

1. 2. Exact solutions. In order to solve the coupled partial differential equations of the parabolic type we have, at present, at our disposal a variety of numerical methods [97, 42] as well as fast electronic computers. With the aid of these, it becomes possible to obtain almost arbitrarily close approximations to the exact solutions with a tolerable expenditure of time. The properties of the fluid can be conceived as quantities that vary with position, and arbitrary boundary conditions can be prescribed. It is possible to obtain similar solutions if the external velocity, the blowing velocity, as well as the temperature on the wall, are prescribed in a definite manner. In such cases, the system of partial differential equations reduces to a system of ordinary differential equations, and the latter can be integrated numerically. There exist such numerical results for incompressible wedge-flows (inclusive of stagnation flow [102, 29]), compressible flow over a flat plate at zero incidence, and supersonic boundary layers on wedges and cones [110]. The diagram in Fig. 14.21 illustrates, by way of example,the laminar velocity, temperature, and concentration boundary layer on a cone with helium injection.

A method designed to calculate laminar, hypersonic, binary-mixture boundary layers was given by J. Steinheuer [91] who applied it to the example of cooling by ablation with the aid of pyrolizing teflon.

All of the numerical calculations mentioned so far neglect the terms which stem from thermal diffusion, that is the terms which have been underlined in eqn. (14.32). Such a simplification is sometimes permissible as far as the computation of skin friction and heat-transfer rate is concerned. Experiments show that the equilibrium temperature on an adiabatic wall does not decrease in the presence of thermal diffusion, but calculations based on this simplified scheme always predict such a decrease.

Exact calculations on two-substance boundary layers which occur in flows with evaporation or sublimation present us with considerable difficulties. The distribution of velocity of the evaporating substance (i. e. of the velocity of blowing) and of temperature at the phase boundary in the flow direction can no longer be prescribed arbitrarily. Both distributions arise spontaneously as a result of the coupled heat and mass transfer and neither is known *a priori*. In this domain, W. Splettstoesser [90] calculated a large number of solutions in which the evaporation rate as well as the locally satisfied energy balance have been evaluated on the basis of eqn. (14.53).

F. Eisfeld [21] published solutions for flows of binary mixtures that arise in the presence of the adiabatic evaporation of a film of carbon dioxide with a special mathematical form assumed for the law of evaporation. In this work, he discovered that the process of adiabatic evaporation of a plane film leads to self-similar solutions, *cf*. Chap. VIII. In such cases, it turns out that the local rate of evaporation must follow a $1/\sqrt{x}$-type law. This is the distribution of the normal velocity in suction or blowing on a flat plate at zero incidence that leads to self-similar solutions, as illustrated in Fig. 14.12. The temperature and concentration at the surface of the film turns out to be uniform.

1.3. Approximate solutions. It is possible to simplify the problem by assuming that the Prandtl number, P, and the Schmidt number, $S_c = \nu/D_{12}$ are equal to unity and that the viscosity is a linear function of temperature. With these assumptions, C. R. Faulders [25] calculated the shearing stress at the wall when a light gas is injected; he considered cases of varying molecular mass ratio with respect to the base gas. More general cases of external-velocity and injection-velocity distributions can be analyzed with the aid of the integral equations [109].

2. Experimental results

Most experimental investigations into the problem of injecting a foreign gas into a laminar supersonic boundary layer concentrate almost exclusively on the measurement of the equilibrium temperature on an adiabatic wall.

When the boundary layer contains several components, the exact calculation becomes tedious because the flux of each component depends on the fluxes of all the others. A considerable simplification occurs when a model with constant properties combined with multicomponent diffusion coefficients referred to a specified state is adopted. Such a model gives very good agreement with exact solutions even in the case of strongly variable properties [97].

Considerably more numerous experiments have been performed with turbulent boundary layers (*cf*. Chap. XXII). The details of the process of ablation are only partly understood and for this reason, a calculation of ablation heat-transfer must still be based on crude semi-empirical equations [3a].

References

[1] Ackeret, J.: Das Rotorschiff und seine physikalischen Grundlagen. Vandenhoeck und Rupprecht, Göttingen, 1925.

[2] Ackeret, J.: Grenzschichtabsaugung. Z. VDI *70*, 1153—1158 (1926).

[3] Ackeret, J., Ras, M., and Pfenninger, W.: Verhinderung des Turbulentwerdens einer Reibungsschicht durch Absaugung. Naturwissenschaften, 622 (1941); see also Helv. phys. Acta *14*, 323 (1941).

[4] MacAdams, C.: Recent advances in ablation. ARS J. *29*, 625—632 (1959).

[5] Attinello, S.: Auftriebserhöhung durch Grenzschichtsteuerung. Interavia *10*, 925—927 (1955).

[6] Baron, J.R., and Scott, P.E.: Some mass transfer results with external flow pressure gradients. JASS *27*, 625—626 (1960).

[7] Betz, A.: Die Wirkungsweise von unterteilten Flügelprofilen. Berichte und Abh. Wiss. Gesellschaft f. Luftfahrt, No. 6 (1922); NACA TM 100 (1922).

[8] Betz, A.: Beeinflussung der Reibungsschicht und ihre praktische Verwertung. Schriften dt. Akad. f. Luftfahrtforschung No. 49 (1939).

[9] Betz, A.: History of boundary layer control research in Germany. In: Boundary layer and flow control (G. V. Lachmann, ed), *I*, 1—20, London 1961.

[10] Braslow, A.L., Burrows, D.L., Tetervin, N., and Visconti, F.: Experimental and theoretical studies of area suction for the control of the laminar boundary layer. NACA Rep. 1025 (1951).

[11] Brown, B.: Exact solutions of the laminar boundary layer equations for a porous plate with variable fluid properties and a pressure gradient in the main stream. Proc. First US Nat. Congr. Appl. Mech., 843—852 (1951).

[12] Brown, W.B., and Donoughe, P.L.: Table of exact laminar boundary layer solutions when the wall is porous and fluid properties are variable. NACA TN 2479 (1951).

[12a] Chang, P.K.: Control of flow separation. Hemisphere Publishing Corporation, Washington DC (1976).

[13] Carrière, P., and Eichelbrenner, E.A.: Theory of flow reattachment by a tangential jet discharging against a strong adverse pressure gradient. In: Boundary layer and flow control (G.V. Lachmann, ed.), *I*, 209—231, London, 1961.

[14] Clarke, J.H., Menkes, H.R., and Libby, P.A.: A provisional analysis of turbulent boundary layers with injection. JAS *22*, 255—260 (1955).

[15] Curle, N.: The estimation of laminar skin friction including effects of distributed suction. Aero. Quart. *11*, 1—21 (1960).

[16] Culick, F.E.C.: Integral method for calculating heat and mass transfer in laminar boundary layers. AIAA J. *1*, 783—793 (1963).

[17] Dannenberg, R.E., and Weiberg, J.A.: Effect of type of porous surface and suction velocity distribution on the characteristics of a 10·5 per cent thick airfoil with area suction. NACA TN 3093 (1953).

[18] von Doenhoff, A.E., and Loftin, L.K.: Present status of research on boundary layer control. JAS *16*, 729—740 (1949).

[19] Donoughe, P.L., and Livingood, J.N.B.: Exact solutions of laminar boundary layer equations with constant property values for porous wall with variable temperature. NACA Rep. 1229 (1955).

[20] Dorrance, W.H., and Dore, F.J.: The effect of mass transfer on the compressible turbulent boundary layer skin friction and heat transfer. JAS *21*, 404—410 (1954).

[20a] Eckert, E.R.G.: Thermodynamische Kopplung von Stoff und Wärmeübergang. Forschg. Ing.-Wes. *29*, 147—151 (1963).

[21] Eisfeld, F.: Die Berechnung der Grenzschichten für gekoppelten Wärmeübergang und Stoffaustausch bei Verdunstung eines Flüssigkeitsfilms über einer parallel angeströmten Platte unter Berücksichtigung veränderlicher Stoffbeiwerte. Int. J. Heat Mass Transfer *14*, 1537—1550 (1971).

[22] Emmons, H.W., and Leigh, D.C.: Tabulation of Blasius function with blowing and suction. ARC CP 157 (1954).

[23] Eppler, R.: Praktische Berechnung laminarer und turbulenter Absauge-Grenzschichten. Ing.-Arch. *32*, 221—245 (1963).

[24] Eppler, R.: Gemeinsame Grenzschichtabsaugung für Hochauftrieb und Schnellflug. Jb. WGL 140—149 (1962).

[25] Faulders, C.R.: A note on laminar layer skin friction under the influence of foreign gas injection. JASS *28*, 166—167 (1961).
[26] Favre, A.: Contribution à l'étude expérimentale des mouvements hydrodynamiques à deux dimensions. Thesis University of Paris 1938, 1—192.
[27] Flatt, J.: The history of boundary layer control research in the United States of America. In: Boundary layer and flow control (G.V. Lachmann, ed.), *I*, 122—143, London, 1961.
[28] Flügel, G.: Ergebnisse aus dem Strömungsinstitut der Technischen Hochschule Danzig. Jb. Schiffbautechn. Gesellschaft *31*, 87—113 (1930).
[29] Fox, H., and Libby, P.A.: Helium injection into the boundary layer at an axisymmetric stagnation point. JASS *29*, 921 (1962).
[29a] Gersten, K., and Gross, J.F.: Flow and heat transfer along a plane wall with periodic suction. ZAMP *25*, 399—408 (1974).
[30] Gerber, A.: Untersuchungen über Grenzschichtabsaugung. Rep. Inst. of Aerodynamics,
[31] Goldstein, S.: Low-drag and suction airfoils. JAS *15*, 189—220 (1948).
[32] Gregory, N., and Walker, W.S.: Wind-tunnel tests on the NACA 63 A 009 aerofoil with distributed suction over the nose. ARC RM 2900 (1955).
[33] Gregory, N.: Research on suction surface for laminar flow. In: Boundary layer and flow control (G.V. Lachmann, ed.), *II*, 924—960, London, 1961.
[34] Gross, J.F., Hartnett, J.P., Masson, D.J., and Gazley, C., Jr.: A review of binary boundary layer characteristics. J. Heat Mass Transfer *3*, 198—221 (1961).
[35] Head, M.R.: The boundary layer with distributed suction. ARC RM 2783 (1955).
[36] Head, M.R.: History of research on boundary layer control for low drag in the U. K. In: Boundary layer and flow control (G.V. Lachmann, ed.), *I*, 104—121, 1961.
[37] Holstein, H.: Messungen zur Laminarhaltung der Grenzschicht an einem Flügel. Lilienthal-Bericht S 10, 17—27 (1940).
[38] Holzhauser, C.A., and Bray, R.S.: Wind-tunnel and flight investigations of the use of leading edge area suction for the purpose of increasing the maximum lift coefficient of a 35° swept-wing airplane. NACA Rep. 1276 (1956).
[39] Hurley, D.G., and Thwaites, B.: An experimental investigation of the boundary layer on a porous circular cylinder. ARC RM 2829 (1955).
[40] Iglisch, R.: Exakte Berechnung der laminaren Reibungsschicht an der längsangeströmten ebenen Platte mit homogener Absaugung. Schriften dt. Akad. d. Luftfahrtforschung, 8 B, No. 1 (1944); NACA RM 1205 (1949).
[41] Jones, M., and Head, M.R.: The reduction of drag by distributed suction. Proc. Third Anglo-American Aeronautical Conference, Brighton, 199—230 (1951).
[41a] Kay, J.M.: Boundary layer along a flat plate with uniform suction. ARC RM 2628 (1948).
[42] Kordulla, W., and Will, E.: Tangentiales Ausblasen von Helium in laminaren Hyperschall-grenzschichten. ZFW *22*, 295—307 (1974).
[43] Lachmann, G.V.: Boundary layer control. J. Roy. Aero. Soc. *59*, 163—198 (1955); see also Aero. Eng. Rev. *13*, 37—51 (1954) and Jb. WGL 132—144 (1954).
[44] Lachmann, G.V. (ed.): Boundary layer and flow control, *I* and *II*. Pergamon Press, London, 1961.
[45] Lew, H.G.: On the compressible boundary layer over a flat plate with uniform suction. Reissner Annivers. Vol. Contr. Appl. Mech. Ann Arbor/Mich. 43—60 (1949).
[46] Lew, H.G., and Mathieu, R.D.: Boundary layer control by porous suction. Dep. Aero. Eng. Pennsylvania State Univ. Rep. No. 3 (1954).
[47] Lew, H.G., and Vanucci, J.B.: On the laminar compressible boundary layer over a flat plate with suction or injection. JAS *22*, 589—597 (1955).
[48] Libby, P.A., Kaufmann, L., and Harrington, R.P.: An experimental investigation of the isothermal laminar boundary layer on a porous flat plate. JAS *19*, 127 (1952).
[49] Libby, P.A., and Pallone, A.: A method for analyzing the heat insulating properties of the laminar compressible boundary layer. JAS *21*, 825—834 (1954).
[50] Libby, P.A., and Cresci, R.J.: Experimental investigation of the down-stream influence of stagnation point mass transfer. JASS *28*, 51 (1961).
[50a] Libby, P.A.: Heat and mass transfer at a general three-dimensional stagnation point. AIAA J. *5*, 507—517 (1967).
[50b] Libby, P.A.: Laminar flow at a three-dimensional stagnation point with large rates of injection. AIAA J. *14*, 1273—1279 (1976).
[51] Low, G.M.: The compressible laminar boundary layer with fluid injection. NACA TN 3404 (1955).

[52] Meredith, F.W., and Griffith, A.A.: in: Modern developments in fluid dynamics. Oxford University Press, 2, 534, Oxford, 1938.

[53] Mickley, H.S., Rose, R.C., Squires, A.L., and Stewart, W.E.: Heat mass, and momentum transfer for flow over a flat plate with blowing or suction. NACA TN 3208 (1954).

[54] Miles, E.G.: Sucking away boundary layers. Flight 35, 180 (1939).

[55] Morduchow, M.: On heat transfer over a sweat-cooled surface in laminar compressible flow with pressure gradient. JAS 19, 705—712 (1952).

[56] Ness, N.: Foreign gas injection into a compressible turbulent boundary layer on a flat plate. JASS 28, 645—654 (1961).

[57] Nickel, K.: Eine einfache Abschätzung für Grenzschichten. Ing.-Arch. 31, 85—100 (1962).

[58] Pankhurst, R.C., Raymer, W.G., and Devereux, A.N.: Wind-tunnel tests of the stalling properties of an 8 per cent thick symmetrical section with nose suction through a porous surface. ARC RM 2666 (1953).

[59] Pankhurst, R.C.: Recent British work on methods of boundary layer control. Proc. Symp. at Nat. Phys. Lab. (1955).

[60] Pappas, C.C.: Effect of injection of foreign gases in the skin friction and heat transfer on the turbulent boundary layer. JAS Paper, 59—78 (Jan. 1959).

[60a] Pechau, W.: Ein Näherungsverfahren zur Berechnung der kompressiblen laminaren Grenzschicht mit kontinuierlich verteilter Absaugung. Ing.-Arch. 32, 157—186 (1963).

[61] Pfenninger, W.: Untersuchung über Reibungsverminderung an Tragflügeln, insbesondere mit Hilfe von Grenzschichtabsaugung. Rep. Inst. Aerodynamics, ETH Zürich, No. 13 (1946); see also JAS 16, 227—236 (1949); NACA TM 1181 (1947).

[62] Pfenninger, W., and Bacon, J.W.: About the development of swept laminar suction. In: Boundary layer and flow control (G.V. Lachmann, ed.) II, 1007—1032, London, 1961.

[63] Pfenninger, W., and Groth, E.: Low drag boundary layer suction experiments in flight on a wing glove of an F-94 A airplane with suction through a large number of fine slots. In: Boundary layer and flow control (G.V. Lachmann, ed.) II, 987—999, London, 1961.

[63a] Pientka, K.: Theoretische Untersuchung der laminaren Zweistoffgrenzschichtströmung längs einer benetzten Platte bei nicht-adiabater Verdunstung. Diss. Braunschweig 1977.

[64] Poisson-Quinton, Ph.: Récherches théoriques et expérimentales sur le controle de circulation par soufflage appliqué aux ailes d'avions. ONERA Publication, Note Technique No. 37 (1956); see also Jb. WGL 1956, 29—51 (1957).

[65] Poisson-Quinton, Ph., and Lepage, L.: Survey of French research on the control of boundary layer and circulation. In: Boundary layer and flow control (G.V. Lachmann, ed.) I, 21—73, London, 1961.

[66] Poppleton, E.D.: Boundary layer control for high lift by suction of the leading-edge of a 40 degree swept-back wing. ARC RM 2897 (1955).

[67] Prandtl, L.: The mechanics of viscous fluids. In: Aerodynamic theory (W. F. Durand, ed.) III, 34—208 (1935).

[68] Preston, J.H.: The boundary layer flow over a permeable surface through which suction is applied. ARC RM 2244 (1946).

[69] Pretsch, J.: Grenzen der Grenzschichtbeeinflussung. ZAMM 24, 264—267 (1944).

[70] Raspet, A.: Boundary layer studies on a sailplane. Aero. Eng. Rev. 11, 6, 52 (1952).

[71] Regenscheit, B.: Eine neue Anwendung der Absaugung zur Steigerung des Auftriebes eines Tragflügels. F. B. 1474 (1941).

[72] Regenscheit, B.: Absaugung in der Flugtechnik. Jb. WGL 1952, 55—63 (1953).

[73] Richards, E.J., Walker, W., and Greening, J.: Tests of a Griffith aerofoil in the 13 · 9 ft tunnel. ARC RM 2148 (1954).

[74] Ringleb, F.: Computation of the laminar boundary layer with suction. JAS 19, 48—54 (1952).

[75] Rheinboldt, W.: Zur Berechnung stationärer Grenzschichten bei kontinuierlicher Absaugung mit unstetig veränderlicher Absaugegeschwindigkeit. J. Rat. Mech. Analysis 5, 539—596 (1956).

[76] Rubesin, M.W.: An analytical estimation of the effect of transpiration cooling on the heat-transfer and skin-friction characteristics of a compressible, turbulent boundary layer. NACA TN 3341 (1954).

[77] Schlichting, H.: Die Grenzschicht an der ebenen Platte mit Absaugung und Ausblasen. Luftfahrtforschung 19, 293—301 (1943).

[78] Schlichting, H.: Die Grenzschicht mit Absaugung und Ausblasen. Luftfahrtforschung 19, 179—181 (1942).

[79] Schlichting, H., and Bussmann, K.: Exakte Lösungen für die laminare Reibungsschicht mit Absaugung und Ausblasen. Schriften dt. Akad. d. Luftfahrtforschung 7 B, No. 2 (1943).

[80] Schlichting, H.: Die Beeinflussung der Grenzschicht durch Absaugung und Ausblasen. Jb. dt. Akad. d. Luftfahrtforschung 90—108 (1943/44).

[81] Schlichting, H.: Ein Näherungsverfahren zur Berechnung der laminaren Reibungsschicht mit Absaugung. Ing.-Arch. 16, 201—220 (1948); NACA TM 1216 (1949).

[82] Schlichting, H.: Absaugung in der Aerodynamik. Jb. WGL 1956, 19—29 (1957); see also: L'aspiration de la couche limite en technique aéronautique. Technique et Science Aéronautique Part 4, 149—161 (1956).

[83] Schlichting, H.: Einige neuere Ergebnisse über Grenzschichtbeeinflussung. Advances in Aeronautical Sciences. II, Proc. First Internat. Congr. in the Aeronautical Sciences in Madrid 1958, London, 563—586 (1959).

[84] Schrenk, O.: Versuche mit Absaugflügeln. Luftfahrtforschung 12, 10—27 (1935).

[85] Schrenk, O.: Tragflügel mit Grenzschichtabsaugung. Luftfahrtforschung 2, 49 (1928); see also ZFM 22, 259 (1931); Luftfahrtforschung 12, 10 (1935); Luftwissen 7, 409 (1940); also NACA TM 974 (1941).

[86] Sinhar, K.D.P.: The laminar boundary layer with distributed suction on an infinite yawed cylinder. ARC CP 214 (1956).

[87] Smith, A.M., and Roberts, H.E.: The jet airplane utilizing boundary layer air for propulsion. JAS 14, 97—109 (1947).

[88] Smith, M.H.: Bibliography on boundary layer control. Literature Search No. 6, Library Bulletin. The James Forrestal Research Center, Princeton Univ. (1955).

[88a] Stuart, J.T.: On the effect of uniform suction on the steady flow due to a rotating disk. Quart. J. Mech. Appl. Math. 7, 446—457 (1954).

[89] Smith, A.M.O., and Jaffe, N.A.: General method for solving the laminar nonequilibrium boundary layer equations of a dissociating gas. AIAA J. 4, 611—620 (1966).

[90] Splettstösser, W.: Untersuchung der laminaren Zweistoffgrenzschichtströmung längs eines verdunstenden Flüssigkeitsfilms. Diss. Braunschweig 1974. Wärme- und Stoffübertragung 8, 71—86 (1975).

[91] Steinheuer, J.: Berechnung der laminaren Zweistoff-Grenzschicht in der hypersonischen Staupunktströmung mit temperaturabhängigen Stoffbeiwerten. Diss. Braunschweig 1970; ZAMM 51, 209—223 (1971).

[92] Stuart, J.T.: On the effects of uniform suction on the steady flow due to a rotating disk. Quart. J. Mech. Appl. Math. 7, 446—457 (1954).

[93] Stüper, J.: Flight experiments and tests on two airplanes with suction slots. NACA TM 1232 (1950). Engl. transl. of ZWB Forschungsbericht No. 1821 (1943).

[94] Thwaites, B.: The production of lift independently of incidence. J. Roy. Aero. Soc. 52, 117—124 (1948).

[95] Thwaites, B.: Investigations into the effects of continuous suction on laminar boundary layer flow under adverse pressure gradients. ARC RM 2514 (1952).

[96] Thwaites, B.: On the momentum equation in laminar boundary layer flow. A new method of uniparametric calculation. ARC RM 2587 (1952).

[97] Taitel, Y., and Tamir, A.: Multicomponent boundary layer characteristics. Use of the reference state. Int. J. Heat Mass Transfer 18, 123—129 (1975).

[98] Torda, T.P.: Boundary layer control by distributed surface suction or injection. Biparametric general solution. J. Math. Phys. 32, 312—314 (1954).

[99] Trilling, L.: The incompressible boundary layer with pressure gradient and suction. JAS 17, 335—341 (1950).

[100] Truckenbrodt, E.: Die laminare Reibungsschicht an einer teilweise mitbewegten längsangeströmten ebenen Platte. Abh. Braunschweig. Wiss. Ges. 4, 181—195 (1952).

[101] Truckenbrodt, E.: Ein einfaches Näherungsverfahren zum Berechnen der laminaren Reibungsschicht mit Absaugung. Forschg. Ing.-Wes. 22, 147—157 (1956).

[102] Watson, E.J.: The asymptotic theory of boundary layer flow with suction. ARC RM 2619 (1952).

[103] Wieghardt, K.: Zur Berechnung ebener und drehsymmetrischer Grenzschichten mit kontinuierlicher Absaugung. Ing.-Arch. 22, 368—377 (1954).

[104] Williams, J.: A brief review of British research on boundary layer control for high lift. In: Boundary layer and flow control (G.V. Lachmann, ed.), I, 74—103, London, 1961.

[105] Wortmann, F.X.: Progress in the design of low drag aerofoils. In: Boundary layer and flow control (G.V. Lachmann, ed.), II, 748—770, London, 1961.

[106] Wuest, W.: Entwicklung einer laminaren Grenzschicht hinter einer Absaugestelle. Ing.-Arch. *17*, 199—206 (1949).

[107] Wuest, W.: Asymptotische Absaugegrenzschichten an längsangeströmten zylindrischen Körpern. Ing.-Arch. *23*, 198—208 (1955).

[108] Wuest, W.: Survey of calculation methods of laminar boundary layers with suction in incompressible flow. In: Boundary layer and flow control (G. V. Lachmann, ed.), *II*, 771—800, London, Pergamon Press, 1961.

[109] Wuest, W.: Laminare Grenzschichten bei Ausblasen eines anderen Mediums (Zweistoffgrenzschichten). Ing.-Arch. *31*, 125—143 (1962).

[110] Wuest, W.: Kompressible laminare Grenzschichten bei Ausblasen eines anderen Mediums. ZFW *11*, 398—409 (1963).

[111] Yuan, S. W.: Heat transfer in laminar compressible boundary layer on a porous flat plate with fluid injection. JAS *16*, 741—748 (1949).

[112] Young, A. D.: Note on the velocity and temperature distributions attained with suction on a flat plate of infinite extent in compressible flow. Quart. J. Mech. Appl. Math. *1*, 70—75 (1948).

[113] Young, A. D., and Zamir, M.: Similar and asymptotic solutions of the incompressible boundary layer equations with suction. Aero. Quart. *18*, 103—120 (1967).

CHAPTER XV

Non-steady boundary layers†

a. General remarks on the calculation of non-steady boundary layers

The examples of solutions of the boundary-layer equations which have been considered until now referred to steady motion. They are by far the most important cases encountered in practical applications. Nevertheless, in this chapter we propose to consider several examples of motions which depend on time, i. e. of non-steady boundary layers.

The most common examples of non-steady boundary layers occur when the *motion* is *started from rest* or when it is periodic. When motion is started from rest both the body and the fluid have zero velocities up to a certain instant of time. The motion begins at that instant and we can consider either that the body is dragged through the fluid at rest or that the body is at rest and that the external fluid motion varies with time. In this latter case an initially very thin boundary layer is formed near the body, and the transition from the velocity of the body to that in the external flow takes place across it. Immediately after the start of the motion the flow in the whole fluid space is irrotational and potential with the exception of a very thin layer near the body. The thickness of the boundary layer increases with time, and it is important to investigate at which instant separation (reverse flow) first occurs as the boundary layer continues to build up. One such example was already considered in Sec. V 4; it was the exact solution of the Navier-Stokes equations for the flow near a wall which is accelerated impulsively from rest and moves in a direction parallel to itself. Also, the start of the flow in a pipe (Sec. V 6) belongs to the same category.

Further examples of non-steady boundary layers occur when either the body performs a periodic motion in a fluid at rest or when the body is at rest and the fluid executes a periodic motion. The motion of a fluid near a wall which oscillates in its own plane (Sec. V 7) affords an example of this type of problem.

1. Boundary-layer equations. The fundamental equations for non-steady boundary layers have already been deduced in Sec. VII a. In the general case when the flow is compressible and non-steady but two-dimensional, we must resort to the following equations for the velocity and temperature fields (*cf.* eqns. (12.50 a to e)):

---

† I am indebted to Professor K. Gersten who revised this chapter for the Fifth Edition of this book.

$$\frac{\partial \varrho}{\partial t} + \frac{\partial (\varrho\, u)}{\partial x} + \frac{\partial (\varrho\, v)}{\partial y} = 0 \; , \tag{15.1}$$

$$\varrho \left(\frac{\partial u}{\partial t} + u\, \frac{\partial u}{\partial x} + v\, \frac{\partial u}{\partial y} \right) = - \frac{\partial p}{\partial x} + \frac{\partial}{\partial y} \left(\mu\, \frac{\partial u}{\partial y} \right), \tag{15.2}$$

$$\varrho\, c_p \left(\frac{\partial T}{\partial t} + u\, \frac{\partial T}{\partial x} + v\, \frac{\partial T}{\partial y} \right) = \frac{\partial}{\partial y} \left(k\, \frac{\partial T}{\partial y} \right) + \mu \left(\frac{\partial u}{\partial y} \right)^2 + \frac{\partial p}{\partial t} + u\, \frac{\partial p}{\partial x} \; , \tag{15.3}$$

$$p = \varrho\, R\, T \; , \tag{15.4}$$

$$\mu = \mu (T) \; . \tag{15.5}$$

The boundary conditions are:

$$y = 0: \quad u = U_w(t) \; , \quad v = 0 \; , \qquad T = T_w(x,t)$$
$$y = \infty: \quad u = U(x,t) \; , \qquad\qquad T = T_\infty (x,\, t).$$

Here, $U_w(t)$ denotes the velocity of the wall if it is in motion, and $U(x, t)$ refers to the non-viscous external motion; the latter is related to the pressure by the equation,

$$- \frac{1}{\varrho_\infty}\, \frac{\partial p}{\partial x} = \frac{\partial U}{\partial t} + U\, \frac{\partial U}{\partial x} \; , \tag{15.6}$$

which follows immediately when the viscous terms are omitted from eqn. (15.2). Generally speaking, it will be convenient to choose a system of coordinates linked with the stationary, external flow. As far as incompressible flows are concerned, these different systems of coordinates are equivalent (cf. [27]). The definition of the point of separation in non-steady flow is also closely related to the choice of the system of coordinate axes (cf. [33]). In what follows we shall consider that separation occurs at the point where $(\partial u/\partial y)_w$ vanishes in a system of coordinates linked with the solid surface.

In complete analogy with steady boundary layers, it is possible to derive integral relations from the differential equations of non-steady boundary-layer flows. These are:

$$U\, \frac{\partial}{\partial t} \int_0^\infty (\varrho - \varrho_\infty)\, \mathrm{d}y + \frac{\partial}{\partial t}\, (\varrho_\infty U \delta_1) + \frac{\partial U}{\partial x}\, \varrho_\infty\, U\, \delta_1 + \frac{\partial}{\partial x}\, (\varrho_\infty U^2 \delta_2) = \tau_0 \tag{15.7}$$

and

$$c_p\, T_\infty\, \frac{\partial}{\partial t} \int_0^\infty (\varrho_\infty - \varrho)\, \mathrm{d}y + \left(\varrho_\infty\, c_p\, \frac{\partial T_\infty}{\partial t} - \frac{\partial p}{\partial t} \right) \delta_1 + \varrho_\infty\, U\, \frac{\partial U}{\partial t}\, \delta_H +$$
$$\tag{15.8}$$
$$+ \frac{\partial}{\partial x}\, (\varrho_\infty\, c_p\, T_\infty\, U \delta_H) + \varrho_\infty\, U^2\, \frac{\partial U}{\partial x}\, \delta_H = \int_0^\infty \mu \left(\frac{\partial u}{\partial y} \right)^2 \mathrm{d}y - k \left(\frac{\partial T}{\partial y} \right)_{y=0} .$$

Here δ_1 denotes the displacement thickness, δ_2 the momentum thickness and δ_H the enthalpy thickness defined earlier in eqns. (13.74), (13.75) and (13.77), respectively. Further, the quantities $U(x, t)$, $\varrho_\infty (x, t)$ and $T_\infty (x, t)$ refer to the frictionless external flow. In the special case of a stationary flow, we recover the relations known to us from eqns. (13.80) and (13.82). When the flow is incompressible, these relations simplify to:

$$\frac{\partial}{\partial t}(U\delta_1) + U\frac{\partial U}{\partial x}\delta_1 + \frac{\partial}{\partial x}(U^2\delta_2) = \frac{\tau_0}{\varrho},\tag{15.9}$$

and

$$\frac{\partial\delta_1}{\partial t} + \frac{1}{U^2}\frac{\partial}{\partial t}(U^2\delta_2) + U\frac{\partial\delta_3}{\partial x} + 3\,\delta_3\frac{\partial U}{\partial x} =$$

$$= \frac{2}{\varrho U^2}\int_0^\infty \mu\left(\frac{\partial u}{\partial y}\right)^2\mathrm{d}y.\tag{15.10}$$

When the flow is steady, eqn. (15.9) becomes identical with eqn. (8.35), whereas eqn. (15.10) transforms into eqn. (8.32).

We begin our study with the analysis of non-steady boundary layers in an incompressible fluid. Section XVf will contain some solutions of the boundary-layer equations for compressible non-steady flows.

2. The method of successive approximations. The integration of the non-steady boundary layer equations (15.1) to (15.3) can be carried out in most cases by a process of successive approximations, the method being based on the following physical reasoning: In the first instant, after the motion had started from rest, the boundary layer is very thin and the viscous term $\nu(\partial^2 u/\partial y^2)$ in eqn. (15.2) is very large, whereas the convective terms retain their normal values. The viscous term is then balanced by the non-steady acceleration $\partial u/\partial t$ together with the pressure term in which, at first, the contribution of $\partial U/\partial t$ is of major importance. Selecting a system of coordinates which is at rest with respect to the body and assuming that the fluid moves with respect to the body at rest, we can make the assumption that the velocity is composed of two terms

$$u(x,y,t) = u_0(x,y,t) + u_1(x,y,t).\tag{15.11}$$

Under these conditions the first approximation, u_0, satisfies the linear differential equation

$$\frac{\partial u_0}{\partial t} - \nu\frac{\partial^2 u_0}{\partial y^2} = \frac{\partial U}{\partial t}\tag{15.12}$$

with the boundary conditions $y = 0: u_0 = 0$; $y = \infty: u_0 = U(x,t)$. The equation for the second approximation, u_1, is obtained with reference to eqn. (15.2) in which the convective terms are calculated from u_0 and in which the convective pressure-term can now be taken into account. Hence we have

$$\frac{\partial u_1}{\partial t} - \nu\frac{\partial^2 u_1}{\partial y^2} = U\frac{\partial U}{\partial x} - u_0\frac{\partial u_0}{\partial x} - v_0\frac{\partial u_0}{\partial y},\tag{15.13}$$

with the boundary conditions $u_1 = 0$ at $y = 0$ and $u_1 = 0$ at $y = \infty$. This, too, is a linear equation. In addition to eqns. (15.12) and (15.13) we have the continuity equation for u_0, v_0 and u_1, v_1. Higher-order approximations u_2, u_3, ... can be obtained in a similar manner. The same method can be applied to the study of periodic boundary layers. However, the complexity of the method of successive approximations increases rapidly as higher approximations are considered.

3. C. C. Lin's method for periodic external flows. An alternative method has been devised by C. C. Lin [28] who modelled it on the approach employed for the study of turbulent flows, to be described in Chap. XVIII. It can be used for the solution of problems involving periodic motions in the free stream and relies on forming suitable averages of the quantities under investigation and on a linearization of the equation which describes the oscillatory component of the velocity in the boundary layer. On the other hand, the full equation which describes the *mean flow* is retained.

If the free-stream velocity $U(x, t)$ has an oscillating component, it can be written

$$U(x, t) = \overline{U}(x) + U_1(x, t) , \tag{15.14}$$

where the bar denotes an average value with respect to time over one period. Hence the average of the periodic component, $U_1(x, t)$, vanishes. Thus

$$\overline{U}_1(x, t) = 0 . \tag{15.15}$$

The velocity components u and v in the boundary layer and the pressure p are also separated into mean and periodic components:

$$\left.\begin{aligned} u(x, y, t) &= \bar{u}(x, y) + u_1(x, y, t) , \\ v(x, y, t) &= \bar{v}(x, y) + v_1(x, y, t) , \\ p(x, t) &= \bar{p}(x) \quad + p_1(x, t) , \end{aligned}\right\} \tag{15.16}$$

with

$$\bar{u}_1 = \bar{v}_1 = \bar{p}_1 = 0 . \tag{15.17}$$

Substituting from eqn. (15.14) into eqn. (15.6) and taking averages, we obtain

$$\overline{U} \frac{d\overline{U}}{dx} + \overline{U_1 \frac{\partial U_1}{\partial x}} = - \frac{1}{\varrho} \frac{\partial \bar{p}}{\partial x} \tag{15.18}$$

which, subtracted from eqn. (15.6) gives

$$\frac{\partial U_1}{\partial t} + \overline{U} \frac{\partial U_1}{\partial x} + U_1 \frac{\partial \overline{U}}{\partial x} + U_1 \frac{\partial U_1}{\partial x} - \overline{U_1 \frac{\partial U_1}{\partial x}} = - \frac{1}{\varrho} \frac{\partial p_1}{\partial x} . \tag{15.19}$$

Similarly, eqn. (15.2) will yield

$$\bar{u} \frac{\partial \bar{u}}{\partial x} + \bar{v} \frac{\partial \bar{u}}{\partial y} = \overline{U} \frac{d\overline{U}}{dx} + v \frac{\partial^2 \bar{u}}{\partial y^2} + F(x, y) \tag{15.20}$$

where

$$F(x, y) = \overline{U_1 \frac{\partial U_1}{\partial x}} - \overline{\left(u_1 \frac{\partial u_1}{\partial x} + v_1 \frac{\partial u_1}{\partial y} \right)} \tag{15.21}$$

and

$$\frac{\partial u_1}{\partial t} + \left(\bar{u} \frac{\partial u_1}{\partial x} + \bar{v} \frac{\partial u_1}{\partial y} \right) + \left(u_1 \frac{\partial \bar{u}}{\partial x} + v_1 \frac{\partial \bar{u}}{\partial y} \right) +$$

$$+ \left(u_1 \frac{\partial u_1}{\partial x} + v_1 \frac{\partial u_1}{\partial y} \right) - \overline{\left(u_1 \frac{\partial u_1}{\partial x} + v_1 \frac{\partial u_1}{\partial y} \right)} = \tag{15.22}$$

$$= \underline{\frac{\partial U_1}{\partial t}} + \overline{U} \frac{\partial U_1}{\partial x} + U_1 \frac{\partial \overline{U}}{\partial x} + U_1 \frac{\partial U_1}{\partial x} - \overline{U_1 \frac{\partial U_1}{\partial x}} + \underline{v \frac{\partial^2 u_1}{\partial y^2}} .$$

The essential simplification of the theory consists in retaining only the three under-lined terms in eqn. (15.22), which is thereby linearized and reduces to

$$\frac{\partial u_1}{\partial t} = \frac{\partial U_1}{\partial t} + \nu \frac{\partial^2 u_1}{\partial y^2} \, . \tag{15.23}$$

By estimating orders of magnitude it can be shown that the preceding approximation is a valid one if the ratio of the so-called "ac" boundary-layer thickness,

$$\delta_0 = \sqrt{\frac{2 \, \nu}{n}} \, , \tag{15.24}$$

formed with the frequency n of the oscillation, is small compared with the steady-state boundary-layer thickness δ which would exist if $U(x, t)$ were equal to $U(x)$. Hence, for the approximation to be valid we must have

$$\left(\frac{\delta_0}{\delta}\right)^2 \ll 1 \, , \tag{15.25}$$

which, in practice, restricts the theory to very high frequencies. It will be recalled that the quantity δ_0, eqn. (15.24), occured in the solution to the problem of an oscillating plate which has been considered in Sec. V a 7.

Equation (15.23) which is linear and related to the so-called heat-conduction equation (5.17) describes the oscillating component u_1 of the boundary-layer profile and can be solved in terms of the given oscillating component U_1 of the potential flow alone, because the process of linearization has made it independent of the mean motion. The normal components of the flow can be calculated from the equation of continuity (15.1) which can be split into an average part

$$\frac{\partial \bar{u}}{\partial x} + \frac{\partial \bar{v}}{\partial y} = 0 \, , \tag{15.26}$$

and an oscillating part

$$\frac{\partial u_1}{\partial x} + \frac{\partial v_1}{\partial y} = 0 \, . \tag{15.27}$$

Having solved for the oscillation $u_1(x, y, t)$, $v_1(x, y, t)$ we can return to eqn. (15.21) and calculate the function $F(x, y)$ which appears in eqn. (15.20). The latter now describes the mean motion $\bar{u}(x, y)$.

It should be noted that the equation for the mean flow, eqn. (15.20), has a form which is identical with the steady-state version of the boundary-layer equation. The only difference consists in the appearance of the additional term $F(x, y)$; this now plays the same part as the term $\bar{U} \cdot d\bar{U}/dx$ which originates in the pressure gradient. Both terms represent known functions in the differential equation. The only difference consists in the fact that the mean pressure gradient $\bar{U} \cdot d\bar{U}/dx$ is "impressed" on the boundary layer and is independent of the transverse coordinate y, whereas the additional term $F(x, y)$ depends on it.

Owing to the existence of oscillatory components, the average flow is different from that which would be obtained if the potential velocity $U(x, t)$ were averaged from the outset. The difference is clearly brought into evidence by the appearance of the function $F(x, y)$; it has its origin in the non-linearity of the differential equation.

It will be stated later in Chaps. XVIII and XIX that the essential characteristic of a steady turbulent stream consists in the fact that on the mean velocity of flow there is superimposed a random, three-dimensional, quasi-periodic oscillation. Consequently, problems involving turbulent free streams exhibit the same features as those being discussed now; they involve changes in direction as well as in the magnitude of the free-stream velocity U. In most cases it is customary to neglect the free-stream oscillation and to calculate as if the flow were steady and as if the potential velocity were given by $\bar{U}(x)$ instead of $U(x, t)$. This is equivalent to omitting the additional term $F(x, y)$ in eqn. (15.20) and necessarily leads to an average velocity profile which is different from $\bar{u}(x,y)$. The preceding remarks show clearly that the order in which the two operations, averaging and solving the equations, are performed is not immaterial and affects the final result.

4. Expansion into a series when a steady stream is perturbed slightly. Very often, problems in non-steady boundary layers involve an essentially steady flow on which there is superimposed a small non-steady perturbation. If it is assumed that the perturbation is small compared with the steady basic flow, it is possible to split the equations into a non-linear boundary-layer equation for the steady perturbation. A well-known example is that for which the external stream has the form

$$U(x,t) = \bar{U}(x) + \varepsilon \, U_1(x,t) + \ldots , \tag{15.28}$$

where ε denotes a very small number. The most important special case when the external perturbation is purely harmonic was studied exhaustively by M. J. Lighthill [27]. The same type of linearization can be employed when the temperature at the wall is represented by the expression

$$T_w(x,t) = \bar{T}_w(x) + \varepsilon \, T_{w1}(x,t) \tag{15.29}$$

or when the wall itself performs small, non-steady, perturbing motions (oscillating bodies).

In such cases we start with the assumption that the solutions for the dynamic as well as for the thermal boundary layer are of the following forms:

$$\left. \begin{aligned} u(x,y,t) &= u_0(x,y) + \varepsilon \, u_1(x,y,t) + \varepsilon^2 \, u_2(x,y,t) + \ldots , \\ v(x,y,t) &= v_0(x,y) + \varepsilon \, v_1(x,y,t) + \varepsilon^2 \, v_2(x,y,t) + \ldots , \\ T(x,y,t) &= T_0(x,y) + \varepsilon \, T_1(x,y,t) + \varepsilon^2 \, T_2(x,y,t) + \ldots . \end{aligned} \right\} \tag{15.30}$$

The postulated forms from eqns. (15.30) are introduced into eqns. (15.1) to (15.3) and the resulting terms are ordered with respect to the powers of ε. From the requirement that the differential expressions which multiply each power of ε must vanish singly, we obtain a cascade of differential equations. We list them for the case when $\varrho = \text{const}$, when the external flow is of the form of eqn. (15.28), and when the wall temperature is given by eqn. (15.29):

Equations for zeroth order (steady basic flow):

$$\left. \begin{aligned} \frac{\partial u_0}{\partial x} + \frac{\partial v_0}{\partial y} &= 0 , \\ u_0 \frac{\partial u_0}{\partial x} + v_0 \frac{\partial u_0}{\partial y} &= \bar{U} \frac{d\bar{U}}{dx} + v \frac{\partial^2 u_0}{\partial y^2} , \\ u_0 \frac{\partial T_0}{\partial x} + v_0 \frac{\partial T_0}{\partial y} &= a \frac{\partial^2 T_0}{\partial y^2} , \end{aligned} \right\} \tag{15.31}$$

with the boundary conditions

$$y = 0 : \quad u_0 = v_0 = 0 \; ; \; T_0 = \bar{T}_w(x) \; ,$$

$$y = \infty : \quad u_0 = \bar{U}(x) \quad ; \; T_0 = T_\infty \; .$$

Equations of first order (purely non-steady):

$$\frac{\partial u_1}{\partial x} + \frac{\partial v_1}{\partial y} = 0 \; ,$$

$$\frac{\partial u_1}{\partial t} + u_0 \frac{\partial u_1}{\partial x} + u_1 \frac{\partial u_0}{\partial x} + v_0 \frac{\partial u_1}{\partial y} + v_1 \frac{\partial u_0}{\partial y} =$$

$$= \frac{\partial U_1}{\partial t} + \bar{U} \frac{\partial U_1}{\partial x} + U_1 \frac{d\bar{U}}{dx} + \nu \frac{\partial^2 u_1}{\partial y^2} \; ,$$

$$\frac{\partial T_1}{\partial t} + u_0 \frac{\partial T_1}{\partial x} + u_1 \frac{\partial T_0}{\partial x} + v_0 \frac{\partial T_1}{\partial y} + v_1 \frac{\partial T_0}{\partial y} = a \frac{\partial^2 T_1}{\partial y^2} \; ,$$

(15.32)

with the boundary conditions

$$y = 0 : \quad u_1 = v_1 = 0 \; ; \; T_1 = T_{w1}(x,t) \; ,$$

$$y = \infty : \quad u_1 = U_1(x,t) \; ; \; T_1 = 0 \; .$$

Equations of second order (steady and non-steady terms):

$$\frac{\partial u_2}{\partial x} + \frac{\partial v_2}{\partial y} = 0 \; ,$$

$$\frac{\partial u_2}{\partial t} + u_0 \frac{\partial u_2}{\partial x} + u_1 \frac{\partial u_1}{\partial x} + u_2 \frac{\partial u_0}{\partial x} + v_0 \frac{\partial u_2}{\partial y} +$$

$$+ v_1 \frac{\partial u_1}{\partial y} + v_2 \frac{\partial u_0}{\partial y} = U_1 \frac{\partial U_1}{\partial x} + \nu \frac{\partial^2 u_2}{\partial y^2} \; ,$$

$$\frac{\partial T_2}{\partial t} + u_0 \frac{\partial T_2}{\partial x} + u_1 \frac{\partial T_1}{\partial x} + u_2 \frac{\partial T_0}{\partial x} + v_0 \frac{\partial T_2}{\partial y} +$$

$$+ v_1 \frac{\partial T_1}{\partial y} + v_2 \frac{\partial T_0}{\partial y} = a \frac{\partial^2 T_2}{\partial y^2} \; ,$$

(15.33)

with the boundary conditions

$$y = 0 : \quad u_2 = v_2 = T_2 = 0 \; ,$$

$$y = \infty : \quad u_2 = T_2 = 0 \; .$$

The equations of higher orders have corresponding structures. The preceding systems of equations can be solved one after the other, it being noted that all, except those of zeroth order, are linear. If equations (15.1) to (15.3) were to possess exact solutions of the postulated form (15.30) up to order ε^n, then, generally speaking, the solutions arrived at by the preceding scheme would differ from the exact solution by terms of order ε^{n+1}.

The application of this method to the calculation of periodic boundary layers will be discussed in Sec. XVe3. A similar series expansion, but in terms of powers of

$$\frac{x^k}{U^{k+1}} \frac{\partial^k U}{\partial t^k} \; , \quad \text{and} \quad \frac{1}{T_w - T_\infty} \left(\frac{x}{U} \right)^k \frac{\partial^k T_w}{\partial t^k} \; ,$$

(15.34)

was employed by F. K. Moore [31], S. Ostrach [35], F. K. Moore and S. Ostrach [32] and E. M. Sparrow [50] (cf. Sec. XVf2).

5. Similar and semi-similar solutions. When we studied the theory of steady, two-dimensional boundary layers (see Sec. VIII b), we described as similar that class of solutions for which the dependence on the two variables x and y could be reduced to that on a single variable η by the application of a suitable similarity transformation. In an analogous manner, we say that a solution of a non-steady two-dimensional problem belongs to the class of similar solutions when the three independent variables x, y, t can be reduced to a single variable η. H. Schuh [46] and Th. Geis [10] have indicated all such solutions for which a reduction to a single variable is possible, that is, such as are of the form

$$u(x,y,t) = U(x,t) \cdot H(\eta) , \quad \text{with} \quad \eta = \frac{y}{N(x,t)} . \tag{15.35}$$

For example, external flows of the form $U(x,t) = mx/t$ and the cases when $U(x,t) = Ct^n$ mentioned in Sec. XV c belong to this class. The similar solutions for an external stream of the form $U(x,t) = x/(a+bt)$, where a and b are constants, were analyzed by K. T. Yang [71].

If a transformation can be found which reduces the three independent variables x, y, t to two, we say that the resulting solution is semi-similar [21]. In particular, when the variables are reduced to y and x/t, the solutions are also called pseudo-steady (*cf.* [7]). A solution of this type was discovered by I. Tani [56] for the case when the external flow is given by $U(x,t) = U_0 - x/(T-t)$, with U_0 and T denoting constants. A wider class of semi-similar solutions was considered by H. A. Hassan [19]; see also ref. [21].

6. Approximate solutions. Attempts to solve the complete set of equations for the general case when the external flow, $U(x,t)$, is an arbitrary function of the variables would lead to very great difficulties. For this reason, one must often resort to approximate methods, for example to analogs of the Kármán-Pohlhausen procedure, discussed in Chap. X. Such procedures have been developed in detail for incompressible, non-steady boundary layers by H. Schuh [46], L. A. Rozin [42], and K. T. Yang [72]. Reference [72] deals also with thermal boundary layers. The integral relations given in eqns. (15.9) and (15.10) form here the starting point. Since the process of integrating over the boundary-layer thickness eliminates only one variable, y, one is still left with a partial differential equation.

b. Boundary-layer formation after impulsive start of motion

We now propose to analyze the first phases of the motion after it has been started from rest. The problem can be simplified considerably, as suggested by H. Blasius [8], if it is assumed that the body is accelerated very rapidly, the fluid being at rest, or, in other words, that it is started impulsively. Thus the body assumes its full velocity discontinuously and the velocity remains constant afterwards. In the system of coordinates which is, as assumed before, linked with the body, the potential flow is defined by the conditions

$$\left.\begin{array}{ll} t \leq 0 : & U(x,t) = 0, \\[2mm] t > 0 : & U(x,t) = U(x), \end{array}\right\} \tag{15.36}$$

where $U(x)$ denotes the potential flow about the body in the steady state. In this particular case we have $\partial U/\partial t = 0$, and equation (15.12) of the first approximation becomes simply

$$\frac{\partial u_0}{\partial t} - \nu \frac{\partial^2 u_0}{\partial y^2} = 0 \; , \qquad (15.37)$$

with $u_0 = 0$ for $y = 0$, and $u_0 = U(x)$ for $y = \infty$. This equation is identical with that for one-dimensional heat conduction. It was solved in Sec. V 4 for the case of a plate started impulsively in its own plane, while the fluid was at rest at a large distance from it. It was then possible to introduce a new dimensionless variable (*similarity transformation*):

$$\eta = \frac{y}{2 \sqrt{\nu t}} \; . \qquad (15.38)$$

In this manner we obtain the solution in the form

$$u_0(x, y, t) = U(x) \times \zeta_0{}'(\eta) = U(x) \; \text{erf} \; \eta \; . \qquad (15.39)$$

This is the first approximation both for the two-dimensional and for the axi-symmetrical case. Further, if the potential velocity is independent of x, i. e. if $U = U_0 = \text{const}$ (flat plate at zero incidence), eqn. (15.39) constitutes the exact solution of eqn. (15.2), since the convective terms in eqn. (15.13) vanish together with the pressure term so that $u_1 \equiv 0$. However, the solution arrived at in this way does not constitute the complete solution to the problem and applies only sufficiently far downstream where the influence of the edge is negligible and where the flow behaves as if the plate were infinitely long. Strictly speaking, the complete solution must also satisfy the condition that $u(0, y, t) = 0$ for all values of y and t. The complete solution is given in ref. [54].

In the general case, when the external flow $U(x, t)$ depends on the space coordinate, it is necessary to make a distinction between the two-dimensional and the axisymmetrical cases.

1. Two-dimensional case. We shall begin by considering the two-dimensional case. For this case we assume a power series in time for the stream function stipulating that it has the form

$$\psi(x, y, t) = 2 \sqrt{\nu t} \; \left\{ U \zeta_0(\eta) + t \, U \frac{dU}{dx} \zeta_1(\eta) + \cdots \right\} . \qquad (15.40)$$

Hence, the velocity components $u = \partial\psi/\partial y$ and $v = -\partial\psi/\partial x$ become:

$$\left. \begin{aligned} u &= U \zeta_0{}' + t \, U \frac{dU}{dx} \zeta_1{}' + \cdots \\[2mm] -v &= 2 \sqrt{\nu t} \left\{ \frac{dU}{dx} \zeta_0 + t \left[\left(\frac{dU}{dx} \right)^2 + U \frac{d^2 U}{dx^2} \right] \zeta_1 + \cdots \right\} . \end{aligned} \right\} \qquad (15.41)$$

Inserting these expressions into eqn. (15.12) we obtain the differential equation of the first approximation:

$$\zeta_0{}''' + 2 \, \eta \, \zeta_0{}'' = 0 \; , \qquad (15.42)$$

with the boundary conditions $\zeta_0 = \zeta_0' = 0$ at $\eta = 0$, and $\zeta_0' = 1$ at $\eta = \infty$. Equation (15.42) is identical with eqn. (5.21) and the solution for ζ_0' is indicated in eqn. (15.39). The function ζ_0' is shown plotted in Fig. 15.1.

Combining eqn. (15.13) with (15.40) we obtain the differential equation for the second approximation $\zeta_1(\eta)$ in the form:

$$\zeta_1''' + 2\,\eta\,\zeta_1'' - 4\,\zeta_1' = 4\,(\zeta_0'^2 - \zeta_0\zeta_0'' - 1)\,,$$

with the boundary conditions $\zeta_1 = \zeta_1' = 0$ at $\eta = 0$ and $\zeta_1' = 0$ at $\eta = \infty$. The solution derived by H. Blasius is:

$$\zeta_1' = -\frac{3}{\sqrt{\pi}}\,\eta\,\exp(-\eta^2)\,\mathrm{erfc}(\eta) + \frac{1}{2}\,(2\,\eta^2 - 1)\,\mathrm{erfc}^2(\eta) + \frac{2}{\pi}\,\exp(-2\,\eta^2) +$$

$$+\frac{1}{\sqrt{\pi}}\,\eta\,\exp(-\eta^2) + 2\,\mathrm{erfc}(\eta) - \frac{4}{3\pi}\,\exp(-\eta^2) +$$

$$+\left(\frac{3}{\sqrt{\pi}} + \frac{4}{3\pi^{3/2}}\right)\{\eta\,\exp(-\eta^2) - \frac{\sqrt{\pi}}{2}\,(2\,\eta^2 + 1)\,\mathrm{erfc}(\eta)\}\,. \qquad (15.43)$$

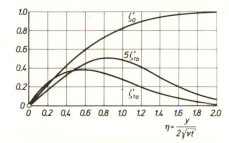

Fig. 15.1. The functions ζ_0' and $\zeta_1' = \zeta_{1a}'$ and ζ_{1b}' for the velocity distribution in the nonsteady boundary layer, eqns. (15.41) and (15.50), for impulsive motion

The function ζ_1' is shown plotted (as function ζ_{1a}') in Fig. 15.1. The initial slopes of the two functions, required for the calculation, of separation are given by

$$\zeta_0''(0) = \frac{2}{\sqrt{\pi}} = 1\cdot128\,; \quad \zeta_1''(0) = \frac{2}{\sqrt{\pi}}\left(1 + \frac{4}{3\pi}\right) = 1\cdot607. \qquad (15.44)$$

An exact expression for the next term of the expansion of the stream function in terms of time was obtained by S. Goldstein and L. Rosenhead [14]. E. Boltze [9] previously derived a less accurate solution when he considered the axially symmetrical problem (see succeeding section).

The question of the position of the point of separation can be answered with the aid of the second approximation. In this connexion we shall consider the cases of the circular and the elliptic cylinder. The condition for the point of separation is given by $\partial u/\partial y = 0$ for $y = 0$, which leads to the following condition for the time of separation t_s:

$$\zeta_0''(0) + \zeta_1''(0)\,t_s\,\frac{dU}{dx} = 0\,,$$

as seen from eqn. (15.41). With the values (15.44) this becomes

$$1 + \left(1 + \frac{4}{3\pi}\right) \frac{dU}{dx} t_s = 0 \,. \tag{15.45}$$

Equation (15.45) allows us to calculate the instant at which separation begins at a given place. Separation occurs only at points where dU/dx is negative. The point of earliest separation occurs at a place where the absolute value of dU/dx is largest. It does not follow that this coincides with the downstream stagnation point as will be demonstrated on the example of the elliptic cylinder.

Example : *Circular cylinder*

For the circular cylinder of radius R in a stream of velocity U_∞ we obtain :

$$U(x) = 2\,U_\infty \sin \frac{x}{R} \quad \text{and} \quad \frac{dU}{dx} = 2\,\frac{U_\infty}{R} \cos \frac{x}{R} \,,$$

where x denotes the arc measured from the upstream stagnation point. The abolute value of the gradient dU/dx has its maximum at the downstream stagnation point, and separation occurs at a time

$$t_s = \frac{R/U_\infty}{2\left(1 + \frac{4}{3\pi}\right)} \,, \tag{15.46}$$

as seen from eqn. (15.45). The distance covered until separation begins is $s_s = t_s\,U_\infty$, so that

$$s_s = \frac{R}{2\left(1 + \frac{4}{3\pi}\right)} = 0{\cdot}351\,R \,.$$

The boundary layer in the neighbourhood of the downstream stagnation point of a circular cylinder has been calculated by I. Proudman and K. Johnson [35a] for the case of sudden acceleration; they solved the problem on the basis of the Navier-Stokes equations. *Cf.* M. Katagiri [25a].

Example: *Elliptic cylinder* [16, 59]

Let the semi-axes of the elliptic cylinder be a and b, and let $k = b/a$ be their ratio, no assumption about their relative magnitude being made, so that $a \gtrless b$. The equation of the ellipse can be written as $x^2/a^2 + y^2/b^2 = 1$. Introducing the angular coordinate ϕ, defined by $x/a = \cos \phi$ and $y/b = \sin \phi$, and assuming that the cylinder is started impulsively with a velocity U_∞ in a direction parallel to the axis a, we can write for the velocity distribution along the contour of the ellipse:

$$\frac{U(s)}{U_\infty} = \frac{1 + k}{\sqrt{1 + k^2 \cot^2 \phi}} \,,$$

and for the velocity gradient

$$\frac{a}{U_\infty} \frac{dU}{ds} = \frac{(1 + k)\,k^2 \cos \phi}{(\sin^2 \phi + k^2 \cos^2 \phi)^2} \,.$$

It is easy to verify that the maximum value of the velocity gradient coincides with the downstream stagnation point if $k^2 < 4/3$. For $k^2 > 4/3$ the maximum value of the gradient occurs at $\phi = \phi_m$, where

$$\cos^2 \phi_m = \frac{1}{3\,(k^2 - 1)} \,.$$

The maximum values of the gradient become

$$
\begin{aligned}
k^2 \leq \frac{4}{3} &: \quad \frac{b}{U_\infty}\left(\frac{dU}{ds}\right)_m = \frac{1+k}{k} \\
k^2 \geq \frac{4}{3} &: \quad \frac{b}{U_\infty}\left(\frac{dU}{ds}\right)_m = \frac{3\sqrt{3}}{16}\,\frac{k^3\,(1+k)}{\sqrt{k^2-1}}.
\end{aligned}
\tag{15.47}
$$

Inserting the values from eqns. (15.47) into eqn. (15.45), we find that the time elapsed until the onset of separation is

$$
\begin{aligned}
t_s\,\frac{U_\infty}{a} &= \frac{k^2}{\left(1+\frac{4}{3\pi}\right)(1+k)} && \text{for } k^2 \leq \frac{4}{3}\;; \\[2mm]
t_s\,\frac{U_\infty}{b} &= \frac{16\,\sqrt{k^2-1}}{\left(1+\frac{4}{3\pi}\right)3\sqrt{3}\,k^3\,(1+k)} && \text{for } k^2 \geq \frac{4}{3}.
\end{aligned}
\tag{15.48}
$$

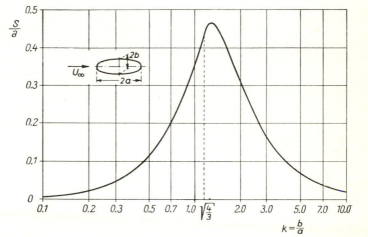

Fig. 15.2.   Distance s traversed by elliptic cylinder until the onset of separation in the case of impulsive acceleration from rest

The distance s traversed by the elliptic cylinder until the onset of separation, and given by $s = t_s\,U_\infty$, is seen plotted in Fig. 15.2 in terms of the ratio of the axes $k = b/a$. The place where separation occurs first is given by

$$
y_s = 0 \quad \text{for} \quad k^2 \leq \frac{4}{3}\;,
$$

$$
\frac{y_s^2}{b^2} = 1 - \frac{1}{3\,(k^2-1)} \quad \text{for} \quad k^2 \geq \frac{4}{3}.
$$

For $k = 1$ eqn. (15.48) transforms into eqn. (15.46) for the circular cylinder. Beginning with this value the time t_s for the onset of separation decreases with increasing $k = b/a$, and the position of the point of separation moves from the end of axis a towards the end of axis b. In the limit $b/a \to \infty$, i.e. for a plate at right angles to the direction of motion, we have $t_s = 0$ and $y_s = b$. Hence the onset of separation is immediate for the case of a flat plate perpendicular to the direction of motion, and it takes place at the edge.

The formation of the boundary layer on a rotating cylinder started impulsively was calculated by W. Tollmien [60] by an analogous method in his Goettingen thesis presented in 1924. In this case separation is suppressed on that side of the cylinder where the tangential velocity has the same direction as the velocity of flow.

The process of acceleration for an elliptic cylinder at an angle of incidence has been treated in a paper by H. J. Lugt [28a]. In it, the author succeeded in calculating the formation of the starting vortices at Reynolds numbers in the range $\mathsf{R} = Vd/\nu = 15$ to 200. We wish to refer the reader also to a paper by D. Dumitrescu and M. D. Cazacu [9a] which discusses the same problem but for a flat plate at an angle of incidence. See also Fig. 4.2 for the plate at right angles to the stream.

2. Axially symmetrical problem. The process of boundary-layer formation about an axially symmetrical body accelerated impulsively was investigated by E. Boltze [9] in his Goettingen thesis. We consider the boundary layer on a body of revolution whose shape is defined by $r(x)$, Fig. 11.6, and which is set in motion at $t = 0$. The acceleration is impulsive, and the cylinder moves in the direction of its axis. The relevant equations are now eqns. (15.2) and (11.27b), and the solution can again be represented as a sum of a first approximation, u_0, and a second approximation, u_1, defined by eqns. (15.12) and (15.13) respectively. In view of the changed form of the continuity equation we introduce a different stream function, namely

$$u = \frac{1}{r}\frac{\partial \psi}{\partial y} ; \quad v = -\frac{1}{r}\frac{\partial \psi}{\partial x} ,$$

and we assume it to be of the form

$$\psi(x, y, t) = 2\sqrt{\nu t}\left\{r\, U\, \zeta_0(\eta) + t\left[r\, U\frac{dU}{dx}\zeta_{1a}(\eta) + U^2\frac{dr}{dx}\zeta_{1b}(\eta)\right] + \dots\right\}. \quad (15.49)$$

Hence

$$\frac{u}{U} = \zeta_0' + t\left[\frac{dU}{dx}\zeta_{1a}' + \frac{U}{r}\frac{dr}{dx}\zeta_{1b}'\right]. \quad (15.50)$$

The variable η has the same meaning as in the two-dimensional problem, eqn. (15.38). The differential equation for ζ_0 resulting from eqn. (15.12) is identical with equation (15.42) for the two-dimensional problem, as already mentioned. For the second approximation in the expansion in terms of time we now obtain from eqn. (15.13) the following differential equations, defining ζ_{1a} and ζ_{1b}:

$$\left.\begin{array}{l}
\zeta_{1a}''' + 2\,\eta\,\zeta_{1a}'' - 4\,\zeta_{1a}' = 4(\zeta_0'^2 - 1 - \zeta_0\,\zeta_0'') , \\[2mm]
\zeta_{1b}''' + 2\,\eta\,\zeta_{1b}'' - 4\,\zeta_{1b}' = -4\,\zeta_0\,\zeta_0'' ,
\end{array}\right\} \quad (15.51)$$

with the boundary conditions

$$\eta = 0 : \quad \zeta_{1a} = \zeta_{1a}' = 0 ; \quad \zeta_{1b} = \zeta_{1b}' = 0 ;$$

$$\eta = \infty : \quad \zeta_{1a}' = 0 ; \quad \zeta_{1b}' = 0 .$$

The equation for ζ_{1a} is identical with that for ζ_1 of the two-dimensional problem, and the equation for ζ_{1a} was solved numerically by E. Boltze [9]. The character of ζ_{1a}' and ζ_{1b}' can be ascertained from Fig. 15.1. The initial slope of ζ_{1b}' is $\zeta_{1b}''(0) = 0\cdot169$.

In accordance with eqn. (15.50) the onset of separation is defined by the condition $(\partial u/\partial y)_{y=0} = 0$ which gives

$$\zeta_0''(0) + t_s \left[\frac{dU}{dx} \, \zeta_{1a}''(0) + \frac{U}{r} \frac{dr}{dx} \, \zeta_{1b}''(0) \right] = 0$$

or, with the preceding numerical values of $\zeta_0''(0)$, $\zeta_{1a}''(0) = \zeta_1''(0)$ and $\zeta_{1b}''(0)$,

$$1 + t_s \left[\frac{dU}{dx} \left(1 + \frac{4}{3\pi} \right) + 0.150 \, \frac{U}{r} \frac{dr}{dx} \right] = 0. \tag{15.52}$$

E. Boltze calculated two further terms of the expansion for the stream function in eqn. (15.49).

Example: *Sphere*

By way of example E. Boltze computed the process of boundary-layer formation on a sphere which is started impulsively from rest. Denoting the radius of the sphere by R and the free-stream velocity by U_∞, we have in this case

$$r = R \sin \frac{x}{R} \, ; \qquad U(x) = \frac{3}{2} \, U_\infty \sin \frac{x}{R}.$$

The beginning of separation now follows from eqn. (15.52), or

$$1 + t_s \, \frac{3}{2} \, \frac{U_\infty}{R} \, 1.573 \cos \frac{x}{R} = 0.$$

Separation sets in at the stagnation point downstream, i. e. at a place where $\cos (x/R) = -1$, so that $\frac{3}{2} \, t_s \, U_\infty / R = 1/1.573 = 0.635$. Taking into account the two further terms of the expansion for the stream function calculated by E. Boltze, we obtain the more accurate value 0.589 for this constant. Thus, the instant of separation for a sphere started impulsively becomes

$$t_s = 0.392 \, \frac{R}{U_\infty}. \tag{15.53}$$

The distance covered in that time is $s_s = U_\infty t_s = 0.392 \, R$, or, in round figures, 40 per cent of the radius of the sphere. The point of separation moves from $\phi = \pi$, at first rapidly, and later slowly, towards $\phi \approx 110°$ which is its position in steady flow, and reaches it only after an infinite time. Fig. 15.3 represents the pattern of streamlines and the velocity distribution for an intermediate instant, which corresponds to a distance of $0.6 \, R$ covered by the sphere. This corresponds to a time of 0.6 sec, with a radius of $R = 10$ cm (about 4 in) and a velocity $U_\infty = 10$ cm/sec (about 0.33 ft/sec). The streamlines are seen plotted in Fig. 15.3 in which the linear scale of the thickness of the boundary layer has been exaggerated for the sake of clarity. For water with $\nu = 0.01 \times 10^{-4}$ m^2/sec (about 0.1×10^{-4} ft^2/sec) the magnification factor is about 30. The magnitudes of the velocities in the closed vortex are very small and the velocity gradient and the circulation are greatest outside the streamline $\psi = 0$ at the point of separation.

The idealized process of instantaneous acceleration assumed in the preceding theory is a good approximation to actual cases if the time of acceleration is small compared with the time which elapses before separation sets in.

The process of the formation of a boundary layer on a rotating disk was studied by K. H. Thiriot [58] in his thesis presented to the University of Goettingen. He considered the case of a disk accelerated impulsively in a fluid at rest to a uniform angular velocity, as well as the case of a disk rotating with the fluid and suddenly arrested in its motion. The ultimate state of motion for the first case is the solution for a disk rotating in a fluid at rest given by W. G. Cochran and discussed in Sec. V 11.

The final state of motion for the second problem is given by the solution due to U. T. Boedewadt and discussed in Sec. X a. It concerns the rotation of the fluid body over a fixed plane. A generalization of all these cases has been discussed by K. H. Thiriot [57] in a further paper, when he considered the case of a disk rotating with the fluid body and impulsively accelerated, or decelerated, so that its angular velocity is changed by a small quantity compared with that of the fluid. It is noteworthy that a stationary boundary layer is then formed in the neighbourhood of the rotating disk. The details of the growth of a boundary layer on a disk started impulsively were computed by S. D. Nigam [34].

E. M. Sparrow and J. L. Gregg [51] solved the problem of a disk which rotates with a non-uniform augular velocity; C. R. Illingworth [25] and Y. D. Wadhwa [64] treated the problem of the growth of a boundary layer on a rotating body of revolution. The case considered by H. Wundt [70], namely that of a yawed cylinder accelerated impulsively, constitutes another example of a three-dimensional, non-steady boundary layer. Additional solutions for three-dimensional, non-steady boundary layers can be found in refs. [20, 21, 22, 52, and 53]:

W. Wuest [69] obtained solutions for three-dimensional non-steady boundary layers on bodies which perform non-steady motions at right angles to the main flow. One example considered was that of a cylinder in steady cross-flow which is made to perform axial periodic oscillations. The case of a wedge which oscillates harmonically in a direction parallel to its leading edge, also considered, contains as special cases those of a flat plate and stagnation flow.

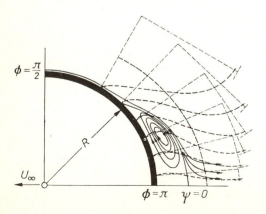

Fig. 15.3. Boundary layer on the downstream side of a sphere accelerated impulsively after the onset of separation; from Boltze [9]. The sphere has traversed a distance of $0{\cdot}6\,R$

c. Boundary-layer formation in accelerated motion

The process of boundary-layer formation in two-dimensional flow for the case of uniform acceleration of the body has been calculated by H. Blasius. The results are very similar to those for an impulsive start of the motion. The potential velocity of the body is now given in the form

$$
\begin{aligned}
t \leq 0 : \quad & U(x,t) = 0 , \\
t > 0 : \quad & U(x,t) = t \times w(x) .
\end{aligned}
\Bigg\} \tag{15.54}
$$

It is again possible to assume a series in terms of successive approximation, as given by eqn. (15.11). These approximations then satisfy eqns. (15.12) and (15.13). Assuming an expansion of the stream function in terms of time of the form

$$
\psi(x, y, t) = 2 \sqrt{\nu t} \left\{ t\, w\, \zeta_0(\eta) + t^3\, w\, \frac{dw}{dx} \, \zeta_1(\eta) + \cdots \right\}
$$

and

$$
u(x, y, t) = U \left(\zeta_0' + t^2 \frac{dw}{dx} \zeta_1' + \cdots \right), \tag{15.55}
$$

it is possible to deduce the following differential equations for $\zeta_0(\eta)$ and $\zeta_1(\eta)$:

$$
\begin{aligned}
& \zeta_0''' + 2\,\eta\,\zeta_0'' - 4\,\zeta_0' = -4 \\
& \zeta_1''' + 2\,\eta\,\zeta_1'' - 12\,\zeta_1' = -4 + 4(\zeta_0'^2 - \zeta_0\,\zeta_0'')
\end{aligned}
\Bigg\} , \tag{15.56}
$$

with the boundary conditions

$$
\begin{aligned}
\eta = 0 : \quad & \zeta_0 = \zeta_0' = 0 , \quad \zeta_1 = \zeta_1' = 0 , \\
\eta = \infty : \quad & \zeta_0' = 1 , \quad \zeta_1' = 0 .
\end{aligned}
$$

The solution for the function ζ_0' given by H. Blasius is of the form:

$$
\zeta_0' = 1 + \frac{2}{\sqrt{\pi}} \, \eta \exp(-\eta^2) - (1 + 2\,\eta^2)\, \mathrm{erfc}\,\eta . \tag{15.57}
$$

Blasius was also able to give a solution for ζ_1' in closed form. The initial slopes which are required for the calculation of separation are:

$$
\zeta_0''(0) = \frac{4}{\sqrt{\pi}} = 2\cdot257; \quad \zeta_1''(0) = \frac{31}{15\sqrt{\pi}} - \frac{256}{225\sqrt{\pi^3}} = 0\cdot427 \, \frac{4}{\sqrt{\pi}} = 0\cdot964.
$$

The beginning of separation in this case is given by eqn. (15.55), and when only the first two terms of the expansion are used we obtain

$$
\zeta_0''(0) + t_s^2 \frac{dw}{dx} \zeta_1''(0) = 0
$$

or, with the preceding numerical values of $\zeta_0''(0)$ and $\zeta_1''(0)$:

$$
1 + 0\cdot427\, t_s^2 \frac{dw}{dx} = 0 ,
$$

so that

$$
t_s^2 \frac{dw}{dx} = -2\cdot34 .
$$

This expression can also be written in the form:

$$
1 + 0\cdot427\, t_s \frac{dU}{dx} = 0 .
$$

Upon comparing with eqn. (15.45), it is seen that for equal values of dU/dx separation occurs earlier when the motion is started impulsively than when the acceleration is uniform.

H. Blasius calculated two further terms of the expansion, and with their aid the equation for t_s is obtained in the following modified form:

$$1 + 0.427 \frac{dw}{dx} t_s{}^2 - 0.026 \left(\frac{dw}{dx}\right)^2 t_s{}^4 - 0.01 w \frac{d^2w}{dx^2} t_s{}^4 = 0 .$$

For the case of a cylinder which is placed symmetrically with respect to the direction of flow the last term vanishes at the downstream stagnation point, and we obtain

$$t_s{}^2 \frac{dw}{dx} = - 2.08 . \qquad\qquad (15.58)$$

Example: *Circular cylinder*

For the case of a circular cylinder we have

$$U(x, t) = t\, w(x) = 2\, b\, t \sin \frac{x}{R} ,$$

where b denotes the constant acceleration. Hence

$$w(x) = 2\, b \sin \frac{x}{R} ; \qquad \frac{dw}{dx} = \frac{2\, b}{R} \cos \frac{x}{R} .$$

The point at which separation occurs first coincides, in this case too, with the downstream stagnation point, $\cos (x/R) = -1$. Thus from eqn. (15.58) we obtain

$$t_s{}^2 = 1.04 \frac{R}{b} .$$

The distance covered by the cylinder until separation begins is given by $s = \frac{1}{2} b\, t_s{}^2$. which then becomes $s = 0.52\, R$, and is also greater than that for the case of impulsive motion. The argument in Sec. XVb, concerning the point at which separation first occurs, remains valid in the present case. The pattern of streamlines for the case under consideration is given in Fig. 15.4, which is based on Blasius's work. This pattern corresponds to time $T = t \sqrt{b/R} = 1.58$, the distance covered by the cylinder being equal to $1.25\, R$. Assuming $R = 10$ cm (about 4 in), $b = 0.1$ cm/sec^2 (about 0.04 in/sec^2 = 0.0033 ft/sec^2), we obtain $\sqrt{b/R} = 0.1$ sec^{-1}, and the time elapsed since the beginning of the motion is $t = 15.8$ sec. Figure 15.4 shows the shape of the resulting boundary layer, the linear scale having been increased in the same way as in Fig. 15.3. For water with $v = 0.01 \times 10^{-4}$ m^2/sec (about 0.1×10^{-4} ft^2/sec) the linear factor is equal to about $\sqrt{10}$.

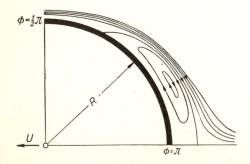

Fig. 15.4. Boundary layer on the downstream side of circular cylinder during start with uniform acceleration after the beginning of separation (Blasius)

Velocity: $U(t) = b \times t$;

Pattern at time $T = t \sqrt{b/R} = 1.58$;

Separation first occurs at $T_s = t_s \sqrt{b/R} = 1.02$

H. Goertler [15] extended the theoretical calculation of the process of boundary-layer formation during acceleration assuming a potential flow of the form $U(x, t) = w(x)t^n$, with $n = 0$, $1, 2, 3, 4$. For $n = 0$ and $n = 1$ he obtained the preceding cases of impulsive and uniform acceleration respectively. H. Goertler gave explicit expressions for the first term in the expansion of the stream function in powers of time for the values $n = 0$ to 4. The second term was evaluated at the wall together with its initial slope so that the instant at which separation begins and the distance covered, e. g. by a cylinder, can be computed. In this connexion a paper by E. J. Watson [65] may also be consulted.

d. Experimental investigation of the starting process

The process of boundary-layer formation can be studied with the aid of the previously discussed analytical methods but it cannot be carried very much beyond the beginning of separation. The flow pattern outside the boundary layer becomes markedly changed after the onset of separation, particularly on the downstream side of blunt bodies such as a circular cylinder. Consequently, calculations based on the theoretical pressure distribution derived from potential theory give an inaccurate representation of the further course of the process. The photographs in Fig. 15.5 illustrate the development of the flow pattern around a circular cylinder. Figure 15.5a shows that a potential frictionless flow-pattern does exist during the first instants after starting. Figure 15.5b represents the moment when separation has just begun at the downstream stagnation point, and in Fig. 15.5c the point of separation has already moved a considerable distance upstream. The streamline through the point of separation encloses a region where the flow velocities are very small. The vorticity is largest outside this streamline; it forms a vortex sheet which curls up as the pattern continues to develop and forms two concentrated vortices, Fig. 15.5d. In the free stream behind these vortices it is possible to discern the existence of a stagnation point which coincides with the junction of the two streamlines through the points of separation. Figure 15.5e shows that the vortices continue to grow. They become unstable with the course of time and are carried away from the body by the external flow, Fig. 15.5f. In the steadystate the motion oscillates and the pressure distribution around the body differs considerably from that stipulated by potential-flow theory.

The phenomena under consideration have been investigated in more detail on a circular cylinder by M. Schwabe [47], who measured, in particular, the pressure distribution around the cylinder during the process of acceleration from rest. Pressure-distribution curves around the cylinder contour for several phases of the process are given in Fig. 15.6. The distance between the cylinder and the stagnation point in the free stream behind the two vortices is denoted here by d. It is seen that the measured pressure distribution is very close to that in potential flow in the early stages of the process but deviates progressively more from it as time advances. H. Rubach [43] attempted to describe this type of flow about a circular cylinder with the aid of potential theory, assuming the existence of two symmetrical point-vortices downstream from the body at a position roughly corresponding to that in Fig. 15.5e. It is, however, necessary to remark here that the resemblance to a pattern with two such vortices is only transitory. Very extensive experimental investigations of the wake formed behind a circular cylinder in the range of Reynolds numbers $5 < R < 40$ have recently been performed by M. Coutanceau and R. Bouard

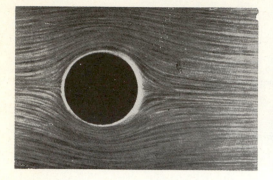

Fig. 15.5a

Fig. 15.5b

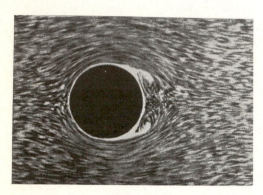

Fig. 15.5c

Fig. 15.5d

Fig. 15.5e

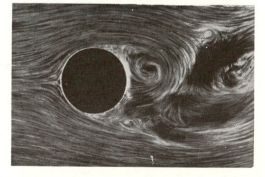

Fig. 15.5f

Fig. 15.5 a to f.   Formation of vortices in flow past a circular cylinder after acceleration from rest
(L. Prandtl)

[9b, c]; the preceding two papers cover the steady as well as the nonsteady case. Reference [9c] establishes the limits of the Reynolds number range in which the "twin" vortices, shown in Figs. 15.5d and 15.5c, can exist and adhere to the body.

Separation: The process of separation is much more difficult to describe in the case of non-steady laminar boundary layers and in the case of moving walls than for steady flows along a solid, stationary wall. In the latter case, separation is determined by the simple condition that the shearing stress at the wall must vanish: $\tau_0 = \mu(\partial u/\partial y)_0 = 0$. It was shown in a paper by W. Sears and D.P. Telionis [47a], as already intimated in earlier papers by F.K. Moore [33] and N. Rott [38], that in non-steady flows separation occurs when the shearing stress at an internal stagnation point vanishes. Thus, for *separation*

$$u = 0 \text{ and } \partial u/\partial y = 0 \text{ in the interior.}$$

This condition is known as the *Moore-Rott-Sears criterion*. Physically, this condition describes a blow-up of the laminar boundary layer. Such a separating, non-steady, two-dimensional boundary layer exhibits, to a certain extent, the same character as a three-dimensional boundary layer formed in the angle between a flat plate and a squat body mounted on it. In this case, shown in Figs. 11.20 and 11.21, the flow forms a separation surface; see also refs. [47b, c].

An extensive review on the unsteady flow around blunt bodies with many excellent flow pictures has been given by S. Taneda [56a].

In conclusion, it may be worth mentioning that these separation processes occur on a much reduced scale in the case of slender bodies, such as e. g. slender elliptical cylinders, whose longer axes are parallel to the direction of flow, or of aerofoils, Consequently, the experimental pressure distribution around such bodies agrees, in most cases, very closely with that given by potential theory (see also Fig. 1.11).

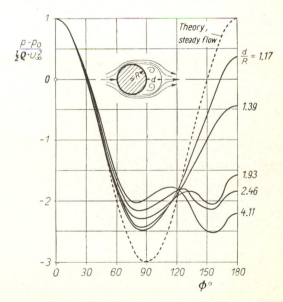

Fig. 15.6. Pressure distribution measured around a circular cylinder during the starting process, after M. Schwabe [47]

e. Periodic boundary-layer flows

1. Oscillating cylinder in fluid at rest. In order to give an example of a periodic boundary-layer flow we now propose to calculate the boundary layer on a body which performs a reciprocating, harmonic oscillation of small amplitude in a fluid at rest. This is an extension of the problem of the boundary layer on a flat plate performing harmonic oscillations in its plane which was already discussed in Sec. V 7.

It will be shown in this section that small oscillations of a body in a fluid at rest induce characteristic secondary flows whose nature is such that a *steady* motion is imparted to the whole fluid in spite of the fact that the motion of the body is purely periodic. Effects of this kind occur, e. g., when dust patterns are created in a Kundt tube and are of some importance in acoustics.

Suppose that the potential velocity distribution for the cylindrical body which we shall now consider is given by $U_0(x)$. The potential flow in the case of periodic oscillations with a circular frequency n is then given by

$$U(x,t) = U_0(x) \cos(nt) . \qquad (15.59)$$

We shall now assume a system of coordinates linked with the solid body. Thus eqns. (15.1) and (15.2) may be applied, the pressure distribution being given by eqn. (15.6). The boundary conditions are $u = 0$ for $y = 0$ and $u = U$ for $y = \infty$.

It is possible to attempt to solve this problem by the method which was used in the case of acceleration from rest, i. e. by calculating successive approximations for the velocity-distribution function as defined in eqn. (15.11) and with the aid of eqns. (15.12) and (15.13).

This method appears to be admissible if

$$\left| U \frac{\partial U}{\partial x} \right| \ll \left| \frac{\partial U}{\partial t} \right| .$$

Now $U \, \partial U/\partial x \sim U_m{}^2/d$ where d denotes a linear dimension of the body (e. g. the diameter of the cylinder). On the other hand $\partial U/\partial t \sim U_m \times n$, where U_m denotes the maximum velocity of the body. Thus we have

$$U \frac{\partial U}{\partial x} \bigg/ \frac{\partial U}{\partial t} \sim \frac{U_m}{nd} .$$

The maximum velocity U_m is proportional to $n \times s$, where s is the amplitude, so that

$$U \frac{\partial U}{\partial x} \bigg/ \frac{\partial U}{\partial t} \sim \frac{s}{d} \ll 1 .$$

The preceding argument shows that the proposed method of solution may be used in cases when the amplitude of oscillation is small compared with the dimensions of the body.

The calculation was performed by H. Schlichting [44] (see also ref. [36]). Since the differential equations are linear, it seems convenient to adopt here the complex notation and to write eqn. (15.59) in the form

$$U(x,t) = U_0(x) \, e^{int} ,$$

with the convention that only the real parts of the complex quantities in question have physical meaning attached to them. Introducing a dimensionless coordinate defined by

$$\eta = y \sqrt{\frac{n}{\nu}}, \tag{15.60}$$

and assuming that the first approximation to the stream function, ψ_0, is of the form

$$\psi_0 (x, y, t) = \sqrt{\frac{\nu}{n}} \, U_0(x) \, \zeta_0(\eta) \, e^{int},$$

and hence

$$u_0(x,y,t) = U_0(x) \, \zeta'_0 \, e^{int}; \qquad v_0(x,y,t) = - \frac{dU_0}{dx} \sqrt{\frac{\nu}{n}} \, \zeta_0 \, e^{int}, \tag{15.61}$$

we obtain from eqn. (15.12) the following differential equation for $\zeta_0(\eta)$:

$$i \, \zeta_0' - \zeta_0''' = i \,,$$

with the boundary conditions $\zeta_0 - \zeta_0' = 0$ at $\eta = 0$ and $\zeta_0' = 1$ at $\eta = \infty$. The solution is

$$\zeta_0' = 1 - \exp \{ - (1 - i) \, \eta / \sqrt{2} \}.$$

Reverting to the real notation[†] we obtain the function

$$u_0(x,y,t) = U_0(x) \, [\cos{(nt)} - \exp{(- \eta / \sqrt{2})} \cos{(nt - \eta / \sqrt{2})}] \tag{15.62}$$

which represents the first approximation to the velocity-distribution function. This is the same solution as that for the oscillating flat plate in eqn. (5.26a)[‡].

If the second approximation $u_1(x, y, t)$ is now calculated from eqn. (15.13), it is seen that the convective terms on the right-hand side of the equation will contribute terms with $\cos^2{nt}$. These, in turn, can be reduced to terms with $\cos{2nt}$, $\sin{2nt}$ and steady-state, i. e. time-independent terms. Taking into account these circumstances we can express the stream function of the second approximation in the form

$$\psi_1(x, y, t) = \sqrt{\frac{\nu}{n}} \, U_0(x) \, \frac{dU_0}{dx} \, \frac{1}{n} \, \{ \zeta_{1a}(\eta) \, e^{2\,int} + \zeta_{1b}(\eta) \},$$

and hence

$$u_1 \, (x, y, t) = U_0(x) \, \frac{dU_0}{dx} \, \frac{1}{n} \, \{ \zeta_{1a}' \, e^{2\,int} + \zeta'_{1b} \},$$

where ζ_{1a} denotes the periodic and ζ_{1b} the steady-state contribution of the second approximation, respectively. As seen from eqn. (15.13) these two functions satisfy the following differential equations:

[†] This is necessary for the correct calculation of the convective terms on the right-hand side in eqn. (15.13).

[‡] It should be noted that here, as distinct from Sec. V a 7, the system of coordinates is linked with the body; furthermore, the dimensionless coordinate η differs from that used there by a factor $\sqrt{2}$.

$$2 \, \mathrm{i} \, \zeta_{1a}{}' - \zeta_{1a}{}''' = \tfrac{1}{2} \, (1 - \zeta_0{}'^2 + \zeta_0 \, \zeta_0{}'') \, ,$$

$$- \zeta_{1b}{}''' = \tfrac{1}{2} - \tfrac{1}{2} \, \zeta_0{}' \, \overline{\zeta_0{}'} + \tfrac{1}{4} \, (\zeta_0 \overline{\zeta_0}{}'' + \overline{\zeta_0} \, \zeta_0{}'') \, ,$$

where the bar over the symbols denotes the respective conjugate complex quantities.

The normal and tangential components of the periodic contribution must vanish at the wall, whereas at a large distance from it only the tangential component vanishes. Putting $\eta' = \eta/\sqrt{2}$ we obtain

$$\zeta_{1a}{}' = - \frac{\mathrm{i}}{2} \exp \left[-(1+\mathrm{i}) \, \sqrt{2} \, \eta' \right] + \frac{\mathrm{i}}{2} \exp \left[-(1+\mathrm{i}) \, \eta' \right] - \frac{\mathrm{i}-1}{2} \, \eta' \exp \left[-(1+\mathrm{i})\eta' \right] .$$

Regarding the steady-state contribution it is found that only the boundary conditions at the wall can be satisfied, and that at a large distance from it is possible to make the tangential component finite but not zero. Thus

$$\zeta_{1b}{}' = - \frac{3}{4} + \frac{1}{4} \exp \left(- 2 \, \eta' \right) + 2 \sin \eta' \exp \left(- \eta' \right) +$$

$$+ \frac{1}{2} \cos \eta' \exp \left(- \eta' \right) - \frac{\eta'}{2} \left(\cos \eta' - \sin \eta' \right) \exp \left(- \eta' \right)$$

so that

$$\zeta_{1b}{}' (\infty) = - \tfrac{3}{4} \, .$$

The second approximation is seen to contain a steady-state term which does not vanish at a large distance from the body, i. e. outside the boundary layer. Its magnitude is given by

$$u_2 \, (x, \infty) = - \frac{3}{4 \, n} \, U_0 \, \frac{\mathrm{d} U_0}{\mathrm{d} x} \, . \tag{15.63}$$

The preceding argument has thus led us to the remarkable result that a potential flow which is periodic with respect to time induces a steady, secondary ('streaming') motion at a large distance from the wall as a result of viscous forces. Its magnitude, given by eqn. (15.63), is independent of the viscosity. The steady-state component of the velocity is such that fluid particles are seen to flow in the direction of decreasing amplitude of that component of the potential velocity which is parallel to the wall.

An example of such a motion, viz. the pattern of streamlines of the steady flow about a circular cylinder which oscillates in a fluid at rest, is shown in Fig. 15.7. Figure 15.8 contains a photograph of the flow pattern about a cylinder which performs an oscillatory motion in a tank filled with water. The camera with which the photograph was taken moved with the cylinder and the surface of the water was covered with fine metallic particles which rendered the motion visible. The particles show up as wide bands in the picture owing to the long exposure time and to their reciprocating motion. The fluid particles flow towards the cylinder from above and from below, and move away in both directions parallel to the reciprocating motion of the cylinder. This is in good agreement with the theoretical pattern of streamlines shown in Fig. 15.7. Similar photographs were also published by E. N. Andrade [1], who induced standing sound waves about a circular cylinder and rendered the resulting secondary flow visible by the injection of smoke.

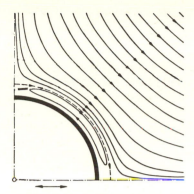

Fig. 15.7. Pattern of streamlines of the steady secondary motion in the neighbourhood of an oscillating circular cylinder

Fig. 15.8. Secondary flow in the neighbourhood of an oscillating circular cylinder. The camera moves with the cylinder. The metallic particles which serve to render the flow visible show up as wide bands owing to the long exposure time and to their reciprocating motion, after Schlichting [44]

It is important to notice here that the first approximation, u_0 in eqn. (15.62), shows that the different layers in the fluid oscillate with different phase shifts compared with the forcing oscillations, and that their amplitudes decrease outwards from the wall. The same features were exhibited by the solutions discussed in Chap. V. The first approximation, u_0, as well as the solutions in Chap. V were obtained from differential equations which did not contain the convective terms

$$u \frac{\partial u}{\partial x}, \qquad v \frac{\partial u}{\partial y}, \qquad \text{and} \qquad U \frac{\partial U}{\partial x}.$$

It can, therefore, be stated that y-dependent phase shifts and amplitudes decaying with distance from the wall are caused exclusively by the action of viscosity. On the other hand, in the second approximation, u_1, there appears a term which is not periodic and which represents steady streaming superimposed on the oscillatory motion. Hence, it can also be stated that secondary flow has its origin in the convective terms and is due to the interaction between inertia and viscosity. It should be borne in mind that simplifications in which the convective terms have been omitted lead to solutions which are free from streaming and may, therefore, give a misleading representation of the flow. Streaming does, in general, appear only when the solution is carried to at least the second-order approximation.

The phenomena under consideration offer a simple explanation of Kundt's dust patterns which are used to demonstrate the existence of standing sound waves in a tube. The sound waves in question are longitudinal ones and the maxima of their amplitudes are located at points of maximum amplitude in the standing waves (Fig. 15.9). Thus a secondary flow is induced in the pipe and its velocity near the wall is directed from the point of maximum amplitude to the nodes. At a large distance from the wall the velocity must, evidently, change sign to satisfy the continuity requirement. This induces 'streaming' effects, the shifting of the particles of dust, and causes them to form little heaps at the nodes.

It is clear from the preceding description that the quantity of dust used to produce Kundt patterns is of great importance. A large quantity of dust will become agitated and may reach the region of inner flow when vibrations of the tube are excited. Consequently it may not be possible to cause the dust to move away from the points of maximum amplitude. If, however, only a small quantity of it is taken, the influence of the flow near the wall will be stronger and the points of maximum amplitude will soon become free of dust. Problems connected with steady motion which accompany oscillations have been treated in greater detail in publications on acoustics, cf. [68].

An analogous investigation of the flow about an axially symmetric ellipsoid which oscillates about its axis of symmetry in a fluid at rest was carried out by A. Gosh [17]; cf. also D. Roy [40, 41].

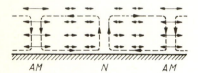

Fig. 15.9. Explanation of the formation of Kundt's dust patterns

AM = amplitude maximum;
N = node of oscillation

2. C. C. Lin's theory of harmonic oscillations. In the preceding section we have considered typical examples of oscillations involving fluids at rest. Problems in which the oscillation is superimposed on a stream are much more important in applications, but also much more difficult to analyse. A certain insight into this type of process can be obtained with the aid of C. C. Lin's theory [28] described in Sec. XV a.

If the external oscillation is described by the function

$$U(x, t) = \overline{U}(x) + U_1(x) \sin n t , \tag{15.64}$$

we can find from eqn. (15.23) that the oscillating component of the longitudinal velocity u is given by

$$u_1(x,y,t) = U_1(x) \{ \sin n t - [\exp(-y/\delta_0)] \cdot [\sin(n t - y/\delta_0)] \} . \tag{15.65}$$

It is noteworthy that the phase shift of the longitudinal perturbation component, $u_1(x, y, t)$, with respect to the external flow again depends on the transverse coordinate, y. The transverse component, $v_1(x, y, t)$, can be obtained with the aid of the continuity equation (15.27), and it, too, exhibits the typical phase shift. Having secured expressions for $u_1(x, y, t)$ and $v_1(x, y, t)$, we can calculate the apparent pressure gradient $F(x, y)$ from eqn. (15.21). This assumes the form

$$F(x, y) = \frac{1}{2} U_1 \frac{dU_1}{dx} \overline{F} \left(\frac{y}{\delta_0} \right) , \tag{15.66}$$

where

$$F\left(\frac{y}{\delta_0}\right) = \exp(-y/\delta_0)\,[(2+y/\delta_0)\cos(y/\delta_0) - (1-y/\delta_0)\sin(y/\delta_0) -$$
$$- \exp(-2\,y/\delta_0)]\,. \tag{15.67}$$

A diagram of this function is seen plotted in Fig. 15.10. The expression (15.66) shows that deviations between the true mean velocity profile $\bar{u}$ and the quasi-steady velocity profile u_s which would exist if we were to assume $F(x, y) = 0$, depend essentially on the amplitude $U_1(x)$ of the oscillation and on its variation dU_1/dx along the flow. In particular, even a large amplitude of oscillation will produce no change in the velocity profile if it remains constant along the flow, i. e. if $U_1 = \text{const}$. From the diagram in Fig. 15.10 it can be deduced that the largest relative modification of the velocity profile occurs near the wall, because $\bar{F}(y/\delta_0)$ has the largest value $\bar{F}(0) = 1$ there. Since the fluid particles nearest to the wall move under relatively small accelerations, the additional pressure gradient will produce the greatest changes near the wall.

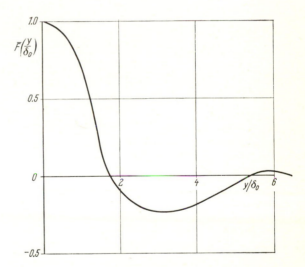

Fig. 15.10. Plot of the function $\bar{F}(y/\delta_0)$ from eqn. (15.67) for a single, harmonic component in the external stream

If there were a spectrum of harmonics of frequencies kn ($k = 1, 2, \ldots$), i. e. for a free-stream velocity

$$U(x,t) = \bar{U}(x) + \sum_k U_{1k}(x)\,\sin(knt)\,, \tag{15.68}$$

we would obtain simply

$$F(x, y) = \sum_k \tfrac{1}{2}\,U_{1k}\frac{dU_{1k}}{dx}\,\bar{F}\left(\frac{y}{\delta_{0k}}\right) \tag{15.69}$$

with

$$\delta_{0k} = \sqrt{\frac{2\,\nu}{k\,n}}\,.$$

From what has been said before it is clear that the position of the point of laminar separation is affected by the external oscillations and that the point of separation must oscillate itself. Finally, C. C. Lin's method leads to the valuable conclusion that the fundamental oscillation induces higher harmonics in the boundary-layer oscillation.

3. External flow with small, harmonic perturbation. The case when the external flow performs small, harmonic oscillations has been treated in a number of publications. The method employed was that of a series expansion in the perturbation parameter described in Sec. XV a 3. We assume that the external flow is of the form

$$U\,(x,t) = \overline{U}\,(x) + \varepsilon\,U_1\,(x)\,e^{int}\;,\tag{15.70}$$

and note that, for it, most investigations restrict themselves to the calculation of the first approximation, that is, of the functions u_1, v_1, and T_1 from eqn. (15.30). M. J. Lighthill [27] formulated an approximate method for the solution of eqn. (15.32) for arbitrary forms of the function $\overline{U}\,(x)$ and $U_1(x)$. The particular case when both functions can be represented in the form of power series has been considered by E. Hori [24], whereas N. Rott and M. L. Rosenzweig [39] examined the example when the two functions $\overline{U}(x)$ and $\overline{U}_1(x)$ are simple powers of x. The example of stagnation flow studied by M. B. Glauert [13] and N. Rott [39] as well as the flow along a flat plate at zero incidence discussed by A. Gosh [17] and S. Gibbelato [11, 12] constitute sub-cases of the latter. Finally, A. Gosh [17] and P. G. Hill and A. H. Stenning [23] performed experimental measurements on non-steady boundary layers.

If the external flow is of the form

$$U\,(x,t) = c x^m\,(1\,+\,\varepsilon\,e^{int}) = \overline{U}\,(1\,+\,\varepsilon\,e^{int})\tag{15.71}$$

then eqns. (15.31) lead to the familiar differential equations for similar solutions, eqns. (9.8) and (9.8a), namely,

$$f''' + \frac{m+1}{2}\,ff'' + m\,(1 - f'^2) = 0\;,\tag{15.72}$$

$$\frac{1}{P}\,\theta'' + \frac{m+1}{2}\,f\,\theta' = 0\tag{15.73}$$

with

$$u_0 = c x^m\,f'\,(\eta)\quad\text{and}\quad\frac{T_0 - T_\infty}{T_w - T_\infty} = \theta\,(\eta)\;,$$

where

$$\eta = y\,\sqrt{\frac{\overline{U}}{\nu\,x}}\;.$$

Assuming in eqns. (15.32) that

$$u_1 = \varepsilon\,e^{int}\,\overline{U}\,\Phi_\eta\,(\xi,\,\eta)\;,\tag{15.74}$$

$$\frac{T_1 - T_\infty}{T_{\mathrm{w}} - T_\infty} = \varepsilon\,e^{int}\,\Theta\,(\xi,\,\eta)\tag{15.75}$$

with

$$\xi = \frac{i\,nx}{\overline{U}}\;,\tag{15.75a}$$

we are led to the following differential equations for the auxiliary functions $\Phi(\xi,\eta)$ and $\Theta\,(\xi,\,\eta)$:

$$\Phi_{\eta\eta\eta} + \frac{m+1}{2}\,f\,\Phi_{\eta\eta} - (\xi + 2\,mf')\Phi_\eta + \frac{m+1}{2}\,f''\,\Phi - (1 - m)\,f'\,\xi\,\Phi_{\eta\xi} +$$

$$+\,(1 - m)\,f''\,\xi\,\Phi_\xi + \xi + 2\,m = 0\;,\tag{15.76}$$

$$\frac{1}{P}\,\Theta_{\eta\eta} + \frac{m+1}{2}\,f\,\Theta_{\eta} - (1-m)\,f'\,\xi\,\Theta_{\xi} - \xi\,\Theta = -\,\frac{m+1}{2}\,\Phi\,\theta' - $$

$$- (1-m)\,\xi\,\Phi_{\xi}\,\theta' \tag{15.77}$$

with the boundary conditions

$$\eta = 0 \;:\;\; \Phi = \Phi_{\eta} = \Phi_{\xi} = \Theta = 0\,,$$
$$\eta = \infty \;:\;\; \Phi_{\eta} = 1\,;\;\; \Theta = 0\,. \tag{15.78}$$

The preceding differential equations are, normally, solved in the form of series expansions, first for small values of ξ and then for large values of ξ. Assuming that

$$\Phi\,(\xi,\eta) = \sum_{k=0}^{\infty} \xi^k \,\Phi_k\,(\eta)\,;\;\;\; \Theta\,(\xi,\eta) = \sum_{k=0}^{\infty} \xi^k \,\theta_k\,(\eta) \tag{15.79}$$

for small values of ξ, we are led to ordinary differential equations for the functions $\Phi_k(\eta)$ and $\theta_k(\eta)$. The derivatives at $\eta = 0$ serve to calculate the shearing stress at the wall as well as the local Nusselt number. In this manner we can derive that

$$\frac{1}{2}\,c_f'\,\sqrt{R_x} = \frac{\tau_0}{\mu\,\overline{U}\,\sqrt{\dfrac{\overline{U}}{\nu x}}} = f''\,(0) + \varepsilon\,e^{int} \sum_{k=0}^{\infty} \xi^k \,\Phi_k{}''\,(0)\,, \tag{15.80}$$

and that

$$\frac{N_x}{\sqrt{R_x}} = -\,\{\theta'\,(0) + \varepsilon\,e^{int} \sum_{k=0}^{\infty} \xi^k \,\theta_k{}'\,(0)\}\,. \tag{15.81}$$

According to F. K. Moore [31] (see also A. Gosh [17] and S. Gibbelato [12]), the case of the flat plate at zero incidence is represented by the expression:

$$\frac{1}{2}\,c_f'\,\sqrt{R_x} = 0{\cdot}332 + \varepsilon\,e^{int} \left\{0{\cdot}498 + 0{\cdot}470 \left(\frac{n\,x}{U_\infty}\right)^2 + \ldots + i\left(0{\cdot}849\,\frac{n\,x}{U_\infty} + \ldots\right)\right\} \tag{15.82}$$

and

$$\frac{N_x}{\sqrt{R_x}} = 0{\cdot}296 + \varepsilon\,e^{int} \left[0{\cdot}148 + 0{\cdot}125 \left(\frac{n\,x}{U_\infty}\right)^2 + \ldots \right.$$

$$\left. \ldots - i\left(0{\cdot}021\,\frac{n\,x}{U_\infty} + \ldots\right)\right]\,, \;\; (P = 0{\cdot}72)\,. \tag{15.83}$$

Substituting $n = 0$, we recover the quasi-steady solution, which signifies that at every instant the solution behaves like the steady solution for the instantaneous external velocity. The appearance of an imaginary term at $n \neq 0$ means that the boundary layer suffers a phase shift with respect to the external flow, the shift being different for velocity and temperature. Whereas the maxima in shearing stress lead the maxima in the external flow (in the limit $n\,x/U_\infty \to \infty$ the phase angle tends to $45°$), the maxima in temperature lag behind them (in the limit $n\,x/U_\infty \to \infty$ the phase angle tends to $90°$). In addition, it turns out that at large values of $n\,x/U_\infty$ the amplitude of the shearing-stress oscillation increases without bound, whereas that of the heat flux slowly decays to zero as $n\,x/U_\infty$ is made to increase.

When the solution of the system of equations (15.33) is carried to second order, it is found that the functions $u_2(x,y,t)$, $v_2(x,y,t)$, and $T_2(x,y,t)$ contain a harmonic part of double frequency and a supplementary, steady part which is independent of time. The latter modifies the basic flow and can be interpreted as a secondary flow in complete analogy with that encountered in the solutions of the preceding section.

For stagnation flow, we have $U_1(x) = \text{const}$, and it is found that then u_2, v_2 and all higher-order terms vanish, as demonstrated by M. B. Glauert [13]. Consequently, the basic flow augmented by the terms u_1 and v_1 constitutes an exact solution, one, moreover, which is also exact for the complete Navier-Stokes equations (cf. also ref. [67]). By a suitable transformation of variables, the preceding case can be made to yield the solutions for stagnation flow on an oscillating wall first given in refs. [13, 67, 2]. A solution for the case of an infinite flat plate with suction and periodic external flow obtained by J. T. Stuart [52], and extended by J. Watson [66] is intimately related to the former. The flow along a flat plate at zero incidence whose external flow is perturbed by a travelling wave was treated in detail by J. Kestin, P. F. Maeder and H. E. Wang [26]. The case of three-dimensional flow in the neighbourhood of a circular cylinder which oscillates in the direction of its axis was solved by W. Wuest [69].

4. Oscillating flow through a pipe. The case of the flow of a fluid through a pipe under the influence of a periodic pressure difference affords another example of an oscillating flow in the boundary layer. This type of flow occurs, e. g., under the influence of a reciprocating piston, and its theory was given by Th. Sexl [48] and S. Uchida [63]. It will now be assumed that the pipe is very long and circular in cross-section. We shall denote the coordinate in the direction of the axis of the pipe by x, denoting the radial distance from it by r. Under the previous assumptions the flow may be taken to be independent of x. When the axial velocity component, u, ceases to depend on x, the other velocity component must vanish together with the convective terms parallel to the tube axis. Thus the Navier-Stokes equation (3.36) assumes the form

$$\frac{\partial u}{\partial t} = -\frac{1}{\varrho}\frac{\partial p}{\partial x} + \nu\left(\frac{\partial^2 u}{\partial r^2} + \frac{1}{r}\frac{\partial u}{\partial r}\right),\tag{15.84}$$

which is exact as it implies no additional simplifications. The boundary condition is $u = 0$ for $r = R$ at the wall. We shall assume that the pressure gradient caused by the motion of the piston is harmonic and is given by

$$-\frac{1}{\varrho}\frac{\partial p}{\partial x} = K\cos nt\,,\tag{15.85}$$

where K denotes a constant. It is, again, convenient to use complex notation and to put

$$-\frac{1}{\varrho}\frac{\partial p}{\partial x} = K\,e^{int},$$

attributing physical significance only to the real part.

Assuming that the velocity function has the form $u(r, t) = f(r)\,e^{int}$, and referring to eqn. (15.84), we obtain the following differential equation for the function $f(r)$:

$$f''(r) + \frac{1}{r}f'(r) - \frac{i\,n}{\nu}f(r) = -\frac{K}{\nu}$$

whose solution is given by

$$u(r,t) = -i\frac{K}{n}\,e^{int}\left\{1 - \frac{J_0\left(r\sqrt{\frac{-in}{\nu}}\right)}{J_0\left(R\sqrt{\frac{-in}{\nu}}\right)}\right\}.\tag{15.86}$$

Here J_0 denotes the Bessel function of the first kind and of zero order. Owing to the linearity of eqn. (15.84), the solutions obtained in eqn. (15.86) can be superimposed for different frequencies. A full discussion of this equation for arbitrary values of n is somewhat tedious owing to the presence of the Bessel function with a complex argument, but the two limiting cases of very large and very small circular frequencies, n, respectively prove to be extremely simple.

Expanding the Bessel function in a series and retaining only the quadratic terms we obtain an expression which is valid for the case of very small values of the dimensionless group $\sqrt{n/\nu}\, R$ (very slow oscillations):

$$u(r,t) = -\,\mathrm{i}\,\frac{K}{n}\,\mathrm{e}^{\mathrm{i}nt}\left\{1 - \frac{1 + \frac{\mathrm{i}n}{4\,\nu}r^2}{1 + \frac{\mathrm{i}n}{4\,\nu}R^2}\right\}, \tag{15.87}$$

or, returning to the real notation,

$$u(r,t) = \frac{K}{4\,\nu}\,\mathrm{e}^{\mathrm{i}nt}\,(R^2 - r^2) = \frac{K}{4\,\nu}\,(R^2 - r^2)\cos\,(nt)\,.$$

The velocity distribution is seen to be in phase with the exciting pressure distribution, the amplitude being a parabolic function of the radius as was the case in steady flow.

Using the asymptotic expansion of the Bessel function $J_0(z) \to \sqrt{2/\pi z}\ \mathrm{e}^{\mathrm{i}}\,\mathrm{i}^{-1/2}$ we obtain an expression for very large values of $\sqrt{n/\nu}\, R$:

$$u(r,t) = -\,\frac{\mathrm{i}\,K}{n}\,\mathrm{e}^{\mathrm{i}nt}\left\{1 - \sqrt{\frac{R}{r}}\,\exp\left[-(1+\mathrm{i})\sqrt{\frac{n}{2\,\nu}}\,(R-r)\right]\right\},$$

or, in the real notation, $\tag{15.88}$

$$u(r,t) = \frac{K}{n}\left\{\sin nt - \sqrt{\frac{R}{r}}\,\exp\left(-\sqrt{\frac{n}{2\,\nu}}\,(R-r)\right)\sin\left[nt - \sqrt{\frac{n}{2\,\nu}}\,(R-r)\right]\right\}.$$

The second term is quickly damped out as the distance from the wall, $R - r$, increases, provided that $\sqrt{n/\nu}\, R$ is large. Consequently at a large distance from the wall only the first term is important; it is seen to be independent of that distance. This solution has a form typical for boundary layers because at a large distance from the wall the fluid moves as if it were frictionless and, moreover, its phase is shifted by half a period with respect to the exciting force.

The sketch in Fig. 15.11 represents the velocity profile for an intermediate frequency ($\sqrt{n/\nu}\, R = 5$) of oscillation at different instants of one period. When a comparison is made between the velocity profiles and the diagram of the variation of the pressure gradient with time, plotted at the bottom, it emerges that the flow on the axis of the pipe lags behind that in the layers near the wall. It should be noted that in line with our remarks in the preceding section, the present solution is free from secondary flow because the non-linear inertia terms did not appear in the differential equation (15.84). On the other hand, the characteristic phase shifts and amplitude decays can be clearly discerned (cf. ref. [27]).

The preceding type of flow was investigated experimentally by E. G. Richardson and E. Tyler [37] who measured the mean with respect to time of the velocity squared, to be denoted by $\overline{u^2}$. In the case of fast oscillations we obtain from eqn. (15.88) the expression

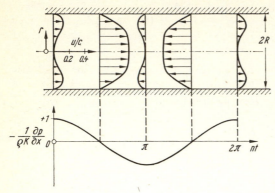

Fig. 15.11. Velocity distribution in oscillating pipe flow at different instants of one period, after S. Uchida [63].

Pressure gradient: $-\dfrac{\partial p}{\partial x} = \varrho\, K \cos(nt)$

$$k = \sqrt{\frac{n}{\nu}}\quad R = 5; \qquad c = \frac{K\, k^2}{8\, n} = 3\cdot125\,\frac{K}{n}$$

$$\overline{u^2}\,(r) = \frac{K^2}{2\,n^2}\left\{ 1 - 2\,\sqrt{\frac{R}{r}}\,\exp\left[-\sqrt{\frac{n}{2\,\nu}}\,(R-r)\right]\cos\left[\sqrt{\frac{n}{2\,\nu}}\,(R-r)\right] + \right.$$

$$\left. + \frac{R}{r}\,\exp\left[-2\,\sqrt{\frac{n}{2\,\nu}}\,(R-r)\right]\right\}.$$

If the distance from the wall $y = R - r$ is small compared with the pipe radius R, the ratio R/r can be replaced by unity. Thus, introducing the dimensionless distance from the wall $\eta = (R-r)\sqrt{n/2\,\nu} = y\sqrt{n/2\,\nu}$, we have

$$\frac{\overline{u^2}\,(y)}{K^2/2\,n^2} = 1 - 2\cos\eta\,\exp(-\eta) + \exp(-2\,\eta)\,. \tag{15.89}$$

The variation of this mean is seen plotted against η in Fig. 15.12. The maximum value does not coincide with the axis of the pipe (large distance), but occurs near the wall at $\eta = y\sqrt{n/2\,\nu} = 2\cdot28$. This value agrees very well with measurement (E. G. Richardson's [37]. "annular effect"). In this connexion the reader is also referred to M. Z. Krzywoblocki's calculations for compressible fluids (Chap. XI [61]).

Recently R. B. Kinney and his collaborators [26a, b; 44a] succeeded in calculating the unsteady viscous flow around a lifting aerofoil. This includes the process of development of the vortices which start at the leading and trailing edges.

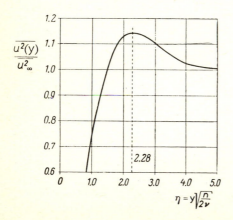

Fig. 15.12. Variation of the mean with respect to time of the velocity squared for periodic pipe flow (E. G. Richardson's [37] "annular effect")

y = distance from the wall of the pipe;
$u_\infty^2 = K^2/2\,n^2$ = mean with respect to time of the velocity squared at large distance from wall

f. Non-steady, compressible boundary layers

The contemporary, fast development of supersonic aerodynamics created a growing interest in compressible, non-steady boundary layers. Boundary layers of this kind arise, for example, behind shock waves or expansion trains in shock tubes or similar installations used in aerodynamic research. A good knowledge of non-steady, compressible boundary layers is also required for the calculation of the drag and amount of heat transferred by a fast-moving body which may be decelerated or accelerated in its flight, and whose surface temperature may vary with time owing to aerodynamic heating. In what follows, we shall examine two simple examples of non-steady, laminar, compressible boundary layers. The first example will discuss the formation of a boundary layer behind a travelling, normal shock wave. The second example will concern itself with a flat plate at zero incidence in non-uniform motion and with a varying surface temperature. Readers who wish to undertake a deeper study of non-steady, compressible boundary layers are referred to the summarizing reviews by E. Becker [6] and K. Stewartson [56].

For the sake of simplicity, we shall restrict ourselves to the considerations of a perfect gas whose specific heats and Prandtl number are constants and whose viscosity is proportional to absolute temperature ($\omega = 1$ in eqn. (13.4a)). The two-dimensional dynamic and thermal boundary layers are now determined by eqns. (15.1) to (15.5) together with the boundary conditions listed along with them. The equation of continuity can be satisfied by the introduction of a stream function $\psi(x, y, t)$, and the velocity components are then related to it by the equations

$$u = \frac{\varrho_0}{\varrho} \frac{\partial \psi}{\partial y} \; ; \quad v = -\frac{\varrho_0}{\varrho} \left(\frac{\partial \psi}{\partial x} + \frac{\partial \bar{y}}{\partial t} \right) , \tag{15.90}$$

where the new transverse coordinate is defined as

$$\bar{y} = \int_0^y \frac{\varrho}{\varrho_0} \, \mathrm{d}y \; ; \tag{15.91}$$

it may be described as an "equivalent, incompressible distance from the wall". The symbol ϱ_0 represents a convenient reference density (here reference may be made to Sec. XIIId 1.1).

1. Boundary layer behind a moving normal shock wave. The first problem of interest is defined in Fig. 15.13; it concerns the boundary layer which is formed behind a normal shock wave moving at a constant velocity U_S into a fluid at rest and at a state described by the subscript 0. The state of the gas behind the shock wave, but outside the boundary layer, will be denoted by the subcript ∞. We shall simplify the problem by assuming that the parameters in the external flow behind the shock wave are independent of x and t. This is equivalent to neglecting the effect of the growing boundary layer on the external flow which must be expected to make itself felt in a shock tube. It turns out that the problem so formulated leads to a set of similar profiles and reduces it to one in the single variable

$$\eta = \frac{\bar{y}}{\sqrt{v_0 \left(t - \frac{x}{U_S} \right)}} = \int_0^y \frac{(\varrho/\varrho_0) \, \mathrm{d}y}{\sqrt{v_0 \left(t - \frac{x}{U_S} \right)}} , \tag{15.92}$$

which replaces the original, three variables x, y, t. Assuming that the stream function is of the form

$$\psi(x,y,t) = U_\infty \sqrt{\nu_0\left(t - \frac{x}{U_S}\right)} f(\eta) ,\qquad (15.93)$$

we can describe the velocity distribution in the boundary layer by the equation

$$u = U_\infty f'(\eta) .\qquad (15.94)$$

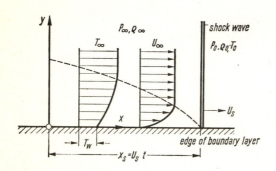

Fig. 15.13. Formation of a boundary layer behind a normal shock wave moving with a velocity U_S

Substitution of the above form for the stream function together with the corresponding form

$$T = T_\infty\, \theta(\eta)\qquad (15.95)$$

for the temperature distribution into eqns. (15.1) to (15.5) allows us to derive the following, ordinary differential equations for the functions $f(\eta)$ and $\theta(\eta)$. These are:

$$f''' + \frac{1}{2}\left(\eta - \frac{U_\infty}{U_S}f\right)f'' = 0 ,\qquad (15.96)$$

$$\frac{1}{P}\theta'' + \frac{1}{2}\left(\eta - \frac{U_\infty}{U_S}f\right)\theta' = -\frac{U_\infty^{\,2}}{c_p\,T_\infty}f''^{\,2} .\qquad (15.97)$$

The requisite boundary conditions can be recorded as:

$$\eta = 0:\; f = f' = 0,\; \theta = \frac{T_w}{T_\infty} ;$$
$$\eta = \infty:\; f' = 1,\;\; \theta = 1 .$$
$$(15.98)$$

The solutions for $u/U_\infty = f'(\eta)$ from eqn. (15.96) are seen plotted in Fig. 15.14a. The parameter U_∞/U_S for the family of curves characterizes the strength of the shock wave. The highest possible value for U_∞/U_S is $(U_\infty/U_S)_{max} = 2/(\gamma+1)$ and corresponds to an infinitely strong shock; with $\gamma = 1\cdot4$, this yields $(U_\infty/U_S)_{max} = 0\cdot83$. Negative values of U_∞/U_S correspond to fictitious, non-steady, continuous expansion fans, each imagined concentrated in a single front. In the particular case when $U_\infty/U_S = 0$ we are led to the so-called Rayleigh problem (Stokes's first problem,

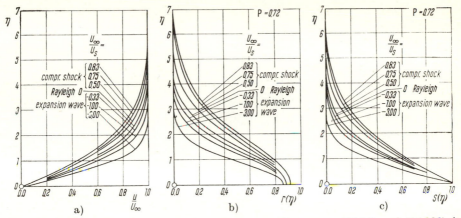

Fig. 15.14. Velocity and temperature distributions from eqns. (15.94) and (15.106) in the laminar boundary layer behind a normal shock wave of constant velocity, after H. Mirels [29] The parameter U_∞/U_S characterizes the strength of the wave

Sec. V a 4) which deals with the impulsive start of a flat wall. It is seen from Fig. 15.14a that the thickness of a boundary layer behind a normal shock exceeds that for the so-called Rayleigh problem. This means that upon the lapse of a certain time $t - x/U_S$ after the passage of the shock wave, the boundary layer at a given position has grown thicker than on an impulsively started plate after the same period of time has lapsed from start. The opposite is true for expansion waves.

The solutions for the linear differential equation (15.97) for $\theta(\eta)$ can be represented in the form of a linear combination of two basic solutions, defined as follows:

$$\frac{T - T_\infty}{T_\infty} = \theta(\eta) - 1 =$$

$$= \frac{\gamma - 1}{2} M_\infty^2 r(\eta) - \left\{ \frac{\gamma - 1}{2} M_\infty^2 r(0) + 1 - \frac{T_w}{T_\infty} \right\} s(\eta). \tag{15.99}$$

The functions $r(\eta)$ and $s(\eta)$ are solutions of the following ordinary differential equation:

$$\frac{1}{P} r'' + \frac{1}{2} \left(\eta - \frac{U_\infty}{U_S} f \right) r' = -2 f''^2, \tag{15.100}$$

$$\frac{1}{P} s'' + \frac{1}{2} \left(\eta - \frac{U_\infty}{U_S} f \right) s' = 0, \tag{15.101}$$

together with the boundary conditions

$$\left. \begin{array}{llll} \eta = 0 : & r' = 0, & s = 1; \\[2mm] \eta = \infty : & r = 0, & s = 0. \end{array} \right\} \tag{15.102}$$

The solutions for $P = 0.72$ have been plotted in Figs. 15.14b and c. The numerical value $r(0)$ is a measure of the recovery temperature, T_a, that is, of the temperature at the surface of an adiabatic wall. In this case, we have $\theta'(0) = 0$, and hence $s(\eta) = 0$. It follows from eqn. (15.99) that the adiabatic wall temperature is

$$T_a = T_\infty \left\{ 1 + \frac{\gamma - 1}{2} \, M_\infty^2 \, r(0) \right\}. \tag{15.103}$$

When $P = 1$, we have $r(0) = 1$, and the adiabatic wall temperature becomes identical with the stagnation temperature [cf. eqn. (13.17)]. When the Prandtl number of the gas differs little from unity, it is possible, according to H. Mirels [29], to employ the approximation that

$$r(0) = P^\alpha,$$

with

$$\alpha = 0.39 - \frac{0.02}{1 - (U_\infty/U_S)} \quad \text{for} \quad \frac{U_\infty}{U_S} > 0 \quad \text{(compression waves)} \tag{15.104}$$

$$\alpha = 0.50 - \frac{0.13}{1 - (U_\infty/U_S)} \quad \text{for} \quad \frac{U_\infty}{U_S} < 0 \quad \text{(expansion waves)}. \tag{15.105}$$

Thus, finally, the temperature distribution becomes,

$$T - T_\infty = \frac{\gamma - 1}{2} \, M_\infty^2 \, T_\infty \, r(\eta) + (T_w - T_a) \, s(\eta). \tag{15.106}$$

For the skin-friction coefficient,

$$c_f' = \frac{\tau_w}{\frac{1}{2} \, \varrho_w \, U_\infty^2},$$

we find

$$c_f' \sqrt{R} = 2 \, f''(0), \tag{15.107}$$

and the local Nusselt number is

$$N = \frac{q}{T_w - T_a} \, \frac{U_\infty^2 \left(t - \frac{x}{U_S} \right)}{k_w} = R \, s'(0), \tag{15.108}$$

where

$$R = \frac{U_\infty^2}{\nu_w} \left(t - \frac{x}{U_S} \right). $$

Once again, according to H. Mirels [29], when the Prandtl number is near to unity, it is possible to resort to the following approximations:

$$c_f' \sqrt{R} = 1.128 \, \sqrt{1 - \beta \frac{U_\infty}{U_S}}, \tag{15.109}$$

$$N = \frac{1}{2} \, c_f' \, R \, P^\lambda, \tag{15.110}$$

where for compression waves $(U_\infty/U_S > 0)$

$$\beta = 0.346$$

$$\left.\lambda = 0.35 + \frac{0.15}{1 - (U_\infty/U_S)} \, , \right\} \tag{15.111}$$

and for expansion waves $(U_\infty/U_S < 0)$

$$\beta = 0.375$$

$$\left.\lambda = 0.48 + \frac{0.02}{1 - (U_\infty/U_S)} \, . \right\} \tag{15.112}$$

The boundary-layer thickness exceeds the so-called Rayleigh value when the wave is compressive; this causes the shearing stress, the skin-friction coefficient, and the Nusselt number to become smaller compared with their Rayleigh values. The opposite is true for expansion waves. In the special case when $\mathsf{P} = 1$, the heat-transfer formulae reduce to the simple Reynolds analogy

$$\mathsf{N} = \tfrac{1}{2} c_f \mathsf{R} \, , \quad (\mathsf{P} = 1) \tag{15.113}$$

known to the reader as eqn. (12.55).

The preceding problem which discussed the boundary layer behind a shock wave of constant velocity constitutes an idealized special case in that it can be reduced to a steady problem by the felicitous choice of a coordinate system in which the shock wave is at rest. More general solutions of the same problem have been treated in the works of E. Becker [3, 4, 6, 7] and H. Mirels and J. Hamman [30].

2. Flat plate at zero incidence with variable free-stream velocity and surface temperature. In our second example we consider the compressible boundary layer on a flat plate when the free-stream velocity, $U_\infty(t)$, as well as the temperature at the surface, $T_w(t)$, vary in the course of time. The stream function ψ from eqn. (15.90), and the temperature distribution

$$\theta = \frac{T - T_\infty}{T_w - T_\infty}$$

are now determined by the equations

$$\psi_{\bar{y}t} + \psi_{\bar{y}}\,\psi_{x\bar{y}} - \psi_x\,\psi_{\bar{y}\bar{y}} = \dot{U}_\infty + \nu_\infty\,\psi_{\bar{y}\bar{y}\bar{y}} \, , \tag{15.114}$$

$$\theta_t + \theta\,\frac{\dot{T}_w}{T_w - T_\infty} + \psi_{\bar{y}}\,\theta_x - \psi_x\,\theta_{\bar{y}} = \frac{\nu_\infty}{\mathsf{P}}\left\{\theta_{\bar{y}\bar{y}} + \frac{\mu_\infty/k_\infty}{T_w - T_\infty}\,\psi^2_{\bar{y}\bar{y}}\right\}, \tag{15.115}$$

in which the pressure-gradient term has been deleted. The variable $\bar{y}$ has been defined in eqn. (15.91), and $\dot{U}_\infty$ and $\dot{T}_w$ denote the derivatives of free-stream velocity and surface temperature with respect to time, respectively. In order to arrive at solutions, the following series expansions are postulated:

$$\psi = \sqrt{\nu_\infty \, U_\infty \, x} \, \{F(\eta) + \zeta_0 \, f_0(\eta) + \zeta_1 \, f_1(\eta) + \dots \} \qquad (15.116)$$

$$\theta = \theta_0(\eta) + \beta_1 \, \theta_1(\eta) + \beta_2 \, \theta_2(\eta) + \dots + \zeta_0 \, h_0(\eta) + \zeta_1 \, h_1(\eta) + \zeta_2 \, h_2(\eta) + \dots$$

$$+ \frac{U_\infty^2}{2 \, c_p \, (T_w - T_\infty)} \, \{ S(\eta) + \zeta_0 \, s_0(\eta) + \zeta_1 \, s_1(\eta) + \dots \} \, . \qquad (15.117)$$

Here

$$\eta = \frac{\bar{y}}{2 \, x} \sqrt{\frac{U_\infty \, x}{\nu_\infty}}$$

defines a new, dimensionless coordinate, and the following abbreviations have been
employed:

$$\left. \begin{array}{l} \zeta_0 = \dfrac{\dot{U}_\infty}{U_\infty} \left(\dfrac{x}{U_\infty} \right), \quad \zeta_1 = \dfrac{\ddot{U}_\infty}{U_\infty} \left(\dfrac{x}{U_\infty} \right)^2, \quad \dots ; \quad \text{etc.} \\[3mm] \beta_1 = \dfrac{\dot{T}_w}{T_w - T_\infty} \left(\dfrac{x}{U_\infty} \right), \quad \beta_2 = \dfrac{\ddot{T}_w}{T_w - T_\infty} \left(\dfrac{x}{U_\infty} \right)^2, \quad \dots \quad \text{etc.} \end{array} \right\} \qquad (15.118)$$

The preceding forms are substituted into the differential equations for the boundary
layer and it is found that the functions $F(\eta)$, $f_0(\eta)$, ... satisfy ordinary differential
equations. Solutions for them when $P = 0.72$ have been given in refs. [35, 49].
The functions $F(\eta)$, $\theta_0(\eta)$ and $S(\eta)$ are identical with the solutions for the steady
problem with U_∞ interpreted as the instantaneous velocity (quasi-steady flow).
The remaining terms describe the departures from the quasi-steady solution.

The ratio of the shearing stress at the wall, τ_w, to that for quasi-steady flow,
τ_{ws}, is given by

$$\frac{\tau_w}{\tau_{ws}} = 1 + \frac{x}{U_\infty} \left\{ 2.555 \, \frac{\dot{U}_\infty}{U_\infty} - 1.414 \, \frac{\ddot{U}_\infty}{U_\infty} \left(\frac{x}{U_\infty} \right) + \dots \right\} . \qquad (15.119)$$

Correspondingly, the ratio of heat fluxes at the wall for $P = 0.72$ (*cf.* [50]) is
described by

$$\frac{q}{q_s} = 1 + \frac{x}{U_\infty} \left\{ 2.39 \, \frac{\dot{T}_w}{T_w - T_{as}} + \dots \right.$$

$$- \frac{\dot{U}_\infty}{U_\infty} \left[0.0692 \, \frac{T_w - T_\infty}{T_w - T_{as}} - 0.0448 \, \frac{T_\infty - T_{as}}{T_w - T_{as}} \right] + \dots \right\} . \qquad (15.120)$$

The adiabatic-wall temperature of quasi-static flow used as a reference should be
calculated from the formula

$$T_{as} = T_\infty + 0.848 \, \frac{U_\infty^2}{2 \, c_p} \, . \qquad (15.121)$$

When the method just described is being used, it should be realized, as pointed
out by H. Tsuji [62], that the expressions for $\zeta_0, \zeta_1, \dots, \beta_1, \beta_2, \dots$ are, generally
speaking, interdependent even for prescribed forms of $U_\infty(t)$ and $T_w(t)$. See also the
paper by H.D. Harris and A.D. Young [73].

The theory of laminar, non-steady boundary layers has been developed considerably in the last years. Information on this phase can be found in three volumes of conference proceedings. The first, edited by E.A. Eichelbrenner, reports on the IUTAM Symposium "Recent Research on Unsteady Boundary Layers", Quebec 1972 [74]. The second, edited by R.B. Kinney [75], concerns a symposium on "Unsteady Aerodynamics" held in 1975 at the University of Arizona. The third is devoted to an AGARD meeting held in 1977 [76]. A review paper by N. Riley may also merit comparison [37a].

References

[1] Andrade, E.N.: On the circulation caused by the vibration of air in a tube. Proc. Roy. Soc. A *134*, 447—470 (1931).

[2] Arduini, C.: Strato limite incompressibile laminare nell'intorno del punto di ristagno di un cilindro indefinito oscillante. L'Aerotecnica *41*, 341—346 (1961).

[3] Becker, E.: Das Anwachsen der Grenzschicht in und hinter einer Expansionswelle. Ing.-Arch. *25*, 155—163 (1957).

[4] Becker, E.: Instationäre Grenzschichten hinter Verdichtungsstössen und Expansionswellen. ZFW *7*, 61—73 (1959).

[5] Becker, E.: Die laminare inkompressible Grenzschicht an einer durch laufende Wellen deformierten ebenen Wand. ZFW *8*, 308—316 (1960).

[6] Becker, E.: Instationäre Grenzschichten hinter Verdichtungsstössen und Expansionswellen. Progress in Aero. Sci. *1* (A. Ferry, D. Küchemann, and L.H. Sterne, ed.), 104—173, London, 1961.

[7] Becker, E.: Anwendung des numerischen Fortsetzungsverfahrens auf die pseudostationäre, kompressible laminare Grenzschicht in einem Stosswellenrohr. ZFW *10*, 138—147 (1962).

[7a] Berger, E., and Wille, R.: Periodic flow phenomena. Annual Review of Fluid Mech. *4*, 313—340 (1972).

[8] Blasius, H.: Grenzschichten in Flüssigkeiten mit kleiner Reibung. Z. Math. Phys. *56*, 1—37 (1908).

[9] Böltze, E.: Grenzschichten an Rotationskörpern in Flüssigkeiten mit kleiner Reibung. Diss. Göttingen 1908.

[9a] Dumitrescu, D., and Cazacu, M.D.: Theoretische und experimentelle Betrachtungen über die Strömung zäher Flüssigkeiten um eine Platte bei kleinen und mittleren Reynoldszahlen. ZAMM *50*, 257—280 (1970).

[9b] Coutanceau, M., and Bouard, R.: Experimental determination of the main features of the viscous flow in the wake of a circular cylinder in uniform translation. Part 1: Steady flow. JFM *79*, 231—256 (1977).

[9c] Coutanceau, M., and Bouard, R.: Experimental determination of the main features of the viscous flow in the wake of a circular cylinder in uniform translation. Part 2: Unsteady flow. JFM *79*, 257—272 (1977).

[10] Geis, Th.: Bemerkung zu den "ähnlichen" instationären laminaren Grenzschichtströmungen. ZAMM *36*, 396—398 (1956).

[11] Gibellato, S.: Strato limite attorno ad una lastra piana investita da un fluido incompressibile clotato di una velocità che e somma di una parte constante e di una parte alternata. Atti della Accademia delle Scienze di Torino *89*, 180—192 (1954—1955) and *90*, 13—24 (1955—1956).

[12] Gibellato, S.: Strato limite termico attorno a una lastra piana investita da una corrente lievemente pulsante di fluido incompressibile. Atti della Accademia delle Scienze di Torino *91*, 152—170 (1956—1957).

[13] Glauert, M.B.: The laminar boundary layer on oscillating plates and cylinders. JFM *1*, 97—110 (1956).

[14] Goldstein, S., and Rosenhead, L.: Boundary layer growth. Proc. Cambr. Phil. Soc. *32*, 392—401 (1936).

[15] Görtler, H.: Verdrängungswirkung der laminaren Grenzschicht und Druckwiderstand. Ing.-Arch. *14*, 286—305 (1944).

[16] Görtler, H.: Grenzschichtentstehung an Zylindern bei Anfahrt aus der Ruhe. Arch. d. Math. *1*, 138—147 (1948).

[17] Gosh, A.: Contribution à l'étude de la couche limite laminaire instationnaire. Publications Scientifiques et Techniques du Ministère de l'Air No. 381 (1961).

[18] Gribben, R. J.: The laminar boundary layer on a hot cylinder fixed in a fluctuating stream. J. Appl. Mech. *28*, 339—346 (1961).

[19] Hassan, H. A.: On unsteady laminar boundary layers. JFM *9*, 300—304 (1960); see also JASS *27*, 474—476 (1960).

[20] Hayasi, N.: On similar solutions of the unsteady quasi-two-dimensional incompressible laminar boundary-layer equations. J. Phys. Soc. Japan *16*, 2316—2329 (1961).

[21] Hayasi, N.: On semi-similar solutions of the unsteady quasi-two-dimensional incompressible laminar boundary-layer equations. J. Phys. Soc. Japan *17*, 194—203 (1962).

[22] Hayasi, N.: On the approximate solution of the unsteady quasi-two-dimensional incompressible laminar boundary-layer equations. J. Phys. Soc. Japan *17*, 203—212 (1962).

[23] Hill, P. G., and Stenning, A. H.: Laminar boundary layers in oscillatory flow. J. Basic Engg. *82*, 593—608 (1960).

[24] Hori, E.: Unsteady boundary layers (4 reports). Bulletin of JSME *4*, 664—671 (1961); *5*, 57—64 (1962); *5*, 64—72 (1962); *5*, 461—470 (1962).

[25] Illingworth, C. R.: Boundary layer growth on a spinning body. Phil. Mag. *45* (7), 1—8 (1954).

[25a] Katagiri, M.: Unsteady boundary-layer flows past an impulsively started circular cylinder. J. Phys. Soc. Japan *40*, 1171—1177 (1976).

[26] Kestin, J., Maeder, P. F., and Wang, W. E.: On boundary layers associated with oscillating streams. Appl. Sci. Res. A *10*, 1—22 (1961).

[26a] Kinney, R. B., and Cielak, Z. M.: Analysis of unsteady viscous flow past an airfoil. Part I: Theoretical developments. AIAA J. *15*, 1712—1717 (1977); Part II: Numerical formulation and results. AIAA J. *16*, 105—110 (1978); see also: AGARD C. P. 227, 26/1 to 26/14 (1978).

[27] Lighthill, M. J.: The response of laminar skin friction and heat transfer to fluctuations in the stream velocity. Proc. Roy. Soc. A *224*, 1—23 (1954).

[28] Lin, C. C.: Motion in the boundary layer with a rapidly oscillating external flow. Proc. 9th Intern. Congress Appl. Mech. Brussels 1957, *4*, 155—167.

[28a] Lugt, H. J., and Haussling, H. J.: Laminar flow past an abruptly accelerated cylinder at 45° incidence. JFM *65*, 711—734 (1974).

[28b] Mehta, V. B., and Lavan, Z.: Starting vortex, separation bubbles and stall: A numerical study of laminar unsteady flow around an airfoil. JFM *67*, 227—256 (1975).

[29] Mirels, H.: Boundary layer behind shock or thin expansion wave moving into stationary fluid. NACA TN 37 12 (1956).

[30] Mirels, H., and Hamman, J.: Laminar boundary layer behind strong shock moving with non-uniform velocity. Physics of Fluids *5*, 91—95 (1962).

[31] Moore, F. K.: Unsteady, laminar boundary layer flow. NACA TN 2471 (1951).

[32] Moore, F. K., and Ostrach, S.: Average properties of compressible laminar boundary layer on a flat plate with unsteady flight velocity. NACA TN 3886 (1956).

[33] Moore, F. K.: On the separation of the unsteady laminar boundary layer. IUTAM-Symposium, Boundary layers, Freiburg 1957 (H. Görtler, ed.), 296—311, Berlin, 1958. posium. Boundary-layers research, Freiburg 1957 (H. Görtler, ed.), 296—311, Berlin, 1958.

[34] Nigam, S. D.: Zeitliches Anwachsen der Grenzschicht an einer rotierenden Scheibe bei plötzlichem Beginn der Rotation. Quart. Amer. Math. *9*, 89—91 (1951).

[35] Ostrach, S.: Compressible laminar boundary layer and heat transfer for unsteady motions of a flat plate. NACA TN 3569 (1955).

[35a] Proudman, I., and Johnson, K.: Boundary layer growth near a rear stagnation point. JFM *12*, 161—168 (1962).

[36] Lord Rayleigh: On the circulation of air observed in Kundt's tubes and on some allied acoustical problems. Phil. Trans. Roy. Soc. London *175*, 1—21 (1884).

[37] Richardson, E. G., and Tyler, E.: The transverse velocity gradient near the mouths of pipes in which an alternating or continuous flow of air is established. Proc. Phys. Soc. London *42*, 1—15 (1929).

[37a] Riley, N.: Unsteady laminar boundary layers. SIAM Review *17*, 274—297 (1975).

[38] Rott, N.: Unsteady viscous flow in the vicinity of a stagnation point. Quart. Appl. Math. *13*, 444—451 (1956).

[39] Rott, N., and Rosenzweig, M. L.: On the response of the laminar boundary layer to small fluctuations of the free-stream velocity. JASS *27*, 741—747, 787 (1960).

[39a] Rott, N.: Theory of time-dependent laminar flows. Princeton University Series, High Speed Aerodynamics and Jet Propulsion. Princeton University Press, Vol. *IV*, 395—438 (1964).

[40] Roy, D.: Non-steady periodic boundary layer. J. Appl. Math. Phys. *12*, 363—366 (1961).

[41] Roy, D.: On the non-steady boundary layer. ZAMM *42*, 252—256 (1962).

[42] Rozin, L.A.: An approximation method for the integration of the equations of a non-stationary laminar boundary layer in an incompressible fluid. NASA Techn. Transl. *22* (1960).

[43] Rubach, H.: Über die Entstehung und Fortbewegung des Wirbelpaares bei zylindrischen Körpern. Diss. Göttingen 1914; VDI-Forschungsheft *185* (1916).

[44] Schlichting, H.: Berechnung ebener periodischer Grenzschichtströmungen. Phys. Z. *33*, 327—335 (1932).

[44a] Schmall, R.A., and Kinney, R.B.: Numerical study of unsteady viscous flow past a lifting plate. AIAA J. *12*, 1566—1573 (1974).

[45] Schuh, H.: Calculation of unsteady boundary layers in two-dimensional laminar flow. ZFW *1*, 122—131 (1953).

[46] Schuh, H.: Über die "ähnlichen" Lösungen der instationären laminaren Grenzschicht-gleichungen in inkompressibler Strömung. Fifty years of boundary-layer research (H. Görtler and W. Tollmien, ed.), Braunschweig, 147—152.

[47] Schwabe, M.: Über Druckermittlung in der nichtstationären ebenen Strömung. Diss. Göttingen, 1935; Ing.-Arch. *6*, 34—50 (1935); NACA TM 1039 (1943).

[47a] Sears, W.R., and Telionis, D.P.: Boundary layer separation in unsteady flow. SIAM J. Appl. Math. *28*, 215—235 (1975).

[47b] Telionis, D.P., and Tsahalis, D.Th.: Unsteady turbulent boundary layers and separation. AIAA J. *14*, 468—474 (1976).

[47c] Tsahalis, D.Th.: Laminar boundary-layer separation from an upstream moving wall. AIAA J. *15*, 561—566 (1975).

[48] Sexl, Th.: Über den von E.G. Richardson entdeckten "Annulareffekt". Z. Phys. *61*, 349 (1930); see also: Tollmien, W.: Handbuch der Exper. Physik *IV*, Part I, 281—282 (1931).

[49] Sparrow, E.M., and Gregg, J.L.: Nonsteady surface temperature effects on forced convection heat transfer. JAS *24*, 776—777 (1957).

[50] Sparrow, E.M.: Combined effects of unsteady flight velocity and surface temperature on heat transfer. Jet Propulsion *28*, 403—405 (1958).

[51] Sparrow, E.M., and Gregg, J.L.: Flow about an unsteady rotating disc. JASS *27*, 252—257 (1960).

[52] Squire, L.C.: Boundary layer growth in three dimensions. Phil. Mag. *45* (7), 1272—1283 (1954).

[53] Squire, L.C.: The three-dimensional boundary layer equations and some power series solutions. ARC RM 3006 (1955).

[54] Stewartson, K.: The theory of unsteady laminar boundary layers. Adv. Appl. Mech. *6*, 1—37 (1960).

[55] Stuart, J.T.: A solution of the Navier-Stokes and energy equations illustrating the response of skin friction and temperature of an infinite plate thermometer to fluctuations in the stream velocity. Proc. Roy. Soc. A *231*, 116—130 (1955).

[55a] Stuart, J.T.: Unsteady boundary layers. L. Rosenhead (ed.): Laminar boundary layers. Clarendon Press, Oxford 1963, pp. 349—408.

[56] Tani, I.: An example of unsteady laminar boundary layer flow. Inst. Univ. of Tokyo, Rep. No. 331 (1958); see also: IUTAM-Symposium, Boundary-layer research, Freiburg 1957 (H. Görtler, ed.), 347, Berlin, 1958.

[56a] Taneda, S.: Visual study of unsteady separated flows around bodies. Progress in Aerospace Sciences (D. Küchemann, ed.), Pergamon Press, London, Vol. *XVII*, 287—348 (1977).

[57] Thiriot, K.H.: Untersuchungen über die Grenzschicht einer Flüssigkeit über einer rotierenden Scheibe bei kleiner Winkelgeschwindigkeitsänderung. ZAMM *22*, 23—28 (1942).

[58] Thiriot, K.H.: Grenzschichtströmung kurz nach dem plötzlichen Anlauf bzw. Abstoppen eines rotierenden Bodens. ZAMM *30*, 390—393 (1950); see also Diss. Göttingen, 1940; ZAMM *20*, 1—13 (1940).

[59] Tollmien, W.: Grenzschichten. Handbuch der Exper.-Physik *IV*, Part I, 274 (1931).

[60] Tollmien, W.: Die zeitliche Entwicklung der laminaren Grenzschicht am rotierenden Zylinder. Diss. Göttingen 1924; see also: Handbuch der Exper.-Physik *IV*, Part I, 277 (1931).

[61] Trimpi, R. L., and Cohen, N. B.: An integral solution to the flat plate laminar boundary layer flow existing inside and after expansion waves moving into quiescent fluid with particular application to the complete shock tube flow. NACA TN 3944 (1957).

[62] Tsuji, H.: Note on the solution of the unsteady laminar boundary layer equations. JAS 20, 295—296 (1953).

[63] Uchida, S.: The pulsating viscous flow superposed on the steady laminar motion of incompressible fluid in a circular pipe. ZAMP 7, 403—422 (1950).

[64] Wadhwa, Y. D.: Boundary layer growth on a spinning body; accelerated motion. Phil. Mag. 3 (8), 152—158 (1958).

[65] Watson, E. J.: Boundary layer growth. Proc. Roy. Soc. A 231, 104—116 (1955).

[66] Watson, J.: A solution of the Navier-Stokes-equations, illustrating the response of a laminar boundary layer to a given change in the external stream velocity. Quart. J. Mech. Appl. Math. 11, 302—325 (1958).

[67] Watson, J.: The two-dimensional laminar flow near the stagnation point of a cylinder which has an arbitrary transverse motion. Quart. J. Mech. Appl. Math. 12, 175—190 (1959).

[68] Westervelt, P. J.: The theory of steady rotational flow generated by a sound field. J. Acoust. Soc. Amer. 25, 60—67 (1953).

[69] Wuest, W.: Grenzschichten an zylindrischen Körpern mit nichtstationärer Querbewegung. ZAMM 32, 172—178 (1952).

[70] Wundt, H.: Wachstum der laminaren Grenzschicht an schräg angeströmten Zylindern bei Anfahrt aus der Ruhe. Ing.-Arch. 23, 212—230 (1955).

[71] Yang, K. T.: Unsteady laminar boundary layers in an incompressible stagnation flow. J. Appl. Mech. 25, 421—427 (1958).

[72] Yang, K. T.: Unsteady laminar boundary layers over an arbitrary cylinder with heat transfer in an incompressible flow. J. Appl. Mech. 26, 171—178 (1959).

[73] Young, A. D., and Harris, H. D.: A set of similar solutions of the compressible laminar boundary layer equations for the flow over flat plate with unsteady wall temperature. ZFW 15, 295—301 (1967).

[74] Eichelbrenner, E. A. (ed.): Recent research on unsteady boundary layers. IUTAM Symposium 1971, I and II, Presse de l'Université Laval, Quebec, 1972.

[75] Kinney, R. B. (ed.): Unsteady aerodynamics. Proc. Symp. Univ. of Arizona, March 18—20, 1975, I and II (1975).

[76] AGARD-CP-227: Unsteady Aerodynamics. Papers presented at the Fluid Dynamics Panel Symposium, at Ottawa, Canada, September 1977 (1978).

Part C. Transition

CHAPTER XVI

Origin of turbulence I

Some experimental results; foundations of the stability theory and experimental verification for the boundary layer on the flat plate

Introduction. This and the succeeding chapter are devoted to a presentation of the complex of problems which relate to transition from laminar to turbulent flow. The first *experimental* results on this problem were obtained by O. Reynolds in the eighties of the preceding century (Reynolds's dye experiment), as mentioned in Sec. II c and illustrated in Fig. 2.22. The principal theoretical idea for the analysis of this problem was conceived by O. Reynolds and Lord Rayleigh; this is to the effect that transition constitutes a problem in the stability of laminar flows (Reynolds's hypothesis). The *theoretical* investigations into this problem were crowned by a decisive breakthrough which occurred half a century later after decades of futile efforts. This was accomplished in about 1930 in the form of the stability theory formulated by Prandtl's school in Goettingen. In the time interval between 1930 and 1970, the whole body of knowledge regarding transition was successfully enlarged, both experimentally and theoretically, after the above theory had been brilliantly confirmed with the aid of very careful experiments performed in 1940 by H. L. Dryden and his collaborators.

The last two decades saw the appearance of a large number of summaries of the field. These were, in the order of their appearance: 1959, H. L. Dryden [20a]; 1959, H. Schlichting [79]; 1961, W. Tollmien and D. Grohne [102]; 1963, J.T. Stuart [91]; 1964, S. F. Shen [85a]; 1969, I. Tani [96]; 1969, M. V. Morkovin [61a]; 1976, E. Reshotko [70a]. In most recent times, the contemporary status of research in this field was the subject of a conference of the AGARD Fluid Dynamics Panel on "Laminar-turbulent transition". The conference was held in Copenhagen in May 1977, and its proceedings are available as AGARD Conference Proceedings No. 224 [1a].

a. Some experimental results on transition from laminar to turbulent flow

1. Transition in pipe flow. Very often the flows of real fluids differ from the laminar flows considered in the preceding chapters. They exhibit a characteristic feature which is termed *turbulence*. When the Reynolds number is increased, internal flows and boundary layers formed on solid bodies undergo a remarkable transition from the laminar to the turbulent regime. The origin of turbulence and the accom-

panying transition from laminar to turbulent flow is of fundamental importance
for the whole science of fluid mechanics. The incidence of turbulence was first recognized in relation to flows through straight pipes and channels. In a flow at very
low Reynolds number through a straight pipe of uniform cross-section and smooth
walls, every fluid particle moves with a uniform velocity along a straight path.
Viscous forces slow down the particles near the wall in relation to those in the external
core. The flow is wellordered and particles travel along neighbouring layers (laminar
flow), Fig. 2.22a. However, observation shows that this orderly pattern of flow
ceases to exist at higher Reynolds numbers, Fig. 2.22b, and that strong mixing
of all the particles occurs. This mixing process can be made visible in a flow through
a pipe, as first shown by O. Reynolds [71], by feeding into it a thin thread of liquid
dye. As long as the flow is laminar the thread maintains sharply defined boundaries
all along the stream. As soon as the flow becomes turbulent the thread diffuses into
the stream and the fluid appears uniformly coloured at a short distance downstream.
In this case there is superimposed on the main motion in the direction of the axis
of the pipe a subsidiary motion at right angles to it which effects mixing. The pattern
of streamlines at a fixed point becomes subjected to continuous fluctuations and the
subsidiary motion causes an exchange of momentum in a transverse direction because
each particle substantially retains its forward momentum while mixing is taking
place. As a consequence, the velocity distribution over the cross-section is considerably more uniform in turbulent than in laminar flow. The measured velocity distribution for these two types of flow is shown in Fig. 16.1, where the mass flow is the
same for both cases. In laminar flow, according to the Hagen-Poiseuille solution
given in Chap. I, the velocity distribution over the cross-section is parabolic (see
also Fig. 1.2), but in turbulent flow, owing to the transfer of momentum in the transverse direction, it becomes considerably more uniform. On closer investigation it
appears that the most essential feature of a turbulent flow is the fact that at a given
point in it, the velocity and the pressure are not constant in time but exhibit very
irregular, high-frequency fluctuations, Fig. 16.17. The velocity at a given point
can only be considered constant on the average and over a longer period of time
(quasi-steady flow).

    The first systematic investigation into these two fundamentally different patterns
of flow were conducted by O. Reynolds [71]. O. Reynolds was also the first to
investigate in greater detail the circumstances of the transition from laminar to
turbulent flow. The previously mentioned dye experiment was used by him in this
connexion, and he discovered the law of similarity which now bears his name, and
which states that transition from laminar to turbulent flow always occurs at nearly
the same Reynolds number $\bar{w}\,d/\nu$, where $\bar{w} = Q/A$ is the mean flow-velocity
(Q = volume rate of flow, A = cross-sectional area). The numerical value of the
Reynolds number at which transition occurs (critical Reynolds number) was

Fig. 16.1. Velocity distribution in pipe;   a) laminar;   b) turbulent

established as being approximately

$$R_{crit} = \left(\frac{\overline{w}\,d}{\nu}\right)_{crit} = 2\,300\;. \tag{16.1}$$

Accordingly, flows for which the Reynolds number $R < R_{crit}$, are supposed to be laminar, and flows for which $R > R_{crit}$, are expected to be turbulent. The numerical value of the critical Reynolds number depends very strongly on the conditions which prevail in the initial pipe length as well as in the approach to it. Even Reynolds thought that the critical Reynolds number increases as the disturbances in the flow before the pipe are decreased. This fact was confirmed experimentally by H. T. Barnes and E. G. Coker [1b], and later by L. Schiller [80] who reached critical values of the Reynolds number of up to 20,000. V. W. Ekman [24] succeeded in maintaining laminar flow up to a critical Reynolds number of 40,000 by providing an inlet which was made exceptionally free from disturbances. The upper limit to which the critical Reynolds number can be driven if extreme care is taken to free the inlet from disturbances is not known at present. There exists, however, as demonstrated by numerous experiments, a lower bound for R_{crit} which is approximately at 2000. Below this value, the flow remains laminar even in the presence of very strong disturbances.

Transition from laminar to turbulent flow is accompanied by a noticeable change in the law of resistance. In laminar flow, the longitudinal pressure gradient which maintains the motion is proportional to the first power of the velocity (*cf.* Sec. Id); by contrast, in turbulent flow this pressure gradient becomes nearly proportional to the square of the mean velocity of flow. The increase in the resistance to flow has its origin in the turbulent mixing motion. This change in the law of pipe friction can be inferred from Fig. 20.1.

Detailed investigations of the process of transition reveal that in a certain range of Reynolds numbers around the critical the flow becomes "intermittent" which means that it alternates in time between being laminar or turbulent. The variation of the velocity of flow with time in this range is shown graphically in Fig. 16.2 which represents the results of measurements performed by J. Rotta [75] in 1956 at different distances along a pipe radius. The velocity plots demonstrate that periods of laminar and turbulent flow succeed each other in a random sequence. At positions closer to the centre-line, the velocity in the laminar intervals exceeds the temporal mean value of the velocity of flow in the turbulent intervals; at positions closer to the pipe wall, conditions are reversed. Since during the experiments care was taken to maintain a constant rate of flow over long intervals of time, it is concluded that in the region of intermittent flow the velocity distribution alternates between a corresponding developed laminar distribution, and a corresponding fully developed turbulent distribution, as shown in Fig. 16.1a and 16.1b respectively. The physical nature of this flow can be aptly described with the aid of the intermittency factor γ, which is defined as that fraction of time during which the flow at a given position remains turbulent. Hence $\gamma = 1$ corresponds to continuous turbulent flow, and $\gamma = 0$ denotes continuous laminar flow. The intermittency factor is shown plotted in Fig. 16.3 for various Reynolds numbers in terms of the axial distance x. At a constant Reynolds number, the intermittency factor increases continuously with the distance. The Reynolds numbers cover the

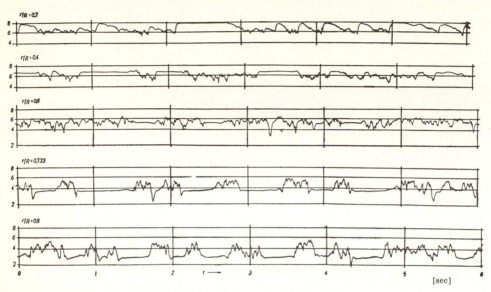

Fig. 16.2. Variation of flow velocity in a pipe in the transition range at different distances *r* from pipe axis, as measured by J. Rotta [75]

Reynolds number $R = \overline{w}d/\nu = 2550$; axial distance $x/d = 322$; $\overline{w} = 4\cdot27$ m/sec ($= 14\cdot0$ ft/sec); velocities given in m/sec. These velocity plots, obtained with the aid of a hot-wire anemometer, demonstrate the intermittent nature of the flow in that periods of laminar and turbulent flow succeed each other in time

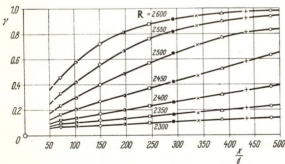

Fig. 16.3. Intermittency factor γ for pipe flow in the transition range in terms of the axial distance x for different Reynolds numbers R, as measured by J. Rotta [75]

Here $\gamma = 1$ denotes continuously turbulent, and $\gamma = 0$ continuously laminar flow

range from $R = 2300$ to 2600 over which transition is completed. At Reynolds numbers near the lower limit, the process of transition to fully developed turbulent which the flow extends over very large distances measured in thousands of diameters. Measurements of this kind have been recently amplified by J. Meseth [60].

2. Transition in the boundary layer on a solid body. As already stated, the flow in a boundary layer can also undergo transition, a fact which was discovered much later than transition in a pipe. The whole field of flow about a body immersed in a stream, and in particular, the force exerted on it, are strongly dependent on whether the flow in the boundary layer is laminar or turbulent. Transition in a boundary layer on a solid body in a stream is affected by many parameters, the most important ones being the pressure distribution in the external flow, the nature of the wall (roughness) and the nature of the disturbances in the free flow (intensity of turbulence).

Blunt Bodies: A particularly remarkable phenomenon connected with transition in the boundary layer occurs with blunt bodies, for example spheres or circular cylinders. It is seen from Figs. 1.4 and 1.5 that the drag coefficient of a sphere or cylinder decreases abruptly at Reynolds numbers $R = V D/\nu$ of about 3×10^5. This abrupt drop in the drag coefficient, noticed first by G. Eiffel [23] in relation to spheres, is a consequence of transition in the boundary layer. Transition causes the point of separation to move downstream which considerably decreases the width of the wake. The truth of this explanation was demonstrated by L. Prandtl [41] who mounted a thin wire hoop somewhat ahead of the equator of a sphere. This causes artificially the boundary layer to become turbulent at a lower Reynolds number and produces the same drop in drag as occurs when the Reynolds number is made to increase. The smoke photographs in Fig. 2.24 and 2.25 show clearly the extent of the wake on a sphere: in the sub-critical flow regime the wake is wide and the drag is large, and in the super-critical regime it is narrow and the drag is small. The latter flow regime was here created with the aid of Prandtl's 'tripping wire'. These experiments show conclusively that the jump in the drag curve of a sphere is due to a boundary-layer effect and is caused by transition.

Flat plate: The process of transition on a flat plate at zero incidence is somewhat simpler to understand than that on a blunt body. The process of transition in the boundary layer on a flat plate was first studied by J. M. Burgers [6], B. G. van der Hegge Zijnen [41] and later by M. Hansen and, in greater detail, by H. L. Dryden [16, 17, 18]. According to Chap. VII, the boundary-layer thickness on a flat plate increases in proportion to $\sqrt{x}$, where x denotes the distance from the leading edge. Near the leading edge the boundary layer is always laminar†, becoming turbulent further downstream. On a plate with a sharp leading edge and in a normal air stream (i. e. of intensity of turbulence $T \approx 0.5\%$) transition takes place at a distance x from it, as determined by

$$R_{x, crit} = \left(\frac{U_\infty x}{\nu} \right)_{crit} = 3.5 \times 10^5 \text{ to } 10^6 .$$

On a flat plate, in the same way as in a pipe, the critical Reynolds number can be increased by providing for a disturbance-free external flow (very low intensity of turbulence), cf. Sec. XII d 2.

Transition is easiest to perceive by a study of the velocity distribution in the boundary layer. As seen from Fig. 2.23, transition is shown prominently by a sudden increase in the boundary-layer thickness. In a laminar boundary layer the dimensionless boundary-layer thickness, $\delta/\sqrt{\nu x/U_\infty}$, remains constant and equal, approximately, to 5. The dimensionless boundary-layer thickness is seen plotted against the length Reynolds number $R_x = U_\infty x/\nu$ in Fig. 2.23 already mentioned; at $R_x > 3.2 \times 10^5$ a sudden increase in the boundary-layer thickness is clearly visible. Transition also involves a noticeable change in the shape of the velocity-

---

† Except for leading-edge separation which may occur on a flat plate of finite thickness — if no precautions have been taken to suppress it, as explained later.

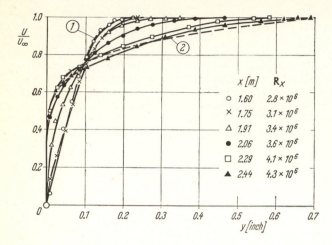

Fig. 16.4. Velocity profiles in a boundary layer on a flat plate in the transition region, as measured by Schubauer and Klebanoff [83]

(1) laminar, Blasius profile; (2) turbulent, $^1/_7$-th power law, $\delta = 17$ mm ($= 1\cdot36$ in), external velocity $U_{\infty} = 27$ m/sec (89 ft/sec); turbulence intensity $T = 0\cdot03\%$

x [m]	R_X
○ 1.60	2.8×10^6
× 1.75	3.1×10^6
△ 1.91	3.4×10^6
● 2.06	3.6×10^6
□ 2.29	4.1×10^6
▲ 2.44	4.3×10^6

distribution curve. The changes in the velocity profiles in the transition region have been plotted in Fig. 16.4. They are based on measurements performed by G. B. Schubauer and P. S. Klebanoff [83] in a stream of very low turbulence intensity and it is seen that in this case the transition region extends over a range of Reynolds numbers from about $R_x = 3 \times 10^6$ to 4×10^6. In this range, the boundary-layer profile changes from that of fully developed laminar flow, as calculated by Blasius, to fully developed turbulent flow (see Chap. XXI). The process of transition involves a large decrease in the shape factor $H_{12} = \delta_1/\delta_2$, as seen from Fig. 16.5; here δ_1 denotes the displacement thickness and δ_2 is the momentum thickness. In the case of a flat plate, the shape factor decreases from $H_{12} \approx 2\cdot6$ in the laminar regime to $H_{12} \approx 1\cdot4$ in the turbulent regime.

This change in the velocity distribution in the transition region can be utilized for the convenient determination of the point of transition, or, rather, of the transition region. The principle is explained with the aid of Fig. 16.6. A total-head tube or a Pitot tube is moved parallel to the wall at a distance which corresponds to the maximum difference between the velocities in the laminar and turbulent regimes. On being moved downstream across the transition region, the tube shows a fairly sudden increase in the total or dynamic pressure.

Transition on a flat plate also involves a large change in the resistance to flow, in this case in the skin friction. In laminar flow the skin friction is proportional to the 1·5 power of velocity, eqn. (7.33), whereas in turbulent flow the power increases to about 1·85, as shown a long time ago by W. Froude [29] who performed towing experiments with plates at very high Reynolds numbers. In this connexion the reader may also wish to consult Fig. 21.2.

More recent experiments performed by H. W. Emmons [25], and G. B. Schubauer and P. S. Klebanoff [83] have shown that in the case of a flat plate the process of transition is also intermittent and consists of an irregular sequence of laminar and turbulent regions. As explained in Fig. 16.7, at a given point in the boundary layer there occurs suddenly a small turbulent area ('turbulent spot'), irregular

$$H_{12} = \frac{\delta_1}{\delta_2}$$

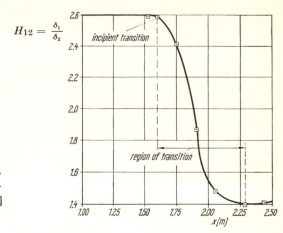

Fig. 16.5. Change in the shape factor $H_{12} = \delta_1/\delta_2$ of the boundary layer for a flat plate in the transition region as measured by Schubauer and Klebanoff [83] quoted from [65]

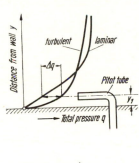

$w_1 = 14.1\ m/sec$

Fig. 16.6. Explanation of the method of determining the position of the point of transition with the aid of a total-head tube or a Pitot tube

in shape, which then travels downstream in a wedge-shaped region, as shown. Such turbulent spots appear at irregular intervals of time and at different, randomly distributed points on the plate. In the interior of the wedge-like domain the flow is predominantly turbulent, whereas in the adjoining regions it alternates continuously between being laminar and turbulent. In this connexion see also ref. [13]. A paper by M. E. McCormick [57a] deals with the problem of the origin of such turbulent spots. It turns out that an artificially created turbulent spot does not persist when the Reynolds number has a value lower than $R_{\delta1} = 500$; this is consistent with the value of the critical Reynolds number calculated with the aid of the linear stability theory, eqn. (16.22). Very detailed experimental investigations of turbulent spots, and in particular of the velocity distribution in them, have been carried out by J. Wygnanski et al. [108].

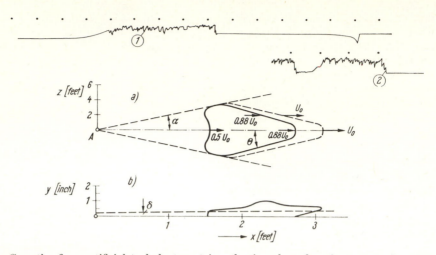

Fig. 16.7.   Growth of an artificial turbulent spot in a laminar boundary layer on a flat plate
at zero incidence as measured by G. B. Schubauer and P. S. Klebanoff [83], quoted from [20]
(a) plan; (b) side view of a turbulent spot artificially created at A when it is at a distance of about 2·4 ft from the
point of origin. The point A is at a distance of 2·3 ft from the leading edge. $\alpha = 11\cdot3°$, $\theta = 15\cdot3°$, δ = boundary layer
thickness, $U_\infty \approx 10$ m/sec (velocity). (1) and (2) represent oscillograms obtained with the aid of a hot-wire anemometer when an artificial or natural turbulent spot swims by. Time interval between marker traces is $^1/_{60}$ sec

Slender bodies: It has been established that the pressure gradient along a wall
exerts an important influence on the position of the point of transition in the boundary layer. In ranges of decreasing pressure (accelerated flow), the boundary layer
remains, generally speaking, laminar, whereas even a very small pressure increase
almost always brings transition with it. Making use of this fact, it is always possible
to reduce the skin friction on slender bodies (aerofoils, streamline bodies) by displacing the point of transition as far downstream as possible; this is achieved by a
suitable choice of shape or profile and of the corresponding pressure distribution.
The skin friction of bodies possessing such long laminar initial lengths of boundary
layer (laminar profiles) can be reduced to as little as half or less of what it would
be on a more normal shape.

The position of the point of transition and hence the magnitude of the skin
friction can be strongly affected by other means also, for example by sucking away
the boundary layer.

b. Principles of the theory of stability of laminar flows

1. Introductory remarks. Efforts to clarify and to explain theoretically the
remarkable process of transition just described were initiated many decades ago;
they have led to success only in the last thirty years. These theoretical investigations are based on the assumption that laminar flows are affected by certain small
disturbances; in the case of pipe flow these disturbances may originate, for example,
in the inlet, whereas in the case of a boundary layer on a solid body placed in a
stream they may also be due to wall roughness or to irregularities in the external

flow. The theory endeavours to follow up in time the behaviour of such disturbances when they are superimposed on the main flow, bearing in mind that their exact form still remains to be determined in particular cases. The decisive question to answer in this connexion is whether the disturbances increase or die out with time. If the disturbances decay with time, the main flow is considered stable; on the other hand, if the disturbances increase with time the flow is considered unstable, and there exists the possibility of transition to a turbulent pattern. In this way a *theory of stability* is created, and its object is to predict the value of the critical Reynolds number for a prescribed main flow. The basis of the theory of stability can be traced to O. Reynolds [72] who supposed that the laminar pattern, being a solution of the differential equations of fluid dynamics, always represents a possible type of flow, but becomes unstable above a definite limit (precisely above the critical Reynolds number) and changes into the turbulent pattern.

Much work has been done on the mathematical foundations of Reynolds's hypothesis during many decades, first by O. Reynolds himself and later, notably, by Lord Rayleigh [70]. These efforts, which led to very complicated calculations, have remained for a long time devoid of achievement. About 1930, L. Prandtl and his collaborators succeeded in attaining the initial object of predicting the value of the critical Reynolds number in a satisfactory way. The experimental verification of the stability theory came some ten years later when H. L. Dryden with his co-workers were able to obtain brilliant agreement between theory and experiment. Comprehensive accounts of the theory of stability were given by H. Schlichting [78, 79], by R. Betchov and W. O. Criminale [4], and by C. C. Lin [55].

2. Foundation of the method of small disturbances. The theory of stability of laminar flows decomposes the motion into a mean flow (whose stability constitutes the subject of the investigation) and into a disturbance superimposed on it. Let the mean flow, which may be regarded as steady, be described by its Cartesian velocity components U, V, W and its pressure P. The corresponding quantities for the non-steady disturbance will be denoted by u', v', w', and p', respectively. Hence, in the resultant motion the velocity components are

$$u = U + u', \quad v = V + v', \quad w = W + w', \tag{16.2}$$

and the pressure is

$$p = P + p'. \tag{16.3}$$

In most cases it is assumed that the quantities related to the disturbance are small compared with the corresponding quantities of the main flow.

The investigation of the stability of such a disturbed flow can be carried out with the aid of either of two different methods. The first method (*energy method*) consists merely in the calculation of the variation of the energy of the disturbances with time. Conclusions are drawn depending on whether the energy decreases or increases as time goes on. The theory admits an arbitrary form of the superimposed motion and demands only that it should be compatible with the equation of continuity. The energy method was developed mainly by H. A. Lorentz [57] and did not prove successful; we shall, therefore, refrain from considering it in detail.

The second method accepts only flows which are consistent with the equations of motion and analyzes the manner in which they develop in the flow, by reference to the appropriate differential equations. This is the *method of small disturbances*. This second method has led to complete success and will, for this reason, be described with some detail.

We shall now consider a two-dimensional incompressible mean flow and an equally two-dimensional disturbance. The resulting motion, described by eqns. (16.2) and (16.3), satisfies the two-dimensional form of the Navier-Stokes equations as given in eqns. (4.4a, b, c). We shall further simplify the problem by stipulating that the mean velocity U depends only on y, i. e., $U = U(y)$, whereas the remaining two components are supposed to be zero everywhere, or $V \equiv W \equiv 0$†. We have encountered such flows earlier, describing them as *parallel flows*. In the case of a channel with parallel walls or a pipe, such a flow is reproduced with great accuracy at a sufficient distance from the inlet section. The flow in the boundary layer can also be regarded as a good approximation to parallel flow because the dependence of the velocity U in the main flow on the x-coordinate is very much smaller than that on y. As far as the pressure in the main flow is concerned, it is obviously necessary to assume a dependence on x as well as on y, i. e., $P(x,y)$, because the pressure gradient $\partial P/\partial x$ maintains the flow. Thus we assume a mean flow with

$$U(y) ; \quad V \equiv W \equiv 0 ; \quad P(x,y) . \tag{16.4}$$

Upon the mean flow we assume superimposed a two-dimensional disturbance which is a function of time and space. Its velocity components and pressure are, respectively,

$$u'(x,y,t) , \quad v'(x,y,t) , \quad p'(x,y,t) . \tag{16.5}$$

Hence the resultant motion, according to eqns. (16.2) and (16.3), is described by

$$u = U + u' ; \quad v = v' ; \quad w = 0 ; \quad p = P + p' . \tag{16.6}$$

It is assumed that the mean flow, eqn. (16.4), is a solution of the Navier-Stokes equations, and it is required that the resultant motion, eqn. (16.6), must also satisfy the Navier-Stokes equations. The superimposed fluctuating velocities from eqn. (16.5) are taken to be "small" in the sense that all quadratic terms in the fluctuating components may be neglected with respect to the linear terms. The succeeding section will contain a more detailed description of the form of the disturbance. Now, the task of the stability theory consists in determining whether the disturbance is amplified or whether it decays for a given mean motion; the flow is considered unstable or stable depending on whether the former or the latter is the case.

Substituting eqns. (16.6) into the Navier-Stokes equations for a two-dimensional, incompressible, non-steady flow, eqns. (4.4a, b, c), and neglecting quadratic terms in the disturbance velocity components, we obtain

$$\frac{\partial u'}{\partial t} + U \frac{\partial u'}{\partial x} + v' \frac{dU}{dy} + \frac{1}{\varrho} \frac{\partial P}{\partial x} + \frac{1}{\varrho} \frac{\partial p'}{\partial x} = \nu \left(\frac{d^2 U}{dy^2} + \nabla^2 u' \right) ,$$

$$\frac{\partial v'}{\partial t} + U \frac{\partial v'}{\partial x} + \frac{1}{\varrho} \frac{\partial P}{\partial y} + \frac{1}{\varrho} \frac{\partial p'}{\partial y} = \nu \nabla^2 v' , \quad \frac{\partial u'}{\partial x} + \frac{\partial v'}{\partial y} = 0 ,$$

† There are reasons to suppose, as shown by G. B. Schubauer and P. S. Klebanoff [83], that these components are always present in real flows, particularly in flows past flat plates. Their magnitude is negligible for most purposes, but they seem to play a part, not yet fully elucidated, in the process of transition; see also footnote on p. 468.

where ∇^2 denotes the Laplacian operator $\partial^2/\partial x^2 + \partial^2/\partial y^2$.

If it is considered that the mean flow itself satisfies the Navier-Stokes equations, the above equations can be simplified to

$$\frac{\partial u'}{\partial t} + U \frac{\partial u'}{\partial x} + v' \frac{dU}{dy} + \frac{1}{\varrho} \frac{\partial p'}{\partial x} = \nu \nabla^2 u' , \tag{16.7}$$

$$\frac{\partial v'}{\partial t} + U \frac{\partial v'}{\partial x} \qquad + \frac{1}{\varrho} \frac{\partial p'}{\partial y} = \nu \nabla^2 v' , \tag{16.8}$$

$$\frac{\partial u'}{\partial x} + \frac{\partial v'}{\partial y} = 0 . \tag{16.9}$$

We have obtained three equations for u', v' and p'. The boundary conditions specify that the turbulent velocity components u' and v' vanish on the walls (no-slip condition). The pressure p' can be easily eliminated from the two equations, (16.7) and (16.8), so that together with the equation of continuity there are two equations for u' and v'. It is possible to criticize the assumed form of the mean flow, eqn. (16.4), on the ground that the variation of the component U of the velocity with x as well as the normal component V have been neglected. In this connexion, however, J. Pretsch [44] proved that the resulting terms in the equations are unimportant for the stability of a boundary layer (see also S. J. Cheng [7]).

3. The Orr-Sommerfeld equation. The mean laminar flow in the x-direction with a velocity $U(y)$ is assumed to be influenced by a disturbance which is composed of a number of discrete partial fluctuations, each of which is said to consist of a wave which is propagated in the x-direction. As it has already been assumed that the perturbation is two-dimensional, it is possible to introduce a stream function $\psi(x,y,t)$ thus integrating the equation of continuity (16.9). The stream function representing a single oscillation of the disturbance is assumed to be of the form

$$\psi(x, y, t) = \phi(y) \, e^{i(\alpha x - \beta t)} . \tag{16.10}\dagger$$

Any arbitrary two-dimensional disturbance is assumed expanded in a Fourier series; each of its terms represents such a partial oscillation. In eqn. (16.10) α is a real quantity and $\lambda = 2\pi/\alpha$ is the wavelength of the disturbance. The quantity β is complex,

$$\beta = \beta_r + i\beta_i ,$$

where β_r is the circular frequency of the partial oscillation, whereas β_i (amplification factor) determines the degree of amplification or damping. The disturbances are damped if $\beta_i < 0$ and the laminar mean flow is stable, whereas for $\beta_i > 0$ instability sets in. Apart from α and β it is convenient to introduce their ratio

$$c = \frac{\beta}{\alpha} = c_r + i\,c_i . \tag{16.11}$$

---

† The convenient complex notation is used here. Physical meaning is attached only to the real part of the stream function, thus

$$\mathrm{Re}(\psi) = e^{\beta_i t} \, [\phi_r \cos(\alpha x - \beta_r t) - \phi_i \sin(\alpha x - \beta_r t)]$$

where $\phi = \phi_r + i\phi_i$ is the complex amplitude.

Here c_r denotes the velocity of propagation of the wave in the x-direction (phase velocity) whereas c_i again determines the degree of damping, or amplification, depending on its sign. The amplitude function, ϕ, of the fluctuation is assumed to depend on y only because the mean flow depends on y alone. From eqn. (16.10) it is possible to obtain the components of the perturbation velocity

$$u' = \frac{\partial \psi}{\partial y} = \phi'(y)\, e^{i(\alpha x - \beta t)}, \tag{16.12}$$

$$v' = -\frac{\partial \psi}{\partial x} = -i\,\alpha\,\phi(y)\, e^{i(\alpha x - \beta t)}. \tag{16.13}$$

Introducing these values into eqns. (16.7) and (16.8), we obtain, after the elimination of pressure, the following, ordinary, fourth-order, differential equation for the amplitude $\phi(y)$:

$$(U - c)\,(\phi'' - \alpha^2\,\phi) - U''\,\phi = -\frac{i}{\alpha\,R}\,(\phi'''' - 2\,\alpha^2\,\phi'' + \alpha^4\,\phi). \tag{16.14}$$

This is the fundamental *differential equation for the disturbance (stability equation)* which forms the point of departure for the stability theory of laminar flows. It is commonly referred to as the *Orr-Sommerfeld equation*. Equation (16.14) has been cast in dimensionless form in that all lengths have been divided by a suitable reference length b or δ (width of channel or boundary-layer thickness), and velocities have been divided by the maximum velocity U_m of the main flow. The primes denote here differentiation with respect to the dimensionless coordinates y/δ or y/b, and

$$R = \frac{U_m\, b}{\nu} \quad \text{or} \quad R = \frac{U_m\, \delta}{\nu}$$

denotes the Reynolds number which is a characteristic of the mean flow. The terms on the left-hand side of eqn. (16.14) are derived from the inertia terms, and those on the right-hand side from the viscous terms in the equations of motion. By way of example, the boundary conditions for a boundary-layer flow demand that the components of the perturbation velocity must vanish at the wall ($y = 0$) and at a large distance from it (free stream). Thus:

$$\left.\begin{array}{llll} y = 0: & u' = v' = 0: & \phi = 0, & \phi' = 0; \\ y = \infty: & u' = v' = 0: & \phi = 0, & \phi' = 0. \end{array}\right\} \tag{16.15}$$

At this stage it is possible to raise the objection that disturbances superimposed on a two-dimensional flow pattern need not be two-dimensional, if a complete analysis of the question of stability is to be achieved. This objection was removed by H. B. Squire [87] who proved, by assuming disturbances which were periodic also in the z-direction, that a two-dimensional flow pattern becomes unstable at a higher Reynolds number when the disturbance is assumed three-dimensional than when it is supposed to be two-dimensional. In this sense two-dimensional disturbances are "more dangerous" for two-dimensional flows than three-dimensional disturbances. Hence the value of the critical Reynolds number, or, more precisely, of the lowest limit of stability, is obtained by considering two-dimensional disturbances.

4. The eigenvalue problem. The problem of stability has now been reduced to an eigenvalue problem of eqn. (16.14) with the boundary conditions (16.15).

When the mean flow $U(y)$ is specified, eqn. (16.14) contains four parameters, namely α, R, c_r and c_i. Of these the Reynolds number of the mean flow is likewise specified and, further, the wavelength $\lambda = 2\pi/\alpha$ of the disturbance is to be considered given. In this case the differential equation (16.14), together with the boundary conditions (16.15), furnish one eigenfunction $\phi(y)$ and one complex eigenvalue $c = c_r + i c_i$ for each pair of values α, R. Here c_r represents the phase velocity of the prescribed disturbance whereas the sign of c_i determines whether the wave is amplified ($c_i > 0$) or damped ($c_i < 0$)†. For $c_i < 0$ the corresponding flow (U, R) is stable for the given value of α, whereas $c_i > 0$ denotes instability. The limiting case $c_i = 0$ corresponds to neutral (indifferent) disturbances.

The result of such an analysis for a prescribed laminar flow $U(y)$ can be represented graphically in an α, R diagram because every point of this plane corresponds to a pair of values of c_r and c_i. In particular, the locus $c_i = 0$ separates the region of stable from that of unstable disturbances. This locus is called the *curve of neutral stability* (Fig. 16.8). The point on this curve at which the Reynolds number has its smallest value (tangent parallel to the α-axis) is of greatest interest since it indicates that value of the Reynolds number below which all individual oscillations decay, whereas above that value at least some are amplified. This smallest Reynolds number is the *critical Reynolds number* or *limit of stability* with respect to the type of laminar flow under consideration.

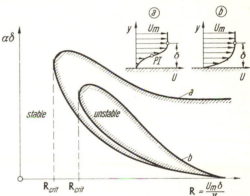

Fig. 16.8. Curves of neutral stability for a two-dimensional boundary layer with two-dimensional disturbances

(a) "non-viscous" instability; in the case of velocity profiles of type a *with* point of inflexion *PI*, the curve of neutral stability is of type a

(b) "viscous" instability; in the case of velocity profiles of type b *without* point of inflexion, the curve of neutral stability is of type b

The asymptotes for the curve of neutral stability a at R → ∞ are obtained from the "frictionless" stability equation (16.16)

The experimental evidence concerning transition from laminar to turbulent flow referred to previously leads us to expect that, at small Reynolds numbers for which laminar flow is observed, all wavelengths would produce only stable disturbances, whereas at larger Reynolds numbers, for which turbulent flow is observed, unstable disturbances ought to correspond to at least some wavelengths. However, it is necessary to remark at this point that the critical Reynolds number calculated from stability considerations cannot be expected to be equal to the Reynolds

---

† On the other hand, it is also possible to regard R and the circular frequency β_r as fixed. In this case the eigenvalue problem determines a corresponding value of α (the wavelength) and the coefficient of amplification, β_i. These were the conditions satisfied by the experiments carried out by H. L. Dryden and his collaborators, as described in Sec. XVId, when an artificial disturbance of a definite frequency was superimposed on a laminar flow with the aid of a suitably excited strip.

number observed at the point of transition. If attention is fixed on the flow in the boundary layer along a wall, then the theoretical critical Reynolds number indicates the point on the wall at which amplification of some individual disturbances begins and proceeds downstream of it. The transformation of such amplified disturbances into turbulence takes up some time, and the unstable disturbance has had a chance to travel some distance in the downstream direction. It must, therefore, be expected that the observed position of the point of transition will be downstream of the calculated, theoretical limit of stability, or, in other words, that the experimental critical Reynolds number exceeds its theoretical value. This remark, evidently, applies to Reynolds numbers based on the current length as well as to those based on the boundary-layer thickness. In order to distinguish between these two values it is usual to call the theoretical critical Reynolds number (limit of stability) the *point of instability* whereas the experimental critical Reynolds number is called the *point of transition* †.

The stability problem, briefly described in the preceding paragraphs, leads to extremely difficult mathematical considerations. Owing to these, success in the calculation of the critical Reynolds number eluded the workers in this field for several decades, in spite of the greatest efforts directed towards this goal. Consequently, in what follows we shall be unable to provide a complete presentation of the stability theory and will be forced to restrict ourselves to giving an account of the most important results only.

5. General properties of the Orr-Sommerfeld equation. Since from experimental evidence the limit of stability $c_i = 0$ is expected to occur for large values of the Reynolds number, it is natural to simplify the equation by omitting the viscous terms on the right-hand side of it, as compared with the inertia terms, because of the smallness of the coefficient $1/R$. The resulting differential equation is known as the *frictionless stability equation*, or *Rayleigh's equation:*

$$(U - c)\,(\phi'' - \alpha^2\,\phi) - U''\,\phi = 0 \,. \qquad (16.16)$$

It is important to note here that of the four boundary conditions (16.15) of the complete equation it is now possible to satisfy only two, because the frictionless stability equation is of the second order. The remaining boundary condition to be satisfied is the vanishing of the normal components of velocity near the wall of a channel, or, in boundary-layer flow, their vanishing at the wall and at infinity. Thus, in the latter case, we have

$$y = 0: \quad \phi = 0; \quad y = \infty: \quad \phi = 0 \,. \qquad (16.17)$$

The omission of the viscous terms constitutes a drastic simplification, because the order of the equation is reduced from four to two, and this may result in a loss of important properties of the general solution of the complete equation, as compared with its simplified version. Here we may repeat the remarks noted previously in Chap. IV in connexion with the transition from the Navier-Stokes equations of a viscous fluid to those for a frictionless fluid.

---

† As already explained in Sec. XVIa, recent experimental results (H. W. Emmons [25], and Schubauer and Klebanoff [83]) indicate that there is no well-defined point of transition but that the process of transition from laminar to fully developed turbulent flow extends over a finite distance.

The majority of earlier papers on the theory of stability used the frictionless equation (16.16) as their point of departure. In this manner, evidently, no critical Reynolds number can be obtained but it is possible to answer the question of whether a given laminar flow is stable or not†. The complete equation (16.14) was analyzed much later when, after many failures, critical Reynolds numbers were at last successfully evaluated.

Starting with the above frictionless stability equation, eqn. (16.16), Lord Rayleigh [70] succeeded in deriving several important, general theorems concerning the stability of laminar velocity profiles. The validity of these has later been confirmed also for the case when the effect of viscosity is taken into account.

Theorem I: The first important, general theorem of this kind, the so-called *point-of-inflexion criterion*, asserts that velocity profiles which possess a point of inflexion are unstable.

Lord Rayleigh was able only to prove that the existence of a point of inflexion constitutes a *necessary condition* for the occurrence of instability. Much later, W. Tollmien [100] succeeded in showing that this constitutes also a *sufficient condition* for the amplification of disturbances. The point-of-inflexion criterion is of fundamental importance for the theory of stability because it provides — except for a correction due to the omission of the influence of viscosity — a first, rough classification of all laminar flows. From the practical point of view, this criterion is important owing to the direct connexion between the existence of a point of inflexion and the presence of a pressure gradient. In the case of convergent-channel flow with a favourable pressure gradient, as seen from Fig. 5.15, the velocity profiles are very full and possess no points of inflexion. In contradistinction, in a divergent channel with an adverse pressure gradient, the velocity profiles are pointed and points of inflexion are present. Identical differences in the geometrical form of the velocity profiles occur in the laminar boundary layer on a body immersed in a stream. According to boundary-layer theory, the velocity profiles in the interval where the pressure decreases are free from points of inflexion, whereas those in the interval where the pressure increases always possess them, see Sec. VIIc. Hence, the point-of-inflexion criterion becomes equivalent to a statement about the effect of the pressure gradient in the external flow on the stability of the respective boundary layers. As applied to boundary-layer flow, it amounts to this: a favourable pressure gradient stabilizes the flow, whereas an adverse pressure gradient enhances instability. It follows that the position of the point of minimum pressure on a body placed in a stream is decisive for the position of the point of transition, and, roughly speaking, we can say that the position of the point of minimum pressure determines that of the point of transition and causes the latter to lie close behind the former.

The influence of viscosity on the solution of the stability equation, which has been neglected up to this point, changes the preceding conclusions only very slightly. The preceding instability of velocity profiles with points of inflexion is usually referred to as "frictionless instability" because the laminar mean flow proves to be unstable even without taking into account the effect of viscosity on the oscillating motion. In the diagram of Fig. 16.8, the case of frictionless instability corresponds to the curve of type a. Even at $R = \infty$ there exists already a certain unstable range

---

† With the reservation that the influence of viscosity on the disturbance itself has been left out of consideration.

of wavelengths; in the direction of decreasing Reynolds numbers, this range is separated from the stable range by the curve of neutral stability.

In contrast with the preceding case, *viscous instability* is associated with a curve of neutral stability of shape *b*, also shown in Fig. 16.8, and with boundary-layer profiles possessing no point of inflexion. At Reynolds numbers tending to infinity, the range of unstable wavelengths is contracted to a point and domains of unstable oscillations are seen to exist only for finite Reynolds numbers. Generally speaking, the amount of amplification is much larger in the case of frictionless instability than in the case of viscous instability.

The existence of viscous instability can be discovered only in connexion with a discussion of the full Orr-Sommerfeld equation; it constitutes, therefore, the more difficult analytical case. The simplest case of flow, namely that along a flat plate with zero pressure gradient belongs to the kind for which only viscous instability does occur; it was successfully tackled only comparatively recently.

Theorem II: The second important general theorem states that the velocity of propagation of neutral disturbances ($c_i = 0$) in a boundary layer is smaller than the maximum velocity of the mean flow, i. e. that $c_r < U_m$.

This theorem was also first proved by Lord Rayleigh [70], albeit under some restrictive assumptions; it was proved again by W. Tollmien [100] for more general conditions. It asserts that in the interior of the flow there exists a layer where $U - c = 0$ for neutral disturbances. This fact, too, is of fundamental importance in the theory of stability. The layer for which $U - c = 0$ corresponds, namely, to a singular point of the frictionless stability equation (16.16). At this point ϕ'' becomes infinite if U'' does not vanish there simultaneously. The distance $y = y_K$ where $U = c$ is called the *critical layer* of the mean flow. If $U_K'' \neq 0$, then ϕ'' tends to infinity as

$$\frac{U''_K}{U'_K} \frac{1}{y - y_K}$$

in the neighbourhood of the critical layer where it is permissible to put $U - c = U_K'(y - y_K)$ approximately; consequently the *x*-component of the velocity can be written as

$$u' = \phi' \sim \frac{U''_K}{U'_K} \ln(y - y_K) . \tag{16.18}$$

Thus, according to the frictionless stability equation, the component u' of the velocity which is parallel to the wall becomes infinite if the curvature of the velocity profile at the critical layer does not vanish simultaneously. This mathematical singularity in the frictionless stability equation points to the fact that the effect of viscosity on the equation of motion must not be neglected in the neighbourhood of the critical layer. The inclusion of the effect of viscosity removes this physically absurd singularity of the frictionless stability equation. The analysis of the effect of this so-called viscous correction on the solution of the stability equation plays a fundamental part in the discussion of stability.

The two theorems due to Lord Rayleigh show that the curvature of the velocity profile affects stability in a fundamental way. Simultaneously it has been demonstrated that the calculation of velocity profiles in laminar boundary layers must proceed with very high accuracy for the investigation of stability to be possible: it is not enough to evaluate $U(y)$ with a sufficient degree of accuracy but its second derivative $d^2 U/dy^2$ must also be accurately known.

A summary of the solutions of Rayleigh's equation, presented from the mathematician's point of view, was prepared by P. G. Drazin and L. N. Howard [15a].

c. Results of the theory of stability as they apply to the boundary layer on a flat plate at zero incidence

1. Some older investigations into stability. The earlier investigations undertaken as a continuation of Lord Rayleigh's work limited themselves at first to the consideration of Couette flow, i. e. to the case of linear velocity distribution in a flow between two parallel walls, Fig. 1.1. The very exhaustive discussion of the case which included the full effect of viscosity provided by A. Sommerfeld [86], R. von Mises [61] and L. Hopf [45] led to the conclusion that this type of flow remains stable at all Reynolds numbers and at all wavelengths. For a time, after this negative result had been obtained, it was thought that the method of small oscillations was unsuitable for the theoretical solution of the problem of transition. It transpired later that this view was not justified, because Couette flow is a very special and restricted example. Moreover, as shown earlier, the curvature of the velocity profile plays an essential physical role in the flow, and it is not permissible to leave it out of account.

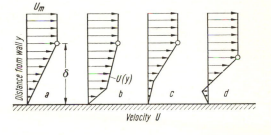

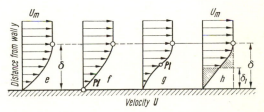

Fig. 16.9. Velocity profiles illustrating the study of the stability of laminar boundary layers

$U(y)$ = velocity distribution; U_m = velocity in the free stream; δ = boundary layer thickness; δ_1 = displacement thickness; PI = point of inflexion of the velocity profile. For $R \to \infty$ the profiles of type a, b, e, and f are stable; profiles of type c, d, g are unstable; profiles of type e exist in a favourable pressure gradient; type f corresponds to constant pressures; type g exists in an adverse pressure gradient

In the year 1921 L. Prandtl [67] reverted to the attempt to examine the problem of stability by theoretical methods. In order to consider the stability of a laminar boundary layer on a flat plate without undue mathematical complications, velocity profiles with straight segments were used, Figs. 16.9a, b, c, d in the same way as was done previously by Lord Rayleigh. A calculation performed by O. Tietjens [98] on the basis of the frictionless stability equation showed that in the case of boundary-layer profiles, the existence of convex corners, Figs. 16.9a, b, ensures stability, whereas concave corners. Figs. 16.9c, d, always lead to instability. This investigation made it plausible to suppose that velocity profiles with points of inflexion, Fig. 16.9g, are unstable. The truth of this supposition was later demonstrated by W. Tollmien [100], as already stated in Sec. XVIb, Theorem I.

In order to obtain a limit of stability expressed in terms of a Reynolds number for unstable velocity profiles (Figs. 16.9c and d), the largest viscous terms appearing in the complete stability equation (16.14) were taken into account, and it was expected that they will promote damping. The influence of viscosity on the disturbances extended here only over a very small region of the whole velocity profile, being located in the immediate neighbourhood of the wall, in order to satisfy the no-slip condition. The calculations performed by O. Tietjens led to the very unexpected result that the introduction of a small value of viscosity into the equations did not produce damping but amplification for all Reynolds numbers, and for all wavelengths of the disturbances. Moreover, this result was obtained not only for unstable velocity profiles (Figs. 16.9c, d) but also for the profiles of type a and b in Fig. 16.9, which have been shown to be stable when viscosity was neglected.

An interim review describing progress achieved between the years 1920 and 1930 was given by L. Prandtl [67a] on the occasion of the annual GAMM meeting (German Society for Applied Mathematics and Mechanics) in Bad Elster, 1931.

2. Calculation of the curve of neutral stability. A satisfactory explanation of the above paradox was supplied by W. Tollmien [99] in the year 1929. He demonstrated that the influence of viscosity on disturbances must be taken into account not only in the immediate neighbourhood of the wall, as supposed by O. Tietjens, but that, in addition, it must be accounted for also in the neighbourhood of the critical layer, where the velocity of wave propagation of the disturbances becomes equal to the velocity of the main flow and where, as shown in Sec. XVI b 5, the component u' becomes infinite according to the simplified, frictionless theory, the curvature of the profile being different from zero. The existence of viscosity causes large changes in this *critical layer*, while it is also evident that in reality u' remains finite there. However, the influence of viscosity becomes evident only if the curvature of the velocity profile is not left out of account. These considerations demonstrated that it was necessary to study the behaviour of small disturbances with respect to curved velocity profiles ($d^2U/dy^2 \neq 0$), and with viscosity taken into account both in the neighbourhood of the wall and in the critical layer. This programme was carried out by W. Tollmien in the paper quoted earlier, and as a result, he was able to find a limit of stability (critical Reynolds number) for the example of the flow in the boundary layer on a flat plate at zero incidence which agreed well with experiments.

In order to integrate the Orr-Sommerfeld equation (16.14), which is of 4th order, it is necessary to establish a fundamental system of solutions for it. For $y \to \infty$ and with $U(y) = U_m = \mathrm{const}$, this is

$$\left. \begin{array}{ll} \phi_1 = \mathrm{e}^{-\alpha y}; & \phi_2 = \mathrm{e}^{+\alpha y}; \\ \phi_3 = \mathrm{e}^{-\gamma y}; & \phi_4 = \mathrm{e}^{+\gamma y}; \end{array} \right\} \tag{16.19}$$

where

$$\gamma^2 = \alpha^2 + \mathrm{i}\, \mathsf{R}(\alpha - \beta). \tag{16.20}$$

Generally speaking, for neutral oscillations we find that

$$|\gamma| \gg |\alpha|, \tag{16.20a}$$

and, consequently, ϕ_1 and ϕ_2 represent the slowly varying solutions, whereas ϕ_3 and ϕ_4 become the fast varying solutions. The pair of solutions ϕ_1, ϕ_2 satisfies both the frictionless disturbance equation (Rayleigh's equation) and the viscous, Orr-Sommerfeld equation, eqns. (16.16) and (16.14), as $y \to \infty$. By contrast, the pair of solutions ϕ_3, ϕ_4 satisfies only the viscous disturbance equation. For this reason ϕ_1, ϕ_2 are referred to as the frictionless solutions, whereas ϕ_3, ϕ_4 are called the viscous solutions.

When we proceed to calculate the constants in the general solution

$$\phi = C_1 \phi_1 + C_2 \phi_2 + C_3 \phi_3 + C_4 \phi_4,$$

we notice that the solutions ϕ_2 and ϕ_4 must be dropped. This is due to the fact that the boundary conditions (16.15) require the vanishing of ϕ and ϕ' at $y \to \infty$. Thus

$$\phi = C_1 \phi_1 + C_3 \phi_3, \tag{16.21}$$

with the boundary condition that $\phi = \phi' = 0$ at $y = 0$. The non-viscous solution ϕ_1 does not satisfy the no-slip condition at the wall ($y = 0$) because $\phi_1' \neq 0$ there. Furthermore, at the critical layer given by $U - c = 0$ we discover that $\phi_1' \to \infty$, as explained earlier. It follows that the contribution from friction becomes particularly large at those two locations, and that the required particular solution $\phi_3(y)$, as well as the general solution $\phi(y)$, vary with y at a fast rate there. As a consequence, it becomes very tedious to calculate the characteristic function $\phi(y)$ and the eigenvalue $c = c_r + i c_i$, whether analytically or numerically, for a given pair of values of α and R. When numerical methods are used, the special difficulties stem from the fact that the highest derivative in the Orr-Sommerfeld equation, ϕ'''', is multiplied by the very small factor $1/R$. Mathematically speaking, the large difference between the course of the function $\phi(y)$ at the wall and at the critical layer as depicted by the frictionless (Rayleigh) equation and the equation containing friction (Orr-Sommerfeld) stems from the fact that the order of the differential equation is reduced from four to two when the viscous terms are deleted in it.

An attempt to calculate numerically the characteristic functions $\phi(y)$ of the Orr-Sommerfeld equation (16.14) for a large set of prescribed pairs of values of the reciprocal wavelength, α, and Reynolds number, R, puts enormous demands on the capacity of a computer. This explains why O. Tietjens [98] and W. Heisenberg [42], who attacked this problem in the twenties, failed to achieve success. At the end of the twenties, Tollmien reverted to this problem and found no other way but to fall back on a very tedious analytic procedure. Nevertheless, these time-consuming analytic methods proved eminently successful[†]. Details of these calculations can be found in the original papers of W. Tollmien [99, 100, 101] and D. Grohne [38]. There is no need to summarize this work here, because the calculations have been rendered

---

[†] Tollmien's analytic investigations [99] produced, among others, the physically important result, that the velocity components u' of the disturbance experiences a phase shift of angle π upon crossing the critical layer. This shift is due to the viscosity; cf. the calculations in [39a].

obsolete by modern numerical methods based on large, efficient electronic computers. The first successful numerical solution of the Orr-Sommerfeld equation was published in the year 1962 by E. F. Kurtz and S. H. Crandall [51], that is thirty years after the publication of W. Tollmien's [99] original results! This work was improved in 1970 in two papers by R. Jordinson [47, 48]. Important introductory work was performed by M. R. Osborne [62] and L. H. Lee and W. C. Reynolds [53]. The peculiar difficulties of the numerical evaluation of the characteristic solutions and the eigenvalues of the Orr-Sommerfeld equation have been again discussed shortly afterwards in the work of J. M. Gersting and D. F. Jankowski [30] and A. Davie [12]. Furthermore, R. Betchov and W. O. Criminale gave a summary account of the difficulties associated with the numerical solution of the Orr-Sommerfeld equation in their book [4], basing it on R. E. Kaplan's [48a] MIT thesis. In this connexion the reader may also wish to consult Chap. 5 of F. M. White's book [107].

As a sequel to ref. [47], the effect of a slight streamwise change in the basic flow was studied a number of times [2, 4a, 31, 46a, 84a, 106]. As already pointed out by J. Pretsch [69], this effect is small.

It may be useful to point out at this stage in our description that the stability analysis of a flow field in a boundary layer is, generally speaking, more difficult than of that through a channel. This is due to the fact that one of the boundary conditions for boundary-layer flow is at infinity, whereas both boundaries of a channel are located at finite distances. This is aggravated by the circumstance that the velocity profile $U(y)$ of the main flow in a boundary layer is not an exact solution of the Navier-Stokes equations, in contrast with a channel (e. g. Hagen-Poiseuille flow). Finally, we wish to recall that the Orr-Sommerfeld equation itself was derived on the assumption that the main flow $U(y)$ does not change in the direction of the stream†. This assumption is satisfied in channel flow but not in a boundary layer. All these circumstances combine to render the stability analysis of a boundary layer fundamentally more difficult than for channel flow.

3. Results for the flat plate. As a first application, W. Tollmien [99] employed his method to the investigation of the stability of the boundary layer on a flat plate at zero incidence. The velocity profile of such a boundary layer (Blasius profile) is shown in Fig. 7.7. The profiles at different stations along the plate are similar which means that they can be made to coincide when they are plotted against $y/\delta(x)$. Here $\delta(x)$ denotes the boundary-layer thickness which has been

---

† The extended form of the Orr-Sommerfeld equation (16.14) which contains the additional terms introduced by the lack of parallelism in the basic flow can be found in a paper by W. S. Saric and A. H. Nayfeh [84a]. There are six additional terms. Two terms are introduced by the change in the x-direction of the amplitude of the disturbance, two terms are added by the transverse component of the velocity in the basic flow, one more term is due to the change in the wavelength of the disturbance in the x-direction and, finally, the sixth term corresponds to higher-order terms in boundary-layer theory (see Chap. IX). The presence of suction or blowing gives rise to further terms. An investigation of the numerical solutions of the so modified Orr-Sommerfeld equation for various velocity profiles of the Falkner-Skan series, [2, 31, 106], failed to include all additional terms in most cases. For this reason it is difficult to make a comparison between such solutions as well as between them and the solutions of the "simplified" Orr-Sommerfeld equation. However, in most cases the change in the limit of stability due to lack of parallelism turns out to be small. Numerical examples have been given by F. C. T. Shen et al. [85b].

shown in eqn. (7.35) to be given by $\delta = 5 \cdot 0 \sqrt{\nu\, x/U_\infty}$. The velocity profile possesses a point of inflexion at the wall and corresponds to the one shown in Fig. 16.9f. Thus, in the light of the point-of-inflexion criterion which we stated in Sec. XVIb 5, it is seen that this profile lies just on the border-line between profiles with no point of inflexion, which are stable according to the frictionless theory, and profiles with a point of inflexion, which are unstable.

The results of stability calculations performed in accordance with the method described in the preceding section are shown in Figs. 16.10 and 16.11 as well as Table 16.1. The statepoints along the curves themselves represent neutral disturbances; the region embraced by the curve corresponds to unstable disturbances, and that outside it contains stable points. The two branches of the curve of neutral stability tend towards zero at very large Reynolds numbers. The smallest Reynolds number, for which one indifferent disturbance still exists, represents the critical Reynolds number and is given by

$$\left(\frac{U_\infty \delta_1}{\nu}\right)_{crit} = \mathsf{R}_{crit} = 520 \quad \text{(point of instability)}\dagger. \qquad (16.22)$$

Table 16.1. Wavelength $\alpha\,\delta_1$ and frequency $\beta_r\,\delta_1/U_\infty$ of neutral disturbances in terms of the Reynolds number R for the boundary layer on a flat plate at zero incidence (Blasius profile). Theory after W. Tollmien [99]; numerical calculations by R. Jordinson [47] and D.R. Houston, both for parallel flow. See Figs. 16.10 and 16.11

		Lower branch		Upper branch	
$R = \dfrac{U_\infty \delta_1}{\nu}$		$\alpha\,\delta_1$	$\dfrac{\beta_r\,\delta_1}{U_\infty}$	$\alpha\,\delta_1$	$\dfrac{\beta_r\,\delta_1}{U_\infty}$
1×10^6		0·017	0·0010	0·102	0·0153
5×10^5		0·021	0·0015	0·111	0·0178
2×10^5		0·027	0·0024	0·126	0·0224
1×10^5		0·032	0·0035	0·136	0·0257
5×10^4		0·040	0·0050	0·140	0·0267
2×10^4		0·053	0·0083	0·176	0·0393
1×10^4		0·066	0·0122	0·212	0·0534
5×10^3		0·085	0·0184	0·254	0·0724
4×10^3		0·092	0·0211	0·269	0·0975
3×10^3		0·103	0·0253	0·288	0·0895
2×10^3		0·122	0·0330	0·315	0·105
1000		0·171	0·0551	0·351	0·130
800		0·195	0·067	0·356	0·137
603		0·241	0·090	—	—
600		—	—	0·347	0·138
558		0·261	0·099	—	—
530		0·281	0·110	0·324	0·130
$R_{crit} = 520$		0·302	0·120	—	—

---

† In his first paper of 1929, W. Tollmien indicated a value of 420; a subsequent calculation by H. Schlichting, performed in 1933 [76] yielded the value of 575. The numerical value of 520 quoted here was taken from a contemporary investigation carried out by Jordinson [47] in 1970.

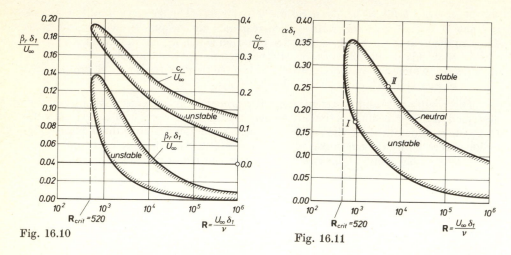

Fig. 16.10

Fig. 16.11

Fig. 16.10. Curves of neutral stability for the disturbance frequency β_r and the wave velocity c_r as a function of Reynolds number for the boundary layer on a flat plate at zero incidence (Blasius profile). Theory according to W. Tollmien [99]; numerical calculations by R. Jordinson [47]; see also Table 16.1

Fig. 16.11. Curves of neutral stability for the disturbance wavelength $\alpha\,\delta_1$ as a function of Reynolds number for the boundary layer on a flat plate at zero incidence (Blasius profile). Theory according to W. Tollmien [99]; numerical calculations by R. Jordinson [47]; see also Table 16.1. The amplitude distribution for disturbances I and II is given in Fig. 16.20

This is the point of instability for the boundary layer on a flat plate. It is remarkable that only a comparatively narrow range of wavelengths and frequencies is "dangerous" for the laminar boundary layer. On the one hand, there is a *lower* limit for the Reynolds number, on the other, there is an *upper* limit for the characteristic magnitudes of the disturbances. Once the latter are exceeded no instability is caused. The numerical values are:

$$\frac{c_r}{U_\infty} = 0.39 \; ; \quad \alpha\,\delta_1 = 0.36; \quad \frac{\beta_r\,\delta_1}{U_\infty} = 0.15 \, .$$

It is noteworthy that the wavelength is very large compared with the boundary layer thickness. The smallest unstable wavelength is

$$\lambda_{min} = \frac{2\,\pi}{0.36}\,\delta_1 = 17.5\,\delta_1 \approx 6\,\delta \, .$$

A detailed comparison between the preceding theoretical results and experiment will be given in the next section. Here we shall only remark that the position where the boundary layer becomes first unstable according to theory (point of instability) must always be expected to lie upstream of the experimentally observed point of transition because actual turbulence is created along the path from the point of instability to the point of transition owing to the amplification of the unstable disturbances. This condition is satisfied in the case under consideration. We have already said in Sec. XVI a that according to older measurements the point of transition occurs at $(U_\infty x/\nu)_{crit} = 3 \cdot 5 \times 10^5$ to 10^6. Using the value of $\delta_1 = 1 \cdot 72 \sqrt{\nu x/U_\infty}$ from eqn. (7.37) we can find that this corresponds to a critical Reynolds number

$$\left(\frac{U_\infty \delta_1}{\nu}\right)_{crit} = 950 \quad \text{(point of transition)},$$

which is considerably larger than the value of 520 which we quoted earlier for the point of instability.

The distance between the point of instability and the point of transition depends on the *degree of amplification* and the kind of disturbances present in the external stream (intensity of turbulence), but the actual mechanism of amplification can be obtained from the study of the magnitudes of the parameters in the interior of the curve of neutral stability, $\beta_i > 0$. Calculations of this kind were first performed by H. Schlichting [76] for the flat plate; they have been repeated by S. F. Shen [85].

In order to gain a clearer insight into the mechanics of the oscillating motion, H. Schlichting [77] determined the eigenfunctions $\phi(y)$ for several neutral disturbances. This enabled him to draw the pattern of streamlines of the disturbed motion for the neutral oscillations. An example of such a pattern can be found in Fig. 16.14.

The diagram in Fig. 16.12 illustrates the amplification of unstable disturbances in the boundary layer on a flat plate. The diagram, based on a recent calculation performed by H. G. Ombrewski et al. [63], extends over a wide range of Reynolds numbers. It turns out that the maximum amplification rate does not place itself at a very high Reynolds number $(R \to \infty)$ but is located in the moderate range of $R = 10^3$ to 10^4. This is due to the fact that the curve of neutral stability for a flat plate is of

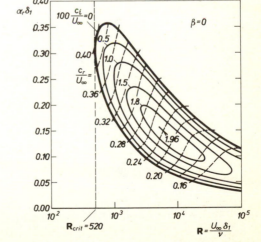

Fig. 16.12. Curves of constant *temporal* amplification for the boundary layer on a flat plate at zero incidence over a wide range of Reynolds numbers, after H. G. Ombrewski et al. [63]

the type of "viscous instability", curve (b) in Fig. 16.8, which shows no amplification at very large Reynolds numbers. Figure 16.13 depicts curves of constant amplification in the lower range of Reynolds numbers. The curves $c_i =$ const in Fig. 16.12 represent the amplification *in time* of the unstable disturbances, because $\beta_i = c_i \alpha$. By contrast, in this diagram, with $\alpha_i =$ const, the curves describe the *spatial* amplification of a disturbance being propagated downstream†.

In later times, J. T. Stuart [9, 90] and D. Grohne [34] made an attempt to determine the course of the amplification of unstable disturbances taking into account the effect of the *non-linear* terms in the equations. In this connexion it is important to realize that the amplification of the unstable disturbances causes the mean flow to contract quite considerably. This, in turn, causes a change in the transfer of energy from the main motion to the oscillating motion, since it is proportional to dU/dy. The main effect of this is that at a later stage the unstable disturbances no longer amplify in proportion to $\exp{(\beta_i t)}$ but tend to a finite value which is independent of the initial value.

The distance between the point of transition and the point of neutral stability depends considerably on the *turbulence intensity* in addition to its dependence on amplification (see also Sec. XVI d).

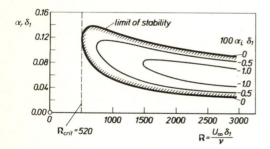

Fig. 16.13. Curves of constant *spatial* amplification for the boundary layer on a flat plate at zero incidence in the lower range of Reynolds numbers as calculated by R. Jordinson [47]

---

† The stream function of the disturbance introduced in eqn. (16.10) describes its amplification in time when $\alpha = \alpha_r$ is real and $\beta = \beta_r + i\beta_i$ is complex, because

$$\psi(x, y, t) = \Phi(y) \exp{(\beta_i t)} \exp{[i(\alpha_r x - \beta_r t)]}.$$

On the other hand, if we regard $\alpha = \alpha_r + i\alpha_i$ as complex in eqn. (16.10), and take $\beta = \beta_r$ as being real, we conclude that the *local* amplification is described by

$$\psi(x, y, t) = \Phi(y \exp{(-\alpha x_i)} \exp{[i(\alpha_r x - \beta_r t)]}.$$

It follows that the temporal amplification of the unstable disturbances is described by $\beta_i > 0$, whereas the local amplification is described by $\alpha_i < 0$. Neutral stability corresponds to $\beta_i = 0$ and $\alpha_i = 0$ indicating that it is the same for both time and place dependent amplification. A more detailed study of the relation between temporal and local amplification (or damping) of disturbances is contained in two papers by M. Gaster [32, 33]. Compare here the paper by A. R. Wazzan [104a]; non-Newtonian fluids are discussed by J. R. A. Pearson [64a].

More than a decade was to elapse before an experimental verification of the above theory of stability could be obtained. This was brilliantly achieved by G. B. Schubauer and H. K. Skramstad [82] and we shall give an account of their work in the succeeding section. At a time when these experimental results were already known, C. C. Lin [54] repeated all the calculations required in the development of the theory; his calculations agreed at all essential points with those due to W. Tollmien and H. Schlichting.

Navier-Stokes equations: At a much later time, H. Fasel [28 b] calculated the temporal amplification of artificially induced, periodic disturbances. He used a numerical method and employed the full Navier-Stokes equations. The results agree in all essential points with those reported earlier, that is with the results deduced from the linear stability theory based on the Orr-Sommerfeld equation; see also [33a] and [57 d].

d. Comparison of the theory of stability with experiment

1. Older measurements of transition. The preceding results were the first solutions of the theory of small disturbances which led to the evaluation of a critical Reynolds number of the same order of magnitude as that measured experimentally. In accordance with the theory, small disturbances which fall within a certain range of frequency and wavelength are amplified, whereas disturbances of smaller or larger wavelengths are damped, provided that the Reynolds number exceeds a certain limiting value. The theory shows that disturbances whose wavelengths are large and equal to a multiple of the boundary-layer thickness are particularly "dangerous". It is further assumed that the amplification of disturbances eventually effects the transition from laminar to turbulent flow. The process of amplification represents, so to say, the link between the stability theory and the experimentally established fact of the existence of transition.

Some time before the first successes of the theory of stability had been achieved, L. Schiller [81] carried out extensive experimental investigations into the phenomenon of transition, particularly as it occurs in pipes. These led to the development of a semi-empirical theory of transition which was based on the premise that, essentially, transition is due to finite disturbances which originate in the inlet to the pipe, or, in the case of boundary-layer flow, in the external free stream. These ideas were further developed theoretically, particularly by G. I. Taylor [97]†.

The decision as to which of the two theories should be adopted had to be left to experiment. Even before the stability theory was established, transition on a flat plate had been investigated experimentally and in detail by J. M. Burgers [6],

---

† "In the late thirties the prevailing view was probably as expressed by G.I. Taylor [97] in 1938 who was of the opinion that stability theory had little or no connexion with boundary layer transition. Only the Germans, who propounded the theory and reported qualitative agreement with the flow-visualization experiment of Prandtl (1933) in the Goettingen watertunnel, gave any credence to the theory. They were right: The early experiments could not detect Tollmien-Schlichting waves, because they were overwhelmed by the very noisy background turbulence in the wind-tunnel of those days" (quoted from F.M. White [107], 1974).

B. G. van der Hegge Zijnen [41] and M. Hansen. These measurements led to the
result that the critical Reynolds number was contained in the range

$$\left(\frac{U_\infty\, x}{\nu}\right)_{crit} = 3\cdot5 \text{ to } 5 \times 10^5 \,.$$

Soon after, H. L. Dryden [16, 17] and his collaborators undertook a very thorough
and careful investigation of this type of flow. During the course of these investigations
extensive data on the velocity distribution were carefully plotted with the aid of
hot-wire anemometers in terms of space coordinates and time. However, the selective amplification of disturbances predicted by the theory could not yet be detected.

   At about the same time, experiments carried out at Goettingen on a flat plate
in a water channel yielded a qualitative confirmation of the theory of stability.
The photographs in Fig. 16.15 depict a turbulent region which originated from a
disturbance of long wavelength. The similarity between these photographs and the
theoretical pattern of streamlines of a neutral disturbance shown in Fig. 16.14 is
irrefutable.

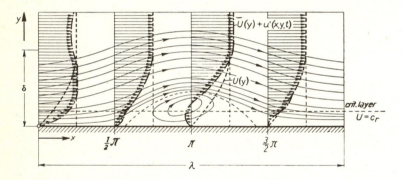

Fig. 16.14.   Patterns of streamlines and velocity distribution for a neutral disturbance in the
boundary layer on a flat plate at zero incidence (disturbance I in Fig. 16.11)

$U(y)$ = mean flow; $U(y) + u'(x, y, t)$ = disturbed velocity distribution; $U_\infty\, \delta_1/\nu = 893$ = Reynolds number;
$\lambda = 40\, \delta_1$ = wavelength of disturbance; $c_r = 0.35\, U_\infty$ = wave propagation velocity;

$$\int_0^\delta \sqrt{\overline{u'^2}}\, dy = 0\cdot172\, U_\infty\, \delta = \text{intensity of disturbance}$$

   In discussing transition it is necessary to introduce one very important parameter which measures the "degree of disturbance" in the external stream. Its importance was first recognised when measurements of the drag of spheres were performed in different wind tunnels. In this connexion it was discovered that the critical
Reynolds number of a sphere, that is that value of the Reynolds number which
corresponds to the abrupt decrease in the drag coefficient shown in Fig. 1.5, depends
very markedly on the strength of the disturbances in the free stream. This can be
measured quantitatively with the aid of the time-average of the oscillating, turbulent velocities as they occur, for example, behind a screen (see also Sec. XVIIIf).

Denoting this time-average of the three components by $\overline{u'^2}, \overline{v'^2}, \overline{w'^2}$, we define the *intensity* (or *level*) *of turbulence* of a stream as

$$\mathsf{T} = \sqrt{\tfrac{1}{3}(\overline{u'^2} + \overline{v'^2} + \overline{w'^2})}\big/ U_\infty,$$

where U_∞ denotes the mean velocity of the flow. In general, at a certain distance from the screens or honeycombs, the turbulence in a wind tunnel becomes *isotropic*, i. e. one for which the mean oscillations in the three components are equal:

$$\overline{u'^2} = \overline{v'^2} = \overline{w'^2} \,.$$

In this case it is sufficient to restrict oneself to the oscillation u' in the direction of flow, and to put

$$\mathsf{T} = \sqrt{\overline{u'^2}}\big/ U_\infty \,.$$

Fig. 16.15. Flow along a flat plate; turbulence originating from a disturbance of long wavelength after L. Prandtl [68]

The photographs were taken with the aid of a slow motion-picture camera, which travelled on a trolley along with the flow; consequently, the camera is trained on the same group of vortices all the time. The flow is made visible by sprinkling aluminium dust on the water surface

This simpler definition of turbulence intensity is very often used in practice even in cases when the turbulence is not isotropic. Measurements in different wind tunnels show that the critical Reynolds number of a sphere depends very strongly on the turbulence intensity, T, the value of R_{crit} increasing fast as T decreases. In older wind tunnels (constructed before 1940), the intensity of turbulence was of the order of 0·01.

2. Verification of the theory of stability by experiment. In 1940, H. L. Dryden, assisted by G. B. Schubauer and H. K. Skramstad of the National Bureau of Standards in Washington, undertook a new and extensive experimental programme of investigation into the phenomenon of transition from laminar to turbulent flow [82]. In the meantime it became accepted that the intensity of turbulence exerts a decisive influence on the process of transition. Consequently, a special wind tunnel was constructed for these investigations, and by the use of a large number of suitable screens and a very large contraction ratio, the intensity of turbulence was reduced to the extremely low, and never previously attained, value

$$\mathsf{T} = \sqrt{\overline{u'^2}} \,/\, U_\infty = 0.0002 \;.$$

The stream was then used for the thorough investigation of the laminar boundary layer on a flat plate at zero incidence, when it was discovered that at very low turbulence intensities, i. e. of the order of $\mathsf{T} < 0.001$, the previously established value of the critical Reynolds number of $R_{crit} = 3.5$ to 5×10^5 was increased to

$$\left(\frac{U\,x}{\nu} \right)_{crit} \approx 2.8 \times 10^6 \;,$$

see Fig. 16.16. It was further discovered, as revealed by Fig. 16.16, that a decrease in the intensity of turbulence causes the critical Reynolds number to increase, at first quite fast; after a value of about $\mathsf{T} = 0.001$ has been attained, a critical Reynolds number of $R_{crit} = 2.8 \times 10^6$ is reached and retained at lower turbulence intensities. This demonstrates the existence of an upper limit of the critical Reynolds number on a flat plate. A measured point obtained earlier by A. A. Hall and G. S. Hislop [39] fits quite well into the graph of Fig. 16.16.

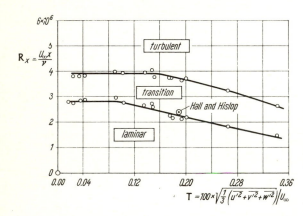

Fig. 16.16. Influence of intensity of turbulence on critical Reynolds number on flat plate at zero incidence, as measured by Schubauer and Skramstad [82]

All the measurements which we are about to discuss were carried out at an intensity of turbulence of $\mathsf{T}' = 0.0003$. Velocities were measured with the aid of a hot-wire anemometer and a cathode-ray oscillograph. The measurement consisted in the determination of the variation of the velocity with time at several stations along the plate; these were undertaken first under normal conditions (i. e. in the presence of natural disturbances), and then with artificially produced disturbances. Such artificial disturbances of a definite frequency were created with the aid of a thin metal strip which was placed at a distance of 0·15 mm from the wall and in which oscillations were excited electromagnetically. The existence of amplified sinusoidal disturbances could be clearly demonstrated even in the presence of *natural oscillations* (i. e. with no excitation), see Fig. 16.17. Owing to the extremely low intensity of turbulence, there are hardly any irregular oscillations left in the boundary layer, but, as the point of transition is approached, there appear almost purely sinusoidal oscillations; their amplitude is at first small and increases rapidly in the downstream direction. A short distance ahead of the point of transition, oscillations of very high amplitude make their appearance. At the point of transition, these regular oscillations break down and are suddenly transformed into the irregular patterns of high frequency which are characteristic of turbulent motion.

Fig. 16.17. Oscillogram of the u'-component of fluctuations caused by random ("natural") disturbances in the laminar boundary layer on a flat plate in a stream of air. Measurements on transition from laminar to turbulent flow due to Schubauer and Skramstad [82]

Distance from wall: 0·57 mm; free-stream velocity $U_\infty = 24$ m/sec, interval between time marks: $^1/_{30}$ sec

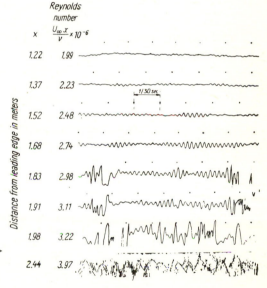

The measurements under consideration also threw light on the question as to why such amplified sinusoidal oscillation escaped detection during earlier experiments. It turns out, namely, that transition is caused directly by the random disturbances and is not preceded by a selective amplification of sinusoidal oscillations if the intensity of turbulence is increased, as already mentioned, from the above value of $T = 0·0003$ to $T = 0·01$, i. e. to a value commonly encountered in previous measurements.

During the experiments with *artificial disturbances* a thin metal strip extending over a width of about 30 cm, 0·05 mm thick and 2·5 mm deep was placed at a distance of 0·15 mm from the wall and was excited by a magnetic field induced with the aid of an alternating current. In this way it was possible to induce two-dimensional disturbances of prescribed frequency, as stipulated by the theory. This gave rise to amplified, damped and neutral oscillations simultaneously. They were again measured with the aid of a hot-wire anemometer. Results of such measurements are shown plotted in Fig. 16.18. The experimental points, which are joined by a broken line, represent measured neutral oscillations. The theoretical curve of neutral stability from Fig. 16.12 has been drawn for comparison, and the agreement is seen to be very good.

In order to gain more insight into the mechanism of transition, measurements of the amplitude of the u'-component were carried out for several neutral distur- bances at varying distance from the wall. Fig. 16.19 shows oscillograms of the sinu- soidal motion for the component u'. Each oscillogram contains two simultaneous curves, one of which was always taken at the same distance from the wall, the other having been taken at various distances. The variation of the amplitude of the u' oscillation over the boundary layer width is shown in Fig. 16.20. The diagram represents the results obtained by Schubauer and Skramstad and refers to the neutral disturbances marked I and II in Fig. 16.11. There is good agreement with the theory due to H. Schlichting [77].

Very careful experiments of this kind have been performed more recently by J. A. Ross et al. [74] who had at their disposal a wind tunnel of the very low turbulence intensity $T = 0.0003$. They report equally good agreement between theory and ex- periment.

We have already remarked earlier that the experimental verification of the stability theory was first made possible when a stream of very low turbulence inten- sity could be produced. The older experiments which were performed at a turbulence intensity of $T = 0.01$ confirmed the expectation that the observed point of transition lies downstream of the point of instability predicted by theory. However, the dis- tance between the points of instability and transition depends to a marked degree on turbulence intensity. It is to be expected that this distance should decrease as the intensity of turbulence is increased because in the presence of high turbulence a small amount of amplification suffices to produce turbulence from the unstable dis- turbances. The graph due to P. S. Granville [36] and shown in Fig. 16.21 illustrates this point in relation to the boundary layer on a flat plate. The difference between the Reynolds numbers formed with the momentum thickness at the points of tran- sition and instability, namely

$$\left(\frac{U_\infty \delta_2}{\nu} \right)_{tr} - \left(\frac{U_\infty \delta_2}{\nu} \right)_i$$

has been used as a measure of this distance and has been plotted against the inten- sity of turbulence. For the point of instability of a flat plate the value

$$\left(\frac{U_\infty \delta_2}{\nu} \right)_i = \frac{1}{2 \cdot 6} \left(\frac{U_\infty \delta_1}{\nu} \right)_i = \frac{520}{2 \cdot 6} = 200$$

was used.

The following diagram correlates G. B. Schubauer's and H. K. Skramstad's [82] measurements performed at very low turbulence intensities as well as the older measurements due to Hall and Hislop and performed at higher turbulence intensities.

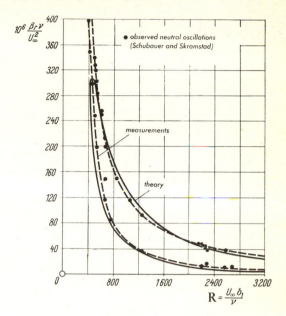

$$10^6 \frac{\beta_r \nu}{U_\infty^2}$$

● observed neutral oscillations
(Schubauer and Skramstad)

measurements

theory

$$R = \frac{U_\infty \delta_1}{\nu}$$

Fig. 16.18. Curves of neutral stability for neutral frequencies of disturbances on a flat plate at zero incidence. Measurements due to Schubauer and Skramstad [82]. Theory due to Tollmien [99]

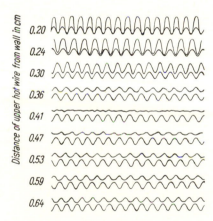

Distance of upper hot wire from wall in cm

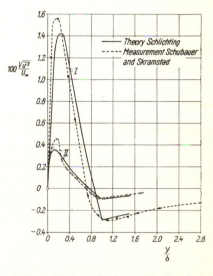

$$100 \frac{\sqrt{\overline{u'^2}}}{U_\infty}$$

Theory Schlichting
Measurement Schubauer
and Skramstad

$$\frac{y}{\delta}$$

Fig. 16.19. Measurements on oscillations in the laminar boundary layer performed by Schubauer and Skramstad [82]

Simultaneous recording of velocity with the aid of two hot-wire anemometers placed at a distance of 30 cm behind the strip. The lower curve corresponds to a hot wire placed at a distance of 1·4 mm from the wall; the upper curve corresponds to a hot wire placed at varying distances from the wall as indicated. The strip was placed at a distance of 90 cm behind the leading edge of the plate. Frequency 70 sec⁻¹. Velocity $U_\infty = 13$ m/sec

Fig. 16.20. Variation of amplitude of the u'-fluctuation for two neutral disturbances in a laminar boundary layer on flat plate at zero incidence. Measurements due to Schubauer and Skramstad [82.] Theory due to Schlichting [77]

The curves labelled I and II correspond to the two neutral disturbances I and II in Fig. 16.11

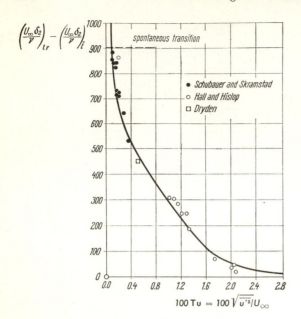

$$\left(\frac{U_\infty \delta_2}{\nu}\right)_{tr} - \left(\frac{U_\infty \delta_2}{\nu}\right)_{i}$$

spontaneous transition

• Schubauer and Skramstad
○ Hall and Hislop
□ Dryden

$$100\,\mathrm{T}u = 100\,\sqrt{\overline{u'^2}}/U_\infty$$

Fig. 16.21. Measurements on transition of a flat plate, after P. S. Granville [36]. Difference between the Reynolds numbers at the points of transition and instability in terms of turbulence intensity. As turbulence intensity increases, the point of transition moves closer to the point of instability

All experimental points trace a *single* curve. The point of transition does not coincide with the point of instability until very high turbulence intensities of about T = 0·02 to 0·03 have been reached; *cf.* [1].

Other velocity profiles: We now proceed to describe briefly investigations into the stability of other velocity profiles; a more detailed account is given in Chap. XVII.

The theorem on the instability of velocity profiles possessing a point of inflexion, which is due to Lord Rayleigh and W. Tollmien, had already been subjected to an experimental verification by G. Rosenbrook [73]; he was also able to report complete agreement between theoretical predictions and experiment.

A paper by S. Hollingdale [43] contains a contribution to the study of the stability of velocity profiles in the wake of a solid body. The stability of laminar jets was studied by N. Curle [10]. Mention may, finally, be made of the work of A. Michalke and H. Schade [58], T. Tatsumi [96a], L. N. Howard [46], and C. W. Clenshaw and D. Elliot [11]. The last reference established a limit of stability of $\mathrm{R}_{crit} = 6\cdot5$ for a plane jet, the Reynolds number being formed with the jet width at half height.

A linear stability analysis for the two-dimensional case of channel flow was first published by C. C. Lin [54]. He found that the critical Reynolds number, referred to the maximum velocity U_m and the half-width of the channel, b, had the value

$$\mathrm{R}_{crit} = \left(\frac{U_m\,b}{\nu}\right)_{crit} = 5314\,.$$

A subsequent, and more careful calculation by L. H. Thomas [103] confirmed this result. The non-linear stability theory was applied to this case by K. Stewartson [88]

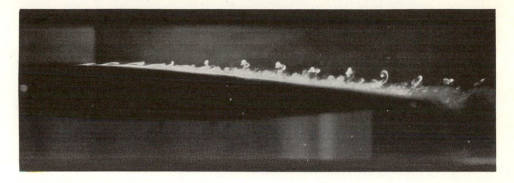

Fig. 16.22. Smoke picture of the flow in the boundary layer on an aerofoil in the presence of periodic disturbances, after H. Bergh [5]

Free-stream velocity $U_\infty = 4$ m/sec;   frequency of disturbance $\beta_r = 145$ 1/sec

and J. T. Stuart [90]; see also R. G. DiPrima et al. [14] and J. T. Stuart [91]. A summary of this problem area was given by A. Michalke. M. Ikeda [46 b] is also of interest.

A clear idea of the details of the mechanism of amplification can be formed by studying the smoke pictures of the zone of transition in the boundary layer on an airfoil taken by H. Bergh [5] and reproduced in Fig. 16.22. The artificial disturbances were produced with the aid of a loudspeaker; they are seen to induce in the boundary layer a succession of amplified, regular waves, their amplitude increasing in the downstream direction. See also [1].

Three-dimensional flows. The experimental evidence adduced so far shows that transition is started as a result of the amplification of two-dimensional disturbances. The growth of such disturbances was investigated in great detail by G. B. Schubauer and H. K. Skramstad [82], G. B. Schubauer and P. S. Klebanoff [83], as well as by I. Tani [92, 95]. It turned out that the amplification of the unstable plane waves always produces a distinctly three-dimensional flow structure. After the amplitude of the wave has reached a certain magnitude there sets in a period of strong and non-linear amplification of the disturbance. This process is accompanied by a transfer of energy in the transverse direction and this distorts the original, two-dimensional character of the base flow. Thus, the breakdown of laminar flow and, hence, the birth of turbulence appear to be a consequence of the development of the unstable disturbances in three dimensions. This is accompanied by the appearance, to a certain extent in the boundary layer too, of vortices whose axes lie in the direction of the flow.

Further light on this problem can be thrown by a study of the work of G. B. Schubauer, P. S. Klebanoff and K. D. Tidstrom [84, 49, 50]. H. Goertler and H. Witting [35] as well as C. C. Lin, D. J. Benny and H. P. Greenspan [56, 3, 37].

The experimental results reported in this chapter show such complete agreement with the theory of stability of laminar flows that the latter may now be regarded as a verified component of fluid mechanics. The hypothesis — that the process of transition from laminar to turbulent flow is the consequence of an instability in the laminar flow, enunciated by O. Reynolds, is hereby completely vindicated. It certainly represents a *possible* and *observable* mechanism of transition. The question as to whether it paints a complete picture of the process and whether it constitutes the *only* mechanism encountered in nature is still at present an open one. The latter questions now occupy the attention of many research workers.

e. Effect of oscillating free stream on transition

After it had been discovered with the aid of the experiment described earlier that the intensity of turbulence of the external stream, that is that the presence of an *irregular* time-dependent fluctuation in the free stream, exerted a strong influence on transition, it was natural to undertake studies on the effect of *regular* fluctuation in the free stream on transition. The effect of a superimposed fluctuation of small amplitude ($\varepsilon \ll 1$) in an external stream $U(x, t)$ of the form

$$U(x, t) = U_0(x) + \varepsilon\, U_1(x) \cos n\, t$$

on the structure of a laminar boundary layer was discussed in Secs. XV a 3 and XV e 3.

Since the Reynolds number at transition decreases considerably as the intensity of turbulence increases, it is plausible to suppose that a similar effect should occur as the amplitude $\Delta U = \varepsilon U_1$ of the periodic external stream is made to increase. The effect of an oscillation superimposed on the external stream on the transition of a laminar boundary layer was clarified experimentally by J. H. Obremski and A. A. Fejer [63a] as well as by J. A. Miller and A. A. Fejer [60a]. These investigations concentrated attention, in the first place, on the boundary layer on a flat plate (Blasius profile). In this case, the velocity distribution in the external stream is

$$U(x, t) = U_\infty + \Delta U \cos n\, t.$$

Here U_∞ is the time-average of the free-stream velocity which is independent of x, ΔU is the amplitude of the temporal fluctuation in the external stream, and n denotes its circular frequency. The measurements reported in [63a] were performed in an incompressible stream with

$$U_\infty = 20 \text{ to } 40 \text{ m/sec, with } \Delta U/U_\infty = 0 \cdot 014 \text{ to } 0 \cdot 29,$$

and with frequencies of $n = 4$ to 62 sec^{-1}.

These very careful experimental investigations yielded the following essential results:

(a) The critical Reynolds number of the start of transition, $\mathsf{R}_{x,tr} = U_\infty\, x_{tr}/\nu$ depends only on the amplitude $\Delta U/U_\infty$ of the external fluctuation.

(b) The dimensionless transition length, that is the distance between the start of transition and its completion, $\mathsf{R}_{x,t} - \mathsf{R}_{x,tr}$ depends only on the frequency of the external oscillation[†].

(c) The record showing the variation of velocity with time demonstrates that the line of transition is characterized by a regular and intermittent transition. The measurements led to the conclusion that transition can be described by the following "non-steady" Reynolds number:

$$\mathsf{R}_{NS} = \Delta U \times L/2\,\pi\,\nu.$$

Since the characteristic length of the external, oscillating stream is $L = U_\infty/n$, it is possible to express the "non-steady" Reynolds number in the form

$$\mathsf{R}_{NS} = \frac{U_\infty^2\,(\Delta U/U_\infty)}{2\,\pi\,\nu\,n}.$$

Here $\Delta U/U_\infty$ is the dimensionless amplitude of the impressed oscillation and $n\nu/U_\infty^2$ is its dimensionless frequency. The measurements showed that the Reynolds number $\mathsf{R}_{x,\,tr} = U_\infty\, x_{tr}/\nu$ at

---

† Start of transition at $\mathsf{R}_{x,tr} = U_\infty\, x_{tr}/\nu$ — lower curve in Fig. 16.16. Completion of transition at $\mathsf{R}_{x,tr} = U_\infty\, x_T/\nu$ — upper curve in Fig. 16.16. Over the distance from x_{tr} to x_T it is observed that the intermittency factor increases from $\gamma = 0$ to $\gamma = 1$; this is interpreted by the statement that in this zone we observe "transitional turbulence".

the point of transition was always considerably reduced compared with that for stationary flow when the "non-steady" Reynolds number was large, i. e. when $R_{NS} > 27000$. In these experiments, transition on the flat plate in stationary flow started at $R_{x,tr} = 1\cdot8 \times 10^6$. According to Fig. 16.16, this value of $R_{x,tr}$ corresponds, approximately, to a turbulence intensity of $T = 0\cdot28\%$ in the external stream.

So far, a satisfactory theory of stability for boundary layers in the presence of an external oscillating stream does not exist [11a]. Observation of intermittent turbulence in the presence of a free-stream oscillation shows that its frequency β_r is of the same order of magnitude as that of natural, neutral disturbances of the Tollmien-Schlichting type from stability theory; see also Fig. 16.18. The frequency, n, of the oscillating flow investigated here was smaller by a factor of about 100 than that in the natural, neutral disturbances.

A review of the process of transition in the presence of free-stream oscillations was recently published by R.J. Loehrke, M.V. Morkovin, and A.A. Fejer [48b].

f. Concluding remark

At the end of this chapter we wish to present, by way of summary, the process of transition in the boundary layer on a flat plate in the presence of an external flow of low turbulence intensity. As seen from Fig. 16.23, the flow goes through the following stages, starting with the leading edge:

(1)   Stable laminar flow following the leading edge.

(2)   Unstable, laminar flow with two-dimensional Tollmien-Schlichting waves.

(3)   Development of unstable, laminar, three-dimensional waves and vortex formation.

(4)   Bursts of turbulence in places of very high local vorticity.

(5)   Formation of turbulent spots in places when the turbulent velocity fluctuations are large.

(6)   Coalescence of turbulent spots into a fully developed turbulent boundary layer.

In most cases, the transition from turbulent spots to fully developed turbulence is associated with the formation of a separation bubble, as already mentioned in connexion with Fig. 10.13. At the present time, only stages (1), (2) and (3) are amenable to a theoretical analysis. The complete clarification of the remaining stages will require much additional theoretical research work.

(1) Stable flow
(2) Unstable Tollmien-Schlichting waves
(3) Three-dimensional waves and vortex formation
(4) Bursting of vortices
(5) Formation of turbulent spots
(6) Fully developed turbulent flow

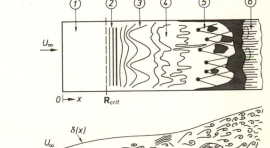

Fig. 16.23. Idealized sketch of transition zone in the boundary layer on a flat plate at zero incidence after F. M. White [107]

References

[1] Arnal, D., Julien, J. C., and Michel, R.: Analyse expérimentale et calcul de l'appartition et du développement de la transition de la couche limite. AGARD CP 224, 13—1 to 13—17 (1977).

[1a] AGARD-CP-224: Laminar-turbulent transition. Papers presented at the Fluid Dynamics Panel Symposium at Technical University of Denmark, Lyngby, Denmark, 2—4 May 1977 (1977).

[1b] Barnes, H.T., and Coker, E.G.: The flow of water through pipes. Proc. Roy. Soc. London 74, 341—356 (1905).

[2] Barry, M.D.J., and Ross, M.A.S.: The flat plate boundary layer. Part 2: The effect of increasing thickness on stability. JFM 43, 813—818 (1970).

[3] Benny, D.J.: A non-linear theory for oscillations in a parallel flow. JFM 10, 209—236 (1961).

[4] Betchov, R., and Criminale, W.O.: Stability of parallel flows. Academic Press, 1967.

[4a] Bouthier, M.: Stabilité linéaire des écoulements presque parallèles. I. Journal de Mécanique 11, 599—621 (1972). II. La couche limite de Blasius. Journal de Mécanique 12, 75—95 (1973).

[5] Bergh, H.: A method for visualizing periodic boundary layer phenomena. IUTAM Symposium Boundary-layer research (H. Görtler, ed.), Berlin, 1958, 173—178.

[6] Burgers, J.M.: The motion of a fluid in the boundary layer along a plane smooth surface. Proc. First Intern. Congress for Appl. Mech. 113, Delft, 1924.

[7] Cheng, S.J.: On the stability of laminar boundary layer flow. Quart. Appl. Math. 11, 346—350 (1953).

[8] Corner, D., Barry, M.D.J., and Ross, M.A.S.: Non-linear stability theory of the flat plate boundary layer. ARC CP No. 1296, (1974).

[9] Couette, M.: Etudes sur le frottement des liquides. Ann. Chim. Phys. 21, 433—510 (1890).

[10] Curle, N.: Hydrodynamic stability in unlimited fields of viscous flow. Proc. Roy. Soc. London A 238, 489—501 (1957).

[11] Clenshaw, C.W., and Elliott, D.: A numerical treatment of the Orr-Sommerfeld equation in the case of a laminar jet. Quart. J. Mech. Appl. Math. 13, 300—313 (1960).

[11a] Davis, S.: The stability of periodic flows. Annual Review Fluid Mech. 8, 57—74 (1976).

[12] Davie, A.: A simple numerical method for solving Orr-Sommerfeld problems. Quart. J. Mech. Appl. Math. 26, 401—411 (1973).

[13] Dhawan, S., and Narasimha, R.: Some properties of boundary layer flow during transition from laminar to turbulent motion. JFM 3, 418—436 (1958).

[14] Di Prima, R.C., Eckhaus, W., and Sege, L.A.: Non-linear wave number interactions in near critical two-dimensional flows. JFM 49, 705—744 (1971).

[15] Doetsch, H.: Untersuchungen an einigen Profilen mit geringem Widerstand im Bereich kleiner c_a-Werte. Jb. dt. Luftfahrtforschung I, 54—57 (1940).

[15a] Drazin, P.G., and Howard, L.N.: Hydrodynamic stability of parallel flow of inviscid fluid. Adv. Appl. Mech. 9, 1—89 (1966).

[16] Dryden, H.L.: Boundary layer flow near flat plates. Proc. Fourth Intern. Congress for Appl. Mech. Cambridge, England, 1934, 175.

[17] Dryden, H.L.: Airflow in the boundary layer near a plate. NACA Rep. 562 (1936).

[18] Dryden, H.L.: Turbulence and the boundary layer. JAS 6, 85—100 and 101—105 (1939).

[19] Dryden, H.L.: Some recent contributions to the study of transition and turbulent boundary layers (Papers presented at the Sixth Internat. Congress for Appl. Mech., Paris, Sept. 1946; NACA TN 1168 (1947); see also: Recent advances in the mechanics of boundary layer flow. Advances Appl. Mech. New York I, 2—40 (1948).

[20] Dryden, H.L.: Recent investigation of the problem of transition. ZFW 4, 89—95 (1956).

[20a] Dryden, H.L: Transition from laminar to turbulent flow. High Speed Aerodynamics and Jet Propulsion 5, 3—74, Princeton and Oxford, 1959.

[21] Dryden, H.L.: Recent advances in the mechanics of boundary layer flow. Adv. Appl. Mech. 1, 2—40 (1948).

[22] Dubs, W.: Über den Einfluss laminarer und turbulenter Strömung auf das Röntgenbild von Wasser und Nitrobenzol. Ein röntgenographischer Beitrag zum Turbulenzproblem. Helv. phys. Acta 12, 169—228 (1939).

[23] Eiffel, G.: Sur la résistance des sphères dans l'air en mouvement. Comptes Rendus 155, 1597—1599 (1912).

[24] Ekman, V.W.: On the change from steady to turbulent motion of liquids. Ark. f. Mat. Astron. och Fys. *6*, No. 12 (1910).

[25] Emmons, H.W., and Bryson, A.E.: The laminar-turbulent transition in a boundary layer. Part I: JAS *18*, 490—498 (1951); Part II: Proc. First US National Congress Appl. Mech. 859—868 (1952).

[26] Fage, A.: Fluid motion transition from laminar to turbulent flow in a boundary layer. Phys. Soc. Rep. on Progress in Physics *6*, 270 (1939).

[27] Fage, A.: Experiments on the breakdown of laminar flow. JAS *7*, 513—517 (1940).

[28] Fage, A., and Preston, J.H.: Experiments on transition from laminar to turbulent flow in the boundary layer. Proc. Roy. Soc. London A *178*, 201—227 (1941).

[28a] Fage, A.: Transition in the boundary layer caused by turbulence. ARC RM 1896 (1942).

[28b] Fasel, H.: Investigation of the stability of boundary layers by a finite difference model of the Navier-Stokes equations. JFM *78*, 355—383 (1976); see also: Diss. Stuttgart 1974.

[29] Froude, W.: Experiments on the surface friction. Brit. Ass. Rep. 1872.

[30] Gersting, J.M., and Jankowski, D.F.: Numerical methods for Orr-Sommerfeld problems. Intern. J. Numerical Methods in Engineering *4*, 195—206 (1972).

[31] Gaster, M.: On the effect of boundary layer growth on flow stability. JFM *66*, 465—480 (1974).

[32] Gaster, M.: A note on the relation between temporally-increasing and spatially-increasing disturbances in hydrodynamic stability. JFM *14*, 222—224 (1962).

[33] Gaster, M.: The role of spatially growing waves in the theory of hydrodynamic stability. Progress in Aero. Sciences (D. Küchemann, ed.), *6*, 251—270 (1965).

[33a] Gaster, M., and Jordinson, R.: On the eigenvalues of the Orr-Sommerfeld equation. JFM *72*, 121 133 (1975).

[34] Grohne, D.: Ein Beitrag zur nicht-linearen Stabilitätstheorie von ebenen Laminarströmungen. ZAMM *52*, 256—257 (1972).

[35] Görtler, H., and Witting, H.: Theorie der sekundären Instabilität der laminaren Grenzschichten. IUTAM Symposium, Boundary-layer Research (H. Görtler, ed.) 110—126, Berlin, 1958.

[36] Granville, P.S.: The calculations of viscous drag of bodies of revolution. Navy Department, The David Taylor Model Basin, Report 849 (1953).

[37] Greenspan, H.P., and Benny, D.J.: On shear-layer instability, breakdown and transition, JFM *15*, 133—153 (1963).

[38] Grohne, D.: Über das Spektrum bei Eigenschwingungen ebener Laminarströmungen. ZAMM *34*, 344—357 (1954).

[39] Hall, A.A., and Hislop, G.S.: Experiments on the transition of the laminar boundary layer on a flat plate. ARC RM 1843 (1938).

[39a] Habermann, R.: Nonlinear perturbations of the Orr-Sommerfeld equation asymptotic expansion of the logarithmic phase shift across the critical layer. SIAM J. Math. Analysis *7*, 70—81 (1976).

[40] Hamel, G.: Zum Turbulenzproblem. Nachr. Ges. Wiss. Göttingen, Math. Phys. Klasse 261—270 (1911).

[41] van der Hegge Zijnen, B.G.: Measurements of the velocity distribution in the boundary layer along a plane surface. Thesis Delft 1924.

[42] Heisenberg, W.: Über Stabilität und Turbulenz von Flüssigkeitsströmen. Ann. d. Phys. *74*, 577—627 (1924).

[43] Hollingdale, S.: Stability and configuration of the wakes produced by solid bodies moving through fluids. Phil. Mag. VII *29*, 209—257 (1940).

[44] Holstein, H.: Über die äussere und innere Reibungsschicht bei Störungen laminarer Strömungen. ZAMM *30*, 25—49 (1950).

[45] Hopf, L.: Ann. d. Phys. *44*, 1 (1914) and *59*, 538 (1919); see also: Summary report by F. Noether, ZAMM *1*, 125—138 (1921).

[46] Howard, L.N.: Hydrodynamic stability of a jet. J. Math. Phys. *37*, 283—298 (1959).

[46a] Kachanov, S., Kozlov, V.V., and Levchenko, I.A.: Uchenyie zapiski TSAGI *VI*, 5, 137—140 (1975).

[46b] Ikeda, M.: Finite disturbances and growing vortices in a two-dimensional jet. JFM *80*, 401—421 (1977).

[47] Jordinson, R.: The flat plate boundary layer. Part 1. Numerical integration of the Orr-Sommerfeld equation. JFM *43*, 801—811 (1970). See also: Ph.D. Thesis Edinburgh Univ. 1968.

[48] Jordinson, R.: Spectrum of eigenvalues of the Orr-Sommerfeld equation for Blasius flow. Phys. Fluids *14*, 2535—2537 (1971).

[48a] Kaplan, R.E.: The stability of laminar incompressible boundary layers in the presence of compliant boundaries. Ph.D. Thesis, Massachusetts Inst. of Technology, Aero-Elastic and Structures Research Laboratory, ASRL TR 116—1 (1964).

[48b] Loehrke, R.J., Morkovin, M.V., and Fejer, A.A.: Review. Transition in nonreversing oscillating boundary layers. J. Fluids Eng. Trans. ASME Series I, *97*, 534—549 (1975).

[49] Klebanoff, P.S., and Tidstrom, K.D.: Evolution of amplified waves leading to transition in a boundary layer with zero pressure gradient. NASA TN D-195, (1959)

[50] Klebanoff, P.S., Tidstrom, K.D., and Sargent, L.M.: The three-dimensional nature of boundary-layer instability. JFM *12*, 1—34 (1962).

[51] Kurtz, E.F., and Crandall, S.H.: Computer-aided analysis of hydrodynamic stability. J. Math. Phys. *44*, 264—279 (1962).

[52] Lewis, G.W.: Some modern methods of research in the problems of flight. The 1939 (27th) Wilbur Wright Memorial Lecture. J. Roy. Aero. Soc. London *43*, 769—802 (1939).

[53] Lee, L.H., and Reynolds, W.C.: On the approximate and numerical solution of Orr-Sommerfeld problems. Quart. J. Mech. Appl. Math. *20*, 1—22 (1967).

[54] Lin, C.C.: On the stability of two-dimensional parallel flows. Quart. Appl. Math. *3*, 117—142 (July 1945); *3*, 213—234 (Oct. 1945); *3*, 277—301 (Jan. 1946).

[55] Lin, C.C.: The theory of hydrodynamic stability. Cambridge Univ. Press, 1955.

[56] Lin, C.C., and Benny, D.J.: On the instability of shear flows. Proc. Symp. Appl. Math. *13*, Hydrodynamic Instability 1—24 (1962).

[57] Lorentz, H.A.: Abhandlung über theoretische Physik *I*, 43—71, Leipzig, 1907; new version of earlier paper: Akad. v. Wet. Amsterdam *6*, 28 (1897); see also: Prandtl, L.: The mechanics of viscous fluids, in Durand, W.F.: Aerodynamic theory *III*, 34—208 (1935).

[57a] McCormick, M.E.: An analysis of the formation of turbulent patches in the transition boundary layer. J. Appl. Mech. *35*, 216—219 (1968).

[57b] Lessen, M.: see Chap. IX.

[57c] Lessen, M., and Ko, S.H.: On the low Reynolds number stability characteristics of the laminar incompressible half jet. Phys. of Fluids *12*, 404—407 (1969).

[57d] Mack, L.M.: A numerical study of the temporal eigenvalue spectrum of the Blasius boundary layer. JFM *73*, 497—520 (1976).

[57e] Mack, L.M.: Transition prediction and linear stability theory. AGARD-CP-224, 1—1 to 1—22 (1977).

[57f] Mattingly, G.E., and Criminale, W.O.: The stability of an incompressible two-dimensional wake. JFM *51*, 233—272 (1972).

[58] Michalke, A., and Schade, H.: Zur Stabilität von freien Grenzschichten. Ing.-Arch. *33*, 1—23 (1963).

[59] Michalke, A.: The instability of free shear layers. Progress in Aerospace Sciences (D. Küchemann, ed.) *12*, 213—239 (1972).

[60] Meseth, J.: Experimentelle Untersuchung der Übergangszonen zwischen laminaren und turbulenten Strömungsgebieten intermittierender Rohrströmung. Mitteilungen aus dem Max-Planck-Institut für Strömungsforschung und der Aerodynamischen Versuchsanstalt, Göttingen, No. 58 (1974).

[60a] Miller, J.A., and Fejer, A.A.: Transition phenomena in oscillating boundary layer flows. JFM *18*, 438—449 (1964).

[61] von Mises, R.: Kleine Schwingungen und Turbulenz. Jahresber. Dt. Math. Verein. 241—248 (1912).

[61a] Morkovin, M.V.: On the many faces of transition, in: Viscous drag reduction (C.S. Wells, ed.). Plenum Press, New York, 1969, pp. 1—31; see also: Critical evaluation of transition from laminar to turbulent shear layer with emphasis on hypersonically travelling bodies. Air Force Flight Dynamics Lab., Wright-Patterson Air Force Base, Ohio, TR 68—149 (1969).

[62] Osborne, M.R.: Numerical methods for hydrodynamic stability problems. SIAM J. Appl. Math. *15*, 539—557 (1967).

[63] Ombrewski, H.G., Morkovin, M.V., and Landahl, M.: A portfolio of stability characteristics of incompressible boundary layers. AGARDograph No. 134 (1969).

[63a] Obremski, H.J., and Fejer, A.A.: Transition in oscillating boundary layer flows. JFM *29*, 93—111 (1967).

[64] Orr, W. M. F.: The stability or instability of the steady motions of a perfect liquid and of a viscous liquid. Part I: A perfect liquid; Part II: A viscous liquid. Proc. Roy. Irish Acad. *27*, 9—68 and 69—138 (1907).

[64a] Pearson, J. R. A.: Instability of non-Newtonian flow. Ann. Rev. Fluid Mech. *8*, 163—181 (1976).

[65] Persh, J.: A study of boundary-layer transition from laminar to turbulent flow. US Naval Ordnance Lab. Rep. 4339 (1956).

[66] Prandtl, L.: Über den Luftwiderstand von Kugeln. Nachr. Ges. Wiss. Göttingen, Math. Phys. Klasse, 177—190 (1914); see also Coll. Works *II*, 597—608.

[67] Prandtl, L.: Bemerkungen über die Entstehung der Turbulenz. ZAMM *I*, 431—436 (1921) and Phys. Z. *23*, 19—25 (1922); see also Coll. Works *II*, 687—696.

[67a] Prandtl, L.: Über die Entstehung der Turbulenz. ZAMM *11*, 407—409 (1931).

[68] Prandtl, L.: Neuere Ergebnisse der Turbulenzforschung. Z. VDI *77*, 105—114 (1933); see also: Coll. Works, *II*, 105—114.

[69] Pretsch, J.: Die Stabilität einer ebenen Laminarströmung bei Druckgefälle und Druckanstieg. Jb. dt. Luftfahrtforschung *I*, 58—75 (1941).

[70] Lord Rayleigh: On the stability of certain fluid motions. Proc. Math. Soc. London *11*, 57 (1880) and *19*, 67 (1887); Scientific Papers *I*, 474—487 and *III*, 17; see also Scientific Papers *IV*, 203 (1895) and *VI*, 197 (1913).

[70a] Reshotko, E.: Boundary layer stability and transition. Ann. Rev. Fluid Mech. *8*, 311—349 (1976).

[71] Reynolds, O.: On the experimental investigation of the circumstances which determine whether the motion of water shall be direct or sinuous, and the law of resistance in parallel channels. Phil. Trans. Roy. Soc. *174*, 935—982 (1883); see also Coll. Papers *II*, 51.

[72] Reynolds, O.: On the dynamical theory of incompressible viscous fluids and the determination of the criterion. Phil. Trans. Roy. Soc. A *186*, 123—164 (1895); see also Coll. Papers *II*.

[73] Rosenbrook, G.: Instabilität der Gleitschichten im schwach divergenten Kanal. ZAMM *17*, 8—24 (1937). Diss. Göttingen 1937.

[74] Ross, J. A., Barnes, F. H., Burns, J. G., and Ross, M. A. S.: The flat plate boundary layer. Part 3. Comparison of theory and experiment. JFM *43*, 819—832 (1970).

[75] Rotta, J.: Experimenteller Beitrag zur Entstehung turbulenter Strömung im Rohr. Ing.-Arch. *24*, 258—281 (1956).

[76] Schlichting, H.: Zur Entstehung der Turbulenz bei der Plattenströmung. Nachr. Ges. Wiss. Göttingen, Math. Phys. Klasse 182—208 (1933); see also ZAMM *13*, 171—174 (1933).

[77] Schlichting, H.: Amplitudenverteilung und Energiebilanz der kleinen Störungen bei der Plattenströmung. Nachr. Ges. Wiss. Göttingen, Math. Phys. Klasse, Fachgruppe I, *1*, 47—78 (1935).

[78] Schlichting, H.: Über die Theorie der Turbulenzentstehung. Summary Report, Forschg. Ing.-Wes. *16*, 65—78 (1950).

[79] Schlichting, H.: Entstehung der Turbulenz. Handbuch der Physik (S. Flügge, ed.), *8/1*, 351—450, Springer-Verlag, 1959. See also: Handbook of Fluid Dynamics (V. L. Streeter, ed.), McGraw-Hill, 1961.

[80] Schiller, L.: Untersuchungen über laminare und turbulente Strömung. Forschg. Ing.-Wes. Heft 428 (1922), or ZAMM *2*, 96—106 (1922), or Physikal. Z. *23*, 14—19 (1922).

[81] Schiller, L.: Neue quantitative Versuche zur Turbulenzentstehung. ZAMM *14*, 36—42 (1934).

[82] Schubauer, G. B., and Skramstad, H. K.: Laminar boundary layer oscillations and stability of laminar flow. National Bureau of Standards Research Paper 1772. Reprint of confidential NACA Rep. dated April 1943 (later published as NACA War-time Rep. W-8) and JAS *14*, 69—78 (1947); see also NACA Rep. 909.

[83] Schubauer, G. B., and Klebanoff, P. S.: Contributions on the mechanics of boundary layer transition. NACA TN 3489 (1955) and NACA Rep. 1289 (1956); see also Proc. Symposium on Boundary Layer Theory, NPL, England, 1955.

[84] Schubauer, G. B., and Klebanoff, P. S.: Mechanism of transition at subsonic speeds. IUTAM Symposium, Boundary-layer Research (H. Görtler, ed.), 84—107, Berlin, 1958.

[84a] Saric, W. S., and Nayfeh, A. H.: Nonparallel stability of boundary layer flows. Phys. of Fluids *18*, 945—950 (1975).

[85] Shen, S. F.: Calculated amplified oscillations in plane Poiseuille and Blasius flows. JAS *21*, 62—64 (1954).

[85a] Shen, S. F.: Stability of laminar flows. High Speed Aerodynamics and Jet Propulsion *4*, 719—853, Princeton and Oxford, 1964.

[85b] Shen, F. C. T., Chen, T. S., and Huang, L. M.: The effects of mainflow radial velocity on the stability of developing laminar pipe flow. J. Appl. Mech., Trans. ASME Ser. E *43*, 209—212 (1976).

[86] Sommerfeld, A.: Ein Beitrag zur hydrodynamischen Erklärung der turbulenten Flüssigkeitsbewegungen. Atti del 4. Congr. Internat. dei Mat. *III*, 116—124, Roma, 1908.

[87] Squire, H. B.: On the stability of three-dimensional distribution of viscous fluid between parallel walls. Proc. Roy. Soc. London A *142*, 621—628 (1933).

[88] Stewartson, K.: A non-linear instability theory for a wave system in plane Poiseuille flow. JFM *48*, 529—545 (1971).

[89] Stuart, J. T.: On the effects of the Reynolds stress on hydrodynamic stability. ZAMM-Sonderheft 32—38 (1956).

[90] Stuart, J. T.: Non-linear effects in hydrodynamic stability. Proc. Xth Intern. Congress of Appl. Mech., Stresa, 63—97 (1960).

[91] Stuart, J. T.: Hydrodynamic stability, in: Laminar boundary layers (L. Rosenhead, ed.), 482—579, Clarendon Press, Oxford, 1963; see also: Appl. Mech. Rev. *18*, 523—531 (1965).

[92] Tani, I.: Some aspects of boundary layer transition at subsonic speeds. Advances in Aeronautical Sciences (Th. v. Kármán, ed.), *3*, 143—160, Pergamon Press, New York and London, 1962.

[93] Tani, I.: Einige Bemerkungen über den laminar-turbulenten Umschlag in Grenzschichtströmungen. ZAMM *53*, T 25—T 32 (1973).

[94] Tani, I.: Low speed flows involving bubble separation. Progress Aero. Sci. *5*, 70—103 (1964).

[95] Tani, I.: Review of some experimental results on boundary layer transition. Phys. of Fluids Suppl. *10*, 11—16 (1967).

[96] Tani, I.: Boundary layer transition. Annual Review of Fluid Mechanics *1*, 169—196 (1969).

[96a] Tatsumi, A., and Katukani, T.: The stability of a two-dimensional laminar jet. JFM *4*, 261—275 (1958).

[97] Taylor, G. I.: Some recent developments on the study of turbulence. Proc. of the Fifth Intern. Congress for Appl. Mech., New York, 294 (1938); see also: Statistical theory of turbulence. V. Effect of turbulence on boundary layer. Proc. Roy. Soc. London A *156*, 307—317 (1936); see also: Scientific Papers *II*, 356—364.

[98] Tietjens, O.: Beiträge zur Entstehung der Turbulenz. Diss. Göttingen 1922; ZAMM *5*, 200—217 (1925).

[99] Tollmien, W.: Über die Entstehung der Turbulenz. 1. Mitt. Nachr. Ges. Wiss. Göttingen, Math. Phys. Klasse 21—44 (1929); Engl. transl. in NACA TM 609 (1931).

[100] Tollmien, W.: Ein allgemeines Kriterium der Instabilität laminarer Geschwindigkeitsverteilungen. Nachr. Ges. Wiss. Göttingen, Math. Phys. Klasse, Fachgruppe I, *1*, 79-114 (1935); Engl. transl. in NACA TM 792 (1936).

[101] Tollmien, W.: Asymptotische Integration der Störungsdifferentialgleichung ebener laminarer Strömungen bei hohen Reynoldsschen Zahlen. ZAMM *25/27*, 33—50 and 70—83 (1947).

[102] Tollmien, W., and Grohne, D.: The nature of transition. In: Boundary Layer and Flow Control (G. V. Lachmann, ed.). Vol. *2*, 602—636, Pergamon Press, London, 1961.

[103] Thomas, L. H.: The stability of plane Poiseuille flow. The Physical Rev. *86*, 812 (1952).

[104] Townsend, H. C. H.: Note on boundary layer transition. ARC RM 1873 (1939).

[104a] Wazzan, A. R.: Spatial stability of Tollmien-Schlichting waves. Progress in Aerospace Sciences (D. Küchemann, ed.) *16*, 99—127 (1975).

[105] Wieselsberger, C.: Der Luftwiderstand von Kugeln. ZFM *5*, 140—145 (1914).

[106] Wazzan, A. R., Taghavi, H., and Keltner, G.: Effect of boundary layer growth on stability of incompressible flat plate boundary layer with pressure gradient. Phys. of Fluids *17*, 1655—1670 (1974).

[107] White, F. M.: Viscous fluid flow. McGraw-Hill, New York, 1974.

[108] Wygnanski, J., Sokolov, M., and Frieman, D.: On a turbulent "spot" in a laminar boundary layer. JFM *78*, 785—819 (1976).

CHAPTER XVII

Origin of turbulence II †

Effect of pressure gradient, suction, compressibility, heat transfer, and roughness on transition

Introductory remark

The results described in Chap. XVI have demonstrated, in principle, the applicability of the method of small disturbances to the study of the phenomenon of transition from laminar to turbulent flow. We may, therefore, expect that this theory should also supply us with information concerning the other parameters which exert an important influence on transition, in addition to the single one, the Reynolds number, discussed so far. We have already briefly reported in Sec. XVIb that the pressure gradient in the external flow has a great influence on the stability of the boundary layer, and hence on transition, in the sense that a favourable pressure gradient stabilizes the flow and an adverse pressure gradient renders it less stable. Body forces, such as the centrifugal force in a curved stream and the buoyancy in a non-homogeneous stream, are very important for transition. In more modern times problems connected with boundary-layer control through suction or blowing and their effect on transition have become important (*cf*. Chap. XIV). Suction exerts a stabilizing effect, but blowing promotes instability. In the case of flows which occur at very high speeds, when the fluid must be regarded as being compressible, the presence of heat transfer from or to the wall (heating or cooling) affects transition to an important degree. The transfer of heat from the fluid to the wall has a highly stabilizing effect but if the flow of heat is from the wall to the fluid the effect is reversed. Finally, problems connected with the influence of roughness on transition are of great practical importance.

The present chapter will contain a review of all these diverse problems, and we shall begin with the study of the effect of pressure gradients because of its great importance in practical applications; in this connexion the reader may consult the two summary reviews, the one by I. Tani [238] written in 1969 and the other by E. Reshotko [194a] published in 1976. The former concentrates on incompressible flows, whereas the latter emphasizes compressible flow and heat transfer. We wish once again to draw the reader's attention to the somewhat older, very excellent, review of this problem area prepared by J. T. Stuart [227a].

---

† I am very much indebted to Dr. L.M. Mack of the Jet Propulsion Laboratory of the California Institute of Technology, Pasadena, California for the completely rewritten, present version of Section XVIIe of this chapter.

a. Effect of pressure gradient on transition in boundary layer along smooth walls

The boundary layer on a flat plate at zero incidence whose stability was investigated in Chap. XVI has the peculiar characteristic that its velocity profiles at different distances from the leading edge are similar to each other (cf. Chap. VII). In this case similarity results from the absence of a pressure gradient in the external flow. On the other hand, in the case of a cylindrical body of arbitrary shape when the pressure gradient along the wall changes from point to point, the resulting velocity profiles are not, generally speaking, similar to each other. In the ranges where the pressure decreases downstream, the velocity profiles have no points of inflexion and are of the type shown in Fig. 16.9e whereas in regions where the pressure increases downstream they are of the type shown in Fig. 16.9g and do possess points of inflexion. In the case of a flat plate all velocity profiles have the same limit of stability, namely $R_{crit} = (U_\infty \delta_1/\nu)_{crit} = 520$; in contrast with that, in the case of an arbitrary body shape, the individual velocity profiles have markedly different limits of stability, higher than for a flat plate with favourable pressure gradients, and lower with adverse pressure gradients. Consequently, in order to determine the position of the point of instability for a body of a given, prescribed shape, it is necessary to perform the following calculations:

1. Determination of the pressure distribution along the contour of the body for frictionless flow. 2. Determination of the laminar boundary layer for that pressure distribution. 3. Determination of the limits of stability for these individual velocity profiles. The problem of determining the pressure distribution belongs to potential theory which supplies convenient methods of computation as, for example, described by T. Theodorsen and J. E. Garrick [242] and F. Riegels [193]. Convenient methods for the calculation of laminar boundary layers were given in Chap. X. The third step, the stability calculation, will now be discussed in detail.

It is known from the theory of laminar boundary layers, Chap. VII, that, generally speaking, the curvature of the wall has little influence on the development of the boundary layer on a cylindrical body; this is true as long as the radius of curvature of the wall is much larger than the boundary-layer thickness, which amounts to saying that the effect of the centrifugal force may be neglected when analyzing the formation of a boundary layer on such bodies. Hence the boundary layer is seen to develop in the same way as on a flat wall, but under the influence of that pressure gradient which is determined by the potential flow past the body. The same applies to the determination of the limit of stability of a boundary layer with a pressure gradient which is different from zero.

In contrast with the case of a flat plate, where the external flow is uniform at $U_\infty = $ const, we now have to contend with an external stream whose velocity, $U_m(x)$, is a function of the length coordinate. The velocity $U_m(x)$ is related to the pressure gradient dp/dx through the Bernoulli equation

$$\frac{dp}{dx} = - \varrho \, U_m \, \frac{dU_m}{dx} \, . \tag{17.1}$$

In spite of the dependence of the external velocity on the length coordinate, it is possible, as shown by J. Pretsch [177], to analyze the stability of laminar flows with a pressure gradient in the same way as in its absence (Chap. XVI); it is again possible to work with a mean flow whose velocity $U(y)$ depends only on the transverse coordinate y. The influence of the pressure gradient on stability manifests itself through the form of the velocity profile given by $U(y)$. We have already said in Sec. XVIb that the limit of stability of a velocity profile depends strongly on its shape, profiles with a point of inflexion possessing considerably lower limits of stability than those without one (point-of-inflexion criterion). Now, since the pressure gradient controls the curvature of the velocity profile in accordance with eqn. (7.15)

$$\mu \left(\frac{\mathrm{d}^2 U}{\mathrm{d}y^2}\right)_{wall} = \frac{\mathrm{d}p}{\mathrm{d}x},$$

(17.2)

the strong dependence of the limit of stability on the shape of the velocity profile amounts to a large influence of the pressure gradient on stability. It is, therefore, true to say that accelerated flows ($\mathrm{d}p/\mathrm{d}x < 0$, $\mathrm{d}U_m/\mathrm{d}x > 0$, favourable pressure gradient) are considerably more stable than decelerated flows ($\mathrm{d}p/\mathrm{d}x > 0$, $\mathrm{d}U_m/\mathrm{d}x < 0$, adverse pressure gradient).

The strong influence of the pressure gradient on stability and on the amplification of small disturbances predicted by the present theory was confirmed experimentally by G. B. Schubauer and H. K. Skramstad, Sec. XVId. The graphs in Fig. 17.1 represent oscillograms of the velocity oscillations on a flat wall with a pressure gradient. The upper half of the diagram shows that a pressure drop which amounts to 10 per cent of the dynamic pressure causes a complete damping out of the oscillations, whereas the pressure increase which succeeds it and which amounts to only 5 per cent of the dynamic pressure, causes not only strong amplification but produces transition at once. (In this connexion attention is drawn to the reduced scale of the last two oscillograms!)

In the evaluation of stability, it appears to be convenient to express the influence of a pressure gradient by the use of a shape factor of the velocity profile and to

Fig. 17.1. Oscillogram of velocity fluctuations in laminar boundary layer with pressure gradient, as measured by G. B. Schubauer and H. K. Skramstad. Decreasing pressure produces damping; increasing pressure causes strong amplification and produces transition

Distance of measuring station from wall — 0·5 mm.
Velocity $U_\infty = 29$ m/sec

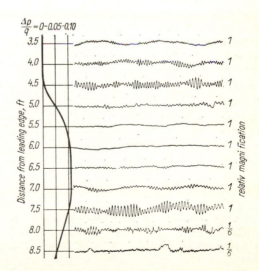

stipulate, for the sake of simplicity, a one-parameter family of laminar velocity profiles. An example of such a one-parameter family of velocity profiles, which, moreover, constitute exact solutions of the boundary-layer equations, is represented by Hartree's wedge flows. Their free-stream velocity is given by

$$U_m(x) = u_1\, x^m \, , \qquad\qquad (17.2\,\text{a})$$

and the associated velocity profiles can be found plotted in Fig. 9.1. Here m denotes the shape factor of the profiles and the wedge angle is $\beta = 2\,m/(m+1)$. When $m < 0$ (increasing pressure), the velocity profiles have a point of inflexion; when $m > 0$ (decreasing pressure), there is no point of inflexion. As early as 1941, J. Pretsch [178, 179] carried out the stability calculation for a series of profiles of this one-parameter family. Later, in 1969, these calculations were considerably extended by H. G. Ombrewski ([63] of Chap. XVI); he evaluated not only the critical Reynolds number but also the amplification rate of the unstable disturbances. The calculations reveal a stronger dependence of the critical Reynolds number on the shape factor m than did earlier work. The diagram of Fig. 17.2 describes one result of these calculations, namely the curves of constant amplification for the velocity profiles associated with the external flow given by eqn. (17.2a) with $m = -0.049$ which corresponds to $\beta = -0.1$; cf. A. R. Wazzan [104] of Chap. XVI.

K. Pohlhausen's approximate method described in Chap. X is the most convenient one for the calculation of laminar velocity profiles and it is, therefore, useful to investigate the stability of the associated velocity profiles. The shape of the velocity profiles is determined by the shape factor

$$\varLambda = \frac{\delta^2}{\nu}\,\frac{dU_m}{dx} \, . \qquad\qquad (17.3)$$

The family of velocity profiles was shown in Fig. 10.4. The shape factor $\varLambda$ assumes values between $\varLambda = +12$ and -12, the latter value corresponding to separation;

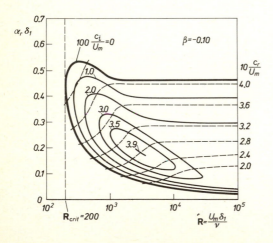

Fig. 17. 2. Curves of constant *temporal* amplification for the boundary layer in retarded flow with free-stream velocity given by $U_m = C\,x^m$ calculated for a wide range of Reynolds numbers after [63] of Chap. XVI

$m = \beta/(2-\beta) = 0.048$

$\beta = -0.1$

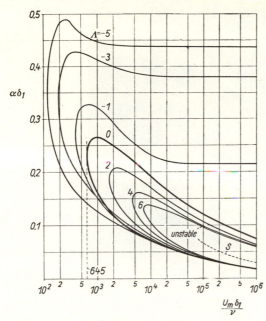

Fig. 17.3. Curves of neutral stability for laminar boundary-layer profiles with pressure decrease $(\varLambda > 0)$ and pressure increase $(\varLambda < 0)$. The shape factor of the velocity profile is defined as

$$\varLambda = \frac{\delta^2}{\nu}\,\frac{dU_m}{dx}\,,\ \text{see also Fig. 10.5}$$

Fig. 17.4.   Critical Reynolds number of boundary-layer velocity profiles with pressure gradient as a function of the shape factor $\varLambda$

at the forward stagnation point it is equal to $\varLambda = +7.05$, and at the point of minimum pressure we have $\varLambda = 0$. For $\varLambda > 0$ the pressure decreases; $\varLambda < 0$ corresponds to an increase in pressure. The velocity profiles for $\varLambda < 0$ each possess a point of inflexion.

H. Schlichting and A. Ulrich [200] carried out stability calculations for this family of velocity profiles. The curves of neutral stability are shown in Fig. 17.3. Both branches of the curves of neutral stability for all velocity profiles with a decreasing pressure $(\varLambda > 0)$ tend to zero as $\mathsf{R} \to \infty$, just as was the case for the flat plate, $\varLambda = 0$. On the other hand the upper branches of curves corresponding to profiles with adverse pressure gradients $(\varLambda < 0)$ tend to an asymptote which differs from zero so that even for $\mathsf{R} \to \infty$ there exists a finite region of wavelengths at which disturbances are always amplified. The velocity profiles in the region of favourable pressure gradient $(\varLambda > 0)$ as well as the profile for constant pressure $(\varLambda = 0)$ belong to the type of "viscous" instability (curve b in Fig. 16.8), whereas the profiles in the range of adverse pressure gradient $(\varLambda < 0)$ are of the type characteristic of "frictionless" instability (curve a in Fig. 16.8). It is seen from Fig. 17.3 that the

Fig. 17.5. Shadograph picture of reverse transition from turbulent to laminar flow in a boundary layer in supersonic flow round a corner at $\mathsf{M} = 3$, after J. Sternberg [215]

Fig. 17.6. Schematic representation of the flow in the boundary layer in supersonic flow around a corner, after J. Sternberg, *cf.* Fig. 17.5

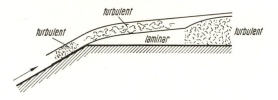

region of unstable wavelengths enclosed by the curve of neutral stability is much greater for boundary layers with adverse pressure gradients than for accelerated flows. The dependence of the critical Reynolds number on the shape factor Λ which follows from Fig. 17.3 has been plotted in Fig. 17.4[†]. It varies with the value of the shape factor Λ, and hence with the pressure gradient, very strongly. The diagram in Fig. 17.2 contains, additionally, curves of constant amplification $c_i/U_m = $ const for a velocity profile associated with the small rate of pressure rise corresponding to $\beta = -0 \cdot 1$. A comparison with Fig. 16.12 convinces us that the slow rate of pressure increase considerably increases the amplification rate.

---

† The value $\mathsf{R}_{crit} = 645$ given here for $\Lambda = 0$ differs somewhat from the value 520 given previously in Fig. 16.11. This is due to the difference between the exact Blasius velocity profile used previously and an approximate one employed for the preparation of Fig. 17.2.

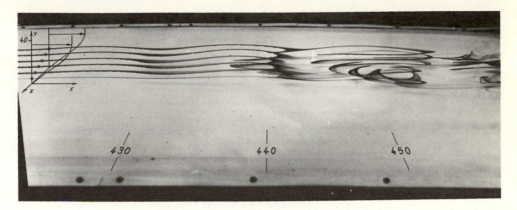

Fig. 17.7. Illustrating the instability of two-dimensional Tollmien-Schlichting waves in the laminar boundary layer along a plane wall and in the presence of an adverse pressure gradient. Photograph of wave-like streak lines in a water channel obtained with the aid of the tellurium method by F. X. Wortmann [257, 258]; disturbance created artificially by an oscillating strip ($3 \times 800 \times 0.03$ mm). The strip is located at a station where $R_1 = 750$; the streak lines are created at $R_1 = 950$ (left border of figure). The rolling up of streak lines downstream is a consequence of the instability of the perturbation waves. The figures denote distances in cm

Numerical data for laminar boundary layer at the center of the figure:
Velocity outside boundary layer, $U_m = 9.1$ cm/sec
Displacement thickness of boundary layer, $\delta_1 = 1.4$ cm
Wavelength of disturbance, $\lambda = 18$ cm
Reynolds number, $R_1 = U_m \, \delta_1/\nu = 1250$
Shape factor, $\delta_1/\delta_2 = H_{12} = 2.9$
Polhausen parameter from eqn. (10.21), $\Lambda = -8$

The photograph of Fig. 17.7, taken by F. X. Wortmann [255, 256] in a water channel conveys a clear impression of unstable oscillations in a laminar boundary layer. The picture was obtained by the tellurium method [256]. The artifical disturbances were generated with the aid of an oscillating strip placed near the wall, in a manner similar to that employed by Schubauer and Skramstad and described in Chap. XVI. The pressure rise along the wall is so small that the Pohlhausen parameter from eqn. (17.3) has the value $\Lambda = -8$. At the station where the disturbance is generated the local Reynolds number has the value $R_{\delta_1} = 750$, and the dimensionless wavelength of the disturbance is $\alpha_1 \, \delta_1 = 2 \, \pi \, \delta_1/\lambda = 0.48$. This point is located far in the unstable field of Fig. 17.3. The instantaneous snapshot of the streak lines in Fig. 17.7 shows the final phase of the two-dimensional development of the disturbance about 20 wavelengths downstream of the oscillating strip. This disturbance amplifies in complete agreement of theory with experiment. The fluctuation, which is still two-dimensional near the left edge of the picture becomes distorted in its middle by the oncoming longitudinal vortices. At the right edge of the picture it is already possible to discern "turbulent cores". This confirms our remarks concerning three-dimensional disturbances given at the end of this chapter.

On several occasions we have stressed the fact that a *pressure increase* along a boundary layer strongly favours transition to turbulent flow in it. Conversely, a strong *pressure decrease*, such as may be created behind sharp edges in supersonic flow, can cause a turbulent boundary layer to become laminar. Interesting observations of this kind were made by J. Sternberg [215] who employed a cylinder provided with a conical fore-body. Figure 17.5 shows a shadograph of the flow along the conical fore-body at a Mach number $M = 3$. The boundary layer turns turbulent at the tripping wire provided for the purpose. Further downstream, behind the corner formed at the junction of the two bodies, the turbulent boundary layer turns laminar again, Fig. 17.6. This phenomenon is explained by the circumstance that the large favourable pressure gradient at the shoulder impresses a very strong acceleration on the flow and this, in turn, extinguishes the turbulence, in a way reminiscent of the effect of a strong contraction placed ahead of the test section of a wind tunnel. Qualitative indications on this process can be found in a paper by W. P. Jones and E. E. Launder [103]. According to these authors, relaminarization (extinction of turbulence) occurs in incompressible streams when the dimensionless acceleration parameter satisfies the inequality

$$K = \frac{\nu}{U_m^2} \cdot \frac{dU_m}{dx} > 3 \times 10^{-6} \text{ (relaminarization)}.$$

Introducing Pohlhausen's shape factor Λ from eqn. (10.21), and using eqn. (17.3), we can translate the preceding condition to read

$$\Lambda > 3 \times 10^{-6} R_\delta^2,$$

where $R_\delta = U_m \delta / \nu$ denotes the Reynolds number referred to the thickness of the turbulent boundary layer. It is necessary to stress that this is a purely empirical criterion. A more detailed investigation was carried out by R. Narasimha and K. R. Sreenivasan [168]; see also the earlier paper by V. C. Patel and M. R. Head [189].

The transition from a turbulent to a laminar flow pattern in a tube of circular cross-section was investigated in detail experimentally by M. Sibulkin as early as 1962. In particular, this investigation extended to a study of the attenuation of longitudinal turbulent fluctuations and discovered that this is stronger near the wall than in the center of the pipe.

The preceding results will enable us to calculate in the following section the position of the point of instability for the case of two-dimensional flow past a body of arbitrary shape.

b. Determination of the position of the point of instability for prescribed body shape

The determination of the position of the point of transition for prescribed body shapes (in two-dimensional flow) becomes very easy if use is made of the results contained in Figs. 17.3 and 17.4. The essential advantage of the method to be described here consists in the fact that no further laborious calculations are required, the tedious part of the work having been completed once and for all when computing the diagrams in Fig 17.3.

We begin with the evaluation of the laminar boundary layer from the potential velocity distribution $U_m(x)/U_\infty$, which is regarded as known, by the use of Pohlhausen's approximate method outlined in Chap. X. Such a calculation furnishes values of the shape factor Λ and the displacement thickness δ_1 in terms of the length of arc x, measured from the forward stagnation point. On proceeding along the laminar boundary layer from the forward stagnation point in a downstream direction at an assumed constant body Reynolds number $U_\infty l/\nu$ (l — length of body), it is noticed that, at the beginning, the limit of stability, $(U_m \delta_1/\nu)_{crit}$, is very high owing to the sharp pressure decrease. On the other hand the boundary layer is thin and consequently the local Reynolds number $U_m \delta_1/\nu$ is certain to be smaller than the critical value, $(U_m \delta_1/\nu)_{crit}$, and the boundary layer is stable. Further downstream the rate of pressure decrease becomes smaller and is followed by a pressure increase behind the point of minimum pressure so that the local limit of stability, $(U_m \delta_1/\nu)_{crit}$, decreases in the downstream direction, whereas the boundary-layer thickness and, with it, the local Reynolds number, $(U_m \delta_1/\nu)$, increase. At a certain point the two become equal:

$$\frac{U_m \delta_1}{\nu} = \left(\frac{U_m \delta_1}{\nu}\right)_{crit} \quad \text{(point of instability)}, \tag{17.4}$$

and from that point onwards the boundary layer is unstable. The point defined by eqn. (17.4) will be referred to as the *point of instability* and its position does, evidently, depend on the body Reynolds number, $(U_\infty l/\nu)$, because the local boundary-layer thickness is influenced by it.

The calculation of the position of the point of instability in terms of the Reynolds number sketched in the preceding paragraph can be conveniently performed with the aid of the diagrams in Fig. 17.8. It will be developed in more detail for

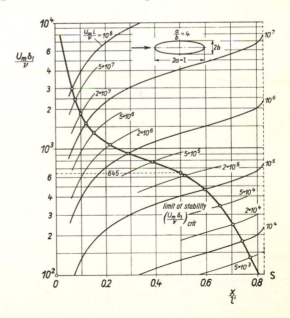

Fig. 17.8. Calculation of the position of the point of instability in terms of the Reynolds number $U_\infty l/\nu$ for an elliptic cylinder with slenderness ratio $\alpha/b = 4$

$2\,l' = $ circumference

the example of an elliptic cylinder whose major axis, a, is related to its minor axis, b, by the ratio $a/b = 4$. The flow will be assumed parallel to the major axis. The potential velocity-distribution function for such a cylinder was already given in Fig. 10.9, and the results of the calculations pertaining to the boundary layer are shown in Figs. 10.10 and 10.10b. From the variation of the shape factor $\varLambda$ with x, Fig. 10.11 b, and with the aid of Fig. 17.4. it is now possible to plot the variation of the local critical Reynolds number, $R_{crit} = (U_\infty \delta_1/\nu)_{crit}$, as shown by the curve marked *limit of stability* in Fig. 17.8. From the calculation of the laminar boundary layer we can also take the variation of the dimensionless displacement thickness $(\delta_1/l)\,(\sqrt{U_\infty\, l/\nu})$, as shown in Fig. 10.10a. For a given body Reynolds number $U_\infty\, l/\nu$, it is now possible to evaluate the local Reynolds number, $U_m\, \delta_1/\nu$, based on the displacement thickness, since

$$\frac{U_m\,\delta_1}{\nu} = \left(\frac{\delta_1}{l}\,\sqrt{\frac{U_\infty l}{\nu}}\right)\sqrt{\frac{U_\infty l}{\nu}}\,\frac{U_m}{U_\infty}\,, \tag{17.5}$$

where the value of $U_m(x)/U_\infty$ is known from the potential velocity function. The curves of $U_m\,\delta_1/\nu$ in terms of the arc length, x/l', have also been drawn in Fig. 17.8 for various values of the Reynolds number $U_\infty\, l/\nu$. The points of intersection of these curves with the *limit of stability* give the position of the point of instability for the respective value of the Reynolds number†. The points of instability for a family of elliptic cylinders of slenderness ratios $a/b = 1, 2, 4, 8$ are shown in Fig. 17.9. It is remarkable that the shift of the point of instability with an increasing Reynolds number is very small for the case of a circular cylinder. This shift becomes more pronounced as the slenderness ratio is increased.

The position of the point of instability for an aerofoil can be easily calculated in a similar manner. In this connexion it is particularly important to determine the dependence on the angle of incidence in addition to that on the Reynolds number. The results of such calculations for the case of a symmetrical Zhukovskii aerofoil at varying angles of incidence and lift coefficients are shown in Fig. 17.10. It is seen that, as the angle of incidence increases, the minimum of pressure on the suction side becomes more and more prominent and moves forward, whereas that on the pressure side becomes flatter and moves to the rear. This causes the point of instability to move upstream on the suction side and downstream on the pressure side as the angle of incidence is increased. Simultaneously the point of instability on the suction side closes up towards the point of minimum pressure for all Reynolds numbers because of the steep course of the curve near the minimum; the opposite effect occurs on the pressure side, where the curves are flat near the minimum causing the points of instability to diverge. In any case the diagram in Fig. 17.10 displays very clearly the dominating influence of the pressure distribution on the position of the point of instability and hence on that of the point of transition. Even at high Reynolds numbers the point of instability (and hence the point of transition) hardly moves in front of the point of minimum pressure, whereas behind the point

---

† The curves $U_m\,\delta_1/\nu$ for various values of $U_\infty\,l/\nu$ can be drawn from each other by translating them in a direction parallel to the axis of ordinates, if a logarithmic scale is used for the latter. This is a very convenient simplification to use when a graphical method is employed.

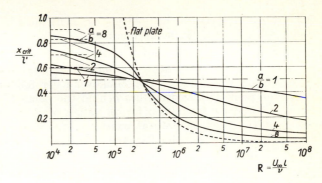

Fig. 17.9. Position of points of instability for elliptic cylinders of slenderness ratio $a/b = 1, 2, 4, 8, \infty$ (flat plate) plotted against the body Reynolds number R

2 l' — circumference; S — point of laminar separation; M — point of minimum pressure

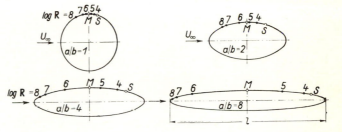

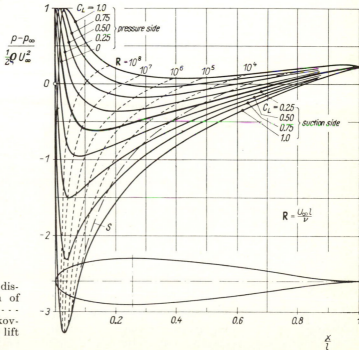

S = position of point of laminar separation

Fig. 17.10. Pressure distribution ——, position of point of instability · · · · on a symmetrical Zhukovskii aerofoil at varying lift coefficients

of minimum pressure instability and, consequently, transition sets in almost at once even at low Reynolds numbers.

Figure 17.11 shows, further, the position of the point of instability, as determined experimentally for a NACA aerofoil, which possessed an almost identical pressure distribution with that of the Zhukovskii aerofoil under consideration. It is seen that the point of transition lies behind the point of instability but in front of the point of laminar separation for all values of Reynolds number and lift coefficient as expected from theoretical considerations. Secondly, the shift of the point of transition with a varying Reynolds number and lift coefficient follows that of the point of instability. Results of systematic calculations on the position of the point of transition for aerofoils of varying thickness and camber can be found in a report by K. Bussmann and A. Ulrich [10].

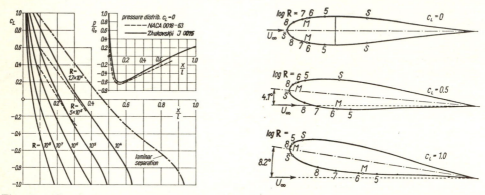

Fig. 17.11. Position of point of instability and of point of transition as a function of lift coefficient and Reynolds number. ———— theoretical point of instability: J 0015; - - - - measured point of transition: NACA 0018

$\bar{S}$ = stagnation point; M = point of minimum pressure; S = point of laminar separation

As a rough guide in approximate calculations it is possible to deduce the rule that the point of transition almost coincides with the point of minimum pressure of the potential flow in the range of Reynolds numbers from 10^6 to 10^7. At very large Reynolds numbers the point of transition may lie a short distance in front of that position and it may move a considerable distance behind it at small Reynolds numbers, particularly when the pressure gradient, whether positive or negative, is small. On the other hand, it will be noted that the point of instability always lies in front of the point of laminar separation irrespective of the value of the Reynolds number. Thus we can establish the rule that the point of instability lies behind the point of minimum pressure but in front of the point of laminar separation, at all except very large Reynolds numbers.

The precise distance between the point of transition and the point of instability depends on the rate of amplification of the unstable disturbances and on the intensity of turbulence in the free stream. In turn, the rate of amplification is strongly influenced by the pressure gradient. R. Michel [150] discovered a remarkably simple, purely empirical, relationship between the rate of amplification and the distance between the theoretical position of the point of instability and the experimentally determined position of the point of transition. More recently, A. M. O. Smith [211] succeeded in confirming this relation on the basis of the stability theory. As it enters the region of instability, Fig. 17.3, every unstable disturbance which travels downstream suffers an amplification which is proportional to $\exp(\beta_i t)$, or to

$$\exp(\int \beta_i \, dt) \,, \tag{17.6}$$

if β_i depends on time. Here the integral should extend over the range of unstable disturbances which is traversed by the disturbance after it had entered the region of instability. The amplification diagrams of $\beta_i = \text{const}$ (of the kind shown in Fig. 16.13) which are associated with different pressure gradients have been evaluated by J. Pretsch [179]. In 1957, A.M.O. Smith utilized these diagrams and performed a large number of calculations for aerofoils and bodies of revolution for which experimental determinations of the point of transition were available. He calculated the amplification rate from eqn. (17.6) extending the integration over the path from the theoretical limit of stability to the experimental point of transition. The result of his calculations is shown in Fig. 17.12. The result of these calculations which related to many different measurements performed at very low turbulence intensities in the free stream and with very smooth surfaces leads to the conclusion that the amplification rate of unstable disturbances, integrated along the path from the point of instability to the point of transition, has a value of

$$\exp(\int \beta_i \, dt) = \exp 9 = 8103. \tag{17.7}$$

This discovery was confirmed, at about the same time, by J.L. van Ingen [94]. See also a paper by R. Michel [166].

    In modern times this discovery was confirmed by many measurements [104] which indicate an amplification factor of about $\exp 10 = 22{,}026$.

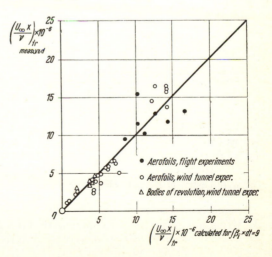

Fig. 17.12. Determination of the amplification rate $\exp(\int \beta_i \, dt)$ for unstable disturbances extended over the path from the theoretical limit of stability to the experimental point of transition, after A. M. O. Smith [211]

The distance between the point of instability and the point of transition can be represented in the form of the difference between the Reynolds numbers formed with the aid of the momentum thickness at these two points, as was already done in Fig. 16.21, that is, as $(U\delta_2/\nu)_{tr} - (U\delta_2/\nu)_i$. Fig. 17.14 shows a plot of this quantity in terms of the mean Pohlhausen parameter $\bar{K}$ and is based on the values found by P. S. Granville [75]. Here we have

$$\bar{K} = \frac{1}{x_{tr} - x_i} \int_{x_i}^{x_{tr}} \frac{\delta_2{}^2}{\nu} \frac{dU_m}{dx} \, dx = \frac{1}{x_{tr} - x_i} \int_{x_i}^{x_{tr}} K(x)\,dx \,, \tag{17.8}$$

and K is defined in eqn. (10.27).

The measurements which have been taken into account in the above calculation were all performed at very low turbulence intensities (free-flight measurements and measurements in low-turbulence wind tunnels). The diagram in Fig. 17.13 shows that the results due to many experimenters arrange themselves satisfactorily on a single

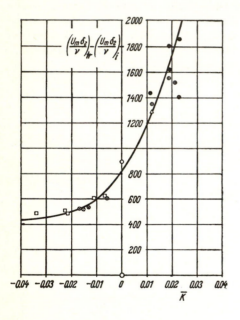

Fig. 17.13. Measurements on the point of transition in boundary layers with pressure gradient, after Granville [82]. Difference between the Reynolds numbers at the point of transition, $R_{\delta_2,tr} = (U_m \delta_2/\nu)_{tr}$, and at the point of instability, $R_{\delta_2,i} = (U_m \delta_2/\nu)_i$ as a function of the mean pressure gradient $\bar{K}$ from eqn. (17.8). $\bar{K} > 0$ corresponds to accelerated and $\bar{K} < 0$ to decelerated flows

O   Flat plate, Schubauer and Skramstad [203]
⊗   NACA aerofoil 0012, von Doenhoff [30]
●   Suction side   } Aerofoil NACA 65(215)—114, Bras-
⊕   Pressure side   } low and Visconti [7]
□   Aerofoil of 8% thickness ratio, B. M. Jones [98]
     [7, 30, 203] measurements in low-turbulence wind tunnel;
     [75] free-flight measurements

curve. The difference $(U\delta_2/\nu)_{tr} - (U\delta_2/\nu)_i$ is considerably larger for favourable pressure gradients ($\bar{K} > 0$) than for adverse ones ($\bar{K} < 0$). At constant pressure ($\bar{K} = 0$) this difference attains a value of about 800 which agrees well with that given in Fig. 16.21 for a flat plate at very small turbulence intensity. In this connexion see a note by E. R. van Driest and C. B. Blumer [50].

Laminar aerofoils: The stability calculations summarized in Figs. 17.9 and 17.10 demonstrate very convincingly that the pressure gradient has a decisive influence on stability and transition in complete agreement with measurements. The design of *laminar aerofoils* is based on the same circumstance. The small skin friction of such aerofoils is achieved by designing for long stretches of laminar boundary layer. This aim is achieved by moving the point of maximum thickness, and hence the point of minimum pressure, a considerable distance towards the trailing edge. In any case the desired shift in the position of pressure minimum can only be attained in a certain narrow range of angles of incidence.

Very extensive measurements on laminar aerofoils were carried out during the Second World War in the United States [1]. H. Doetsch [31] published the first experimental results on laminar aerofoils as early as in 1939, but B. M. Jones [98] had previously observed remarkably long stretches of laminar boundary layers during experiments in flight. Laminar aerofoils are widely used in the construction of gliders. Research of fundamental importance on aerofoils for glider wings have been performed by F. X. Wortmann; they are known as F. X. aerofoils and are characterized in [2]. Fig. 17.14 shows the amount of saving in drag that can be attained with laminar aerofoils. The saving due to the "laminar effect" reaches values of 30 to 50 per cent. of the drag of normal aerofoils in the interval of Reynolds number $R = 2 \times 10^6$ to 3×10^7. At very large Reynolds numbers, say $R > 5 \times 10^7$, the laminar effect is lost because the point of transition on the aerofoil shifts suddenly forward as demanded by the stability theory. The pressure distribution curves for some of the aerofoils are shown in Fig. 17.15. The measured

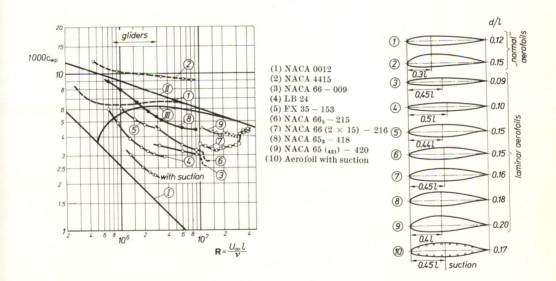

Fig. 17. 14. Skin-friction coefficients of laminar aerofoils and of "normal" aerofoils after [1] and [96]. LB 24 — Japanese laminar aerofoil after I. Tani [234]. F X 35 — 153 — laminar aerofoil after F. X. Wortmann [257]. Aerofoil with suction after W. Pfenninger, [61] of Chap. XIV. Curves (I), (II), (III) represent the skin friction of a flat plate at zero incidence in laminar, fully turbulent, and transitional flow

position of the point of transition is shown in addition for aerofoil R 2525. It is seen that transition occurs shortly after the pressure minimum in complete agreement with the theoretical results in Fig. 17.10. Figure 17.16 shows, further, plots of drag coefficients in terms of the lift coefficient for three aerofoils of equal thickness but varying camber. It should be noted that by increasing the camber it is possible to cause a shift in the region of very small drag in the direction of higher values of lift, but even so, the region of reduced drag still extends over a definite width only. Needless to say, in the case of laminar aerofoils the interaction between the external stream and the boundary layer is very important; methods for the calculation of such effects have been developed by R. Eppler [60]. At this point it is necessary to remark that certain circumstances cause considerable difficulties in the practical application of laminar aerofoils. Principally these are due to the great demands on the smoothness of the surfaces in order to exclude premature transition owing to roughness. In this connexion we wish to draw the reader's attention to a paper by L. Speidel [212] on laminar aerofoils placed in a harmonically disturbed free stream.

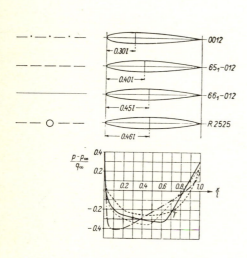

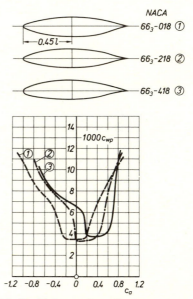

Fig. 17.15. Pressure distribution for laminar aerofoils at zero incidence ($c_a = 0$). Aerofoils 0012, 65_1—012, 66_1—012 from ref. [1]; aerofoil R 2525, after Doetsch [31]

T = position of point of transition for $R = 3 \cdot 5 \times 10^6$

Fig. 17.16. Coefficients of profile drag, c_{wp}, plotted against lift coefficient, c_a, for three laminar aerofoils with varying camber, $R = 9 \times 10^6$, from ref. [7]. The region of small drag moves towards higher lift coefficients, c_a, as camber increases

The discussion in this section may be summarized as follows:

1. The theory of stability shows that the pressure gradient exerts an overwhelming influence on the stability of the laminar boundary layer; a decrease in pressure in the downstream direction has a stabilizing effect, whereas increasing pressure leads to instability.

2. In consequence, the position of the point of maximum velocity of the potential velocity distribution function (= point of minimum pressure) influences decisively the position of the point of instability and of the point of transition. It can be assumed, as a rough guiding rule, that at medium Reynolds numbers ($R = 10^6$ to 10^7) the point of instability coincides with the point of minimum pressure and that the point of transition follows shortly afterwards.

3. As the angle of incidence of an aerofoil is increased at a constant Reynolds number, the points of instability and transition move forwards on the suction side and rearwards on the pressure side.

4. As the Reynolds number is increased at constant incidence the points of instability and transition move forwards.

5. At very high Reynolds numbers and with a flat pressure minimum, the point of instability may, under certain circumstances, slightly precede the point of minimum pressure.

6. Even at low Reynolds numbers ($R = 10^5$ to 10^6) the points of instability and transition precede the point of laminar separation; under certain circumstances the laminar boundary layer may become separated and may re-attach as a turbulent boundary layer.

Flexible wall: Another effective method of stabilizing a laminar boundary layer is to make the wetted wall flexible. In connexion with the observed astonishing swimming performance of porpoises [90], it has been suggested that these animals have a very small skin-friction coefficient because the boundary layer on them remains laminar even at very large Reynolds numbers owing to the flexibility of their skin. In order to put this hypothesis to the test, M. O. Kramer [110] performed measurements of drag on elastic circular cylinders placed in a stream parallel to their axes. Indeed, reductions of the order of 50% in drag, compared with rigid cylinders, have been observed in the range of Reynolds numbers $R = 3 \times 10^6$ to 2×10^7.

Furthermore, T. B. Benjamin [4] and M. T. Landahl [120] instituted comprehensive theoretical analyses on the stability of boundary layers on flexible plates with the aid of the method explained in Sec. XVIc. These revealed that in addition to the Tollmien-Schlichting waves which occur in a form modified by the flexibility of the wall, there appear modified elastic waves in the wall itself. Such elastic waves are created owing to the presence of the flow outside the wall. Furthermore, there appear waves of the Kelvin-Helmholtz type, rather like those observed on free shear layers. The first effect — the modification of the Tollmien-Schlichting waves by the flexibility of the wall — may, taken by itself, explain the drastic displacement of the point of neutral stability in the upstream direction. However, the three effects which depend on the internal friction in the wall counteract each other to a certain extent. For this reason, we would expect only a small overall effect. Thus, M. O. Kramer's experimental results appear to be confirmed by the stability theory only qualitatively but not quantitatively. The supposition that M. O. Kramer's results could perhaps be explained by the influence of wall flexibility on the fully developed turbulent boundary layer induced G. Zimmermann [259] to undertake a theoretical investigation into this problem. He came to the conclusion that the flexibility of the wall could lead to a reduction of the shearing stress on the wall of the order of 10 per cent, at least in the presence of a fluid of high density such as water. In the absence of a complete theory of turbulence, it is impossible to view these results as more than estimates. The paper, [259], contains references to additional contributions which concern themselves with the effect of wall flexibility on the stability and turbulence of boundary-layer flows.

c. Effect of suction on transition in a boundary layer

It has already been pointed out in Chap. XIV that the application of suction to a laminar boundary layer is an effective means of reducing drag. The effect of suction is to stabilize the boundary layer in a way similar to the effect of the pressure gradient discussed in the preceding section, and the reduction in drag is achieved by preventing transition from laminar to turbulent flow. A more detailed analysis reveals that the influence of suction is due to two effects. First, suction reduces the boundary-layer thickness and a thinner boundary layer is less prone to become turbulent. Secondly, suction creates a laminar velocity profile which possesses a higher limit of stability (critical Reynolds number) than a velocity profile with no suction.

So far only continuous suction can be treated mathematically and several solutions of such cases have already been given in Chap. XIV. In connexion with the problem of maintaining laminar flow it is important to estimate the quantity of fluid to be removed. It is possible to obtain any desired reduction in boundary-layer thickness, and hence to keep the Reynolds number below the limit of stability, provided that enough fluid is sucked away. However, a large suction volume is uneconomical because a large proportion of the saving in power due to the reduction in drag is then used to drive the suction pump. It is, therefore, important to determine the *minimum suction volume* which is required in order to maintain laminar flow. The saving in drag achieved through suction is greatest when this minimum value is used because any higher suction volume will lead to a thinner boundary layer and to an increase in shearing stress at the wall.

As shown in Chap. XIV the solution of the boundary-layer equations with suction included is particularly simple for the case of a flat plate at zero incidence with uniform suction (velocity of suction denoted by $- v_0$†). It will be recalled that the velocity profile, and hence the boundary-layer thickness, become independent of the current coordinate from a certain distance from the leading edge onwards. As shown in eqn. (14.7) the displacement thickness of this *asymptotic suction profile* is given by

$$\delta_1 = \frac{v}{- v_0} \, . \tag{17.9}$$

K. Bussmann and H. Muenz [9] carried out an investigation into the stability of this profile (Fig. 14.6) on the lines of the method explained in Chap. XVI. As seen from eqn. (14.6), the velocity profile is decribed by the equation

$$u(y) = U_\infty \left[1 - \exp(v_0 \, y/v) \right]$$

for which the critical Reynolds number has the very large value of

$$\left(\frac{U_\infty \delta_1}{v} \right)_{crit} = 70{,}000 \, . \tag{17.10}$$

Thus, the critical Reynolds number of the asymptotic suction profile is seen to

---

† Here $v_0 < 0$ denotes suction, and $v_0 > 0$ denotes blowing.

be more than 130 times larger than on a flat plate at zero incidence and in the absence of a pressure gradient or suction. This value demonstrates the highly-stabilizing effect of suction. Furthermore, the preceding argument shows that laminar flow is maintained not only owing to the reduction in boundary-layer thickness but also, and in particular, owing to a large increase in the limit of stability of the velocity profile. The curve of neutral stability for the asymptotic suction profile is shown in Fig. 17.17. ($\xi = \infty$). It should be noted that the limit of stability is increased as compared with the case with no suction and that, in addition, the range of unstable disturbance wavelengths circumscribed by the curve of neutral stability is reduced considerably.

The preceding results allow us now to find an answer to the important question of how much fluid must be removed in order to maintain laminar flow. Assuming, by way of simplification, that the asymptotic profile already exists at the leading edge of the flat plate to which uniform suction has been applied, we conclude that a laminar boundary layer which is stable along the whole plate exists if the value of the displacement thickness Reynolds number is smaller than the limit of stability given by eqn. (17.10). Hence,

$$\text{condition of stability}: \frac{U_\infty \delta_1}{\nu} < \left(\frac{U_\infty \delta_1}{\nu}\right)_{crit} = 70{,}000 \;.$$

Using the value of δ_1 for the asymptotic profile from eqn. (17.9) we have

$$\text{condition of stability}: \frac{(-v_0)}{U_\infty} = c_Q > \frac{1}{70{,}000} \;. \tag{17.11}$$

According to this result the boundary layer would be stable if the volume coefficient of suction had the extremely low value of $1/70{,}000 = 1\cdot4 \times 10^{-5}$.

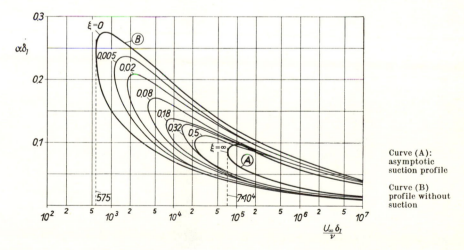

Fig. 17.17. Curves of neutral stability for the velocity profiles on a flat plate at zero incidence with uniform suction; $\xi = \left(\dfrac{-v_0}{U_\infty}\right)^2 \dfrac{U_\infty x}{\nu} = c_Q^2\,R_x$ denotes the dimensionless inlet length

It might be remarked here that a more accurate calculation would presumably lead to a higher value of the volume coefficient. This is due to the fact that the asymptotic velocity profile, on whose existence the above calculation was based, develops only at a certain distance from the leading edge. The velocity profiles between that point and the leading edge are of different shapes, changing gradually from the Blasius form with no suction at short distances behind the leading edge to the above asymptotic form. The profile shapes in this initial, starting length for the laminar boundary layer with suction have been plotted in detail in Fig. 14.8. All these velocity profiles have lower limits of stability than the asymptotic one, and it follows that the quantity of fluid to be removed over the initial length must be larger than the value given in eqn. (17.11), if laminar flow is to be maintained.

In order to analyze this matter in greater detail it is necessary to repeat the stability calculation for the series of velocity profiles in the starting length taking suction into account. These profiles constitute a one-parameter family of curves as shown in Fig. 14.8, the parameter being given by

$$\xi = c_Q{}^2 \times \frac{U_\infty x}{\nu},$$

and changing from $\xi = 0$ at the leading edge to $\xi = \infty$ for the asymptotic profile. In practice, however, it may be assumed that the starting length ends with $\xi = 4$. The resulting critical Reynolds numbers have been computed by A. Ulrich [243] and are given in Table 17.1; the corresponding curves of neutral stability have been plotted in Fig. 17.17. The amplification of unstable disturbances for the asymptotic profile has been calculated by J. Pretsch [180]. The highest degree of amplification obtained in this calculation was about 10 times smaller than that for

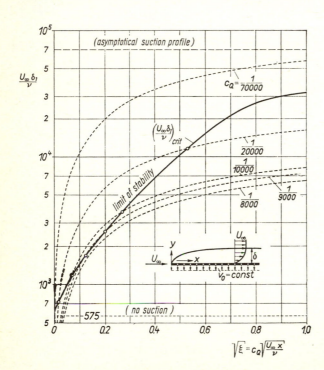

Fig. 17.18. Determination of critical value of volume coefficient for maintenance of laminar flow through suction for boundary layer on flat plate

Table 17.1.   Dependence of critical Reynolds number of velocity profiles with suction on dimensionless suction volume factor ξ, after Ulrich [243]

$\xi = c_Q^2 \dfrac{U_\infty x}{\nu}$	0	0·005	0·02	0·08	0·18	0·32	0·5	∞
$\left(\dfrac{U_\infty \delta_1}{\nu}\right)_{crit}$	575	1120	1820	3940	7590	13500	21900	70000

the flat plate (Blasius flow) in Fig. 16.13. With the results of this calculation it is now easy to determine the volume coefficient of suction which is sufficient to ensure stability over the starting length. It can be obtained from Fig. 17.18.in which the limit of stability from Table 17.1 and the variation of the dimensionless displacement thickness

$$\frac{U_\infty \delta_1}{\nu} = \frac{-v_0\,\delta_1}{\nu}\,\frac{1}{c_Q}$$

for a prescribed value of $c_Q = (-v_0)/U_\infty$ have been plotted against the dimensionless length coordinate. Here $(-v_0)\,\delta_1/\nu$ is known in terms of ξ from the calculation of the boundary layer, Table 14.1. It is seen from Fig. 17.19 that the limit of stability is not crossed at any point over the whole length only if the volume coefficient is kept at a value larger than 1/8,500. Hence, the critical value of the volume coefficient becomes

$$c_{Q\ crit} = 1·2 \times 10^{-4}. \tag{17.12}$$

We are now in a position to answer the question which was left open in Chap. XIV, namely, that concerning the actual decrease in the drag on a flat plate at zero incidence whose boundary layer is kept laminar by suction. Figure 14.9 contained a plot of the coefficient of skin friction under these conditions expressed in terms of the Reynolds number with the volume coefficient c_Q appearing as a parameter. If the curve which corresponds to c_{Qcrit} from eqn. (17.12) is now plotted in the diagram, it is possible to deduce the variation of the coefficient of skin friction for a flat plate under conditions of *optimum suction*, as shown in Fig. 17.19. The distance between the curve marked 'optimum suction' and that marked 'turbulent' corresponds to the saving in drag effected by the application of suction.

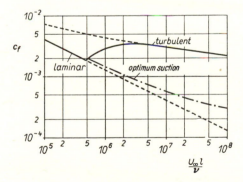

Fig. 17.19.   Coefficient of skin friction of a flat plate at zero incidence. *Optimum suction* denotes smallest volume coefficient $c_{Q\ crit} = 1·2 \times 10^{-4}$ which just suffices to maintain laminar flow

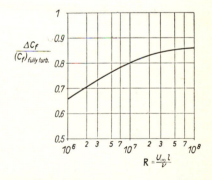

Fig. 17.20.   Relative saving in drag on flat plate at zero incidence with suction maintaining laminar flow at *optimum suction* from Fig. 17.19

$\Delta c_f = c_{f\ turb} - c_{f\ laminar\ with\ suction}$

The relative saving in drag calculated with respect to turbulent drag increases somewhat as the Reynolds number is increased, Fig. 17.20. It varies from 65 to 85 per cent. in a range of Reynolds numbers $R = 10^6$ to 10^8. Experimental results concerning boundary-layer control have already been discussed in Chap. XIV. These theoretical results concerning the saving of drag due to suction have found an excellent confirmation in the experiments carried out in flight and in a wind tunnel [89, 99, 105], see also Fig. 14.19.

The effect of suction on the limit of stability together with that of a pressure gradient can be represented graphically by plotting the critical Reynolds number against the shape factor $H_{12} = \delta_1/\delta_2$ of the boundary layer profile, as was done in Fig. 17.21. The critical Reynolds numbers for a flat plate with zero pressure gradient and uniform suction (Iglisch profiles, Fig. 14.8), those for a plate with suction described by $v_0 \sim 1/\sqrt{x}$ (Bussmann profiles, Fig. 14.12) as well as those for the case with no suction but with a pressure gradient (Hartree profiles) place themselves well on a single curve. For the asymptotic suction profile we have $H_{12} = 2$ and for the plate with no suction the value is $H_{12} = 2 \cdot 59$.

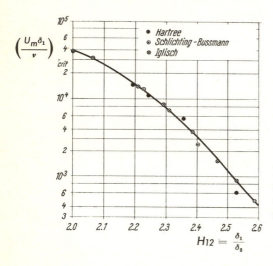

Fig. 17.21. The critical Reynolds number for laminar velocity profiles with suction, and with a pressure gradient plotted as a function of the shape factor $H_{12} = \delta_1/\delta_2$

A paper by H. Krueger [111] contains calculations of the critical Reynolds number for several examples involving *wing sections*. W. Wuest [247, 248] proved analytically that the stabilizing effect of several single slits placed one behind the other is markedly smaller than that produced by uniform suction.

d. Effect of body forces on transition

1. Boundary layer on convex walls (centrifugal forces). There are cases when transition from laminar to turbulent flow is materially affected by external forces impressed on the boundary layer. The flow in the annulus between two rotating concentric cylinders affords an example of such a case. When the inner cylinder is at rest and the outer cylinder rotates uniformly, the velocity in the annulus increases practically linearly from zero at the inner wall to the peripheral velocity of the outer wall. A fluid particle from an outer layer opposes a tendency to being moved inwards because its centrifugal force exceeds that on a particle nearer the axis of the cylinder and shows a tendency to being thrown outwards. Equally, motion outwards is made more difficult because the centrifugal force acting on an inner particle is smaller than that on a particle further away from the axis so that, consequently, particles are acted upon by what might be termed a 'centripetal lift'. Hence, it can be appreciated that transverse motions which are characteristic of turbulent flow are impeded by centrifugal forces. Thus, in this case, the centrifugal forces have a stabilizing effect.

All stability calculations described so far were confined to flat plates. H. Goertler [83] generalized Tollmien's stability criterion for profiles with a point of inflexion to include the influence of wall curvature, this case being of great practical importance. Tollmien's theorem for flat walls which states that in the limiting case of very large Reynolds number velocity profiles with a change in sign of d^2U/dy^2 become unstable, see Sec. XVI b, must be modified by stating that a change in the sign of the expression

$$\frac{d^2U}{dy^2} + \frac{1}{R}\frac{dU}{dy}$$

causes frictionless instability in the case of curved walls. Here R denotes the radius of curvature of the wall with $R > 0$ denoting a convex and $R < 0$ denoting a concave region of the wall. According to this criterion, the two-dimensional disturbances become unstable on convex walls a short distance ahead of the pressure minimum, whereas on concave walls this occurs a short distance behind it. However, on the whole, the influence of wall curvature is very small if the ratio of boundary-layer thickness, δ, to the radius of curvature, R, satisfies the conditions that $\delta/|R| \ll 1$. With concave walls a different kind of instability, namely that with respect to certain three-dimensional disturbances, to be discussed in Sec. f of this chapter, becomes of far greater importance.

Couette flow: The stability of laminar flow between two concentric, rotating cylinders (Couette flow) is governed to a large extent by the centrifugal forces. The velocity distributions which occur in this case were given in Fig. 5.4 on the basis of exact solutions of the Navier-Stokes equations; they covered various values of the ratio of radii $\varkappa = r_1/r_2$, and concerned two basic cases: (I) inner cylinder rotates, outer cylinder at rest; (II) outer cylinder rotates, inner cylinder at rest. In Case I (inner cylinder rotating), the layers at the rotating inner wall experience larger centrifugal forces than those near the outer wall. For this reason, the present case turns out to be highly unstable; it was investigated very early by G.I. Taylor who assumed the existence of three-dimensional disturbances. G.I. Taylor discovered the existence of a secondary flow in the form of ring-shaped vortices in excellent agreement between theory and experiment, see Sec. XVIIf and Figs. 17.32—17.34. In Case II (outer cylinder rotating) the larger centrifugal forces occur in the fluid layers at the outer wall, which has a highly stabilizing effect on the flow. The stability of such flows with respect to two-dimensional disturbances was thoroughly investigated theoretically by F. Schultz-Grunow [204c]; he was able to demonstrate the inherent stability of this arrangement. Experimental investigations by the same author [204 b] confirmed this conclusion; see also the more recent papers by F. Schultz-Grunow [204 d].

2. The flow of non-homogeneous fluids (stratification). The influence of vertical *density variations* on the stability of flow past a flat horizontal wall is in a sense related to the case of centrifugal forces occurring in a homogeneous fluid flowing along a curved wall. When the density decreases upwards, the arrangement is stable, and it becomes unstable when the density variation is reversed. In the latter case there is instability, even without flow, when the fluid is heated from below. The fluid then becomes unstable in that the horizontal layers of fluid become honeycombed into regular hexagonal eddy patterns [5, 97, 190]. In the case of flow with stable density stratification, turbulent mixing in the vertical direction is impeded because heavier particles must be lifted and lighter particles must be depressed against hydrostatic forces. Turbulence can even be completely suppressed if the density gradient is strong enough, the phenomenon being of some importance in certain meteorological processes. It is, for example, possible to observe that on cool summer evenings damp meadows are blanketed in sharply outlined mists with a gentle wind blowing. This is a sign that the wind ceased to be turbulent so that layers of air slide over each other in laminar motion and without turbulent mixing. The cause of this phenomenon lies in the pronounced temperature gradient which is formed in the air as the earth cools in the evening and prevents mixing of the warmer, and therefore lighter, upper layers of the atmosphere with·the colder and heavier layers near the ground. The "falling off" of the wind which can sometimes be observed towards the evening is due to the same effect. The wind prevails in all its force at higher altitudes but the suppression of turbulence near the ground on cooling greatly reduces its speed. Furthermore, the streaming of sweet over salt water which occurs, e. g., in the Kattegat, as well as the remarkable stability of Bjerknes's polar fronts, when the cold masses of air form a wedge under the warm air, belong to this group of phenomena.

L. Prandtl [173] analyzed the phenomena connected with density gradients as well as with the previously discussed flows over curved surfaces involving the influence of centrifugal forces with the aid of an energy method. He has shown that the stability of stratified flows depends on the stratification parameter

$$\mathsf{R}_i = -\frac{g}{\varrho}\frac{\mathrm{d}\varrho}{\mathrm{d}y} \bigg/ \left(\frac{\mathrm{d}U}{\mathrm{d}y}\right)^2_w \tag{17.13}$$

known as the Richardson number, in addition to the usual dependence on Reynolds number. Here g denotes the acceleration due to gravity, ϱ the density, and the positive direction of y is measured vertically upwards. The subscript w refers to the value of the velocity gradient at the wall, and $\mathsf{R}_i = 0$ corresponds to a homogeneous fluid, $\mathsf{R}_i > 0$ denoting stable, and $\mathsf{R}_i < 0$ unstable stratification. The energy method used by L. F. Richardson [192] and L. Prandtl has shown that turbulence may be expected to disappear at $\mathsf{R}_i > 2$. G. I. Taylor [240] refined Prandtl's reasoning and obtained $\mathsf{R}_i \geq 1$ as the limit of stability. H. Ertel [56] supplied a thermodynamic justification for this criterion.

G. I. Taylor [240] and S. Goldstein [69] were the first ones to apply the method of small disturbances to this problem. Assuming a continuous density distribution and a linear velocity profile in an infinite fluid they found the limit of stability to be at $R_i = \frac{1}{4}$. The influence of viscosity and of curvature in the velocity profile have been neglected in this connexion. H. Schlichting [199] investigated the stability of flows with density stratification with the aid of Tollmien's theory. The calculation was based on the assumption of a Blasius profile for a flat plate with a density gradient in the boundary layer and constant density outside it. It was found that the critical Reynolds number increased rapidly as the Richardson number increased, Fig. 17.22, changing its value from $R_{crit} = 645$ for $R_i = 0$ (homogeneous flow) to $R_{crit} = \infty$ for $R_i = 1/24 = 0.042$. Thus for

$$R_i > 0.042 \quad \text{(stable)} \tag{17.14}$$

the flow remains stable everywhere on the flat plate. It is seen that the present limit of stability is considerably smaller than that given by previous theories.

A comparison between theory and the experimental results due to H. Reichardt [174] is given in Fig. 17.22; measurements were performed in a special

Fig. 17.22. Critical Reynolds number for the boundary layer on a flat plate at zero incidence in flow with density gradient as a function of the Richardson number R_i

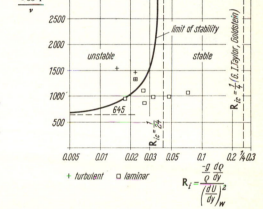

rectangular channel in Goettingen. The air was passed through the channel whose upper wall was heated with the aid of steam, the bottom wall being cooled with water. It is seen that all observed laminar flows fall within the stable region, whereas all turbulent flows fall within the unstable region. Agreement is, therefore, excellent.

G. I. Taylor [239] observed turbulent flow in ocean currents at considerably higher values of the Richardson number and it appears that this phenomenon is due to the absence of walls. Recently, J. T. Stuart [227a] investigated theoretically the effect of a magnetic field on transition. It turned out that the critical Reynolds number increases considerably for the case of laminar flow between two parallel flat walls when the lines of the magnetic field are parallel to the walls.

e. Effects due to heat transfer and compressibility†

1. Introductory remark. The theoretical and experimental results concerning transition described in the preceding sections are valid only for flows at moderate speeds (incompressible flow). The effect of the compressibility of the fluid on transition has recently been exhaustively investigated under the stimulus from aeronautical engineering. In the case of compressible flows, apart from the Mach number, it is necessary to take into account one additional, important parameter which is connected with the rate of heat transferred between the fluid and the wall. When the fluid is incompressible, heat can be exchanged between the wall and the fluid only if the temperature of the wall is higher or lower than that of the fluid flowing past it. In the case of a compressible fluid, the heat evolved in the boundary layer produces an additional, important influence, as already shown in Chap. XIII. In either case a thermal boundary layer develops in addition to the velocity boundary layer and plays its part in the determination of the instability of a small disturbance. The theoretical and experimental considerations which we are about to discuss will show that for the subsonic flow of a gas, heat transfer from the boundary layer to the wall exerts a stabilizing influence, while heat transfer from the wall to the gas has the opposite effect. Both of these are reversed for the flow of a liquid. For supersonic flow, a new type of unstable disturbance is possible which responds to the transfer of heat in an entirely different manner.

2. The effect of heat transfer in incompressible flow. Some of the main features of the effect of the transfer of heat from the wall to the fluid on the stability of a laminar boundary layer can be readily recognized even in the case when the flow is incompressible. We shall, therefore, explain it first in this simplified form. The first experimental investigations on the influence of heat transfer on transition were performed some time ago by W. Linke [131]. W. Linke measured the drag of a vertical heated plate placed in a horizontal stream in a range of length Reynolds number $R = 10^5$ to 10^6, and observed that heating caused it to increase by a large amount. He concluded from this increase, quite correctly, that the heating of the plate caused the transition Reynolds number to decrease.

With the aid of the point-of-inflexion criterion which was discussed in Chapter XVI it is easy to show that there is a stabilizing or destabilizing effect due to the transfer of heat when $T_w \neq T_\infty$. This effect is a consequence of the dependence of the viscosity μ of the fluid on the temperature T. When the temperature-dependence of the viscosity is taken into account, the curvature of the profile $U(y)$ of the main flow at the wall is given for the case of a flat plate at zero incidence, according to eqn. (13.6), by

$$\left(\frac{\mathrm{d}^2 U}{\mathrm{d} y^2}\right)_w = -\frac{1}{\mu_w}\left(\frac{\mathrm{d}\mu}{\mathrm{d} y}\right)_w \left(\frac{\mathrm{d} U}{\mathrm{d} y}\right)_w. \tag{17.15}$$

Now, if the wall is hotter than the fluid in the free stream, we have $T_w > T_\infty$ and the temperature gradient at the wall is negative: $(\mathrm{d}T/\mathrm{d}y)_w < 0$. Since for a gas the viscosity increases with temperature according to eqn. (13.3), we must have $(\mathrm{d}\mu/\mathrm{d}y)_w$

---

† This section was prepared by Dr. L. M. Mack of the Jet Propulsion Laboratory of the California Institute of Technology at Pasadena, California.

< 0. Since the velocity gradient is positive at the wall it follows from eqn. (17.15) that

$$T_w > T_\infty \text{ implies } \left(\frac{\mathrm{d}^2 U}{\mathrm{d} y^2}\right)_w > 0. \tag{17.16}$$

Thus for a heated wall the curvature of the velocity profile at the wall is positive, and it follows immediately that a point of inflexion ($\mathrm{d}^2 U/\mathrm{d} y^2 = 0$) must exist within the boundary layer because the curvature is vanishingly small but negative at $y = \infty$ (*cf.* Fig. 7.4). This means that the transfer of heat from the wall to a gas flowing past it renders the boundary layer unstable by the criterion given in Chap. XVI in a manner analogous to a pressure increase in the downstream direction. On the contrary, cooling the wall renders the boundary layer more stable by increasing the curvature of the velocity profile at the wall and acts like a favourable pressure gradient.

A numerical calculation by T. Cebeci and A. M. O. Smith [22] for air confirmed the decrease in the critical Reynolds number for the onset of instability of a heated flat plate, and a similar decrease in the transition Reynolds number was observed in the experiment of H. W. Liepmann and G. H. Fila [129] on a vertical flat plate at zero incidence.

Since the viscosity of a liquid decreases as the temperature is increased, the effects of heating and cooling should be reversed according to eqn. (17.15). An investigation by A. R. Wazzan, T. Okamura and A. M. O. Smith [249, 250, 251] for water confirmed this expectation. The critical Reynolds number for the onset of instability is shown in Fig. 17.23 for heated and cooled walls along with the maximum dimensionless amplification factor $(\beta_i \delta_1/U_\infty)_{max}$ and the ratio of the dimensionless displacement thickness $\delta_{1\,i}\sqrt{U_\infty/x\,\nu_\infty}$ to its value $1\cdot721$ for the unheated wall.

There is a strong stabilizing effect as the wall temperature is increased from its initial value of $15\cdot6\,°C$ ($60\,°F$), but then further heating is de-stabilizing. Although the dimensionless amplification factor is constant for $T_w > 60\,°C$, the dimensional $(\beta_i)_{max}$ increases in inverse proportion to δ_1. The results for a cooled wall show the expected de-stabilizing effect for liquids. In the theory of A. R. Wazzan, the only influence of heat transfer, other than on the mean velocity profile, is through the temperature dependence of the viscosity. A more complete theory by R. L. Lowell and

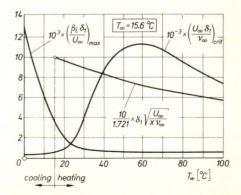

Fig. 17.23. Effect of wall temperature on instability and displacement thickness of water boundary layer on a flat plate, after A. R. Wazzan, T. Okamura and A. M. O. Smith [250]

T_w = wall temperature
T_∞ = free-stream temperature

E. Reshotko [146] included the temperature and density fluctuations, but led to almost identical numerical results. A stability experiment performed by A. Strazisar, J. M. Prahl and E. Reshotko [227] verified the predicted shift of the minimum critical Reynolds number with a small amount of heating.

Free convection: Transition of a free-convection boundary layer on a vertical heated flat plate was first related to the amplification of small disturbances by E. R. G. Eckert and E. Soehngen [57, 59]. The reader is referred to articles by B. Gebhardt [79, 80, 81] for a comprehensive review of this field in which much progress has been made in explaining observed transition phenomena by means of accurate numerical calculations based on the method of small disturbances.

Whereas for the vertical heated plate the instability originates from progressing waves of the Tollmien-Schlichting type, on the inclined heated plate standing unstable vortices with axes along the direction of flow have been observed; these are of the Taylor-Goertler type, see [147, 228, 81].

The stability of a free convective stream on a heated vertical plate was investigated by P. R. Nachtsheim [167] who employed the method of small disturbances. The velocity and temperature distributions were those of Figs. 12.23 and 12.24, respectively. Velocity profiles with a strong point of inflexion, such as those in Fig. 12.24, are intrinsically characterized by a low limit of stability. The inclusion of temporal temperature fluctuations on top of the velocity fluctuations produces an additional strong destabilizing effect of the main flows, because this mechanism transfers energy from the main motion to the disturbance. The calculation leads to two coupled differential equations which now replace the Orr-Sommerfeld equation (16.14). One of them refers to velocity and the other to temperature. These two equations contain the Prandtl number and the Grashof number in addition to the Reynolds number. In this connexion the reader should consult the papers by E. Eckert et al. [59], A. Szewczyk [229] and T. Benjamin [15] which contain also experimental results.

3. The effect of compressiblity. Of the numerous transition phenomena which have been encountered in supersonic and hypersonic boundary layers, we shall focus on the effects of Mach number and heat transfer on the zero pressure gradient boundary layers which are formed on flat plates or on cones at zero angle of incidence. We proceed first to a summary of the principal results which have been obtained with the method of small disturbances, and will then show how the theory can account for some of the experimental observations. Many of the theoretical results to be presented are taken from a detailed study by L. M. Mack [153] of compressible stability theory, and in the absence of a specific reference this work is to be understood as the source.

The first analysis of the stability of compressible laminar boundary layers was given by D. Kuechemann [112] on the basis of the neglect of viscosity on the motion of the disturbance. The temperature gradient and curvature of the velocity profile were first included in the frictionless analysis by L. Lees and C. C. Lin [122]. They classified the disturbances, which were assumed of the same periodic form as in eqn. (16.10) into three categories called subsonic, sonic, and supersonic, depending on whether the phase velocity c_r is greater than, equal to, or less than $U_\infty - a_\infty$, where a is the speed of sound. In particular, L. Lees and C. C. Lin proved that a sufficient condition for the existence of an unstable subsonic disturbance is

Fig. 17. 24. Effect of Mach number on phase velocities of two-dimensional neutral disturbances and displacement thickness of adiabatic flat-plate boundary
layers

c_s = phase velocity of neutral subsonic disturbance
c_0 = phase velocity of neutral sonic disturbance

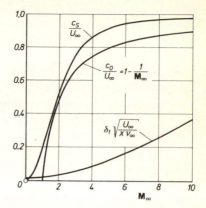

$$\left[\frac{d}{dy}\left(\varrho\,\frac{dU}{dy}\right)\right]_{y_s} = 0, \tag{17.17}$$

provided that $U(y_s) > U_\infty - a_\infty$.

This theorem is the extension to compressible flow of Theorem I of Sec. XVI b,
and y_s is the compressible counterpart of the inflexion point in incompressible flow.
It can conveniently be referred to as the generalized inflexion point. With a generalized inflexion point, there is a neutral subsonic disturbance with $c_r = c_s = U(y_s)$ and
also a neutral sonic disturbance when $M_\infty > 1$ with phase velocity $c_r = c_0 = U_\infty - a_\infty$
and $\alpha = 0$. Neutral supersonic disturbances are possible in certain flows, but no
general conditions for their existence have been given. Figure 17.24 shows the dimensionless phase velocities c_s/U_∞ and c_0/U_∞ of the neutral subsonic and sonic disturbance
as functions of M_∞ for a family of adiabatic flat-plate boundary layers. The mean
boundary layer profiles which were used in the calculation of c_s, and will be used
throughout this Section, are accurate numerical solutions of the compressible laminar
boundary layer equations for air with both the viscosity coefficient and Prandtl
number functions of temperature, and with a free-stream stagnation temperature of
311 K at $M_\infty = 5\cdot1$ where $T = 50$ K. At higher Mach numbers, T_∞ remains at 50 K.
These temperature conditions are characteristic of supersonic and hypersonic wind
tunnels. Since $c_s > c_0 > 0$ in Fig. 17.24, all of the boundary layers of this family
satisfy the conditions of the extended theorem and are unstable to frictionless disturbances. The movement of the generalized inflexion point to larger y/δ with increasing M_∞ is similar to the movement of the inflexion point with increasing adverse
pressure gradient in incompressible flow. Figure 17.24 also gives the dimensionless
displacement thickness $\delta_1 \sqrt{U_\infty/x\,\nu_\infty}$ as a function of M_∞ for the family of adiabatic
boundary layers. L. Lees and C. C. Lin were able to prove that the wave number of the
neutral subsonic disturbance is unique as in incompressible flow, provided that the
mean flow relative to the phase velocity is everywhere subsonic, i. e. $\hat{M}^2 < 1$ throughout the boundary layer, where $\hat{M} = (U - c_r)/a$ is the local relative Mach number.
Although their proof that eqn. (17.17) is a sufficient condition for the instablity had
the same restriction, it appears from extensive numerical calculations that eqn. (17.17)
is a true sufficient condition even when $\hat{M}^2 > 1$. On the contrary, L.M. Mack [152] show-

ed by numerical calculations that with a region in the boundary layer where $\hat{M}^2 > 1$ there are an infinite number of neutral wave numbers, or modes, with the same phase velocity c_s. The multiple modes are a result of the change in the governing differential equation for, say, the pressure oscillation from elliptic when $\hat{M}^2 < 1$ to hyperbolic when $\hat{M}^2 > 1$. The first mode is the same as in incompressible flow, and was first computed for compressible flow by L. Lees and E. Reshotko [142]. The additional, or higher, modes have no incompressible counterparts. With $c_r = c_s$, $\hat{M}_w^2$ first reaches unity at $M_\infty = 2\cdot2$, and the upper boundary of the region of supersonic relative flow is at $y/\delta = 0\cdot16$, $0\cdot43$, $0\cdot59$, for $M_\infty = 3$, 5, 10, respectively.

The multiple neutral disturbances with phase velocity c_s are not the only ones possible when $\hat{M}_w^2 > 1$. There are also multiple neutral disturbances with $U_\infty \leq c_r \leq U_\infty + a_\infty$. These disturbances do not depend on the boundary layer having a generalized inflexion point. Furthermore, there are always adjacent amplified disturbances of the *same type with phase velocities* $c_r < U_\infty$. *Consequently, the compressible boundary layer is unstable to frictionless disturbances regardless of any other features of the velocity and temperature profiles as long as there is a region where* $\hat{M}^2 > 1$.

A limiting factor in the amplification of first-mode disturbances is that c_r must lie between c_0 and c_s. Anything that increases the difference $c_s - c_0$ also increases the amplification factor β_i. As shown by Fig. 17.24, this difference can be quite small. The constraint imposed by c_0, which unlike c_s is unrelated to the boundary-layer profile, can only be removed by considering a more general form of disturbance than has been used up to this point. With

$$u'(x, y, z, t) = f(y) \exp\left[i\left(\alpha x + \gamma z - \beta t\right)\right] \tag{17.18}$$

in place of eqn. (16.12), the wave normal is inclined at the angle $\tilde{\psi} = \tan^{-1}(\gamma/\alpha)$ to the x-direction. It has been shown by C.C. Lin [144] that if the coordinate system is rotated about the y-axis until the x-axis is coincident with the wave normal, then the three-dimensional frictionless equations are identical to the two-dimensional equations (except for an additional momentum equation in the new z-direction which is uncoupled from the other equations). Therefore, the preceding considerations are still valid for a three-dimensional disturbance, but with the governing Mach number $\tilde{M}_\infty = M_\infty \cos\tilde{\psi}$ instead of M_∞. Consequently, the phase velocity of a three-dimensional neutral sonic disturbance is $c_0/U_\infty = 1 - 1/\tilde{M}_\infty$, and c_0 decreases as $\tilde{\psi}$ increases.

As a result of $c_s - c_0$ increasing with $\tilde{\psi}$ the maximum amplification factor of a three-dimensional first-mode disturbance is larger than for a two-dimensional disturbance, as is shown in Fig. 17.25. In this figure, the maximum dimensionless amplification factor $(\beta_i \delta_1/U_\infty)_{max}$, where the maximum is with respect to α for two-dimensional disturbances and with respect to both α and $\tilde{\psi}$ for three-dimensional disturbances, is plotted as a function of M_∞ for the family of adiabatic boundary layers. The results for second-mode two-dimensional disturbances are also shown in Fig. 17.25. In contrast to the first mode, the most unstable second-mode disturbances are two-dimensional. The second mode depends only on the extent in the y-direction of the supersonic relative flow region as determined by c_r and the Mach number $\tilde{M}_\infty$. This region has its maximum extent for a two-dimensional disturbance.

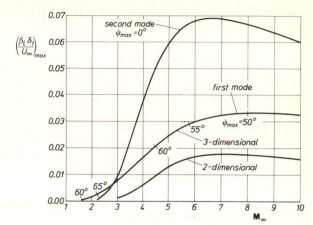

Fig. 17. 25. Effect of Mach number on maximum amplification factor of first- and second-mode disturbances as given by frictionless theory for adiabatic flat-plate boundary layers, after L. M. Mack [153]

L. Lees and C.C. Lin [122] also developed a theory of viscous disturbances along the same general lines as the incompressible asymptotic theory presented in Chapter XVI. This theory was later extended by L. Lees [124], D.W. Dunn and C.C. Lin [43] and E. Reshotko [142]. However, when it became possible to obtain the eigenvalues of the compressible stability equations with great accuracy on a digital computer, as was done first by W.B. Brown [14] and then by L.M. Mack [151], it was found that the asymptotic theory is only valid up to slightly supersonic Mach numbers. For this reason, only those results which have been obtained by numerical integration of the complete viscous stability equations will be mentioned here.

As the Mach number increases, three Mach number regions with different instability characteristics can be distinguished for the adiabatic flat-plate boundary layers. In Fig. 17.26 the ratio $(\beta_i)_{max}/(\beta_i)_{max,inc}$ at $R = U_\infty x/\nu_\infty = 2\cdot25 \times 10^6$, where $(\beta_i)_{max,inc} = 0\cdot00432\, U_\infty/\delta_1$ is the amplification factor at the same Reynolds number for incompressible flow, is given as a function of M_∞ for two-dimensional second-mode disturbances. In the first region, up to about $M_\infty = 2\cdot5$, only first-mode disturbances are of importance. The maximum amplification factor of two-dimensional disturbances decreases sharply, but for $M_\infty > 1$, three-dimensional disturbances are the most unstable. In the second region, from $M_\infty = 2\cdot5$ to $5\cdot0$, the increasing frictionless instability shown in Fig. 17.25 begins to make its influence felt at lower Reynolds numbers.

Fig. 17. 26. Effect of Mach number on maximum amplification factor of first- and second-mode disturbances as given by viscous theory for adiabatic flat-plate boundary layers, after L. M. Mack [153]

$R = U_\infty x/\nu_\infty = 2\cdot25 \times 10^6$
$\tilde{\nu}_{max}$ = most unstable wave angle to nearest 5°

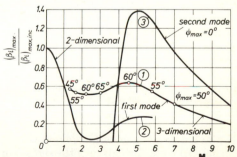

By $M_\infty = 3\cdot5$, the maximum amplification factor of both three- and two-dimensional disturbances occurs at $R = \infty$. It is in this second region, where the instability assumes an essentially frictionless nature, that an unstable band of frequencies associated with the second mode first appears for $R = 2\cdot25 \times 10^6$. In the third region, $M_\infty > 5$, the amplification factors decrease steadily in proportion to the increase in δ_1 shown in Fig. 17.24.

For the low speed flow of a gas, we have already discussed the destabilizing effect of a heated wall and the stabilizing effect of a cooled wall. L. Lees [124] calculated similar effects for compressible air boundary layers, and, in addition, predicted the possibility of completely stabilizing supersonic boundary layers by cooling. Although this prediction and subsequent calculations of the cooling required for complete stabilization by M. Bloom [11] and E. R. van Driest [50] were based on the asymptotic theory of two-dimensional disturbances and took no account of the higher modes, the more recent computer calculations have verified that sufficient cooling will indeed completely stabilize, or nearly so, both two- and three-dimensional first-mode disturbances over a wide Mach number range. Fig. 17.27 shows, from the frictionless theory, the ratio of $(\beta_i)_{max}$ to its value for the adiabatic wall, $(\beta_i)_{max,ad}$ as a function of T_w/T_{ad}, the ratio of the wall temperature to the adiabatic wall temperature. The stabilization of three-dimensional first-mode disturbances is clearly seen, with the stabilization decreasing with increasing Mach number for the same temperature ratio. On the contrary, second-mode disturbances, far from being stabilized by cooling, are de-stabilized. The reason for this different behaviour is, once again, that the generalized inflexion point, which is strongly influenced by cooling, has no importance for the unstable higher modes. The important quantity, the extent of the supersonic relative-flow region, is little influenced by cooling.

Similar results to those shown in Fig. 17.27 are obtained from the viscous theory, except that for $M_\infty > 3$ less cooling is required for stabilization at any finite Reynolds number than is given by the frictionless theory. In this connexion see also a paper by E. Reshotko [194b].

Experimental results: The first experiment to demonstrate the existence of laminar instability waves in supersonic flow was performed by J. Laufer and T. Vrebalovich [139]. A later experiment by J. M. Kendall [114] went further and provided a quantitative verification of supersonic stability theory at $M_\infty = 4\cdot5$. The success of J. M. Kendall's experiment derived in part from the operation of his wind tunnel with laminar boundary layers on the walls. The absence of the usual large acoustic disturbances that originate from supersonic turbulent boundary layers made possible increased accuracy in the measurement of the growth of small artificial disturbances produced by a glow discharge between electrodes embedded in the surface of the flat plate and skewed at an angle $\tilde{\psi}$ to the z-direction. A comparison of theoretical and experimental amplification factors is given in Fig. 17.28, where the ordinate is the dimensionless spatial amplification factor $-\alpha_i \delta_1$. The spatial amplification factor is related to the rms amplitude A of any oscillating flow variable by

$$\frac{1}{A}\frac{dA}{dx} = -\alpha_i. \tag{17.18a}$$

If a hot-wire anemometer follows the peak rms disturbance amplitude downstream, the logarithmic derivative of the signal amplitude can be interpreted as $-\alpha_i$. Although

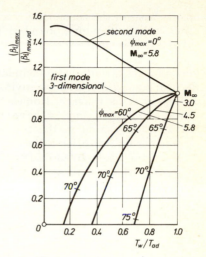

Fig. 17. 27. Effect of wall cooling on maximum amplification factor of first- and second-mode disturbances as given by frictionless theory for flat-plate boundary layers

T_w = wall temperature
T_{ad} = adiabatic will temperature
$\tilde{\psi}_{max}$ = most unstable wave angle to nearest 5°

the theoretical α_i can be obtained from β_i with $\alpha_i = \beta_i/(\partial\beta_r/\partial\alpha)$, where α is the group velocity, as was originally done by H. Schlichting [76] Chap. XVI (see also M. Gaster [78]), it is now more convenient to calculate α_i directly by letting the wave numbers α and γ in eqn. (17.18) be complex numbers and the frequency β a real number.

Figure 17.28 shows fine agreement of the measurements with the theory for a $\tilde{\psi} = 55°$ first-mode disturbance and two-dimensional second-mode disturbance. There is an appreciable discrepancy only in the maximum amplification rate of the second mode, but the frequencies of maximum amplification agree very well.

We are able to turn now to a consideration of the effects of Mach number and cooling on the transition Reynolds number of zero pressure gradient boundary layers

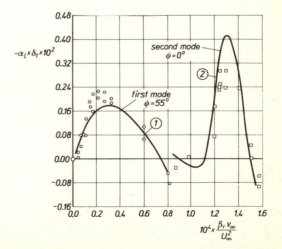

Fig. 17. 28. Comparison of experimental and theoretical spatial amplification factors for the adiabatic flat-plate boundary layer at $M_\infty = 4\cdot5$, $U_\infty x/\nu_\infty = 2\cdot4 \times 10^6$. Experimental data obtained from hot-wire anemometer with disturbances produced artificially by glow-discharge device skewed at an angle $\tilde{\psi}$ equal to theoretical wave angle

○, □ experiment: J. M. Kendall [114]
— theory: L. M. Mack [153]

on the basis of the theoretical results just presented. An important point to keep in mind is that although a boundary layer has definite instability properties, its transition Reynolds number depends not only on these properties but also on the type and intensity of the disturbances present in the flow. The only facility in which it is convenient to study adiabatic boundary layers is the supersonic wind tunnel which has its own special disturbance environment. Below $M_\infty = 3$, transition measurements differ widely for different tunnels. For $M_\infty > 3$, both J. Laufer [135] and E. R. van Driest and J. C. Boison [34] have shown that turbulence from the supply section does not affect transition in the test section. Instead, the primary disturbance source responsible for transition is the acoustic radiation from the turbulent boundary layers on the tunnel walls. In addition to the effect on transition measurements of differences in the disturbance environment, there is also the problem of defining and measuring the transition Reynolds number in a consistent manner. An instructive comparison of five different methods of measuring transition has been given by J. L. Potter and J. D. Whitfield [187]. The method of small disturbances can properly be applied only to the calculation of a start-of-transition Reynolds number.

The numerous wind-tunnel transition data for $M_\infty > 3$ accumulated by S. R. Pate and C. J. Schueler [183] for flat plates and by S. R. Pate [181] for cones, formed the basis of their correlations based solely on parameters of the acoustic radiation. These data, together with measurements at $M_\infty < 3$ by J. Laufer and J. E. Marte [138] on cones, by D. Coles [24] on a flat plate, a single observation at $M_\infty = 1\cdot6$ by J. M. Kendall of laminar flow on a flat plate at $R = 4\cdot3 \times 10^5$ in the same tunnel used by D. Coles, and measurements on cones in transonic tunnels by N. S. Dougherty and F. W. Seinle [49], suggest the following pattern for the Mach number dependence of the transition Reynolds number in a good wind tunnel: An initial increase for $M_\infty > 1$ with a peak, perhaps rather broad, between $M_\infty = 1\cdot5$ and $2\cdot0$, followed by a decline, and then, starting somewhere between $M_\infty = 3$ and 5, a monotonic increase which continues to at least $M_\infty = 16$ according to measurements in a helium tunnel [157]. It is of particular interest that these three Mach number regions correspond roughly to the three regions discussed previously in connexion with Fig. 17.27. A more direct connexion with stability theory was made by L. M. Mack [154] by means of a simplified calculation of the start of transition on a flat plate based solely on a critical amplitude A, of the most amplified single-frequency disturbance, as given by eqn. (17.18 a). The results of this calculation along with some experimental flat-plate data [24, 45] are shown in Fig. 17.29. With A_0 the value of A at the neutral-stability point, the upper curve results from assuming that A_0 is independent of Mach number, and the lower curve results from assuming $A_0 \propto M_\infty^2$ for $M_\infty > 1\cdot3$. It is the lower curve which corresponds to transition in a wind tunnel, where J. Laufer [136] has determined that from $M_\infty = 1\cdot6$ to $5\cdot0$ the free-stream rms disturbance amplitude varies essentially as M_∞^2. The general similarity of this curve to the measurements fully supports the view that transition in supersonic boundary layers results from the amplification of particular flow disturbances in accordance with the method of small disturbances.

In an experimental investigation of the effect on transition of a flow parameter such as Mach number, it is necessary to keep the unit Reynolds number U_∞/ν_∞ constant, as with the measurements given in Fig. 17.29. The dependence of the transition Reynolds number on unit Reynolds number has been noted in various types of wind tunnels by several investigators [12, 185, 195, 116] as well as in ballistic ranges [184].

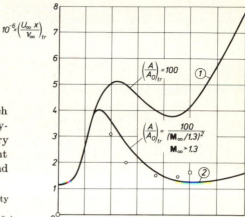

Fig. 17. 29. Comparison of the effect of Mach number on experimental start-of-transition Reynolds numbers of adiabatic flat-plate boundary layers with calculations based on two different amplitude criteria (D. Coles [25], R. E. Deem and J. S. Murphy [45], L. M. Mack [154])

A_0 = initial disturbance amplitude at neutral-stability points

$(A/A_0)_{tr}$ = amplitude ratio at transition, from eqn. (17.18a)

Fig. 17. 30. Comparison of the experimental effect of unit Reynolds number on the start-of-transition Reynolds number of an adiabatic flat plate with a theoretical calculation based on stability theory and the measured properties of free-stream disturbances; measurements: D. Coles [25]; calculations: L. M. Mack [155]

— calculated
measured, M_∞
    ○ 1·97
    □ 2·57
    △ 3·70
    ◇ 4·54

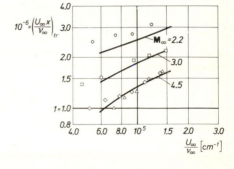

Stability theory offers an explanation of this effect in those instances where some gross transition parameter such as roughness or leading-edge bluntness has not been overlooked. As discussed by E. Reshotko [194], the boundary layer frequency response and the spectrum of the flow disturbances shift relative to one another as the unit Reynolds number changes. Because of this shift, the initial amplitudes of the individual frequency components of the disturbance in the boundary layer must change and so, too, must the transition Reynolds number. A calculation by L. M. Mack [155] of the unit Reynolds effect on the start-of-transition Reynolds number at three Mach numbers, the results of which are compared in Fig. 17.30 with the measurements of D. Coles [24], indicates that this explanation is correct for a smooth flat plate. The latter calculation was more realistic than the one leading to Fig. 17.29 because it took into account the power spectrum of the free-stream disturbances as measured by J. Laufer [137], and also the influence of unit Reynolds number on the intensity of the free-stream disturbances. The initial disturbances in the boundary layer were assumed proportional to the free-stream disturbances.

The destabilizing effect of heating on transition was confirmed for a flat plate at $M_\infty = 2\cdot4$ by R. W. Higgins and C. C. Pappas [91] and for a body of revolution at $M_\infty = 1\cdot6$ by K. R. Czarnecki and A. R. Sinclair [25]. The latter experiment also showed the predicted stabilizing effect of cooling, as did the experiment of J. R. Jack and N. S. Diaconis [100] on two bodies of revolution at $M_\infty = 3\cdot12$. Of the numerous flight experiments with cooled boundary layers in which laminar flow has been observed at high length Reynolds numbers, we may cite the experiment of J. Sternberg [215] on a cone at $M_e = 2\cdot7$, where M_e is the Mach number at the edge of the boundary layer. The surface temperature history definitely showed the presence of laminar flow at $R = 40 \times 10^6$ with $T_w/T_{ad} = 0\cdot51$, as compared to a wind-tunnel measurement [51] of $R = 12 \times 10^6$ for the end of transition at the same Mach number with $T_w/T_{ad} = 0\cdot65$.

The results of an investigation by E. R. van Driest and J. C. Boison [34] on a cone at $M_e = 1\cdot9$, $2\cdot7$ and $3\cdot7$ are shown in Fig. 17.31. The increase in the transition Reynolds number with cooling is clearly seen, as is the reduction of the stabilization effect with increasing Mach number. The latter trend continues to higher Mach numbers, as was shown by the small stabilization effect found on flat plates at $M_\infty = 6\cdot0$ and $6\cdot5$, respectively, by A. M. Cary [21] and D. V. Maddalon [156]. These effects of cooling are in complete accord with the behaviour of first-mode disturbances shown in Fig. 17.27. However, in two experiments by N. S. Diaconis, J. R. Jack and R. J. Wisniewski [47, 101], at $M_\infty = 3\cdot12$, cooling beyond the region of stabilization shown in Fig. 17.31 resulted in a decrease rather than an increase in the transition Reynolds number. This phenomenon has been called "transition reversal" because it is contrary to the expected trend. It has also been observed in shock tunnels by B. E. Richards and J. L. Stollery [195, 196] on a flat plate at $M_\infty = 8\cdot2$, and by K. F.

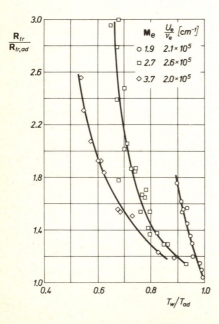

Fig. 17.31. Experimental transition data obtained on a 10°-cone at zero angle of incidence showing stabilizing effect of wall cooling at three Mach numbers in a supersonic wind tunnel, after E. R. van Driest and J. C. Boison [34]

○ Coles
□ Deem and Murphy $\Big\}$ $u_\infty/\nu_\infty = 1\cdot2 \times 10^5$ cm^{-1}
— calculated, Mack

Stetson and G. H. Rushton [226] on a cone at $M_e = 5\cdot5$; by N. W. Sheetz [223] on highly cooled cones in a ballistic range from $M_\infty = 3$ to 9; and by G. G. Mateer [159] on cones at $M_e = 5\cdot0$ and $6\cdot6$ in a conventional wind tunnel. The origin of this phenomenon has not been established, although somewhat similar effects have been obtained by adding small roughness elements [34]. A complicating factor is that the reversal does not appear in many experiments which cover an equally wide range of surface temperature [35, 21, 156].

In the hypersonic flow regime, three wind-tunnel experiments [45, 58, 221] at $M_\infty = 10$, the first one on a flat plate and the other two on cones, showed almost no effect of cooling on transition. When eqn. (17.18a) is integrated for all frequencies for which second-mode disturbances are unstable, the resulting maximum amplitude ratio $(A/A_0)_{max}$ at a fixed Reynolds number is found to be almost independent of T_w/T_{ad}. This result suggests that transition in hypersonic wind tunnels may be a consequence of unstable second-mode disturbances, a view which has experimental support. J. L. Potter and J. D. Whitfield [172] were the first to observe well-defined rope-like periodic disturbances immediately preceding transition in a hypersonic boundary layer. Similar observations have since been made by M. C. Fisher and L. M. Weinstein [64, 66] and A. Demetriades [46]. Many of the features of these non-linear disturbances, in particular a wavelength of about $2\,\delta$ are remarkably close to the theoretical properties of unstable second-mode disturbances. Finally, J. M. Kendall [115] measured the spectrum of naturally occurring disturbances well before transition in the boundary layer of a cooled cone ($T_w/T_{ad} = 0\cdot6$) at $M_e = 7\cdot7$. He found a pronounced maximum at a frequency within 7 per cent of the theoretical frequency of the most amplified second-mode disturbance at the comparable flat-plate length Reynolds number ($= 1/3$ of the cone Reynolds number).

Numerous experiments have been performed on other aspects of transition in supersonic and hypersonic boundary layers to which the method of small disturbances has not yet been applied. An extensive review of this subject has been provided by M. V. Morkovin [160]. For coverage of specific transition effects, the reader is referred to the following groups of references: [13, 186, 226] on the effect of leading-edge or nose bluntness; [48, 159, 224, 226] on the effect on a body of revolution of an angle of incidence; [23, 45, 102, 182] on the effect of sweep angle; and [48, 65, 158, 253] on the effect of ablation.

f. Stability of a boundary layer in the presence of three-dimensional disturbances

1. Flow between concentric rotating cylinders. In all examples discussed so far the basic flow under consideration was two-dimensional and its stability was investigated on the assumption that the disturbance superimposed on it was also two-dimensional. Moreover the disturbance was assumed to be in the form of a plane wave which progressed in the direction of the main flow. As far as flows along a flat plate are concerned, this scheme leads to the lowest limit of stability because, as noticed by H. B. Squire (Sec. XVI b 3), three-dimensional disturbances will always lead to a higher limit of stability.

When flows along curved walls are considered, it is found that a different kind of instability must be taken into account. The case of flow between two rotating concentric cylinders of which the inner cylinder is in motion and the outer cylinder is at rest affords an example of an unstable stratification caused by centrifugal forces. The fluid particles near the inner wall experience a higher centrifugal force and show a tendency to being propelled outwards. The stability of this type of flow was first investigated by Lord Rayleigh [191] who assumed that the *fluid* was *non-viscous*. He found that the flow becomes unstable when the peripheral velocity, u, decreases with the radius, r, more strongly than $1/r$, that is, when

$$u(r) = \frac{\text{const}}{r^n} \text{ with } n > 1 \quad \text{(unstable)} . \tag{17.19}$$

The case of a *viscous fluid* was first investigated in detail by G. I. Taylor [149] who used the framework of a linear theory for this purpose. When a certain Reynolds number has been exceeded, there appear in the flow vortices, now known as Taylor vortices, whose axes are located along the circumference and which rotate in alternately opposite directions. Figure 17.32 contains a schematic representation of this motion which is characterized by the fact that the annulus between the two cylinders is completely filled by these ring-like vortices. The conditions for the flow to become unstable can be expressed with the aid of a characteristic number known as the *Taylor number*, T_a, of the form

$$\mathsf{T}_a = \frac{U_i d}{\nu} \sqrt{\frac{d}{R_i}} \geq 41\cdot3 \quad \text{(viscous instability)} , \tag{17.20}$$

where d denotes the width of the gap, R_i the inner radius, and U_i the peripheral velocity of the inner cylinder. G. I. Taylor's stability criterion is in excellent agree-

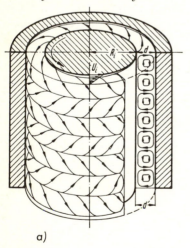

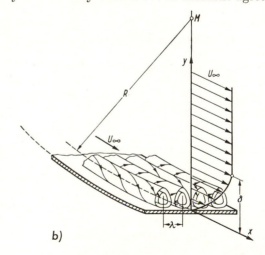

a)   b)

Fig. 17.32.

(a) Taylor vortices between two concentric cylinders. ınner cylinder rotating, outer cylinder at rest; d — width of annular gap; h — height of cylinder, eqn. (17.21).

(b) Goertler vortices in the boundary layer on a concave wall
$U(y)$ — base flow
δ — boundary layer thickness
λ — wavelength of disturbance

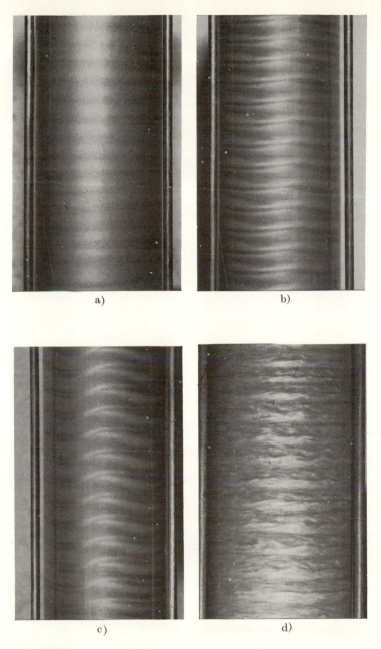

a)   b)

c)   d)

Fig. 17.33. Photographs of Taylor vortices from Fig. 17.32a for flow between concentric rotating cylinders, after F. Schultz-Grunow and H. Hein [204]. a) $R = 94.5$; $T_a = 41.3$: laminar, onset of vortex formation; b) $R = 322$; $T_a = 141$: still laminar; c) $R = 868$; $T_a = 387$: still laminar; d) $R = 3960$; $T_a = 1715$: turbulent

ment with measurements. This can be inferred very clearly from the pictures of such
Taylor vortices obtained by F. Schultz-Grunow and H. Hein [204], several of which
have been reproduced in Fig. 17.33. In their experimental arrangement in which the
gap had the dimension of $d = 4$ mm, and the inner radius was $R_i = 21$ mm, the
vortices appeared at a peripheral velocity, U_i, which corresponds to a Reynolds
number $R = U_i d/v = 94.5$, Fig. 17.33a. It is noteworthy that the flow remained
laminar at the much higher Reynolds numbers of $R = 322$ ($T_a = 141$) and $R = 868$
($T_a = 387$), Figs. 17.33b, c. Turbulent flow did not become developed until a Reynolds
number $R = 3960$ ($T_a = 1715$) had been reached, Fig. 17.33d. It should be stressed
emphatically that the first appearance of neutral vortices at the limit of stability
in accordance with eqn. (17.20) and the persistence of amplified vortices at higher
Taylor numbers does not in any way imply that the flow has become turbulent. On
the contrary, even if the limit of stability is exceeded by a large margin, the flow
remains well ordered and laminar. Turbulent flow does not become developed until
Taylor, and therefore, Reynolds numbers vastly exceeding the limit of stability
are attained.

J. T. Stuart [218] succeeded in computing the flow pattern of the unstable
laminar flow in the presence of Taylor vortices and with the non-linear terms in
the equation of motion retained. He discovered the existence of equilibrium between

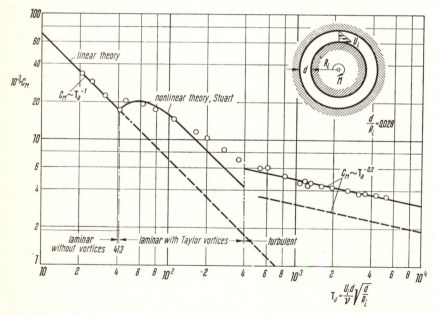

Fig. 17.34.   Flow between two concentric cylinders; torque coefficient for inner cylinder in terms
of the Taylor number, T_a.

Inner cylinder rotates, outer cylinder at rest.
Relative gap size $d/R_i = 0.028$
Measurements due to G. I. Taylor [241].
Linear theory from eqn. (17.22). Non-linear theory due to J. T. Stuart [218]

the transfer of energy from the base flow to the secondary flow and the viscous energy dissipation in the secondary flow. The transfer of energy from the base flow to the secondary flow causes a large increase in the torque required to rotate the inner cylinder. The diagram in Fig. 17.34 contains a comparison between the theoretically derived and the experimentally measured values of the torque coefficient, C_M. The latter is defined as

$$C_M = \frac{M_i}{\frac{1}{2}\pi\varrho\,U_i^2\,R_i^2\,h}\,, \tag{17.21}$$

with h as the height of the cylinder. The linear theory with small relative gaps, d/R_i, yields

$$C_M = A\left(\frac{U_i\,d}{\nu}\right)^{-1} = A\sqrt{\frac{d}{R_i}}\,\mathsf{T}_a^{-1} \quad \text{(linear theory)}\,. \tag{17.22}$$

In addition to the curve which corresponds to this linear theory, and which leads to a torque coefficient $C_M = 0\cdot67/\mathsf{T}_a$ for $d/R_i = 0\cdot028$, the diagram contains the curve provided by J. T. Stuart's non-linear theory as well as one given by a theory for turbulent flow; the latter leads to the formula that $C_M \sim \mathsf{T}_a^{-0.2}$. In all, we may discern three regimes of flow, each circumscribed by the Taylor number in the following way:

$$\mathsf{T}_a < 41.3: \text{ laminar Couette flow,}$$
$$41.3 < \mathsf{T}_a < 400: \text{ laminar flow with Taylor vortices,}$$
$$\mathsf{T}_a > 400: \text{ turbulent flow.}$$

Agreement between theory and experiment is excellent in the first two ranges[†]. An extension of Taylor's theory can be found in a study by K. Kirchgaessner [106]. A detailed experimental investigation of Couette flow, particularly in transition, was carried out in 1965 by D. Coles [29].

Effect of an axial velocity: The preceding stability calculations have been extended by H. Ludwieg [132, 133] to include the case when the two cylinders are also axially displaced with respect to each other. Let $u(r)$ denote the tangential velocity, and let $w(r)$ denote the axial velocity. If we now introduce the dimensionless velocity gradients

$$\tilde{u} = \frac{r}{u}\frac{du}{dr} \quad \text{and} \quad \tilde{w} = \frac{r}{u}\frac{dw}{dr}\,,$$

we can write the stability criterion for a non-viscous fluid in the form

$$(1-\tilde{u})\,(1-\tilde{u}^2) - (\tfrac{5}{3}-\tilde{u})\,\tilde{w}^2 > 0 \quad \text{(stable)}\,. \tag{17.23}$$

---

† The experimental results displayed in Fig. 17.34 demonstrate further that an increase in the Taylor number, that is, that an increase in the Reynolds number at a constant value of d/R_i, causes a transition from cellular to turbulent flow. When the flow is turbulent ($\mathsf{T}_a > 400$), we have $C_M \sim \mathsf{T}_a^{-0.2}$, and hence, at constant d/R_i also $C_M \sim (U_i\,d/\nu)^{-0.2} = \mathsf{R}^{-0.2}$. The same result was discovered by H. Reichardt ([26] in Chap. XIX) when he studied the case of linear Couette flow between flat parallel walls. It is remarkable that the same dependence of the torque coefficient on Reynolds number exists for a disk rotating in a fluid at rest, eqn. (21.36).

Fig. 17.35. Unstable flow with vortices in the shape of concentric, rotating cylinders with axial motion superimposed. Experiments due to H. Ludwieg [134]

$$\tilde{u} = 37; \qquad \tilde{w} = 1{\cdot}58$$

This inequality contains Rayleigh's criterion from eqn. (17.19) as a special case and results when $w = 0$ is assumed here; we then find that $1 + \tilde{u} > 0$. The stability calculation which led to eqn. (17.23) took into account disturbances which were not necessarily axially symmetric; the latter turned out to be the "most dangerous" ones and determined the limit of stability implied by the inequality (17.23). Figure 17.35 shows an example of an unstable flow which contains vortices in the shape of spirals. H. Ludwieg's theory has been compared with experimental results [134] in Fig. 17.36. Every base flow investigated experimentally is represented by a point in the $\tilde{u}$, $\tilde{w}$ plane. The open and full circles characterize stable and unstable flow, respectively, it being noted that vortices were observed for the latter. It is seen that H. Ludwieg's stability criterion from eqn. (17.23) is fully confirmed by experiments.

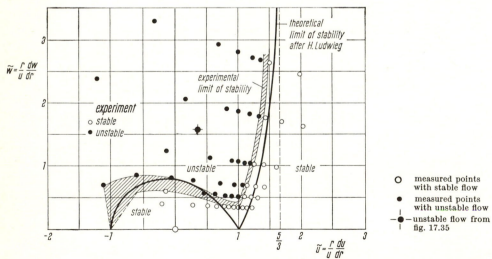

Fig. 17.36. Experimental verification of the stability theory for flow between two concentric, rotating cylinders with axial motion superimposed, after H. Ludwieg [134]

$\mathsf{R} = (R_0 - R_i)^2\, \omega_i / \nu = 650$

Full curve: limit of stability, according to eqn. (17.23)

Shaded area: experimentally determined limit of stability

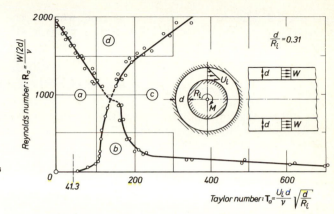

(a) laminar flow
(b) laminar flow with Taylor vortices
(c) turbulent flow with vortices
(d) turbulent flow

Fig. 17.37. Ranges of laminar and turbulent flow in annulus between two concentric cylinders; inner cylinder rotates, outer cylinder at rest in pressure of axial flow; plot in terms of Taylor number T_a and Reynolds number R_a; measurements by J. Kaye and E. C. Elgar [119] $W =$ axial velocity

The flow through the annulus between two concentric cylinders, with the inner cylinder rotating and the outer cylinder at rest, on which an axial velocity component is superimposed is of great practical importance. Such flow patterns occur in the hydrodynamic lubrication of journal bearing as well as in the air cooling of electric generators. A knowledge of the flow pattern is required for the calculation of the torque of the bearing as well as of the cooling effectiveness of the generator. The diagram in Fig. 17.37, drawn on the basis of measurements performed by J. Kaye and E.C. Elgar [119], allows us to determine the prevailing flow regime — laminar or turbulent — for an annulus with a given axial stream. This is determined by two characteristic numbers, the Taylor number T_a defined in eqn. (17.20) and by a Reynolds number formed with the axial velocity W and the width, d, of the annulus, viz.

$$R_a = \frac{W\,(2\,d)}{\nu}\,. \tag{17.23a}$$

Experiment points to the existence of four zones:

(a)   at low Reynolds number with $T_a < 41\cdot3$; laminar, streamline flow;

(b)   Taylor number in the range $41\cdot3 < T_a < 300$ with low to moderate Reynolds numbers R_a; laminar flow with a system of Taylor vortices;

(c)   large values of the Taylor number $T_a > 150$; "orderly" turbulent flow with vortices;

(d)   moderate Taylor number $T_a < 150$ and large Reynolds numbers $R_a > 100$; "disorderly", fully developed turbulent flow.

The boundaries between these four zones depend strongly on the parameter d/R_i, that is on the ratio between the width of the annulus and the inner cylinder radius.

A theoretical delineation of these boundaries is not now possible.

Sphere: O. Sawatzki and J. Zierep [230] performed similar investigations to those depicted in Fig. 17.34; they used two concentric spheres with gap ratios

$d/R_i < 0\cdot 2$. The outer sphere was at rest, whereas the inner sphere rotated. The character of the flow in such a spherical annulus is also determined by the Taylor number from eqn. (17.20) and the Reynolds number formed with the annulus width, d, and the peripheral velocity, U_i, that is by

$$\mathsf{R}_i = \frac{U_i\, d}{\nu}\ .$$

In the range of validity of linear theory, that is before the appearance of Taylor vortices of the kind shown in Fig. 17.32, the torque acting on the inner sphere is given by the torque coefficient

$$C_m = \frac{16\,\pi/3}{\mathsf{R}_i}\ ,$$

where

$$C_m = \frac{M_i}{\tfrac{1}{2}\,\varrho\, U_i^2\, R_i^3}\ ,$$

with M_i denoting the torque, and R_i the inner radius.

Whereas in the preceding case with rotating concentric cylinders the *entire* flow field is either laminar or turbulent, depending on the values of the Taylor and Reynolds numbers, the case of the sphere is more complex, because different flow regimes can occur simultaneously side by side. As the Reynolds number is increased, Taylor vortices, and hence also transition conditions, appear first near the plane of the equator whereas the flow near the poles remains laminar. See also references [161, 162, 163].

2. Boundary layers on concave walls. A similar kind of instability with respect to three-dimensional disturbances occurs in flows along concave walls. In a boundary layer formed on a convex wall the centrifugal forces exert a stabilizing effect whose magnitude however, is numerically small, as was already shown in Sec. XVIId. In contrast with that, the de-stabilizing effect of centrifugal forces on concave walls leads to a type of instability which resembles the pattern of Taylor vortices shown in Fig. 17.32a. The existence of the latter effect was first demonstrated by H. Goertler [74]. Considering a basic flow in the x-direction given by $U(x)$ (y-distance from the wall, x-measured at right angles to flow direction in the plane of the wall, Fig. 17.32b) it is assumed that there is superimposed on it a three-dimensional disturbance of the form

$$\left.\begin{aligned}
u' &= u_1\,(y)\,\{\cos\,(\alpha z)\}\ \mathrm{e}^{\beta t},\\[4pt]
v' &= v_1\,(y)\,\{\cos\,(\alpha z)\}\ \mathrm{e}^{\beta t},\\[4pt]
w' &= w_1\,(y)\,\{\sin\,(\alpha z)\}\ \mathrm{e}^{\beta t}.
\end{aligned}\right\} \qquad (17.24)$$

Here β is real and denotes the amplification factor, whereas $\lambda = 2\,\pi/\alpha$ represents the wavelength of the disturbance at right angles to the principal flow direction. The vortices have the shape shown in Fig. 17.32b, their axes being parallel to the basic flow direction. The present problem is concerned with standing waves (cellular vortices) which are known as Taylor-Goertler vortices. They are of the same kind as the Taylor vortices from Fig. 17.32a.

The calculation of the amplification of these three-dimensional vortices with time based on the method of small disturbances leads to an eigenvalue problem in a manner similar to that discussed in connexion with two-dimensional disturbances (Chap. XVI). The influence of viscosity was taken into account in the investigation under discussion. The first, approximate, solution of this very difficult eigenvalue problem was published in 1940 by H. Goertler [72]. Later, in 1973, F. Schultz-Grunow [204a] formulated a more accurate theory in that he took into account all terms of first order of smallness. The diagram in Fig. 17.38 contains his numerical results. It is seen that the minimum of the limit of stability occurs between $R_\delta = U_0 \delta/\nu = 4$ to 6 when the relative curvature in $\delta/R = 0.02$ to 0.10.

The phenomenon of transition in boundary layers on bodies placed in external streams involving both convex and concave walls was investigated experimentally by F. Clauser [17] and H. W. Liepmann [127, 128]. Some of the results provided by H. W. Liepmann are illustrated in Fig. 17.39. The plot in Fig. 17.39a confirms the theoretical prediction that the effect of curvature on the critical Reynolds number in the case of convex walls is very slight and that it is smaller for concave than for convex walls. Figure 17.39b shows a plot of the parameter

$$\frac{U_\infty \delta_{2tr}}{\nu} \sqrt{\frac{\delta_{2tr}}{R}}$$

against δ_{2tr}/R. This parameter describes the effect of curvature on transition in the boundary layer; it corresponds to the Taylor number of eqn. (17.20) for flows between rotating cylinders illustrated in Fig. 17.34. The graph of Fig. 17.39b shows that transition occurs for

$$\frac{U_\infty \delta_{2tr}}{\nu} \sqrt{\frac{\delta_{2tr}}{R}} > 7. \tag{17.25}$$

This numerical value is considerably larger than the corresponding limit of stability which places itself at 0.4, as seen from Fig. 17.38. In this connexion it is necessary to note that the boundary layer thickness δ must there be eliminated in favour of the momentum thickness $\delta_2 = 0.047 \delta\dagger$.

According to H. L. Dryden [38], the numerical value in (17.25) depends, in addition, on the intensity of turbulence; its value is contained between 6 and 9, where the lower limit corresponds to an external intensity of turbulence $T = 0.003$, the higher value corresponding to a considerably lower intensity.

---

† Here it is necessary to bear in mind that the location of the zone of transition must be expected to lie considerably further downstream than the limit of stability. This is due to the fact that unstable disturbances, whether coupled with stationary waves or with progressing waves of the Tollmien-Schlichting type (Sec. XVIb), must have time to amplify or, in other words, until the amplification factor has had a chance to attain its appropriate value. On the other hand, in the case of a Couette experiment with the inner cylinder rotating and the outer stationary, Fig. 17.33, we must expect that the experimentally observed, critical Reynolds number at which Taylor vortices first appear ought to have a value which is very close to that given by theory. This is due to the fact that at a constant rate of rotation the process of amplification of disturbances occurs at a constant Reynolds number. Thus, the amplification factor attains its appropriate value provided that the experiment lasts long enough; see here Fig. 17.33.

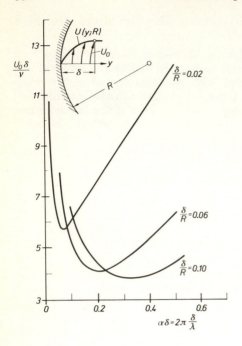

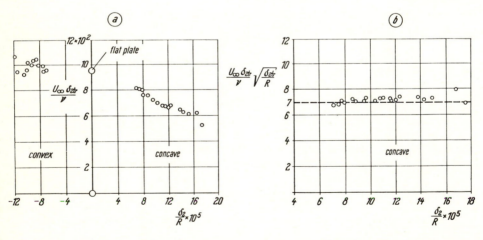

Fig. 17.38. Limit of stability ($\beta = 0$) for the boundary layer on a concave, curved wall in terms of the ratio δ/R of boundary-layer thickness to radius of curvature, after F. Schultz-Grunow [204a]

boundary-layer thickness $\delta = (2\,\nu\,x/U_0)^{1/2}$
radius of wall curvature R

Fig. 17.39. Measurements of the point of transition on slightly concave walls, after H. W. Liepmann [127, 128]; (a) critical Reynolds number $\dfrac{U_\infty \delta_{2tr}}{\nu}$ versus $\dfrac{\delta_2}{R}$; (b) the characteristic quantity $\dfrac{U_\infty \delta_{2tr}}{\nu} \sqrt{\dfrac{\delta_{2tr}}{R}}$ versus $\dfrac{\delta_2}{R}$

δ_2 = momentum thickness; R = radius of curvature of wall

A very thorough experimental investigation of transition along a concave, curved wall was recently carried out by H. Bippes [16] who employed models dragged along a water channel. These experiments throw light on the origin of longitudinal vortices like those in Fig. 17.32 b. In this connexion see the papers by F. X. Wortmann [256] and H. Goertler and H. Hassler [83].

Recently H. Goertler drew attention to the fact that the same type of instability can occur near the forward stagnation point of a bluff body in a stream. The necessary condition that the streamlines must be concave in the direction of increasing velocity is here present. So far, the calculations performed by H. Goertler [47] and G. Haemmerlin [86] for the case of two-dimensional stagnation flow represented in Fig. 5.9 have shown the existence of unstable disturbances, but no limit of stability in the form of a Reynolds number can be given. Experiments performed by N. A. V. Piercy and E. G. Richardson [170, 171] suggest that the flow in the neighbourhood of the forward stagnation point on a circular cylinder does indeed become unstable. Reviews of three-dimensional effects in the theory of stability were given by H. Goertler [74]. See also the more recent papers by J. Kestin et al. [118] and W. Sadeh et al. [232, 233].

The considerations contained in the present section together with those in Chap. XVI and Secs. XVII a, b lead to the following picture of transition in the boundary layer of a solid body (e. g. an aerofoil); transition on flat and convex walls is governed by the instability of travelling, two-dimensional Tollmien-Schlichting waves whereas that on concave walls is governed by the stationary Taylor-Goertler vortices.

3. Stability of three-dimensional boundary layers. The details of the process of transition in a three-dimensional boundary layer appear to be entirely different from those associated with the two-dimensional flows considered earlier. One example of this type of transition is afforded by the case of a disk rotating in a fluid at rest

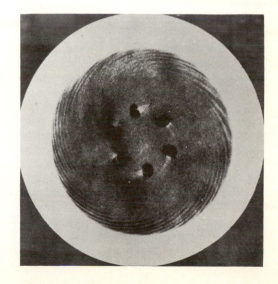

Fig. 17.40.   Photograph illustrating transition in the boundary layer on a disk rotating in a fluid at rest after N. Gregory, J. T. Stuart and W. S. Walker [77]. Direction of rotation is counter-clockwise; speed $n = 3200$ rpm radius of disk $= 15$ cm

Stationary vortices are seen forming in an annular region of inner radius $R_i = 8 \cdot 7$ cm and outer radius $R_0 = 10 \cdot 1$. The inner radius constitutes the limit of stability, with
$$\mathbf{R}_i = R_i\, \omega^2/\nu = 1 \cdot 9 \times 10^5.$$
Transition occurs at the outer radius where
$$\mathbf{R}_0 = R_0\, \omega^2/\nu = 2 \cdot 8 \times 10^5$$

for which the details of the laminar layer are known from Sec. V b. A photograph illustrating the process of transition on a rotating disk and taken by N. Gregory, J. T. Stuart and W. S. Walker [77] is reproduced in Fig. 17.40. The photograph shows that in an annular region there appear stationary vortices which assume the shape of logarithmic spirals. The inner radius of this region marks the onset of instability and transition occurs at the outer radius. The inner radius corresponds to a Reynolds number of $R_i = R_i{}^2 \, \omega/\nu = 1\cdot9 \times 10^5$ and at the outer radius we have $R_o = R_o{}^2 \, \omega/\nu = = 2\cdot8 \times 10^5$. J. T. Stuart complemented the experimental work with an analytic study of the stability of such a motion. In it, he assumed the existence of three-dimensional, periodic disturbances whose forms included as special cases the progressing Tollmien-Schlichting waves as well as the stationary, three-dimensional Taylor-Goertler vortices. The results of his calculations showed qualitative agreement with the experimental results of Fig. 17.40.

Another case of this kind occurs on a yawed flat plate in supersonic flow when the associated laminar boundary layer becomes unstable. As shown experimentally by J. J. Ginoux [84], the boundary layer develops longitudinal vortices which produce transition.

g. The influence of roughness on transition

1. Introductory remark. The problem which we are about to examine in this section, namely the question of how the process of transition depends on the roughness of the solid walls, is one of considerable practical importance; so far, however, it has not been possible to analyze it theoretically. The problem under consideration has gained in importance in the recent past, particularly since the advent of laminar aerofoils in aeronautical applications. The very extensive experimental material collected up to date includes information on the effect of cylindrical (two-dimensional roughness elements), point-like (three-dimensional, single roughness elements) and distributed roughness elements. Many of the investigations include additional data on the influence of pressure gradients, turbulence intensity or Mach number.

Generally speaking, the presence of roughness favours transition in the sense that under otherwise identical conditions transition occurs at a lower Reynolds number on a rough wall than on a smooth wall. That this should be so follows clearly from the theory of stability: the existence of roughness elements gives rise to additional disturbances in the laminar stream which have to be added to those generated by turbulence and already present in the boundary layer. If the disturbances created by roughness are bigger than those due to turbulence, we must expect that a lower degree of amplification will be sufficient to effect transition. On the other hand, if the roughness elements are very small, the resulting disturbances should lie below the 'threshold' which is characteristic of those generated by the turbulence of the free stream. In this case, the presence of roughness would be expected to have no effect on transition. The preceding considerations show complete agreement with experiment. When the roughness elements are very large, transition will occur at the points where they are present themselves, as is, for example, the case with the tripping wire on the sphere shown in Fig. 2.25. In this connexion the reader may wish to consult the paper by J. Stueper [220].

The earlier papers which addressed themselves to this problem, namely those by L. Schiller [202], I. Tani, R. Hama and S. Mituisi [235], S. Goldstein [67], and A. Fage and J. H. Preston [62], assumed that the point of transition was located at the position of roughness elements, when they were large, or that their presence had no influence at all when they were small. However, A. Fage has shown that the point of transition moves continuously upstream as the height of the roughness elements is increased until it ultimately reaches the position of the roughness elements themselves. Consequently, in discussing the influence of roughness on transition, it is necessary to provide answers to the following three questions:

1. What is the maximum height of roughness elements below which no influence on transition exists? (Critical height of roughness elements in laminar flow).

2. What is the (larger) limiting height of a roughness element which causes transition to occur at the element itself?

3. How is it possible to describe the position of the point of transition in the range intermediate between these two limits?

2. Single, cylindrical roughness elements. A single, cylindrical (or two-dimensional) roughness element usually takes the form of, say, a wire which is attached to the wall at right angles to the stream direction. For this type of roughness element, S. Goldstein deduced from older measurements, that the *critical height*, k_{crit}, i. e. the height which just does not affect transition, can be represented by

$$\frac{u_k^* \, k_{crit}}{\nu} = 7 \, . \tag{17.26}$$

Here $u_k^* = \sqrt{\tau_{ok}/\varrho}$ denotes the friction velocity and τ_{ok} is the shearing stress at the wall in the laminar boundary layer at the position of the roughness element. According to I. Tani and his co-workers [235] the minimum height for which transition occurs at the element itself can be found from the relation $u_k^* \, k_{crit}/\nu = 15$, whereas A. Fage and J. H. Preston [62] quote

$$\frac{u_k^* \, k_{crit}}{\nu} = 20 \, . \tag{17.27}$$

The preceding characteristic values apply to circular wires. In the case of flat and cupped cross-sections or for grooves the values are considerably larger, whereas for sharp elements they become smaller.

H. L. Dryden [39] provided an argument of a dimensional nature which leads to an empirical law for the determination of the position of the point of transition, x_{tr}, in terms of the height k of the roughness element and its position, x_k. Dryden discovered that in incompressible flow all experimental points for the case when transition does not occur at the roughness element itself, i. e. when $x_{tr} > x_k$, arrange themselves on a *single* curve in a plot of the Reynolds number $R_{tr} = U \, \delta_{1tr}/\nu$ formed with the displacement thickness δ_{1tr} of the boundary layer at the point of transition against the ratio k/δ_{1k}, where δ_{1k} denotes the displacement thickness at the position of the roughness element, Fig. 17.41. The diagram in Fig. 17.41 contains an auxiliary scale of $R_{xtr} = U \, x_{tr}/\nu$†.

---

† The two Reynolds numbers on the axis of ordinates are related through the equation

$$R_{tr} = \frac{U\delta_{1tr}}{\nu} = 1 \cdot 72 \, \sqrt{\frac{U x_{tr}}{\nu}} = 1 \cdot 72 \, \sqrt{R_{xtr}} \, .$$

As the height k is increased, the position of the point of transition x_{tr} moves closer to the roughness element which means that the curves in Fig. 17.41 are traversed from left to right. The experimental points begin to deviate from this curve upwards as soon as the point of transition has reached the roughness element, i. e. when $x_{tr} = x_k$. They then lie along the family of straight lines which contain x_k/k as a parameter and is given by

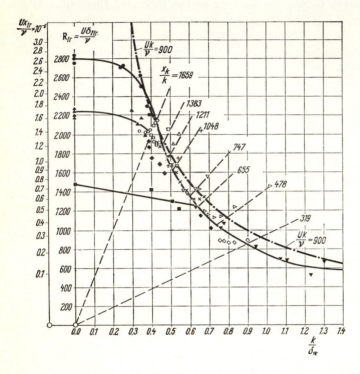

Fig. 17.41.   The critical Reynolds number for laminar boundary layer as a function of the ratio of height k of roughness element to the displacement thickness of the boundary layer at the position of the roughness element, δ_{1k}, for single, two-dimensional roughness elements in incompressible flow

The measurements are satisfactorily interpolated by eqn. (17.28)

$$R_{tr0} = \frac{U\delta_{tr0}}{\nu} \text{ and } R_{xtr0} = \frac{Ux_{tr}}{\nu}$$ denote the critical Reynolds numbers for a smooth flat plate.

- - - calculated from eqn. (17.28)

○ □ ◺ △ ▽ ▷ × + $x_{tr} > x_k$; $R_{tr0} = 1.7 \times 10^6$;
  $p = $ const, after [63];

▲ $R_{tr0} = 1.7 \times 10^6$; $p = $ const;
  after I. Tani et al. [237];

● $R_{xtr0} = 2.7 \times 10^6$; $p = $ const;
  after I. Tani et al. [237]

◀ $R_{xtr0} = 2.7 \times 10^6$;

  Pressure drop
  $$\frac{p_1 - p_{tr}}{\frac{1}{2}\varrho U_1^2} = 0.2 \text{ to } 0.8, \text{ according to } [237]$$

▼ $p = $ const; after G. B. Schubauer and
  H. K. Skramstad [203]

■ $R_{xtr0} = 6 \times 10^5$; $p = $ const;
  after I. Tani et al. [237]

Full points refer to experimental results at $x_{tr} > x_k$

$$\frac{U\delta_{1tr}}{\nu} = 3 \cdot 0 \; \frac{k}{\delta_{1k}} \; \frac{x_k}{k} \; ; \qquad (17.28)$$

it is also shown in Fig. 17.41. According to Japanese measurements [237], the hyperbola-like branch of the curves in Fig. 17.41 possesses universal validity, both for flows with different, weak pressure gradients, and with different intensities of turbulence. Increased turbulence causes merely an earlier deviation of the curve to the left, in the direction of the turbulence-dependent critical Reynolds number of a flat plate, $(R_{x tr})_{k=0} = R_{x tr0}$. An analysis coupled with additional measurements have led K. Kraemer [109] to conclude that a wire at an arbitrary position is fully effective if

$$\frac{U k}{\nu} \geq 900 \;. \qquad (17.29)$$

The graph of this equation has been included in Fig. 17.41; the equation agrees well with experiments. It should be noted, however, that even in the case of a "fully effective" tripping wire, there remains a certain minimum distance between the position, x_{tr}, of the transition point and the position, x_k, of the wire itself. According to K. Kraemer, this distance is given by

$$\frac{U (x_{tr} - x_k)}{\nu} = 2 \times 10^4 \;. \qquad (17.30)$$

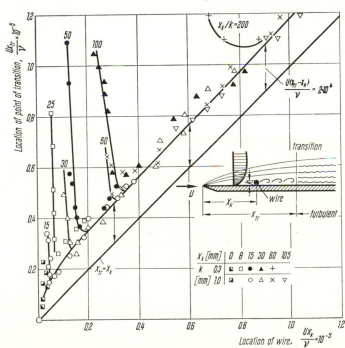

Fig. 17.42. Distance between point of transition, x_{tr}, and the position of the tripping wire, x_k, for "fully effective" operation, plotted from eqn. (17.30), after K. Kraemer [109]

The corresponding curve is represented in Fig. 17.42.

According to H. L. Dryden [39, 40], it is possible to take into account the variation in the turbulence intensity by plotting the ratio of the critical Reynolds number for a rough wall to that for a smooth wall, namely $(R_{x\,tr})_{rough}/(R_{x\,tr})_{smooth}$, as a function of k/δ_{1k}, Fig. 17.43. When plotted in this system of coordinates, the results of measurements with different intensities of turbulence fall on a single curve which means that the ratio $(R_{x\,tr})_{rough}/(R_{x\,tr})_{smooth}$ is a function of the single parameter k/δ_{1k}. The three questions posed at the end of the last section can now be easily answered with the aid of the three graphs of Figs. 17.41, 17.42 and 17.43.

Very detailed experiments concerning the influence of a two-dimensional, discrete roughness element (wire) on transition were performed recently by P. S. Klebanoff and K. D. Tidstrom [107a]; these were conceived as a continuation of earlier work [107b]. In particular, measurements were made in the disturbed boundary layer close behind the roughness element. These measurements lead to the conclusion that the behaviour of the disturbed boundary layer can best be understood if the effect of roughness is conceived as that of a strong, wavelike disturbance which vigorously destabilizes the boundary layer and thus has the same effect as an increase in the turbulence intensity of the free stream.

The influence of roughness on transition is considerably smaller in compressible than in incompressible flow. This fact can be deduced from Fig. 17.44 which refers to a flat plate at zero incidence and which, as far as the results for compressible flow are concerned, is based on the measurements performed by P.F. Brinich [12]. The measurements were performed with the aid of cylindrical roughness elements of circular cross-section at a Mach number $M = 3\cdot1$; when plotted in the coordinates of Fig. 17.44, they arrange themselves in a family of curves which cover the shaded area in the diagram, but which still strongly depend on the position x_k of the roughness element. The curve for incompressible flow, shown in Fig. 17.44 for the purpose of comparison, illustrates the fact that at high Mach numbers the boundary layer can 'tolerate' a considerably larger roughness element than in incompressible flow. According to the graph, the critical height of a roughness element is some 3 to 7 times larger than in incompressible flow. Experiments performed by R. H. Korkegi [108] at the even higher Mach number of $M = 5\cdot8$ showed that at such large Mach numbers a tripping wire produces no turbulence at all. On the other hand, the blowing of air into the boundary layer seems to be effective in promoting transition even in compressible flow.

In recent times, E. R. van Driest and C. B. Blumer [37] undertook a series of measurements on a cone with its axis parallel to the stream at a Mach number $M_\infty = 2\cdot7$; this was a continuation of earlier work [36]. In addition to varying the diameter of the circular tripping wire, the experimenters also varied its position on the cone as well as the rate of heat transfer to it.

3. Distributed roughness. There exist only scant results concerning measurements on transition in the presence of distributed roughness [19]. A paper by E. G. Feindt [63] contains a brief description of an investigation into the influence of a pressure gradient and grain size, k_S, in the presence of sand roughness. The measurements were performed in a convergent and a divergent channel of circular cross-section

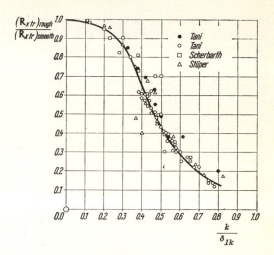

Fig. 17.43. Ratio of the critical Reynolds number on a flat plate at zero incidence with a single roughness element to that of a smooth plate, after Dryden [39]
$$R_{x\,tr} = U\,x_{tr}/\nu$$

k—height of roughness element δ_{1k} — displacement thickness of the boundary layer at the roughness element. Measurement due to Tani [235]

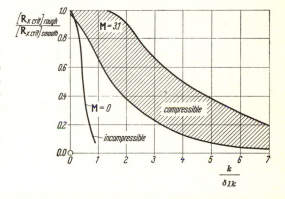

Fig. 17.44. Influence of single, two-dimensional roughness elements on the critical Reynolds number on a flat plate in compressible flow, as measured by P. F. Brinich [8]

k — height of single roughness element;
δ_{1k} — displacement thickness of boundary layer at roughness element

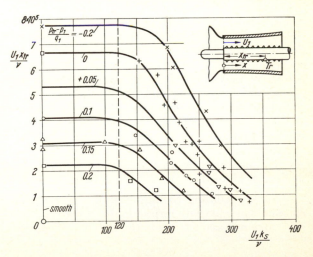

Fig. 17.45. Influence of pressure gradient and surface roughness on sand-covered wall on position of point of transition for incompressible flow, as measured by E. G. Feindt [63]

k_s — sand-grain size. The sand roughness has no influence on transition when $U_1 k_s/\nu < 120$

with a cylinder covered with sand placed axially in them. The walls of the channels were smooth and their slope controlled the pressure gradient. The graph in Fig. 17.45 represents the relation between the critical Reynolds number $U_1 x_{tr}/\nu$ formed with the coordinate of the point of transition and the Reynolds number $U_1 k_S/\nu$ formed with the sand grain size, k_S, for different pressure gradients measured by E. G. Feindt. The values for a smooth wall ranged from $U_1 x_{tr}/\nu = 2 \times 10^5$ to 8×10^5 at different pressure gradients, corresponding to the strongly stabilizing or de-stabilizing influence of the pressure gradient. It is seen that the measurements under consideration lead to the conclusion that when $U_1 k_S/\nu$ increases, at first there is no change in the critical Reynolds number. The critical Reynolds number decreases steeply, but only after the value

$$\frac{U_1 \, k_S}{\nu} = 120 \tag{17.31}$$

has been exceeded. Hence, this value determines the critical roughness and answers Question 1 posed earlier. Roughness is seen to exert an influence comparable to that of the pressure gradient at values exceeding this limit.

h. Axially symmetrical flows

The most important case of an axially symmetrical flow is that existing in a straight pipe, i. e. when the velocity profile is parabolic. This case was investigated very early by Th. Sexl [205] who was unable to discover any instability; he was equally unable, however, to prove the existence of stability for all Reynolds numbers. Some time later, J. Pretsch [177] succeeded in proving that the analysis of the stability of these parabolic velocity profiles can be reduced to that of plane Couette flow (i. e. pure shear flow). Since the latter is stable at all Reynolds numbers, the same is seen to be true about the parabolic velocity profiles in a pipe. The same conclusion was reached by G. M. Corcos and J. R. Sellars [18], by C. L. Pekeris [169] as well as by several contemporary investigators [54, 66a, 148, 175]; it was finally confirmed by Th. Sexl and K. Spielberg [206]. This fact is surprising for two reasons. First, because flows in pipes do undergo transition. In fact, as the reader will recall, the earliest experiments on transition have been performed by O. Reynolds on pipes. Secondly, it is difficult to visualize the fact that parabolic velocity profiles in channels can (Sec. XVIc) — but parabolic velocity profiles in pipes cannot — be made unstable by very small disturbances. For these reasons attempts are being made to investigate this matter still further, both analytically and experimentally. In this connexion it may be noted that R. J. Leite [125] failed to observe any amplification of small, axi-symmetrical disturbances travelling downstream in a circular pipe at Reynolds numbers as high as $R = 13000$. Th. Sexl and K. Spielberg [206] established that in relation to axially symmetrical flows Squire's theorem, mentioned in Sec. XVIb 3, no longer holds, and that symmetrical plane waves are, therefore, no longer more critical for the flow, than disturbances of a three-dimensional nature.

Since according to the linear theory fully developed Hagen-Poiseuille flow is completely stable with respect to rotationally symmetric as well as skew disturbances, it became necessary to study the stability of the flow in the *inlet length* (see Chap. XI) of the pipe with respect to such disturbances. This was done by T. Sarpkaya who performed measurements himself [197b] and who wrote a summary paper on the subject [197a] as well as on the theoretical results of various other authors. Theoretical calculations always lead to the discovery of a limit of stability which exceeds the experimentally measured critical Reynolds number by a wide margin.

The stability of Hagen-Poiseuille flow in a pipe with parabolic velocity distribution with respect to infinitesimal *three-dimensional* disturbances was first studied by M. Lessen and P. J. Singh [126a], who considered disturbances of azimuthal periodicity unity. No instability was found for Reynolds numbers ranging up to 16000. The results were confirmed for modes of unity and higher order azimuthal periodicity up to Reynolds numbers of 50,000 by H. Salwen and C. E. Grosch [197c] using a different method of solution of the eigenvalue problem.

The stability of the *swirling wake* model for a trailing vortex, where the axial flow has the wake velocity distribution and the swirling flow has the decaying vortex distribution, was studied by M. Lessen, P. J. Singh and F. L. Paillet [126b, 126c] for azimuthally periodic disturbances of periodicity unity and higher. It was found that a small amount of swirl could reduce the minimum critical Reynolds number by as much as 50 per cent. At larger amounts of swirl, the higher azimuthal modes were more destabilized than the lower ones and at still larger amounts of swirl, the flow was stabilized.

Older experiments (Chap. XX in [53] and [54]) as well as more recent results of the application of stability theory for axial flows with a swirl component [132, 133] suggest that even a very small tangential velocity component considerably reduces the stability of Hagen-Poiseuille flow. For this reason, P. A. Mackrodt [164, 165] studied the stability of laminar pipe flow with superimposed *rigid* rotation by way of modelling the flow of real interest. In this case, the disturbance moves along a spiral down the axis of the pipe. The result of this calculation is depicted in Fig. 17.46. The limit of stability (neutral disturbances) is shown here in the plane R_ϕ, R_z, where $R_\phi = \omega R^2/\nu$ is the Reynolds number formed with the peripheral velocity ωR, whereas $R_x = U_0 R/\nu$ refers to the maximum velocity U_0.

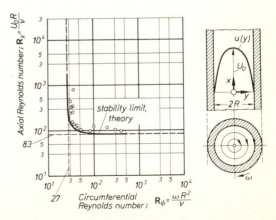

Fig. 17.46. Stability limit of Hagen-Poiseuille flow in a circular pipe with solid-body rotation, after P. A. Mackrodt [164, 165]; $R_{xcrit} = 83$ after T.J. Pedley [188]; ○ observed unstable flow

The measured points represent observed neutral disturbance vortices at the boundary between damping and amplification. The agreement between theory and experiment is very good. The theory confirms the supposition that small velocity components in the tangential directions cause Hagen-Poiseuille flow to become unstable.

J. Rotta, whose work was discussed in detail in Sec. XVIa, performed measurements on the intermittency factor of large disturbances propagated downstream in the inlet section of a pipe. Similar experiments were performed by E. R. Lindgren [130] who made the disturbance visible by the use of polarized light and a bi-refringent, weak solution of bentonite. E. R. Lindgren was able to show that even strong initial disturbances decay in the inlet length when the Reynolds number of the flow (based on the pipe diameter) is small. At Reynolds numbers from about $R = 2600$ upwards the process of transition begins. It is characterized by an amplification of the initial disturbances and by the appearance of self-sustaining turbulent flashes which emanate from fluid layers near the wall along the tube.

The preceding peculiarities of laminar flows through pipes force us to re-consider the relation between the theory of small disturbances and transition and, in particular, to pose the question as to whether transition can *always* be said to be due to an amplification of *small* disturbances. No conclusive answer to this question can at present be given without further work on the behaviour of small, three-dimensional disturbances. In this connexion it should also be remembered that the limit of stability for *plane* Poiseuille flow which lies at $R_{crit} = 5314$ as stated on p. 480, considerably exceeds the critical Reynolds number for transition observed in channels. This is inconsistent with the theory which asserts that the limit of stability must always occur at a lower Reynolds number than transition itself. However, at the present stage of knowledge, and in the face of the present interest in the subject, judgement must be reserved until further results become available.

The stability of a laminar boundary layer on a body of revolution was also investigated by J. Pretsch [176]; in this connexion consult a paper by P. S. Granville [82]. In cases when the ratio of boundary-layer thickness to curvature is very small compared with unity, the resulting stability equation for the axially symmetrical case becomes identical with that for the plane case. Hence, all results obtained for the latter can be extended to apply to the former without reservation.

References

[1] Abbott, J.H., von Doenhoff, A.E., and Stivers, L.S.: Summary of airfoil data. NACA Rep. 824 (1954).

[2] Althaus, D.: Stuttgarter Profilkatalog. Inst. Aerodynamik of Stuttgart Univ. (1972).

[3] ARC RM 2499; Transition and drag measurements on the Boulton Paul sample of laminar flow wing construction. Part I: by J.H. Preston and N. Gregory; Part II: by K.W. Kimber; Part III: Joint Discussion.

[3a] Beasley, J. A.: Calculation of the laminar boundary layer and prediction of transition on a sheared wing. ARC RM 3787 (1976); RAE TR-73156 (1974).

[4] Benjamin, T.B.: Effects of a flexible boundary on hydrodynamic stability. JFM 9, 513—532 (1961).

[5] Bénard, H.: Les tourbillons cellulaires dans une mappe liquide. Rev. Gen. Sci. Pure Appl. 11, 1261—1271 and 1309—1328 (1900).

[6] Bloom, M.: The effect of surface cooling on laminar boundary layer stability. JAS 18, 635—636 (1951).

[7] Braslow, A.L., and Visconti, F.: Investigation of boundary layer Reynolds number for transition on an NACA 65(215)—114 airfoil in the Langley two-dimensional low-turbulence pressure tunnel. NACA TN 1704 (1948).

[8] Brinich, P.F.: Boundary layer transition at Mach 3·12 with and without single roughness element. NACA TN 3267 (1954).

[9] Bussmann, K., and Münz, H.: Die Stabilität der laminaren Reibungsschicht mit Absaugung. Jb. dt. Luftfahrtforschung I, 36—39 (1942).

[10] Bussmann, K., and Ulrich, K.: Systematische Untersuchungen über den Einfluss der Profilform auf die Lage des Umschlagpunktes. Preprint Jb. dt. Luftfahrtforschung 1943 in Techn. Berichte 10, 9 (1943).

[11] Bloom, M.: The effect of surface cooling on laminar boundary layer stability. JAS 18, 635—636 (1951).

[12] Brinich, P.F.: Boundary layer transition at Mach 3·12 with and without single roughness elements. NACA TN 3267 (1954).

[13] Brinich, P.F., and Sands, N.: Effect of bluntness on transition for a cone and a hollow cylinder at Mach 3·1. NACA TN 3979 (1957).

[14] Brown, W.B.: Exact solution of the stability equations for laminar boundary layers in compressible flow. Boundary Layer and Flow Control (G.V. Lachmann, ed.), Vol. 2, 1033—1048, Pergamon Press, New York, 1961.

[15] Brooke-Benjamin, T.: Wave formation in laminar flow down on inclined plane. JFM 2, 554—574 (1957).

[16] Bippes, H.: Experimentelle Untersuchung des laminar-turbulenten Umschlages an einer parallel angeströmten konkaven Wand. Diss. T.U. Berlin, 1972; Sitzungsber. Heidelberger Akademie der Wiss. Math. Naturw. Klasse, 1972, Springer, Berlin, pp. 103—180; see also: Bippes, H., and Görtler, H.: Acta Mechanica 14, 251—267 (1972).

[17] Clauser, L.M., and Clauser, F.: The effect of curvature on the transition from laminar to turbulent boundary layer. NACA TN 613 (1937).

[18] Corcos, G.M., and Sellars, J.R.: On the stability of fully developed flow in a pipe. JFM 5, 97—112 (1959).

[19] Czarnecki, K.R., Robinson, R.B., and Hilton, Jr., J.H.: Investigation of distributed surface roughness on a body of revolution at a Mach numer of 1·61. NACA TN 3230 (1954).

[20] Czarnecki, K.R., and Sinclair, A.R.: An investigation of the effects of heat transfer on boundary-layer transition on a parabolic body of revolution (NACA RM 10) at a Mach number of 1·61. NACA Rep. 1240 (1955).

[21] Cary, A.M.Jr.: Turbulent boundary layer heat transfer and transition measurements for cold-wall conditions at Mach 6. AIAA J. 6, 958—959 (1968).

[22] Cebeci, T., and Smith, A.M.O.: Investigation of heat transfer and of suction for tripping laminar boundary layers. J. Aircraft 5, 450 (1968).

[22a] Cebeci, T., and Keller, H.B.: Stability calculations for a rotating disk. AGARD-CP-224, 7—1 to 7—9 (1977).

[23] Chapman, G.T.: Some effects of leading-edge sweep in boundary layer transition at supersonic speeds. NASA TN D-1075 (1961).

[24] Coles, D.: Measurements of turbulent friction on a smooth flat plate in supersonic flow. JAS 21, 433—448 (1954).

[25] Czarnecki, K.R., and Sinclair, A.R.: An investigation of the effects of heat transfer on a parabolic body of revolution (NACA RM-10) at a Mach number of 1·61. NACA Rep. 1240 (1955).

[26] Colak-Antic, P.: Hitzdraht-Messungen des laminar turbulenten Umschlages bei freier Konvektion. Jb. WGL 1964, 171—176 (1965).

[27] Colak-Antic, P.: Dreidimensionale Stabilitätserscheinungen des laminar turbulenten Umschlages bei freier Konvektion längs einer vertikalen geheizten Platte. Sitzungsberichte Heidelberger Akademie der Wiss. Jb. 1962/64, 315—416, Heidelberg (1964).

[28] Crowder, H.J., and Dalton, C.: Stability of Poiseuille flow in a pipe. J. Comput. Phys. 7, 12—31 (1971).

[29] Coles, D.: Transition in circular Couette flow. JFM 21, 385—425 (1965).

[29a] DiPrima, R.C., and Stuart, J.T.: Nonlocal effects in the stability of flow between eccentric rotating cylinders. JFM 54, 393—415 (1972).

[29b] DiPrima, R.C., and Stuart, J.T.: The nonlinear calculation of Taylor-vortex flow between eccentric rotating cylinders. JFM 67, 85—111 (1975).

[30] von Doenhoff, A. E.: Investigation of the boundary layer about a symmetrical airfoil in a wind tunnel of low turbulence. NACA Wartime Rep. L 507 (1940).

[31] Doetsch, H.: Untersuchungen an einigen Profilen mit geringem Widerstand im Bereich kleiner c_a-Werte. Jb. dt. Luftfahrtforschung I, 54—57 (1940).

[32] van Driest, E. R.: Cooling required to stabilize the laminar boundary layer on a flat plate. JAS 18, 698—699 (1951).

[33] van Driest, E. R.: Calculation of the stability of the laminar boundary layer in a compressible fluid on a flat plate with heat transfer. JAS 19, 801—812 (1952).

[34] van Driest, E. R., and Boison, J. C.: Experiments on boundary layer transition at supersonic speeds. JAS 24, 885—899 (1957).

[35] van Driest, E. R., and McCauley, W. D.: Boundary layer transition on a 10 degree cone at Mach number 2·81 as affected by extreme cooling. JAS 24, 780—781 (1957).

[36] van Driest, E. R., and Blumer, C. B.: Boundary layer transition at supersonic speeds. Three-dimensional roughness effects (spheres). JASS 29, 909—916 (1962).

[37] van Driest, E. R., and Blumer, C. B.: Boundary layer transition: Free-stream turbulence and pressure gradient effects. AIAA J. 1, 1303—1306 (1963).

[38] Dryden, H. L.: Recent advances on the mechanics of boundary layer flow (R. v. Mises, and Th. v. Kármán, ed.). Advances in Appl. Mech 1, 1—40, New York (1948).

[39] Dryden, H. L.: Review of published data on the effect of roughness on transition from laminar to turbulent flow. JAS 20, 477—482 (1953).

[40] Dryden, H. L.: Effects of roughness and suction on transition from laminar to turbulent flow. Publ. Scient. et Techn. de Ministère de l'Air Paris (SDJT) 49—60 (1954).

[41] Dryden, H. L.: Transition from laminar to turbulent flow at subsonic and supersonic speeds Proc. Conference on High-Speed Aeronautics, New York, 1955, 41—74.

[42] Dryden, H. L.: Recent investigations on the problem of transition. ZFW 4, 89—95 (1956).

[43] Dunn, W. D., and Lin, C. C.: On the stability of the boundary layer in a compressible fluid. JAS 22, 455—477 (1955); see also JAS 20, 577 (1953) and 19, 491 (1952).

[44] Dunning, R. W., and Ulmann, E. F.: Effects of sweep and angle of attack on boundary layer transition on wings at Mach number 4·04. NACA TN 3473 (1955).

[45] Deem, R. E., and Murphy, J. S.: Flat plate boundary layer transition at hypersonic speeds. AIAA Paper 65—128 (1965).

[46] Demetriades, A.: Hypersonic viscous flow over a slender cone. Part III. Laminar instability and transition. AIAA Paper 74—535 (1974).

[47] Diaconis, N. S., Jack, J. R., and Wisniewski, R. J.: Boundary layer transition at Mach 3·12 by cooling and nose blunting. NACA TN 3928 (1957).

[48] Dicristina, V.: Three-dimensional boundary-layer transition on a sharp 8° cone at Mach 10. AIAA J. 8, 852—856 (1970).

[49] Dougherty, Jr., N. S., and Steinle, Jr., F. W.: Transition Reynolds number comparisons in several major transonic tunnels. AIAA Paper 74—627 (1974).

[50] van Driest, E. R., and Blumer, C. B.: Boundary-layer transition at supersonic speeds: Roughness effects with heat transfer. AIAA J. 4, 603—607 (1968).

[51] van Driest, E. R., and McCauley, W. D.: The effect of controlled three-dimensional roughness on boundary layer transition at supersonic speeds. JASS 27, 261—271 (1960).

[53] Dunn, D. W., and Lin, C. C.: On the stability of the laminar boundary layer in a compressible fluid. JAS 22, 455—477 (1955).

[54] Davey, A., and Drazin, P. G.: The stability of Poiseuille flow in a pipe. JFM 36, 209—218 (1969).

[56] Ertel, H.: Thermodynamische Begründung des Richardsonschen Turbulenzkriteriums. Meteorol. Z. 56, 109 (1939).

[57] Eckert, E. R. G., and Soehngen, E.: Interferometric studies on the stability and transition to turbulence of a free-convection boundary layer. Proc. of the General Discussion on Heat Transfer, Sept. 1951, publ. by Inst. Mech. Eng. London and ASME.

[58] Everhart, P. E., and Hamilton, H. H.: Experimental investigation of boundary layer transition on a cooled 7·5° total-angle cone at Mach 10. NASA TN D 4188 (1967).

[59] Eckert, E. R. G., Soehngen, E., and Schneider, P. J.: Studien zum Umschlag laminar-turbulent der freien Konvektionsströmung an einer senkrechten Platte. Fifty Years of Boundary-layer Research (W. Tollmien and H. Görtler, ed.), 407—412, 1955.

[60] Eppler, R.: Ergebnisse gemeinsamer Anwendung von Grenzschicht- und Profiltheorie. ZFW 8, 247—260 (1960).

[61] Eppler, R.: Laminarprofile für Reynoldszahlen grösser als $4 \cdot 10^6$. Ing.-Arch. 38, 232—240 (1969).

[62] Fage, A., and Preston, J. H.: On transition from laminar to turbulent flow in the boundary layer. Proc. Roy. Soc. A *178*, 201−227 (1941).

[63] Feindt, E. G.: Untersuchungen über die Abhängigkeit des Umschlages laminar-turbulent von der Oberflächenrauhigkeit und der Druckverteilung. Diss. Braunschweig 1956; Jb. 1956 Schiffbautechn. Gesellschaft *50*, 180−203 (1957).

[63a] Frenkiel, F. N., Landahl, M. T., and Lumley, L.: Structure of turbulence and drag reduction. IUTAM Symp., Washington, D. C., 7−12 June 1976, The Physics of Fluids *20*, No. 10, Part II, p. S 1−S 292, 1977; also B. A. Thom in Proc. Intern. Congress in Rheology, North Holland, Amsterdam, Sect. II, p. 135.

[64] Fischer, M.C.: Turbulent bursts and rings on a cone in helium at Ma = 7·6. AIAA J. *10*, 1387−1389 (1972).

[65] Fischer, M.C.: An experimental investigation of boundary layer transition on a 10° half-angle cone at Mach 6·9. NASA D-5766 (1970).

[66] Fischer, M.C., and Weinstein, L.M.: Cone transitional boundary-layer structure. AIAA J. *10*, 699−701 (1972).

[66a] Garg, V.K., and Rouleau, W.T.: Linear spatial stability of pipe Poiseuille flow. JFM *54*, 113−127 (1972).

[67] Goldstein, S.: A note on roughness. ARC RM 1763 (1936).

[68] Goldstein, S.: The stability of viscous fluids between rotating cylinders. Proc. Cambr. Phil. Soc. *33*, 41−61 (1937).

[69] Goldstein, S.: On the stability of superposed streams of fluids of different densities. Proc. Roy. Soc. A *132*, 523 (1939).

[70] Goldstein, S.: Low-drag and suction airfoils. 11th Wright Brothers Lecture). JAS *15*, 189−215 (1948).

[71] Görtler, H.: Über den Einfluss der Wandkrümmung auf die Entstehung der Turbulenz. ZAMM *20*, 138−147 (1940).

[72] Görtler, H.: Über eine dreidimensionale Instabilität laminarer Grenzschichten an konkaven Wänden. Nachr. Wiss. Ges. Göttingen, Math. Phys. Klasse, Neue Folge *2*, No. 1 (1940); see also ZAMM *21*, 250−252 (1941).

[73] Görtler, H.: Dreidimensionale Instabilität der ebenen Staupunktströmung gegenüber wirbelartigen Störungen. Fifty Years of Boundary-layer Research (W. Tollmien and H. Görtler, ed.), 303−314, Braunschweig, 1955.

[74] Görtler, H.: Dreidimensionales zur Stabilitätstheorie laminarer Grenzschichten. ZAMM *35*, 362/63 (1955).

[75] Granville, P.S.: The calculation of viscous drag of bodies of revolution. Navy Department. The David Taylor Model Basin. Rep. No. 849 (1953).

[76] Gregory, N., and Walker, S.: The effect on transition of isolated surface excrescences in the boundary layer. ARC RM *13*, 436 (1950).

[77] Gregory, N., Stuart, J.T., and Walker, W.S.: On the stability of three-dimensional boundary layers with application to the flow due to a rotating disk. Phiel. Trans.Roy. Soc. London A *248*, 155−199 (1955).

[78] Gaster, M.: A note on the relation between temporally-increasing and spatially-increasing disturbances in hydrodynamic stability. JFM *14*, 222−224 (1962).

[78a] Gaster, M.: On the flow along swept leading edges. Aero. Quart. *18*, 165−184 (1967).

[79] Gebhart, B.: Instability, transition and turbulence in bouyancy-induced flows. Annual Review of Fluid Mechanics (M. Van Dyke, ed.) *5*, 213−246 (1973).

[80] Gebhart, B.: Natural convection flows and stability. Advances in Heat Transfer *9*, 273−348 (1973).

[81] Gebhart, B.: Natural convection flow, instability and transition. Trans. ASME Ser. C *91*, 293−309 (1969).

[82] Granville, P.S.: The prediction of transition from laminar to turbulent flow in boundary layers on bodies of revolution. Rep. No. 3900 of the Naval Ship Research and Development Center, Bethesda, Maryland, 1974.

[83] Görtler, H., and Hassler, H.: Einige neue experimentelle Beobachtungen über das Auftreten von Längswirbeln in Staupunktströmungen. Schiffstechnik *20*, 67−72 (1973).

[84] Ginoux, J.J.: Instabilité de la couche limite sur ailes en flèche. ZFW *15*, 302−305 (1967).

[85] Hämmerlin, G.: Über das Eigenwertproblem der dreidimensionalen Instabilität laminarer Grenzschichten an konkaven Wänden. Diss. Freiburg 1954; J. Rat. Mech. Anal. *4*, 279−321; see also ZAMM *35*, 366−367 (1955).

[86] Hämmerlin, G.: Zur Instabilitätstheorie der ebenen Staupunktströmung. Fifty Years of Boundary-layer Research (W. Tollmien and H. Görtler, ed.), 315—327 (1955).

[87] Harrin, E.N.: A flight investigation of laminar and turbulent boundary layers passing through shock waves at full-scale Reynolds-numbers. NACA TN 3056 (1953).

[88] Hausamann, W.: Flugwehr und Technik, Zürich 4, 179 (1942).

[89] Head, M.R.: The boundary layer with distributed suction. ARC RM 2783 (1955).

[90] Hertel, H.: Struktur, Form, Bewegung. Series: Biologie und Technik. Krausskopf-Verlag, Mainz, 190—195, 1963.

[91] Higgins, R.W., and Pappas, C.C.: An experimental investigation of the effect of surface heating on boundary layer transition on a flat plate in supersonic flow. NACA TN 2351 (1951).

[92] Holstein, H.: Messungen zur Laminarhaltung der Reibungsschicht. Lilienthal-Bericht S 10 17—27 (1940).

[93] Huang, L.M., and Chen, T.S.: Stability of developing pipe flow subjected to non-axisymmetrical disturbances. JFM 83, 183—193 (1974), see also Phys. Fluids 17, 245—247 (1974).

[94] Van Ingen, J.L.: A suggested semi-empirical method for the calculation of the boundary layer transition region. Techn. Univ. Dep. of Aeronautics, Delft, Report V.T.H. 74 (1956).

[95] Jack, J.R., and Diaconis, N.S.: Variation of boundary-layer transition with heat transfer on two bodies of revolution at a Mach number of 3·12. NACA TN 3562 (1955).

[96] Jacobs, E.N., and Sherman, A.: Airfoil section characteristics as affected by variations of the Reynoldsnumber. NACA TR 586 (1937).

[97] Jeffreys, H.: The instability of a layer of fluid heated below. Phil. Mag. 2, 833—844 (1926); see also Proc. Roy. Soc. A 118, 195—208 (1928).

[98] Jones, B.M.: Flight experiments on the boundary layer. Wright Brothers Lecture. JAS 5, 81—102 (1938); also Aircraft Eng. 10, 135—141 (1938).

[99] Jones, B.M., and Head, M.R.: The reduction of drag by distributed suction. Proc. Third Anglo-American Aero. Conference, Brighton 199—230 (1951).

[100] Jack, J.R., and Diaconis, N.S.: Variation of boundary-layer transition with heat transfer on two bodies of revolution at a Mach number of 3·12. NACA TN 3562 (1955).

[101] Jack, J.R., Wisniewski, R.J., and Diaconis, N.S.: Effects of extreme surface cooling on boundary layer transition. NACA TN 4094 (1957).

[102] Jillie, D.W., and Hopkins, E.J.: Effects of Mach-number, leading-edge bluntness and sweep on boundary-layer transition on a flat plate. NASA TN D-1071 (1961).

[103] Jones, W.P., and Launder, B.E.: The prediction of laminarization with a two-equation model of turbulence. J. Heat and Mass Transfer 15, 301—314 (1972); see also JFM 56, 337—351 (1972).

[104] Jaffe, N.A., Okamura, T.T., and Smith, A.M.O.: Determination of spatial amplification factors and their application to predicting transition. AIAA J. 8, 301—308 (1970).

[105] Kay, J.M.: Boundary layer flow along a flat plate with uniform suction. ARC RM 2628 (1948).

[106] Kirchgässner, K.: Die Instabilität der Strömung zwischen zwei rotierenden Zylindern gegenüber Taylorwirbeln für beliebige Spaltbreiten. ZAMP 12, 14—30 (1961).

[107] Kirchgässner, K.: Einige Beispiele zur Stabilitätstheorie von Strömungen an konkaven und erwärmten Wänden. Ing.-Arch. 31, 115—124 (1962).

[107a] Klebanoff, P.S., and Tidstrom, K.D.: Mechanism by which a two-dimensional roughness element induces boundary layer transition. Phys. of Fluids 15, 1173—1188 (1972).

[107b] Klebanoff, P.S., Tidstrom, K.D., and Sargent, J.: The three-dimensional nature of boundary layer instability. JFM 12, 1—34 (1962); see also JAS 22, 803—804 (1955).

[108] Korkegi, R.H.: Transition studies and skin-friction measurements on an insulated flat plate at a Mach number of 5·8. JAS 23, 97—102 (1956).

[109] Krämer, K.: Über die Wirkung von Stolperdrähten auf den Grenzschichtumschlag. ZFW 9, 20—27 (1961).

[110] Kramer, M.O.: Boundary layer stabilization by distributed damping. J. Amer. Soc. Naval Eng. 72, 25—33 (1960).

[111] Krüger, H.: Über den Einfluss der Absaugung auf die Lage der Umschlagstelle an Tragflügelprofilen. Ing.-Arch. 19, 384—387 (1951).

[112] Küchemann, D.: Störungsbewegungen in einer Gasströmung mit Grenzschicht. ZAMM 18, 207—222 (1938); Diss. Göttingen 1938; see also note by H. Görtler, ZAMM 23, 179—183 (1943).

[113] Kuethe, A.M.: On the character of the instability of the laminar boundary layer near the nose of a blunt body. JAS 25, 338—339 (1958).

[114] Kendall, J.M.: Supersonic boundary layer stability experiments. Proc. Boundary Layer Transition Study Group Meeting (W.D. McCauley, ed.), II, Aerospace Corp., Cal., 1967.

[115] Kendall, J.M.: Wind tunnel experiments relating to supersonic and hypersonic boundary-layer transition. AIAA J. 13, 290—299 (1975).

[116] Koschmieder, E.L.: Taylor vortices between eccentric cylinders. Phys. of Fluids 19, 1—4 (1976).

[117] Krogmann, P.: An experimental investigation of laminar and transitional heat transfer to a sharp slender cone at $Ma_\infty = 5$, including effects of angle of attack and circumferential heat transfer. Diss. Braunschweig 1975; AIAA Paper 74—628 (1974); see also ZFW 1, 101—115 (1977).

[117a] Küchemann, D.: Störungsbewegung in einer Gasströmung mit Grenzschicht. Diss. Göttingen 1938. ZAMM 18, 207—222 (1938).

[118] Kestin, J., and Wood, R.T.: On the stability of two-dimensional stagnation flow. JFM 44, 461—479 (1970).

[119] Kaye, J., and Elgar, E.C.: Modes of adiabatic and diabatic fluid flow in an annulus with inner rotating cylinder. Trans. ASME 80, 753—765 (1958).

[120] Landahl, M.T.: On the stability of a laminar incompressible boundary layer over a flexible surface. JFM 13, 609—632 (1962).

[121] Laufer, J., and Vrebalovich, Th.: Stability and transition of a supersonic laminar boundary layer on a flat plate. JFM 9, 257—299 (1960).

[122] Lees, L., and Lin, C.C.: Investigation of the stability of the laminar boundary layer in a compressible fluid. NACA TN 1115 (1946).

[123] Lees, L.: The stability of the laminar boundary layer in a compressible flow. NACA TN 1360 (1947) and NACA Rep. 876 (1947).

[124] Lees, L.: Comments on the "Effect of surface cooling on laminar boundary-layer stability". JAS 18, 844 (1951).

[125] Leite, R.J.: An experimental investigation of the stability of Poiseuille flow. JFM 5, 81—96 (1959).

[126] Lessen, M., and Gangwani, S.T.: Effect of wall small amplitude waviness on the stability of the laminar boundary layer. Phys. Fluids 19, 510—513 (1976).

[126a] Lessen, M., and Singh, P.J.: The stability of axisymmetric free shear layers. JFM 60, 433—457 (1973).

[126b] Lessen, M., Singh, P.J., and Paillet, F.L.: Stability of a trailing line vortex. Part I: Incisvid theory. JFM 65, 753—763 (1974).

[126c] Lessen, M., and Paillet, F.L.: Stability of a trailing line vortex. Part II: Viscous theory. JFM 65, 769—779 (1974).

[127] Liepmann, H.W.: Investigations on laminar boundary layer stability and transition on curved boundaries. ARC RM 7802 (1943).

[128] Liepmann, H.W.: Investigation of boundary layer transition on concave walls. NACA Wartime Rep. W-87 (1945).

[129] Liepmann, H.W., and Fila, G.H.: Investigations of effect of surface temperature and single roughness elements on boundary layer transition. NACA TN 1196 (1947) and NACA Rep. 890 (1947).

[130] Lindgren, E.R.: Liquid flow in tubes I, II and III. Archiv för Fysik 15, 97 (1959); 15, 3 (1959) and 103 (1959).

[131] Linke, W.: Über den Strömungswiderstand einer beheizten ebenen Platte. Luftfahrtforschung 19, 157—160 (1942).

[132] Ludwieg, H.: Stabilität der Strömung in einem zylindrischen Ringraum. ZFW 8, 135—140 (1960).

[133] Ludwieg, H.: Ergänzung zu der Arbeit "Stabilität der Strömung in einem zylindrischen Ringraum". ZFW 9, 359—361 (1961).

[134] Ludwieg, H.: Experimentelle Nachprüfung der Stabilitätstheorien für reibungsfreie Strömungen mit schraubenlinienförmigen Stromlinien. ZFW 12, 304—309 (1964).

[135] Laufer, J.: Factors affecting transition Reynolds numbers on models in supersonic wind tunnels. JAS 21, 497—498 (1954).

[136] Laufer, J.: Aerodynamic noise in supersonic wind tunnels. JASS 28, 685—692 (1961).

[137] Laufer, J.: Some statistical properties of the pressure field radiated by a turbulent boundary layer. Phys. Fluids 7, 1191—1197 (1964).

[138] Laufer, J., and Marte, J.E.: Results and a critical discussion of transition-Reynolds-number measurements on insulated cones and flat plates in supersonic wind tunnels. Jet Propulsion Lab., Pasadena, Calif., Rep. 20—96 (1955).

[139] Laufer, J., and Vrebalovich, T.: Stability and transition of a supersonic laminar boundary layer on a flat plate. JFM 9, 257—299 (1960).

[140] Lees, L.: The stability of the laminar boundary layer in a compressible flow. NACA TN 1360 (1947) and NACA Rep. 876 (1947).

[141] Lees, L., and Lin, C.C.: Investigation of the stability of the laminar boundary layer in a compressible fluid. NACA TN 1115 (1946).

[142] Lees, L., and Reshotko, E.: Stability of the compressible laminar boundary layer. JFM 12, 555—590 (1962).

[143] Liepmann, H.W., and Fila, G.: Investigations of effect of surface temperature and single roughness elements on boundary-layer transition. NACA TN 1196 (1947) and NACA Rep. 890 (1947).

[144] Lin, C.C.: The theory of hydrodynamic stability (Chap. 5). Cambridge University Press, 1955.

[145] Linke, W.: Über den Strömungswiderstand einer beheizten ebenen Platte. Luftfahrt-forschung 19, 157—160 (1942).

[146] Lowell, R.L., and Reshotko, E.: Numerical study of the stability of a heated, water boundary layer. Div. Fluid. Thermal and Aero. Sci., Case Western Reserve Univ., Cleveland, Ohio, Rep. FTAS—TR 73—93 (1974).

[147] Lloyd, J.R., and Sparrow, E.M.: On the instability of natural convection flow on inclined plates. JFM 42, 465—470 (1970).

[148] Lessen, M., Sadler, S., and Liu, T.Y.: Stability of pipe Poiseuille flow. Phys. Fluids 11, 1404—1409 (1968).

[149] Meksyn, D.: Stability of viscous flow over concave cylindrical surfaces. Proc. Roy. Soc. A 203, 253—265 (1950).

[150] Michel, R.: Etude de la transition sur les profiles d'aile; établissement d'un critère de détermination du point de transition et calcul de la trainée de profil en incompressible. ONERA Rapport 1/1578 A (1951).

[151] Mack, L.M.: Computation of the stability of the laminar compressible boundary layer. Methods in Computational Physics (B. Alder, ed.) 4, 247—299, Academic Press, 1965.

[152] Mack, L.M.: The stability of the compressible laminar boundary layer according to a direct numerical solution. AGARDOgraph 97, Part 1, 483—501 (1965).

[153] Mack, L.M.: Boundary layer stability theory. Jet Propulsion Lab., Pasadena, Calif., Rep. 900—277 (1969).

[154] Mack, L.M.: Linear stability theory and the problem of supersonic boundary-layer transition. AIAA J. 13, 278—289 (1975).

[155] Mack, L.M.: A numerical method for the prediction of high-speed boundary-layer transition using linear theory. Proc. Conf. on Aerodynamic Analyses Requiring Advanced Computers, NASA Sp-347 (1975).

[156] Maddalon, D.V.: Effect of varying wall temperature and total temperature on transition Reynolds number at Mach 6·8. AIAA J. 7, 2355—2357 (1969).

[157] Maddalon, D.V., and Henderson, T.A.: Hypersonic transition studies on a slender cone at small angles of attack. AIAA J. 6, 176—177 (1968).

[158] Marvin, J.G., and Akin, C.M.: Combined effects of mass addition and nose bluntness on boundary-layer transition. AIAA J. 8, 857—863 (1970).

[159] Mateer, G.G.: Effect of wall cooling and angle of attack on boundary-layer transition on sharp cones at $M_\infty = 7\cdot4$. NASA TN D-6908 (1972).

[160] Morkovin, M.V.: Critical evaluation of transition from laminar to turbulent shear layers with emphasis on hypersonically travelling bodies. Air Force Flight Dynamics Lab., Wright-Patterson Air Force Base, Ohio, TR 68—149 (1969).

[161] Munson, B.R., and Joseph, D.D.: Viscous incompressible flow between concentric rotating spheres. Part I: Basic flow. JFM 49, 289—304 (1971).

[162] Munson, B.R., and Joseph, D.D.: Viscous incompressible flow between concentric rotating spheres. Part II: Hydrodynamic stability. JFM 49, 305—318 (1971).

[163] Munson, B.R., and Menguturk, M.: Viscous incompressible flow between concentric rotating spheres. Part III: Linear stability and experiments. JFM 69, 705—719 (1975).

[164] Mackrodt, P.A.: Stabilität von Hagen-Poiseuille-Strömungen mit überlagerter starrer Rotation. Mitt. Max-Planck-Institut für Strömungsforschung and AVA No. 55, Göttingen (1971); see also ZAMM 53, T 111—T 112 (1973).

[165] Mackrodt, P.A.: Stability of Hagen-Poiseuille flow with superimposed rigid rotation. JFM *73*, 153—164 (1976).

[166] Michel, R.: Détermination du point de transition et calcul de la trainée des profiles d'ailes en incompressible. ONERA Publ. No. 58 (1952).

[167] Nachtsheim, P.R.: Stability of the free convection boundary layer flow. NACA TN D 2089 (1963).

[168] Narasimha, R., and Sreenivasan, K.R.: Relaminarization in highly accelerated turbulent boundary layers. JFM *61*, 417—447 (1973).

[169] Pekeris, C.L.: Stability of the laminar flow through a straight pipe in infinitesimal disturbances which are symmetrical about the axis of the pipe. Proc. Nat. Acad. Sci. Washington *34*, 285 (1948).

[170] Piercy, N.A.V., and Richardson, E.G.: The variation of velocity amplitude close to the surface of a cylinder moving through a viscous fluid. Phil. Mag. *6*, 970—976 (1928).

[171] Piercy, N.A.V., and Richardson, E.G.: The turbulence in front of a body moving through a viscous fluid. Phil. Mag. *9*, 1038—1041 (1930).

[172] Potter, J.L., and Whitfield, J.D.: Effects of slight nose bluntness and roughness on boundary layer transition in supersonic flows. JFM *12*, 501—535 (1962).

[173] Prandtl, L.: Einfluss stabilisierender Kräfte auf die Turbulenz. Lectures on aerodynamics and related fields, Aachen, 1929, 1—10; Berlin, 1930; see also Coll. Works *II*, 778—785.

[174] Prandtl, L., and Reichardt, H.: Einfluss von Wärmeschichtung auf die Eigenschaften einer turbulenten Strömung. Dt. Forschung No. 21, 110—121 (1934); see also Coll. Works *II*, 846—854.

[175] Prandtl, L.: Bericht über neuere Untersuchungen über das Verhalten der laminaren Reibungsschicht, insbesondere den laminar-turbulenten Umschlag. Mitt. dt. Akad. Luftfahrtforschung *2*, 141 (1942).

[176] Pretsch, J.: Über die Stabilität der Laminarströmung um eine Kugel. Luftfahrtforschung *18*, 341—344 (1941).

[177] Pretsch, J.: Über die Stabilität der Laminarströmung in einem geraden Rohr mit kreisförmigem Querschnitt. ZAMM *21*, 204—217 (1941).

[178] Pretsch, J.: Die Stabilität einer ebenen Laminarströmung bei Druckgefälle und Druckanstieg. Jb. dt. Luftfahrtforschung *I*, 58—75 (1941).

[179] Pretsch, J.: Die Anfachung instabiler Strömungen in einer laminaren Reibungsschicht. Jb. dt. Luftfahrtforschung *I*, 54—71 (1942).

[180] Pretsch, J.: Umschlagbeginn und Absaugung. Jb. dt. Luftfahrtforschung *I*, 1—7 (1942).

[181] Pate, S.R.: Measurements and correlations of transition Reynolds numbers on sharp slender cones at high speeds. AIAA J. *9*, 1082—1090 (1971).

[182] Pate, S.R., and Groth, E.E.: Boundary-layer transition measurements on swept wings with supersonic leading edge. AIAA J. *4*, 737—738 (1966).

[183] Pate, S.R., and Schueler, C.J.: An investigation of radiated aerodynamic noise effects on boundary-layer transition in supersonic and hypersonic wind tunnels. AIAA J. *7*, 450—457 (1969).

[184] Potter, J.L.: Observations on the influence of ambient pressure on boundary-layer transition. AIAA J. *6*, 1907—1911 (1968).

[185] Potter, J.L., and Whitfield, J.D.: Effects of unit Reynolds number, nose bluntness and roughness on boundary layer transition. AGARD Rep. 256 (1960).

[187] Potter, J.L., and Whitfield, J.D.: Boundary-layer transition under hypersonic conditions. AGARDograph 97, Part 3, 1—61 (1965).

[188] Pedley, T.J.: On the instability of viscous flow in a rapidly rotating pipe. JFM *35*, 97—115 (1969).

[189] Patel, V.C., and Head, M.R.: Reversion of turbulent to laminar flow. JFM *34*, 371—392 (1968).

[190] Lord Rayleigh: On convection currents in a horizontal layer of fluid when the higher temperature is on the underside. Phil. Mag. *32*, 529 (1916) or Scientific Papers *6*, 432—446.

[191] Lord Rayleigh: On the dynamics of revolving fluids. Proc. Roy. Soc. A *93*, 148—154 (1916); reprinted in Scientific Papers, *6*, 447—453.

[192] Richardson, L.F.: The supply of energy from and to atmospheric eddies. Proc. Roy. Soc. A *97*, 354—373 (1926).

[193] Riegels, F.: Das Umströmungsproblem bei inkompressiblen Potentialströmungen. Ing.-Arch. *16*, 373—376 (1948) and *17*, 94—106 (1949).

[194] Reshotko, E.: Stability theory as a guide to the evaluation of transition data. AIAA J. 7, 1086—1091 (1969).

[194a] Reshotko, E.: Boundary layer stability and transition. Annual Review of Fluid Mechanics (M. Van Dyke, ed.) 8, 311—349 (1976).

[194b] Reshotko, E.: Transition reversal and Tollmien-Schlichting instability. Phys. of Fluids 6, 335—342 (1963).

[195] Richards, B. E., and Stollery, J. L.: Transition reversal on a flat plate at hypersonic speeds. AGARDograph 97, Part 1, 477—489 (1965).

[196] Richards, B. E., and Stollery, J. L.: Further experiments on transition reversal at hypersonic speeds. AIAA J. 4, 2224—2226 (1966).

[197] Sato, H., and Kuriki, K.: The mechanism of transition in the wake of a thin flat plate placed parallel to a uniform flow. JFM 11, 321—352 (1961).

[197a] Sarpkaya, T.: A note on the stability of developing laminar pipe flow subjected to axisymmetric and non-axisymmetric disturbances. JFM 68, 345—351 (1975).

[197b] Sarpkaya, T.: Evolution of small disturbances in the laminar transition region of Hagen-Poiseuille flow. Ann. Rep. Nat. Sci. Foundation. N U Hydro Rep. No. 027. TS (1966).

[197c] Salwen, H., and Grosch, C. E.: Stability of Poiseuille flow in a pipe of circular cross section. JFM 54, 93-112 (1972).

[198] Schlichting, H.: Über die Stabilität der Couette-Strömung. Ann d. Phys. V, 905—936 (1932).

[199] Schlichting, H.: Turbulenz der Wärmeschichtung. ZAMM 15, 313—338 (1935); see also Proc. Fourth Int. Congr. Appl. Mech. 245, Cambridge, 1935.

[200] Schlichting, H., and Ulrich, A.: Zur Berechnung des Umschlages laminar-turbulent. Jb. dt. Luftfahrtforschung I, 8—35 (1942). Detailed presentation in Report of the Lilienthal-Gesellschaft S 10, 75—135 (1940).

[201] Schlichting, H.: Die Beeinflussung der Grenzschicht durch Absaugung und Ausblasen. Jb. dt. Akad. d. Luftfahrtforschung 90—108 (1943/44).

[202] Schiller, L.: Handbuch der Experimental-Physik IV, Part 4, 1—207, Leipzig, 1932.

[203] Schubauer, G. B., and Skramstad, H. K.: Laminar boundary layer oscillations and stability of laminar flow. National Bureau of Standards Research Paper 1772 (1943); JAS 14, 69—78 (1947); see also NACA Rep. 909 (1947).

[204] Schultz-Grunow, F., and Hein, H.: Beitrag zur Couette-Strömung. ZFW 4, 28—30 (1956).

[204a] Schultz-Grunow, F., and Behbahani, D.: Boundary layer stability at longitudinally curved walls. ZAMP 24, 499—506 (1973) and ZAMP 26, 493—495 (1975).

[204b] Schultz-Grunow, F.: Zur Stabilität der Couette-Strömung. ZAMM 39, 101—110 (1959).

[204c] Schultz-Grunow, F.: The stability of Couette flow with respect to two-dimensional perturbations. In W. Fiszdon (ed.): Fluid Dynamics Transaction 3, 83—93, Warszawa, 1967.

[204d] Schultz-Grunow, F.: Exakte Zugänge zu hydrodynamischen Problemen. 18. Ludwig Prandtl Memorial Lecture, ZFW 23, 175—183 (1975).

[205] Sexl, Th.: Zur Stabilitätsfrage der Poiseuilleschen und der Couette-Strömung. Ann. Phys. (4) 83, 835—848 (1927).

[206] Sexl, Th., and Spielberg, K.: Zum Stabilitätsproblem der Poiseuille-Strömung. Acta Phys. Austriaca 12, 9—28 (1958).

[207] Shapiro, N. M.: Effects of pressure gradient and heat transfer on the stability of the compressible laminar boundary layers. JAS 23, 81—83 (1956).

[208] Shen, S., and Persh, J.: The limiting wall temperature ratios required for complete stabilization of laminar boundary layers with blowing. JAS 23, 286—287 (1956).

[209] Silverstein, A., and Becker, J. V.: Determination of boundary layer transition on three symmetrical airfoils in the NACA full-scale wind tunnel. NACA TR 637 (1938).

[210] Smith, A. M. O.: On the growth of Taylor-Görtler vortices along highly concave walls. Quart. Appl. Math. 13, 233—262 (1955).

[211] Smith, A. M. O.: Transition pressure gradient and stability theory. Paper presented at the IX. Intern. Congress of Appl. Mech. 4, 234—244, Brussels, 1957; see also JASS 26, 229—245 (1959).

[212] Speidel, L.: Beeinflussung der laminaren Grenzschicht durch periodische Störungen der Zuströmung. ZFW 5, 270—275 (1957).

[213] Stalder, J. R., Rubesin, M. W., and Tendeland, T. H.: A determination of the laminar-transitional- and turbulent-boundary-layer temperature-recovery factor on a flat plate in supersonic flow. NACA TN 2077 (1950).

[214] Stender, W.: Laminarprofil-Messungen des NACA, eine Auswertung zur Gewinnung all-gemeiner Erkenntnisse über Laminarprofile. Luftfahrttechnik 2, 218—227 (1956).

[215] Sternberg, J.: A free-flight investigation of the possibility of high Reynolds-number super-sonic laminar boundary layers. JAS 19, 721—733 (1952).

[216] Sternberg, J.: The transition from a turbulent to a laminar boundary layer. Ballistic Research Laboratories Rep. 906 (1954). Aberdeen Proving Ground, Maryland, USA.

[217] Stuart, J.T.: On the stability of viscous flow between parallel planes in the presence of a coplanar magnetic field. Proc. Roy. Soc. London A 221, 189—206 (1954).

[218] Stuart, J.T.: On the nonlinear mechanics of hydrodynamic stability. JFM 4, 1—21 (1958).

[219] Stuart, J.T.: On three-dimensional non-linear effects in the stability of parallel flows. Advances in Aeronautical Sciences (Th. v. Kármán, ed.) 3, 121—142. Pergamon Press, New York/London, 1962.

[220] Stüper, J.: Der Einfluss eines Stolperdrahtes auf den Umschlag der Grenzschicht an einer ebenen Platte. ZFW 4, 30—34 (1956).

[221] Sanator, R.J., De Carlo, J.P., and Torillo, D.J.: Hypersonic boundary layer transition data for a cold wall slender cone. AIAA J. 3, 758—760 (1965).

[222] Schlichting, H.: Zur Entstehung der Turbulenz bei der Plattenströmung. Nachr. Ges. Wiss. Göttingen, Math. Phys. Klasse, 182—208 (1933).

[223] Sheetz, N.W., Jr.: Boundary-layer transition on cones at hypersonic speeds. Proc. Symposium on Viscous Drag Reduction (S.G. Spangler and C.S. Wells, Jr., ed.), Plenum Press (1969).

[223a] Sibulkin, M.: Transition from turbulent to laminar pipe flow. Phys. of Fluids 5, 280—284 (1962).

[224] Steinbeck, P.C.: Effects of unit Reynolds number, nose bluntness, angle of attack, and roughness on transition on a 5° half-angle cone at Mach 8. NASA TN D-4961 (1969).

[225] Sternberg, J.: A free-flight investigation of the possibility of high Reynolds-number super-sonic laminar boundary layers. JAS 19, 721—733 (1952).

[226] Stetson, K.F., and Rushton, G.H.: Shock tunnel investigation of boundary layer tran-sition at M = 5·5. AIAA J. 5, 899—909 (1967).

[227] Strazisar, A., Prahl, J.M., and Reshotko, E.: Experimental study of the stability of heated laminar boundary layers in water. Report FTAS/TR 75—113, Dept. of Fluid Thermal and Aero. Sci. Case Western Reserve Univ. (1975).

[227a] Stuart, J.T.: Hydrodynamic stability. In: Rosenhead, L. (ed.): Laminar boundary layers, pp. 492—579, Clarendon Press, Oxford, 1963.

[228] Sparrow, E.M., and Husar, R.B.: Longitudinal vortices in natural convection flow on inclined plates. JFM 37, 251—253 (1969).

[229] Szewczyk, A.: Stability and transition of the free-convection layer along a vertical flat plate. Int. J. Heat and Mass Transfer 5, 903—914 (1962).

[230] Sawatzki, O., and Zierep, J.: Das Stromfeld im Spalt zwischen zwei konzentrischen Kugel-flächen, von denen die innere rotiert. Acta Mechanica 9, 13—25 (1970); see also ZAMM 50, 205—208 (1970) and Eighth Symposium on Naval Research, SRC-179, 275—287 (1970).

[231] Sexl, Th.: Über dreidimensionale Störungen der Poiseuilleschen Strömung. Ann. Phys. 83, 835 (1927).

[232] Sadeh, W.S., Sutera, S.P., and Maeder, P.F.: Analysis of vorticity amplification in the flow approaching a two-dimensional stagnation point. ZAMP 21, 699—716 (1970).

[233] Sadeh, W.S., Sutera, S.P., and Maeder, P.F.: An investigation of vorticity amplification in stagnation flow. ZAMP 21, 717—742 (1970).

[234] Tani, I., and Mituisi, S.: Contributions to the design of aerofoils suitable for high speeds. Aero. Res. Inst. Tokyo, Imp. Univ. Rep. 198 (1940).

[235] Tani, I., Hama, R., and Mituisi, S.: On the permissible roughness in the laminar boundary layer. Aero. Res. Inst. Tokyo, Imp. Univ. Rep. 199 (1940).

[236] Tani, I., and Hama, T.: Some experiments on the effects of a single roughness element on boundary layer transition. JAS 20, 289—290 (1953).

[237] Tani, I., Juchi, M., and Yamamoto, K.: Further experiments on the effect of a single roughness element on boundary layer transition. Rep. Inst. Sci. Technol. Tokyo, Univ. 8 (Aug. 1954).

[238] Tani, I.: Boundary layer transition. Annual Review of Fluid Mech. 1, 169—196 (1969).

[238a] Tatsumi, T.: Stability of the laminar inlet-flow prior to the formation of Poiseuille regime. Part I: J. Phys. Soc. Japan 7, 489—495 (1952). Part II: J. Phys. Soc. Japan 7, 495—502 (1952).

[238b] Tani, I., and Sato, H.: Boundary layer transition by roughness elements. J. Phys. Soc. Japan *11*, 1284—1291 (1956); see also IXe Congres International de Mécanique Appliquée, Actes, *IV*, 86—93 (1957).

[239] Taylor, G. I.: Internal waves and turbulence in a fluid of variable density. Rapp. Proc. Verb. Cons. Internat. pour l'Exploration de la Mer. LXXVI Copenhagen, 35—42 (1931).

[240] Taylor, G. I.: Effects of variation in density on the stability of superposed streams of fluid. Proc. Roy. Soc. A *132*, 499—523 (1931).

[241] Taylor, G. I.: Stability of a viscous liquid contained between two rotating cylinders. Phil. Trans. A *223*, 289—343 (1923); see also Proc. Roy. Soc. A *151*, 494—512 (1935) and *157*, 546—564 and 565—578 (1936).

[242] Theodorsen, T., and Garrick, J.: General potential theory of arbitrary wing section. NACA TR 452 (1933).

[243] Ulrich, A.: Theoretische Untersuchungen über die Widerstandsersparnis durch Laminar-haltung mit Absaugung. Schriften dt. Akad. d. Luftfahrtforschung *8* B, No. 2 (1944).

[244] Wendt, F.: Turbulente Strömung zwischen zwei rotierenden koaxialen Zylindern. Diss. Göttingen 1934. Ing.-Arch. *4*, 577—595 (1933).

[245] Wijker, H.: On the determination of the transition point from measurements of the static pressure along a surface. Holl. Ber. A 1210 (1951).

[246] Wijker, H.: Survey of transition point measurements at the NLL, mainly for twodimensional flow over a NACA 0018 profile. Holl. Ber. A 1269 (1951).

[247] Wuest, W.: Näherungsweise Berechnung und Stabilitätsverhalten von laminaren Grenz-schichten mit Absaugung durch Einzelschlitze. Ing.-Arch. *21*, 90—103 (1953).

[248] Wuest, W.: Stabilitätsmindernde Einflüsse der Absaugegrenzschichten. ZFW *4*, 81—84 (1956).

[249] Wazzan, A. R., Okamura, T., and Smith, A. M. O.: The stability of water flow over heated and cooled flat plates. J. Heat Transfer *90*, 109—114 (1968).

[250] Wazzan, A. R., Okamura, T., and Smith, A. M. O.: The stability in incompressible flat plate laminar boundary layer in water with temperature dependent viscosity. Proc. Sixth Southeastern Seminar on Thermal Sciences, Raleigh, N. C., 184—202 (1970).

[251] Wazzan, A. R., Okamura, T., and Smith, A. M. O.: The stability and transition of heated and cooled incompressible laminar boundary layers. Proc. Fourth Int. Heat Transfer Conf. (U. Grigull and E. Hahne, ed.), *2*, FC 1·4, Elsevier Publ. Co., Amsterdam, 1970.

[252] Wazzan, A. R., Okamura, T., and Smith, A. M. O.: Stability of laminar boundary layers at separation. Phys. of Fluids, *10*, 2540—2545 (1967).

[253] Wilkins, M. E., and Tauber, M. E.: Boundary layer transition on ablating cones at speeds up to 7 km/sec. AIAA J. *4*, 1344—1348 (1966).

[254] Wieghardt, K.: Theoretische Strömungslehre. Teubner, Stuttgart, 1965.

[255] Wortmann, F. X.: Experimentelle Untersuchungen laminarer Grenzschichten bei instabi-ler Schichtung. Proc. Eleventh Int. Congress of Appl. Mech. München 1964, 815—825 (1965).

[256] Wortmann, F. X.: Längswirbel in instabilen laminaren Grenzschichten. Der Ingenieur *83*, L 52—L 60 (1971).

[257] Wortmann, F. X.: Visualization of transition. JFM *38*, 473—480 (1969).

[258] Wortmann, F. X.: The incompressible fluid motion downstream of two-dimensional Tollmien-Schlichting waves. AGARD Conf. Proc. 224, 12—1 to 12—8 (1977).

[259] Zimmermann, G.: Wechselwirkungen zwischen turbulenten Wandgrenzschichten und flexiblen Wänden. Bericht 10/1974 of the Max-Planck-Institut Göttingen, 1974.

Part D. Turbulent boundary layers

CHAPTER XVIII

Fundamentals of turbulent flow

a. Introductory remarks

Most flows which occur in practical applications are turbulent, the term denoting a motion in which an irregular fluctuation (mixing, or eddying motion) is superimposed on the main stream. Several photographs of turbulent flows in an open water channel are shown in Figs. 18.1a, b, c, and d in order to illustrate the type of motion under consideration, the pattern having been made visible by sprinkling the free surface with powder. The velocity of flow was the same in all pictures, but the camera was moved at different speeds along the axis of the channel. It is easy to deduce from each picture whether the longitudinal velocity of the fluid particles was smaller than or exceeded that of the camera, and their appearance gives a very impressive idea of the complexity of turbulent flow.

The fluctuation which is superimposed on the principal motion is so hopelessly complex in its details that it seems to be inaccessible to mathematical treatment, but it must be realized that the resulting mixing motion is very important for the course of the flow and for the equilibrium of forces. The effects caused by it are as if the viscosity were increased by factors of one hundred, ten thousand, or even more. At large Reynolds numbers there exists a continuous transport of energy from the main flow into the large eddies. However, energy is dissipated preponderantly by the small eddies, and the process occurs in a narrow strip inside the boundary layer, in the neighbourhood of the wall, as shown in detail in ref. [25].

Mixing is responsible for the large resistance experienced by turbulent flow in pipes, for the drag encountered by ships and aeroplanes and for the losses in turbines and turbocompressors, yet, on the other hand, turbulence enables us to achieve greater pressure increases in diffusers or along aeroplane wings and compressor blades. These devices would all show separation if the flow were laminar and free of turbulence, and consequently the degree of energy recovery in a diffuser would be small, and wings and blades would operate in an unsatisfactory manner.

In the succeeding chapter we shall discuss problems involving *fully developed turbulent motion*. In this connexion we are forced to restrict ourselves to the consideration of time-averages of turbulent motion, because a complete theoretical formulation has so far proved impossible, owing to the complexity of turbulent fluctuations.

Fig. 18.1a.   Camera velocity 12·15 cm/sec

Fig. 18.1b.   Camera velocity 20 cm/sec

Fig. 18.1c.   Camera velocity 25 cm/sec

Fig. 18.1d.   Camera velocity 27·6 cm/sec

Figs. 18.1a, b, c, d.   Turbulent flow in a water channel 6 cm wide, photographed with varying camera speeds. Photographs taken by Nikuradse [39] and published by Tollmien [57]

In following this path it has at least proved possible to establish certain theoretical principles which allow us to introduce a measure of order into the experimental material. Moreover in many cases it proved possible to predict these mean values under the assumption of certain plausible hypotheses and so to obtain good agreement with experiment. The following chapters will give an account of such a semi-empirical theory of turbulent flow†.

The present chapter will be devoted to the study of the influence of fluctuations on the mean flow. The succeeding chapter will be concerned with the semi-empirical assumptions used in the calculation of turbulent motion; most of them are linked with the concept of *mixing length* due to Prandtl. The remaining chapters will then deal with specific groups of turbulent motions on this basis and will include the flow through pipes, along plates, in turbulent boundary layers with pressure gradients, and free turbulent flow, i. e. the flow in jets and wakes where no restricting walls are present. Contributions to the treatment of detailed problems can be found in Conference Proceedings [3, 17, 17a, 18].

b. Mean motion and fluctuations

Upon close investigation it appears that the most striking feature of turbulent motion consists in the fact that the velocity and pressure at a fixed point in space do not remain constant with time but perform very irregular fluctuations of high frequency (see Fig. 16.17). The lumps of fluid which perform such fluctuations in the direction of flow and at right angles to it do not consist of single molecules as assumed in the kinetic theory of gases; they are macroscopic fluid balls of varying small size. It may be noted, by way of example, that although the velocity fluctuation in channel flow does not exceed several per cent., it nevertheless has a decisive influence on the whole course of the motion. The fluctuations under consideration may be visualized by realizing that certain bigger portions of the fluid have their own intrinsic motion which is superimposed on the main flow. Such *fluid balls* or *lumps* are clearly visible in the photographs, Figs. 18.1b, c, d. The size of such fluid balls, which continually agglomerate and disintegrate, determines the *scale of turbulence*; their size is determined by the external conditions associated with the flow, that is, for example, by the mesh of a screen or honeycomb through which the stream had passed. Several quantitative measurements of the magnitudes associated with such fluctuations will be given in Sec. XVIII d.

In natural winds these fluctuations manifest themselves very clearly in the form of squalliness and often attain a magnitude of 50 per cent. of the mean wind speed. The size of turbulence elements in the atmosphere can be judged, for example, by observing the eddying of a corn field.

---

† Several workers, in particular J. M. Burgers, Th. von Kármán and G. I. Taylor quite early developed a theory which exceeds these limits and which is based on statistical concepts. However, this theory has not so far been able to solve the fundamental problem mentioned earlier. We do not propose to consider this statistical theory of turbulence in the remainder of this book and refer the reader to the comprehensive reviews by G. K. Batchelor [1], A. A. Townsend [62], J.O. Hinze [20], S. Corrsin [7], C.C. Lin [35, 36], J.C. Rotta [46, 42], P. Bradshaw [2], D.C. Leslie [34], M. Rosenblatt and C. Van Atta [45], H. Tennekes and J.L. Lumley [56].

It has already been pointed out in Chap. XVI that in describing a turbulent flow in mathematical terms it is convenient to separate it into a *mean motion* and into a *fluctuation*, or *eddying motion*. Denoting the time-average of the u-component of velocity by $\bar{u}$ and its velocity of fluctuation by u', we can write down the following relations for the velocity components and pressure

$$u = \bar{u} + u'; \quad v = \bar{v} + v'; \quad w = \bar{w} + w'; \quad p = \bar{p} + p', \qquad (18.1\,\mathrm{a, b, c, d})$$

as indicated in eqn. (16.2). When the turbulent stream is compressible (Chap. XXIII), it is necessary to include fluctuations in the density, ϱ, and in the temperature, T and to put

$$\varrho = \bar{\varrho} + \varrho'; \quad T = \bar{T} + T' . \qquad (18.1\,\mathrm{e, f})$$

The time-averages are formed at a fixed point in space and are given, e. g., by

$$\bar{u} = \frac{1}{t_1} \int\limits_{t_0}^{t_0+t_1} u \, \mathrm{d}t . \qquad (18.2)$$

In this connexion it is understood that the mean values are taken over a sufficiently long interval of time, t_1, for them to be completely independent of time. Thus, by definition, the time-averages of all quantities describing the fluctuations are equal to zero:

$$\overline{u'} = 0; \quad \overline{v'} = 0; \quad \overline{w'} = 0; \quad \overline{p'} = 0; \overline{\varrho'} = 0; \overline{T'} = 0 . \qquad (18.3)$$

The feature which is of fundamental importance for the course of turbulent motion consists in the circumstance that the fluctuations u', v', w' influence the mean motion $\bar{u}$, $\bar{v}$, $\bar{w}$ in such a way that the latter exhibits an apparent increase in the resistance to deformation. In other words the presence of fluctuations manifests itself in an apparent increase in the viscosity of the fundamental flow. This increased *apparent viscosity* of the mean stream forms the central concept of all theoretical considerations of turbulent motion. We shall begin, therefore, by endeavouring to obtain a closer insight into these relations.

It is useful to list here several rules of operating on mean time-averages, as they will be required for reference. If f and g are two dependent variables whose mean values are to be formed and if s denotes any one of the independent variables x, y, z, t then the following rules apply:

$$\bar{\bar{f}} = \bar{f}; \qquad \overline{f + g} = \bar{f} + \bar{g},$$

$$\overline{\bar{f} \cdot g} = \bar{f} \cdot \bar{g},$$

$$\overline{\frac{\partial f}{\partial s}} = \frac{\partial \bar{f}}{\partial s}; \qquad \overline{\int f \, \mathrm{d}s} = \int \bar{f} \, \mathrm{d}s .$$
$$\left.\begin{array}{c} \\ \\ \\ \\ \\ \end{array}\right\} \qquad (18.4)$$

c. Additional, "apparent" turbulent stresses

Before deducing the relation between the mean motion and the apparent stresses caused by the fluctuations we shall give a physical explanation which will illustrate their occurrence. The argument will be based on the momentum theorem.

Let us now consider an elementary area dA in a turbulent stream whose velocity components are u, v, w. The normal to the area is imagined parallel to the x-axis and the directions y and z are in the plane of dA. The mass of fluid passing through this area in time dt is given by $dA \cdot \varrho u \cdot dt$ and hence the flux of momentum in the x-direction is $dJ_x = dA \cdot \varrho\, u^2 \cdot dt$; correspondingly the fluxes in the y and z-directions are $dJ_y = dA \cdot \varrho\, u\, v \cdot dt$ and $dJz = dA \cdot \varrho\, u\, w \cdot dt$, respectively. Remembering that the density is constant we can calculate the following time-averages for the fluxes of momentum per unit time:

$$\overline{dJ_x} = dA \cdot \varrho\, \overline{u^2}; \quad \overline{dJ_y} = dA \cdot \varrho\, \overline{u\, v}; \quad \overline{dJ_z} = dA \cdot \varrho \cdot \overline{u\, w}.$$

By eqn. (18.1) we find that, e. g.,

$$u^2 = (\bar{u} + u')^2 = \bar{u}^2 + 2\, \bar{u}\, u' + u'^2 \; ;$$

applying the rules in eqns. (18.3) and (18.4) we find that

$$\overline{u^2} = \bar{u}^2 + \overline{u'^2} \, ,$$

and that, similarly,

$$\overline{u \cdot v} = \bar{u} \cdot \bar{v} + \overline{u'\, v'} \, ; \quad \overline{u \cdot w} = \bar{u} \cdot \bar{w} + \overline{u'\, w'} \, .$$

Hence, the expressions for the momentum fluxes per unit time become

$$\overline{dJ_x} = dA \cdot \varrho\, (\bar{u}^2 + \overline{u'^2}) \, , \quad \overline{dJ_y} = dA \cdot \varrho\, (\bar{u} \cdot \bar{v} + \overline{u'\, v'}) \, ,$$

$$\overline{dJ_z} = dA \cdot \varrho\, (\bar{u} \cdot \bar{w} + \overline{u'\, w'}) \, .$$

These quantities, denoting the rate of change of momentum, have the dimension of forces on the elementary area dA, and upon dividing by it we obtain forces per unit area, i. e. stresses. Since the flux of momentum per unit time through an area is always equivalent to an equal and opposite force exerted on the area by the surroundings, we conclude that the area under consideration, which is normal to the x-axis, is acted upon by the stresses $-\varrho(\bar{u}^2 + \overline{u'^2})$ in the x-direction, $-\varrho(\bar{u}\,\bar{v} + \overline{u'\, v'})$ in the y-direction and $-\varrho(\bar{u}\,\bar{w} + \overline{u'\, w'})$ in the z-direction. The first of the three is a normal stress and the latter two are shearing stresses. It is thus seen that the superposition of fluctuations on the mean motion gives rise to three additional stresses

$$\sigma_x' = -\varrho\, \overline{u'^2}; \quad \tau_{yx}' = -\varrho\, \overline{u'\, v'}; \quad \tau_{zz}' = -\varrho\, \overline{u'\, w'} \tag{18.5}$$

acting on the elementary surface. They are termed "apparent" or *Reynolds stresses* of turbulent flow and must be added to the stresses caused by the steady flow as explained earlier in connexion with laminar flow. Corresponding expressions apply in the case of elementary areas normal to the two remaining axes y and z. They lorm together a complete *stress tensor of turbulent flow*. Equations (18.5) were first deduced by O. Reynolds [43] from the equations of motion of fluid dynamics (see also the next section).

It is easy to visualize that the time-averages of the mixed products of velocity fluctuations, such as e. g. $\overline{u' v'}$ do, in fact, differ from zero. The stress component $\tau_{xy}' = \tau_{yx}' = - \varrho\, \overline{u' v'}$ can be interpreted as the transport of x-momentum through a surface normal to the y-axis. Considering, for example, a mean flow given by $\bar{u} = \bar{u}(y)$, $\bar{v} = \bar{w} = 0$ with $d\bar{u}/dy > 0$, Fig. 18.2, we can see that the mean product $\overline{u' v'}$ is different from zero: The particles which travel upwards in view of the turbulent

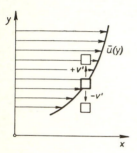

Fig. 18.2. Transport of momentum due to turbulent velocity fluctuation

fluctuation ($v' > 0$) arrive at a layer y from a region where a smaller mean velocity $\bar{u}$ prevails. Since they do, on the whole, preserve their original velocity $\bar{u}$, they give rise to a negative component u' in a layer y. Conversely, the particles which arrive from above the layer ($v' < 0$) give rise to a positive u' in it. On the average, therefore, a positive v' is "mostly" associated with a negative u' and a negative v' is "mostly" associated with a positive u'. We may thus expect that the time-average $\overline{u' v'}$ is not only different from zero but also negative. The shearing stress $\tau_{xy}' = - \varrho\, \overline{u' v'}$ is positive in this case and has the same sign as the relevant laminar shearing stress $\tau_l = \mu\, du/dy$. This fact is also expressed by stating that there exists a *correlation* between the longitudinal and transverse fluctuation of velocity at a given point.

d. Derivation of the stress tensor of apparent turbulent friction from the Navier-Stokes equations

Having illustrated the origin of the additional forces caused by turbulent fluctuation with the aid of a physical argument we shall now proceed to derive the same expression in a more formal way and directly from the Navier-Stokes equations. The object of the succeeding argument is to derive the equations of motion which must be satisfied by the time-averages of the velocity components $\bar{u}, \bar{v}, \bar{w}$ and of the pressure $\bar{p}$. The Navier-Stokes equations (3.32) for incompressible flow can be rewritten in the form

$$\varrho \left\{ \frac{\partial u}{\partial t} + \frac{\partial (u^2)}{\partial x} + \frac{\partial (uv)}{\partial y} + \frac{\partial (uw)}{\partial z} \right\} = - \frac{\partial p}{\partial x} + \mu\, \nabla^2 u, \qquad (18.6\text{a})$$

$$\varrho \left\{ \frac{\partial v}{\partial t} + \frac{\partial (vu)}{\partial x} + \frac{\partial (v^2)}{\partial y} + \frac{\partial (vw)}{\partial z} \right\} = - \frac{\partial p}{\partial y} + \mu \nabla^2 v, \qquad (18.6\,\mathrm{b})$$

$$\varrho \left\{ \frac{\partial w}{\partial t} + \frac{\partial (wu)}{\partial x} + \frac{\partial (wv)}{\partial y} + \frac{\partial (w^2)}{\partial z} \right\} = - \frac{\partial p}{\partial z} + \mu \nabla^2 w, \qquad (18.6\,\mathrm{c})$$

$$\frac{\partial u}{\partial x} + \frac{\partial v}{\partial y} + \frac{\partial w}{\partial z} = 0 . \qquad (18.6\,\mathrm{d})$$

where ∇^2 denotes Laplace's operator. We now introduce the hypotheses regarding the decomposition of velocity components and pressure into their time-averages and fluctuation terms from eqn. (18.1) and form time-averages in the resulting equations term by term, taking into account the rules from eqn. (18.4). Since $\overline{\partial u'/\partial x} = 0$ etc. the equation of continuity becomes

$$\frac{\partial \bar{u}}{\partial x} + \frac{\partial \bar{v}}{\partial y} + \frac{\partial \bar{w}}{\partial z} = 0 . \qquad (18.7)$$

From eqns. (18.7) and (18.6d) we obtain also that

$$\frac{\partial u'}{\partial x} + \frac{\partial v'}{\partial y} + \frac{\partial w'}{\partial z} = 0 .$$

It is seen that the time-averaged velocity components and the fluctuating components each satisfy the incompressible equation of continuity.

Introducing the assumptions from eqn. (18.1) into the equations of motion (18.6a, b, c) we obtain expressions similar to those given in the preceding section. Upon forming averages and considering the rules in eqn. (18.4) it is noticed that the quadratic terms in the mean values remain unaltered because they are already constant in time. The terms which are linear in the turbulent components such as e. g. $\partial u'/\partial t$ and $\partial^2 u'/\partial x^2$ vanish in view of eqn. (18.3). The same is true of the mixed terms such as e. g. $\bar{u} \cdot u'$, but the quadratic terms in the fluctuating components remain in the equations. Upon averaging they assume the form $\overline{u'^2}, \overline{u'\,v'}$ etc. Hence, if the averaging process is carried out on eqns. (18.6), and if simplifications arising from the continuity equation (18.7) are introduced, the following system of equations results

$$\varrho \left(\bar{u} \frac{\partial \bar{u}}{\partial x} + \bar{v} \frac{\partial \bar{u}}{\partial y} + \bar{w} \frac{\partial \bar{u}}{\partial z} \right) = - \frac{\partial \bar{p}}{\partial x} + \mu \nabla^2 \bar{u} - \varrho \left[\frac{\partial \overline{u'^2}}{\partial x} + \frac{\partial \overline{u'\,v'}}{\partial y} + \frac{\partial \overline{u'\,w'}}{\partial z} \right] \quad (18.8\,\mathrm{a})$$

$$\varrho \left(\bar{u} \frac{\partial \bar{v}}{\partial x} + \bar{v} \frac{\partial \bar{v}}{\partial y} + \bar{w} \frac{\partial \bar{v}}{\partial z} \right) = - \frac{\partial \bar{p}}{\partial y} + \mu \nabla^2 \bar{v} - \varrho \left[\frac{\partial \overline{u'v'}}{\partial x} + \frac{\partial \overline{v'^2}}{\partial y} + \frac{\partial \overline{v'\,w'}}{\partial z} \right] \quad (18.8\,\mathrm{b})$$

$$\varrho \left(\bar{u} \frac{\partial \bar{w}}{\partial x} + \bar{v} \frac{\partial \bar{w}}{\partial y} + \bar{w} \frac{\partial \bar{w}}{\partial z} \right) = - \frac{\partial \bar{p}}{\partial z} + \mu \nabla^2 \bar{w} - \varrho \left[\frac{\partial \overline{u'\,w'}}{\partial x} + \frac{\partial \overline{v'w'}}{\partial y} + \frac{\partial \overline{w'^2}}{\partial z} \right] \quad (18.8\,\mathrm{c})$$

The quadratic terms in turbulent velocity components have been transferred to the right-hand side for a reason which will soon become apparent. Eqns. (18.8) together with the equation of continuity, eqn. (18.7), determine the problem under consideration. The left-hand sides of eqns. (18.8) are formally identical with the steady-state Navier-Stokes equations (3.32), if the velocity components u, v, w are replaced by their time-averages, and the same is true of the pressure and friction terms on the right-hand side. In addition the equations contain terms which depend on the turbulent fluctuation of the stream.

Comparing eqns. (18.8) with eqns. (3.11) it is seen that the additional terms on the right-hand side of eqns. (18.8) can be interpreted as components of a stress tensor. By eqn. (3.10a) the resultant surface force per unit area due to the additional terms is seen to be

$$P = i\left(\frac{\partial \sigma'_x}{\partial x} + \frac{\partial \tau'_{xy}}{\partial y} + \frac{\partial \tau'_{xz}}{\partial z}\right) + j\left(\frac{\partial \tau'_{xy}}{\partial x} + \frac{\partial \sigma'_y}{\partial y} + \frac{\partial \tau'_{yz}}{\partial z}\right) + k\left(\frac{\partial \tau'_{xz}}{\partial x} + \frac{\partial \tau'_{yz}}{\partial y} + \frac{\partial \sigma'_z}{\partial z}\right).$$

Carrying the analogy with eqns. (3.11) still further we can rewrite eqns. (18.8) in the form

$$\begin{aligned}
\varrho\left(\bar{u}\frac{\partial \bar{u}}{\partial x} + \bar{v}\frac{\partial \bar{u}}{\partial y} + \bar{w}\frac{\partial \bar{u}}{\partial z}\right) &= -\frac{\partial \bar{p}}{\partial x} + \mu\nabla^2\bar{u} + \left(\frac{\partial \sigma'_x}{\partial x} + \frac{\partial \tau'_{xy}}{\partial y} + \frac{\partial \tau'_{xz}}{\partial z}\right) \\
\varrho\left(\bar{u}\frac{\partial \bar{v}}{\partial x} + \bar{v}\frac{\partial \bar{v}}{\partial y} + \bar{w}\frac{\partial \bar{v}}{\partial z}\right) &= -\frac{\partial \bar{p}}{\partial y} + \mu\nabla^2\bar{v} + \left(\frac{\partial \tau'_{xy}}{\partial x} + \frac{\partial \sigma'_y}{\partial y} + \frac{\partial \tau'_{yz}}{\partial z}\right) \\
\varrho\left(\bar{u}\frac{\partial \bar{w}}{\partial x} + \bar{v}\frac{\partial \bar{w}}{\partial y} + \bar{w}\frac{\partial \bar{w}}{\partial z}\right) &= -\frac{\partial \bar{p}}{\partial z} + \mu\nabla^2\bar{w} + \left(\frac{\partial \tau'_{xz}}{\partial x} + \frac{\partial \tau'_{yz}}{\partial y} + \frac{\partial \sigma'_z}{\partial z}\right)
\end{aligned}\right\} \quad (18.9)$$

and upon comparing eqns. (18.9) with (18.8) we can see that the components of the stress tensor due to the turbulent velocity components of the flow are:

$$\begin{pmatrix} \sigma'_x & \tau'_{xy} & \tau'_{xz} \\ \tau'_{xy} & \sigma'_y & \tau'_{yz} \\ \tau'_{xz} & \tau'_{yz} & \sigma'_z \end{pmatrix} = -\varrho\begin{pmatrix} \overline{u'^2} & \overline{u'v'} & \overline{u'w'} \\ \overline{u'v'} & \overline{v'^2} & \overline{v'w'} \\ \overline{u'w'} & \overline{v'w'} & \overline{w'^2} \end{pmatrix}. \quad (18.10)$$

This stress tensor is identical with the one obtained in eqn. (18.5) with the aid of the momentum equation.

From the preceding argument it can be concluded that the components of the mean velocity of turbulent flow satisfy the same equations, i. e. eqns. (18.9), as those satisfied by laminar flow, except that the laminar stresses must be increased by additional stresses which are given by the stress tensor in eqn. (18.10). These additional stresses are known as *apparent*, or *virtual stresses of turbulent flow* or *Reynolds stresses*. They are due to turbulent fluctuation and are given by the time-mean values of the quadratic terms in the turbulent components. Since these stresses are added to the ordinary viscous terms in laminar flow and have a similar influence on the course of the flow, it is often said that they are caused by *eddy viscosity*. The total stresses are the sums of the viscous stresses from eqn. (3.25a,b) and of these apparent stresses, so that, e. g.,

$$\left.\begin{aligned}
\sigma_x &= -p + 2\mu\frac{\partial \bar{u}}{\partial x} - \varrho\overline{u'^2}, \\
\tau_{xy} &= \mu\left(\frac{\partial \bar{u}}{\partial y} + \frac{\partial \bar{v}}{\partial x}\right) - \varrho\overline{u'v'}, \dots.
\end{aligned}\right\} \quad (18.11)$$

Generally speaking, the apparent stresses far outweigh the viscous components and, consequently, the latter may be omitted in many actual cases with a good degree of approximation.

Boundary-layer equations: At this stage it may be useful briefly to outline the form of the boundary-layer equations for turbulent flow. In the case of *two-dimensional* flows ($\bar{w} \equiv 0$) eqns. (18.7) and (18.8a, b, c), modified by the boundary-layer approximations as outlined in Chap. VII, lead to

$$\frac{\partial \bar{u}}{\partial x} + \frac{\partial \bar{v}}{\partial y} = 0, \tag{18.12}$$

$$\bar{u}\,\frac{\partial \bar{u}}{\partial x} + \bar{v}\,\frac{\partial \bar{u}}{\partial y} = -\frac{1}{\varrho}\frac{\partial \bar{p}}{\partial x} + \frac{\partial}{\partial y}\left\{ v\,\frac{\partial \bar{u}}{\partial y} - \overline{u'\,v'} \right\}. \tag{18.13}$$

(two-dimensional, turbulent boundary layer)

Due to the boundary-layer simplifications, the term

$$+ \frac{\partial}{\partial x}\,\overline{(v'^2 - u'^2)}$$

which is generated by the normal stresses, can be neglected. A comparison with the equations for the laminar boundary layer, eqns. (7.10) and (7.11), leads to the following rules:

(a)   The velocity components and the pressure, u, v, and p, are to be replaced by their time-averages $\bar{u}$, $\bar{v}$, $\bar{p}$.

(b)   The inertia terms and the pressure term remain unchanged, whereas the viscous term $v\,\partial^2 u/\partial y^2$ must be replaced by

$$\frac{\partial}{\partial y}\left(v\,\frac{\partial \bar{u}}{\partial y} - \overline{u'\,v'} \right).$$

This is equivalent to stating that the laminar viscous force per unit volume $\partial \tau_l/\partial y$ must be replaced by

$$\frac{\partial}{\partial y}\,(\tau_l + \tau_t)$$

where $\tau_l = \mu\,\partial u/\partial y$ is the laminar shearing stress from Newton's law, and $\tau_t = -\varrho\,\overline{u'v'}$ is the apparent turbulent stress from Reynolds's hypothesis.

Boundary conditions: The boundary conditions to be satisfied by the mean velocity components in eqns. (18.9) are the same as in ordinary laminar flow, namely they all vanish at solid walls (no-slip condition). Moreover, all turbulent components must vanish at the walls and they are very small in their immediate neighbourhood. It follows, therefore, that all components of the tensor of apparent stresses vanish at the solid walls and the only stresses which act near them are the viscous stresses of laminar flow as they, generally speaking, do not vanish there. Furthermore it is seen that in the immediate neighbourhood of a wall the apparent stresses are small compared with the viscous stresses, and it follows that in every turbulent flow there exists a very thin layer next to the wall which, in essence, behaves like one in laminar motion. It is known as the *laminar sub-layer* and its velocities are so small that the viscous forces dominate over the inertia forces. Thus, no turbulence can exist in it. The laminar sub-layer joins a transitional layer in which the velocity fluctuations are so large that they give rise to turbulent shearing stresses which are, in turn, comparable with the viscous stresses. At still larger distances from the wall the turbulent stresses eventually completely outweigh the viscous stresses. This is the actual turbulent boundary layer. The thickness of the laminar sub-layer is so small, in most cases, that it is impossible, or very difficult, to observe it under experimental conditions. Nevertheless, it is of decisive importance for the flow under consideration because it is the seat of phenomena by which the shearing at the wall and hence the viscous drag are determined. We shall revert to this point later in the book.

Equations (18.9) and (18.10) constitute the starting point for the mathematical treatment of turbulent-flow problems, or, more precisely, for the calculation of the time-averages of the magnitudes which describe the flow. The time-averaged values of the turbulent velocity components can be interpreted as the components of a stress tensor but it must be borne in mind that such an interpretation does not in itself lead to very much. Equations (18.9) and (18.10) cannot be used for a rational evaluation of the mean flow as long as the relation between the mean and the turbulent components is not known. Such a relation can only be obtained empirically and forms the essential contents of all the hypotheses concerning turbulence which will be discussed in the succeeding chapter.

e. Some measurements on fluctuating turbulent velocities

In experimental work on turbulent flow it is usual to measure only the mean values of pressure and velocity because they are the only quantities which can be measured conveniently. The measurement of the turbulent, fluctuating components $u', v', \ldots$ themselves, or of their mean values such as $\overline{u'^2}, \overline{u'v'}, \ldots$ is rather difficult and requires elaborate equipment. Reliable measurements of the fluctuation-velocity components have been obtained with the aid of hot-wire anemometers. The measurement of the mean values is quite sufficient for most practical applications, but only through the actual measurement of the fluctuating components is it possible to gain a deeper understanding of the mechanism of turbulent flow. We now propose to give a short account of some experimental work on the measurement of the fluctuating-velocity components in order to present a more vivid physical picture of the phenomena and in order to give some justification to the preceding mathematical argument.

H. Reichardt [41] carried out such measurements in a wind tunnel with a rectangular test section 1 m wide and 24·4 cm high. The variation of the mean velocity over the height of the tunnel, $\bar{u}(y)$, is seen plotted in Fig. 18.3; measurements were made in the central section of the tunnel. It is seen to be a typical turbulent velocity profile with a steep increase near the wall and a fairly uniform velocity near the centre-line. The maximum velocity was $U = 100$ cm/sec. The same diagram contains also plots of the root-mean-square values of the longitudinal and transverse components $\sqrt{\overline{u'^2}}$ and $\sqrt{\overline{v'^2}}$ respectively. The transverse fluctuation does not vary greatly over the height of the channel and its average value is about 4 per cent. of U, but the longitudinal turbulent component exhibits a pronounced steep maximum of

$0.13\ U$ very close to the wall. It is clearly seen from the diagram that both turbulent components decrease to zero at the wall, as stated earlier. Figure 18.4 shows a plot of the mean value of the product $-\overline{u'\,v'}$, which is equal to the turbulent shearing stress except for a factor ϱ. The value of $-\overline{u'\,v'}$ falls to zero in the centre of the test section for reasons of symmetry, whereas its maximum occurs near the wall showing that turbulent friction has its largest value there. The broken line τ/ϱ shows the variation of shearing stress which was obtained from the measured pressure distribution and independently of the measurement of velocity. The two curves nearly coincide over the major portion of the height of the test section, and this may be interpreted as a good check on the measurements; it also shows that almost all of the shearing stress is due to turbulence. The two curves under consideration diverge near the wall, the curve of $-\overline{u'\,v'}$ decreasing to zero, because turbulent fluctuations die out near the wall. The difference between the two curves gives laminar friction. Finally Fig. 18.4 contains values of the *correlation coefficient*, ψ, between the longitudinal and transverse fluctuations at the same point; it is defined by

$$\psi = \frac{\overline{u'\,v'}}{\sqrt{\overline{u'^2}}\cdot\sqrt{\overline{v'^2}}}.\qquad(18.12)$$

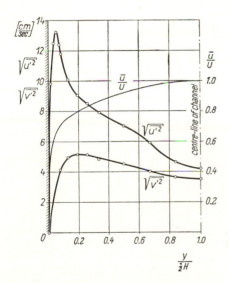

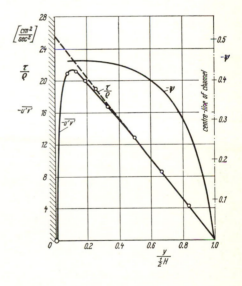

Fig. 18.3. Measurement of fluctuating turbulent components in a wind tunnel, at maximum velocity $U = 100$ cm/sec after Reichardt [41]

Root-mean-square of longitudinal fluctuation $\sqrt{\overline{u'^2}}$, transverse fluctuation $\sqrt{\overline{v'^2}}$, mean velocity $\bar{u}$

Fig. 18.4. Measurement of fluctuating components in a channel, after Reichardt [41]

The product $\overline{u'\,v'}$, the shearing stress τ/ϱ, and the correlation coefficient ψ

The correlation coefficient ψ ranges over values up to $\psi = -0.45$.

More extensive measurements on the turbulent fluctuations have also been performed in the boundary layer of a flat plate at zero incidence. Figure 18.5 reproduces some of the results obtained by P. S. Klebanoff [25] in a boundary layer on a flat plate associated with a stream of the very low turbulence intensity of 0.02 % (cf. Secs. XVI d and XVIII f), at a Reynolds number $R_x = U_\infty x/\nu = 4.2 \times 10^6$. The profile of the temporal mean of the velocity, $\bar{u}$, exhibits a shape which is very much like that in a channel, Fig. 18.3. The variation of the longitudinal fluctuation $\sqrt{\overline{u'^2}}$ with its pronounced maximum in close proximity of the wall as well as the flatter course of the curve of transverse oscillation at right angles to the wall, $\sqrt{\overline{v'^2}}$, closely resemble those obtained in the channel, Fig. 18.3. It is remarkable that in the boundary layer on a flat plate, Fig. 18.5, the transverse oscillation parallel to the wall, $\sqrt{\overline{w'^2}}$, also attains considerable values, values which, moreover, exceed those attained by $\sqrt{\overline{v'^2}}$. The turbulent shearing stress $-\overline{u'v'}/U_\infty^2$ measured close to the wall agrees with $\tau/\rho\, U_\infty^2 \approx 0.0015$, the local value of the skin-friction coefficient $\frac{1}{2}\, c_f'$ in the diagram of Fig. 21.10. A comparison between Figs. 18.3 and 18.4 for the channel and Fig. 18.5 for the boundary layer reveals that the turbulent fluctuations are very similar in both cases. They provide a justification for the application of the laws of turbulent flow deduced from the study of flows through channels and pipes to the description of the flow in a boundary layer. We shall make use of this possibility in Chap. XXI.

G. B. Schubauer and P. S. Klebanoff [50] performed also very careful measurements of the fluctuations of the turbulent velocity components and of the correlation coefficient in a turbulent boundary layer on a flat wall with a *favourable* and an *adverse* pressure gradient.

J. Laufer [32] performed extensive measurements on the fluctuating components in *pipe flow*. Earlier measurements performed by P. S. Klebanoff and Z. W. Diehl [24] on an artificially thickened boundary layer on a flat plate demonstrated that it behaves substantially like an ordinary boundary layer with a correspondingly increased inlet length. Detailed results on turbulent flow through a channel can be found described in a paper by J. Laufer [30]. A subsequent paper by J. C. Laurence [33] contains the results of his investigations of the intensity of turbulence in free jets.

The investigations into turbulent oscillations in the boundary layer of a flat plate described in ref. [25] have shown, further, that the turbulence at the outer edge of the boundary layer is intermittent and resembles in this respect the flow in the inlet length of a pipe described in Sec. XVI a and Figs. 16.2 and 16.3. Oscillograms of the oscillating turbulent velocity components demonstrate that the position of the fairly sharp boundaries between the highly turbulent flow in the boundary

---

† It may be remarked here in passing that the existence of an apparent shearing stress due to velocity fluctuations always implies a correlation between the turbulent velocity components in two different directions. Such a correlation also exists in the case of the disturbances which have been investigated in connexion with the theory of stability of laminar motion; see ref. [58]

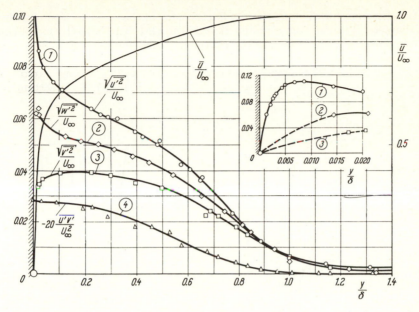

Fig. 18.5. Variation of the fluctuating turbulent velocity components in the boundary layer on a flat plate at zero incidence, as measured by P. S. Klebanoff [25]

Curve (1), longitudinal oscillation: $\sqrt{\overline{u'^2}}$

Curve (2), transverse oscillation parallel to wall: $\sqrt{\overline{w'^2}}$

Curve (3), transverse oscillation at right angles to wall: $\sqrt{\overline{v'^2}}$

Curve (4), turbulent shearing stress: $\overline{u'\,v'} = -\,\tau_t/\varrho$;

$\bar{u}$ denotes mean velocity

layer and the nearly turbulence-free external stream fluctuates strongly with time. The variation of the intermittency factor γ over the width of the boundary layer is shown plotted in Fig. 18.6. The value $\gamma = 1$ signifies that the flow is turbulent at all times and $\gamma = 0$ corresponds to a flow which remains laminar. It can be inferred from the diagram that the boundary layer is intermittent in that respect from $y = 0.5\,\delta$ to $y = 1.2\,\delta$. Turbulent jets and wakes behave in a similar manner (cf. [25a, 26, 28, 63]).

Apart from the distribution of the velocity fluctuations, it is necessary to provide additional data in order to characterize a turbulent stream. A quantitative statement regarding the spatial structure of turbulence can be obtained by simultaneously observing the velocity fluctuations at two neighbouring points 1 and 2 in the flow field. These allow us to determine the correlation function

$$R = \frac{\overline{u_1'\,u_2'}}{\sqrt{\overline{u_1'^2}}\cdot\sqrt{\overline{u_2'^2}}} \,. \tag{18.13}$$

first introduced by G. I. Taylor [52]. The diagram in Fig. 18.7 displays a typical correlation function for the longitudinal fluctuations in the cross-section of a circular pipe of diameter d taken from a publication due to G. I. Taylor [54]. One of the hot wires was placed in the centre and the other at a variable distance r. At $r = 0$ the two fluctuations u_1' and u_2' are identical, and for this reason we find that $R(0) = 1$ there. As r is increased, the correlation function decreases in value very rapidly, even assuming, in this particular example, small negative values over a certain range.

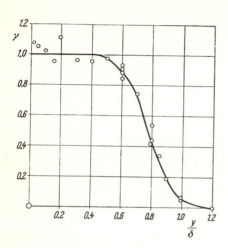

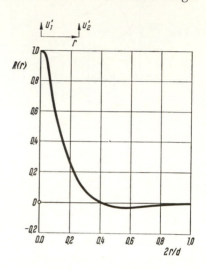

Fig. 18.6. Variation of the intermittency factor γ in a turbulent boundary layer on a flat plate at zero incidence, as measured by P. S. Klebanoff [25]

Fig. 18.7. Correlation function for the turbulent fluctuation of the longitudinal component of velocity, u_1', measured in the centre of a pipe and related to the fluctuating component u_2' at distance r. Measurements due to L. F. G. Simmons and reported by G. I. Taylor [54]

This is explained by the requirement of continuity according to which, as we know, the rate of flow through any cross-section remains constant in time. The integral of the correlation function R, that is, the quantity

$$L = \int_0^{1/2\,d} R(r)\,\mathrm{d}r\,. \qquad (18.14)$$

yields a characteristic length of the structure of the turbulence in the flow. This length, which is also known as the *scale of turbulence*, establishes a measure of the extent of the mass which moves as a unit and gives an idea of the average size of the turbulent eddies ("balls of turbulence"). In the example under discussion it was found that $L \approx 0\cdot 14\ (\tfrac{1}{2}\,d)$.

If the second velocity, u_2' in eqn. (18.13), is measured at the same location but at a different instant of time (u_1' at instant t_1 and u_2' at instant $t_2 = t_1 + t$), we obtain the so-called *autocorrelation function*. The provision of space-time correlations, that is, of observations of two velocity components, each measured at a different location in space and at a different instant in time, allows us to gain a good deal of insight. As an example, we reproduce in Fig. 18.8 such space-time correlations

Distance from wall:
$y/\delta = 0.24$
Boundary-layer thickness:
$\delta = 16.8$ mm

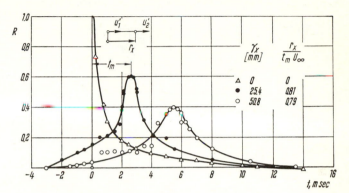

Fig. 18.8. Space-time correlations of velocity fluctuations in the turbulent boundary layer on a flat plate, as measured by A. J. Favre, J. J. Gaviglio and R. J. Dumas [16]

obtained by A. J. Favre and his coworkers [16] in the turbulent boundary layer on a flat plate. The temporal displacement, t_m, of the maximum of each curve is imposed by the passage of turbulent eddies; the eddies move with an approximate velocity which is equal to $0.8\,U_\infty$. The decrease in the maxima is the result of a process which can be visualized as follows: With the passage of time, the turbulent eddies lose their individuality through mixing with the surrounding turbulent fluid. Concurrently, new eddies continuously spring into being.

An alternative description of the structure of turbulence is obtained when a *frequency analysis* of the motion is provided instead of a correlation function. This leads us to the concept of the *spectrum* of a turbulent stream. Let n denote the frequency and $F(n)\,dn$ the fractional content of the root-mean-square value, $\overline{u'^2}$, of the longitudinal fluctuation which belongs to the frequency interval from n to $n + dn$. The function $F(n)$, which represents the density of the distribution of $\overline{u'^2}$ in n, is known as the *spectral distribution* of $\overline{u'^2}$. By definition, we must have

$$\int_0^\infty F(n)\,dn = 1 . \tag{18.15}$$

The spectral function $F(n)$ can be interpreted as the Fourier transform of the autocorrelation function†. The spectra displayed in Fig. 18.9 were obtained by P. S. Klebanoff [25] in the turbulent boundary layer formed on a flat plate. Except for the measurement at the outer edge of the boundary layer ($y/\delta = 1$), the highest value of $F(n)$ always occurs at the lowest measured frequency. As the frequency, n, is raised, the function $F(n) \sim n^{-5/3}$ in concordance with the theory developed by

---

† This was first recognized by G. I. Taylor [47].

A. N. Kolmogorov, C. F. von Weizsaecker [64] and W. Heisenberg. As the frequency becomes even larger, $F(n)$ decreases under the action of kinematic viscosity at a faster rate still. According to W. Heisenberg's theory [19a], at very high values of frequency we should observe that $F(n) \sim n^{-7}$. The two theoretical laws are represented in Fig. 18.9 by the two straight lines labelled (1) and (2), respectively.

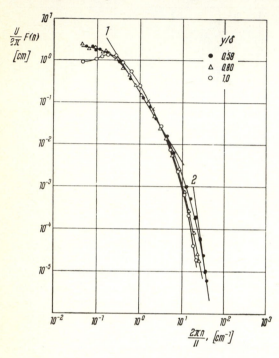

Fig. 18.9.   Frequency spectrum of the longitudinal fluctuation in the turbulent boundary layer on a flat plate measured by P. S. Klebanoff [25] Curve (1): $F \sim n^{-5/3}$ Curve (2): $F \sim n^{-7}$ Theory due to W. Heisenberg [19a]

J. Maréchal [36a] performed detailed measurements on the frequency spectrum in flows with homogeneous turbulence. In particular, he investigated the effect of a strong two-dimensional contraction of the stream.

The character of the spectral distribution of the energy of fluctuation leads us directly to the idea that a turbulent stream does not contain eddies of a definite size only. On the contrary, there must exist eddies of very many different sizes. At very large Reynolds numbers, these eddies may differ by orders of magnitude one from another. Recent papers deal with the effect of turbulence intensity on the development of a turbulent boundary layer [5, 19, 22]. R. Emmerling [15] made a detailed investigation of the wall pressure of a turbulent boundary layer making use of an optical method. He thus discovered that the flow develops zones of high-amplitude pressure fluctuations which occur at irregular time intervals. The extrema in wall-pressure fluctuations in these zones move at an instantaneous convective velocity of 40 to 80% of mean stream and in the direction of the mean stream. The wave form of such fluctuations changes slowly with time. See also the papers by W. K. Blake [1a], M. K. Bull [3a], S. J. Kline et al. [26], and P. J. Mulhearn [38a], as well as the reviews by W. W. Willmarth [66, 67].

f. Energy distribution in turbulent streams

The subsidiary, oscillatory motion continually drains away energy from the mean, principal motion through the intermediary of the turbulent stresses. Ultimately, this energy is completely dissipated into heat owing to the action of viscosity. If we introduce the velocities from eqn. (18.1) into the expression for the dissipation function from eqn. (12.8), we can isolate a group of terms which depend solely on the velocity gradients of the mean motion. This part is known as the *direct dissipation function*. The remainder which corresponds to the energy dissipated by virtue of the existence of fluctuations is known as the *turbulent dissipation function*. According to eqn. (12.8), the turbulent dissipation function is given by the expression

$$\varepsilon = \nu \left[2 \overline{\left(\frac{\partial u'}{\partial x}\right)^2} + 2 \overline{\left(\frac{\partial v'}{\partial y}\right)^2} + 2 \overline{\left(\frac{\partial w'}{\partial z}\right)^2} + \overline{\left(\frac{\partial u'}{\partial y} + \frac{\partial v'}{\partial x}\right)^2} + \right.$$
$$\left. + \overline{\left(\frac{\partial u'}{\partial z} + \frac{\partial w'}{\partial x}\right)^2} + \overline{\left(\frac{\partial v'}{\partial z} + \frac{\partial w'}{\partial y}\right)^2} \right]. \tag{18.16}$$

The direct dissipation function makes a non-negligible contribution only in the neighbourhood of solid walls. In the remainder of the field, the turbulent dissipation function predominates over the former.

The expression in eqn. (18.16) assumes a much simpler form when the turbulence is homogeneous and isotropic. A turbulent field is termed homogeneous when the statistical distributions are the same at every point in space; it is termed isotropic when the distributions remain invariant with respect to arbitrary rotations and reflexion of the coordinate axes. Taking into account symmetry properties and the requirement of continuity, it is possible to reduce the right-hand side of eqn. (18.16) to a *multiple* of a *single* term, for example, to a multiple of $\overline{(\partial u'/\partial x)^2}$. This enabled G. I. Taylor [53] to simplify eqn. (18.16) to the form

$$\varepsilon = 15 \, \nu \, \overline{\left(\frac{\partial u'}{\partial x}\right)^2}. \tag{18.17}$$

Strictly speaking, isotropic turbulence does not exist in nature. A turbulent field which approximates in its structure the hypothetical case of isotropy can be produced by passing a parallel stream through a wire screen as is done in wind tunnels. Departures from isotropy become much larger in pipe flow, boundary layers, etc. Nevertheless, the notion of isotropic turbulence acquires wider applicability if it is restricted to distribution functions of velocity differences instead of to those formed with respect to velocities themselves. Following A. N. Kolmogorov†, we consider correlation functions of the form

$$B = \overline{(u_2' - u_1')^2},$$

(*cf.* eqn. (18.13) and Fig. 18.7), and designate the turbulence as "locally isotropic" when the correlation function remains invariant with respect to rotations and reflexions of the coordinate system in a restricted domain, that is in a restricted range of distances r between points 1 and 2.

It is found that such local isotropy exists in any turbulent flow in a sufficiently small interval $r \ll L$, where L has been defined in eqn. (18.14) on condition that the Reynolds number

$$\mathbf{R} = \frac{\sqrt{\overline{u'^2}} \, L}{\nu}$$

of the turbulence has become large enough. It exists even in shear flows, such as flows in pipes, boundary layers, etc., in which there are present large shearing stresses, but regions in close proximity of walls and boundaries are excepted. The regions of locally isotropic turbulence extend precisely over ranges where the gradients of the fluctuations ($\partial u'/\partial x$ etc.) assume large values.

---

† The works of A. N. Kolmogorov are now accessible in a German as well as in an English translation [18a, 18b].

For this reason, equation (18.17) enjoys very wide applicability. A dimensional argument which was first advanced by A. N. Kolmogorov and later, independently, by C. F. von Weizsaecker [64] and W. Heisenberg [19a], leads to the establishment of further details regarding the form of the correlation functions for small distances r or of the form of the spectral function for large frequencies. We must, however, refrain from pursuing this subject here, except for referring the reader once more to Fig. 18.9.

The circumstances which are essential for the understanding of turbulence are these: The turbulent stresses are created predominantly by the large eddies, that is, by eddies whose size is of order L. As a consequence of the instability of the flow, there appear eddies of smaller and smaller size until, ultimately, the gradients $\partial u'/\partial x$ etc become so steep in the smallest eddies that they produce in them a transformation of mechanical energy into heat. The mechanical power which is transferred from the mean motion to the large eddies by the action of turbulent stresses is independent of viscosity; it cascades step by step to ever smaller eddies until a size is reached when it is dissipated. This mechanism is responsible for the fact that the skin friction as well as the distribution of mean velocities depend very little on the Reynolds number in spite of the fact that all losses in energy are due to the viscosity.

g. Wind-tunnel turbulence

The relative magnitude of the longitudinal and transverse fluctuations of velocity is a very important variable in wind-tunnel measurements; it determines the degree to which measurements performed on a model can be applied to the full-scale structure as well as how measurements performed in different tunnels can be compared among themselves. We have already mentioned in Sec. XVI d that, in particular, transition from laminar to turbulent flow strongly depends on the magnitude of the oscillating velocity component. The whole development of the turbulent boundary layer and the location of the separation point, as well as the rate of heat transfer depend on the intensity of turbulence in the free stream (*cf.* Sec. XII g). The magnitude of the fluctuations in a given tunnel is determined by the mesh of its screens, grids or honeycombs. At a certain distance from the screens there is *isotropic turbulence* which means that the mean velocity fluctuations in the three coordinate directions are equal to each other:

$$\overline{u'^2} = \overline{v'^2} = \overline{w'^2}\,.$$

In such cases the *level, degree,* or *intensity of turbulence* can be described by the quantity $\sqrt{\overline{u'^2}}/U_\infty$ which is then identical with

$$\sqrt{\tfrac{1}{3}\,(\overline{u'^2} + \overline{v'^2} + \overline{w'^2})}\,\Big/\,U_\infty\,.$$

The degree of turbulence of a wind tunnel expressed as $\sqrt{\overline{u'^2}}/U_\infty$ can be reduced to values as low as 0·1 per cent., if a sufficient number of fine-mesh screens or honeycombs is used, see Fig. 16.16†.

---

† H. L. Dryden and G. B. Schubauer [10] undertook extensive measurements on the effect of placing fine-mesh screens in a wind tunnel on its turbulence level. The addition of a single screen reduces the intensity of turbulence in the ratio $1/\sqrt{1+c}$, where c denotes the resistance coefficient of the screen; hence when n screens are used, the reduction in turbulence intensity is in the ratio $\{1/(1+c)\}^{n/2}$. Consequently, for a given pressure loss the reduction in turbulence intensity is greater when a large number of screens of small resistance is chosen in preference to a single screen of large resistance. According to ref. [10], the addition of a contraction cone to the tunnel brings with it a great reduction of the absolute value of the longitudinal, oscillating component. On the other hand, the transverse components either remain constant or even increase.

The experimentally verified fact that the critical Reynolds number of a sphere for which the drag coefficient decreases steeply (Fig. 1.5) depends strongly on the degree of turbulence of the wind tunnel, is of great practical importance. The value of the critical Reynolds number† is of the order of $(V D/\nu)_{crit} = 1 \cdot 5$ to 4×10^5 and decreases with an increasing intensity of turbulence. This fact is evident on physical grounds because a high intensity of turbulence in the free stream leads to transition at low Reynolds numbers so that the point of separation is shifted downstream causing the wake to decrease, and this, in turn, reduces drag. On the other hand, free-flight measurements on a sphere performed by C. B. Millikan and A. L. Klein [37] gave the surprising result that in the free atmosphere the critical Reynolds number of the sphere is independent of the structure of turbulence which varies with the weather. Free-flight measurements gave a critical Reynolds number of $R_{crit} = 3 \cdot 85 \times 10^5$, which is larger than that for most wind tunnels, although measurements in low-turbulence tunnels approach the value in free flight. The fact that the critical Reynolds number measured in free flight is independent of the weather is explained by the circumstance that the turbulent eddies in the atmosphere are so large that they cannot affect the phenomena in the thin boundary layer on a sphere. In any case, these measurements lead to the conclusion that it is necessary to design wind tunnels of low turbulence intensity if model measurements are to be applicable to the design of full-scale aircraft. This is particularly important when measurements are performed on low-drag aerofoils whose boundary layers remain laminar over long stretches (laminar aerofoils, Sec. XVII b). The characteristics of such aerofoils can be successfully measured only in low-turbulence tunnels, i. e. in tunnels whose intensity of turbulence is extremely small ($T \approx 0 \cdot 0005$; cf. [14]). Reference [6] contains a summary of measurements on the intensity of turbulence of a large number of wind tunnels.

Since the direct measurement of the velocity fluctuation $\sqrt{\overline{u'^2}}/U_\infty$ is quite difficult, attempts were made to regard the critical Reynolds number on a sphere as the parameter which describes the intensity of turbulence in a wind-tunnel. This critical Reynolds number for a sphere can be determined either by measuring the drag, as suggested by H. L. Dryden [8, 9], or by measuring the pressure difference between the forward stagnation point and a point at the rear of the sphere, as suggested by S. Hoerner [21]‡. The latter method was extensively used by R. C. Platt [40]. H. L. Dryden and A. M. Kuethe [8] correlated the critical Reynolds number for a sphere with the mean longitudinal fluctuation, Fig. 18.10, and discovered that these two quantities satisfy a unique functional relationship. The value of the critical Reynolds number, $R_{crit} = 3 \cdot 85 \times 10^5$, measured on a sphere in free flight corresponds to a vanishingly small intensity of turbulence, $T \to 0$.

Bodies of other than spherical shapes also exhibit some influence of the intensity of turbulence on their drag. This has been demonstrated by the measurements on flat plates placed at right angles to the stream which have been performed by G. B. Schubauer and H. L. Dryden [49].

---

† The critical Reynolds number of a sphere is defined as that for which the drag coefficient assumes the conventional value of $c_D = 0 \cdot 3$.

‡ To the conventional value of $c_D = 0 \cdot 3$ there corresponds a pressure difference between the forward stagnation point and a point at the rear of the sphere of $\Delta p = 1 \cdot 22\, q$, where q denotes the dynamic pressure in the free stream.

Detailed investigations which were carried out by G. I. Taylor [53] and H. L. Dryden [11] led to the conclusion that the drag in a stream cannot be adequately described by specifying the magnitude of the fluctuation of the velocity components alone, because it is also affected by the structure of the turbulent stream. On the basis of a theory of turbulence developed by himself, G. I. Taylor proposed that the critical Reynolds number of a sphere depends on the parameter

$$\frac{\sqrt{\overline{u'^2}}}{U_\infty} \left(\frac{D}{L}\right)^{1/5} ,$$

where L is the scale of turbulence, that is, the integral of the correlation function defined in eqn. (18.14), and D is the diameter of the sphere. H. U. Meier et al. [36b] investigated the influence of the scale of turbulence, L, on the turbulent boundary layer at low turbulence intensity. They obtained maximum values of the wall shear stress when the scale is of the order of the boundary-layer thickness.

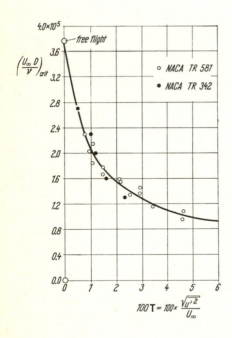

Fig. 18.10.   Relation between the critical Reynolds number of a sphere and the intensity of turbulence of the tunnel, after H. L. Dryden and A. M. Kuethe [8, 10]

The scale of turbulence, L, in a wind tunnel is determined by the mesh of the screens and by the pitch of the directional blades. Since small eddies dissipate their energy faster than the large ones, the mean value of the scale of turbulence, L, increases away from the screens. There exist numerous theoretical and experimental investigations into the development of turbulence behind screens; in this connexion, the works of G. K. Batchelor [1], S. Corrsin [7], G. Charney [5], H. L. Dryden [11], G. D. Huffman [22] and J. E. Green [19], Th. von Kármán [23], C. C. Lin [35, 36], G. I. Taylor [53], and W. Tollmien [60, 61], should be consulted.

References

[1] Batchelor, G.K.: The theory of homogeneous turbulence. Cambridge, 1953, reprint 1970.

[1a] Blake, W.K.: Turbulent boundary layer wall pressure fluctuations on smooth and rough wall. JFM *44*, 637—660 (1970).

[2] Bradshaw, P.: An introduction to turbulence and its measurement. Pergamon Press, 1971.

[3] Bowden, K.F., Frenkiel, F.N., and Tani, I. (ed.): Boundary layers and turbulence. Proc. IUGG/IUTAM Symp. Kyoto 1966, Phys. Fluids Suppl. (1967).

[3a] Bull, M.K.: Wall pressure fluctuations associated with subsonic turbulent boundary flow. JFM *28*, 719—754 (1967).

[4] Burgers, J.M.: A mathematical modell illustrating the theory of turbulence. Advances in Appl. Mech. Vol. *I* (R. von Mises and Th. von Kármán, ed.), New York, 1948.

[5] Charney, G., Comte-Bellot, G., and Mathieu, J.: Development of a turbulent boundary layer on a flat plate in an external turbulent flow. AGARD Conf. Proc. *93*, 27.1—27.10 (1971).

[6] Cooper, R.D., and Tulin, M.P.: Turbulence measurements with the hot-wire anemometer. AGARDograph No. 12 (1955).

[7] Corrsin, S.: Turbulence, experimental methods. Handb. Physik (S. Flügge, ed.), Vol. *VIII/2* Springer-Verlag, Berlin/Göttingen/Heidelberg, 1963.

[8] Dryden, H.L., and Kuethe, A.M.: Effect of turbulence in wind-tunnel measurements. NACA Rep. 342 (1929).

[9] Dryden, H.L.: Reduction of turbulence in wind tunnels. NACA Rep. 392 (1931).

[10] Dryden, H.L., Schubauer, G.B., Mock, W.C., and Skramstad, H.K.: Measurements of intensity and scale of wind-tunnel turbulence and their relation to the critical Reynolds number of spheres. NACA Rep. 581 (1937).

[11] Dryden, H.L.: Turbulence investigations at the National Bureau of Standards. Proc. Fifth Intern. Congress of Appl. Mech., p. 362 (1938).

[12] Dryden, H.L.: Turbulence and the boundary layer. JAS *6*, 85—100 (1939).

[13] Dryden, H.L., and Schubauer, G.B.: The use of damping screens for the reduction of wind-tunnel turbulence. JAS *14*, 221—228 (1947).

[14] Dryden, H.L., and Abbott, J.H.: The design of low turbulence wind tunnels. NACA TN 1755 (1948).

[15] Emmerling, R.: Die momentane Struktur des Wanddruckes einer turbulenten Strömung. Mitteilungen aus dem Max-Planck-Institut für Strömungsforschung und der Aerodynamischen Versuchsanstalt, No. 56 (1973); see also: Emmerling, R., Meier, G.E.A., and Dinkelacker, A.: Investigation of the instantaneous structure of the wall pressure under a turbulent boundary layer flow. AGARD Conf. Proc. No. 131 on Noise Mechanisms, 24.1—24.12 (1974).

[16] Favre, A.J., Gaviglio, J.J., and Dumas, R.J.: Space-time double correlations and spectra in a turbulent boundary layer. JFM *2*, 313—342 (1957); Further space-time correlations of velocity in a turbulent boundary layer. JFM *3*, 344—356 (1958).

[17] Favre, A.J.: La mécanique de la turbulence. Edited by Centre National de la Recherche Scientifique, No. 108, Paris, 1962.

[17a] Fiedler, H. (ed.): Structure and mechanisms of turbulence, Vol. I and II. Proceedings, Berlin, 1977. Lecture Notes in Physics, Vol. 75 and 76, Springer Verlag, 1978.

[18] Frenkiel, F.N.: Turbulence in Geophysics. Publ. by the American Geophysical Union, Washington, D. C., 1962.

[18a] Friedlander, S.K., and Topper, L. (ed.): Turbulence: Classic papers on statistical theory. Interscience Publ., New York, 1961.

[18b] Goering, H. (ed.): Sammelband zur statistischen Theorie der Turbulenz. Akademie-Verlag, Berlin, 1958.

[19] Green, J.E.: On the influence of free stream turbulence on a turbulent boundary layer, as it relates to wind tunnel. AGARD Rep. No. 602 (1973).

[19a] Heisenberg, W.: Zur statistischen Theorie der Turbulenz. Z. Phys. *124*, 628—657 (1948).

[20] Hinze, J.O.: Turbulence. McGraw-Hill, New York, 2nd ed. 1975.

[21] Hoerner, S.: Versuche mit Kugeln betreffend Kennzahl, Turbulenz und Oberflächenbeschaffenheit. Luftfahrtforschung *12*, 42 (1934).

[22] Huffmann, G.D., Zimmermann, D.R., and Benett, W.A.: The effect of free stream turbulence level on turbulent boundary layer behaviour. AGARDograph 164, 89—115 (1972).

[23] von Kármán, Th.: Progress in the statistical theory of turbulence. Proc. Nat. Acad. Sci. Washington *34*, 530 (1948); see also Coll. Works *IV*, 362—371.

[24] Klebanoff, P.S., and Diehl, Z.W.: Some features of artificially thickened fully developed turbulent boundary layers with zero pressure gradient. NACA Rep. 1110 (1952).

[25] Klebanoff, P.S.: Characteristics of turbulence in a boundary layer with zero pressure gradient. NACA Rep. 1247 (1955).

[25a] Kim, H.T., Kline, S.J., and Reynolds, W.C.: The production of turbulence near a smooth wall in a turbulent boundary layer. JFM 50, 133—160 (1971).

[26] Kline, S.J., Reynolds, W.C., Schaub, F.A., and Runstadler, P.W.: The structure of turbulent boundary layers. JFM 30, 741—773 (1967).

[27] Kovasznay, L.S.G.: Turbulent measurements. Sec. F. of Physical Measurements in Gasdynamics and Combustion. High Speed Aerodynamics and Jet Propulsion. Vol. IX (W.R. Ladenburg, ed.), Princeton University Press, 1954, 213—285.

[28] Kovasznay, L.S.G., Kibens, V., and Blackwelder, R.F.: Largescale motion in the intermittent region of a turbulent boundary layer. JFM 41, 283—325 (1970).

[29] Kovasznay, L.S.G.: The turbulent boundary layer. Annual Review of Fluid Mech. 2, 95—112 (1970).

[30] Laufer, J.: Investigation of turbulent flow in a two-dimensional channel. NACA Rep. 1053 (1951).

[31] Laufer, J.: New trends in experimental turbulence research. Annual Review of Fluid Mech. 7, 307—326 (1975).

[32] Laufer, J.: The structure of turbulence in fully developed pipe flow. NACA Rep. 1174 (1954).

[33] Laurence, J.C.: Intensity, scale, and spectra of turbulence in mixing region of free subsonic jet. NACA Rep. 1292 (1956).

[34] Leslie, D.C.: Developments in the theory of turbulence. Clarendon Press, Oxford, 1973.

[35] Lin, C.C.: Statistical theories of turbulence. High Speed Aerodynamics and Jet Propulsion Vol. V, Sec. C, 196—253 (1959), Princeton.

[36] Lin, C.C., and Reid, W.H.: Turbulent flow, theoretical aspects. Handb. Physik (S. Flügge, ed.) Vol. VIII/2, Springer-Verlag, Berlin/Göttingen/Heidelberg, 1963.

[36a] Maréchal, J.: Etude expérimentale de la déformation plane d'une turbulence homogène. J. Mécanique 11, 263—294 (1972).

[36b] Meier, H.U., and Kreplin, H.P.: The influence of turbulent velocity fluctuations and integral length scale of low speed windtunnel flow on the boundary layer development. AIAA 10th Aerodynamic Testing Conference, San Diego, Cal., April 1978. Conf. Proc. No. 783, 232—238 (1978).

[37] Millikan, C.B., and Klein, A.L.: The effect of turbulence. Aircraft Eng. 169 (1933).

[38] Motzfeld, H.: Frequenzanalyse turbulenter Schwankungen. ZAMM 18, 362—366 (1938).

[38a] Mulhearn, P.J.: On the structure of pressure fluctuations in turbulent shear flow. JFM 71, 801—813 (1975).

[39] Nikuradse, J.: Kinematographische Aufnahme einer turbulenten Strömung. ZAMM 9, 495—496 (1929).

[40] Platt, R.C.: Turbulence factors of NACA wind tunnel as determined by sphere tests. NACA Rep. 558 (1936).

[41] Reichardt, H.: Messungen turbulenter Schwankungen. Naturwissenschaften 404 (1938); see also ZAMM 13, 177—180 (1933) and ZAMM 18, 358—361 (1939).

[42] Rotta, J.C.: Turbulente Strömungen. B. G. Teubner, Stuttgart, 1972.

[43] Reynolds, O.: On the dynamical theory of incompressible viscous fluids and the determination of the criterion. Phil. Trans. Roy. Soc. 186, A 123—164 (1895) and Sci. Papers I, 355.

[44] Ribner, H.S., and Tucker, M.: Spectrum of turbulence in a contracting stream. NACA Rep. 1113 (1953).

[45] Rosenblatt, M., and van Atta, C. (ed.): Statistical models and turbulence. Proc. Symp. Univ. California, San Diego (La Jolla), 1971. In: Lecture Notes in Physics 12, Springer-Verlag, 1972.

[46] Rotta, J.C.: Turbulent boundary layers in incompressible flow. Progress in Aeronautical Sciences 2, 1—219 (A. Ferri, D. Küchemann and L.H.G. Sterne, ed.), Pergamon Press, Oxford, 1962.

[47] Taylor, G.I.: The spectrum of turbulence. Proc. Roy. Soc. A 164, 476—490 (1938).

[47a] Schlichting, H.: Neuere Untersuchungen über die Turbulenzentstehung. Naturwissenschaften 22, 376—381 (1934).

[48] Schlichting, H.: Amplitudenverteilung und Energiebilanz der kleinen Störungen bei der Plattenströmung. Nachr. Ges. d. Wiss. Göttingen, Math. Phys. Klasse, Fachgruppe I, 1, 47—78 (1935).

[49] Schubauer, G.B., and Dryden, H.L.: The effect of turbulence on the drag of flat plates. NACA Rep. 546 (1935).

[50] Schubauer, G.B., and Klebanoff, P.S.: Investigation of separation of the turbulent boundary layer. NACA Rep. 1030 (1951).

[51] Simmons, L.F.G., and Salter, C.: An experimental determination of the spectrum of turbulence. Proc. Roy. Soc. A *165*, 73—89 (1938).

[52] Taylor, G.I.: Statistical theory of turbulence. Parts 1—4. Proc. Roy. Soc. London A *151*, 421—478 (1935).

[53] Taylor, G.I.: Statistical theory of turbulence. Part 5, Effect of turbulence on boundary layer. Theoretical discussion of relationship between scale of turbulence and critical resistance of spheres. Proc. Roy. Soc. London A *151*, 307—317 (1936); see also JAS *4*, 311—315 (1937).

[54] Taylor, G.I.: Correlation measurements in turbulent flow through a pipe. Proc. Roy. Soc. A *157*, 537—546 (1936).

[55] Taylor, G.I.: The spectrum of turbulence. Proc. Roy. Soc. London A *164*, 476—490 (1938).

[56] Tennekes, H., and Lumley, J.L.: A first course in turbulence. The MIT Press, 1972.

[57] Tollmien, W.: Turbulente Strömungen. Handb. der Experimentalphysik, Vol. *4*, Part I, 291—339 (1931).

[58] Tollmien, W.: Über die Korrelation der Geschwindigkeitskomponenten in periodisch schwankenden Wirbelverteilungen. ZAMM *15*, 96—100 (1935).

[59] Tollmien, W., and Schäfer, M.: Zur Theorie der Windkanalturbulenz. ZAMM *21*, 1—17 (1941).

[60] Tollmien, W.: Fortschritte der Turbulenzforschung. Zusammenfassender Bericht. ZAMM *33*, 200—211 (1953).

[61] Tollmien, W.: Abnahme der Windkanalturbulenz nach dem Heisenbergschen Austauschansatz als Anfangswertproblem. Wiss. Z. T. H. Dresden *2*, 443—448 (1952/53).

[62] Townsend, A.A.: The structure of turbulent shear flow. Cambridge University Press 2nd ed. 1976.

[63] Thomas, R.M.: Conditional sampling and other measurements in a plane turbulent wake. JFM *57*, 549—582 (1973).

[64] von Weizsäcker, C.F.: Das Spektrum der Turbulenz bei grossen Reynoldsschen Zahlen. Z. Phys. *124*, 614—627 (1948).

[65] Wieghardt, K.: Über die Wirkung der Turbulenz auf den Umschlagpunkt. ZAMM *20*, 58—59 (1940).

[66] Willmarth, W.W.: Pressure fluctuations beneath turbulent boundary layers. Annual Review of Fluid Mech. *7*, 13—38 (1975).

[67] Willmarth, W.W.: Structure of turbulence in boundary layers. Advances in Appl. Mech. Academic Press, New York, *15*, 159—254 (1975).

CHAPTER XIX

Theoretical assumptions for the calculation of turbulent flows

a. Fundamental equations

It is not very likely that science will ever achieve a complete understanding of the mechanism of turbulence because of its extremely complicated nature. The main variables which are of practical interest are the mean velocities, but so far no rational theory which would enable us to determine them by calculation has been formulated. For this reason many attempts have been made to create a mathematical basis for the investigation of turbulent motion with the aid of semi-empirical hypotheses. The empirical assumptions advanced in the past have been developed into more-or-less complete theories, but none of them succeeded in fully analyzing even a single case of turbulent flow. It is necessary to supplement the original hypothesis with additional hypotheses which vary from case to case, and the form of certain functions, or at least certain numerical values, must be derived experimentally. The aim which underlies such empirical theories of turbulence is to deduce the still missing fundamental physical ideas from results of experimental measurements.

The turbulent mixing motion is responsible not only for an exchange of momentum, but it also enhances the transfer of heat and mass in fields of flow associated with non-uniform distributions of temperature or concentration. The methods for the calculation of turbulent flow, temperature, and concentration fields developed so far are based on empirical hypotheses which endeavour to establish a relationship between the Reynolds stresses produced by the mixing motion and the mean values of the velocity components together with suitable hypotheses concerning heat and mass transfer. The momentum equations for the mean motion, eqn. (18.8), as well as the differential equation for temperature (not quoted in Chap. XVIII) cannot acquire a form which is suitable for being integrated unless assumptions of this kind have been introduced beforehand.

J. Boussinesq [7, 8] was the first to work on the problem stated in the preceding section. In analogy with the coefficient of viscosity in Stokes's law for laminar flow

$$\tau_l = \mu \, \frac{\partial u}{\partial y} \, ,$$

he introduced a *mixing coefficient*, A_τ, for the Reynolds stress in turbulent flow by putting

$$\tau_t = - \varrho \, \overline{u' \, v'} = A_\tau \, \frac{\mathrm{d}\bar{u}}{\mathrm{d}y} \, . \tag{19.1}$$

The turbulent mixing coefficient, A_τ, corresponds to the viscosity, μ, in laminar flow and is, therefore, often called "apparent" or "virtual" (also "eddy") viscosity.

The assumption in equation (19.1) has the great disadvantage that the eddy viscosity, A_τ, is not a property of the fluid like μ, but depends itself on the mean velocity $\bar{u}$. This can be recognized if it is noted that viscous forces in turbulent flow are approximately proportional to the square of the mean velocity rather than to its first power as in laminar flow. According to equation (19.1), this would imply that A_τ is approximately proportional to the first power of the mean velocity.

Often, use is made of the apparent (virtual or eddy) kinematic viscosity $\varepsilon_\tau = A_\tau/\varrho$ which is analogous to the kinematic viscosity $\nu = \mu/\varrho$. If this is done, the equations for the shearing stress are rewritten

$$\tau_l = \varrho \cdot \nu \frac{du}{dy}$$

and

$$\tau_t = \varrho \cdot \varepsilon_\tau \frac{d\bar{u}}{dy} . \tag{19.2}$$

It is now possible to introduce into the Navier-Stokes equations for the mean flow, eqns. (18.9), the boundary-layer simplifications. In the case of the velocity boundary layer these will be similar to the considerations discussed in Sec. VII a in connexion with laminar boundary layers. In the case of two-dimensional, incompressible, turbulent flow, with due regard being given to equation (19.1), we obtain the following system of differential equations:

$$\bar{u} \frac{\partial \bar{u}}{\partial x} + \bar{v} \frac{\partial \bar{u}}{\partial y} = -\frac{1}{\varrho} \frac{d\bar{p}}{dx} + \frac{\partial}{\partial y} \left[(\nu + \varepsilon_\tau) \frac{\partial \bar{u}}{\partial y} \right] \tag{19.3a}$$

$$\frac{\partial \bar{u}}{\partial x} + \frac{\partial \bar{v}}{\partial y} = 0 , \tag{19.3b}$$

which should be compared with eqns. (18.12) and (18.13). The preceding set of equations corresponds to equations (7.10) and (7.11) for laminar flow, and the boundary conditions for the velocity components are identical with those in the laminar case, eqn. (7.12).

b. Prandtl's mixing-length theory

The hypotheses in eqns. (19.1) and (19.2) cannot be used for the calculation of actual examples if nothing is known about the dependence of A_τ on velocity. In order to develop the preceding method (initiated by Boussinesq) it is necessary to find empirical relations between the coefficients and the mean velocity. In discussing these, we shall confine ourselves in the present section to the velocity field in incompressible flow because the latter is then independent of the temperature field. The calculation of compressible-flow fields and of temperature fields, and, in particular, of the rates of heat transfer in turbulent motion, will be taken up in detail in Chap. XXIII.

In 1925 L. Prandtl [21] made an important advance in this direction. In developing his hypothesis we shall refer to the simplest case of parallel flow in which the velocity varies only from streamline to streamline. The principal direction of flow is assumed parallel to the x-axis and we have

$$\bar{u} = \bar{u}(y) ; \qquad \bar{v} = 0 ; \qquad \bar{w} = 0 .$$

The preceding type of flow is realized in a rectangular channel for which the results of measurement on turbulent velocity components were given in Figs. 18.3 and 18.4. In the present case only the shearing stress

$$\tau_{xy}' = \tau_t = -\varrho \, \overline{u' v'} = A_\tau \frac{d\bar{u}}{dy} \tag{19.4}$$

remains different from zero.

With L. Prandtl we can now visualize the following simplified mechanism of the motion: as the fluid passes along the wall in turbulent motion, fluid particles coalesce into lumps which move bodily and which cling together for a given traversed length, both in the longitudinal and in the transverse direction, retaining their momentum parallel to x. It will now be assumed that such a lump of fluid, which comes, say, from a layer at $(y_1 - l)$ and has a velocity $\bar{u}(y_1 - l)$, is displaced over a distance l in the transverse direction, Fig. 19.1. This distance l is known as *Prandtl's mixing length* †. As the lump of fluid retains its original momentum, its velocity in the new lamina at y_1 is smaller than the velocity prevailing there. The difference in velocities is then

$$\Delta u_1 = \bar{u}(y_1) - \bar{u}(y_1 - l) \approx l \left(\frac{d\bar{u}}{dy}\right)_1 .$$

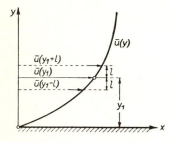

Fig. 19.1.   Explanation of the mixing-length concept

The last expression is obtained by developing the function $u(y_1 - l)$ in a Taylor series and neglecting all higher-order terms. In this transverse motion we have $v' > 0$. Similarly a lump of fluid which arrives at y_1 from the lamina at $y_1 + l$ possesses a velocity which exceeds that around it, the difference being

$$\Delta u_2 = \bar{u}(y_1 + l) - \bar{u}(y_1) \approx l \left(\frac{d\bar{u}}{dy}\right)_1 .$$

† The term *mixture length* has also been used.

Here $v' < 0$. The velocity differences caused by the transverse motion can be regarded as the turbulent velocity components at y_1. Hence we can calculate the time-average of the absolute value of this fluctuation, and we obtain

$$\overline{|u'|} = \tfrac{1}{2}(|\Delta u_1| + |\Delta u_2|) = l \left|\left(\frac{d\overline{u}}{dy}\right)_1\right|. \tag{19.5}$$

Equation (19.5) leads to the following physical interpretation of the mixing length l. The mixing length is that distance in the transverse direction which must be covered by an agglomeration of fluid particles travelling with its original mean velocity in order to make the difference between its velocity and the velocity in the new lamina equal to the mean transverse fluctuation in turbulent flow. The question as to whether the lump of fluid completely retains the velocity of its original lamina as it moves in a transverse direction, or whether it partly assumes the velocity of the crossed lamina and continues beyond it in a transverse direction, is here left entirely open. Prandtl's concept of a mixing length is analogous, up to a certain point, with the mean free path in the kinetic theory of gases, the main difference being that the latter concerns itself with the microscopic motion of molecules, whereas the present concept deals with the macroscopic motion of large agglomerations of fluid particles†.

It may be imagined that the transverse velocity fluctuation originates in the following way: Consider two lumps of fluid meeting in a lamina at a distance y_1, the slower one from $(y_1 - l)$ preceding the faster one from $(y_1 + l)$. In these circumstances the lumps will collide with a velocity $2\,u'$ and will diverge sideways. This is equivalent to the existence of a transverse velocity component in both directions with respect to the layer at y_1. If the two lumps appear in the reverse order they will move apart at a velocity $2\,u'$ and the empty space between them will be filled from the surrounding fluid, again giving rise to a transverse velocity component in the two directions at y_1. This argument implies that the transverse component v' is of the same order of magnitude as u' and we put

$$\overline{|v'|} = \text{const} \cdot \overline{|u'|} = \text{const} \cdot l \frac{d\overline{u}}{dy}. \tag{19.6}$$

In order to find an expression for the shearing stress from eqn. (19.1) it is necessary to investigate the mean value $\overline{u'\,v'}$ a little closer. It follows from the preceding

---

† In analogy with eqn. (19.5) we can write

$$u' = l' \frac{d\overline{u}}{dy}, \tag{19.5a}$$

for the variation of the longitudinal, turbulent component u' with time. Here l' denotes a length which varies with time and which may assume both positive and negative values. Hence, from eqn. (19.2) we obtain

$$\tau_t = -\varrho\,\overline{v'\,l'}\,\frac{d\overline{u}}{dy} = \varrho\varepsilon_\tau\frac{d\overline{u}}{dy}, \tag{19.5b}$$

and the virtual kinematic viscosity becomes

$$\varepsilon_\tau = -\overline{v'\,l'}. \tag{19.5c}$$

representation that the lumps which arrive at layer y_1 with a positive value of v' (upwards from below in Fig. 19.1) give rise "mostly" to a negative u' so that their product $u'\,v'$ is negative. The lumps with a negative value of v' (downwards from above in Fig. 19.1) are "mostly" associated with a positive u' and the product $u'\,v'$ is again negative. The qualifying word "mostly" in the above context expresses the fact that the appearance of particles for which u' has the opposite sign to the above is not completely excluded but is, nevertheless, much less frequent. Thus, the temporal average $\overline{u'\,v'}$ is different from zero, and negative. Hence, we assume,

$$\overline{u'\,v'} = -\, c\, \overline{|u'|} \cdot \overline{|v'|}\,. \tag{19.6a}$$

with $0 < c < 1$ $(c \neq 0)$. Nothing is known about the numerical factor c but, in essence, it appears to be identical with the correlation factor defined in eqn. (18.12). The experimental results plotted in Fig. 18.4 give some idea as to its behaviour. Combining eqns. (19.5) and (19.6) we now obtain

$$\overline{u'\,v'} = -\, \text{const} \cdot l^2 \left(\frac{d\bar{u}}{dy}\right)^2\,.$$

It should be noted that the constant in the above equation is different from that in eqn. (19.6), as the former also contains the factor c from eqn. (19.6a). The constant can now be included with the still unknown mixing length, and we may write

$$\overline{u'\,v'} = -\, l^2 \left(\frac{d\bar{u}}{dy}\right)^2\,. \tag{19.6b}$$

Consequently, the shearing stress from eqn. (19.1) can be written as

$$\tau_t = \varrho\, l^2 \left(\frac{d\bar{u}}{dy}\right)^2\,. \tag{19.6c}$$

Taking into account that the sign of τ_t must change with that of $d\bar{u}/dy$, it is found that it is more correct to write

$$\tau_t = \varrho\, l^2 \left|\frac{d\bar{u}}{dy}\right| \frac{d\bar{u}}{dy}\,. \tag{19.7}$$

This is *Prandtl's mixing-length hypothesis*. It will be shown later that it is very useful in the calculation of turbulent flows.

Comparing eqn. (19.7) with the Boussinesq hypothesis in eqn. (19.1), we find the following expressions for the virtual viscosity

$$A_\tau = \varrho\, l^2 \left|\frac{d\bar{u}}{dy}\right| \tag{19.8a}$$

and for the virtual kinematic viscosity from eqn. (19.2)

$$\varepsilon_\tau = l^2 \left|\frac{d\bar{u}}{dy}\right|\,. \tag{19.8b}$$

It is known from experimental evidence that turbulent drag is roughly proportional to the square of velocity and the same result is obtained from eqn. (19.7) if the mixing length is assumed to be independent of the magnitude of velocity. The mixing length, unlike viscosity in Stokes's law, is still not a property of the fluid, but it is, at least, a purely local function.

In numerous cases it is possible to establish a simple relation between the mixing length, l, and a characteristic length of the respective flow. For example, in flows along smooth walls l must vanish at the wall itself, because transverse motions are inhibited by its presence. In flows along rough walls the mixing length near the wall must tend to a value of the same order of magnitude as the solid protrusions.

Prandtl's equation (19.7) has been successfully applied to the study of *turbulent motion along walls* (pipe, channel, plate, boundary layer), and to the problem of so-called *free turbulent flow*. The latter term refers to flow without solid walls, such as the mixing of a jet with the surrounding still air. Examples of such applications will be given in Chaps. XX, XXI, and XXIV. R. A. M. Galbraith et al. [13a] provided good experimental support for the utility of the mixing-length concept.

c. Further assumptions for the turbulent shearing stress

Prandtl's equation (19.7) for shearing stress in turbulent flow is still unsatisfactory in that the apparent, kinematic viscosity ε, eqn. (19.7b), vanishes at points where $d\bar{u}/dy$ is equal to zero, i. e. at points of maximum or minimum velocity. This is certainly not the case because turbulent mixing does not vanish at points of maximum velocity (centre of channel). The latter view is confirmed by Reichardt's measurements on turbulent fluctuations, Fig. 18.3, which show that in the centre of the channel the longitudinal and transverse fluctuations both differ from zero.

In order to counter these difficulties L. Prandtl [23] established a considerably simpler equation for the apparent kinematic viscosity. It is valid only in the case of free turbulent flow and was derived from extensive experimental data on free turbulent flow due to H. Reichardt [24]. In setting up this new hypothesis L. Prandtl assumed that the dimensions of the lumps of fluid which move in a transverse direction during turbulent mixing are of the same order of magnitude as the width of the mixing zone. It will be recalled that the previous hypothesis implied that they were small compared with the transverse dimensions of the region of flow. The virtual kinematic viscosity, ε, is now formed by multiplying the maximum difference in the time-mean flow velocity with a length which is assumed to be proportional to the width, b, of the mixing zone. Thus,

$$\varepsilon_\tau = \varkappa_1 \, b \, (\bar{u}_{max} - \bar{u}_{min}) . \tag{19.9}†$$

Here, $\varkappa_1$ denotes a dimensionless number to be determined experimentally. It follows from eqn. (19.9) that ε remains constant over the whole width of every cross-section, whereas the previous hypothesis (19.7b) implied that it varied even if the mixing length were assumed to be constant. From eqns. (19.9) and (19.1) we obtain that the

---

† On comparing this equation with eqn. (19.5c), it is seen that according to the present hypothesis the transverse fluctuation v' is proportional to $\bar{u}_{max} - \bar{u}_{min}$, and that the mixing length l' is proportional to the width b. An alternative hypothesis which relates to the apparent kinematic viscosity ε_τ and is very similar to that in eqn. (19.9) was formulated by H. Reichardt [24].

turbulent shearing stress is given by

$$\tau_t = \varrho\, \varkappa_1\, b\, (\bar{u}_{max} - \bar{u}_{min}) \frac{d\bar{u}}{dy}. \tag{19.10}$$

Examples of the application of this hypothesis are also given in Chap. XXIV.

According to T. Cebeci and A. M. O. Smith [11a], a calculation procedure based on a combination of the mixing-length method expressed by eqn. (19.7) and an assumption of the kind represented by eqn. (19.10) has withstood the test of time. The inner part of the boundary layer $(0 \leq y \leq y_k)$ is described by the van Driest [12] mixing-length formula

$$l = \varkappa\, y \left[1 - \exp\left(-\frac{y \sqrt{\tau_0/\varrho}}{\nu\, A}\right)\right], \tag{19.11}$$

(see also eqn. (20.15b)). Employing eqn. (19.8) in the region $0 \leq y \leq y_k$, we find that

$$\varepsilon_i = \left\{\varkappa\, y \left[1 - \exp\left(-\frac{y \sqrt{\tau_0/\varrho}}{\nu\, A}\right)\right]\right\}^2 \cdot \frac{d\bar{u}}{dy}. \tag{19.12}$$

As far as the outer portion of the boundary layer is concerned, the hypothesis to be used is suggested by eqn. (19.9). In this range we assume that

$$\varepsilon_a = \varkappa_2\, U\, \delta_1\, \gamma, \tag{19.13}$$

where the originally constant ε_a is now multiplied by the intermittency factor γ. In turn, and in accordance with the measurements of P. Klebanoff (see Fig. 18.6), the intermittency factor is approximated by the relation

$$\gamma = [1 + 5\cdot 5\, (y/\delta)^6]^{-1}. \tag{19.14}$$

The constants are given by the values $\varkappa = 0\cdot 4$, $A = 26$ and $\varkappa_2 = 0\cdot 017$. The value of y_k which delineates the range of validity of eqn. (19.12) from eqn. (19.13) is obtained from the requirement that the apparent kinematic viscosities must assume equal values there. Thus we put

$$\varepsilon_i = \varepsilon_a \quad \text{for} \quad y = y_k. \tag{19.15}$$

See also ref. [9b].

A result, similar to the one contained in eqn. (19.7), has been obtained by G. I. Taylor [32] on the basis of his vorticity transport theory. In Prandtl's theory, the assumption is made that the mean velocity $\bar{u}$ remains constant during the transverse motion of a lump of fluid; Taylor's theory substitutes for this the hypothesis that the rotation, that is that $d\bar{u}/dy$ remains constant. This yields the equation

$$\tau' = \frac{1}{2}\, \varrho\, l_w^2\, \left|\frac{d\bar{u}}{dy}\right|\frac{d\bar{u}}{dy}, \tag{19.15a}$$

which differs from eqn. (19.7) merely by the factor 1/2. This means that the mixing length of G. I. Taylor's vorticity-transfer theory is larger by a factor $\sqrt{2}$ than that in L. Prandtl's momentum-transfer theory. Thus $l_w = \sqrt{2}\, l$. On the basis of his considerations, G. I. Taylor concluded that the diffusion of temperature differences and vorticity in the mixing zone behind a cylindrical rod occur in conformity with identical laws. This is in essential agreement with experiments, and the explanation turns on the fact that here the axes of the vortices arrange themselves principally at right angles to the main stream and to the direction of the velocity gradient. By contrast, in a flow field in the proximity of a solid wall there predominate vortices whose axes are parallel to the flow direction. For this reason, the temperature field becomes similar to the velocity field directly.

d. Von Kármán's similarity hypothesis

It would be very convenient to possess a rule which allowed us to determine the dependence of mixing length on space coordinates. Th. von Kármán [17] made an attempt to establish such a rule assuming that turbulent fluctuations are similar at all point of the field of flow (*similarity rule*), i. e. that they differ from point to point only by time and length scale factors. A velocity which is characteristic of the turbulent, fluctuating motion can be formed with the aid of the turbulent shearing stress by defining it, with the aid of eqn. (19.1), as follows:

$$v_* = \sqrt{\frac{|\tau_t|}{\varrho}} = \sqrt{|\overline{u'\, v'}|} \, . \tag{19.16}$$

The quantity v_* is called the *friction velocity* and is a measure of the intensity of turbulent eddying and of the correlation which exists between the fluctuating components in the x and y directions. For the similarity rule under consideration we imagine a two-dimensional mean flow in the x direction, such that $\bar{u} = \bar{u}(y)$ and $\bar{v} = 0$ (parallel flow), and an auxiliary motion which is also two-dimensional. In this case it is possible to show that the rule that

$$l \sim \frac{d\bar{u}/dy}{d^2\bar{u}/dy^2} \tag{19.17}$$

constitutes a necessary condition to secure compatibility between the similarity hypothesis and the vorticity transport equation (4.10).

Introducing an empirical dimensionless constant $\varkappa$, von Kármán made the assumption that the mixing length satisfies the equation:

$$l = \varkappa \left| \frac{d\bar{u}/dy}{d^2\bar{u}/dy^2} \right| . \tag{19.18}$$

In accordance with the above hypothesis, the mixing length, l, is independent of the magnitude of velocity, being a function of the velocity distribution only. The mixing length becomes a purely local function as already required earlier, and the constant $\varkappa$ in eqn. (19.18), can only be determined empirically. It is a universal dimensionless constant which must have the same value for all turbulent flows, provided that the assumptions made previously are satisfied (parallel flow).

Introducing, finally, eqn. (19.18) into eqn. (19.6c), we find that the turbulent shearing stress is

$$\tau_t = \varrho\, \varkappa^2\, \frac{(d\bar{u}/dy)^4}{(d^2\bar{u}/dy^2)^2} \, . \tag{19.19}$$

A. Betz [4] gave a very lucid derivation of eqn. (19.18). In later times von Kármán's hypothesis has been extended to include compressible turbulent flows too [20a]; see the paper by C.C. Lin et al. and the observations made by G. Hamel [9] and O. Bjorgum [6].

e. Universal velocity-distribution laws

Both von Kármán's law of turbulent friction, eqn. (19.19), and Prandtl's law, eqn. (19.7), are easily applied to the problem of finding the velocity distribution in a rectangular channel. Since this universal law is of fundamental importance for the considerations in the succeeding chapters, and since it is applicable to circular channels as well, we shall devote a little space to its derivation.

The channel will be assumed to have a width $2h$ and the x-axis will be placed along its centre-line, the coordinate y being measured from the latter. We shall assume a constant pressure gradient along the axis, putting $\partial \bar{p}/\partial x = C$†. Since $-\partial \bar{p}/\partial x + \partial \tau/\partial y = 0$, the shearing stress is a linear function of the width of the channel, i. e.

$$\tau = \tau_0 \frac{y}{h} \, , \tag{19.20}$$

where τ_0 denotes the stress at the wall.

1. Von Kármán's velocity-distribution law. Applying von Kármán's similarity rule, eqn. (19.19), to eqn. (19.20), we obtain

$$\frac{\tau_0}{\varrho} \frac{y}{h} = \varkappa^2 \frac{(du/dy)^4}{(d^2u/dy^2)^2} \, .$$

Integrating twice and determining the constants of integration from the condition that $u = u_{max}$ at $y = 0$, we have

$$u = u_{max} + \frac{1}{\varkappa} \sqrt{\frac{\tau_0}{\varrho}} \left\{ \ln \left[1 - \sqrt{\frac{y}{h}} \right] + \sqrt{\frac{y}{h}} \right\} .$$

Introducing the frictional velocity at the wall, $v_{*0} = \sqrt{\tau_0/\varrho}$, we can rewrite the last equation in dimensionless form

$$\frac{u_{max} - \bar{u}}{v_{*0}} = -\frac{1}{\varkappa} \left\{ \ln \left[1 - \sqrt{\frac{y}{h}} \right] + \sqrt{\frac{y}{h}} \right\} ; \tag{19.21}$$

(y=distance from centre-line). This is the form in which the universal velocity-distribution law was deduced by von Kármán [17]. It is shown as curve (2) in Fig. 19.2. The predicted velocity-distribution curve has a kink near the centre-line of the channel which is due to the fact that the requirement of similarity cannot be satisfied here, because, in accordance with eqn. (19.18), the mixing length becomes equal to zero at the centre. At the wall, for $y = h$, eqn. (19.21) leads to an infinitely large velocity which is explained by the fact that molecular friction has been neglected in comparison with the

---

† From this point onwards we shall omit the bar above the symbol to denote time-averages as confusion with time-dependent quantities is no longer possible.

apparent, turbulent friction. This assumption breaks down in the neighbourhood of the wall where the turbulent boundary layer goes over into the laminar sub-layer. We are thus led to additional considerations which will be given later. In what follows we shall, therefore, exclude from consideration a small region near the centre-line and a small region near the wall. It is particularly remarkable that the universal velocity-distribution law in its form given by eqn. (19.21) does not contain either the roughness or the Reynolds number explicitly†. The velocity-distribution law in eqn. (19.21), also known as the *velocity-defect law*, can be expressed in words as follows: The velocity-distribution curves for a rectangular channel can be made to coincide if the difference $u_{max} - u$ is made dimensionless with the aid of the friction velocity at the wall, v_{*0}, and plotted against y/h. This result, which turns out to be valid for circular pipes as well, will be compared with experimental measurements in Sec. XX c.

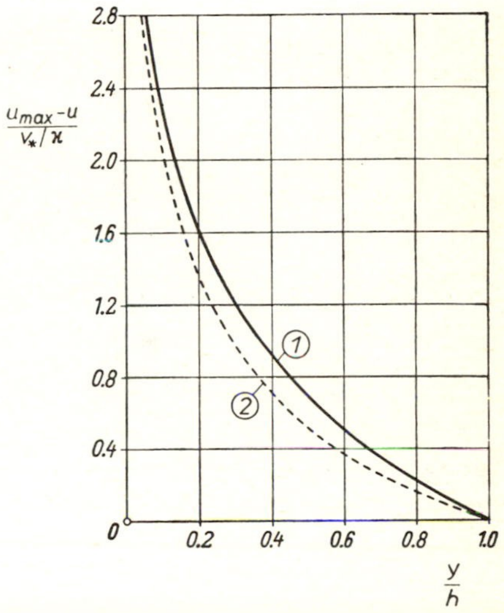

Fig. 19.2. Universal velocity distribution law for turbulent channel flow, after von Kármán and Prandtl [17, 21]

Curve (1) corresponds to eqn. (19.28);
Curve (2) corresponds to eqn. (19.21).
y — distance from wall

2. Prandtl's velocity-distribution law. A similar velocity-distribution law can also be deduced from Prandtl's hypothesis for the turbulent shearing stress, eqn. (19.7). In the process of deriving the relevant expression we shall gain more insight into the conditions which prevail in the immediate neighbourhood of the wall and which we were forced to exclude from the preceding argument. We shall consider a turbulent stream along a smooth flat wall and we shall denote the distance from the wall by the symbol y, with $u(y)$ denoting the velocity. In the neighbourhood of the wall we shall assume proportionality between mixing length and wall distance, so that

$$l = \varkappa y \,. \tag{19.22}$$

Here $\varkappa$ denotes a dimensionless constant which must be deduced from experiment. This assumption is reasonable, because the turbulent shearing stress at the wall is

---

† They are, of course, contained implicitly in the wall stress τ_0.

zero owing to the disappearance of the fluctuations. Hence, according to Prandtl's assumption, the turbulent shearing stress becomes

$$\tau = \varrho \, \varkappa^2 \, y^2 \left(\frac{du}{dy}\right)^2 .$$ (19.23)

At this stage Prandtl introduced an additional, far-reaching assumption, namely that the shearing stress remains constant, i. e. that $\tau = \tau_0$, where τ_0 denotes the shearing stress at the wall. Introducing once more the friction velocity

$$v_{*0} = \sqrt{\frac{\tau_0}{\varrho}} ,$$ (19.24)

we obtain

$$v_{*0}^2 = \varkappa^2 \, y^2 \left(\frac{du}{dy}\right)^2$$ (19.25)

or

$$\frac{du}{dy} = \frac{v_{*0}}{\varkappa \, y} .$$ (19.26)

On integrating we have

$$u = \frac{v_{*0}}{\varkappa} \ln y + C .$$ (19.27)

Here the constant of integration, C, must be determined from the condition at the wall and serves to fit the turbulent velocity distribution to that in the laminar sub-layer. However, even without determining C it is possible to deduce from eqn. (19.27) a law analogous to that in eqn. (19.21). In spite of the fact that eqn. (19.27) is valid only in the neighbourhood of the wall, because of the assumption that $\tau = $ const., we shall attempt to use it for the whole region, i. e. up to $y = h$. Since at $y = h$ we have $u = u_{max}$, we obtain

$$u_{max} = \frac{v_{*0}}{\varkappa} \ln h + C ,$$

and hence, by forming the velocity difference, we deduce

$$\frac{u_{max} - u}{v_{*0}} = \frac{1}{\varkappa} \ln \frac{h}{y} ; \quad (y = \text{distance from wall}) .$$ (19.28)

This universal velocity-defect law due to Prandtl is shown plotted as curve (1) in Fig. 19.2. In the preceding argument we succeeded in deriving a universal velocity-distribution law from Prandtl's law of friction in complete analogy with that in eqn. (19.21), which was obtained from von Kármán's similarity rule. The only difference is in the form of the functions of y/h which appear on the right-hand side of eqns. (19.21) and (19.28) respectively. On reflexion this will not appear incomprehensible, if we take into account the difference in the assumption concerning the shearing stress. Von Kármán assumed a linear shearing-stress distribution, the mixing length being $l \sim u'/u''$. On the other hand, Prandtl assumed a constant shearing stress and $l \sim y$. Figure 19.2 contains a comparison between these two laws. A further comparison with experiment is deferred to Chap. XX.

It may be worth noting in passing that it is possible to obtain the simple result that $l = \varkappa y$ from the velocity-defect law (19.27), together with von Kármán's equation for mixing length. This may be easily verified by the reader. Finally, it should be noted that the preceding argument proves that the coefficients $\varkappa$ in eqns. (19.22) and (19.18) are identical.

We shall now revert to the problem of determining the constant of integration, C, in eqn. (19.27). As already mentioned, the constant should be determined from the condition that the turbulent velocity distribution must join the laminar velocity distribution in the immediate neighbourhood of the wall where the laminar and turbulent shearing stresses are of the same order of magnitude. We determine the constant of integration C from the condition that $u = 0$ at a certain distance y_0 from the wall. In this manner

$$u = \frac{v_{*0}}{\varkappa} (\ln y - \ln y_0) \,. \tag{19.29}$$

The distance y_0 is of the order of magnitude of the thickness of the laminar sub-layer. Using a dimensional argument we find that this distance y_0 is proportional to the ratio ν/v_{*0} of the kinematic viscosity, ν, and the friction velocity, v_{*0}, as its dimension is that of a length. We may thus put

$$y_0 = \beta \frac{\nu}{v_{*0}} \,, \tag{19.30}$$

where β denotes a dimensionless constant. Substituting β into eqn. (19.29) we obtain

$$\frac{\bar{u}}{v_{*0}} = \frac{1}{\varkappa} \left(\ln \frac{y v_{*0}}{\nu} - \ln \beta \right) \tag{19.29 a}$$

which is the dimensionless, logarithmic, universal velocity-distribution law, and asserts that the velocity, referred to the friction velocity v_{*0}, is a function of the dimensionless wall distance, $y v_{*0}/\nu$. The latter is a kind of Reynolds number based on the wall distance, y, and on the friction velocity at the wall. Equation (19.29 a) contains the two empirical constants, $\varkappa$ and β. In accordance with the previous reasoning we may expect that the constant $\varkappa$ is independent of the nature of the wall (whether smooth or rough) and that it is, moreover, a universal constant of turbulent flow. Experimental results, to be discussed in greater detail in the succeeding chapter, give a value of $\varkappa = 0.4$. The second constant, β, depends on the nature of the wall surface; relevant numerical values will be given in Chap. XX.

Introducing the abbreviations

$$\frac{u}{v_{*0}} = \phi \,, \tag{19.31}$$

$$\frac{y v_{*0}}{\nu} = \eta \,, \tag{19.32}$$

we can shorten eqn. (19.29 a) to read

$$\phi(\eta) = A_1 \ln \eta + D_1 \,, \tag{19.33}$$

where

$$A_1 = \frac{1}{\varkappa} = 2.5; \qquad D_1 = -\frac{1}{\varkappa} \ln \beta \,. \tag{19.34}$$

The universal velocity-distribution law, eqn. (19.33), which has now been derived for the case of a flat wall (rectangular channel) retains its fundamental importance for flows through circular pipes, as will be seen in the next chapter. We may now state, in anticipation, that it leads to good agreement with experiment.

In concluding this chapter it may be worth stressing once again that the two universal velocity-distribution laws in eqns. (19.21) and (19.27) were obtained for turbulent flow, and took into account, apart from the small sub-layer near the wall, only turbulent shearing stresses, and it should be realized that such an assumption is satisfied at large Reynolds numbers only. Consequently the velocity-distribution law, particularly that in eqn. (19.33), must be regarded as an asymptotic law applicable to very large Reynolds numbers. For smaller Reynolds numbers, when laminar friction exerts some influence outside the very thin sub-layer, experiment leads to a power law of the form

$$\phi(\eta) = C \, \eta^n \tag{19.35}$$

or

$$\frac{u}{v_{*0}} = C \left(\frac{y \, v_{*0}}{\nu} \right)^n,$$

where the exponent n is approximately equal to $\frac{1}{7}$, but varies somewhat with the Reynolds number. This point will also be taken up again in the succeeding chapter.

The case of so-called Couette flow between two parallel flat plates which are displaced relative to each other (Fig. 1.1) constitutes a very simple example of a flow in which the shearing stress remains constant. The shearing stress τ remains

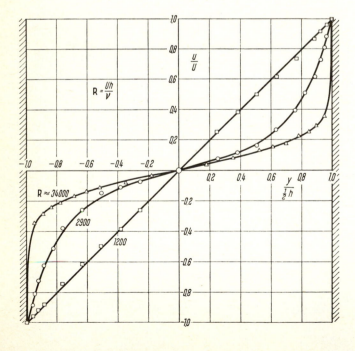

Fig. 19.3. Velocity profiles in parallel Couette flow between two parallel plates moving in opposite directions, after H. Reichardt [25, 26]

At R = 1200 the flow is laminar; at R = 2900 and 34,000 the flow is turbulent rigorously constant in turbulent as well as in laminar flow, and is equal to that at the wall, τ_0. H. Reichardt [25, 26] carried out an extensive investigation of this case; some of his results can be inferred from Fig. 19.3 which shows several velocity profiles observed in Couette flow. The flow remains laminar as long as the Reynolds number $R < 1500$ and the velocity distribution is then linear to a good degree of approximation. When the Reynolds number R exceeds the value 1500 the flow is turbulent. The turbulent velocity profiles are very flat near the centre and become very steep near the walls. A profile of this kind is to be expected in turbulent flow if it is remembered that the shearing stress consists of a laminar contribution

$$\tau_l = \mu \left(\frac{du}{dy} \right)$$

and a turbulent contribution

$$\tau_t = A_\tau \left(\frac{d\bar{u}}{dy} \right)$$

due to turbulent mixing. Hence

$$\tau = \tau_0 = (\mu + A_\tau) \frac{d\bar{u}}{dy} ,$$

where A_τ denotes the mixing coefficient defined in eqn. (19.1). In this manner the velocity gradient turns out to be proportional to $1/(\mu + A)$. Since A varies from zero at the wall to its maximum in the centre of the channel, the velocity profile must become steep at the wall and flat at the centre, as confirmed by the plots in Fig. 19.3. The turbulent mixing coefficient increases with an increasing Reynolds number and the curvature of the velocity profile becomes, correspondingly, more pronounced; compare the paper by A. A. Szeri [31a].

f. Further development of theoretical hypotheses

The calculation of turbulent flows on the basis of the different semi-empirical hypotheses discussed previously, and carried out in detail in the succeeding chapters, is not satisfactory in so far as it is still impossible to analyze different kinds of turbulent flow with the aid of the same hypothesis concerning turbulent friction. For example, Prandtl's hypothesis on the mixing length, eqn. (19.7), fails completely in the case of so-called isotropic turbulence as it exists behind a screen of fine mesh, because in this case the velocity gradient of the basic flow is equal to zero everywhere. The hypotheses for the calculation of developed turbulent flow, discussed in Secs. XIXb and c, have been considerably extended by L. Prandtl [22] in an attempt to derive a universally valid system of equations (turbulent flow near wall, free turbulent flow, isotropic turbulence).

Energy equation: L. Prandtl based his new development on the consideration of the kinetic energy of turbulent fluctuation, $E = \frac{1}{2} \varrho(\overline{u'^2} + \overline{v'^2} + \overline{w'^2})$, and calculated the change of the energy of the subsidiary motion with time, DE/Dt, for a particle which moves with the basic stream. This is composed of three terms: of the decrease in energy due to internal friction in the motion of the lumps of fluid, of the transfer of energy from the basic motion to the subsidiary motion — this term being proportional to $(dU/dy)^2$ — and, finally, of the transfer of kinetic energy from the more turbulent to the less turbulent zones. The energy balance between these three terms leads to a differential equation for the energy of the turbulent subsidiary motion which must be added to the system of differential equations for the mean motion; it has the form

$$\frac{DE}{Dt} = \frac{\partial E}{\partial t} + \bar{u} \frac{\partial E}{\partial x} + \bar{v} \frac{\partial E}{\partial y} = \tag{19.36}$$

$$= - c \underbrace{\frac{E^{3/2}}{L}}_{\text{dissipation}} + \underbrace{\frac{\tau_t}{\varrho} \frac{\partial \bar{u}}{\partial y}}_{\text{production}} + \underbrace{\frac{1}{y^j} \frac{\partial}{\partial y} \left(y^j \, k_q \, E^{1/2} \, L \frac{\partial E}{\partial y} \right)}_{\text{diffusion}} .$$

Here $j = 0$ for two-dimensional mean flows, $j = 1$ for axially symmetric mean flows (y-radial distance from axis). L. Prandtl referred to the preceding as to the first fundamental equation. A second equation relates the turbulent shearing stress with the velocity gradient of the mean flow and is analogous to the old mixing equation (19.2), but also contains the energy of the turbulent subsidiary motion, that is

$$\tau_t = \varrho \, k \, E^{1/2} \, L \, \partial \bar{u}/\partial y \, . \tag{19.37}$$

The two equations — (19.36) and (19.37) — contain the three free constants c, k, k_q which must be derived by a reference to experimental results. The length scale L is a local function which represents, essentially, the mixing length of eqn. (19.7). The definition of this quantity can, however, also be based on an integral of the correlation function of the velocity components measured at two points (see J.C. Rotta [29], p. 177 ff).

If the structure of turbulence does not change along a streamline, as is the case with the logarithmic law of the wall, and if it is assumed that there is no diffusion of turbulent energy, it is possible to equate the first two terms on the right-hand side of eqn. (19.36) and to show that

$$E = \frac{k}{c} \, L^2 \left(\frac{\partial \bar{u}}{\partial y} \right)^2 \tag{19.38}$$

in view of eqn. (19.37). Putting $\varrho \, E = |\tau_t| \, k/c$ we recognize that eqn. (19.38) reduces to the mixing-length formula of eqn. (19.7) with $l = L$. Finally, if we equate τ_t from eqns. (19.37) and (19.7) we derive that

$$c = k^3 \, . \tag{19.39}$$

The preceding relation is used as a definition of k and this leaves only two adjustable constants, namely c and k_q. Various investigations appear to indicate that $c = 0 \cdot 165$ and $k_q = 0 \cdot 6$ are the appropriate numerical values. Thus, finally, equations (19.38) and (19.39) yield

$$\tau_t = c^{2/3} \, \varrho \, E \approx 0 \cdot 3 \, \varrho \, E \, . \tag{19.40}$$

In the presence of homogeneous turbulence, such as that found behind a screen, only the first of the above three turbulent energy terms is present and for this reason the turbulence behind the screen decays downstream. In the case of channel flow all three terms are present, but the third term (transfer of kinetic energy from more to less turbulent regions) is significant only near the wall where, owing to the vigorous creation of new turbulence by wall shear, there exists a zone of particularly high turbulence (*cf.* Fig. 18.3), and near the centre, where no turbulence is created and where the flow is, therefore, much less turbulent. G.S. Glushko [13] extended the preceding method to include the effects of the Reynolds number and performed calculations for the turbulent boundary layer on a flat plate. The calculation covered the region of transition into a viscous sublayer as well as transition from laminar to turbulent flow. I.E. Beckwith and D.M. Bushnell [5] repeated the same calculations and extended their applicability to boundary layers with variable pressure gradients. They examined the effect of such modifications on the numerical values of the empirical constants.

Bradshaw's method: In the method proposed by P. Bradshaw, D.H. Ferriss and N.P. Atwell [10] — which is designed principally for the calculation of turbulent boundary layers — the expression for the turbulent shearing stress from eqn. (19.37) is replaced by a linear dependence

$$\tau_t = 2 \, a_1 \, \varrho \, E \tag{19.41}$$

on turbulent energy; this corresponds to eqn. (19.40) with $a_1 = 0 \cdot 15$. Thus the energy equation (19.36) has been transformed into a differential equation for the turbulent shearing stress. In the case of two-dimensional mean flows, we obtain

$$\bar{u} \, \frac{\partial}{\partial x} \left(\frac{\tau_t}{2 \, a_1 \, \varrho} \right) + \bar{v} \, \frac{\partial}{\partial y} \left(\frac{\tau_t}{2 \, a_1 \, \varrho} \right) =$$

$$= -\frac{(\tau_t/\varrho)^{2/3}}{L} + \frac{\tau_t}{\varrho} \, \frac{\partial \bar{u}}{\partial y} - \left(\frac{\tau_{max}}{\varrho} \right)^{1/2} \frac{\partial}{\partial y} \left(G \, \frac{\tau_t}{\varrho} \right) \, . \tag{19.42}$$

The length scale L is assumed to be a function of y/δ. Thus,

$$L = \delta \, f_1 \, (y/\delta).$$ (19.43)

In contrast to Prandtl who thought of the diffusion of turbulent energy as the migration of energy from regions where its value is high to those where it is low, Bradshaw and collaborators assume that the flux of diffusion of turbulent energy is proportional to $(\tau_{max})^{1/2} \, \tau_t$. Here τ_{max} denotes the maximum value of the shearing stress in the interval $0{\cdot}25 \, \delta \leqslant y \leqslant \delta$. The function G in eqn. (19.42) is defined as

$$G = (\tau_{max}/\varrho \, U^2)^{1/2} \, f_2 \, (y/\delta),$$ (19.44)

where f_2 is a second function of y/δ. The two functions, f_1 and f_2 ,have been plotted in Fig. 19.4. Bradshaw's method has found favour for the calculation of turbulent boundary layers on smooth and rough surfaces; its validity has been extended to include compressible flows as well as suction and blowing.

Fig. 19.4. Empirical functions f_1 and f_2 which occur in P. Bradshaw's [10] method of calculating turbulent boundary layers; see eqns. (19.42) to (19.44)

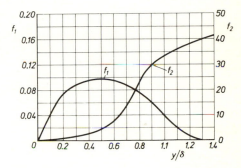

Differential equation for the length scale: The application of the differential equations (19.36) and (19.37) for the calculation of a velocity field with prescribed initial and boundary conditions cannot be performed unless some statement is made concerning the length scale L in the same way as in the case with the mixing length l in eqn. (19.7). For this reason, J.C. Rotta [28, 29] complemented eqns. (19.36) and (19.37) with a third equation which serves to calculate L. Numerical calculations [32 a] have demonstrated that the preceding system of equations leads to satisfactory agreement with experiment without requiring any additional assumptions. This is true for many specific configurations, including channel flow, pipe flow, two-dimensional and axially symmetric jets, free jet surface and two-dimensional wake. In modern times many additional calculation methods have been proposed; they are in essence, similar to those discussed here. Recent developments in this field have been summarized by P. Bradshaw [9a], B. E. Launder and D. B. Spalding [20], J.C. Rotta [30], G. L. Mellor and H. J. Herring [20b] and W.C. Reynolds [26a].

Curvature of streamlines: The curvature of streamlines in turbulent flow produces unexpectedly large changes in the structure of the turbulence. In most cases, such changes are one order of magnitude more important than is the effect of the pressure gradient normal to the boundary layer. In the case of ordinary aerofoils, the effects on momentum and heat transfer are considerable, becoming extremely large in the case of turbine and compressor blades. The present-day status of our knowledge of these curvature effects, and our ability to master them by calculation, have been discussed by P. Bradshaw [11].

References

[1] Batchelor, G.K.: Energy decay and self-preserving correlation functions in isotopic turbulence. Quart. Appl. Math. *6*, 97—116 (1948).

[2] Batchelor, G.K., and Townsend, A.A.: The nature of turbulent motion at large wave numbers. Proc. Roy. Soc. London A *199*, 238—255 (1949).

[3] Batchelor, G.K.: The theory of homogeneous turbulence. Cambridge University Press, 1953.

[4] Betz, A.: Die von Kármánsche Ähnlichkeitsüberlegung für turbulente Vorgänge in physikalischer Auffassung. ZAMM *11*, 397 (1931).

[5] Beckwith, I.E., and Bushnell, D.M.: Detailed description and results of a method for computing mean and fluctuating quantities in turbulent boundary layers. NASA TN D-4815 (1968).

[6] Bjorgum, O.: On the steady turbulent flow along an infinitely long smooth and plane wall. Universitet i. Bergen, Arbok, Naturvitenskapelig rekke No. 7, (1951).

[7] Boussinesq, J.: Essai sur la théorie des eaux courantes. Mém. prés. Acad. Sci. *XXIII*, 46, Paris (1877).

[8] Boussinesq, J.: Theorie de l'écoulement tourbillonant et tumultueux des liquides dans les lits rectilignes à grande section (tuyaux de conduite et cannaux découverts), quand cet écoulement s'est régularisé en un régime uniforme, c'est-à-dire, moyennement pareil a travers toutes les sections normales du lit. Comptes Rendus de l'Académie des Sciences *CXXII*, p. 1290—1295 (1896).

[9] Hamel, G.: Streifenmethode und Ähnlichkeitsbetrachtungen zur turbulenten Bewegung. Abhandl. preuss. Akad. Wiss., Math. Naturwiss. Klasse, Nr. 8 (1943).

[9a] Bradshaw, P.: The understanding and prediction of turbulent flow. Aeronautical J. *76*, 403—418 (1972).

[9b] Bradshaw, P.: An improved Van Driest skin friction formula for compressible turbulent boundary layers. AIAA J. *15*, 212—214 (1975).

[10] Bradshaw, P., Ferriss, D.H., and Atwell, N.P.: Calculation of boundary-layer development using the turbulent energy equation. JFM *28*, 593—616 (1967).

[11] Bradshaw, P.: Effects of streamline curvature on turbulent flow. AGARDograph No. 169 (1973).

[11a] Cebeci, T., and Smith, A.M.O.: A finite-difference solution of the incompressible turbulent boundary layer equations by an eddy-viscosity concept. In: Kline, S.J., Morkovin, M.V., Sovran, G., and Cockrell, D.J. (eds.): Computation of turbulent boundary layers. Vol. *1*: Methods, prediction, and flow structure, 346—355 (1973).

[12] Van Driest, E.R.: On turbulent flow near a wall. JAS *23*, 1007—1011 (1956).

[12a] Eckelmann, H.: Experimentelle Untersuchungen in einer turbulenten Kanalströmung mit starken viskosen Wandschichten. Mitt. Max-Planck-Inst. Strömungsforsch. u. Aerodyn. Versuchsanstalt No. 48 (1970).

[12b] Eckelmann, H., Wallace, J.M., and Brodkey, R.: The wall region in turbulent shear flow. JFM *54*, 39—48 (1972).

[12c] Eckelmann, H.: The structure of viscous sublayer and the adjacent wall region in a turbulent channel flow. JFM *65*, 439—459 (1974).

[13] Glushko, G.S.: Turbulent boundary layer on a flat plate in an incompressible fluid. Izv. Ak. Nauk. SSSR, Ser. Mekh. No. 4, 13—23 (1965). Engl. transl. in NASA TTF 10, 080.

[13a] Galbraith, R.A.M., and Head, M.R.: Eddy viscosity and mixing length from measured boundary layer developments. Aero. Quart. *26*, 133—154 (1975).

[14] Glushko, G.S.: Differential equation for turbulence scale and prediction of turbulent boundary layer on a flat plate. In: Turbulentnye techeniya (M.D. Millionschchikov, ed.). Moscow, Nauka, 1970 [in Russian].

[15] Glushko, G.S.: Transition in the turbulent flow regime in a boundary layer on a flat plate for different turbulence scales of free stream. Izv. Ak. Nauk. SSSR, Mekh. Zhidkosti i gaza, No. 3, 68—70 (1972).

[16] Jevlev, V.M.: Turbulent motion of high temperature continuous media. Moscow, Nauka, 1975 [in Russian].

[17] von Kármán, Th.: Mechanische Ähnlichkeit und Turbulenz. Nachr. Ges. Wiss. Göttingen, Math. Phys. Klasse, 58 (1930) and Proc. 3rd. Intern. Congress Appl. Mech., Stockholm, Part I, 85 (1930); NACA TM 611 (1931); also Coll. Works *II*, 337—346.

[18] von Kármán, Th.: Progress in the statistical theory of turbulence. Proc. Nat. Acad. Sci. Washington, *34*, 530—539 (1948); also Coll. Works *IV*, 363—371.

[19] Kolmogorov, A. N.: Equations of turbulent motion of an incompressible fluid. Izv. Ak. Nauk. SSSR. Seria fizicheskaya IV (1942), No. 1–2, pp. 56–58 [in Russian].

[20] Launder, B. E., and Spalding, D. B.: Mathematical models of turbulence. Academic Press, London, 1972.

[20a] Lin, C. C., and Shen, S. F.: Studies of von Kármán's similarity theory and its extension to compressible flow. NACA TN 2542 (1951).

[20b] Mellor, G. L., and Herring, H. J.: A survey of the mean turbulent field closure models. AIAA J. *11*, 590–599 (1973).

[21] Prandtl, L.: Über die ausgebildete Turbulenz. ZAMM *5*, 136–139 (1925) and Proc. 2nd. Intern. Congr. Appl. Mech., Zürich 1926, 62–75; also Coll. Works *II*, 736–751.

[22] Prandtl, L.: Über ein neues Formelsystem der ausgebildeten Turbulenz. Nachr. Akad. Wiss. Göttingen 6–19 (1945); also Coll. Works *II*, 874–888.

[23] Prandtl, L.: Bemerkungen zur Theorie der freien Turbulenz. ZAMM *22*, 241–243 (1942); also Coll. Works *II*, 869–873.

[24] Reichardt, Gesetzmässigkeiten der freien Turbulenz. VDI-Forschungsheft 414, 1st ed., Berlin, 1942; 2nd ed., Berlin, 1951.

[25] Reichardt, H.: Über die Geschwindigkeitsverteilung in einer geradlinigen turbulenten Couette-Strömung. ZAMM-Sonderheft *36*, 26–29 (1956); see also Rep. No. 9 of the Max-Planck-Inst. für Strömungsforschung, Göttingen (1954).

[26] Reichardt, H.: Gesetzmässigkeiten der geradlinigen turbulenten Couette-Strömung. Max-Planck-Inst. für Strömungsforschung and Aerodyn. Versuchsanstalt, Göttingen, Rep. No. 22 (1959).

[26a] Reynolds, W. C.: Computation of turbulent flows. Ann. Rev. Fluid Mech. (M. van Dyke, ed.) *8*, 183–208 (1976).

[27] Rotta, J. C.: Über eine Methode zur Berechnung turbulenter Scherströmungsfelder. ZAMM *50*, T 204–T 205 (1970).

[28] Rotta, J. C.: Recent attempts to develop a generally applicable calculation method for turbulent shear flow layers. AGARD CP No. 93 (1972).

[29] Rotta, J. C.: Turbulente Strömungen. B. G. Teubner, Stuttgart, 1972.

[30] Rotta, J. C.: Turbulent shear layer prediction on the basis of the transport equations for the Reynolds stresses. Proc. 13th Int. Congr. Theor. Appl. Mech. Moscow 1972. Springer Verlag, 1973, pp. 295–308.

[31] Schmidt, W.: Der Massenaustausch in freier Luft und verwandte Erscheinungen. Hamburg, 1925.

[31a] Szeri, A. A., Yates, C. C., and Hai, S. M.: Flow development in a parallel plate channel. J. Lubrication Technology, Trans. ASME Ser. F *98*, 145–156 (1976).

[32] Taylor, G. I.: The transport of vorticity and heat through fluids in turbulent motion. Appendix by A. Fage and V. M. Faulkner. Proc. Roy. Soc. London A *135*, 685–705 (1932); see also Phil. Trans. A *215*, 1–26 (1915).

[32a] Vollmers, H., and Rotta, J. C.: Similar solutions of the mean velocity, turbulent energy and length scale equation. AIAA J. *15*, 714–720 (1977); see also: Ähnliche Lösungen der Differentialgleichungen für gemittelte Geschwindigkeiten, Turbulenzenergie und Turbulenzlänge. DLR–FB 76–24 (1976).

[33] von Weizsäcker, C. F.: Das Spektrum der Turbulenz bei grossen Reynoldsschen Zahlen. Z. Phys. *124*, 614–627 (1948).

CHAPTER XX

Turbulent flow through pipes

a. Experimental results for smooth pipes

The case of turbulent flow through pipes was investigated very thoroughly in the past because of its great practical importance [24, 31, 48, 49, 59, 61, 62, 71, 72]†. Moreover, the results arrived at are important not only for pipe flow; they also contribute to the extension of our fundamental knowledge of turbulent flow in general. Methods of dealing with other turbulent flows, such as the flow along a flat plate or a streamline body, could be devised only on the basis of the detailed experimental results obtained with pipe flow.

When a fluid is allowed to enter a circular pipe from a large container, the velocity distribution in the cross-sections of the *inlet length* varies with the distance from the initial cross-section. In sections close to that at entrance the velocity distribution is nearly uniform. Further downstream the velocity distribution changes, owing to the influence of friction, until a fully developed velocity profile is attained at a given cross-section and remains constant downstream of it. The variation of the velocity profile in the inlet length of a pipe in *laminar* flow was described in Sec. XIb (Fig. 11.8). Its length is approximately $l_I = 0.03\, d \cdot \mathsf{R}$ so that for $\mathsf{R} = 5{,}000$ to $10{,}000$ it ranges from 150 to 300 pipe-diameters. The inlet length in *turbulent* flow is considerably shorter than in laminar flow. According to the measurements performed by H. Kirsten [33] its length is about 50 to 100 diameters, but J. Nikuradse [45] determined that the fully formed velocity profile exists already after an inlet length of 25 to 40 diameters; the reader may also consult ref. [75].

In what follows we shall concern ourselves mainly with fully developed turbulent flow through a straight pipe of circular cross-section. The radial coordinate measured outwards from the axis will be denoted by y' and we shall consider a fluid cylinder of length L and radius y' in *fully developed* turbulent flow. The cylinder is not acted upon by any inertia forces, so that in accordance with eqn. (1.9) we can write down the condition of equilibrium between the force due to shearing stress τ on the circumference, and the pressure difference $p_1 - p_2$‡ on the end faces in the form

$$\tau = \frac{p_1 - p_2}{L}\, \frac{y'}{2}\,, \tag{20.1}$$

---

† The following description is largely based on the experimental results reported by J. Niku-radse [45, 46].

‡ From this point onwards we shall omit the bar above the symbol to denote time-averages because confusion with time-dependent quantities is no longer possible (as we already did on p. 586).

the relation being equally valid for laminar and turbulent motion. In the present analysis τ denotes the sum of laminar and turbulent shearing stress. Thus the shearing stress distribution over a cross-section is linear, and its largest value, τ_0, occurs at the wall, where

$$\tau_0 = \frac{p_1 - p_2}{L} \frac{R}{2} . \tag{20.2}$$

It is seen that the shearing stress at the wall, τ_0, can be determined directly by measuring the pressure gradient along the pipe.

The relationship between the pressure gradient and the rate of flow $Q = \pi R^2 \bar{u}$ could be determined theoretically for the case of laminar flow and the result agreed with experiment. In the case of turbulent flow such a relationship can only be obtained empirically[†], because attempts to perform a purely theoretical analysis of turbulent flow, even for one particular case, have so far been entirely unsuccessful. This relation is usually given by so-called *laws of friction* or *laws of resistance*. Available papers contain a large number of empirical equations for the law of friction in pipes. Furthermore, the older equations were often given in a form which depended on the respective system of units and did not satisfy Reynolds's law of similitude. In order to make use of dimensionless variables it is now common to use the dimensionless coefficient of resistance, λ, and to define it (see also eqn. (5.10)) as

$$\frac{p_1 - p_2}{L} = \frac{\lambda}{d} \frac{\varrho}{2} \bar{u}^2 , \tag{20.3}$$

where $d = 2R$ denotes the diameter of the cross-section. Comparing eqn. (20.2) with (20.3) we can deduce the relation

$$\tau_0 = \frac{1}{8} \lambda \varrho \bar{u}^2 , \tag{20.4}$$

which will be required later. In 1911 H. Blasius [5] made a critical survey of the then existing and already numerous experimental results and arranged them in dimensionless form in accordance with Reynolds's law of similarity. He was able to establish the following empirical equation:

$$\lambda = 0 \cdot 3164 \left(\frac{\bar{u} d}{\nu} \right)^{-\frac{1}{4}} = 0 \cdot 3164 / R^{0 \cdot 25} , \tag{20.5}$$

which is valid for the frictional resistance of *smooth pipes* of circular cross-section and is known as the *Blasius resistance formula*. Here $\bar{u} d / \nu = R$ denotes the Reynolds number calculated with the mean flow velocity $\bar{u}$ and the diameter of the pipe. According to this result the dimensionless coefficient of resistance in a pipe is a function of the Reynolds number only. It is found that the Blasius formula is valid in the range of Reynolds numbers $R = \bar{u} d / \nu \leq 100,000$. Consequently, the pressure drop in turbulent flow in that range is seen to be proportional to $\bar{u}^{7/4}$. At the time when Blasius established eqn. (20.5), measurements for higher Reynolds numbers were

---

[†] The mean velocity of flow through a pipe, $\bar{u}$, will be defined as $\bar{u} = Q/\pi R^2$, and U will denote the maximum velocity in the cross-section.

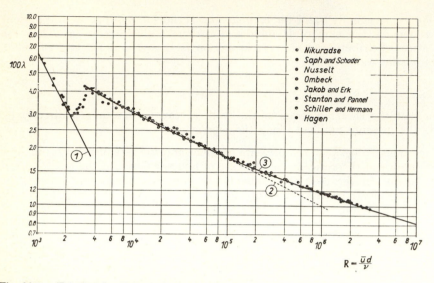

Fig. 20.1.   Frictional resistance in a smooth pipe

Curve (1) from eqn. (5.11), after Hagen-Poiseuille for laminar flow; curve (2) from eqn. (20.5), after Blasius [5] for turbulent flow; curve (3) from eqn. (20.30), after Prandtl [52] for turbulent flow

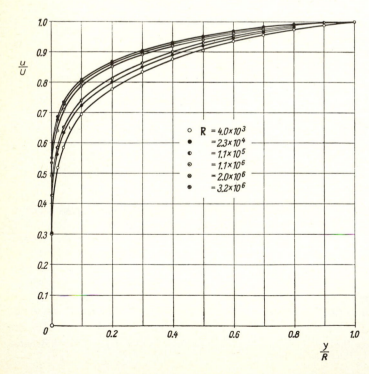

Fig. 20.2.   Velocity distribution in smooth pipes for varying Reynolds number, after Nikuradse [45]

not available. In Fig. 20.1 the Blasius formula, eqn. (20.5), is seen compared with experimental results; it reproduces them very accurately for Reynolds numbers up to R = 100,000. However, points obtained at R > 100,000 deviate considerably upwards, as seen from Fig. 20.1 in relation to experimental values reported by J. Nikuradse [38].

J. Nikuradse carried out a very thorough experimental investigation into the law of friction and velocity profiles in smooth pipes in a very wide range of Reynolds numbers $4 \times 10^3 \leq R \leq 3\cdot2 \times 10^6$. Velocity profiles for several Reynolds numbers are seen plotted in Fig. 20.2. They are given in dimensionless form in that u/U has been plotted against y/R. It will be noticed that the velocity profile becomes fuller as the Reynolds number increases. It is possible to represent it by the empirical equation

$$\frac{u}{U} = \left(\frac{y}{R}\right)^{\frac{1}{n}}, \tag{20.6}$$

where the exponent n varies slightly with the Reynolds number. The plots in Fig. 20.3 show that the assumption of a simple $1/n$-th-power law agrees well with experiment as the graphs of $(u/U)^n$ against y/R, fall on straight lines, when a suitable choice for n has been made. The value of the exponent n is $n = 6$ at the lowest Reynolds number $R = 4 \times 10^3$; it increases to $n = 7$ at $R = 100 \times 10^3$ and to $n = 10$ at the highest Reynolds number, $R = 3240 \times 10^3$, attained in this investigation.

We shall note here for further reference the expression for the ratio of the mean to the maximum velocity, $\bar{u}/U$, which can be easily derived from eqn. (20.6). It is found that

$$\frac{\bar{u}}{U} = \frac{2n^2}{(n+1)\,(2n+1)}, \tag{20.7}$$

and the respective numerical values are given in Table 20.1.

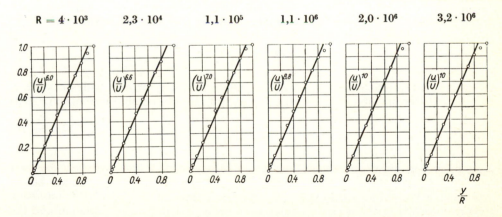

Fig. 20.3.   Velocity distribution in smooth pipes. Verification of the assumption in eqn. (20.6)

Table 20.1.   Ratio of mean to maximum velocity in pipe flow in terms of the exponent n of the velocity distribution, according to eqn. (20.6)

n	6	7	8	9	10
$\bar{u}/U$	0·791	0·817	0·837	0·852	0·865

b. Relation between law of friction and velocity distribution

The equation for the velocity distribution (20.6) is related to Blasius's law of friction in eqn. (20.5) and this relation, first discovered by L. Prandtl [51], is of fundamental importance in the theory of turbulent flow; it allows us to draw conclusions from pipe experiments which are valid for the flat plate [32]; use of them will be made in Chap. XXI.

On substituting the value of λ from eqn. (20.5) into eqn. (20.4) we obtain the following expression for the shearing stress at the wall:

$$\tau_0 = 0.03955\,\varrho\,\bar{u}^{7/4}\,\nu^{1/4}\,d^{-1/4}.$$

Introducing the radius R instead of the diameter d it is necessary to divide the numerical factor in the above equation by $2^{1/4} = 1.19$. Thus we obtain.

$$\tau_0 = 0.03325\,\varrho\,\bar{u}^{7/4}\,\nu^{1/4}\,R^{-1/4} = \varrho\,v_*^{\,2}, \tag{20.8}$$

where $v_* = \sqrt{\tau_0/\varrho}$ denotes the friction velocity introduced in Chap. XIX. If we split $v_*^{\,2}$ into $v_*^{\,7/4}$ and $v_*^{\,1/4}$, we obtain

$$\left(\frac{\bar{u}}{v_*}\right)^{\frac{7}{4}} = \frac{1}{0.03325}\left(\frac{v_*\,R}{\nu}\right)^{\frac{1}{4}} \quad \text{or} \quad \left(\frac{\bar{u}}{v_*}\right) = 6.99\left(\frac{v_*\,R}{\nu}\right)^{\frac{1}{7}},$$

and if we eliminate the mean velocity $\bar{u}$ with the aid of the maximum velocity U by putting $\bar{u}/U = 0.8$ which, as seen from Table 20.1, corresponds approximately to an exponent $n = 7$, i. e. to a Reynolds number $R = 10^5$, we have

$$\frac{U}{v_*} = 8.74\left(\frac{v_*\,R}{\nu}\right)^{\frac{1}{7}}. \tag{20.9}$$

It is now natural to assume that this equation is valid for any wall distance y, and not only for the pipe axis (wall distance $y = R$). Hence, we obtain from eqn. (20.9)

$$\frac{u}{v_*} = 8.74\left(\frac{y\,v_*}{\nu}\right)^{\frac{1}{7}}. \tag{20.10}$$

The preceding argument shows that the $\frac{1}{7}$-*th-power velocity-distribution law* can be derived from Blasius's resistance formula. It has already been shown before that such a law agrees with experiment over a certain range of Reynolds numbers, and it is seen that there exists a relation between Blasius's law of friction and the n-th-power velocity-distribution law. Introducing the abbreviations $u/v_* = \phi$ and

$y\,v_*/\nu = \eta$, which were already used in eqns. (19.31) and (19.32), we can transform eqn. (20.10) to

$$\phi = 8{\cdot}74\,\eta^{1/7}\,.\tag{20.11}†$$

Thus we have once more deduced eqn. (19.35) which was first obtained from considerations of similarity, except that the numerical values of the constants C and n, which then remained undertermined, are now known from the law of pipe friction. Equation (20.11) has been compared with J. Nikuradse's experiments in Fig. 20.4, curve (4). It is seen that the $\frac{1}{7}$-th-power law agrees well with experiment up to a Reynolds number of about $R = 100{,}000$. No better agreement can be expected because Blasius's equation (20.5) from which it was derived is valid only to that limit, Fig. 20.1.

In order to obtain better agreement it would be necessary to introduce a smaller exponent into Blasius's equation, say $\frac{1}{5}$ or $\frac{1}{6}$ instead of $\frac{1}{4}$. Performing corresponding calculations it is found that the exponent $\frac{1}{7}$ in the velocity-distribution law would have to be replaced by $\frac{1}{8}$, $\frac{1}{9}$, etc., respectively, in agreement with measured values. The relation $\phi = C \times \eta^{1/10}$ has been plotted as curve (5) in Fig. 20.4, and it is seen that it does, in fact, reproduce the experimental values at higher Reynolds numbers with a good measure of agreement, but that the fit is inferior at lower Reynolds numbers.

(1) $\phi = \eta$ laminar;

(2) transition from laminar
     to turbulent;
     after Reichardt [55]

(3) eqn. (20.14), turbulent,
     all Reynolds numbers;

(4) eqn. (20.11), turbulent,
     $R < 10^5$;

(5) $\phi = 11{\cdot}5\,\eta^{1/10}$

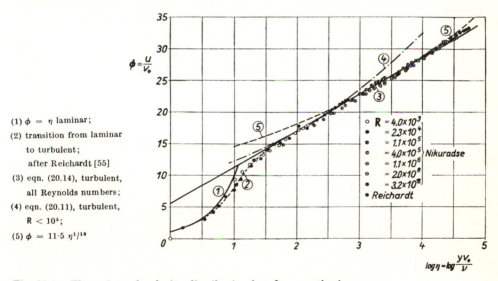

Fig. 20.4.   The universal velocity-distribution law for smooth pipes

† Upon generalizing for other exponents, we obtain, with K. Wieghardt [82a], $u/v_* = C(n) \times (y\,v_*/\nu)^{1/n}$ and the following numerical values:

n	7	8	9	10
$C(n)$	8·74	9·71	10·6	11·5

For future reference we now propose to write down an expression for the friction velocity v_* from eqn. (20.10). We obtain

$$v_* = 0 \cdot 150 \; u^{\frac{7}{8}} \left(\frac{\nu}{y} \right)^{\frac{1}{8}}$$

and

$$\tau_0 = \varrho \, v_*{}^2 = 0 \cdot 0225 \, \varrho \, u^{\frac{7}{4}} \left(\frac{\nu}{y} \right)^{\frac{1}{4}} \tag{20.12}$$

or

$$\tau_0 = 0 \cdot 0225 \, \varrho \, U^{\frac{7}{4}} \left(\frac{\nu}{R} \right)^{\frac{1}{4}} . \tag{20.12a}$$

This can also be written in dimensionless form as

$$c_f' = \frac{\tau_0}{\frac{1}{2} \varrho \, U^2} = 0 \cdot 045 \left(\frac{U R}{\nu} \right)^{-1/4} , \tag{20.12b}$$

where c_f' denotes the local skin-friction coefficient. This relation, which is equivalent to the one in eqn. (20.5), is known as Blasius's law of skin friction in pipe flow. This relation will be used later.

c. Universal velocity-distribution laws for very large Reynolds numbers

The fact that the exponent in the law of pipe resistance as well as in the expression for velocity distribution decreases with increasing Reynolds numbers suggests that both must tend asymptotically to some expressions which are valid for very high Reynolds numbers and which must contain the logarithm of the independent variable, as it is the limit of a polynomial for very small values of the exponent. A detailed examination of experimental results for very large Reynolds numbers shows that such logarithmic laws do, in fact, exist. Physically such asymptotic laws are characterized by the fact that laminar friction becomes completely negligible compared with turbulent friction. The great advantage of such logarithmic laws, as compared with the $1/n$-th-power laws, consists in their being asymptotic expressions for very large Reynolds numbers; they may, therefore, be extrapolated to arbitrarily large values beyond the range covered by experiment. On the other hand, when the $1/n$-th-power laws are used the value of the exponent n changes, as the range of Reynolds numbers is extended.

Such an asymptotic logarithmic law has already been given in eqn. (19.33) for the case of flow along a flat plate. It was deduced from Prandtl's equation (19.7) for turbulent shearing stress under the assumption that the mixing length is proportional to the distance from the wall, $l = \varkappa y$, and was valid for small wall distances y. This equation has the form:

$$\phi = A_1 \ln \eta + D_1 \tag{20.13} \dagger$$

---

† H. Reichardt [55] indicated a refined expression for the velocity distribution. It covers the whole range of distances, from the wall of the pipe at $y = 0$ to the centre-line at $y = R$, i. e., it is also true for the laminar sub-layer, to which eqn. (20.13) does not apply. It is also valid in the neighbourhood of the centre-line, where measured velocity-distribution curves show systematic deviations from eqn. (20.13). In particular, the transition region shown as curve (2) in Fig. 20.4 is well reproduced by the formula. This universal velocity-distribution law was deduced with the aid of theoretical estimations and very careful measurements of the turbulent mixing coefficient A defined by eqn. (19.1). Compare also a paper by W. Szablewski [74].

with $A_1 = 1/\varkappa$ and $D_1 = -(1/\varkappa) \cdot \ln \beta$ as free constants. We shall apply this equation without change to pipe flow. Comparing it with the measurements performed by J. Nikuradse, as shown by curve (3) in Fig. 20.4, it is seen that excellent agreement is obtained not only for points near the wall but for the whole range up to the axis of the pipe. The numerical values of the constants are found to be

$$A_1 = 2\cdot5; \qquad D_1 = 5\cdot5 .$$

This gives the following values of $\varkappa$ and β:

$$\varkappa = 0\cdot4; \qquad \beta = 0\cdot111 .$$

Hence the universal velocity-distribution law for very large Reynolds numbers has the form†

or
$$\left. \begin{aligned} \phi &= 2\cdot5 \ln \eta + 5\cdot5 \\ \phi &= 5\cdot75 \log \eta + 5\cdot5 . \end{aligned} \right\} \quad \text{(smooth)} \qquad (20.14)$$

By a reasoning similar to the one given in the preceding section it is possible to arrive at a corresponding universal asymptotic resistance formula from the above universal velocity-distribution equation.

Equation (20.14), being one for turbulent flow, is valid only in regions where the laminar shearing stress can be neglected in comparison with the turbulent stress. In the immediate neighbourhood of the wall, where the turbulent shearing stress decreases to zero and laminar stresses predominate, deviations from this law must be expected. H. Reichardt [54, 55] extended this kind of measurement to include very small distances from the wall in a flow in a channel. Curve (2) in Fig 20.4 represents the transition from the laminar sub-layer (cf. Sec. XVIII c) to the turbulent boundary layer. The curve denoted by (1) in the above diagram corresponds to laminar flow for which $\tau_0 = \mu \, u/y$. With $\tau_0 = \varrho \, v_*{}^2$ we obtain $u/v_* = y \, v_*/\nu$, or

$$\phi = \eta \quad \text{(laminar)} . \qquad (20.14\,\mathrm{a})$$

From this it can be seen that for values $y \, v_*/\nu < 5$ the contribution from turbulent friction may be completely neglected compared with laminar friction. In the range $5 < y \, v_*/\nu < 70$ both contributions are of the same order of magnitude, whereas for $y \, v_*/\nu > 70$ the laminar contribution is negligible compared with turbulent friction. Thus:

$$\left. \begin{aligned} \frac{y\,v_*}{\nu} &< 5: & &\text{purely laminar friction} \\[2mm] 5 < \frac{y\,v_*}{\nu} &< 70: & &\text{laminar-turbulent friction} \\[2mm] \frac{y\,v_*}{\nu} &> 70: & &\text{purely turbulent friction} . \end{aligned} \right\} \qquad (20.15)$$

---

† In the following equations ln denotes the natural logarithm and log the logarithm to base 10.

Hence the thickness of the laminar sub-layer is seen to be equal to

$$\delta_l \approx 5 \frac{\nu}{v_*}. \tag{20.15a}$$

We now propose to compare the experimental results on velocity-distribution measurements in pipe flow with the alternative universal equation, which was deduced in Chap. XIX in the form $(U-u)/v^* = f(y/R)$. It will be recalled that it followed both from von Kármán's similarity theory and from Prandtl's assumption about the shearing stress, together with the relation $l = \varkappa y$ for the mixing length. In the first case we obtained eqn. (19.21) and in the second case eqn. (19.28) was obtained.

The universal velocity distribution can be calculated over the whole range, that is from the wall ($y = 0$) to the zone of fully developed turbulence, by the application of a suitably chosen function of y for the mixing length. Such a relation was developed by E. R. van Driest [12] on the basis of an argument which leans on Stokes's solution for a flat plate which oscillates in its own plane (Stokes's second problem, Sec. V7); it has the form

$$l = \varkappa y \left[1 - \exp\left(-\eta/A\right)\right]. \tag{20.15b}$$

The constant has the value $A = 26$. For very large values of y we retrieve the relation $l = \varkappa y$, but in the overlapping region between laminar and turbulent friction the values of l are smaller. In order to integrate the equation

$$\tau_0 = \tau_l + \tau_t = \varrho \left(\nu + l^2 \left|\frac{du}{dy}\right|\right) \frac{du}{dy}$$

for a constant value of τ_0 we first solve for du/dy and obtain

$$u = 2 \, v_* \int_0^{\eta} \frac{d\eta'}{1 + [1 + 4 \, (v_* \, l/\nu)^2]^{1/2}}.$$

Next, we introduce the expression for l from eqn. (20.15b) and arrive at a velocity distribution which is in good agreement with the experimental results displayed in Fig. 20.4. The expression due to E. R. van Driest was successfully applied to the calculation of turbulent boundary layers with suction and blowing [58] as well as to compressible flows[†].

Law of the wall: For sufficiently large distances from the wall (fully developed turbulent layer), the velocity distribution is represented by the logarithmic law quoted in eqn. (20.14). In the zone near the wall (laminar sublayer), the linear law

---

[†] In a recent publication, W. C. Reynolds (Annual Reviews of Fluid Mechanics, Vol. 8, p. 187) reports that the numerical value $A = 26$ of van Driest's constant is suitable only for boundary layers with a zero pressure gradient and only along an impermeable wall. In the presence of blowing or suction or of a pressure gradient, the constant A can assume considerably different values. With a favourable pressure gradient or with blowing, the values may exceed $A = 26$ by a wide margin

of eqn. (20.14a) is applicable. In the transition (mixed laminar-turbulent) zone, the velocity distribution is represented by Reichardt's law, curve (2) in Fig. 20.4. These three laws taken together are now referred to as the "law of the wall"; van Driest's integral represents another form of this law of the wall. Important considerations on this topic can be found in a summary article by F. H. Clauser [5a]; see also the papers by D. B. Spalding [68a] and G. Kleinstein [33a]. Further discussions are given in refs. [5a] of Chap. XXI and [46, 47] of Chap. XXIII.

Since the simpler assumption on the mixing length, $l = \varkappa y$, does not seem to be suitable for the whole pipe diameter, it appears preferable to deduce the dependence of mixing length on distance directly from experiment and then to apply Prandtl's hypothesis

$$\tau = \varrho \, l^2 \left(\frac{du}{dy}\right)^2 \tag{20.16}$$

to calculate the velocity distribution from a linear shearing-stress distribution

$$\tau = \tau_0 \left(1 - \frac{y}{R}\right). \tag{20.17}$$

It is now possible to calculate the variation of mixing length with y/R directly from eqns. (20.16) and (20.17) together with the measured velocity distribution $u(y)$. This calculation was carried out by J. Nikuradse [45] who obtained the remarkable result shown in Fig. 20.5; it represents the variation of mixing length over the diameter of the pipe for the case of smooth pipes and it seen that it is independent of the Reynolds number, when values below 10^5 are excluded. This function can be represented by the empirical relation

$$\frac{l}{R} = 0.14 - 0.08 \left(1 - \frac{y}{R}\right)^2 - 0.06 \left(1 - \frac{y}{R}\right)^4. \tag{20.18}$$

In the neighbourhood of the wall this equation can be simplified to

$$l = 0.4 \, y - 0.44 \, \frac{y^2}{R} + \ldots \tag{20.18a}$$

which shows that Prandtl's hypothesis, $l = \varkappa y$, is confirmed for small distances from the wall, with

$$\varkappa = 0.4 . \tag{20.19}$$

It can be shown, further, that the variation of mixing length with wall distance given in eqn. (20.18) remains valid for rough and not only for smooth pipes. Figure 20.6 represents the results of J. Nikuradse's [46] measurements on pipes artificially roughened with sand of different grain size, and the preceding statement is seen to be confirmed. Furthermore, it may now be expected that the velocity distribution calculated from the mixing length from eqn. (20.18) will be valid for rough as well as for smooth pipes.

For the sake of simplicity the expression for the mixing length can be written as

$$l = \varkappa \, y \cdot f\left(\frac{y}{R}\right) \tag{20.20}$$

where $f(y)/R \to 1$ for $y/R \to 0$. Introducing $v_* = \sqrt{\tau_0/\varrho}$ and combining eqn. (20.17) with eqn. (20.16) we obtain the following differential equation for the velocity distribution

$$\frac{du}{dy} = \frac{1}{l}\sqrt{\frac{\tau}{\varrho}} = \frac{v_*}{\varkappa}\frac{\sqrt{1-\dfrac{y}{R}}}{yf\left(\dfrac{y}{R}\right)}$$

whence, by integration

$$u = \frac{v_*}{\varkappa}\int\limits_{y_0/R}^{y/R}\frac{\sqrt{1-\dfrac{y}{R}}\,d\left(\dfrac{y}{R}\right)}{\dfrac{y}{R}\,f\left(\dfrac{y}{R}\right)}. \tag{20.21}$$

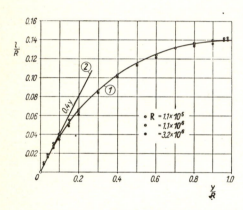

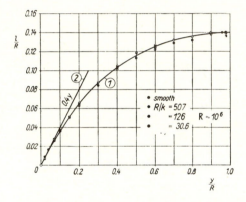

Fig. 20.5. Variation of mixing length over pipe diameter for smooth pipes at different Reynolds numbers

Curve (1) from eqn. (20.18)

Fig. 20.6. Variation of mixing length over pipe diameter for rough pipes

Curve (1) from eqn. (20.18)

Here the lower limit of integration at y_0, where the velocity is equal to zero, is of the order of the thickness of the laminar sub-layer and, therefore, proportional to ν/v_* as seen from eqn. (20.15a). Thus $y_0/R = F_1(v_* \, R/\nu)$. The maximum velocity U in the centre of the pipe can be deduced from eqn. (20.21) and becomes

$$U = \frac{v_*}{\varkappa}\int\limits_{y_0/R}^{1}\frac{\sqrt{1-\dfrac{y}{R}}\,d\left(\dfrac{y}{R}\right)}{\dfrac{y}{R}\,f\left(\dfrac{y}{R}\right)}. \tag{20.21a}$$

By eqns. (20.21) and (20.21 a) we now have

$$U - u = v_* \, F\left(\frac{y}{R}\right).$$
(20.22)

Thus we have again been led to the universal velocity-distribution law, eqn. (19.21) and eqn. (19.28). The essential generalization which has now been achieved consists in the fact that the universal law in eqn. (20.22) is valid for rough as well as for smooth pipes, the function $F(y/R)$ being the same in both cases. Equation (20.22) asserts that curves of velocity distribution plotted over the pipe radius contract into a single curve for *all* values of Reynolds number and for *all* degrees of roughness, if $(U-u)/v_*$ is plotted in terms of y/R, Fig. 20.7. It may be noted that the above form of the velocity-distribution law was first deduced by T. E. Stanton [72]. An explicit expression for $F(y/R)$ could be obtained by evaluating the integral in eqn. (20.21); it is, however, simpler to make use of the already known form of the velocity-distribution law for smooth pipes as given in eqn. (20.14). Hence, in a way similar to eqns. (20.9) and (20.10), we have

$$U - u = 2{\cdot}5 \, v_* \ln \frac{R}{y} = 5{\cdot}75 \, v_* \log \frac{R}{y}$$

and so

$$\frac{U - u}{v_*} = 5{,}75 \log \frac{R}{y}.$$
(20.23)

This equation is seen plotted in Fig. 20.7 as curve (1) and compared with experimental results for smooth and rough pipes. It contains the empirical constant, $\varkappa$, whose

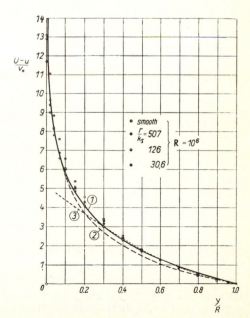

Fig. 20.7. Universal velocity-distribution law for smooth and rough pipes

Curve (1) from eqn. (20.23), Prandtl;
curve (2) from eqn. (20.24), von Kármán;
curve (3) from eqn. (20.25), Darcy

numerical value $\varkappa = 0.4$ was already given in eqn. (20.19). The agreement between theory and experiment is very good.

The universal velocity-distribution law can be deduced also from von Kármán's similarity law, eqn. (19.21), whence we obtain

$$\frac{U-u}{v_*} = -\frac{1}{\varkappa}\left\{ \ln\left[1 - \sqrt{1 - \frac{y}{R}}\right] + \sqrt{1 - \frac{y}{R}}\right\}, \tag{20.24}$$

with y denoting the distance from the wall. This equation, shown as curve (2) in Fig. 20.7, also agrees well with the experimental values, if $\varkappa = 0.36$ is chosen. Figure 20.7 contains an additional curve (3) which is based on H. Darcy's [9] empirical equation. Darcy deduced it in 1855 from his very careful measurements on velocity distribution and in our present notation it can be written as

$$\frac{U-u}{v_*} = 5.08\left(1 - \frac{y}{R}\right)^{\frac{3}{2}}. \tag{20.25}$$

Darcy's formula gives good agreement at all points except those near the wall with $y/R < 0.25$.

It is worth pointing out here that both universal velocity-distribution laws, eqns. (20.23) and (20.24), have been obtained, as seen from the argument in the preceding chapter, for two-dimensional flow in a channel. The fact that they nevertheless agree well with the experimental results for the case of pipe flow with axial symmetry can be taken as proof that there is a far-reaching similarity between the velocity distribution in the two-dimensional and axially symmetrical cases. It will be recalled that in laminar flow the velocity distribution is parabolic in both cases.

Starting with G. I. Taylor's vorticity-transfer theory it is also possible to deduce a universal velocity-distribution law of the form of eqn. (20.22) but, evidently, with a function $\dot{F}(y/R)$ which differs from those appearing in Prandtl's or von Kármán's calculations. A comparison between the results obtained from G. I. Taylor's vorticity-transfer theory and from L. Prandtl's momentum-transfer theory is contained in papers by S. Goldstein [20] and G. I. Taylor [76]. No unequivocal decision in favour of either of the two theories could, however, be obtained.

Good insight into the physical aspects of flow through a pipe can be gained by considering the variation of the apparent kinematic viscosity, ε_τ, over the cross-section of the pipe which is seen plotted in Fig. 20.8 on the basis of J. Nikuradse's experimental results. Starting with $\tau = \varrho\, \varepsilon_\tau(du/dy)$ from eqn. (19.2) we can insert the value of τ from eqn. (20.17), and hence we can obtain the variation of ε_τ from the measured velocity distribution. The apparent kinematic viscosity is independent of the Reynolds number, just as was the case with the mixing length. However, the type of variation of ε_τ is much more complex than that of l, Fig. 20.5. The maximum of ε_τ falls half-way between the wall and the axis, and on the axis ε_τ becomes very small but does not reduce to zero. Considering the diagram in Fig. 20.8 it must be conceded that it would be much more difficult to find a plausible hypothesis describing the variation in ε_τ than was the case with the mixing length, l. This circumstance was advanced earlier (Sec. XIX b) as a reason in favour of introducing the mixing length into the equations rather than the apparent viscosity, and this view is seen to be borne out by experimental results. In order to compare laminar with turbulent friction, Fig. 20.8 contains respective values for Poiseuille flow. The latter are, of course, identical with ν. Since $v_* = (\sqrt{\tfrac{1}{8}\lambda})\,\bar{u}$, as seen from eqn. (20.4), and since for laminar flow $\lambda = 64/R$, as seen from eqn. (5.11), we find that

$$\frac{\varepsilon_{lam}}{v_*\,R} = \frac{\nu}{v_*\,R} = \frac{1}{\sqrt{2\,R}}\,.$$

This shows that turbulent friction is much larger than laminar friction. particularly at high Reynolds numbers.

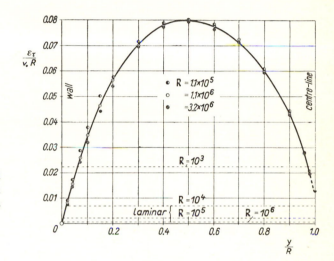

Fig. 20.8.   Dimensionless virtual kinematic viscosity for smooth pipes plotted from Nikuradse's [45] experimental results

d. Universal resistance law for smooth pipes at very large Reynolds numbers

Retracing the steps in the argument of Sec. XX b, which has led us from Blasius's resistance formula to the $\tfrac{1}{7}$-th-power velocity-distribution law, we can now derive a new pipe-resistance formula from the universal logarithmic velocity-distribution law. The logarithmic velocity-distribution law in eqn. (20.23) was derived under the assumption that laminar friction was negligible compared with turbulent friction which meant that it could be extrapolated to arbitrarily large Reynolds numbers. The same may now be expected to be true of the resistance law about to be deduced. The following reasoning will be based on a paper by L. Prandtl [52].

Upon integrating eqn. (20.23) over the cross-sectional area we obtain the mean velocity of flow

$$\bar{u} = U - 3\cdot75\,v_*\,. \qquad (20.26)$$

Nikuradse's experiments show that the constant 3·75 must be adjusted slightly, so that

$$\bar{u} = U - 4\cdot07\,v_*\,. \qquad (20.27)$$

From eqn. (20.4), we obtain

$$\lambda = 8 \left(\frac{v_*}{u} \right)^2 ,$$ (20.28)

and from the universal velocity-distribution law, eqn. (20.14), we have

$$U = v_* \left\{ 2 \cdot 5 \ln \frac{R v_*}{\nu} + 5 \cdot 5 \right\}$$

which combined with eqn. (20.26) gives

$$\overline{u} = v_* \left\{ 2 \cdot 5 \ln \frac{R v_*}{\nu} + 1 \cdot 75 \right\} .$$ (20.29)

We can introduce the Reynolds number from

$$\frac{R v_*}{\nu} = \frac{1}{2} \frac{\overline{u} d}{\nu} \frac{v_*}{\overline{u}} = \frac{\overline{u} d}{\nu} \frac{\sqrt{\lambda}}{4 \sqrt{2}} ,$$

so that we obtain from eqns. (20.28) and (20.29)

$$\lambda = \frac{8}{\left\{ 2{,}5 \ln \left(\frac{\overline{u} d}{\nu} \sqrt{\lambda} \right) - 2 \cdot 5 \ln 4 \sqrt{2} + 1 \cdot 75 \right\}^2} = \frac{1}{\left\{ 2 \cdot 035 \log \left(\frac{\overline{u} d}{\nu} \sqrt{\lambda} \right) - 0 \cdot 91 \right\}^2}$$

or

$$\frac{1}{\sqrt{\lambda}} = 2 \cdot 035 \log \left(\frac{\overline{u} d}{\nu} \sqrt{\lambda} \right) - 0 \cdot 91 .$$

According to this result the universal law of friction for a smooth pipe should give a straight line if $1/\sqrt{\lambda}$ is plotted against $\log (R \sqrt{\lambda})$. This feature agrees extremely well with experiment, as seen from Fig. 20.9, where the results of measurements of

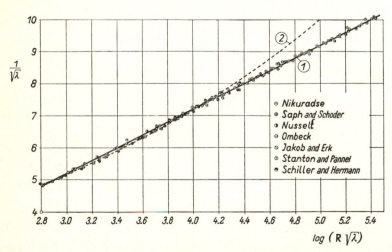

Fig. 20.9.   Universal law of friction for a smooth pipe
Curve (1) from eqn. (20.30), Prandtl; curve (2) from eqn. (20.5), Blasius

many authors have been plotted. The numerical coefficients for the averaged curve passing through the experimental results differ only very little from the preceding, derived values. The straight line (1) passing through the experimental points in Fig. (20.9) can be represented by the equation

$$\frac{1}{\sqrt{\lambda}} = 2 \cdot 0 \log\left(\frac{\bar{u}\,d}{\nu}\sqrt{\lambda}\right) - 0 \cdot 8 \qquad \text{(smooth)}. \qquad (20.30)$$

This is *Prandtl's universal law of friction for smooth pipes*. It has been verified by J. Nikuradse's [45] experiments up to a Reynolds number of $3 \cdot 4 \times 10^6$ and the agreement is seen to be excellent. From its derivation it is clear that it may be extrapolated to arbitrarily large Reynolds numbers, and it may be stated that measurements with higher Reynolds numbers are, therefore, not required. Values computed from eqn. (20.30) are given in Table 20.2. The universal law of friction is represented by curve (3) in Fig. 20.1.

Table 20.2. Coefficient of resistance for smooth pipes in terms of the Reynolds number; see also Fig. 20.9

$R = \dfrac{\bar{u}\,d}{\nu}$	Prandtl, eqn. (20.30) λ	Blasius, eqn. (20.5) λ
10^3	(0·0622)	(0·0567)
$2 \cdot 10^3$	(0·0494)	(0·0473)
$5 \cdot 10^3$	0·0374	0·0376
10^4	0·0309	0·0316
$2 \cdot 10^4$	0·0259	0·0266
$5 \cdot 10^4$	0·0209	0·0212
10^5	0·0180	0·0178
$2 \cdot 10^5$	0·0156	0·0150
$5 \cdot 10^5$	0·0131	—
10^6	0·0116	(0·0100)
$2 \cdot 10^6$	0·0104	—
$5 \cdot 10^6$	0·0090	—
10^7	0·0081	(0·0056)

The universal equation agrees well with Blasius's equation (20.5) up to $R = 10^5$, but at higher values Blasius's equation deviates progressively more from the results of measurement, whereas eqn. (20.30) maintains good agreement.

The flow of gases through smooth pipes at very high velocities was investigated by W. Froessel [18]. The variation in pressure along a pipe for different mass flow rates is represented in Fig. 20.10. The numbers shown against the curves indicate the fraction of maximum mass flow through a nozzle of equal diameter and with equal stagnation pressure. The curves which fall off to the right refer to subsonic flow, whereas the increasing curves apply to supersonic flow. The latter curves include jumps to higher pressures and subsonic flow effected by a shock. The coefficients of resistance are not markedly different from those in incompressible flow, as seen from Fig. 20.11. The straight line marked (1), shown for comparison, corresponds to eqn. (20.30). K. Oswatitsch and M. Koppe [50] gave a theoretical analysis of the flow of a compressible fluid in a pipe; their results agree well with Froessel's experiments.

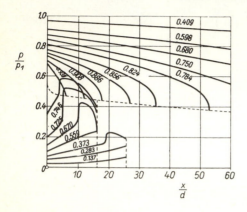

Fig. 20.10. Pressure distribution along a pipe in compressible flow, after Froessel [18]

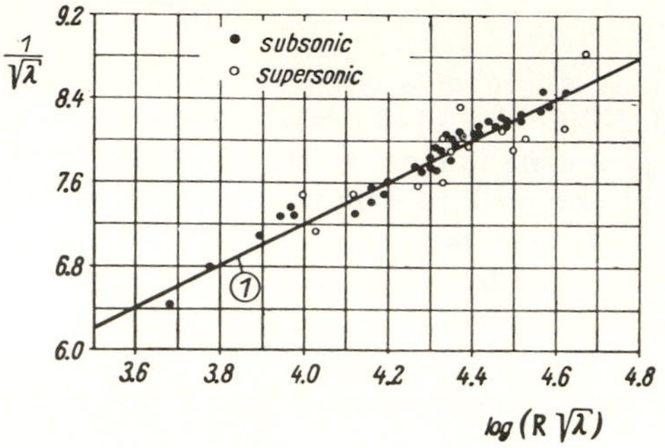

Fig. 20.11. Law of resistance for smooth pipes in compressible flow. Curve (1) from eqn. (20.30)

e. Pipes of non-circular cross-section

Turbulent flow through pipes of non-circular cross-section was investigated by L. Schiller [60] and J. Nikuradse [44], who have determined the law of friction and the velocity distribution for pipes of rectangular, triangular, and trapezoidal cross-section, as well as for circular pipes with notches. It is convenient to introduce a coefficient of resistance to flow which is referred to the *hydraulic diameter* d_h:

$$\frac{p_1 - p_2}{L} = \frac{1}{2} \frac{\lambda}{d_h} \varrho \, \overline{u}^2 \, ,$$

where

$$d_h = \frac{4A}{C}$$

and A denotes the cross-sectional area, C denoting the wetted perimeter. In the case of a circular cross-section the hydraulic diameter is equal to the diameter of the circle.

Figure 20.12 contains plots of λ against R for a series of cross-sectional shapes. For turbulent flow the results are well represented by the law for circular pipes, curve 2. In the laminar region, however, the experimental data, when reduced to the hydraulic diameter, do not fall on a single curve. The deviations depend on the shape of the duct cross-section. The applicability of the concept of the hydraulic

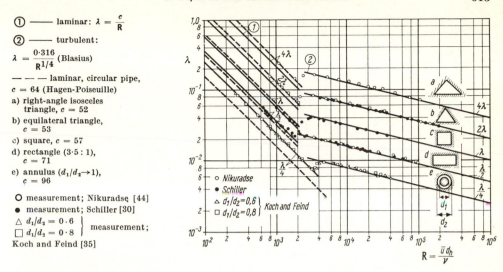

① —— laminar: $\lambda = \dfrac{c}{R}$

② —— turbulent:

$\lambda = \dfrac{0.316}{R^{1/4}}$ (Blasius)

— — — laminar, circular pipe,
$c = 64$ (Hagen-Poiseuille)

a) right-angle isosceles
   triangle, $c = 52$

b) equilateral triangle,
   $c = 53$

c) square, $c = 57$

d) rectangle (3·5 : 1),
   $c = 71$

e) annulus ($d_1/d_2 \rightarrow 1$),
   $c = 96$

○ measurement; Nikuradse [44]

● measurement; Schiller [30]

△ $d_1/d_2 = 0.6$ ⎫
□ $d_1/d_2 = 0.8$ ⎬ measurement;
Koch and Feind [35]

$R = \dfrac{\bar{u}\,d_h}{\nu}$

Fig. 20.12. Resistance formula for smooth pipes of non-circular cross-section

diameter for turbulent flow has been verified by experiments [42] up to a Mach number $M = 1$. The velocity-distribution curves in such pipes are particularly noteworthy. The curves of constant velocity for rectangular and triangular cross-section obtained by J. Nikuradse [43, 44] are shown in Figs. 20.13 and 20.14. In all cases the velocities at the corners are comparatively very large; this stems from the fact that in all straight pipes of non-circular cross-section there exist secondary flows. These are such that the fluid flows towards the corner along the bisectrix of the angle and then outwards in both directions. The secondary flows continuously transport momentum from the centre to the corners and generate high velocities there. Schematic

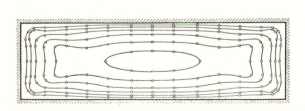

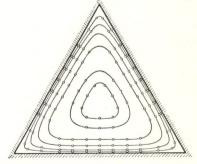

Fig. 20.13. Curves of constant velocity for a pipe of rectangular cross-section, after Nikuradse [43]

Fig. 20.14. Curves of constant velocity for a pipe of equilateral triangular cross-section, after Nikuradse [44]

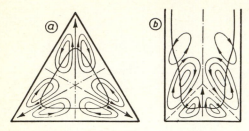

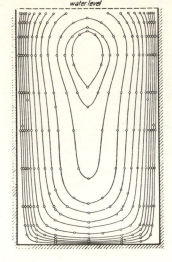

Fig. 20.15. Secondary flows in pipes of triangular and rectangular cross-section (schematic)

Fig. 20.16. Curves of constant velocity for a
rectangular open channel, after Nikuradse [43]

diagrams of secondary flows in triangular and rectangular pipes are shown in Fig. 20.15.
It is seen that the secondary flow in the rectangular cross-section which proceeds
from the wall inwards in the neighbourhood of the ends of the larger sides and
of the middle of the shorter sides creates zones of low velocity. They appear very
clearly in the picture of curves of constant velocity in Fig. 20.13. Such secondary
flows come into play also in open channels, as evidenced by the pattern of curves
of constant velocity in Fig. 20.16. The maximum velocity does not occur near the
free surface but at about one fifth of the depth down, and the flow in the free surface
is not at all two-dimensional as might have been expected. When the cross-section
of the channel contains a narrow region, transition does not occur simultaneously
over the whole of the flow. For example, in the region within an acute angle of a
triangular cross-section, the flow remains laminar to very large Reynolds numbers,
whereas in the bulk it had turned turbulent long ago. Such a state of affairs is seen
illustrated with the aid of Fig. 20.17 which represents the results of measurements

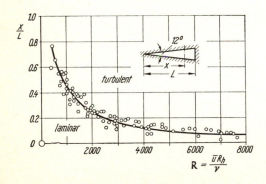

Fig. 20.17. Boundary between laminar
and turbulent flow in an acute, triangular channel, determined visually by the
use of smoke injection, after E. R. G.
Eckert and T. E. Irvine [13]

R_h — hydraulic radius $= d_h/2$

performed by E. R. G. Eckert and T. E. Irvine [13]. At a Reynolds number of R = 1000, the flow remains laminar over 40 per cent. of the height of the triangle, the region of laminar flow decreasing as the Reynolds number is increased.

E. Meyer [38] investigated the pressure and velocity distribution in a flow through a straight channel with a cross-section whose shape varied but whose cross-sectional area remained constant. He used a channel in which a circular cross-section was gradually transformed into a rectangle with its sides in the ratio 1 : 2. Transition was effected in both directions over two different lengths, and it was discovered that the pressure loss in the portion with transition from circle to rectangle considerably exceeded that in the opposite direction.

f. Rough pipes and equivalent sand roughness

Most pipes used in engineering structures cannot be regarded as being hydraulically smooth, at least at higher Reynolds numbers. The resistance to flow offered by rough walls is larger than that implied by the preceding equations for smooth pipes. Consequently, the laws of friction in rough pipes are of great practical importance, and experimental work on them began very early. The desire to explore the laws of friction of rough pipes in a systematic way is frustrated by the fundamental difficulty that the number of parameters describing roughness is extraordinarily large owing to the great diversity of geometric forms. If we consider, for example, a wall with identical protrusions we must come to the conclusion that its drag depends on the density of distribution of such roughnesses, i. e. on their number per unit area as well as on their shape and height and, finally, also on the way in which they are distributed over its surface. It took, therefore, a long time to formulate clear and simple laws which describe the flow of fluids through rough pipes. L. Hopf [25] made a comprehensive review of the numerous earlier experimental results and found two types of roughness in relation to the resistance formula for rough pipes and open channels. The first kind of roughness causes a resistance which is proportional to the square of the velocity; this means that the coefficient of resistance is independent of the Reynolds number and corresponds to relatively coarse and tightly spaced roughness elements such as for example coarse sand grains glued on the surface, cement, or rough cast iron. In such cases the nature of the roughness can be expressed with the aid of a single roughness parameter k/R, the so-called *relative roughness*, where k is the height of a protrusion and R denotes the radius or the hydraulic radius of the cross-section. From considerations of similitude we may conclude that in this case the resistance coefficient depends on the relative roughness only. The actual relation can be determined experimentally by performing measurements on pipes or channels of differing hydraulic radii but of the same absolute roughness. Such measurements were carried out by K. Fromm [17] and W. Fritsch [16], who found that for geometrically similar roughnesses λ is proportional to $(k/R_h)^{0.314}$.

The second type of resistance formula occurs when the protrusions are more gentle or when a small number of them is distributed over a relatively large area, such as those in wooden or commercial steel pipes. In such cases the resistance coefficient depends both on the Reynolds number and on the relative roughness.

From the physical point of view it must be concluded that the ratio of the height of protrusions to the boundary-layer thickness should be the determining factor. In particular, the phenomenon is expected to depend on the thickness of the laminar sub-layer δ_l, so that k/δ_l must be regarded as an important dimensionless number which is characteristic of the kind of roughness. It is clear that roughness will cause no increase in resistance in cases where the protrusions are so small (or the boundary layer is so thick) that they are all contained within the laminar sublayer, i. e. if $k < \delta_l$, and the wall may be considered hydraulically smooth. This is similar to the absence of the influence of roughness on resistance in Hagen-Poiseuille flow. Recalling our considerations in Sec. XIX e, we find that the thickness of the laminar sub-layer is given by $\delta_l = \text{const} \cdot \nu/v_*$ and that the dimensionless roughness factor is

$$k/\delta_l \sim k \, v_*/\nu \, .$$

This is a Reynolds number which is formed with the grain size of roughness and the friction velocity v_*.

Very systematic, extensive, and careful measurements on rough pipes have been carried out by J. Nikuradse [46]†, who used circular pipes covered on the inside as tightly as possible with sand of a definite grain size glued on to the wall. By choosing pipes of varying diameters and by changing the size of grain, he was able to vary the relative roughness k_S/R from about $1/500$ to $1/15$. The regularities of behaviour discovered during the course of these measurements can be correlated with those for smooth pipes in a simple manner.

We shall begin by describing Nikuradse's measurements and we shall then show that the relation between the resistance formula and the velocity distribution, which we found earlier in the case of smooth pipes, can be extended to the case of rough pipes in a natural way.

Resistance formula: Figure 20.18 represents the law of friction for pipes roughened with sand. In the region of laminar flow all rough pipes have the same resistance as a smooth pipe. The critical Reynolds number is equally independent of roughness, and in the turbulent region there is a range of Reynolds numbers over which pipes of a given relative roughness behave in the same way as smooth pipes. The rough pipe can, therefore, be said to be *hydraulically smooth* in this range and λ depends on R alone. Beginning with a definite Reynolds number the magnitude of which increases as k_S/R decreases, the resistance curve for a rough pipe deviates from that for a smooth pipe and reaches the region of the quadratic resistance law at some higher value of Reynolds number, where λ depends on k_S/R only. Hence it is necessary to consider *three regimes*:

1. *Hydraulically smooth‡ regime:*

$$0 \leq \frac{k_S v_*}{\nu} \leq 5: \quad \lambda = \lambda \, (\text{R}) \, .$$

---

† In what follows we shall use the symbol k_S to denote the grain size in Nikuradse's sand roughness, reserving the symbol k for any other kind of roughness.

‡ The numerical values of $k_S v_*/\nu$ quoted here will be derived later from the velocity distribution law. They are valid only for roughnesses obtained with sand.

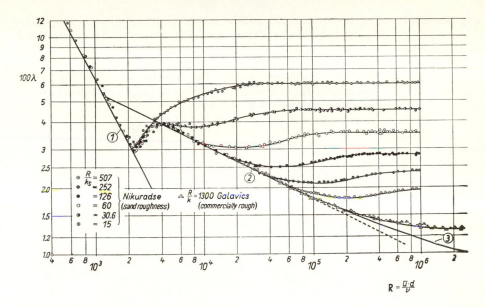

Fig. 20.18. Resistance formula for rough pipes

Curve (1) from eqn. (5.11), laminar; curve (2) from eqn. (20.5), turbulent, smooth; curve (3) from eqn. (20.30), turbulent, smooth

The size of the roughness is so small that all protrusions are contained within the laminar sub-layer.

2. *Transition regime:*

$$5 \leq \frac{k_s v_*}{v} \leq 70: \quad \lambda = \lambda \left(\frac{k_s}{R}, \mathsf{R} \right).$$

Protrusions extend partly outside the laminar sub-layer and the additional resistance, as compared with a smooth pipe, is mainly due to the form drag experienced by the protrusions in the boundary layer.

3. *Completely rough regime:*

$$\frac{k_s v_*}{v} > 70: \quad \lambda = \lambda \left(k_s/R \right).$$

All protrusions reach outside the laminar sub-layer and by far the largest part of the resistance to flow is due to the form drag which acts on them. For this reason the law of resistance becomes quadratic.

Velocity distribution: The velocity gradient near a rough wall is less steep than that near a smooth one, as can be seen from Fig. 20.19, in which the velocity ratio u/U has been plotted against the distance ratio y/R for a smooth and for several rough pipes, all having been measured within the range of validity of the square resistance law. Expressing the velocity distribution function again by a power formula of the type of eqn. (20.6) we obtain exponents of $\frac{1}{4}$ to $\frac{1}{5}$. The variation of mixing length over the cross-section calculated from these curves has already been plotted in Fig. 20.6 from which it is seen that it is exactly the same for rough and for smooth pipes. It can be represented by the empirical equation (20.18). In particular, in the neighbourhood of the wall we have $l = \varkappa\, y = 0.4\, y$.

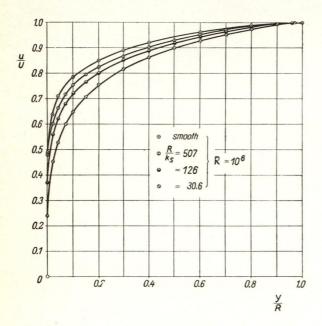

Fig. 20.19. Velocity distribution in rough pipes, after Nikuradse [46]

It follows, therefore, that the logarithmic law for velocity distribution, eqn. (19.29), remains valid for rough pipes, except that the constant of integration, y_0, must be given a different numerical value. Furthermore, it is natural to make it proportional to the roughness height k_S, i. e. to put $y_0 = \gamma\, k_S$, so that eqn. (19.29) now becomes

$$\frac{u}{v_*} = \frac{1}{\varkappa}\left(\ln\frac{y}{k_S} - \ln\gamma\right),\tag{20.31}$$

the constant γ still depending on the nature of the particular roughness. Comparing this equation with J. Nikuradse's measurements, we find that they can, in fact, be represented by an equation of the form:

$$\frac{u}{v_*} = 2.5\ln\frac{y}{k_S} + B,\tag{20.32}$$

where the constant $2 \cdot 5 = 1/\varkappa = 1/0 \cdot 4$, whereas B assumes different values for the three ranges of roughness discussed previously. In the range of the completely rough regime, we have $B = 8 \cdot 5$, so that in this region

$$\frac{u}{v_*} = 5 \cdot 75 \log \frac{y}{k_s} + 8 \cdot 5 \quad \text{(completely rough)} . \tag{20.32a}$$

The corresponding straight line is seen to agree well with the results of measurement, Fig. 20.20. Generally speaking B is a function of the roughness Reynolds number $v_* k_S/v$. The value which corresponds to hydraulically smooth flow follows at once from eqns. (20.32) and (20.14), and is

$$B = 5 \cdot 5 + 2 \cdot 5 \ln \frac{v_* k_S}{v} \quad \text{(hydraulically smooth)} . \tag{20.33}$$

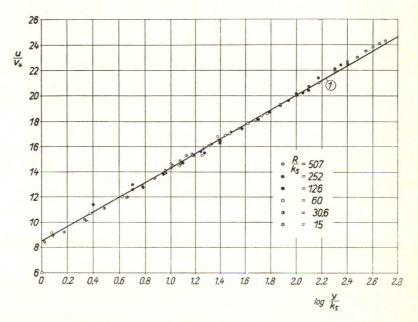

Fig. 20.20. Universal velocity distribution in rough pipes
Curve (1) from eqn. (20.32a)

The values of B in the transition region from hydraulically smooth flow to completely rough flow are shown plotted against $v_* k_S/v$ in Fig. 20.21; the points are seen to arrange themselves exceedingly well on one curve.

Writing eqn. (20.32) for the axis of the pipe $y = R$, $u = U$ and forming the difference $U - u$, we obtain the velocity-defect equation (20.23):

$$\frac{U-u}{v_*} = 2 \cdot 5 \ln \frac{R}{y} = 5 \cdot 75 \log \frac{R}{y} , \tag{20.23}$$

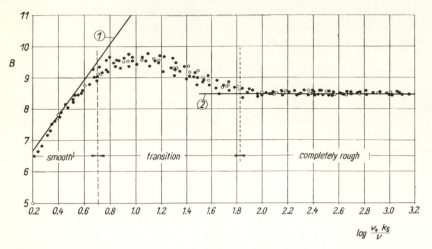

Fig. 20.21.   Roughness function B in terms of $v_* \, k_S/\nu$, for Nikuradse's sand roughness
Curve (1): hydraulically smooth, eqn. (20.33); curve (2): $B = 8 \cdot 5$; completely rough

once more. It has been found to apply to smooth pipes in connexion with Fig. 20.7. In order to see more clearly the connexion between the velocity distributions for smooth and rough pipes, it is useful to re-plot the results for rough pipes in the form of a relation between the dimensionless velocity $u/v_* = \phi$ and the Reynolds number $y \, v_*/\nu = \eta$, as was done in eqn. (20.13) and Fig. 20.4 in relation to smooth pipes. Writing eqn. (20.32a) for the rough pipe in the form

$$\frac{u}{v_*} = 5 \cdot 75 \log \frac{y v_*}{\nu} + D_1 \quad \text{(completely rough) ,} \tag{20.33a}$$

and comparing it with eqns. (20.33a) and (20.32a), we obtain

$$D_1 = 8 \cdot 5 - 5 \cdot 75 \, \log \frac{k_s v_*}{\nu} \quad \text{(completely rough) .} \tag{20.33b}$$

This velocity distribution is seen plotted in Fig. 20.22, after N. Scholz [65]; it represents the velocity distribution for smooth pipes as well as that for rough pipes, in accordance with eqn. (20.33a). The diagram consists of a family of parallel straight lines with $v_* \, k_S/\nu$ playing the part of a parameter. The value of $v_* \, k_S/\nu = 5$ corresponds to hydraulically smooth walls, the range between $v_* \, k_S/\nu = 5$ to 70 corresponds to transition from the hydraulically smooth to the completely rough regime, and for $v_* \, k_S/\nu > 70$ the flow is completely rough, as mentioned previously. In particular, the diagram shows clearly that the laminar sub-layer which reaches as far as $y \, v_*/\nu = 5$ in hydraulically smooth pipes, has no importance for completely rough walls.

Relation between resistance formula and velocity distribution: This type of relation exists for rough pipes also and can be deduced in the same manner as was done

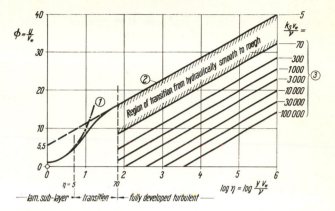

Fig. 20.22. Universal velocity profile for turbulent flows through pipes which is valid for smooth as well as for rough walls, after N. Scholz [65]

(1) smooth, laminar sublayer, $\phi = \eta$
(2) smooth, turbulent, eqn. (20.14)
(3) rough, turbulent, eqn. (20.33a)
   with D_1 from eqn. (20.33b)

in Sec. XX d for the case of smooth pipes. The relation is simplest for the *completely rough regime*. We begin by calculating the mean velocity from eqn. (20.23) in the same way as in eqn. (20.26):

$$\bar{u} = U - 3{\cdot}75\, v_* . \tag{20.34}$$

Substituting $U = v_*\,(2{\cdot}5 \ln R/k_S + 8{\cdot}5)$ from eqn. (20.32a), we have

$$\bar{u}/v_* = 2{\cdot}5 \ln (R/k_S) + 4{\cdot}75 \quad \text{or} \quad \lambda/8 = (v_*/\bar{u})^2 = [2{\cdot}5 \ln (R/k_S) + 4{\cdot}75]^{-2} ,$$

i. e.

$$\lambda = [2 \log (R/k_S) + 1{\cdot}68]^{-2} ,$$

which is the quadratic resistance formula for completely rough flow. It was first derived by Th. von Kármán (Chap. XIX [17]) from the similarity law. A comparison with J. Nikuradse's experimental results (Fig. 20.23) shows that closer agreement can be obtained, if the constant $1{\cdot}68$ is replaced by $1{\cdot}74$. Hence the *resistance formula for the completely rough regime* becomes

$$\lambda = \frac{1}{\left(2 \log \dfrac{R}{k_S} + 1{\cdot}74\right)^2} . \tag{20.35}\dagger$$

The experimental results lie very close to a straight line in a plot of $1/\sqrt{\lambda}$ against $\log (R/k_S)$ and it is worth noting that eqn. (20.35) may be applied to pipes with non-

---

† An equation which correlates the whole transition region from hydraulically smooth to completely rough flow was established by Colebrook and White [6]:

$$\frac{1}{\sqrt{\lambda}} = 1{\cdot}74 - 2 \log \left(\frac{k_S}{R} + \frac{18{\cdot}7}{R\,\sqrt{\lambda}} \right). \tag{20.35a}$$

For $k_S \to 0$ this equation transforms into eqn. (20.30), valid for hydraulically smooth pipes. For $R \to \infty$, it transforms into eqn. (20.35) for the completely rough regime. In the transition region eqn. (20.35a) plots λ against R in a way which resembles the curve labelled "commercially rough" in Figs. 20.18 and 20.25.

circular cross-sectional areas if R is replaced by the hydraulic radius $R_h = 2\,A/C$
(A = area; C = wetted perimeter).

It is also easy to derive the relation between the resistance law and the velocity distribution in the *transition region*. From eqn. (20.32) we have

$$B = \frac{u}{v_*} - 2{\cdot}5 \ln \frac{y}{k_S} = \frac{U}{v_*} - 2{\cdot}5 \ln \frac{R}{k_S}\,.$$

On the other hand, from eqn. (20.34) we obtain

$$\frac{U}{v_*} = \frac{\bar{u}}{v_*} + 3{\cdot}75 = \frac{2\sqrt{2}}{\sqrt{\lambda}} + 3{\cdot}75\,,$$

and the preceding equation gives

$$B\left(\frac{v_* k_S}{\nu}\right) = \frac{u}{v_*} - 2{\cdot}5 \ln \frac{y}{k_S} = \frac{2\sqrt{2}}{\sqrt{\lambda}} + 3{\cdot}75 - 2{\cdot}5 \ln \frac{R}{k_S}\,. \qquad (20.36)$$

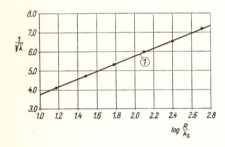

Fig. 20.23.   Resistance formula of sand-roughened pipes in completely rough regime

Curve (1) from eqn. (20.35)

The last equation determines the value of the resistance coefficient λ if the constant B is known from the velocity distribution. On the other hand, eqn. (20.36) can be used to determine the constant B as a function of $v_* k_S/\nu$ either from the velocity distribution or from the resistance formula. The plot in Fig. 20.21 agrees well with the results from either of these methods and proves that the calculation of the velocity distribution from the resistance formula is permissible for rough pipes too.

The limits between the three regimes, namely those of hydraulically smooth flow, the transitional regime, and the completely rough regime, which have been given earlier, can now be taken directly from Fig. 20.21. We have

$$\text{hydraulically smooth:} \qquad \frac{v_* k_S}{\nu} < 5\,,$$

$$\text{transition:} \qquad 5 < \frac{v_* k_S}{\nu} < 70\,, \qquad (20.37)$$

$$\text{completely rough:} \qquad \frac{v_* k_S}{\nu} > 70\,.$$

These limits are in complete accord with the velocity distribution in the boundary layer in the immediate neighbourhood of a smooth wall measured by H. Reichardt and plotted in Fig. 20.4. The limit of the hydraulically smooth regime $v_* \, k_S/\nu = 5$ gives the thickness of the laminar sub-layer and coincides with the limit of the range in which the Hagen-Poiseuille, purely laminar, velocity-distribution law retains its validity. The limit of $v_* \, k_S/\nu = 70$ for the transitional regime also coincides with the point where the measured velocity distribution goes over tangentially into the logarithmic formula (20.14) in fully turbulent friction.

S. Goldstein [19] succeeded in deducing the limit of $v_* \, k_S/\nu = 5$ for the hydraulically smooth regime from the criterion that at that point a von Kármán vortex street is about to begin to form on an individual protrusion. According to measurements on circular cylinders performed by F. Homann this occurs at a Reynolds number of 60 to 100, where the Reynolds number is formed with the diameter and the free-stream velocity (Fig. 1.6). In a more recent investigation J. C. Rotta [58] found that the thickness of the laminar sub-layer is smaller for a rough wall than for a smooth one to which eqn. (20.15a) was found to apply.

g. Other types of roughness

The roughness obtained by Nikuradse with sand can be said to be of maximum density, because the grains of sand were glued to the wall as closely to each other as possible. In many practical applications the roughness density of the walls is considerably smaller, and such roughnesses can no longer be described by the indication of the height of a protrusion, k, or by the relative measure k/R only. It is convenient to arrange such roughnesses on a scale of *standard roughness* and to adopt Nikuradse's sand roughness for correlation because it has been investigated in a very large range of values of R and k_S/R. The correlation is simplest in the completely rough regime when, according to what was said previously, the resistance coefficient is given by eqn. (20.35). It is convenient to correlate any given roughness with its *equivalent sand roughness* and to define it as that value which gives the actual coefficient of resistance when inserted into eqn. (20.35).

H. Schlichting [63] determined experimentally these values of equivalent sand roughness for a large number of roughnesses arranged in a regular fashion. The special experimental channel used for this purpose had a rectangular cross-section with three smooth side-walls and one long, interchangeable side-wall whose roughness was varied to suit the experiment. By measuring the velocity distribution in the central cross-section it is possible to determine the shearing stress on the rough wall with the aid of the logarithmic formula and hence, also, the equivalent sand roughness. In order to do that it is only necessary to determine the constant B in the universal equation $u/v_* = 5{\cdot}75 \, \log \, (y/k) + B$ for a given value of k. On comparing with eqn. (20.32a) we obtain the equivalent sand roughness from the equation

$$5{\cdot}75 \log k_S/k = 8{\cdot}5 - B \,. \qquad (20.38)$$

Some results of such measurements are seen summarized in Fig. 20.24. Similar measurements were carried out by V. L. Streeter [73] and H. Moebius [39] on pipes which had been made artificially rough by cutting threads of different forms into them.

Generally speaking, pipes which are regarded as *smooth in* engineering *practice* cannot be taken to be *hydraulically smooth*. An example of this discrepancy is given

No	item	dimensions	D [cm]	d [cm]	k [cm]	k_s [cm]	photographs
1	spheres	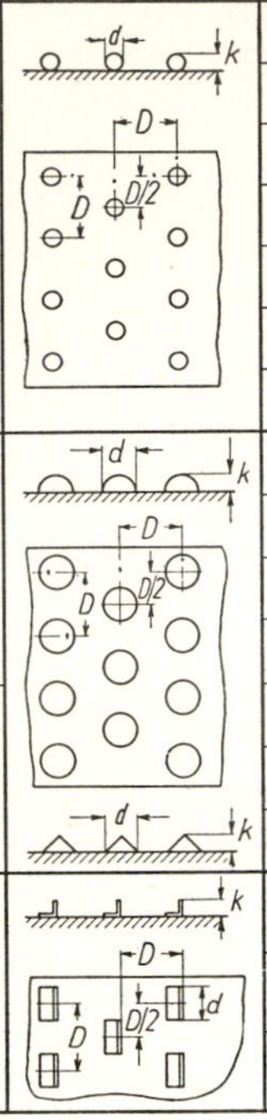	4	0.41	0.41	0.093	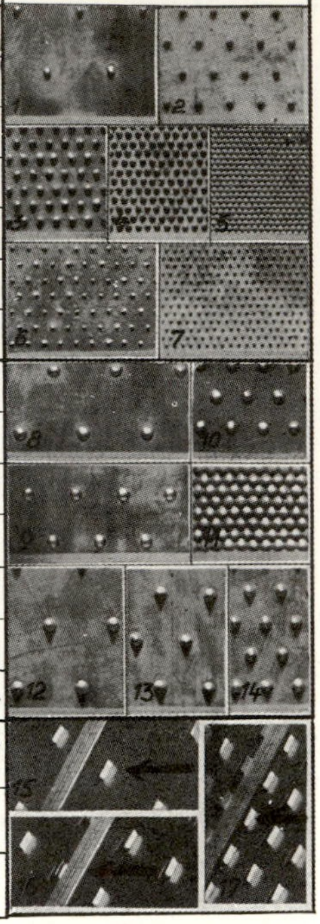
2			2	0.41	0.41	0.344	
3			1	0.41	0.41	1.26	
4			0.6	0.41	0.41	1.56	
5			densest arrgt.	0.41	0.41	0.257	
6			1	0.21	0.21	0.172	
7			0.5	0.21	0.21	0.759	
8	spherical segments		4	0.8	0.26	0.031	
9			3	0.8	0.26	0.049	
10			2	0.8	0.26	0.149	
11			densest arrgt.	0.8	0.26	0.365	
12	cones		4	0.8	0.375	0.059	
13			3	0.8	0.375	0.164	
14			2	0.8	0.375	0.374	
15	"short" angles		4	0.8	0.30	0.291	
16			3	0.8	0.30	0.618	
17			2	0.8	0.30	1.47	

Fig. 20.24. Results of measurements on regular roughness patterns after H. Schlichting [63]

k — actual height of protrusion; k_s — equivalent sand roughnesses

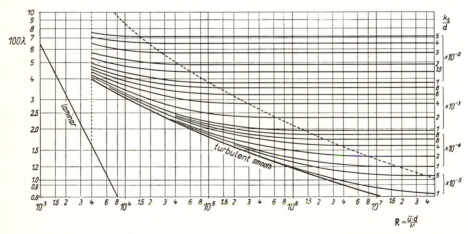

$$R = \frac{\bar{u} \cdot d}{\nu}$$

Fig. 20.25. Resistance of commercially rough pipes after L. F. Moody [40]

k_s — equivalent sand roughness, to be determined in particular cases from the auxiliary graph in Fig. 20.26. The broken line indicates the boundary of the completely rough regime where the quadratic law of friction applies in Fig. 20.18, where results of the measurements carried out by B. Bauer and F. Galavics [3] on a "commercially smooth" steel pipe with a flow of hot water are seen plotted together with Nikuradse's values for pipes roughened with sand.

The difficulty in applying the above calculations to practical cases lies in the fact that the value of roughness to be ascribed to a given pipe is not known. Very extensive experimental results on the resistance of commercially rough pipes are contained in a paper by L. F. Moody [40]. Fig. 20.25 shows that the graph of λ against R for different values of k_S/d is in essence identical with J. Nikuradse's diagram in Fig. 20.18. The individual values of equivalent relative sand roughness k_S/d can be obtained from the auxiliary graph in Fig. 20.26 where pipes are seen to have been arranged in the order of values on Nikuradse's equivalent sand-roughness scale. This follows from the fact that the values of λ in terms of k_S/d agree with Nikuradse's values from Fig. 20.18 in the completely rough regime. The transition from hydraulically smooth conditions at small Reynolds numbers to complete roughness at large Reynolds numbers occurs much more gradually in such commercial pipes than in Nikuradse's artificially roughened ones.

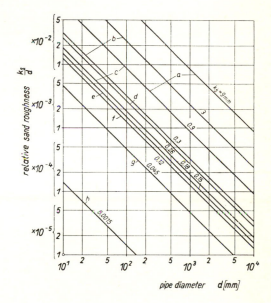

Fig. 20.26. Auxiliary diagram for the evaluation of equivalent relative sand roughness for commercial pipes, after L. F. Moody [40]

a) riveted steel
b) reinforced concrete
c) wood
d) cast iron
e) galvanized steel
f) bitumen-coated steel
g) structural and forged steel
h) drawn pipes

It is sometimes impossible to fit commercially rough surfaces satisfactorily into the scale of sand roughness. A peculiar type of roughness giving very large values of the resistance coefficient was discovered in the water duct in the valley of the Ecker [68, 82]. This pipe had a diameter of 500 mm and after a long period of usage it was noticed that the mass flow decreased by more than 50 per cent. Upon examination it was found that the walls of the duct were covered with a rib-like deposit only 0·5 mm high, the ribs being at right angles to the flow direction. Thus the geometrical roughness had the small value of $k/R = 1/500$, but the effective sand roughness showed values of $k_S/R = 1/40$ to $1/20$, as calculated from the resistance coefficient which was, in turn, determined with the aid of the measured values of mass flow. It appears, therefore, that rib-like corrugations lead to much higher resistances than sand roughness of the same absolute dimension. Extensive experiments on the increase in the resistance found in commercial ducts, for example in mine shafts, can be found described in a paper by E. Huebner [26].

Further details concerning the resistance offered to flow by rough walls, particularly those due to single protrusions, will be given in Chap. XXI in connexion with the discussion on the resistance of flat plates.

h. Flow in curved pipes and diffusers

Curved pipes: The preceding considerations concerning pipe flow are valid only for straight pipes. In curved pipes there exists a secondary flow because the particles near the flow axis which have a higher velocity are acted upon by a larger centrifugal force than the slower particles near the walls. This leads to the emergence of a secondary flow which is directed outwards in the centre and inwards (i. e. towards the centre of curvature) near the wall, Fig. 20.27.

The influence of curvature is stronger in laminar than in turbulent flow. C. M. White [80] and M. Adler [2] carried out experiments on laminar flow. The turbulent case was investigated experimentally by H. Nippert [47] and H. Richter [56]. Theoretical calculations for the laminar case were carried out by W. R. Dean [10] and M. Adler [2]. The characteristic dimensionless variable, which determines the influence of curvature *in the laminar case*, is the *Dean number*

$$\mathsf{D} = \frac{1}{2}\,\mathsf{R}\,\sqrt{\frac{R}{r}} = \frac{\bar{u}R}{\nu}\,\sqrt{\frac{R}{r}}\,,\tag{20.39}$$

where R is the radius of the cross-section and r is the radius of curvature.

The measurements carried out by M. Adler for the values: $r/R = 50$, 100, and 200, demonstrated the existence of a large increase in the resistance to flow caused by the curvature for $\mathsf{R}\sqrt{R/r} > 10^{1/2}$. According to his calculations the resistance coefficient λ for laminar flow in a curved pipe is given by

$$\frac{\lambda}{\lambda_0} = 0\cdot1064\left[\mathsf{R}\,\sqrt{\frac{R}{r}}\right]^{1/2},\tag{20.40}$$

where λ_0 denotes the coefficient of resistance of a straight pipe, eqn. (20.30). Measurements indicate, however, that the above equation only has asymptotic validity, and

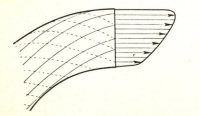

Fig. 20.27. Flow in a curved pipe, after Prandtl [52]

may be used for values of the parameter $\sqrt{R/r}$ exceeding about $10^{2\cdot8}$. The results of measurements are approximated with a higher degree of precision by the following empirical equation, first given by L. Prandtl [53]:

$$\frac{\lambda}{\lambda_0} = 0\cdot37 \; \mathsf{D}^{0\cdot36}. \qquad (20.41)$$

This equation gives good agreement with experimental results in the range

$$10^{1\cdot6} < \mathsf{R} \, (R/r)^{1/2} < 10^{3\cdot0} \, .$$

C. M. White [81] has found that the resistance coefficient for *turbulent flow* in a curved pipe can be represented by the equation

$$\frac{\lambda}{\lambda_0} = 1 + 0\cdot075 \; \mathsf{R}^{1/4} \left(\frac{R}{r} \right)^{1/2} \qquad (20.42) \dagger$$

whose form indicates clearly that the Dean number can no longer serve as a suitable independent variable. In more recent times, H. G. Cuming [8] carried out an investigation into the phenomenon of secondary flow in curved pipes.

In the laminar case Ito [26a] extended the validity of eqn. (20.40) to lower Dean numbers. According to his calculations the resistance coefficient λ is given by

$$\lambda/\lambda_0 = 0\cdot103 \; K^{1/2} \, (1 + 3\cdot945 \; K^{-1/2} + 7\cdot782 \; K^{-1} + 9\cdot097 \; K^{-3/2} + \ldots),$$

in which $K = 2 \, \mathsf{D}$. If the numerical coefficient outside the parentheses is replaced by $0\cdot101$, the equation gives good agreement with experimental results in the range of $K > 30$.

In the turbulent case Ito [27] has shown theoretically that the ratio of the resistance coefficients, λ/λ_0, may be expressed in terms of the dimensionless variable $\mathsf{R} \, (R/r)^2$. The experimental results of Ito [28] can be represented with sufficient accuracy by the equations mentioned in the footnote.

In flow through a bend or elbow there is not only some loss of energy within the bend itself, but a part of the loss produced by the bend takes place in the straight pipe following it. Extensive measurements of the loss coefficients for smooth pipe bends and a correlation of results were given by H. Ito [29]. Theoretical results are reported by W. M. Collins et al. [5b].

In flow through a radially rotating straight pipe, a secondary flow similar to that found in a curved pipe is set up by the action of a Coriolis force; it gives rise to a large increase in resistance. Extensive measurements and theoretical calculations on this subject were carried out by H. Ito and K. Nanbu [30].

---

† H. Ito [27] gives:

$$\lambda \left(\frac{r}{R} \right)^{1/2} = 0\cdot029 + 0\cdot304 \left[\mathsf{R} \left(\frac{R}{r} \right)^2 \right]^{-0\cdot25} ; \quad 300 > \mathsf{R} \, (R/r)^2 > 0\cdot034$$

and
$$\frac{\lambda}{\lambda_0} = \left[\mathsf{R} \left(\frac{R}{r} \right)^2 \right]^{0\cdot05} ; \quad \mathsf{R}(R/r)^2 > 6 \, .$$

These differ somewhat from, but are in general agreement with, C. M. White's equation above.

Extensive measurements and theoretical calculations on frictional losses in turbulent flow have also been carried out by R. W. Detra [11] who included curved pipes of noncircular cross-section in his investigations. It is found that the resistance offered by an elliptic pipe is greater when the major axis of the ellipse lies in the plane of curvature than when it is perpendicular to it.

E. Becker [4] studied secondary flows in a rectangular channel of constant curvature in which the radial extent of the cross-section is much larger than its height. The formation of dead-water areas and separation in a 90° rectangular bend provided with a sharp entrance was investigated by D. Haase [21].

Diffusers: J. Ackeret [1], H. Sprenger [69, 70] as well as S. J. Kline and his collaborators [15, 34, 41] performed a large number of experimental investigations into the characteristics of straight and curved diffusers. One of the most important results of these researches is the establishment of the fact that the thickness of the turbulent boundary layer at inlet exerts a very large influence on the efficiency of pressure recovery. This is defined as

$$\eta_D = \frac{p_2 - p_1}{\frac{1}{2}\,\varrho(\bar{u}_1{}^2 - \bar{u}_2{}^2)}\,. \tag{20.43}$$

Here p denotes the static pressure and $\bar{u}$ the mean velocity over the cross-section, whereas subscripts 1 and 2 refer to conditions at inlet and exit, respectively. The

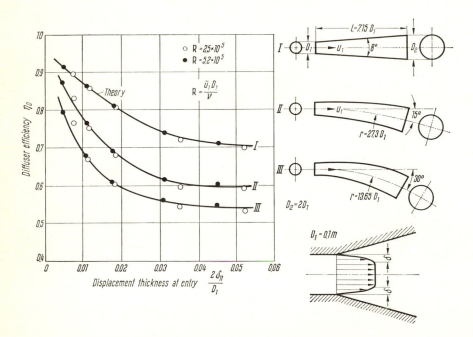

Fig. 20.28. Diffuser efficiency in straight and curved diffusers with circular cross-sections. Efficiency defined in eqn. (20.43). Diagram represents dependence on ratio of boundary-layer thickness to radius of cross-section at entry; after J. Ackeret [1, 1a] and H. Sprenger [70]

diagram in Fig. 20.28 represents the variation of efficiency with the ratio $2\,\delta_{11}/D_1$ for diffusers with circular cross-sections. The symbol δ_{11} denotes the displacement thickness at inlet, and D_1 is the corresponding diffuser diameter. For straight diffusers $\eta_D = 0\cdot9$ when the displacement thickness at inlet, δ_{11}, constitutes $0\cdot5\,\%$ of the radius $\frac{1}{2}\,D_1$ and falls to $\eta_D = 0\cdot7$ when δ_{11} increases to $5\,\%$ of $\frac{1}{2}\,D_1$. It is further recognized that the efficiency decreases strongly as the angle of deflexion increases. It has been also established that the shape of the cross-section at inlet plays an important part when the diffuser is curved. For example, when the cross-section is circular at inlet and elliptic at exit, the efficiency η_D is considerably lower if the major axis of the ellipse is placed in the plane of curvature. If the major axis is turned to be at right angles to the plane of curvature, the efficiency η_D increases. In the former case there arises a much more vigorous secondary flow which leads to increased losses. J. Ackeret succeeded in theoretically calculating the efficiency of a straight diffuser whose cross-sections are circular. This was done with the aid of the equations for turbulent boundary layers contained in Chap. XXII. It is seen from Fig. 20.28 that this calculation agrees well with measurements. Systematic calculations on boundary layers in straight diffusers were also performed by H. Schlichting and K. Gersten [64]. These lead to the conclusion that there exists an optimum included angle of divergence, $2\,\alpha$, for diffusers of equal area ratio (inlet to exit) and equal Reynolds number at entry. The efficiency, η_D, then attains a maximum. The optimum angle lies between $2\,\alpha = 3\,°$ and $8\,°$ and decreases as the Reynolds number is increased.

In this connexion attention is drawn to the experimental studies on diffusers performed by F. A. L. Winternitz and W. J. Ramsay [83] and by J. S. Sobey [68 b].

A summary review of the mechanics of flow in diffusers was prepared by D. J. Cockerell and E. Markladd [7].

i. Non-steady flow through a pipe

The problem of pulsating flow, i. e. of a steady mean flow on which there is superimposed a periodic pulsation, was investigated by F. Schultz-Grunow [66]. The experimental arrangement consisted of a pipe which was fed with water at a constant head and whose end section was rhythmically increased and decreased in area. The velocity profiles for the periods of acceleration and deceleration, respectively, differ markedly from one another. They are very similar to steady-flow profiles through a gradually convergent pipe or channel during the periods of acceleration, and during periods of deceleration they resemble steady-state profiles in a divergent channel (diffuser), as explained in Chap. XXII, where such profiles will be found plotted in detail. In certain circumstances both reversal of flow and separation near the wall may occur during periods of deceleration. The time-averaged value of the coefficient of resistance does not differ considerably from its steady-state value when the pulsations are gradual and slow. In a further paper, F. Schultz-Grunow [67] described a practical method for the measurement of the rate of discharge in pulsating flow.

j. Drag reduction by the addition of polymers

In turbulent flow, the pressure drop in a pipe can be considerably reduced relative to eqn. (20.30) by the addition of small quantities of polymer particles. In laminar flow, similar additives leave the pressure drop practically unchanged. The extent of drag reduction depends on the molecular weight of the polymer and on its concentration. The graph of Fig. 20.29 describes the resistance coefficient, λ, as a function of the Reynolds number $R = \bar{u}d/\nu_p$ — where ν_p is the kinematic viscosity of the solution — for different values of the concentration c of the solution. The measurements were performed by R. W. Paterson and F. H. Abernathy [50a]. As the concentration increases, the reduced resistance coefficient tends to a curve of maximum reduction, curve (3), indicated by P. S. Virk [77]. The diagram of Fig. 20.29 displays points obtained at two measuring stations along the pipe. The difference in the values of the resistance coefficient at those two sections is explained by the fact that the polymer molecules are torn apart in the turbulent stream resulting in an effective decrease in concentration in the downstream direction. In spite of intensive research into this phenomenon (cf. the paper by M. T. Landahl [36]), no satisfactory explanation for its occurrence has yet been advanced. Nevertheless, experiments demonstrate unmistakably that the reduction in drag is linked to changes in the structure of the turbulence. The process is best illustrated with the aid of the experimentally determined velocity-distribution laws.

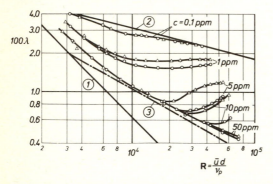

ν_p = kinematic viscosity of the solution
O = measuring station at 214 d from inlet
$\triangle$ = measuring station at 1541 d from inlet
(1) laminar flow with $\lambda = 64/R$
(2) turbulent flow, Newtonian fluid, eqn. (20.5)
(3) asymptote for maximum drag reduction after
    P. S. Virk [77], eqn. (20.45)

Fig. 20.29. Resistance coefficient, λ, of smooth pipes in turbulent flow of polymer solutions as a function of Reynolds number as measured by R. W. Paterson and F. H. Abernathy [50]. Solutions of polyethylene oxide of given concentrations c; 1 ppm designates 1 g of polymer per 10^6 g of water

According to P. S. Virk [78, 79], it is necessary to distinguish three velocity-distribution zones:

(1) The laminar sublayer $(0 \leq \eta \leq 10)$; this corresponds to curve (1) in Fig. 20.4.

(2) The fully developed turbulent zone. Here the distribution follows the law according to eqn. (20.13) with $A_1 = 2.5$ regardless of the physical properties of the solution. The constant D_1 varies strongly with the concentration.

(3) The zone described as "elastic". This zone places itself between the laminar sublayer and the fully developed turbulent region [77]. Here the velocity is represented by the logarithmic law

$$\phi = 11{\cdot}7 \ln \eta - 17\,.\tag{20.44}$$

This law is valid for all intents and purposes as far as the centre of the pipe for sufficiently high concentrations. In analogy with eqn. (20.30), we can integrate it to derive the resistance formula

$$\frac{1}{\sqrt{\bar{\lambda}}} = 9{\cdot}5 \log\left(\frac{\bar{u}\,d}{\nu_p}\right) - 19{\cdot}3 \quad \text{(high polymer concentrations)}.\tag{20.45}$$

The preceding equation can be regarded as an asymptotic law for the largest drag reduction attainable. Experiments by P. S. Virk [78] have shown, further, that the effect of surface roughness is suppressed to a large extent in flows of polymer solutions. This problem area has recently been reported extensively at an International Symposium [15a]; compare also the review by N. S. Berman [4a].

References

[1] Ackeret, J.: Grenzschichten in geraden und gekrümmten Diffusoren. IUTAM-Symposium Freiburg/Br. 1957 (H. Görtler, ed.), Berlin, 1958, 22—37.

[1a] Ackeret, J.: Aspects of internal flow. Fluid mechanics of internal flow (G. Sovran, ed.). Elsevier Publishing Company, Amsterdam/London/New York, 1967, 1—24.

[2] Adler, M.: Strömung in gekrümmten Rohren. ZAMM *14*, 257—275 (1934).

[3] Bauer, B., and Galavics, F.: Experimentelle und theoretische Untersuchungen über die Rohrreibung von Heizwasserleitungen. Mitt. d. Fernheizkraftwerkes d. ETH Zürich 1936; see also: F. Galavics: Schweizer Archiv *5*, 12, 337 (1939).

[4] Becker, E.: Beitrag zur Berechnung von Sekundärströmungen. ZAMM *36*, special issue, 3—8 (1956); see also: Mitt. Max-Planck-Inst. für Strömungsforschung *13* (1956).

[4a] Berman, N.S.: Drag reduction by polymers. Ann. Rev. Fluid Mech. (M. van Dyke, ed.) *10*, 47—64 (1978).

[5] Blasius, H.: Das Ähnlichkeitsgesetz bei Reibungsvorgängen in Flüssigkeiten. Forschg. Arb. Ing.-Wes. No. 134, Berlin (1913).

[5a] Clauser, F.H.: The turbulent boundary layer. Adv. Appl. Mech. *4*, 1—51 (1956). Academic Press, New York.

[5b] Collins, W.M., and Dennis, S.C.R.: The steady motion of a viscous fluid in a curved tube. Quart. J. Mech. Appl. Math. *28*, 133—156 (1975).

[6] Colebrook, C.F.: Turbulent flow in pipes with particular reference to the transition region between the smooth and rough pipe laws. J. Institution Civil Engineers, 1939; see also: Engineering hydraulics (H. Rouse, ed.). Chap. VI, Steady flow in pipes and conduits, by V.L. Streeter. New York, 1950.

[7] Cockrell, D.J., and Markladd, E.: A review of incompressible diffuser flow. Aircraft Eng. *35*, 286—292 (1963).

[8] Cuming, H.G.: The secondary flow in curved pipes. ARC RM 2880 (1955).

[9] Darcy, H.: Recherches expérimentales relatives aux mouvements de l'eau dans tuyaux. Mem. Prés. à l'Académie des Sciences de l'Institut de France *15*, 141 (1858).

[10] Dean, W.R.: The streamline motion of a fluid in a curved pipe. Phil. Mag. (7) *4*, 208 (1927) and *5*, 673 (1928).

[11] Detra, R.W.: The secondary flow in curved pipes. Inst. Aerodyn. ETH Zürich Rep. No. 20 (1953).

[12] Van Driest, E.R.: On turbulent flow near a wall. JAS *23*, 1007—1011 (1956).

[13] Eckert, E.R.G., and Irvine, T.F.: Flow in corners with non-circular cross-sections. Trans. ASME 709—718 (1956); see also JAS *22*, 65—66 (1955).

[14] Eckert, E. R. G., and Irvine, T. F.: Incompressible friction factor, transition and hydro-
     dynamic entrance-length studies of ducts with triangular and rectangular cross sections.
     Paper presented at Fifth Midwestern Conf. on Fluid Mech. 1957.

[15] Fox, R. W., and Kline, S. J.: Flow regimes in curved subsonic diffusers. J. Basic Eng., Trans.
     ASME *84*, Series D, 303—312 (1962).

[15a] Frenkiel, F. N., Landahl, M. T., and Lumley, J. L. (ed.): Structure of turbulence and drag
     reduction. IUTAM Symposium, Washington D. C., 7—12 June 1976, The Physics of Fluids,
     20, No. 10, Part II, S 1—S 292 (1977); see also B. A. Tom in Proc. Intern. Congr. Rheology,
     North-Holland, Amsterdam, 1949, Sec. II, 135.

[16] Fritsch, W.: Einfluss der Wandrauhigkeit auf die turbulente Geschwindigkeitsverteilung
     in Rinnen. ZAMM *8*, 199—216 (1928).

[17] Fromm, K.: Strömungswiderstand in rauhen Rohren. ZAMM *3*, 339—358 (1923).

[18] Frössel, W.: Strömung in glatten, geraden Rohren mit Über- und Unterschallgeschwindig-
     keit. Forschg. Ing.-Wes. *7*, 75—84 (1936).

[19] Goldstein, S.: A note on roughness. ARC RM 1763 (1963).

[20] Goldstein, S.: The similarity theory of turbulence, and flow between planes and through
     pipes. Proc. Roy. Soc. A *159*, 473—496 (1937).

[21] Haase, D.: Strömung in einem 90°-Knie. Ing.-Arch. *22*, 282—292 (1954).

[22] For additional references see: H. W. Hahnemann: Der Strömungswiderstand in Rohr-
     leitungen und Leitungselementen. Forschg. Ing.-Wes. *16*, 113—119 (1950).

[23] Hawthorne, W. R.: Secondary circulation in fluid flow. Proc. Roy. Soc. London A *206*, 374
     (1951).

[24] Hermann, R.: Experimentelle Untersuchungen zum Widerstandsgesetz des Kreisrohres bei
     hohen Reynoldsschen Zahlen und grossen Anlauflängen. Diss. Leipzig; Akad. Verlagsgesell-
     schaft, Leipzig, 1930.

[25] Hopf, L.: Die Messung der hydraulischen Rauhigkeit. ZAMM *3*, 329—339 (1923).

[26] Hübner, E.: Über den Druckverlust in Rohren mit Einbauten. Forschg. Ing.-Wes. *19*, 1—16
     (1953).

[26a] Ito, H.: Laminar flow in curved pipes. ZAMM *49*, 653—663 (1969).

[27] Ito, H.: On the pressure loss of turbulent flow through curved pipes. Mem. Inst. High
     Speed Mech., Tohoku Univ., Sendai, Japan, *7*, 63—76 (1952).

[28] Ito, H.: Friction factors for turbulent flow in curved pipes. Trans. ASME, Series D, *81*
     (J. Basic Eng.), 123—134 (1959); in more detail: Mem. Inst. High Speed Mech., Tohoku
     Univ., Sendai, Japan, *14*, 137—172 (1958/59).

[29] Ito, H.: Pressure losses in smooth pipe bends. Trans. ASME, Series D, *82* (J. Basic Eng.),
     131—143 (1960).

[30] Ito, H., and Nanbu, K.: Flow in rotating straight pipes of circular cross section. Trans.
     ASME, Series D, *93* (J. Basic Eng.), 383—394 (1971).

[31] Jakob, M., and Erk, S.: Der Druckabfall in glatten Rohren und die Durchflussziffer von
     Normaldüsen. Forschg. Arb. Ing.-Wes. No. 267, Berlin (1924).

[32] von Kármán, Th.: Über laminare und turbulente Reibung. ZAMM *1*, 233—252 (1921); see
     also Coll. Works *II*, 70—97.

[33] Kirsten, H.: Experimentelle Untersuchungen der Entwicklung der Geschwindigkeitsvertei-
     lung der turbulenten Rohrströmung. Diss. Leipzig 1927.

[33a] Kleinstein, G.: Generalized law of the wall and eddy-viscosity model for wall boundary
     layer. AIAA J. *5*, 1402—1407 (1967).

[34] Kline, S. J., Abbott, D. E., and Fox, R. W.: Optimum design of straight-walled diffusers.
     J. Basic Eng., Trans. ASME, Series D, *81*, 305—320 (1959).

[35] Koch, R., and Feind, K.: Druckverlust und Wärmeübergang in Ringspalten. Chemie-Ing.-
     Techn. *30*, 577—584 (1958).

[36] Landahl, M. T.: Drag reduction by polymer addition. In: Proc. 13th Int. Congr. Theor.
     Appl. Mech., Moscow, Aug. 1972 (E. Becker and G. K. Mikhailov, ed.), Springer-Verlag,
     177—199 (1973).

[37] Lumley, J. L.: Drag reduction by additives. Annual Review of Fluid Mech. *1*, 367—384
     (1969).

[37a] Lumley, J. L.: Drag reduction in turbulent flow by polymer additives. J. Polym. Sci.
     Macromol. Rev. *7*, 263—290 (1978).

[38] Meyer, E.: Einfluss der Querschnittsverformung auf die Entwicklung der Geschwindigkeits-
     und Druckverteilung bei turbulenten Geschwindigkeitsverteilungen in Rohren. VDI-For-
     schungsheft 389 (1938).

[39] Möbius, H.: Experimentelle Untersuchungen des Widerstandes und der Geschwindigkeits-verteilung in Rohren mit regelmässig angeordneten Rauhigkeiten bei turbulenter Strömung. Phys. Z. *41*, 202—225 (1940).

[40] Moody, L. F.: Friction factors for pipe flow. Trans. ASME *66*, 671—684 (1944).

[41] Moore, C. A., Jr., and Kline, S. J.: Some effects of vanes and of turbulence in two-dimensional wide-angle subsonic diffusers. NACA TN 4080 (1958).

[42] Naumann, A.: Druckverlust in Rohren nichtkreisförmigen Querschnittes bei hohen Ge-schwindigkeiten. ZAMM *36*, special issue, 25 (1956); see also Allg. Wärmetechnik *7*, 32—41 (1956).

[43] Nikuradse, J.: Untersuchungen über die Geschwindigkeitsverteilung in turbulenten Strö-mungen. Diss. Göttingen 1926; VDI-Forschungsheft 281 (1926).

[44] Nikuradse, J.: Turbulente Strömung in nichtkreisförmigen Rohren. Ing.-Arch. *1*, 306—332 (1930).

[45] Nikuradse, J.: Gesetzmässigkeit der turbulenten Strömung in glatten Rohren. Forschg. Arb. Ing.-Wes. No. 356 (1932).

[46] Nikuradse, J.: Strömungsgesetze in rauhen Rohren. Forschg. Arb. Ing.-Wes. No. 361 (1933).

[47] Nippert, H.: Über den Strömungswiderstand in gekrümmten Kanälen. Forschg. Arb. Ing.-Wes. No. 320 (1929).

[48] Nusselt, W.: Wärmeübergang in Rohrleitungen. Forschg. Arb. Ing.-Wes. No. 89, Berlin (1910).

[49] Ombeck, H.: Druckverlust strömender Luft in geraden zylindrischen Rohrleitungen. Forschg. Arb. Ing.-Wes. No. 158/159, Berlin (1914).

[50] Oswatitsch, K.: Grundlagen der Gasdynamik. Wien, 1976; also: Gasdynamics. Engl. transl. by G. Kuerti, Academic Press, 1956.

[50a] Paterson, R. W., and Abernathy, F. H.: Turbulent flow drag reduction and degradation with dilute polymer solutions. JFM *43*, 689—710 (1970).

[51] Prandtl, L.: Über den Reibungswiderstand strömender Luft. Ergebnisse AVA Göttingen, 1st Series, 136 (1921); see also Coll. Works *II*, 620—626.

[52] Prandtl, L.: The mechanics of viscous fluids. In: W. F. Durand: Aerodynamic Theory, *III*, 142 (1935); see also summary by L. Prandtl: Neuere Ergebnisse der Turbulenzforschung. Z. VDI *77*, 105—114 (1933); see also Coll. Works *II*, 819—845.

[53] Prandtl, L.: Führer durch die Strömungslehre. 3rd ed., 159, Braunschweig, 1949. Also: Essentials of fluid dynamics. Engl. transl. by W. M. Deans, Blackie, 1952.

[54] Reichardt, H.: Die Wärmeübertragung in turbulenten Reibungsschichten. ZAMM *20*, 297—328 (1940).

[55] Reichardt, H.: Vollständige Darstellung der turbulenten Geschwindigkeitsverteilung in glatten Leitungen. ZAMM *31*, 208—219 (1951).

[56] Richter, H.: Der Druckabfall in gekrümmten glatten Rohrleitungen. Forschg. Arb. Ing.-Wes. No. 338 (1930).

[57] Rotta, J. C.: Das in Wandnähe gültige Geschwindigkeitsgesetz turbulenter Strömungen. Ing.-Arch. *18*, 277—280 (1950).

[58] Rotta, J. C.: Control of turbulent boundary layers by uniform injection and suction of fluid. Jb. DGLR, 91—104 (1970).

[59] Saph, V., and Schoder, E. H.: An experimental study of the resistance to the flow of water in pipes. Trans. Amer. Soc. Civ. Eng. *51*, 944 (1903).

[60] Schiller, L.: Über den Strömungswiderstand von Rohren verschiedenen Querschnitts- und Rauhigkeitsgrades. ZAMM *3*, 2—13 (1923).

[61] Schiller, L.: Rohrwiderstand bei hohen Reynoldsschen Zahlen. Lectures on aerodynamics and related fields, 69, Berlin, 1930.

[62] Schiller, L.: Strömung in Rohren. Handb. Exper. Physik, *IV*, Part 4, 1—210, Leipzig, 1931.

[63] Schlichting, H.: Experimentelle Untersuchungen zum Rauhigkeitsproblem. Ing.-Arch. *7*, 1—34 (1936). Engl. transl. in Proc. Soc. Mech. Eng. USA (1936); see also: Werft, Reederei, Hafen 99 (1936), and Jb. Schiffbautechn. Ges. 418 (1936).

[64] Schlichting, H., and Gersten, K.: Berechnung der Strömung in rotationssymmetrischen Diffusoren mit Hilfe der Grenzschichttheorie. ZFW 9, 135—140 (1961).

[65] Scholz, N.: Strömungsvorgänge in Grenzschichten. VDI-Ber. *6*, 7—12 (1955).

[66] Schultz-Grunow, F.: Pulsierender Durchfluss durch Rohre, Forschg. Ing.-Wes. *11*, 170—187 (1940).

[67] Schultz-Grunow, F.: Durchflussmessverfahren für pulsierende Strömungen. Forschg. Ing.-Wes. *12*, 117—126 (1941).

[68] Seiferth, R., ana Krüger, W.: Überraschend hohe Reibungsziffer einer Fernwasserleitung.
      Z. VDI *92*, 189 (1950).
[68a] Spalding, D.B.: A single formula for the "law of the wall". J. Appl. Mech. *28*, 455—458
      (1961).
[68b] Sobey, J.S.: Inviscid secondary motions in a tube of slowly varying ellipticity. JFM *73*,
      621—639 (1976).
[69] Sprenger, H.: Messungen an Diffusoren. VDI-Ber. *3*, 10—110 (1955); see also ZAMP *7*,
      372—374 (1956).
[70] Sprenger, H.: Experimentelle Untersuchungen an geraden und gekrümmten Diffusoren.
      Mitt. Inst. Aerodyn. ETH Zürich No. 27 (1959).
[71] Stanton, T.E.: The mechanical viscosity of fluids. Proc. Roy. Soc. London A *85*, 366 (1911).
[72] Stanton, T.E., and Pannel, J.R.: Similarity of motion in relation of the surface friction
      of fluids. Phil. Trans. Roy. Soc. A *214*, 199 (1914); see also Proc. Roy. Soc. London A *91*,
      46 (1915).
[73] Streeter, V.L.: Frictional resistance in artificially roughened pipes. Proc. Amer. Soc. Civil
      Eng. *61*, 163—186 (1935).
[74] Szablewski, W.: Berechnung der turbulenten Strömung im Rohr auf der Grundlage der
      Mischungsweghypothese. ZAMM *31*, 131—142 (1951).
[75] Szablewski, W.: Der Einlauf einer turbulenten Rohrströmung. Ing.-Arch. *21*, 323—330
      (1953).
[76] Taylor, G.I.: Flow in pipes and between parallel planes. Proc. Roy. Soc. London A *159*,
      496—506 (1937).
[76a] Toms, B.A.: Some observations on the flow of linear polymer solutions through straight
      tubes at large Reynolds numbers. Proc. 1st Int. Congr. Rheol. *2*, 135—141 (1948), Amster-
      dam, North Holland.
[77] Virk, P.S.: An elastic sublayer model for drag reduction by dilute solutions of linear macro-
      molecules. JFM *45*, 417—440 (1971).
[78] Virk, P.S.: Drag reduction in rough pipes. JFM *45*, 225—246 (1971).
[79] Virk, P.S., Mickley, H.S., and Smith, K.A.: The ultimate asymptote and mean flow
      structure in Toms's phenomenon. J. Appl. Mech., Trans. ASME, Series E, *37*, 488—493
      (1970).
[80] White, C.M.: Streamline flow through curved pipes. Proc. Roy. Soc. London A *123*, 645—
      663 (1929).
[81] White, C.M.: Fluid friction and its relation to heat transfer. Trans. Inst. Chem. Eng. *10*,
      66 (1932).
[82] Wiederhold, W.: Über den Einfluss von Rohrablagerungen auf den hydraulischen Druck-
      abfall. Gas- u. Wasserfach *99*, 634 (1949).
[82a] Wieghardt, K.: Turbulente Grenzschichten. Göttinger Monographie, Part B *5* (1946).
[83] Winternitz, F.A.L., and Ramsay, W.J.: Effects of inlet boundary layer on pressure re-
      covery energy conversion and losses in conical diffusers. J. Roy. Aero. Soc. *61*, 116—124
      (1957).

CHAPTER XXI

Turbulent boundary layers at zero pressure gradient; flat plate; rotating disk; roughness

It might be surmised that it would be possible to perform calculations on the turbulent boundary layer along a flat plate, or along any shape for that matter, from eqns. (19.3a) and (19.3b) and by the same general methods as those applied to laminar boundary layers, having first established an expression for the magnitude of the viscous forces with the aid of one of the hypotheses which have been discussed in Chap. XIX. So far, however, this scheme has met with no success owing to insurmountable difficulties, nothing being known about the zone of transition from the turbulent boundary layer to the laminar sub-layer which exists in the immediate neighbourhood of the wall, and the laws of friction in the sublayer are also unknown. From this point of view conditions are more favourable as far as problems of so-called *free turbulent flows* are concerned (Chap. XXIV). These include turbulent motions in which no solid boundaries exist, such as when a jet of fluid is mixed with the surrounding atmosphere at rest, or when the wake behind a body diffuses into the stream. Such cases can be solved with the aid of the differential equations together with the empirical laws of turbulent friction. As far as the other problems of turbulent flow are concerned no successful scheme for the integration of the equations of motion has yet been advanced. The only methods available at the present time for the mathematical treatment of turbulent boundary layers are approximate methods of the type used in laminar boundary-layer theory. These are based, principally, on the momentum integral equation which has been used successfully in the study of laminar boundary layers too.

The simplest case of a turbulent boundary layer occurs on a flat plate at zero incidence; it is, furthermore, of great practical importance. It occurs, for example, in the calculation of the skin-friction drag on ships, on lifting surfaces and aeroplane bodies in aeronautical engineering, and on the blades of turbines and rotary compressors. The flat plate at zero incidence is simpler to consider, because the pressure gradient along the wall is zero so that the velocity outside the boundary layer is constant. In some of the above examples the pressure gradient may differ from zero but, just as was the case with laminar flow, the skin friction in such instances is not materially different from that on a flat plate, provided that there is no separation. The study of the flat plate is thus the basis for the calculation of the skin-friction drag for all body shapes which do not suffer appreciably from separation. The next chapter will contain an extension of this study to the case of a turbulent boundary layer with a definite pressure gradient. In many practical cases (ships, aeroplanes) the Reynolds numbers

$R = U_\infty \, l/\nu$ (U_∞ — free-stream velocity; l — length of plate) are so large that they cannot be subjected to measurement in a laboratory. Moreover, even at moderate Reynolds numbers it is much more difficult to carry out measurements in the boundary layer on a plate than in that inside a pipe. It is, therefore, very advantageous that it is possible to calculate the skin friction on a plate from the extensive data available for pipes by the use of a method due to L. Prandtl [40] and Th. von Kármán [30]. This calculation of the skin-friction drag on a plate can be carried out both for smooth and for rough walls. A good summary of this work was given by F. R. Hama [23].

a. The smooth flat plate

The approximate method to be applied to this problem is based on the momentum integral equation of boundary-layer theory as given in eqn. (8.32) of Chap. VIII, the velocity profile over the boundary-layer thickness being approximated by a suitable empirical equation. The momentum equation then provides a relation between the *characteristic parameters* of the boundary layer, i.e. between displacement thickness, momentum thickness and shearing stress at the wall.

In the following argument we shall assume at first that the boundary layer is turbulent already at the leading edge ($x = 0$) and we shall choose a system of co-ordinates as shown in Fig. 21.1, b denoting the width of the plate. The boundary-layer thickness $\delta(x)$ increases with x and on translating the data for a pipe into those for a plate we notice that the maximum velocity, U, of the former corresponds to the free-stream velocity, U_∞, of the latter, the radius, R, of the pipe corresponding to the boundary-layer thickness, δ.

At this stage we introduce with L. Prandtl the fundamental assumption that the velocity distribution in the boundary layer on a plate is identical with that inside a circular pipe. This assumption cannot, certainly, be exact, because the velocity distribution in a pipe is formed under the influence of a pressure gradient, whereas on a plate the pressure gradient is zero. However, small differences in the velocity distribution are unimportant, because the drag is calculated from the integral of momentum. Furthermore, the experimental results obtained by M. Hansen [23a] and J. M. Burgers [6] prove that this assumption is well satisfied at least in the

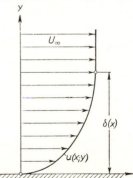

Fig. 21.1. Turbulent boundary layer
on a flat plate at zero incidence

range of moderately large Reynolds numbers ($U_\infty \, l/\nu < 10^6$). They both found that the velocity profile in the boundary layer on a plate can be described fairly well by a power formula of the form of eqn. (20.6), as found for a pipe. We shall revert once more to this problem (p. 643), when we shall discuss some systematic deviations between the velocity profiles in pipes and on plates at larger Reynolds numbers.

The skin-friction drag $D(x)$ of a flat plate of lenght x on one side satisfies the follo- wing relation as seen from eqns. (10.1) and (10.2) in Chap. X:

$$D(x) = b \int_0^x \tau_0(x')\, \mathrm{d}x' = b\, \varrho \int_0^{\delta(x)} u(U_\infty - u)\, \mathrm{d}y\,. \tag{21.1}$$

Here $\tau_0(x)$ denotes the shearing stress at a distance x from the leading edge, and the second integral is evaluated at x over the boundary-layer thickness. Intro- ducing the momentum thickness δ_2, defined by $\delta_2\, U_\infty{}^2 = \int_0^\delta u(U_\infty - u)\, \mathrm{d}y$ in eqn. (8.31), we can rewrite eqn. (21.1) as follows:

$$D(x) = b\, \varrho\, U_\infty{}^2\, \delta_2(x)\,. \tag{21.2}$$

From eqns. (21.1) and (21.2) we obtain the local shearing stress:

$$\frac{1}{b}\frac{\mathrm{d}D}{\mathrm{d}x} = \tau_0(x) = \varrho\, U_\infty{}^2\, \frac{\mathrm{d}\delta_2}{\mathrm{d}x}\,. \tag{21.3}$$

Equation (21.3) is identical with the momentum-integral equation of boundary- layer theory, eqn. (8.32), in the case of uniform potential flow $U(x) = U_\infty = \mathrm{const.}$

We shall now perform the calculation of the drag on a flat plate on the assump- tion of a $\frac{1}{7}$-th-power law for the velocity profile which is true for moderate Reynolds numbers, and we shall then confine ourselves to quoting the results for the logarith- mic law which is valid for arbitrarily large Reynolds numbers, Fig. 20.4, because the complete calculation for this case is fairly tedious.

1. Resistance formula deduced from the $\frac{1}{7}$-th-power velocity distribution law. In accordance with the preceding argument and with eqn. (20.6) it is seen that the $\frac{1}{7}$-th-power law of velocity distribution in a pipe leads to the following velocity distribution in the boundary layer on a flat plate

$$\frac{u}{U_\infty} = \left(\frac{y}{\delta}\right)^{\frac{1}{7}}, \tag{21.4}$$

where $\delta = \delta(x)$ denotes the boundary-layer thickness which is a function of distance, x, and is to be determined in the course of the calculation. The assumption in eqn. (21.4) implies that the velocity profiles along a flat plate are *similar*, i. e. that all velocity profiles plot as one curve of u/U_∞ versus y/δ.

The equation for shearing stress at the wall is also taken over directly from the circular pipe, eqn. (20.12a):

$$\frac{\tau_0}{\varrho U_\infty{}^2} = 0\cdot 0225 \left(\frac{\nu}{U_\infty\, \delta}\right)^{\frac{1}{4}}\,. \tag{21.5}$$

From eqns. (8.30) and (8.31), together with eqn. (21.4) we can calculate the displa- cement thickness, δ_1, and the momentum thickness, δ_2:

$$\delta_1 = \frac{\delta}{8} ; \qquad \delta_2 = \frac{7}{72}\delta . \tag{21.6}†$$

From eqns. (21.3) and (21.6) we have

$$\frac{\tau_0}{\varrho U_\infty^2} = \frac{7}{72}\frac{d\delta}{dx} , \tag{21.7}$$

so that on comparing eqns. (21.5) and (21.7) we obtain

$$\frac{7}{72}\frac{d\delta}{dx} = 0.0225\left(\frac{\nu}{U_\infty\delta}\right)^{\frac{1}{4}}$$

which is the differential equation for $\delta(x)$. Integration from the initial value $\delta = 0$ at $x = 0$ gives

$$\delta(x) = 0.37\, x\left(\frac{U_\infty x}{\nu}\right)^{-\frac{1}{5}} \tag{21.8}$$

and hence

$$\delta_2(x) = 0.036\, x\left(\frac{U_\infty x}{\nu}\right)^{-\frac{1}{5}}. \tag{21.9}$$

The boundary-layer thickness is seen to increase with the power $x^{4/5}$ of the distance, whereas in laminar flow we had $\delta \sim x^{1/2}$. The total skin-friction drag on a flat plate of length l and width b wetted on one side is, by eqn. (21.2), given by

$$D = 0.036\,\varrho\, U_\infty^2\, b\, l\, (U_\infty l/\nu)^{-1/5} .$$

The drag on a plate in turbulent flow is seen to be proportional to $U_\infty^{9/5}$ and $l^{4/5}$ compared with $U_\infty^{3/2}$ and $l^{1/2}$, respectively, for laminar flow, eqn. (7.33). Introducing dimensionless coefficients for the local and the total skin friction by putting

$$c_f' = \frac{\tau_0}{\frac{1}{2}\varrho\, U_\infty^2} , \quad \text{or} \quad c_f = \frac{D}{\frac{1}{2}\varrho\, U_\infty^2\, bl} ,$$

we obtain from eqns. (21.3) and (21.2) that

$$c_f' = 2\,\frac{d\delta_2}{dx} , \qquad c_f = 2\,\frac{\delta_2(l)}{l} . \tag{21.10}$$

Hence, from eqn. (21.9), we can write $c_f' = 0.0576\,(U_\infty x/\nu)^{-1/5}$ and $c_f = 0.072\,(U_\infty l/\nu)^{-1/5}$. The last equation is in very good agreement with experimental results for plates whose boundary layers are turbulent from the leading edge onwards, if the numerical constant 0.072 is changed to 0.074. Thus

$$c_f = 0.074\,(R_l)^{-1/5} ; \qquad 5\times 10^5 < R_l < 10^7 . \tag{21.11}$$

The resistance formula (21.11) is seen plotted as curve (2) in Fig. 21.2. The range of validity of this formula is restricted to $U_\infty\,\delta/\nu < 10^5$ in accordance with the limitation on Blasius's pipe resistance formula. By eqn. (21.8) this corresponds to $U_\infty\,l/\nu < 6\times 10^6$. Since for $R_l < 5\times 10^5$ the boundary layer on a plate is fully

---

† In the general case of a power law $u/U = (y/\delta)^{1/n}$ we have:

$$\frac{\delta_1}{\delta} = \frac{1}{1+n} , \qquad \frac{\delta_2}{\delta} = \frac{n}{(1+n)(2+n)} .$$

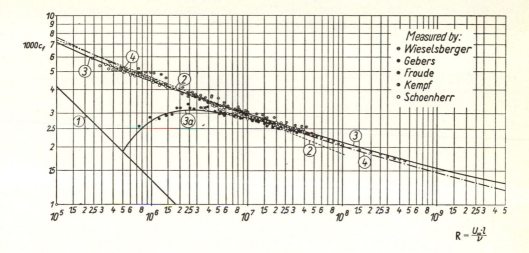

$$R = \frac{U_\infty l}{\nu}$$

Fig. 21.2. Resistance formula for smooth flat plate at zero incidence; comparison between theory and measurement

Theoretical curves: curve (1) from eqn. (7.34), laminar, Blasius; curve (2) from eqn. (21.11), turbulent, Prandtl; curve (3) from eqn. (21.16), turbulent, Prandtl-Schlichting; curve (3a) from eqn. (21.16a), laminar-to-turbulent transition; curve (4) from eqn. (21.19b), turbulent, Schultz-Grunow

laminar, it is possible to specify the following range of validity for eqn. (21.11): $5 \times 10^5 < R_l < 10^7$, using round numbers. Introducing the necessary corrections for the numerical coefficients we obtain the following expression for the local coefficient of skin friction

$$\frac{\tau_0}{\varrho\, U_\infty^2} = \frac{1}{2}\, c_f' = 0\cdot0296\, (R_x)^{-1/5} = 0\cdot0128 \left(\frac{U_\infty \delta_2}{\nu}\right)^{-\frac{1}{4}}. \qquad (21.12)$$

Equation (21.11), as already mentioned, is valid on the assumption that the boundary layer is turbulent from the leading edge onwards. In reality, the boundary layer will be laminar to begin with, and will change to a turbulent one further downstream. The position of the point of transition will depend on the intensity of turbulence in the external flow and will be defined by the value of the critical Reynolds number which ranges over $(U_\infty x/\nu)_{crit} = R_{crit} = 3 \times 10^5$ to 3×10^6 (see Sec. XVI a). The existence of the laminar section causes the drag to decrease and, following L. Prandtl, the decrease can be estimated if it is assumed that behind the point of transition the turbulent boundary layer behaves as if it were turbulent from the leading edge. Thus, from the drag of a wholly turbulent boundary layer it is necessary to subtract the turbulent drag of the length up to the point of transition at x_{crit} and to add the laminar drag for the same length. Thus, the decrease becomes $\Delta D = - (\varrho/2)\, U_\infty^2\, b\, x_{crit}\, (c_{ft} - c_{fl})$, where c_{ft} and c_{fl} denote the coefficient of turbulent and laminar skin friction, respectively, for the total drag at the section where transition occurs, i.e. at R_{crit}. Hence the correction for c_f is

$$\Delta c_f = - (x_{crit}/l)\, [c_{ft} - c_{fl}] = - (R_{x,crit}/R_l)[c_{ft} - c_{fl}] .$$

Putting $\Delta c_f = -A/R_l$, we find that the value of the constant A is determined by the position of the point of transition R_{crit}, namely

$$A = R_{x,crit}\,(c_{ft} - c_{fl}).$$

Consequently, the coefficient of total skin friction including the effect of the laminar initial length becomes

$$c_f = \frac{0 \cdot 074}{\sqrt[5]{R_l}} - \frac{A}{R_l}\,, \quad 5 \times 10^5 < R_l < 10^7\,. \tag{21.13}$$

Taking c_{ft} from eqn. (21.11) and $c_{fl} = 1 \cdot 328\ R_x^{-1/2}$ from the Blasius formula, eqn. (7.34), we obtain the following values for A:

$R_{x,crit}$	3×10^5	5×10^5	10^6	3×10^6
A	1050	1700	3300	8700

2. Resistance formula deduced from the logarithmic velocity-distribution law. The Reynolds numbers which occur in practical applications in connexion with flat plate problems considerably exceed the range of validity of eqn. (21.13)†, and it is, therefore, necessary to find a resistance formula which would be valid for much higher Reynolds numbers. In principle such a formula can be derived in the same way as before, except that the universal logarithmic velocity-profile equation should be used instead of the $\tfrac{1}{7}$-th-power formula, in analogy with eqns. (20.13) and (20.14) for pipe flow. Since the universal logarithmic formula, as shown earlier, may be extrapolated to arbitrarily large Reynolds numbers in the case of pipe flow, we may expect to obtain a resistance formula for the plate which would also lend itself to extrapolation to arbitrarily large Reynolds numbers. In any case, it is again implied that pipe flow and boundary-layer flow on a flat plate exhibit identical velocity profiles (see also p. 643).

The derivation is not so simple for the logarithmic law as it was for the $\tfrac{1}{7}$-th-power formula. This is mainly due to the fact that the application of the logarithmic law to the flat plate does not lead to similar profiles any longer. We shall, therefore, refrain from reproducing here the details of the calculation, referring the reader to L. Prandtl's original paper [40].

The logarithmic formula for pipe flow was derived in eqn. (20.14) in the form

$$\phi = A_1 \log \eta + D_1 \tag{21.14}$$

where

$$\phi = \frac{\bar{u}}{v_*} \text{ and } \eta = \frac{y\,v_*}{v}\,,$$

---

† In large and fast aeroplanes the Reynolds numbers of the wing are of the order of $R_l = 5 \times 10^7$; a large, modern fast steamer reaches about $R_l = 5 \times 10^9$; see also Table 21.3, p. 661.

with

$$v_* = \sqrt{\tau_0/\varrho}$$

denoting the characteristic velocity formed with the wall shearing stress τ_0. In the case of pipe flow considered in Chap. XX, the constants were indicated to have the numerical values $A_1 = 5{\cdot}75$ and $D_1 = 5{\cdot}5$. However, extensive experimental investigations (see Fig. 21.3) have demonstrated that the velocity profiles in the two cases under consideration, in a pipe and on a flat plate, are somewhat different and it becomes necessary to modify the numerical values to

$$A_1 = 5{\cdot}85 \; ; \quad D_1 = 5.56 \; . \tag{21.15}$$

The calculation leads to a fairly cumbersome set of equations for the local and total coefficients of skin friction in terms of the length Reynolds number $R_l = U_\infty l/\nu$. In the process, a formula for the dimensionless boundary-layer thickness $v_* \, \delta/\nu = \eta_\delta$ is also obtained. The numerical results are shown in Table 21.1 and the graph of c_f versus R_l has been plotted in Fig. 21.2 as curve (3).

Since the exact formulae from which the resistance law represented by Table 21.1 has been evaluated is exceedingly inconvenient, H. Schlichting fitted the relation between c_f and R_l from Table 21.1 into an empirical equation of the form

$$c_f = \frac{0{\cdot}455}{(\log R_l)^{2{\cdot}58}} \; . \tag{21.16} \dagger$$

In order to make an allowance for the laminar initial length, it is required to make the same deduction as before, eqn. (21.13). Thus

$$\boxed{c_f = \frac{0{\cdot}455}{(\log R_l)^{2{\cdot}58}} - \frac{A}{R_l} \; ,} \tag{21.16a}$$

where the value of the constant A depends on the position of the point of transition as specified in the Table on p. 601. This is the *Prandtl-Schlichting skin-friction formula for a smooth flat plate at zero incidence*. It is valid in the whole range of Reynolds numbers up to $R_l = 10^9$ and it agrees with eqn. (21.13) up to $R_l = 10^7$. It is seen plotted as curve (3a) in Fig. 21.2 where $A = 1700$ was chosen, corresponding to transition at $R_x = 5 \times 10^5$. Blasius's curve for laminar flow corresponding to $c_f = 1{\cdot}328 \, R_l^{-1/2}$ is also shown for comparison, curve (1).

A very similar theoretical calculation for the skin friction of a flat plate was devised by Th. von Kármán [29]. K. E. Schoenherr [50] made use of von Kármán's scheme and derived from it the expression

$$\frac{1}{\sqrt{c_f}} = 4{\cdot}13 \log (R_l \, c_f) \; . \tag{21.17}$$

Results of numerous experimental measurements are seen plotted together with these theoretical curves in Fig. 21.2. The measurements performed by C. Wiesels-

---

† The results for the coefficient of local skin friction, c_f', in Table 21.1 can also be fitted into an empirical equation as follows

$$c_f' = (2 \log R_x - 0{\cdot}65)^{-2{\cdot}3} \; .$$

Table 21.1.   Resistance formula for flat plate computed from the logarithmic velocity profile in eqns. (21.14) and (21.15); see curve (3) in Fig. 21.2

$\left(\dfrac{v_* \, \delta}{\nu}\right) \cdot 10^{-3} = \eta_\delta \cdot 10^{-3}$	$R_l \cdot 10^{-6}$	$c_f' \cdot 10^3$	$c_f \cdot 10^3$
0·200	0·107	5·51	7·03
0·353	0·225	4·54	6·04
0·500	0·355	4·38	5·48
0·707	0·548	4·03	5·05
1·00	0·864	3·74	4·59
1·30	1·20	3·53	4·33
2·00	2·07	3·22	3·92
3·00	3·43	2·97	3·57
5·00	6·43	2·69	3·23
7·07	9·70	2·53	3·02
12·0	18·7	2·30	2·71
20·0	34·3	2·11	2·48
28·3	51·8	2·00	2·34
50·0	102	1·83	2·12
100	229	1·65	1·90
170	425	1·53	1·75
283	768	1·42	1·63
500	1476	1·32	1·50

berger [67] on cloth-covered glazed plates lie somewhat above the turbulent curve (2), which would indicate that there was no substantial laminar length in his experiments and that the roughness was small. The measurements due to F. Gebers [19], which range from $R_l = 10^6$ to 3×10^7, fall on the transition curve (3a), eqn. (21.16a), at the lower end of the range. At the higher Reynolds numbers his results lie on curve (3) from eqn. (21.16). The measurements reported by K. E. Schoenherr [50] also show good agreement with theory. The highest Reynolds numbers have been achieved by G. Kempf [31]† who attained values of up to $R_l = 5 \times 10^8$. They show excellent agreement with the theoretical curve from eqn. (21.16a). Extensive measurements have been described by D. W. Smith and J. H. Walker [56]; they cover the range $10^6 < R_l < 4.5 \times 10^7$ and agree well with those due to G. Kempf [31] and F. Schultz-Grunow [53], but place themselves somewhat below the graph of eqn. (21.17). It is noteworthy that D. W. Smith and J. H. Walker measured skin-friction coefficients with the aid of a Pitot tube placed at the surface, and that this method has recently been used by many investigators with great success. In this connexion the work of J. H. Preston [43], R. A. Dutton [11], G. E. Gadd [18], P. Bradshaw and N. Gregory [5] as well as J. F. Naleid and M. J. Thompson [37] may be consulted. On summing up, it is possible to state that the preceding results have been confirmed by measurement over the whole range of Reynolds numbers.

---

† He measured only local frictional coefficients. L. Prandtl evaluated the corresponding total values by integration, see Reports AVA Goettingen, part IV.

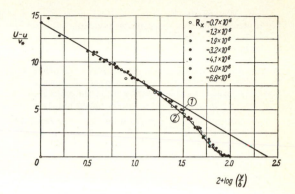

Fig. 21.3. Velocity distribution in
the boundary layer on a flat plate
at zero incidence, after Schultz-
Grunow [53]

Curve (1) logarithmic law of pipe flow. In
the outer portion the velocity distribution
on a plate is seen to deviate markedly
from that inside a circular tube. Curve (2)
was used by Schultz-Grunow as a basis
for the calculation of the boundary layer
of a plate and led to eqns. (21.19a) and
(21.19b)

3. Further refinements. As already stated, the preceding method of calculation is based
on the assumption that the velocity profiles in the boundary layer on a plate and inside a pipe
are identical, if the maximum velocity U and the radius R of the circular tube are replaced by
the free-stream velocity U_∞ and the boundary-layer thickness δ of the plate. This assumption
was checked by F. Schultz-Grunow [53] on the basis of very careful measurements on the
boundary layer on a plate. The investigation showed that the velocity profile in the outer portion
of the boundary layer of a plate deviates systematically upwards from the logarithmic velocity
distribution law of a circular pipe. The results of his measurements on a plate are given in Fig. 21.3.
They can be well represented by a velocity-defect law

$$\frac{U_\infty - u}{v_*} = f_1\left(\frac{y}{\delta}\right),\tag{21.18}$$

as already found in the case of a pipe, eqn. (20.23). It is seen that the loss of momentum on a
plate is somewhat smaller than that given by the logarithmic pipe formula and, consequently,
the drag must be smaller than that obtained by the direct application of pipe formulae. The
departure of the velocity distribution from the universal logarithmic velocity distribution (law
of the wall), eqn. (21.14), represented by eqn. (21.18) and Fig. 21.3, was determined empirically
by F. Schultz-Grunow†; it represents the so-called law of the wake introduced by D. Coles [8a]
According to F. Schultz-Grunow, the law of the wake is independent of the Reynolds number
and possesses, therefore, in some measure the properties of a universal law. For more details
concerning the law of the wake the reader should consult D. Coles's original paper [8a].

F. Schultz-Grunow repeated the derivation of the resistance formula from the preceding
system of equations and with the aid of the function $f_1(y/\delta)$ which he found by measurement.
The result can be represented by the following interpolation formulae

$$c_f' = 0.370\,(\log R_x)^{-2.584}\tag{21.19a}$$

$$c_f = 0.427\,(\log R_l - 0.407)^{-2.64}.\tag{21.19b}$$

The last equation has been plotted in Fig. 21.2 as curve (4), and it will be noticed that the deviation from the Prandtl-Schlichting curve (3) is small.

The different methods for the calculation of turbulent skin friction have been critically
examined by L. Landweber [33].

---

† On applying the pipe formula we have
$$f_1(y/\delta) = A\ln(\delta/y) = 2.5\ln(\delta/y),$$
which leads to a straight line in Fig. 21.3. The points near the wall are seen to fall on this
straight line; the portion of the curve which corresponds to the outer portion of the boundary
layer deviates strongly downwards from the straight line.

K. Wieghardt [65] advanced an explanation for the difference between the velocity profile in a pipe and that on a plate, pointing out that the influence of turbulence at the outer edge of the boundary layer differs in the two cases. In the case of a plate a low degree of turbulence in the external stream gives rise to velocity fluctuations which are practically zero at the outer edge of the boundary layer, whereas in the centre of the pipe they would have an appreciable magnitude because of the influence of the other side. To the smaller intensity of turbulence on a plate there corresponds a steeper increase in velocity and hence a thinner total boundary layer. He was also able to show that the velocity profile on a plate becomes very close to that in pipe flow if the external flow is made highly turbulent.

J. Nikuradse [38] also conducted a very comprehensive series of experiments on flat plates. He found that in the range of large Reynolds numbers of $R_x = 1 \cdot 7 \times 10^6$ to 18×10^6 the velocity profiles are similar, if u/U is plotted against y/δ_1, where δ_1 denotes the displacement thickness. The universal velocity-distribution law $u/U = f(y/\delta_1)$ turns out to be independent of the Reynolds number. The local and total coefficients of skin friction have been calculated from the measured velocity profiles with the aid of the momentum theorem.

The following interpolation formulae were obtained for the velocity distribution, displacement thickness, momentum thickness, and coefficients of skin friction, respectively:

$$\frac{u}{U_\infty} = 0 \cdot 737 \left(\frac{y}{\delta_1} \right)^{0 \cdot 1315} ,$$

$$\frac{U_\infty \delta_1}{\nu} = 0 \cdot 01738 \, R_x^{\,0 \cdot 861} ,$$

$$H_{12} = \frac{\delta_1}{\delta_2} = 1 \cdot 30 .$$

$$c_f{}' = 0 \cdot 02296 \, R_x^{-0 \cdot 139} ,$$

$$c_f = 0 \cdot 02666 \, R_l^{-0 \cdot 139} .$$

In connexion with the calculation of skin friction on a plate, the paper by V. M. Falkner [15] may also be consulted. In a paper by D. Coles [8a] the velocity profiles are represented by a linear combination of two universal functions, one of which is called the law of the wake, the other being the law of the wall as already mentioned.

Measurements performed by H. Motzfeld [36] concerned themselves with the turbulent boundary layer on a wavy wall. H. Schlichting [46] gave some estimates concerning turbulent boundary layers with suction and blowing. When homogeneous (that is, continuously and uniformly distributed) suction is applied, the asymptotic boundary-layer thickness remains constant in the same manner as for a laminar boundary layer. However, in the turbulent case the boundary layer is much more sensitive to changes in the suction flow-rate than in the laminar. Very extensive measurements performed in turbulent boundary layers on porous flat walls by A. Favre, R. Dumas and E. Verollet [16] show that the application of suction exerts a strong influence on the turbulent motion.

4. Effect of finite dimensions; boundary layers in corners. When a flat plate of finite span is placed in a stream which flows in the direction of its length, it is found that near the side-edge the boundary layer is no longer two-dimensional, as it is along the centre-line of the plate. Experiments performed by J. W. Elder [13] demonstrated that near the edges there arise secondary flows which are similar to those observed in pipes of non-circular cross-section (cf. Sec. XXe). This causes a large increase in the local skin-friction coefficient along the edges. According to Elder's measurements, and remarkably enough, this additional drag, always averaged over the span, turns out to be independent of the length Reynolds number, R_l, or the width of the plate. However, the region situated very close to the leading edge of the plate forms an exception, the local skin-friction coefficient varying irregularly in the flow direction and at right angles to it. Still according to Elder's measuremants, the increase in drag is given by

$$\Delta c_f = 3 \cdot 62 \times 10^{-4} - \frac{30}{R_l} . \tag{21.20}$$

The second term in this equation accounts for the rapidly decaying effect of the leading edge (on this detail the reader may also refer to A. A. Townsend [64]).

A similar effect arises when two plates aligned with the flow are made to form a concave corner. The interaction between the two boundary layers for the case of a rectangular corner was studied by K. Gersten [20] who indicates the existence of an additional drag of magnitude

$$\varDelta D = \tfrac{1}{2}\, \varrho\, U^2{}_\infty\, l^2\, \varDelta c_f\,, \tag{21.21}$$

where, according to K. Gersten, the interaction contribution is

and

$$\varDelta c_f = -\, \frac{5\cdot 76}{R_l} \qquad \text{in laminar flow}\,,$$

$$\varDelta c_f = -\, \frac{0\cdot 0052}{R_l{}^{2/5}} \qquad \text{in turbulent flow}\,. \tag{21.21a}$$

The supplementary drag has turned out to be negative, which means that the drag of two plates which are wetted only on the inner side of the corner and which are joined at right angles, is smaller than the drag of a flat plate of equal total area.

E. Eichelbrenner [12] examined the case of a corner of arbitrary angle.

5. Boundary layers with suction and blowing. *Measurement:* In this section we shall introduce brief remarks concerning turbulent boundary layers on a flat plate with suction and blowing which may serve as an extension of the considerations of Chap. XIV on laminar boundary layers with suction. The first theoretical study of this topic was made as early as 1942 by H. Schlichting [46, 47]. In modern times experimental as well as theoretical studies have been performed by J.C. Rotta [44].

Some of Rotta's experimental results are shown graphically in Fig. 21.4. This is a diagram showing the variation of the momentum thickness $\delta_2(x)$ along a porous flat plate with homogeneous suction and blowing at various values of the suction velocity, v_w, at the wall. The external velocity was $U_\infty = 20$ to 30 m/sec and the normal wall velocity ranged from $v_w = -0\cdot 10$ m/sec (suction) to $0\cdot 13$ m/sec (blowing). The volume coefficient varied from $c_Q = v_w/U_\infty = -0\cdot 005$ to $+0\cdot 005$ and was, thus, very small†. These measurements confirmed the well-known fact that the rate of boundary-layer thickness growth in the downstream direction increases as the blowing

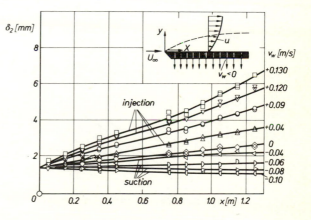

Fig. 21.4. Turbulent boundary layer on a flat plate with uniform suction or injection: momentum thickness δ_2, according to eqn. (7.38), along the plate; measurements by J. C. Rotta [44]

---

† Suction and blowing started at a short distance from the leading edge rather than at the leading edge itself.

rate increases. For $c_Q = -0\cdot005$ the boundary-layer thickness reaches a constant value downstream and constitutes an asymptotic boundary layer in the sense of Sec. XIVb.

The study of turbulent boundary layers with suction has many applications. Among them we may mention that the introduction of a foreign gas into the boundary layer through a porous wall or through slots constitutes a very effective means of film or transpiration cooling. This reduces the rate of heat transfer from the hot, streaming gas to the solid body, as is done for gas-turbine blades. Similarly, this is a means of reducing the rate of heat flow from the boundary layer rendered very hot by kinetic heating on a body flying at a hypersonic velocity to its wall. Blowing can also produce a considerable reduction in drag. A very good review of such applications was published by L. O. F. Jeromin [28].

Theory: In order to calculate the asymptotic turbulent boundary layer on a flat plate with homogeneous suction, we observe from eqn. (18.13) that the normal velocity $v = v_w$ is constant over the whole thickness of the layer. Hence, we can integrate the equation of motion in the x-direction with respect to the normal y-direction, and thus obtain

$$v_w u = v \frac{\partial u}{\partial y} - \frac{\tau_w}{\varrho} - \overline{u'\,v'} .$$ (21.22)

Introducing the friction velocity $v_* = (\tau_w/\varrho)^{1/2}$ and taking into account the fact that at large distances from the wall (i.e. outside the laminar sublayer) it is possible to neglect the viscous shear $v(\partial u/\partial y)$ with respect to the turbulent stress $-\overline{u'v'}$, we derive from eqn. (21.22) that

$$v_w u + v_*^2 = -\overline{u'\,v'}$$ (21.23)

With Prandtl's mixing-length assumption

$$-\overline{u'\,v'} = l^2 \left(\frac{du}{dy}\right)^2$$

from eqn. (19.6c), and putting $l = \varkappa\,y$, we deduce from eqn. (21.23) that

$$(\varkappa y)^2 \left(\frac{du}{dy}\right)^2 = v_w u + v_*^2 .$$ (21.24)

Here $\varkappa = 0\cdot4$ denotes von Kármán's constant. The preceding equation immediately proves that the velocity distribution can be given the following dimensionless form:

$$\frac{u}{v_*} = f\left(\eta, \frac{v_w}{v_*}\right) .$$ (21.25)

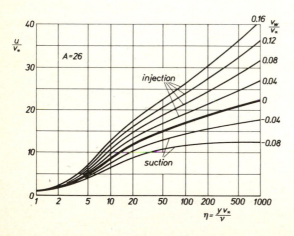

Fig. 21.5. Turbulent boundary layer on a flat plate with uniform suction or injection: theoretical velocity distribution according to eqn. (21.26) after J. C. Rotta [44]

$A = 26$, Van Driest's constant, eqn. (19.11)

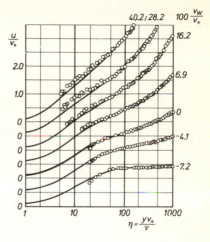

Fig. 21.6. Turbulent boundary layer on a flat plate with uniform suction or injection: velocity distribution in the boundary layer according to eqn. (21.26) for different values of the suction parameter v_w/v_* after J. C. Rotta [44]

○ experiment
— calculation

Here $\eta = y\,v_*/\nu$ is the dimensionless distance from the wall from eqn. (19.32). The integration of eqn. (21.23) gives

$$\frac{u}{v_*} = \frac{1}{\varkappa}\ln\eta + C + \frac{1}{4}\frac{v_w}{v_*}\left(\frac{1}{\varkappa}\ln\eta + C\right)^2. \tag{21.26}$$

Equation (21.26) can be regarded as a generalization of the universal velocity distribution law for impermeable turbulent boundary layers, eqn. (19.33), to the case of pervious walls with either suction or blowing. In order to include in our considerations the existence of a laminar sublayer, it is pertinent to introduce E. R. van Driest's [10] damping term, eqn. (19.11). The result of such a calculation is shown in Fig. 21.5. A comparison with the experiments of J. C. Rotta is given in Fig. 21.6. The agreement is satisfactory if a suitable value is chosen for the adjustable constant C.

Experimental investigations on turbulent boundary layers with injection of the same or another gas through porous walls into a *compressible* stream at Mach numbers up to M = 3·6 have been performed by L. C. Squire [59]. Calculations show that the assumption of Prandtl's mixing length here too leads to satisfactory results.

b. The rotating disk

1. The "free" disk. The flow in the neighbourhood of a rotating disk is of great practical importance, particularly in connexion with rotary machines. It becomes turbulent at larger Reynolds numbers, R = $U\,R/\nu > 3 \times 10^5$, in the same way as the flow about a plate. Here R denotes the radius and $U = \omega\,R$ is the tip velocity of the disk. The character of this kind of flow was described in Sec. V 11, which contained the complete solution for the laminar case when the disk rotates in an infinitely extended body of fluid ("free" disk). Owing to friction, the fluid in the immediate neighbourhood of the disk is carried by it and then forced outwards by the centrifugal acceleration. Thus the velocity in the boundary layer has a radial and a tangential component, and the mass of fluid which is driven outwards by centrifugal forces is replaced by an axial flow. Making a simple estimation of the balance of viscous and centrifugal forces in laminar flow it was possible to show that the boundary-layer thickness δ is proportional to $\sqrt{\nu/\omega}$, and hence, independent of the radius, and that the torque, M, which is proportional to $\mu R^3 U/\delta$, must be given by an expression of the form $M \sim \varrho U^2 R^3 (U R/\nu)^{-1/2}$. The exact solution for the laminar case showed, further, that the dimensionless torque coefficient, defined as

$$C_M = \frac{2 M}{\frac{1}{2} \varrho \omega^2 R^5} \tag{21.27}$$

for a disk wetted on both sides, is given by eqn. (5.56), and is equal to

$$C_M = 3 \cdot 87 \; \mathsf{R}^{-1/2} \quad \text{(laminar)} , \tag{21.28}$$

where $\mathsf{R} = R^2\omega/\nu$ is the Reynolds number, Fig. 5.14.

It is now proposed to make the same estimation for the turbulent case basing it on the same resistance formula for turbulent flow as was used in the case of the flat plate, i. e., in the simplest case, on the $\frac{1}{7}$-th-power law for the velocity distribution. A fluid particle which rotates in the boundary layer at a distance r from the axis is acted on by a centrifugal force per unit volume of magnitude $\varrho r \omega^2$. The centrifugal force on a volume of area $dr \times ds$ and height δ becomes $\varrho r \omega^2 dr \times ds$. The shearing stress τ_0 forms an angle θ with the tangential direction and its radial component must balance the centrifugal force. Hence we have $\tau_0 \sin \theta \, dr \times ds = \varrho r \omega^2 \delta dr \times ds$ or

$$\tau_0 \sin \theta = \varrho r \omega^2 \delta .$$

On the other hand, the tangential component of shearing stress can be expressed with the aid of eqn. (21.5) which was used in the case of a flat plate, replacing U_∞ by the tangential velocity $r \omega$. Thus

$$\tau_0 \cos \theta \sim \varrho (\omega r)^{7/4} (\nu/\delta)^{1/4} .$$

Equating τ_0 in these two expressions, we find that

$$\delta \sim r^{3/5} (\nu/\omega)^{1/5} .$$

It is seen that in the turbulent case the boundary-layer thickness increases outwards in proportion to $r^{3/5}$ and does not remain constant as in the laminar case. Further, the torque becomes $M \sim \tau_0 R^3 \sim \varrho R \omega^2 (\nu/\omega)^{1/5} R^{3/5} R^3$ so that

$$M \sim \varrho U^2 R^3 \left(\frac{\nu}{U R}\right)^{1/5} ,$$

Th. von Kármán [30] investigated the turbulent boundary layer on a rotating disk with the aid of an approximate method based on the momentum equation and similar to the one applied in the preceding section to the study of the flat plate. The variation of the tangential velocity component through the boundary layer was assumed to obey the $\frac{1}{7}$-th-power law. The viscous torque for a disk wetted on both sides was shown to be equal to

$$2 M = 0 \cdot 073 \, \varrho \, \omega^2 \, R^5 \, (\nu/\omega \, R^2)^{1/5} \tag{21.29}$$

and the torque coefficient defined in eqn. (21.27) becomes

$$C_M = 0\cdot146 \ \mathsf{R}^{-1/5} \quad \text{(turbulent)} . \tag{21.30}$$

This equation has been plotted in Fig. 5.14 as curve (2). It shows very good agreement with the experimental results due to W. Schmidt and G. Kempf† for $\mathsf{R} > 3 \times 10^5$. The numerical factor in the equation for the boundary-layer thickness which was left undetermined becomes

$$\delta = 0\cdot526 \ r \ (\nu/r^2 \ \omega)^{1/5}, \tag{21.31}$$

and the volume of flow in the axial direction is given by

$$Q = 0\cdot219 \ R^3 \ \omega \ \mathsf{R}^{-1/5} , \tag{21.32}$$

as compared with eqn. (5.57) for laminar flow.

An approximate calculation based on the logarithmic velocity-distribution law $u/v_* = A_1 \ln(y \, v_*/\nu) + D_1$ was performed by S. Goldstein [21], who found the following formula for the torque:

$$\frac{1}{\sqrt{C_M}} = 1\cdot97 \ \log (\mathsf{R} \ \sqrt{C_M}) + 0\cdot03 \quad \text{(turbulent)} . \tag{21.33}$$

It is noteworthy that this equation has the same form as the universal pipe-resistance formula, eqn. (20.30). The numerical factors have been adjusted to obtain the best possible agreement with experimental results. This equation is seen plotted as curve (3) in Fig. 5.14. On this topic see also P. S. Granville [22].

2. The disk in a housing. The disk in turbines or rotary compressors mostly revolve in very tight housings in which the width of the gap, s, is small compared with the radius, R, of the disk, Fig. 21.7. Consequently, it was found necessary to investigate the case of a disk rotating in a housing.

Laminar flow. The relations become particularly simple when the flow is laminar, $\mathsf{R} < 10^5$, and when the gap is very small. If the gap, s, is smaller than the boundary-layer thickness the variation of the tangential velocity across the gap becomes linear in the manner of Couette-flow. Hence, the shearing stress at a distance r from the axis is equal to $\tau = r\omega \mu/s$ and the torque of the viscous forces on one side of a disk is given by

$$M = 2\pi \int_0^R \tau \, r^2 \, \mathrm{d}r = \frac{\pi}{2} \ \frac{\omega \, \mu \, R^4}{s} .$$

Consequently for both sides we have

$$2 \, M = \pi \, \omega \, R^4 \, \mu/s ,$$

and the torque coefficient from eqn. (21.27) becomes

---

† See refs. [16] and [31] in Chap. V.

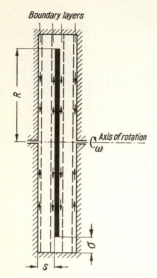

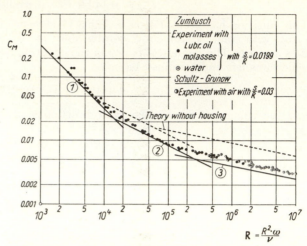

Fig. 21.7. Explanation of symbols for the problem of a disk rotating in a housing

Fig. 21.8.   Viscous drag of disk rotating in a housing

Curve (1), from eqn. (21.34), laminar; curve (2), from eqn. (21.35), laminar; curve (3), from eqn. (21.36), turbulent. Theory with no housing (free disk) see Fig. 5.14

$$C_M = 2\pi \frac{R}{s} \frac{1}{R} \quad \text{(laminar)} . \tag{21.34}$$

This equation is seen plotted as curve (1) in Fig. 21.8 for a value of $s/R = 0.02$. It shows very good agreement with the experimental values due to O. Zumbusch (see ref. [54]).

C. Schmieden [49] investigated the influence of the width σ of the lateral spacing of a disk in a cylindrical housing, Fig. 21.7, on the assumption of very small Reynolds numbers (creeping motion). The Navier-Stokes equations can be simplified because of the very low Reynolds numbers (see Sec. IVd) and the solution for the moment coefficient appears in the form $C_M = K/R$, in analogy with eqn. (21.34). The constant K depends on the two dimensionless ratios s/R and σ/R. In the case of very small values of σ/R (< 0.1) the values of C_M are markedly larger than those in eqn. (21.34), whereas for large values of σ/R eqn. (21.34) retains its validity ($K = 2\pi R/s$).

The flow pattern in the case of larger gaps differs considerably from the above simple scheme. This latter case was investigated theoretically and experimentally by F. Schultz-Grunow [54]. If the gap is a multiple of the boundary-layer thickness, then an additional boundary layer will be formed on the housing, Fig. 21.7. The fluid in the boundary layer on the rotating disk is centrifuged outwards, and this is compensated by a flow inwards in the boundary layer on the housing at rest. There is no appreciable radial component in the intermediate layer of fluid which rotates with about half the angular velocity of the disk. F. Schultz-Grunow investigated this flow both for the laminar and for the turbulent case. The expression for the torque is of the same form as for the free disk in eqn. (5.56), only the numerical factor has a different value. The frictional moment of a disk in laminar flow and wetted on both sides becomes $2\,M = 1\cdot334\,\mu\,R^4\,\omega\,\sqrt{\omega/\nu}$, and hence the coefficient

$$C_M = 2\cdot67\ R^{-1/2} \quad \text{(laminar)} .\tag{21.35}$$

This equation is seen plotted as curve (2) in Fig. 21.8. It agrees with measured values up to about $R = 2 \times 10^5$ and connects fairly well with eqn. (21.34).

Turbulent flow. For Reynolds numbers $R > 3 \times 10^5$ the flow around a disk rotating in a housing becomes turbulent as usual. This case was also solved by F. Schultz-Grunow who used an approximate method based on the scheme of Fig. 21.7. The tangential velocity was assumed to obey the $\frac{1}{7}$-th-power law and it was shown that the core revolves with about half the angular velocity in this case too. The moment coefficient was shown to be equal to

$$C_M = 0\cdot0622\ (R)^{-1/5} \quad \text{(turbulent)} .\tag{21.36}$$

This equation has been plotted in Fig. 21.8 as curve (3). Compared with measurement it leads to values which are too small by about 17 per cent., and this must be attributed to the crude assumptions made in the calculation.

It is particularly noteworthy that, apart from the case of very small gaps, eqn. (21.34), the moment of viscous forces is completely independent of the width of the gap, as seen from eqns. (21.35) and (21.36). Comparing the frictional moment on a "free" disk and on one rotating in a housing, eqns. (21.35) and (21.36) as against eqns. (21.28) and (21.30), it is seen that the moment on a free disk is greater than that on a disk in a housing, Fig. 21.8. This fact can be explained by the existence of the core which moves at half the angular velocity. This decreases the transverse gradient of the tangential velocity to approximately one half of what it would be on a free disk and, consequently, the drag is also smaller than on a "free" disk.

The flow process depicted in Fig. 21.7 in which the boundary layer on the rotating disk flows outwards and that on the casing flows inwards was later investigated experimentally by J. Dailey and R. Nece [8b]; their measurements covered the wide range of gap widths $s/R = 0\cdot01$ to $0\cdot20$, and a range of Reynolds numbers $R = R^2\omega/\nu = 10^3$ to 10^7 and included both laminar and turbulent flows. The results shown in Fig. 21.8 concerning the torque have been largely confirmed.

Heat transfer: The rate of heat transferred from a heated rotating disk to the cooler casing at rest is important in the design of gas turbines. The temperature field which develops in the gap between the disk and the casing is strongly influenced by the complex flow pattern which prevails in it; in turn, this has a large influence on the flux of heat from disk to housing. The simpler case of a rotating "free" disk was investigated some time ago by K. Millsaps and K. Pohlhausen [34a], see also Sec. XIId and Fig. 5.11. Theoretical and experimental information concerning the disk in a housing in laminar as well as in turbulent flow can be found in a thesis by R. Caly [6a] presented to Aachen University. Caly made measurements of both the velocity as well as the temperature boundary layer and included the case of a narrow gap with a *single* boundary layer and that of a wide gap with *two* separate boundary layers, one on the inner and one on the outer wall. In most cases good agreement between theory and measurement of heat flow was obtained.

c. The rough plate

1. The resistance formula for a uniformly rough plate. In most practical applications connected with the flat plate (e. g. ships, lifting surfaces of an aircraft, turbine blades) the wall cannot be considered hydraulically smooth. Consequently, the flow past a rough plate is of as much practical interest as that through a rough pipe.

The relative roughness k/R of the pipe is now replaced by the quantity k/δ, where δ denotes the boundary-layer thickness. The essential difference between the flow through a rough pipe and that over a rough plate consists in the fact that the relative roughness k/δ decreases along the plate when k remains constant because δ increases downstream, whereas in a pipe k/R remains constant. This circumstance causes the front of the plate to behave differently from its rearward portion as far as the influence of roughness on drag is concerned. Assuming, for the sake of simplicity, that the boundary layer is turbulent from the leading edge onwards, we find completely rough flow over the forward portion, followed by the transition regime and, eventually, the plate may become hydraulically smooth if it is sufficiently long. The limits between these three regions are determined by the dimensionless roughness parameter $v_* \, k_S/\nu$ as given in eqn. (20.37) for sand roughness.

The result of the calculation for pipes can be transposed to the case of rough plates in exactly the same way as for smooth plates in complete analogy with the detailed description given in Sec. XXI a. Such calculations were carried out by L. Prandtl and H. Schlichting [41] with the use of Nikuradse's results on pipes roughened with sand (Sec. XXf). The calculation was based on the logarithmic velocity-distribution law for rough pipes in the form of eqn. (20.32), whence $u/v_* = 2 \cdot 5 \ln (y/k_S) + B$. The dependence of the roughness function B on the roughness parameter $v_* \, k_S/\nu$ is given by the plot in Fig. 20.21. The calculation, which is essentially the same as in Sec. XXI a, must be carried out separately for the transition and completely rough regimes respectively. For the details of this method reference should be made to the original paper.

The result can be represented in two graphs, Figs. 21.9. and 21.10, in which the coefficient of total skin-friction drag, c_f, and the local coefficient, c_f', have been plotted against the Reynolds number $\mathsf{R} = U_\infty \, l/\nu$ with the relative roughness l/k_S as a parameter. In the case of the local coefficient, $U_\infty \, x/\nu$ and x/k_S are used. In addition the diagrams contain curves of $U_\infty \, k_S/\nu = \text{const}$, which can be computed at once from the previous ones. The two families of curves have the following significance: if the velocity on a given plate is changed, l/k_S remains constant, and the coefficient of skin friction varies along a curve $l/k_S = \text{const}$. If, on the other hand, the length of the plate is changed, $U_\infty \, k_S/\nu$ remains constant, and the coefficient of skin friction varies along a curve $U_\infty \, k_S/\nu = \text{const}$. Both graphs have been computed on the assumption that the turbulent boundary layer begins right at the leading edge. The broken curve shown in the diagrams corresponds to the limit of complete roughness and it may be noted that a given relative roughness

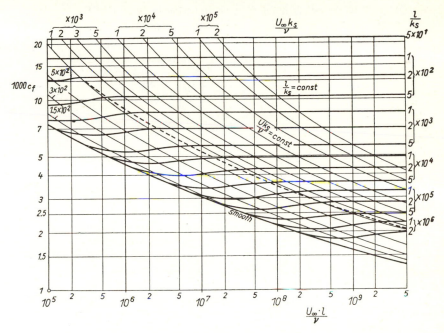

Fig. 21.9. Resistance formula of sand-roughened plate; coefficient of *total* skin friction

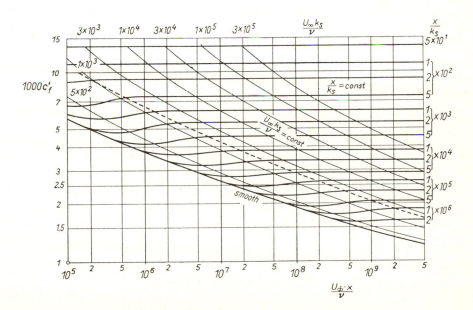

Fig. 21.10. Resistance formula of sand-roughened plate; coefficient of *local* skin friction

causes the coefficient of skin friction to increase only if the Reynolds number exceeds a certain value, in complete similarity with pipe flow (see Sec. XXI d).

In the completely rough regime it is possible to make use of the following interpolation formulae for the coefficients of skin friction in terms of relative roughness:

$$c_f' = \left(2\cdot87 + 1\cdot58 \log \frac{x}{k_S} \right)^{-2\cdot5}, \tag{21.37}$$

$$c_f = \left(1\cdot89 + 1\cdot62 \log \frac{l}{k_S} \right)^{-2\cdot5}, \tag{21.38}$$

which are valid for $10^2 < l/k_S < 10^6$.

In order to use these diagrams for roughness other than the sand roughness assumed here, it is necessary to determine the equivalent sand roughness as explained in Sec. XX g.

In the calculation of the drag on ships it is important to consider plates with very small roughness (painted metal plates) as well as smooth plates covered with single protuberances, such as rivet heads, welded seams, joints, etc. F. Schultz-Grunow [52] carried out a large number of measurements on such surfaces in the open channel of the Institute in Goettingen mentioned in Sec. XX g. Additional comprehensive data on roughnesses occuring in shipbuilding can also be found in several papers by G. Kempf [32]. According to these measurements it is possible to use an average value of equivalent sand roughness of $k_S = 0\cdot3$ mm ($= 0\cdot012$ in approx.) for newly launched ships. At the high Reynolds numbers which occur in ships this causes an increase in resistance of 34 to 45 per cent. due to roughness, as compared with hydraulically smooth walls. Roughness due to weeds adhering to ships' hulls has a particularly detrimental effect on resistance. Increases in resistance of 50 per cent., as compared with normal conditions, may well occur under such circumstances. The roughness of surfaces is also important in turbines, turbo-compressors and similar engines. The smoothness of normally manufactured blade surfaces is not sufficient to secure hydraulically smooth conditions [17a, 57]; see also p. 660.

Camouflage paints used on aeroplane surfaces can be well fitted into the scale of equivalent sand roughness as proved by the investigations carried out by A. D. Young [69], during which equivalent sand roughnesses of $k_S = 0\cdot003$ to $0\cdot2$ mm ($0\cdot001$ to $0\cdot01$ in approx.) have been measured. They are equal to about $1\cdot6$ times the size of the mean geometrical protrusions, i. e., $k_S = 1\cdot6 \, k$. In this connexion it is noteworthy that the increase in resistance due to roughness in the subsonic range of flow is independent of the Mach number.

W. Paeschke [39] demonstrated that the laws of friction, in flows along rough walls, which have emerged from these experimental investigations can be applied to the motion of natural winds over the surface of the earth. The effective roughness of surfaces covered with different kinds of vegetation could be determined by measurements of the velocity distribution of the wind in the layer just above the surface of the earth. Equation (20.32) $[u/v_* = 2\cdot5 \ln (y/k) + B]$, which has been deduced from pipe-flow results and which represents the velocity profile over a rough surface, was confirmed, and the value of $B = 5$ was found when the physical height of the vegetation is used as the roughness parameter k. In accordance with eqn. (20.38), this is the same as taking the equivalent sand roughness to be $k_S = 4\,k$.

2. Measurements on single roughness elements. K. Wieghardt [66] carried out a large number of measurements on roughness in the special tunnel in Goettingen. The tunnel, operated on air, had smooth walls and a rectangular cross-section measuring 140×40 cm (about $4{\cdot}5 \times 1{\cdot}3$ ft) and was 6 m (about 20 ft) long. The drag was measured with the aid of a balance which was attached to a rectangular test plate (50×30 cm, or $1{\cdot}65 \times 1{\cdot}00$ ft approximately). The test plate was accommodated in a recess in the lower wall ($1{\cdot}4 \times 6$ m or $4{\cdot}5 \times 20$ ft approximately) of the tunnel and it was free to move over a short distance. The difference between the drag on the test plate with and without the roughness elements gave the increase in drag, $\varDelta D$, due to roughness. Generally speaking this increase consists of two terms. The first term is the form drag due to roughness itself and the second is due to the fact that the presence of roughness elements changes the velocity profile in its neighbourhood and hence the shearing stress on the wall as, for example, in the region of back flow behind a fillet or ledge. The ratio of the height of the roughness element to the boundary-layer thickness, k/δ, is an important parameter for the application of such results to actual conditions on a ship's hull or an aeroplane. Its value was varied by setting up the same roughness elements at different places along the wall of the tunnel. From the point of view of practical applications it is also important to define a suitable dimensionless coefficient with the aid of the additional drag. K. Wieghardt used one defined by

$$C_D = \frac{\varDelta D}{\overline{q}\,A}. \tag{21.39}$$

where $\varDelta D$ denotes the measured additional drag, A the largest frontal area of the roughness element perpendicular to the direction of flow, and $\overline{q}$ is the stagnation pressure averaged over the height of the roughness element, or

$$\overline{q} = \frac{1}{k} \int_0^k \frac{1}{2}\,\varrho\,u^2\,(y)\,\mathrm{d}y = \frac{1}{k}\,\frac{1}{2}\,\varrho\,U^2 \int_0^k \left(\frac{y}{\delta}\right)^{\frac{2}{7}}\,\mathrm{d}y\,.$$

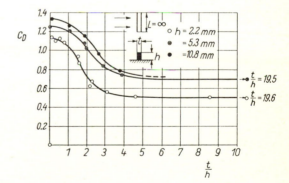

Fig. 21.11 Resistance coefficient for rectangular ribs, as measured by Wieghardt [66]

Here $u(y)$ denotes the velocity distribution on the smooth wall, that is, e. g., $u/U = (y/\delta)^{1/7}$. A large variety of roughness elements was subjected to test, including rectangular ribs arranged at right angles or at an acute angle to the stream, shaped fillets of triangular and circular cross-sections, sheet metal joints, single rivet heads and rows of rivets, cavities in the wall and others. Some of the results for rectangular ribs at right angles to the stream are seen plotted in Fig. 21.11. The value of the coefficient C_D decreases with increasing t/h (t — width, h — height). Holes and cavities in the surface lead to increased values of the resistance coefficient because the external flow causes the fluid in the cavity to take part in the motion.

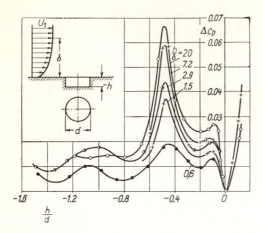

Fig. 21.12.   Resistance coefficient of circular cavities of varying depth in a flat wall, as measured by Wieghardt [66]

Figure 21.12 presents the increase in drag caused by circular cavities shown in the sketch (diameter d and depth h). Since the definition of q adopted previously loses its sense in this case, the drag was made dimensionless with reference to the stagnation pressure outside the boundary layer, $\Delta c_D = \Delta D / \frac{1}{4} q \pi d^2$. The increase in drag is smaller for smaller values of the ratio of the depth of the cavity, h, to the boundary-layer thickness, δ. It is noteworthy that all curves have a common maximum at $h/d \approx -0.5$. Further, small local maxima occur at $-h/d \approx 0.1$ and 1.0. The minima between them occur at $-h/d \approx 0.2$, 0.8, and 1.35. Depending on the depth of the cavity it may sometimes happen that regular vortex patterns are formed in it, leading to the different values of drag. As seen from the symmetry of the curves about $h/d = 0$, shallow cavities of up to $-d/h = 0.1$ give the same increase in drag as corresponding small protuberances.

Roughness in the form of rifling or ridges on a plate cut normal to the flow direction have been the subject of modern studies by A. E. Perry et al. [39a].

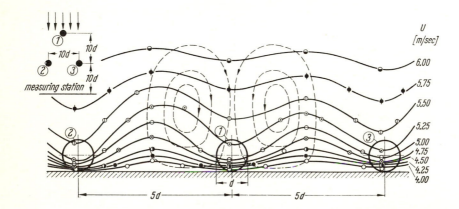

Fig. 21.13.   Curves of constant velocity in the flow field behind a row of spheres (full lines), as measured by H. Schlichting [45], and accompanying it the secondary flow (broken lines) in the boundary layer behind sphere (1), as calculated by F. Schultz-Grunow [55a]. In the neighbourhood of the wall, the velocity behind the spheres is larger than that in the gaps. The spheres produce a "negative wake effect" which is explained by the existence of secondary flow

Diameter of spheres $d = 4$ mm

The flow pattern which exists behind an obstacle placed in the boundary layer near a wall differs markedly from that behind an obstacle placed in the free stream. This circumstance emerges clearly from an experiment performed by H. Schlichting [45] and illustrated in Fig. 21.13. The experiment consisted in the measurement of the velocity field behind a row of spheres placed on a smooth flat surface. The pattern of curves of constant velocity clearly shows a kind of *negative wake effect*. The smallest velocities have been measured in the free gaps in which no spheres are present over the whole length of the plate; on the other hand, the largest velocities have been measured behind the rows of spheres where precisely the smaller velocities would be expected to exist. W. Jacobs [26] carried out a more detailed investigation of this peculiar effect. According to a remark made by F. Schultz-Grunow [55], the reason for such behaviour seems to be connected with the existence of secondary flow of a kind which is similar to that on a lift-generating body. The streamlines of this secondary flow have been sketched in Fig. 21.13. The existence of this effect was confirmed by D. H. Williams and A. F. Brown [68] who performed measurements on an aerofoil provided with rows of rivets.

There is in existence a very extensive literature concerning the roughness of aerofoils [9, 24, 25].

3. Transition from a smooth to a rough surface. W. Jacobs [27] investigated the flow pattern near a wall which consisted of a smooth section followed by a rough one, or vice versa. The problem is of some interest in meteorology and occurs when a wind passes from sea to land, or from land to sea, flowing past surfaces whose roughnesses differ considerably from each other. It is noticed that the velocity profile which corresponds to the downstream section of the wall forms only at a certain distance behind the boundary between the two sections. The variation of shearing stress calculated from the measured velocity profile with the aid of Prandtl's hypothesis, i. e., $\tau = \varrho\, l^2\, (du/dy)^2$, is seen plotted in Figs. 21.14 and 21.15. The diagrams show one remarkable feature, namely that the shearing stress at the wall assumes its new value which corresponds to fully developed flow immediately behind the boundary between the two sections. This result is important, e. g., when it is desired to calculate the drag on a plate which consists of a smooth and a rough section. In the zone of transition the variation of shearing stress at right angles to the wall, $\tau\,(y)$, has a form which is intermediate between the linear functions characteristic of fully developed flow over a rough and a smooth wall respectively. The shearing stress function $\tau\,(y)$ obtained from measurement can be interpolated with the aid of the empirical relation

$$\tau\,(x, y) = \left\{ \tau_s - (\tau_s - \tau_r)\, e^{-11.6\,\frac{y}{x}} \right\} \frac{h - y}{h} \quad \text{(smooth → rough)} , \qquad (21.40)$$

which is shown dotted in Figs. 21.14 and 21.15. Here τ_r and τ_s denote the shearing stresses on the rough and smooth wall, respectively, both for fully developed flow, x is the distance along the wall measured from the border line between the two portions of the plate, y is the distance from the wall, and h denotes the height of the channel. For the reverse order of transition (rough → smooth) the same formula may be used, except that τ_s and τ_r must be interchanged.

The influence of a pressure gradient on the transition from a smooth to a rough surface has been investigated by W. H. Schofield [51] and R. A. Antonia [1a]. Severe local pressure fluctuations have been observed by P. J. Mulhearn [36a] downstream from such an abrupt change.

d. Admissible roughness

The amount of roughness which is considered "admissible" in engineering applications is that maximum height of individual roughness elements which causes no increase in drag compared with a smooth wall. The practical importance of determining the amount of admissible roughness for a given set of circumstances is very great, because it determines the amount of labour which is worth spending in manufacturing a given surface. The answer to this question is essentially different depending on whether the flow under consideration is laminar or turbulent.

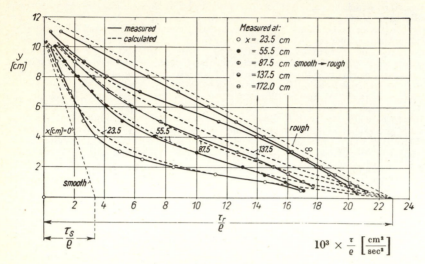

Fig. 21.14. Variation of shearing stress in the boundary layer on passing from a *smooth* to a *rough* portion of wall, as measured by W. Jacobs [27]

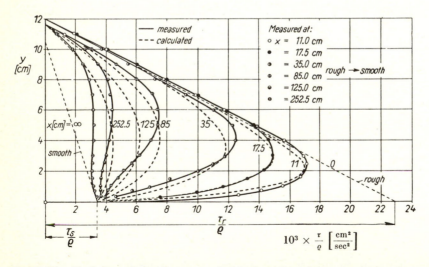

Fig. 21.15. Variation of shearing stress in the boundary layer on passing from a *rough* to a *smooth* portion of wall, measured by W. Jacobs [27]

In the case of *turbulent boundary layers* roughness has no effect, and the wall is hydraulically smooth if all protuberances are contained within the laminar sub-layer. As mentioned before, the thickness of the latter is only a small fraction of the boundary-layer thickness. In connexion with pipe flow it was found that the condition for a wall to be hydraulically smooth is given by eqn. (20.37) which stated that the dimensionless roughness Reynolds number†

$$\frac{v_* k}{v} < 5 \quad \text{(hydraulically smooth)}, \tag{21.41}$$

where $v_* = \sqrt{\tau_0/\varrho}$ denotes the friction velocity. This result can be considered valid also for the flat plate at zero incidence. However, from the practical point of view it seems more convenient to specify a value of relative roughness k/l. Referring to the diagram in Fig. 21.9, which represents the resistance formula for a plate, we can obtain the admissible value of k/l from the point at which a given curve $l/k = \text{const.}$ deviates from the curve for a smooth wall. It is seen that the admissible value of k/l decreases as the Reynolds number $U_\infty l/v$ is increased. Rounded-off values from Fig. 21.9 are listed in Table 21.2. They can be summarized by the following simple formula:

$$\frac{U_\infty k_{adm}}{v} = 100, \tag{21.42}$$

whose approximate validity can also be deduced directly from Fig. 21.9.

Table 21.2. Admissible height of protuberances in terms of the Reynolds number

$R_l = \dfrac{U_\infty l}{v}$	10^5	10^6	10^7	10^8	10^9
$\left(\dfrac{k}{l}\right)_{adm}$	10^{-3}	10^{-4}	10^{-5}	10^{-6}	10^{-7}

This formula gives only one value of k_{adm} for the whole length of the plate. Since, however, the boundary-layer thickness is smaller near the leading edge, the admissible value of k is smaller upstream than further downstream. A formula which takes this circumstance into account is obtained when $v_*^2/U_\infty^2 = \tau_0/\varrho\,U_\infty^2 = \frac{1}{2}\,c_f'$ is introduced, c_f' denoting the local coefficient of skin friction, as given in Table 21.1. Thus we obtain

$$\frac{U_\infty k_{adm}}{v} < \frac{7}{\sqrt{c_f'}}. \tag{21.43}$$

---

† The estimates performed in this section make no distinction between the equivalent sand height, k_S, and the actual height, k, of a protuberance.

For small Reynolds numbers $R_x < 10^6$ eqns. (21.42) and (21.43) give practically the same results, whereas at larger Reynolds numbers eqn. (21.43) gives somewhat greater values. We are, thus, justified in retaining the simpler equation (21.42) because there is no danger of finding values of k_{adm} which are too high. Equation (21.42) states that the *admissible height of roughness elements is independent of the length of the plate*; it is determined solely by the velocity and by the kinematic viscosity in accordance with the condition

$$k_{adm} \leq 100 \, \frac{\nu}{U_\infty} \, . \tag{21.44}$$

It follows that the absolute values of admissible roughness for a model and its original are equal if the velocity and kinematic viscosity are the same in both cases. For long bodies this may lead to extremely small admissible roughnesses as compared, with their linear dimensions, see Table 21.3.

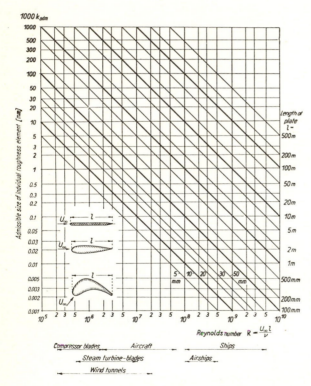

Fig. 21.16. Admissible roughness k_{adm} for rough plates at zero incidence and aircraft wings from eqn. (21.44)

For practical applications it is still more convenient to relate the admissible value of roughness directly to the length of the plate, l, or more generally, to the length, l, of the body under consideration, (e. g. length of ship's hull, wing chord, blade chord in turbines or rotary compressors), because this leads to a more graphic

Table 21.3. Examples on the calculation of admissible roughness from Fig. 21.16

Item	Description	Length l		Velocity w			Kinematic viscosity $10^6 \times \nu$		$R = \dfrac{wl}{\nu}$	Admissible roughness k_{adm}	
		m (ft)		km/h	m/sec	ft/sec	m²/sec	ft²/sec		mm	in
Ship's hull	large fast	250 (820)		56 30 knots	15	49	1·0	10	4×10^9	0·007	0·00028
	small slow	50 (165)		18 10 knots	5	165	1·0	10	3×10^8	0·02	0·0008
Airship	—	250 (820)		120	33	1100	15	150	5×10^8	0·05	0·002
Aeroplane (wing)	large fast	4† (13)		600	166	545	15	150	5×10^7	0·01	0·0004
	small slow	2 (6·5)		200	55	180	15	150	8×10^6	0·025	0·001
Compressor blades	slow	0·1 (0·33)		—	150	490	15	150	1×10^6	0·01	0·0004
Model wings	small	0·2 (0·65)		144	40	130	15	150	5×10^5	0·05	0·002
Steam turbine blades	high pressure $t = 300$ C (~ 550 F)	10 mm (0·4 in)		—	200	650	0·4	4	5×10^6	0·0002	0·000008
	high pressure $t = 500$ C (~ 950 F)	10 mm (0·4 in)		—	200	650	0·8	8	$2·5 \times 10^6$	0·0005	0·00002
	low pressure	100 mm		—	400	1300	8	80	5×10^6	0·002	0·00008

† Chord.

measure for the required surface smoothness. To achieve this, equation (21.44) may be rewritten as

$$k_{adm} \leq l \times \frac{100}{\mathsf{R}_l}, \tag{21.45}$$

where $\mathsf{R}_l = U_\infty \, l/\nu$. The diagram in Fig. 21.16 may be used to facilitate calculations with the aid of eqn. (21.45). The diagram contains a plot of admissible sizes of protuberances against Reynolds number, with the characteristic length as a parameter. The ranges of Reynolds numbers encountered in various engineering applications (ship, airship, aircraft, compressor blades, steam turbine blades) have been shown at the bottom of the diagram for convenience of reference. In addition, Table 21.3 gives a summary of several examples which have been computed with the aid of Fig. 21.16. In the case of *ships' hulls* admissible roughnesses are of the order of several hundredths of one millimetre (several tenths of one thousandth to several thousandths of an inch); such values cannot be attained in practice and it is always necessary to allow for a considerable increase in drag due to roughness. The same is true of *airships*. As far as *aircraft surfaces* are concerned, it is seen that admissible roughness dimensions lie between 0·01 and 0·1 mm (0·0004 and 0·004 in). With very careful preparation of the surface it is possible to meet these demands. In the case of *model aircraft* and *compressor blades* which require the same order of smoothness, i. e. 0·01 to 0·1 mm (0·0004 to 0·004 in), hydraulically smooth surfaces can be obtained without undue difficulty. The Reynolds numbers encountered in *steam turbines* are comparatively large because the pressures are comparatively high[†] in spite of the small linear dimensions, and admissible roughness values are, consequently, very small. The required values of between 0·0002 to 0·002 mm (10^{-5} to 10^{-4} in) can hardly be attained on newly manufactured blades. They are certainly exceeded after a period of operation due to corrosion and scaling. It may now be remarked that the preceding considerations apply to tightly spaced protuberances which correspond to sand roughness. In the case of widely spaced obstacles and in the case of wall waviness the admissible values are somewhat larger.

The influence of roughness on the losses in a steam turbine stage depends to a great extent on the pressure drop across it, i. e. on the degree of reaction of the stage. This point emerges clearly from Fig. 21.17 which represents the results of measurements performed by L. Speidel [58] on turbine cascades with varying sand roughness. The diagram contains a plot of the loss coefficient $\zeta_t = \Delta g / \frac{1}{2} \, \varrho \, w_2^2$, where Δg denotes the mean value of the loss in total head averaged over one pitch; the loss has been made dimensionless with reference to the total head at exit (w_2 denotes the leaving velocity). The increase in the value of the loss coefficient ζ_t as β_1 is increased is caused by an increase in w_1 with β_1, as may be verified with reference to the velocity triangle. The broken straight lines in Fig. 21.17 represent the rate at which ζ_t increases on the assumption that the boundary layer is turbulent all along the blade. For lower

---

[†] For values of the kinematic viscosity of superheated steam consult Escher Wyss Reports, vol. X, No. 1, p. 3 (1937), or NBS-NACA Tables of Thermodynamic Properties of Gases, Washington, 1954. See also J. Kestin and J. Whitelaw, Trans. ASME (A), J. Basic Engineering, *88* (1966) 82—104. Most recent values can be taken from Mechanical Engineering, *88*, 79 (1976).

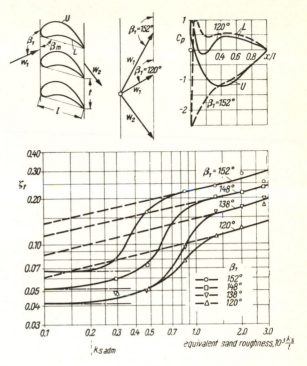

Fig. 21.17. Loss coefficients for turbine blades provided with sand roughness, as measured by L. Speidel [58]

$\zeta_t = \Delta g / \frac{1}{2} \varrho\, w_2^2$;
Δg — loss in total head. Solidity $t/l = 0.67$;
mean blade angle $\beta_m = 72°$;
Reynolds number $R = w_2\, l/\nu = 5 \times 10^5$;
$C_p = (p - p_2)/\frac{1}{2}\varrho\, w_2^2$

roughness values the measured points fall considerably below these straight lines. It has been ascertained during the investigation that this behaviour results from the existence of long stretches of laminar boundary layers; as the roughness is increased the length of the laminar portion of the boundary layer decreases. The estimate of the amount of permissible roughness from eqn. (21.45) leads here to a value of $k_s/l = 0.2 \times 10^{-3}$ at $R = w_2\, l/\nu = 5 \times 10^5$. This limit has been marked in Fig. 21.17 and it is seen that it agrees well with the experimental results. Reference is also made to papers by K. Bammert and K. Fiedler [2, 3].

The height of a protuberance which causes transition in a laminar boundary layer will be called *critical height* or *critical roughness* (*cf.* Sec. XVII g). Roughness affects the resistance offered by the wall by moving the point of transition in an upstream direction, and, depending on the shape of the body, the drag may be either increased or decreased. The drag is *increased* by such a shift in the point of transition when the drag of the body is predominantly due to skin friction (for example an aerofoil). It may be *decreased* under certain circumstances if the drag of the body is mainly due to form drag (e. g. circular cylinder). In accordance with some Japanese measurements [62], this critical value of the roughness is given by

$$\frac{v_* \, k_{crit}}{\nu} = 15 \, . \tag{21.46}$$

We shall now calculate the value of k_{crit} for a wing of length $l = 2$ m (about 6·5 ft) in air ($\nu = 14 \times 10^{-6}$ m²/sec) at a velocity $U_\infty = 83$ m/sec $= 300$ km/hr (about 185 mph). We have $R_l = U_\infty l/\nu \approx 10^7$. Consider a point on the wing at $x = 0\cdot 1 \; l$, i. e. at $R_x = U_\infty x/\nu \approx 10^6$. The boundary layer can remain laminar as far as this point owing to the existence of a negative pressure gradient. The shearing stress at the wall for a laminar boundary layer is given by eqn. (7.32) and is $\tau_0/\varrho = 0\cdot332 \; U_\infty^2 \sqrt{\nu/U_\infty x} =$ $= 0\cdot332 \times 6900 \times 10^{-3}$ m²/sec² $= 2\cdot29$ m²/sec². Hence $v_* = \sqrt{\tau_0/\varrho} = 1\cdot52$ m/sec. Inserting into eqn. (21.46) we have

$$k_{crit} = 15 \, \frac{\nu}{v_*} = \frac{15}{1\cdot52} \times 0\cdot14 \times 10^{-4} \, \text{m} = 0\cdot14 \, \text{mm (about 0·0056 in) .}$$

This shows that the critical size of a protuberance which causes transition is about ten times larger than the value of about 0·02 mm (0·0008 in) in the turbulent boundary layer, as calculated in Table 21.3, for the case in hand (small aeroplane). The laminar boundary layer "can stand" much larger roughness than the turbulent boundary layer. K. Scherbarth [48] carried out experiments on the behaviour of laminar boundary layers on walls provided with single obstacles (rivet heads). It was ascertained that behind the obstacle there forms a wedge-like turbulent disturbed region whose angle of spread is about 14° to 18°.

The very extensive measurements carried out by E. G. Feindt [17] have led to a refinement of the criterion for the critical height given in eqn. (21.46) as mentioned in Sec. XVIIg.

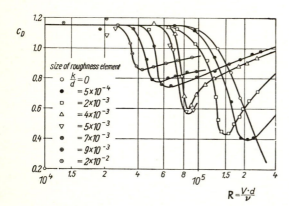

Fig. 21.18. Drag on circular cylinders at varying roughness, after Fage and Warsap [14]

The influence of roughness on form drag can be summarized as follows: bodies with sharp edges, such as e. g. a flat plate at right angles to the stream, are quite insensitive to surface roughness, because the point of transition is determined by the edges. On the other hand, the drag of bluff bodies, such as circular cylinders, is very sensitive to roughness. The value of the critical Reynolds number for which the drag shows a sudden drop (Fig. 1.4) depends to a marked degree on the roughness of the surface. According to measurements, [1, 14] as shown in Fig. 21.18, the critical Reynolds number decreases with increasing relative roughness k/R ($d = 2 \; R = $ diameter of cylinder). The boundary layer appears to be disturbed by roughness to such a degree that transition occurs at considerably lower Reynolds numbers than is the case with smooth cylinders. Roughness has, therefore, the same effect as Prandtl's tripping wire (Fig. 2.25), namely, it does reduce drag in a certain range of Reynolds numbers. In any case the drag in the supercritical range of Reynolds numbers is always larger for the rough than for the smooth cylinders; see here [60].

References

[1] Ackeret, J.: Schweiz. Bauzeitung *108*, 25 (1936).
[1a] Antonia, R.A., and Wood, D.H.: Calculation of a turbulent boundary layer downstream of a small step change in surface roughness. Aero. Quart. *26*, 202—210 (1975).
[2] Bammert, K., and Fiedler, K.: Der Reibungsverlust von rauhen Turbinenschaufeln. Brennstoff-Wärme-Kraft *18*, 430—436 (1966).
[2a] Banner, M.L., and Melville, W.K.: On the separation of air flow over water waves. JFM *77*, 825—842 (1976).
[3] Bammert, K., and Fiedler, K.: Hinterkanten- und Reibungsverlust in Turbinenschaufelgittern. Forschg. Ing.-Wes. *32*, 133—141 (1966).
[4] Blenk, H., and Trienes, H.: Strömungstechnische Beiträge zum Windschutz. Grundlagen der Landtechnik. VDI-Verlag, No. 8, 1956.
[5] Bradshaw, P., and Gregory, N.: The determination of local turbulent skin friction from observations in the viscous sub-layer. ARC RM 3202 (1961).
[6] Burgers, J.M.: The motion of a fluid in the boundary layer along a plane smooth surface. Proc. First Intern. Congress Appl. Mech. 121, Delft (1924).
[6a] Caly, R.: Der Wärmeübergang an einer im geschlossenen Gehäuse rotierenden Scheibe. Thesis Aachen 1966.
[7] Chapmann, D.R., and Kester, R.H.: Measurements of turbulent skin friction in cylinders in axial flow at subsonic and supersonic velocities. JAS *20*, 441—448 (1953).
[8] Coles, D.: The problem of the turbulent boundary layer. ZAMP *5*, 181—202 (1954).
[8a] Coles, D.: The law of the wake in the turbulent boundary layer. JFM *1*, 191—226 (1956).
[8b] Daily, J., and Nece, R.: Chamber dimension effects on induced flow and friction resistance of enclosed rotating disks. J. Basic Eng., Trans. ASME. Series D, *82*, 217—232 (1960).
[9] Doetsch, H.: Einige Versuche über den Einfluss von Oberflächenstörungen auf die Profileigenschaften, insbesondere auf den Profilwiderstand im Schnellflug. Jb. dt. Luftfahrtforschung *1*, 88—97 (1939).
[10] Van Driest, E.R.: On turbulent flow near a wall. JAS *23*, 1007—1011 (1956).
[11] Dutton, R.A.: The accuracy of measurement of turbulent skin friction by means of surface Pitot tubes and the distribution of skin friction on a flat plate. ARC RM 3058 (1957).
[12] Eichelbrenner, E.: La couche-limite turbulente à l'intérieur d'un dièdre. Rech. Aéro. Paris No. 83, 3—8 (1961).
[13] Elder, J.W.: The flow past a flat plate of finite width. JFM *9*, 133—153 (1960).
[14] Fage, A., and Warsap, J.H.: The effects of turbulence and surface roughness on the drag of circular cylinders. ARC RM 1283 (1930).
[15] Falkner, V.M.: The resistance of a smooth flat plate with turbulent boundary layer. Aircraft Engineering *15* (1943).
[16] Favre, A., Dumas, R., and Verollet, E.: Couche limite sur paroi plane poreuse avec aspiration. Publications Scientifiques et Techniques du Ministère de l'Air, No. 377 (1961).
[17] Feindt, E.G.: Untersuchungen über die Abhängigkeit des Umschlages laminar-turbulent von der Oberflächenrauhigkeit und der Druckverteilung. Diss. Braunschweig 1956. Jb. Schiffbautechn. Ges. *50*, 180—203 (1957).
[17a] Forster, V.T.: Performance loss of modern stream-turbine plant due to surface roughness. The Inst. of Mech. Eng., Preprint, London, 1967.
[18] Gadd, G.E.: A note on the theory of the Stanton tube. ARC RM 3147 (1960).
[19] Gebers, F.: Ein Beitrag zur experimentellen Ermittlung des Wasserwiderstandes gegen bewegte Körper. Schiffbau *9*, 435—452 and 475—485 (1908); also: Das Ähnlichkeitsgesetz für den Flächenwiderstand in Wasser geradlinig fortbewegter polierter Platten. Schiffbau *22*, 687—930 (1920/21), continuations.

[20] Gersten, K.: Die Grenzschichtströmung in einer rechtwinkeligen Ecke. ZAMM *39*, 428—429 (1959); see also: Corner interference effects. AGARD Rep. 299 (1959).

[21] Goldstein, S.: On the resistance to the rotation of a disk immersed in a fluid. Proc. Cambr. Phil. Soc. *31*, Part 2, 232 (1935).

[22] Granville, P.S.: The torque and turbulent boundary layer of rotating disks with smooth and rough surfaces, and in drag-reduction polymer solutions. J. Ship Research *17*, 181—195 (1973).

[23] Hama, F.R.: Boundary layer characteristics for smooth and rough surfaces. Transaction of the Society of Naval Architects and Marine Engineers *62*, 333—358 (1954).

[23a] Hansen, M.: Die Geschwindigkeitsverteilung in der Grenzschicht an einer eingetauchten Platte. ZAMM *8*, 185—199 (1928); NACA TM 585 (1930).

[24] Hood, M.J.: The effects of some common surface irregularities on wing drag. NACA TN 695 (1939).

[25] Jacobs, E.N.: Airfoil section characteristics as affected by protuberances. NACA Rep. 446 (1932).

[26] Jacobs, W.: Strömung hinter einem einzelnen Rauhigkeitselement. Diss. Göttingen 1938, Part I, Ing.-Arch. *9*, 343—355 (1938).

[27] Jacobs, W.: Umformung eines turbulenten Geschwindigkeitsprofils. Diss. Göttingen 1938, Part II, ZAMM *19*, 87—100 (1939).

[28] Jeromin, L.O.F.: The status of research in turbulent boundary layers with fluid injection. Progress in Aero. Sciences *10*, 65—190 (1970).

[29] von Kármán, Th.: Mechanische Ähnlichkeit und Turbulenz. Proc. IIIrd Intern. Congr. of Appl. Mech. *85*, Stockholm 1931, and Hydromechanische Probleme des Schiffsantriebes, Hamburg, 1932; see also: JAS *1*, 1 (1932); NACA TM 611 (1931); see also Coll. Works *II*, 322—346.

[30] von Kármán, Th.: Über laminare und turbulente Reibung. ZAMM *1*, 233—252 (1921); NACA TM 1092 (1946); see also Coll. Works *II*, 70—97.

[31] Kempf, G.: Neue Ergebnisse der Widerstandsforschung. Werft, Reederei, Hafen *10*, 234 und 247 (1929).

[32] Kempf, H.: Über den Einfluss der Rauhigkeit auf den Widerstand von Schiffen. Jb. Schiffbautechn. Ges. *38*, 159 and 233 (1937); and: The effect of roughness on the resistance of ships. Engineering, London *143*, 417 (1937); see also: Trans. Inst. Nav. Architects *79*, 109 and 137 (1937).

[33] Landweber, L.: Der Reibungswiderstand der längsangeströmten ebenen Platte. Jb. Schiffbautechn. Ges. *46*, 137—150 (1952).

[34] Liepmann, H.W., and Fila, G.H.: Investigations of effects of surface temperature and single roughness elements on boundary layer transition. NACA Rep. 890 (1947).

[34a] Millsaps, K., and Pohlhausen, K.: Heat transfer by laminar flow from a rotating plate. JAS *19*, 120—126 (1952).

[35] Mottard, E.J., and Loposer, J.D.: Average skin friction drag coefficient from tank tests of a parabolic body of revolution (NACA RM-10). NACA Rep. 1161 (1954).

[36] Motzfeld, H.: Die turbulente Strömung an welligen Wänden. Diss. Göttingen 1935, ZAMM *17*, 193—212 (1937).

[36a] Mulhearn, P.J.: Turbulent boundary layer wall pressure fluctuations downstream from an abrupt change in surface roughness. Physics of Fluids *19*, 796—801 (1976).

[37] Naleid, J.F., and Thompson, M.J.: Pressure-gradient effects on the Preston tube in supersonic flow. JASS *28*, 940—944 (1961).

[38] Nikuradse, J.: Turbulente Reibungsschichten an der Platte. Published by ZWB, R. Oldenbourg, München und Berlin, 1942.

[39] Paeschke, W.: Experimentelle Untersuchungen zum Rauhigkeits- und Stabilitätsproblem in der bodennahen Luftschicht. Diss. Göttingen 1937. Summary in: Beiträge zur Physik der freien Atmosphäre *24*, 163 (1937); see also: Z. Geophysik *13*, 14 (1937).

[39a] Perry, A.E., and Schofield, W.H.: Rough wall turbulent boundary layers. JFM 37, 383—413 (1969).

[40] Prandtl, L.: Über den Reibungswiderstand strömender Luft. Ergebnisse AVA Göttingen, IIIrd Series (1927) and: Zur turbulenten Strömung in Rohren und längs Platten. Ergebnisse AVA Göttingen, IVth Series (1932); First mention in 1st Series 136 (1921); see also Coll. Works *II*, 620—626 and 632—648.

[41] Prandtl, L., and Schlichting, H.: Das Widerstandsgesetz rauher Platten. Werft, Reederei, Hafen 1—4 (1934); see also Coll. Works *II*, 648—662.

[42] Prandtl, L.: The mechanics of viscous fluids, in: W. F. Durand: Aerodynamic theory, *III* 34—208 (1935).

[43] Preston, J. H.: The determination of turbulent skin friction by means of Pitot tubes. J. Roy. Aero. Soc. *58*, 109—121 (1954).

[44] Rotta, J. C.: Control of turbulent boundary layers by uniform injection or suction of fluid. Jb. 1970 Dt. Gesellschaft für Luft- und Raumfahrt, ed. by H. Blenk and W. Schulz, Braunschweig, 1971, pp. 91—104; see also ZAMM *46*, T 213—T 215 (1966).

[45] Schlichting, H.: Experimentelle Untersuchungen zum Rauhigkeitsproblem. Ing.-Arch. *7*, 1—34 (1936); NACA TM 823 (1937).

[46] Schlichting, H.: Die Grenzschicht an der ebenen Platte mit Absaugung und Ausblasen. Luftfahrtforschung *19*, 293—301 (1942).

[47] Schlichting, H.: Die Grenzschicht mit Absaugen und Ausblasen. Luftfahrtforschung *19*, 179—181 (1942).

[48] Scherbarth, K.: Grenzschichtmessungen hinter einer punktförmigen Störung in laminarer Strömung. Jb. dt. Luftfahrtforschung *I*, 51—53 (1942).

[49] Schmieden, C.: Über den Widerstand einer in einer Flüssigkeit rotierenden Scheibe. ZAMM *8*, 460—479 (1928).

[50] Schoenherr, K. E.: Resistance of flat surfaces moving through a fluid. Trans. Soc. Nav. Arch. and Mar. Eng. *40*, 279 (1932).

[51] Schofield, W. H.: Measurements in adverse pressure gradient turbulent boundary layers with a step change in surface roughness. JFM *10*, 573—593 (1975).

[52] Schultz-Grunow, F.: Der hydraulische Reibungswiderstand von Platten mit mässig rauher Oberfläche, insbesondere von Schiffsoberflächen. Jb. Schiffbautechn. Ges. *39*, 176—198 (1938).

[53] Schultz-Grunow, F.: Neues Widerstandsgesetz für glatte Platten. Luftfahrtforschung *17*, 239 (1940); also NACA TM 986 (1941).

[54] Schultz-Grunow, F.: Der Reibungswiderstand rotierender Scheiben in Gehäusen. ZAMM *15*, 191—204 (1935); see also: H. Föttinger: ZAMM *17*, 356—358 (1937) and K. Pantell: Forschg. Ing.-Wes. *16*, 97—108 (1950).

[55] Schultz-Grunow, F.: Der Mechanismus des Widerstandes von Einzelrauhigkeiten. ZAMM *36*, 309 (1956).

[55a] Schultz-Grunow, F.: Die Entstehung von Längswirbeln in Grenzschichten. ZAMM *38*, 85—95 (1958).

[56] Smith, D. W., and Walker, J. H.: Skin friction measurements in incompressible flow. NASA TR R-26 (1959).

[57] Sörensen, E.: Wandrauhigkeitseinfluss bei Strömungsmaschinen. Forschg. Ing.-Wes. *8*, 25 (1937).

[58] Speidel, L.: Einfluss der Oberflächenrauhigkeit auf die Strömungsverluste in ebenen Schaufelgittern. Forschg. Ing.-Wes. *20*, 129—140 (1954).

[59] Squire, L. C.: Eddy viscosity distributions in compressible turbulent boundary layers with injection. Aero. Quart. *22*, 169—182 (1971).

[60] Szeckenyi, E.: Supercritical Reynolds number simulation for two-dimensional flow over circular cylinders. JFM *70*, 529—542 (1975).

[61] Szablewski, W.: Berechnung der turbulenten Strömung längs der ebenen Platte. ZAMM *31*, 309—324 (1951).

[62] Tani, I., Hama, J., and Mituisi, S.: On the permissible roughness in the laminar boundary layer. Aero. Res. Inst. Tokyo, Rep. 199 (1940).

[63] Tillmann, W.: Neue Widerstandsmessungen an Oberflächenstörungen in der turbulenten Grenzschicht. Forschungshefte für Schiffstechnik No. 2 (1953).

[64] Townsend, A. A.: The turbulent boundary layer. Boundary-layer Research, IUTAM Symposium Freiburg/Br. 1957 (H. Görtler, ed.), 1—15 (1958).

[65] Wieghardt, K.: Über die turbulente Strömung im Rohr und längs der Platte. ZAMM *24* 294—296 (1944).

[66] Wieghardt, K.: Erhöhung des turbulenten Reibungswiderstandes durch Oberflächenstörungen. Techn. Berichte *10*, Heft 9 (1943); see also: Forschungshefte für Schiffstechnik *1*, 65—81 (1953).

[67] Wieselsberger, C.: Untersuchungen über den Reibungswiderstand von stoffbespannten Flächen. Ergebnisse AVA Göttingen, 1st Series, 120—126 (1921).

[68] Williams, D. H., and Brown, A. F.: Experiments on a riveted wing in the compressed air tunnel. ARC RM 1855 (1938).

[69] Young, A. D.: The drag effects of roughness at high subcritical speeds. J. Roy. Aero. Soc. *18*, 534 (1950).

CHAPTER XXII

The incompressible turbulent boundary layer with pressure gradient †

In the present chapter we shall discuss the behaviour of a turbulent boundary layer in the presence of a positive or negative pressure gradient along the wall, thus providing an extension of the subject matter of the preceding chapter in which the boundary layer on a flat plate with no pressure gradient was considered. The present case is particularly important for the calculation of the drag of an aeroplane wing or a turbine blade as well as for the understanding of the processes which take place in a diffuser. Apart from skin friction we are interested in knowing whether the boundary layer will separate under given circumstances and if so, we shall wish to determine the point of separation. The existence of a negative and, in particular, of a positive pressure gradient exerts a strong influence on the formation of the layer just as was the case with laminar layers. At the present time these very complicated phenomena are far from being understood completely but there are in existence several semi-empirical methods of calculation which lead to comparatively satisfactory results.

In the year 1962, J.C. Rotta [85] prepared a comprehensive and careful review of this vast field of knowledge. In order to develop methods of calculating incompressible, turbulent boundary layers with pressure gradients it is necessary to derive from experiment relations which go beyond those employed for pipes and flat plates at zero incidence. For this reason we shall begin by giving a short account of some experimental results.

a. Some experimental results

Early systematic experiments on two-dimensional flows with pressure drop and pressure rise in convergent and divergent channels with flat walls have been carried out by F. Doench [28], J. Nikuradse [71], H. Hochschild [45], R. Kroener [57] and J. Polzin [76]. Measurements on circular diffusers, and particularly on the efficiency of the process of energy transformation, are described in papers by F. A. L. Winternitz and W. J. Ramsay [123]. These experiments demonstrate that the shape of the velocity profile depends very strongly on the pressure gradient. Figure 22.1 shows the velocity profiles which were measured by J. Nikuradse during his experiments with

† The new version of this chapter was prepared by Professor E. Truckenbrodt whose assistance I hereby gratefully acknowledge.

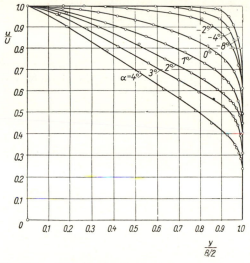

Fig. 22.1. Velocity distribution in *convergent and divergent* channels with flat walls, as measured by J. Nikuradse [71]

α — half included angle; B — width of channel

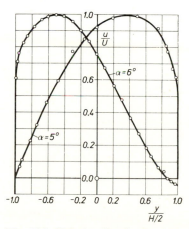

Fig. 22.2. Velocity distribution in a *divergent* channel of half included angle $\alpha = 5°$ and $\alpha = 6°$, as measured by J. Nikuradse [71]. The lack of symmetry in the velocity distribution signifies incipient separation

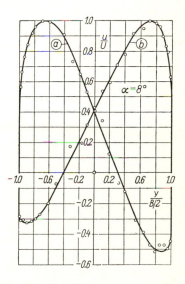

Fig. 22.3. Velocity distribution in a *divergent* channel of half included angle $\alpha = 8°$, measured by J. Nikuradse [71]. Reverse flow is completely developed. The flow oscillates at longer intervals between patterns (a) and (b)

slightly convergent or divergent channels. The half included angle of the channels ranged over the values $\alpha = -8°, -4°, -2°, 0°, 1°, 2°, 3°, 4°$. The boundary-layer thickness in a convergent channel is much smaller than that at zero pressure gradient, whereas in a divergent channel it becomes very thick and extends as far as the centre-line of the channel. For semi-angles up to $4°$ in a divergent channel the velocity profile is fully symmetrical over the width of the channel and shows no features associated with separation. On increasing the semi-angle beyond $4°$ the shape of the velocity profile undergoes a fundamental change. The velocity profiles for channels with $5°$, $6°$ and $8°$ of divergence, respectively, shown in Figs. 22.2 and 22.3, cease to be symmetrical. With a $5°$ angle of divergence, Fig. 22.2, no back-flow can yet be discerned, but separation is about to begin on one of the channel walls. In addition the flow becomes unstable so that, depending on fortuitous disturbances, the stream adheres alternately to the one or the other wall of the channel. Such an instability is characteristic of incipient separation. J. Nikuradse observed the first occurrence of separation at an angle between $\alpha = 4.8°$ and $5.1°$. At an angle of $\alpha = 6°$, Fig. 22.2, the lack of symmetry in the velocity profile is even more pronounced, and the reversal of the flow indicates the start of separation. At $\alpha = 8°$ the width of the region of

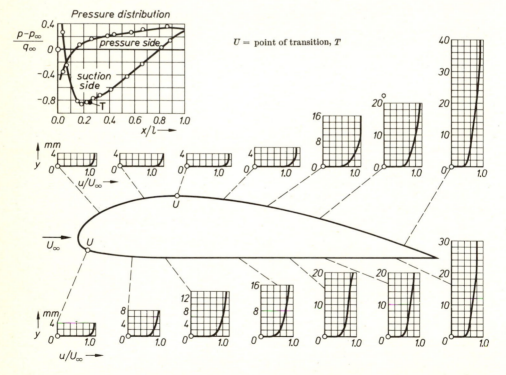

Fig. 22.4. Boundary layer on wing aerofoil, as measured by Stueper [105]; measurements in flight; lift coefficient $c_L = 0.4$; Reynolds number $R = 4 \times 10^6$; chord $l = 1800$ mm. The boundary layer is turbulent all along the *pressure side* owing to adverse pressure gradient; on the *suction side* it is laminar upstream of pressure minimum and turbulent downstream from it

reversed flow is considerably larger than for $\alpha = 6°$, and frequent oscillation of the stream from one side to the other is observed, the phenomenon being absent at $\alpha = 5°$ and $6°$. However, the duration of one particular flow configuration is sufficiently long for a full set of readings to be obtained. As the angle of divergence is increased, the region of reverse flow becomes wider, and the beats are more frequent.

The diagram in Fig. 22.4 shows an example of a turbulent boundary layer formed on an aerofoil and measured by J. Stueper [105] in free flight. In the case represented here, the boundary layer on the pressure side is turbulent from the leading edge onwards, because here the pressure rises over the whole width of the wing. On the suction side, the point of transition places itself a short distance behind the pressure minimum in agreement with the description given in Sec. XVII b. The fact that the boundary layer has become turbulent is inferred from the sudden increase in its thickness.

Very thorough experimental investigations into the behaviour of turbulent boundary layers with pressure gradients have been later performed by G. B. Schubauer and P. S. Klebanoff [97], by J. Laufer [58], and by F. H. Clauser [21]. The first two of the above papers contain, in particular, results of measurements on turbulent fluctuations and on the correlation coefficients which were defined in Chap. XVIII. The last paper contains extensive results of measurements on shearing stresses. The calculations described in the following sections can evidently apply only to flows which adhere completely to the walls, that is, to cases which are similar to the one shown in Figs. 22.1 and 22.4.

b. The calculation of two-dimensional turbulent boundary layers

1. General remarks. To this day, all methods for the calculation of turbulent boundary layers with pressure gradients have been later performed by G. B. and tangential stress components created by the turbulent fluctuations as well as the thus released energy losses cannot be calculated by purely theoretical means. Furthermore, it is still necessary to introduce here empirical relations of the type of Prandtl's famous mixing-length formula invented in 1925, because the statistical theory of turbulence has yet to produce a replacement for it. It is astonishing that Prandtl's hypothesis, half a century after its discovery, still plays a very important role in the literature on the calculation of turbulent boundary layers. Most contemporary methods are approximate; they make use of the momentum and energy equations of the velocity layer (as distinct from the thermal layer which will not be discussed in this section) and of certain relations that follow from them. The corresponding relations for laminar boundary layers were derived in Chaps. X and XI.

The procedures for the calculation of turbulent boundary layers available today can be divided into two classes: methods based on *integral forms* of the principal equations and methods based on *differential* equations. The former can be traced to work that was done by Th. von Kármán in 1921. In this procedure, the partial differential equations are reduced to a system of ordinary differential equations in that an analytic integration in the transverse direction is first performed, *cf.* Chaps. VIII and XIII. In the other class of cases, the partial differential equations are integrated directly by the application of numerical methods, such as the method of finite differences outlined in Sec. IXi, or by finite elements. It is evident that the amount of work involved when differential equation methods are used is substantially larger than in the case of integral methods. The former require the use of a very large digital computer equipped with a large memory, whereas the latter can be done on a small calculator or, even, with the aid of a slide rule.

In the following paragraphs we shall confine ourselves to the description of methods which result merely in the calculation of time-averaged values of such variables of the turbulent flow as the velocity, the local shearing stress and the region of separation, because we subscribe to the view that only such mean values are of real interest to the engineer. Thus we refrain from calculating all those quantities that result from fluctuations, for example the correlation coefficients, the intensity of turbulence and its scale. Readers interested in these aspects are referred to more specialized publications, e. g. [10, 81].

Research into turbulent boundary layers was considerably advanced by the Stanford University Conference organized by S. J. Kline in 1968. The results achieved at the time have been published in two large volumes edited by S. J. Kline, M. V. Morkovin, G. Sovran, D. J. Cockrell, D. E. Coles and E. A. Hirst [54]. In the appended [79] "morphology" prepared by W. C. Reynolds, the reader will find a description of 20 integral and 8 differential methods and characterized according to their respective physical basis (status as of 1967). They differ, principally, in the empirical closure functions which are introduced in order to make the system of equations solvable. In addition, the conference had at its disposal 33 sets of experimental data which served as testing material for the computational algorithms. About ten years later, W. C. Reynolds [81] provided once again a summary review of the very large number of computational schemes; this appeared in his contribution to the Annual Reviews of Fluid Mechanics of 1976 (*cf.* the same author's 1974 contribution in *Chemical Engineering* [80]). In 1974 there appeared the book by F. M. White [119] which describes 20 integral and 11 differential procedures. It is difficult, and we shall not attempt, to select a "best method" from among the very large number proposed so far.

A summary of many of these methods, principally *integral* ones, was prepared earlier by A. Walz [116] and J. C. Rotta [86, 87]. A review of *differential* methods is contained in P. Bradshaw's contributions [9, 12, 13, 14]. Further, the book by T. Cebeci and A. M. O. Smith [20] and two earlier papers by the same authors [18, 19], contain good reviews of many calculational procedures. The two earlier reviews by L. S. G. Kovasznay [56] and F. H. Clauser [21a] may also be consulted.

We shall also refrain from describing in detail several of these numerous methods. Instead, we shall concentrate on a single one of the many and bring it to a point where the reader can work with it directly. For this purpose we have selected the *integral* method developed by E. Truckenbrodt [111]. The first version of this method was published in 1952; it has now been brought up to date in the light of the present physical understanding of the subject [114]. The procedure is convenient to handle and constitutes one of the best integral methods from the point of view of accuracy.

2. Truckenbrodt's integral method. Before we proceed with the description of the details of E. Truckenbrodt's [114] method, we find it helpful for its understanding to preface it with a few historical remarks. As already mentioned earlier, all computational algorithms for turbulent boundary layers rely on certain empirical relations.

As time progressed, and, in particular, since the middle of the thirties, the empirical basis, and hence also the semi-empirical and theoretical computational procedures, underwent a process of continuous improvement.

The first method for the calculation of turbulent boundary layers with pressure gradients was formulated by E. Gruschwitz [41] in 1931. The experimental data on which this method was based were later improved by A. Kehl [53]. At about the same time A. Buri [15] published a similar procedure. H.C. Garner [35] developed a method based on the work of A. E. von Doenhoff and N. Tetervin [27] that turned out to be superior to the first one mentioned above from the point of view of numerical convenience. In 1952, E. Truckenbrodt [111] formulated a simple quadrature based on the experimental results of K. Wieghardt [120], H. Ludwieg and W. Tillmann [60], as well as J.C. Rotta [82, 83]. The method was suitable for two-dimensional as well as for rotationally symmetric flow. This method was improved in 1974 on the basis of later insights [114]. It is this version that we now propose to discuss in some detail.

Characteristic numbers: In order to provide a description of the essential behaviour of a velocity layer it is necessary to know its thickness and to have an indication of the velocity distribution in the boundary layer. Since the *boundary-layer thickness*, $\delta(x)$, across which the dissipative layer merges with the frictionless external flow $U(x)$, so that $u(x, y = \delta) = U(x)$, cannot be defined with any accuracy, it is convenient to operate with the quantities defined earlier in eqns. (8.30), (8.31), and (8.34). These include the following

$$\delta_1(x) = \int_0^\delta (1 - u/U)\, dy \qquad \text{(displacement thickness)}, \qquad (22.1\,\text{a})$$

$$\delta_2(x) = \int_0^\delta (1 - u/U)\,(u/U)\, dy \qquad \text{(momentum thickness)}, \qquad (22.1\,\text{b})$$

$$\delta_3(x) = \int_0^\delta [1 - (u/U)^2]\,(u/U)\, dy \qquad \text{(energy thickness)}. \qquad (22.1\,\text{c})$$

These quantities can be made dimensionless by introducing appropriate *Reynolds numbers* formed with the external velocity. Thus, we may use,

$$R_2 = \delta_2\, U/\nu; \quad R_3 = \delta_3\, U/\nu; \quad \text{etc.} \qquad (22.2\,\text{a, b})$$

The velocity profile depends strongly on the external pressure gradient, expressed through the derivative dU/dx, and is characterized by a number of *shape factors*. These are also made dimensionless for preference and can be defined in the form of ratios of thicknesses from eqns. (22.1). It is customary to use the contractions

$$H_{12} = \delta_1/\delta_2; \quad H_{23} = \delta_2/\delta_3; \quad H_{32} = \delta_3/\delta_2; \quad \text{etc.} \qquad (22.3\,\text{a, b, c})$$

For the moment we refrain from reproducing here the shape factors used by E. Gruschwitz [41] and A. Buri [15]. The fact that the shape factors defined in eqns. (22.3) constitute useful quantities for the description of velocity profiles has been known for a long time; this has been corroborated by the summaries presented at the Stanford Conference [54]. Beyond this, measurements indicate that turbulent velocity profiles can be described approximately by a *one-parameter family of curves*. This means that the shape factors H_{12} and $H_{32} = 1/H_{23}$ are related to each other uniquely, as evidenced by the graph of Fig. 22.5. This fact is expressed by a relation $H_{12} = f(H_{32})$, if a slight, residual dependence on the Reynolds number is neglected. Guided by the preceding observation, E. Truckenbrodt [114] introduced† the *modified shape factor*

$$H = \exp\left\{ \int_{(H_{32})_\infty}^{H_{32}} \frac{dH_{32}}{(H_{21}-1)\,H_{32}} \right\} = \exp\left\{ -\int_{(H_{23})_\infty}^{H_{23}} \frac{dH_{23}}{(H_{12}-1)\,H_{23}} \right\}. \quad (22.4\text{a, b})$$

The reference value $(H_{23})_\infty = (1/H_{32})_\infty$ has been chosen as the lower limit of integration because it represents an average value for flows without a pressure gradient. In the case of turbulent boundary layers we choose $(H_{12})_\infty = 1\cdot3$. The numerical evaluation of the relation in eqn. (22.4) can be undertaken on the basis of a relation indicated by H. Fernholz [33]. The result is seen plotted in Fig. 22.6.

In the case of flow with zero pressure gradient we find that $H = H_\infty = 1$ by definition (mean value in the case of turbulent flow). Flows with adverse pressure gradients (pressure rising in the downstream direction) are characterized by $H_S \leqslant H < 1$, whereas for accelerated flows (pressure decreasing) we find that $1 < H \leqslant H_0$, where H_S denotes the shape factor for the velocity profile with incipient separation, and H_0 denotes the shape factor of a two-dimensional stagnation-flow profile. According to K. Wieghardt, the shape factors H_{12} and H_{32} are related to each other by the equation

$$H_{12} = H_{32}/(3\,H_{32} - 4),$$

on the assumption of so-called power-law profiles. Substituting this expression into eqn. (22.4a) and integrating with respect to H_{32}, we can derive the following expression for the modified shape factor in a turbulent boundary layer:

$$H = \frac{(H_{32})_\infty}{H_{32}}\left(\frac{2 - (H_{32})_\infty}{2 - H_{32}}\right)^{1/2} = 0\cdot5442\,H_{23}\left(\frac{H_{23}}{H_{23} - 0\cdot5049}\right)^{1/2}. \quad (22.5\text{a, b})\,{}^{+}_{+}$$

The numerical values indicated in the literature for turbulent boundary layers for which *separation* can occur vary considerably. J.C. Rotta [86] recommends $4\cdot05 > (H_{12})_S > 4\cdot0$ or $H_S \approx 0\cdot723$, whereas A. Walz [116] proposes the values $1\cdot50 < (H_{32})_S < 1\cdot57$, or $0\cdot736 < H_S < 0\cdot761$. According to A.A. Townsend [110a] (*cf.* Stratford [104]) a vanishing shear stress occurs for $(H_{12})_S = 2\cdot274$ or $H_S = 0\cdot784$ in the case of profiles created by an external flow with $U(x) \sim x^p$ with $p = -0\cdot234$. The various shape factors for incipient separation have been indicated in Fig. 22.6. The values of the modified factor H_S fluctuate much less than those of $(H_{12})_S$ and $(H_{23})_S$. Reference [114] indicates that *separation* can occur for

$$H_S \leqslant 0\cdot723 \qquad \text{(separation)}. \qquad (22.6)$$

The range $H_S = 0\cdot723 \leqslant H \leqslant 0\cdot761 = H'_S$
describes velocity profiles that are prone to separate.

---

† Reference [111] employs the modified shape factor $L = \ln H$.

‡ The numerical constants in eqn. (22.5 b) have been adjusted to represent the experiments available at the time.

$$H_{32} = \frac{\delta_3}{\delta_2}$$

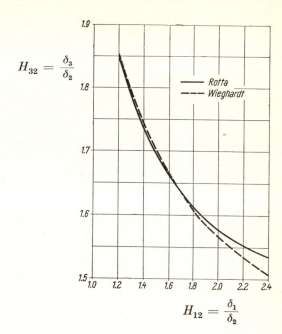

—— Rotta
---- Wieghardt

$$H_{12} = \frac{\delta_1}{\delta_2}$$

Fig. 22.5. The ratio of boundary layer thicknesses $H_{32} = \delta_3/\delta_2$ plotted against $H_{12} = \delta_1/\delta_2$, after J.C. Rotta [84] and K. Wieghardt [120]. See also H. Fernholz [33]

H = modified shape factor from eqn. (22.4)
H_∞ = reference state; approximately flow at constant pressure
H_S = separated boundary layer
$H_S \leqslant H \leqslant H_S'$ = boundary layer prone to separation
(1) = H_{23} in terms of H
(1a) determined numerically from eqn. (22.4a) after [23a, 46a], $H_{12} = 1 + 1\cdot48\,(2 - H_{32})$ $+ 104\,(2 - H_{32})^{6\cdot7}$
(1b) calculated with the aid of eqn. (22.5)
(2) = H_{12} in terms of H
(2a) and (2b) are analogous to (1a) and (1b)

Fig. 22.6. Shape factor of a turbulent boundary layer

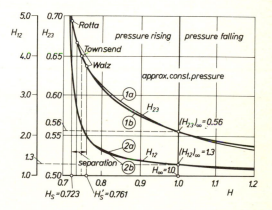

3. Basic equations. In order to calculate the boundary-layer thickness and the shape factor, the latter characterizing the velocity profile, it is necessary to have two equations. These are: the momentum integral equation (8.32) and the energy equation (8.35). In the second equation it is necessary to introduce on the right-hand side a still unspecified expression for the dissipation work associated with the shearing stresses of the turbulent stream.

As the basic equations for *momentum thickness* $\delta_2(x)$ and for *energy thickness* $\delta_3(x)$, we obtain†:

$$\frac{d\delta_2}{dx} + (2 + H_{12})\frac{\delta_2}{U}\frac{dU}{dx} = c_T, \qquad (22.7\,\text{a})$$

and

$$\frac{d\delta_3}{dx} + 3\frac{\delta_3}{U}\frac{dU}{dx} = 2\,c_D, \qquad (22.7\,\text{b})$$

respectively. Here c_T is the skin-friction coefficient and c_D is the dissipation coefficient. The preceding two coefficients related to the shearing stress depend strongly on the Reynolds number, R, according to eqn. (22.2), and on the shape factor H in conformity with eqn. (22.4). The following power-laws for their description have withstood the test of time:

$$c_T = \frac{\tau_0}{\varrho\,U^2} = \frac{\alpha(H)}{R_2^a}; \quad c_D = -\int_0^\delta \frac{u}{U}\frac{\partial}{\partial y}\left(\frac{\tau}{\varrho\,U^2}\right)dy = \frac{\beta(H)}{R_3^b}. \qquad (22.8\,\text{a, b})$$

The expressions contain the factors $\alpha(H)$ and $\beta(H)$ which are unique functions of the shape factor and a specified power of the local Reynolds numbers R_2 or R_3. The diagrams in Figs. 22.7a and b represent the quantities a and b as well as $\alpha' = \alpha/\alpha_\infty$ and $\beta' = \beta/\beta_\infty$ together with α_∞ and β_∞ (denoting values for zero-gradient flow) as functions of H. The respective formulae are quoted in the captions. It is seen that β' varies slowly with H, whereas α' assumes the value $\alpha'(H = H_S) = 0$ at separation and then increases fast with increasing H.

Equations (22.8a, b) are now substituted into eqns. (22.7a, b) and this leads us to the modified forms of the momentum and energy equations for $\delta_2(x)$ and $\delta_3(x)$, respectively.

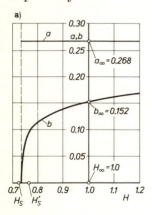

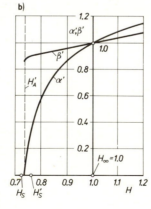

$a = 0.268$

$b = 0.2317\,H_{32} - 0.2644 - 0.87 \times 10^5 (2 - H_{32})^{20}$

$\alpha' = \alpha/\alpha_\infty$ with $\alpha_\infty = 0.0157$

$\beta' = \beta/\beta_\infty$ with $\beta_\infty = 0.0053$

$\alpha = 0.0245\,(1 - 2.007\log H_{12})^{1.705}$

$\beta = [0.00481 + 0.0822(H_{32} - 1.5)^{4.81}]\,H_{32}^b$

Fig. 22.7. Shearing stresses in turbulent boundary layer corresponding to eqn. (22.8) after [33], [120]; a) exponents a and b in terms of H; b) factors α and β in terms of H

---

† The above equations neglect the effect of the normal components $\overline{\varrho u'^2}$ and $\overline{\varrho v'^2}$ of the tensor of Reynolds stresses from eqn. (18.10). Among others, references [85, 87] contain indications on how to modify these equations if this simplification is not acceptable.

In order to complete the evaluation it is still necessary to know the *shape-factor function H(x)*. It is shown on p. 487 of [113] that eqns. (22.7a, b) lead to the working forms

$$\delta_2 \frac{dH_{32}}{dx} - (H_{12} - 1) H_{32} \frac{\delta_2}{U} \frac{dU}{dx} = 2 c_D - H_{32} c_T \tag{22.9a}$$

$$\delta_3 \frac{dH_{23}}{dx} + (H_{12} - 1) H_{23} \frac{\delta_3}{U} \frac{dU}{dx} = c_T - 2 H_{23} c_D \tag{22.9b}$$

for the calculation of $H_{32}(x)$ and of $H_{23}(x)$. The shape factors can now be calculated either by the use of a coupled pair of equations, namely (22.7a) and (22.9a) or (22.7b) and (22.9b). To distinguish between these two possibilities we speak of the *momentum method* in the first case and of the *energy method* in the second case. In most procedures use is made of the momentum method, whereas E. Truckenbrodt [111, 114] favours the energy method. The latter choice has been made for two reasons: (a) The left-hand side of eqn. (22.7b), unlike that of eqn. (22.7a), does not depend explicitly on the shape factor. Thus instead of eqn. (22.7b) we can also write

$$\frac{1}{U^3} \frac{d}{dx} (U^3 \delta_3) = 2 c_D. \tag{22.10}$$

(b) The dissipation factor c_D on the right side of eqn. (22.7b) must be calculated from eqn. (22.8b) by performing a quadrature extended over the boundary-layer thickness $0 \leq y \leq \delta(x)$, whereas the skin-friction coefficient c_T on the right-hand side of eqn. (22.7a) depends only on the local shearing stress at the wall, eqn. (22.8a). This signifies that the dissipation work depends much less on the shape factor than the shearing stress at the wall. This is confirmed by the graphs of $\alpha'(H)$ and $\beta'(H)$ in Fig. 22.7b. Thus, in the energy method, the coupling between the equations determining the boundary-layer thickness (energy equation) and the equation which determines the shape factor turns out to be much weaker than in the momentum method.

Reference [114] shows how the basic equations for the boundary-layer thicknesses, eqns. (22.7a, b), can be transformed into equations that determine the local Reynolds numbers defined in eqn. (22.2). Similarly, this reference shows how the basic equations for the shape factors, eqns. (22.9a, b), can be transformed into equations for the modified shape factor defined in eqn. (22.4). In this manner, we obtain

$$\frac{dR}{dx} + m \frac{R}{U} \frac{dU}{dx} = U(x) \Phi (R, H), \tag{22.11a}$$

$$\frac{d}{dx} \left(\frac{H}{U} \right) = \frac{\psi (R, H)}{R(x)}. \tag{22.11b}$$

Table 22.1 summarizes the expression for the contractions m, Φ and ψ. The quantities R, m, Φ and ψ are provided with subscript 2 for the momentum method, and with subscript 3 for the energy method.

4. Quadrature for the calculation of plane turbulent boundary layers. Under certain simplifying assumptions it is possible still further to simplify the system of equations (22.11a, b). In this manner it is possible to derive explicit expressions for $R(x)$ and $H(x)$ by quadrature for an arbitrary velocity variation, $U(x)$, in the outer flow, that

Table 22.1.   Summary of the quantities which occur in the equations for the calculation of the dimensionless momentum thickness, R_2, of the dimensionless energy thickness, R_3, and of the shape factor; see eqns. (22.11a, b)

	Momentum method (subscript 2)	Energy method (subscript 3)
R	$R_2 = U\,\delta_2/\nu$	$R_3 = U\,\delta_3/\nu$
m	$m_2 = H_{12} + 1 = m_2\,(H)$	$m_3 = 2 = \text{const}$
$\Phi(R, H)$	$\Phi_2 = c_T/\nu = \alpha\,(H)/\nu\;R_2^a$	$\Phi_3 = 2\,c_D/\nu = 2\,\beta\,(H)/\nu\;R_3^b$
$\psi(R, H)$	$\psi_2 = -\dfrac{H}{\nu}\;\dfrac{c_T - 2\,H_{23}\,c_D}{H_{12} - 1}$	$\psi_3 = \dfrac{H}{\nu}\;\dfrac{2\,c_D - H_{32}\,c_T}{H_{12} - 1}$
	$= -\dfrac{H}{\nu}\;\dfrac{\alpha - 2\,\beta\,H_{23}^{1+b}\,R_2^{a-b}}{(H_{12} - 1)\,R_2^a}$	$= \dfrac{H}{\nu}\;\dfrac{2\,\beta - \alpha\,H_{32}^{1+a}\,R_3^{b-a}}{(H_{12} - 1)\,R_3^b}$

is for zero, adverse, or favourable pressure gradients. The integration is here extended only over certain powers of $U(x)$. We now proceed to derive such approximate, explicit formulae and to show how a suitable choice of approximate procedures can lead to a step-by-step improvement.

Analogy with laminar boundary layer (momentum method): In a manner analogous with K. Pohlhausen's approximate method for the laminar boundary layer outlined in Chap. X, A. Buri [15] assumes that in the case of a turbulent boundary layer the ratio $H_{12} = \delta_1/\delta_2$ as well as the skin-friction coefficient, $c_T = \tau_0/\varrho U^2$, continue to be functions of a single shape factor†

$$\Gamma = \frac{\delta_2}{U}\,\frac{dU}{dx}\,R_2^a; \quad (a = 1/4). \tag{22.12}$$

He introduces the following functions, cf. eqn. (22.8a):

$$H_{12} = \frac{\delta_1}{\delta_2} = f_1\,(\Gamma), \qquad c_T = \frac{\tau_0}{\varrho\,U^2} = \frac{f_2(\Gamma)}{R_2^a}. \tag{22.13a, b}$$

These relations are now substituted into the momentum equation (22.7a) to obtain

$$\frac{d}{dx}\,(\delta_2\,R_2^a) = F(\Gamma), \tag{22.14}$$

where $F(\Gamma)$ is a universal function given by

$$F(\Gamma) = (1 + a)\,f_2(\Gamma) - [2 + a + (1 + a)\,f_1(\Gamma)]\,\Gamma.$$

---

† The laminar counterpart of the shape factor Γ is

$$K = \frac{\delta_2^2}{\nu}\,\frac{dU}{dx} = \frac{\delta_2}{U}\,\frac{dU}{dx}\,R_2$$

from eqn. (10.27).

Using his own and J. Nikuradse's [71] measurements, A. Buri established that $F(\Gamma)$ can be represented by the linear relation

$$F(\Gamma) = \Gamma_\infty - n\,\Gamma = \Gamma_\infty - n\,\frac{\delta_2}{U}\frac{dU}{dx}\,R_2^a \qquad (22.15)$$

with a good degree of approximation. For accelerated and retarded streams he found the numerical values $0{\cdot}01475 \leq \Gamma_\infty \leq 0{\cdot}0175$ and $3{\cdot}94 \leq n \leq 4{\cdot}15$ for $a = 0{\cdot}25$. If we now introduce eqn. (22.15) into eqn. (22.14), we can obtain the integral with respect to x in closed form. The result, expressed in terms of the Reynolds number formed with the momentum thickness, is

$$[R_2(x)]^{1+a} = \frac{1}{\nu'}\frac{E_2(x)}{[U(x)]^e}, \qquad (22.16)$$

where $\nu' = \nu/\Gamma_\infty$ and

$$E_2(x) = E_2(x_1) + \int_{x_1}^{x} U^n\,dx. \qquad (22.16\,\mathrm{a})$$

The numerical values for the exponents a, e and n as well as for the modified kinematic viscosity, ν', are listed in Table 22.2. The constant of integration is

$$E_2(x_1) = \nu'\,[U(x)]^e\,[R_2(x_1)]^{1+a}.$$

Analogy with turbulent boundary layer on a flat plate: Whereas along a flat plate at zero incidence (constant-pressure flow) the velocity of the external flow remains constant, $U(x) = U_\infty = \mathrm{const}$, the general case is characterized by a variable outer velocity $U(x) \neq \mathrm{const}$, that is $dU/dx \neq 0$. We shall suppose that for $U(x) \neq \mathrm{const}$ the value assumed by the shape factor is the same as that for a hypothetical $U_\infty = U(x)$ which implies that $H(x) = H_\infty = 1$†. It follows that $H_{12} = (H_{12})_\infty = 1{\cdot}3$ and, according to Table 22.1, we must have $m = \mathrm{const}$ for the momentum as well as for the energy method. Correspondingly $\alpha = \alpha_\infty$, $\beta = \beta_\infty$, $a = a_\infty$ and $b = b_\infty$, see Fig. 22.7. As far as the external flow is concerned, we always substitute the actual dis-

Table 22.2.   Summary of numerical constants that occur in the explicit equations for the calculation of momentum and energy thickness; see eqns. (22.16), (22.17), and (22.19); for b see Fig. 22.7a; take β from Fig. 22.7b

| Analogy | Momentum method | | | Energy method |
	Laminar boundary layer	Turbulent boundary layer on flat plate (subscript ∞)		Self-similar solution
m	—	$H_{12} + 1 = 2{\cdot}30$	2	$(H = \mathrm{const})$
c	$a = 0{\cdot}25$	$a = 0{\cdot}268$	$b = 0{\cdot}152$	b from Fig. 22.7a
e	$2{\cdot}94 < e < 3{\cdot}15$	$m\,(1+a) = 2{\cdot}92$	$2\,(1+b) = 2{\cdot}30$	b from Fig. 22.7a
n	$3{\cdot}94 < n < 4{\cdot}15$	$1 + m\,(1+a) = 3{\cdot}92$	$3 + 2\,b = 3{\cdot}30$	b from Fig. 22.7a
ν'	$57\nu < \nu' < 68\,\nu$ $68\,\nu > \nu/\Gamma_\infty > 57$	$\dfrac{\nu}{(1+a)\,\alpha} \approx 50\,\nu$	$\dfrac{\nu}{2\,(1+b)} \approx 78\,\nu$	β from Fig. 22.7b

---

† The fact that, strictly speaking, $H(x) \approx 1{\cdot}0$ for a flat plate is here ignored.

tribution $U(x)$. Since the value of the shape factor, $H(x) = 1$, has already been assigned, the only quantity that we need to calculate is the local Reynolds number determined by eqn. (22.11a). Since $m = \text{const}$, we can contract the two terms on the left-hand side and solve the problem by performing two integrations, one each for $R_2(x)$ and $R_3(x)$. In contracted form these are

$$\{R(x)\}^{1+c} = \frac{1}{\nu'} \frac{E(x)}{[U(x)]^e}, \text{ where } E(x) = E(x_1) + \int_{x_1}^{x} U^n \, dx. \qquad (22.17)$$

The relations and numerical values to be used for the exponents i, e and n, as well as for the modified kinematic viscosity ν' are listed in Table 22.2, separately for the momentum and for the energy method. The constant of integration is

$$E(x_1) = \nu \, \{U(x_1)\}^e \, \{R(x_1)\}^{1+i}.$$

Differentiating eqn. (22.17) with respect to x and taking into account the contractions defined in Tables 22.1 and 22.2 we can demonstrate consistency with eqn. (22.11a). In the case of the momentum method, eqn. (22.17) becomes identical with eqn. (22.16) if we put $i = a$, $R = R_2$ and $E = E_2$.

Comparing the numerical data of Table 22.2, we find far-reaching agreement. In spite of considerable differences in the assumptions for the shape factor and for the shearing stress at the wall we discover that the two explicit equations for the calculation of the momentum thickness are equivalent. The following specific numerical values can be recommended:

$$R \equiv R_2: \quad a = 0 \cdot 268, \quad e = 3, \quad n = 4, \quad \nu' = 50 \, \nu. \qquad (22.18)$$

The energy method is discussed below.

Analogy with self-similar solutions (energy method): Self-similar solutions in boundary-layer theory are generally described as *equilibrium flows* when they occur in turbulent motion. They are characterized by the fact that the velocity profiles u/U at varying positions x become similar for certain velocity distributions $U(x)$ of the outer flow. This means that the shape factor $H(x)$ remains constant with x, that is that $dH/dx = 0$. Figure 22.7 implies that all quantities which depend on x in general must become constant for such equilibrium boundary layers.

We now substitute in eqn. (22.10) the expression for c_D from eqn. (22.8b) and note that the integration with respect to x can be performed in closed form with $b = \text{const}$ and $\beta' = \text{const}$. The Reynolds number formed with the energy thickness is thus given by

$$\{R_3(x)\}^{1+b} = \frac{1}{\nu'} \frac{E_3(x)}{[U(x)]^e} \text{ where } E_3(x) = E_3(x_1) + \beta' \int_{x_1}^{x} U^n \, dx. \qquad (22.19)$$

The numerical values for b, e, n, and ν' are to be selected in accordance with the relations in Table 22.2 and Fig. 22.7a.

In the special case of separation-prone flows for which $H_S \leq H \leq H'_S$, we find, for example, that $1 + b'_S = 1 \cdot 094$ and $1 + b_\infty = 1 \cdot 152$. These two values differ by about 5%. Such a discrepancy can be disregarded in view of the uncertainties inherent in such approximate methods. In other words, this signifies that it is possible to perform calculations using numerical values based on the flat-plate analogy. The quantity $\beta' = \beta/\beta_\infty$ that appears in eqn. (22.19) also depends only weakly on the shape factor; by way of approximation, we let it be $\beta' = 1$. Thus, the calculation of the energy thickness with the aid of eqn. (22.19) can be based on the following numerical values:

$$R \equiv R_3: \quad b = 0\cdot152; \quad e = 2\cdot3; \quad n = 3\cdot3; \quad \nu' = 78\,\nu; \quad \beta' = 1. \qquad (22.20)$$

With these assumption, eqn. (22.19) transforms into eqn. (22.17) bearing in mind that $i = b$, $R = R_3$, and $E = E_3$, as expected.

Reference [114] shows that eqn. (22.11a) suffices by itself to solve the problem when the energy method is used. By contrast, when the momentum method is used, the coupling between eqns. (22.11a) and (22.11b) cannot be disregarded†. The latter leads us to the trivial result that $R_2 = H_{23}R_3$ in view of the definition $H_{23} = \delta_2/\delta_3$. At the level of approximation considered so far, the momentum method turns out to be identical with the energy method. Nevertheless, the two procedures differ essentially from one another in that the momentum method employs the two basic equations (22.11a) and (22.11b), whereas the energy method gets by with eqn. (22.11a) alone. As far as the development of further approximations in the form of simple integrals is concerned, we have exhausted the potential inherent in the momentum method. In the energy method, eqn. (22.11b) is used to derive a formula for the shape factor by closed-form integration, as we are about to show.

Integration method due to E. Truckenbrodt: E. Truckenbrodt [111, 114] worked out an approximate method for the explicit integration of the equations of turbulent boundary layers which serves to obtain the boundary-layer thickness (energy thickness) as well as the (modified) shape factor. The first version of the method [111] has proved to be practicable for calculations in engineering applications. It was, therefore, thought useful to modify it in the light of more recent discoveries. Employing the classification introduced above, we describe this as an *energy method*. The method can be used for two-dimensional as well as for axially symmetric flows, *cf.* Sec. XXIX c1.

The method is based on eqn. (22.11a) which is used to calculate the Reynolds number formed with the energy thickness. Hence we put $R = R_3$, $m = 2$ and $\Phi = \Phi_3 = (2/\nu)\,\beta\,R_3^{\,b}$ in accordance with Table 22.1. If we assume that $b = $ const and $\beta(x)$ is known, we can integrate eqn. (22.11a) with respect to x and obtain

$$\{R_3(x)\}^{1+b} = \frac{1}{\nu'}\frac{E_3(x)}{[U(x)]^e} \quad \text{where} \quad E_3(x) = E_3(x_1) + \int_{x_1}^{x} \beta'\,U^n\,\mathrm{d}x. \qquad (22.21)$$

The numerical values in eqn. (22.20) are valid up to $\beta'(x) = \beta(x)/\beta_\infty$. However, an inspection of Fig. 22.7b shows that β' does not deviate much from the value $1\cdot0$, and we may calculate with $\beta' = 1\cdot0$ by way of approximation. In this case, eqn. (22.21) transforms into eqn. (22.19). Thus, if no great demands of accuracy are made on the values of the Reynolds number, we obtain

$$R_3(x) = \left(\frac{1}{\nu'}\,\frac{E_3(x_1) + \int_{x_1}^{x} U^{3+2b}\,\mathrm{d}x}{\{U(x)\}^{2(1+b)}}\right)^{1/(1+b)} \qquad (22.22\,\text{a})$$

† By way of amplification, we mention that the result derived in Table 6 of [114] is also valid when $a = $ const and $b = $ const.

where the following numerical values (*cf.* Table 22.2 — energy method) have been employed:

$$b = 0 \cdot 152; \quad v' = 80 \, v; \quad \text{with } E_3(x_1) = v' \, \{[U(x_1)]^2 \, R_3(x_1)\}^{1+b}. \tag{22.22 b}$$

This explicit formula contains only the external free-stream velocity $U(x)$ which may be known from potential theory or from measurement. The position $x = x_1$ constitutes the starting point for the calculation.

Apart from the velocity $U(x_1)$, the constant of integration $E_3(x_1)$ contains also the energy thickness $\delta_3(x_1)$. If the station x_1 coincides with the point of transition the energy thickness should be calculated over the laminar boundary layer in the range $0 \leq x \leq x_1$. Here $x = 0$ denotes the start of the boundary layer; for example, the leading edge of a plate or the stagnation point of a blunt body. It was shown in [114] that eqn. (22.22a) is also valid for laminar boundary layers when $b = 1$, $v_l' = v/4 \beta_\infty = 0 \cdot 917 v$ and $E_3(x_1 = 0) = 0$ should be specified. In this case, with a laminar starting length, the constant of integration becomes

$$E_3(x_1) = \left[\frac{1}{v'} \frac{\int\limits_0^{x_1} U^5 \, dx}{\{U(x)\}^4} \right]^{1/2} \quad \text{(point of transition).} \tag{22.23}$$

If the boundary layer already is turbulent at $x = x_1$, it is necessary to substitute into eqn. (22.22) for $E_3(x_1)$ the local value $R_3(x_1) = \delta_3(x_1) \, U(x_1)/v$.

In many practical applications it is not enough to know the behaviour of the boundary-layer thickness, here the energy thickness $\delta_3(x)$. This is the case with separation-prone or separated boundary layers. If, for example, it is necessary to make a statement about the possibility of separation, it is necessary to know the velocity parameters along the wall. All methods discussed in Sec. XXIIb1 provide procedures for the calculation of some shape factor in addition to that of a boundary-layer thickness, such as the momentum thickness $\delta_2(x)$ discussed there. The shape factors are defined differently in different methods and different differential equations are specified for their calculation. A review and intercomparison was given by J.C. Rotta [85].

The differential equation (22.9a, b) for the shape factors $H_{32}(x)$ and $H_{23}(x)$ were obtained by the coupling of the momentum-integral and energy-integral equations (22.7a, b). The preceding differential equations determine the shape factor in a unique way provided that one-parameter velocity profiles $H_{12} = f(H_{32})$ or $H_{12} = f(H_{23})$ are postulated and approximate expressions for the shear-stress coefficients c_T and c_D are substituted from eqn. (22.8a, b). The determining equation (22.11b) for the shape factor can be written in terms of the *modified shape factor* $H = f(H_{23})$ proposed by E. Truckenbrodt. Together with eqn. (22.11a), this relation forms a system of simultaneous differential equations for the Reynolds number formed with the energy thickness, $R_3(x)$, and for the shape factor $H(x)$. According to Table 22.1 we must put $m = 2 = $ const for the energy method discussed here. The forms of the functions $\Phi_3(R_3, H)$ and $\psi_3(R_3, H)$ are to be taken from the same table. Reference [114] summarizes eqns. (22.11a) and (22.11b) as follows:

$$H(x) = \frac{U(x) G'(x)}{E'(x)}, \tag{22.24}$$

with

$$E'(x) = E(x_1) + \int_{x_1}^{x} \beta' \, U^n \, \mathrm{d}x$$

and

$$G'(x) = G(x_1) + \int_{x_1}^{x} \gamma' \, U^n \, \mathrm{d}x,$$

where e, n and γ' are listed in Table 22.2 (energy method). The correction function $\gamma'(x) = \gamma'(R_3, H)$ can be calculated with the aid of $a(H)$, $b(H)$, $\alpha(H)$, $\beta(H)$, $H_{12}(x)$, $H_{32}(H)$, as well as $R_3(x)$ and $H(x)$.

The correction γ' differs by a larger or smaller amount from the value $1 \cdot 0$ in the case of a turbulent boundary layer, and cannot be determined with an adequate degree of reliability. By way of approximation, we assume $\gamma'(x) = \text{const} = 1 \cdot 0$ and introduce a new quantity $c = \text{const} = 4 \cdot 0$ in order further to simplify the analytic solution. The quantity c has been so determined as to achieve optimum agreement between available measurements [54] and theoretical results; see also [114]. The modified shape factor is obtained from the equation

$$H(x) = U(x) \, G(x) \, [N(x)]^{-1/c} \tag{22.25}$$

which is the result of some algebraic transformations not reproduced here. Here the *influence functions* of the external velocity distribution are defined as

$$\left. \begin{aligned} G(x) &= G(x_1) + \int_{x_1}^{x} U^{2\,(1+b)} \, \mathrm{d}x; \\ N(x) &= N(x_1) + c \int_{x_1}^{x} U^{2\,(1+b)+c} \, G^{c-1} \, \mathrm{d}x. \end{aligned} \right\} \tag{22.26}$$

The initial values, i.e. the constants of integration are

$$\left. \begin{aligned} G(x_1) &= \nu' \, [H(x_1) \, \{U(x_1)\}^{1+2b} \, \{R_3(x_1)\}^{1+b}]; \\ N(x_1) &= [U(x_1) \, G(x_1)/H(x_1)]^c. \end{aligned} \right\} \tag{22.26a}$$

We take the numerical constants as

$$b = 0 \cdot 152; \quad c = 4 \cdot 0; \quad \nu' = 78 \, \nu. \tag{22.26b}$$

The integral expression (22.25) for the calculation of the shape factor contains only the external velocity distribution $U(x)$, as was the case with the corresponding integral expression (22.22) for the calculation of the Reynolds number. The determination of the influence function $N(x)$ requires the performance of a double integration with respect to x. The position $x = x_1$ once again represents the starting point of the calculation.

The constants of integration $G(x_1)$ and $N(x_1)$ contain the shape factor $H(x_1)$ in addition to the velocity $U(x_1)$ and the Reynolds number $R_3(x_1)$. If the position x_1 coincides with the point of transition, it is necessary to require that the energy thickness of the laminar boundary layer must be equal to that of the turbulent boundary layer in accordance with eqn. (22.23). On the other hand, the shape factor may change its value at the point of transition. The numerical values of the shape factor lie in the range $1 \cdot 0 \geq H \geq H_S = 0 \cdot 723$.

The theory of the origin of turbulence presented in Chap. XVII leads to the conclusion, which agrees with measurements, that transition from laminar to turbulent flow in the boundary layer occurs at a place which lies a small distance downstream from the velocity maximum of the external stream. For this reason, and by way of approximation, it is permissible to base the calculation at the point of transition on the value that corresponds to an external flow with a zero pressure gradient. According to the definition of H in eqn. (22.4) the latter is equal for laminar and turbulent boundary layers, namely

$$H(x_1) = 1 \quad \text{(transition)}. \tag{22.27}$$

If the boundary layer is already turbulent at $x = x_1$, it is necessary to employ the corresponding local values $G(x_1)$ and $N(x_1)$.

5. Application of the method. The approximate method described in the preceding paragraphs can be applied with ease because only simple integrations are required. Such detailed calculations have been performed for all experimental data (33 sets) collected in [54]; in particular, using eqn. (22.22), the calculations yielded the variation $R_3(x)$ of the Reynolds number formed with the energy thickness as well as the corresponding variation $H(x)$ of the modified shape factor after eqn. (22.25)†. In this manner, the practical calculations included very diverse external flow regimes and so covered a wide range of applications. Figure 22.8 illustrates the comparison between theory and measurement for an aerofoil in an adverse pressure gradient. Similar comparisons for other measurements are shown in Figs. 22.9a, b‡. The latter diagram contains a comparison of calculated and measured values of the Reynolds number and of the shape factor for the measuring station located furthest downstream. Deviations from the straight line constitute a measure of the quality of the approximate method. The comparison for $\log R_3$ contained in Fig. 22.9a is satisfactory, particularly if account is taken of the fact that excessive demands on the accuracy of calculated values of the Reynolds number are of no great practical significance.

According to theory, the six sets of measurements illustrated in Fig. 22.9b for which $H < H_S$ exhibit incipient separation. Measurements have confirmed this, and Ref. [114] contains a more detailed discussion of this circumstance. The sets of measurements designated Ident 1500 and Ident 2600 show particularly large discrepancies between theory and measurement. The case Ident 1500 represents a reattached boundary layer behind a ledge. It is understandable that the preceding method is not quite satisfactory in this case as far as the calculation of the Reynolds number and of the shape factor is concerned. Case Ident 2600 refers to a so-called equilibrium boundary layer formed under an external stream with $U(x) \sim x^{-0.255}$. A. A. Townsend investigated a similar boundary layer, namely one with $U(x) \sim x^{-0.234}$. He obtained the value $H = 0.748$ for the shape factor which differs considerably from the measured value $H = 0.823$. The approximate method yields $H = 0.731$. At the present time it is not possible to explain the reason for these discrepancies. To conclude, we wish to draw the reader's attention to the fact that the simple

---

† We have introduced corrections for three-dimensional effects in order to account for a possible convergence or divergence of streamlines. The correction was based on the method of J.C. Rotta [86]. *Cf.* remark on p. 676

‡ The diagrams include the cases of axially symmetric flows discussed in Sec. XXII d 1.

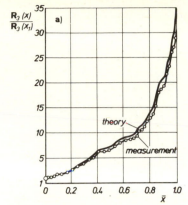

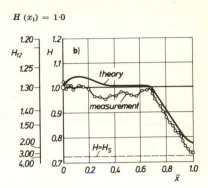

Fig. 22.8. Turbulent boundary layer on a wing aerofoil in adverse pressure gradient [54]; case Ident $2100: \bar{x} = (x - x_1)/(x - x_N)$ where x_1 = initial measuring station (start of measurement), x_N = final measuring station (end of measurement). Measured points by G. B. Schubauer and P. S. Klebanoff. Theory — full line — after eqns. (22.22) and (22.25). a) Reynolds number; b) shape factors H_{12} and H

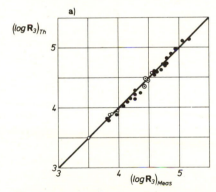

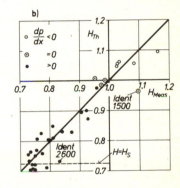

Fig. 22.9. Turbulent boundary layer data taken from 33 sets of measurements with different velocity distributions in the free stream; plotted points refer to end station at x_N. Measurements (subscript $Meas$) after [54]. Theory (subscript Th) as in eqns. (22.22) and (22.25). a) Reynolds number R_3; b) shape factor H

assumption regarding the coefficient c_D from eqn. (22.8b) for dissipated work is only conditionally valid because it describes merely its variation with the local Reynolds number and shape factor. A more accurate calculation would have to include the effect of the upstream portion of the boundary layer on c_D (*cf.* here the investigation in [86]).

In cases when the external velocity can be assumed to be proportional to a power of x, say $U(x) \sim x^p$ with p = const, the application of our method becomes very simple. Let us assume that the turbulent boundary layer starts at $x = 0$ without a laminar inlet portion so that the constants of integration in eqns. (22.22a) and

(22.26) vanish. The required integrals can be written in closed form, and, according to [114], we obtain

$$R_3(x) = \left(\frac{\beta}{r}\frac{U\,x}{\nu}\right)^{1/(1+b)} \tag{22.28a}$$

and

$$H(x) = (s/r)^{1/c}, \tag{22.28b}$$

with $b = 0\cdot152$, $\bar{\beta} = 2\,(1 + b)\,\beta_\infty = 0\cdot0127$, $c = 4\cdot0$, $r = 1 + (3 + 2\,b)\,p$ and $s = 1 + 2\,(1 + b)\,p$. For a given value of p the shape factor is $H(x) = \text{const}$. This means that for $U(x) \sim x^p$ we are dealing with a self-similar solution (equilibrium boundary layer). The case $p = 0$ represents a flat plate at zero incidence with $U(x) = U_\infty = \text{const}$.

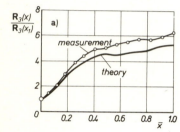

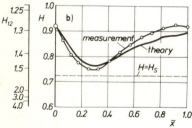

Fig. 22.10. Turbulent boundary layer on a body of revolution with initially strong pressure rise and transition to constant pressure [54]; case Ident 4000: $\bar{x} = (x - x_1)/(x - x_N)$, where x_1 = initial measuring station (start of measurement), x_N = final measuring station (end of measurement). Measured points by Moses (case 5) Theory (full line) after eqns. (22.1) and (22.2). a) Reynolds number R_3; b) shape factor H_{12}

6. Remarks on the behaviour of turbulent boundary layers in the presence of a pressure gradient. The application of the method described in Chap. XXIIb4 to turbulent boundary layers leads us to the calculation of the variation along the flow of the Reynolds number $R_3(x)$ formed with the energy thickness $\delta_3(x)$, and of that of the modified shape factor $H(x)$. Additional quantities pertaining to the boundary layer can be obtained by adding the relations depicted in Figs. 22.6 and 22.7.

The Reynolds stresses do not change much along streamlines in relatively short turbulent boundary layers in the presence of strong pressure gradients. R. G. Deissler [25] demonstrated that the assumption of a constant shearing stress can lead to good agreement between calculation and measurement; he also succeeded in calculating heat-transfer coefficients for turbulent boundary layers [26] by the use of the same method.

Boundary layer thickness: When the values of $H(x)$ are known, the diagram in Fig. 22.6 yields the relation $H_{12}(x) = H_{12}[H(x)]$ and $H_{23}[H(x)]$. In turn, employing the definitions given in eqns. (22.3b, c), we can calculate the displacement thickness and the momentum thickness

$$\delta_1(x) = H_{12}(x)\,H_{23}(x)\,\delta_3(x)\,, \tag{22.29a}$$

$$\delta_2(x) = H_{23}(x)\,\delta_3(x)\,, \tag{22.29b}$$

respectively. For *equilibrium boundary layers* for which $H_{12}(x) = $ const and $H_{23}(x) = $ const, we obtain

$$\delta_1(x) \sim \delta_2(x) \sim \delta_3(x) \sim x^{(1-bp)/(1+b)},$$

as seen from eqn. (22.28a).

Total drag: The form drag of a body in a stream consists of skin friction and pressure drag. The skin friction is the integral of shearing stresses taken over the surface of the body. Even in cases without separation it is necessary to add the *pressure drag* to skin friction. The origin of the pressure drag lies in the fact that the boundary layer exerts a displacement action on the external stream. The streamline of the potential flow are displaced from the contour of the body by an amount equal to the displacement thickness. This modifies somewhat the pressure distribution on the body surface. In contrast with potential flow (d'Alembert's paradox), the resultant of this pressure distribution modified by friction no longer vanishes but produces a pressure drag which must be added to skin friction. The two together give *form drag*. The calculation of form drag which is determined by the momentum thickness at the trailing edge will be discussed in detail in Chap. XXV.

Non-separating boundary layers: The pressure drag remains small only if separation can be avoided. This can be achieved by the proper design of the *shape* of the body. The self-similar laminar flows discussed in Chaps. VIII and IX afford examples of flows which do not lead to separation in the presence of an adverse pressure gradient. When the external flow follows the power law $U(x) \sim x^p$, separation occurs in laminar flow for values of $p_S \leqslant -0.09$. The corresponding value in turbulent flow is obtained from eqn. (22.28b) by substituting in it $H = H_S \leqslant 0.723$. This gives $p_S \leqslant -0.27$, whereas A. A. Townsend indicates the value $p_S \leqslant -0.234$. This signifies that a turbulent boundary layer can sustain a considerably larger adverse pressure gradient without separating than does a laminar boundary layer. Self-similar solution give a hint on how to arrange the *pressure distribution* in order to sustain the largest possible adverse pressure gradient without separation. A pressure distribution that starts with a large and continues with a decreasing adverse pressure gradient generates a thinner boundary layer and makes it possible to sustain a large total pressure increase than a uniform gradient would. This fact was confirmed experimentally by G. B. Schubauer and W. G. Spangenberg [97] and by B. S. Stratford [104]. A critical review of different methods of calculating the position of the point of separation is contained in [17].

Re-attaching boundary layers: More recent contributions concerning the particularly interesting case when a separated shear layer re-attaches itself to the wall and develops further as a boundary layer in the downstream direction are contained in the paper by P. Bradshaw and F. Y. F. Wong [14] as well as P. Wauschkuhn and V. Vasanta Ram [117]. The discussion relates to a boundary layer which has separated at a backward-facing step. The essential difference between such a boundary layer and a "normal" boundary layer, for example on a flat plate or an aerofoil, consists in the fact that its turbulence structure has become strongly disturbed by the prior separation. Such a perturbation in structure makes it very difficult to formulate a procedure for calculation. P. Wauschkuhn and V. Vasanta Ram [117] report measurements of wall shear stress, mean-velocity distribution and Reynolds stress in the re-attached layer and describe comparisons with several evaluation procedures.

7. Turbulent boundary layers with suction and injection. The possibility of influencing the flow in a boundary layer by blowing or suction is of some practical importance, particularly with a view to increasing the maximum lift of aerofoils. The procedure for calculating laminar boundary layers with suction was given in Sec. XIVb; the corresponding method for a turbulent boundary layer was discussed in Sec. XXIa.

A procedure for the calculation of a turbulent boundary layer with homogeneous suction and blowing on a flat plate at zero incidence was first formulated by H. Schlichting [90]. Experimental investigations and a comparison between them and theory were discussed in Sec. XXIa. The preceding procedures were extended by W. Pechau [75] and R. Eppler [32] to include the case of an arbitrary velocity distribution $-v_0(x)$ of suction velocity. The results obtained by these methods are discussed in [92, 94]. They contain further calculations performed with the aid of this procedure; they illustrate the effect of the magnitude and position of the suction zone on the minimum suction flow required to eliminate separation on aerofoils. It turns out that the optimum arrangement is to concentrate the suction zone in a narrow region on the suction side of the aerofoil and to place it at a short distance behind the nose. This is understandable, because the largest local adverse pressure gradients occur in that region when the angles of incidence are large. The required minimum suction rates, as described by the suction coefficients $c_{Q,\,min}$, are of the order of 0·002 to 0·004. A. Raspet [78] performed flight measurements on wings provided with suction at the nose.

Another effective method to *increase maximum lift*, particularly in wings with a large flap-deflection angle, consists in the *injection* of a thin jet of air of large velocity close to the nose of the flap, Fig. 22.11. This device imparts a considerable amount of energy to the turbulent boundary layer and causes it to adhere to the wing. The gain in lift achieved by this method can be estimated by comparing the pressure distributions of the flap wing with and without separation, respectively. According to J. Williams [122], the effectiveness of the jet can be judged with reference to the dimensionless momentum coefficient

$$c_\mu = \frac{\varrho_j\,v^2_j\,s}{\frac{1}{2}\,\varrho\,U_\infty{}^2\,l}\,,\qquad(22.30)$$

where v_j denotes the velocity of the jet and s represents its width. F. Thomas [109, 110] performed extensive measurements on the effectiveness of injection for the increase in the lift of flap wings. He was also able to formulate a procedure which allows us to calculate the value of the momentum coefficient required to avoid separation by injection through a slit into a turbulent boundary layer. In addition, F. Thomas [109] performed detailed measurements in the turbulent boundary layer behind an injection slit. Similarly, investigations were performed by P. Carrière and E. A. Eichelbrenner[16] on the question of the return of a separated boundary layer in a large adverse pressure gradient through the application of a tangential jet.

H. Schlichting [91] gave a short summary of investigations into the problem of increasing the maximum lift of wings by suitably controlling the boundary layer.

If a different gas is injected into a turbulent boundary layer, we are again faced with a *binary layer*, as was the case with laminar flow (Sec. XIV c), in which the concentration varies throughout the flow field. Various physical hypotheses have been proposed in order to be in a position to analyze the process of injection into a turbulent boundary layer. D. L. Turcotte [115] assumes that the process of mixing is essentially complete in the laminar sublayer and derives in this manner an approximate formula for the shearing stress at the wall for the case of an incompressible fluid. The formula was extended to include compressible boundary layers; its form is:

$$\frac{\tau_w}{\tau_{w0}} = \exp\left\{-6\cdot94\left(\frac{\varrho_w}{\varrho_1}\,\frac{v_w}{u_1}\,\sqrt{\frac{2\,\varrho_w}{\varrho_1\,c_{f0}}}\right)\left(1+\sqrt{\frac{\tau_{w0}}{\tau_w}}\right)\right\}$$

In this equation, the subscript w refers to the wall, the subscript 0 relates to the case without injection and the subscript 1 describes the free stream. The validity of the preceding equation has been confirmed by measurements performed by several authors on plates and cones at Mach numbers ranging from 0 to 4·3.

Extensive measurements on the effect of the injection of an other gas on the shearing stress at the wall in boundary layers formed on cones in compressible flow have been reported upon by C. C. Pappas and A. F. Okuno [73].

M. W. Rubesin and C. C. Pappas [88] proposed a mixing-length theory for the calculation of the effects of the injection of a foreign gas into a turbulent boundary layer. This was applied to the calculation of the rate of heat transferred from the wall, and the corresponding results for the injection of helium and hydrogen are shown in Fig. 22.12; they have been plotted along with experimental results for comparison. The latter show an even larger decrease in heat transfer rates than predicted by the theory. By contrast, the recovery factor seems to be affected but little by the injection of a lighter gas, in a turbulent as well as in a laminar boundary layer.

Experiments in which a heavy gas (freon) was blown into a turbulent boundary layer of air yielded approximately identical velocity profiles as those in which air was discharged, even though the density ratio of the gases between the wall and outer edge of the boundary layer was as high as 4. Except for the case of an adverse pressure gradient or of very vigorous blowing, the phenomena can be described quite well with the aid of Prandtl's mixing-length theory.

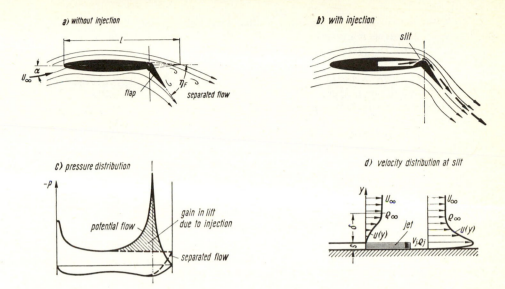

Fig. 22.11. Flat wing with injection through a slit at the nose of the flap for the purpose of increasing maximum lift; a) separated flow, without injection; b) adhering flow with injection; c) pressure distribution; d) velocity distribution in the boundary layer

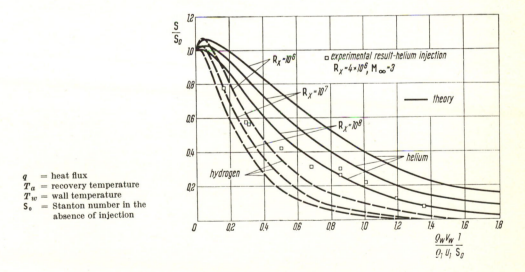

q = heat flux
T_a = recovery temperature
T_w = wall temperature
S_0 = Stanton number in the absence of injection

Fig. 22.12. Heat-transfer rates for a binary boundary layer on a flat plate at zero incidence with the injection of hydrogen or helium into air in a turbulent boundary layer, after M. W. Rubesin and C. C. Pappas [88]. Comparison between theory and measurement for the Stanton number $S = q/\varrho_1 u_1 c_{p1} (T_a - T_w)$

8. Boundary layers on cambered walls. Two-dimensional boundary layers on curved walls have been investigated by H. Wilcken [121] (see also A. Betz [4]). If the wall is *concave* the faster particles are pressed against it by centrifugal forces and slower particles are deflected away from it. Thus the process of turbulent mixing which takes place between faster and slower fluid particles is accentuated and the intensity of turbulence is increased. The reverse is true of *convex* walls in the neighbourhood of which the faster particles are forced away from the wall, the slower particles being pressed towards it, and turbulent mixing is impeded. Consequently with equal pressure gradients, the thickness of a turbulent boundary layer on a concave wall is greater than, and that on a convex wall is smaller than, the thickness on a flat plate. H. Schmidbauer [96] extended Gruschwitz's method to include the case of convex walls. Further results were provided by G. L. Mellor [101a, 101b] and R. N. Meroney and P. Bradshaw [65a] and B. R. Ramaprian and B. G. Shivaprasad [77a].

c. Turbulent boundary layers on aerofoils: maximum lift

A very comprehensive survey of the problem of high-lift of aerofoils has recently been given by A. M. O. Smith [101]. In the following, we propose to deal with the theoretical aspects of calculating the maximum lift of aerofoils.

It is well known that the maximum lift of an aerofoil is associated with the separation of the boundary layer on its suction side. Thus the theoretical prediction of the maximum lift must deal with the pressure distribution of an aerofoil section with partly separated flow and with the interaction between this pressure distribution and the boundary layer. This problem has been attacked by K. Jacob [47]; see also the summary article by G. K. Korbacher [55]. Figure 22.13 refers to a profile at the rather large angle of incidence of $\alpha = 10.7°$, and presents some theoretical and experimental results for the pressure distribution. The pressure distributions (a) and (b) for the two Reynolds numbers, $R = 0.4 \times 10^5$ and 4.2×10^5, differ considerably: for the low Reynolds number the flow on the suction side of the profile is nearly fully separated; at the higher Reynolds number, the flow is only partly separated, S being the point of separation. Both pressure distributions are characterized by a rather long stretch of nearly constant pressure on the suction side of the aerofoil. In the separated aera in terms of the potential flow theory, these pressure distributions are calculated by assuming that there exists a region of "dead air" on the suction side with approximately constant pressure at its boundaries. With a surface singularity method such a region can be simulated by an outflow region produced by a certain distribution of sources on the aft part of the suction side of the profile. Realizing this, the main problem now is to determine how the Reynolds number influences separation. This is achieved with the aid of boundary-layer theory in the following way: in the potential-flow calculation the location of the point of separation is treated as a free parameter. The determination of this parameter is achieved by combining the calculation of the pressure distribution of the potential flow with separation with the calculation of the laminar or turbulent boundary layer generated by this pressure distribution. An "adequate flow" demands that the point of separation of the boundary layer must coincide with the point of separation of the potential flow with a dead-air region; the required result is achieved by iteration. In this way the point of separation can be located. The calculation brings to bear the influence of the Reynolds number, because the location of the point of separation of a turbulent bound-

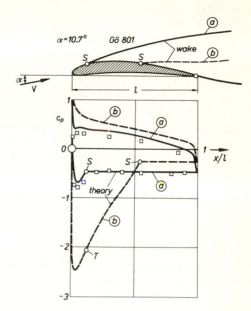

(a) R = 0·4 × 10⁵, α = 10·7°
$\qquad\qquad\quad$ α = 8°
(b) R = 4·2 × 10⁵, α = 10·7

S = separation;
T = transition

Fig. 22.13. Pressure distribution on an aerofoil in separated flow, after K. Jacob [47], at two different Reynolds numbers $R = Vl/\nu$

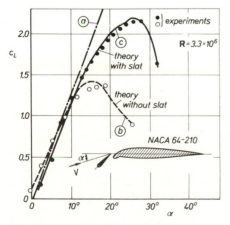

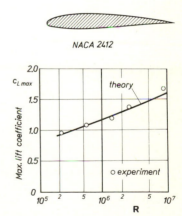

Fig. 22.14. Lift coefficient c_L against angle of incidence α for an aerofoil with a slat. Theory by K. Jacob and D. Steinbach [48], measurements by W. Baumert [3]

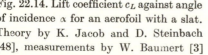

(a) theory, assuming inviscid flow *with* slat
(b) theory, assuming viscous flow *without* slat
(c) theory, assuming viscous flow *with* slat

Fig. 22.15. Maximum lift coefficient of an aerofoil $c_{L\,max}$ as a function of Reynolds number, after K. Jacob [47]

ary layer depends on the Reynolds number. Figure 22.13 shows that for the pressure distribution of the profile Gö 801 there exists rather good agreement between experiment and the theory under consideration.

The theory was extended to multi-element aerofoil systems with separation [48]. Additional results, especially on the lift, are presented in Fig. 22.14. The diagram demonstrates that the curve of the lift coefficient versus the angle of incidence $C_L(\alpha)$, and especially the maximum lift coefficient C_{Lmax}, for an aerofoil NACA 64-210 with a slat is considerably improved by the slat. The agreement between theory and experiment is quite satisfactory here, too. Finally, Fig. 22.15 shows the dependence of the maximum lift coefficient, C_{Lmax}, of the profile NACA 2412 on the Reynolds number, R. The increase in the maximum lift coefficient with increasing Reynolds number, which is oberved in experiments, is well confirmed by the theory.

Calculations of maximum lift of wings in *laminar* flow have been performed by G. H. Goradia et al. [37, 38].

d. Three-dimensional boundary layers

General remarks: The physical nature of a three-dimensional boundary layer is characterized by the fact that the direction of the velocity in the interior of the boundary layer deviates considerably from that in the outer flow. This is brought about by a pressure gradient that acts at an angle to the main flow. As a result, there occur vigorous secondary motions, *cf.* Fig. 11.1 in Chap. XI. A good example of such a flow pattern is contained in the measurements performed by R. C. Sachdeva and J. H. Preston [89] in the boundary layer on a ship's hull.

There exists a summary account describing the calculation of three-dimensional, incompressible boundary layers prepared by J. C. Cooke and M. G. Hall [23]; it deals predominantly with laminar boundary layers. A comprehensive monograph on turbulent three-dimensional boundary layers was published by J. F. Nash and V. C. Patel [70]. The analytic calculation of a general case, for example that of the boundary layers on swept or delta wings, is still very difficult, even though numerous proposals of such methods exist. Here we may mention, for example, the work of N. A. Cumpsty and M. R. Head [24], J. C. Cooke [22], P. Bradshaw [7], L. F. East [29], R. Michel et al. [66], and A. Elsenaar and B. van den Berg [31] and F. M. White et al. [118a]. The present status of research in this field was reviewed by Fanneloep at a symposium held in Trondheim in 1975 [30a]. In what follows, we shall describe several simpler examples of three-dimensional turbulent boundary layers. The state of the theory is, however, still unsatisfactory†.

1. Boundary layers on bodies of revolution. C. B. Millikan [67] was the first to calculate a turbulent boundary layer on a body of revolution, the method having been based on the momentum integral equation. The relevant momentum equation was given in eqn. (11.39). Using our present notation, we can write it as

$$\frac{d\delta_2}{dx} + \delta_2 \left(\frac{2 + H_{12}}{U} \frac{dU}{dx} + \frac{1}{R} \frac{dR}{dx} \right) = \frac{\tau_0}{\varrho\, U^2} \,. \tag{22.31}$$

Here $R(x)$ denotes the radius of the local cross-section of the body of revolution.

---

† "It is curious and depressing that the more sophisticated the theory, the poorer is the agreement with experiments" (F. M. White [119] p. 549).

At the aft portion of a body of revolution the two derivatives, dU/dx and dR/dx, become negative. It follows from the preceding equation that the momentum thickness $\delta_2(x)$ increases and becomes very large there. This may create circumstances which nullify the main assumption of boundary-layer theory, namely that $\delta_2 \ll R$. As a consequence, the calculation near the butt of the body of revolution may become erroneous and the position of the region of separation cannot be determined reliably. According to F.M. White [119], equation (22.31) remains usable when the local Reynolds number satisfies the condition that

$$\frac{U(x)\,R(x)}{\nu} > 1000\,.$$

P.S. Granville [39] formulated a multiparameter procedure for the calculation of turbulent boundary layers on rotationally symmetric bodies placed in an axially directed stream. The method hinges on the calculation of momentum thickness and of a shape factor and can be used for the aft portion of the body where the boundary layer thickness is of the same order of magnitude as the local radius of the body.

In a manner similar to that used for two-dimensional boundary layers, Truckenbrodt [111, 114] was able to show that the use of the energy integral equation leads to an explicit integral formula for the calculation of the *energy thickness*. If x denotes the current arc length measured along a meridian, and $R(x)$ the radius of a section normal to the axis of symmetry, then the extension of eqn. (22.22a) for the Reynolds number formed with the energy thickness can now be written

$$\mathsf{R}_3(x) = \left[\frac{1}{\nu'} \, \frac{E_3(x_1) + \int_{x_1}^{x} R^{1+b}\,U^{3+2b}\,dx}{\{R(x)\}^{1+b}\,\{U(x)\}^{2(1+b)}} \right]^{1/(1+b)} \tag{22.32}$$

The numerical constants b and ν' should be taken from eqn. (22.22b) and the constant of integration is

$$E_3(x_1) = \nu'\,[R(x)\,\{U(x_1)\}^2\,\mathsf{R}_3(x_1)]^{1+b}\,.$$

In the more recent formulation [114], the equation for the modified *shape factor* in the axially symmetric case contains the function describing the variation of the body radius. This is in contrast with the earlier formulation [111] according to which the modified shape factor was the same for bodies of revolution and two-dimensional bodies. The generalized form of eqn. (22.25) is now

$$H(x) = U(x)\,G(x)\,\{N(x)\}^{-1/c}\,, \tag{22.32a}$$

where the *influence functions* for the radius and external velocity distributions are

$$G(x) = G(x_1) + \int_{x_1}^{x} R^{1+b}\,U^{2\,(1+b)}\,dx\,; \qquad N(x) = N(x_1) + c \int_{x_1}^{x} R^{1+b}\,U^{2\,(1+b)+c}\,G^{c-1}\,dx\,.$$

The constants of integration are

$$\left. \begin{aligned} G(x_1) &= \nu'\,[H(x_1)\,\{R(x_1)\}^{1+b}\,\{U(x_1)\}^{1+2b}\,\{\mathsf{R}_3(x_1)\}^{1+b}]; \\ N(x_1) &= [U(x_1)\,G(x_1)/H(x_1)]^c\,. \end{aligned} \right\} \tag{22.33}$$

The numerical constants follow from eqn. (22.26b).

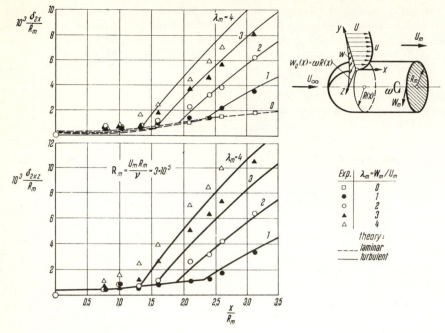

Fig. 22.16. Momentum thicknesses δ_{2x} and δ_{2xz} on a rotating body of revolution placed in an axial stream, after O. Parr [74]

δ_{1x} and δ_{2x} from eqn. (11.50) rotation parameter $\lambda_m = W_m/U_m$
Reynolds number $R = U_m R_m/\nu = 3 \times 10^5$

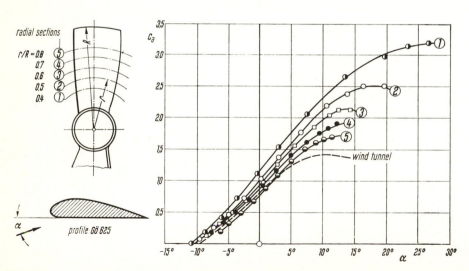

Fig. 22.17. Local lift coefficients, c_a, at various radial sections on a rotating propeller according to measurements performed by H. Himmelskamp [44]

The diagrams in Fig. 22.10 show a comparison between theory and measurement in a flow past an axially symmetric body; the diagrams plot the Reynolds number formed with the energy thickness and the modified shape factor.

In order to take into account corrections due to three-dimensionality caused by the possible convergence or divergence of streamlines, J.C. Rotta [86] proposes to base the calculation on an effective radius $R(x)$. Numerical values for $R(x)$ are summarized in [86] for all measurements catalogued in [54]; compare here the measurements by W.W. Willmarth et al. [122a] and A.M.O. Smith [104a].

2. Boundary layers on rotating bodies. The calculation of laminar boundary layers on rotating bodies placed in an axial stream was discussed in Sec. XIc. The method of calculation which makes use of momentum integral equations, formulated for the meridional and circumferential directions respectively, has been extended by E. Truckenbrodt [112] to include the turbulent case. He was, moreover, fortunate to succeed in giving convenient integrals for the calculation of the parameters of the boundary layer. Experimental and further theoretical investigations into the boundary layer on rotating streamline bodies were carried out by O. Parr [74]. In this case, the boundary layer grows rapidly with the rotation parameter $\lambda = \omega R/U_m$; here ω denotes the angular velocity, R the largest radius of the body, and U_m is the axial reference velocity. The turbulent boundary layer on a rotating body of revolution placed in an axial stream can be calculated with the aid of the system of equations (11.45) to (11.48), in which the shearing stress must be assumed to vary with the rotation parameter. The diagram in Fig. 22.16 compares the calculated and measured values of the momentum thicknesses δ_{2x} and δ_{2xz}, as reported by O. Parr [74] for a cylindrical body provided with a spherical nose. The agreement is good. The region of transition from laminar to turbulent flow moves forward as the rotation parameter increases; its position coincides with the point at which the momentum thicknesses increase abruptly. See also Sec. XIb2.

A method for the calculation of three-dimensional boundary layers on stationary bodies as well as on rotating ones, such as propellers or blades of rotary compressors and turbines, was indicated by A. Mager [61]; comparative measurements are contained in ref. [62]. H. Himmelskamp [44] carried out measurements in the boundary layer on a rotating airscrew and determined local lift coefficients of the blade from measurements of pressure distributions. Some of his results are seen reproduced in Fig. 22.17; they are given in the form of plots of the local lift coefficient, c_a, at various radial sections, in terms of the angle of incidence, α. Corresponding measurements on a stationary blade placed in a wind tunnel are also shown for comparison. Figure 22.17 shows that markedly increased lift coefficients are obtained near the hub, and the effect can be traced to separation being delayed to larger angles of incidence. For example, the section closest to the hub has a maximum lift coefficient of 3·2 compared with 1·4 on the stationary blade. The displacement of separation towards larger angles of incidence is explained by the appearance of an additional acceleration which acts in the flow direction and which is created by Coriolis forces; it has the same effect as a favourable pressure gradient. In addition, but to a lesser extent, the centrifugal forces acting in the boundary layer carried with the blade exert a beneficial influence with respect to separation. Fluid particles in the boundary layer are acted upon by a centrifugal force which is proportional to the radius. Consequently, less fluid is transported to each blade from the center than away from it and outwards, and the boundary layer is thinner than would be the case in two-dimensional flow about the same shape. A. Betz [5] gave some theoretical arguments on this point. F. Gutsche [42] made the flow on a propeller blade visible by painting the former with a dye. Centrifugal forces also exert a large influence on the proceess of transition. H. Muesmann [68] showed in his thesis that, other things being equal, transition occurs on a rotating propeller blade at a considerably lower Reynolds number than on one which is stationary.

Fig. 22.18. Convergent and divergent boundary layers; system of
coordinates;
a) divergent, $a + x > 0$;
b) convergent, $a + x < 0$

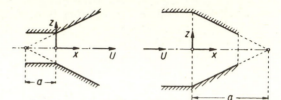

3. Convergent and divergent boundary layers.

The methods for the calculation of turbulent boundary layers which were described in Sec. XXIIb have been extended by A. Kehl [53] to include cases when the streamlines either converge or diverge sideways, Fig. 22.18. Boundary layers of this type occur in a diffuser or in a nozzle and also near the bow or the stern of a body of revolution. In this connexion the measurements due to Gruschwitz have been extended to $R = U\,\delta_2/\nu = 3 \times 10^4$ and his method of calculation has been generalized to include this case as well. Assuming a system of coordinates as shown in Fig. 22.18, x and z being chosen in the plane of the wall, y being measured at right angles to it, we notice that along the streamline which coincides with the axis, i. e. along that for which $w = 0$, the same equation of motion is satisfied as that for the two-dimensional case, eqn. (8.29). On the other hand the equation of continuity changes to

$$\frac{\partial u}{\partial x} + \frac{\partial v}{\partial y} + \frac{\partial w}{\partial z} = 0\,.$$

The momentum integral equation (22.7a) obtains an additional term which stems from the convergence or divergence of the streamlines, as the case may be. Retracing the steps in the derivation of the momentum equation, which was given in Sec. VIIIe in connexion with eqn. (8.29), we see that the integral of the second term in the first equation of motion becomes

$$\int_0^h v\,\frac{\partial u}{\partial y}\,dy = -U\,\frac{\partial}{\partial x}\int_0^h u\,dy + u\int_0^h \frac{\partial u}{\partial x}\,dy - U\int_0^h \frac{\partial w}{\partial z}\,dy + \int_0^h u\,\frac{\partial w}{\partial z}\,dy\,.$$

The two last terms on the right-hand side are due to the divergence of the flow in the z-direction. Taking into account the divergence of the streamlines, we have

$$\frac{w}{u} = \frac{z}{x + a} \qquad \text{and} \qquad \left(\frac{\partial w}{\partial z}\right)_{z=0} = \frac{u}{x + a}\,.$$

Hence the two additional terms become

$$-\frac{1}{x+a}\int\limits_0^h u\,(U-u)\,\mathrm{d}y = -\frac{\delta_2}{x+a}\,U^2\,,$$

and the momentum integral equation must be supplemented with the additional term $\delta_2\,U^2/(x+a)$. Consequently, the momentum integral equation for the plane of symmetry which replaces eqn. (22.7a) is now

$$\frac{\mathrm{d}\,\delta_2}{\mathrm{d}x} + \delta_2\left(\frac{1}{a+x} + \frac{H_{12}+2}{U}\,\frac{\mathrm{d}U}{\mathrm{d}x}\right) = \frac{\tau_0}{\varrho\,U^2}\,. \tag{22.34}$$

For divergent streamlines we have $a+x > 0$, and $a+x < 0$ corresponds to a convergent stream, Fig. 22.18. It follows at once from eqn. (22.34) that the increase in the momentum thickness proceeds at a smaller rate in the case of divergent, and at a larger rate in the case of convergent streamlines, both compared with the two-dimensional case. This result is also to be expected from physical considerations.

Boundary layer in a corner: The flow in a turbulent boundary layer formed in a rectangular corner between two flat walls was investigated theoretically and experimentally by K. Gersten [36] (see also Sec. XXIa 4). The related problem of the structure of a turbulent boundary layer at the junction of a cylindrical body and a flat plate was first considered by J.P. Johnston [49, 50, 51] and later, and more thoroughly, by H. G. Hornung and P.N. Joubert [46]; see also S. G. Rubin [88a] and M. Shafir and S. G. Rubin [99a].

References

[1] Ackeret, J.: Zum Entwurf dicht stehender Schaufelgitter. Schweiz. Bauzeit. 103 (1942).
[2] Baker, R. J., and Launder, B. E.: The turbulent boundary layer with foreign gas injection. Part I: Measurements in zero pressure gradient. Part II: Predictions and measurements in severe streamwise pressure gradients. Int. J. Heat and Mass Transfer 17, 275—306 (1974).
[3] Baumert, W., and Enghardt, K.: Dreikomponentenmessungen an einem Rechteckflügel mit Vorflügel und abgesenkter Nase. DFVLR Bericht 71—C—29 (1971).
[4] Betz, A.: Über turbulente Reibungsschichten an gekrümmten Wänden. Lectures on aerodynamics and related subjects, Aachen 1929. Verlag Springer, Berlin, 1930, 10—18.
[5] Betz, A.: Höchstauftrieb von Flügeln an umlaufenden Rädern. ZFW 9, 97—99 (1961).
[6] Bienert, P.: Strömungsbild einer turbulenten Ablösung. ZFW 16, 141—147 (1968).
[7] Bradshaw, P.: Calculation of three-dimensional turbulent boundary layers. JFM 46, 417—445 (1971).
[8] Bradshaw, P.: The understanding and prediction of turbulent flow. Aero. J. 76, 413—418 (1972).
[9] Bradshaw, P.: Turbulence research — progress and problems. Proc. of the 1976 Heat Transfer and Fluid Mech. Institute (A. A. McKillop, ed.), Stanford Univ. Press, 1976.
[10] Bradshaw, P. (ed.): Turbulence. Springer Verlag, Berlin/Heidelberg/New York, 1976.
[11] Bradshaw, P., and Ferriss, D. H.: Calculation of boundary layer development using the turbulent energy equations: Compressible flow on adiabatic walls. JFM 46, 83—110 (1971).
[12] Bradshaw, P., and Ferriss, D. H.: Applications of a general method of calculating turbulent shear layers. J. Basic Eng., Trans. ASME Series D, 94, 345—354 (1972).
[13] Bradshaw, P., Ferriss, D. H., and Atwell, N. P.: Calculation of boundary layer development using the turbulent energy equation. JFM 28, 593—616 (1967).
[14] Bradshaw, P., and Wong, F. Y. F.: The reattachment and relaxation of a turbulent shear layer. JFM 52, 113—135 (1972).
[15] Buri, A.: Eine Berechnungsgrundlage für die turbulente Grenzschicht bei beschleunigter und verzögerter Strömung. Diss. Zürich 1931.

[16] Carrière, P., and Eichelbrenner, E.A.: Theory of flow reattachment by a tangential jet discharging against a strong adverse pressure gradient. Boundary layer and flow control (G.V. Lachmann, ed.), Vol. *I*, 209—231, 1961.

[17] Cebeci, T., Mosinskis, G.J., and Smith, A.M.O.: Calculation of separation points in incompressible turbulent boundary layers. J. Aircr. *9*, 618—624 (1972).

[18] Cebeci, T. and Smith, A.M.O.: A finite-difference solution of the incompressible turbulent boundary layer equations by an eddy viscosity concept. AFOSR-IFP, Stanford Conference on Computation of Turbulent Boundary Layers, Vol. *I*, 346—355 (1968).

[19] Cebeci, T., and Smith, A.M.O.: A finite-difference method for calculating compressible laminar and turbulent boundary layers. J. Basic Eng., Trans. ASME, Series D, *92*, 523—535 (1970).

[20] Cebeci, T., and Smith, A.M.O.: Analysis of turbulent boundary layers. Academic Press, New York, 1974.

[21] Clauser, F.H.: Turbulent boundary layers in adverse pressure gradients. JAS *21*, 91—108 (1954).

[21a] Clauser, F.H.: The turbulent boundary layer. Adv. Appl. Mech. *4*, 1—51 (1965).

[22] Cooke, J.C.: Boundary layers over infinite yawed wings. Aero. Quart. *11*, 333—347 (1960).

[23] Cooke, J.C., and Hall, M.G.: Boundary layer in three dimensions. Progress in Aeronautical Sciences *2*, 222—282 (1962).

[24] Cumpsty, N.A., and Head, M.R.: The calculation of three-dimensional turbulent boundary layers. Part I: Flow over the rear of an infinite swept wing. Aero. Quart. *18*, 55—84 (1967). Part II: Attachment-line flow on an infinite swept wing. Aero. Quart. *18*, 150—164 (1967). Part III: Comparison of attachment-line calculations with experiment. Aero. Quart. *20*, 99—113 (1969). Part IV: Comparison of measurements with calculations on the rear of a swept wing. Aero. Quart. *21*, 121—132 (1970).

[25] Deissler, R.G.: Evolution of a moderately short turbulent boundary layer in a severe pressure gradient. JFM *64*, 763—774 (1974).

[26] Deissler, R.G.: Evolution of the heat transfer and flow in moderately short turbulent boundary layers in severe pressure gradients. J. Heat and Mass Transfer *17*, 1079—1085 (1974).

[27] von Doenhoff, A.E., and Tetervin, N.: Determination of general relations for the behavior of turbulent boundary layers. NACA Rep. 772 (1943).

[28] Dönch, F.: Divergente und konvergente Strömungen mit kleinen Öffnungswinkeln. Diss. Göttingen 1925. Forschungsarbeiten VDI No. 292 (1926).

[29] East, L.F.: Measurements of the three-dimensional incompressible turbulent boundary layer on the surface of a slender delta wing by the leading edge vortex. ARC RM 3768 (1973).

[30] East, L.F., and Hoxey, R.P.: Low-speed three-dimensional turbulent boundary layer data, Part I. RAE Techn. Rep. 69041 (1969).

[30a] East, L.F. (ed.): Computation of three-dimensional boundary layers. Symposium Euromech 60, Trondheim, 1975, FFA TN AE 1211 (1975). See article by Fannelöp, T.K., and Kragstad, P.A.: Three-dimensional turbulent boundary layers in external flows. Also JFM *71*, 815—826 (1975).

[31] Elsenaar, A., van den Berg, B., and Lindhout, J.F.P.: Three-dimensional separation of an incompressible turbulent boundary layer on an infinite swept wing. AGARD Conf. Proc. No. 168, Flow Separation, 34—1 to 34—15 (1975).

[32] Eppler, R.: Praktische Berechnung laminarer und turbulenter Absauge-Grenzschichten. Ing.-Arch. *32*, 221—245 (1963).

[33] Fernholz, H.H.: Halbempirische Gesetze zur Berechnung turbulenter Grenzschichten nach der Methode der Integralbedingungen. Ing.-Arch. *33*, 384—395 (1964).

[34] Fernholz, H.H.: Experimentelle Untersuchung einer inkompressiblen turbulenten Grenzschicht mit Wandreibung nahe Null in einem längsangeströmten Kreiszylinder. ZFW *16*, 401—406 (1968).

[35] Garner, H.C.: The development of turbulent boundary layers. ARC RM 2133 (1944).

[36] Gersten, K.: Corner interference effects. AGARD Rep. No. 299 (1959).

[37] Goradia, S.H., and Colwell, G.T.: Analysis of high-lift wing systems. Aero. Quart. *26*, 88—108 (1975).

[38] Goradia, S.H., and Lyman, V.: Laminar stall prediction and estimation of $C_{L\,max}$. J. Aircr. *11*, 528—536 (1974).

[39] Granville, P.S.: Similarity-law entrainment method for thick axisymmetric turbulent boundary layers in pressure gradients. David Taylor Naval Ship Research and Development Center, Bethesda, MD. Rep. No. 4525 (1975).

[40] Gruschwitz, E.: Die turbulente Reibungsschicht in ebener Strömung bei Druckabfall und Druckanstieg. Ing.-Arch. *2*, 321—346 (1931); summary in ZFW *23*, 308 (1932).

[41] Gruschwitz, E.: Turbulente Reibungsschichten mit Sekundärströmungen. Ing.-Arch. *6*, 355—365 (1935).

[42] Gutsche, F.: Versuche an umlaufenden Flügelschnitten mit angerissener Strömung. Jb. Schiffbautechn. Ges. *41*, 188—226 (1940).

[43] Head, M. R.: Entrainment in the turbulent boundary layer. ARC RM 3152 (1960).

[44] Himmelskamp, H.: Profiluntersuchungen an einem umlaufenden Propeller. Diss. Göttingen 1945. Max-Planck-Inst. für Strömungsforschung, Göttingen, Rep. No. 2 (1950).

[45] Hochschild, H.: Versuche über Strömungsvorgänge in erweiterten und verengten Kanälen. Forschungsarbeiten VDI No. 114 (1910).

[46] Hornung, H. G., and Joubert, P. N.: The mean velocity profile in three-dimensional turbulent boundary layers. JFM *15*, 368—384 (1963).

[47] Jacob, K.: Berechnung der abgelösten inkompressiblen Strömung um Tragflügelprofile und Bestimmung des maximalen Auftriebs. ZFW *17*, 221—230 (1969).

[48] Jacob, K., and Steinbach, D.: A method for prediction of lift for multi-element airfoil systems with separation. AGARD CP 143, V/STOL-Aerodynamics, 12—1 to 12—16 (1974).

[49] Johnston, J. P.: On the three-dimensional turbulent boundary layer generated by secondary flow. Trans. ASME, Ser. D, J. Basic Eng. *82*, 233—248 (1960).

[50] Johnston, J. P.: The turbulent boundary layer at a plane of symmetry in a three-dimensional flow. Trans. ASME, Ser. D, J. Basic Eng. *82*, 622—628 (1960).

[51] Johnston, J. P.: Measurements in a three-dimensional turbulent boundary layer induced by a forward facing step. JFM *42*, 823—844 (1970).

[52] Johnston, J. P., and Wheeler, A. J.: An assessment of three-dimensional turbulent boundary layer prediction methods. Trans. ASME, Ser. I, J. Fluids Eng. *95*, 415—421 (1973).

[53] Kehl, A.: Untersuchungen über konvergente und divergente turbulente Reibungsschichten. Diss. Göttingen 1942; Ing.-Arch. *13*, 293—329 (1943).

[54] Kline, S. J., Morkovin, M. V., Sovran, G., Cockrell, D. J., Coles, D. E., and Hirst, E. A. (eds.): Proc. AFOSR-IFP-Stanford Conference 1968. Computation of turbulent boundary layers, Vol. I and II. Stanford Univ. Press, 1969.

[55] Korbacher, G. K.: Aerodynamics of powered high-lift systems. Ann. Rev. Fluid Mech. *6*, 319—358 (1974).

[56] Kovasznay, L. S. G.: The turbulent boundary layer. Ann. Rev. Fluid Mech. (M. van Dyke, ed.) *2*, 95—112 (1970).

[57] Kröner, R.: Versuche über Strömungen in stark erweiterten Kanälen. Forschungsarbeiten VDI No. 222 (1920).

[58] Laufer, J.: Investigation of turbulent flow in a two-dimensional channel. NACA Rep. 1053 (1951).

[59] Ludwieg, H.: Ein Gerät zur Messung der Wandschubspannung turbulenter Reibungsschichten. Ing.-Arch. *17*, 207—218 (1949).

[60] Ludwieg, H., and Tillmann, W.: Untersuchungen über die Wandschubspannung in turbulenten Reibungsschichten. Ing. Arch. *17*, 288—299 (1949); summary of both papers ZAMM *29*, 15—16 (1949). Engl. transl. in NACA TM 1285 (1950).

[61] Mager, A.: Generalization of boundary-layer momentum-integral equations to three-dimensional flows including those of rotating system. NACA Rep. 1067 (1952).

[62] Mager, A., Mahoney, I. I., and Budinger, R. E.: Discussion of boundary layer characteristics near the wall of an axial-flow compressor. NACA Rep. 1085 (1952).

[63] Mellor, G. L.: The effects of pressure gradients on turbulent flow near a smooth wall. JFM *24*, 255—274 (1966).

[64] Mellor, G. L., and Gibson, D. M.: Equilibrium turbulent boundary layers. JFM *24*, 225—253 (1966).

[65] Mellor, G. L., and Herring, H. J.: A survey of the mean turbulent field closure methods. AIAA J. *11*, 590—599 (1973).

[65a] Meroney, R. N., and Bradshaw, P.: Turbulent boundary layer growth over a longitudinally curved surface. AIAA J. *13*, 1448—1453 (1975).

[66] Michel, R., Quemard, C., and Cousteix, J.: Méthode pratique de prévision des couches limites turbulentes bi- et tri-dimensionelles. Recherche Aérosp. No. 1972—1, 1—14 (1972).

[67] Millikan, C. B.: The boundary layer and skin friction for a figure of revolution. Trans. ASME, J. Appl. Mech. *54*, 2—29 (1932).

[68] Muesmann, H.: Zusammenhang der Strömungseigenschaften des Laufrades eines Axial-gebläses mit denen eines Einzelflügels. Diss. Braunschweig 1958; ZFW 6, 345—362 (1958).

[69] Nash, J.F.: The calculation of three-dimensional turbulent boundary layers in incompressible flow. JFM 37, 625—242 (1969).

[70] Nash, J.F., and Patel, V.C.: Three-dimensional turbulent boundary layers. S.B.C. Technical Books (Scientific & Business Consultants, Inc., Atlanta, Georgia), 1972.

[71] Nikuradse, J.: Untersuchungen über die Strömungen des Wassers in konvergenten und divergenten Kanälen. Forschungsarbeiten VDI No. 289 (1929).

[72] Orzag, S.A., and Israeli, M.: Numerical simulation of viscous incompressible flows. Ann. Rev. Fluid Mech. 6, 281—318 (1974).

[73] Pappas, C.C., and Okuno, A.F.: Measurements of skin friction of the compressible turbulent boundary layer on a cone with foreign gas injection. JASS 27, 321—331 (1960).

[74] Parr, O.: Untersuchungen der dreidimensionalen Grenzschicht an rotierenden Drehkörpern bei axialer Anströmung. Diss. Braunschweig 1962; Ing.-Arch. 32, 393—413 (1963).

[75] Pechau, W.: Ein Näherungsverfahren zur Berechnung der ebenen und rotationssymmetrischen turbulenten Grenzschicht mit beliebiger Absaugung oder Ausblasung. Jb. WGL 1958, 82—92 (1959).

[76] Polzin, J.: Strömungsuntersuchungen an einem ebenen Diffusor. Ing.-Arch. 11, 361—385 (1940).

[77] Pretsch, J.: Zur theoretischen Berechnung des Profilwiderstandes. Jb. dt. Luftfahrtforschung I, 61—81 (1938).

[77a] Ramaprian, B.R., and Shivaprasad, B.G.: Mean flow measurements in turbulent boundary layers along midly curved surfaces. AIAA J. 15, 189—196 (1977).

[78] Raspet, A., Cornish, J.J., and Gryant, G.D.: Delay of the stall by suction through distributed perforations. Aero. Eng. Rev. 11, 6, 52—60 (1952).

[79] Reynolds, W.C.: A morphology of the prediction methods (of turbulent boundary layers). Article in [54] Vol. I, pp. 1—15 (1969).

[80] Reynolds, W.C.: Recent advances in the computation of turbulent flow. Advances in Chemical Engineering 8, 193—246 (1974), ed. by T.B. Brew et al., Academic Press.

[81] Reynolds, W.C.: Computation of turbulent flow. Ann. Rev. Fluid Mech. 9, 183—208 (1976).

[82] Rotta, J.: Beitrag zur Berechnung der turbulenten Grenzschichten. Ing.-Arch. 19, 31—41 (1951) and Max-Planck-Inst. für Strömungsforschung Göttingen Rep. No. 1 (1950).

[83] Rotta, J.: Schubspannungsverteilung und Energiedissipation bei turbulenten Grenzschichten. Ing.-Arch. 20, 195—207 (1952).

[84] Rotta, J.: Näherungsverfahren zur Berechnung turbulenter Grenzschichten unter Benutzung des Energiesatzes. Max-Planck-Inst. für Strömungsforschung Göttingen Rep. No. 8 (1953).

[85] Rotta, J.: Turbulent boundary layers in incompressible flow. Progress in Aero. Sci. 2, 1—219 (1962), ed. by A. Ferri, D. Küchemann and L.H.G. Sterne, Pergamon Press, Oxford, 1962.

[86] Rotta, J.: Vergleichende Berechnungen von turbulenten Grenzschichten mit verschiedenen Dissipationsgesetzen. Ing.-Arch. 38, 212—222 (1969).

[87] Rotta, J.: Turbulente Strömungen. Stuttgart, 1972.

[88] Rubesin, M.W., and Pappas, C.C.: Analysis of the turbulent boundary-layer characteristics on a flat plate with distributed light-gas injection. NACA TN 4149 (1958).

[88a] Rubin, S.G.: Incompressible flow along a corner. JFM 26, 97—110 (1966).

[89] Sadcheva, R.D., and Preston, J.H.: investigation of turbulent boundary layers on a ship model. Schiffstechnik 23, 1—45 (1976).

[90] Schlichting, H.: Die Grenzschicht an der ebenen Platte mit Absaugung und Ausblasen. Luftfahrtforschung 19, 293—301 (1942).

[91] Schlichting, H.: Einige neuere Ergebnisse über Grenzschichtbeeinflussung. Proc. First Int. Congr. Aero. Sci. Madrid; Adv. in Aero. Sci. II, 563—586, Pergamon Press, London, 1959.

[92] Schlichting, H., and Pechau, W.: Auftriebserhöhung von Tragflügeln durch kontinuierlich verteilte Absaugung. ZFW 7, 113—119 (1959).

[93] Schlichting, H.: Three-dimensional boundary layer flow. Intern. Assoc. Hydraulic Research, IXth Congr., Dubrovnik, 1262—1290, (1961).

[94] Schlichting, H.: Aerodynamische Probleme des Höchstauftriebes. Lecture at Third Int. Congr. Aero. Sci. (ICAS) Stockholm, Sweden, 1962; ZFW 13, 1—14 (1965).

[95] Schlichting, H.: Einige neuere Ergebnisse aus der Aerodynamik des Tragflügels (Tenth Prandtl Memorial Lecture 1966). Jb. WGLR 1966, 11—32 (1967).

[96] Schmidbauer, H.: Verhalten turbulenter Reibungsschichten an erhaben gekrümmten Wänden. Diss. München 1934; see also Luftfahrtforschung *13*, 161 (1936); Engl. transl. in NACA TM 791 (1936).

[97] Schubauer, G. B., and Klebanoff, P. S.: Investigation of separation of the turbulent boundary layer. NACA Rep. 1030 (1951).

[98] Schubauer, G. B., and Spangenberg, W. G.: Forced mixing in boundary layers. JFM *8*, 10—32 (1960).

[99] Schwarz, F., and Wuest, W.: Flugversuche am Baumuster Do 27 mit Grenzschichtabsaugung zur Steigerung des Höchstauftriebes. ZFW *12*, 108—120 (1964).

[99a] Shafir, M., and Rubin, S. G.: The turbulent boundary layer near a corner. J. Appl. Mech. (Trans. ASME, Ser. E). *43*, 567—570 (1976).

[100] Shanebrook, J. R., and Summer, W. J.: A small cross flow theory for the three-dimensional compressible turbulent boundary layer on adiabatic walls. AIAA J. *11*, 950—954 (1973).

[101] Smith, A. M. O.: High-lift aerodynamics. 37th Wright Brothers Lecture 1974. J. of Aircr. *12*, 501—530 (1975).

[101a] So, R. M. C., and Mellor, G. L.: Experiments on turbulent boundary layers on concave wall. Aero. Quart. *26*, 25—40 (1975).

[101b] So, R. M. C., and Mellor, G. L.: Experiments on convex curvature effect in turbulent boundary layers. JFM *60*, 43—62 (1973).

[102] Speidel, L., and Scholz, N.: Untersuchungen über die Strömungsverluste in ebenen Schaufelgittern. VDI-Forschungsheft 464 (1957).

[103] Squire, H. B., and Young, A. D.: The calculation of profile drag of airflow. ARC RM 1838 (1938).

[104] Stratford, B. S.: Prediction of separation of the turbulent boundary layer. JFM *5*, 1—16 (1959); An experimental flow with zero skin friction throughout its region of pressure rise. JFM *5*, 17—35 (1959).

[104a] Smith, A. M. O.: Stratford's turbulent separation criterion for axially symmetric flow. ZAMP *28*, 928—938 (1977).

[105] Stüper, J.: Untersuchung von Reibungsschichten am fliegenden Flugzeug. Luftfahrtforschung *11*, 26—32 (1934); see also NACA TM 751 (1934).

[106] Szablewski, W.: Turbulente Strömung in konvergenten Kanälen. Ing.-Arch. *20*, 37—45 (1952).

[107] Szablewski, W.: Turbulente Strömungen in divergenten Kanälen (mittlerer und starker Druckanstieg). Ing.-Arch. *22*, 268—281 (1954).

[108] Szablewski, W.: Wandnahe Geschwindigkeitsverteilung turbulenter Grenzschichtströmungen mit Druckanstieg. Ing.-Arch. *23*, 295—306 (1955).

[109] Thomas, F.: Untersuchungen über die Erhöhung des Auftriebes von Tragflügeln mittels Grenzschichtbeeinflussung durch Ausblasen. Diss. Braunschweig 1961; ZFW *10*, 46—65 (1962).

[110] Thomas, F.: Untersuchungen über die Grenzschicht an einer Wand stromabwärts von einem Ausblasspalt. Abhandl. Wiss. Ges. Braunschweig *15*, 1—17 (1963).

[111] Truckenbrodt, E.: Ein Quadraturverfahren zur Berechnung der laminaren und turbulenten Reibungsschicht bei ebener und rotationssymmetrischer Strömung. Ing.-Arch. *20*, 211—228 (1952).

[112] Truckenbrodt, E.: Ein Quadraturverfahren zur Berechnung der Reibungsschicht an axial angeströmten rotierenden Drehkörpern. Ing.-Arch. *22*, 21—35 (1954).

[113] Truckenbrodt, E.: Strömungsmechanik. Springer, Berlin/Heidelberg/New York, 1968.

[114] Truckenbrodt, E.: Neuere Erkenntnisse über die Berechnung von Strömungsgrenzschichten mittels einfacher Quadraturformeln. Part I: Ing.-Arch. *43*, 9—25 (1973); Part II: Ing.-Arch. *43*, 136—144 (1974).

[115] Turcotte, D. L.: A sublayer theory for fluid injection into the incompressible turbulent boundary layer. JASS *27*, 675—678 (1960).

[116] Walz, A.: Strömungs- und Temperaturgrenzschichten. Braun, Karlsruhe, 1966.

[117] Wauschkuhn, P., and Vasanta Ram, V.: Die turbulente Grenzschicht hinter einem Ablösungsgebiet. ZFW *23*, 1—9 (1975).

[118] White, F. M.: A new integral method for analyzing the turbulent boundary layer with arbitrary pressure gradient. J. Basic Eng., Trans. ASME, Ser. D, *91*, 371—378 (1969).

[118a] White, F. M., and Lessmann, R. C.: A three-dimensional integral method for calculating incompressible turbulent skin friction. Trans. ASME, Ser. I, *97*, 550—557 (1975).

References to this chapter continued on p. 728.

CHAPTER XXIII

Turbulent boundary layers in compressible flow †

a. General remarks

It has been demonstrated in Sec. XIII a that the presence of high velocities in the boundary layer gives rise to such large temperature differences that it becomes necessary to take into account the effect of temperature on the properties of the fluid in addition to that of the changes in its volume. Beyond this, it is found that the transfer of heat plays an essential part in the behaviour of a compressible boundary layer; its presence leads to the appearance of a strong interaction between the velocity field and the temperature field‡.

1. Turbulent heat transfer. When a liquid or a gas of non-uniform temperature is caused to move turbulently, it is found that the turbulent mixing motion creates in it temperature fluctuations in addition to the more familiar velocity fluctuations. In analogy with eqn. (18.1) for velocity fluctuations, we may represent the fluctuating temperature

$$T = \bar{T} + T' \qquad (23.1)$$

in the form of the sum of a temporal average, $\bar{T}$, and a pure fluctuation, T'. These fluctuations give rise to a supplementary heat flux which is analogous to the flux of momentum evolved by the velocity fluctuations. In order to show this more clearly, we assume, as we did earlier in Sec. XVIII b, that through a surface element, dA, whose normal points in the x-direction, there flows a mass of fluid $dA \, \varrho u \, dt$ during time dt. The enthalpy of this mass per unit volume is $\varrho \, c_p \, T$, and the convective flux in the x-direction has a value $dQ_x = dA \, \varrho \, u \, c_p \, T$. If we now introduce the expression for u from eqn. (18.1) and that for T from eqn. (23.1) and form the temporal average of the heat flux, we shall obtain

$$\overline{dQ_x} = dA \, \varrho \, c_p (\bar{u} \, \bar{T} + \overline{u' \, T'}) \, .$$

It is seen that the presence of velocity and temperature fluctuations generates the supplementary heat flux $dA \, \varrho \, c_p \, \overline{u' \, T'}$ in the x-direction. Corresponding expressions are obtained for the supplementary fluxes of heat in the directions y and z. We

---

† I am indebted to Dr. J.C. Rotta for the text of this chapter which is new.
‡ A comprehensive summary of the theory of turbulent boundary layers in compressible flows is given in the book of S.S. Kutateladze and A. I. Leont'ev [50].

conclude, therefore, that the three components of the additional heat flux (quantity of heat per unit area and time) are:

$$q_x{'} = \varrho\, c_p\, \overline{u'\, T'} \; ; \quad q_y{'} = \varrho\, c_p\, \overline{v'\, T'} \; ; \quad q_z{'} = \varrho\, c_p\, \overline{w'\, T'} \; . \tag{23.2}$$

It has been assumed here that there exists a statistical correlation between the velocity and temperature fluctuations. The existence of such a correlation in the presence of a gradient $d\bar{T}/dy$ of the mean temperature can be demonstrated in the same way as that used earlier to demonstrate the existence of the correlation $\overline{u'\, v'}$. The argument advanced in the last paragraph of Sec. XVIII b retains its force if $\bar{T}$ is substituted for $\bar{u}$ and T' for u'. In such circumstances there will arise a correlation $\overline{v'\, T'}$. It follows further from this argument that the simultaneous existence of the gradients $d\bar{u}/dy$ and $d\bar{T}/dy$ must impose a strong correlation between u' and T'. This conclusion has been confirmed by measurements with hot-wire anemometers in compressible [47] and incompressible boundary layers formed on a heated wall [41, 42]. According to measurements performed by A. L. Kistler [47]. the correlation coefficient

$$\frac{\overline{u'\, T'}}{\sqrt{\overline{u'^2}} \times \sqrt{\overline{T'^2}}}$$

attains values of 0·6 to 0·8 in a compressible boundary layer.

2. The fundamental equations for compressible flow. Temperature fluctuations together with the pressure fluctuations mentioned earlier in Sec. XVIII b produce density fluctuations. For this reason it is assumed that the density

$$\varrho = \bar{\varrho} + \varrho' \tag{23.3}$$

is also equal to the sum of a time-average, $\bar{\varrho}$. and a density fluctuation, ϱ'. The fluctuations in temperature, pressure, and density are related through the equation of state of the gas, eqn. (12.20). When the gas is treated as perfect, and when the fluctuations are small, we may put

$$\frac{\varrho'}{\varrho} \approx \frac{p'}{p} - \frac{T'}{T} \, , \tag{23.4}$$

to a first approximation. In addition to the turbulent transfer of heat, the presence of density fluctuations constitutes the second important new phenomenon which occurs in compressible, turbulent streams. Evidently their presence may not be neglected when expressions for the tensor of apparent stresses (Sec. XVIII c) is derived. Formally, when eqn. (23.3) is taken into account, eqns. (18.5) must be replaced by the following additional terms due to turbulence

$$\left. \begin{aligned} \sigma'_x &= -\,\bar{\varrho}\,\overline{u'^2} - 2\,\bar{u}\,\overline{\varrho'\,u'} - \overline{\varrho'\,u'^2} \, , \\ \tau'_{xy} &= -\,\bar{\varrho}\,\overline{u'\,v'} - \bar{u}\,\overline{\varrho'\,v'} - \bar{v}\,\overline{\varrho'\,u'} - \overline{\varrho'\,u'\,v'} \, , \\ \tau'_{xz} &= -\,\bar{\varrho}\,\overline{u'\,w'} - \bar{u}\,\overline{\varrho'\,w'} - \bar{w}\,\overline{\varrho'\,u'} - \overline{\varrho'\,u'\,w'} \, . \end{aligned} \right\} \tag{23.5}$$

Here $\overline{\varrho'\,u'}$, $\overline{\varrho'\,v'}$, and $\overline{\varrho'\,w'}$ play the part of the components of a turbulent flux of mass in the three directions: x, y, z. On averaging, the equation of continuity for a compressible stream, eqn. (3.30), leads to

$$\frac{\partial(\bar{\varrho}\,\bar{u})}{\partial x} + \frac{\partial(\bar{\varrho}\,\bar{v})}{\partial y} + \frac{\partial(\bar{\varrho}\,\bar{w})}{\partial z} + \frac{\partial\,\overline{\varrho'u'}}{\partial x} + \frac{\partial\,\overline{\varrho'v'}}{\partial y} + \frac{\partial\,\overline{\varrho'w'}}{\partial z} = 0 \,. \tag{23.6}$$

Regarding the density fluctuations, it is possible to say at first that $\varrho'/\bar{\varrho}$ is hardly likely to exceed $u'/\bar{u}$. Since, now, $u'/\bar{u} \ll 1$, it appears possible to neglect the last term in each of eqns. (23.5) with respect to the first. Further simplifications result when attention is confined to boundary layers in which $\bar{v} \ll \bar{u}$. J. C. Rotta [80] demonstrated that in such cases it is possible altogether to eliminate the density fluctuations from the equations for boundary layers if, as is customary, the normal stresses themselves are neglected. First we notice that $\bar{v}\,\overline{\varrho'\,u'} \ll \bar{u}\,\overline{\varrho'\,v'}$ in the equation for τ'_{xy} in (23.5), so that only two terms need be retained. Furthermore, since $\partial\,\overline{\varrho'\,u'}/\partial x \ll \partial\overline{\varrho'\,v'}/\partial y$, the continuity equation (23.6), written for a boundary layer which is two-dimensional on the average, acquires the form

$$\frac{\partial(\bar{\varrho}\,\bar{u})}{\partial x} + \frac{\partial(\bar{\varrho}\,\bar{v})}{\partial y} + \frac{\partial\,\overline{\varrho'\,v'}}{\partial y} = 0 \,. \tag{23.6a}$$

The boundary-layer equation is derived from eqn. (12.50b) in that eqn. (12.50a), multiplied with u, is added, with eqns. (18.1) and (23.3) substituted; the result is then averaged in accordance with eqn. (18.4). When the above-mentioned terms are neglected, the following, final form for the boundary-layer equation is obtained:

$$\bar{\varrho}\,\bar{u}\,\frac{\partial\bar{u}}{\partial x} + (\bar{\varrho}\,\bar{v} + \overline{\varrho'\,v'})\,\frac{\partial\bar{u}}{\partial y} = -\frac{d\bar{p}}{dx} + \frac{\partial}{\partial y}\left(\mu\,\frac{\partial\bar{u}}{\partial y}\right) - \frac{\partial(\bar{\varrho}\,\overline{u'\,v'})}{\partial y} \,. \tag{23.7}$$

It is noted that in eqns. (23.6a) and (23.7) the density fluctuation appears only in the form of $\overline{\varrho'\,v'}$ added to $\bar{\varrho}\,\bar{v}$. It is, therefore, convenient to re-introduce the original expression for the mass flux $\overline{\varrho\,v} = \bar{\varrho}\,\bar{v} + \overline{\varrho'\,v'}$ in the y-direction, and to define the turbulent, apparent stress as

$$\tau'_{xy} = -\bar{\varrho}\,\overline{u'\,v'} \,.$$

In any case, the exact value of the mean velocity component at right angles to the wall, $\bar{v}$, remains undetermined, being of little interest anyway. The energy equation (12.19) can be treated in like manner. Introducing the turbulent heat flux

$$\bar{q}_y = c_p\,\bar{\varrho}\,\overline{v'\,T'} \,,$$

we obtain the following set of equations which describe the processes in compressible, turbulent boundary layers:

$$\frac{\partial\overline{\varrho\,u}}{\partial x} + \frac{\partial\overline{\varrho\,v}}{\partial y} = 0 \,, \tag{23.8a}$$

$$\overline{\varrho\,u}\,\frac{\partial\bar{u}}{\partial x} + \overline{\varrho\,v}\,\frac{\partial\bar{u}}{\partial y} = -\frac{d\bar{p}}{dx} + \frac{\partial}{\partial y}\left(\mu\,\frac{\partial\bar{u}}{\partial y}\right) + \frac{\partial\tau'_{xy}}{\partial y} \,, \tag{23.8b}$$

$$c_p\left(\overline{\varrho\,u}\,\frac{\partial\bar{T}}{\partial x} + \overline{\varrho\,v}\,\frac{\partial\bar{T}}{\partial y}\right) = \frac{\partial}{\partial y}\left(k\,\frac{\partial\bar{T}}{\partial y}\right) - \frac{\partial q_y'}{\partial y} + \overline{\mu\,\Phi} + \bar{u}\,\frac{d\bar{p}}{dx} \,. \tag{23.8c}$$

Here, the term $\overline{\mu\Phi}$ represents the mean value of the dissipation, and for it, the following approximation may be employed:

$$\overline{\mu\,\Phi} = \left(\mu\,\frac{\partial\bar{u}}{\partial y} + \tau'_{xy}\right)\frac{\partial\bar{u}}{\partial y}\ .\tag{23.8d}$$

The set must be augmented by the approximate form of the equation of state for mean values:

$$\bar{p} = \bar{\varrho}\,R\,\bar{T}\ .\tag{23.9}$$

The preceding system of equations for compressible, turbulent boundary layers replaces equations (12.50a) to (12.50d) for corresponding laminar flow. The boundary conditions remain unchanged (*cf.* Chap. XII).

In order to explore the details of turbulent motion in compressible media, it is necessary to undertake measurements with hot wires. This is made difficult by the need to uncouple the effect of temperature and velocity fluctuations within a single signal. The problems which arise in this way form the subject of the publications [49, 65] by L. S. G. Kovasznay and M. V. Morkovin, respectively. Leaving apart the appearance of density and temperature fluctuations, it is found that the flow remains, in its general outline, the same as in an incompressible fluid. However, as the Mach number is increased, the velocity fluctuations lose in intensity, as demonstrated by the experimental results due to A. L. Kistler [47] and shown in Fig. 23.1. The effect of density fluctuations which go beyond those included in eqns. (23.8a) to (23.8c) have been investigated by J. C. Rotta [80].

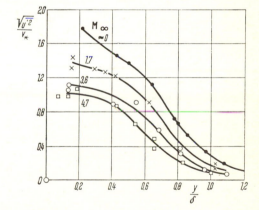

Fig. 23.1. Distribution of turbulent velocity fluctuations in the boundary layer on a flat plate placed at zero incidence in a supersonic stream. Measurements due to A. L. Kistler [47] and P. S. Klebanoff [48]

In order to render the system of equations (23.8a) to (23.8d) more amenable to practical calculations, it is possible, as was done in Chap. XIX, to introduce into it empirical assumptions for momentum and heat transport. Equation (19.1) for the apparent shearing stress $\tau_t = \tau'_{xy}$ is usually taken over unchanged. As far as the turbulent heat flux is concerned, it is customary to give it a form reminiscent of Fourier's law of thermal conduction, eqn. (12.2), according to which we have

$$q_l = -\,k\,\frac{\partial T}{\partial y}\qquad\text{(laminar)}\ ,$$

and to postulate that

$$q_t = - c_p A_q \frac{\partial T}{\partial y} \qquad \text{(turbulent)}. \qquad (23.10)$$

It is true that the exchange mechanisms for momentum and heat are similar; nevertheless, they are not identical. The exchange coefficients A_τ and A_q for momentum and heat, respectively, have, therefore, different values in general. Taking into account eqns. (19.1) and (23.10), together with eqn. (23.8d), we can transform the system of equations (23.8a) to (23.8d) to the form:

$$\frac{\partial \bar{\varrho}\, \bar{u}}{\partial x} + \frac{\partial \overline{\varrho\, v}}{\partial y} = 0 , \qquad (23.11\,\text{a})$$

$$\bar{\varrho}\, \bar{u}\, \frac{\partial \bar{u}}{\partial x} + \overline{\varrho\, v}\, \frac{\partial \bar{u}}{\partial y} = - \frac{d\bar{p}}{dx} + \frac{\partial}{\partial y} \left[(\mu + A_\tau)\, \frac{\partial \bar{u}}{\partial y} \right] , \qquad (23.11\,\text{b})$$

$$c_p \left(\bar{\varrho}\, \bar{u}\, \frac{\partial \overline{T}}{\partial x} + \overline{\varrho\, v}\, \frac{\partial \overline{T}}{\partial y} \right) = \frac{\partial}{\partial y} \left[(k + c_p A_q)\, \frac{\partial \overline{T}}{\partial y} \right] +$$

$$+ (\mu + A_\tau) \left(\frac{\partial \bar{u}}{\partial y} \right)^2 + \bar{u}\, \frac{d\bar{p}}{dx} . \qquad (23.11\,\text{c})$$

3. Relation between the exchange coefficients for momentum and heat. We have stressed in the past that the occurrence of a fluctuating motion in a turbulent flow causes momentum to be exchanged vigorously between the layers of different velocities. It also causes an increase in the transfer of heat and mass when temperature or concentration gradients are present. For this reason, there exists an intimate connexion between heat and momentum transfer in general. In particular, we must expect the existence of a relation between the heat flux and the shearing stress at the wall itself. The existence of such an analogy between heat and momentum transfer was first discovered by O. Reynolds [76], and for this reason we now speak of Reynolds's analogy (*cf.* Sec. XIIe 3). This analogy enables us to make statements concerning the transfer of heat from the known laws of drag in a turbulent boundary layer. The exchange coefficients for momentum and heat — A_τ and A_q — both have the dimension of a viscosity, μ (kg/m sec or lb/m sec in absolute systems), so that in addition to the molecular Prandtl number $\mathsf{P} = \mu\, c_p/k$, it is convenient to introduce a corresponding, dimensionless, *turbulent Prandtl number*

$$\mathsf{P}_t = \frac{A_\tau}{A_q} . \qquad (23.12)$$

Thus by definition,

$$\frac{q_t}{\tau_t} = - \frac{c_p}{\mathsf{P}_t} \frac{\partial \overline{T}/\partial y}{\partial \bar{u}/\partial y}. \qquad (23.13)$$

The total rate of heat transferred assumes the form

$$q = - c_p \left(\frac{\mu}{\mathsf{P}} + \frac{A_\tau}{\mathsf{P}_t} \right) \frac{\partial \overline{T}}{\partial y} . \qquad (23.14)$$

The turbulent Prandtl number can be determined with the aid of simultaneous determinations of velocity and temperature profiles; unfortunately, the level of

Fig. 23.2. Ratio of the turbulent transfer coefficients A_q/A_τ over the length of a radius in turbulent pipe flow, after H. Ludwieg [55]

Reynolds number $R = 3 \cdot 2 \times 10^5$ to $3 \cdot 7 \times 10^5$

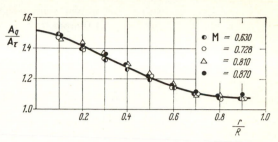

confidence with which the results of such measurements can be accepted is low owing to the difficulties of measuring local temperatures in flows in general, and to the uncertainties in the values of the gradients $d\bar{u}/dy$ and $d\bar{T}/dy$. It turns out that P_t varies with the distance from the wall. In an investigation performed by H. Ludwieg [55] it was found, as shown in Fig. 23.2, that the ratio $A_q/A_\tau = 1/P_t$ varies from about unity at the wall $(r/R=1)$ to about $1\cdot5$ in the centre of a pipe $(r/R=0)$ and is independent of the Mach number. Similar results were reported by D. S. Johnson [42] who made measurements in a boundary layer on a heated wall. According to these, the ratio A_q/A_τ increases from about unity at the wall to approximately 2 at the edge of the boundary layer. A. Fage and V. M. Falkner (*cf.* ref. [97]) and H. Reichardt [72] measured a value of 2, the former in the wake behind a circular cylinder, and the latter in a free jet, both in an incompressible stream. According to the preceding measurements the ratio A_q/A_τ is smaller in a boundary layer than in a free stream owing to the influence of the wall on the boundary layer. It seems, therefore, plausible to assume that the ratio A_q/A_τ has a value of unity at the wall (according to Ludwieg the value is $1\cdot08$ giving $P_t \approx 0\cdot9$) and increases to a value of 2 $(P_t = 0\cdot5)$ away from the wall. In practice, frequently, a constant value of $A_q/A_\tau=1$ $(P_t=1)$ or of $1\cdot3$ (Reichardt, giving $P_t = 0\cdot769$) is assumed. It must, however be pointed out that the manner in which the turbulent Prandtl number varies across a boundary layer has not been determined beyond doubt, and that there exist experimental results which are in conflict with the preceding ones, as reported in the summaries by J. Kestin and P. D. Richardson [45, 46].

The relation between heat transfer and skin friction was utilized by H. Ludwieg [54] in order to measure the shearing stress at a wall by measuring the amount of heat transferred from a small resistance element built into the wall and heated to a temperature exceeding that of the stream.

The analogy between heat and momentum transfer in a free jet is discussed in Sec. XXIVe.

b. Relation between velocity and temperature distribution†

1. The transfer of heat from a flat plate. It was shown in Chap. XII that the velocity and temperature profiles are identical in the case of laminar flow past a flat plate at zero indicence on condition that frictional heat is neglected and that the Prandtl number is equal to unity. The same can be asserted in relation to turbulent flow, on condition that $P_t = 1$ as well as $P = 1$. This implies physically that it is

† References to papers on turbulent boundary layers with suction and injection in *compressible* flow have been indicated in Sec. XXIa5.

assumed that the same mechanism causes the exchange of momentum as well as of heat. Since the velocity and temperature profiles are identical, we can then write, that

$$q(x) = \frac{k}{\mu} \frac{T_w - T_\infty}{U_\infty} \tau_0(x) . \tag{23.15}$$

The preceding equation can be easily re-arranged to the form

$$\mathsf{N}_x = \tfrac{1}{2} \mathsf{R}_x c_f' \qquad (\text{Reynolds, } \mathsf{P} = \mathsf{P}_t = 1) \tag{23.16}$$

described earlier as the Reynolds analogy. It is seen that the relation of direct proportionality between the Nusselt number and the coefficient of skin friction which was derived in Chapter XII for the case of laminar flow past a flat plate at zero incidence, (*cf.* equation (12.56b) remains valid in the turbulent case. Equation (23.16) retains its validity in the presence of compressibility, just as was the case with laminar flow, on condition that the Nusselt number is now formed with the temperature difference $T_w - T_a$.†

As already mentioned before, the principal difficulty in studying turbulent boundary layers and turbulent heat transfer problems stems from the fact that the eddy or exchange coefficients A_τ and A_q are not properties of the fluid, unlike the viscosity μ or the thermal conductivity k, but that they depend on the distance from the wall inside the boundary layer. At a sufficiently large distance from the wall they assume values which are many times larger than the molecular coefficients μ and k, so much so, in fact, that in most cases the latter can be neglected with respect to the former. By contrast, in the immediate neighbourhood of the wall, i. e. in the laminar sub-layer, the eddy coefficients vanish because in it turbulent fluctuations and hence turbulent mixing are no longer possible. Nevertheless, the rate of heat transfer between the stream and the wall depends precisely on the phenomena in the laminar sub-layer and so on the molecular coefficients μ and k. It is fortunate that eqn. (23.16) remains valid throughout, regardless of the existence of a laminar sub-layer, because when $\mathsf{P} = 1$, as shown in Section XIIg, the velocity and tempe-rature distribution in the laminar sub-layer remain identical. The assumption that $\mathsf{P}_t = 1$ in turbulent boundary layers leads, as a rule, to useful results; by contrast, the Prandtl number in the laminar sub-layer can differ appreciably from unity, as is the case, for example, with liquids (Table 12.1). When this is the case, eqn. (23.16) loses its validity. Extensions of the Reynolds analogy to cases when $\mathsf{P} \neq 1$ have been formulated by many authors, among them L. Prandtl [70], G. I. Taylor [96] and Th. von Kármán [44], and R. G. Deissler [20, 21, 22, 23].

L. Prandtl assumed that $\mathsf{P}_t = 1$ and divided the boundary layer into two zones: the laminar sub-layer in which the eddy coefficients vanish, and the turbu-

---

† Frequently, instead of the Nusselt number use is made of the so-called Stanton number

$$S = \frac{\alpha}{\varrho \, c_p \, U_\infty} = \frac{\mathsf{N}_x}{\mathsf{R}_x \, \mathsf{P}} .$$

If this is preferred, the Reynolds analogy from eqn. (23.16) becomes

$$S = \tfrac{1}{2} c_f' .$$

The remaining equations can be easily transformed to replace N by S.

lent, external boundary layer, in which the molecular coefficients μ and k can be neglected. Under these assumptions, eqns. (19.1) and (23.14), written for the laminar sub-layer will lead to the form

$$\frac{q}{\tau} = -\frac{k}{\mu}\frac{dT}{du} \, ,$$

whereas in the turbulent layer they will lead to

$$\frac{q}{\tau} = -c_p \frac{dT}{du} \, .$$

Remembering that at the wall $u = 0$, assuming that the temperature at the wall is constant and equal to T_w, and denoting the velocity and temperature, respectively, at the outer edge of the laminar sub-layer by u_l and T_l, and in the free stream by U_∞, T_∞, Prandtl introduced the assumption that the ratio q/τ remains constant across the whole width of the boundary layer†. Integration over the laminar sub-layer will then lead to

$$\frac{q}{\tau} = -\frac{k}{\mu}\frac{T_l - T_w}{u_l} = -\frac{k}{\mu U_\infty}\frac{T_l - T_w}{(u_l/U_\infty)} \, . \qquad (23.17)$$

Similarly, integration over the turbulent zone will lead to

$$\frac{q}{\tau} = -c_p \frac{T_l - T_\infty}{u_l - U_\infty} \, .$$

Equating the two right-hand sides we obtain

$$\mathsf{P}\,(T_w - T_\infty) = -\frac{U_\infty}{u_l}\left[1 + \frac{u_l}{U_\infty}(\mathsf{P}-1)\right](T_l - T_w) \, .$$

Hence, the local coefficient of heat transfer becomes

$$\alpha = \frac{q}{T_w - T_\infty} = -\frac{\mathsf{P}}{1 + (u_l/U_\infty)(\mathsf{P}-1)} \cdot \frac{(u_l/U_\infty)\,q}{T_l - T_w} \, .$$

On introducing eqn. (23.17) we have

$$\alpha = \frac{1}{1 + (u_l/U_\infty)(\mathsf{P}-1)} \cdot \frac{c_p\,\tau}{U_\infty} \, .$$

We can express this result in terms of the Nusselt number and are led in this way to the extension of the Reynolds analogy which was derived independently by L. Prandtl and G. I. Taylor:

$$\mathsf{N}_x = \frac{\frac{1}{2}c_f'\,\mathsf{R}_x\,\mathsf{P}}{1 + (u_l/U_\infty)(\mathsf{P}-1)} \qquad (\text{Prandtl-Taylor, } \mathsf{P}_t = 1)\,. \qquad (23.18)$$

---

† This condition is satisfied exactly for the flat plate at constant temperature because then $u/U_\infty = (T - T_w)/(T_\infty - T_w)$, if $\mathsf{P} = 1$, see eqn. (13.13).

In order to apply the preceding equation to particular cases it is still necessary to make a suitable assumption about the ratio of the mean velocity at the outer edge of the laminar sub-layer to that in the free stream †. In the particular case when $P = 1$, the Prandtl-Taylor equation (23.18) reduces to Reynolds's equation (23.16).

In deriving the Prandtl-Taylor equation (23.18) it was supposed that the boundary layer could be sharply divided into a turbulent layer and a laminar sub-layer. In actual fact one merges into the other in a continuous way and it is possible to discern the existence of an intermediate, or buffer layer in which the magnitudes of the molecular and turbulent exchange are comparable. Th. von Kármán [44] subdivided the boundary layer into three zones and derived a similar formula for the relation between the coefficients of heat transfer and skin friction. This is of the form

$$N_x = \frac{\frac{1}{2} R_x P c_f'}{1 + 5 \sqrt{\frac{1}{2} c_f'} \left\{ (P - 1) + \ln \left[1 + \frac{5}{6} (P - 1) \right] \right\}} \qquad \text{(von Kármán, } P_t = 1). \qquad (23.19)$$

Von Kármán's equation (23.19) also reduces to Reynolds's equation (23.16) in the special case when $P = 1$. The relation between the local Nusselt number N_x and the Reynolds number R_x is seen plotted in Fig. 23.3 for the case of a flat plate and for three values of the Prandtl number, namely $P = 10$, 1 and 0.01. The curves (b) and (c) represent, respectively, the plots from equations (23.19) and (19.42) for $P_t = 1$.

The analogy relations between the rate of heat transfer and skin friction in turbulent flow are of great practical importance because their application is not restricted to flows past flat plates. They can be used for arbitrary turbulent flows and thus enjoy much more general applicability. The latter statement has been confirmed by numerous measurements.

The equations under consideration have turned out to be valid when applied to the calculation of heat transfer from slender bodies in parallel streams, that is in cases when the pressure gradients outside the bodies are not unduly large. It was also possible to show that the analogy carries over to compressible flows when it remains independent of the Mach number. All of the forms of the Reynolds analogy quoted earlier remain approximately valid when they are applied to internal flows in circular pipes. It is then necessary to replace the current length, x, in the expression for the Nusselt and Reynolds numbers by the diameter D of the pipe, and the velocity

---

† In the case of turbulent flow in a pipe, the ratio of the velocity at the outer edge of the laminar sub-layer to the velocity U on the axis is given by

$$u_l/U = 5 \sqrt{\tau/\varrho U^2} = 5 \sqrt{\tfrac{1}{2} c_f} \;,$$

as shown in Chap. XX, eqn. (20.15a). With this approximation Prandtl's equation becomes

$$N_x = \frac{\frac{1}{2} R_x P c_f'}{1 + 5 \sqrt{\tfrac{1}{2} c_f'} (P - 1)} \;.$$

Referring c_f' to the mean pipe velocity $\bar{u}$, we would have

$$\frac{u_l}{\bar{u}} = 5 \sqrt{\frac{1}{2} c_f'} \;.$$

and temperature of the external stream, respectively, must be replaced by the mean velocity and the mean temperature of the fluid in the pipe.

In all preceding derivations we have assumed that the turbulent Prandtl number $P_t = 1$. In other words, it has been assumed that the eddy coefficients for momentum and thermal energy transfer are equal. It is, however, known from measurements that the value of this ratio differs from unity. The case when $P_t \neq 1$ in heat transfer was extensively investigated by H. Reichardt [73]. According to this work, the Nusselt number is represented by the relation

$$N_x = \frac{\frac{1}{2}c_f' \, R_x \, P}{P_t + \sqrt{\frac{1}{2}c_f'} \left\{ (P - P_t)\,a + A \right\}} \qquad \text{(Reichardt)} . \qquad (23.20)$$

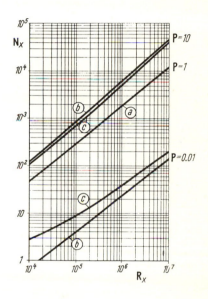

Fig. 23.3. Variation of Nusselt number with Reynolds number at different Prandtl numbers in the case of turbulent heat transfer on a flat plate (Reynolds analogy)

(a) Reynolds, eqn. (23.16)
(b) L. Prandtl and G. I. Taylor, eqn. (23.18)
(c) Th. von Kármán, eqn. (23.19)

It has been assumed that $P_t = 1$,

$c_f' = 0.0592 \, R_x^{-0.2}$, eqn. (21.12),

$u_l/U_\infty = 5 \sqrt{\dfrac{1}{2}c_{f'}}$

The quantity a, which makes an allowance for the transfer of heat through the sublayer, depends on the ratio of the two Prandtl numbers, P/P_t, and is given by

$$a = \frac{v_*}{v} \int\limits_0^\infty \frac{dy}{\left(1 + \dfrac{A_\tau}{\mu}\right)\left[1 + \left(\dfrac{P}{P_t}\right)\left(\dfrac{A_\tau}{\mu}\right)\right]} . \qquad (23.21)$$

H. Reichardt evaluated this integral assuming a smooth variation of A_τ, that is of the velocity, from the wall up to the region of fully developed turbulent flow. The numerical results are seen reproduced in Table 23.1. The quantity A which also appears in eqn. (23.20) is a function of the turbulent Prandtl number, P_t, and is slightly affected by changes in $\sqrt{c_f'/2}$; however, eqn. (23.20) is not, generally speaking, very sensitive to it. According to calculations performed by J. C. Rotta [81], it is possible to approximate it by

$$A \approx 4\,(1 - P_t) .$$

Table 23.1. The constants a and b for the calculation of the coefficient of heat transfer from eqn. (23.20) and of the recovery factor from eqn. (23.27), after H. Reichardt [73] and J. C. Rotta [81].

P/P_t	a	b
0·5	10·22	123·8
0·72	9·55	108·1
1·44	8·25	82·2
2·0	7·66	71·6
5·0	6·04	47·5
10	5·05	34·3
20	4·10	24·5
30	3·61	20·1
100	2·47	10·9
200	1·98	7·7
1000	1·17	3·4

The temperature distribution in turbulent boundary layers on flat plates in the presence of an arbitrarily varying, turbulent Prandtl number, P_t, was studied by E. R. von Driest [28] and J. C. Rotta [81]. In the latter reference it is shown that only the values which the turbulent Prandtl number, P_t, assumes close to the wall determine the rate of heat transfer and the temperature distribution; consequently, the details of the variation of P_t away from the wall are less important. The variation of P_t with distance from the wall is brought to bear only through the intermediary of the quantity A when in the remainder the value of P_t at the wall is substituted. A suitable value for this seems to be $P_t \approx 0.9$. J. R. Taylor [98] performed such calculations for boundary layers with variable pressure and temperature along the wall.

2. The transfer of heat from rough surfaces. It has been demonstrated in Secs. XXf and XXIc that rough surfaces develop considerably larger values of skin friction in turbulent flow than do smooth ones. The same is true of the coefficient of heat transfer. Normally, however, the percentage increase in the rate of heat transfer is smaller than that in skin friction. This is understandable, because a part of the turbulent shearing stresses can be transmitted to the wall through pressure forces exerted on protuberances; but there exists no analogue for this mechanism in heat transfer. Experimental investigations on the transfer of heat to a rough tube were carried out, among others, by W. Nunner [66] and D. F. Dipprey and R. H. Sabersky [25]. The latter authors made measurements at different values of the Prandtl number. Theoretical considerations due to D. F. Dipprey and R. H. Sabersky [25] as well as to P. R. Owen and W. R. Thomson [67] are based on the hypothesis that the effect of roughness on the mechanism of exchange is confined to the regions located in the proximity of the wall. Starting with this hypothesis, it is possible to derive an equation which has the same structure as eqn. (23.20), and differs only in that the term $(P - P_t)$ must be replaced by a quantity, β, which is a function of the Prandtl number, P, and of the roughness. In the particular case when $P_t = 1$, we obtain

$$N_x = \frac{\tfrac{1}{2} c_f' \, R_x \, P}{1 + \sqrt{\tfrac{1}{2} c_f'} \; \beta \, (v_* \, k/v \, ; P)} \qquad \text{(Dipprey, Sabersky, Owen, Thomson; } P_t = 1)\,. \quad (23.22)$$

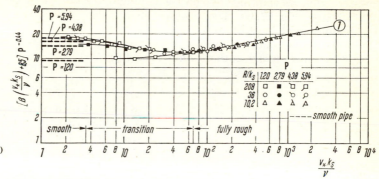

Fig. 23.4. The roughness function $(\beta + 8\cdot5)$ $P^{-0\cdot44}$ as a function of $v_* k_s/\nu$ for sand roughness at various Prandtl numbers, from the measurements by D. F. Dipprey and R. H. Sabersky [25]

Dipprey and Sabersky quote that

$$\beta\left(\frac{v_* k_s}{\nu}\; ; \mathsf{P}\right) = 5\cdot19 \left(\frac{v_* k_s}{\nu}\right)^{0\cdot2} \mathsf{P}^{0\cdot44} - 8\cdot5 . \tag{23.23}$$

This correlation is based on their own experimental results with sand roughness and is valid in the completely rough regime when $v_* k_s/\nu > 70$. The graph of the function β together with the experimental results over the whole range of roughness Reynolds number $v_* k_s/\nu$ is shown in Fig. 23.4. Owen and Thomson correlated experimental results from various sources, including those from refs. [25] and [66], and concluded that

$$\beta\left(\frac{v_* k_s}{\nu}\; ; \mathsf{P}\right) = 0\cdot52 \left(\frac{v_* k_s}{\nu}\right)^{0\cdot45} \mathsf{P}^{0\cdot8} . \tag{23.24}$$

Procedures for the calculation of heat-transfer rates in turbulent flows with non-isothermal surfaces have been worked out by D. B. Spalding [88], and J. Kestin and coworkers [36, 45, 46]. Extensive measurements under such conditions were performed by W. C. Reynolds, W. M. Kays, and S. J. Kline [77].

3. Temperature distribution in compressible flow. In order to understand the laws which govern the temperature distribution in compressible flows, the reader may wish first to refer to the relevant considerations for laminar boundary layers which were advanced in Sec. XIIIb. When the pressure remains constant and $\mathsf{P} = \mathsf{P}_t = 1$, the temperature distribution satisfies eqns. (13.12), and eqn. (13.13) in the general case with heat transfer, both owing to the evolution of frictional heat. When $\mathsf{P} \neq \mathsf{P}_t \neq 1$, it is possible to evaluate the recovery temperature on an (adiabatic) wall by the use of eqn. (13.19), i. e. by

$$T_a = T_\infty \left(1 + r\,\frac{\gamma-1}{2}\,\mathsf{M}_\infty{}^2\right) . \tag{23.25}$$

The recovery factor, r, is somewhat larger in turbulent flow than it was in laminar flow, experiments showing that on the average its value places itself between 0·875 and 0·89 (see Fig. 17.31). The diagram in Fig. 23.5 reproduces L. M. Mack's [56] comparison of values of the recovery factor, r, measured on cones at different Mach numbers and at different Reynolds numbers. In order to estimate the effect of Prandtl number, many authors quote the formula

$$r = \sqrt[3]{\mathsf{P}} \tag{23.26}$$

which yields $r = 0.896$ at $\mathsf{P} = 0.72$. It is also possible to obtain this estimate theoretically, in a manner analogous to that used for the calculation of the coefficient of heat transfer. For this purpose it is necessary to start with the energy equation (23.11c) and to include the effects of the molecular and of the turbulent transfer mechanisms in accordance which the hypothesis contained in eqn. (23.14). Proceeding in this way, J. C. Rotta [81] obtained the equation:

$$r = \mathsf{P}_t + \tfrac{1}{2} c_f{}' \, (\mathsf{P} - \mathsf{P}_t) \, b + B \sqrt{\tfrac{1}{2} c_f{}'} \; . \tag{23.27}$$

The quantity b is a function of the ratio P/P_t and accounts for, like the quantity a in eqn. (23.20), the processes taking place in the laminar sublayer. It is given by the integral

$$b = \frac{2}{v} \int\limits_0^\infty \frac{u \, dy}{\left(1 + \dfrac{A_\tau}{\mu}\right)\left[1 + \left(\dfrac{\mathsf{P}}{\mathsf{P}_t}\right)\!\left(\dfrac{A_\tau}{\mu}\right)\right]} \; . \tag{23.28}$$

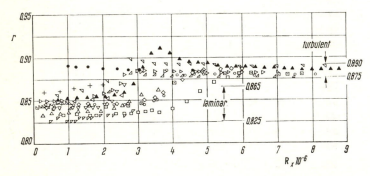

Fig. 23.5. Recovery factors in terms of Reynolds number from measurements on cones at Mach numbers $\mathsf{M}_\infty = 1\cdot2$ to $\mathsf{M}_\infty = 6\cdot0$ after L. M. Mack [56]

	Wind tunnel		M_∞	Type of cone		Wind tunnel		M_∞	Type of cone
●	Ames	1 × 3 ft No. 1	1,97	10° hollow; steel	▷	Aberdeen		2,18	10° wood
□	Lewis	8 × 6 ft	1,98	10° hollow; steel	×	GALCIT 5 × 5 in.		6,0	20° ceramic
△	Lewis	18 × 18 in.	1,94	10° hollow; steel	○	Ames	1 × 3 ft No. 1	2,0	20° hollow; steel
▽	Ames	6 × 6 ft	1,9	10° hollow; steel	▲	JPL	18 × 20 in.	4,50	5° fibreglass
◁	Ames	10 × 14 in.	4,48	10° hollow; steel	◁	JPL	18 × 20 in.	1,63	13° lucite
▽	Ames	2 × 2 ft	1,21	10° hollow; steel	+	JPL	18 × 20 in.	4,50	13° lucite
▷	Lewis	2 × 2 ft	3,93	10° hollow; steel	◇	JPL	12 × 12 in.	1,63	13° lucite
◁	Ames	1 × 3 ft No. 2	3,00	10° hollow; steel	✳	JPL	12 × 12 in.	2,54	13° lucite

Numerical values have been included in Table 23.1. The factor B depends on P_t and somewhat on $\sqrt{c_f'/2}$. According to Rotta, we may take

$$B = 7\,(1 - P_t)\,.$$

When the turbulent Prandtl number varies over the thickness of the boundary layer, it is necessary to insert into eqn. (23.24) the value assumed by it at the wall. When the Prandtl number, P, as well as the turbulent Prandtl number, P_t, differ from unity, it is worth noting that, normally, eqn. (13.21) given in Chap. XIII for laminar boundary layers constitutes a usable approximation for the temperature distribution in a compressible turbulent boundary layer. B. Schultz-Jander [95] developed a procedure for the calculation of temperature distributions in turbulent compressible boundary layers.

c. Influence of Mach number; laws of friction

To date, the calculation of turbulent boundary layers in incompressible flow has not developed to a point where it could be classed as being more than a semi-empirical theory. It is, therefore, not surprising that the same remark applies to the calculation of compressible turbulent boundary layers. In the case of incompressible turbulent boundary layers a starting point is provided by the hypotheses which were discussed in the preceding chapters, namely by Prandtl's mixing-length hypothesis, by von Kármán's similarity rule or by Prandtl's universal velocity-distribution law. The authors of numerous contemporary papers have endeavoured to create a semi-empirical theory of compressible turbulent boundary layers by transposing these hypotheses and by adapting them to the compressible case. This necessitated the introduction of additional *ad hoc* hypotheses. In the absence of detailed investigations into the mechanics of compressible turbulent flows, the transposition of the semi-empirical theories of turbulent flows from the incompressible to the compressible case involves a good deal of arbitrariness.

From the practical point of view, the difficulties increase because, on the one hand, there are two additional parameters — the Mach number, M_∞, of the free stream and the temperature, T_w, of the solid surface — which influence the flow, and, on the other hand, the available experimental results are not entirely free of contradictions. Three methods should be singled out from among the numerous proposals for handling the problem, because they have been employed particularly frequently:

(1) Introduction of a reference temperature for the density and viscosity of the gas.

(2) Application of Prandtl's mixing length hypothesis or of von Kármán's similarity hypothesis.

(3) Transformation of the coordinates.

Over and above, the literature of the subject contains expositions of methods which cannot be classified under any one of the three preceding headings. In an impressive comparison, D. R. Chapman and R. H. Kester [11] brought to the fore the large differences which result when different methods are used to calculate skin friction (*cf.* [30]). An extensive comparison between twenty different computational schemes and existing, experimental results was carried out by D. B. Spalding and S. W. Chi [89].

1. The flat plate at zero incidence. The guiding idea of the methods of class (1) is the hypothesis that the laws of incompressible flow remain valid in the compressible case on condition that the values of density, ϱ, and viscosity, μ, are taken at a suitably chosen reference temperature, T^*. Th. von Kármán [43] was the first one to utilize this possibility and selected the temperature at the wall as his reference temperature. Starting with the law of friction for a flat plate at zero incidence in incompressible flow embodied in eqn. (21.17), von Kármán obtained the following equation for the skin-friction coefficient in the compressible case:

$$\frac{0.242}{\sqrt{c_f}} \left\{1 + \frac{\gamma - 1}{2} \, \mathsf{M}_\infty^2\right\}^{-1/2} = \log{(\mathsf{R}_l \, c_f)} - \frac{1}{2} \log{\left\{1 + \frac{\gamma - 1}{2} \, \mathsf{M}_\infty^2\right\}}, \qquad (23.29)$$

where $\mathsf{M}_\infty = U_\infty/c_\infty$ denotes the Mach number of the free stream. The preceding equation is valid only for an adiabatic wall; in it, the viscosity function was assumed in the form $\mu/\mu_0 = \sqrt{T/T_0}$. Various attempts have been made to improve the method of the reference temperature by choosing a value T^* which lies between the highest and the lowest values of temperature, T, encountered within the boundary layer. E. R. G. Eckert [29, 30] proposed to place the reference temperature at

$$T^* = T_0 + 0.5 \, (T_w - T_1) + 0.22 \, (T_a - T_1), \qquad (23.30)$$

where T_1 denotes the temperature at the edge of the boundary layer, T_w is the surface temperature at the wall and T_a represents the recovery (i. e. adiabatic wall) temperature. Eckert's formula includes the case with heat transfer. The introduction of a reference temperature constitutes the simplest way of accounting for the influence of Mach number and heat transfer on skin friction and leads to results which are often adequate in engineering applications. For this reason, M. H. Bertram [2] carried out a large programme of calculations of skin-friction coefficients covering a wide range of Mach numbers and temperature ratios.

The idea of applying Prandtl's mixing-length hypothesis was taken up by E. R. van Driest [27]. He stipulated that $l = \varkappa \, y$, as given in eqn. (19.22). The effect of compressibility is brought to bear by allowing the density to vary thus causing the boundary-layer thickness to change too. He obtained explicit formulae for turbulent skin friction on a flat plate, with and without heat transfer, which account for the influence of the Reynolds and Mach numbers simultaneously. For the case of an adiabatic wall the formula for the coefficient of total skin friction has the form:

$$\frac{0.242}{\sqrt{c_f}} \, (1 - \lambda^2)^{1/2} \, \frac{\sin^{-1}\lambda}{\lambda} = \log{(\mathsf{R}_l \, c_f)} + \frac{1 + 2\,\omega}{2} \log{(1 - \lambda^2)} \qquad (23.31)$$

where

$$1 - \lambda^2 = \frac{1}{1 + \frac{\gamma - 1}{2} \, \mathsf{M}_\infty^2} \, . \qquad (23.32)$$

and $\mathsf{M}_\infty = U_\infty/c_\infty$ denotes the free-stream Mach number. The symbol ω denotes the exponent in the viscosity law $\mu/\mu_0 = (T/T_0)^\omega$ from eqn. (13.4). This equation differs from (23.29) by the factor $(\sin^{-1}\lambda)/\lambda$ on the left-hand side and by the appearance of the exponent ω of the viscosity law. For $\mathsf{M}_\infty \to 0$ eqn. (23.31) transforms into

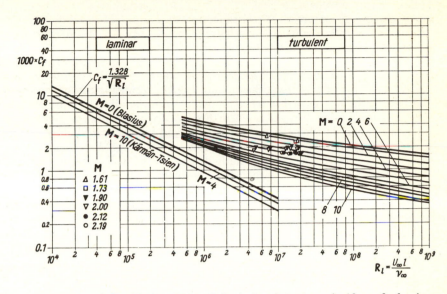

Fig. 23.6. Coefficient of total skin friction for an adiabatic flat plate at zero incidence for laminar and turbulent boundary layer. Theoretical curves for turbulent flow from eqn. (23.31), after E. R. van Driest [27]; $\gamma = 1 \cdot 4$, $\omega = 0 \cdot 76$, P $= 1$

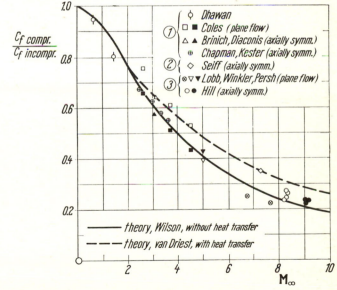

——— Theory due to Wilson [102] for an adiabatic wall and zero pressure gradient; the ratio T_w/T_∞ varies between $1 \cdot 8$ for M $= 2$ and $21 \cdot 0$ for M $= 10$

— — — Theory due to van Driest [27], with heat transfer, zero pressure gradient

Measurements:

(1) adiabatic wall, zero pressure gradient

(2) with heat transfer, zero pressure gradient

(3) with heat transfer, $T_w/T_\infty = 8 \cdot 0$, favourable pressure gradient

Fig. 23.7. Skin friction coefficient of a flat plate at zero incidence as a function of the Mach number for a turbulent boundary layer; comparison between theory and measurement; $R_x \approx 10^7$, from [38]

von Kármán's incompressible resistance formula, eqn. (21.17). Fig. 23.6 gives a plot of eqn. (23.31) and a comparison with experimental results. The measure of agreement between theory and experiment is not satisfactory in all cases, but in this connexion it must be pointed out that measurements at high Mach numbers are somewhat uncertain. R. E. Wilson [102] carried out similar calculations, but based them on von Kármán's similarity hypothesis, eqn. (19.19). Limiting himself to the case of an adiabatic wall, he derived a result which is quite similar to eqn. (23.31). Further experimental results are contained in Fig. 23.7 which shows a plot of the ratio of the skin-friction coefficients in compressible and incompressible flow in terms of the Mach number, covering a range which includes very high Mach numbers. The graph contains two theoretical curves; the first one due to R. E. Wilson [102] presupposes an adiabatic wall, and the second one, derived by E. R. van Driest [27], includes the effect of heat transfer. The measurements were performed by several workers [7, 14, 38, 53, 87] and show good agreement with theory. Additional information concerning the influence of heat transfer on skin friction is contained in Fig. 23.8 which was also based on van Driest's calculations [27]. The diagram shows that the skin friction on an adiabatic wall is somewhat smaller than is the case when heat flows from the fluid to the wall.

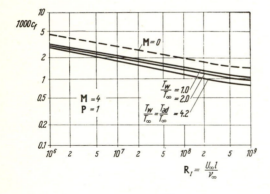

Fig. 23.8.   Skin friction coefficient for a flat plate at zero incidence in turbulent flow *with* heat transfer as a function of Reynolds number for different values of the temperature ratio T_w/T_∞, after E. R. van Driest [27]

The curve for $\dfrac{T_w}{T_\infty} = \dfrac{T_{ad}}{T_\infty} = 4\cdot 2$ corresponds to the case with zero heat transfer
$M = 4, \ P = 1$

Coordinate transformation: The coordinate transformation described in Sec. XIIId and valid for laminar flow can also be carried through formally when applied to the differential equations for compressible turbulent boundary layers. The Reynolds stress τ'_{xy} is transformed to

$$\tilde{\tau}'_{xy} = \frac{1}{b}\left(\frac{c_0}{c_1}\right)^2 \frac{p_0}{p_1}\, \tau'_{xy} \ ,\tag{23.33}$$

and with this substitution, the momentum equation (23.8b) acquires the form:

$$\tilde{u}\,\frac{\partial \tilde{u}}{\partial \tilde{x}} + \tilde{v}\,\frac{\partial \tilde{u}}{\partial \tilde{y}} = \tilde{u}_1\,\frac{\partial \tilde{u}_1}{\partial \tilde{x}}\,(1+S) + \nu_0\,\frac{\partial^2 \tilde{u}}{\partial \tilde{y}^2} + \frac{1}{\varrho_0}\,\frac{\partial \tilde{\tau}'_{xy}}{\partial \tilde{y}} \ .\tag{23.34}$$

The symbols used here are identical with those defined for eqns. (13.24) to (13.41). With the mathematical possibility of transforming the equations for compressible flow into a form identical with that for incompressible flow, many authors (e. g. B. A. Mager [57], D. Coles [15], L. Crocco [16], D. A. Spence [91, 92]) coupled a physical hypothesis, according to which the velocity profiles in the transformed plane retain the same form as that valid for incompressible flow. Consequently, the law of friction as well as other relations remain valid when the transformed parameters are substituted into them. This conclusion, which is certainly valid for laminar flows, does not necessarily carry over to turbulent flows because the transformation of coordinates cannot be applied to the equations which describe the fluctuating motion. This leads to contradictions with respect to all theories which are based on Boussinesq's assumption embodied in eqn. (19.1). These include theories which utilize Prandtl's mixing-length hypothesis or von Kármán's similarity hypothesis. If we accept the physically plausible assumption that the eddy kinematic viscosity ε_τ defined in eqn. (19.2) is independent of density, we are faced with the fact that a transformation to the incompressible form ceases to be possible. However, a transformation to

$$\tilde{\tau}'_t = \varrho_1 \, \tilde{\varepsilon} \, \frac{\partial \tilde{u}}{\partial \tilde{y}} \left(\frac{\varrho}{\varrho_1} \right)^2 \tag{23.35}$$

can still be carried out. In this case the new eddy kinematic viscosity, ε_τ, is related to the original quantity, ε, through the equation

$$\tilde{\varepsilon} = \frac{1}{b} \left(\frac{c_0}{c_1} \right)^2 \varepsilon_\tau.$$

Now, it is known that the density ratio ϱ/ϱ_1 varies considerably with the distance, y, from the wall when the Mach number is large. Consequently, one of two conclusions forces itself upon us. If we assume that the velocity profiles remain unchanged compared with the incompressible case, we find that the distribution of ε has changed. If, however, we admit that ε remains unaltered, we end up with modified velocity profiles. The statements concerning the effect of Mach number on the velocity profiles in the original coordinates which can be made on the basis of the two preceding schemes turn out to be exactly opposite. This observation throws a good deal of light on the whole complex of problems which arise when the laws obtained in the incompressible case are translated to apply to the compressible case.

Further details: The effect of Mach number on the velocity profile is brought to bear through the increase in temperature in the direction of the wall. Since it is possible to suppose that the pressure, p, is independent of y, it is found that the density distribution in the boundary layer is described by

$$\frac{\varrho}{\varrho_\infty} = \frac{T_\infty}{T} . \tag{23.36}$$

As the Mach number increases outside an adiabatic wall, it is seen that the density must decrease very strongly at small values of y and this must cause the boundary-layer thickness to increase considerably. On the other hand, an increase in the Mach number effects an increase in viscosity and a decrease in the skin-friction coefficient. This, in turn, causes the laminar sub-layer to increase strongly. An example of the

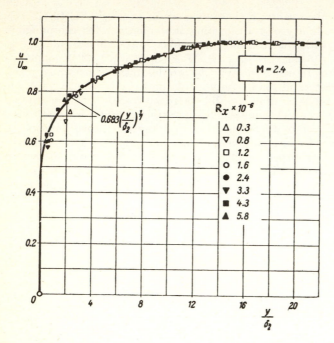

Fig. 23.9. Measurements on velocity distribution in turbulent boundary layer on flat plate at zero incidence at supersonic velocity, after R. M. O'Donnell [26]

$M_\infty = 2\cdot4$; δ_2 — momentum thickness from eqn. (13.75); $T_w = T_a$

Incompressible theory:

$$\frac{u}{U_\infty} = 0\cdot716 \left(\frac{u}{\delta_2}\right)^{1/7} ;$$

Compressible theory:

$$\frac{u}{U_\infty} = 0\cdot683 \left(\frac{u}{\delta_2}\right)^{1/7}$$

velocity profile in a compressible turbulent boundary layer is given in Fig. 23.9 which contains a plot of u/U_∞ in terms of y/δ_2 for $M_\infty = 2\cdot4$ as measured by R. M. O'Donnell [26]. Here, δ_2 represents the momentum thickness defined in eqn. (13.75).

In the adopted system of coordinates, the points for different Reynolds numbers arrange themselves well on a single curve. The theoretical curve shown on the graph deviates from the corresponding curves for incompressible flow much less than was the case with laminar flow, Fig. 13.10. As expected, the boundary-layer thickness increases with Mach number; this is brought into evidence in Fig. 23.10 which displays velocity profiles up to $M_\infty = 9\cdot9$. It is worth noting in this connexion that

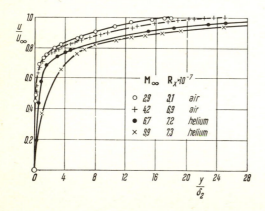

Fig. 23.10. Velocity distribution in a turbulent boundary layer on a flat plate in supersonic flow at various Mach numbers, as measured by F. W. Matting, D. R. Chapman, J. R. Nyholm, and A. G. Thomas [58]

$T_w = T_a$

the momentum thickness from eqn. (13.75) becomes smaller compared with the boundary-layer thickness, δ, as the Mach number is increased, because the density decreases in the direction of the wall.

The diagram in Fig. 23.11 contains a logarithmic plot of the velocity ratio u/v_* against $\eta = y\, v_*/v_w$ of the type encountered in Fig. 20.4, in which the values of density, ϱ, and kinematic viscosity, ν, have been taken at the wall temperature. It is noted that the characteristic shape familiar from incompressible flow persists at higher Mach numbers, but quantitative departures make their appearance. This follows from the experimental results plotted in the figure and due to R. K. Lobb, E. M. Winkler, and J. Persh [53]. In the presence of heat transfer, there appears to exist a large influence on the sub-layer, whereas the curves in the fully developed

Fig. 23.11. Universal velocity distribution law in a turbulent boundary layer on the wall of a channel at supersonic flow with heat transfer, as measured by R. K. Lobb, E. M. Winkler, and J. Persh [53]

Properties referred to wall temperature, hence:

$v_* = \sqrt{\tau_0/\varrho_w}\,$; $\quad \eta = y\, v_*/v_w$

Curves (1) and (2): Theoretical laws for incompressible flow
Curve (1): Laminar sub-layer, $u/v_* = \eta$
Curve (2): universal logarithmic law

$\qquad u/v_* = 5\cdot5 + 5\cdot75 \log \eta$

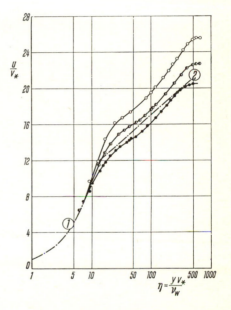

	M_∞	$\dfrac{T_a - T_w}{T_a}$	$R_2 \times 10^{-4}$	M_τ	β_q
●	5·75	0·108	1·16	0·117	0·0074
◖	5·79	0·238	1·24	0·114	0·0162
○	5·82	0·379	1·14	0·116	0·0273

region become nearly parallel. Theoretical investigation into the proper generalization of the universal velocity-distribution law to include compressibility effects have been published by R. G. Deissler [21, 23] and J. C. Rotta [78]. According to these, the velocity distribution in the proximity of the wall is influenced by two parameters, namely by the Mach number M_τ and the heat-flux number, β_q, each defined by the respective equation, as follows:

$$\mathsf{M}_\tau = \frac{v_*}{c_w} = \mathsf{M}_\infty \sqrt{\frac{1}{2}\, c_f{}'}$$

and

$$\beta_q = -\frac{q_w}{\varrho_w\, c_{pw}\, T_w\, v_*} = \frac{s}{\sqrt{\frac{1}{2}\, c_f{}'}} \frac{T_a - T_w}{\sqrt{T_w\, T_\infty}} \right\}$$

(23.37)

where c_w denotes the velocity of sound at the wall, S is the Stanton number, and c_f' the local skin-friction coefficient. Calculations performed by J. C. Rotta [78] under certain simplifying assumptions yielded results which were qualitatively correct; however, the effect of β_q on the laminar sub-layer turns out to be larger in experiments than that which can be reflected in the calculations. The measurements undertaken by H. U. Meier [60, 61, 62] give an indication of the corresponding temperature distributions. The evaluation of these results showed that the turbulent Prandtl number increases across the sublayer and reaches a value exceeding unity; this means that the factor A_q for heat transfer to the wall decreases faster than the corresponding eddy coefficient A_τ for momentum transfer. According to H. U. Meier and J.C. Rotta [63], it is possible to describe this state of affairs theoretically by transposing Prandtl's mixing-length hypothesis (Chap. XIX) to the transport of heat. Thus, eqn. (23.14) is transformed into

$$q_t = - c_p \, \varrho \, l_q^2 \, \left| \frac{d\bar{u}}{dy} \right| \frac{d\bar{T}}{d\,y} \ .$$

The mixing length l_q for heat transfer differs as to magnitude from that for momentum transfer, l in eqn. (19.7). In analogy with E. R. van Driest's equation (20.15b), it is assumed that in the neighbourhood of the wall we may put

$$l_q = \varkappa_q \, y \, [1 - \exp{(- \, y \, \sqrt{\varrho \, \tau_w}/\mu \, A_1)}]. \tag{23.37a}$$

The dimensionless constants $\varkappa_q$ and A_1 have different values than $\varkappa$ and A in eqn. (20.15b). The turbulent Prandtl number, as defined in eqn. (23.12), becomes

$$\mathsf{P}_t = (l/l_q)^2 .$$

The variation of P_t across the boundary layer was computed by H. U. Meier [64]. Figure 23.12 allows us to conclude that measured total-temperature distributions are reproduced quite well by calculations based on J.C. Rotta's [78] law of the wall for compressible boundary layers. The diagrams represent the ratio T_0/T_∞ of total

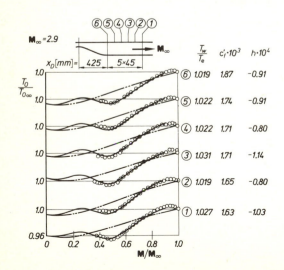

Fig. 23.12. Total temperature T_0 in the turbulent boundary layer on a flat wall and in the presence of a weak heat flux at supersonic velocity, after H. U. Meier et al. [62]

Mach number $\mathsf{M}_\infty = 2 \cdot 9$
Reynolds number $R/cm = 0 \cdot 8 \times 10^5 \ cm^{-1}$
measurements by H. U. Meier [60]
– – – theory as in eqn. (23.37a) with $(\varkappa/\varkappa_q)^2 = 0.9$; $A/A_1 = 1 \cdot 3$
Dimensionless heat transfer coefficient
$h = q_w/c_p \, T_w \, \varrho \, U$
Local skin-friction coefficient $c_f' = \tau_w/\frac{1}{2} \, \varrho_w \, U^2$

temperatures as functions of the Mach-number ratio M/M_∞. Here

$$T_0 = T + \bar{u}^2/2\,c_p\,.$$

When the rate of heat transfer is small $(q_w \approx 0)$, the temperature increases from the wall outwards and reaches a maximum which is followed by a decrease to a minimum and an ultimate increase.

When the wall is *rough* the influence of the Mach number on skin friction is even greater. According to H.W. Liepmann and F.E. Goddard [37, 52], the ratio $c_{f\,compr}/c_{f\,inc}$ for the completely rough regime becomes proportional to the density ratio ϱ_w/ϱ_∞, and hence

$$\frac{c_{f\,compr}}{c_{f\,inc}} = \frac{1}{1 + r\,\dfrac{\gamma-1}{2}\,M^2_\infty} \tag{23.38}$$

where r denotes the recovery factor.

2. Variable pressure. In practical applications, it is frequently necessary to perform calculations for turbulent boundary layers in compressible flows with varying pressure. The need is particularly acute in the design of convergent-divergent nozzles for supersonic wind tunnels, because the displacement effect of the boundary layer in them must be known fairly accurately. As was the case with incompressible flow, the known approximate procedures are based on the integral momentum equation; in some cases, the energy integral equation has also been employed. The two integral equations in question have been already given as eqns. (13.80) and (13.87) for adiabatic walls. As far as turbulent boundary layers are concerned, these are written:

momentum-integral equation —

$$\frac{d\delta_2}{dx} + \frac{\delta_2}{U}\,\frac{dU}{dx}\,(2 + H_{12} - M^2) = \frac{\tau_0}{\varrho_1\,U^2}\,, \tag{23.39}$$

energy-integral equation (kinetic energy) —

$$\frac{d\delta_3}{dx} + \frac{\delta_3}{U}\,\frac{dU}{dx}\left(3 + \frac{2\,\delta_H}{\delta_3} - M^2\right) = \frac{2}{\varrho_1\,U^3}\int_0^\delta \tau\,\frac{\partial u}{\partial y}\,dy\,; \tag{23.40}$$

they are valid for $P = 1$, and are not restricted to adiabatic walls. Here, δ_3 denotes the energy thickness, eqn. (13.76), δ_H represents an enthalpy thickness, eqn. (13.77), and $H_{12} = \delta_1/\delta_2$.

A number of authors, including G. W. Englert [31], E. Reshotko and M. Tucker [75], N.B. Cohen [12] and D.A. Spence [92], applied the Illingworth-Stewartson transformation with respect to the momentum-integral equation (23.39) and thus reduced it to its incompressible form. A. Walz [100] reduced the two equations (23.39) and (23.40) to a relatively convenient form from the point of view of numerical computation and encompassed the required universal functions in a set of tables of numerical values.

J.C. Rotta [84] described a similar procedure for two-dimensional and axisymmetric flows as well as for the calculation of a body of revolution in subsonic and supersonic flow [105]. The agreement between calculations and measurement is satisfactory up to a Mach number of $M_\infty = 2$. The deviations which occur at $M_\infty = 2\cdot4$ and $2\cdot8$ are explained, partially, by the fact that the curvature of the streamlines in conjunction with the variations of density exerts an unexpectedly large influence on the development of the boundary layer — an effect not accounted for in the calculation. The reasons for this effect of streamline curvature were investigated by J.C. Rotta [82]; a contribution to this problem was also made by P. Bradshaw [4]. Methods of finite differences have also been adapted to deal with turbulent boundary layers in compressible streams. T. Cebeci and A.M.O. Smith [9] developed a method based on mixing theory (see Sec. XIXc) whose validity has been extended to include three-dimensional boundary layers [10]. The method due to P. Bradshaw (see Sec. XIXf) that makes use of the equation for kinetic energy has also been extended to apply to compressible flows [6]. P. Bradshaw [5] reached the conclusion that the volumetric dilatation exerts a deep influence on the structure of the turbulence in the boundary layer. Agreement between measurement and calculation could be considerably improved by the introduction of an additional term in eqn. (19.42). A method of integration for three-dimensional compressible boundary layers was developed by P.D. Smith [94]; a proposal in this matter was made by J. Cousteix [9a]; compare also D. Arnal et al. [1a] and J. Cousteix et al. [9b].

References

[1] Anon.: Compressible turbulent boundary layers. A symposium held at Langley Research Center, Hampton, Virginia. December 10—11, 1968; NASA SP 216 (1969).

[1a] Arnal, D., Cousteix, J., and Michel, R.: Couche limite se développant avec gradient de pression positif dans un écoulement extérieur turbulent. Rech. Aerosp. Paris, 1976, *1*, 13—26 (1976).

[2] Bertram, M.H.: Calculations of compressible average turbulent skin friction. NASA TR R—123 (1962).

[3] Bourne, D.E., and Davies, D.R.: On the calculation of eddy viscosity and of heat transfer in a turbulent boundary layer on a flat surface. Quart. J. Mech. Appl. Math. *11*, 223—234 (1958).

[4] Bradshaw, P.: Effects of streamline curvature on turbulent flow. AGARDograph No. 169 (1973).

[5] Bradshaw, P.: The effect of mean compression or dilatation on the turbulence structure of supersonic boundary layers. JFM *63*, 449—464 (1974).

[6] Bradshaw, P., and Ferriss, D.H.: Calculation of boundary-layer development using the turbulent energy equation: compressible flow on adiabatic walls. JFM *46*, 83—110 (1971).

[7] Brinich, P.F., and Diaconis, N.S.: Boundary layer development and skin friction at Mach-number 3·05. NASA TN 2742 (1952).

[8] Burggraf, O.R.: The compressibility transformation and the turbulent boundary layer equation. JASS *29*, 434—439 (1962).

[9] Cebeci, T., and Smith, A.M.O.: A finite-difference method for calculating compressible laminar and turbulent boundary layers. Trans. ASME Ser. D, J. Basic Eng. *92*, 523—535 (1970).

[9a] Cousteix, J.: Analyse théorique et moyens de prévision de la couche limite turbulente tridimensionelle. ONERA Publ. No. 157 (1974).

[9b] Cousteix, J., and Houdeville, R.: Epaississement et séparation d'une couche limite turbulente soumise en interaction avec un choc oblique. Rech. Aerosp. Paris, 1976, *1*, 1—11 (1976).

[10] Cebeci, T.: Kaups, K., Ramsey, J., and Moser, A.: Calculation of three-dimensional compressible turbulent boundary layers on arbitrary wings. Douglas Aircr. Co., Report No. MDC J. 6866 1—40, (1975).

[11] Chapman, D.R., and Kester, R.H.: Measurements of turbulent skin friction on cylinders in axial flow at subsonic and supersonic velocities. JAS *20*, 441—448 (1953).

[12] Cohen, N.B.: A method for computing turbulent heat transfer in the presence of a streamwise pressure gradient for bodies in high-speed flow. NASA Memo. 1—2—59 L (1959).

[13] Coles, D.: Measurements of turbulent friction on a smooth flat plate in supersonic flow. JAS *21*, 433—448 (1954).

[14] Coles, D.: Measurements in the boundary layer on a smooth flat plate in supersonic flow. I. The problem of the turbulent boundary layer. Cal. Inst. Techn. Jet Propulsion Lab. Rep. 20—69 (1953); II. Instrumentation and experimental techniques at the Jet Propulsion Laboratory. Cal. Inst. Techn. Jet Propulsion Lab. Rep. 20—70 (1953); III. Measurements in a flat plate boundary layer at the Jet-Propulsion Laboratory. Cal. Inst. Techn. Jet Propulsion Lab. Rep. 20—71 (1953).

[15] Coles, D.: The turbulent boundary layer in a compressible fluid. Phys. Fluids 7, 1403—1423 (1964).

[16] Crocco, L.: Compressible turbulent boundary layer with heat exchange. AIAA J. 1, 2723—2731 (1963).

[17] Culick, F.E.C., and Hill, J.A.F.: A turbulent analog of the Stewartson-Illingworth transformation. JAS 25, 259—262 (1958).

[18] Davies, D.R.: On the calculation of eddy viscosity and heat transfer in a turbulent boundary layer near a rapidly rotating disk. Quart. J. Mech. Appl. Math. 12, 211—221 (1959).

[19] Davies, D.R., and Bourne, D.E.: On the calculation of heat and mass transfer in laminar and turbulent boundary layers. I. The laminar case. II. The turbulent case. Quart. J. Mech. Appl. Math. 9, 457—488 (1956).

[20] Deissler, R.G., and Eian, C.S.: Analytical and experimental investigation of fully developed turbulent flow of air in a smooth tube with heat transfer with variable fluid properties. NACA TN 2629 (1952).

[21] Deissler, R.G.: Analysis of turbulent heat transfer, mass transfer, and friction in smooth tubes at high Prandtl and Schmidt numbers. NACA TR 1210 (1955).

[22] Deissler, R.G., and Taylor, M.F.: Analysis of turbulent flow and heat transfer in non-circular passages. NACA TN 4384 (1959).

[23] Deissler, R.G.: Analysis of turbulent flow and heat transfer on a flat plate at high Mach numbers with variable fluid properties. NACA Techn. Rep. R—17, 1—33 (1959).

[24] Dhawan, S.: Direct measurements of skin friction. NACA Rep. 1121 (1953).

[25] Dipprey, D.F., and Sabersky, R.H.: Heat and momentum transfer in smooth and rough tubes at various Prandtl numbers. Int. J. Heat Mass Transfer 6, 329—353 (1963).

[26] O'Donnell, R.M.: Experimental investigation at Mach number of 2·41 of average skin friction coefficients and velocity profiles for laminar and turbulent boundary layers and assessment of probe effects. NACA TN 2132 (1954).

[27] Van Driest, E.R.: Turbulent boundary layer in compressible fluids. JAS 18, 145—160 (1951).

[28] Van Driest, E.R.: The turbulent boundary layer with variable Prandtl number. Fifty years of boundary-layer research (W. Tollmien and H. Görtler, eds.), Braunschweig, 1955, 257—271.

[29] Eckert, E.R.G.: Engineering relations for friction and heat transfer to surfaces in high velocity flow. JAS 22, 585—587 (1955).

[30] Eckert, E.R.G.: Survey on heat transfer at high speeds. Rep. Univ. of Minnesota, Minneapolis, Minn. (1961).

[31] Englert, G.W.: Estimation of compressible boundary-layer growth over insulated surfaces with pressure gradient. NACA TN 4022 (1957).

[32] Ferrari, C.: Study of the boundary layer at supersonic speeds in turbulent flow: Case of flow along a flat plate. Quart. Appl. Math. 8, 33 (1950).

[33] Ferrari, C.: The turbulent boundary layer in a compressible fluid with positive pressure gradient. Cornell Aeronautical Laboratory CAL/CM—560 (1949); summary: JAS 18, 460—477 (1951).

[34] Ferrari, C.: Comparison of theoretical and experimental results for the turbulent boundary layer in supersonic flow along a flat plate. JAS 18, 555—564 (1951).

[35] Ferrari, C.: Determination of the heat transfer properties of a turbulent boundary layer in the case of supersonic flow when the temperature distribution along the wall is arbitrarily assigned. Fifty years of boundary-layer research, (W. Tollmien and H. Görtler, eds.), Braunschweig, 1955, 364—384.

[36] Gardner, G.O., and Kestin, J.: Calculation of the Spalding function over a range of Prandtl-numbers. Int. J. Heat Mass Transfer 6, 289—299 (1963).

[37] Goddard, F.E.: Effect of uniformly distributed roughness on turbulent skin friction drag at supersonic speeds. JASS 26, 1—15, 24 (1959).

[38] Hill, F.K.: Boundary-layer measurements in hypersonic flow. JAS 23, 35—42 (1956).

[39] Hill, F.K.: Turbulent boundary layer measurements at Mach numbers from 8 to 10. Phys. Fluids 2, 668—680 (1959).

[40] Hoffmann, E.: Der Wärmeübergang bei der Strömung im Rohr. Z. Ges. Kälte-Ind. *44*, 99—107 (1937).

[41] Johnson, D.S.: Velocity, temperature, and heat transfer measurements in a turbulent boundary layer downstream of a stepwise discontinuity in wall temperature. J. Appl. Mech. *24*, 2—8 (1957).

[42] Johnson, D.S.: Velocity and temperature fluctuation measurements in a turbulent boundary layer downstream of a stepwise discontinuity in wall temperature. Trans. ASME J. Appl. Mech. *26*, 325—336 (1959).

[43] von Kármán, Th.: The problem of resistance in compressible fluids. Volta Congress Rome 1935, 222—277; see also Coll. Works *III*, 179—221.

[44] von Kármán, Th.: The analogy between fluid friction and heat transfer. Trans. ASME *61*, 705—710 (1939); see also Coll. Works *III*, 355—367.

[45] Kestin, J., and Richardson, P.D.: Heat transfer across turbulent incompressible boundary layers. Int. J. Heat and Mass Transfer *6*, 147—189 (1963).

[46] Kestin, J., and Richardson, P.D.: Wärmeübertragung in turbulenten Grenzschichten. Forschg. Ing.-Wes. *29*, 93—104 (1963).

[46a] Kestin, J., and Persen, L.N.: The transfer of heat across a turbulent boundary layer at very high Prandtl numbers. Int. J. Heat and Mass Transfer *5*, 355—371 (1962).

[47] Kistler, A.L.: Fluctuation measurements in a supersonic turbulent boundary layer. Phys. Fluids *2*, 290—296 (1959).

[48] Klebanoff, P.S.: Characteristics of turbulence in a boundary layer with zero pressure gradient. NACA TN 3178 (1954); TR 1247, 1135—1153 (1955).

[49] Kovasznay, L.S.G.: The hot-wire anemometer in supersonic flow. JAS *17*, 565—573 (1950).

[50] Kutateladze, S.S., and Leont'ev, A.I.: Turbulent boundary layer in compressible gases. Transl. by D.B. Spalding. Edward Arnold Publishers Ltd., London, 1964.

[51] Lilley, G.M.: An approximation solution of the turbulent boundary layer equation in incompressible and compressible flow. Coll. Aero. Cranfield Rep. 134 (1960).

[52] Liepmann, H.W., and Goddard, F.E.: Note on the Mach number effect upon the skin friction of rough surfaces. JAS *24*, 784 (1957).

[53] Lobb, R.K., Winkler, E.M., and Persh, J.: Experimental investigation of turbulent boundary layers in hypersonic flow. NAVORD Rep. 3880 (1955).

[54] Ludwieg, H.: Ein Gerät zur Messung der Wandschubspannung turbulenter Reibungsschichten. Ing.-Arch. *17*, 207—218 (1949).

[55] Ludwieg, H.: Bestimmung des Verhältnisses der Austauschkoeffizienten für Wärme und Impuls bei turbulenten Grenzschichten. ZFW *4*, 73—81 (1956).

[56] Mack, L.M.: An experimental investigation of the temperature recovery-factor. Jet Propulsion Laboratory, Calif. Inst. Techn., Pasadena, Rep. 20—80 (1954).

[57] Mager, A.: Transformation of the compressible turbulent boundary layer. JAS *25*, 305—311 (1958).

[58] Matting, F.W., Chapman, D.R., Nyholm, J.R., and Thomas, A.G.: Turbulent skin friction at high Mach numbers and Reynolds numbers in air and helium. NASA TR R—82 (1961).

[59] McLafferty, G.H., and Babber, R.E.: The effect of adverse pressure gradients on the characteristics of turbulent boundary layers in supersonic streams. JASS *29*, 1—10, 18 (1962).

[60] Meier, H.U.: Experimentelle und theoretische Untersuchungen von turbulenten Grenzschichten bei Überschallströmung. Mitt. MPI Strömungsforschg. u. Aerodyn. Versuchsanst. Nr. 49, 1—116, (1970); Diss. Braunschweig 1970.

[61] Meier, H.U., Lee, R.E., and Voisinet, R.L.P.: Vergleichsmessungen mit einer Danberg-Temperatursonde und einer kombinierten Druck-Temperatursonde in turbulenten Grenzschichten bei Überschallströmung. ZFW. *22*, 1—10 (1974).

[62] Meier, H.U., Voisinet, R.L.P., and Gates, D.F.: Temperature distributions using the law of the wall for compressible flow with variable turbulent Prandtl numbers. AIAA 7th Fluid and Plasma Dynamics Conf., Palo Alto, Calif. 1974, AIAA Paper No. 74—596 (1974).

[63] Meier, H.U., and Rotta, J.C.: Temperature distributions in supersonic turbulent boundary layers. AIAA J. *9*, 2149—2156 (1971).

[64] Meier, H.U.: Investigation of the heat transfer mechanism in supersonic turbulent boundary layers. Wärme- und Stoffübertragung *8*, 159—165 (1975).

[65] Morkovin, M.V.: Effects of compressibility on turbulent flows. Colloques Int. CNRS No. 108, 367—380, Mécanique de la turbulence, Marseille, 1962.

[66] Nunner, W.: Wärmeübergang und Druckabfall in rauhen Rohren. VDI-Forsch. 455 (1956).

[67] Owen, P.R., and Thomson, W.R.: Heat transfer across rough surfaces. JFM *15*, 321—334 (1943).

[68] Pappas, C.C.: Measurement of heat transfer in the turbulent boundary layer on a flat plate in supersonic flow and comparison with skin friction results. NACA TN 3222 (1954).

[69] Persen, L.N.: A note on the basic equations of turbulent boundary layers and the heat transfer through such layers. ZFW *15*, 311—314 (1967).

[70] Prandtl, L.: Eine Beziehung zwischen Wärmeaustausch und Strömungswiderstand der Flüssigkeiten. Phys. Z. *11*, 1072—1078 (1910); see also Coll. Works *II*, 585—596.

[71] Reichardt, H.: Die Wärmeübertragung in turbulenten Reibungsschichten. ZAMM *20*, 297—328 (1940); NACA TM 1047 (1943).

[72] Reichardt, H.: Impuls- und Wärmeaustausch bei freier Turbulenz. ZAMM *24*, 268—272 (1944).

[73] Reichardt, H.: Der Einfluss der wandnahen Strömung auf den turbulenten Wärmeübergang. Rep. Max-Planck-Inst. für Strömungsforschung No. 3, 1—63 (1950).

[74] Reichardt, H.: Die Grundlagen des turbulenten Wärmeüberganges. Arch. Wärmetechn. *2*, 129—142 (1951).

[75] Reshotko, E., and Tucker, M.: Approximate calculation of the compressible turbulent boundary layer with heat transfer and arbitrary pressure gradient. NACA TN 4154 (1957).

[76] Reynolds, O.: On the extent and action of the heating surface for steam boilers. Proc. Manchester Lit. Phil. Soc. *14*, 7—12 (1874).

[77] Reynolds, W.C., Kays, W.M., and Kline, S.J.: Heat transfer in the turbulent incompressible boundary layer. I. Constant wall temperature. NASA Memo. 12—1—58 W (1958); II. Step wall temperature distribution. NASA Memo. 12—2—58 W (1958); III. Arbitrary wall temperature and heat flux. NASA Memo. 12—3—58 W (1958); IV. Effect of location of transition and prediction of heat transfer in a known transition region. NASA Memo. 12—4—58 W (1958).

[78] Rotta, J.C.: Über den Einfluss der Machschen Zahl und des Wärmeübergangs auf das Wandgesetz turbulenter Strömungen. ZFW *7*, 264—274 (1959).

[79] Rotta, J.C.: Turbulent boundary layers with heat transfer in compressible flow. AGARD Rep. 281 (1960).

[80] Rotta, J.C.: Bemerkung zum Einfluss der Dichteschwankungen in turbulenten Grenzschichten bei kompressibler Strömung. Ing.-Arch. *32*, 187—190 (1963).

[81] Rotta, J.C.: Temperaturverteilungen in der turbulenten Grenzschicht an der ebenen Platte. Int. J. Heat Mass Transfer *7*, 215—228 (1964).

[82] Rotta, J.C.: Effect of streamwise wall curvature on compressible turbulent boundary layers. IUTAM Symp. Kyoto, Japan, 1966. Phys. Fluids *10*, S 174—S 180 (1967).

[83] Rotta, J.C.: Eine Beziehung zwischen den örtlichen Reibungsbeiwerten turbulenter Grenzschichten bei kompressibler und inkompressibler Strömung. ZFW *18*, 195—201 (1970).

[84] Rotta, J.C.: FORTRAN IV — Rechenprogramm für Grenzschichten bei kompressiblen ebenen und achsensymmetrischen Strömungen. DLR FB 71—51, 1—82 (1971).

[85] Rubesin, M.W.: A modified Reynolds analogy for the compressible turbulent boundary layer on a flat plate. NACA TN 2917 (1953).

[86] Schubauer, G.B., and Tchen, C.M.: Turbulent flow. High Speed Aerodynamics and Jet Propulsion *V*, 75—195, Princeton (1959).

[87] Seiff, A.: Examination of the existing data on the heat transfer of turbulent boundary layers at supersonic speeds from the point of view of Reynolds analogy. NACA TN 3284 (1954).

[88] Spalding, D.B.: Heat transfer to a turbulent stream from a surface with a step-wise discontinuity in wall temperature. International developments in heat transfer (Proc. Conf. organized by ASME at Boulder, Colorado, 1961), Part II, 439—446.

[89] Spalding, D.B., and Chi, S.W.: The drag of a compressible turbulent boundary layer on a smooth flat plate with and without heat transfer. JFM *18*, 117—143 (1964).

[90] Spence, D.A.: Velocity and enthalpy distributions in the compressible turbulent boundary layer on a flat plate. JFM *8*, 368—387 (1960).

[91] Spence, D.A.: Some applications of Crocco's integral for the turbulent boundary layer. Proc. 1960 Heat Transfer Fluid Mech. Inst., Stanford Univ. 62—76 (1960).

[92] Spence, D.A.: The growth of compressible turbulent boundary layers on isothermal and adiabatic walls. ARC RM 3191 (1961).

[93] Stratford, B.S., and Beavers, G.S.: The calculation of the compressible turbulent boundary layer in an arbitrary pressure gradient. A correlation of certain previous methods. ARC RM 3207 (1959).

[94] Smith, P.D.: An integral prediction method for three-dimensional compressible turbulent boundary layers. ARC RM 3739, 1—54 (1974).

[95] Schultz-Jander, B.: Heat transfer calculations in turbulent boundary layers using integral relations. Acta Mechanica 21, 301—312 (1975).

[96] Taylor, G.I.: Conditions at the surface of a hot body exposed to the wind. ARC RM 272 (1919).

[97] Taylor, G.I.: The transport of vorticity and heat through fluids in turbulent motion. Appendix by A. Fage and V. M. Falkner. Proc. Roy. Soc. 135, 685 (1932); see also Phil. Trans. A 215, 1 (1915).

[98] Taylor, J.R.: Temperature and heat flux distributions in incompressible turbulent equilibrium boundary layers. Int. J. Heat Mass Transfer 15, 2473—2488 (1972).

[99] Tucker, M.: Approximate turbulent boundary layer development in plane compressible flow along thermally insulated surfaces with application to supersonic-tunnel contour correction. NACA TN 2045, 78 (1950).

[100] Walz, A.: Näherungstheorie für kompressible turbulente Grenzschichten. ZAMM-Sonderheft 36, 50—56 (1956).

[101] Walz, A.: Über Fortschritte in Näherungstheorie und Praxis der Berechnung kompressibler laminarer und turbulenter Grenzschichten mit Wärmeübergang. ZFW 13, 89—102 (1965).

[102] Wilson, R.E.: Turbulent boundary layer characteristics at supersonic speeds — Theory and experiment. JAS 17, 585—594 (1950).

[103] Winkler, E.M.: Investigation of flat plate hypersonic turbulent boundary layers with heat transfer. J. Appl. Mech. 83, 323—329 (1961).

[104] Winkler, E.M., and Cha, M.H.: Investigation of flat plate hypersonic turbulent boundary layers with heat transfer at a Mach number of 5·2 (U). NAVORD Rep. 6631 (1959).

[105] Winter, K.G., Rotta, J.C., and Smith, K.G.: Untersuchungen der turbulenten Grenzschicht an einem taillierten Drehkörper bei Unter- und Überschallströmung. DLR FB 65—52, 1—71 (1965); see also ARC RM 3633, 1—75 (1970).

[106] Young, A.D.: The drag effects of roughness at high subcritical speeds. J. Roy. Aero. Soc. 18, 534 (1950).

Additional references to Chap. XXII, page 701

[119] White, F.M.: Viscous fluid flow. McGraw-Hill, New York, 1974, 725 pp.

[120] Wieghardt, K.: Turbulente Grenzschichten. Göttinger Monographie, Part B 5 (1945/46).

[121] Wilcken, H.: Turbulente Grenzschichten an gewölbten Wänden. Ing.-Arch. 1, 357—376 (1930).

[122] Williams, J.: British research on boundary layer and flow control for high lift by blowing. ZFW 6, 143—160 (1958).

[122a] Willmarth, W.W., Winkel, R.E., Sharma, L.K., and Bogar, T.J.: Axially symmetric turbulent boundary layers on cylinders: mean velocity profiles and wall pressure fluctuations JFM 76, 35—64 (1976).

[123] Winternitz, F.A.L., and Ramsay, W.J.: Effects of inlet boundary layer on the pressure recovery in conical diffusers. Mech. Eng. Res. Lab., Fluid Mech. Div., East Kilbride, Glasgow, Rep. No. 41 (1956).

[124] Young, A.D.: The calculation of the total and skin friction drags of bodies of revolution at 0° incidence. ARC RM 1874 (1939).

[125] Young, A.D., and Majola, O.O.: An experimental investigation of the turbulent boundary layer along a streamwise corner. Turbulent Shear Flow, AGARD Conf. Proc. 93, 12—1 to 12—9 (1971).

[126] Young, A.D.: Some special boundary-layer problems. 20th Prandtl Memorial Lecture, ZFW 1, 401—414 (1977).

CHAPTER XXIV

Free turbulent flows; jets and wakes

a. General remarks

In the preceding chapters we have considered turbulent flows along solid walls and we propose to continue the study of turbulent streams with the discussion of several examples of so-called *free turbulent flow*. Turbulent flows will be termed free if they are not confined by solid walls. We shall discern three kinds of turbulent free flows, Fig. 24.1: free jet boundaries, free jets, and wakes.

A jet boundary occurs between two streams which move at different speeds in the same general direction. Such a surface of discontinuity in the velocity of flow is unstable and gives rise to a zone of turbulent mixing downstream of the point, where the two streams first meet. The width of this mixing region increases in a downstream direction, Fig. 24.1 a.

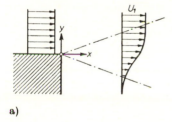

a)

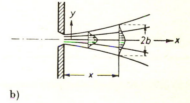

b)

Fig. 24.1.   Examples of free turbulent flows;
a) jet boundary, b) free jet, c) wake

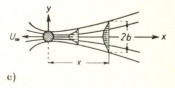

c)

A free jet is formed when a fluid is discharged from a nozzle or orifice, Fig. 24.1b. Disregarding very small velocities of flow, it is found that the jet becomes completely turbulent at a short distance from the point of discharge. Owing to turbulence, the emerging jet becomes partly mixed with the surrounding fluid at rest. Particles of fluid from the surroundings are carried away by the jet so that the mass-flow increases in the downstream direction. Concurrently the jet spreads out and its velocity decreases, but the total momentum remains constant. A comprehensive account of the problems of free jets was given by S.I. Pai [26]. See also the book by G.N. Abramovich [1].

A wake is formed behind a solid body which is being dragged through fluid at rest, Fig. 24.1c, or behind a solid body which has been placed in a stream of fluid. The velocities in a wake are smaller than those in the main stream and the losses in the velocity in the wake amount to a loss of momentum which is due to the drag on the body. The spread of the wake increases as the distance from the body is increased and the differences between the velocity in the wake and that outside become smaller.

Qualitatively such flows resemble similar flows in the laminar region (Chaps. IX and XI), but there are large quantitative differences which are due to the very much larger turbulent friction. Free turbulent flows are much more amenable to mathematical analysis than turbulent flows along walls because turbulent friction is much larger than laminar friction in the whole region under consideration. Consequently, laminar friction may be wholly neglected in problems involving free turbulent flows, which is not the case in flows along solid walls. It will be recalled that in the latter case, by contrast, laminar friction must always be taken into account in the immediate neighbourhood of the wall (i. e. in the laminar sub-layer), and that causes great mathematical difficulties.

Furthermore, it will be noted that problems in free turbulent flow are of a *boundary-layer* nature, meaning that the region of space in which a solution is being sought does not extend far in a transverse direction, as compared with the main direction of flow, and that the transverse gradients are large. Consequently it is permissible to study such problems with the aid of the boundary-layer equations. In the two-dimensional incompressible flow these are

$$\frac{\partial u}{\partial t} + u \frac{\partial u}{\partial x} + v \frac{\partial u}{\partial y} = \frac{1}{\varrho} \frac{\partial \tau}{\partial y}, \tag{24.1}$$

$$\frac{\partial u}{\partial x} + \frac{\partial v}{\partial y} = 0. \tag{24.2}$$

Here τ denotes the turbulent shearing stress. The pressure term has been dropped in the equation of motion because in all problems to be considered it is permissible to assume, at least to a first approximation, that the pressure remains constant. In the case of wakes this assumption is satisfied only from a certain distance from the body onwards.

In order to be in a position to integrate the system of equations (24.1) and (24.2), it is necessary to express the turbulent shearing stress in terms of the parameters of the main flow. At present such an elimination can only be achieved with the aid of semi-empirical assumptions. These have already been discussed in Chap. XIX. In this connexion it is possible to make use of Prandtl's mixing length theory, eqn. (19.7):

$$\tau_t = \varrho \, l^2 \left| \frac{\partial u}{\partial y} \right| \frac{\partial u}{\partial y}, \tag{24.3}$$

or of its extension

$$\tau_t = \varrho \, l^2 \frac{\partial u}{\partial y} \sqrt{\left(\frac{\partial u}{\partial y}\right)^2 + l_1^2 \left(\frac{\partial^2 u}{\partial y^2}\right)^2},$$ (24.4)

where the mixing lengths l and l_1 are to be regarded as purely local functions[†]. They must be suitably dealt with in each particular case. Further, it is possible to use Prandtl's hypothesis in eqn. (19.10), namely

$$\tau_t = \rho \, \varepsilon_\tau \frac{\partial u}{\partial y} = \varrho \, \varkappa_1 \, b \, (u_{max} - u_{min}) \frac{\partial u}{\partial y},$$ (24.5)

where b denotes the width of the mixing zone and $\varkappa_1$ is an empirical constant. Moreover

$$\varepsilon_\tau = \varkappa_1 \, b \, (u_{max} - u_{min})$$ (24.5a)

is the virtual kinematic viscosity, assumed constant over the whole width and, hence, independent of y. In addition it is possible to use von Kármán's hypothesis, eqn. (19.19) and that due to G. I. Taylor, eqn. (19.15a).

When either of the assumptions (24.3), (24.4) or (24.5) is used it is found that the results differ from each other only comparatively little. The best measure of agreement with experimental results is furnished by the assumption in eqn. (24.5) and, in addition, the resulting equations are more convenient to solve. For these reasons we shall express a preference for this hypothesis. Nevertheless, some examples will be studied with the aid of the hypotheses in eqns. (24.3) and (24.4) in order to exhibit the differences in the results when different hypotheses are used. Moreover, the mixing length formula, eqn. (24.3), has rendered such valuable service in the theory of pipe flow that it is useful to test its applicability to the type of flow under consideration. It will be recalled that, among others, the universal logarithmic velocity distribution law has been deduced from it.

b. Estimation of the increase in width and of the decrease in velocity

Before proceeding to integrate eqns. (24.1) and (24.2) for several particular cases we first propose to make estimations of orders of magnitude. In this way we shall be able to form an idea of the type of law which governs the increase in the width of the mixing zone and of the decrease in the 'height' of the velocity profile with increasing distance x. The following account will be based on one first given by L. Prandtl [27].

When dealing with problems of turbulent jets and wakes it is usually assumed that the mixing length l is proportional to the width of jet, b, because in this way we are led to useful results. Hence we put

$$\frac{l}{b} = \beta = \text{const}.$$ (24.6)

[†] This extension was not discussed in Chap. XIX because it is used only very rarely.

In addition, the following rule has withstood the test of time: The rate of increase of the width, b, of the mixing zone with time is proportional to the transverse velocity v':

$$\frac{Db}{Dt} \sim v' .$$

Here D/Dt denotes, as usual, the substantive derivative, so that $D/Dt = u\, \partial/\partial x + v\, \partial/\partial y$. According to a previous estimate, eqn. (19.6), we have $v' \sim l\, \partial u/\partial y$, and thus

$$\frac{Db}{Dt} \sim l\frac{\partial u}{\partial y} .$$

Further, the mean value of $\partial u/\partial y$ taken over half the width of the jet may be assumed to be approximately proportional to u_{max}/b. Consequently,

$$\frac{Db}{Dt} = \text{const} \times \frac{l}{b}\, u_{max} = \text{const} \times \beta\, u_{max} . \qquad (24.7)$$

Jet boundary: With the use of the preceding relations we shall now estimate the rate at which the width of the mixing zone which accompanies a free jet boundary increases with the distance, x. For the jet boundary we have

$$\frac{Db}{Dt} \sim u_{max}\frac{db}{dx} . \qquad (24.8)$$

On comparing eqns. (24.8) and (24.7) we obtain

$$\frac{db}{dx} = \text{const} \times \frac{l}{b} = \text{const} ,$$

or

$$b = \text{const} \times x$$

which means that the width of the mixing zone associated with a free jet boundary is proportional to the distance from the point where the two jets meet. The constant of integration which must, strictly speaking, appear in the above equation can be made to vanish by a suitable choice of the origin of the coordinate system.

Two-dimensional and circular jet: Equation (24.8) remains valid in the case of a two-dimensional and of a circular jet, u_{max} denoting now the velocity at the centre-line. Thus in such cases we also have

$$b = \text{const} \times x . \qquad (24.9)$$

The relation between u_{max} and x can be obtained from the momentum equation. Since the pressure remains constant the integral of the x-component of momentum taken over the whole cross-sectional area must remain constant and independent of x, i. e.

$$J = \varrho \int u^2\, dA = \text{const} .$$

In the case of a *two-dimensional jet* we have $J' = \text{const} \times \varrho\, u^2_{max}\, b$, where J' denotes momentum per unit length, and hence $u_{max} = \text{const} \times b^{-1/2}\sqrt{J'/\varrho}$. In view of eqn. (24.9) we have, further,

$$u_{max} = \text{const} \times \frac{1}{\sqrt{x}}\sqrt{\frac{J'}{\varrho}} \quad \text{(two-dimensional jet)} . \qquad (24.10)$$

In the case of a *circular jet* the momentum is

$$J = \text{const} \times \varrho\, u^2_{max}\, b^2$$

and hence

$$u_{max} = \text{const} \times \frac{1}{b}\sqrt{\frac{J}{\varrho}} .$$

In view of eqn. (24.9) we now have

$$u_{max} = \text{const} \times \frac{1}{x}\sqrt{\frac{J}{\varrho}} \quad \text{(circular jet)} . \qquad (24.11)$$

Two-dimensional and circular wake: Instead of eqn. (24.8) we now have

$$\frac{Db}{Dt} = U_\infty \frac{db}{dx} ,$$

and eqn. (24.7) is replaced by

$$\frac{Db}{Dt} = \text{const} \times \frac{l}{b}\, u_1 = \text{const} \times \beta\, u_1 .$$

where $u_1 = U_\infty - u$. On equating the two expressions, we obtain

$$U_\infty \frac{db}{dx} \sim \frac{l}{b}\, u_1 = \beta\, u_1$$

or

$$\frac{db}{dx} \sim \beta\, \frac{u_1}{U_\infty} \quad \text{(two-dimensional and circular wake)} . \qquad (24.12)$$

The calculation of momentum in problems involving wakes differs from that for the case of jets, because now there is a direct relationship between momentum and the drag on the body. As already mentioned, eqn. (9.26), the momentum integral is

$$D = J = \varrho \int u\,(U_\infty - u)\, dA ,$$

provided that the control surface has been placed so far behind the body that the static pressure has become equal to that in the undisturbed stream. At a large distance behind the body $u_1 = U_\infty - u$ is small compared with U_∞ so that we may put $u(U_\infty - u) = (U_\infty - u_1)\, u_1 \approx U_\infty\, u_1$. Thus for two-dimensional and circular wakes

$$J = D \approx \varrho\, U_\infty \int u_1\, dA . \qquad (24.13)$$

Two-dimensional wake: Let h denote the height of the cylindrical body and d its diameter; its drag will then be $D = \frac{1}{2}\, c_D\, \varrho\, U_\infty^2\, h\, d$ and the momentum, eqn. (24.13), is $J \sim \varrho\, U_\infty\, u_1\, h\, b$. Equating, according to eqn. (24.13), we have

$$\frac{u_1}{U_\infty} \sim \frac{c_D\, d}{2\, b} . \qquad (24.14)$$

Inserting eqn. (24.12) for the rate of increase in width, we obtain

$$2\,b\,\frac{db}{dx} \sim \beta\,c_D\,d$$

or

$$b \sim (\beta\,x\,c_D\,d)^{1/2} \quad \text{(two-dimensional wake)} . \tag{24.15}$$

Inserting this value into eqn. (24.14) we find that the rate at which the 'depression' in the velocity curve decreases downstream is represented by

$$\frac{u_1}{U_\infty} \sim \left(\frac{c_D d}{\beta x}\right)^{\frac{1}{2}} \quad \text{(two-dimensional wake)} . \tag{24.16}$$

In other words, the width of a two-dimensional wake increases as $\sqrt{x}$ and the velocity decreases as $1/\sqrt{x}$.

Circular wake: Denoting the frontal area of the body by A we can write its drag as $D = \frac{1}{2}\,c_D\,A\,\varrho\,U_\infty^2$ and the momentum, eqn. (24.13), becomes $J \sim \varrho\,U_\infty\,u_1\,b^2$. Equating D and J, we obtain

$$\frac{u_1}{U_\infty} \sim \frac{c_D A}{b^2} . \tag{24.17}$$

Inserting this value into eqn. (24.12), we find that the increase in width is given by

$$b^2\,\frac{db}{dx} \sim \beta\,c_D\,A$$

or

$$b \sim (\beta\,c_D\,A\,x)^{1/3} \quad \text{(circular wake)} . \tag{24.18}$$

Inserting eqn. (24.18) into (24.17) we find for the decrease in the depression in the velocity profile the expression

$$\frac{u_1}{U_\infty} \sim \left(\frac{c_D A}{\beta^2 x^2}\right)^{\frac{1}{3}} \quad \text{(circular wake)} . \tag{24.19}$$

Table 24.1.   Power laws for the increase in width and for the decrease in the centre-line velocity in terms of distance x for problems of free turbulent flow

	laminar		turbulent	
	width b	centre-line velocity u_{max} or u_1	width b	centre-line velocity u_{max} or u_1
Free jet boundary	$x^{1/2}$	x^0	x	x^0
Two-dimensional jet	$x^{2/3}$	$x^{-1/3}$	x	$x^{-1/2}$
Circular jet	x	x^{-1}	x	x^{-1}
Two-dimensional wake	$x^{1/2}$	$x^{-1/2}$	$x^{+1/2}$	$x^{-1/2}$
Circular wake	$x^{1/2}$	x^{-1}	$x^{+1/3}$	$x^{-2/3}$

Thus, for a circular wake we find that the width of the wake increases in proportion to $x^{1/3}$ and that the velocity decreases in proportion to $x^{-2/3}$.

The power-laws for the width and for the velocity in the centre have been summarized in Table 24.1. The corresponding laminar cases which were partly considered in Chaps. IX and XI have been added for completeness.

c. Examples

The preceding estimates give in themselves a very good idea of the essential features encountered in problems involving free turbulent flows. We shall, however, now go one step further and shall examine several particular cases in much greater detail deducing the complete velocity distribution function from the equations of motion. In order to achieve this result it is necessary to draw on one of the hypotheses in eqns. (24.3) to (24.5). The examples which have been selected here for consideration all have the common feature that the *velocity profiles* which occur in them are *similar* to each other. This means that the velocity profiles at different distances x can be made congruent by a suitable choice of a velocity and a width scale factor.

1. The smoothing out of a velocity discontinuity. As our first example we shall consider the problem of the smoothing out of a velocity discontinuity which was first treated by L. Prandtl [27]. At time $t = 0$ there are two streams moving at two different velocities, U_1 and U_2 respectively, their boundary being at $y = 0$ (Fig. 24.2). As already mentioned, the boundary across which the velocity varies discontinuously is unstable and the process of turbulent mixing smoothes out the transition so that it becomes continuous. The width of the zone over which this continuous transition from velocity U_1 to velocity U_2 takes place increases with increasing time. We are here concerned with a problem in non-steady parallel flow for which

$$u = u(y, t) \; ; \quad v = 0 \, . \tag{24.20}$$

The convective terms in eqn. (24.1) vanish identically. Making use of Prandtl's mixing theory, eqn. (24.3), we can transform eqn. (24.1) to give

$$\frac{\partial u}{\partial t} = l^2 \left| \frac{\partial u}{\partial y} \right| \frac{\partial^2 u}{\partial y^2} \, . \tag{24.21}$$

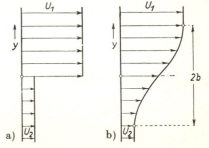

Fig. 24.2.   The smoothing out of a velocity discontinuity, after Prandtl [27]; a) Initial pattern $(t = 0)$, b) Pattern at later instant

The width of the mixing zone, b, increases with time and $b = b(t)$; the mixing length is assumed to be proportional to b in the same way as before so that $l = \beta\, b$. Assuming that the velocity profiles are similar, we may put

$$u \sim f(\eta)$$

with $\eta = y/b$ and $b \sim t^p$. The exponent p in the expression for the width can be determined from the condition that in eqn. (24.21) the acceleration and frictional terms must be proportional to equal powers of time, t. Thus $\partial u/\partial t$ is proportional to t^{-1}, whereas the right-hand side is proportional to $t^{2p-3p} = t^{-p}$, so that $p = 1$.

In this manner we obtain the following assumptions for the problem in hand:

$$b = B\,t\,; \qquad \eta = \frac{y}{b} = \frac{y}{B\,t}\,.$$

The velocity u is best assumed to be of the form

$$u = \tfrac{1}{2}\,(U_1 + U_2) + \tfrac{1}{2}\,(U_1 - U_2)\,f\,(\eta) \tag{24.22}$$

or

$$u = U_m + A\,f\,(\eta)\,,$$

with $U_m = \tfrac{1}{2}(U_1 + U_2)$ and $A = \tfrac{1}{2}(U_1 - U_2)$. In order to make sure that at the edges of the mixing zone, i. e. at $y = \pm\,b$, the velocity becomes equal to U_1 and U_2 respectively, we must put $f = \pm\,1$ at $\eta = \pm\,1$. Inserting the value from eqn. (24.22) into eqn. (24.21) we obtain the following differential equation for $f(\eta)$:

$$\eta\,f' + \frac{\beta^2\,A}{B}\,f'\,f'' = 0\,.$$

The equation has one solution $f' = 0$, i. e. $f = \mathrm{const}$, which represents the trivial case of a constant velocity. If, however, f' differs from zero, we may divide through, whence we find

$$\eta + \frac{\beta^2\,A}{B}\,f'' = 0\,.$$

Upon integration, we have

$$f\,(\eta) = c_0\,\eta^3 + c_1\,\eta\,,$$

with $c_0 = -\,B/6\,\beta^2\,A$. The above solution satisfies the condition $f(0) = 0$ so that the constants c_0 and c_1 can be determined from the condition $f(\eta) = 1$ and $f'(\eta) = 0$ at $y = b$, i. e. at $\eta = 1$. Hence,

$$c_0 = -\tfrac{1}{2}\,, \qquad c_1 = \tfrac{3}{2}\,.$$

Introducing these values into eqn. (24.22) we obtain the solution in its final form

$$u\,(y,\,t) = \frac{1}{2}\,(U_1 + U_2) + \frac{1}{2}\,(U_1 - U_2)\left[\frac{3}{2}\left(\frac{y}{b}\right) - \frac{1}{2}\left(\frac{y}{b}\right)^3\right] \tag{24.23}$$

with

$$b = \tfrac{3}{2}\,\beta^2\,(U_1 - U_2)\,t\,. \tag{24.24}$$

The velocity distribution from eqn. (24.23) is seen plotted in Fig. 24.2. It has the remarkable property that the velocity in the mixing region does not go over into the two free-stream velocities asymptotically. Transition occurs at a finite distance $y = b$ with a discontinuity in $\partial^2 u / \partial y^2$. This is a general property of all solutions obtained on the basis of Prandtl's hypothesis (24.3) for the shearing stress in turbulent flow. It constitutes what may be called an esthetical deficiency of this hypothesis. The improved hypotheses (24.4) or (24.5) are free of this blemish.

The quantity $\beta = l/b$ is the only empirical constant which appears in the solution; it can be determined solely from experimental data.

2. Free jet boundary. The conditions at a free jet boundary are closely related to those in the preceding example. With reference to Fig. 24.1a we shall consider the more general case when at $x = 0$ there is a meeting of two streams whose constant velocities are U_1 and U_2, respectively, it being assumed that $U_1 > U_2$. Downstream of the point of encounter the streams will form a mixing zone whose width b increases proportionately to x, Fig. 24.1a. The first solution to the problem under consideration was given by W. Tollmien [52], who made use of Prandtl's mixing length hypothesis for turbulent shear, eqn. (24.3). We shall review here the mathematically simpler solution due to H. Goertler [18] who based it on Prandtl's hypothesis in eqn. (24.5). Since the virtual kinematic viscosity ε is independent of y, eqns. (24.1) and (24.5) give

$$u \frac{\partial u}{\partial x} + v \frac{\partial u}{\partial y} = \varepsilon_\tau \frac{\partial^2 u}{\partial y^2} . \tag{24.25}$$

Putting $b = c\,x$ we obtain the following expression for the virtual kinematic viscosity, eqn. (24.5a), which is applicable to our case:

$$\varepsilon_\tau = \varkappa_1\, c\, x (U_1 - U_2) . \tag{24.26}$$

In view of the similarity of the velocity profiles u and v are functions of y/x. Putting $\xi = \sigma\, y/x$ we can integrate the equation of continuity by the adoption of a stream function $\psi = x\, U\, F(\xi)$ where $U = \frac{1}{2}(U_1 + U_2)$. Then $u = U\, \sigma\, F'(\xi)$ and eqn. (24.25) leads to the following differential equation for $F(\xi)$:

$$F''' + 2\,\sigma^2\, F\, F'' = 0 , \tag{24.27}$$

where $\sigma = \frac{1}{2}(\varkappa_1\, c\, \lambda)^{-1/2}$ and $\lambda = (U_1 - U_2)/(U_1 + U_2)$. The boundary conditions are $\xi = \pm \infty : F'(\xi) = 1 \pm \lambda$. The differential equation (24.27) is identical with Blasius' equation for the flat plate at zero incidence, eqn. (7.28), but the present boundary conditions are different. H. Goertler solved eqn. (24.27) by assuming a power-series expansion of the form

$$\sigma\, F\,(\xi) = F_0\,(\xi) + \lambda\, F_1\,(\xi) + \lambda^2\, F_2\,(\xi) + \cdots \tag{24.28}$$

with $F_0 = \xi$. Substituting (24.28) into (24.27) and arranging in ascending powers of λ, we obtain a system of differential equations which is solved by recursion. The first of the differential equations is of the form

$$F_1''' + 2\,\xi\, F_1'' = 0 , \tag{24.29}$$

with the boundary conditions $F'_1(\xi) = \pm 1$ at $\xi = \pm \infty$. The solution of (24.29) is given by the error function

$$F_1'(\xi) = \operatorname{erf} \xi = \frac{2}{\sqrt{\pi}} \int_0^{\xi} e^{-z^2}\, dz .$$

The contributions of the succeeding terms of the series in eqn. (24.28) are not significant. Hence the solution becomes

$$u = \frac{U_1 + U_2}{2} \left\{ 1 + \frac{U_1 - U_2}{U_1 + U_2} \operatorname{erf} \xi \right\} \tag{24.30}$$

with

$$\xi = \sigma \frac{y}{x} . \tag{24.30a}$$

Figure 24.3 compares the theoretical solution with H. Reichardt's [29] measurements for the case when $U_2 = 0$ and agreement is seen to be very good. The quantity σ is the only empirical constant left free to be adjusted from experiment. According to the measurements performed by H. Reichardt the width $b_{0.1}$ of the mixing zone, measured between stations where $(u/U_1)^2 = 0.1$ (corresponding to $\xi = -0.345$) and $(u/U_1)^2 = 0.9$ (corresponding to $\xi = 0.975$) has the value $b_{0.1} = 0.098\,x$, which yields $\sigma = 13.5$. The virtual kinematic viscosity becomes $\varepsilon = 0.014\,b_{0.1} \times U_1$.

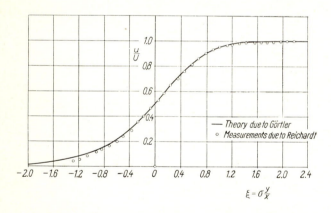

Fig. 24.3. Velocity distribution in the mixing zone of a jet; $\sigma = 13.5$

Blunt body: The process of turbulent mixing that occurs in the wake behind a blunt body was explored in detail by M. Tanner [49]. The results are displayed in Fig. 24.4. At each edge behind a blunt two-dimensional body or around the sharp circular edge behind cylindrical bodies there form mixing zones of the kind sketched in the figure. The velocity distribution across such a zone is of the same shape as that in Fig. 24.3; it can be described by eqn. (24.30). The similarity parameter σ from eqn. (24.30a) strongly depends on the angle ϕ of the two-dimensional wedge or axially symmetric cone. This dependence is represented graphically by Fig. 24.4. The parameter σ decreases considerably as the wedge angle ϕ is increased. For $\phi = 180°$ (plate at right angles to the flow direction) the value of σ is only one half of that for $\phi = 0$ (free jet). This signifies that in the wake the angle of spread of the mixing

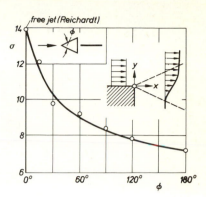

Fig. 24.4. Turbulent mixing zone in the wake close behind a squat wedge-like body as investigated by M. Tanner [49]. The similarity parameter σ from eqn. (24.30a) represented as a function of the wedge angle ϕ

zone behind a flat plate at right angles to the flow is about double of that in a free stream. However, this is true only for the case when a flat splitter plate is placed in the wake to prevent the formation of a von Kármán vortex street.

W. Szablewski [46, 47, 48] extended these calculations, as well as those given in Sec. XXIVc1, to cases when there is a large difference in the densities of the two streams, but a small difference in their velocities. It turns out that the widths of the mixing zones are affected only very slightly by this difference in density. Nevertheless, as the difference in the densities is increased, the zone of mixing becomes displaced in the direction of the less dense jet. The preceding results can also be applied when the two jets differ in their chemical concentrations. P. B. Gooderum, G. P. Wood and H. J. Brevoort [17] carried out an experimental investigation into the conditions at the free boundary of a supersonic jet. The results showed that the mixing zone is somewhat narrower and the level of turbulence is somewhat smaller than in incompressible flow.

3. Two-dimensional wake behind a single body. Two-dimensional wakes were first investigated by H. Schlichting [35] in his thesis presented to Goettingen University. The investigation was based on Prandtl's mixing length hypothesis, eqn. (24.3). A solution for the same problem which was based on Prandtl's hypothesis in eqn. (24.5) was later given by H. Reichardt [29] and H. Goertler [18]. We shall now give a short account of both solutions in order to illustrate the fact that the two results do not differ much one from the other.

In the case of a wake, the velocity profiles become similar only at large distances downstream from the body, there being no similarity at smaller distances. We shall restrict ourselves to the consideration of large distances x so that the velocity difference

$$u_1 = U_\infty - u \tag{24.31}$$

is small compared with the free stream velocity U_∞. At large distances the static pressure in the wake is equal to the static pressure in the free stream. Consequently, the application of the momentum theorem to a control surface which encloses the body, assumed to be a cylinder of height h, gives

$$D = h \varrho \int\limits_{y=-\infty}^{+\infty} u\, (U_\infty - u)\, \mathrm{d}y = h \varrho \int\limits_{y=-\infty}^{+\infty} u_1\, (U_\infty - u_1)\, \mathrm{d}y \,.$$

Neglecting $u_1{}^2$, we obtain

$$D = h \varrho\, U_\infty \int\limits_{y=-\infty}^{+\infty} u_1\, \mathrm{d}y \,.$$

Substituting $D = \tfrac{1}{2} c_D\, d\, h\, \varrho\ U_\infty^2$, where d denotes the thickness of the cylinder, we obtain

$$\int\limits_{y=-\infty}^{+\infty} u_1\, \mathrm{d}y = \tfrac{1}{2} c_D\, d\, U_\infty \,. \qquad (24.32)$$

As deduced in Sec. XXIVb, the width and the velocity difference vary in a manner to give $b \sim x^{1/2}$ and $u_1 \sim x^{-1/2}$.

Shearing stress hypothesis from eqn. (24.3): Since the term $v\, \partial u / \partial y$ in eqn. (24.1) is small, we obtain

$$- U_\infty \frac{\partial u_1}{\partial x} = 2\, l^2\, \frac{\partial u_1}{\partial y}\, \frac{\partial^2 u_1}{\partial y^2} \,. \qquad (24.33)$$

It is assumed that the mixing length l is constant over the width b and proportional to it, i. e. that $l = \beta\, b\, (x)$. In view of the similarity of the velocity profiles the ratio $\eta = y/b$ is introduced as the independent variable. In agreement with the power laws for the width and for the depth of depression in the velocity profile we make the assumptions:

$$b = B\, (c_D\, d\, x)^{1/2} \qquad (24.34)$$

$$u_1 = U_\infty \left(\frac{x}{c_D d}\right)^{-\frac{1}{2}} f\, (\eta) \,. \qquad (24.35)$$

Inserting into eqn. (24.33), we are led to the following differential equation for the function $f(\eta)$:

$$\frac{1}{2}\, (f + \eta\, f') = \frac{2\beta^2}{B}\, f'\, f''$$

with the boundary conditions $u_1 = 0$ and $\partial u_1 / \partial y = 0$ at $y = b$, i. e. $f = f' = 0$ at $\eta = 1$. Integrating once, we obtain

$$\frac{1}{2}\, \eta\, f = \frac{\beta^2}{B}\, f'^2 \,,$$

where the constant of integration has been made equal to zero in view of the boundary condition. Repeated integration yields

$$f = \frac{1}{9}\, \frac{B}{2\,\beta^2}\, (1 - \eta^{3/2})^2 \,.$$

Now it only remains to determine the constant of integration B from the momentum integral (24.32); we thus obtain $B = \sqrt{10}\,\beta$†, and the final solution becomes

$$b = \sqrt{10}\,\beta\,(x\,c_D\,d)^{1/2} \tag{24.36}$$

$$\frac{u_1}{U_\infty} = \frac{\sqrt{10}}{18\,\beta}\left(\frac{x}{c_D d}\right)^{-\frac{1}{2}}\left\{1 - \left(\frac{y}{b}\right)^{\frac{3}{2}}\right\}^2. \tag{24.37}$$

It is noticed that the resulting width has a finite magnitude, the same feature having been observed in connexion with the solution for the smoothing out of a velocity discontinuity, for which the same assumption for shearing stress had been used. At the edge, $y = b$, there is again a discontinuity in the curvature of the velocity profile. Moreover, in the centre at $y = 0$ the second derivative $\partial^2 u/\partial y^2$ even becomes infinitely large and the velocity profile exhibits here a sharp kink.

The results of this theoretical calculation, eqn. (24.37) have been compared with Schlichting's measurements [35] in Fig. 24.5. The measurements were performed in the wake behind a circular cylinder, and the theoretical curve is shown as curve (1). It is seen that there is an excellent measure of agreement. The single free constant, the constant β in eqns. (24.36) and (24.37), must, again, be determined on the basis of measured values. The value of β can be deduced from Fig. 24.6 in which the width of the wake has been plotted against the distance, x, from the body. The measured points have been obtained by H. Reichardt [29] and H. Schlichting [35] in the wakes behind circular cylinders of different diameters d. According to these, $b_{1/2} = \frac{1}{4}(x\,c_D\,d)^{1/2}$, where $b_{1/2}$ denotes half the width at half depth. Since

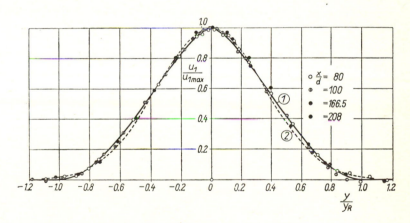

Fig. 24.5.   Velocity distribution in a two-dimensional wake behind a circular cylinder. Comparison between theory and measurement after Schlichting [35]

Theory: curve (1) corresponds to eqn. (24.37); curve (2) corresponds to eqn. (24.39)

---

† It will be noted that $\displaystyle\int_{-1}^{+1}(1 - \eta^{3/2})^2\,\mathrm{d}\eta = \tfrac{9}{10}$.

$b_{1/2} = 0.441\ b$, we have $0.441\ \sqrt{10}\ \beta = \frac{1}{4}$ and thus

$$\beta = \frac{l}{b} = 0.18\ .$$

The preceding solution constitutes an approximation for large distances x; measurements show that it is valid for $x/c_D\ d > 50$. In the case of smaller distances it is possible to calculate additional terms for the velocity, the terms being proportional to x^{-1} and $x^{-3/2}$, respectively.

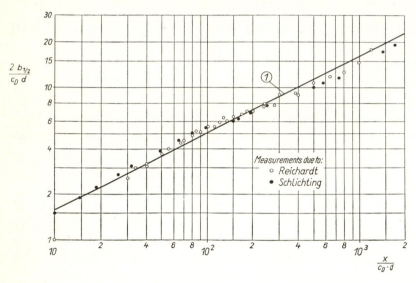

Fig. 24.6.   Two-dimensional wake. Increase in width of wake behind circular cylinder
Curve (1): $b_{1/2} = \frac{1}{4}\ (x\ c_D\ d)^{1/2}$

Shearing stress hypothesis from eqn. (24.5): From eqns. (24.1) and (24.5) we now obtain

$$U_\infty \frac{\partial u_1}{\partial x} = \varepsilon_\tau \frac{\partial^2 u_1}{\partial y^2}\ . \qquad (24.38)$$

The virtual kinematic viscosity is here $\varepsilon_\tau = \varkappa_1\ u_{1max}\ b$ and, hence, constant and equal to ε_0, say. Consequently, the differential equation for u_1 is identical with that for a laminar wake, eqn. (9.30), except that the laminar kinematic viscosity ν must be replaced by ε_0. Thus we can simply copy the solution which was found in Chap. IX. Denoting $\eta = y\ \sqrt{U_\infty/\varepsilon_0 x}$, we obtain from eqns. (9.31) and (9.34) that

$$u_1 = U_\infty\ C\left(\frac{x}{d}\right)^{-\frac{1}{2}}\ \exp\left(-\frac{1}{4}\ \eta^2\right)\ .$$

The constant C follows from the momentum integral, and is

$$C = \frac{c_D}{4\sqrt{\pi}} \sqrt{\frac{U_\infty d}{\varepsilon_0}} \, ,$$

so that finally

$$\frac{u_1}{U_\infty} = \frac{1}{4\sqrt{\pi}} \sqrt{\frac{U_\infty c_D d}{\varepsilon_0}} \left(\frac{x}{c_D d}\right)^{-\frac{1}{2}} \exp\left(-\frac{1}{4}\,\eta^2\right). \tag{24.39}$$

The value of half the width at half the depth is $b_{1/2} = 1{\cdot}675 \sqrt{\varepsilon_0/U_\infty c_D d} \; (x\, c_D\, d)^{1/2}$. Comparing with the preceding measured value of $b_{1/2}$ it is found that the empirical quantity ε_0 has the value

$$\frac{\varepsilon_0}{U_\infty c_D d} = 0{\cdot}0222 \, .$$

Taking into account that $U_\infty c_D d = 2{\cdot}11 \times 2\, b_{1/2}\, u_{1m}$, we have

$$\varepsilon_0 = 0{\cdot}047 \cdot 2\, b_{1/2}\, u_{1m} \, .$$

The preceding solution shows that the velocity distribution in the wake can be represented by Gauss's function. The alternative solution from eqn. (24.39) is seen plotted in Fig. 24.5 as curve (2). The difference between this solution and that in eqn. (24.37) is very small.

W. Tollmien [53] solved the same problem on the basis of von Kármán's hypothesis from eqn. (19.19). In the neighbourhood of the points of inflexion in the velocity profile, where $\partial^2 u/\partial y^2 = 0$, it has proved necessary to make additional assumptions. Extensive experiments, which were carried out by A. A. Townsend [54] in the wake of a cylinder and which were concerned with turbulent fluctuations at Reynolds numbers near 8000, showed that at a distance equal to about 160 to 180 diameters the turbulent microstructure is not yet fully developed. Furthermore, oscillograms taken in the stream demonstrate that the flow is fully turbulent only around the centre, and fluctuates between laminar and turbulent in the neighbourhood of the outer boundaries of the wake. Measurements on circular cylinders at very large Reynolds numbers were described in Chap. II; cf. H. Pfeil [26b].

Circular wakes have been investigated by Miss L. M. Swain [41] who based the calculation on the hypothesis in eqn. (24.3). She obtained the same expression for velocity as in the two-dimensional case, eqn. (24.37), but the power laws for the width and for the centre-line velocity were found to be different, namely $b \sim x^{1/3}$ and $u_{1max} \sim x^{-2/3}$, as already shown in Table 24.1.

Until recently, it has been accepted that the velocity distribution in a wake becomes independent of the shape of the body far enough behind it, and is therefore of a universal form. This belief was put to the test in a series of experiments performed by H. Reichardt and R. Ermshaus [31] and related to wakes behind bodies of revolution. It turned out that in each individual case the velocity profiles remain similar at varying distances behind the body. Nevertheless, the profiles behind bluff bodies (plates, cones with a ratio diameter/height $= 1$) tend to be fuller than those behind slender ones (for example a cone with a ratio diameter/height $= 1/4$ to $1/6$). Differences of this kind have not been observed in two-dimensional wakes.

4. The wake behind a row of bars. The wake behind a row, or cascade, of bodies, such as that behind a row which is composed of a very large number of cylindrical bars whose pitch is equal to λ, Fig. 24.7, is closely related to the wake behind a single body. The present case was investigated both theoretically and experimentally by R. Gran Olsson [19]. At a certain distance from the row, the width of the wake cast by a single element of the row is equal to the pitch, i. e. $b = \lambda$. The velocity difference $u_1 = U_\infty - u$ is here also small compared with U_∞, and eqn. (24.1) can be simplified to

$$- U_\infty \frac{\partial u_1}{\partial x} = \frac{1}{\varrho} \frac{\partial \tau}{\partial y}. \tag{24.40}$$

The calculation for the case in hand becomes very simple when the more general mixing length hypothesis from eqn. (24.4) is used. The first step consists in the determination of the exponent in the power function for the decrease of u_1 with x. On putting $u_1 \sim x^p \, f(y)$, we have $\partial u_1 / \partial x \sim x^{p-1}$. The right-hand side of eqn. (24.40) becomes proportional to $\partial \tau / \partial y \sim (\partial u / \partial y) \cdot (\partial^2 u / \partial y^2) \sim x^{2p}$, because the mixing length, being proportional to the width, is constant. Thus $p - 1 = 2\,p$ and it follows that $p = -1$, or, that the velocity difference u_1 decreases in proportion to x^{-1}.

In the case of fully developed flow the velocity distribution must be expected to be a periodic function in y, whose period is equal to λ. Thus we assume

$$u_1 = U_\infty \, A \left(\frac{x}{\lambda} \right)^{-1} \cos \left(2\pi \, \frac{y}{\lambda} \right).$$

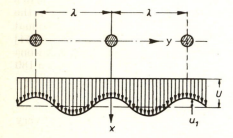

Fig. 24.7. Flow pattern behind a row of bars. Explanatory sketch

The point $y = 0$ has here been made to coincide with the centre of one depression in the velocity distribution, and A is a free constant whose value is still to be determined. We now form the expression for the shearing stress τ from eqn. (24.4) with $l = \text{const}$ and assume that $l_1 = \lambda / 2 \, \pi$, which seems permissible. The result is a very simple expression of the form

$$\frac{1}{\varrho} \frac{\partial \tau}{\partial y} = l^2 \left(\frac{x}{\lambda} \right)^{-2} U_\infty^2 \, A^2 \left(\frac{2\pi}{\lambda} \right)^3 \cos \left(2\pi \, \frac{y}{\lambda} \right).$$

Inserting this expression into eqn. (24.40), we obtain $A = (\lambda/l)^2/8\,\pi^3$ and hence the final solution

$$u_1 = \frac{U_\infty}{8\pi^3} \left(\frac{\lambda}{l} \right)^2 \frac{\lambda}{x} \cos \left(2\pi \, \frac{y}{\lambda} \right). \tag{24.41}$$

According to the measurements performed by R. Gran Olsson, this equation is valid for $x/\lambda > 4$.

Behind a row of circular bars for which $\lambda/d = 8$ the magnitude of the mixing length is given by

$$\frac{l}{\lambda} = 0 \cdot 103 .$$

R. Gran Olsson also studied the case with τ from eqn. (24.3) which implies $l_1 = 0$; with this assumption the calculation becomes much more cumbersome. H. Goertler [18] solved the same problem with the aid of assumption (24.5) for τ and found that the solution was identical with eqn. (24.41)†. A second approximation for smaller distances from the cascade was deduced by G. Cordes [7].

Cascades with a very narrow spacing between the bars are often used in wind tunnels to obtain a locally uniform velocity distribution. But often several jets close in on each other, and this process prevents the velocity from becoming uniform. J. G. von Bohl [5] made a more detailed study of such phenomena and performed experiments on several rows of parallel, polygonal bars varying the solidity m, i. e. the ratio of that portion of the cross-section which is filled by bars to the total channel cross-section over the values $m = 0 \cdot 308$, $0 \cdot 462$ and $0 \cdot 615$. When the value of m is small the single jets remain parallel; the closing-in of jets occurs at about $m = 0 \cdot 37$ to $0 \cdot 46$.

5. The two-dimensional jet. The turbulent two-dimensional jet was first calculated by W. Tollmien [52] who used Prandtl's mixing length hypothesis, eqn. (24.3). In the present section we shall, however, give a short account of the simpler solution based on Prandtl's second hypothesis, eqn. (24.5), which was given by H. Reichardt [29] and H. Goertler [18]. Measurements of the velocity distribution were performed by E. Foerthmann [11] and H. Reichardt [29].

The rate of increase in the width of the jet, $b \sim x$, and that of the decrease in the centre-line velocity, $U \sim x^{-1/2}$, have already been given in Table 24.1. Equations (24.1) and (24.5) lead to the differential equation

$$u \frac{\partial u}{\partial x} + v \frac{\partial u}{\partial y} = \varepsilon_\tau \frac{\partial^2 u}{\partial y^2} \tag{24.42}$$

which must be combined with the equation of continuity. The virtual kinematic viscosity is given by

$$\varepsilon_\tau = \varkappa_1 \, b \, U ,$$

where U denotes the centre-line velocity. Denoting the centre-line velocity and the width of the jet at a fixed characteristic distance s from the orifice by U_s and b_s, re-

---

† With $\varepsilon_\tau = K \lambda (u_{max} - u_{min})$, we have $u = \dfrac{U_\infty}{8 \, \pi^2 K} \dfrac{\lambda}{x} \cos \left(2\pi \dfrac{y}{\lambda} \right)$ or, on comparing with eqn. (24.41), $K = \pi (l/\lambda)^2 = 0 \cdot 103^2 = 0 \cdot 0333$. Thus the virtual kinematic viscosity becomes $\varepsilon_\tau = 0 \cdot 0333 \, \lambda (u_{max} - u_{min})$.

spectively, we may write

$$U = U_s \left(\frac{x}{s}\right)^{-\frac{1}{2}} ; \qquad b = b_s \frac{x}{s} .$$

Consequently,

$$\varepsilon_\tau = \varepsilon_s \left(\frac{x}{s}\right)^{\frac{1}{2}} \quad \text{with} \quad \varepsilon_s = \varkappa_1 \, b_s \, U_s .$$

Further, we put

$$\eta = \sigma \frac{y}{x} ,$$

where σ denotes a free constant. The equation of continuity is integrated by the use of a stream function ψ, which is assumed to be of the form

$$\psi = \sigma^{-1} \, U_s \, s^{1/3} \, x^{1/2} \, F(\eta) .$$

Thus

$$u = U_s \left(\frac{x}{s}\right)^{-\frac{1}{2}} F' ; \qquad v = \sigma^{-1} \, U_s \, s^{1/2} \, x^{-1/2} \left(\eta \, F' - \frac{1}{2} F\right) .$$

On substituting into eqn. (24.42) we obtain the following differential equation for $F(\eta)$:

$$\frac{1}{2} F' + \frac{1}{2} F F'' + \frac{\varepsilon_s}{U_s s} \sigma^2 \, F''' = 0 ,$$

with the boundary conditions $F = 0$ and $F' = 1$ at $\eta = 0$, and $F' = 0$ at $\eta = \infty$. Since ε_s contains the free constant $\varkappa_1$, we may put

$$\sigma = \frac{1}{2} \sqrt{\frac{U_s \, s}{\varepsilon_s}} . \tag{24.43}$$

This substitution simplifies the preceding differential equation which can now be integrated twice, whence we obtain

$$F^2 + F' = 1 . \tag{24.44}$$

This is exactly the same equation as that for the two-dimensional laminar jet, eqn. (9.42). Its solution is $F = \tanh \eta$ so that the velocity is $u = U_s \, (x/s)^{-1/2} (1 - \tanh^2 \eta)$. The characteristic velocity can be expressed in terms of the constant momentum per unit length: $J = \varrho \int_{-\infty}^{+\infty} u^2 \, dy$. Hence $J = \frac{4}{3} \varrho \, U_s^2 \, s/\sigma$. With $J/\varrho = K$ (kinematic momentum), we obtain the final form of the solution:

$$\left.\begin{array}{l} u = \dfrac{\sqrt{3}}{2} \sqrt{\dfrac{K\sigma}{x}} (1 - \tanh^2 \eta) , \\[2mm] v = \dfrac{\sqrt{3}}{4} \sqrt{\dfrac{K}{x\sigma}} \left\{2\eta \, (1 - \tanh^2 \eta) - \tanh \eta\right\} . \end{array}\right\} \tag{24.45}$$

The value of the single empirical constant σ was determined experimentally by

H. Reichardt [29] who found that $\sigma = 7\cdot67$. Fig. 24.8 contains a comparison between the theoretical curve from eqn. (24.45) with the measurements due to E. Foerthmann, curve (2). The theoretical curve obtained by W. Tollmien [52] on the basis of

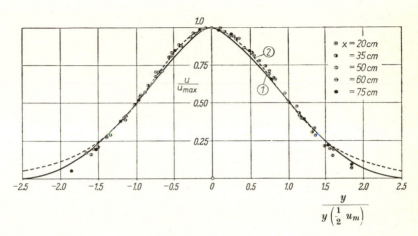

Fig. 24.8. Velocity distribution in a two-dimensional, turbulent jet. Measurements due to Foerthmann [11]

Theory: curve (1) due to Tollmien [52]; curve (2) from eqn. (24.45)

Prandtl's mixing-length hypothesis, curve (1), has also been shown for comparison. The first theoretical curve shows a slightly superior agreement with measurement as it is fuller near its maximum.

From the given numerical value of σ we obtain $\varepsilon_\tau = \dfrac{1\cdot125}{4\,\sigma}\, b_{1/2}\, U$, or

$$\varepsilon_\tau = 0\cdot037\, b_{1/2}\, U \ ,$$

where $b_{1/2}$ again denotes half the width at half depth.

A generalization of this problem consisting in a study of turbulent mixing undergone between a two-dimensional jet with a co-directional external stream was explored by S. Yamaguchi [60]. See also S. Mohammadian [24a] and H. Pfeil et al. [26a].

6. The circular jet. Experimental results on circular jets were given by W. Zimm [61] and P. Ruden [33] as well as by H. Reichardt [29] and W. Wuest [59]. Some results of measurements on circular jets are also contained in the series of reports published by the Aerodynamic Institute in Goettingen [62].

The first theoretical treatment of a circular jet was given by W. Tollmien [52] who based his study on Prandtl's mixing-length theory. In this case, as well as in the preceding one, the assumption for shearing stress given in eqn. (24.5) leads to a considerably simpler calculation. According to Table 24.1 the width of the jet is proportional to x and the centre-line velocity $U \sim x^{-1}$. Thus the virtual kinematic viscosity becomes

$$\varepsilon_\tau = \varkappa_1\, b\, U \sim x^0 = \text{const} = \varepsilon_0$$

which means that it remains constant over the whole of the jet, as it was in the two-dimensional wake. Consequently, the differential equation for the velocity distribution becomes formally identical with that for the laminar jet, it being only necessary to replace the kinematic viscosity, ν, of laminar flow by the virtual kinematic viscosity, ε_0, of turbulent flow. It is, therefore, possible to carry over the solution for the laminar, circular jet, eqns. (11.15) to (11.17). Introducing, once more, the constant kinematic momentum, K, as a measure of the strength of the jet†, we obtain

$$
\left.
\begin{aligned}
u &= \frac{3}{8\pi}\frac{K}{\varepsilon_0 x}\frac{1}{\left(1+\frac{1}{4}\eta^2\right)^2}, \\[2mm]
v &= \frac{1}{4}\sqrt{\frac{3}{\pi}}\frac{\sqrt{K}}{x}\frac{\eta-\frac{1}{4}\eta^3}{\left(1+\frac{1}{4}\eta^2\right)^2}, \\[2mm]
\eta &= \frac{1}{4}\sqrt{\frac{3}{\pi}}\frac{\sqrt{K}}{\varepsilon_0}\frac{y}{x}.
\end{aligned}
\right\}
\qquad (24.46)
$$

The empirical constant is now equal to $\sqrt{K}/\varepsilon_0$. According to the measurement performed by H. Reichardt the width of the jet is given by $b_{1/2} = 0{\cdot}0848\, x$. With $\eta = 1{\cdot}286$ at $u = \frac{1}{2} u_m$ we have $b_{1/2} = 5{\cdot}27\, x\, \varepsilon_0/\sqrt{K}$, and hence

$$\frac{\varepsilon_0}{\sqrt{K}} = 0{\cdot}0161.$$

On the other hand we have

$$\sqrt{K} = 1{\cdot}59\, b_{1/2}\, U$$

so that

$$\varepsilon_0 = 0{\cdot}0256\, b_{1/2}\, U$$

where, as before, $b_{1/2}$ denotes half the width at half depth.

The diagram in Fig. 24.9 contains a comparison between measured velocity distribution points and the theoretical results from eqns. (24.46) shown as curve (2). Curve (1) provides a further comparison with the theory due to W. Tollmien [52]. The mixing length theory leads here also to a velocity distribution curve which is somewhat too pointed near the maximum, whereas eqns. (24.46) give excellent agreement over the whole width. The pattern of stream-lines is shown plotted in Fig. 24.10. It is seen that the jet draws in at its boundary fluid from the surrounding mass at rest so that the mass of fluid carried by the jet increases in a downstream

---

† We have $K = 2\pi \int\limits_0^\infty u^2\, y\, dy$.

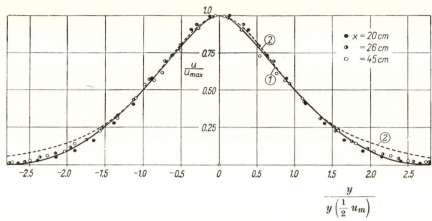

Fig. 24.9.   Velocity distribution in a circular, turbulent jet. Measurements due to Reichardt [29]
Theory: curve (1) due to Tollmien [52]; curve (2) from eqns. (24.46)

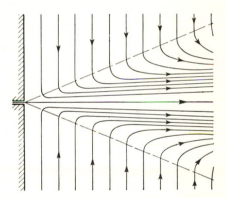

Fig. 24.10. Pattern of streamlines
in a circular, turbulent free jet

direction. The mass of fluid carried at a distance x from the orifice can be calculated
from eqn. (11.18). Inserting the above value for ε_0, we obtain

$$Q = 0.404 \sqrt{K\,x}. \qquad (24.47)$$

Calculations on the velocity and temperature distributions in two-dimensional
and circular jets have also been carried out by L. Howarth [21], both on the basis
of L. Prandtl's and of G. I. Taylor's assumption concerning turbulent mixing. The
mechanism which governs the mixing of a jet issuing from a circular nozzle with
the fluid in a large pipe was studied experimentally by K. Viktorin [55]. The
experiments covered a range of values of the velocity ratio in the pipe to that in
the jet of from 0 to 4. Compared with the mixing of a free jet with the surrounding fluid it is noticed that the pressure increases in the direction of flow in a manner

which resembles the phenomena near a sudden increase in cross-sectional area and sometimes described as Carnot's loss. A theoretical calculation based on Prandtl's mixing length hypothesis showed that the velocity distribution behaves in the same way as in a circular wake (width $\sim x^{-1/3}$, centre-line velocity $\sim x^{-2/3}$).

When a jet of finite width emerges into a uniform stream, the uniform velocity distribution becomes transformed near the mouth of the nozzle into the preceding profile. The case in hand was studied by A. M. Kuethe [24] and H. B. Squire and J. Trouncer [38]. Turbulent jets issuing into a parallel flow whose velocity is U_∞ differ from wakes formed behind single bodies essentially only in that the sign of u_1 in eqn. (24.31) is opposite; in a jet $u > U_\infty$, whereas in a wake $u < U_\infty$. In particular, at a large distance from the orifice where we have $|u_1| \ll U_\infty$, we find that the laws which describe the spreading of the jet are the same as those given in Table 24.1 for the two-dimensional or axially symmetric wake. Now, H. Reichardt [32] discovered on the basis of experiments that the distribution of total pressure, that is, of excess momentum

$$\varrho\, u_e{}^2 = \varrho\,(u^2 - U_\infty{}^2)$$

approximates a Gaussian distribution over the width of the jet, except in the proximity of the orifice itself. It can, therefore, be represented by the formula

$$\frac{u^2 - U_\infty{}^2}{u_m{}^2 - U_\infty{}^2} = \exp\left\{-\,(\ln 2)\,(y/b)^2\right\}.$$

The measure of width, b, was so chosen as to satisfy the condition that

$$\frac{u^2 - U^2}{u_m{}^2 - U^2} = \frac{1}{2} \quad \text{for} \quad y = b\,.$$

A paper by J. F. Keffer and W. D. Baines [22] treats the case of a turbulent jet under the influence of an external stream directed at right angles to it. A paper by R. Wille [58] summarizes experimental investigations on free jets.

Buoyant jets: Theoretical predictions of momentum jets and forced plumes discharged into a homogeneous or stratified infinite ambient atmosphere of different density depend on buoyancy forces. These forces have a vital effect on the diffusion mechanism of a jet, although the basic theory with its boundary layer assumptions is generally retained. The approach is the method of integral balance laws of mass, momentum and energy and the emerging system of differential equations is completed by a so-called entrainment hypothesis. The literature of this subject is vast [1b, 20a, 24b]. There is experimental evidence, however, that the usual boundary layer assumption can be relaxed in buoyant jets and plumes, at least at larger distances from the nozzle. In this connexion the paper by W. Schneider [36a] may also be mentioned.

7. The two-dimensional wall jet. A two-dimensional jet is formed when a fluid jet of large lateral width issues from a narrow opening and flows along a wall on one side, the other side mixing with an expanse of fluid at rest. Thus, the velocity distribution acquires the nature of a boundary layer near the wall but becomes that of a free jet at a larger distance from it, see Fig. 24.11. In most practical cases, the flow is turbulent. The earliest experiments on such a configuration were performed in 1934 by E. Foerthmann [11]. Later measurements are due to A. Sigalla [37] and P. Barke [3]. E. Foerthmann discovered that the velocity profiles are self-similar, disregarding the immediate downstream distance from the slit; they can be described by the equation

$$u \sim x^{-1/2} \, f(y/x). \tag{24.48}$$

This demonstrates that the velocity maximum decreases as $x^{-1/2}$ and that the width of the velocity profile increases as x, where x denotes a fictitious distance from the exit slit.

Evaluating the distribution of shearing stresses, E. Foerthmann found that the mixing length follows a law of the form

$$l = 0 \cdot 068 \, b, \tag{24.49}$$

where b denotes the width of the wall jet.

The preceding results were confirmed by the measurements undertaken by A. Sigalla [37]. It was ascertained that the local shearing stress is

$$\frac{\tau_0}{\frac{1}{2} \varrho \, u_m^2} = 0 \cdot 0565 \left(\frac{u_m \, \delta_1}{\nu} \right)^{-1/4}, \tag{24.48a}$$

where u_m is the maximum velocity of the wall jet, and δ_1 is the corresponding distance from the wall.

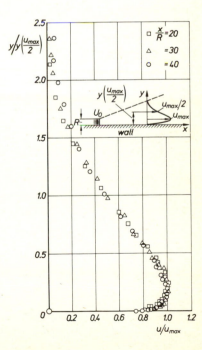

Fig. 24.11. Velocity distribution in a two-dimensional wall jet after the measurements of N.V.C. Swamy et al. [44]. The similarity law stated in eqn. (24.48) is satisfied well

The first attempt to describe the circumstances of a wall jet by theory was undertaken by M. B. Glauert [16]. The former was considerably improved by E. A. Eichelbrenner et al. [13]. The semi-empirical theory succeeded for the first time in predicting the separation of a wall jet. Subsequently, J. S. Gartshore and B. G. Newman [14] established an integral-momentum method which was based on very extensive measurements. These included wall jets with injection. The calculation made it possible to determine the numerical value of the momentum coefficient that is necessary to avoid separation of the wall jet. Further experimental results can be found in the papers by P. Bradshaw and M. T. Gee [4] as well as of V. Kruka and S. Eskinazi [23]. The account by F. Thomas [51] describes experiments concerning the mixing of a turbulent, two-dimensional jet bounded by a wall on one side with an external flow on the other.

Two-dimensional jets on highly convex, curved walls exhibit the well-known *Coanda effect*, that is the adherence of the jet over wide distances along the wall in the flow direction. Experimental and theoretical investigations into the pattern created by a plane jet flowing along the contour of a circular cylinder have been carried out by J. Gersten [15]. F. A. Dvorak [10] deals with the calculation of turbulent boundary layers on highly convex, curved walls, paying special attention to wall jets flowing along curved walls. Wall jets are employed in practice for boundary layer control and in film cooling; compare also B. G. Newman [25a], A. Metral [24c, 24d] and D. W. Young [60a].

Three-dimensional wall jets with a finite ratio of the two sides have been recently studied experimentally by P. M. Sforza and G. Herbst [42], by B. G. Newman et al. [25], by N. V. C. Swamy and B. L. Gowda [43], as well as by N. V. C. Swamy and P. Bandyopadhyay [44]. These measurements revealed a very fast rate of spreading of the jet in the spanwise direction and the existence of a very different fictitious origin for the growth of the width of the jet in the parallel as opposed to the normal wall direction.

e. Diffusion of temperature in free turbulent flow

The process of turbulent mixing causes a transfer of the properties of the fluid in a direction at right angles to the main stream. On the one hand the mixing motion causes *momentum* to flow away from the main stream, on the other hand, particles suspended in the fluid (floating particles of dust, chemical additives) are directed into the stream, and in addition there is a transfer of heat, that is a diffusion of a temperature field. The intensity of the transfer of a given property in turbulent motion is usually described by a suitable coefficient. Denoting the coefficient for momentum transfer by A_τ and that for heat by A_q, we can define them (see Sec. XXIIIa) by writing

$$\tau = A_\tau \frac{\mathrm{d}u}{\mathrm{d}y} \; ; \qquad q = c_p A_q \frac{\mathrm{d}T}{\mathrm{d}y} \, .$$

Here u and $c_p T$ denote momentum and heat per unit mass, respectively, and τ and q denote the flux of momentum and heat (= quantity of heat per unit area and time) respectively. In this connexion u and T denote temporal means. Since the mechanisms for the transfer of momentum and heat are not identical the values of A_τ and A_q are, generally speaking, different. However, according to Prandtl's mixing-length theory the mechanisms of the transfer of momentum and heat in free turbulent flows are identical which means that A_τ and A_q are assumed equal to each other. The measurement performed by A. Fage and V. M. Falkner [50] in the wake behind a row of heated bars have shown that the temperature profile is wider than the velocity profile and that, by way of approximation, we may assume $A_q = 2\,A_\tau$. This result agrees with G. I. Taylor's theory which was discussed in Sec. XIX c, and according to which turbulent mixing motion causes an exchange of vorticity rather than momentum. The problem of the diffusion of temperature in free turbulent flows was also considered by H. Reichardt [30], who made both theoretical and experimental contributions. The theoretical work is closely related to that described in the preceding section. First, empirical relations have been deduced for the temperature profile from experimental results in the same way as was done previously for the velocity (momentum) distribution, hypotheses on turbulent flow having been avoided. On the basis of an argument which we shall omit here, Reichardt succeeded in deriving a remarkable relation between the temperature and the velocity distribution. This is given by

$$\frac{T}{T_{max}} = \left(\frac{u}{u_{max}}\right)^{A_\tau/A_q} . \tag{24.50}$$

Here, the subscript *max* refers to the maximum values, and the scales for u and T must be so arranged as to render the points for which $u = 0$ and $T = 0$ coincident. Reichardt's experimental results for the two-dimensional jet (Fig. 24.13) and for the two-dimensional wake show good agreement with the law $T/T_{max} = (u/u_{max})^{1/2}$ which implies $A_q/A_\tau = 2$ in agreement with G. I. Taylor's theory [50]. Measurements on the temperature distribution in a heated, circular turbulent jet have been performed by S. Corrsin and M. S. Uberoi [8], as well as by J. O. Hinze and B. G. van der Hegge Zijnen [20]. The temperature distribution behind a plane row of bars was also measured by R. Gran Olsson as was reported in a paper already quoted [19].

Mixing of coaxial turbulent jets issuing with different velocities and temperatures in a pipe: Theoretical and experimental investigations concerning the mixing of coaxial turbulent jets issuing with different velocities and temperatures in a pipe were carried out by S. R. Ahmed [1] in the incompressible case, Fig. 24.12. In this case, the inner jet moved with a moderately larger temperature and velocity than did the outer jet. The most important parameter which governs the mixing process turns out to be the ratio $U^* = u_{HO}/u_{SO}$ of the two jets, where u_{HO} und u_{SO} denote the velocities in the inlet cross-section of the inner and outer jet, respectively. The equalization of momentum as well as of temperature depends on this parameter.

The diagram in Fig. 24.12a represents the variation along the pipe axis of the ratio u_H/u_{HO} determined theoretically by S. R. Ahmed [1] in terms of the velocity ratio $U^* = u_{HO}/u_{SO}$. The diagram in Fig. 24.12b depicts the same quantity in terms of the temperature ratio $\Theta^* = \Theta_{HO}/\Theta_{SO}$. Both diagrams contain experimental points for comparison. Here Θ_{HO} and Θ_{SO} denote the temperature of the inner or outer jet in the pipe inlet. The agreement between theory and measurement is good. The influence of the temperature ratio Θ^* on the mixing process is insignificant in the range covered by the measurements.

Natural convection, i. e. the diffusion of temperature in a stream created by thermal buoyancy, was investigated by W. Schmidt [36] who considered the following

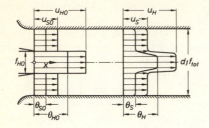

Exp.	U^*	F^*	Θ^*
●	1·36	0·40	1·12
◑	2·34	0·40	1·09
○	3·65	0·40	1·10
▲	1·19	0·63	1·08
△	2·01	0·63	1·20

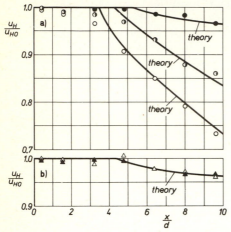

Fig. 24.12. The mixing of coaxial turbulent jets of different velocities and temperatures in a pipe, after S. R. Ahmed [1a]. Variations of the velocity along the axis of the pipe a) for various velocity ratios $U^* = u_{HO}/u_{SO}$ at a constant value of the temperature ratio Θ^*; b) for various values of the temperature ratio $\Theta^* = \theta_{HO}/\theta_{SO}$ at a constant value of the velocity. $F^* = f_{HO}/f_{SO}$ denotes the area ratio of the inner jet to the whole jet

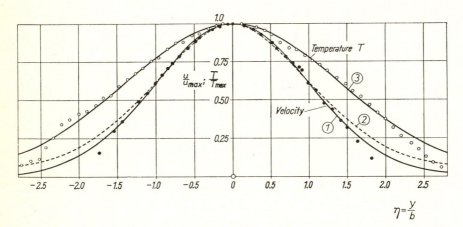

Fig. 24.13. Temperature and velocity distribution in a two-dimensional jet. Measurements are due to Reichardt [29]

Curve (1): $u/u_{max} = \exp(-\frac{1}{2}\eta^2)$;

curve (2): $u/u_{max} = 1 - \tanh^2(\eta)$, eqn. (24.45);

curve (3): $T/T_{max} = \exp(-\frac{1}{4}\eta^2) = (u/u_{max})^{1/2}$, eqn. (24.50)

two cases: 1. two-dimensional flow above a linear source of heat placed on a horizontal floor and 2. axi-symmetrical flow above a point-source. In both cases the width of the velocity and temperature profile increases in direct proportion to the height above the floor, x. In the two-dimensional case the velocity remains constant at all heights, whereas the temperature decreases as x^{-1}. In the axially symmetrical case the velocity is proportional to $x^{-1/3}$, the temperature being proportional to $x^{-5/2}$. The two-dimensional case was treated theoretically on the basis of Prandtl's mixing-length theory (transport of momentum) as well as on the basis of G. I. Taylor's vorticity transport theory. The axially symmetrical case could be investigated only with the aid of Prandtl's theory because G. I. Taylor's theory breaks down in this case. Measurements performed for the axi-symmetrical case confirm the theoretical calculations. The diffusion of temperature behind a point-source and behind a linear source placed in the boundary layer on a flat plate were investigated experimentally by K. Wieghardt [56]. In the case of the point-source it is found that the transfer of heat is much larger sideways than at right angles to the wall. The paper contains equations which allow us to transpose the experimental results to similar cases. In this connexion the paper by B. Frost [12] may also be consulted. Temperature fluctuations in a turbulent wake have been measured by D. W. Schmidt and W. J. Wagner [45].

References

[1] Abramovich, G. N.: The theory of turbulent jets (Translation from the Russian). MIT Press, Cambridge, Mass., 1963.

[1a] Ahmed, S. R.: Die Vermischung von koaxialen und turbulenten Strahlen verschiedener Geschwindigkeit und Temperatur in einem Rohr. Diss. Braunschweig 1970. VDI-Forschungsheft 547, 18—30 (1971).

[1b] Albertson, M. L., Dai, Y. B., Jenson, R. A., and Rouse, H.: Diffusion of submerged jets. Trans. Am. Soc. Civil Engrs. 115, 639—696 (1950).

[2] Anderlik, E.: Math. termeszett. Ertes 52, 54 (1935).

[2a] Antonia, R. A., and Bilger, R. W.: The heated round jet in a coflowing stream. AIAA J 14, 1541—1547 (1976).

[3] Barke, P.: An experimental investigation of a wall-jet. JFM 2, 467—472 (1957).

[4] Bradshaw, P., and Gee, M. T.: Turbulent wall jets with and without an external stream. ARC RM 3252, 1—48 (1962).

[5] von Bohl, J. G.: Das Verhalten paralleler Luftstrahlen. Ing.-Arch. II, 295—314 (1940).

[6] Bourque, C., and Newmann, B. G.: Re-attachment of a two-dimensional incompressible jet to an adjacent flat plate. Aero. Quart. 11, 201—232 (1960).

[7] Cordes, G.: Statische Druckmessung in turbulenter Strömung. Ing.-Arch. 8, 245—270 (1937)

[8] Corrsin, S., and Uberoi, M. S.: Further experiments on the flow and heat transfer in a heated turbulent air jet. NACA TN 998 (1950).

[9] Davies, D. R.: The problem of diffusion into a turbulent boundary layer from a plane area source bounded by two straight perpendicular edges. Quart. J. Mech. Appl. Math. 7, 468—471 (1954).

[10] Dvorak, F. A.: Calculation of turbulent boundary layers and wall jets over curved surfaces. AIAA J. 11, 517—524 (1973).

[11] Förthmann, E.: Über turbulente Strahlausbreitung. Diss. Göttingen 1933; Ing.-Arch. 5, 42—54 (1934); NACA TM 789 (1936).

[12] Frost, B.: Turbulence and diffusion in the lower atmosphere. Proc. Roy. Soc. A 186, 20 (1946).

[13] Eichelbrenner, E. A., and Dumargue, P.: Le problème du "jet pariétal" plan en régime turbulent pour un écoulement extérieur de vitesse U_e constante. J. Mécanique I, 109—122 and I, 123—134 (1962).

[14] Gartshore, J.S., and Newman, B.G.: The turbulent wall jet in an arbitrary pressure gradient. Aero. Quart. *20*, 25—56 (1969).

[15] Gersten, K.: Flow along highly curved surfaces. Lecture at EUROMECH *I*, Berlin 1965; see also [57].

[16] Glauert, M.B.: The wall-jet. JFM *1*, 625—643 (1956).

[17] Gooderum, P.B., Wood, G.P., and Brevoort, M.J.: Investigation with an interferometer of the turbulent mixing of a free supersonic jet. NACA Rep. 968 (1950).

[18] Görtler, H.: Berechnung von Aufgaben der freien Turbulenz auf Grund eines neuen Näherungsansatzes. ZAMM *22*, 244—254 (1942).

[19] Gran Olsson, R.: Geschwindigkeits- und Temperaturverteilung hinter einem Gitter bei turbulenter Strömung. ZAMM *16*, 257—267 (1936).

[20] Hinze, J.O., and van der Hegge Zijnen, B.G.: Transfer of heat and matter in the turbulent mixing zone of an axially symmetric jet. Proc. 7th Intern. Congr. Appl. Mech. *2*, Part I, 286—299 (1948).

[20a] Hirst, E.: Buoyant jet discharged into quiescent stratified ambients. J. Geophys. Res. *76*, 7375—7384 (1971).

[21] Howarth, L.: Concerning the velocity and temperature distributions in plane and axially symmetrical jets. Proc. Cambr. Phil. Soc. *34*, 185—203 (1938).

[22] Keffer, J.F., and Baines, W.D.: The round turbulent jet in a cross-wind. JFM *15*, 481—496 (1963).

[23] Kruka, V., and Eskinazi, S.: The wall jet in a moving stream. JFM *20*, 555—579 (1964).

[24] Kuethe, A.M.: Investigations of the turbulent mixing regions formed by jets. J. Appl. Mech. *2*, 87—95 (1935).

[24a] Mohammadian, S., Sailey, M., and Peerless, J.: Fluid mixing with unequal free-stream turbulence intensities. J. Fluids Eng. Trans. ASME I, *98*, 229—235 (1976).

[24b] List, E.H., and Imberger, J.: Turbulent entrainment in buoyant jets and plumes. J. Hydr. Div. ASCE *99*, HY 9, 1461—1474 (1973).

[24c] Metral, A.: Sur un phénomène de déviation des veines fluides et ses applications. Effet Coanda, Cabinet Technique du Ministère de l'Air (1938).

[24d] Metral, A., and Zerner, F.: L'effet Coanda. Publication Scientifiques et Techniques du Ministère de l'Air, No. 218 (1948). MOS. TIB/T 4027 (1953).

[25] Newman, B.G., Patel, R.P., Savage, S.B., and Tjio, H.K.: Three-dimensional wall jet originating from a circular orifice. Aero. Quart. *23*, 188—200 (1972).

[25a] Newman, B.G.: The deflection of plane jets by adjacent boundaries — Coanda effect. G.V. Lachmann (ed.): Boundary Layer and Flow Control. Pergamon Press. Vol. 1, 232—264 (1961).

[26] Pai, S.I.: Fluid dynamics of jets. New York, 1954.

[26a] Pfeil, H., and Eifler, J.: Zur Frage der Schubspannungsverteilung für die ebenen freien turbulenten Strömungen. Forschg. Ing.-Wes. *41*, 105—136 (1975).

[26b] Pfeil, H., and Eifler, J.: Messungen im turbulenten Nachlauf des Einzelzylinders. Forschg. Ing.-Wes. *41*, 137—145 (1975).

[27] Prandtl, L.: The mechanics of viscous fluids. In W.F. Durand (ed.): Aerodynamic Theory, *III*, 16—208 (1935); see also Proc. IInd Intern. Congress Appl. Mech. Zürich 1926.

[28] Reichardt, H.: Über eine neue Theorie der freien Turbulenz. ZAMM *21*, 257—264 (1941).

[29] Reichardt, H.: Gesetzmässigkeiten der freien Turbulenz. VDI-Forschungsheft 414 (1942), 2nd ed. 1951.

[30] Reichardt, H.: Impuls- und Wärmeaustausch in freier Turbulenz. ZAMM *24*, 268—272 (1944).

[31] Reichardt, H., and Ermshaus, R.: Impuls- und Wärmeübertragung in turbulenten Windschatten hinter Rotationskörpern. Int. J. Heat Mass Transfer *5*, 251—265 (1962).

[32] Reichardt, H.: Turbulente Strahlausbreitung in gleichgerichteter Grundströmung. Forschg. Ing.-Wes. *30*, 133—139 (1964).

[33] Ruden, P.: Turbulente Ausbreitung im Freistrahl. Naturwissenschaften *21*, 375—378 (1933).

[34] Sawyer, R.A.: The flow due to a two-dimensional jet issuing parallel to a flat plate. JFM *9*, 543—560 (1960).

[35] Schlichting, H.: Über das ebene Windschattenproblem. Diss. Göttingen 1930; Ing.-Arch. *1*, 533—571 (1930).

[36] Schmidt, W.: Turbulente Ausbreitung eines Stromes erhitzter Luft. ZAMM *21*, 265—278 and 351—363 (1941).

[36a] Schneider, W.: Über den Einfluß der Schwerkraft auf anisotherm, turbuiente Freistrahlen. Abh. Aerod. Inst. T.H. Aachen, No. 22, 59—65 (1975).

[37] Sigalla, S.: Measurements of skin friction in plane turbulent wall jet. J. Roy. Aero. Soc. 62, 873—877 (1958).

[38] Squire, H.B., and Rouncer, J.T.: Round jets in a general stream. ARC RM 1974 (1944).

[39] Squire, H.B.: Reconsideration of the theory of free turbulence. Phil. Mag. 39, 1—20 (1948).

[40] Squire, H.B.: Jet flow and its effect on aircraft. Aircarft Engineering 22, 62—67 (1950).

[41] Swain, L.M.: On the turbulent wake behind a body of revolution. Proc. Roy. Soc. London A 125, 647—659 (1929).

[42] Sforza, P.M., and Herbst, G.: A study of three-dimensional incompressible turbulent wall jet. AIAA J. 8, 276—283 (1970).

[43] Swamy, N.V.C., and Gowda, B.H.L.: Characteristics of three-dimensional wall jets. ZFW 22, 314—323 (1974).

[44] Swamy, N.V.C., and Bandyopadhyay, P.: Mean and turbulence characteristics of three-dimensional wall jets. JFM 71, 541—562 (1975).

[45] Schmidt, D.W., and Wagner, W.J.: Measurements of the temperature fluctuations in turbulent wakes. ZFW 22, 10—14 (1974).

[46] Szablewski, W.: Zur Theorie der turbulenten Strömung von Gasen stark veränderlicher Dichte. Diss. Göttingen 1947; Ing.-Arch. 20, 67—72 (1952).

[47] Szablewski, W.: Zeitliche Auflösung einer ebenen Trennungsfläche der Geschwindigkeit und Dichte. ZAMM 35, 464—468 (1955).

[48] Szablewski, W.: Turbulente Vermischung zweier ebener Luftstrahlen von fast gleicher Geschwindigkeit und stark unterschiedlicher Temperatur. Ing.-Arch. 20, 73—80 (1952).

[49] Tanner, M.: Einfluss des Keilwinkels auf den Ähnlichkeitsparameter der turbulenten Vermischungszone in inkompressibler Strömung. Forschg. Ing.-Wes. 39, 121—125 (1973).

[50] Taylor, G.I.: The transport of vorticity and heat through fluids in turbulent motion. Appendix by A. Fage and V.M. Falkner. Proc. Roy. Soc. A 135, 685—705 (1932).

[51] Thomas, F.: Untersuchungen über die Grenzschicht an einer Wand stromabwärts von einem Ausblasespalt. Abhandl. Wiss. Ges. Braunschweig 15, 1—17 (1963).

[52] Tollmien, W.: Berechnung turbulenter Ausbreitungsvorgänge. ZAMM 6, 468—478 (1926); NACA TM 1085 (1945).

[53] Tollmien, W.: Die von Kármánsche Ähnlichkeitshypothese in der Turbulenz-Theorie und das ebene Windschattenproblem. Ing.-Arch. 4, 1—15 (1933).

[54] Townsend, A.A.: Momentum and energy diffusion in the turbulent wake of a cylinder. Proc. Roy. Soc. London A 197, 124—140 (1949).

[55] Viktorin, K.: Untersuchungen turbulenter Mischvorgänge. Forschg. Ing.-Wes. 12, 16—30 (1941); NACA TM 1096 (1946).

[56] Wieghardt, K.: Über Ausbreitungsvorgänge in turbulenten Reibungsschichten. ZAMM 28, 346—355 (1948).

[57] Wille, R., and Fernholz, H.: Report on the First Mechanics Colloquium on the Coanda effect. JFM 23, 801—819 (1965).

[58] Wille, R.: Beiträge zur Phänomenologie der Freistrahlen (Third Otto-Lilienthal-Lecture 1962). ZFW 11, 222—233 (1963).

[59] Wuest, W.: Turbulente Mischvorgänge in zylindrischen und kegeligen Fangdüsen. Z. VDI 92, 1000—1001 (1950).

[60] Yamaguchi, S.: Turbulente Vermischung eines ebenen Strahles in gleichgerichteter Aussenströmung. Ing.-Arch. 35, 172—180 (1966).

[60a] Young, D.W., and Zonars, D.: Wind tunnel tests of the Coanda wing and nozzle. USAF Techn. Report 6199 (1950).

[61] Zimm, W.: Uber die Strömungsvorgänge im freien Luftstrahl. VDI-Forschungsheft 234 (1921).

[62] Reports of the AVA Göttingen, Ergebnisse der Aerodynamischen Versuchsanstalt Göttingen. R. Oldenbourg, München, Vol. 2, 69—77 (1923).

CHAPTER XXV

Determination of profile drag

a. General remarks

The total drag on a body placed in a stream of fluid consists of *skin friction* (equal to the integral of all shearing stresses taken over the surface of the body) and of *form* or *pressure drag* (integral of normal forces). The sum of the two is called *total* or *profile drag*. The skin friction can be calculated with some accuracy by the use of the methods of the preceding chapters. The form drag which does not exist in frictionless subsonic flow, is due to the fact that the presence of the boundary layer modifies the pressure distribution on the body as compared with ideal flow, but its computation is very difficult. Consequently, reliable data on total drag must, in general, be obtained by measurement. In more modern times methods of estimating the amount of profile drag have, nevertheless, been established. We shall discuss them briefly in Sec. d of the present chapter.

In many cases the determination of total drag by weighing lacks in accuracy because, when measurements are performed, for example, in a wind tunnel, the drag on the suspension wires is too large compared with the force to be measured. In some cases even, such as in free flight experiments, its direct determination becomes impossible. In such cases the method of determining profile drag from the velocity distribution in the wake (Pitot traverse method), which has already been described in Chap. IX, becomes very useful. Moreover, it is often the only practicable way of performing this kind of measurement. In principle it can only be used in two-dimensional and axially symmetrical cases, but we shall restrict ourselves to the consideration of the two-dimensional case.

The formula in eqn. (9.27) which was deduced in Chap. IX and which serves to determine the magnitude of drag from the velocity distribution in the wake is valid only for comparatively large distances from the body. According to it the total drag on a body† is given by the expression:

$$D = b \varrho \int_{y=-\infty}^{+\infty} u \, (U_\infty - u) \, \mathrm{d}y \; . \tag{25.1}$$

Here b denotes the length of the cylindrical body in the direction of the axis of the cylinder, U_∞ is the free-stream velocity, and $u(y)$ denotes the velocity distribution

---

† In Chap. IX the total drag on a body was denoted by $2\,D$ (for the two sides of the plate); in this chapter the symbol D is used for it.

in the wake. The integral must be taken at such a large distance from the body that the static pressure at the measuring section becomes equal to that in the undisturbed stream. In practical cases, whether in a wind tunnel or in free flight measurements, it is necessary to come much closer to the body. Consequently it becomes necessary to take into account the contribution from the pressure term and eqn. (25.1) must be modified. This correction term has an appreciable value when measurements are performed close to the body (e. g. at distances less than one chord in the case of aerofoils) and it is, therefore, important to have a comparatively accurate expression for it. The correction term was first calculated by A. Betz [4] and later by B. M. Jones [26]. At present most measurements are being evaluated with the aid of the formula due to Jones because of its comparative simplicity. Nevertheless, we propose to discuss Betz's formula as well because its derivation exhibits several very interesting features.

b. The experimental method due to Betz

With reference to Fig. 25.1 we select a control surface around the body as shown. In the entry cross-section I in front of the body the flow is lossless, its total pressure being g_∞. The total pressure in cross-section II behind the body is $g_2 < g_\infty$. The remaining cross-sections of the control surface are imagined placed far enough from the body for the flow in them to be undisturbed. In order to satisfy the condition of continuity, the velocity u_2 in cross-section II must in some places exceed the undisturbed velocity U_∞. Applying the momentum theorem to the control surface gives the following expression for the drag on a cylinder of length b:

$$D = b \left\{ \int_{y=-\infty}^{+\infty} (p_1 + \varrho u_1{}^2)\, \mathrm{d}y - \int_{-\infty}^{+\infty} (p_2 + \varrho u_2{}^2)\, \mathrm{d}y \right\}. \tag{25.2}$$

In order to adapt this equation to the evaluation of experimental results it is necessary to transform the above integrals so that they need only be evaluated over that section of the velocity curve which includes the depression of plane II in the profile. The total pressures satisfy the conditions:

$$
\left.
\begin{array}{lll}
\text{at infinity:} & \quad g_\infty = p_\infty + \dfrac{1}{2}\varrho\, U_\infty{}^2 & \\[2mm]
\text{at cross-section I:} & \quad g_\infty = p_1 + \dfrac{1}{2}\varrho\, u_1{}^2 & \\[2mm]
\text{at cross-section II:} & \quad g_2 = p_2 + \dfrac{1}{2}\varrho\, u_2{}^2 \,. &
\end{array}
\right\} \tag{25.3}
$$

Thus, eqn. (25.2) becomes

$$D = b \left\{ \int_{-\infty}^{+\infty} (g_\infty - g_2)\, \mathrm{d}y + \frac{1}{2}\varrho \int_{-\infty}^{+\infty} (u_1{}^2 - u_2{}^2)\, \mathrm{d}y \right\}. \tag{25.4}$$

The first integral already has the desired form, because the total pressure is equal to g_∞ everywhere outside the depression. In order to transform the second integral in the same way we introduce a hypothetical flow $u_2'(y)$ in cross-section II which is identical with u_2 everywhere outside the depression but which differs from u_2 in the region of the depression in that the total pressure for u_2' is equal to g_∞. Thus

$$g_\infty = p_2 + \frac{1}{2}\varrho\,u_2'^2. \tag{25.5}$$

Since the actual flow u_1, u_2 satisfies the equations of continuity, the mass flow of the hypothetical flow u_1, u_2' is too large across section II. This is equivalent to the existence of a source which is located, essentially, at the body and whose strength is

$$Q = b\int (u_2' - u_2)\,\mathrm{d}y. \tag{25.6}$$

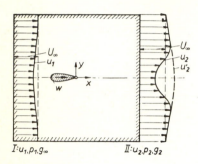

Fig. 25.1. Determination of profile drag by the method due to Betz [4]

A source which exists in a frictionless parallel stream of velocity U_∞ suffers a thrust equal to

$$R = -\varrho\,U_\infty\,Q. \tag{25.7}$$

We now apply the momentum theorem from eqn. (25.4) to the hypothetical flow, i. e. we assume a velocity u_1 in section I, and a velocity u_2' in section II. Since $g_2' = g_\infty$ and since the resultant force is equal to R from eqn. (25.7), we obtain

$$-\varrho\,U_\infty\,Q = b\,\frac{1}{2}\varrho\int (u_1{}^2 - u_2'^2)\,\mathrm{d}y.$$

Subtracting this value from eqn. (25.4) we have

$$D + \varrho\,U_\infty\,Q = b\left\{\int (g_\infty - g_2)\,\mathrm{d}y + \frac{1}{2}\varrho\int (u_2'^2 - u_2{}^2)\,\mathrm{d}y\right\}. \tag{25.8}$$

In view of eqn. (25.6) we have now

$$D = b\left\{\int (g_\infty - g_2)\,\mathrm{d}y + \frac{1}{2}\varrho\int (u_2'^2 - u_2{}^2)\,\mathrm{d}y - \varrho\,U_\infty\int (u_2' - u_2)\,\mathrm{d}y\right\}.$$

Each of the above integrals need only be evaluated over the wake since outside it $u_2' = u_2$. Since $u_2'^2 - u_2{}^2 = (u_2' - u_2)(u_2' + u_2)$, the above can be transformed to

$$D = b\left\{\int (g_\infty - g_2)\,\mathrm{d}y + \frac{1}{2}\varrho\int (u_2' - u_2)(u_2' + u_2 - 2\,U_\infty)\,\mathrm{d}y\right\}. \tag{25.9}$$

In order to determine the drag, D, it is necessary to measure the total pressure, g_2, and the static pressure, p_2, over the cross-section II behind the body. Thus we also obtain g_∞ as it is equal to g_2 outside the depression. The hypothetical velocity u_2' is defined in eqn. (25.5) from which it can be calculated.

In cases when the static pressure over the measuring station equals that in the undisturbed stream, i. e. when $p_2 = p_\infty$, we also have $u_2' = U_\infty$ and eqn. (25.9) transforms back into eqn. (25.1).

Defining a dimensionless coefficient of drag by writing

$$D = c_D\, b\, l\, q_\infty\,, \tag{25.9a}$$

where $q_\infty = \frac{1}{2}\varrho\, U_\infty^2$ denotes the dynamic pressure of the oncoming stream and $b \times l$ is the reference area, we can rewrite eqn. (25.9) to read:

$$c_D = \int \frac{g_\infty - g_2}{q_\infty}\, \mathrm{d}\!\left(\frac{y}{l}\right) +$$
$$+ \int \left(\sqrt{\frac{g_\infty - p_2}{q_\infty}} - \sqrt{\frac{g_2 - p_2}{q_\infty}} \right)\left(\sqrt{\frac{g_\infty - p_2}{q_\infty}} + \sqrt{\frac{g_2 - p_2}{q_\infty}} - 2 \right) \mathrm{d}\!\left(\frac{y}{l}\right). \tag{25.10}$$

This is the most convenient form for the evaluation of experimental results.

c. The experimental method due to Jones

Some time later, B. M. Jones [26] indicated a similar method for the determination of profile drag. The final formula due to Jones is somewhat simpler than that due to A. Betz.

The cross-section II (Fig. 25.2) in which measurements are performed is located behind the body at a short distance from it; the static pressure p_2 at the measuring station is still markedly different from the static pressure in the undisturbed stream. Cross-section I is placed so far behind the body that $p_1 = p_\infty$. Applying eqn. (25.1) to cross-section I, we obtain

$$D = b\,\varrho \int u_1\,(U_\infty - u_1)\, \mathrm{d}y_1\,. \tag{25.11}$$

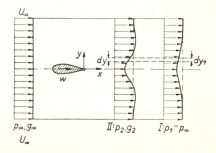

Fig. 25.2. Determination of profile drag
by the method due to B. M. Jones [26]

In order to confine the determination of u_1 to the use of results obtained from measurements in cross-section II, we first apply the equation of continuity along a streamtube

$$\varrho \, u_1 \, dy_1 = \varrho \, u_2 \, dy \,. \tag{25.12}$$

Hence

$$D = b \, \varrho \int u_2 \, (U_\infty - u_1) \, dy \,. \tag{25.13}$$

Secondly, according to B. M. Jones [26], we make the assumption that the flow proceeds from section II to section I without losses, i. e. that the total pressure remains constant along every stream-line between the stations I and II:

$$g_2 = g_1 \,. \tag{25.14}$$

Introducing the total pressures

$$p_\infty + \frac{1}{2} \, \varrho \, U_\infty{}^2 = g_\infty \,; \quad p_\infty + \frac{1}{2} \, \varrho \, u_1{}^2 = g_1 = g_2 \,; \quad p_2 + \frac{1}{2} \, \varrho \, u_2{}^2 = g_2$$

we see from eqn. (25.13) that

$$D = 2b \int \sqrt{g_2 - p_2} \left(\sqrt{g_\infty - p_\infty} - \sqrt{g_2 - p_\infty} \right) dy \,, \tag{25.15}$$

where the integral extends over cross-section II. In this case, as in the previous one, the integrand differs from zero only across the disturbed portion of the velocity profile. Introducing a dimensionless coefficient, in the same way as in eqn. (25.9a), and taking into account that $g_\infty - p_\infty = q_\infty$, we have

$$\boxed{c_D = 2 \int \sqrt{\frac{g_2 - p_2}{q_\infty}} \left(1 - \sqrt{\frac{g_2 - p_\infty}{q_\infty}} \right) d \left(\frac{y}{l} \right)} \,. \tag{25.16}$$

Jones's preceding equation also transforms into the simple equation (25.1) in cases when the static pressure at the measuring station is equal to the undisturbed static pressure, $p_2 = p_\infty$.

A. D. Young [75] indicated a transformation of Jones's formula which simplifies the evaluation of the integral in eqn. (25.16). The resulting equation contains an additive correction term apart from the integral of the total pressure loss taken over the depression in the velocity profile. The correction term depends on the shape of the velocity profile in the measuring station, but it can be computed once and for all. A critical appraisal of this method is contained in a note by G. I. Taylor [67].

The preceding two experimental methods have been used very frequently for the determination of profile drag both in flight and in wind tunnel measurements, [6, 12, 16, 29, 38, 39, 61, 62, 69, 70], and have led to very satisfactory results. H. Doetsch [6] demonstrated that both the Betz and the Jones formulae can be used when the distance between the measuring station behind the aerofoil and the aerofoil itself is as short as 5 per cent. chord. In this case the correction term in Betz's formula amounts to about 30 per cent. of the first term. Both methods are particularly suitable when the influence of surface roughnesses on profile drag is being determined as well as to the determination of the very small drag of laminar aerofoils.

A. D. Young [71] extended the applicability of Jones's method to compressible flows, Retracing the steps in that derivation, we apply the continuity equation for compressible flow.

$$\varrho_1 \, u_1 \, dy_1 = \varrho_2 \, u_2 \, dy_2 \, , \tag{25.17}$$

and deduce the following formula for drag:

$$D = b \int_{-\infty}^{+\infty} \varrho_2 \, u_2 \, (U_\infty - u_1) \, dy_2 \, . \tag{25.18}$$

Here, again, it is necessary to express u_1 in terms of the quantities measured in plane II. In the realm of compressible flow it is necessary to replace Jones's assumption that $g_1 = g_2$ by the assumption that the entropy remains constant along a streamline from plane II to plane I. This leads to the isentropic relation

$$\frac{p_2}{\varrho_2{}^\gamma} = \frac{p_1}{\varrho_1{}^\gamma} \, . \tag{25.19}$$

If, now, the stagnation pressure measured by the Pitot tube in compressible flow is denoted by g, we have

$$g = p_0 = \frac{\varrho_0}{\varrho} \left(\frac{\gamma-1}{2} \frac{\varrho}{2} \, w^2 + p \right) , \tag{25.20}$$

and it can be verified that eqn. (25.19) also leads to the assumption $g_1 = g_2$. The velocity u_1 can be determined from the Bernoulli equation for compressible flow, namely

$$u_1{}^2 = \frac{2\gamma}{\gamma-1} \frac{p_1}{\varrho_1} \left[\left(\frac{g_2}{p_1} \right)^{\frac{\gamma-1}{\gamma}} - 1 \right] . \tag{25.21}$$

In order to solve the problem in principle, it is only necessary to express the velocity u_2 in terms of the measured pressures g_2 and p_2 in plane II. A measurement of the total and static pressures in plane II is again sufficient for the determination of the drag of the body. However, the complicated relation between velocities and pressures in the compressible Bernoulli equation leads to a very cumbersome equation. For this reason, A. D. Young expanded the velocities u_1 and u_2 into series of the form

$$u^2 = \frac{2\gamma}{\gamma-1} \frac{p}{\varrho} \left[\frac{\gamma-1}{\gamma} \frac{g-p}{p} - \frac{\gamma-1}{2\gamma^2} \left(\frac{g-p}{p} \right)^2 + \cdots \right] . \tag{25.22}$$

In this manner, the terms in eqn. (25.15) derived by Jones for the incompressible case can now be separated, and the remaining terms can be arranged in a power series in terms of the Mach number. Thus

$$c_D = c_{D,\,i} + A_1 \, \mathsf{M}_\infty^2 + A_2 \, \mathsf{M}_\infty^4 + \cdots \, , \tag{25.23}$$

where $c_{D,\,i}$ denotes the drag coefficient for the incompressible case, as given by eqn. (25.16), and the coefficients A_1, A_2, $\ldots$ represent certain integrals which can be calculated from the measured data in plane II. Restricting oneself to low Mach numbers, and hence to two terms in the expansion (25.23), one obtains

$$c_D = 2 \int_{y=-\infty}^{+\infty} \sqrt{\frac{g_2 - p_2}{q_\infty}} \left(1 - \frac{g_2 - p_\infty}{q_\infty} \right) \left\{ 1 + \frac{\mathsf{M}_\infty^2}{8} \left[3 \, \frac{p_2 - p_\infty}{q_\infty} + 3 - \right. \right.$$

$$\left. \left. - 2\gamma - 2 \, \frac{g_2 - p_\infty}{q_\infty} - (2\gamma - 1) \, \sqrt{\frac{g_2 - p_\infty}{q_\infty}} \right] \right\} d\left(\frac{y}{l} \right) \tag{25.24}$$

where

$$q_\infty = g_\infty - p_\infty \, .$$

The additional term which depends on the Mach number provides a negative contribution to the drag coefficient. It is possible to evaluate this additional term once and for all if a suitable assumption is made for the shape of the depression in the velocity profile in the wake; this was also done by A. D. Young.

d. Calculation of profile drag

Methods which can be used for the calculation of profile drag and which are based on the same principles as the above experimental methods, have been devised by J. Pretsch [40] and H. B. Squire and A. D. Young [64]. These are tied in with the calculation of boundary layers, as described in Chap. XXII. However, in order to be in a position to calculate pressure drag it is necessary in each case to make use of certain additional, empirical relations. See also H. Goertler [19].

We now propose to give a short description of H. B. Squire's and A. D. Young's method of calculation taking into account some more recent results. We shall begin by transforming eqn. (25.1), which relates the drag on a body with the velocity profile in the wake behind the body. Introducing the momentum thickness $\delta_{2\infty}$ from eqn. (8.31) and the drag coefficient from eqn. (25.9a), we can rewrite it as

$$c_D = 2 \frac{\delta_{2\infty}}{l} . \tag{25.25}$$

Here

$$\delta_{2\infty} = \int\limits_{y=-\infty}^{\infty} \frac{u}{U_\infty} \left(1 - \frac{u}{U_\infty}\right) \mathrm{d}y$$

denotes the momentum thickness of the wake at a large distance from the body. On the other hand, the calculation described in Chap. XXII permits us to evaluate the momentum thickness at the trailing edge, for which the symbol δ_{21} will be used. The essence of Squire's method consists in relating these two quantities, $\delta_{2\infty}$ and δ_{21}, in such a way as to permit the calculation of drag from eqn. (25.25) when the momentum thickness at the trailing edge of the body is known from a boundary-layer calculation.

The momentum integral equation of boundary-layer theory, eqn. (22.6), is valid also for the wake behind a body with the only difference that the shearing stress τ_0 must be equated to zero. Thus we have

$$\frac{\mathrm{d}\delta_2}{\mathrm{d}x} + (H + 2)\, \delta_2 \frac{U'}{U} = 0, \tag{25.26}$$

where $H = \delta_1/\delta_2$ and $U' = dU/dx$†. The symbol x denotes now the distance from the trailing edge of the body measured along the centre-line of the wake. The last equation can also be written in the form

$$\frac{1}{\delta_2} \frac{\mathrm{d}\delta_2}{\mathrm{d}x} = -(H + 2) \frac{\mathrm{d}}{\mathrm{d}x} \left(\ln \frac{U}{U_\infty}\right) .$$

---

† The shape factor δ_1/δ_2 will now be denoted by H, for simplicity, rather than by H_{12}, as before.

Integrating over x from the trailing edge of the body (subscript 1) to a station sufficiently far downstream, so as to have $U = U_\infty$ and $p = p_\infty$, we obtain

$$\left[\ln \delta_2\right]_\infty^1 = -\left[(H+2)\ln \frac{U}{U_\infty}\right]_\infty^1 + \int_\infty^1 \ln \frac{U}{U_\infty}\frac{dH}{dx}\,dx\,.$$

At a large distance behind the body we have $H = 1$, and consequently

$$\ln \frac{\delta_{21}}{\delta_{2\infty}} + (H_1 + 2)\ln \frac{U_1}{U_\infty} = \int_{H=1}^{H=H_1} \ln \frac{U}{U_\infty}\,dH\,.$$

Here $H_1 = \delta_1/\delta_2$ denotes the value of the shape factor $H = \delta_{11}/\delta_{21}$ at the trailing edge which is known from the calculation of the boundary layer. This equation gives the required relation between $\delta_{2\infty}$ and δ_{21}, provided that U_1/U_∞ and the value of the integral on the right-hand side are known. First we find that

$$\delta_{2\infty} = \delta_{21}\left(\frac{U_1}{U_\infty}\right)^{H_1+2}\exp\left(\int_1^{H_1}\ln \frac{U_\infty}{U}\,dH\right)\,. \tag{25.27}$$

In order to be in a position to evaluate the integral, it is necessary to know the relation between the static pressure in the wake, which determines the value of U, and the velocity distribution in the wake which, in turn, determines the value of the shape factor H. The magnitude of $\ln(U_\infty/U)$ decreases monotonically along the wake, starting with the value $\ln(U_\infty/U_1)$ at the trailing edge until it reaches zero at a large distance. Simultaneously H decreases from the value H_1 at the trailing edge, until it reaches unity at a large distance. H. B. Squire established an empirical relation between $\ln(U_\infty/U)$ and H. According to experiment:

$$\frac{\ln(U_\infty/U)}{H-1} = \frac{\ln(U_\infty/U_1)}{H_1-1} = \text{const}\,,$$

so that

$$\int_1^{H_1}\ln \frac{U_\infty}{U}\,dH = \frac{H_1-1}{2}\ln \frac{U_\infty}{U_1}\,.$$

On substituting into eqn. (25.27), we obtain

$$\delta_{2\infty} = \delta_{21}\left(\frac{U_1}{U_\infty}\right)^{(H_1+5)/2}$$

or, with the rounded-off value of $H_1 = 1\cdot4$:

$$\delta_{2\infty} = \delta_{21}\left(\frac{U_1}{U_\infty}\right)^{3.2}$$

On substituting this value into eqn. (25.25) we obtain an expression for the coefficient of total drag in the form

$$\boxed{c_{D\,tot} = 2\frac{\delta_{21}}{l}\left(\frac{U_1}{U_\infty}\right)^{3\cdot2}}\,. \tag{25.28}$$

The coefficient of profile drag can be evaluated from the above equation, if the momentum thickness at the trailing edge is known from the boundary-layer calculation and if, in addition, the ideal, potential velocity at the trailing edge, U_1, is known. The latter can be found, for example, from a reading of the static pressure at the trailing edge. According to a method proposed by H. B. Helmbold [22] the determination of U_1/U_∞ can also proceed as follows: We begin by evaluating the momentum thickness at the trailing edge, δ_{21}/l, from eqn. (22.17) using the value $n = 4$. This value is then substituted into eqn. (25.28), and in the resulting formula U_1/U_∞ is raised to the power $+ 0.2$. Thus this factor can be approximated by the value of unity, because U_1/U_∞ itself does not differ much from unity, and the value of the coefficient of profile drag for one side ($\mathsf{R} = U_\infty l/\nu$) can be found from eqn. (25.28) to be

$$c_{D\,tot} = \frac{0.074}{\mathsf{R}^{1/5}} \left\{ \int\limits_{x_t/l}^{1} \left(\frac{U}{U_\infty} \right)^{3.5} \mathrm{d}\left(\frac{x}{l} \right) + C \right\}^{0.8} \qquad (25.29)\ \dagger$$

with

$$C = 61.6\ \mathsf{R}^{1/4} \left(\frac{\delta_{2t}}{l} \right)^{5/4} \left(\frac{U_t}{U_\infty} \right)^{3.75} . \qquad (25.30)$$

The subscript t refers to the point of transition and the value of the constant C can be determined from the condition that the laminar and turbulent momentum thicknesses must be equal to each other at the point of transition, $\delta_{2t} = \delta_{2turb} = \delta_{2lam}$. The value of δ_{2lam} can be found from eqn. (10.37). For uniform potential flow with $U = U_\infty$, eqn. (25.29) transforms to the corresponding expression for the flat plate at zero incidence, eqn. (21.11), if, in addition, we put $C = 0$ for fully developed turbulent flow.

E. Truckenbrodt [68] transformed eqn. (25.29) replacing the potential velocity distribution by the coordinates of the aerofoil section thus, evidently, effecting a considerable simplification.

H. B. Squire and A. D. Young [64] evaluated a number of examples by the use of a different method. We shall now describe some of them, referring to Fig. 25.3, which contains a *resumé* of these results. The thickness of the aerofoils was varied from $d/l = 0$ (flat plate) to $d/l = 0.25$ and the Reynolds numbers $\mathsf{R} = U_\infty l/\nu$ ranged from 10^6 to 10^8. It is found that the profile drag is very sensitive to the position of the point of transition from laminar to turbulent flow. This latter parameter was varied from $x_t/l = 0$ to 0.4. The increase in profile drag with thickness is, essentially, due to an increase in form drag. Fig. 25.4 shows the relation between form and profile drag. Analogous calculations were performed by J. Pretsch [40] in relation to von Kármán-Trefftz aerofoils. The measure of agreement between calculation and experiment depends decisively on the assumed position of the point of transition. It will be recalled from Chap. XVII that the position of the point of transition is largely dependent on the pressure gradient of the respective

---

† L. Speidel [63] tested the validity of this simple equation against a very large number of actual examples.

potential flow. It was pointed ouf there, Fig. 17.10, that as a first approximation it is possible to assume that the point of transition coincides with the point of minimum pressure provided that the Reynolds number is large, say $R \approx 10^7$. Numerical values obtained on the basis of this assumption show satisfactory agreement with measured values.

The preceding method was first generalized to include axially symmetrical cases by A. D. Young [72]. The method proposed by N. Scholz [58] has been considerably developed and can be, both for two-dimensional and for axially sym-

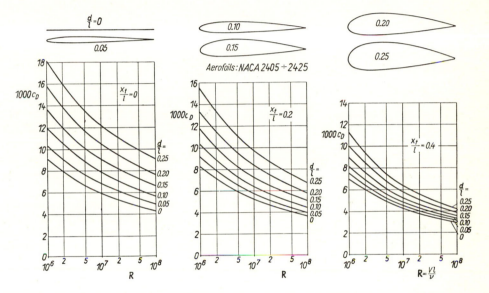

Fig. 25.3.   Profile drag in terms of Reynolds number as evaluated by Squire and Young [64]
x_t denotes the position of the point of transition

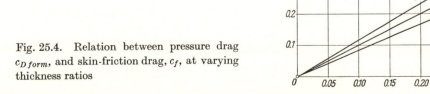

Fig. 25.4.   Relation between pressure drag $c_{D\,form}$, and skin-friction drag, c_f, at varying thickness ratios

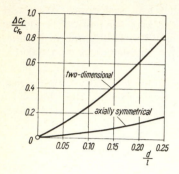

Fig. 25.5. Increase in the coefficient of profile drag plotted in terms of relative thickness, as calculated by Scholz [58]

Total or profile drag $c_{D\,tot} = c_{D\,form} + c_f$

metrical cases, applied to rough walls (equivalent sand roughness) as well. From
a very large number of calculated examples on aerofoils (two-dimensional case)
and bodies of revolution, it proved possible to deduce relations to describe the influence of thickness on profile drag. These are shown plotted in Fig. 25.5. The
difference $\Delta c_f = c_f - c_{f0}$ denotes the increase in the coefficient of skin friction,
related to the wetted surface, as against its value for a flat plate at zero incidence,
c_{f0}. The curve for the two-dimensional case agrees fairly well with the results
shown plotted in Fig. 25.3 for the case of a fully turbulent boundary layer ($x_t/l = 0$).
In this connexion the paper by P. S. Granville [18] may also be consulted.

These calculations give an indication about the effect of friction on lift. The
displacement of the external streamlines caused by the boundary layer modifies
the pressure distribution on an aerofoil and causes the experimental value to become
lower than that given by potential theory. This loss of lift was calculated by K.
Kraemer for the range of angles of incidence below the stalling angle.

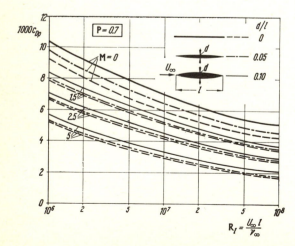

Fig. 25.6. Drag coefficients for biconvex profiles in supersonic flow and
with fully developed turbulent boundary layer, after A. D. Young and
S. Kirkby [76, 77]; no heat transfer

Prandtl number $P = 0.7$. The drag coefficient must be augmented by the contribution
from wave drag, eqn. (25.31)

The preceding method of evaluating total drag by the application of the momentum equation was extended by A. D. Young and S. Kirkby [76] to the case of supersonic flow. Some results of their calculations for biconvex profiles of varying thickness ratio and at zero incidence are shown plotted in Fig. 25.6. The drag coefficient c_{Dtot} includes pressure drag and skin friction and must be augmented by the contribution of the wave drag which exists in supersonic, ideal flow[†]. For biconvex profiles, according to the linearized theory, the latter is given by

$$c_{D\,wave} = \frac{16}{3}\left(\frac{d}{l}\right)^2 \frac{1}{(M^2-1)^{1/2}}\,. \tag{25.31}$$

According to the results shown in Fig. 25.6, the influence of the thickness of the aerofoil on drag is very small, particularly in the supersonic range. The influence of the Mach number is approximately of the same order as in the case of a flat plate at zero incidence.

e. Losses in the flow through cascades

1. General remarks. The numerical calculations of the total drag of a single aerofoil which were explained in Sec. XXVd have been extended by H. Schlichting and N. Scholz [46, 49] to include the case of a row, or cascade, of aerofoils and can, therefore, be applied to the flow through blades. When discussing axial turbine or turbo-compressor stages, it is customary to simplify the problem by taking a co-axial cylindrical section through the stationary and moving row of blades and to develop the resulting pattern on to a plane. The pattern of aerofoils thus obtained is known as a *two-dimensional cascade*. The arrangement of blades in a cascade is usually described by specifying the *solidity ratio t/l* and the *mean blade angle* or *angle of stagger*, β_m, Fig. 25.7. In contrast to the case of flow past a single aerofoil, the application of potential theory to the case of flow past a cascade leads to the conclusion that, generally speaking, there exists a difference in the pressure in front of and behind the cascade. The pressure decreases downstream, when the cascade transforms pressure into velocity (turbine blading). When the cascade performs the reverse (compressor blading), the pressure increases in the direction of flow. This change in pressure, together with the shape of the blade, determines the pressure distribution around the contour of the blade and hence also the structure of the boundary layer. Figure 25.7 shows a plot of the pressure distribution and of the position of the points of separation on two different turbine blades. In the case of profile 9, Fig. 25.7a, the point of separation is located closely behind the pressure minimum for an angle of inflow $\beta_1 = 90°$. However, this occurs only at the low Reynolds number $R = 10^5$, at $R = 10^6$ there is no separation on either side of the blade in this case. As far as profile 15, represented in Fig. 25.7b, is concerned, the point of separation on the pressure side lies downstream of, but very close to, the point of minimum pressure for both angles of inflow. The point of separation on the suction side is very close to the trailing edge in either case.

---

† Here, the pressure drag represents the change in wave drag due to the displacement effect.

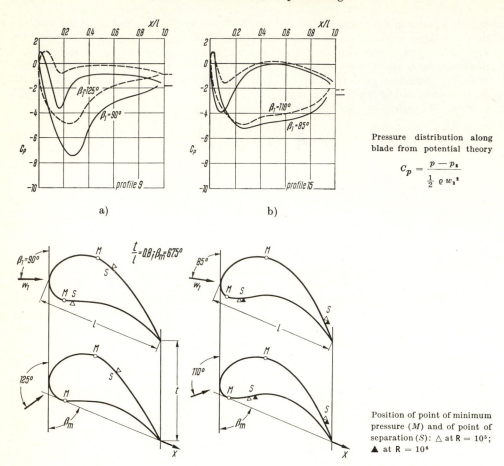

Pressure distribution along blade from potential theory

$$C_p = \frac{p - p_2}{\frac{1}{2}\varrho\, w_2^2}$$

Position of point of minimum pressure (M) and of point of separation (S): $\triangle$ at $R = 10^5$; $\blacktriangle$ at $R = 10^6$

Fig. 25.7. Pressure distribution and position of point of separation of a turbulent boundary layer on the blade of a turbine cascade for two different angles of inflow, after F. W. Riegels [44]

Blade angle $\beta_m = 67\cdot5^\circ$; Solidity ratio $t/l = 0\cdot8$

The work in ref. [46] shows how to employ the method outlined in Sec. XXVd in order to calculate the losses of a two-dimensional cascade at varying angles of inflow. N. Scholz and L. Speidel [60] systematized such calculations and compared them with experimental results.

The velocity distribution immediately behind the exit plane of the cascade shows strong depressions which stem from the boundary layers of the individual blades. Turbulent mixing causes these velocity differences to smooth out further downstream, thus giving rise to an additional loss of energy. The amount of *loss due to mixing* can be evaluated with the aid of the momentum theorem. When determining the total loss in the flow through cascades, it is necessary to take this mixing loss into account in addition to the loss of energy in the boundary layers of the individual blades. Thus a calculation of losses in a cascade consists of the following three partial calculations: 1. Determination of the ideal, potential pressure distribution around the contour of the blades. 2. Calculations of the (laminar or turbulent) boundary layer at a blade. 3. Determination of the losses due to mixing in the wake behind the cascade.

The total amount of losses associated with a cascade is best specified by indicating the difference Δg in the total pressures between the undisturbed flow in front of the cascade and the "smoothed out" actual flow far behind it. Thus

$$\Delta g = g_1 - g_2' = p_1 + \tfrac{1}{2} \varrho \, w_1{}^2 - (p_2' + \tfrac{1}{2} \varrho \, w_2'^2) \,, \tag{25.32}$$

where p_2' and w_2' denote the pressure and velocity in the real (i. e. affected by losses) flow far behind the cascade, respectively. These should be distinguished from the values p_2 and w_2, respectively, which refer to ideal (lossless) flow. It is convenient to render the total loss Δg dimensionless with reference to the dynamic head formed with the axial velocity component $w_{ax} = w_1 \sin \beta_1 = w_2 \sin \beta_2$, as it determines the mass of fluid which passes through the cascade. For reasons of continuity its value must be the same in front of as behind the cascade. We then introduce the following coefficient:

$$\zeta_t = \frac{\Delta g}{\tfrac{1}{2} \varrho \, w_{ax}{}^2} \,. \tag{25.33} \dagger$$

Some results of the systematic investigations on cascades, carried out at the Braunschweig Engineering University [60], also [49], are shown in Fig. 25.8. These represent a comparison between measured and calculated values of the loss coefficient. All blades were derived from the aerofoil NACA 8410. The variable parameters

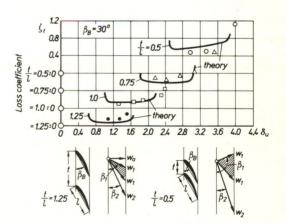

Fig. 25.8. Loss coefficient ζ_t from eqn. (25.33) in terms of the deflexion coefficient $\delta_d = \Delta w_d / w_{ax}$ for turbine cascades with different solidity ratios t/l, after [49]. Measurements and calculations by N. Scholz and L. Speidel [60]

Blade profile: NACA 8410
Reynolds number $\mathsf{R} = w_s l/\nu = 5 \times 10^5$

$\dagger$ In the design of steam turbines it is usual to employ a *velocity coefficient*, ψ, which is defined as the ratio of the real exit velocity to its value in ideal flow, so that $\psi = w'_2/w_2$. Consequently, the two coefficients satisfy the relation $\zeta_t = (1-\psi^2)/\sin^2 \beta_2$.

included the solidity ratio t/l (= 0·5, 0·75, 1·0 and 1·25); the blade angle was $\beta_S =$ 30° (turbine cascade). The loss coefficient defined in eqn. (25.33) is seen plotted in terms of the deflexion coefficient or deflexion ratio

$$\delta_d = \Delta w_d / w_{ax} \, ,$$

where Δw_d denotes the transverse component of velocity (i. e. velocity in circumferential direction) created by the cascade. If we first center our attention on the middle range of the polars (adhering boundary layers), we notice a steep increase in the loss coefficient which occurs as the solidity ratio decreases. The reason for it lies in the fact that the number of blades per unit of length of the circumference is larger when the pitch is small than when the pitch is larger. To a first approximation the loss coefficient is proportional to the number of blades. At the right and left edge of the polar we observe a sudden and large increase in the loss coefficient. This is due to flow separation on the pressure side (left end of curve) or on the suction side (right end of polar) of the blade. In the latter case, an increase in the flow angle causes the admissible load on the blade to be exceeded. It is remarkable that the polar curves displace themselves in the direction of larger angles of deflexion as the solidity ratio decreases.

The measurements and the calculations were carried out for a Reynolds number $R = w_2 \, l/\nu = 5 \times 10^5$. The calculations were performed on the assumption that the boundary layer was turbulent all along the blades. In the experimental arrangement the boundary layers were made turbulent by the provision of tripping wires near the leading edges. The calculated and measured values of the loss coefficient show very good agreement with each other. Further examples and comparisons between theory and experiment are given in [47, 63].

Wake: A very detailed experimental investigation of the flow in a turbulent wake behind a cascade of blades is described in a paper by R. Raj and B. Lakshminarayana [42]. Measurements included determinations of the velocity distribution, intensity of turbulence, and of the apparent Reynolds stresses in the wake at different distances from the cascade. It has transpired that the wakes are not symmetric up to a distance (3/4) l behind the blades in cascades which turn the flow. The decrease in velocity downstream from the cascade exit section is considerably slower than at a flat plate, behind a circular cylinder or downstream from a single aerofoil at zero incidence.

Jet flap: The augmentation of the turning angle $\Delta\beta = \beta_1 - \beta_2$ of compressor cascades by a jet flap has been investigated by U. Stark [64a].

2. Influence of Reynolds number: The changes in the aerodynamic coefficients of a cascade produced by a change in the Reynolds number are important when it becomes necessary to apply the results of tests on models to the design of a full-scale turbomachine. This effect is exerted principally on the loss coefficient, and the problem can be found discussed in a sizeable number of publications [5, 41, 65]. From the physical point of view, the effect of Reynolds number on the loss coefficient of a two-dimensional cascade is analogous to that of the skin friction of a single aerofoil. because in either case the effect originates in the boundary layer. The losses suffered by the cascade stem mainly from the boundary layer if the pressure distribution along a blade in a cascade is such that no important separations occur. They are then affected by the Reynolds number in about the same way as the skin-friction coefficient of a flat plate at zero incidence and are proportional to $R^{-1/2}$ for laminar flow, becoming proportional to $R^{-1/5}$ in turbulent flow. In both cases, the Reynolds number is formed with the blade length, l. The dependence of the loss coefficient on Reynolds number in the absence of separation can be determined by calculation with the aid of a method proposed by K. Gersten [15]. A result of this kind is seen displayed in Fig. 25.9. The diagram describes the variation in the loss coefficient,

$$\zeta_{t2} = \frac{\varDelta g}{\frac{1}{2}\varrho\, w_2^2}\,, \tag{25.34}$$

of a cascade consisting of thick, strongly cambered blades, over a considerable range of Reynolds numbers, that is from $R_2 = w_2 l/\nu = 4 \times 10^4$ to 4×10^5. Here $\varDelta g$ denotes the loss in stagnation pressure and w_2 is the exit velocity. In order to provide a comparison with measurements, the diagram contains a theoretical curve which takes into account separation losses computed with the aid of Ref. [60]. As far as the position of the point of transition is concerned, the calculation was based on the experimentally verified circumstance that the boundary layer on the pressure side of a blade remained laminar as far as the trailing edge, whereas that on the suction side underwent transition at the point of minimum pressure. The diagram in Fig. 25.9 demonstrates that there exists excellent agreement between calculation and measurement.

The magnitude of the losses is strongly influenced by the position of the point of transition. As the Reynolds number is increased, the point of transition moves forward and this lengthens the turbulent portion of the boundary layer and causes the losses to increase. The forward movement of the point of transition is enhanced by increased roughness [13] or by an increased turbulence intensity [8], as one would expect to find in a turbomachine. At very low Reynolds numbers the boundary layer can separate before transition has occurred in it thus causing a large increase in the

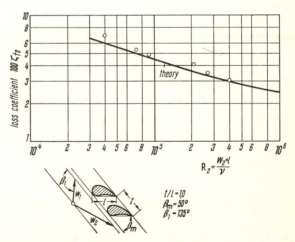

Fig. 25.9.   Loss coefficient of a turbine cascade, eqn. (25.34), in terms of the Reynolds number R_2, after K. Gersten [15]

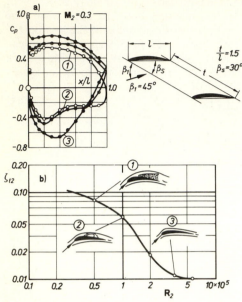

(1) $R_2 = 0.5 \times 10^5$
(2) $R_2 = 1.0 \times 10^5$
(3) $R_2 = 4.0 \times 10^5$

a) Pressure distribution for various Reynolds numbers at $M_2 = 0.3$

b) Loss coefficient ζ_{t2} from eqn. (25.34) as a function of the Reynolds number R_2

Fig. 25.10. Aerodynamic coefficients of a turbine cascade as a function of the Reynolds number as measured by H. Schlichting and A. Das [52, 53]

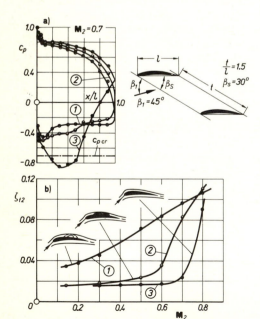

(1) $R_2 = 1.0 \times 10^5$
(2) $R_2 = 2.0 \times 10^5$
(3) $R_2 = 4.0 \times 10^5$

a) Pressure distribution for various Reynolds numbers at $M_2 = 0.7$

b) Loss coefficient ζ_{t2} from eqn. (25.34) as a function of the Mach number M_2 for various values of the Reynolds number

Fig. 25.11. Aerodynamic coefficients of a turbine cascade as a function of Mach number as measured by H. Schlichting and A. Das [52, 53]

loss coefficient under certain circumstances. This large increase in the loss coefficient at low Reynolds numbers is illustrated in Fig. 25.10b which refers to a turbine cascade. At larger Reynolds numbers, $R_2 = 5 \times 10^5$, the transition is spontaneous and the losses are small. At moderate Reynolds numbers, $R_2 = 1 \times 10^5$, there is laminar separation followed by turbulent re-attachement. Thus under the boundary layer there forms a so-called separation bubble and the loss coefficient increases considerably. At very low Reynolds numbers, $R_2 = 0.5 \times 10^5$, the laminar layer separates and stays separated to the end of the blade. The losses increase by a large amount once more.

The details of the separation of the boundary layer are once again mirrored in the pressure distributions plotted in Fig. 25.10a for three values of the Reynolds number. The extent of the separation bubble depends strongly on the Reynolds number and on the intensity of turbulence of the oncoming stream. See [8, 20, 28, 37, 43, 57, 60], and the paper by R. Kiock [30]. *Cf.* W. B. Roberts [43].

In conjunction with our discussion of the effect of the Reynolds number, it is necessary to stress that under certain circumstances the surface roughness can have a large influence on the losses. In addition to enhancing transition, roughness can also directly increase the losses. This occurs when the protuberances exceed a certain admissible value; see [3, 56].

3. Effect of Mach number: The preceding results concerning the loss coefficient of cascades refer to incompressible flows ($M < 0.3$). The effect of compressibility can be said to set in at $M > 0.4$. An example of this effect is shown in Fig. 25.11b. The plot represents the loss coefficient for a cascade producing a small angle of turn in a subsonic flow. The Mach number M_2 is the independent variable and the three curves refer to three different Reynolds numbers. The pressure distribution for $M = 0.7$, Fig. 25.11a, shows that at $R_2 = 4 \times 10^5$ the loss coefficient increases sharply as the Mach number is increased. The sharp increase occurs as a result of shock formation in regions where the local value of the velocity of sound, $c_{p,\,crit}$, has been exceeded in the flow. For the two lower Reynolds numbers, $R_2 = 1.0 \times 10^5$ and $R_2 = 2.0 \times 10^5$, the pressure distribution points to a separated flow. The results displayed in Figs. 25.10 and 25.11 demonstrate that the Mach number exerts a deep influence on the flow through cascades in the range of Reynolds numbers from $R = 10^4$ to 10^5, in addition to the large effect of the Reynolds number itself. The preceding measurements were performed in the high-speed cascade wind tunnel in Brunswick [54] in which the Reynolds number and the Mach number can be varied independently.

The diagram of Fig. 25.12 illustrates the effect of the Mach number on the loss coefficient of a cascade that produces a large angle of turn in the flow. The cascade was designed for incompressible flow. The loss coefficient remains nearly constant at the value $\zeta_{t2} \approx 0.03$ up to $M_2 = 0.7$; it increases sharply as the Mach number is further increased. The reason for this behaviour is clear from Fig. 25.13 in which it is possible to discern the existence of shock waves on the suction side of the blade. These cause separation of the boundary layer.

The effect of the Mach number and of the turbulence intensity on the loss coefficient of cascades has been studied in two theses presented to the Engineering University at Braunschweig by J. Bahr [2] and H. Hebbel [21], respectively. Reference [50] may also be consulted on this point.

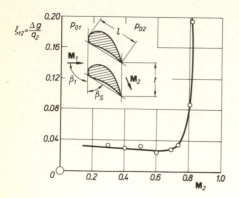

Fig. 25.12. Loss coefficient of a turbine cascade, ζ_{t2} from eqn. (25.34) in terms of the Mach number M_2 after O. Lawaczeck [34]

$\Delta g = p_{01} - p_{02}$ = total pressure loss

$q_2 = \frac{1}{2} \varrho_2 w_2^2$ = stagnation pressure

blade angle: $\beta_S = 56°$; solidity ratio: $t/l = 0.81$
angle at inlet: $\beta_1 = 90°$; Reynolds number:
$R_2 = 6 \times 10^5$

Fig. 25.13. Transonic flow through a turbine cascade. Photograph obtained with the aid of Schlieren method by O. Lawaczeck and H.J. Heinemann [32]. Exposure 20×10^{-9} sec. The strong shock waves on the suction side of the aerofoil cause separation and hence large losses, see also Fig. 25.12

$\beta_S = 70°$;
$t/l = 0.5$
$\beta_1 = 65°$
$\beta_2 = 36°$
$M_1 = 0.40$
$M_2 = 0.85$
$R_2 = 8 \times 10^5$

In modern times, the development of steam turbines of increased power density has caused the outer blade sections of the low-pressure stages to operate in the transonic velocity regime. This made it necessary to undertake systematic investigations into the behaviour of transonic turbine blades. Here the Mach number of the approaching stream is lower than unity ($M_1 < 1$), whereas that at exit exceeds it ($M_2 > 1$); cf. [31]. References [33, 34] contain an account of transonic flow across cascade with a large angle of turn.

H. Haas and H. Maghon [20a] give a comprehensive account of the practical applications of these research results on flow through cascades as they relate to modern developments in steam and gas turbines.

References

[1] Abbott, J.H., von Doenhoff, A.E., and Stivers, L.S.: Summary of airfoil data. NACA Rep. 824 (1945).

[2] Bahr, J.: Untersuchungen über den Einfluss der Profildicke auf die kompressible ebene Strömung durch Verdichtergitter. Diss. Braunschweig 1962. Forschg. Ing.-Wes. *30*, 14—25 (1964).

[3] Bammert, K., and Milsch, R.: Boundary layers on rough compressor blades. ASME Paper No. 72—GT—48. Gas Turbine Conference, San Francisco, 1972.

[4] Betz, A.: Ein Verfahren zur direkten Ermittlung des Profilwiderstandes. ZFM *16*, 42—44 (1925).

[5] Davis, H., Kottas, H., and Moody, A.M.: The influence of Reynolds number on the performance of turbo-machinery. Trans. ASME *73*, 499—509 (1951).

[6] Doetsch, H.: Profilwiderstandsmessungen im grossen Windkanal der DVL. Luftfahrtforschung *14*, 173—178 and 370—372 (1937).

[7] Dowlen, E.M.: A shortened method for the calculation of aerofoil profile drag. J. Roy. Aero. Soc. *56*, 109—116 (1952).

[8] Dunham, J.: Predictions of boundary layer transition on turbomachinery blades. AGARD AG—164 (1972).

[9] Eastman, N., Jacobs, E.N., and Sherman, A.: Aerofoil section characteristics as affected by variation of the Reynolds number. NACA RM 586 (1937).

[10] Evans, R.L.: Stream turbulence effects on the turbulent boundary layer in a compressor cascade. British ARC Rep. 34 587 (1973).

[11] Fage, A., Falkner, V.M., and Walker, W.S.: Experiments on a series of symmetrical Joukowsky sections. ARC RM 1241 (1929).

[12] Fage, A.: Profile and skin-friction airfoil drags. ARC RM 1852 (1938).

[13] Feindt, E.G.: Untersuchungen über die Abhängigkeit des Umschlages laminar—turbulent von der Oberflächenrauhigkeit und der Druckverteilung. Diss. Braunschweig 1956; Jb. Schiffbautechn. Ges. *50*, 180—205 (1956).

[14] Gersten, K.: Experimenteller Beitrag zum Reibungseinfluss auf die Strömung durch ebene Schaufelgitter. Abhandl. Wiss. Ges. Braunschweig *7*, 93—99 (1955).

[15] Gersten, K.: Der Einfluss der Reynolds-Zahl auf die Strömungsverluste in ebenen Schaufelgittern. Abhandl. Wiss. Ges. Braunschweig *11*, 5—19 (1959).

[16] Goett, H.J.: Experimental investigation of the momentum method for determining profile drag. NACA Rep. 660 (1939).

[17] Göthert, B.: Widerstandsbestimmung bei hohen Unterschallgeschwindigkeiten aus Impulsverlustmessungen. Jb. dt. Luftfahrtforschung *I*, 148—155 (1941).

[18] Granville, P.S.: The calculation of the viscous drag of bodies of revolution. David W. Taylor Model Basin, Rep. 849 (1953).

[19] Görtler, H.: Verdrängungswiderstand der laminaren Grenzschicht und Druckwiderstand. Ing.-Arch. *14*, 286—305 (1943/44).

[20] Gaster, M.: The structure and behaviour of laminar separation bubbles. British ARC Rep. 28 226 (1966).

[20a] Haas, H., and Maghon, H.: Mögliche Wirkungsgradverbesserungen bei Dampf- und Gasturbinen. Conference on technologies for more efficient utilization of energy in electric power stations at Kernforschungsanlage Jülich, Febr. 1976. Jülich-Conf. *19*, 74—80 (1976).

[21] Hebbel, H.: Über den Einfluss der Mach-Zahl und der Reynolds-Zahl auf die aerodynamischen Beiwerte von Turbinenschaufelgittern bei verschiedener Turbulenz der Strömung Diss. Braunschweig 1962; Forschg. Ing.-Wes. *30*, 65—77 (1964).

[22] Helmbold, H.B.: Zur Berechnung des Profilwiderstandes. Ing.-Arch. *17*, 273—279 (1949).

[23] Hebbel, H.: Über den Einfluss der Mach-Zahl und der Reynolds-Zahl auf die aerodynamischen Beiwerte von Verdichter-Schaufelgittern bei verschiedener Turbulenz der Strömung. Forschg. Ing.-Wes. *33*, 141—150 (1967).

[24] Horlock, J.H., and Lakshminarayana, B.: Secondary flows: Theory, experiment, and application in turbomachinery aerodynamics. Annual Review of Fluid Mechanics (M. Van Dyke, ed.) *5*, 247—280 (1973).

[25] Horlock, J.H., Shaw, R., Pollhard, D., and Lewkowicz, A.: Reynolds number effects in cascades and axial flow compressors. Trans. ASME, J. Eng. Power *86*, 236—242 (1964).

[26] Jones, B.M.: The measurement of profile drag by the pitot traverse method. ARC RM 1688 (1936).

[27] Jaumotte, A.L., and Devienne, P.: Influence du nombre de Reynolds sur les pertes dans les grilles d'aubes. Technique et Science Aéronautique 5, 227—232 (1956).

[28] Horton, H.P.: A semi-empirical theory for the growth and bursting of laminar separation bubbles. British ARC Rep. CP No. 1073 (1969).

[29] Jones, B.M.: Flight experiments on boundary layers. JAS 5, 81—101 (1938); see also Engineering 145, 397 (1938) and Aircraft Eng. 10, 135—141 (1938).

[30] Kiock, R.: Einfluss des Turbulenzgrades auf die aerodynamischen Beiwerte von ebenen Verzögerungsgittern. Diss. Braunschweig 1971; Forschg. Ing.-Wes. 39, 17—28 (1973).

[31] Lawaczeck, O., and Amecke, J.: Probleme der transsonischen Strömung durch Turbinen-Schaufelgitter. VDI-Forschungsheft 540 (1970).

[32] Lawaczeck, O., and Heinemann, H.J.: Von Kármán vortex streets in the wakes of subsonic and transonic cascades. Paper AGARD Meeting on Unsteady Phenomena in Turbumachinery, Monterey, Cal., Sept. 1975. AGARD CP—177, 28—1 to 28—13 (1976).

[33] Lehthaus, F.: Berechnung der transsonischen Strömung durch ebene Turbinengitter nach dem Zeitschritt-Verfahren. Diss. Braunschweig 1977. VDI-Forschungsheft 586, 5—24 (1978).

[34] Lawaczeck, O.: Halbempirisches Berechnungsverfahren für ebene transsonische Turbinen-profile mit Plattenprofilen. VDI-Forschungsheft 586, 25—36 (1978).

[35] Lawson, T.V.: An investigation into the effect of Reynolds number on a cascade of blades with parabolic arc camberline. British NGTE Memo. M 1975 (1953).

[36] Lock, C.N.H., Hilton, W.F., and Goldstein, S.: Determination of profile drag at high speeds by a pitat traverse method. ARC RM 1971 (1946).

[37] Ntim, B.A.: A theoretical and experimental investigation of separation bubbles. Ph.D. Thesis Univ. of London 1969.

[38] Piercy, N.V.A., Preston, J.H., and Whitehead, L.G.: Approximate prediction of skin friction and lift. Phil. Mag. 26, 791—815 (1938).

[39] Pfenninger, W.: Vergleich der Impulsmethode mit der Wägung bei Profilwiderstands-messungen. Rep. Inst. of Aerodynamics ETH Zürich, No. 8 (1943).

[40] Pretsch, J.: Zur theoretischen Berechnung des Profilwiderstandes. Jb. dt. Luftfahrt-forschung 1938, I, 61—81; Engl. transl. NACA TM 1009 (1942).

[41] Rhoden, H.G.: Effects of Reynolds number on the flow of air through a cascade of compressor blades. ARC RM 2919 (1956).

[42] Raj, R., and Lakshminarayana, B.: Characteristics of the wake behind a cascade of air-foils. JFM 61, 707—730 (1973).

[43] Roberts, W.B.: The effect of Reynolds number and laminar separation on axial cascade performance. Trans. ASME, Ser. A, J. Eng. Power 97, 261—274 (1975).

[44] Riegels, F.W.: Fortschritte in der Berechnung der Strömung durch Schaufelgitter. ZFW 9, 2—15 (1961).

[45] Schäffer, H.: Untersuchungen über die dreidimensionale Strömung durch axiale Schaufel-gitter mit zylindrischen Schaufeln. Diss. Braunschweig 1954. Forschg. Ing.-Wes. 21, 9—19 and 41—49 (1955).

[46] Schlichting, H., and Scholz, N.: Über die theoretische Berechnung der Strömungsverluste eines ebenen Schaufelgitters. Ing.-Arch. 19, 42—65 (1951).

[47] Schlichting, H.: Ergebnisse und Probleme von Gitteruntersuchungen. ZFW 1, 109—122 (1953).

[48] Schlichting, H.: Berechnung der reibungslosen inkompressiblen Strömung für ein vor-gegebenes ebenes Schaufelgitter. VDI-Forschungsheft 447 (1955).

[49] Schlichting, H.: Anwendung der Grenzschichttheorie auf Strömungsprobleme der Turbo-maschinen. Siemens Z. 33, 429—438 (1959); see also: Application of boundary layer theory in turbomachinery. Trans. ASME Ser. D, J. Basic Eng. 81, 543—551 (1959).

[50] Schlichting, H.: Neuere Untersuchungen über Schaufelgitterströmungen. Siemens Z. 37, 827—837 (1963).

[51] Surugue, J. (ed.): Boundary layer effects in turbomachines. AGARDograph No. 164 (1972).

[52] Schlichting, H., and Das, A.: Über einige grundlegende Fragen auf dem Gebiet der Aero-dynamik der Turbomaschinen. L'Aerotecnica 46, 179—194 (1966).

[53] Schlichting, H., and Das, A.: On the influence of turbulence level on the aerodynamic losses of axial turbomachines. In: Flow research on blading (L.S. Dzung, ed.), Elsevier, Amsterdam, 1970, 243—274.

[54] Scholz, N., and Hopkes, U.: Der Hochgeschwindigkeitsgitterwindkanal der Deutschen Forschungsanstalt für Luftfahrt Braunschweig. Forschg. Ing.-Wes. *25*, 133—147 (1959). See also Schlichting, H.: The variable density high speed cascade wind tunnel of the Deutsche Forschungsanstalt für Luftfahrt Braunschweig. AGARD Rep. 91 (1956).

[55] Scholz, N.: Aerodynamik der Schaufelgitter. Vol. I, Braun, Karlsruhe, 1965. Revised Engl. translation by A. Klein: Aerodynamics of cascades. AGARDograph No. 220, AGARD, Paris, 1977.

[56] Speidel, L.: Einfluss der Oberflächenrauhigkeit auf die Strömungsverluste in ebenen Schaufelgittern. Forschg. Ing.-Wes. *20*, 129—140 (1954).

[57] Seyb, N.J.: Determination of cascade performance with particular reference to the prediction of boundary layer parameters. British ARC Rep. 27 214 (1965).

[58] Scholz, N.: Über eine rationelle Berechnung des Strömungswiderstandes schlanker Körper mit beliebig rauher Oberfläche. Jb. Schiffbautechn. Ges. *45*, 244—259 (1951).

[59] Scholz, N.: Strömungsuntersuchungen an Schaufelgittern. VDI-Forschungsheft 442 (1954).

[60] Scholz, N., and Speidel, L.: Systematische Untersuchungen über die Strömungsverluste von ebenen Schaufelgittern. VDI-Forschungsheft 464 (1957).

[61] Schrenk, M.: Über die Profilwiderstandsmessung im Fluge nach dem Impulsverfahren. Luftfahrtforschung *2*, 1—32 (1928); NACA TM 557 and 558 (1930).

[62] Serby, J.E., Morgan, M.B., and Cooper, E.R.: Flight tests on the profile drag of 14% and 25% thick wings. ARC RM 1826 (1937).

[63] Speidel, L.: Berechnung der Strömungsverluste von ungestaffelten ebenen Schaufelgittern. Diss. Braunschweig 1953; Ing.-Arch. *22*, 295—322 (1954).

[64] Squire, H.B., and Young, A.D.: The calculation of the profile drag of aerofoils. ARC RM 1838 (1938).

[64a] Stark, U.: A theoretical investigation of the jet flap compressor cascade in incompressible flow. Diss. Braunschweig 1971; Trans. ASME, J. Eng. Power. *94*, 249—260 (1972).

[65] Stuart, D.J.K.: Analysis of Reynolds number effects in fluid flow through two-dimensional cascades. ARC RM 2920 (1956).

[66] Tani, I.: Low speed flows involving bubble separations. Progress in Aeronautical Sciences, (D. Küchemann, ed.) Pergamon Press, *5*, 70—103 (1964).

[67] Taylor, G.I.: The determination of drag by the pitot traverse method. ARC RM 1808 (1937).

[68] Truckenbrodt, E.: Die Berechnung des Profilwiderstandes aus der vorgegebenen Profilform. Ing.-Arch. *21*, 176—186 (1953).

[69] Wanner, A., and Kretz, P.: Druckverteilungs- und Profilwiderstandsmessungen im Flug an den Profilen NACA 23012 und Göttingen 549. Jb. dt. Luftfahrtforschung *I*, 111—119 (1941).

[70] Weidinger, H.: Profilwiderstandsmessungen an einem Junkers-Tragflügel. Jb. WGL 1926, 112; NACA TM 428 (1927).

[71] Young, A.D.: Note on the effect of compressibility on Jones's momentum method of measuring profile drag. ARC RM 1881 (1939).

[72] Young, A.D.: The calculation of the total and skin friction drags of bodies of revolution at 0° incidence. ARC RM 1947 (1939).

[73] Young, A.D., Winterbottom, B.A., and Winterbottom, N.E.: Note on the effect of compressibility on the profile drag of aerofoils at subsonic Mach numbers in the absence of shock waves. ARC RM 2400 (1950).

[74] Young, A.D.: Note on momentum methods of measuring profile drags at high speeds. ARC RM 1963 (1946).

[75] Young, A.D.: Note on a method of measuring profile drag by means of an integrating comb. ARC RM 2257 (1948).

[76] Young, A.D., and Kirkby, S.: The profile drag of biconvex wing sections at supersonic speeds. Fifty years of boundary-layer research (W. Tollmien and H. Görtler, eds.), Braunschweig, 1955, 419—431.

[77] Young, A.D.: The calculation of the profile drag of aerofoils and bodies of revolution at supersonic speeds. WGL Jb. 1953 (H. Blenk, ed.), 66—75 (1954).

Bibliography

A. Reviews organized in serial publications

A 1. Annual Review of Fluid Mechanics, Annual Review Inc., Palo Alto, Cal.

Vol. 1 (1969)
Goldstein, S.: Fluid mechanics in the first half of this century.
Turner, J.S.: Buoyant plumes and thermals.
Brown, S.N., and Stewartson, K.: Laminar separation.
Yih, Chia-Shun: Stratified flows.
Melcher, J.R., and Taylor, G.I.: Electrohydrodynamics: A review of the role of interfacial shear stresses.
Kennedy, J.F.: The formation of sediment ripples, dunes, and antidunes.
Tani, I.: Boundary-layer transition.
Ffowcs Williams, J.E.: Hydrodynamic noise.
Jones, R.T.: Blood flow.
Phillips, O.M.: Shear-flow turbulence.
Van Dyke, M.: Higher-order boundary-layer theory.
Levich, V.G., and Krylov, C.S.: Surface-tension-driven phenomena.
Sherman, F.S.: The transition from continuum to molecular flow.
Hawthorne, W.R., and Novak, R.A.: The aerodynamics of turbo-machinery.
Lumley, J.L.: Drag reduction by additives.
Zel'Divich, Y.B., and Raizer, Y.P.: Shock waves and radiation.
Lighthill, M.J.: Hydromechanics of aquatic animal propulsion.

Vol. 2 (1970)
Loitsianskii, L.G.: The development of boundary-layer theory in the USSR.
Emmons, H.W.: Critique of numerical modeling of fluid-mechanics phenomena.
Veronis, G.: The analogy between rotating and stratified fluids.
Newman, J.N.: Applications of slender-body theory in ship hydrodynamics.
Kovasznay, L.S.G.: The turbulent boundary layer.
Lick, W.: Nonlinear wave propagation in fluids.
Brenner, H.: Rheology of two-phase systems.
Philip, J.R.: Fluid in porous media.
Hendershott, M., and Munk, W.: Tides.
Monin, A.S.: The atmospheric boundary layer.
Phillips, N.A.: Models for weather prediction.
Robinson, A.R.: Boundary layers in ocean circulation models.
Spreiter, J.R., and Alksne, A.U.: Solar-wind flow past objects in the solar system.
Rich, J.W., and Treanor, Ch.E.: Vibrational relaxation in gas-dynamic flows.
Marble, F.E.: Dynamics of dusty gases.

Vol. 3 (1971)

Busemann, A.: Compressible flow in the thirties.

Jaffrin, M.Y., and Shapiro, A.H.: Peristaltic pumping.

Hunt, J.C.R., and Shercliff, J.A.: Magnetohydrodynamics at high Hartmann number.

Friedmann, H.W., Linson, L.M., Patrick, R.M., and Petschek, H.E.: Collisionless shocks in plasmas.

Vincenti, W.G., and Traugott, S.C.: The coupling of radiative transfer and gas motion.

Rivlin, R.S., and Sawyers, K.N.: Nonlinear continuum mechanics of viscoelastic fluids.

Willmarth, W.W.: Unsteady force and pressure measurements.

Williams, F.A.: Theory of combustion in laminar flows.

Fung, Y.C., and Zweifach, B.W.: Microcirculation: Mechanics of blood flow in capillaries.

Rohsenow, W.M.: Boiling.

Wehausen, J.V.: The motion of floating bodies.

Hayes, W.D.: Sonic boom.

Cox, R.G., and Mason, S.G.: Suspended particles in fluid flow through tubes.

Korobeinikov, V.P.: Gas dynamics of explosions.

Stuart, J.T.: Nonlinear stability theory.

Mikhailov, V.V., Neiland, V.Y., and Sychev, V.V.: The theory of viscous hypersonic flow.

Vol. 4 (1972)

Villat, H.: As luck would have it — a few mathematical reflections.

Harleman, D.R.F., and Stolzenbach, K.D.: Fluid mechanics of heat disposal from power generation.

Turcotte, D.L., and Oxburgh, E.R.: Mantle convection and the new global tectonics.

Long, R.R.: Finite amplitude disturbances in the flow of inviscid rotating and stratified fluids over obstacles.

Jahn, Th.L., and Votta, J.J.: Locomotion of protozoa.

Roberts, P.H., and Soward, A.M.: Magnetohydrodynamics of the earth's core.

Becker, E.: Chemically reacting flows.

Hall, M.G.: Vortex breakdown.

Hunter, C.: Self-gravitating gaseous disks.

Wu, Th. Yao-tsu: Cavity and wake flows.

Barenblatt, G.I., and Zel'dovich, Y.B.: Self-similar solutions as intermediate asymptotics.

Berger, E., and Wille, R.: Periodic flow phenomena.

Hoult, D.P.: Oil spreading on the sea.

van Wijngaarden, L.: One-dimensional flow of liquids containing small gas bubbles.

Milgram, J.H.: Sailing vessels and sails.

Ashley, H., and Rodden, W.P.: Wing-body aerodynamic interaction.

Howard, L.N.: Bounds on flow quantities.

Vol. 5 (1973)

Flügge, W., and Flügge-Lotz, I.: Ludwig Prandtl in the nineteen-thirties: Reminiscences.

Penner, S.S., and Jerskey, T.: Use of lasers for local measurement of velocity components, species, and temperatures.

Oppenheim, A.K., and Soloukhin, R.I.: Experiments in gasdynamics of explosions.

Fischer, H.B.: Longitudinal dispersion and turbulent mixing in open-channel flow.

Wegener, P.P., and Parlange, J.-Y.: Spherical-cap bubbles.

Mollo-Christensen, E.: Intermittency in large-scale turbulent flows.

Nieuwland, G.Y., and Spee, B.M.: Transonic airfoils: Recent developments in theory, experiment, and design.

Fay, J.A.: Buoyant plumes and wakes.

Acosta, A.J.: Hydrofoils and hydrofoil craft.

Saibel, E.A., and Macken, N.A.: The fluid mechanics of lubrication.

Gebhart, B.: Instability, transition, and turbulence in buoyancy-induced flows.

Horlock, J.H., and Lakshminarayana, B.: Secondary flows: Theory, experiment, and application in turbomachinery aerodynamics.

McCune, J.E., and Kerrebrock, J.L.: Noise from aircraft turbomachinery.

Ferri, A.: Mixing-controlled supersonic combustion.

Eichelbrenner, E.A.: Three-dimensional boundary layers.

Werle, H.: Hydrodynamic flow visualization.

Kogan, M.N.: Molecular gas dynamics.

Nickel, K.: Prandtl's boundary-layer theory from the viewpoint of a mathematician.

Vol. 6 (1974)

Taylor, G.I.: The interaction between experiment and theory in fluid mechanics.

Miles, J.W.: Harbor seiching.

Turner, J.S.: Double-diffusive phenomena.

Streeter, V.L., and Wylie, E.B.: Waterhammer and surge control.

van Atta, Ch.W.: Sampling techniques in turbulence measurements.

Phillips, O.M.: Nonlinear dispersive waves.

Truesdell, C.: The meaning of viscometry in fluid dynamics.

Panofsky, H.A.: The atmospheric boundary layer below 150 meters.

Roberts, P.H., and Donelly, R.J.: Superfluid mechanics.

Batchelor, G.K.: Transport properties of two-phase materials with random structure.

Benton, E.R., and Clark, jr., A.: Spin-up.

Orzag, St.A., and Israeli, M.: Numerical simulation of viscous incompressible flows.

Korbacher, G.K.: Aerodynamics of powered high-lift systems.

Vol. 7 (1975)

Burgers, J.M.: Some memories of early work in fluid mechanics at the Technical University of Delft.

Willmarth, W.W.: Pressure fluctuations beneath turbulent boundary layers.

Palm, E.: Nonlinear thermal convection.

Lomax, H., and Steger, L.: Relaxation methods in fluid mechanics.

Wieghardt, K.: Experiments in granular flow.

Christiansen, W.H., Russell, D.A., and Hertzberg, A.: Flow lasers.

Widnall, Sh.E.: The structure and dynamics of vortex filaments.

Tien, C.L.: Fluid mechanics of heat pipes.

Koh, R.C.Y., and Brooks, N.H.: Fluid mechanics of waste-water disposal in the ocean.

Goldsmith, H.L., and Skalak, R.: Hemodynamics.

Ladyzhenskaya, O.A.: Mathematical analysis of Navier-Stokes equations for incompressible liquids.

Maxworthy, T., and Browand, F.K.: Experiments in rotating and stratified flows: Oceanographic application.

Laufer, J.: New trends in experimental turbulence research.

Raichlen, F.: The effect of waves on rubble-mound structures.

Csamady, G.T.: Hydrodynamics of large lakes.

Vol. 8 (1976)

Rouse, H.: Hydraulics' latest golden age.

Bird, R.B.: Useful non-Newtonian models.

Peterlin, A.: Optical effects in flow.

Davis, St.H.: The stability of time-periodic flows.

Cermak, J.E.: Aerodynamics of buildings.

Fischer, H.B.: Mixing and dispersion in estuaries.

Hill, J.C.: Homogeneous turbulent mixing with chemical reaction.

Pearson, J.R.A.: Instability in non-Newtonian flow.

Reynolds, W.C.: Computation of turbulent flows.

Comte-Bellot, G.: Hot-wire anemometry.

Wooding, R.A., and Morel-Seytoux, H.J.: Multiphase fluid flow through porous media.

Inman, D.L., Nordstrom, Ch.E., and Flick, R.E.: Currents in submarine canyons: An air-sea-land interaction.

Reshotko, E.: Boundary-layer stability and transition.

Libby, P.A., and Williams, F.A.: Turbulent flows involving chemical reactions.

Rusanov, V.V.: A blunt body in a supersonic stream.

Vol. 9 (1977)

Jones, R.T.: Recollections from an earlier period in American aeronautics.

Pipkin, A.C., and Tanner, R.I.: Steady non-viscometric flows of viscoelastic liquids.

Bradshaw, P.: Compressible turbulent shear layers.

Davidson, J.F., Harrison, D., and Guedes de Carvalho, J.R.F.: On the liquidlike behavior of fluidized beds.

Tani, I.: History of boundary-layer theory.

Williams, III, J.C.: Incompressible boundary-layer separation.

Plesset, M.S., and Prosperetti, A.: Bubble dynamics and cavitation.

Holt, M.: Underwater explosions.

Zel'dovich, Y.B.: Hydrodynamics of the universe.

Pedley, T.J.: Pulmonary fluid dynamics.

Canny, M.J.: Flow and transport in plants.

Spielman, Ll.A.: Particle capture from low-speed laminar flows.

Saville, D.A.: Electrokinetic effects with small particles.

Brennen, Ch., and Winet, H.: Fluid mechanics of propulsion by cilia and flagella.

Hütter, U.: Optimum wind-energy conversion systems.

Shen, Shan-fu: Finite-element methods in fluid mechanics.

Ffowcs Williams, J.E.: Aeroacoustics.

Belotserkovskii, S.M.: Study of the unsteady aerodynamics of lifting surfaces using the computer.

Vol. 10 (1978)

Binnie, A.M.: Some notes on the study of fluid mechanics in Cambridge.

Tuck, E.O.: Hydrodynamic problems of ships in restricted waters.

Bird, G.A.: Monte Carlo simulation of gasflows.

Berman, N.S.: Drag reduction by polymers.

Ryzhov, O.S.: Viscous transonic flows.

Griffith, W.C.: Dust explosions.

Leith, C.E.: Objective methods for weather prediction.

Callander, R.A.: River meandering.

Dickinson, R.E.: Rossby waves — Long-period oscillations of oceans and atmospheres.

Jenkins, J.T.: Flows of nematic liquid crystals.

Leibovich, S.: The structure of vortex breakdown.

Laws, E.M., and Livesey, J.L.: Flow through screens.

Sherman, F.S., Imberger, J., and Corcos, G.M.: Turbulence and mixing in stably stratified waters.

Patterson, G.S., Jr.: Prospects for computational fluid mechanics.

Taub, A.H.: Relativistic fluid mechanics.

Reethof, G.: Turbulence-generated noise in pipe flow.

Ashton, G.D.: River ice.

Mei, Chiang C.: Numerical methods in water-wave diffraction and radiation.

Keller, H.B.: Numerical methods in boundary-layer theory.

Busse, F.H.: Magnetohydrodynamics of the earth's dynamo.

A 2. Advances in Applied Mechanics, Academic Press, New York
    (only contributions to fluid mechanics listed)

Vol. 1 (1948), ed. by R. von Mises and Th. von Kármán,

Dryden, H.L.: Recent advances in the mechanics of boundary layer flow. p. 2—40.

Burgers, J.M.: A mathematic model illustrating the theory of turbulence. p. 171—199.

von Mises, R., and Schiffer, M.: On Bergman's integration method in two-dimensional compressible fluid flow. p. 249—285.

Vol. 2 (1951), ed. by R. von Mises and Th. von Kármán

von Kármán, Th., and Lin, C.C.: On the statistical theory of isotropic turbulence. p. 2—19.

Kuerti, G.: The laminar boundary layer in compressible flow. p. 23—92.

Polubarinova-Kochina, P.Y.: Theory of filtration of liquids in porous media. p. 154—225.

Vol. 3 (1953), ed. by R. von Mises and Th. von Kármán

Carrier, G.F.: Boundary layer problems in applied mechanics. p. 1—19.

Zaldastani, O.: The one-dimensional isentropic fluid flow. p. 21—59.

Frenkiel, F.N.: Turbulent diffusion: Mean concentration distribution in a flow field of homogeneous turbulence. p. 62—107.

Ludloff, H.F.: On aerodynamics of blasts. p. 109—144.

Guderley, G.: On the presence of shocks in mixed subsonic-supersonic flow patterns. p. 145—184.

Rosenhead, L.: Vortex systems in wakes. p. 185—195.

Vol. 4 (1956), ed. by H.L. Dryden and Th. von Kármán

Clauser, F.H.: The turbulent boundary layer. p. 2—51.

Moore, F.K.: Three-dimensional boundary layer theory. p. 160—228.

Vol. 5 (1958), ed. by H.L. Dryden and Th. von Kármán

Fabri, J., and Siestrunck, R.: Supersonic air ejectors. p. 1—34.

Van De Vooren, A.I.: Unsteady airfoil theory. p. 36—89.

Frieman, E.A., and Kulsrud, R.M.: Problems in hydromagnetics. p. 195—231.

Wegener, P.P., and Mack, L.M.: Condensation in supersonic and hypersonic wind tunnels. p. 307—447.

Vol. 6 (1960), ed. by H.L. Dryden and Th. von Kármán

Stewartson, K.: The theory of unsteady laminar boundary layers. p. 1—37.

Ludwig, G., and Heil, M.: Boundary layer theory with dissociation and ionization. p. 39—118.

Chester, W.: The propagation of shock waves along ducts of varying cross section. p. 120—152.

Oswatitsch, K.: Similarity and equivalence in compressible flow. p. 153—271.

Wille, R.: Karman vortex streets. p. 273—287.

Vol. 7 (1962), ed. by H.L. Dryden and Th. von Kármán

Mirels, H.: Hypersonic flow over slender bodies associated with power-law shocks. p. 2—54.

Raymond, H., and Robert, P.H.: Some elementary problems in magneto-hydrodynamics. p. 216—319.

Vol. 8 (1964), ed. by H.L. Dryden and Th. von Kármán,
Sears, W.R., and Resler, E.L.: Magneto-aerodynamics flow past bodies. p. 1—68.
Markovitz, H., and Coleman, B.: Incompressible second order fluids. p. 69—101.
Ribner, H.S.: The generation of sound by turbulent jets. p. 104—182.
Rumyantsev, V.V.: Stability of motion of solid bodies with liquid filled cavities by Lyapunov's method. p. 184—232.
Moiseev, N.N.: Introduction to the theory of oscillations of liquid-containing bodies. p. 233—289

Vol. 9 (1966), ed. by G.G. Cherny et al.
Drazin, P.G., and Howard, L.N.: Hydrodynamic stability of parallel flow of inviscid fluid. p. 1—89.
Moiseev, N.N., and Petrev, A.A.: The calculation of free oscillations of a liquid in a motionless container. p. 91—154.

Vol. 10 (1967), ed. by G.G. Cherny et al.
Lick, W.: Wave propagation in real gases. p. 1—72.
Paria, G.: Magneto-elasticity and magneto-thermoelasticity. p. 73—112.

Vol. 11 (1971), ed. by Chia-Shun Yih
Yao, Th., and Tsu, W.V.: Hydrodynamics of swimming fishes and cetaceans. p. 1—63.
Fung, Y.C.: A survey of the blood flow problem. p. 65—130.
Sichel, M.: Two-dimensional shock structure in transonic and hypersonic flow. p. 131—207.

Vol. 12 (1972), ed. by Chia-Shun Yih
Harper, J.F.: The motion of bubbles and drops through liquids. p. 59—129.
Germain, P.: Shock waves, jump relations and structure. p. 131—194.
Liu, V.C.: Interplanetary gas dynamics. p. 195—237.

Vol. 13 (1973), ed. by Chia-Shun Yih
Veronis, G.: Large scale ocean circulation. p. 2—92.
Wehausen, J.V.: Wave resistance of ships. p. 93—245.
Kuo, H.L.: Dynamics of quasigeostrophic flows and instability theory. p. 248—330.

Vol. 14 (1974), ed. by Chia-Shun Yih
Stewartson, K.: Multistructured boundary layers on flat plates and related bodies. p. 145—239.
Joseph, D.D.: Response curves for plane Poiseuille flow. p. 241—278.
Cowin, S.C.: The theory of polar fluids. p. 279—347.

Vol. 15 (1975), ed. by Chia-Shun Yih
Vanoni, V.A.: River dynamics. p. 1—87.
Tuck, T.O.: Matching problems involving flow through small holes. p. 89—158.
Willmarth, W.W.: Structure of turbulence in boundary layers. p. 159—259.

Vol. 16 (1976), ed. by Chia-Shun Yih
Batchelor, G.K.: G.I. Taylor as I knew him. p. 1—8.
Peregrine, D.H.: Interaction of water waves and currents. p. 10—117.
Moffatt, H.K.: Generation of magnetic fields by fluid motion. p. 120—181.
Yih, Chia-Shun: Instability of surface and internal waves. p. 369—419.

Vol. 17 (1977), ed. by Chia-Shun Yih
Long, R.R.: Some aspects of turbulence in geophysical systems. p. 2—90.
Ogilvie, T.F.: Singular-perturbation problems in ship hydrodynamics. p. 92—188.

A 3. Progress in Aeronautical Sciences, Pergamon Press, London
(only contributions to fluid mechanics listed)

Vol. I (1961), ed. by A. Ferri, D. Küchemann and L.H.G. Sterne

Maskell, E.C.: On the principles of aerodynamic design. p. 1—7.

Legendre, R.: Calcul des profils d'aubes pour turbomachines transsoniques. p. 8—25.

Fenain, M.: La théorie des écoulements à potentiel homogène et ses applications au calcul des ailes en régime supersonique. p. 26—103.

Becker, E.: Instationäre Grenzschichten hinter Verdichtungsstössen und Expansionswellen. p. 104—173.

Goldworthy, F.A.: On the dynamics of ionized gas. p. 174—205.

Warren, C.H.E., and Randall, D.C.: The theory of sonic bangs. p. 238—274.

Vol. II (1962), ed. by A. Ferri, D. Küchemann, and L.H.G. Sterne

Rotta, J.: Turbulent boundary layers in incompressible flow. p. 1—219.

Cooke, J.C., and Hall, M.G.: Boundary layers in three-dimensions. p. 221—282.

Vol. III (1962), ed. by A. Ferri, D. Küchemann, and L.H.G. Sterne

Bagley, J.A.: Some aerodynamic principles for the design of swept wings. p. 1—83.

Sacks, A.H., and Burnell, J.A.: Ducted propellers — a critical review of the state of the art. p. 85—135.

Cox, R.N.: Experimental facilities for hypersonic research. p. 137—178.

Panofsky, H.A., and Press, H.: Meteorological and aeronautical aspects of atmospheric turbulence. p. 179—232.

Vol. V (1964), ed. by D. Küchemann and L.H.G. Sterne

Bradshaw, P., and Pankhurst, R.C.: The design of low speed wind tunnels. p. 1—69.

Tani, I.: Low speed flow involving bubble separation. p. 70—103.

Teipel, I.: Ergebnisse der Theorie schallnaher Strömungen. p. 104—142.

Germain, P.: Écoulements transsoniques homogènes. p. 143—273.

Esterma, J., and Roshko, A.: Rarified gas dynamics. p. 274—294.

Vol. VI (1965), ed. by D. Küchemann and L.H.G. Sterne

Smolderen, J.J.: The evolution of the equations of gas flow at low density. p. 1—132.

Gaster, M.: The role of spatially growing waves in the theory of hydrodynamics stability. p. 251—270.

Küchemann, D.: Hypersonic aircraft and their aerodynamic problems. p. 271—353.

Vol. VII (1966), ed. by D. Küchemann

Roy, M.: On the rolling-up of the conical vortices above a delta wing. p. 1—5.

Legendre, R.: Vortex sheets rolling-up along leading-edges of delta wings. p. 7—33.

Smith, J.H.B.: Theoretical work on the formation of vortex sheets. p. 35—51.

Hall, M.G.: The structure of concentrated vortex cores. p. 53—110.

Rott, N., and Lewellen, W.S.: Boundary layers and their interactions in rotating flows. p. 111—144.

Morton, B.R.: Geophysical vortices. p. 145—194.

Wille, R.: On unsteady flows and transient motions. p. 195—207.

Vol. VIII (1967), ed. by D. Küchemann

Hess, J.L., and Smith, A.M.O.: Calculation of potential flow about arbitrary bodies. p. 1—138.

Lock, R.C., and Bridgewater, J.: Theory of aerodynamic design of swept-winged aircraft at transonic and supersonic speeds. p. 139—228.

Wuest, W.: Boundary layers in rarefied gas flow. p. 295—352.

Vol. IX (1968), ed. by D. Küchemann

Vries, D.G., and Beatrix, C.: Les procédés généraux de mesure des caratéristiques. p. 1—39.

Chushkin, P.J.: Numerical method of characteristics for three-dimensional supersonic flows. p. 41—122.

Michel, R.: Caractéristiques thermiques des couches limites et calcul pratique des transferts de chaleur en hypersonique. p. 123—214.

Broadbent, E.G.: A review of fluid mechanical and related problems in MHD generators. p. 215—327.

Küchemann, D., and Weber, J.: An analysis of some performance aspects of various types of aircraft designed to fly over different ranges at different speeds. p. 329—465.

Vol. X (1970), ed. by D. Küchemann

Jeromin, L.O.F.: The status of research in turbulent boundary layers with fluid injection. p. 65—189.

Fenain, M.: Calcul numérique des ailes en régime supersonique stationnaire ou instationnaire. p. 191—259.

Enselme, M.: Contribution au calcul des caractéristiques aérodynamiques d'un aéronef en écoulement supersonique stationnaire ou instationnaire. p. 261—336.

Vol. XI (1970), ed. by D. Küchemann

Dutton, J.A.: Effects of turbulence on aeronautical systems. p. 67—109.

Coupry, G.: Problémes du vol d'un avion en turbulence. p. 111—181.

Burnham, J.: Atmospheric turbulence at the cruise altitudes of supersonic transport aircraft. p. 183—234.

Green, J.E.: Interactions between shock waves and turbulent boundary layers. p. 235—340.

Vol. XII (1972), ed. by D. Küchemann

Jaffe, N.A., and Smith, A.M.O.: Calculation of laminar boundary layers by means of a differential-difference method. p. 49—212.

Michalke, A.: The instability of free shear layers. p. 213—239.

Smith, J.H.B.: Remarks on the structure of conical flow. p. 241—272.

Chevallier, J.P., and Taillet, J.: Récent progrès dans les techniques de mesure en hypersonique. p. 273—358.

Vol. XIII (1972), ed. by D. Küchemann

Barantsey, R.G.: Some problems of gas solid surface interaction. p. 1—80.

Drummond, A.M.: Performance and stability of hypervelocity aircraft flying on a minor circle. p. 137—221.

Vol. XIV (1973) ed. by D. Küchemann

Lukasiewicz, J.: A critical review of development of experimental methods in high-speed aerodynamics. p. 1—26.

Gonor, A.L.: Theory of hypersonic flow about a wing. p. 109.—175

Tanner, M.: Theoretical prediction of base pressure for steady base flow. p. 177—225.

Fuchs, H.V., and Michalke, A.: Introduction to aerodynamic noise theory. p. 227—297.

Vol. XV (1974), ed. by D. Küchemann

Rasmussen, H.: Application of variational methods in compressible flow calculations. p. 1—35.

Lichtfuss, H.J., and Starken, H.: Supersonic cascade flow. p. 37—149.

Hacker, T., and Oprisiv, C.: A discussion of the roll-coupling problem. p. 151—180.

Bütefisch, K.A., and Venemann, D.: The electron beam technique in hypersonic rarefied gas dynamics. p. 217—255.

Vol. XVI (1975), ed. by D. Küchemann

Wazzan, A.R.: Spatial stability of Tollmien-Schlichting waves. p. 99—127.

Clements, R.R., and Maul, D.J.: The representation of sheets of velocity by discrete vortices. p. 129—146.

Chue, S.H.: Pressure probes for fluid measurements. p. 147—223.

Tanner, M.: Reduction of base drag. p. 369—384.

Carriere, P., Sirieix, M., and Delery, J.: Méthodes de calcul des écoulement turbulents décollées et supersonique. p. 385—429.

Vol. XVII (1976/77), ed. by D. Küchemann

Broadbent, E.G.: Flows with heat addidion. p. 93—107.

Jones, D.S.: The mathematical theory of noise shielding. p. 149—229.

Broadbent, E.G.: Noise shielding for aircraft. p. 231—268.

Glass, J.J.: Shock waves on earth and in space. p. 269—286.

Taneda, S.: Visual study of unsteady separated flows around bodies. p. 287—348.

A 4. Advances in Aeronautical Sciences

Vol. I and Vol. II: Proceedings of the First International Congress in Aeronautical Sciences, Madrid, 8—13 September, 1958. Pergamon Press, London, 1959.

Vol. III and Vol. IV: Proceedings of the Second International Congress in Aeronautical Sciences, Zürich, 12—16 September, 1960. Pergamon Press, London, 1962.

Vol. V: Proceedings of the Third Congress of the International Council of the Aeronautical Sciences, Stockholm, 27—31 August, 1962. Spartan Books, Washington D.C., 1964.

Vol. VI: Proceedings of the Fourth Congress of the International Council of the Aeronautical Sciences, Paris, 24—28 August, 1964. Spartan Books, Washington D.C., 1965.

Two Volumes Aerospace Proceedings 1966: Proceedings of the Fifth Congress of the International Council of the Aeronautical Sciences, London, 12—16 September, 1966, ed. by The Royal Aeronautical Society and McMillan, London, 1967.

B.   Handbooks, Collected Papers, Applied Mechanics Congresses

B 1. Handbooks

Princeton University Series on High Speed Aerodynamics and Jet Propulsion, Princeton University Press, 1955—1964, Vol. I to XII

Vol. I (1955), ed. by F.D. Rossini: Thermodynamics and physics of matter.

Vol. II (1956), ed. by B. Lewis, R.N. Pease, H.S. Taylor: Combustion processes.

Vol. III (1958), ed. by H.W. Emmons: Fundamentals of gas dynamics.

Vol. IV (1964), ed. by F.K. Moore: Theory of laminar flows.
Contributions by:
Moore, F.K.: Introduction.
Lagerstrom, P.A.: Laminar flow theory.
Mager, A.: Three-dimensional laminar boundary layers.
Rott, N.: Theory of time-dependent laminar flows.
Moore, F.K.: Hypersonic boundary layer theory.
Ostrach, S.: Laminar flow with body forces.
Shen, S.F.: Stability of laminar flows.

Vol. V (1959), ed. by C.C. Lin: Turbulent flows and heat transfer.
Contributions by:
Dryden, H.L.: Transition from laminar to turbulent flow.
Schubauer, G.B.: Turbulent flow.

Lin, C.C.: Statistical theories of turbulence.

Yachter, M., and Mayer, E.: Conduction of heat.

Deissler, R.G., and Sabersky, R.H.: Convective heat transfer and friction in flow of liquids.

Van Driest, E.R.: Convective heat transfer in gases.

Yuan, S.W.: Cooling by protective fluid films.

Penner, S.S.: Physical basis of thermal radiation.

Hottel, H.C.: Engineering calculations of radiant heat exchange.

Vol. VI (1954), ed. by W.R. Sears: General theory of high speed aerodynamics.

Vol. VII (1957), ed. by A.F. Donovan and H.R. Lawrence: Aerodynamic components of aircraft at high speeds.

Vol. VIII (1961), ed. by A.F. Donovan, H.R. Lawrence, F.E. Goddard, and R.R. Gilruth: High speed problems of aircraft and experimental methods.

Vol. IX (1954), ed. by R.W. Ladenburg, B. Lewis, R.N. Pease, and H.S. Taylor: Physical measurements in gas dynamics and combustion.

Vol. X (1964), ed. by W.R. Hawthorne: Aerodynamics of turbines and compressors.

Vol. XI (1960), ed. by W.R. Hawthorne and W.T. Olson: Design and performance of gas turbine power plants.

Vol. XII (1959), ed. by O.E. Lancaster: Jet propulsion engines.

Handbuch der Physik, ed. by S. Flügge, Springer Verlag, Berlin/Göttingen/Heidelberg.

Vol. VIII/1 (1959), Strömungsmechanik I
Oswatitsch, K.: Physikalische Grundlagen der Strömungslehre. p. 1—124.
Serrin, J.: Mathematical principles of classical fluid mechanics. p. 125—263.
Howarth, L.: Laminar boundary layers. p. 264—350.
Schlichting, H.: Entstehung der Turbulenz. p. 351—450.

Vol. VIII/2 (1963), Strömungsmechanik II
Berker, R.: Intégration des équations du mouvement d'un fluide visqueux incompressible. p. 1—384.
Weissinger, J.: Theorie des Tragflügels bei stationärer Bewegung in reibungslosen, inkompressiblen Medien. p. 385—437.
Lin, Chia-Chiao, and Reid, H.: Turbulent flow, theoretical aspects. p. 438—523.
Corrsin, S.: Turbulence, experimental methods. p. 524—590.
Schaaf, S.A.: Mechanics of rarefied gases. p. 591—624.
Scheidegger, A.E.: Hydrodynamics of porous media. p. 625—662.

Vol. IX/1 (1960), Strömungsmechanik III
Schiffer, M.: Analytical theory of subsonic and supersonic flows. p. 1—161.
Cabannes, H.: Théorie des ondes de choc. p. 162—224.
Meyer, R.E.: Theory of characteristics of inviscid gas dynamics. p. 225—282.
Timman, R.: Linearized theory of unsteady flow of a compressible fluid. p. 283—310.
Gilbarg, D.: Jets and cavities. p. 311—445.
Wehausen, V., and Laitone, E.V.: Surface waves. p. 446—778.

Handbook of Fluid Dynamics, 27 Sections, ed. by V.L. Streeter, Section 9: Schlichting, H.: Boundary layer theory. p. 9,1—9,68. McGraw-Hill, New York, 3rd. ed. 1966.

Handbuch der Experimentalphysik. Vol. 4, Part I, ed. by W. Wien, and F. Harms, Leipzig, 1931. Contributions by Tollmien, W.: Grenzschicht-Theorie. p. 239—287; Turbulente Strömungen. p. 289—339.

B 2. Collected Works

Prandtl, L.: Gesammelte Abhandlungen zur angewandten Mechanik, Hydro- und Aerodynamik. 3 Volumes, ed. by W. Tollmien, H. Schlichting, and H. Görtler. Springer Verlag, 1961.

von Kármán, Th.: Collected works of Theodore von Kármán. 4 Volumes (1902—1951). Butterworth, London, 1956; Supplement Volume (1952—1963), von Kármán Institute Rhode St. Genèse, Belgium, 1975.

Taylor, G.I.: The scientific papers of Sir Geoffrey Ingram Taylor. 4 Volumes, ed. by G.K. Batchelor. Cambridge University Press, 1958—1971.

Taylor, G.I.: Surveys in mechanics. The G.I. Taylor 70th anniversary volume, ed. by G.K. Batchelor, and R.M. Daies. Cambridge, 1956.

B 3. Applied Mechanics Congresses

Görtler, H., and Tollmien, W. (ed.): Fünfzig Jahre Grenzschichtforschung. Eine Festschrift in Originalbeiträgen. Vieweg, Braunschweig, 1955, 499 pp.

Görtler, H. (ed.): Grenzschichtforschung. IUTAM-Symposium, Freiburg/Breisgau, 1957. Springer Verlag, 1958, 411 pp.

Mécanique de la Turbulence, Marseille, 28 August—2 September 1961. Colloques Internationaux du Centre National de la Récherche Scientifique, No. 108, Paris, 1962, 470 pp.

Proceedings of the 10th International Congress of Applied Mechanics, Stresa, Italy, September 1960, ed. by F. Rolla and W.F. Koiter. Elsevier Publishing Co., Amsterdam/New York, 1962, 370 pp.

Proceedings of the 11th International Congress of Applied Mechanics, München, Germany, August 1964, ed. by H. Görtler. Springer Verlag, Berlin, 1966, 1190 pp.

Proceedings of the 12th International Congress of Applied Mechanics, Stanford University, Cal., USA, August 1968, ed. by M. Hetényi and W.G. Vincenti. Springer Verlag, Berlin, 1969, 420 pp.

Proceedings of the 13th International Congress of Applied Mechanics, Moskau University, August 1972, ed. by E. Becker and G.K. Mikhailov. Springer Verlag, Berlin, 1973, 366 pp.

Proceedings of the 14th International Congress of Applied Mechanics, Delft, Holland, August—September 1976, ed. by W.I. Koiter. North-Holland Publishing Co., Amsterdam, 1976, Preprints, 260 pp.

Proceedings of the Boeing Symposium on Turbulence, held at the Boeing Scientific Research Laboratories, Seattle, Washington, USA, 23—27 June, 1969. University Press, Cambridge, 1970; reprinted from the Journal of Fluid Mechanics, Vol. 41, 480 pp.

C.   General Treatises, Surveys, Textbooks
C 1. General Treatises

Howarth, L. (ed.): Modern developments in fluid dynamics. High speed flow. Vol. I, 330 pp. and Vol. II, 475 pp. 2nd ed., Clarendon Press, Oxford, 1956.

Thwaites, B. (ed.): Incompressible aerodynamics. Fluid Motion Memoirs, 636 pp. Clarendon Press, Oxford, 1960.

Lachmann, G.V. (ed.): Boundary layers and flow control. Vol. I and Vol. II, 1360 pp. Pergamon Press, London, 1961.

Rosenhead, L. (ed.): Laminar boundary layers. Fluid Motion Memoirs, 687 pp. Clarendon Press, Oxford, 1963.

Contributions by:

Lighthill, M.J.: Introduction; Real and ideal fluids. p. 1—45.

Lighthill, M.J.: Introduction; Boundary layer theory. p. 46—113.

Whitham, G.B.: The Navier Stokes equations of motion. p. 114—162.

Illingworth, C.R.: Flow at small Reynolds number. p. 163—197.

Jones, C.W., and Watson, E.J.: Two-dimensional boundary layers. p. 198—257.

Gadd, G.E., Jones, C.W., and Watson, E.L.: Approximate methods of solution. p. 258—348.

Stuart, J.T.: Unsteady boundary layers. p. 349—408.

Crabtree, L.F., Küchemann D., and Sowerby, L. Three-dimensional boundary layers. p. 409—491.

Stuart, J.T.: Hydrodynamic stability. p. 492—579.

Pankhurst, R.C., and Gregory, N.: Experimental methods. p. 580—628.

C 2. Surveys

AGARD (= Advisory Group Aeronautical Research and Development)

AGARD Conference Proceedings, C.P. No. 30 (1968). Hypersonic boundary layers and flow fields. Symposium Fluid Dynamics Panel, London, 1968.

AGARD Conference Proceedings, C.P. No. 83 (1971). Facilities and techniques for aerodynamic testing at transonic speeds and high Reynolds numbers. Symposium Fluid Dynamics Panel, Göttingen, 1971.

AGARD Conference Proceedings, C.P. No. 93 (1971). Turbulent shear flow. Symposium Fluid Dynamics Panel, London, 1971.

AGARD Conference Proceedings, C.P. No. 168 (1975). Flow separation. Symposium Fluid Dynamics Panel, Göttingen, 1975.

AGARDograph No. 97 (1965); Parts I, II, III, IV. Recent developments in boundary layer research. Symposium Fluid Dynamics Panel, Naples, 1965.

AGARDograph No. 164 (1972); Surugue, J. (ed.): Boundary layer effects in Turbomachines.

Bradshaw, P. (ed.): Turbulence. 335 pp. Springer Verlag, Berlin/Heidelberg/New York, 1976.
Contributions by:

Bradshaw, P.: Introduction. p. 1—44.

Fernholz, H.H.: External flow. p. 45—107.

Johnston, J.: Internal flow. p. 109—169.

Bradshaw, P., and Woods, J.D.: Geophysical turbulence and buoyant flows. p. 171—192.

Reynolds, W.C., and Cebeci, T.: Calculation of turbulent flows. p. 193—229.

Launder, B.E.: Heat and mass transport. p. 231—287.

Lumley, J.L.: Two-phase and non-Newtonian flows. p. 289—328.

Bradshaw, P. (ed.): Turbulence research. Progress and problems. Proceedings of the 1976 Heat Transfer and Fluid Mech. Inst. Stanford University Press, 1976.

Kline, S.J., Morkovin, M.V., Sovran, G., Cockerill, D.J., Coles, D.E., and Hirst, E.A. (ed.): Proceedings: Computation of turbulent boundary layers. Vol. I and Vol. II: AFOSR-IFP Stanford Conference, Thermosci. Div. Dep. Mech. Eng. Stanford University, 1969. Vol. I, ed. by Kline, Morkovin, Sovran, Cockerill: Methods, predictions, evaluation and flow structure. 590 pp. Vol. II, ed. by Coles and Hirst: Compiled data. 519 pp.

Fiedler, H. (ed.): Structure and mechanisms of turbulence, Vol. I and II. Proceedings, Berlin, 1977. Lecture Notes in Physics, Vol. 75 and 76. Springer Verlag, Berlin 1978.

Schmidt, F.W.: Symposium on Turbulent Shear Flows, 18 Sessions, April 18—20, 1977. The Pennsylvania State University, University Park, Pennsylvania.

C 3. Textbooks

Abramovich, G.N.: The theory of turbulent jets. (Translated from Russian) MIT Press, Cambridge, Mass., 1963.

Batchelor, G.K.: An introduction to fluid dynamics. Cambridge University Press, London/New York, 1967.

Batchelor, G.K.: The theory of homogeneous turbulence. 2nd ed., Cambridge University Press, London, 1970.

Batchelor, G.K.: An introduction to turbulence and its measurements. Pergamon Press, Oxford/New York, 1971.

Betchov, R., and Criminale, W.O.: Stability of parallel flow. Academic Press, New York, 1967.

Betz, A.: Konforme Abbildung. 2nd ed., Springer Verlag, Berlin, 1964.

Bird, R.B., Stewart, W.E., and Lightfoot, E.M.: Transport phenomena. John Wiley, New York, 1960.

Brun, E.A., Martinot-Lagarde, A., and Mathieu, J.: Mécanique des fluides, Vol. I and II. 2nd ed. Dunod, Paris, 1968.

Cebeci, T., and Bradshaw, P.: Momentum transfer in boundary layers. McGraw-Hill, New York, 1977.

Cebeci, T., and Smith, A.M.O.: Analysis of turbulent boundary layers. Academic Press, New York, 1974.

Chandrasekhar, S.: Hydrodynamic and hydromagnetic stability. Clarendon Press, Oxford, 1961.

Chang, P.K.: Separation of flow. Pergamon Press, New York, 1970.

Chang, P.K.: Control of flow separation. Hemisphere Publishing Corporation, Washington, D.C. 1976.

Chapman, A.J.: Heat transfer. 2nd ed., Macmillan, New York, 1967.

Curle, N.: The laminar boundary layer equations. Clarendon Press, Oxford, 1962.

Curle, N., and Davies, H.J.: Modern fluid dynamics. Vol. I: Incompressible flow. Vol. II: Compressible flow. Van Nostrand Reinhold Comp., London/New York, 1968 and 1971.

Dorrance, W.H.: Viscous hypersonic flow. Theory of reacting and hypersonic boundary layers. McGraw-Hill, New York, 1963.

Dryden, H.L., Murnagan, F.P., and Bateman, H.: Hydrodynamics. Reprint, Dover Publications, New York, 1956.

Duncan, W.J., Thom, A.S., and Young, A.D.: An elementary treatise on the mechanics of fluids. Edward Arnold Publ., London, 1st ed. 1960; 2nd ed. 1970.

Eck, B.: Technische Strömungslehre. 7th ed., Springer, Berlin/Heidelberg/New York, 1966.

Eckert, E.R.G.: Einführung in den Wärme- und Stoffaustausch. 3rd ed., Springer, Berlin, 1966.

Eckert, E.R.G., and Drake, jr., R.M.: Heat and mass transfer. McGraw-Hill, New York, 1959.

Eckhaus, W.: Studies in non-linear stability theory. Springer, New York, 1965.

Evans, H.: Laminar boundary layers. Addison-Wesley Publishing Comp., Reading, Mass., 1968.

Favre, A., Kovasznay, L.S.H., Dumas, R., Gaviglio, J., and Coantic, M.: Preface de E.A. Brun: La turbulence en mécanique des fluides; bases théoriques et expérimentales, méthodes statistiques. Gauthier-Villars, Paris, 1976.

Gersten, K.: Einführung in die Strömungsmechanik. Düsseldorf, 1974.

Goldstein, S. (ed.): Modern developments in fluid dynamics. Vols. I and II. University Press, Oxford, 1938.

Goldstein, S.: Lectures in fluid mechanics. In: Lectures in applied mathematics, Vol. II. Interscience Publ., London/New York, 1960.

Gosman, A.O. et al.: Heat and mass transfer in recirculating flows. Academic Press, London/New York, 1969.

Grigull, U. (ed.): Groeber, H., Erk, S., and Grigull, U.: Die Grundgesetze der Wärmeübertragung. 3rd ed. 1955, 3rd printing 1963.

Hansen, A.G.: Fluid mechanics. Wiley Press, New York, 1967.

Hayes, D., and Probstein, F.: Hypersonic flow theory. Academic Press, New York, 1959.

Hinze, J.O.: Turbulence. 2nd ed. McGraw-Hill, New York, 1975.

Hoerner, S.F.: Fluid-dynamic drag. 2nd ed., Washington D.C., 1965.

Kays, W.M.: Convective heat and mass transfer. McGraw-Hill, New York, 1966.

Kaufmann, W.: Technische Hydro- und Aerodynamik. 3rd ed., Springer Verlag, Berlin, 1963. Vol. I: Inviscid flow. 2nd ed., 1966.

Knudsen, J.G., and Katz, D.L.: Fluid dynamics and heat transfer. McGraw-Hill, New York, 1958.

Kuethe, A.M., and Schnetzler, J.D.: Foundations of aerodynamics. John Wiley, New York, 1959.

Kutateladze, S.S., and Leont'ev: Turbulent boundary layers in compressible gases. Translated from Russian by D.P. Spalding. Edward Arnold Publ., London, 1964.

Lamb, H.: Hydrodynamics. 6th ed., Cambridge, 1957. German translation: Lehrbuch der Hydro-
    dynamik. 2nd ed., 1931.

Liepmann, H.W., and Roshko, A.: Elements of gasdynamics. John Wiley, New York, 1957.

Lin, C.C.: The theory of hydrodynamic stability. Cambridge University Press, 1955.

Lin, C.C.: Turbulent flows and heat transfer. Princeton University Press, 1959.

Langlois, W.E.: Slow viscous flow. Macmillan, New York, 1964.

Loitsianski, L.G.: Laminare Grenzschichten. Translated from Russian by W. Szablewski. Aka-
    demie-Verlag, Berlin, 1967.

Leslie, D.C.: The development of the theory of turbulence. Clarendon Press, Oxford, 1973.

Meksyn, D.: New methods in laminar boundary layer theory. Pergamon Press, London, 1961.

Müller, W.: Einführung in die Theorie der zähen Flüssigkeiten. Akademische Verlagsgesellschaft,
    Leipzig, 1932.

Oswatitsch, K.: Grundlagen der Gasdynamik. Springer Verlag, Wien/New York, 1976.

Pai, S.I.: Viscous flow theory. Vol. I: Laminar flow. Vol. II: Turbulent flow. Van Nostrand,
    Princeton, N.J., 1956 and 1957.

Pai, S.I.: Fluid dynamics of jets. Van Nostrand, New York, 1954.

Patankar, S.U., and Spalding, D.B.: Heat and mass transfer in boundary layers. 2nd ed., Inter-
    text Book, London, 1970.

Plate, E.J.: Aerodynamic characteristics of atmospheric boundary layers. Atomic Energy Com-
    mission, 1971.

Prandtl, L., and Tietjens, O.: Hydro- und Aeromechanik, 2 Vols. Springer, Berlin, 1929 and 1931.
    Vol. I: Fundamentals of hydro- and aeromechanics. English translation by L. Rosenhead.
    McGraw-Hill, New York/London, 1924. Vol. II: Applied hydro- and aeromechanics. English
    translation by J.P. Den Hartog. McGraw-Hill, New York/London, 1934.

Prandtl, L.: Führer durch die Strömungslehre. 7th ed., Vieweg, Braunschweig, 1969. English
    translation by W.M. Deans: Essentials in fluid dynamics. Blackie and Son, London, 1952.

Reynolds, A.J.: Turbulent flows in engineering. John Wiley, London, 1974.

Riegels, F.W.: Aerodynamische Profile. Oldenbourg, München, 1958. English translation by
    D.G. Randall: Aerofoil sections. Butterworths, London, 1961.

Rohsenow, W.M., and Choi, H.U.: Heat, mass and momentum transfer. Prentice Hall, Engle-
    wood Cliffs, New Jersey, 1961.

Rotta, J.C.: Turbulente Strömungen. Teubner, Stuttgart, 1972.

Rouse, H.: Advanced mechanics of fluids. John Wiley, New York, 1959.

Sovran, G. (ed.): Fluid mechanics of internal flow. Elsevier, Amsterdam, 1967.

Schmidt, E.: Einführung in die technische Thermodynamik und die Grundlagen der chemischen
    Thermodynamik. 9th ed., Springer, Berlin, 1962. English translation by J. Kestin. Clarendon
    Press, Oxford, 1949; also Dover reprint.

Shapiro, A.H.: The dynamics and thermodynamics of compressible flow. Vols. I and II. Ronald
    Press, New York, 1953 and 1954.

Schlichting, H.: Grenzschichttheorie. 5th ed., Braun, Karlsruhe, 1965.

Schlichting, H., and Truckenbrodt, E.: Aerodynamik des Flugzeuges. Vols. I and II, 2nd ed.,
    Springer, Berlin/Heidelberg/New York, 1967 and 1969.

Schubauer, G.B., and Tcen, C.M.: Turbulent flow. Princeton Aeronautical Paperbacks No. 9,
    Princeton, New Jersey, 1961.

Stewartson, K.: The theory of laminar boundary layers in compressible fluids. Clarendon Press,
    Oxford, 1964.

Townsend, A.A.: The structure of turbulent shear flow. 2nd ed., Cambridge University Press, 1976.

Truitt, R.W.: Hypersonic aerodynamics. Ronald Press, New York, 1959.

Truitt, R.W.: Fundamentals of aerodynamic heating. Ronald Press, New York, 1960.

Truckenbrodt, E.: Strömungsmechanik; Grundlagen und technische Anwendungen. Springer
    Verlag, Berlin/Heidelberg/New York, 1968.

Van Dyke, M.: Perturbation methods in fluid mechanics. 2nd ed., Academic Press, New York, 1975

Wieghardt, K.: Theoretische Strömungslehre, Teubner, Stuttgart, 1965.

White, F.M.: Viscous fluid flow. McGraw-Hill, New York, 1974.

Walz, A.: Strömungs- und Temperaturgrenzschichten. Braun-Verlag, Karlsruhe, 1966. English translation: Boundary layers of flow and temperature, by H. J. Oser, MIT Press, Cambridge, Mass., USA, 1969.

D.   **Bench-mark publications** (in chronological order)

Prandtl, L.: Über Flüssigkeitsbewegung bei sehr kleiner Reibung. Verhandlungen IIIrd Intern. Math. Kongress Heidelberg 1904, 484—491 (1904), Teubner, Leipzig, 1905. English translation: NACA Memo No. 452 (1928). Reprinted in: Vier Abhandlungen zur Hydro- und Aerodynamik, Göttingen, 1927; Coll. Works, Vol. II, 575—584.

Blasius, H.: Grenzschichten in Flüssigkeiten mit kleiner Reibung. Diss. Göttingen 1907. Z. Math. u. Phys. *56*, 1—37 (1908). English translation: NACA TM 1256.

Boltze, E.: Grenzschichten an Rotationskörpern. Diss. Göttingen 1908.

Hiemenz, K.: Die Grenzschicht an einem in den gleichförmigen Flüssigkeitsstrom eingetauchten geraden Kreiszylinder. Diss. Göttingen 1911. Dingl. Polytechn. J. *28*, 321—410 (1911).

Prandtl, L.: Der Luftwiderstand von Kugeln. Nachr. Ges. Wiss. Göttingen, Math. Phys. Klasse, 177—190 (1914); Coll. Works, Vol. II, 597—608.

von Kármán, Th.: Über laminare und turbulente Reibung. ZAMM *1*, 233—252 (1921). NACA TM 1092 (1946).

Pohlhausen, K.: Zur näherungsweisen Integration der Differentialgleichung der laminaren Grenzschicht. ZAMM *1*, 252—268 (1921).

Prandtl, L.: Bemerkungen über die Entstehung der Turbulenz. ZAMM *1*, 431—436 (1921); Coll. Works, Vol. II, 687—696.

Tietjens, O.: Beiträge zur Entstehung der Turbulenz. Diss. Göttingen 1922. ZAMM *5*, 200—217 (1925).

Burgers, J.M.: The motion of a fluid in the boundary layer along a plane smooth surface. Proc. First Intern. Congress Appl. Mech., Delft, 113—128 (1924).

Betz, A.: Ein Verfahren zur direkten Ermittlung des Profilwiderstandes. ZFM *16*, 42 (1925).

Prandtl, L.: Über die ausgebildete Turbulenz. ZAMM *5*, 136—139 (1925); Verhandlungen II. Intern. Kongress Angew. Mechanik, Zürich, 62—75 (1926); Coll. Works, Vol. II, 714—718.

Tollmien, W.: Berechnung turbulenter Ausbreitungsvorgänge. ZAMM *6*, 468—478 (1926).

Prandtl, L.: The generation of vortices in fluids of small viscosity. 15th Wilbur Wright Memorial Lecture, London 1927. J. Roy. Aero. Soc. *31*, 720 (1927). See also: Die Entstehung von Wirbeln in einer Flüssigkeit mit kleiner Reibung. Z. Flugtechn. Motorluftsch. *18*, 489—496 (1927); Coll. Works, Vol. II, 752—777.

Tollmien, W.: Über die Entstehung der Turbulenz. I. Mitteilung Nachr. Ges. Wiss. Göttingen, Math. Phys. Klasse, 21—44 (1929). NACA TM 609 (1931).

Schlichting, H.: Über das ebene Windschattenproblem. Diss. Göttingen 1930. Ing.-Arch. *1*, 533—571 (1930).

Nikuradse, J.: Gesetzmäßigkeiten der turbulenten Strömung in glatten Rohren. Forsch.-Arb. Ing.-Wes. Heft 356 (1932).

Taylor, G.I.: The transport of vorticity and heat through fluids in turbulent motion. Appendix by A. Fage and V.M. Falkner. Proc. Roy. Soc. *135*, 685—705 (1932).

Prandtl, L.: Neuere Ergebnisse der Turbulenzforschung. Z. VDI *77*, 105—114 (1933); Coll. Works, Vol. II, 819—845.

Nikuradse, J.: Strömungsgesetze in rauhen Rohren. Forsch.-Arb. Ing.-Wes. Heft 361 (1933).

Schlichting, H.: Zur Entstehung der Turbulenz bei der Plattenströmung. Nachr. Ges. Wiss. Göttingen, Math. Phys. Klasse, 182—203 (1933); see also ZAMM *13*, 171—174 (1933).

Prandtl, L.: The mechanics of viscous fluids. In W. F. Durand (ed.) Aerodynamics Theory, Vol. III. Springer Verlag, 34—208 (1935).

Tollmien, W.: Ein allgemeines Kriterium der Instabilität laminarer Geschwindigkeitsverteilungen. Nachr. Ges. Wiss. Göttingen, Math. Phys. Klasse, Fachgruppe I, *1*, 79—114 (1935).

Schlichting, H.: Amplitudenverteilung und Energiebilanz der kleinen Störungen bei der Plattenströmung. Nachr. Ges. Wiss. Göttingen, Math. Phys. Klasse, Fachgruppe I, *1*, 47—78 (1935).

Busemann, A.: Gasströmung mit laminarer Grenzschicht entlang einer Platte. ZAMM *15*, 23–25 (1935).

Jones, B.M.: Flight experiments on the boundary layer. First Wright Brothers' Memorial Lecture 1937. J. Aero. Sci. *5*, 81–101 (1938).

von Kármán, Th., and Tsien, H.S.: Boundary layer in compressible fluids. J. Aero. Sci. *5*, 227–232 (1938). See also: Th. von Kármán: Report on the Volta Congress, Rome, 1935.

Görtler, H.: Über eine dreidimensionale Instabilität laminarer Grenzschichten an konkaven Wänden. Nachr. Ges. Wiss. Göttingen, Math. Phys. Klasse, New Series *2*, No. 1 (1940).

Schubauer, G.B., and Skramstad, H.K.: Laminar boundary layer oscillations and stability of laminar flow. J. Aero. Sci. *14*, 69–78 (1947). NACA Rep. 909 (1948).

Tollmien, W.: Asymptotische Integration der Störungsdifferentialgleichung ebener laminarer Strömungen bei hohen Reynolds-Zahlen. ZAMM *25/27*, 33–50 and 70–83 (1947).

Mangler, W.: Zusammenhang zwischen ebenen und rotationssymmetrischen Grenzschichten in kompressiblen Flüssigkeiten. ZAMM *28*, 97–103 (1948).

Truckenbrodt, E.: Ein Quadraturverfahren zur Berechnung der laminaren und turbulenten Reibungsschicht bei ebener und rotationssymmetrischer Strömung. Ing.-Arch. *20*, 211–228 (1952).

Dryden, H.L.: Fifty years of boundary layer theory and experiment. Sciences *121*, 375–380 (1955).

Schlichting, H.: Application of boundary layer theory in turbomachinery. J. Basic Eng. *81*, 543–551 (1959).

Kestin, J.: The effect of free-stream turbulence on heat transfer rates. Advances in Heat Transfer *3*, 1–32 (1966).

Bradshaw, P.: The understanding and prediction of turbulent flow. 6th Reynolds-Prandtl Lecture. J. Roy. Aero. Soc. *76*, 403–418; see also DGLR Jb. 1972, 51–82.

Schlichting, H.: Recent progress in boundary layer research. 36th Wright Brothers' Memorial Lecture 1973. AIAA J. *12*, 427–440 (1974).

Smith, A.M.O.: High lift aerodynamics. 37th Wright Brothers' Memorial lecture 1974. J. Aircraft *12*, 501–530 (1975).

Schlichting, H.: An account of the scientific life of Ludwig Prandtl. Invited Lecture presented at the Symposium on Flow Separation of the AGARD Fluid Dynamics Panel at Göttingen, May 27 to 30, 1975. ZFW *23*, 297–316 (1975).

Tani, J.: History of boundary layer theory. Ann. Review Fluid Mech. *9*, 87–111 (1977).

E. Ludwig Prandtl Memorial Lectures (since 1957)

Betz, A.: Lehren einer fünfzigjährigen Strömungsforschung. ZFW *5*, 97–105 (1957).

Dryden, L.: Gegenwartsprobleme der Luftfahrtforschung. ZFW *6*, 217–233 (1958).

Roy, M.: Über die Bildung von Wirbelzonen in Strömungen mit geringer Zähigkeit. ZFW *7*, 217–227 (1959).

Schmidt, E.: Thermische Auftriebsströmungen und Wärmeübergang. ZFW *8*, 273–284 (1960).

Lighthill, M.J.: A technique for rendering approximate solutions to physical problems uniformly valid. ZFW *9*, 267–275 (1961).

Tollmien, W.: Aspekte der Strömungsphysik 1962. ZFW *10*, 403–413 (1962).

Sauer, R.: Die Aufgabe des Mathematikers in der Aerodynamik. ZFW *11*, 349–357 (1963).

Ackeret, J.: Anwendungen der Aerodynamik im Bauwesen. ZFW *13*, 109–122 (1965).

Busemann, A.: Minimalprobleme der Luft- und Raumfahrt. ZFW *13*, 401–411 (1965).

Schlichting, H.: Einige neuere Ergebnisse aus der Aerodynamik des Tragflügels. DGLR Jb. 1966, 11–32 (1967).

Küchemann, D.: Entwicklungen in der Tragflügeltheorie. DGLR Jb. 1967, 11–22 (1968).

Wittmeyer, H.: Aeroelastomechanische Untersuchungen an dem Flugzeug SAAB 37 „VIGGEN". DGLR Jb. 1968, 11–23 (1969).

Oswatitsch, K.: Möglichkeiten und Grenzen der Linearisierung in der Strömungsmechanik. DGLR Jb. 1969, 11—17 (1970).

Germain, P.: Progressive waves. DGLR Jb. 1971, 11—30 (1972).

Magnus, K.: Fortschritte in der Kinetik von Mehrkörpersystemen. DGLR Jb. 1972, 11—26 (1973).

Becker, E.: Stoßwellen. DGLR Jb. 1973, 11—40 (1974).

Olszak, W.: Gedanken zur Entwicklung der Plastizitätstheorie. ZFW *24*, 123—139 (1976).

Schultz-Grunow, F.: Exakte Zugänge zu hydrodynamischen Problemen. ZFW *23*, 175—183 (1975).

Truckenbrodt, E.: Näherungslösungen der Strömungsmechanik und ihre physikalische Deutung. ZFW *24*, 178—187 (1976).

Young, A.D.: Some special boundary layer problems. ZFW *1*, 401—414 (1977).

Zierep, J.: Instabilitäten in Strömungen zäher, wärmeleitender Medien. ZFW *2*, 143—150 (1978).

Index of Authors

Subject Index

ablation   382, 399
acceleration   47, 423
—, sudden   *see* impulsive motion
acoustics   428, 432
adhesion   *see* no-slip condition
adiabatic compression   268, 327, 763
— temperature   270, 279, 286, 332, 717
— wall   279, 286, 332, 354, 358, 717
aerodynamics   2
—, structural   37
aerofoil   22, 23, 37, 38, 213, 219, 221, 358, 397, 399, 499, 500, 503, 657, 684, 690, 769
—, laminar   *see* laminar aerofoil
—, maximum lift   690;   *see also* Zhukovskii aerofoil; NACA aerofoil
— theory   2
air   8, 269
aircrew   695
d'Alembert's paradox   5, 20
amplification   458, 460, 471, 472, 491, 508, 530
— factor   459, 532
analogy   *see* Reynolds's analogy
angle of inflow   769
— of stagger   769
annular effect   438
approximate methods   109, 112, 127, 158, 201, 214, 219, 239, 304, 352, 392, 635, 671
asymptotic suction profile   385, 506
attack   *see* incidence
autocorrelation function   569
axial symmetry   420;   *see also* body of revolution; cylinder; pipe; sphere

back-flow   *see* reverse flow
bearing   117, 122
Bernoulli's equation   10, 96, 130, 269, 490
Betz's method   759
binary layers   400
Bjerknes's polar front   512
blade angle   769
Blasius series   168, 236, 250, 251, 303
blowing   379, 380, 382, 400, 687
body force   48, 71, 102, 226, 258, 285, 489, 510
body of revolution   235, 686, 692;   *see also* cylinder; pipe; sphere
— — —, rotating   694
boundary layer   2, 3, 24, 78, 127, 131, 635, 668, 702

— —, axially symmetrical   225, 234, 235, 692
— —, compressible   327, 702
— —, concave wall   532
— —, concept   24
— —, control   43, 378, 379, 400, 506, 687
— —, convergent and divergent   696
— —, corner   303, 644, 697
— — equations   44, 128, 130, 151, 163, 201, 282, 284, 285, 330, 409, 676, 704, 706, 730
— — fence   253
— —, higher order   144, 194
— —, nonseparating   687
— —, nonsteady   408, 409
— —, periodic   411, 428, 431, 432
— — profile   *see* velocity distribution
— — re-attachment   688
— —, rotating body   695
— —, similar solutions and self-similar solutions   136, 151, 152, 203, 231, 287, 293, 300, 344, 389, 415, 416, 490, 680, 735
— — simplifications   80, 128, 282;   *see also* boundary layer
— — theory   2, 24
— —, thermal   3, 265, 282, 292, 303, 309, 315
— — thickness   25, 26, 40, 42, 128, 131, 140, 230, 636, 649, 673, 686
— —, three-dimensional   225, 239, 247, 525, 535, 692
— —, turbulent   635, 668, 671, 677, 702, 729
— —, turbulent and compressible   702
bulk viscosity   61
buoyancy   271, 285, 489

camber   690
camouflage paint   654
capillary   12
cascade   769
— flow   248, 313, 660, 744, 769
centrifugal forces   102, 226, 510, 647
channel flow   6, 280, 586, 613;   *see also* Couette flow
— —, divergent or convergent   34, 43, 107, 157, 166, 223, 281, 668
circular jet   230
Coanda effect   792

Abbreviations

The following abbreviations have been used throughout the book

AIAA J.	= *Journal of the American Institute of Aeronautics and Astronautics*, New York, published since 1963 (see JAS and JASS)
ARC	= Aeronautical Research Council, London. Publishes two series of documents, each numbered separately
	ARC RM — Reports and Memoranda
	ARC CP — Current Papers
ARSJ	= Journal American Rocket Society
ASME	= American Society of Mechanical Engineers, New York
J. Appl. Mech.	= *Journal of Applied Mechanics*, being part E of the *Transactions of the ASME* (see above)
J. Heat Transfer	= *Journal of Heat Transfer*, being part C of the *Transactions of the* ASME
AVA	= Aerodynamische Versuchsanstalt, Goettingen, Germany
DFL	= Deutsche Forschungsanstalt für Luft- und Raumfahrt, Braunschweig, Germany (till 1969)
DFVLR	= Deutsche Forschungs- und Versuchsanstalt für Luft- und Raumfahrt, Köln (since 1969)
DVL	= Deutsche Versuchsanstalt für Luft- und Raumfahrt, Köln, Germany (till 1969)
DGLR	= Deutsche Gesellschaft für Luft- und Raumfahrt, Köln
ETH	= Federal Institute of Technology (Eidgenoessische Technische Hochschule) Zurich, Switzerland
Forschg. Ing.-Wes.	= Scientific journal entitled
	Forschung auf dem Gebiete des Ingenieur-Wesens, VDI (German Society of Engineers), Berlin and Duesseldorf (since 1948)
Forschungsheft	= Research supplement to Forschg. Ing.-Wes. (see above)
Ing.-Arch.	= Scientific journal entitled
	Ingenieur-Archiv, Berlin and, since 1947, Berlin and Heidelberg
JAS	= *Journal of the Aeronautical Sciences*, New York, (1932—1958); replaced in 1959 by JASS
JASS	= *Journal of Aero/Space Sciences*, New York (1959—1962); replaced in 1963 by AIAA J.
JFM	= *Journal of Fluid Mechanics*, Cambridge, England
J. Roy. Aero. Soc.	= *Journal of the Royal Aeronautical Society*, London, England

NACA   = The National Advisory Committee for Aeronautics, Washington D. C.
                           [replaced in 1959 by NASA (see below)]. Published three series of
                           documents, each numbered separately:

                           NACA Rep.   — Reports

                           NACA TM   — Technical Memoranda

                           NACA TN   — Technical Notes

NASA   = National Aeronautics and Space Administration (created in 1959 in
                           replacement of NACA)

NGTE   = National Gas Turbine Establishment, Great Britain

ONERA   = Office National d'Études et de Recherches Aérospatiales, Châtillon-sous-
                           Bagneux, France

Proc. Roy. Soc. A   = *Proceedings of the Royal Society, London*, series A

RAE   = Royal Aircraft Establishment, Great Britain

USAF   = United States Air Force

VDI   = Verein Deutscher Ingenieure (German Society of Engineers), Duesseldorf.
                           Publishes: Forschg. Ing.-Wes. with its supplement Forschungsheft
                           (see above)

WGL Jb.   = Jahrbuch der Wissenschaftlichen Gesellschaft für Luftfahrt, 1952—1962;
                           für Luft- und Raumfahrt, 1963—1975 (H. Blenk and W. Schulz, eds.,
                           Vieweg, Braunschweig)

ZAMM   = *Zeitschrift für angewandte Mathematik und Mechanik*, Berlin, Germany

ZAMP   = *Zeitschrift für angewandte Mathematik und Physik*, Basel, Switzerland

ZFM   = *Zeitschrift für Flugtechnik und Motorluftschiffahrt*, Munich and Berlin,
                           Germany

ZFW   = *Zeitschrift für Flugwissenschaft*, Braunschweig, Germany, 1953—1976;
                           from 1977 replaced by *Zeitschrift für Flugwissenschaften und Weltraum-
                           forschung*, Köln.

List of most commonly used symbols

In order not to depart too drastically from the conventions normally employed in papers on the subject, it was found necessary to use the same symbol to denote several different quantities. Thus, for example, λ denotes the resistance coefficient of pipe flow, both laminar and turbulent, and in the theory of stability of laminar boundary layers it denotes the wavelength of a disturbance. Similarly, k denotes thermal conductivity in the theory of thermal boundary layers, and the height of a protuberance in the discussion of the influence of roughness on turbulent flow.

The following is a list of symbols most commonly used in the book.

I. General symbols

$$
\begin{aligned}
A &= \text{wetted area, or frontal area} \\
c &= \text{velocity of sound} \\
d, D &= \text{diameter} \\
g &= \text{acceleration due to gravity} \\
h &= \text{channel width} \\
l, L &= \text{length} \\
p &= \text{pressure (force per unit area)} \\
q &= \tfrac{1}{2} \varrho \, V^2 = \text{dynamic head} \\
r, \phi, z &= \text{cylindrical coordinates} \\
r, R &= \text{radius} \\
u &= \text{mean velocity (in pipe)} \\
U_\infty &= \text{free-stream velocity} \\
U(x) &= \text{velocity in potential flow} \\
u, v, w &= \text{velocity components} \\
\bar{u} &= \text{temporal mean of velocity (pipe or boundary layer)} \\
x, y, z &= \text{cartesian coordinates} \\
V &= \text{free-stream velocity} \\
\varrho &= \text{density (mass per unit volume)} \\
\omega &= \text{angular velocity}
\end{aligned}
$$

II. Viscous flow, turbulence

$$
\begin{aligned}
A_\tau &= \text{eddy viscosity} \\
b &= \text{width of jet or wake} \\
c_D &= \text{drag coefficient} \\
c_f &= \text{skin-friction coefficient} \\
c_f' &= \text{local skin-friction coefficient} \\
D &= \text{drag force} \\
H_{12} &= \delta_1/\delta_2 = \text{first shape factor of velocity profile}
\end{aligned}
$$

$H_{32} = \delta_3/\delta_2 =$ second shape factor of velocity profile

$\mathsf{M} = (v/c) =$ Mach number

$k =$ height of roughness element (protuberance)

$k_S =$ height of grain for equivalent sand roughness

$K =$ shape factor of velocity profile in boundary layer

$l =$ mixing length

$\mathsf{R} = (VL/\nu$ or $\bar{u}d/\nu$ or $U\delta/\nu) =$ Reynolds number

$\mathsf{R}_i =$ Richardson number

$\mathsf{S} =$ Strouhal number

$\mathsf{T} =$ turbulence intensity (also degree or level of turbulence)

$u', v', w' =$ components of turbulent velocity

$\overline{u'^2}, \overline{v'^2}, \overline{u'v'} \ldots =$ temporal means of turbulent velocities

$U =$ maximum velocity at pipe centre

$U_\infty =$ free stream velocity

$v_* = \sqrt{\tau_0/\varrho} =$ friction velocity

$y =$ distance from wall

$\delta =$ boundary-layer thickness

$\delta_1 =$ displacement thickness

$\delta_2 =$ momentum thickness

$\delta_3 =$ energy thickness

$\varepsilon_\tau =$ apparent (virtual) kinematic viscosity in turbulent flow ("eddy viscosity")

$\eta = yv_*/\nu =$ dimensionless distance from wall

$\varkappa =$ empirical constant in turbulent flow; $l = \varkappa y$

$\lambda =$ resistance coefficient of pipe flow

$\Lambda =$ shape factor of velocity profile in laminar boundary layer

$\mu =$ absolute viscosity

$\nu = \mu/\varrho =$ kinematic viscosity

$\tau =$ shearing stress (force per unit area)

$\tau_0, \tau_w =$ shearing stress at a wall

$\phi = u/v_* =$ dimensionless velocity

$\psi =$ stream function

III. Transition from laminar to turbulent flow

$c = \beta/\alpha = c_r + ic_i$

$c_i =$ amplification (or damping) factor

$c_r =$ wave propagation velocity of disturbance

$\mathsf{R}_{crit} = (U_m \delta_1/\nu)_{crit} =$ critical Reynolds number

$\mathsf{T}_a =$ Taylor number

$u', v' =$ velocity components of disturbance

$U(y) =$ velocity profile in boundary layer

$U_m(x) =$ velocity in potential flow

$\beta = \beta_r + i\beta_i$

$\beta_i =$ amplification (or damping) factor

$\beta_r =$ circular frequency of disturbance

$\gamma =$ intermittency factor

$\lambda = 2\pi/\alpha =$ wavelength of disturbance

$\phi(y) =$ amplitude of stream function of disturbance

IV. Thermal and compressible boundary layers

$a = k/\varrho\, c_p$ = thermal diffusivity

$c = \sqrt{(\gamma p/\varrho)}$ = velocity of sound

c_p, c_v = specific heats at constant pressure and volume, respectively

$E = U_\infty^2/c_p\, \varDelta T$ = Eckert number

$G = g\beta l^3\, \varDelta T/v^2$ = Grashof number

h = enthalpy

k = thermal conductivity

$M = V/c$ = Mach number

$N = \alpha l/k$ = Nusselt number

$P = v/a$ = Prandtl number

q = heat flux (quantity of heat per unit area and time)

r = recovery factor

R = gas constant

$R_i = -\{(g/\varrho)\,(d\varrho/dy)\}/(dU/dy)^2{}_W$ = Richardson number

$S = N/RP$ = Stanton number

S_c = Schmidt number

T = temperature

T_a = adiabatic wall temperature (recovery temperature)

T_w = temperature at the wall

α = coefficient of heat transfer

β = coefficient of thermal expansion

$\gamma = c_p/c_v$ = isentropic exponent

δ_T = thickness of thermal boundary layer

$\varDelta T$ = temperature difference

Φ = dissipation function

ω = exponent in viscosity-temperature relation

J.

S

l